Intelligent Production Machines and Systems

First I*PROMS Virtual Conference

Elsevier Internet Homepage

http://www.elsevier.com

Consult the Elsevier homepage for full catalogue information on all books, journals and electronic products and services.

Elsevier Titles of Related Interest

The following journals related to industrial engineering can all be found at: http://www.sciencedirect.com

Computers in Industry
Computers & Industrial Engineering
Advanced Engineering Informatics
Journal of Manufacturing Systems
Journal of Manufacturing Processes
Robotics and Computer-Integrated Manufacturing
Robotics and Autonomous Systems
Automatica
Mechatronics
Engineering Applications of Artificial Intelligence
Knowledge-based Systems
Applied Soft Computing
Neural Networks
Expert Systems with Applications
Fuzzy Sets and Systems
Journal of Network and Computer Applications
Journal of Production Economics
Journal of Product Innovation Management
Journal of Materials Processing Technologies
International Journal of Machine Tools and Manufacture
Journal of Mechanical Working Technology
Mechanics and Machine Theory

To Contact the Publisher
Elsevier welcomes enquiries concerning publishing proposals: books, journal special issues, conference proceedings, etc. All formats and media can be considered. Should you have a publishing proposal you wish to discuss, please contact, without obligation, the commissioning editor responsible for Elsevier's industrial engineering books publishing programme:

Henri van Dorssen
Publisher, Industrial Engineering
Elsevier Limited
The Boulevard, Langford Lane
Kidlington, Oxford
OX5 1GB, UK

Tel.: +44 1865 84 3682
Fax: +44 1865 84 3987
E-mail: h.dorssen@elsevier.com

General enquiries including placing orders, should be directed to Elsevier's Regional Sales Offices – please access the Elsevier internet homepage for full contact details.

Intelligent Production Machines and Systems

First I*PROMS Virtual Conference

4 - 15 July 2005

Edited by:

D T Pham
E E Eldukhri
A J Soroka

Organised by: FP6 I*PROMS Network of Excellence
Sponsored by: The European Commission

2005

ELSEVIER

Amsterdam Boston Heidelberg London New York Oxford
Paris San Diego San Francisco Singapore Sydney Tokyo

Elsevier BV
Radarweg 29, PO Box 211,
1000 AE Amsterdam
The Netherlands

Elsevier Inc
525 B Street, Suite 1900
San Diego, CA 92101-4495
USA

Elsevier Ltd
The Boulevard, Langford Lane, Kidlington,
Oxford OX5 1GB UK

Elsevier Ltd
84 Theobald's Road,
London WC1Z 8RR
UK

First edition 2005

ISBN–10: 0080447309

♾ The paper used in this publication meets the requirements of ANSI/NISO Z39.48-1992 (Permanence of Paper).

Printed in Great Britain

05 06 07 08 09 10 10 9 8 7 6 5 4 3 2 1

IPROMS 2005 Sponsors

Llywodraeth Cynulliad Cymru
Welsh Assembly Government

Preface

Intelligent Production Machines and Systems (IPROMS) employ advanced IT and computational techniques to provide competitive advantage. The 2005 Virtual International Conference on IPROMS took place on the Internet between 4 and 15 July 2005. IPROMS 2005 was an outstanding success. During the Conference, some 4168 registered delegates and guests from 71 countries participated in the Conference, making it a truly global phenomenon.

This book contains the Proceedings of IPROMS 2005. The 107 peer-reviewed technical papers presented at the Conference have been grouped into twelve sections, the last three featuring contributions selected for IPROMS 2005 by Special Sessions chairmen:

- Collaborative and Responsive Manufacturing Systems
- Concurrent Engineering
- E-manufacturing, E-business and Virtual Enterprises
- Intelligent Automation Systems
- Intelligent Decision Support Systems
- Intelligent Design Systems
- Intelligent Planning and Scheduling Systems
- Mechatronics
- Reconfigurable Manufacturing Systems
- Tangible Acoustic Interfaces (Tai Chi)
- Innovative Production Machines and Systems
- Intelligent and Competitive Manufacturing Engineering

Many of the IPROMS 2005 papers were written by partners and associate partners in the I*PROMS EU-funded FP6 Network of Excellence for Innovative Production Machines and Systems, but there were also large numbers of authors external to the Network. In total, IPROMS 2005 authors were from 32 countries across five continents. By attracting contributors and participants globally, IPROMS 2005 has made the first step in establishing the I*PROMS Network as the world's pre-eminent forum for the discussion of research issues in the field of Innovative Manufacturing.

Numerous people and organisations have helped make IPROMS 2005 a reality. We are most grateful to the IPROMS 2005 sponsors, I*PROMS partners, Conference Programme and Organising Committees, Special Session Organisers, Session Chairmen and Facilitators, Authors, Referees, and the I*PROMS Central Coordination Team.

The names of contributors to the success of IPROMS 2005 can be found elsewhere in the Proceedings. Here, we would highlight the much appreciated efforts of the Special Session Organisers, Professor R Teti of the University of Naples Federico II, Professor A Subic and Professor X Yu of the RMIT University in Melbourne and Dr M Yang and Dr Z Wang of Cardiff University.

Finally, our strongest vote of thanks must go to our colleague Vladimir Zlatanov, the technical co-ordinator of IPROMS 2005, who implemented all the IT infrastructure for our Virtual Conference. Without his expertise and dedication, IPROMS 2005would have forever remained virtual.

D.T. Pham, E.E. Eldukhri and A.J. Soroka
MEC, Cardiff University

Introduction by Mr A Gentili, European Commission

IPROMS 2005 is an innovative idea for holding conferences. It allows researchers and industrial practitioners interested in the area of Intelligent Production Machines and Systems to participate free of charge from wherever they are located and whenever they find it convenient. During IPROMS 2005, participants had the opportunity to view presentations, view/download full papers and contribute to the online discussions.

As a sponsor of I*PROMS FP6 Network of Excellence, the European Commission strongly supports the organisation of this event annually. This will enable the Network to disseminate the results of its work globally for the benefit of the wider community. Moreover, it will contribute to the integration of research resources in Europe for an efficient spending of R&D budget, avoiding overlaps in European research activities and exploiting synergies in order to build an effective European Research Area.

Andrea Gentili
Future Manufacturing
European Commission

Programme and Organising Committees

D. Pham (Chair), MEC, Cardiff University, UK
E. Eldukhri (Organising Committee Chair), MEC, ibid
A. Soroka (Programme Committee Chair), MEC, ibid
V. Zlatanov (Technical Co-ordinator), MEC, ibid
S. Dimov, MEC, ibid
M. Packianather, MEC, ibid
A. Thomas, MEC, ibid
B. Peat, MEC, ibid
R. Setchi, Cardiff University, UK
N. Aspragathos, University of Patras, Greece
A. Colombo, Schneider Electric, Germany
B. Denkena, IFW, University of Hannover, Germany
J. Efstathiou, University of Oxford, UK
F. Feenstra, TNO, The Netherlands
B. Grabot, ENIT, France
M. Hoepf, Fraunhofer IPA, Germany
E. Hohwieler, Fraunhofer IPK, Germany
A. Labib, University of Portsmouth, UK
V. Marik, Czech Technical Univ., Czech Republic
F. Meo, Fidia S.p.A, Italy
E. Oztemel, Sakarya University, Turkey
V. Raja, University of Warwick, UK
T. Schlegel, Fraunhofer IAO, Germany
R. Teti, University of Naples Federico II, Italy
A. Thatcher, University of the Witwatersrand, S. Africa
X. Xie, INRIA, France

Special Session Organisers

A. Subic and X. Yu (Innovative Production Machines and Systems)
R. Teti (Intelligent and Competitive Manufacturing Engineering)
M. Yang and Z. Wang (Tangible Acoustic Interfaces - Tai Chi)

Session Chairmen and Co-Chairmen

N. Aspragathos, University of Patras, Greece
E. Brousseau, MEC, Cardiff University, UK
A. Colombo, Schneider Electric, Germany
D. D'Addona, University of Naples Federico II, Italy
E. Eldukhri, MEC, Cardiff University, UK
F. Feenstra, TNO, The Netherlands
B. Grabot, ENIT, France
E. Hohwieler, Fraunhofer IPK, Germany
A. Labib, University of Portsmouth, UK
N. Lagos, MEC, Cardiff University, UK
F. Meo, Fidia S.p.A, Italy
M. Packianather, MEC, Cardiff University, UK
V. Raja, University of Warwick, UK
A. Subic, RMIT University, Australia
A. Soroka, MEC, Cardiff University, UK
R. Teti, University of Naples Federico II, Italy
A. Thomas, MEC, Cardiff University, UK
D. Tsaneva, MEC, Cardiff University, UK
Y. Wang, MEC, Cardiff University, UK
Z. Wang, MEC, Cardiff University, UK
X. Xie, INRIA, France
M. Yang, MEC, Cardiff University, UK
N. Zlatov, MEC, Cardiff University, UK

Referees

A. Afify, MEC, Cardiff University, UK
M. Aksoy, King Saud University, Riyadh, Saudi Arabia
D. Alisantoso, Nanyang Technological University, Singapore
N. Aspragathos, University of Patras, Greece
B. Benhabib, University of Toronto, Canada
E. Brousseau, MEC, Cardiff University, UK
M. Castellani, Universidade Nova Lisboa, Portugal
E. Charles, MEC, Cardiff University, UK
L. Cragg, University of Essex, UK
D. Creighton, Deakin University, Australia
E. Eldukhri, MEC, Cardiff University, UK
F. Feenstra, TNO, The Netherlands
C. Geisert, Fraunhofer IPK, Germany
M. Kraft, Fraunhofer IPK, Germany
N. Lagos, MEC, Cardiff University, UK
J. Lazansky, Czech Technical University, Czech Republic
K. Mahajan, University of Paderborn, Germany
S. Mekid, University of Manchester, UK
F. Meo, FIDIA, Italy
M. Mostafa, Zagazig University, Egypt
D. Müller, Clausthal University of Technology, Germany
S. Nahavandi, Deakin University, Australia
D. Pham, MEC, Cardiff University, UK
G. Putnik, University of Minho, Portugal
V. Raja, University of Warwick, UK
Y. Rao, Huazhong University of Science and Technology, China
T. Rusu, Petru Poni Institute of Macromolecular Chemistry, Romania
G. Ruz, University of Chile, Chile
S. Sagiroglu, Gazi University, Turkey
O. Sauer, Fraunhofer IITB, Germany
R. Setchi, MEC, Cardiff University, UK
A. Soroka, MEC, Cardiff University, UK
T. Strasser, PROFACTOR Produktionsforschungs GmbH, Austria
T. Szecsi, Dublin City University, Ireland
A. Thomas, MEC, Cardiff University, UK
D. Tsaneva, MEC, Cardiff University, UK
S. Walsh, Deakin University, Australia
K. Wang, Norwegian University of Science and Technology, Norway
X. Xie, INRIA, France
M. Yang, MEC, Cardiff University, UK
A. Zacharia, University of Patras, Greece

Facilitators

F. Abdul Aziz, MEC, Cardiff University
A. Afify, MEC, ibid
M. Al Kutubi, MEC, ibid
R. Barton, MEC, ibid
S. Bigot, MEC, ibid
E. Charles, MEC, ibid
A. Ghanbarzadeh, MEC, ibid
Z. Ji, MEC, ibid
E. Koc, MEC, ibid
C. Le, MEC, ibid
H. Liu, MEC, ibid
M. Mahmuddin, MEC, ibid
V. Mineva, MEC, ibid
A. Noyvirt, MEC, ibid
C. Pasantonopoulos, MEC, ibid
S. Rahim, MEC, ibid
M. Ridley, MEC, ibid
G. Ruz, MEC, ibid
S. Sahran, MEC, ibid
B. Sha, MEC, ibid
M. Sourisseau, MEC, ibid
M.Suarez, MEC, ibid
W. Wan, MEC, ibid
Y. Wu, MEC, ibid

I*PROMS Central Coordination and MEC Teams

A. Glanfield
F. Marsh
M. Matthews
C. Rees
M. Takala

Contents

Intelligent Planning and Scheduling Systemsxi

Mechatronics

Reconfigurable Manufacturing Systems

Intelligent Production Machines and Systems
D. T. Pham, E. E. Eldukhri and A. J. Soroka (eds)

The EU FP6 I*PROMS Network of Excellence for Innovative Production Machines and Systems

D T Pham, E E Eldukhri, A Soroka, V Zlatanov, M S Packianather, R Setchi, P T N Pham, A Thomas and Y Dadam
Manufacturing Engineering Centre, Cardiff University, Cardiff CF24 3AA UK

Abstract

This article presents the essence of the I*PROMS Network of Excellence. It gives the rationale for Networks of Excellence, outlines the scope and structure of I*PROMS and summarises its programme of activities. The article also briefly describes the I*PROMS management arrangements and concludes with an assessment of the positive impact the Network has made in its first year of operation.

1. Rationale for Networks of Excellence

Manufacturing is a significant wealth generation sector, accounting for over 20% of the EU's Gross Domestic Product. To compete successfully in the global market, European manufacturing industry needs to be underpinned by well focused advanced production research. Because of the breadth of the field, commercial considerations, and the multi-nationalism of the EU, production research activities within it have been naturally fragmented.

There is potential to co-ordinate pre-competitive research for common benefit. Under its Sixth Framework Programme (FP6), the EU has introduced Networks of Excellence as a new 'instrument' to overcome fragmentation of European research and help shape the conduct of research in Europe. The operation of these networks is based on a joint programme of activities aimed principally at integrating the research activities of the network partners while also advancing knowledge on the topic.

2. The I*PROMS Network of Excellence

The EU FP6 Network of Excellence (NoE) for Innovative Production Machines and Systems (I*PROMS) was inaugurated in October 2004. I*PROMS integrates the production research activities of 30 research centres from 14 countries in Europe:

MEC, Cardiff University (UK) (Co-ordinator)
Profactor (Austria)
Czech Technical University in Prague (Czech Republic)
VTT (Finland)
CETIM (France)
ENIT (France)
INRIA (France)
Robosoft (France)
IAO Fraunhofer Institute (Germany)
IPA Fraunhofer Institute (Germany)
IPK Fraunhofer Institute (Germany)
Schneider Electric (Germany)
TUC (Germany)
University of Hannover (Germany)
University of Patras (Greece)
Dublin City University (Ireland)
CRF (Italy)
FIDIA (Italy),
University of Naples Federico II (Italy)

PIAP (Poland)
University of Minho (Portugal)
Fatronik (Spain)
Tekniker (Spain)
TNO (The Netherlands)
Sakarya University (Turkey)
University of Warwick (UK)
University of Cambridge (UK)
University of Manchester (UK)
University of Newcastle (UK).
University of Oxford (UK)

I*PROMS addresses production research in an integrated manner to help shape this research area and overcome fragmentation. By creating an EU-wide research community concentrating on future manufacturing concepts, processes and systems, I*PROMS acts as the main research hub within the EU for the whole area of production machines and systems.

I*PROMS adopts the knowledge-based 'Autonomous Factory' vision for delivering increased competitiveness for manufacturing in 2020. The Network focuses on intelligent and adaptive production machines and systems that meet dynamic business and value drivers through advanced Information and Communication Technology.

I*PROMS promotes the development of common concepts, tools and techniques enabling the creation and operation of flexible, re-configurable, sustainable, fault-tolerant and eco- and user-friendly production systems. Such systems should react rapidly to changing customer needs, environmental requirements, design inputs, and material/ process / labour availability to manufacture high quality, cost-effective products.

3. The Clusters in I*PROMS

I*PROMS addresses six Manufacturing Challenges, namely, Concurrent Manufacturing, Integration of Human and Technical Resources, Conversion of Information to Knowledge, Environmental Compatibility, Reconfigurable Enterprises, and Innovative Manufacturing Processes and Products. Work on those themes is prosecuted by four interconnected clusters (see Table 1): Advanced Production Machines, Production Automation and Control, Innovative Design Technology and Production Organisation and Management.

The following sections give an outline of the I*PROMS clusters and their scopes. Further information can be found in [1, 2, 3].

3.1. Advanced Production Machines (APM) Cluster

Advanced production machines are the workmen of the factory of the future. These include machines for processing new/nano/smart/high-performance materials, micro-manufacture (MEMS) machines, rapid manufacturing machines (rapid prototyping and rapid tooling), and manufacturing robots (stationary and mobile). In addition to the challenging 'stand-alone' research issues connected to these technologies, this cluster also considers two key issues that require cross-technology co-operation, multi-function machines and reconfigurable machines.

Table 1
Manufacturing Challenges Addressed by I*PROMS Research Clusters

Manufacturing Challenge	**Network Clusters**			
	APM	*PAC*	*IDT*	*POM*
Concurrent Manufacturing	X	X	X	X
Integration of Human and Technical Resources		X	X	X
Conversion of Information to Knowledge		X	X	X
Environmental Compatibility	X		X	
Reconfigurable Enterprises	X	X		X
Innovative Manufacturing Processes and Products	X			

3.2. Production Automation and Control (PAC) Cluster

Production Automation and Control represents the foremen of the factory of the future, responsible for overseeing the machines and for communicating between machines and management. The scope of the PAC cluster includes self-adaptive control, flexible/reconfigurable manufacturing, adaptive quality systems, agent-based distributed architectures, (machine) knowledge management, and human-machine interaction. The major research issue identified for this cluster is collaborative agent-based (or holonic) manufacturing automation, and related to that, self-adaptive control and human-machine interaction. There is considerable interaction between this cluster and the APM cluster in making these machines more 'intelligent' and between this cluster and the Production Organisation and Management cluster in controlling these machines within the overall Autonomous Factory environment. It is the combination of these cluster areas that will ultimately provide the overall flexible/reconfigurable Autonomous Factory.

3.3. Innovative Design Technology (IDT) Cluster

Innovative Design Technology relates to the designers of the factory and products of the future, conceiving novel, customisable, value-added products, and the factory machines and systems required to manufacture them. The coverage of the IDT cluster includes product knowledge management, computer-aided innovation, and advanced computer-aided manufacture. It is with appropriate IDT that concurrency through the product life-cycle, as well as across the extended enterprise, can be achieved, with huge potential impact on time-to-market and product development, manufacturing, maintenance and disposal costs. Two key areas of work are identified: virtual product design technology and design complexity.

3.4. Production Organisation and Management (POM) Cluster

Production Organisation and Management concerns the management of the factory of the future. It covers advanced process control, enterprise/manufacturing simulation, (human) knowledge management, and human-computer interaction at the organisational, enterprise level. This cluster deals with the management of operations ranging from innovation, planning, organising, scheduling and controlling of production processes as well as the management of the interface with the extended enterprise and other supporting functions. Two key themes are identified: Fit Manufacturing enterprises (both individual and extended) and Integration of human and technical resources enabling effective collaboration between man and machine. This includes effective human-computer interaction by means of multimodal/multimedia knowledge-based user interfaces, with active knowledge access and adaptive, dynamic knowledge presentation.

4. I*PROMS Activities

The work of the I*PROMS Network and its clusters is laid out in a comprehensive joint programme of activities (JPA). The JPA comprises the following components:

- A set of integrating activities aimed at structuring the way the participants carry out research on the topics considered. These activities include: defining common research priorities within the scope of I*PROMS; creating and operating joint research platforms; preparing joint bids to government and industry for research grants; creating communication and collaboration facilities; training Network researchers; developing synergistic links with existing research capacities within the EU.
- A programme of jointly executed research to support the Network goals. The research undertaken by the four I*PROMS clusters covers a broad range of topics including new processes for new materials, miniaturisation, mechatronic modules, nanotechnology, modelling and simulation, product life cycle planning, flexible manufacturing systems, process integration, new process control and sensors concepts, intelligent manufacturing process/near-net shape processes, and substitution of harmful substances.
- A set of activities designed to spread excellence. These include: training of manufacturing engineers; promoting mobility and exchange of researchers from outside the network; transferring knowledge to teams external to the network through networking activities; networking with other research networks; transferring technology through web-based advisory services; interacting with national and EC-funded technology transfer projects for industry including SMEs; organising events to increase public awareness and understanding of I*PROMS-related science and technologies; publishing the Network's vision for the factory of the future; developing a programme of industrial workshops, seminars and conferences; collaborating with other related centres of excellence outside of the network to ensure that knowledge and excellence is spread widely.

5. Network Management

I*PROMS is managed through a three-tiered structure. At the top of the structure, there is a Governing Council (GC). The GC is the decision-making and arbitration body of I*PROMS concerned with policy and strategy. The Executive Board (EB) forms the second tier of management. It has powers delegated to it by GC to implement the policy and strategy decisions of the latter in a unified and efficient manner across the whole Network. The day-to-day management of individual clusters is carried out by cluster Executive Committees (ECs) constituting the third tier. ECs are empowered to make operational decisions within constraints agreed by the EB. In addition to the three management tiers, there are two advisory groups, the Scientific Council (SC) and the Industrial Council (IC). The SC is made up of eminent research leaders who provide input to GC and EB on the science and technology direction of I*PROMS. The IC is formed of senior industrialists who help steer the work of I*PROMS from the point of view of its industrial relevance. To ensure I*PROMS operates in a truly integrated manner, strong central co-ordination is provided by a permanent Central Co-ordination Team (CCT) based at the co-ordinating institution. Led by a full-time Network Manager (NM), CCT performs network-wide functions such as internal and external communication, marketing, fund-raising, co-ordination and financial administration. The Team is responsible to the Network Director (ND) representing the Co-ordinator. Measures to achieve tight co-ordination include making the Network Manager the Secretary of GC and EB and a representative of the Co-ordinator is assigned to act as a member of each of the four cluster ECs[1][2].

6. Impact of I*PROMS

I*PROMS has been operating for less than a year. Early indications are that I*PROMS is already making a strong positive impact through its IPROMS 2005 Virtual International Conference on Intelligent Production Machines and Systems. IPROMS 2005 was one of a series of I*PROMS events aimed at spreading excellence within the EU and beyond. The Conference attracted contributions from authors based in 32 countries across five continents. Between 4 and 15 July 2005, some 4168 registered delegates and guests from 71 countries participated in the Conference, making it a truly global phenomenon. Through such high-profile activities, it is hoped that I*PROMS will attract production researchers internationally to assist it in achieving its goal of being the world's preeminent organisation for research in Innovative Production Machines and Systems.

Acknowledgement

I*PROMS is funded by the European Commission under its Framework 6 Programme.

References

[1] Pham D.T., Eldukhri E.E., Peat B., Setchi R., Soroka A.J., Packianather M.S., Thomas A., Dadam Y., Dimov S. Innovative Production Machines and Systems (I*PROMS): A Network of Excellence Funded by the EU Sixth Framework Programme. Industrial Informatics 04, Berlin, (2004) 540-546

[2] Pham D.T., Eldukhri E.E., Setchi R., Soroka A.J., Packianather M.S., Thomas A., Dadam Y., Dimov S. Integrating European Advanced Manufacturing Research: The FP6 I*PROMS Network of Excellence. Intelligent Computation in Manufacturing Engineering 4, (2004), 543-548

[3] Innovative Production Machines and Systems Network of Excellence (I*PROMS NoE) [online]. Available from: http://www.iproms.org

Intelligent Production Machines and Systems
D. T. Pham, E. E. Eldukhri and A. J. Soroka (eds)

A robustness index for scheduling based on sensitivity analysis

M.E. Aydin

Faculty of Business, Computing and Information Systems, London South Bank University, London, SE1 0AA, UK

Abstract

This paper addresses the sensitivity analysis to test the robustness of schedules produced for job shop scheduling problems. The idea is to check the level of tolerance of the schedule against the changing conditions such as the length of processing time for particular operations. We introduced a new dimension of robustness that has not widely been discussed so far.

1. Introduction

Job shop scheduling is considered one of the most important production scheduling problems due to its representative character. On the other hand, it is quite restricted as many assumptions are held for simplification, which abstract the problems from the real life. That is why, a solution produced for a job shop scheduling problem remains not applicable and very fragile even it is well optimised. The main issue here is to produce solutions as robust as possible so that they can be applied to real problems. For this purposes, re-scheduling approaches have arisen whereas most of them remain expensive to apply.

A sensitivity analysis for job shop scheduling (JSS) solutions is proposed and discussed in this paper in order to attack to the robustness problem and investigate a safe and confident base to examine the solutions. Sensitivity analysis is one of very well-known operational research methodologies to test the solutions against "what if" questions. It has been discussed for a flow-shop type scheduling problems quite widely by Hall and Posner [1]. However, sensitivity analysis for JSS problems has not been considered with respect to robustness.

Robustness is such a term that a wide consensus has not reached on, yet. Some authors mean by robustness that a solution is as robust as it meets to the requirements of various performance measures simultaneously [2]. On the other hand, many others mean the tolerability of the solution against the changes in environmental conditions [3]. In our point of view, the robustness is the tolerability of a solution to changing conditions within the environment. Specifically, robustness is the strength of a solution for a particular JSS problem against changing in the length of processing time of a particular operation.

The changes frequently expected within the environments are of machine failure, staff shortages and change of priorities; each constitutes various difficulties in applicability of the solutions. Since JSS problems are very well defined and structured, any release of assumptions will change the problem type to another scheduling problem. If the intention is to keep holding the main assumptions, which are fundamental to keep the problem in the form of JSS, we will only be allowed to work out the processing times, where any change of environmental condition incurs extension or shrinkage in processing times.

The rest of this paper is organised as follows: Basics of sensitivity analysis is presented in Section 2 while job shop scheduling problems are introduced in Section 3. Section 4 discusses a possible sensitivity analysis for job shop scheduling problems. The relation between sensitivity analysis and robustness is presented in Section 5 and conclusion is included in Section 6.

2. Basics of Sensitivity Analysis

In classical operations research, the optimal solutions are subjected to sensitivity analysis so as to understand their strength (robustness) under changing conditions. The conditions might be changed because of the adding or removing constraints, the extension or reduction in resources, or the changing of priorities. In scheduling problems, the most common issues are the changing priorities and/or the emerging shortage in resources. Obviously, it is not easy to suggest a generic method to cover all changes in circumstances and meet all new conditions.

Sensitivity analysis has emerged as a key issue of verifying linear models with respect to post optimality robustness. The subject is expressed with duality and shadow prices of linear optimisation models. Suppose we are given the following linear model:

$$Max\ F(x) \quad = \quad cx \qquad (1)$$

Subject to :

$$Ax = b \qquad (2)$$

$$x \geq 0 \qquad (3)$$

where x is the subject variable, c is the vector of coefficients, A is a matrix of constraints and b is the vector of boundaries of constraints. The tuple of (c,A,b) are considered as the parameters of the model. The concern of sensitivity analysis is to test the flexibility of the optimum solution with the change in either of the parameters. The new tuple is expected to be any form of $\{(c+\Delta c),A,b\}$, $\{c,(A+\Delta\ A),b\}$, $\{c, A,(b+\Delta b)\}$, ..., $\{\ (c+\Delta c),\ (A+\Delta A),\ (b+\Delta b)\}$. That means any change taking place in the parameters separately or concurrently. Therefore the new model turns to be:

$$Max\ F(x) \quad = \quad (c + \Delta c)x \qquad (4)$$

Subject to :

$$(A + \Delta A)x = (b + \Delta b) \qquad (5)$$

$$x \geq 0 \qquad (6)$$

The expectation is to identify the boundaries of each parameter that will not harm the optimality. If any excessive change happens to the parameters, than the optimality will be lost and the model will need re-optimisation. Another issue is the change would take place in the subject variable (x), that will directly lead to remodelling the problem and re-optimisation.

3. Job shop scheduling

Job Shop Scheduling (JSS) problems have been worked for a long time. It is difficult to reach the optimal solution in short time, since the problems have a very wide solution space and there is no guarantee to move to a state better than the one before [9].

We are given a set of jobs *(J)* to be processed in a set of machines (*M*) subject to a number of technological constraints. Each job consists of m operations, O_j, which must be processed on a particular machine, and each job visits each machine once. There is a predefined order of the operations within a job. Because of this order, each operation has to be processed after its predecessor (PJ_j) and before its successor (SJ_j). Each machine processes n operations in an order that is determined during the scheduling time, although there is no such order initially. Therefore, each operation processed on the M_i has a predecessor (PM_i) and a successor (SM_i). A machine can process only one operation at a time. There are no set-up times, no release dates and no due dates.

Each operation has a processing time (p_{ij}) on related machine starting on r_{ij}. The completion time of o_{ij} is therefore: $c_{ij} = r_{ij} + p_{ij}$ where $i = (1,...,m)$, $j=(1,...,n)$ and $r_{ij} = \max(c_{iPJ_j}, c_{PM_ij})$. Each machine and job have particular completion times, which are identified as: $C_{M_i} = c_{in}$ and $C_{J_j} = c_{jm}$ where c_{in} and c_{jm} are the completion time of the last (n^{th}) operation on i^{th} machine and the completion time of the last (m^{th}) operation of jth job, respectively. The overall objective is to minimise the completion time of the whole schedule (makespan), which is the maximum of machine's completion times, $C_{\max} = \max(C_{M_1},....,C_{M_m})$. The representation is done via a disjunctive graph, as it is widely used.

A job-shop-scheduling problem can be represented

by a disjunctive graph. The tasks are the nodes in the graph, each with a single attribute representing the duration. Two dummy nodes are introduced: the start node and the end node, each with duration 0. Each precedence constraint is represented by a directed arc, from a task to its successor. Additional arcs are added from the start node to the first task in each job, and from the last task in each job to the end node. Each resource constraint is represented by a bi-directional arc (or *disjunctive* arc). Selecting one of the two directions for a disjunctive arc imposes an ordering on the two tasks concerned. Selecting an orientation for every disjunctive arc such that there are no cycles in the graph reduces the disjunctive resource constraints to precedence constraints. Given a fully oriented graph, the minimum makespan for that graph can be found by computing the longest path from start node to the end node, where the length of an arc is equal to the duration of the task that starts the arc. The scheduling problem thus reduces to one of finding orientations for all the disjunctive arcs such that the least makespan can be obtained [10]. Local search methods can operate by changing the orientation of some of the disjunctive arcs, and re-computing the minimum makespan. It has been shown [10] that the makespan can only be reduced by changing the orientation of one of the disjunctive arcs on the longest path.

4. Sensitivity analysis for schedules of job-shop environment

Job shop scheduling is a special type of production scheduling, where a set of assumptions are applicable to the problem models in order to keep it simplified.

The idea of classical sensitivity analysis can be applied to job shop schedules. The reason to apply that idea to job shop problems is to examine the flexibility of optimised schedules, and to reveal their tolerability to changing conditions. The likely changeable conditions are machine failure, staff shortages, and the priorities such as due date tightened, lateness in arrivals etc. Each condition changed causes some sort of difficulties in applicability of schedules. Staff shortages incur prolonging processing times, while changes in priorities affect the optimality of schedules. The worst case for job-shop environments is machine failure, as it locks up of the whole schedule. For that reason, we ignore machine failure in job shop environments. Likewise, we suppose the priorities of jobs do not change as the arrival times and due dates are set constant for simplicity.

In job shop schedules, the order of operations on particular machines are the subject variable undertaken. The rest of figures are to be considered as parameters. The parameters to be considered are mainly due dates, arrival times, and processing times. Since job shop scheduling problems are set due dates and arrival times constant and ignorable, the key parameter remaining is processing time and the makespan is adopted as performance measure (fitness). The rest of performance measures such as tardiness are not the case since arrival times and due dates are assumed unchangeable

Since job shop scheduling problems are ill structured, it is difficult to model the problems with a linear programming sense. Although the objective function can be clear enough to formulate in an algebraic function, it is hard to quantify some of the constraints.

The typical issues fall in the intersection area of sensitivity analysis and JSS with respect to robustness will be as follows:

- What if the processing time of a particular operation is increased/ decreased?
- What if the processing time of a critical operation is increased/ decreased?
- What if the processing time of a non-critical operation is increased/ decreased?
- What are the upper and lower bounds of the change in processing time of non-critical operations to keep the optimality?

Based on the sensitivity of the solutions investigated by such questions, a robustness index can be built so as to reveal applicable solutions and offer them for application.

5. Measuring Robustness of Schedules

Since it is a qualitative concept, it is hard to measure robustness. In the following section, a new index is suggested for that purpose.

5.1 Background

The history of scheduling problems goes back to the start of mass production. Since then a numerous literature on the subject has accumulated and many ideas and approaches have been discussed and implemented. Optimisation algorithms have produced better and better results on the job shop scheduling problem in its restricted and reduced, abstract form but the results produced are often very fragile and far from meeting the real needs of applicability [7]. On one hand, the results may be the best, or very close to them, for abstract problems but planning and manufacturing engineers are not able to use them because they are not tolerant to changing circumstances such as machine failure, unexpected staff shortages, etc., it is necessary to produce more adaptable and robust solutions to job shop scheduling problems to provide practitioners with the flexibility needed in the real world as argued by Kuroda *et. al.*[6]. On the other hand, the problem is solved with respect to a particular criterion, but does not provide impressive result in terms of others. For instance, the schedule optimised may provide the best result for makespan, however it does not mean that it will be impressive with respect to mean tardiness, or numbers of tardy jobs, which attract very much interest from industry.

Some research has been done on producing robust and adaptable schedules considering both points mentioned above. Kuroda et al [6] suggest a due-date based flexibility for more robust schedules. Wang [3] built a fuzzy robustness measure based on qualitative possibility theory for project scheduling problems using fuzzy set theory. Hart *et al* [8] have suggested an artificial immune system to provide a robust scheduling system in which they concentrate on the conditions of job arrivals for a dynamic job shop environment. Their idea is to build a number of libraries, each containing a number of genetic strings, each string being a part of a solution for job-shop problems. They evolve the libraries to have a reserve of schedules based on change of arrival conditions and/or due dates. The main benefit of this work is the arrangement of job priorities. Jensen [5] has addressed a neighbourhood-based robustness for job shop scheduling problems. He measures the robustness by using the methodology produced by Jensen and Hensen [4] who have discussed a hamming distance-based fitness landscape of job shop scheduling problems to look for a valley on the landscape that gives better rescheduling conditions. He has examined neighbourhood-based robustness for various fitness functions [5].

5.2 The robustness index

In this paper, robustness is considered as the strength and the tolerability of a schedule against changing environmental conditions. The concept of "robustness" is qualitative by nature and, therefore, it is hard to measure it. The aim is to build a quantitative index based on the sensitivity analysis. As mentioned in the previous section, solutions for job shop scheduling problems are mainly examined with respect to changing conditions such as extension in processing times, machine failure etc. Sensitivity analysis allows examining the solutions with respect to many issues regarding the possible changes. As a result, it reveals whether the solution keeps its applicability or not with respect to feasibility and quality of solution.

A weighted index is developed as a function of the issues considered. Suppose that we are given k number of issues considered in sensitivity analysis. Then, the index can be formulated as a summation function as follows:

$$R = \sum_{i=1}^{k} w_i r_i \qquad (7)$$

where w_i and r_i denote weight and the robustness of the ith issue examined, respectively. R measures the robustness of the whole schedule examined with respect to multiple issues. This function may be a product, or a trigonometric function as well. The idea is to quantify the results gained from the analysis of the solution.

The issues considered can be the extension of processing times as the reflection of delays and staff shortages. It is not meaningful to consider machine failure sort of changes, as it breaches the main assumptions of job shop scheduling problems. Furthermore, because of the main assumption that assumes every job visits every machine, machine failure will convert the feasibility of the schedule to absolute infeasibility. Due to all these reasons, we prefer not to consider machine failure in sensitivity analysis.

On the other hand, the operations with processing time prolonged are two type; critical operations, non-critical operations. Critical operations directly affect on makespan, and any extension on such operations breaks the optimality. However, non-critical ones may not change the optimality as it affects indirectly. So, the robustness of each issue should be reflected to the function, appropriately. The weight of the issues related to the critical operations will be much higher than non-critical ones. If the machine failure is considered in anyway, its weight must be overwhelming.

Once the function set up, the results come out will reflect the robustness of the schedule. In order to have a proper evaluation, it needs to be compared to some tabulated values, which can be generated on problem base. There can be robustness categories such as "low", "middle", and "high" level, which can be determined based on the comparison between the value calculated and the ones tabulated.

6. Conclusion

In this paper, we discuss the possibility of measuring the robustness of schedules, especially job shop scheduling solutions. We introduced a way to build a robustness index based on the issues considered for sensitivity analysis. This will produce a healthier measure for the solutions subject to conditional changes. The research undertaken is going on.

References

[1] N.G. Hall and M. E. Posner, (2004), "Sensitivity analysis for scheduling problems", *Journal of Scheduling,* 7, 49-83.

[2] M.T. Jensen, (2001), "Improving robustness and flexibility of tardiness and total flow-time Job shop using robustness measures", *Journal of Applied Soft Computing*, 1.

[3] J. Wang, (2004), "A fuzzy robust scheduling approach for product development projects", *European Journal of Operational Research*, 152, 180-194.

[4] M. T. Jensen and T. K. Hansen (1999), "Robust solutions to job shop problems" *in Proceedings of the Congress on Evolutionary Computation*, pp:1138-1144, 1999.

[5] M.T. Jensen, (2000), "Neighbourhood based robustness applied to tardiness and total flowtime", in Schoenauer M. *et. al.* PPSN VI, *Lecture Notes in Computing Science*, 1917 (2000), 283-293.

[6] M. Kuroda, H. Shin, and A. Zinnohara, (2002), "Robust scheduling in an advanced planning and scheduling environment", *International Journal of Production Research*, 40/15 (2002), 3655-3668.

[7] M.A.B. Candido, S. K. Khator, and R.M. Barcia, (1998), "A genetic algorithm based procedure for more realistic job shop scheduling problems", *International Journal of Production Research*, 36/12 (1998), 3437-3457.

[8] E. Hart, P. Ross and J. Nelson, (1997), "Producing robust schedules via an artificial immune system" In *Proceedings of the IEEE World Congress on Computational Intelligence*, 1997.

[9] K. R. Baker, , *Introduction to Sequencing and Scheduling*, John Wiley & Son, 1974.

[10] E. Balas, , "Machine sequencing via disjunctive graphs: an implicit enumeration algorithm", *Operations Research*, 17, 941-957, 1969.

Intelligent Production Machines and Systems
D. T. Pham, E. E. Eldukhri and A. J. Soroka (eds)

Computer aided procedure for selecting technological parameters in the drawing process

A.M. Camacho, E.M. Rubio, M.A. Sebastián

[a] *Department of Manufacturing Engineering, National Distance University of Spain (UNED)*
c/ Juan del Rosal, 12, E-28040 Madrid, Spain

Abstract

The ability to predict the performance of a particular manufacturing process is not always easy. Many factors are usually involved, and their influence can be critical in some cases. In drawing process there are several parameters that are related to the necessary energy to perform the process. However, the knowledge about the influence of these parameters is already scarce. Although there are several analysis techniques, the Finite Element Method has become an indispensable tool for analysing metal forming problems. In this study the drawing process has been simulated for different values of reduction, friction, die semiangle and back tension. The results show the influence of these parameters on the energy employed to develop the process.

---

1. Introduction

The main objective in manufacturing is the production of high quality products at minimal cost and time. The ability to predict the performance of a particular manufacturing process is not always easy. Many factors are usually involved, and their influence can be critical in some cases [1].

In metal forming processes, numerous analytical techniques have been developed to improve the engineer's ability to evaluate and to predict several aspects of the metal forming process [2,3,4]. Early methods are based on simple analytical techniques, such as the Homogeneous Deformation Method, the Slab Method, or the Upper Bound Theorem.

Later (early 1970), the Finite Element Method became a prevalent analysis technique for metal forming applications [5,6] due to the advance in computer technology. This method has been implanted in a widespread way in the engineering and manufacturing industry because of its unique capability in describing complex geometries, boundary conditions and realistic materials response.

Drawing process is the most typical stationary metal forming process. Drawing process consists of reducing or changing the cross-section of a preform to obtain products as wires, rods, bars or plates, making them pass through a die by means of a pulling force.

Plate drawing can be analysed under plane strain conditions; while wire, rod and bar drawing can be studied as axisymmetrical problems.

The typical parameters involved in this process are the cross-sectional area reduction, r, the friction coefficient, μ, (that represents the friction along the die-workpiece interface), and the die semiangle, α, (see Fig. 1).

Nevertheless, other factors such as the back tension, σ_{BP}, or the die shape can affect to the mechanics of the process, and to the properties of the final product. All these variables are mainly related to the energy that is necessary to perform the process.

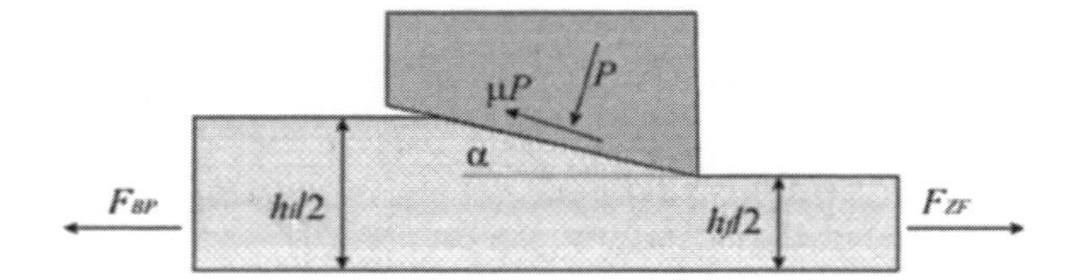

Fig. 1. Technological parameters of the drawing process.

Three are the main contributions to the total energy that is necessary to carry out any metal forming process. These are:

$$E_T = E_{HD} + E_F + E_{ID} \quad (1)$$

where:

E_T, is the total energy to carry out the process

E_{HD} is the energy due to homogeneous deformation

E_F, is the energy due to friction

E_{ID}, is the energy due to internal distortion

In this work, the drawing stress, σ_{zf}, has been calculated for different values of the parameters that were described above. This variable is very useful because it can be assimilated to the specific energy necessary to perform the drawing process.

2. Analytical methods

The Homogeneous Deformation Method (HDM) only takes into account the first contribution, E_{HD}. The HDM provides a drawing load estimate, which is useful in selecting the size of equipment required to form the product [7]. Its analytical expression is:

$$\sigma_{ZF} = 2k \ln\left(\frac{1}{1-r}\right) \quad (2)$$

for plane strain problems, and

$$\sigma_{ZF} = Y \ln\left(\frac{1}{1-r}\right) \quad (3)$$

for axisymmetrical problems.

In these expressions k is the shear yield stress, and Y is the yield stress.

The Slab Method (SM) is based on the force equilibrium of a differential element (slab) in the deformation zone. Figure 2 illustrates an example of the slab method applied to a plate drawing process, assuming a Coulomb friction model.

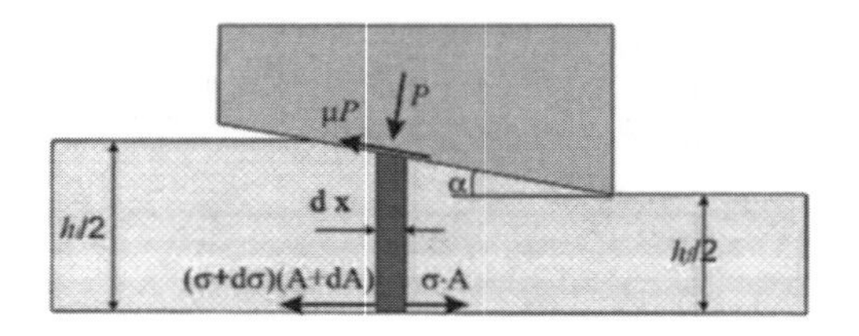

Fig. 2. Analysis of drawing process by the Slab Method.

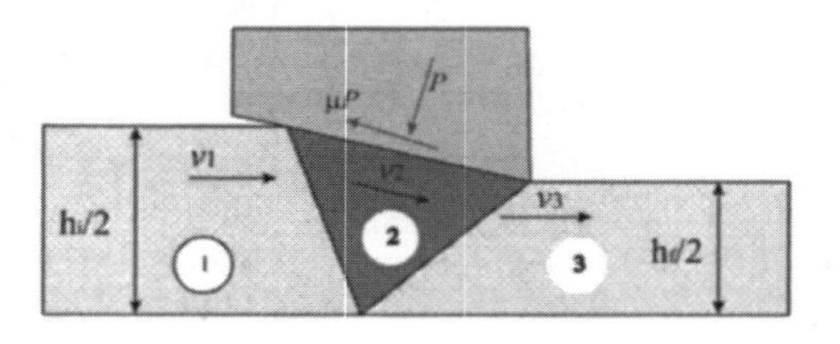

Fig. 3. Analysis of drawing process by the Upper Bound Theorem.

The SM (also called Sachs Method [9]) evaluates the energy due to homogeneous deformation and also the energy due to friction, but the internal distortion is not included in its analytical expression:

$$\sigma_{zf} = \frac{2k(1+B)}{B}\left\{1-\left(\frac{h_f}{h_i}\right)^B\right\}+\left(\frac{h_f}{h_i}\right)^B \sigma_{zi} \quad (4)$$

$$\sigma_{zf} = \frac{Y(1+B)}{B}\left(1-\frac{r_f}{r_i}\right)^{2B}+\left(\frac{r_f}{r_i}\right)^{2B} \sigma_{zi} \quad (5)$$

These are the equations of the SM for plane strain and axisymmetrical problems, respectively; being the factor $B = \mu/\tan\alpha$

Unlike the two previously discussed methods, the Upper Bound Theorem (UBT) is based on an energy principle. The three contributions of the energy are evaluated. In the UBT case, the deformation zone has been modeled by an only triangular rigid zone [10] (Figure 3). Although other suitable models exist, this one fits well to the results found on the classical references about these themes. The election is based on the simplicity of this model and on its robustness versus variations of vertex positions. The applied equation is:

$$\frac{\sigma_{zf}}{2k} = \frac{\overline{AB}\Delta v_{12} + \overline{BC}\Delta v_{23}}{h_f\left[v_3 - \frac{\mu}{sen\alpha + \mu\cos\alpha}v_2\right]} \quad (6)$$

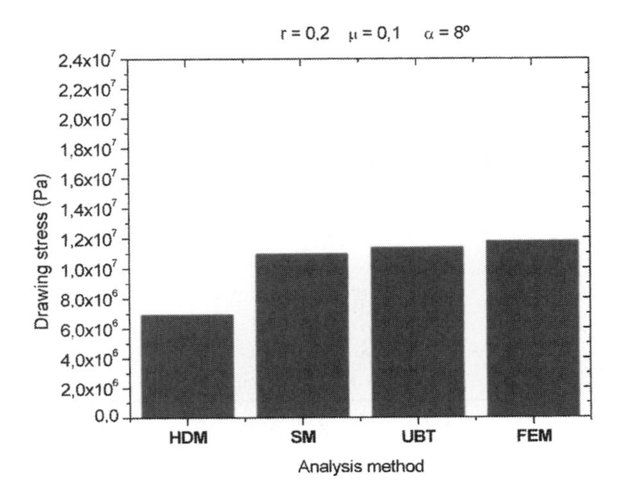

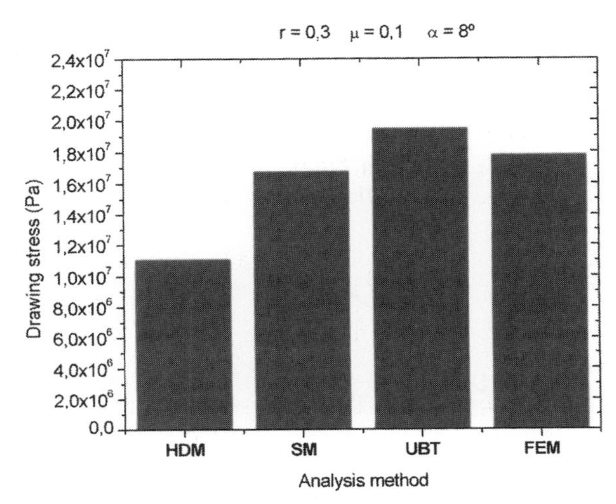

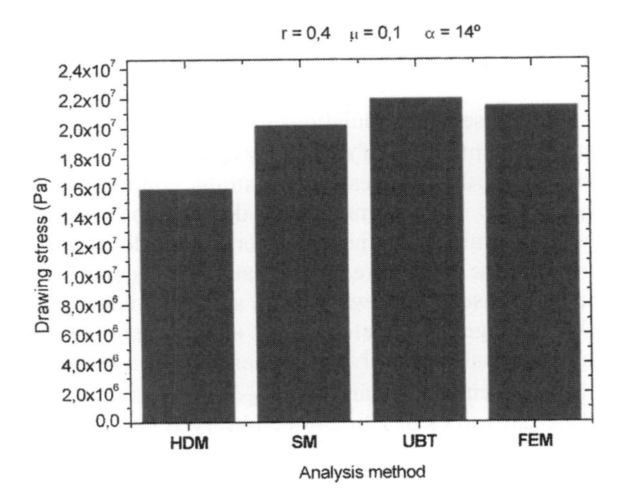

Fig. 4. Results obtained by several methods.

Table 1
Parameters values range

r	α (°)	μ	σ_{BP} (Pa)
0,20 - 0,40	4° - 18°	0 - 0,2	0 - $1{,}67 \cdot 10^7$

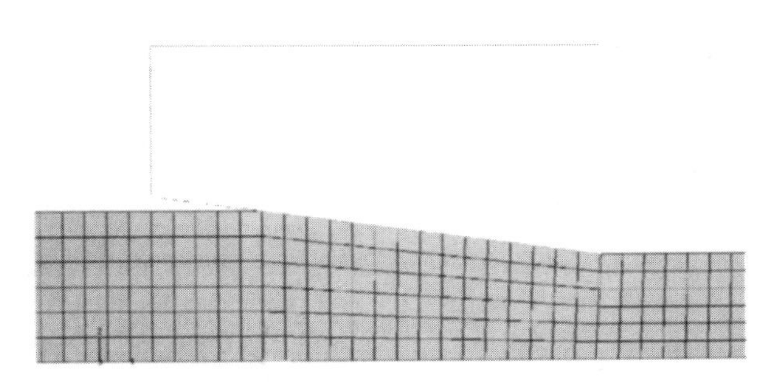

Fig. 5. Detail of the mesh.

Unlike the Finite Element Method (FEM), these methods (HDM, SM and UBT) are basically employed for obtaining the forming load, and its application is generally limited to metal forming processes with simple geometries and idealized material laws. However, nowadays, the application of the UBT is being extended to complex profiles [11].

The results obtained with the methods that have been exposed are shown in Figure 4, for different values of the parameters r, μ and α.. Since FEM is the most accurate one, quantitative differences between FEM and the rest of methods can be observed, what demonstrates the limitations of these methods in the calculation of the drawing force, mainly HDM and SM.

3. Analysis by FEM

The Finite Element Method has become a useful tool for analysing this type of problems. This powerful method take into account the three contributions to the total energy and, besides, phenomena such as the strain hardening and temperature can be considered. Next, this analysis technique is going to be employed to evaluate the influence of certain parameters in the calculation of the drawing force. First, the cross-sectional area reduction, r, that defines the difference between the initial and final height, and it is generally imposed by the specifications of the final product. The following parameter that have been considered is the friction coefficient, μ, and it is controlled by the lubrication mechanism. The die semiangle, α, is a geometrical parameter that is related to the shape of the die.

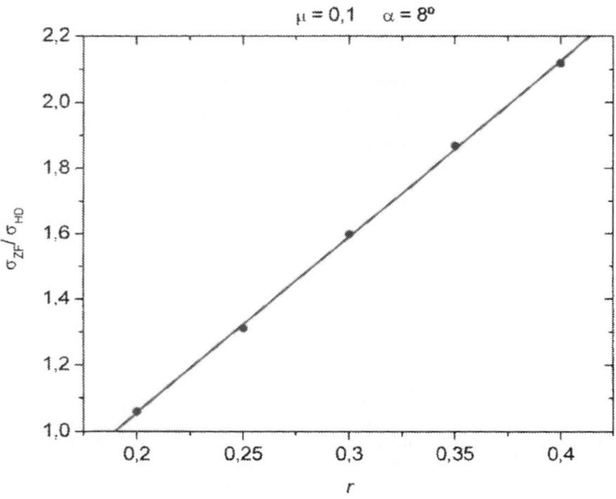

Fig. 6. Influence of the parameter r.

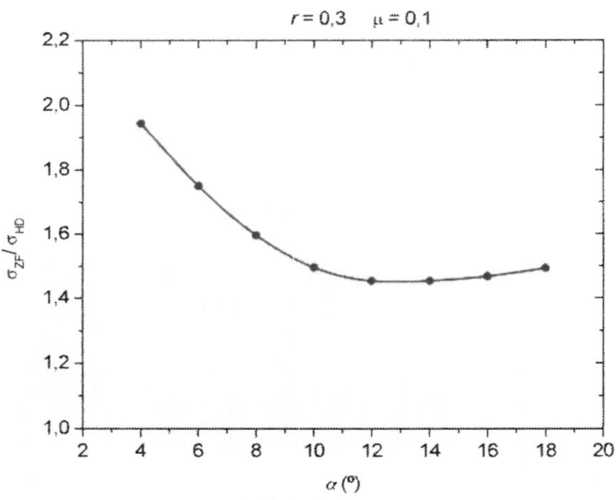

Fig. 8. Influence of the parameter α.

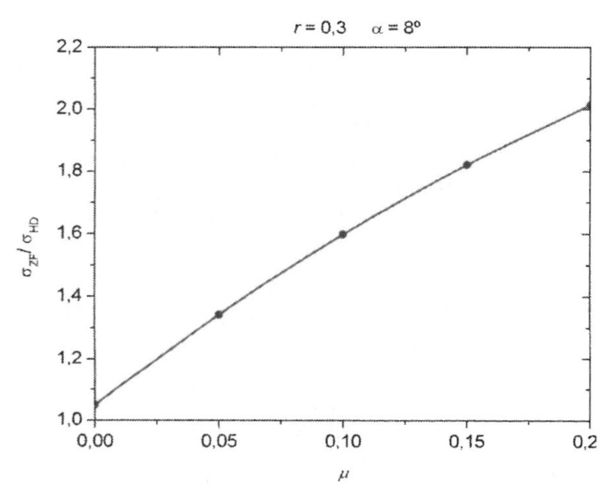

Fig. 7. Influence of the parameter μ..

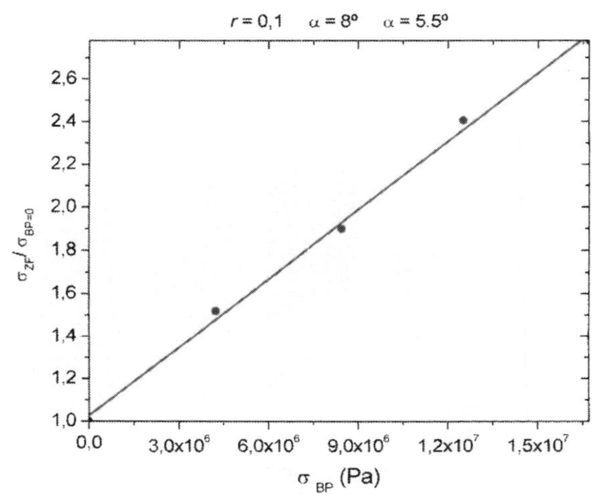

Fig. 9. Influence of the parameter σ_{BP}.

And finally, an external parameter, the back tension, σ_{BP}, that is sometimes applied by external devices, but almost always is imposed involuntarily by the previous capstans. For this purpose, a reference situation is defined by the following conditions: $r = 0,3$, $\mu = 0,1$, $\alpha = 8°$ and $\sigma_{BP} = 0$. Variations of each parameter are imposed for calculating the corresponding values of drawing mean stress. The range of values that have been used in the simulations by FEM appears in Table 1. The numerical analysis has been carried out by a general-purpose finite element software: ABAQUS/Standard [12]. Continuum, 4-node, reduced-integration elements have been used for meshing the workpiece (Fig. 5). The continuum (or solid) elements can be used for complex nonlinear analyses involving contact, plasticity, and large deformations.

4. Results and discussion

First, several simulations have been carried out to study the influence of r, μ and α in the drawing mean stress. In these cases, plane strain conditions are considered. Since homogeneous deformation energy is a lower limit for the necessary energy to develop the process, the results are given in an adimensional way. In this sense, the drawing mean stress is divided into the homogeneous deformation stress, therefore the obtained values are always higher than 1. In Figure 6, the adimensional variable σ_{ZF}/σ_{HD} is shown for different values of r. It can be observed that drawing stress increases with reduction, and the behaviour is practically linear.

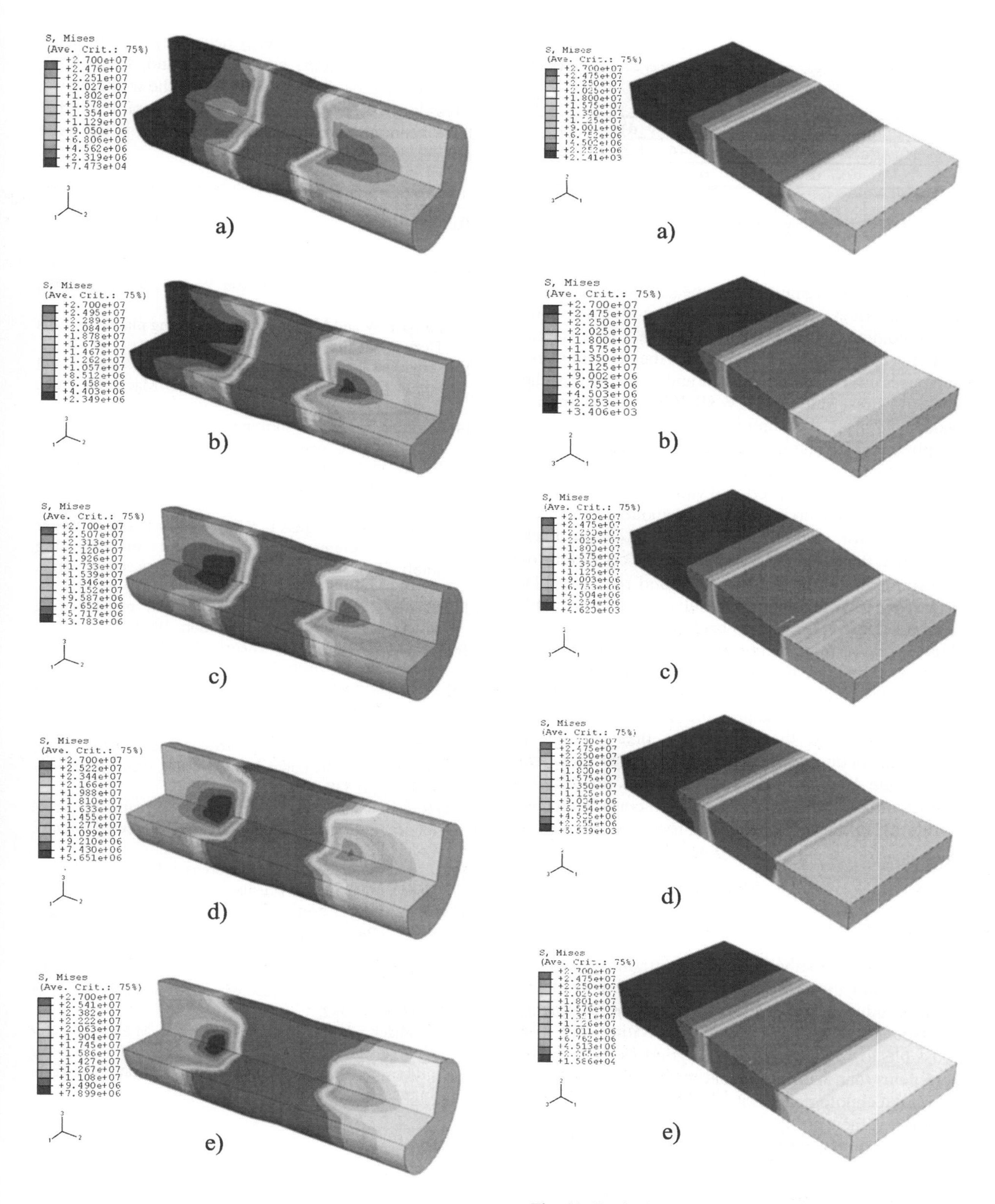

Fig. 10. Equivalent stress contours for $r = 0,1$, $\mu = 0,1$, $\alpha = 5,5°$ and different values of σ_{BP} (see Table 2).

Fig. 11. Equivalent stress contours for $r = 0,3$, $\alpha = 8°$, $\sigma_{BP} = 0$, and different values of μ (see Table 2).

Table 2
Values of μ and σ_{BP}, and correspondence with Figures 10 and 11

	a)	b)	c)	d)	e)
$\sigma_{BP}\cdot 10^7$ (Pa)	0	0,424	0,844	1,25	1,67
μ	0	0,05	0,10	0,15	0,20

Also, the friction between the die workpiece interface, μ, causes an increase of the drawing mean stress (as it can be seen in Fig. 7).

According to Figure 8, the die semiangle, α, has a different effect on the drawing stress than previous technological parameters. In fact, the adimensional drawing stress suffers a decrease as the die semiangle increases until a minimum value and, subsequently, the drawing stress starts to grow slightly. This minimum value defines the optimal die semiangle. The employment of optimal semiangles decreases the required energy to carry out the drawing process, so the knowledge of these critical values is highly useful.

Finally, an axisymmetrical model has been employed for studying the influence of the back tension at the die entry. In this case, the conditions of the simulation are $r = 0{,}1$, $\mu = 0{,}1$ and $\alpha = 5{,}5°$ [13]. Figure 9 shows the obtained results. As it is observed, the back tension (known as back pull effect) causes a high increase of the drawing stress [14]. However, some studies have demonstrated that back tension is desirable for prolonging the tool life, because the back pull effect decreases the pressure in the die [15].

Examples of the simulations that have been developed in this work are presented in Figures 10 and 11. Figure 10 shows equivalent stresses for the five values of back tension that have been considered (see Table 2). In the other hand, in Figure 11, equivalent stresses from $\mu = 0$ to $\mu = 0{,}2$ are shown.

4. Conclusions

Analytical methods have been traditionally used for analysing metal forming processes. Since FEM was first introduced, this powerful method has become an indispensable analysis technique. With FEM it is possible to implement in a program more complex problems than with traditional methods. In this work, several technological parameters have been considered to analyze their influence in the drawing mean stress. It has been demonstrated that parameters such as the reduction, the friction and the back tension increase the drawing stress. On the contrary, optimal die semiangles (where the energy necessary to perform the process is minimum) have been found. Responsive studies could be developed to define the way drawing stress is affected by few changes in the parameters that have been studied, and to specify which is the most critical one.

References

[1] Kalpakjian S. Manufacturing processes for engineering materials (3rd edn). Addison Wesley, Massachusetts, 1997.

[2] Johnson W and Mellor P B. Engineering plasticity. Ellis Horwood, Chichester, 1983.

[3] Slater R A C. Engineering plasticity: theory and application to metal forming processes. Macmillan Press, London, 1977.

[4] Avitzur B. Metal forming: processes and analysis. McGraw-Hill, New York, 1968.

[5] Kobayashi S, Oh S-I and Altan T. Metal forming and the finite-element method. Oxford University Press, New York, 1989.

[6] Rowe G W, Sturgess C E N, Hartley P and Pillinger I. Finite-element plasticity and metalforming analysis. Cambridge University Press, Cambridge, 1991.

[7] Rowe G W. Principles of industrial metalworking processes. Edward Arnold, London, 1977.

[8] Sachs G. Beitrag zur Theorie des Ziehvorganges. Z. Angew. Math. Mech. 7 (1927) 235.

[9] Hoffman O and Sachs G. Introduction to the theory of plasticity for engineers. McGraw-Hill, New York, 1953.

[10] Rubio E M, Sebastián M A and Sanz A. Mechanical solutions for drawing processes under plane strain conditions by the upper-bound. J. Mat. Proc. Tech. 143-144 (2003) 539-545.

[11] Ranatunga V. Three dimensional UBET simulation tool for seamless ring rolling of complex profiles. J. Manuf. Proc. (2004).

[12] Hibbitt D, Karlsson B and Sorensen P. ABAQUS v6.4 User's Manuals. Providence, 2004.

[13] Lueg W and Pomp A. Der einfluss des gegenzuges beim ziehen von stahldraht. Stahl u. Eisen. 63 (1943) 229-236.

[14] Thompson F C. The effect of a backward pull upon the tension required to draw wire. J. Iron Steel Inst. 128 (1933) 369-382.

[15] Camacho A M, Domingo R, Rubio E M and González C. Analysis of the influence of back-pull in drawing process by the finite element method. J. Mat. Proc. Tech. In press (available on line).

Intelligent Production Machines and Systems
D. T. Pham, E. E. Eldukhri and A. J. Soroka (eds)

Intelligent software agents as a basis for collaborative manufacturing systems

Berend Denkena, Alessandro Battino, Peer-Oliver Woelk

Institute of Production Engineering and Machine Tools (IFW), Hannover Centre of Production Technology, University of Hannover, Schoenebecker Allee 2, 30823 Garbsen, Germany

Abstract

Since most of today's industrial products are characterized by high complexity, sophisticated manufacturing processes and numerous customer-specific variants, modern manufacturing is in need of flexible and adaptive concepts for intelligent production. Next-generation manufacturing systems will have to take advantage of innovative information technologies in order to fulfil these demands. Furthermore, companies tend to concentrate on their core competences, since numerous products are too complex to be manufactured by a single company in an economical as well as efficient way today. Thus, manufacturing systems will have to be distributed as well as collaborative, which will increase the demands concerning appropriate information technology even more. In this context, the application of intelligent software agents is one of the most promising approaches for the implementation of collaborative manufacturing systems. This paper will illustrate this approach by means of results of selected research projects.

1. Introduction

Since the last decade, global economy and liberalization are leading to management trends as just-in-time, or reduction of vertical range of manufacture to realize shortened time-to-market and a more flexible consideration of customer' demands. However, modern industrial products are often characterized by a high complexity of design, functionality and necessary manufacturing and assembly processes. Computer systems for the support of planning, scheduling, production control and Supply Chain Management (SCM) tasks have to handle critical paths, bottlenecks and risk of failures within real-time. In the context of design and manufacturing of complex and sophisticated products these tasks involve a high degree of knowledge about the product. Therefore, new information systems are in need of formalized knowledge, which can be handled and processed adequately.

Innovative information infrastructures are needed, which are capable of providing the knowledge at the right place, in the right time, in the right quality and quantity (information logistics in manufacturing). This trend results in more and more knowledge-based manufacturing and new demands concerning the information technology infrastructure to support this new manufacturing paradigm. In particular, new demands arise in context of flexible and adaptive concepts for process planning, production control, and supply chain collaboration in order to meet market requirements. A promising approach to achieve efficient information logistics in collaborative manufacturing is the application of intelligent agents.

This paper will give a brief overview of some adequate applications of intelligent agents and Multi-Agent Systems (MAS) in the production management domain, which are part of the research activities at the Institute of Production Engineering and Machine Tools (IFW). In particular, the following section will give a short introduction to intelligent software agents (section 2) and their application in the manufacturing domain (section 3). The next sections will address the two facets of the collaborative manufacturing: intra-organisational (section 4), and inter-organisational (section 5). For the first facet, a scenario on process planning and production control on shop floor level is described. For the inter-organisational coordination two scenarios are considered: agent-based cooperation of loosely coupled MAS (so-called Multi-MAS or MMAS) on supply-chain level, and agent-based middleware as a collaboration platform for Virtual Enterprises (VE). Finally, a short conclusion is presented.

2. Intelligent software agents

Intelligent software agents represent a modern approach in Artificial Intelligence and deals with software entities, which can act autonomously communicate with other agents, are goal-oriented (pro-active), and use explicit knowledge [1]. They are often deployed for tasks which can be hardly solved monolithically and show a natural distribution [2]. In these domains agents gain great benefit as they are able to solve problems in a distributed manner and to collaborate using agent communication languages. The individual optimization of each agent should lead to a global optimum, which is often associated with an emergent effect. This effect results from the agent's interaction.

Within this paper, intelligent software agents in planning, scheduling, and SCM applications, which are organized in agent communities like MAS, are taken into consideration. Thus, other applications of intelligent agent technology (like mobile agents, lightweight agents on mobile devices or "robots" for information-gathering in the Internet) are not regarded.

3. Agents in the manufacturing application domain

Software agents are not state-of-the-art in the manufacturing application domain yet. Former implementations began usually as joint research projects between industry and universities [3]. Some special applications of agent technology are about to enter the market. Most implementations of MAS are "academic" today. One of the major problems for agent technology on its way from academia to industry are the still missing prove, that its typical advantages like higher flexibility and autonomy are worth its price of the increased effort of developing a MAS application. For example, it is easy to specify the investment costs for MAS development and implementation (including network infrastructure etc.) today, but it is difficult to pre-estimate the future benefits of the gained flexibility and autonomy in term of money. As a consequence, many companies are more or less reserved concerning introduction of agent technology to their IT infrastructure.

In order to point out the possible applications as well as the advantages of agent technology in the manufacturing domain, some research projects deal with realistic or real-world application scenarios for agents. This paper will discuss the approaches and results of two of these research initiatives: the German priority research program "Intelligent software agents and real-world application scenarios" [4] and the European "Multi-agent Business Environment" (MaBE) research project [5].

In this context, it is important to mention that during the last years several successful efforts have been made in areas like standardisation of agent infrastructures (e.g. FIPA [6]) and development methodologies for communication protocols as well as ontology engineering. In the following sections some current development concerning these topics are also presented.

4. Agent-based process planning and scheduling

Within an enterprise and close to the shop-floor level, the information flow can be separated into two major flows: one flow focusing on technological details of the product and another one preserving economical or administrative information, like due dates, priorities, costs, etc. The separated information flows correspond to strong borderlines that exist between process planning, production control and scheduling systems. The traditional approach of separating planning activities (e.g. process planning) from implementing activities (e.g. production control and scheduling) results in a gap between the involved systems. It implies loss of time, information and in consequence loss of quality and prolonged time-to-market. This situation could be overcome by the application of intelligent agents using knowledge explicitly.

Within the IntaPS project, a system of intelligent agents was developed, which is capable of integrated process planning and production control in a very flexible and distributed way. The application of intelligent agents is based on two substantial components, which link together information systems of earlier stages of product development and the resources on the shop floor (see Fig. 1). This link is realised by decentralised planning on shop floor level and by rough level process planning.

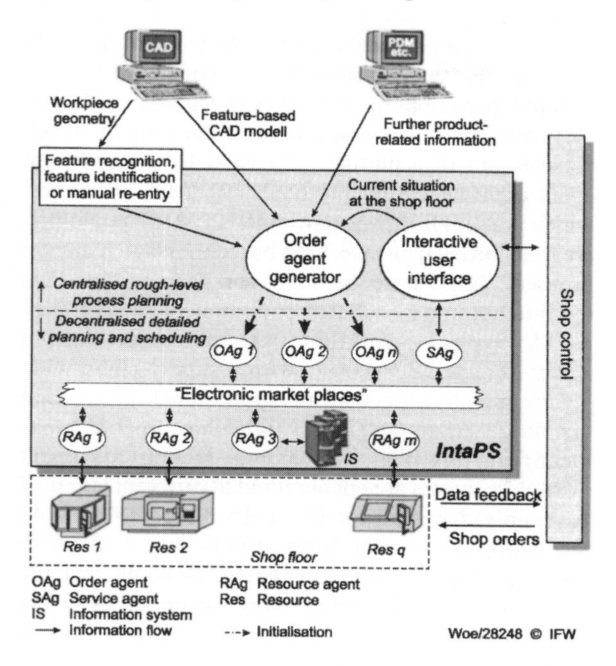

Fig. 1. System architecture of "IntaPS" approach

The implementation of decentralized planning unit on the shop-floor level is based on three different types of agents: resource agents, order agents, and service agents. Each relevant resource of the production system (machines, transportation devices, etc.) and its environment is represented by one resource agent. Resource agents provide local knowledge bases of the associated resources. Order agents represent orders which have to be manufactured. Due to its autonomy and pro-activity, an order agent is able to recognize internal and external disturbances and to react appropriately. Service agents are used for human interaction, transparency and maintenance purposes.

The detailed process planning and scheduling takes place co-operatively within an electronic marketplace. Order agents and resource agents interact according to a "three-phase-model". Starting with a "negotiation phase" required manufacturing skills and due dates as well as capabilities and capacities are communicated. Suitable sequences of manufacturing operations are resulting from auctions between order agents and appropriate resource agents. The optimal sequence of manufacturing operations is accepted as detailed plan.

The second phase is called "verification phase" and ensures the feasibility of the detailed plan. The order agent examines continuously whether its detailed plan is executable under the current conditions. Changing situations (e.g. breakdown of a machine tool) cause order agents to analyze the consequences and to identify those parts of their detailed plans which are affected. If necessary, the order agent enters a "re-negotiation phase" and tenders parts of the detailed plan for a new auction. The re-negotiation phase leads to an improved alternative detailed plan which substitutes the previous plan. Afterwards the verification phase is resumed and lasts until the order is finished.

Centralized rough level planning pre-processes incoming data and generates a rough process plan. These data are geometrical or technological information about the product (e.g. from CAD systems), further organizational information (e.g. from PDM or ERP systems) or the current shop floor situation. Order agents are synthesized with respect to this information and initialized with a rough level process plan. This plan contains only information which the agent is not able to recognize from the environment by itself (e.g. constraints between manufacturing operations). Thus, order agents obtain a maximum scope of data for the allocation of suitable resources and time slots for manufacturing.

One of the current research activities of IntaPS deals with identification of significant performance indicators, which aims at the comparison between agent-based and conventional systems of the job shop scheduling. Thus, the user of the IntaPS MAS will be able to analyse the agent's behaviour and to get advices how to improve the structure of the shop-floor (e.g. to identify bottleneck resources or changes in the product range).

5. Agent-based inter-organisational coordination

Multi-agent systems offer new perspectives compared to conventional, centrally organised architectures also in the scope of SCM. Their structure inherently meets the requirements of decentralised supply chains, whereas conventional SCM systems are often restricted in terms of dynamic behaviour, handling disturbances at supplier sites as well as dealing with highly customised or complex products. In the following, two approaches are presented.

5.1. The Agent.Enterprise approach

Agent.Enterprise is an initiative of the German priority research program "Intelligent Agents in Real-World Business Applications" and deals with realistic application scenarios from the area of manufacturing planning, production control, and SCM. Within these scenarios, the potential advantages of agent technology are analysed and evaluated. *Agent.Enterprise* is an open testbed for agent technology in the manufacturing domain designed to be applied to complex manufacturing supply chains. This testbed is currently under development. Aim of the testbed is to allow for comparative experiments of different research approaches and industrially applied solutions within a common scenario. At the moment, the initial version of the *Agent.Enterprise* testbed, the so-called *Agent.Enterprise NetDemo*, is under development and will be presented to the pubic in mid 2005 [7]. For the testbed, a MMAS is used [8], which incorporates three types of MAS: one for Supply-Chain-wide Planning of orders (SCP-role), one specialized for intra-organizational production planning and scheduling (supplier-role) and one for Supply-Chain Tracking (SCT-role) that is responsible for supply chain event management. The application scenario is a multi-level industrial supply chain for production of agricultural machines. Nevertheless, the basic concepts and underlying architecture of the testbed prototype may be applied to other similar scenarios of complex supply chains in the manufacturing domain.
Each enterprise in the supply chain is represented by three agents within the *Agent.Enterprise* testbed: One SCP-Agent for initial supply chain planning of orders, one agent of the SCT-MAS for tracking and tracing released orders, and one gateway agent. The gateway agent links the internal production planning algorithms of a particular enterprise to the *Agent.Enterprise* supply chain. At the moment, internal production planning is performed by different MAS (developed within the different research projects of the research program), which play the supplier-roles. It is also possible to design a gateway agent which implements a wrapper in order to “wrap” a conventional (non-agent-based) planning system and to link it with the *Agent.Enterprise* supply chain.

An important aspect of *Agent.Enterprise* is the loosely coupling of the MAS: Due to the gateway agent concept, each participating MAS is running without modification to its internal planning algorithms and data structures. The gateway agent performs all necessary actions like translating messages from the *Agent.Enterprise* (neutral) format to the MAS-specific format and vice versa. For this purpose, a system-independent ontology (*Agent.Enterprise* Interface Ontology) has been designed. The mapping between the *Agent.Enterprise* Interface Ontology and the MAS-specific ontologies takes place in the gateway agent. In addition, the interaction between gateway agents of different participating MAS is performed according to the *Agent.Enterprise* Communication Protocol, which consists out of a number of nested FIPA-compliant agent communication protocols [9]. ACL messages, which are sent from one participating MAS to another, are transmitted using a HTTP-based message transport protocol. Thus, these messages can be transmitted via the Internet in a distributed environment easily.

A typical workflow of an order within *Agent.Enterprise* starts with supply-chain-wide planning activities that integrate local production planning. After an initial schedule of orders is identified, this plan is executed. During fulfilment disruptive events are encountered at different levels of the supply chain and are detected by the SCT-MAS. As the disruptive events delay production, replanning activities are triggered by the SCT-agents that are conducted by the enterprise-specific instances of the SCP-MAS. Only if local replanning is not successful a complete replanning on supply chain level is initiated. Interaction among all MAS is visualized within a portal-website that provides the opportunity to initiate new experiments (e.g. placement of new orders) and reset the simulation. A welcome screen of the *Agent.Enterprise NetDemo* presents an overview over the participating MAS and offers links to examine the different MAS in detail. Therefore, each participating MAS is visualized separately in the context of the whole supply chain.

5.2. The MaBE approach

In the European project MaBE an agent-based information and communication infrastructure has been developed in order to enable dynamic and automated combination of services between different users and providers along the whole value chain [10]. The complexity of (dynamic) business environments is addressed through the adoption of the holonic paradigm. In this way a production network is seen as a temporarily existing holon, which is itself composed of sub-holons representing the single enterprises. In the same way, enterprise resources can be seen as organised in a holarchy, so that not only the physical structure is reflected, but also the workflow can be decomposed in different level ranging from strategic down to operative tasks. The agent technology is seen as a mean to implement the holonic paradigm, whereas characteristics such aggregation, inheritance and collaborative behaviour are typical of both concepts.

For resource coordination activities the PROSA reference architecture [11] was adopted. Further information on the use and implementation of this architecture in the MaBE project can be found in [12].

5.2.1. Description of the MaBE middleware

The MaBE software is built on top of the JADE open source platform for agent development [13] (see Fig. 2) and complies the standards delivered by FIPA [6]. Moreover, new standards are going to be set in the fields of security and ontology engineering. For the connection with Jade and 3rd-party software extensions have been created, allowing e.g. ontology services and persistence management. The software architecture is made up of two main layers: the middleware and the application code.

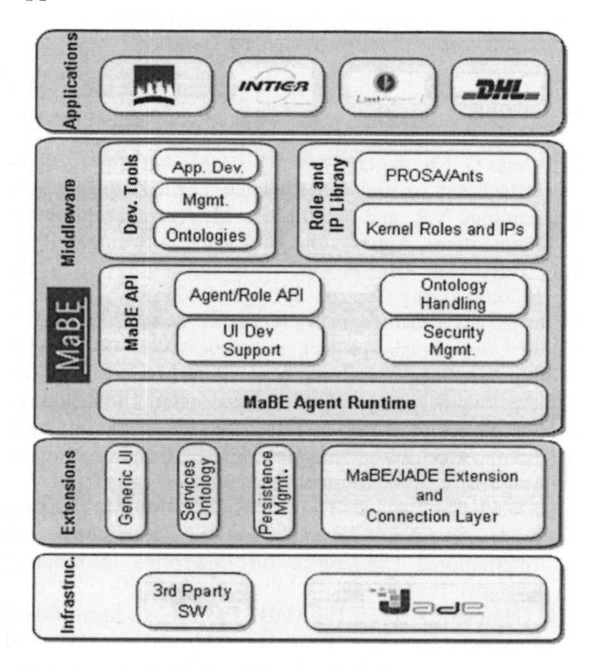

Fig. 2. Architecture of the MaBE agent middleware

In the middleware the general software functionalities are implemented, while the application code is user-specific. The middleware is made up of agent runtime environment, APIs, role and IP library as well as additional tools. In the runtime environment agents play different roles depending on the current necessities. A set of runtime roles with specific functionalities is predefined, which can be used by the roles developed in the application code [14]. Examples of roles are: Creator (creates new entities in MaBE), Mediator (enables communication between entities), Planner (creates sequences of tasks), Information-Source (retrieves information fulfilling given criteria), etc. In addition to the roles handling, APIs support the (dynamical) handling of ontologies, the user interface development and the security management. The role and IP library allows a fast application development through the mapping of application and runtime roles and allows the implementation of the PROSA/ants architecture.

5.2.2. MaBE demonstrator functionalities

One interesting use case taken into consideration in order to test the MaBE approach is related to a German leading group in the heat treatment sector. With the aim of optimizing the use of resources (i.e. facilities and know-how), the nine SMEs constituting the group have been organized in a VE. All companies are connected through a common IT infrastructure. From each company it is possible to access the databases of the whole group. A MaBE demonstrator has been developed aiming at scheduling and automation of time-consuming activities which are still carried out manually. This will allow collaborative, distributed manufacturing: When the capacity in one location is not enough to carry out all orders, some processes can be carried out in other locations. The MaBE software is designed to address coordination tasks on different business levels. In the demonstrator this ability is tested with tasks addressing both, the operative level as well as the coordination of the whole VE. Typical tasks of the operative level are the batching and the scheduling of the orders. In the batching activity an aggregation of order agents takes place on the basis of part treatments that they have to undergo. The aim of the batching is to reach the optimal quantity for a charge, treating together different parts of different orders. The batching is carried out during scheduling, so that charges are recalculated and modified on the basis of the actual capacity of the furnaces. The scheduling is extended outside the single factory of the VE: In case of overload the user is alerted and is able to start a search for capacity in other factories. Collaboration between order agents running in different locations allows allocating orders through a sharing of resources. Moreover, if the reallocation of an order is possible, the logistic provider is involved and the logistic service for transport of parts is booked by the order agent. In this case the logistic provider is seen just as any other resource. Further on, also the whole VE can be seen as a single resource when observed by the customer. This holarchical structure, where each resource is intended as aggregation of resources from a lower level, allows collaboration on different business levels using always the same agent logic.

Another task regarding the inter-organizational collaboration is the knowledge discovery for the process planning activity. Here the intra-organizational collaboration allows a search for part treatments needed to process a particular part in the databases of all locations of the VE. The result of the search is presented as a ranked list, thus a technicians can check the validity of the suggested solutions, select the favourable one, edit the programs on the basis of the current requirements and finally save the data for new orders.

5.2.4. Latest developments in ontology mapping

In MaBE the ontologies are developed using the Protégé tool. A specific ontology is developed for each application, allowing the communication among agents. It is possible to handle ontologies and dynamically update them whenever a modification occurs in the application domain. Moreover the ontologies are exploited in the access to external resources: a functionality allows finding specific data in a database on the basis of the corresponding concepts modelled in the ontology. This plug-in makes use of an XML-file containing the correspondences between the use-case ontology (manually build by the application developer) and an ontology reflecting the structure of the database (automatically extracted). The advantages of this method are that the use-case ontology can be developed without needing to know the structure of the company database, it can be modified and in any case the relative data present in the database can be found by the system without much effort.

6. Conclusions

Despite the fact, that agent-based application are not state-of-the-art in manufacturing industry yet, the concept of intelligent software agents offers an enormous potential for the design of future collaborative manufacturing systems. Within this paper, three different implementations of intelligent agents in the area of intra-enterprise planning and execution as well as SCM are presented. These implementations give interested users the opportunity to learn more about agent-based manufacturing applications and to evaluate their potential to solve real-world problems. In future, tools like MaBE, will offer the basis for further developments. Dealing with collaborative environments most efforts should be invested in the further definition of standards for software agents. Other fields to be deeper investigated are the ontology management and the security issues. Once efficiently addressed these issues, the technology of intelligent software agents will be an "enabling technology" for the design of collaborative next-generation manufacturing systems.

Acknowledgement

The Institute of Production Engineering and Machine Tools (IFW) is partner of the EU-funded FP6 Innovative Production Machines and Systems (I*PROMS) Network of Excellence (www.iproms.org). IntaPS and *Agent.Enterprise* are funded by the Deutsche Forschungsgemeinschaft (DFG). The MaBE project is funded by the European Community.

References

[1] Weiss G. Multiagent Systems – A Modern Approach to Distributed Artificial Intelligence. The MIT Press, 1999.

[2] Jennings N.R. and Wooldridge M. Agent Technology: Foundations, Applications, and Markets. Springer, 1998.

[3] Bussmann S. et al. Multiagent Systems for Manufacturing Control. Springer, 2004.

[4] http://www.realagents.org (accessed 14-04-2005)

[5] http://www.mabe-project.com (accessed 15-04-2005)

[6] http://www.fipa.org/ (accessed 14-04-2005)

[7] http://www.agententerprise.net (accessed 15-04-2005)

[8] Frey D. et al. Integrated Multi-agent-based Supply Chain Management. Workshop on Agent-based computing for Enterprise Collaboration, Linz, 2003.

[9] Stockheim T. et al. How to Build Multi-Multi-Agent Systems: the Agent.Enterprise Approach. 6th International Conference on Enterprise Information Systems. Porto, 2004.

[10] Hämmerle A. et al. The MaBE Project: An Agent-Based Environment for Business Networks. Industrial Applications of Holonic and Multi-Agent Systems. Prague, 2002.

[11] Van Brussel, H. et al. Reference architecture for holonic manufacturing systems: PROSA, Computers In Industry (1998) 37.

[12] Denkena B. et al. A Multi-agent Approach for Production and Logistic Control of a Virtual Enterprise, 3rd International Conference on Advances in Production Engineering, Warsaw, 2004

[13] http://jade.tilab.com/ (accessed 15-04-2005)

[14] Hämmerle, A. et al. A Role-Based Infrastructure for Customised Agent System Development in Supply Networks, IEEE SMC, The Hague, 2004.

Intelligent Production Machines and Systems
D. T. Pham, E. E. Eldukhri and A. J. Soroka (eds)

Modelling and simulation on an RFID based manufacturing system

B. Lu[a], K. Cheng[b]

[a] *School of Mechanical Engineering , Dalian Jiaotong University, Dalian 116028, China*
[b] *School of Technology, Leeds Metropolitan University, Leeds LS1 3HE, UK*
Correspondence to Professor Kai Cheng, email: k.cheng@leedsmet.ac.uk

---

Abstract

RFID is becoming a much more powerful technology and being adopted in supply chains and logistics industries, and now it begins to be applied in manufacturing industry. But how to build a framework of the application and analyse the benefits of application quantitatively is still a challenge. In this paper, RFID technology is critically reviewed with respect to its technological basics. The constraints in a conventional manufacturing system are analysed. A framework for the RFID based manufacturing system is proposed. ProModel software is used as a modelling and simulation tool to simulate the role of RFID in a typical manufacturing system. A case study on gear manufacturing is presented to evaluate the benefits by adopting RFID technology comparing to the conventional manufacturing system in terms of the cost distribution and time utilization. As a result of the improvement the cost of processing was redistributed by the same amount. The study illustrates an 8.39% improvement by implementing RFID technology in the system.

---

1. Introduction

RFID (Radio frequency identification) is becoming a powerful recognized technology with its strong tracking and tracing ability at a much greater level of accuracy, real-time and detail than other alternatives. Currently, most applications of RFID technology are in supply chains and logistics [1], and more and more enterprises are benefiting from adopting this technology [2].

On the other hand, as many manufacturing enterprise environments become more and more information-intensive, and their information systems are sharing the fundamental information of assuming rather than knowing where and what things are [3], which made the manufacturing system lacking of agility and responsiveness. RFID technology provides the solution for closing the gap between the physical flow of materials and the information flow in the production system by communicating in tags and readers.

In this paper, a basic conception of RFID is firstly introduced. After that the problems in a conventional manufacturing system are analyzed. Then a framework for RFID based manufacturing system is proposed with respect to three aspects of machining, assembling as well as packaging in manufacturing industry in the light of enhancing manufacturing intelligence and delivering added-on values. As a new technology introduced to a manufacturing enterprise, traditional methods of description and evaluation such as time-motion studies, arithmetic calculations, and guesswork were simply not sufficient for the task of creating a high complex manufacturing system that was capable of meeting the challenges of the intensely competitive manufacturing market. In this paper, ProModel, a commercial software is therefore used to model an

RFID based generic manufacturing system and to simulate the functions of RFID as defined in terms of production planning dynamically, operation and quality control, product tracking and recall, reusable asset tracking, maintenance and repair, as well as inventory tracking and visibility. At last, a case study is presented on evaluation simulation of RFID application in the mechanical machining process and assembling process to demonstrate the benefits from an RFID based system comparing with the same system without adopting RFID technology. The quantitative analysis based on the simulation yielded many interesting results which are helpful for the manufacturing companies to adopt RFID technology in their production system effectively.

2. RFID conception

Generally, an RFID system consists of the following three components:

(1) A small electronic data carrying device called a transponder, or a tag that is attached to the physical object to be identified. Tags can be active or passive. They can also be classified by their read / write capabilities. The writable tag makes tag can record the information of object through its lifecycle time. Passive tags tend to have the shortest range. The data stored in tags can be encoded by EPC code, which is called the Electronic Product Code [4].

(2) A reader and/or writer with antenna that communicates with the tag by using radio frequency signals without any human intervention. Reader can be used as handhold, dock door reader as well as conveyor belt reader, shown in Figure 1. If tag can be read effectively is a key issue [5]. Reader has an RS232 Com port for connection to PC.

A host system that contains information on the identified object and distributes information to other remote data processing systems. The host system is normally a line of business software application: typically an enterprise resource planning system (ERP), warehouse management system. RFID tags can also provide an automated output from the system allowing dynamic update of the data held on the tag. In all cases, the host system will need software modification to integrate the data provided by the RFID reader/writer. To describe all products related information, Physical Markup Language (PML), an XML format language can be adopted. To find where object data is, the Object Name Service (ONS) is also needed to return the address for a specific EPC. In order to provide data in real time to

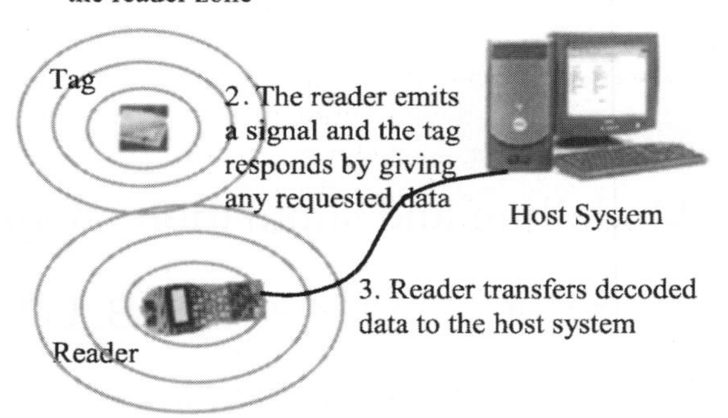
1. Tagged item enters
the reader zone
Tag
2. The reader emits
a signal and the tag
responds by giving
any requested data
Host System
3. Reader transfers decoded
data to the host system
Reader

Fig.1: Working principle of an RFID system

business information system, e.g. ERP system, or to the internet, we need a middle ware, e.g. Savant, to pre-process and manage the data collected by reader [6]. So RFID system is composed of hardware and software, and its work principle can be described in Figure 1.

RFID systems are completely insensitive to dust, moisture, oils, coolants, cuttings, gases, high temperatures and similar problems that can occur in a manufacturing environment. Even particularly dusty or dirty environmental conditions, which would make the use of barcode readers impossible due to the rapid blocking of scanner optics, pose no problems for RFID systems [7].

3. Constraints in conventional manufacturing systems

Currently, the Manufacturing Execution Systems (MES) application does usually not have easy access to detailed information, and therefore, a manufacturer's ERP system has no idea of what is really happening on the shop floor (e.g. subcomponents not being where they were expected, trained people not showing up when they should and machines going down). On the other hand, bar codes have been a major asset on the manufacturing floor. Components and sub-assemblies are bar coded to improve materials handling and increase productivity and maintain inventories at optimum levels. Most of automatic manufacturing systems are based on bar code devices to identify objects by database lookup.

However, barcodes are an optical "line-of-sight" technology where the scanning device must "see" the barcode to be read and interpreted. Barcode labels may be scuffed, damaged, or wrinkled, which makes it difficult to obtain a valid scan. Additionally, barcodes are not really "automatic identification technology," as they are usually read just a few times. They can neither store certain amount of data, nor identify a mobile object timely. Furthermore, in bar code systems, its information can not be rewritten with the change of manufacturing processes.

So barcode based manufacturing systems could not support the production planning dynamically, because production planning addresses the details of resource allocation. The manufacturing information requires a detailed, accurate and timely processing, which will significantly influence the agility of manufacturing system.

In modern production lines the quality of products is tested at inspection points located at a number of stations. When the product is inspected at the end of the production process, it must be possible to unambiguously attribute the quality data gathered earlier to the correct object. But bar code labels are not writable, and can hardly carry these quality information obtained during the production process with the product.

Now in automatic manufacturing lines, some assets such as pallets and containers are tagged with barcode labels, which are used to trigger machining facilities. These assets can be misplaced or read with some mistakes and thus have direct impact on the reliability of the system. Also due to hardly tracing and identifying these assets accurately, sometimes more human intervention or over assets have to be added, which make the manufacturing cost rising. Considering the cost-effectiveness, RFID tags are more and more being used in reusable assets' tracking for their item attendant data service.

Manufacturers have pursued practices such as lean manufacturing and Just In Time (JIT) to obtain the benefits of reduced inventory in manufacturing systems. However, the best laid plans can be impacted when major disruptions to the supply chain occur. Preparation for such events causes companies to build up a buffer of inventory. For instance, each day Boeing sends 4,650 shipments of spare parts to its airline customers worldwide. The airlines have an estimated 44 billion of unused spare parts on their shelves [8]. To trim these inventories for both Boeing and airline is the significant benefits while ensuring parts are always where they need to be to keep the planes flying.

4. A framework for RFID based manufacturing system

In fact, RFID system can be taken as a kind of device for data capturing, storing and transmitting automatically. So to build a RFID based manufacturing system, what kind data need to be concerned is the key issue. Here we address how the RFID could improve the production planning, quality control, maintenance as well as inventory visibility in the light of the analysis of conventional manufacturing systems in above section. Generally, to implement such a function, a basic architecture framework for RFID based manufacturing system can be described as shown in Figure2 [9].

Here the manufacturing process can be considered as three main stages, namely:

(1) Machining of the raw material into parts

(2) Assembly of the parts into a product

(3) Final packaging of the product

Readers and antennae are placed in each station (including inventory). Tags are attached on items to be identified and traced, such as material, semi-finish products, spare parts and facilities. From material to finished product, RFID can make any changes occurred visible and feed back relevant information into the network depending on the needs of the cost-effectiveness of the system.

4.1. Machining stage

In machining stage, all required information from production planning consists of machining specification, material, facility, as well as human labour. The material information associates with the inventory database. When an item goes out of the inventory, reader can update the inventory database automatically. The reader on a machining station can identify whether the item is correct for this process or not. This function is very helpful for the customised or individualised production, because the machining station may cut different part each time according to the change of an order. During the machining process, if any facility does not work, the tag for this machine can be read quickly and the relevant information for this machine can be provided from the maintenance point of view. So the down time can be reduced. After machining, the raw material has been generated into a part to be

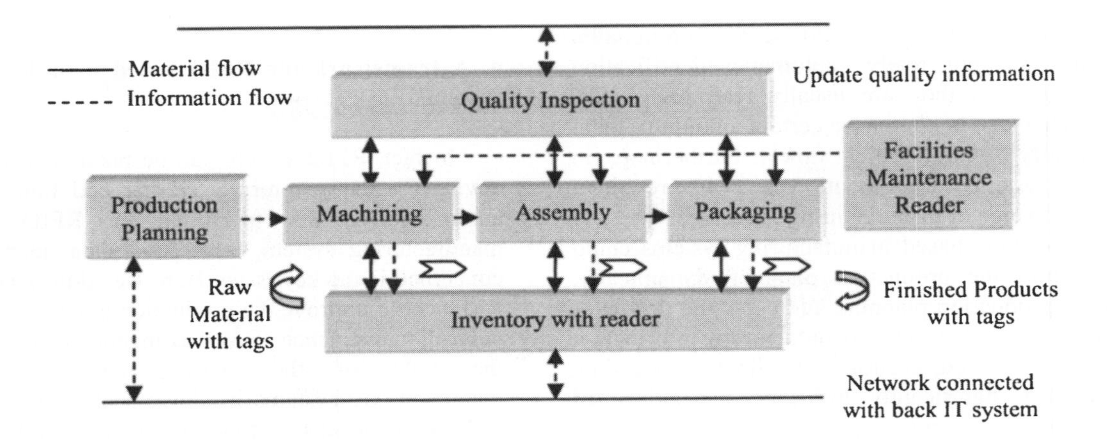

Fig. 2. An architecture for RFID based manufacturing systems

inspected acceptable or not. The quality information can be written by read/write device into tag on this part. Meanwhile the quality data can be uploaded onto the network and used to evaluate the machining process capability. The part can be moved into the inventory or assembly station. If it is moved into inventory, the reader there can record it as a new item and update database automatically.

4.2. Assembly stage

The assembly station can be automatically or manually separated. Similarly, for the customized product, the order may be changed until it to be finally assembled. The reader on assembly station can recognise each item whether it is right for this individualised assembly. So the production process which can be dynamically planed is critical for a customized manufacturing system. Dynamic planning needs inventory visibility and information updating timely. RFID technology will make it possible. During assembly processing, some inspection may be needed and depend the assembly specification. The new quality information can be written into tag on this product, or transmitted to quality information system via network. For instance, by integrating RFID with Golf Car Maker's new assembly line, it has cut the production time down to 46 minutes from 88 minutes per golf car, also improved its ability to customize cars as the customs required, and saved the production cost dramatically.

4.3. Packaging stage

The information need in packaging stage is similar to that in assembly stage, and materials are from up stream and inventory. Due to the products varying significantly, some times packages need different moulds according to the order. Also some specific information (such as ID) for this product used in supply chain can be written on packaging station. Both in assembly stage and in packaging stage, the facilities used may fail for some reasons, but the tags on these machines can help to identify the problem for quick maintenance. Because for RFID based manufacturing systems, all key facilities require tags attached to record the specification for quick maintenance or repair. To make this done is reasonable, for the facility is also the products of machine tool manufacturers. When they have adopted RFID technology, the tag with ID as well as specification must be attached on the machines, which enables them to trace their products for achieving better service, maintenance, quality control and customers' satisfaction, etc.

5. A case study on the machining process and assembling process

5.1. Model construction

Figure 3 shows the simulation for an RFID based mechanical machining and assembling system by using Promodel software [10]. Promodel is a simulation and animation tool to model

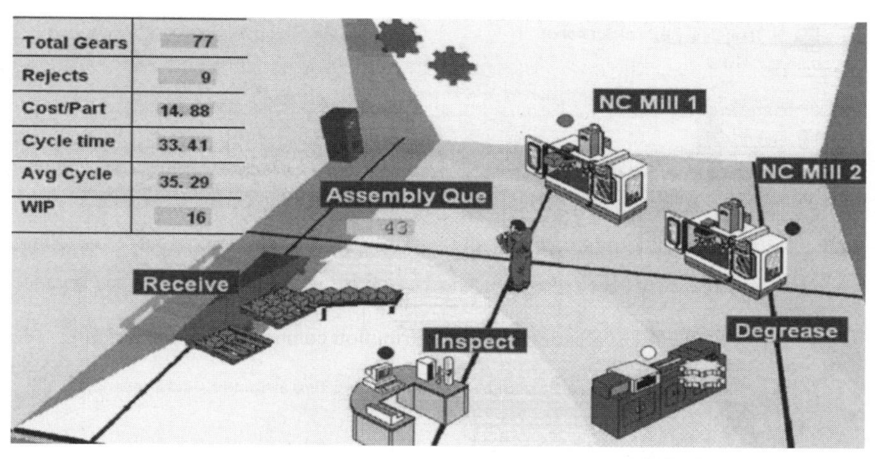

Fig. 3. Simulation on an RFID based manufacturing system

manufacturing system of all types, is completely graphic and object-oriented. In this case, the simulation model is built for a part of gear manufacturing process. Parts tagged are delivered into receive location firstly, there is a reader located to check whether the part is correct for this processing. In this machining unit, there are two NC-Mills, one degrease machine and one inspection station. Part will be sent to NC Mills depending on which machine is available currently. If two machines are ok, then parts will be delivered in turn. After machining, part changed into gear to be transferred to degreasing machine. This machine can process two gears each time. After that gear is sent to inspection station to be test. The test results will be written in to tag on the gear, mean while quality information will be updated on database system. Acceptable gear will be sent to Assembling-Que location waiting for assembling. And reject gear will be sent to rework. During the gear machining processing, any machine is in down time, tag will trigger maintenance system automatically. In Figure 3, each machine attached with light in deferent colors to show the current status of machine. Here blue is in idle/empty, green means in operation, red indicates downtime, pink stands for blocked, and white means in operation for multi-capacity location. In this case, there are seven locations, two NC-Mills set with downtime, and five entities named pallet, blank, gear, reject and assembling, respectively. One dynamic resource used as operator. Simulation facilitates quality control visible in the full process. Communication was improved between material flow and information flow as well as between operators and machines.

5.2. Evaluation based on simulation

To evaluate the benefits from adopting RFID technology is almost every enterprise concerned. But in real life, it is too difficult to evaluate the performance of manufacturing system according to different production strategies. Simulation was used for this study to help understand and evaluate the behavior of the system with RFID and without RFID. Totally nine global variables defined to evaluate the result of simulation. Fig.4 shows the distribution of total cost spent in two scenarios (simulation parameters: replicate 5 times, worm up 5 hours, run 10 hours). For RFID scenario, 96.86% of cost spent on effective products, for normal scenario, only 88.47%. But in reject gears, normal scenario cost 11.53%, and RFID scenario costs only 3.14%. Figure 5 indicates that RFID scenario spent much less time in waiting resources than normal scenario.

As a result of the improvements the cost of processing was redistributed by the same amount. There was an 8.39% improvement by implementing RFID technology in this case. This study brings out the importance of adopting RFID technology only within the concerning quality control function. In fact, minimizing WIP (Work In Process) helps to reduce all waste and abnormalities in the system because excess WIP hides the non-value-adding activities. Also how to evaluate the agility of RFID based manufacturing system is still a challenge. The next step is to increase the usage of the simulation tool in productivity improvement studies.

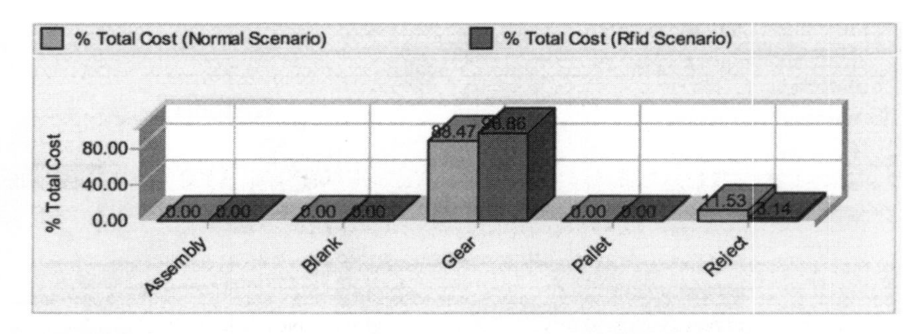

Fig. 4. Evaluation on cost distribution comparing two scenarios

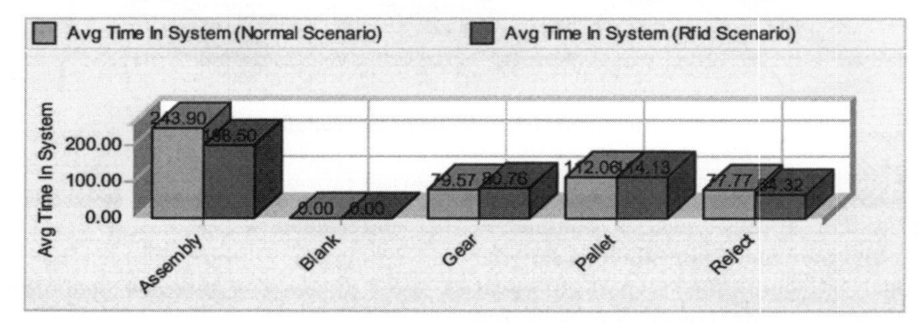

Fig. 5. Evaluation on time spending comparing two scenarios

6. Conclusions

RFID technology is being transferred into manufacturing industry with its strong tracing and tracking ability. This paper established a framework for RFID based manufacturing system, which can help to close the gap between material flow and information flow and to improve the agility of the system. The role of RFID technology in manufacturing system has been verified by using ProModel software. The case study on the gear manufacturing demonstrated how to build and simulate an RFID based manufacturing system and to evaluate the benefits from it. So the outcome of this paper can be used as a methodology in research on how a new technology impact on the business before practice and offering a help in decision making.

Acknowledgements

Authors would like to thank the China Scholarship Council for supporting this research work in the UK.

References

[1] Wada NM. Defense department wants RFID tags on everything but sand. 2003. http://www.changemariner.com/a/001282.htm.

[2] Deal III and Walter F. RFID: A revolution in automatic data recognition. Technology teacher. v 63 n 7. (2004) 23□27.

[3] Tan, J. Turning vision into reality: enabling the supply chain with RFID (2004) http://sap.ittoolbox.com.

[4] Brock DL. The Electronic Product Code (EPC): a naming scheme for physical objects. White Paper of Auto-ID Center, MIT, USA, 2001.

[5] Keskilammi M. Sydänheimo L. Kivikoski M. Radio frequency technology for automated manufacturing and logistics control. Part 1: Passive RFID system and the effects of antenna parameters on operational distance. The international journal of advanced manufacturing technology. v21(2003) 769□774

[6] Thorne A. McFarlane D. Hodges S. The Auto-ID automation laboratory: building tomorrow's systems today, Auto-ID White Paper (2003). http://www.autoidcenter.org/research.asp.

[7] Finkenzeller K. RFID Handbook. (2nd edn). John Wiley & Sons Ltd, Chichester, England, 2003.

[8] Violino B. RFID takes wings, RFID Journal, October, (2004) 13□17.

[9] Lu B. Cheng K. A framework for RFID based manufacturing system, ICAM International Conference on Agility, Helsinki, Finland, 2005

[10] ProModel Corporation. ProModel version 6 user guide. USA, 2004.

Intelligent Production Machines and Systems
D. T. Pham, E. E. Eldukhri and A. J. Soroka (eds)

Production variable management: a software tool based on the connectance concept

Kim Hua Tan
Nottingham University Business School, Jubilee Campus, Nottingham NG8 1JD, England
Tel: 44-1158467749; Fax: 44-1158466341
Website: www.nottingham.ac.uk/business

Abstract

Managing the information associated with a large number of production variables is a complex task. So far, little is available to assist production managers to manage and visualise the information about the variables in such a way that this information could be used to support strategic decision making. This research has sought to address this problem by developing a practical tool that offers a database, and visualisation function, which shows the inter-relationships of the production variables. The tool was developed based on the Connectance Concept. The salient features of the tool are its ability to: a) enable managers to visualise the 'implicit understanding' of the variables cause-effect relationships, b) enable managers to have their knowledge brought into the open and their assumptions challenged, c) see how the decision made on one variable will effect other functional departments in a firm, and d) allow information to be passed, assessed and quantified, so that the ideas and beliefs contained within the model can be altered or modified at will. This paper describes the structure of the tool and its features. The results of application of the tool in companies are briefly presented. This paper concludes by discussing the implications of this research for managers, and identifying directions for future research.

1. Introduction

Do you know the variables that influence the performance of your work centres or factory operations? Do you know how these variables are interrelated and how linked to your operations performance? The answers to these questions can have a dramatic effect on your operations strategies and also on the bottom line performance.

A review of the literature shows that little is available in existing research to assist managers in mapping and managing production variables information in such a way that these valuable information could be used to support managerial decision making. The fishbone diagram, for example, is an existing tool that has been widely applied by managers to study the cause and effect relationships of a problem situation, usually for a small problem within a clear boundary. However, for modelling of complex cause and effect relationships of a production system, the fishbone diagram may not be suitable. There is a range of commercially available software packages built around the modelling techniques of influence diagram and system dynamics. These software tools (i-Think, Stella) automate the application of the techniques and represent all causal relationships of a phenomenon in a manner that is non-ambiguous and probabilistic. However, these techniques have their roots in control theory and could be rather mechanistic [1]. Thus, they may not be suitable to analyze complex problems which involve relationships that are qualitative in nature.

Managing production variable information is difficult because the relationships among variables are complex, and the 'impact' of a variable is likely to be cross functional [2]. For example, insurance requirement has an impact on the production cost but this variable is likely to be under the control of the personnel department. Thus, given such a complex nature, how could managers model, maintain, and visualise the variable relationships? Also, how could they focus their analysis when the model of the variable became too complex? Clearly, if a tool were to be able to support the modelling of the complex variable relationships in a production system, it needs to have the following capabilities:

a) Visualisation – Graphic user interface functions to enable managers to visualise the linkages among variables.
b) Documentation – Database functions to store the complex variables' interrelationships, and support to categorise variables according to different management functions.

In 1984, the late Professor John Burbidge advocated the use of the connectance concept for modelling of production system variables. Burbidge argued that through experience, managers learn the principles which govern the relationships between variables. When driving a car, one learns, for example, that turning the steering wheel in a clockwise direction will cause the car to move to the right, and that pressing the brake will make the vehicle stop. In the same way, by learning something of the effects of changes in input variables on the induced direction of change in output variables, managers can learn how to steer a production system towards achievement of the aims. The working principle of the connectance concept is as follows:

'Providing one does not attempt to specify relationships in quantitative terms, it is possible to make statements about system variable connectance, which are always true, but may not be relevant in all production situations' [3].

Using the connectance concept, Burbidge built a Connectance Model of production variables and intended to use the model for the design of new production systems [4], that was to find the input and system variables required in order to achieve a given combination of output variable requirements. While not necessarily agreeing with Burbidge's assertion that relationships are always true, we believe that the principle of the connectance concept could be used to model a complex production system variable. The developed model could then be used by managers to study how a given direction of change in one variable will induce a particular direction of change in any related variables. The connectance concept, although itself simple, results in large complex models when applied to real production situations. These models are difficult and time consuming to generate manually. Furthermore they are difficult to manipulate manually, and this makes analysis tedious. In order to address these shortcomings we have proposed a computerised tool which facilitates the building of production variables' model. Figure 1 compares the functions of the proposed tool and the Burbidge Connectance Model.

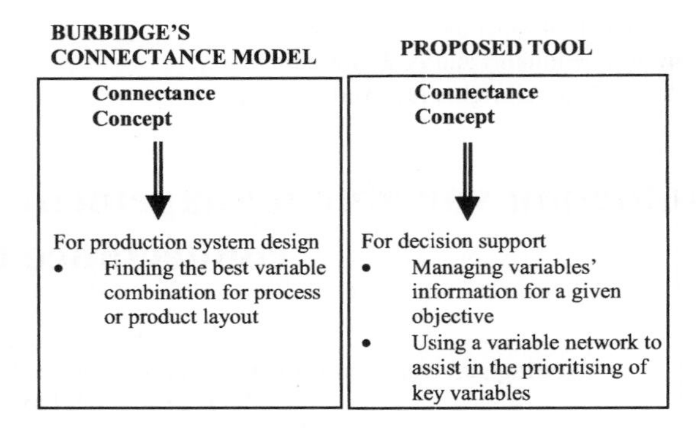

Fig 1. Comparing the Functions of the Proposed Tool and the Burbidge Connectance Model

This paper describes the development and testing of the proposed decision support software tool, we called it **T**ool for **A**ction **P**lan **S**election (TAPS) to assist manager in mapping and model relationships between production variables. The remaining of the paper are organised as follows. The next section explains the structure of TAPS and its features. Then, the results of testing and application of TAPS in companies are presented. This paper concludes by discussing the implications of this research for managers, and identifying directions for future research.

2. Tool For Action Plan Selection (TAPS)

The tool is implemented under the Microsoft Windows operating system using Microsoft's Visual Basic 6.0 programming language. TAPS has four main modules: a) database; b) analysis; c) graphic user interface, and d) AHP analysis

2.1 Database Module

The database module stores the variable information and is based on the original data sheet format as in Burbidge's connectance model. There is one record for each variable, which stores the variable information such as code, name, definitions and its connectance variables. The database module serves as input to the subsequent modules.

To reiterate, the aim of developing the prototype tool is not to build a 'computerised' Burbidge's Connectance Model. The objective is to adopt Burbidge's connectance concept as the main engine for the tool. Since Burbidge's connectance database comprises the most 'comprehensive' production variable information, it is then useful to take Burbidge's database as the main reference. Considering the potentially vast number of variables and the complexity of their connectance, a linear list data structure is used for the database. For a given variable, the connectance information is stored as pointers. Figure 2 illustrates an example of the connectance network for Var 1. In the network, Var 1 has a connectance with Var 2, Var 3 and Var 4; Var 3 has a connectance with Var 4; and Var 4 has connectance with a Var n. Figure 3 shows how the variables information is captured in the database. First, a record table for Var 1 is established and the connectance information of Var 2, Var 3 and Var 4 is stored as pointers. The process then continues for Var 2, followed by Var 3, Var 4 and Var n. In each sequence, the connectance information is stored as a pointer. To accommodate the building of a large production variable model, each variable database is allocated 10 pointers.

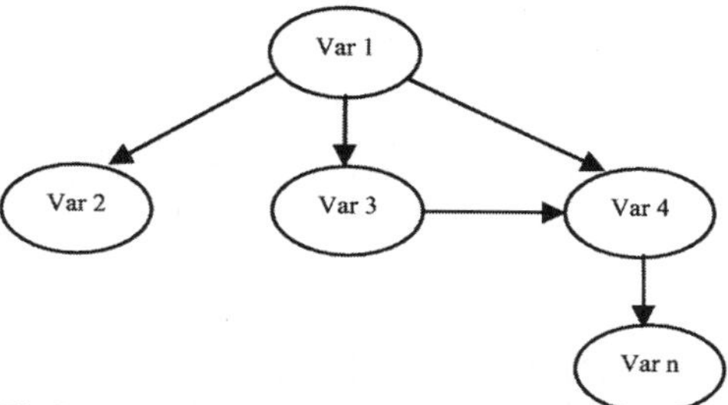

Fig 2: Network Diagram for Var 1

2.2 Analysis Module

The analysis module contains algorithms for accessing the database information and arranges the variable and its connectance variables information in a network diagram. The analysis consists of two main functions: Trace-Down Analysis and Trace-Up Analysis (see Figure 4). A Trace-Down Analysis function is an analysis to determine the effect of a given direction of change in one production system variable on other variables on the network. A Trace-Up Analysis function, however, starts with a desired direction of change in a variable and works back to find which other variables have to be changed in value and in which direction. Example of a trace-up algorithm can be briefly described as follows:

1. Selects a specific variable, such as the capacity of a work centre, interactively from the user interface or directly from the database
2. Initialises a new connectance model with the selected variable.
3. Sets the selected variable as the current variable.
4. Search all parent (connetance) variables of the current variable, and stores all unknown parent variables to the new connectance model in the data array.
5. Moves the index of the current variable to the next datum in the array.
6. If the current index is less than or equal to the total of the variables in the updated connectance model, repeat from 4. Else, the procedure ends.

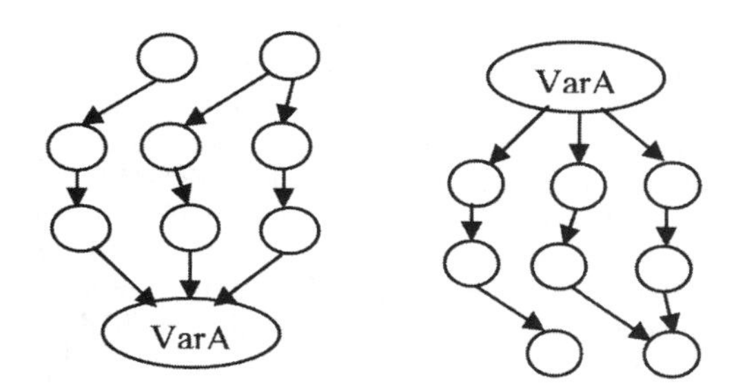

Fig. 4: Trace-up and Trace Down Network for Variable A

2.3 Graphic User Interface (GUI)

The graphic user interface module serves to display the computerised model and contains functions to allow the managers to manipulate the model interactively.

In the network display the variable is displayed as a node with edges (lines) linking it to other nodes with which it has a connectance. Thus, a variable's connectance network is made up of nodes and relations. A network display for the variable 'setup time per operation' (o3302) in the Trace Up Analysis is shown in Figure 5. The variable and its connectance information was extracted from the Burbidge Connectance Model (Burbidge, 1984b). The variable being studied (o3302) is located at the bottom of the network with its immediate connectance at the top. The relationship of the connectance is represented by the direction of the arrow, and the induced variation is given by the sign above the arrow. For example, the arrow from variable 'pre-setting tooling on work centre' (IS205) is pointing to variable 'setup time per operations' (o3302) with a (-) sign above the arrow. This indicate a negative induction: pre-setting tools on a work centre can reduce the setup operations time.

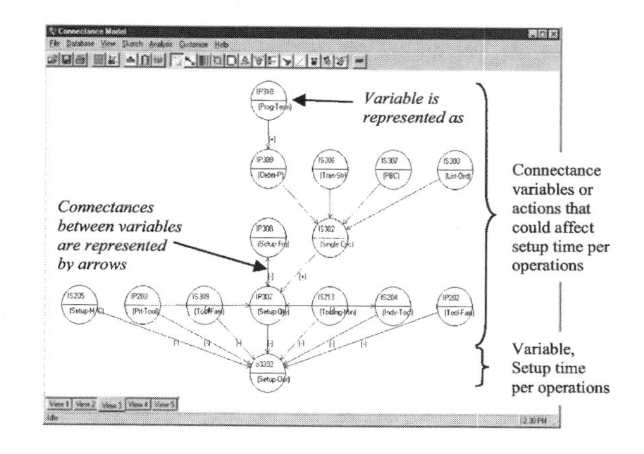

Fig. 5: Sample of Variable's Connectance Network from the Prototype Tool

In the network display, connectances that have 'first order' (Burbidge, 1984b) relationships are displayed on the first row. This is followed by the 'second order' or indirect connectances. For example, variables (IS205; IP203; IS309; IP307; IS213; IS204 and IP202) have first order connectance, they are variables that have a direct knock-on effect on variable (o3302). Variables (IP308 and IS302) have second order or indirect connectance to variable (03302) but they were first order connectance to variable (IP307). Thus, nodes at the higher levels in the network hierarchy have only indirect impact on the variables at the bottom of the hierarchy.

The tool is designed to be user friendly, thus enabling users to update or search the variables database easily. By linking to the database, the tool traces and displays the variable's connectance network in hierarchical form. Other graphic interactivity features in the tool are sketching and node editing functions (size, colour, move, rename etc.) which enable the user to modify the network hierarchy display. The modified connectance network can then be transferred into a new database. In other words, the tool allows the user to build a new connectance database from scratch and modify existing variables' connectance from a database.

2.4 Evaluation Module

In order to allow managers to perform an evaluation process to prioritise variables for further decision analysis, it is desirable for the prototype tool to have an evaluation module. Analytic Hierarchy Process (AHP), a Multi-Attribute Decision Making (MADM) method developed by Saaty in 1987 was added to the prototype tool. The advantage of AHP is its capability to elicit judgements and scale them uniquely using a valid procedure that measures the consistency of these scale values [5].

The built-in AHP function is based on the procedures and eigen vector calculation methods proposed by Saaty [6]. For those selected variables, a pair-wise comparison between two variables v_i and v_j (i=1,2, ..., n) can be quantified by users as a matrix $\mathbf{A} = [a_{ij}]$ which denotes the importance of v_i over v_j. An approximation of the priority vector for the selected variables can then be computed according to the AHP method as follows.

A normalised matrix of **A** by a row vector of the sum of its columns can be given as

$$\mathbf{B} = [b_{ij}] = \left[\frac{a_{ij}}{\sum_{i=1}^{n} a_{ij}} \right] \qquad (1)$$

Then an approximation of the priority vector can be given as

$$\mathbf{x} = [x_i] = \left[\frac{1}{n} \sum_{j=1}^{n} b_{ij} \right] \qquad (2)$$

which provides the priority of those alternative variables. Since the pair-wise comparison is arbitrary, the computation of the consistency ration c_r is also provided. The consistency ratio is a scalar value indicating the consistency of a pair-wise comparison and is calculated by

$$c_R = \frac{c_I}{r_I} \qquad (3)$$

where

$$c_I = \frac{1}{n(n-1)} \sum_{i=1}^{n} \left(\frac{\sum_{j=1}^{n} (a_{ij} x_j)}{x_i} \right) - \frac{n}{n-1} \qquad (4)$$

and r_I depends on the order the pair-wise comparison matrix **A**. Generally, if c_r is less or equal 0.1, the pair-wise comparison can be considered as satisfactory.

Based on the AHP principle, the prototype tool evaluation function enabled users to visualise the problem in a hierarchy tree and see the results of their judgement at each stage of the analysis process (see Figure 6). Moreover, the evaluation function allows managers to state all their assumptions and to make subjective weightings explicit. Thus, the built-in evaluation function helps managers to organise their decision making process and provide a record of how they reached the conclusion they did. To allow managers to answer some of the 'what-if' questions pertaining to their decisions, a sensitivity function was also built into the prototype tool.

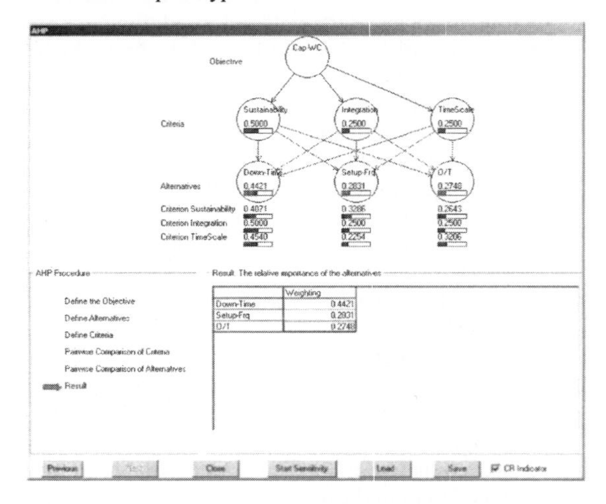

Fig. 6: A Dialog Box for AHP Evaluation

3.0 Applications

The advantage of applying the connectance concept is that models can be created that are consistent with the ways in which managers have experienced the world, and can be transformed in order to answer specific queries. For example, if managers would like to identify the range of action plans to improve capacity utilisation (Cap-WC) on a work centre, they could use the prototype tool to build a 'Cap-WC' connectance diagram of production variables based on their experiences (see Figure 7). The connectance diagram enables managers to represent their knowledge and experience of variable relationships, in a way that can be easily communicated and discussed. Once the variable connectances are understood, the potential actions to achieve a particular objective are revealed. In the example, increasing work centre capacity could be achieved by one or more of the following actions:

- reducing (indicated by a (-) sign) work centre down time (o3314);
- reducing machine set-up frequency (IP308);
- reducing the amount of idle time on work centre (o3311);
- reducing lateness of machine operator (o7303);
- increasing (indicated by a (+) sign) overtime (IP316);
- increasing the number of working shifts (IS310);
- increasing the number of machines (IP205).

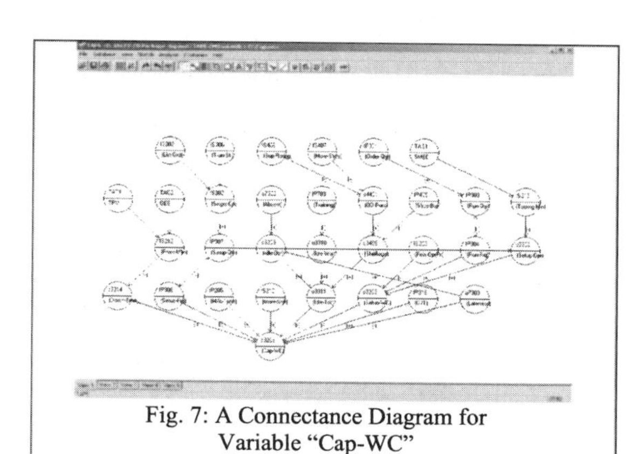

Fig. 7: A Connectance Diagram for Variable "Cap-WC"

The developed prototype tool was tested in five companies in UK: an industrial adhesive manufacturer; a mail delivery service provider; a pharmaceutical firm; a tobacco manufacturer; and a leak detector manufacturer, using a process research based on action research methodology [7]. In each case, a number of workshops were organised to assist managers in applying TAPS to address production operations management problems. Further information on these cases can be found at [8]. The testing cases demonstrated that the prototype tool and its process were feasible. The company participants considered that the tool enabled them to build a connectance model which provides an overview of the issues that they would like to address, and the tool built-in function i.e. filter functions provided them with a structured way to focus their analysis. In general, they pointed that the built-in functions of the software tool was easy to apply.

The participants also felt that the TAPS network structure provided a powerful visual diagram to assist them to get an overall view of a variable and its connectance, which helped them to determine a set of improvement alternatives to work with. The AHP evaluation was well received by the participants. With AHP evaluation module, they pointed that TAPS supports their analysis at every stage (problem definition, ideas generation, and evaluation) of a decision making process that the In general, the company participants agreed that TAPS gave a clear and systematic approach for analysing and diagnosing a particular problem.

4.0 Conclusion

Understanding and managing production variables are a key task for industrial managers. This task is essentially complex as it requires an understanding of the relationships among many variables. The relationships between objectives and variables in a given industrial situation are usually not available to managers in the form of scientific laws. Rather, they are discovered, defined, and labelled by the managers who use an implicit understanding of the operating environment to make sense out of them. We have developed a tool that is capable to enable managers to visualise the 'implicit understanding' of the variables cause-effect relationships. The use of such a visualisation permits managers to have an overall view of the situation.

This paper has discussed the development and application of TAPS. Using TAPS managers can build and modify a variable model interactively and store the information in a database for future reference. Case study results indicate that TAPS is feasible and provides a structured and reliable method for managers to build and understand the complex relationships between production variables. The benefits of the TAPS approach could be summarised as follows:

- Collective understanding – The variable network building process enables everyone to have their knowledge brought into the open and their assumptions challenged. Managers agreed that this process is useful to enhance their understanding of an issue, as well as to facilitate organisational learning.

- Decision support – The built-in filter functions and database enable managers to study variables that have cross-functional relationships. By activating the filter functions, managers could see how the decision made on one variable will effect other functional departments in a firm.
- Facilitate discussion – The variable network helps managers to increase both the depth and breadth of participation in the discussion production management issues. The TAPS approach recognises the importance of assisting the evolution of the managers' ability to deal with the problems confronting them through increasing their understanding of the relevant variables. It provides models of the environment from which a manager can develop insights into the effects of his decisions on progress towards the goals that he wishes to achieve.
- Knowledge Management – The building of a variable network allows information to be passed, assessed and quantified, so that the ideas and beliefs contained within the model can be altered or modified at will.

Results of the case studies showed that the basic structure of this approach seems to be applicable even beyond the manufacturing domain. With TAPS, we are a step closer to provide managers with a useful tool for production operations management.

Acknowledgement
The author would like to thank the University of Nottingham Research Committee New Lecturers Fund for the support of this paper.

References

[1] PIDD, M., 1996, *Tools for Thinking, Modelling in Management Science*, (Chichester: John Wiley and Sons Ltd), pp.211.

[2] TAN, K.H. AND PLATTS, K, 2004, A Connectance Based Approach for Managing Manufacturing Knowledge, *Industrial Management and Data Systems*. Vol.104, No.2, pp. 158-168.

[3] BURBIDGE, J.L., 1984a, *A Production System Variable Connectance Model,* (Bedford: Cranfield Institute of Technology), pp. 1-22.

[4] BURBIDGE, J.L., 1984b, A Classification of Production System Variables. In H. Hubner (ed), *IFIP Production Management Systems: Strategies and Tools for Design*, (North Holland: Elsevier Science Publishers), pp. 3-16.

[5] SAATY, T.L., 1987, The AHP – What It Is and How It Is Used. *Math Modelling*, 9(3-5), pp. 161-176.

[6] SAATY, T.L. (1994). How to Make a Decision: The Analytic Hierarchy Process. *Interfaces*. 24(6), 19-43.

[7] PLATTS, K.W., 1993, A Process Approach To Researching Manufacturing Strategy. *International Journal of Operations and Production Management.* 13(8), pp. 4-17.

[8] TAN, K.H. AND PLATTS, K, 2003, Linking Objectives to Action Plans: A Decision Support Approach Based on the Connectance Concept, *Decision Sciences Journal,* 34(3), pp. 569-593.

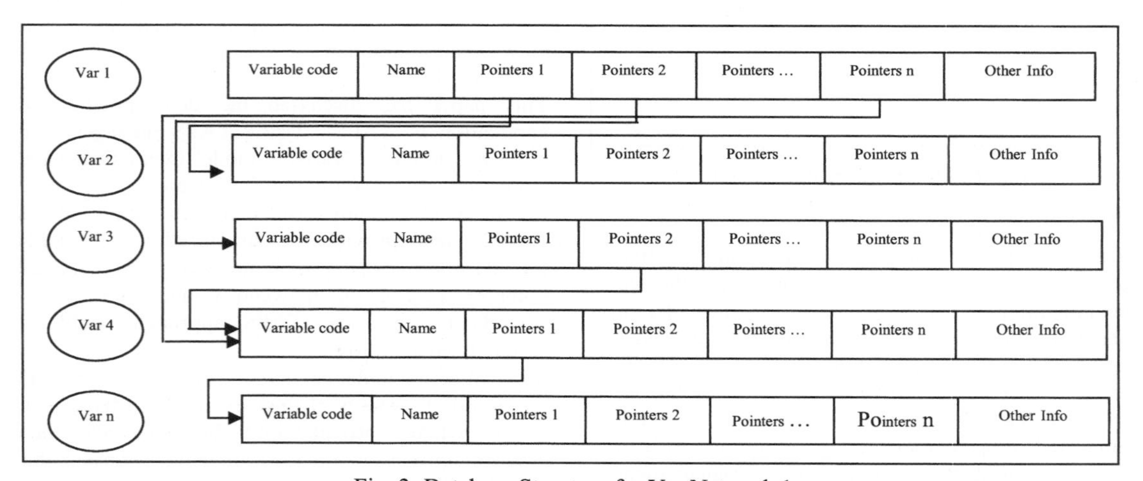

Fig. 3: Database Structure for Var Network 1

Intelligent Production Machines and Systems
D. T. Pham, E. E. Eldukhri and A. J. Soroka (eds)

Toward the Next Generation of Manufacturing Systems. FRABIHO: A Synthesis Model for Distributed Manufacturing

M. Marcos[a], F. Aguayo[b], M.S. Carrilero[a], L. Sevilla[d], J.E. Ares[c], J.R. Lama[b]

[a] *Departamento de Ingenieria Mecanica y Diseño Industrial. Universidad de Cádiz. Escuela Superior de Ingeniería. C/ Chile s/n. E-11003, Cádiz, SPAIN. E-mail: mariano.marcos@uca.es.*
[b] *Escuela Universitaria Politécnica. Universidad de Sevilla. c/ Virgen de África nº 7, E-41011, Sevilla, SPAIN.*
[c] *Departamento de Diseño en la Ingeniería. Universidad de Vigo. Escuela Técnica Superior de Ingenieros Industriales y de Minas. Campus Lagoas Marcosende, Vigo, SPAIN.*
[d] *Departamento de Ingeniería Civil, de Materiales y de Fabricación. Universidad de Málaga. Escuela Técnica Superior de Ingenieros Industriales. Campus de El Ejido, Málaga, SPAIN.*

Abstract

The Next Generation of Manufacturing Systems (NGMS) is being built on the basis of the concepts associated to last generation of enterprise models. These models give rise to the so-called Agile Company, Extended Company and Virtual Company, all them included in a more wide concept which has grown with the Global Company conception. Currently, NGMS contains different distributed manufacturing system models proposed. Some of the most relevant NGMS models have been configured under considerations based on the possibility to establish similitude between a manufacturing system and either social, or biological or fractal systems. However, usually, these models are applied individually without take into consideration the advantages included in the rest of them. In this paper, a synthesis model containing FRActal, BIological and HOlonic (social) considerations has been proposed in order to make good use of the most relevant properties of the previously proposed protomodels. That model has been named as FRABIHO.

---

1. Introduction

New company models to be configured into the Next Generation of Manufacturing Systems (NGMS) are born as a response to the requirements formulated by the new enterprise environments.

According to [1,2], some of the most important requirements are:

a) Complexity and variety of product life-cycle.
b) Decrease of product life-cycle.
c) Increase of capacities and complex technology requirements.
d) Time to market strategies.
e) Clean manufacturing and green product.
f) Product flexibility.
g) Short-time independent enterprises consortia and/or alliances.
h) More relevance of the organisation design in manufacturing systems.

As it has been aforementioned, the above items have promoted the establishment of new company models, which, according to [1-3] can be classified as:

a) Extended company model.
b) Agile company model.
c) Virtual company model.

As it has been reported in [4,5], all of them must be included into the new global manufacturing tendencies, Figure 1.

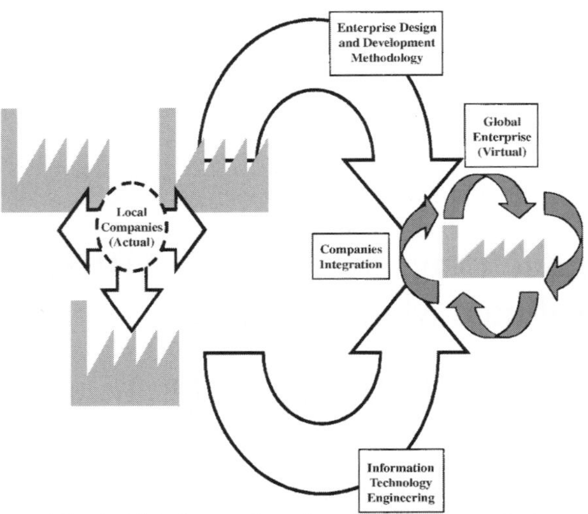

Fig. 1. Global manufacturing through companies' integration

Distributed Manufacturing Systems (DMS) are considered the response to the requirements formulated by these new concepts of companies, acting as a set of manufacturing companies which are distributed local, geographical or virtually.

Precisely, these Manufacturing Systems have given rise to the so-called Next Generation of Manufacturing Systems. Basically, three DMS protomodels have been proposed:

a) Bionic Manufacturing System (BMS).
b) Fractal Manufacturing System (FMS).
c) Holonic Manufacturing System (HMS).

However, these protomodels offer a fractionated research paradigm, without use the synergistic possibilities which would take place through the joint of the advantages of each one of them.

Thus, according to the comments reported by Tharumarajah, Welles and Nemes [3], it can be said that:

a) Low complexity protomodels can be obtained using a FMS model.
b) Self-organised systems can be developed through a BMS protomodel.
c) Self-regularised systems can be used by applying HMS concepts.

In this work, a synthesis model has been proposed in order to both make a good use of the aforementioned advantages and design a model that can be self-organised and self-regularised and, also, can have a low degree of complexity. Nevertheless, before to describe the DMS protomodel, in the next paragraph, a first approximation to Bionic, Holonic and Fractal Manufacturing Systems models is presented.

2. Distributed Manufacturing Systems models

2.1. Fractal Manufacturing Systems

The Fractal Manufacturing System (FMS) is a concept introduced from the Fractal Factory [6]. This concept is based on the establishment of manufacturing cells which can be considering as autonomously cooperating multi-agents. These entities are referred as Basic Fractals Unities (BFU). Each BFU is modelled through modularity criteria including [2,7]:

- Observation
- Analysis
- Resolution
- Organisation
- Report

Fractal architecture is modelled on the basis of a hierarchical structure which is built using BFU as elementary cells [8].

According to this, each fractal unit can represent different cell levels. So, as the entire manufacturing cell as a single machine or element can be named as a fractal [2,8].

2.2. Bionic Manufacturing Systems

The Bionic Manufacturing System (BMS) is a concept which come from considering that a manufacturing system can be modelled through its similarity with a biological systems [1,9]. So, each manufacturing single cell is similar to a biological cell and, because of this, it can be autonomous, cooperative with others, self divisible, and, moreover, its information support acts as the biological genetic code (DNA) [1].

The fundamental BMS cells are called Bio-Modelons [10]. Bio-Modelons are elements and set of elements with static and heuristic properties, from which is built the BMS architecture as a structure where the information control can be made through *enzymatic operators.*

The enzymatic operators define the bio-hierarchy in he BMS model, through the two forms of the information flux control [1]:

- Genetic. The system are described by using analogies with DNA (genetic information variable) and internal biochemical processes.

- Heuristic. BMS is controlled by the information supported, transmitted and filtered through a neurological net.

As it can be appreciated, both points of view contain dynamic and static characteristics.

2.3. Holonic Manufacturing Systems

The Hungarian philosopher Arthur Koestler thought up the term "holon" to describe a basic unit of organisation in biological and social systems in his book "The Ghost in the Machine" [11]. This is the last book in a trilogy where he elaborated the concept of Open Hierarchical Systems (OHS) as a general organiser structure in Nature. The first of the organisation rules of an OHS is based on the consideration that "Parts" and "Wholes" in an absolute sense do not exist in the domain of life. Therefore, the basic elements of a hierarchy in life contain both dependent and autonomous properties. Koestler merged the words "part" and "whole" building the term "holon" as a combination from the Greek "holos" (whole) with the suffix "on", which is suggesting a particle, element or part. Thus, a holon can be considered as a part of a superior holon or a whole that contains other holons. When we are speaking of a holon in the meaning of part it is also called "subholon".

Starting from this point, the corresponding organised system formed by holons is called "Holonic System". The degree of both co-operation and subordination between the different holons that form the system is called "Holarchy".

Holonic Manufacturing Systems (HMS) must be considered as a translation of these concepts into the organiser structure of a manufacturing process [1,12,13]. In this way, the following definitions are given for these kinds of systems [14,15]:

- *Holon:* Basic building block of a manufacturing system which acts in co-operation with other basic elements and autonomously. This building block is thought up for transforming, transporting, storing and/or validating information. Although this information was initially related only with physical objects, at present, it is also associated to each sub-process. Generally, two parts can be distinguished in a holon: an information processing part and, as the case may be, a physical processing part. Of course, in agreement with the concepts introduced in previous sections, a holon can be formed by other holons (subholons).
- *Holarchy:* When a set of co-operative holons is working for achieving a final objective in a manufacturing process, the basic rules for co-operation are defined by the holarchy. Thus, the holarchy is additionally limiting the degree of autonomy of each holon.
- *Holonic Manufacturing Systems:* Holon-based systems, i.e., holarchy, which integrates the full range of activities related to a manufacturing process, including aspects corresponding to design, in its widest conception: production, marketing, etc.

3. DMS protomodel: FRABIHO

As it has been above commented, a synthesis model based on Fractal, Bionic and Holonic considerations has been designed.

The proposed model tries to incorporate the main advantages of the FMS, BMS and HMS models.

This protomodel has been named FRABIHO as an acronym of FRActal, BIonic and HOlonic.

This model tries to be an ontology of both static and dynamic complexity of the natural objects. Therefore, it must include the variety associated to those domains.

FRABIHO has been arranged on the basis of axioms, definitions and immediate propositions, which try to pick up the different analogies with the natural systems employed in the design of manufacturing systems, Table 1.

As a result of combining the varieties of those reference natural objects the following items can be established for FRABIHO protomodel:

a) **FRA**BI**HO**. Holonic model, which contains fractal and holonic varieties. Analogical inspiration at functional levels is articulated by means of holonic procedures. On the other hand, analogical dimension of transposition and imitation of concepts of fractal geometry in the design articulates the fractal behaviour of the model.
b) FRA**BIHO**. Holonic model, which contains bionic and holonic varieties. In this protomodel, the general tasks of design are made under holonic inspiration. On the other

Table 1
Analogies with the natural systems used in the design of manufacturing systems models

FRABIHO							HMS	FMS	BMS
Complexity→ Analogy↓	Functional	Structural	Form	Process	...	Means			
Unconsciousness	-	X	X	X		X			
Inspiration	X	X	X	X		X	X	X	X
Transposition	X	X	X	X		X		X	X
Imitation	X	X	X	X		X		X	X

hand, fine or detailed design is achieved under the imitation and transposition of the reference objects pertaining to biologic domain to the manufacturing domain.

c) **FRABIHO**. Holonic model, which contains fractal, bionic and holonic varieties. It is a synthesis model of the protomodels above described.
d) **FRA**BIHO. Fractal model, which only contains fractal variety.
e) FRA**BI**HO. Bionic model, which only contains bionic variety.

All these can be considered as different variations of the FRABIHO model when they are applied to the different levels of the new enterprise concepts.

When FRABIHO is fully or partially applied, the manufacturing system must be taken as a holon.

Therefore, according to [1,13-14] in order to know, describe and specify this model, a double treatment must be made: WHOLE and ON. These considerations are representative of the holonic conceptions, which give rise to the holonic attributes: autonomy and cooperation. Thus, each single element of a holonic system (holon) can act in collaborative form with the rest of the holons but maintaining its own autonomy.

So, process-product pair and multidimensional complexity must be taken into account and the hierarchy, named holarchy, is the basis of the manufacturing system [16], Fig. 2.

In the NBF notation, the specification of the elements of the FRABIHO model can be made as follows:

<FRABIHO PROTOMODEL> ::=
<<MANUFACTURING SYSTEM AS A HOLON>
<MSYSTEM AS A HOLARCHY>
<HEURISTIC OF HOLONIC MODELLING AND IDENTIFICATION>

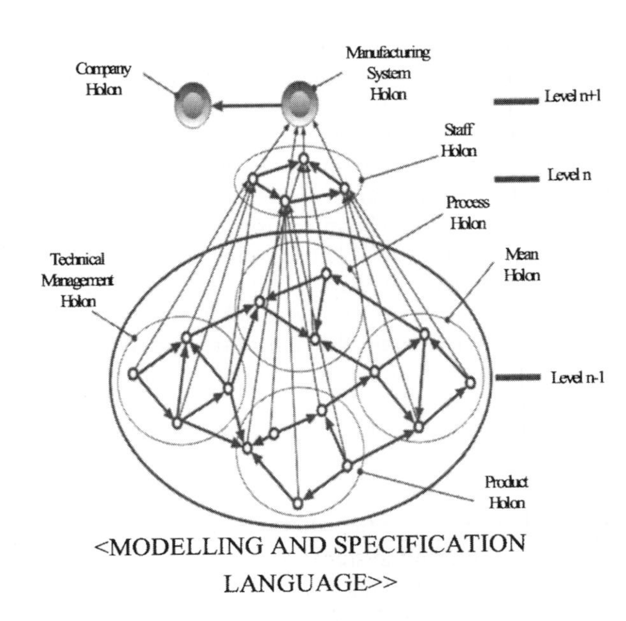

<MODELLING AND SPECIFICATION LANGUAGE>>

Fig. 2. Schematic Interpretation of a Holonic Single Company

4. Modelling FRABIHO

Starting from all previously commented, FRABIHO can be articulated on the basis of the following considerations [1], Fig. 3:

a) Intelligent builders capture all the variety of a manufacturing system when this is considered as a holon. The varieties are captured in all the specification levels.

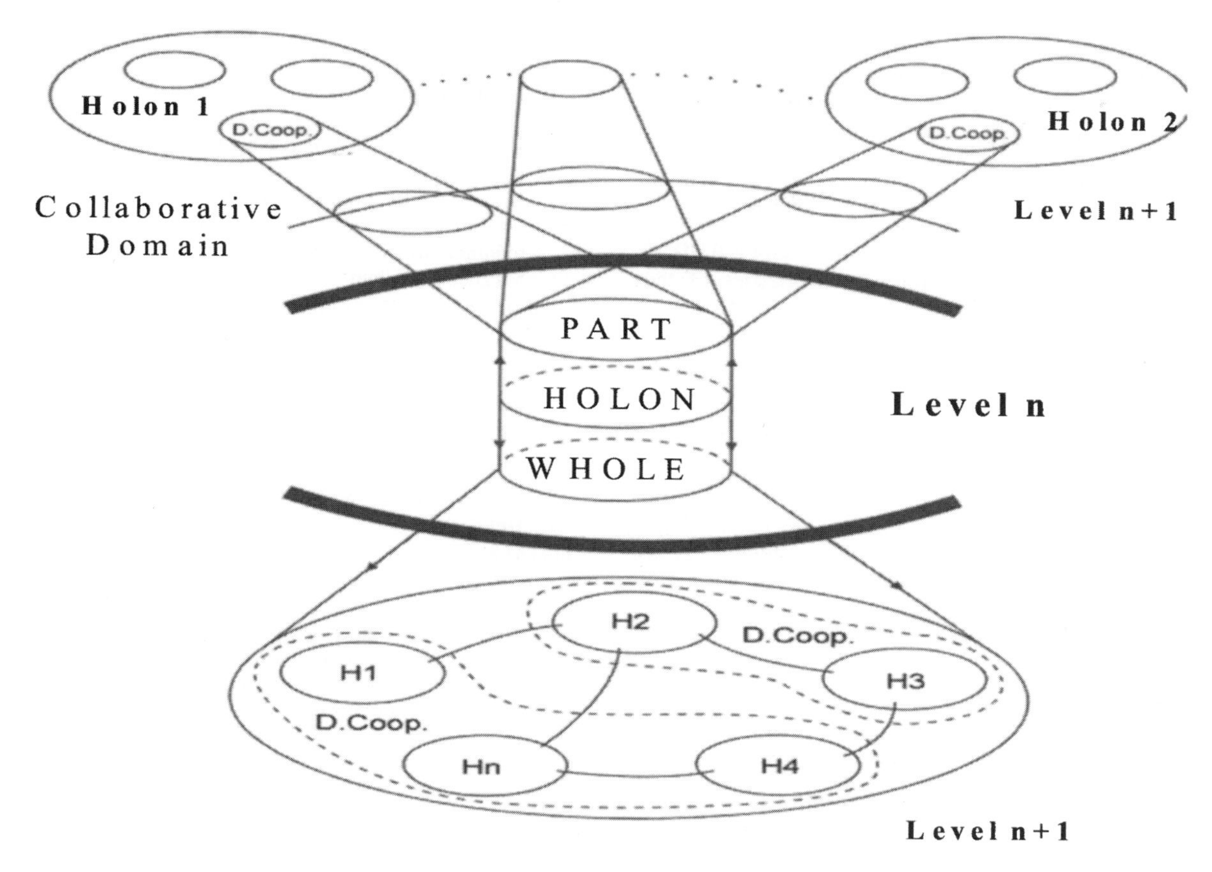

Fig. 3. Main frame of HMS

b) According to the holon definition, the manufacturing system can be taken as either whole or part. Thereby, when manufacturing system is considered as whole, it is available the definition of the internal holarchy and its integration.
c) On the other hand, when the manufacturing system is taken as a part it is available the integration with all the external holarchies.
d) The holonic manufacturing system can be considered as a duality product-process [1]. On this basis, intelligent navigators can move through the general holarchy defined in the preliminary design process.
e) Additionally, intelligent builders can make the detail design and the re-definition of each one of the single or intermediate holons.
f) Finally, the modelling process and the specification of the elements of the holonic system can only be completely achieved by using adequately a modelling language and a correct methodology.

4. Conclusions

New enterprises necessities have given rise to research on modeling new manufacturing systems, which can give a response to the requirements formulated by such enterprises.

The manufacturing systems so emerged are placed into the so-called Next Generation of Manufacturing Systems (NGMS). Most of these models respond to conceptions of Distributed Manufacturing Systems (DMS).

Some of the most relevant DMS models have been configured under considerations based on the possibility to establish similitude between a manufacturing system and either social (Holonic Manufacturing Systems, HMS), or biologic (Bionic

Manufacturing Systems, BMS) or fractal (Fractal Manufacturing Systems, FMS) environments.

However, usually, these protomodels are applied individually without take into account the advantages provided by the rest of them.

In this paper, a synthesis DMS model based on FRActal, BIological and HOlonic attributes has been designed. This protomodel has been called FRABIHO and it contains the most relevant properties of the previously proposed protomodels.

FRABIHO has been articulated on the basis of a set of premises, which give to it, attributes holonic, fractal and bionic from autonomous and collaborative cells that are governed by a hierarchy named holarchy.

Acknowledgements

This work has received financial support by the Andalusian Government (III Plan Andaluz de Investigacion)

References

[1] F. Aguayo, "Design and Manufacturing of products in Holonic Systems. Application to a Holonic Design Module". D. Phil. Thesis, University of Cadiz, Cadiz (Spain), 2003.

[2] A. Molina, J.A. Sánchez, A. Kusiak, Handbook of life cycle engineering. Concepts models y technologies, Ed Kluwer Academic Publishers, London (UK), 1998.

[3] A. Tharumarajah, A.J. Welles, L. Nemes, Comparison of the bionic, fractal and holonic manufacturing system concepts, International Journal Computer Integrated Manufacturing, 9 (3) (1996) 217-226.

[4] F. Aguayo, M.S. Carrilero, J.E. Ares, J.R. Lama, A. González, A. Pastor and M. Marcos, Synthesis models for DMS, 14th International DAAAM Symposium, "Intelligent Manufacturing & Automation: Focus on Reconstruction and Development", Bosnia-Herzegovina, 2003.

[5] R. Bienvenido, F. Aguayo, M.S. Carrilero, J.R. Lama, J. Cano and M. Marcos, Siguiente Generacion de Sistemas de Fabricacion. Un analisis de los distintos marcos paradigmaticos de diseno y desarrollo, XVI Congreso Nacional de Ingenieria Mecanica, Leon (Spain), 2004.

[6] H.J. Warnecke, The Fractal Company: a Revolution in Corporate Culture, Springer, Berlin (Germany), 1993.

[7]K. Ryu, M. Shin, M. Jung, A Methodlogy for Implementing Agent-Based Controllers in the Fractal Manufacturing Systems", 5th Conference on Engineering Design and Automation, Las Vegas (USA), 2001.

[8] K. Ryu, Y. Son and M. Jung, Modelling and specifications of dynamic agents in fractal manufacturing systems, Computers in Industry 52 (2003) 161-182.

[9] G. Langer, Methodology and architecture for holonic multi-cell control system, Ph.D. Thesis, Technical University of Denmark, Copenhagen (Denmark), 1999.

[10] N. Okino, Bionic Manufacturing Systems (BMS). Bio-Modelon based systems design. CAM-I, Annual International Conference, New Orleans (USA), 1980.

[11] A. Koestler, The Ghost in the Machine (1st Edn.), Arkana Books. London (UK), 1967.

[12] A. Tharumarajah, A self-organising view of manufacturing enterprises, Computers in Industry 51(2003) 185-196

[13] R. Bell, S. Rahimifard and K.T.K. Toh, Holonic systems, in: *Handbook of Life Cycle Engineering: Concepts, Models and Technologies*, A. Molina, J.M. Sanchez, A. Kusiak, (Eds), 115-150, Ed. Kluwer Academic Publishers, London (UK), 1998.

[14] J.H. Wyns, *Reference architecture for holonic manufacturing systems*, Ph. D. Thesis, KUL, Leuven, Belgium, 1999.

[15] M.S. Carrilero, F. Aguayo, J.R. Lama, J.E. Ares and M. Marcos, Integracion de modelos Bionicos, Holonicos y Fractales para Fabricación Distribuida, XVI Congreso Nacional de Ingenieria Mecanica, Leon (Spain), 2004.

[16] C.J. Luis, M.S. Carrilero, J.M. González, J.M. Sánchez-Sola and M. Marcos, An approximation to Holonic Manufacturing Systems, Revista del Instituto de la Maqina Herramienta de Espana (Journal of the Spanish MT Institute), IMHE, (1998) 269-274.

Intelligent Production Machines and Systems
D. T. Pham, E. E. Eldukhri and A. J. Soroka (eds)

An Enterprise Collaborative Portal with a Case-Based Reasoning Tool for Business Process Model Creation

D. T. Pham, S. S. Dimov and D. K. Tsaneva

Manufacturing Engineering Centre,
Cardiff University, UK
TsanevaD@cf.ac.uk

Abstract

This paper proposes an approach employing Case-based Reasoning (CBR) techniques for collaborative authoring of business process models by a geographically dispersed team. The approach enables co-ordination of dynamic workflows of the virtual enterprise team. The paper describes an enterprise collaborative portal architecture combining collaboration, graphical process modelling and CBR tools and presents a case study illustrating the proposed approach.

1. Introduction

Business processes change rapidly in the current manufacturing environment, especially when a company is involved in making highly customised products or simultaneously running different projects. In some cases customers require modification of an existing product. This leads to changes in the manufacturing processes and hence to business process re-engineering. For an enterprise to be successful in such environments, it has to establish a corresponding culture for continuous improvements and be supported by responsive business processes. In this aspect, the Case-based Reasoning (CBR) methodology can be applied together with business process modelling techniques to produce new process models from those in the case repository or adapt the existing ones.

Another aspect of contemporary manufacturing enterprises is the need for interdisciplinary teams to collaborate and for their activities to be co-ordinated. When a project team is located at more than one site, it is very difficult to synchronise their tasks and optimise the usage of these distributed resources. Therefore, it is essential that all participants contribute to the creation of the process models in the early stages of the project. In addition, such geographically dispersed teams need a suitable interactive environment to model business processes concurrently. The objective of such collaborative model development is to enable teams to improve their business processes and shorten product lead-times.

This paper proposes an approach for modelling business processes using CBR. The proposed approach enables co-ordination of dynamic workflows in situations where the enterprise team is geographically dispersed. Section 2 reviews current trends in Enterprise Collaborative Portals (ECPs), Business Process Modelling (BPM) and CBR. Section 3 describes the proposed approach and Section 4, the system architecture to support it. Section 5 provides a case study demonstrating an application of the approach. Section 6 concludes the paper.

2. Enterprise Collaborative Portals, Business Process Modelling and Case Based Reasoning

A product development project usually includes a set of work packages or activities that are executed by different people. In the case of complex products developed by a multinational company, projects may involve teams at different locations. At any time, different engineers may work on tasks that are related to one another. To prevent inconsistency and reduce redundant activities, engineers must collaborate closely and project activities must be well co-ordinated. There is a need for an efficient mechanism for carrying out such co-ordination [1].

An Enterprise Collaborative Portal (ECP) can help enterprise users to share information and make it available over the Internet. It gives an opportunity to users to contribute to a project simultaneously throughout the development cycle by accessing and modifying shared information within a controlled environment [2].

A new approach is required in the ECP design, for example, to provide functionality which helps larger groups of people to work together more efficiently [3]. Tight integration with suppliers and customers is also increasingly necessary.

Several reasons for requiring innovation in business process modelling are given in [4]:

- There are always a large number of projects to be managed at any one time;
- Frequent changes in design and delivery requirements during the course of a project call for frequent re-scheduling and re-planning, which are time-consuming;
- Many customers now require detailed information with regard to the time scale and cost breakdown of development projects. There is a need for planning systems allowing customers to tailor the product, and the design tasks, to suit their own lead time and cost requirements.

CBR provides a methodology for capturing problem-solving expertise. The CBR methodology could be applied together with business process modelling techniques to produce new process models from those held in the case base. This should facilitate and speed up the creation of new process models and the reengineering of existing models.

3. Proposed approach

3.1 Collaborative authoring of process models

The issue being addressed is that of process co-ordination in large distributed business environments. A solution will be proposed based on an ECP that supports both decision-making processes and collaborative working practices.

The idea behind developing an enterprise modelling and integration framework is that many components/tasks of business process re-engineering projects are common to most businesses. Thus, proven solutions could be captured, standardised, and re-used instead of being developed from scratch each time. Once standardised and accepted, these frameworks can be supported by models and methodologies, leading to time and cost efficiencies. Such standard models could then be stored in a case repository and then reused by applying intelligent retrieval techniques.

The proposed approach incorporates emerging concepts from the areas of BPM, ECP and CBR. Collaborative portals enable members of distributed teams to discuss any issue associated with the business. To facilitate this, a graphical process-modelling tool is proposed that allows process models to be designed collaboratively. By means of the portal, through a user-friendly interface, simultaneous access by every team member to predefined process templates could be provided. With these templates, users specify the attributes of the processes, for example, cost and duration, and their values, plus events and responsible units. The processes are defined together with their sub-processes and existing interrelationships. In addition, a CBR tool is proposed to assist users in retrieving existing process models from the case repository that could be adapted to the new requirements of the collaborative environment.

3.2 The proposed approach

The proposed approach is presented in Fig. 1. The input information required is any type of process diagram. The approach involves using the Product/process (P/p) graph [5] to represent the modelled process. This particular representation is appropriate particularly because the P/p methodology associated with the P/p graph combines systems concepts with those of traditional engineering disciplines, including rigorous models, measurements and quality management [5].

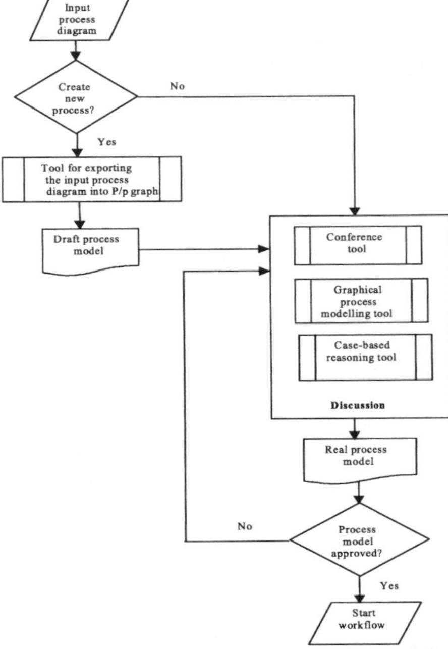

Fig. 1. Algorithm for creating a process model

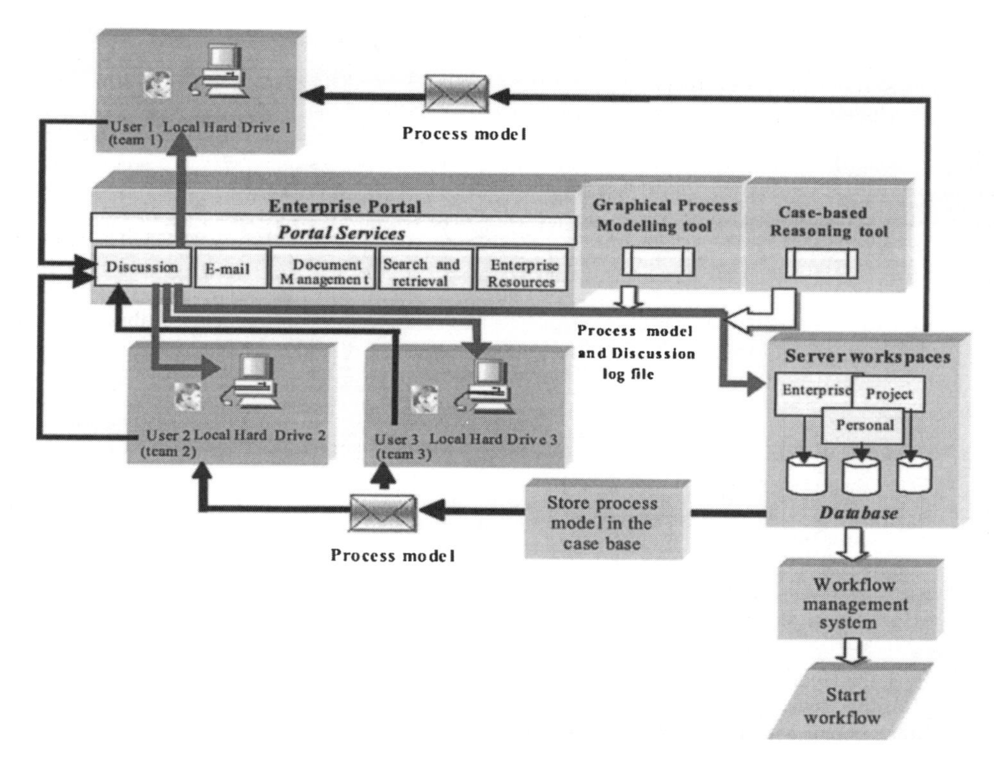

Fig. 2. Proposed system architecture

Input information in the form of a process diagram is used to check whether it is necessary to create a new process model. If the process already exists, it is retrieved from the case base to initiate a discussion among the members of the team concerned. All the participants in the discussion have access to collaboration and Graphical Process Modelling (GPM) tools and the portal services. The CBR tool assists them in identifying the best existing match to specified requirements. When the discussion is completed, the generated process model is added to the case base for future reference and at the same time is sent by email to all concerned parties for final approval. If the generated process model is 'approved, a workflow is activated automatically. Otherwise, a new discussion between the team members is initiated until they reach agreement.

In case it is necessary to create a new business process, the process diagram, once prepared, is converted into an extended P/p graph so that a draft process model can be created. This draft model can then be analysed by all the members of the team employing the collaboration, GMP and CBR tools available at the portal. On completion of this analysis, an executable process model is generated and sent automatically to all participants in the discussion. The process of approval could involve further discussions until the process model satisfies the requirements of all team members. Finally, a workflow based on the generated process model is implemented.

4. System architecture

Several new tools are required to enable automated workflow modelling: Graphical Process Modelling (GPM) tool, Collaboration tool and Case-based reasoning (CBR) tool. These tools communicate with one another and work in collaboration assisted by the services of the enterprise portal.

The architecture of an ECP that includes these tools is shown in (see Fig. 2). The ECP is described only with its server tier (including Portal Services and Server Workspaces) and data tier (databases) for simplicity. The user access tier of the portal could be either via a Web server, or through an Application Programming Interfaces (API) within C, C++, Java or Visual Basic applications.

When a discussion is triggered, all the data exchanged between different teams, for example, Teaml (customer), Team 2 (manufacturer) and Team 3 (supplier), is stored in a log file together with the business process model resulting from it. When the discussion finishes, this log file

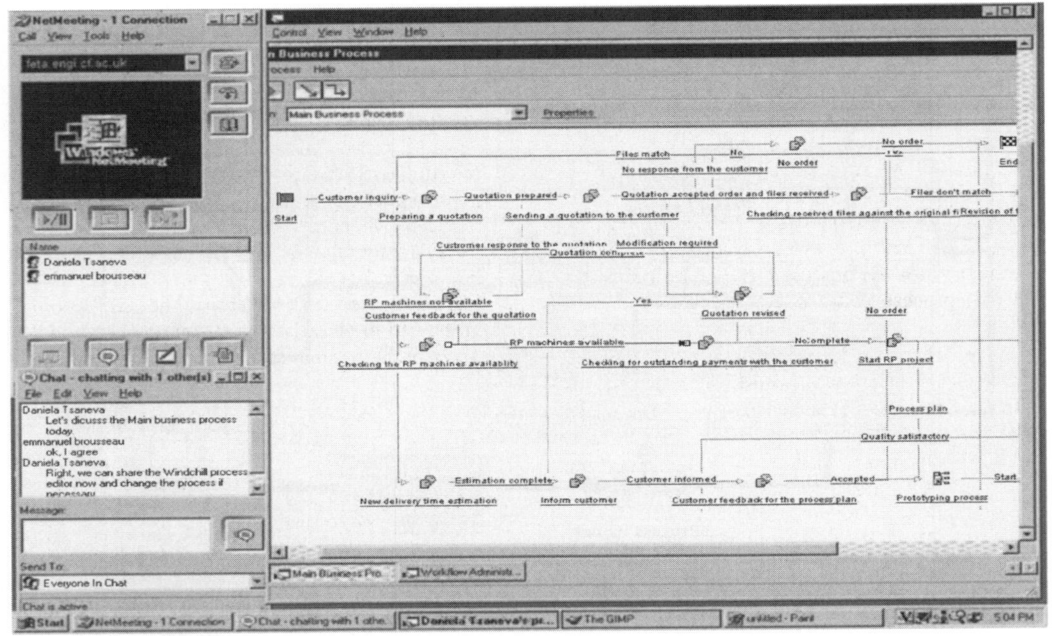

Fig. 3. Collaborative process modelling using NetMeeting Version 3.01 Microsoft Co. TM

(designated as "Process model and Discussion log file" in Fig. 2) is stored, both on the server and on the local disk space for each team member. The GPM tool is employed by the team in collaborative mode to create the process model or introduce changes in an existing model. The approval of the model is carried out via email notification. After approval, a workflow management system takes over the process model execution.

4.1 Graphical process modelling tool

The GPM tool is a very important component of the proposed collaborative environment. It is an application shared by all participants to enable them to visualise business processes during the discussion.

A workflow process editor is employed as the GPM tool for this system. With the proposed approach, the process diagram that is discussed in each session has to be created in advance using this workflow process editor.

After the creation of the draft process diagram, the discussion between the participants in the project could be initiated. The GPM tool is collaboratively applied by the team members to help them visualise changes and discuss the process model until they reach an agreement. The resulting model is saved as a process template and also it is added to the case base for future reference (see Fig. 3).

4.2 Collaboration tool

The collaboration tool employed in this work is an integral part of the ECP. It is based on NetMeeting Version 3.01 (see Fig. 3), which enables distributed product development teams to work together on projects through web-based workspaces. Cross-functional participants from marketing, engineering, procurement, manufacturing, sales and service departments can participate in the product development process in addition to suppliers, manufacturing partners and customers.

4.3 Case-based reasoning tool

The role of the CBR tool is to enable teams working on process modelling tasks to reuse solutions (business process models) developed previously, which could be employed partially or fully to solve new problems [6].

Commercially available systems [7] provide basic business process templates, such as Analysis Process, Change Activity Process, Change Request Process, Change Order Process, Example Auto Routing, Replication Sender, Replication Receive, Review and Submit, that may be adapted to suit specific needs by a workflow designer. These templates, called cases, can be modified individually. Alternatively, multiple cases may be composed into a more complex business process and the assembled process subsequently altered as necessary to meet new business requirements [8].

The case base of the system has to be created so it can be used during the generation of the process model for CBR. In this work, the basic process templates of a commercially available enterprise information management system are used to form a case base. Additionally, several UML basic process patterns [9] are created and stored in there. These UML patterns can be reused in most of the business processes representing basic structures of the process model. The following UML process patterns are added to the case base:

- Process feedback pattern – evaluates the business process results and, based on those results, adjusts the process accordingly to achieve the business process goal;
- Action workflow pattern – represents a tool for analysing and optimising the communication between different parties;
- Basic process structure – provides the basic structure for describing a business process. It shows how to form the business process concept in terms of supplying business resources and goals for the process and the transformation or refinement of input and output resource objects;
- Process layer supply – organises the structure of complex organisations into primary and supporting business processes;
- Time-to-customer – demonstrates how to describe a business with two main processes in order to shorten the lead-time from customer demand to customer satisfaction.

These UML process patterns are represented as workflow templates by a graphical process editor and stored in the case base.

During the creation of the process model, the process designer is provided with a high-level description of the business process. He uses the CBR tool to retrieve past cases that embed characteristics "similar" to the new requirement. After that the process designer will analyse these cases, select relevant cases and composes them into a new solution. This approach is very efficient, leads to time and cost savings, and is very suitable when Business Process Reengineering is required since it deals with existing process models which need to be changed.

The architecture of the CBR tool is presented in Fig. 4. The case base manager provides the basic functionality of the CBR tool, including case base indexing, querying and case retrieval functions. The case base consists of an initial set of models, covering the main UML process patterns and the basic process templates of the Enterprise Information Management System (EIMS) adopted. The CBR process control module provides the link between the process designer, who specifies the queries, and the case base manager which conducts the search for relevant cases. If more than one case matches the search requirements, the cases are ranked in regard to a given matching criterion. The Adaptation engine is used when no similar cases are found or alternatively the retrieved cases need to be adapted and verified, or multiple cases combined to produce a new solution. The GPM tool is employed to visualise the retrieved cases and then to adapt them interactively to new requirements.

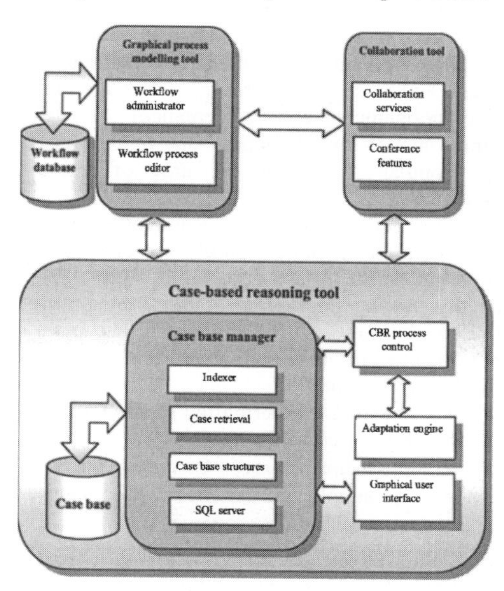

Fig. 4. The case-base reasoning tool architecture

4.4 *Portal services*

The ECP provides a framework for integrating different business applications, such as BPM, workflow automation, product visualisation and collaboration, discussion and conference services. It acts also as a mediator between the business applications and the information provided by the system.

Some of the services supported by the ECP, including meetings, version control, report building, file manager, user login, user profiling, customisation, system notification, calendar, workflow engine, application integration and enterprise resources, are already provided by the commercial system employed in this work. Other services, such as collaboration, are supplied by the Collaboration tool. Additionally, some services, such as channels and the document management and personalisation services, are integrated into the portal as servlets. The portal services interface is based on Java Server Pages (JSP).

5. Case study

A case study concerning the production of a prototype device for training users in golf will be used to demonstrate the proposed approach. This particular case study was chosen because the business processes associated with the rapid prototyping of this product represent a good example of collaboration between the project partners, including customer, supplier and manufacturer. Another reason to choose this case study was that the main business process contains subprocesses and feedback and covers the four main phases (Preparation, Negotiation, Accomplishment and Acceptance) of the Flores model for interaction, one of the most the popular models in the area of action workflow.

The golf training device is assembled from four parts: body, pulley, pulley vee block and squaring bar. The task is to build a prototype of this device employing Rapid Prototyping (RP) technology (SLS DuraForm). The main participants in fulfilling this task are the following three parties:

- the client, ordering the production of the training device – "Company X";
- the manufacturer - "Company Z";
- the supplier of materials – "Company Y".

The team that should create/agree a business process for providing a RP service includes:

- Product engineer from "Company X" who has created the 3D model of the product;
- Production manager from "Company Z" who co-ordinates all the processes within the company;
- RP project manager who is responsible for the project within "Company Z";
- Material supplier from "Company Y".

The client sends his request for prototyping the GTD to the manufacturer. The project manager considers the request and then creates a new project for managing the prototyping business process and the GTD product structure within Windchill [7], which was the EIMS adopted in this case study. The production manager is also responsible for preparing the initial diagrams of the business processes that then will be discussed between all the parties involved. For this case study, extended Product/process graphs constructed in accordance with the P/p methodology [5] are selected as input for the GPM tool.

The project manager then triggers the discussion between the project partners by emailing all participants the CAD model of the product and also the references to the product and project data in Windchill. At the agreed date and time, the discussion commences using the Windows NetMeeting Version 3.01 collaboration tool (see Fig. 3). All participants load the product and project data from the database. The project manager starts the GPM tool and loads the diagram representing the main business process, which has been prepared in advance. Since the GPM tool provides the participants with a shared workspace, all participants can change processes, activities or organisational units and, if required, even delete or add new activities and attributes. The participants can also send remarks to one another, as shown in Fig. 3.

In case a similar previous process is available in the case repository, this process could be retrieved and modified by the project participants using the GPM tool.

During the discussion, it may be required to change the process diagrams several times. For example, the project manager could suggest a new event "Quotation revised" to be added to the process. Also it might be necessary to add other activities such as "Checking received files against the original files", "Checking the RP machine availability" and "Checking for outstanding payments with the customer" before starting the RP subprocess. In addition, the RP project manager could suggest adding quality checks after "Cleaning and post processing", "Joining parts" and "Painting" to the RP subprocess diagram. The product engineer, who represents the client, may propose the activities "Delivery time estimation" and "Process plan revised" to be included in the process diagram to enable the selection of the most appropriate manufacturing route. The material supplier identifies that the activity "Change material" is not always necessary, so he suggests modifying the process by adding a new event "Change material required" in the RP subprocess diagram.

The output from the GPM tool is a workflow diagram that represents the discussed business process (see Fig. 3). The workflow diagram is created using the PTC Windchill Process Administrator ™. The output of the system consists of workflow templates stored in the Windchill database that could be exported into files with neutral format, such as XML.

The data from each session is stored on the server in the common workspace and locally on the computer of every participant in the discussion, together with the generated business process model. The output of the system becomes input information for a workflow system. A workflow is initiated using this process model and then executed.

6. Conclusions

This paper has presented an ECP with a CBR tool for collaborative authoring of business process models. The ultimate idea is to define an innovative approach for business process modelling by employing ECP for internationally distributed team members. The system described in the paper benefits users in several ways:

- **Time savings** – the time to create a new process model or to change an existing model is shortened by using a CBR tool and the ECP features.
- **Cost savings** – communication between the different partners in the team is facilitated by the adoption of ECP technologies.
- **Knowledge enrichment** – this occurs through sharing of ideas and creation of new process models, which are stored for future reference in the case base.

Further work could be performed in the context of BPM and ECP. Tasks to be covered in the future include enriching and reorganising the case base and refining the retrieval and adaptation techniques. In addition, the algorithm for collaborative authoring of dynamic process models could be refined in accordance with the principles of Concurrent Engineering, which will lead to more efficient use of the current resources and reduction of the investment costs related to the business processes involved.

Acknowledgements

The research reported in this paper was carried out within the Objective 2 ERDF project "ITEE", the Objective 1 ERDF project ''Superman'' and the DTI project CoTeam. The case study in the paper includes example information supplied by the Manufacturing Engineering Centre, Cardiff University, Wales, UK. This work was also part funded by the EPSRC Innovative Manufacturing Research Centre and the Network of Excellence "Innovative Production Machines and Systems" (I*PROMS).

References

[1] Huang GQ, Huang J and Mak KL. Agent-based workflow management in collaborative product development on the Internet. CAD, 32 (2), 2000, pp. 133-144

[2] Rezayat M. The Enterprise Web Portal for life-cycle support, CAD, 32 (2), 2000, pp. 85-96

[3] Pham DT, Dimov SS and Tsaneva DK. Enterprise Collaborative Portal for Business Process Modelling. Proc. of Int. Conf. on Perf. Meas., Benchm. and Best Pract. in New Economy, Univ. of Minho, Braga, Portugal, 2003, pp. 517 – 523

[4] Tam S, Lee WB, Chung WWC and Lau HCW. An object-based process planning and scheduling model in a product design environment. Log. Inf. Man., 13 (4), 2000, pp. 191-200

[5] Kaposi A and Myers M. Systems for all. Imperial College Press, London, UK, 2001

[6] Choy KL and Lee WB. Task allocation using case-based reasoning for distributed manufacturing systems. Log. Inf. Man., 13 (3), 2000, pp. 167 - 176

[7] PTC. Windchill Business Administrator's Guide. Windchill Rel. 6.2, Needham, MA, USA, 2001

[8] Madhusudan T and Zhao JL. A Case-Based Framework for Workflow Model Managament, BPM 2003, Spr.-Verlag, Berlin and Heid., LNCS 2678, 2003, pp. 354 – 369

[9] Eriksson H and Penker M. Business modelling with UML: business patterns at work. Wiley, New York, USA, 2000

Intelligent Production Machines and Systems
D. T. Pham, E. E. Eldukhri and A. J. Soroka (eds)

Automatic formation of rules for feature recognition in solid models

S.S. Dimov[a], E.B. Brousseau[a], R.M. Setchi[a]

[a] *The Manufacturing Engineering Centre, Cardiff University, Cardiff CF24 3AA, UK*

Abstract

This paper discusses the application of inductive learning techniques for creation of rule sets that could be utilised for Automatic Feature Recognition (AFR) in 3D solid models. AFR techniques are an important tool for achieving a true integration of Computer-Aided Design (CAD) and Computer-Aided Manufacturing (CAM) processes. In particular, AFR systems allow the identification in CAD models of high-level geometrical entities, features that are semantically significant for manufacturing operations. In this paper, a method is proposed to meet the specific requirements imposed by the utilisation of inductive learning for acquisition of feature recognition rules. The method presented in this study is implemented within a prototype feature recognition system and its capabilities are verified on a benchmarking part.

1. Introduction

The design of products and their consecutive manufacturing requires information at different levels of abstraction. One of the data representation schemes that is widely used to interface Computer-Aided Design (CAD) and Computer-Aided Manufacturing (CAM) processes is the Boundary Representation (B-Rep) scheme. This scheme is very popular in spite of its drawbacks [1]. In particular, the geometrical data stored using the B-Rep scheme cannot be utilised directly for process design because this data lacks high-level geometrical entities that are meaningful from a manufacturing point of view. To bridge this information gap between CAD and CAM, Automatic Feature Recognition (AFR) techniques are applied to identify geometrical entities, features in the CAD model, which are semantically significant in the context of specific downstream manufacturing activities.

For the last twenty-five years, many AFR techniques have been proposed for recognising both simple and interacting features. Such techniques implement approaches based on rules [2, 3] or neural networks [4, 5] for example. However, the focus of the research efforts was on developing techniques for recognising features in the context of a particular manufacturing application, and especially for a range of machining processes. A true CAD/CAM integration requires the development of AFR systems that could be easily applied in different domains. For example, such a system could be used to assess the manufacturability of a given product or part with regard to various production processes.

The objective of this research is to address knowledge acquisition issues associated with the development of AFR systems that could be employed in different application domains. To achieve this, the paper proposes a new method for generating automatically feature recognition rules. In particular, these rules are formed by applying an inductive learning algorithm on training data consisting of feature examples.

The paper is organised as follows. Section 2

presents the proposed method for automatic formation of feature recognition rules. Section 3 illustrates the method for a particular application domain. A prototype AFR system implementing this method is presented in Section 4 and its capabilities are verified on a benchmarking part. Finally, Section 5 presents the main conclusions of this research.

2. Automatic formation of feature recognition rules

2.1. Inductive learning

In this research, it is suggested to generate feature recognition rules by applying an inductive learning algorithm on training data consisting of feature examples. Inductive learning algorithms are a subset of machine learning algorithms. A common characteristic of machine learning techniques is that they identify hidden patterns in training data in order to automatically build classification models for a given application domain. The inductive learning algorithms create models that are represented as rule sets or decision trees. The rule sets include IF-THEN rules that can be readily interpreted by experts and can be used for automatic generation of rule bases for expert systems.

In this study, an inductive learning algorithm that forms classification models represented as rule sets is adopted. In particular, the algorithm utilised is RULES-5 [6]. It belongs to the RULES family of inductive learning algorithms [7]. During the induction process, RULES-5 repeatedly takes an example not covered by the previously created rules and forms a new rule for it until all examples in the training data are covered by the generated rule set.

Like all algorithms for inductive learning, RULES-5 requires the input data to be in a specific format. In particular, the data files presented to such algorithms should contain a collection of objects, each belonging to one of a number of given classes. Each object is described by its class value and by a set of attribute values represented as a vector. Each attribute value in this vector can be either discrete or continuous. The next section discusses the requirements to form training sets for acquisition of feature recognition rules that are suitable for inductive learning.

2.2. Training data creation

2.2.1. Proposed approach

To generate the required training data in the context of this research, the following three steps are proposed (Fig. 1):

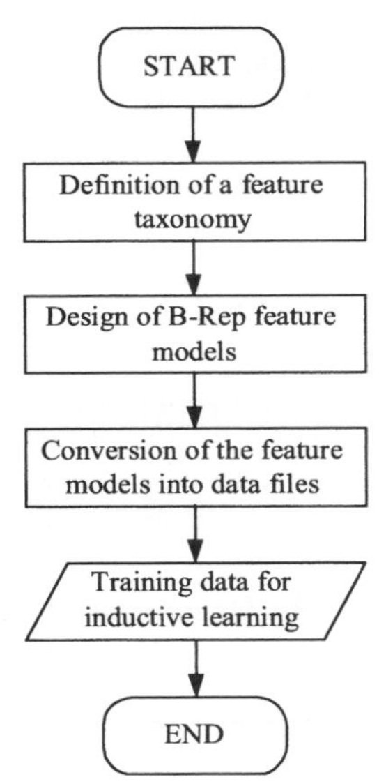

Fig. 1. Training data creation.

- First, a taxonomy that represents the feature classes for a given application domain is defined. For example, the proposed method could be applied to generate rules for recognising machining features that are associated with particular manufacturing methods/machining strategies. Thus, a taxonomy reflecting the specific requirements of this application should be adopted. Such a domain specific classification groups the data available for each feature class and guides the search for pattern recognition rules. However, the proposed method should not be limited to any particular application. For this reason, only the most generic part of a feature taxonomy that could be considered application-independent is adopted in this research. Therefore, the top-level classification of features into protrusions, depressions or surfaces proposed by Gindy [8] is applied because it satisfies this requirement.
- Second, a set of B-Rep models representing examples of features is designed for each class of a given taxonomy. For example, in the machining domain, a feature class could cover all B-Rep models of slot features. A systematic approach is implemented in designing the B-Rep models of features. This is done to achieve some balance between different feature classes represented in the training set. As a result, the weight of all feature classes during the induction process is the

same. The systematic approach adopted in this research for designing the B-Rep models of features is described in more detail in Section 3.1 and an example to illustrate it is provided.

- Third, the B-Rep models of features are converted into data files that are suitable for inductive learning. This conversion is necessary because the models cannot be used directly for inductive learning. In particular, the input data for such algorithms should be in a tabular format. The next section discusses the requirements for converting the B-Rep models of features into training data that is suitable for inductive learning.

2.2.2. Data format

The data files for inductive learning contain a collection of objects. In this research, these objects are called characteristic vectors. Each characteristic vector is composed of attributes that store information about a given B-Rep feature model. Thus, each vector also belongs to the feature class of the respective model. The feature classes determine the taxonomy that is applied to classify features in a particular application domain. Thus, to classify all characteristic vectors belonging to a given class, a coding scheme for storing the information contained in B-Rep feature models should be implemented.

In general, the definition of a coding/representation scheme to encapsulate meaningful data about a specific engineering domain is not a trivial task. In the field of AFR, a systematic approach for defining representation schemes that code feature information from B-Rep models does not exist. Therefore, in this research the following general guidelines are adopted in designing such a scheme:

- The specific characteristics of the application domain should be taken into account in deciding which attributes should be included in a characteristic vector. For example, if machining features are considered, the characteristic vector should be composed of attributes that represent geometrical forms associated with removal volumes.
- A characteristic vector should include as many attributes as possible whilst avoiding the inclusion of irrelevant attributes. In this research, an example of such an attribute is 'the date at which a feature has been created'. Such information is obviously not important for feature recognition tasks. However, this does not necessarily mean dismissing attributes that are considered irrelevant by a domain expert because an interesting characteristic of inductive learning techniques is their capability to generalise and create rules that are not 'obvious' at first for experts.
- The attributes included in a characteristic vector should provide sufficient information to solve a recognition task at a particular level of abstraction. For example, if an AFR system needs to analyse local-level information about faces defining a feature, the characteristic vector should include attributes that represent data associated with feature faces. The following section describes the specific approach adopted in this study for designing characteristic vectors at different levels of abstraction.

2.2.3. Characteristic vector design

In this research, the feature concept proposed by Sakurai and Gossard [9] is adopted. A feature is defined as a single face or a set of contiguous faces, called a face set. Thus, two levels of abstraction should be considered in identifying feature patterns that are representative for a given application domain. Correspondingly, two sets of training data should be formed in order to extract two different sets of rules. The first level of abstraction represents feature faces as single entities. Then, at the second level, the face set that defines a feature is analysed.

In the first training set, each characteristic vector stores information about a single feature face. Thus, several vectors are extracted from each B-Rep feature model. The rules generated from such training data are referred to as the first set of rules. They define feature patterns that are extracted from partial representations of a feature. However, such a coding scheme has some limitations. In particular, it is difficult to identify a set of attributes that does not lead to code duplications, i.e. characteristic vectors that are completely identical, having the same values for all their attributes and, at the same time, belonging to different feature classes. In the machine learning domain, such code duplications are called noise. Thus, such vectors in the training set should be avoided because they introduce ambiguity and could prevent inductive learning algorithms from generating valid rules for some feature classes.

Another problem with such a coding scheme is associated with the intrinsic nature of a feature. In a B-Rep model, any individual feature face represents a low level of information about a feature. Although a feature has been defined previously as a single face or a set of contiguous faces, it is generally the case that a feature is made of more than one face. Thus, this low-level representation scheme should be complemented by another scheme that could represent data at a higher level of abstraction. The training data generated at this second level of abstraction should contain characteristic vectors that include information about the face set defining a feature. Thus, this second training set should include only one vector for each B-Rep feature model. In this research, the rules generated from such training data are referred to as the second set of rules.

3. Illustrative example

This section discusses a possible implementation of the proposed method for automatic formation of feature recognition rules that represent patterns identified in B-Rep feature models.

3.1. Training data creation

3.1.1. Feature taxonomy

In this example, the taxonomy that is adopted categorises features belonging to the machining domain (see Table 1). It is inspired by the classification of machining features proposed by Pham and Dimov [10]. The B-Rep feature models considered in this study are also represented in Table 1. They are constructed using a solid modelling system and thus, they define closed volumes. However, a single face or a face set representing a feature does not necessarily define a closed volume. For this reason, each B-Rep feature model includes feature faces together with faces belonging to a base protrusion.

A systematic approach is implemented to construct the B-Rep feature models shown in Table 1 applying the following three guiding principles:

- First, to balance the importance of each feature class in the training set, the same number of models is created for each class. The importance of this was already highlighted in Section 2.2.1.
- Second, the same base protrusions are used for all classes in order to minimise any influence that they could have on the inductive learning process. Thus, if a given base protrusion is utilised for creating a model for one feature class, it is also utilised for the other classes.
- Third, these base protrusions are different in order to vary the topological and geometrical neighbouring configurations of features. This is required because the number and the type of boundary edges of a feature could differ.

For simplicity, only planar and cylindrical faces are used to construct the feature models shown in Table 1. However, the proposed method is not restricted to these types of geometric entities only. Also, it should be noted that not all faces included in a B-Rep feature model have to be considered in designing a coding scheme. For example, the faces defining the base protrusions in these models should not be taken into account during the coding because they will not provide additional information for the feature recognition process. Even, the vectors created for these faces could introduce a noise in the training sets. That is why the relevant faces in the B-Rep models are selected by end-users to form the characteristic vectors for each feature. The process of designing a feature coding scheme at both levels of abstraction is discussed in the next section.

Table 1
Taxonomy of machining features

Generic feature type	Feature class	B-Rep feature models
Depression	Rectangular pocket (po_re)	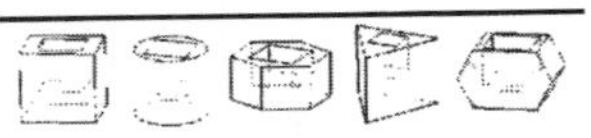
	Obround pocket (po_ob)	
	Blind hole (ho_bl)	
	Through hole (ho_th)	
	Through slot (sl_th)	
	Non-through slot (sl_nt)	
	Step (st)	
Protrusion	Circular protrusion (pr_ci)	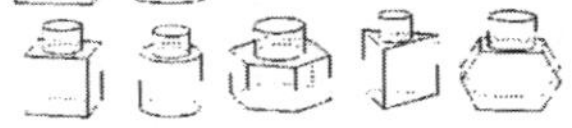
	Rectangular protrusion (pr_re)	

3.1.2. Feature coding schemes

At the first level of abstraction, topological and geometrical data about individual feature faces are utilised in defining a coding scheme. At the second level of abstraction, attributes belonging to a face set defining a feature are utilised. Table 2 shows the attributes that are considered to be of importance in distinguishing one feature from another at this level of abstraction.

Table 2
Attributes included in the characteristic vector at the second level of abstraction

Attribute number	Symbol	Description
1	nFa	The number of feature faces.
2	nPlFa	The number of planar feature faces.
3	nCcClFa	The number of concave cylindrical feature faces.
4	nCvClFa	The number of convex cylindrical feature faces.
5	nCcEd	The number of concave internal edges.
6	nCvEd	The number of convex internal edges.

Table 3
Second set of rules

Rule	Rule description
1	IF nCcEd=8 THEN featureClass = po_re
2	IF nPlFa=3 AND nCcClFa=2 THEN featureClass = po_ob
3	IF nFa = 2 AND nCcClFa=2 THEN featureClass = ho_th
4	IF nPlFa=1 AND nCcEd=2 THEN featureClass = ho_bl
5	IF nCvClFa=2 THEN featureClass = pr_ci
6	IF nPlFa=5 AND nCcEd=0 THEN featureClass = pr_re
7	IF nPlFa=4 THEN featureClass = sl_nt
8	IF nPlFa=3 AND nCcClFa=0 THEN featureClass = sl_th
9	IF nPlFa=2 THEN featureClass = st

3.1.3. Extraction of characteristic vectors

Each of the 45 B-Rep feature models in Table 1 was analysed using an automated procedure to extract the attribute values included in the characteristic vectors at both levels of abstraction. In total, 160 characteristic vectors were created when individual feature faces were considered. The coding at the second level of abstraction resulted in 45 vectors. For both cases, the characteristic vectors formed from the B-Rep models are stored in a text file for further processing by the RULES-5 inductive learning algorithm. In this way, two different training sets are created.

3.2. Rule formation

The RULES-5 algorithm is applied consecutively on both training sets to extract rules that depict feature patterns at both levels of abstraction. In particular, RULES-5 created 15 rules from the first training set that encapsulates information about individual feature faces. The application of the same algorithm on the second training set that includes vectors representing face sets resulted in 9 rules. This set of rules is shown in Table 3. For example, Rule 4 in this set states:

IF nPlFa = 1 AND nCcEd = 2 THEN featureClass = ho_bl.

This means that a face set with one planar face and two concave edges represents a blind hole feature in a B-Rep model.

4. Testing

The proposed method for automatic formation of feature recognition rules was implemented in the learning module of a prototype AFR system. This system is described in more detail in Dimov et al. [11]. It also includes a feature recognition module that employs the rule bases generated applying the proposed method. Both modules of the prototype system were developed using the Java™ programming language. They use as input 3D CAD models in STEP format created using the STEP Application Protocol 203 supported by most commercially available CAD packages. Each module includes several parsers that extract automatically B-Rep data from CAD models in STEP format.

The feature recognition module was applied on the test part shown in Fig. 2, which was used for validation purposes by other researchers [4]. This model contains both simple and intersecting features. According to the taxonomy of machining features defined Section 3.1.1, it is composed of four through slots, one rectangular protrusion, two non-through slots, two rectangular pockets and two other features that are not present in this taxonomy. These two features can be identified as one passage and one corner. The prototype system

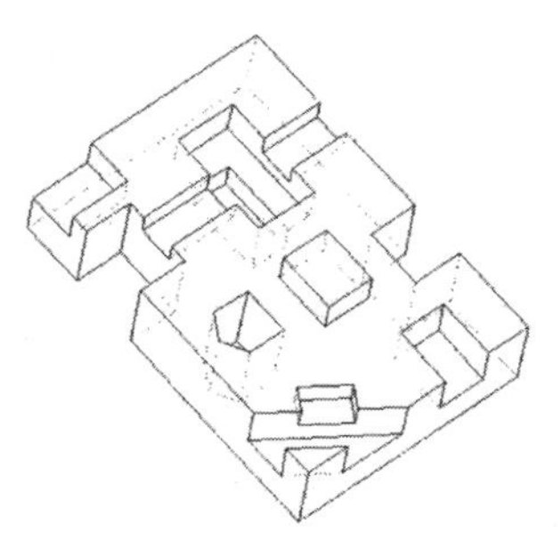

Fig. 2. Test part.

recognised nine of the eleven features present in this part. The corner and the passage could not be recognised because such feature classes were not defined in the taxonomy adopted in this study.

The recognition results obtained for this benchmarking part are similar to those achieved by other researchers. However, by employing the rule bases generated applying the proposed method, the prototype AFR system developed has some advantages over other systems. In particular, it could be deployed in different application domains and the knowledge base that determines its performance could be easily updated to increase its coverage.

5. Conclusions

In this paper, a method for automatic generation of feature recognition rules that is applicable in different application domains is proposed. This is a new method for creating knowledge bases of AFR systems that elevates the knowledge acquisition issues associated with the development of rules-based AFR systems. The two most important characteristics of this method are:

- The application of inductive learning techniques for identification of hidden patterns in sets of feature examples.
- The utilisation of two representation schemes that code feature information at different levels of abstraction to complement and extend the learning capabilities of the method.

This research also suggests that the utilisation of inductive learning techniques for AFR has several advantages:

- It provides a formal mechanism for rule definition and also assures the consistency of the generated rule sets.
- The development of AFR systems for different application domains requires only representative training sets to be formed for each of them. This is a major advantage of the proposed approach due to the domain-dependent nature of features.
- Due to the generalisation capabilities of inductive learning techniques, AFR systems could recognise features that are not present in the training sets.
- The knowledge base of such systems could be extended easily to cover new or user-defined features.

In addition, the following conclusion is also drawn from this research:

- The generation of rule sets at two levels of abstraction offers flexibility in adopting different recognition strategies in AFR systems. The search for features could be carried out by utilising data present in individual faces or face sets.

Acknowledgements

The authors would like to thank the European Commission, the Welsh Assembly Government and the UK Engineering and Physical Sciences Research Council for funding this research under the ERDF Programme 52718 "Support Innovative Product Engineering and Responsive Manufacture" and the EPSRC Programme "The Cardiff Innovative Manufacturing Research Centre". Also, this work was carried out within the framework of the EC Networks of Excellence "Innovative Production Machines and Systems (I*PROMS)" and "Multi-Material Micro Manufacture: Technologies and Applications (4M)".

References

[1] Dimov SS, Setchi RM and Brousseau EB. Automatic feature recognition – a hybrid approach. Fifth IDMME'04, Bath, UK, 5-7 April (2004).
[2] Henderson MR and Anderson DC. Computer recognition and extraction of form features: a CAD/CAM link. Comp. in Industry, 5(4), (1984), pp 329-339.
[3] Dong J and Vijayan S. Feature extraction with the consideration of manufacturing processes. Int. J. Prod. Res., 35(8), (1997), pp 2135-2155.
[4] Nezis K and Vosniakos G. Recognising shape features using a neural network and heuristics. CAD, 29(7), (1997), pp 523-539.
[5] Li WD, Ong SK and Nee AYC. A hybrid method for recognizing interacting machining features. Int. J. Prod. Res., 41(9), (2003), pp 1887-1908.
[6] Pham DT, Bigot S and Dimov SS. Rules-5: a rule induction algorithm for classification problems involving continuous attributes. Proc. IMechE, Part C: J. Mech. Eng. Science, 217(C12), (2003), pp 1273-1286.
[7] Pham DT and Dimov SS. An efficient algorithm for automatic knowledge acquisition. Pattern Recognition, 30(7), (1997), pp 1137-1143.
[8] Gindy NNZ. A hierarchical structure for form features. Int. J. Prod. Res., 27(12), (1989), pp 2089-2103.
[9] Sakurai H and Gossard DC. Recognizing shape features in solid models. IEEE Comp. Graph. & App., 10(5), (1990), pp 22-32.
[10]Pham DT and Dimov SS. An approach to concurrent engineering. Proc. IMechE, Part B: J. Eng. Man., 212(B1), (1998), pp 13-27.
[11]Dimov SS, Brousseau EB and Setchi RM. A hybrid method for feature recognition in CAD models. Proc. IMechE, Part B: J. Eng. Man., submitted for publication.

Intelligent Production Machines and Systems
D. T. Pham, E. E. Eldukhri and A. J. Soroka (eds)

Backlash Error and Effects on the Accuracy of Vertical Machining Center

H.L. Liu, SH.L. Zhang

Engineering college, Zhanjiang Ocean University, Zhanjiang, Guangdong,China, (430074)

Abstract

The position errors in the axis direction of a vertical machine center have been measured using the VM101 linear encored measurement system which has made in HEIDENHAIN cooperation. The setup is simpler and the measurement process is less time consumer compared with the laser interferometer. The character of the backlash error is discussed, its effects on the machine accuracy is presented. Results show that the backlash error has great influence on the position error. The error model for the soft error-compensation based on the Back Propagation Artificial Neural Network (BP-ANN) has developed. The position accuracy is enhanced after the backlash errors are compensated.

---

1. Introduction

The laser interferometer is commonly used in the measurement of the various error components. Geometric errors have various components like linear displacement error (positioning accuracy), straightness and flatness of movement of the axis, spindle inclination angle, square ness error, backlash error etc. The errors can be reduced with the structural improvement of the machine tool through better design and manufacturing practices. However, in most cases, due to physical limitations production and design techniques cannot solely improve the machine tool accuracy. Therefore, identification, characterization and compensation of these error sources are necessary to improve machine tool accuracy cost-effectively. Eung-Suk Lee et al. [1] used the interferometer in measuring 19 of the 21 components on a three-axis CNC milling machine with the exception of angular roll. Ni et al. [2] developed an optical system consisting of a laser interferometer, beam splitters, flat mirrors and dual-axis lateral-effect photo detectors to facilitate simultaneous on-line measurement of five errors on each axis namely linear and vertical straightness, roll, pitch and yaw errors. Shin [3], Ertekin et al [4] also presented the results of accuracy characterization of a CNC Machining Center using laser interferometer. Other techniques such as disc checks [5] and double ball bar (DBB) [6] measurements are also used in the position error measurement. Knapp et al. [7] listed a few tests for assessing the quality of machine tools on a periodic basis. The diagonal displacement test was one of the means adopted to estimate the volumetric accuracy of the machine tool through evaluation of the positioning accuracy and repeatability along the body diagonals. The circular test can also used in assessing the accuracy of the machine tools and the identification of backlash, a software strategy that compensated the backlash was developed [8].

Backlash error can be obtained through kinds of methods, it is either time-costly (laser interferometer) or only the maximum value obtained (DBB) and the effects of backlash error on the machine accuracy are not detail discussed so far. The aim of the investigation presented in this paper was to illustrate the effects of the backlash error on the position accuracy of the machine tools and attracted more attention about the backlash error in the machine production process.

VM101 Comparator System, which has the advantages of easy operation, time-less and also high accuracy was used to measure the axis linear displacement and backlash error in this paper. A simple introduction about the system and measurement process was given in the following section. The third part presents the experimental results. After that, two kinds of models include Back Propagation Artificial Neural Network (BP-ANN) and polynomial function model that can be used in different occasion were presented in the fourth part. The results after backlash error compensated were given in the fifth part. Then some conclusions have made in the final of this paper.

2. Experimental procedures

2.1. The VM101 linear encored measurement System

The VM101 Comparator System which has made in HEIDENHAIN cooperation consists of a precision scanning head, auxiliary carriage, interface electronics and sturdy storage case. The sturdy scale carrier is a U-shaped stainless steel extrusion. The material measure--a precision scale with a phase grating graduation and an 8μm grating period – lies in the neutral stress zone of the U-shaped scale carrier. The scanning head which is usually connected to the spindle of the machine tools runs along the scale without making mechanical contact. It is an exposed incremental linear encoder with high accuracy. Its measuring lengths is 720 mm, accuracy grade is ±1μm at 20 C (±.00004in. at 68 F) [9]. Fig.1 shows the measurement setup on the X-axis.

Fig.1 The X-axis measurement setup

2.2. Measurement process

A new vertical machining centre tape DM4600 was chosen for the tests. Its working volume was 600*400*420(mm). After the workspace was defined, accuracy and repeatability were tested for each axis direction. To minimize the error that might arise due to the temperature change during the measurements because of the axis motions. The machine was put in operation for three hours with a programmed exercise. This routine activated all the axis motions simultaneously before any measurement have taken so that the machine could pass the thermal transition state to reach its equilibrium state.

In order to keep errors resulting from uneven guide ways (Abbe error) at a minimum, the comparator system must be installed as near as possible both to the point of tool contact as well as to the encoder. The sides of the scale carrier should be aligned with the axis to be measured within 0.1mm.

To eliminate the influence of speeds, the machine was programmed to move with a lower feed rate of 50mmpm, after the warm-up cycle, position displacement measurement were taken in both travel directions. The data was captured in equal intervals (20mm) and the measurement cycle CNC part program composes a sequence of moves beginning at one limit of the axis in forward direction (f), extending to the opposite limit, and returning to the beginning position in reverse direction(r). The commanded machine position is nominal axis position, which is determined by the part program being executed. A PC reads the absolute machine positions directly into the database from the VM101.

A sample measurement cycle CNC program for X-axis is given in Table 1.

Table 1 VM101 measurement cycle CNC program

N001 #1=0	N010 X560
N002 G92 X0 Y0 Z0	N011 G04 P10
N003 WHILE #1 LT 28	N012 WHILE #1 LT 20
N004 G91 G01 X20 F50	N013 G91 G01 X-20
N005 G04 P10	N014 G04 P10
N006 #1=#1+1	N015 #1=#1+1
N007 ENDW	N016 ENDW
N008 #1=0	N017 M30
N009 G90 G01 X562	

3. Experimental results

3.1 Axis linear displacement error

The X-axis linear displacement error plots are given in Fig.2. In the legend of the error graphs, Error1-f and Error1-r indicates that errors were measured when the machine was in the first cycle motion for forward (f) and reverse (r) directions after about 3 hours warm-up period. The error plots show that positional errors are generally lower at the beginning of axis travel (home position) and has linearly increasing trend with increasing axis nominal position. It was found that the position error in forward direction was bigger than the position error in the reverse direction at the same point. The maximum displacement error is about 40μm at the end of the axis.

The Y-axis linear displacement error (Fig.3) and the Z-axis linear displacement error (Fig.4) showed similar linear trends with increasing nominal axis position as in X-axis. The Y-axis linear displacement error was smallest among the three axes being tested and the magnitude was approximately 25μm. Maximum linear displacement error of the Z-axis was almost the same value as the end of the X-axis. The slope of the error line of the Y-axis is smallest when the machine is moving in the forward directions. There are an obviously two groups error line on the axis linear displacement error plot. That means there are two different error values at the same location in the axis. The main reason of these results is the backlash error.

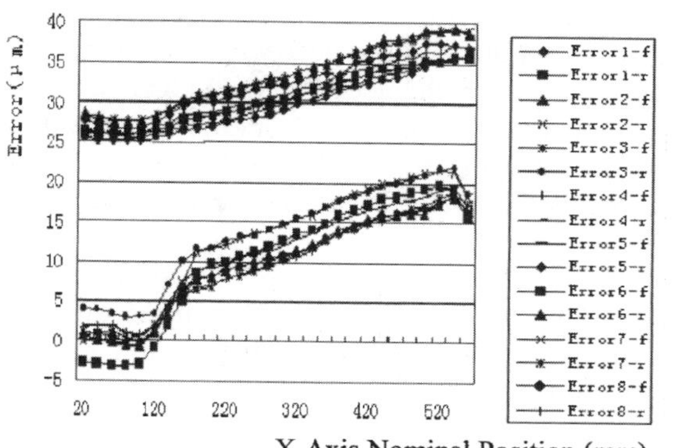

X-Axis Nominal Position (mm)

Fig.2 X-axis linear displacement error

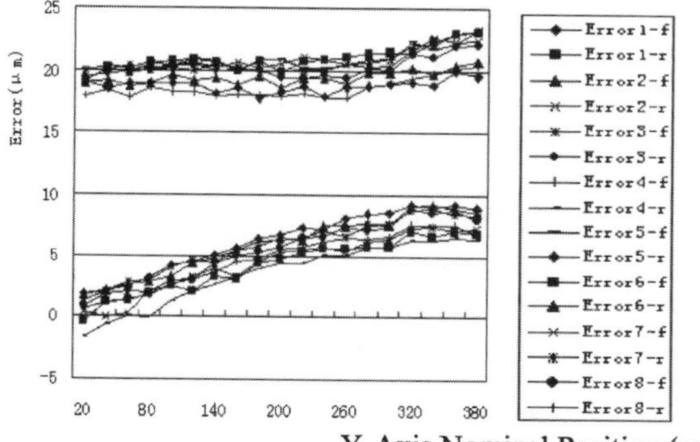

Y-Axis Nominal Position (mm)

Fig.3 Y-axis linear displacement error

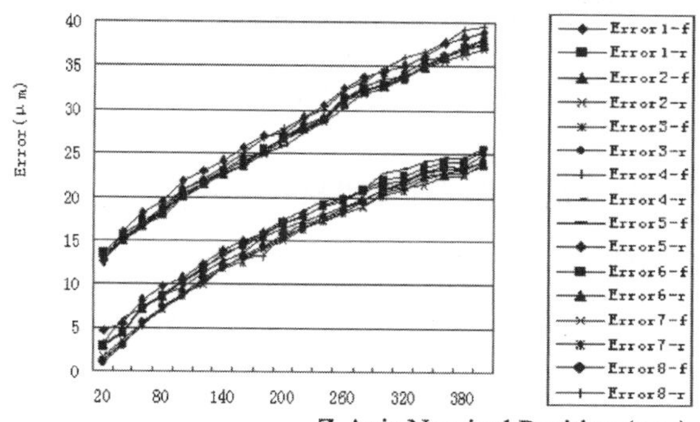

Z-Axis Nominal Position (mm)

Fig.4 Z-axis linear displacement error

3.2 Backlash error

Backlash error is a position dependent error affecting the contouring accuracy. When the axis changes direction from one side to the other, there is a lag before the table starts moving again, that would cause position error-backlash error. The backlash error was the major reason why there have two group error lines on the displacement error plot. It has the great influence on the accuracy of the machine and should be compensated. Fig.5 shows the backlash error of the axes. B_{ex} means the X-axis backlash error, B_{ey} is the Y-axis backlash error and the Z-axis backlash error is B_{ez}.

The backlash error value of the axis was not constant and has the different value at the different axis nominal position. Maximum backlash error is about 27μm at the beginning of travel in X-axis, the backlash error is 18μm and 11μm at the home point of the Y-axis and Z-axis. It leads to the relatively large error of the axis in the forward direction. The Z-axis has the smallest average backlash error among the three axes and obtained maximum B_{ez} at the end of the axis travel with a magnitude of 14μm.

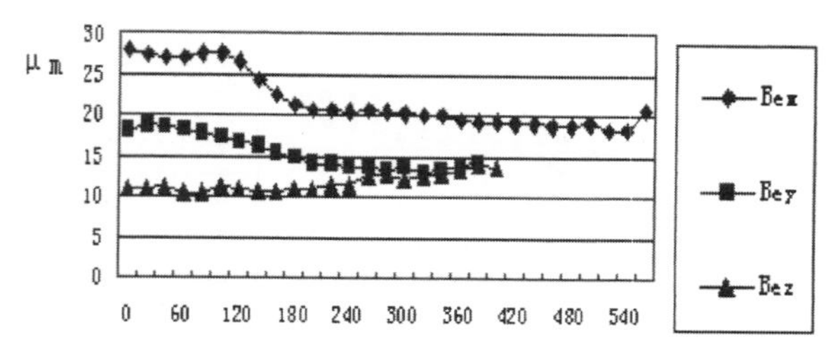

Fig.5 the value of backlash error

4. The models of backlash error

4.1 *The ANN models of backlash error*

The neural net using the back-propagation algorithm has been used to predict the total positioning error at any given thermal state and location of the cutting tool from knowledge of that error at some specified point in the workspace as measured by the laser ball bar [10].the neural network which utilize the learning ability to predicate the backlash error value form the knowledge of the error as measured by the VM101 system was developed. The neural networks developed in this study have three layers — one input layer, one hidden layer, and one output layer. The input layer comprises the units that represent the nominal position of the axis. The output layer has three output units that represent the backlash errors of the three axes. The hidden units of the network are used to extract the underlying features of relationships between the nominal position and the backlash errors of the machine tools. The Fig.6 shows the flow chart of the neural net work.

Training of the networks can proceed only after the entire experimental data is obtained. The training process of the net was shown in Fig.7.

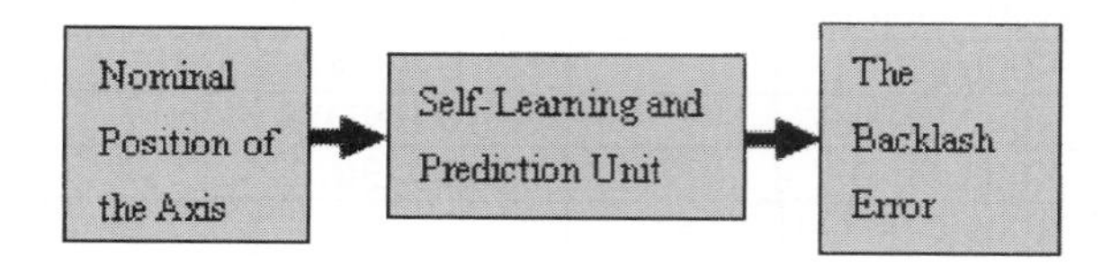

Fig. 6 The flow chart of the neural net

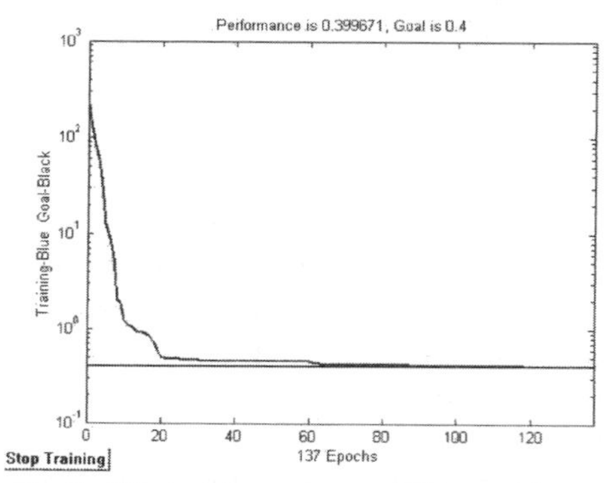

Fig. 7 The training process of the neural network

During learning, the nominal position value of the axis was input into the self-learning and prediction unit, the output of the neural net is the backlash error value. Then the backlash error compensation can be taken. However because the training time is relatively long, the ANN model can only be used in the off-line compensation, the advantage of the net model was very high-nonlinear mapping ability.

4.2 *The polynomial models of backlash error*

For on-line compensation, the backlash errors (B_{ex},B_{ey},B_{ez}) can also be modeled as polynomial functions with the position along each axis. As the equations (1), (2) and (3) expressed.

$$B_{ex}=a_0+a_1x+a_2x^2+a_3x^3 \quad (1)$$

$$B_{ex}=b_0+b_1y+b_2y^2+b_3y^3 \quad (2)$$

$$B_{ex}=c_0+c_1z+c_2z^2+c_3z^3 \quad (3)$$

The error models then can identify using the least-squares criterion method. The calculation process of the backlash error value was very fast, but the drawback of the polynomial model was its relatively lower accuracy.

5. Accuracy after backlash error compensated

After the models have been developed, the backlash error then can be compensated. The procedure can be divided into three stages: calibration of the backlash errors (part 2), modeling of the errors (part 4), and compensation control implementation.

The principle of the error compensation control implementation is as the following steps: the backlash errors on the nominal position of the axis at any given thermal stage are collected and the errors are estimated by the models. Then the resultant error between the commanded position and the real position are calculated. Finally, the compensation singles are sent to the CNC controller to shift the origins of the slide axes to implement the error compensation. The results of the position accuracy after the backlash error compensated using the ANN model were measured using the same procedure as mentioned above. Fig.8, Fig.9 and Fig.10 show the X-axis, Y-axis and Z-axis measurement results.

All three plots show that the backlash errors were well compensated and accuracy of these axes was highly increased. All linear positional errors increasing linearly with respect to axis nominal position and are highest at the end of axis travel range. These results correspond with the most previous work and the assumption that the linear errors components change linearly with position. Machine home position gave best linear accuracy; maximum linear displacement error was about 10μm in the X and Y-axes, the Z-axis maximum linear displacement was about 25μm at the end of the axis position.

The variation of the linear displacement error (position systematic deviation) was about ±2μm at the same position in the axis when the backlash errors were compensated. The repeatability of position of the axes was obtained with a magnitude of about 4μm. This is the limits accuracy through the off-line software for the software-compensation.

The experimental results show that the accuracy was enhanced after the backlash error eliminated. So more attention should be paid with the backlash error during the production and assemble processing of the machine tools.

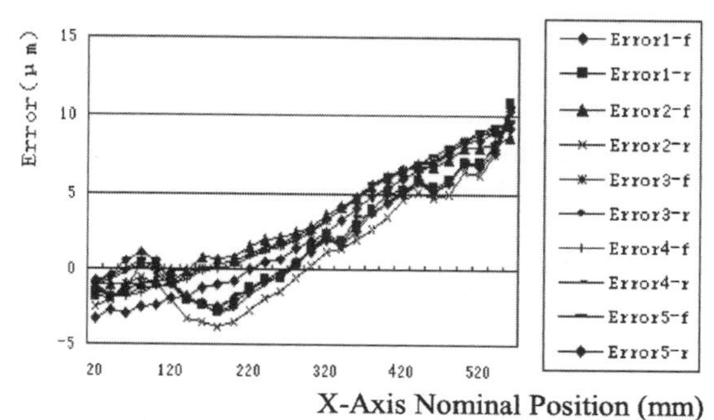

Fig.8 X linear displacement error after compensated

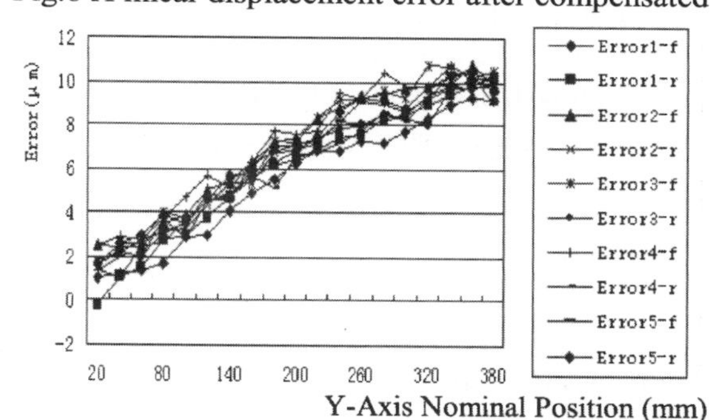

Fig.9 Y linear displacement error after compensated

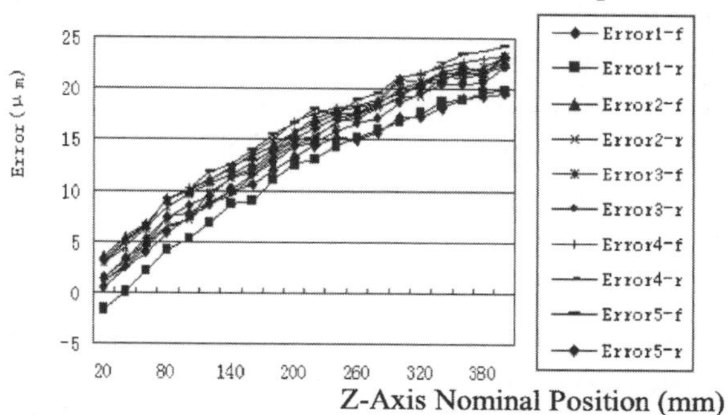

Fig.10 Z linear displacement error after compensated

6. Conclusions

From the results of this research, the following conclusions can be drawn:

- All linear position errors increased almost linearly with respect to axis nominal position and are highest at the end of axis travel range even the backlash error compensated.
- The backlash errors cause relatively large linear displacement error deviation between different directions at the same point in the axis. The backlash errors value is the function of the point in the axis.
- The accuracy of the machine increased after the backlash error compensated, so more attentions should be paid on the backlash errors in the machine tool production procession.

Acknowledgements

This work was financial supported for the research of geometric error measurement and compensation on CNC machine tool by National 863 Subject Plan under grant 2002AA423260.

The authors also would like to express thanks to the HEIDENHAIN cooperation for the free use of the VM101 measurement system.

References

[1] E.S. Lee, S.H. Suh, J.W. Shon, A comprehensivemethod for calibration of volumetric positioning accuracy on CNC machines. The international Journal of Advanced Manufacturing Technology. Vol.14 (1998): 43-49.

[2] J. Ni, S.M. Wu. An on-line measurement technique for machine volumetric error compensation. ASME Trans. Journal of Engineering for Industry. Vol. 115(1993): 85-92.

[3] Yung C. Shin. Characterization of CNC Machining Centers. Journal of Manufacturing Systems. Vol. 10, No. 5(1992): 407-421.

[4] Yalcin M. Ertekin, Anthony Chukwujekwu Okafor. Vertical machining center accuracy characterization using laser interferometer, Part 1. Linear positional errors. Journal of Material Processing Technology. Vol. 105(2000): 394-406.

[5] Knapp W. Test of the three-dimensional uncertainty of machine tools and measuring machine and its relation to the machine errors. Annals of CIRP. Vol .32(1) (1983):459~464

[6] Kakino Y, Ihara Y, Nakatsu Y. The measurement of motion errors of NC machine tools and diagnosis of their origins by using telescoping magnetic ball bar method, Annals of the CIRP.Vol36(1) (1987):377~380

[7] W. Knapp, Interim checks for machine tools. Proc. 3rd Int. Conf on Laser Metrology and Machine Performance－LAMDAMP, 1997: 161-168.

[8] Y.S. Tarng, J.Y. Kao, Y.S. Lim, Identification of and compensation for backlash on the contouring accuracy of CNC machine center. The international Journal of Advanced Manufacturing Technology. Vol. 13(1997): 359-366.

[9] HEIDENHAIN Mounting and Operating Instructions—VM 101 Incremental Comparator System. Germany. 2000.

[10] Srinnivasa, Ziegert, J. C., and Smith, S., "Prediction of Position Errors of a Three Axis Machine Tool Using a Neural Network," Proceedings of the Japan/USA Symposium on Flexible Automation, San Francisco. 1992, 203-209.

Intelligent Production Machines and Systems
D. T. Pham, E. E. Eldukhri and A. J. Soroka (eds)

BM_Virtual Enterprise Architecture Reference Model for Concurrent Engineering processes performance improvement: An experiment

Pithon A.J.C.[a], Putnik G.D.[b]

[a] *Department of Production System, Federal Center of Education, Rio de Janeiro, Brazil*
[b] Department of Production System, *University of Minho [1], Guimarães, Portugal*

Abstract

The BM_Virtual Enterprise Architecture Reference Model (BM_VEARM) is used as an organizational structure for Concurrent Engineering (CE) teams in Agile and/or Virtual Enterprise (A/VE). CE is a resume of a set of principles that includes teamwork, client and supplier involvement, parallel and concurrent processes and continuous improvement of quality. On the other hand, an A/VE is seen as a kind of a short-term networked enterprise, with a dynamic structure, integrated over a universal set of resources, i.e. a global network of independent enterprises. The principal characteristic of the A/VE organizational model presented, in accordance with BM_VEARM, is a broker, which acts as an agent of CE team re-configurability. In order to test the potential of the CE Agile/Virtual teamwork organization according to BM_VEARM, an experiment was organized in which performance of three CE teams, with different organizations, were evaluated: 1) *distributed* CE team (in literature usually called "virtual" CE team); 2) *agile* CE team in accordance with BM_VEARM and 3) *virtual* CE team in accordance with BM_VEARM. The teams had the task to develop a Website. The paper presents basics of the organizational models applied, the experiment organization as well as some experimental results.

1. Introduction

The Concurrent Engineering (CE) and related concepts of teamwork and/or cooperative work has emerged as the knowledge management and organizational tools with the aim to radically cut product development time as well as to radically improve the product quality in order to meet client's requirements better.

To support the cooperative work, and especially the Concurrent Engineering (CE) processes, different organizational models could be applied. These models could be characterized by (physical) distribution of team members, agility and virtuality, besides other characteristics.

The agility together with distributivity with which these groups can be created and reconfigure make possible use of individuals capable to add value to one

[1] The University of Minho is the partner of the EU FP6 Network of Excellence I*PROMS – *Innovative Production Machines and Systems.*

defined task, independently of the place where they are located.

With the objective to compare the performance of distributed CE teams, or, as traditionally defined, "virtual" CE teams, with the performance of CE teams organized in accordance with Virtual Enterprise (VE) concept, in this particular case in accordance with the BM_Virtual Enterprise Architecture Reference Model (BM_VEARM), an experiment is planned and carried through in the Federal Centre of Education (CEFET-RJ) in Rio de Janeiro, Brazil. The participants of the experiment had been the students registered in the last year of the Course of Informatics.

The task consisted of the development of a Website, consisting of several Website pages, adding several links and services, that should serve as a Portal for a Virtual University or as a starting point for navigating the Internet community [1].

This experiment[2] was carried through in the period from October 27th to 30th 2003, which involved three teams working simultaneously. To support the interaction between groups, NetOffice software [2] was used. By being open source software, it was customized in order to take care of the necessities of the task, that is, to allow the broker's function [3].

Initially the experiment was idealized to be applied in the industry. However, due to the problems of technical order and due to needs to satisfy the "timing" of the project development, it was opted, then, to create an environment, where the work of diverse teams could be evaluated. In this way, the experiment presented here contemplates the services, i.e. the domain of services, rather then manufacturing. For each group, i.e. EC team, the same project was considered: to create a Website entitled "Virtual University", to be used as an education environment. Thus, the website is about an Education Portal, to make limitless use of the services and resources offered.

In this context, this paper is organized through three parts: in the first part the basic concepts of CE are presented, as well as its operational model for team work; in the second part, the concept of VE according to BM_VEARM is presented, as well as the way how BM_VEARM is applied for organization of the EC teams (work groups) as agile and virtual organizational model of CE teams. Finally, in the third part, the experiment plan is presented as well as some results.

[2]. At the time, this kind of experiment was never done before in Brazil.

2. Concurrent Engineering Concept

The concept of CE defines that various activities are developed in parallel, simultaneously, interactively, involving professionals from different specialties, covering the entire life-cycle of product development, in opposition to the traditional method of sequencing product development life-cycles phases. Therefore, it becomes possible to "re-feed" an activity by others (the feed-back relation among processes). This new form of work is very beneficial, since it minimizes waste time and resources, which originate from a lack of complete involvement of different sectors, or functions, in all project phases. The new form, also, improves the quality of the product and the quality of the development process [4].

The model for CE processes organization, which has been developed in this study is based on teamwork, also known as the "task force" (Figure 1), which has the (team) leader as the linking element between the members of the team (group) and the company management.

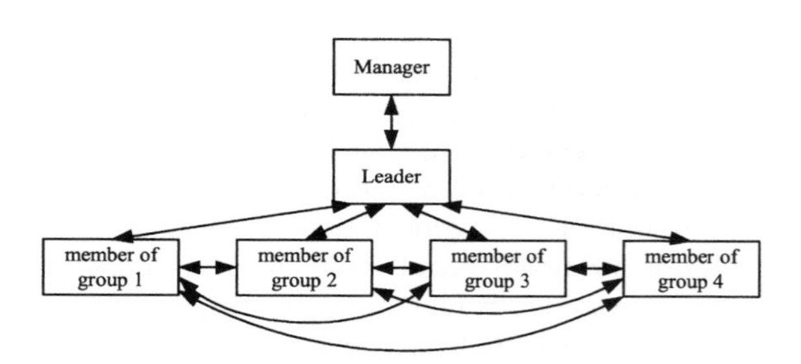

Figure 1 - Management Model of Concurrent Engineering Teams

3. BM_VEARM

The BM_VEARM (figure 2) is defined as a hierarchical structure of multiple levels of inter-enterprise processes control and execution, with broker inserted between two consecutive levels (principal/broker/agent) of the enterprise/ manufacturing process control systems, which ensures integrability, distributivity, agility and virtuality [5].

Integrability (interoperability) is understood to be the capacity of an enterprise to access (interconnect, integrate) existing heterogeneous resources inside and outside the organization. The integration of heterogeneous resources should occur at low cost, which is open system architectures characteristic.

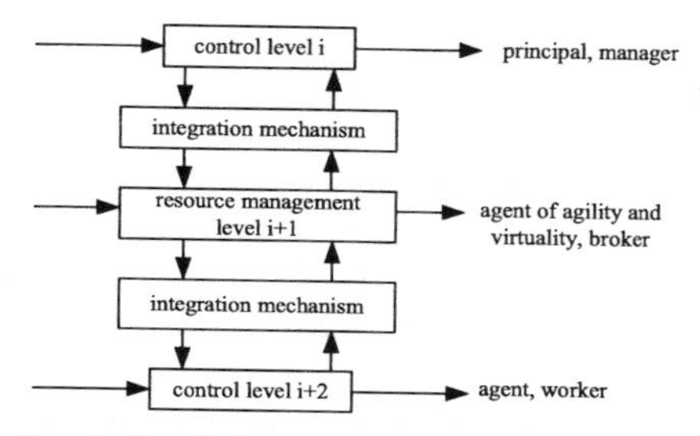

Figure 2 - Elementary hierarchical structure of Virtual Enterprise by BM_VEARM

In the context of virtual enterprises, *distributivity* can be perceived as the capacity the enterprise has to integrate and operate needed resources remotely, i.e. at a distance. The concept of a competitive enterprise implies the ability to access the best resources: simply seeking the cooperation of other enterprises, purchasing components, sub-contracting other companies or creating consortiums, as well as the capacity to manage all business and manufacturing functions, independent of distance, using Wide Area Network (WAN) technologies and corresponding protocols, e.g. Internet.

In the same context (of VE), for an enterprise to be *agile*, in order to answer the abrupt changes of the market, it is necessary that the enterprise has the capability of fast adaptation or fast reconfiguration between two operations (operation off-line).

The *virtuality* has the purpose to further improve the performance of the agile enterprise, that is, it must provide to the system the capacity of reconfiguration along the operation intended (operation on-line), without interruption of the process. The combination of virtuality, agility, distributivity and integrability makes a flexible company with a higher level performance capable to achieve the needed competitiveness on the global market.

In this model, broker is the main element of the agility and the virtuality, being able to reconfigure the organizational structure of the company during an operation in real time (on-line) and/or between two operations (off-line).

Aiming at the integration of the concept of Concurrent Engineering (Figure 1) with the concept of Virtual Enterprise by means of the BM_VEARM model (figure 2), the model of CE teams based on BM_VEARM is conceived (figure 3).

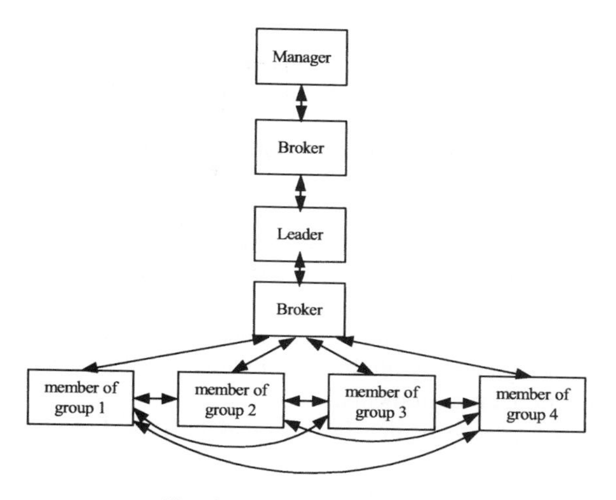

Figure 3 - BM_VEARM Model for CE Group

4. The Experiment

The experiment was carried through by three CE work groups, or teams:

1) *distributed* CE team (usually designated 'virtual teams' in literature);

2) *agile* CE team according to BM_VEARM, and

3) *virtual* CE team according to BM_VEARM.

Thus, the specific objective of this experiment was to observe the behavior and the interaction of the three types of teams, i.e. of the three organizational models for CE teams, and, then, to evaluate the performance of each group/team by the criteria defined in section 4.2.

To make the experiment closer to the real-life conditions pretended, the members of three groups didn't have direct face-to-face contacts or meetings. The management of the communication and work flow among the members of the group, the broker and the leader processes, were carried through using open source software NetOffice, which is advantageous for being accessible from conventional browsers (Internet Explorer, Netscape and Opera), without the necessity of using other tools you should add to participants of the experiment.

4.1. Team work Description

This section describes the profile of the three CE teams that had participated in the experiment as well as the functions/roles by each team member and, still, the concrete model of team organization according to BM_VEARM.

4.1.1. Distributed CE Team

The distributed CE team (figure 4) is formed by the manager; leader; reviser; projector; designer; specialist/programmer of animation and clients: student and professor. The only difference between this group and the traditional CE group are that the CE traditional group works in the face-to-face form and the distributed one works using (through) Web. To the each one of them competes [3].

- *manager*: elaborates the strategy of the Portal, i.e. defines together with the leader which are the Portal's functional requirements;
- *leader*: executes the tasks in details that had been attributed to him by the manager and co-ordinates the members of the group;
- *reviser*: creates and revises texts for the Portal; put texts on the site;
- *projector*: projects the Portal, delineating the objectives, public-target, sections that the Portal must possess, and defines which type of human resources and technology needs each task;
- *designer*: creates and implements the graphical standard of the Portal, in accordance with the specifications defined by the projector;
- *specialist/programmer*: implements the technical/ logical code that defines the operations and functionalities of the pages of the Portal; promotes the integration of diverse individual elements that compose each section;
- *client/student*: represents the public-target, or others, to whom the Portal is destined. The role of this member is to verify if developed software takes care of the students' necessities such as: easy visualization of the pages and the bulletin boards, easy access to the subjects destined to the public-target, etc;
- *client/professor*: analyzes and reviews the implemented pedagogical content to the project; verifies the accessibility to the Portal such as: localization of the contents, navigation on the pages, etc.

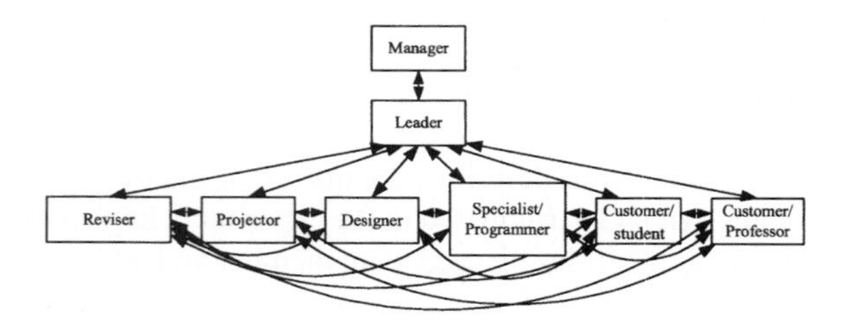

Figure 4. CE Distributed Team work

4.1.2. Agile CE Team According to BM_VEARM

The main characteristic of this group is the figure of broker. It competes to broker, element chosen by the manager, to select necessary resources for execution of the tasks (e.g. to enlist the members of the group). This broker acts as an agent of the agile CE team reconfiguration. In this way, the reconfiguration will be implemented in the 'off-line' way, i.e. when it will have the necessity to substitute some member of the group for another one, broker will go to interrupt the execution of the task, or to wait the interval between switching from one operation to another, and to implement the substitution. The alternative "resources" for reconfiguration of the group structure (i.e. substitution of the group members) are presented in the inferior part of the figure 5.

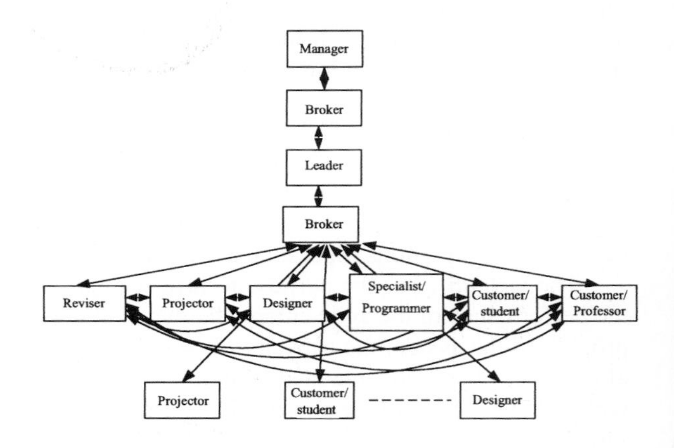

Figure 5. CE Agile Team work according to BM_VEARM

4.1.3. Virtual CE Team According to BM_VEARM

This group, also, has as a main characteristic the broker as the agent of virtual CE team reconfiguration of (figure 6). The performance of broker in this case is different. To reconfigure the group structure, i.e. to substitute a member of the group for another one, it does not need to interrupt the activity that is being developed. This reconfiguration will be realized on-line. The potential of this model consists of the substitution of a member by another one providing that neither the task development being affected by the reconfiguration nor perceived by other members of the group or by another immediate superior level. Broker in this model possess basically the same attributions that was defined for the agile model above but also the broker possesses more 'power' in decisions on reconfiguration. This broker's 'power' is authorized by the Leader or Manager.

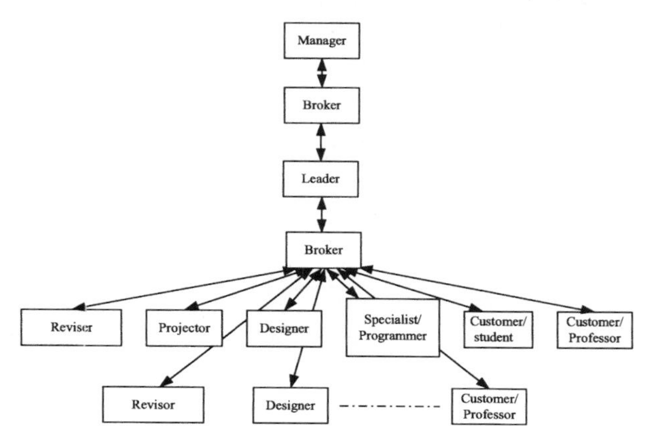

Figure 6. CE Virtual Team work According to BM_VEARM

4.2. Evaluation Criteria

The evaluation was carried through with the objective to survey if the performance of virtual CE team according to BM_VEARM model reached the goal foreseen in the project, i.e. to perform better by various criteria (e.g. to shorten the development time, to improve the product quality, etc.). The verification was based not alone on the comparison of Gantt time diagrams - necessary time to fulfill the task -, as well as on the metrics used in practice for evaluation of web sites, namely the metrics (points) defined for the BigEye Award Program and the requirements of the market.

In conclusion, three performance measures had been evaluated:

1 - Code Quality (product quality criteria)
2 - Load time (product quality criteria), and
3 – Product Development Lead Time (development efficiency criteria)

The lead time is the time used by each team in the construction of each one of the six screens/pages that compose the Site. The constructed screens/pages had been: Homepage, Institution, Classroom, Café, Secretariat and Library.

5. Results

The Site of each one of the three observed groups, was consisted of six screens/pages: Homepage, Institution, Classroom, Café, Secretariat and Library. The criteria used in the evaluation of each screen had been presented in section 4.2. From there, the evaluation of groups has been proceeded.

In this paper, due to limited space, we present only the results for the web pages development lead-times (Table 1).

On the Figure 7, it is presented one of the Gantt diagrams[3]., for the CE team processes lead-times created during the experiment.

Table 1 – Web pages development lead-times [h]

WEBSITE PAGE	TYPE OF CE TEAM		
	VIRTUAL/ DISTRIBUTED	ÁGILE BY BM	VIRTUAL BY BM
HOMEPAGE	41	35	29
INSTITUTION	8,92	5,50	6,42
CLASSROOM	20,92	10,92	32,50
CAFÉ	19,67	19,67	7,92
SECRETARIAT	25,50	23,17	13,17
LIBRARY	49	43,33	30,33
TOTAL	**134,43**	**111,59**	**98,01**

6. Conclusion

The purpose of this work was to demonstrate the performance of the Virtual and Agile CE teams according to BM_VEARM against the traditional model of CE teams organization.

By the experiment results, it can be concluded that the CE Virtual team, with the BM_VEARM based organization, performed better, i.e. it was faster, more efficient, and got better results partially (for partial tasks, i.e. for component web pages of the Portal) and globally (for the total, complete task), when compared with the CE Agile team, with the BM_VEARM based organization, and CE Agile team, with the BM_VEARM based organization, performed better,

[3]. The diagram on the Figure 7 is drastically reduced from its original size. Although the numerical and textual details of the diagram are not readable the authors believed the Figure 7. could serve as an illustration of the documentation and the data based created during the experiment that have served for detailed analysis of the results.

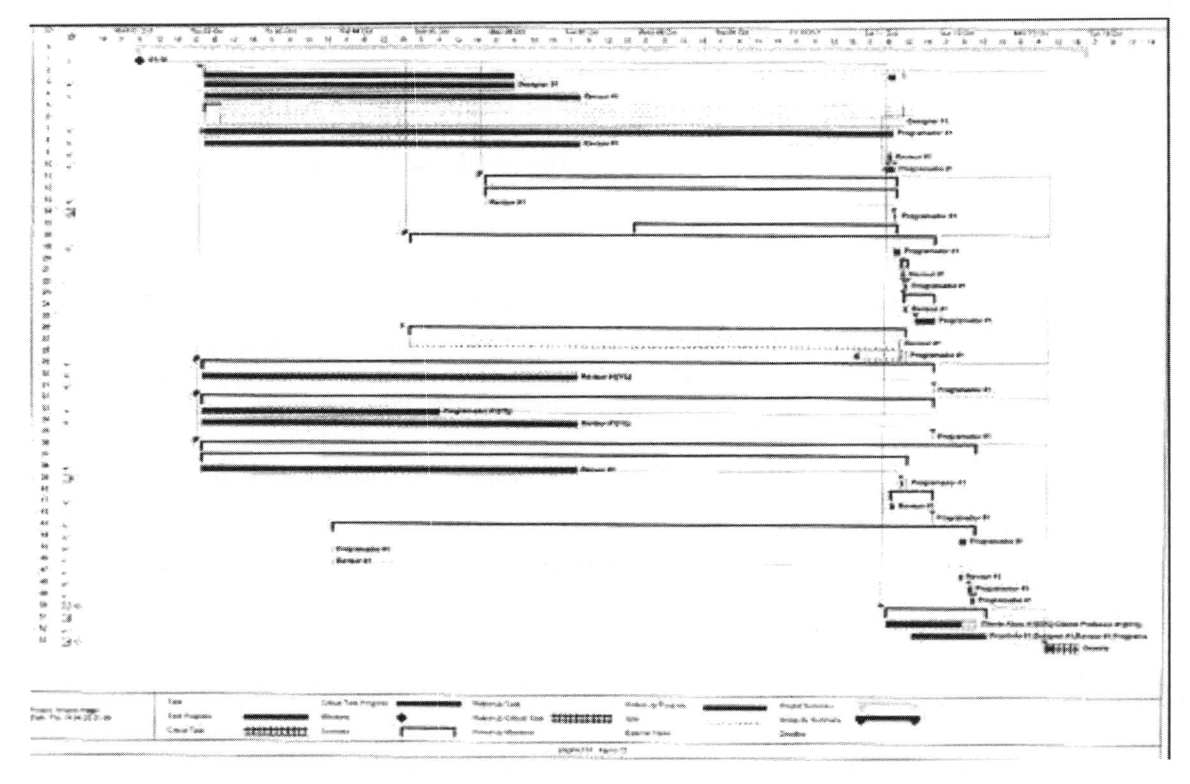

Figure 7. Gant diagram for one CE team processes lead-times

i.e. it was faster, more efficient, and got better results partially and globally, when compared with the 'traditional' CE Distributed (commonly called 'virtual') team.

The same type of results were obtained for two other two criteria; that is, the BM_VEARM based CE teams had the better product (Portal) quality in terms of lesser (page) Load Time and the better Quality Code.

The experiment showed that the BE_VEARM improves the performance of CE teams as it was expected theoretically. The CE Virtual team work based on BM_VEARM can become feasible, if in the application of this model is used adequate group management software as well as the training in teamwork and brokering is provided.

References

[1] Pithon, A.J.C., Putnik, G.D. Concurrent Engineering Based in BM_VEARM for Development of Infrastructure (Portal) for Distance Learning. 10th ISPE Int. Conf. on Concurrent Engineering: The Vision for the Future Generation in Research and Application, R. Jardin-Gonçalves; J. Cha; A. Steiger-Garção. A. A Balkema Publisher, 2003, V.2.:pp 931-936.

[2] NetOffice, *http://www.netoffice.com/*

[3] Pithon, A.J.C. Organization Project for Concurrent Engineering in Virtual Enterprising, Ph.D. Thesis, University of Minho, Portugal, 2004.

[4] Pithon, A.J.C., Putnik, G.D. Team Work for Concurrent Engineering in Agile/Virtual Enterprise by BM_VEARM. Collaborative Business Ecosystems and Virtual Enterprise – Third Conference on Infrastructure for Virtual Enterprise (PROVE'02)L.M. Camarinha-Matos, 1-3 May, Sesimbra, Portugal, Kluwer Academic Publisher, pp 498-496, 2002.

[5] Putnik, G.D. BM_Virtual Enterprise Architecture Reference Model In: A. Gunasekaran (Ed.). Agile Manufacturing Strategy, Elsevier Science Publ. 2001.

Intelligent Production Machines and Systems
D. T. Pham, E. E. Eldukhri and A. J. Soroka (eds)

Holistic Approach for Condition Monitoring and Prediction of Machine Axis

E. Hohwieler[a], C. Geisert[a]

[a] *Fraunhofer Institute for Production Systems and Design Technology, Berlin, Germany*

Abstract

The increasing demands made on operational availability and reliability of machine tools require the adoption of innovative methods to assess the machine's current condition and predict its condition in future. The progression of degradation caused by wear and tear depends strongly on users' product range and the manner of machine operation. Therefore, it is necessary to continuously collect information about the machine's "health status". Currently the condition of the machine axis is typically ascertained using special test routines under defined ambient conditions.

This paper demonstrates a holistic approach for monitoring the condition of the machine axis with simple test routines and additional statistical information. To do this, additional wear relevant data available from the machine control must be logged and included in the computations. This procedure has the advantage that the "load history" of the machine axis can be taken into account for the prediction of future wear progression.

1. Introduction

All machine units which are in relative motion to a mechanically coupled component are subject to friction. This in turn causes abrasion, leading to a reduced accuracy and, in the worst case, to cost-intensive machine breakdown. Thus, ascertaining the "health status" of these wear related components without having to disassemble the machine would allow for efficient implementation of condition based maintenance. This maintenance strategy aims at minimising unscheduled downtime combined with optimised usage of machine components over their entire life expectancy. One component of the machine tool which underlies wear and tear is the machine axis (see Fig. 1). It is subject to highly varying kinds of stress, which depends on many factors, e.g.:

- tool-workpiece-combination,
- kind of machining,
- condition of lubrication.

Although the physical fundamentals have been well established and mathematically modelled, there exists a gap between theoretical knowledge and the ability to predict the real wear status of dynamically stressed components. One reason for this is that the progression of degradation depends on the history of the complex of loads acting on the component. Since machining is in most cases not a static process, it is nearly impossible to record all events that influence wear lifespan.

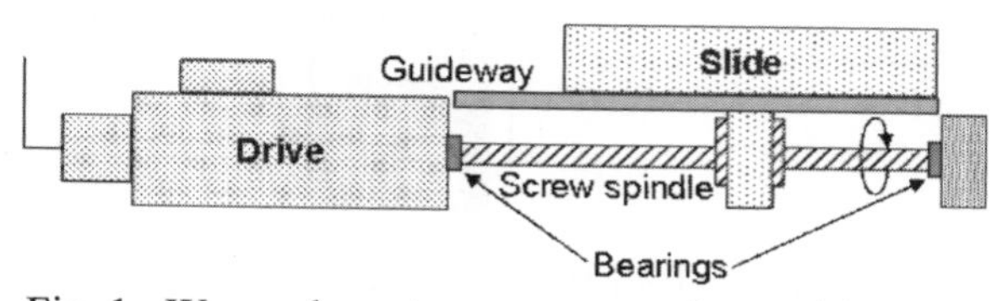

Fig. 1. Wear relevant components of a machine axis.

In this article, a new approach to close this gap is presented. The main idea for condition monitoring of the machine axis is to use intermittent test routines combined with additional background information [1], [2].

2. Research Activities

This chapter shows a small selection of current research activities in the area of Condition Monitoring.

2.1. IMS – Watchdog Agent™

The "Center for Intelligent Maintenance Systems" (IMS) is a common research lab of the University of Michigan and the University of Wisconsin, USA. Researchers of this lab are currently working on the development of new internet-based shared monitoring and diagnosis solutions for Condition Based Maintenance. In addition to the reference architecture for the local data recording and remote-connection of machines, algorithms for condition diagnosis and maintenance prognosis are also being developed. An intelligent Watchdog Agent™ shown in Fig. 2 evaluates machine conditions and predicts time when machine failure will occur. Used methods are:

- Neural Networks,
- Genetic Algorithms,
- Statistic Computing and
- Fuzzy Logic.

Ongoing work deals with the prototypical realisation for the monitoring of spindle systems of a CNC milling machine using auxiliary sensor equipment (vibration, drive current, temperature) [3], [4].

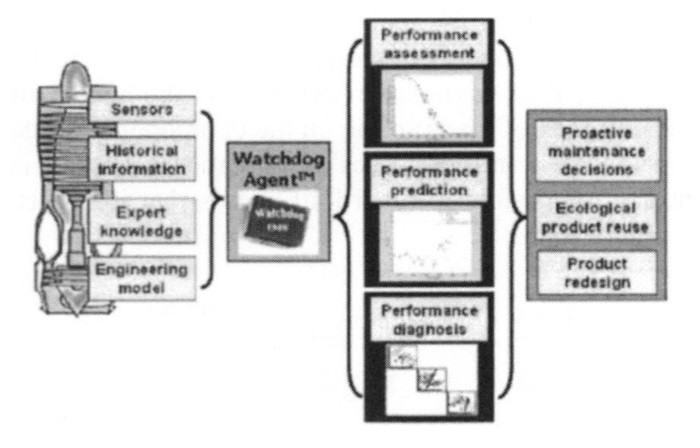

Fig. 2. IMS Watchdog Agent™ [4].

2.2. Live Cycle Unit

During the SFB 281 "Disassembly Factories for the Recovery of Resources in Product and Material Cycles" Product accompanying information systems – so called Life Cycle Units (LCUs) – have been developed at the Technical University of Berlin, see Fig. 3. These LCUs, integrated into products are capable of acquiring, processing, and communicating relevant product and process data for evaluation purposes during the entire product and component life span [5], [6].

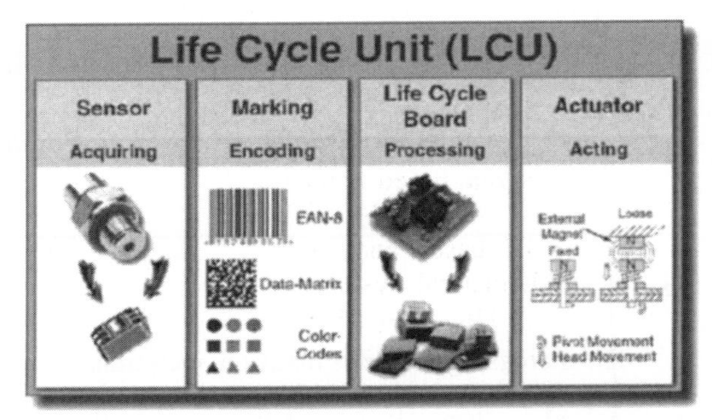

Fig. 3. Elements of the Life Cycle Unit [6].

Within the international research project "Embedded Watchdog Agent/Life Cycle Unit (EWA/LCU) – Monitoring of Standard Components from Machinery" research activities are carried out in close collaboration between the Center for Intelligent Maintenance Systems (Center for IMS). Diagnostic and prognostic algorithms from the Embedded Watchdog Agent (EWA) are integrated into the LCU for performance assessment, performance prediction and performance diagnosis of the monitored products and components [7].

2.3. Intelligent Spindle Unit

The object of the "Intelligent Spindle Unit" project sponsored by BMBF is to increase the availability of the main spindles and machine tools via the development and realisation of an intelligent spindle. Through the integration of sensory technology, the critical operating conditions are identified. In order to ascertain the conditions which may lead to a storage overload, the load, revolutions and temperature changes in the spindle bearings are measured and evaluated [8].

Based on the results of a machine and process

diagnosis study carried out by ISW University Stuttgart, test runs of concrete damage diagnosis are being done as part of a research project sponsored by VDW. Ball screws spindle are being tested on a laboratory construction of machine components using standard signals from the control in order to localise mistakes and to predict breakdowns [9].

3. Test Routines

The above mentioned test routines are used to make a “snapshot” of the present “health status” of the machine axis. In general, two methods of analysis can be used for the generation of characteristic values from measured data:

- model based analysis and
- signal based analysis.

The model based analysis needs an adequate, detailed mathematical system model for parameter identification. This model has to be adapted to the specific nature of each physical system that it is to represent. In contrast, signal-based analysis has a more universal character. It is possible to search for characteristic patterns in the signal that reflect universally valid physical laws. The second method was chosen in this paper, because it can be implemented without detailed knowledge about the system's parameters [10].

3.1. Friction – Indicator for Degradation

Progressive wear modifies the dynamical behaviour of a tribologic system. The roughness of a surface increases and, as a consequence, the friction between the two coupled mechanical components likewise increases. The characteristic curve for the velocity dependent friction, the so called Stribeck-curve, is shown in Fig. 4. The velocity v is represented by the revolution speed of the spindle and can be transformed with Eq. 1.

$$v = \Delta s\, n \tag{1}$$

where Δs is the threw pitch in mm per revolution and n is the revolution speed in revolution per minute.

The Stribeck-curve is subdivided into three velocity dependent sequences. At low speed, after transcending the static friction (also called stiction), dry friction occurs, no lubrication is located between the two surfaces. When the velocity is increased, lubrication fills a gap between the surfaces until they become completely separated. When this stage is reached, the friction rises directly proportional to the velocity.

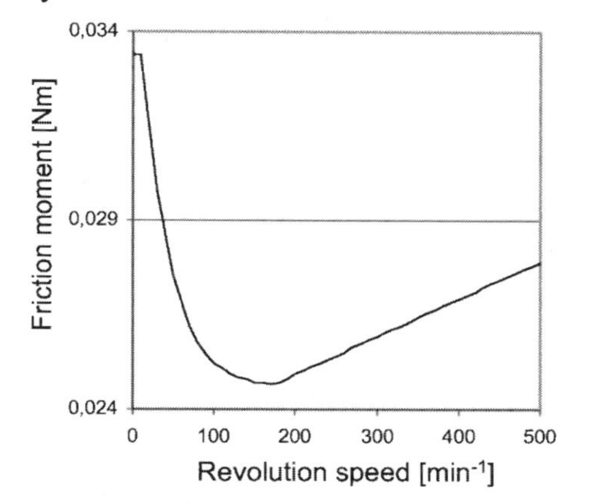

Fig. 4. The Stribeck-curve.

$$F_s = \mu_s F_n = \mu_s m g \tag{2}$$

where F_s is the force needed to overcome the static friction. μ_s is the static coefficient of static friction and F_n the normal force. F_n is the product of the mass m and the acceleration of gravity $g \approx 9.81\ m/s^2$

$$F_v = \mu_v v \tag{3}$$

with F_v as the drive force and v as velocity.

For the development of test routines, there are at least two areas of interest. The first one is assessing when the drive force is high enough to move the axis. This is the point at which the static friction is overpowered (see Eq. 2). This point is very important, because at very slow speed, stick-slip occurs and wear is high.

The second one is the region of viscous friction where the drive force is independent of F_n and increases linearly with the velocity (see Eq. 3). Even if in this region the wear is nearly zero because of the complete separation of the surfaces, μ_v can be seen as an indicator for lubrication quality.

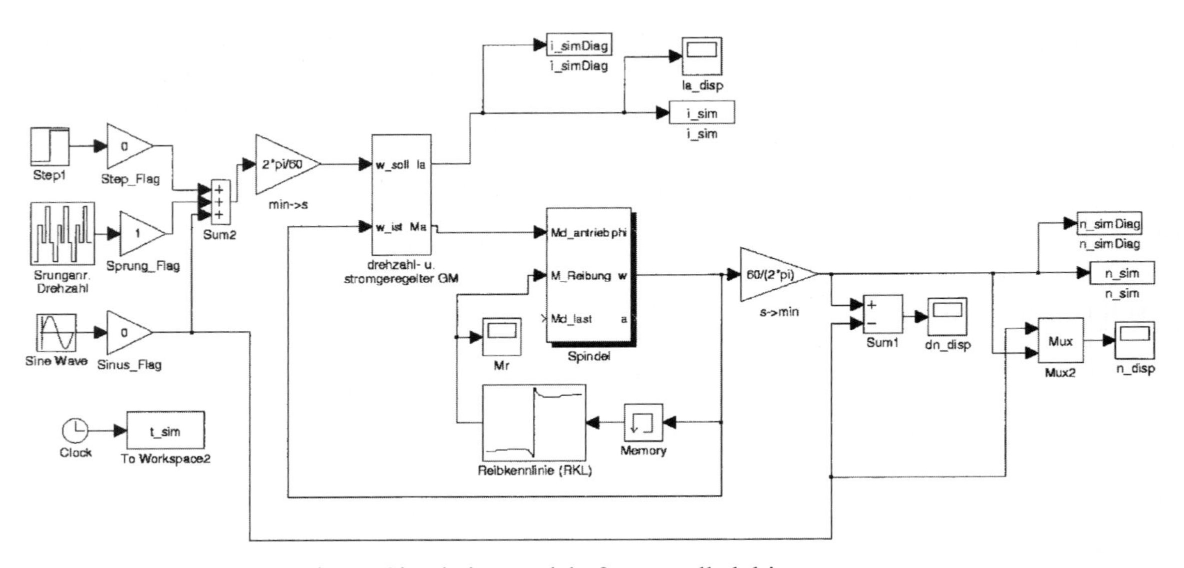

Fig. 5. Simulation model of a controlled drive system.

3.2. NC-Test Cycles – Uncovering of the Indicators

For the estimation of the two parameters μ_s and $\mu_{v,}$ which are indicators of the "health status" of the machine axis, it is necessary to develop special NC-cycles. Each of these cycles must be designed so that it is highlights the desired identifying features for the indicators. Eq. 2 and 3 in section 2.2 evince that the drive force can be used as the information carrier. Unfortunately, ordinary NC-machine tools do not provide drive force data. As an alternative to the force the drive current can be used. This signal is used by the control and is accessible in open control architectures. In a first approximation the drive current i_d is proportional to the drive force F_d (see Eq. 4).

$$i_d = c\, F_d \tag{4}$$

where c denotes the constant in which includes all frictional influences of the drive system [11].

3.2.1. The Stiction Test

It is assumed that the test method which best demonstrates the effect of transcending the stiction is to raise the drive current slowly and synchronously measure the axis position without the active controller. This feature is not available for standard NC-cycle programming. To detect mutations of stiction, a ramp is chosen for the velocity profile and programmed as an NC-cycle by setting the maximal acceleration to a small value[1]. Fig. 6 shows the standardised results of a simulation for this test. While the velocity is still zero in the beginning, the drive current arises very fast due to the controller's influence. After the stiction is exceeded, the axis accelerates and friction decreases. The local maximum of the drive current can be used as a measure of the stiction. The model used represents a controlled drive system created with Simulink®, an application for the modelling and simulation of dynamic systems (Fig. 5) [12], [13].

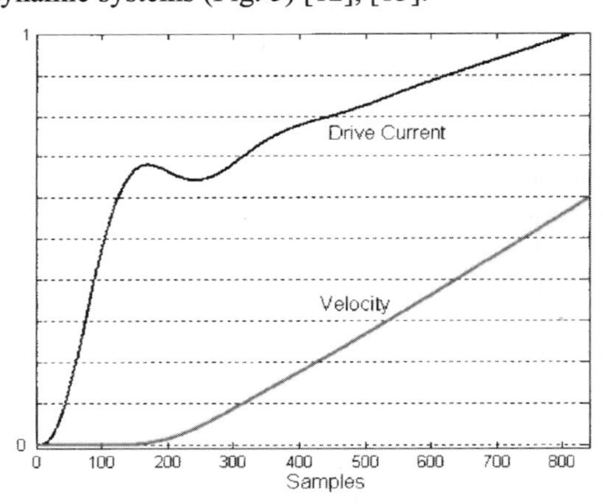

Fig. 6. Simulated drive current and velocity.

[1] Within NC-cycles it is possible to change machine data temporarily.

3.2.2. The Viscosity Test

The viscosity of the lubrication is mainly responsible for the viscous friction. Again, the drive current can be used as a measure of degradation. To test the quality status of the lubrication, the axis is moved at a constant speed within the velocity range where viscous friction occurs. For the estimation of this parameter the sequences of acceleration have to be masked out. After this the mean value is calculated, which represents the measure of the viscosity of the lubrication. Fig. 7 shows the simulated velocity and the drive current for this test. For better illustration both signals were standardised and are presented in the same figure. The simulation was done using the before mentioned model.

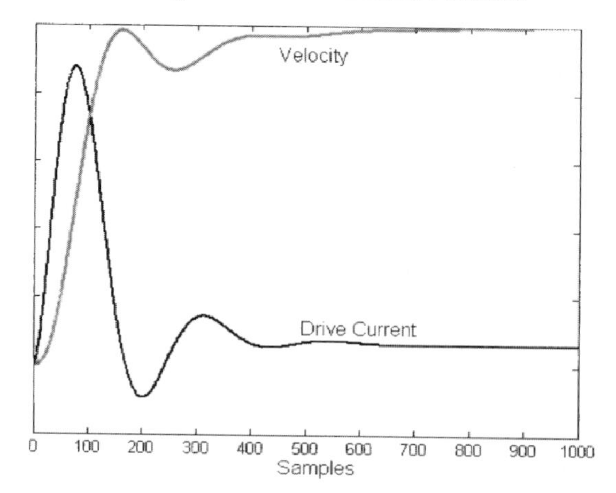

Fig. 7. Simulated drive current and velocity.

3.3. Using Statistical Data

The tests presented here involve taking snapshots of the "health status" of the machine axis. Iterations of these test routines can be used to detect trends within the degradation process. For the prediction of the further progression it is important to know how the machine axis was stressed in more detail.

This can be done by logging additional data along with information about the operation modes of the machine. The result is a kind of "load history" that can be used to make statistical inferences. Among these data rank

- the total number of manufactured workpieces,
- the hours of operation and
- other machine data

which can be accessed via the machine control. Combined with the estimated "health status" parameters a holistic view of the degradation process can be created.

4. Outlook

At present the proposed method is being realised within an industrial project together with a manufacturer of grinding machines. Continuative work has to be done in the sector of data analysis and data interpretation. Here the use of neuro-fuzzy methods appears meaningful and will be implemented in the near future.

Acknowledgement

Fraunhofer Institute for Production Systems and Design Technology is partner of the EU-funded FP6 Innovative Production Machines and Systems (I*PROMS) Network of Excellence.

References

[1] Hohwieler, E. and Geisert, C.: Intelligent Machines Offer Condition Monitoring and Maintenance Prediction Services. In: Roberto Teti (Editor), Proceedings of the 4th CIRP International Seminar on Intelligent Computation in Manufacturing Engineering (CIRP ICME '04), 30 June – 2 July 2004, Sorrento, Italy, pp. 599-604.

[2] Hohwieler, E.; Berger, R.; Geisert, C.: Condition Monitoring Services for e-Maintenance. In: Proceedings of the 7th IFAC Symposium on Cost Oriented Automation, June 7 – 9, 2004, Gatineau/Ottawa, Canada.

[3] Lee, J. et al.: Infotronics Technologies and Predictive Tools for Next-Generation Maintenance Systems. In: Proc. 11th Symposium on Information Control Problems in Manufacturing, Elsevier Ltd., 2004.

[4] Djurdjanovic, D.; Lee, J.; Ni, J.: Watchdog Agent – an infotronics-based prognostics approach for product performance degradation assessment and prediction. In: Advanced Engineering Informatics, Volume 17, Issues 3-4, July-October 2003, pp. 109-125.

[5] Grudzien, W.; Seliger, G.: Life Cycle Unit in Product Life Cycle – Tool for improved maintenance, repair and recycling. In: Proc. 33rd CIRP Intern. Sem. Manufacturing Syst., 2001, pp. 121-125.

[6] Franke, C.: Strategies and Visions for Future Manufacturing in Electronics and Light Assembly Industry. Presentation at: LASSI Summit Seminar, May 2002.

[7] Odry, D.; A. Buchholz: Enabling a Sustainable Cycle Economy with Life Cycle Units. In: Proc. Global Conf. Sustainable Product Development and Life Cycle Eng., Berlin, 2004, pp. 45-50.

[8] Brecher, C. et al.: Neue Technologien verbessern den Service. In: WB – Werkstatt und Betrieb, Carl Hanser-Verlag, September 2004/137, pp. 95-97.

[9] Walther, M.: Maschinendiagnose und Prozessdiagnose – Unterschiede, Gemeinsamkeiten und Schnittstellen. 16. Lageregelseminar, ISW Stuttgart, 18.-19. Juni 2004.

[10] Krüger J.: Methoden zur Verbesserung der Fehlererkennung an Antriebsstrecken. Berlin, Technische Universität, 1998.

[11] Suwalski, I.: Steuerungsintegriertes Prozeßüberwachungssystem für Drehmaschinen. Berlin, Technische Universität, 1999.

[12] Uhlmann, E.; Hohwieler, E.; Becker, F.: New Structures and Methods for Control-Integrated Process Supervision. In: Proceedings of the Second International Workshop on Intelligent Manufacturing Systems, 1999, Leuven, Belgium, pp. 809-815.

[13] Hohwieler, E., Pruschek, P.: Reliability, Process Monitoring and Control; Proceedings MANTYS Conference EMO Milan 2003, "Future Trends of the Machine Tool Industry", Milan, 23. October 2003, pp. 35-45.

Intelligent Production Machines and Systems
D. T. Pham, E. E. Eldukhri and A. J. Soroka (eds)

Overcoming the challenges of product development and engineering changes in a distributed engineering environment

M. Goltz, D. Müller[a], R. Schmitt[b], M. vanden Bosche[c]

[a] *Institut für Maschinenwesen, TU Clausthal, Robert-Koch-Str. 32, 38678 Clausthal-Zellerfeld, Germany*
[b] *Siemens SGP Verkehrstechnik GmbH, VT5 FW6, Eggenberger Str. 31, A-8021 GRAZ, Austria*
[c] *Mission Critical SA, Wijnegemhofstraat 199, B-3071 Erps-Kwerps, Belgium*

Abstract

This paper presents a concept to improve the management of developing complex products in a distributed engineering environment. It will show a product data driven engineering workflow that is able to manage engineering tasks across company boarders. This approach is based on parameters, presenting the smallest product model element, and a user category concept to address tasks to the right persons at the right time. In addition, a concept for an architecture of a web-based distributed workspace will be given.

1. Introduction

Only a small time share of engineering working hours can be considered 'productive'. Approximately 25% are consumed while waiting for decisions or information. In addition an overload of information has to be checked daily, such as notifications about proposed changes in products. The amount of information is increasing rapidly when working in a distributed environment where multiple partners have to work together on a complex product. Product modules not directly related to central competencies of a company are developed and delivered by suppliers and engineering partners. In order to shorten the time for development all partners have to work together from the early stages of product development. This leads to increased needs to exchange and share important and right information among partners.

Today's Product Lifecycle Management (PLM) systems provide good capabilities regarding the management of complex product data within a single company. However, taking into account the variety of systems used at partner sites in an engineering environment one can easily imagine problems regarding the interoperability. It is some kind of a lottery to find the right information at the right place at a time when it is needed. Duplication of data on different sites might be a solution. However, it does not overcome the problem sufficiently because consistency of data can hardly ensured.

2. Product data driven engineering workflow

Within the SIMNET project [1] a workflow management system has been developed with the target to improve the engineering co-operation within the supply chain. For this purpose, the 'classical' document-based engineering workflow functionality of a state-of-the-art system for PLM is enhanced. This enhancement consists in the creation of additional product model controlled functionality, whereby the smallest product model element, i.e. the parameter is used.

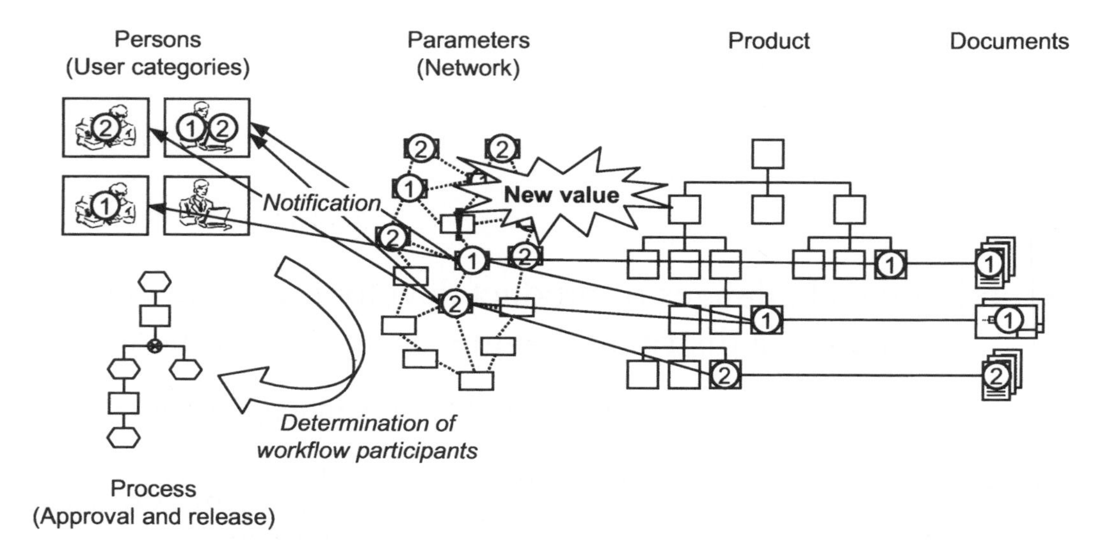

Fig. 1. Example of a parameter-supported engineering change

Parameters describe attributes of product components and relations among them. The approach is based upon observing how engineers think about their tasks. It was found that engineers tend not to view their jobs as dealing with documents, or even as being a process. They view their work as making engineering decisions. These decisions change or determine engineering variables. We refer to these engineering variables as parameters. Parameters represent the specific circumstances in a given engineering situation. They can refer to dimensions as well as forces and movements. Parameters often share complex relationships, which can be represented in terms of functions for example, see Eq. 1:

$$max\_axle\_diameter = f(max\_axle\_load, bear\_distance\_track\_gauge, axle\_material) \quad (1)$$

As people set parameter values, capturing this relationship allows to establish links between an engineering decision and those involved. The relationship between parameters determines the involvement of suppliers and engineering partners. As such, these links can also be captured, represented, and managed through the parameter-based approach.

When introducing this approach, an analysing phase is required to capture the most important parameters. Starting the implementation together with a pilot project enables engineers to create the parameter definitions and instantiate them simultaneously. The relations between the parameters are captured during the performance of the engineering activities. This results in a parameter network that can be used for the support of engineering changes occurring either during the runtime or after the completion of the project.

Each parameter of the network is linked to elements of the product structure as well as to persons. Via the user category scheme it becomes clear which persons have to be informed whenever the status or the value of a particular parameter changes. Thus, the required participants of the approval and release workflow can be identified. A direct link between parameters and most documents[1] is not necessary because the relationship is anyway defined via the items of the product structure. Fig. 1 gives an overview of the existing relations.

Engineering change requests are usually communicated in a parameter-based way. Examples are "The engine power needs to be increased" or "The shaft diameter is not sufficient". The parameter to be changed then automatically becomes the starting point for all further considerations.

As soon as a parameter is subject to a change, its current hardness grade is frozen and the status is set to "in work". The latter is also true for the linked product structure items and related documents. Afterwards, only those persons (or roles) are informed which are assigned to the defined user categories "co-ordinator" and "collaborator" of

[1] There may be working documents that were linked directly to parameters.

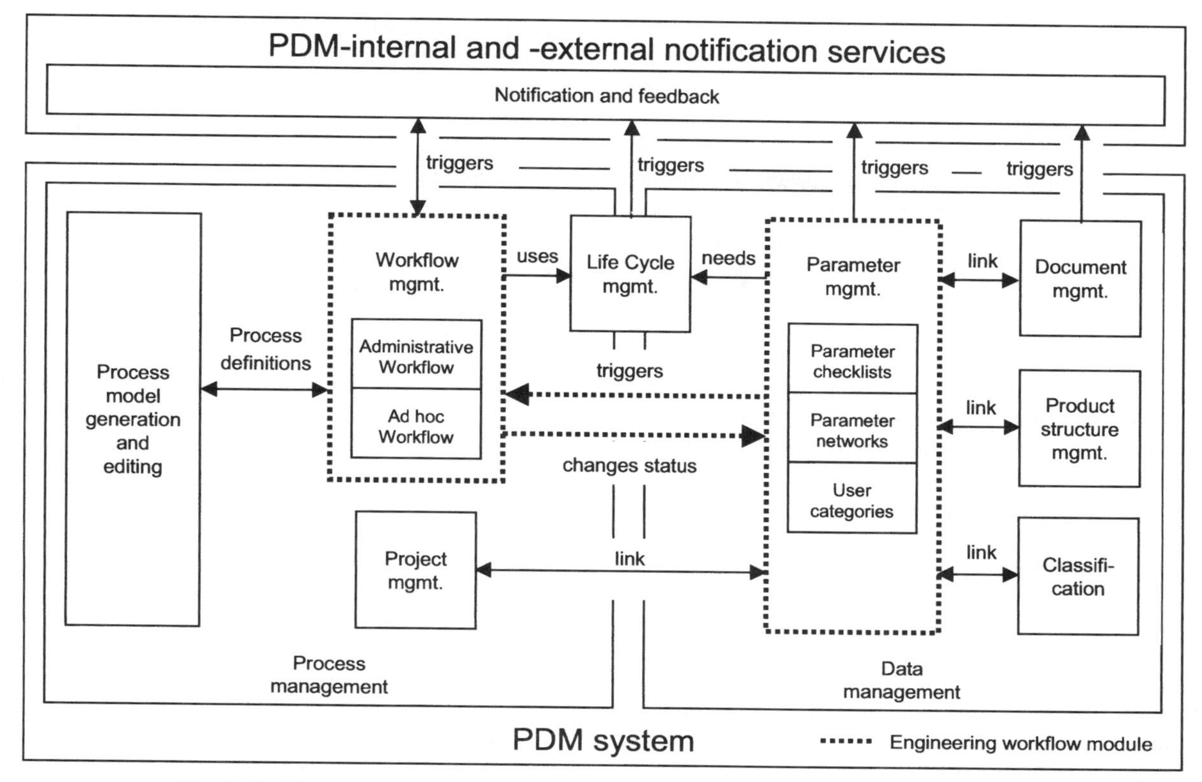

Fig. 2. Architecture for the parameter-based IT support of engineering change processes

- the changed parameter itself and
- all adjacent parameters (which show a first-degree interdependence with the changed parameter).

The system only highlights the existing relation and the possible change propagation between the changed and the neighbouring parameters in the network. The co-ordinator and the collaborators must jointly clarify whether these parameters are actually affected by the change. As long as this clarification is not completed, the neighbouring parameters remain in their current status.

As soon as it becomes clear that an adjacent parameter is affected by the change, its hardness grade is also frozen and the status is set to "in work" (this is also true for related product structure items and documents). In addition, the co-ordinators and collaborators of those parameters with a second-degree interdependence to the initially changed parameters via the affected one are informed (see Fig. 1).

The objective is to cope successfully with the change propagation through the product model by means of a step-by-step identification via the parameters and subsequent interaction of all "co-ordinators" and "collaborators". As soon as the change propagation is clarified to its full extent and a consensus has been found between all people involved in or affected by the change, all parameters set to "in work" during the change process will be jointly set to "in approval" and finally be approved as well as released within a single parameter-based workflow.

The described approach has been implemented into the PLM system of Agile Software Corporation. Fig. 2 illustrates the architecture of the parameter-based IT support of engineering change processes as part of an engineering workflow solution. It furthermore shows the interaction of the new parameter management module with the standard modules of the PLM system. This module has been developed as prototype within the SIMNET project, to verify the approach.

With respect to Fig. 2, the link between parameter and document management is to be considered optional. It depends on whether documentation is created which can be assigned to a specific parameter (e.g. calculation sheets). If documents only describe product structure items such as parts, sub-assemblies, and assemblies, this link is obsolete.

3. Distributed workspace

3.1. How to share information in a distributed environment

During the last years tremendous efforts have been spent to shorten the throughput time in production. Now, these efforts are extended to the development stage of a product, which offers great potential to improve efficiency. Simultaneous and collaborative engineering has been deployed in companies to overcome the time consuming sequential way of working. Experts from other disciplines are integrated into a concurrent engineering team. PLM and ERP systems providing workflow management capabilities support such teams by enabling the exchange of information. However, so far the IT support is limited when it comes to engineering teams built from experts of different companies.

There are two ways to enable access to data that is used by different partners. First way is to grant access to the host system by one partner. All other partners are using a native or web client for the specific PLM system. However, the number of required clients is directly increased by the number of different PLM systems used by the partners.

The second way would be to use an interface between two systems and to replicate data at certain points in time. In this case it is nearly impossible to ensure consistent data, because modification may occur on both sides. In addition, it requires n*(n-1)/2 interfaces to connect all partners of the engineering team.

Therefore, other concepts are required to really enable concurrent design at one product model. Different companies have different views on data, with their own attributes. However, each view does not exist alone. It is somehow connected to the views of partners. Taking that into account the concept of a distributed workspace has been developed. It enables companies to maintain their view on the data created within the project, but without missing the link to other views. That enables a much better transparency regarding the progress of the project and leads to less rework due to missing or outdated information.

3.2. Demands on a distributed workspace

Sharing and exchange of product data across company boarders strikes a very sensible area within a company. Knowledge and information are treated as very valuable goods that require special protection. Therefore, special attention will be put on security issues in terms of:

- granting access to data only for authorised personnel, that includes ensuring that a person can be clearly identified and
- secure data transport independent of the way the data has been accessed (using a web-client, a native PLM client or email).

When accessing data from different systems the distributed workspace has to cope with different PLM systems. To avoid inconsistency it has to be ensured that data uploaded to the distributed workspace is treated the same way it would be handled in the system of origin. Another aspect is related to the parameter management itself. Since the concept presented in chapter 2 leaves a lot of possibilities of usage the mechanisms used in the distributed workspace must not affect the implementation of each company. Partners only have to agree on a common strategy used within their project. Furthermore, it is clearly stated that data stay under the responsibility of the owning company.

3.3 Architecture of the distributed workspace

The distributed workspace is a server on its own hosted by a service provider. Companies with multiple supplying partners are able to create a workspace for each project they are working on. All project participants are able to access the workspace using a web client. Data is exchanged between workspace and the PLM servers using XML standard. The architecture is shown in Fig. 3.

Using the distributed workspace as a central gateway to project relevant data enables efficient handling of information. A user can request data via the web client. The PLM system hosting the requested information checks if access is granted to the user. If so, data is checked out and presented in the web-client.

In addition, the workspace server is used to store the mapping between data coming from different sources, such as a product component handled in a PLM system of the main contractor and at the suppliers site. Changes on one of the sites are communicated via the workspace and users can react accordingly.

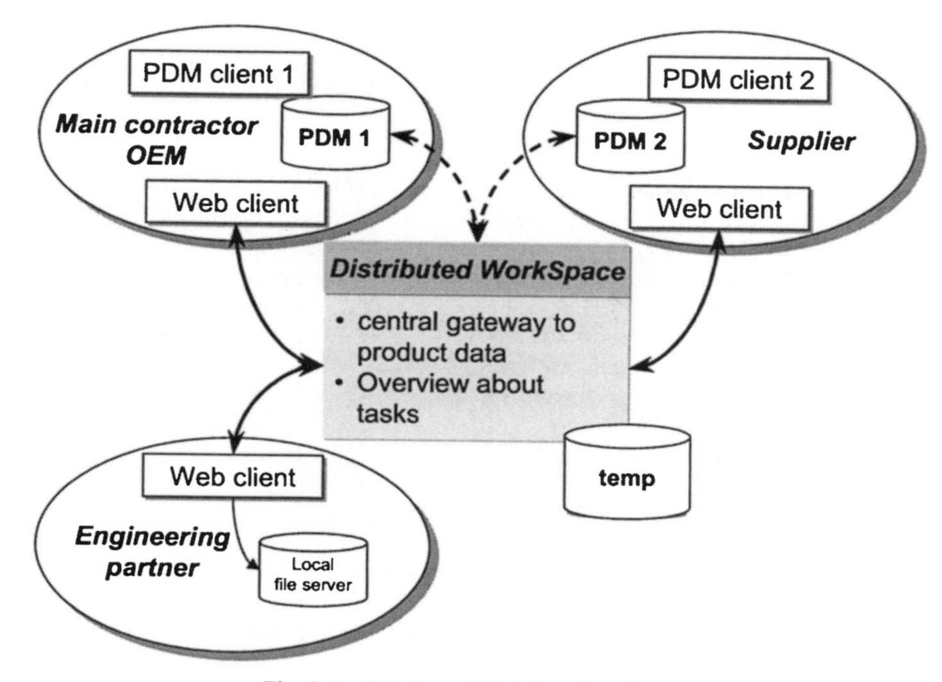

Fig. 3. Architecture of distributed workspace

3.4 Organisation of collaboration

Applying the parameter-based approach as presented before enables engineers to track the ongoing product development from one location. All requests for approval are directed to the workspace in-box of the user. The current view of a parameter is presented by following the link provided in the work item. Attached to the parameter a list of allocated objects can be presented. To retrieve additional information the user does not have to care where to get it from. That is handled by the workspace server. Requests are addressed to all servers that provide parts of the information. All information are compiled into sets of data presentable to the user. Updates done by the user are passed back to the original server. As long as someone is working on the data in the distributed workspace the information is locked for further modification. That ensures consistency of the database, even across companies.

During the runtime of the SIMNET project a distributed workspace has been developed based on a Agile PLM server that hosts only project relevant data and links to original data of the partners systems. The communication among the connected PLM systems has been based on neutral file exchanges.

3.5 Security mechanisms deployed

Such a distributed workspace is based on the interconnections of the participating organisations through the Internet. Within a so called Trusted Virtual Community (TVC), the full spectrum of security services – message encryption, user authentication, service authorisation, integrity and non-repudiation of messages – is required for online interaction (e.g. web) and offline message exchanges (e.g. email).

The SIMNET solution has been based on state of the art security standards – PKI (Public Key Infrastructure), TLS (Transport Layer Security) for online interaction and S/MIME (Secure Multipurpose Internet Mail Extensions) for email – and additional functionalities. Strong security is provided through encryption keys (symmetric/ asymmetric keys). Smart cards will be used to reinforce the security.

All the partners of the TVC are peers, with the exception of a Management Entity (ME), who is independent from any single organisation and is trusted by all the members of the virtual community. The role of the ME is to implement the security policies decided by the community and make sure that the security is correctly enforced, while preserving the autonomy of the members (e.g. with regards to the "visibility" of the data); the ME is also the CA (Certification Authority). Fig. 4 illustrates this.

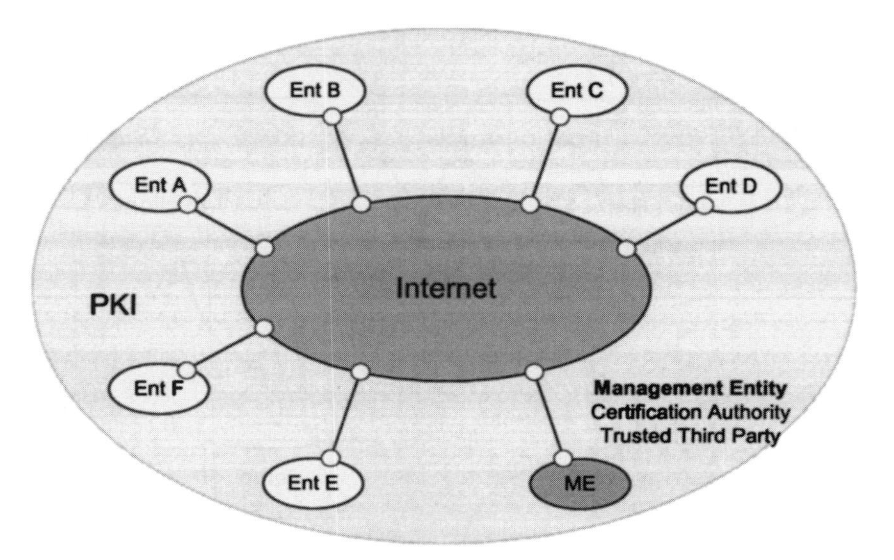

Fig. 4. Security infrastructure

The identification of users, their roles and privileges within the community, and their *Public Key* are managed by a Directory Service, based on the LDAP standard, distributed among the ME and the members. The portion of the directory service under the control of a specific organisation manages the local authorisation rights.

4. Conclusions

Today, the IT support for inter-company teams developing a complex product is still limited to simple file exchanges or access to central PLM systems. The presented parameter-based approach supports such teams in terms of addressing engineering tasks to the right people deploying a user category concept. Relations to components of the product structure and between parameters itself, establishing a parameter network, enable later on an efficient engineering change management.

The presented architecture for a distributed workspace enables access to data needed to execute engineering tasks, even if the data is stored outside the company. The data is protected by state-of-the-art security mechanisms. In addition, the workspace provides communication and a document-free quality assurance in a distributed engineering environment. Major benefits of the shown approach are:

- improved quality of engineering results and
- shortened throughput time regarding engineering change management and product development.

Acknowledgements

[1] The project to which the above described results refer is co-funded by the European Commission under the ESPRIT programme (project No. 26780, *SIMNET - Workflow Management for Simultaneous Engineering Networks,* http://www.imw.tu-clausthal.de/inhalte/forschung/projekte/SIMENT/Welcome.html). The project partners are:

- Siemens SGP Verkehrstechnik GmbH, Graz, Austria (Co-ordinator)
- EIGNER + PARTNER AG, Karlsruhe, Germany
- Knorr-Bremse Systeme für Schienenfahrzeuge GmbH, München, Germany
- Mission Critical SA, Waterloo, Belgium
- BETA at Eindhoven University of Technology, Eindhoven, The Netherlands
- TU Clausthal, Clausthal-Zellerfeld, Germany
- IPS Ingeniería de Productos, Procesos y Sistemas Integrados S.L., Valencia, Spain

The authors wish to acknowledge the Commission for their support. They furthermore wish to acknowledge the SIMNET project partners for their contribution during the development of various ideas and concepts presented in this paper.

[2] The Institute of Mechanical Engineering, Clausthal University of Technology, is partner of the EU-funded FP6 Innovative Production Machines and Systems (I*PROMS) Network of Excellence.

Intelligent Production Machines and Systems
D. T. Pham, E. E. Eldukhri and A. J. Soroka (eds)

A problem solving approach to developing product support systems

R.M. Setchi, N. Lagos

Manufacturing Engineering Centre, Cardiff University, Queen's Buildings, Cardiff CF24 3AA, UK

Abstract

Problem solving has traditionally been one of the principal research areas for artificial intelligence. Yet, although artificial intelligence techniques have been employed in several product support systems, the benefit of integrating product support, artificial intelligence, and problem solving, is still unclear. This paper studies the synergy of these areas and proposes a structured problem solving approach that integrates product support systems and artificial intelligence techniques. The approach includes defining and classifying product support problems, selecting different reasoning techniques for different types of problems, and introducing a multi-modal strategy that combines case- and model-based reasoning. The view that problem solving and product support are interrelated enables the development of product support systems in terms of smaller, more manageable steps. The combination of different reasoning modalities provides a way to overcome the lack of enough documentation resources, which could be a problem in many applications. The prototype system developed illustrates the applicability of the approach.

1. Introduction

Product support is defined as everything needed to support the continued use of a product [1]. Advanced product support takes various forms, ranging from interactive technical manuals (IETMs) [2] to intelligent product manuals (IPMs) [1] and electronic performance support systems (EPSSs) [3].

Efficient support involves answering to users' queries by providing accurate and user-tailored information. Each user query can be represented as a problem, using approaches from problem solving research. The diversity of the problems posed to the system should be an essential consideration throughout its design.

This paper supports the view that a structured problem solving approach can be used throughout the development of a product support system. This approach involves identifying and classifying product support problems and proposing appropriate reasoning methods for each problem type.

The rest of the paper is organised as follows. Section 2 briefly reviews problem solving and artificial intelligence (AI) techniques utilised in product support systems. Section 3 includes definitions of product support problems and solutions, and introduces the proposed problem solving approach. A prototype system is presented in section 4. The final section contains conclusions and directions for further work.

2. Background

2.1. Problem solving and reasoning techniques in AI

Research in AI has greatly benefited from using problem solving techniques [4]. For example, divide and conquer and building block techniques are employed in rule-based reasoning, where expert knowledge is represented using rules. Solving by analogy is utilised in case-based reasoning (CBR), which retains and reuses previous solutions generated by the system. Means-ends analysis is used in model-based reasoning (MBR), which employs general knowledge about the application domain.

These approaches [5, 6] differ in terms of the knowledge, inference cycle and inference process involved, as follows.

- **Knowledge.** Rules and cases represent specific knowledge while models capture general knowledge. For example, an instance of a clutch can be represented using a case with a number of attributes, each of them containing specific knowledge about the type of clutch. The same clutch can be represented using models, which contain generic knowledge such as "clutch is an assembly".
- **Inference cycle.** Rule-base technique follows an iterative cycle, while case- and model-based techniques reuse previous solutions.
- **Inference process.** Rule-based reasoning creates a series of progressive inferences between the problem definition and its solution rather than modifying parts of old solutions, as it is in case-based reasoning.

Each reasoning technique is applicable in different circumstances. Rule-based technique is appropriate for applications that necessitate modular development and uniform knowledge representation. Case-based reasoning is preferred in situations where the domain is not well known but experiences can be easily used to enrich the knowledge of the system. Model-based reasoning is used when the application domain is well known and especially if its causality characteristics are thoroughly understood.

2.2. Reasoning techniques in product support

Studies show that the most common AI technique used is rule-based reasoning, which is primarily employed in troubleshooting. An example is the work of Paul et al. [7], in which a diagnostic system supports the operation of a radar warning receiver.

In addition, a number of researchers have used CBR for diagnosis and help-desk applications. Foo et al. [8] utilise CBR in combination with neural networks for producing a help-desk-support environment, while Auriol et al. [9] use a CBR system in the troubleshooting of a welding robot.

Model-based reasoning has received less attention than these two techniques. An example is the research of Brusilovsky and Cooper [10], who employ models for adapting the interface of a performance support system and creating an 'expert-like' problem solving engine.

Latest attempts focus on the integration of different reasoning techniques. For instance, Pham and Setchi [11] develop adaptive product manuals by combining CBR for interpreting user's requests and rules for adapting the generated documents.

As illustrated above, AI reasoning techniques have been used in product support systems. The research reported in this paper extends these approaches by studying the inference mechanism in a structured manner, which enables the segmentation of the development process into smaller, more manageable steps. Moreover, although most recent research approaches adopt a hybrid reasoning strategy by applying combination of reasoning techniques, the application of multi-modal reasoning in product support systems has not been thoroughly researched. Therefore, the integration of CBR and MBR is a significant part of the presented approach.

3. Problem solving approach to product support

The problem solving process starts with problem identification/recognition [12-14]. In the context of product support, the first step is to identify the required resources and to define the product support problems (PSPs). The problem solving process continues by determining the type of a PSP and the course of actions that should be followed for reaching a product support solution (PSS). A PSS can take several forms, including explanations, instructions, warnings, descriptions, which are delivered through documents.

3.1. Information object cluster

Ideally, product support documents should be constructed using logically linked, semantically distinguishable and classifiable information objects (IOs) [15]. An IO is defined as "a data structure that represents an identifiable and meaningful instance of information in a specific presentation form" [16]. An IO could be a picture that illustrates a part of a product or a textual description of its function. The notion of an Information Object Cluster (IOC) is introduced in this research as a means of organising IOs.

Information Object Cluster is a 2-tuple IOC:=({IO}, S) where {IO} is the set of IOs sharing a common property that are arranged in a structure S.

For example, clutches can be classified in different types according to the number of disks they have. A single-disk clutch has a different structure and operation compared to a double-disk clutch. Assume that four IOs are available, as follows.

- IO1: A textual description of the operation of a single-disk clutch,
- IO2: A textual description of the operation of a double-disk clutch,
- IO3: Image of a single-disk clutch, and
- IO4: Image of a double-disk clutch.

Since all four IOs share the common property of describing the concept of a "clutch", they all belong to the same IOC. Within the IOC, the IOs are structured according to whether they describe a single-disk or a double-disk clutch. IO1 therefore is linked with IO3, and IO2 is linked with IO4. The links are in the form of directed arcs, defining the presentation order. In the case that a user needs a description of the car, other IOCs that correspond to the other subsystems of the car (e.g. brakes) are also needed to produce a different document.

Document (D) is a 2-tuple D := ({IOC}, S) where {IOC} is a set of logically structured IOCs.

3.2. Problem characterisation

There are two main factors that define the structure, content, and presentation of a document generated by a product support system. These are the domain and the environment within which the system is deployed. Domain-specific elements include the products and the tasks that are supported, as well as the documentation resources that describe them. Environmental-specific elements consist of user, location, time and purpose dependent features. A product support system should be able to extract observations from users' queries, related to both types of factors. Therefore, a product support problem could be defined as follows.

Definition 1. A P*roduct Support Problem* (PSP) is a 4-tuple PSP := (MOD, HYP, CON, OBS) where:

- *MOD* is a finite set that includes knowledge about the domain (i.e. supported products and tasks) that is represented with corresponding models. MOD also contains the links between these models and relevant IOCs and IOs.
- *HYP* is a finite set of combinations of elements of MOD. Each combination corresponds to a possible documentation configuration. The arrangement of MOD combinations and documentation configurations will be referred to as hypotheses.
- *CON* is a finite set that includes knowledge about the application environment. It is represented by related models (e.g. user model (UM) or system's specifications model). CON will be referred to as the system's context.
- *OBS* is a finite set that consists of the observations acquired by the current query. OBS should correspond to elements of MOD and CON.

A product support solution is defined as follows.

Definition 2. Given a product support problem PSP := (MOD, HYP, CON, OBS) and a finite set of observations OB such that $OB \subseteq OBS$, then valid hypotheses are represented by VH where $VH \subseteq HYP$ if and only if, $MOD \cup CON \cup VH \vdash OB$. Product support solution (PSS) can be every hypothesis 'h' that $h \in VH$. The best solutions are considered the ones for which OB=OBS.

Product support solutions can be distinguished in terms of their content and presentation.

- Content. The content can include general information (e.g. "any clutch is an assembly") or specific information (e.g. "this clutch has two disks"). It can be generated at run-time, be adapted, or pre-composed.
- Presentation. The presentation can be text- or text and multimedia-based, as well as adaptive, adaptable, and predefined.

3.3. Knowledge reuse for previously solved problems

Previously solved/same PSPs are problems that have already been solved by the system. This means that if the current PSP corresponds to a set of observations $OBS_{current}$, and if $OBS_{current} = OBS_{previous}$, where $OBS_{previous}$ represents a set of observations that have already been processed by the system, then the PSS that relates to $OBS_{previous}$ is reused since the set of current observations is the same as the previous one.

The reasoning technique used should be able to retrieve solutions without repeating the reasoning process, as well as be able to compare the specifics of different problems. Naturally, case-based reasoning is considered as the most appropriate technique. The cases are represented as problem-solution pairs, where the attributes of each case correspond to parts of the domain- and environmental-specific models.

3.4. Knowledge adaptation for similar problems

Similar PSPs are problems for which the current set of observations is not the same as any of the previous ones but which shares similarities with them.

The new PSSs can be produced by removing, adding, replacing, and adapting IOs and IOCs used in previous situations. Assume, for instance, that a $PSS_{previous}$ describes a single-disk clutch. If the current query requires the description of a double-disk clutch then $PSS_{current}$ can be generated by replacing the IOs that describe the single-disk clutch with the IOs that correspond to the double-disk clutch.

Case-based reasoning provides means for adapting solutions according to previous experiences. This feature is highly utilised here. The adapted cases form new problem-solution pairs, which are retained in the case-base.

3.5. Knowledge creation for new problems

New PSPs occur when there are no similarities between earlier observations and new ones, meaning that $OBS_{current} \neq OBS_{previous}$. The PSS in this case requires additional IOCs that are not contained in the document base. For example, assume that a query relates to the description of a transaxle and that all previous PSSs use only an IOC that describes a clutch. If the transaxle IOC does not exist in the document base then means to automatically create it are needed.

The existence of a new problem that has little or no similarity with previous experiences cannot be easily managed by CBR alone. In this case a multi-modal reasoning strategy is needed. General knowledge about the application domains (e.g. product domain), contained in the models, can be utilised to support the case-based reasoning component. In the example discussed, model-based reasoning can be used to identify that a new IOC is required to describe the transaxle. The transaxle concept can then be detected in the product model, and its structure and characteristics acquired. Consequently, a new IOC that describes a transaxle can be generated.

The difference in the solving process of different kinds of problems is depicted in Fig. 1, which is based on the main steps of the case-based reasoning cycle [17].

The stages of the process when a previously solved problem is detected (Fig. 1 (A)) include retrieving relevant cases, reusing them, presenting the best match as a proposed solution, evaluating this solution, and retaining it in the knowledge-base.

If the problem is similar to a previous one (Fig. 1 (B)), an extra stage (i.e. solution adaptation) is needed, where case-based reasoning adaptation techniques can be utilised.

Fig. 1 (C) shows the solving process for a new problem. Model-based reasoning is employed at each stage for identifying new problems, classifying the cases' attributes according to the models, and configuring the generated documents. Enrichment of the models employed is also possible.

The co-relation between the types of a PSP, reasoning technique used, and PSS generated by the product support system, is depicted in Table 1.

Table 1
Co-relation between product support problems (PSPs), reasoning techniques and product support solutions (PSSs)

PSP	**Reasoning Technique**	**PSS**
Previously solved	CBR	**Content**
		Pre-composed and specific
		Presentation
		Text and multimedia, predefined or adaptable
Similar	CBR	**Content**
		Adapted and specific
		Presentation
		Text and multimedia, adaptive and/or adaptable
New	CBR/MBR	**Content**
		Generated and general
		Presentation
		Text and multimedia, adaptive and/or adaptable

4. Implementation

A prototype system, based on the approach proposed, is developed and part of it is depicted in Fig. 2.

The system uses product, task, and user models as described in [18]. The models are created with the Protégé ontology editor. The case-based reasoning tool utilises FreeCBR, for acquisition of cases. The adaptation of the cases is performed using an expert-like engine. The system is implemented in Java; it uses Apache's Tomcat as a standalone web server. The system can be accessed via the Internet.

Fig. 2 shows that the solution provided changes according to the mode used. If the user selects

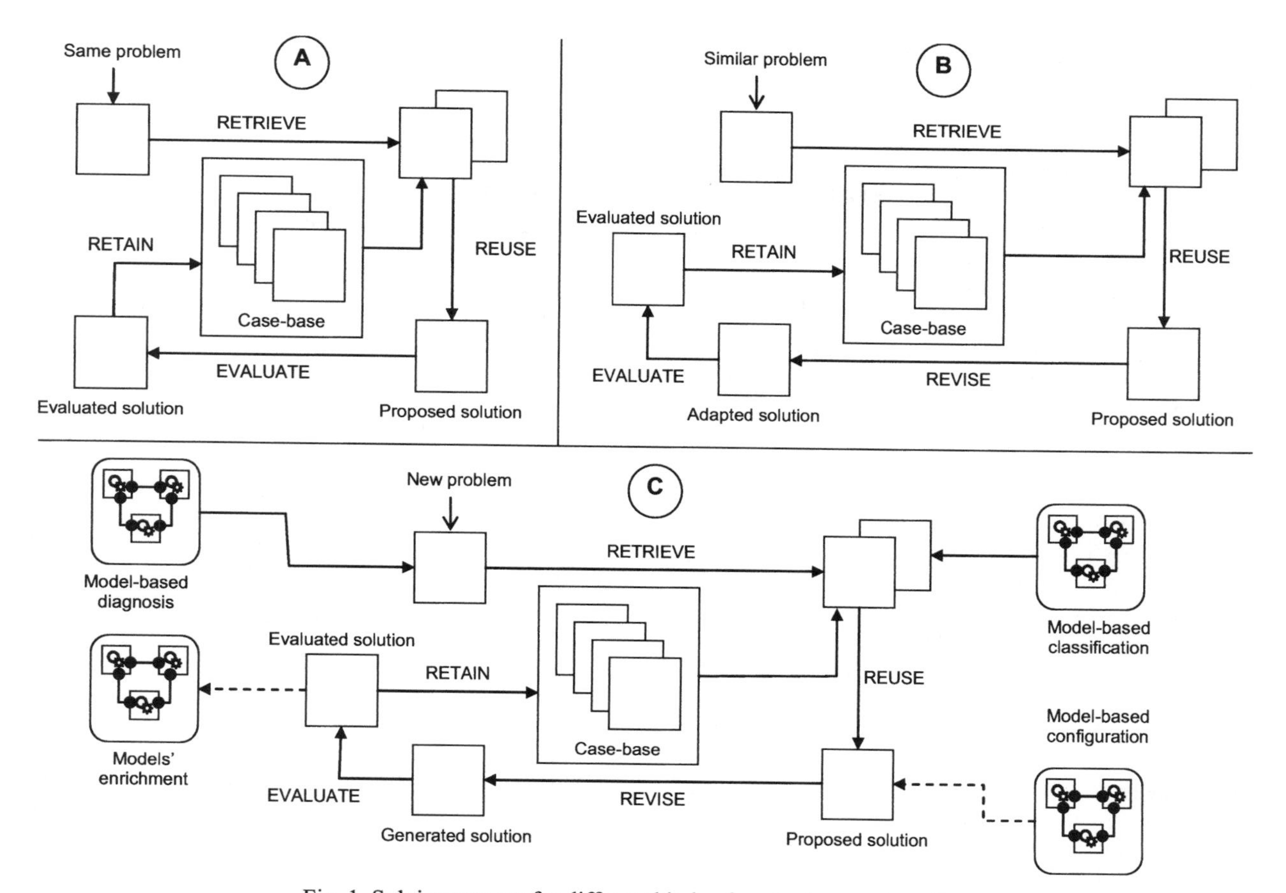

Fig. 1. Solving process for different kinds of product support problems

the browsing mode, the solution given to him/her is a pre-composed document. On the other hand, if the choice is to use the case-based reasoning tool, the user is presented with a list of ranked hypotheses and given the choice to select those that best fit his/her interests. Finally, the user is presented with an adapted solution.

5. Conclusions and future work

The structured approach presented in this paper advances product support by segmenting the development phase of the system's inference mechanism into small manageable steps. These include determining the kind of product support problems that the system is expected to handle, and selecting accordingly the reasoning technique that should be followed. The types of product support problems predetermine the situations with which the product support system should be able to cope with. The multi-modal reasoning strategy proposed enables the system to respond to a variety of user queries since case-based reasoning is utilised in providing accurate and personalised information while model-based reasoning is used to compensate for the lack of documentation resources. The conducted analysis denotes a correlation between problem solving and product support, indicating possible collaboration between the two research communities in terms of approaches and methodologies.

Future work includes the development of a product support knowledge engineering framework, where the proposed approach will be integrated with knowledge-based models.

Acknowledgements

The research described in this paper was performed within the I*PROMS Network of Excellence sponsored by FP6 of the European Community.

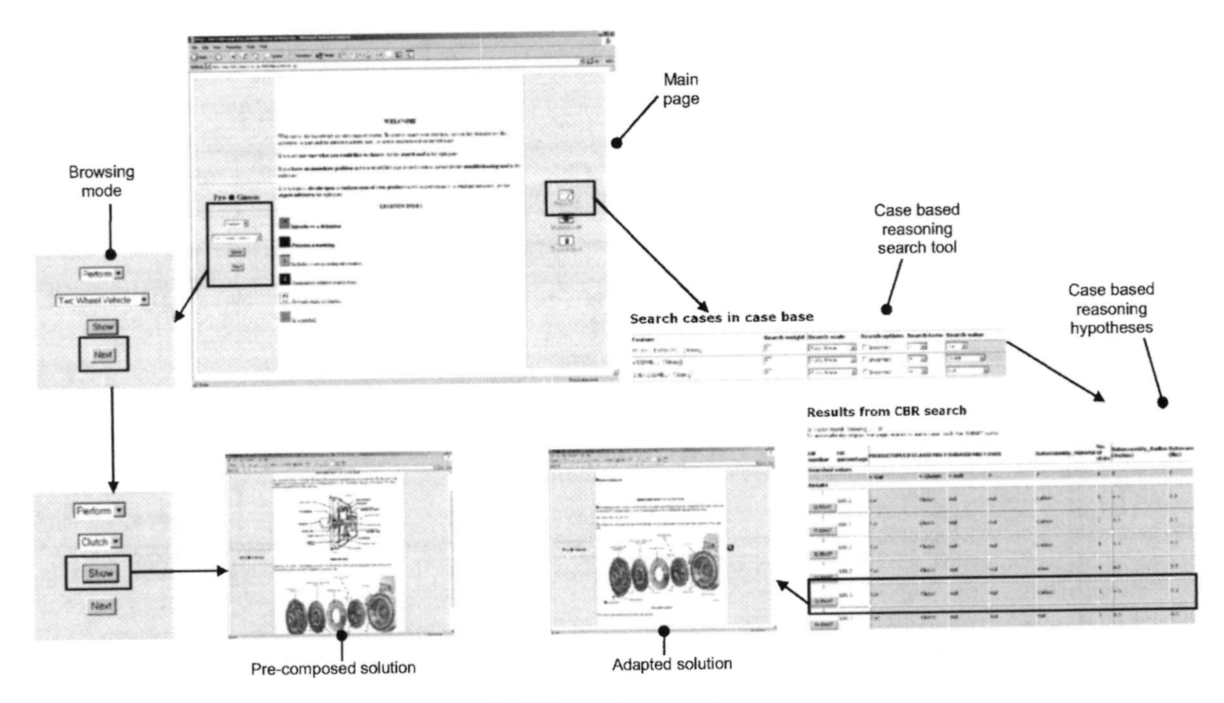

Fig. 2 Illustrative example

References

[1] Pham DT, Dimov SS, Peat BJ. Intelligent Product Manuals. Proc. IMechE. B-214 (2000) 411-419.

[2] Jorgensen EL, Fuller JJ. A Web-based Architecture for Interactive Electronic Technical Manuals (IETMs). ASNE Naval Logistics Conference (1998).

[3] Gery G. Electronic Performance Support Systems: How and Why to Remake the Workplace through Strategic Application of Technology. Boston, Weingarten Publications, 1991.

[4] Dale N, Weems C. Programming and Problem Solving with C++ (4th edition). Jones and Bartlett, Massachusetts, 2005, pp 27-33.

[5] Kolodner J. Case-Based Reasoning. Morgan Kaufmann, San Mateo, 1993, pp 92-97.

[6] Gonzalez AJ, Dankel DD. The Engineering of Knowledge-Based Systems, Theory and Practice. Prentice Hall, US, 1993, pp 263-295.

[7] Paul C, Zeiler G and Nolan M. Integrated Support System for the Self Protection system. IEEE Systems Readiness Technology Conference Proceedings, AUTOTESCON (2003) 155-160.

[8] Foo S, Hui SC and Leong PC. Web-based Intelligent Helpdesk-support Environment. Int. J. Syst. Sc. 33-6 (2002) 389-402.

[9] Auriol E, Crowder RM, McKendrick R, Rowe R and Knudsen T. Integrating Case-based Reasoning and Hypermedia Documentation: an Application for the Diagnosis of a Welding Robot at Odense Steel Shipyard. Eng. App. of AI. 12 (1999) 691-703.

[10] Brusilovsky P, Cooper DW. Domain, Task, and User Models for an Adaptive Hypermedia Performance Support System. Proc. IUI '02, ACM, San Fransisco, California, USA (2002) 23-30.

[11] Pham DT, Setchi RM. Case-based Generation of Adaptive Product Manuals. Proc. IMechE. B-217 (2003) 313-322.

[12] Gunzelmann, G., Anderson, J.R.: Problem solving: Increased Planning with Practice. Cog. Sys. Res. 4 (2003) 57-76.

[13] Gray, P.H.: A Problem solving Perspective on Knowledge Management Practices. Dec. Sup. Sys. 31 (2001) 87-102.

[14] Liao, S.H.: Problem Solving and Knowledge Inertia. Exp. Sys. App. 22 (2002) 21-31.

[15] Pham DT, Dimov SS, Setchi RM, Peat B, Soroka A, Brousseau EB, Huneiti AM, Lagos N, Noyvirt AE, Pasantonopoulos C, Tsaneva DK and Tang Q. Product Lifecycle Management for Performance Support. J. Comp. Inf. Sc. Eng. ASME. 4 (2004) 305-315.

[16] Pham DT., Setchi RM. Authoring Environment for Documentation Development. Proc. IMechE B-215 (2001) 877-882.

[17] Aamodt A, Nygard M. Different Roles and Mutual Dependencies of Data, Information, and Knowledge- An AI perspective on Their Integration. Data & Kn. Eng. 16 (1995) 191-222.

[18] Setchi RM, Lagos N and Dimov SS. Semantic Modelling of Product Support Knowledge. I*PROMS virtual conference (2005).

Intelligent Production Machines and Systems
D. T. Pham, E. E. Eldukhri and A. J. Soroka (eds)

An Object-Oriented Framework for Intelligent Product Manuals

C. Pasantonopoulos, S. S. Dimov, R. M. Setchi

Manufacturing Engineering Centre, Cardiff University, Queen's Buildings, Cardiff CF24 3AA, UK

Abstract

In this paper, an Object-Oriented Framework for the generation of electronic technical manuals is presented. First the problem of documentation generation is discussed. The lack of infrastructure and its effect on the documentation development cycle is then analysed. The need for an Object-Oriented framework that will increase the productivity and will ensure a better information system quality is explained. Furthermore, the technologies enabling the development of systems for generating documentation are outlined, and the system architecture and a case study are described. Finally, the expected benefits of employing such systems are discussed.

1. Introduction

Product documentation has moved from the traditional paper based form into electronic form. As a continuation of this evolution process and through the research conducted on Intelligent Product Manuals [1], interactive electronic technical manuals (IETMs) [2], and electronic performance support systems (EPSSs) [3], the product documentation becomes web-enabled, adaptive [4], with expert system support and thus transforms into a software system.

Software system development is a very demanding process and one that is quite costly and time-consuming. Ways to increase productivity of software developers have always been a major issue within the programming community.

The software team of the documentation-authoring group has to face this software generation process under the trend of rapid customisation of the product and its resulting documentation. Following, rapid rate of change with a total lack of resources for the developer can be a major drawback for the production of high quality systems.

2. Problems on Documentation Systems Development

2.1. Web Enabled Documentation

Web enabled documentation has to be presented through hypertext using the Hypertext Mark-up Language (HTML). It has to reside in a web-server and it inherits all the development aspects of web development. In more advanced documentation solutions, adaptive and personalised documents have to be achieved through the documentation system. Also in many cases it is valuable to offer functionality to the end-user, except for the presentation of the information (e.g. online calendar applications etc). In these cases the web development meets software development and the documentation process becomes part of a software engineering process producing web-applications.

2.2. Information Integration

Software applications such as CAD/CAM (Computer Aided Design/Computer Aided Manufacturing), PDM (Product Data Management) systems and common office applications (Spreadsheet and word processing software) nowadays assist most of the manufacturing processes. All these applications produce data - formatted in one way or another - which can be very valuable to the documentation development process.

One major problem is the easier integration of these generated data with the documentation system in use. The main target is the selection and extraction of reusable text and/or media that can be incorporated within the documentation.

2.3. Ensuring High Quality Information Systems

On top of these problems the documentation team has to exhibit an increased sensitivity on quality assurance matters. Documentation in many cases is legally binded to conform to standards and to provide correct information.

3. An Object-Oriented Framework for the Generation of Technical Documentation

The documentation developers, as it has been identified, require a tool that could ease the burden of generating web-enabled electronic manuals. This shortcoming can be addressed by the creation of an object-oriented [5,6] software framework [7]. In other words, the problems can be reduced by having an infrastructure that would provide the basis for the developers to develop electronic manual applications with minimal effort and time costs.

The first question that has to be answered for the creation of such a framework is what needs to be abstracted? What is it that this framework has to model? Such a framework has to provide basic functions commonly met in documentation systems and allow the developers to extend these resources, customising them to meet the need of their current product documentation development needs.

The main subsystems (Figure 1) that can be identified in this problem domain are the following:

- User Management
- Document Component
- Rule Engine
- PDM integration
- Web Integration
- Text Processing

The User Management subsystem has to handle the users' authentication and user model management issues. By identifying the user and associating him to a user model the system will gain security features and adaptive functionality that can be achieved in accordance with a Rule Based system.

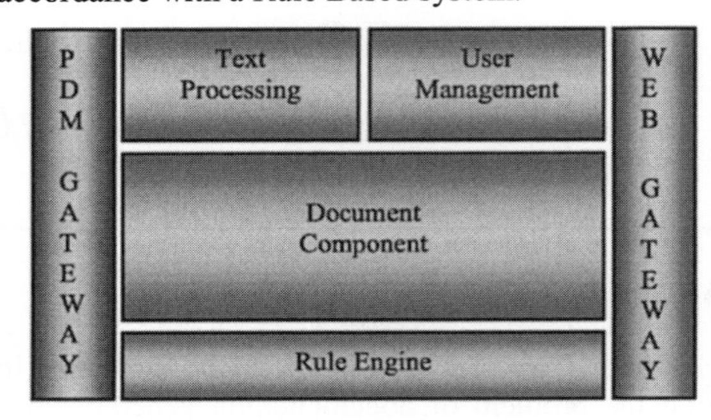

Fig. 1. Identified Subsystems

The Document Component should be able to handle and manage the product data, acting as a placeholder and as a manager, extracting, updating and forwarding the data towards the Web Interface.

A Rule System is needed within the framework to utilise the user model and exert the adaptive strategy to the document components. The rule abstraction can range from being very generic to just being a wrapper for a specific rule-based system according to the needs of the current system.

The PDM integration module must ensure that the PDM is continuously monitored and that the data are extracted and updated at run-time. It has to provide a buffer where selected data can be extracted and isolated from the product data pool for publication as part of the documentation.

The Web Integration component provides the gateway to the web. It must allow for easy integration between the web server and the architecture of the documentation system.

The Text Processing component is concerned with the processing of the textual part of the data. This can assist in uncovering links, textual relations within the application and terms that can be related to external resources.

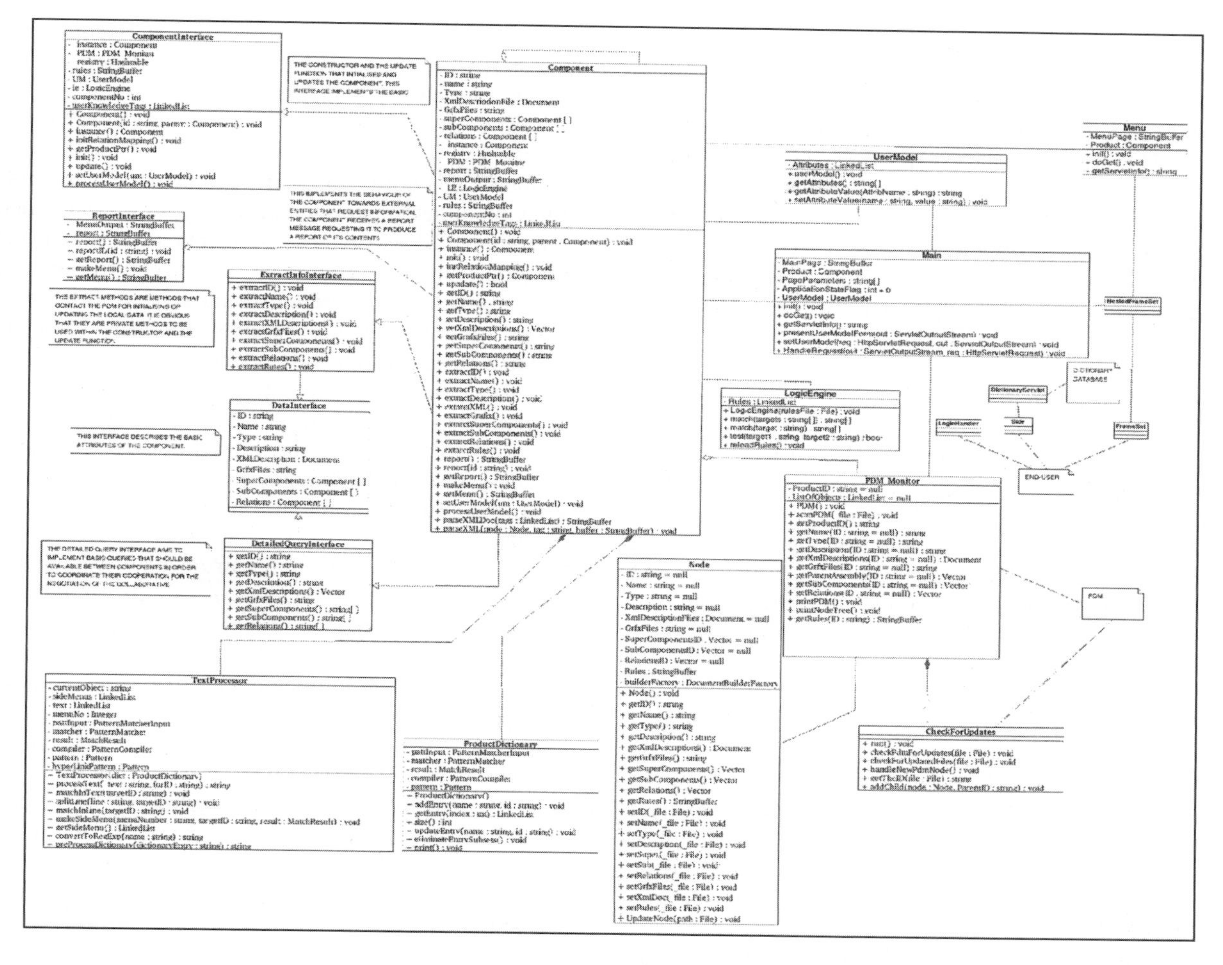

Fig. 2. System Architecture

4. Enabling Object Technologies

There are various Object Oriented technologies that are useful when designing such an application. Some that are strongly advantageous for a documentation system application are Servlet technology, Java Database Connectivity, Java-based XML Parsers, and text pattern-matching tools such as regular expressions.

4.1. Servlet Technologies

Servlets are a new technology from Sun Microsystems. They aim to replace CGI technologies that have a number of shortcomings, including platform dependence and lack of scalability. To address these limitations, Java Servlet technology is used to extend the capabilities of servers of all kinds that host applications, including Web servers. Another main advantage of the Servlets is that they are plain Java classes and can be seemingly integrated to any Java application [8].

4.2. Java Database Connectivity

Java Data Base Connectivity (JDBC) provides cross-DBMS (Database Management System) connectivity to a wide range of SQL (Standard Query Language) databases and access to other tabular data sources, such as spreadsheets or flat files [9]. This makes it a prime candidate for the use of integrating the existing product information sources (such as the PDM and/or product data in spreadsheet format) to the documentation system.

4.3. XML Technology

The eXtensible Mark-up Language (XML) is a text-based mark-up language that is becoming the standard for data interchange [10]. It is used to identify data using tags. Unlike HTML (Hypertext Mark-up

Language) XML is extensible, in the sense that it does not have a standard set of tags but allows the user to define the tags needed and give a descriptive name according to the data that are going to be marked.

The Xerces library can be used for processing XML data using applications written in the Java programming language. Xerces implements the parser standards SAX (Simple API for XML Parsing) and DOM (Document Object Model) and makes the identification, selection, & extraction of data easier [11].

4.4. Text Pattern Matching

Text pattern matching can be very useful as a search tool for text documents. By identifying strings that represent meaningful identifiable terms in a domain, one can automatically discover relations and links between different sets of data. For the text pattern matching the Oro package [12] from the Apache Foundation and its implementation of Regular Expressions can be used.

5. System Architecture

The system architecture as illustrated in Figure 2 can be viewed as a combination of the five distinct subsystems identified earlier. These are the User Management, the Document Component, the Rule Engine, the PDM Integration, the Text Processing and the Web Integration subsystems.

The User Management contains the Login handler and the User Model components. The Login Handler identifies the user according to his username and password. Then the User Model object, according to the username, finds and retrieves the previously set values of the user model for the specific user.

The Document Component integrates different behaviours according to a set of interfaces. The primary behavior is this of the component. In this role the Component is defined according to the Singleton pattern, in order to ensure the instantiation of one and only one root node for the product structure. The rest of the nodes are generated recursively from the root node instantiating components for the complete product structure. The component holds a registry of all the objects participating in the component community to form a pool of references through which the different objects can be looked-up. As a result the component forms a structured community of objects, enables communication and initiates the updating of the data.

The second main behaviour of the component is that of a placeholder for data. As such it plays two main roles, the extractor of data and the responder to queries. The extraction is the part that is associated with the PDM monitor and gets the data, and the detailed query is the interface that ensures that the component can answer specific queries that can be used by the other components.

The Report Interface generates reports of the data contained by the component in the form of HTML Documents. These reports are generated using the Logic Engine and the User Model for selection and adaptation of the data that will be presented to the specific user at hand.

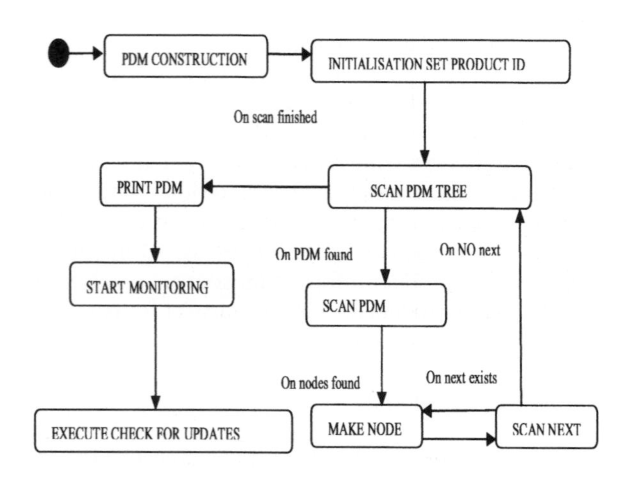

Fig. 3. PDM Integration Subsystem State Machine

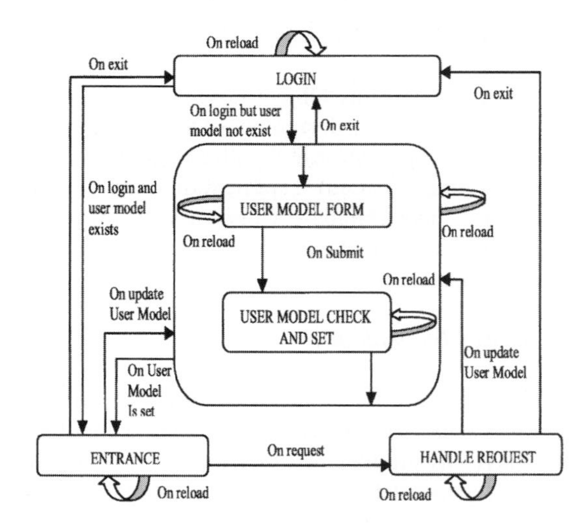

Fig. 4. Web Integration Subsystem State Machine

The Rule Engine component is an abstraction that supports simple IF THEN rules and acts as a wrapper to a proper rule based engine if needed.

The PDM Integration component, described by the state machine in Figure 3, consists of three parts, the PDM Monitor which is the main controller class, the check for updates, which uses a timer and probes the files to check for updates, and the node objects that act as an intermediate placeholder for the data.

The Text Processing subsystem includes the Text Processor and the Product Dictionary. The Product Dictionary keeps a record of all the component names instantiated. Also the names are broken down if they are multiple words and the relevance to existing terms in the dictionary is checked.

In the case that two terms are found to be related, the larger one acts as the main term and the other is noted as a related sub-term.

The text processor uses this dictionary to parse the text content of the components and to automatically create hyperlinks pointing to the related component.

An additional link is created as a star (*). This link shows the lexically related terms and additional links that point to a technical dictionary database.

The Web Integration module comprises of a set of Servlets that feed the output of the components reporting interface to the web. The Servlets also handle the requests received by parsing the parameters and feeding them to the root component. The Main Servlet holds the state management of the web application (Figure 4) by utilising sessions and cookies that identify the user. The Side Servlet holds the lexically related links and the links to the Dictionary Servlet and the Menu Servlet serves the structural menu created by the components of the application.

6. Case Study

In Figures 5 and 6 the resulting output from the web application is shown. In the left frame, the menu that was prepared by the components is presented. In the case that the links are followed a coded request towards the information components of the web application will be generated. In the middle frame the report developed by the currently addressed component is illustrated. The first part of the report is a dynamically generated natural language section describing the structural positioning of the component. The next part is a textual description of the entity represented by the component that has been pre-processed by the text processor and hyperlinks have been added. Next to the link there is the star (*) link that produces the results in the right frame. Then a graphic is presented and finally a selection from the contents of a XML file is included that is based on the processing of the corresponding "user model" and "set of rules" entities. In the right frame, lexically related items and links to dictionary definitions are created according to the text processor results.

In Figure 5 the Entrance State of the web application can be seen. The user model updating form and the login screen are included in the other windows. This application implements a documentation system with a structural menu based on the product structure, customised reports on all the assemblies, subassemblies, parts, and equipped with navigational aids for the documentation user.

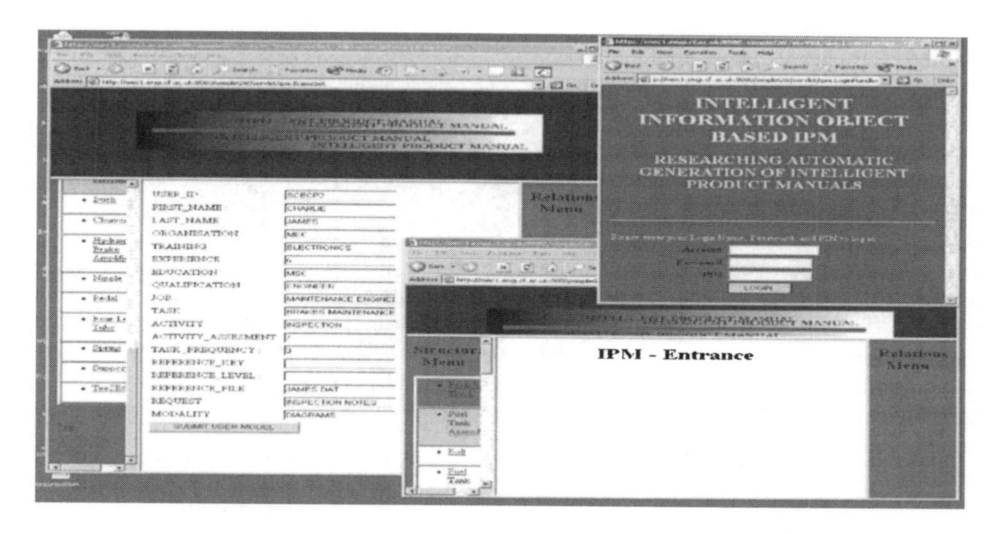

Fig. 5. System Login & User Model Facilities.

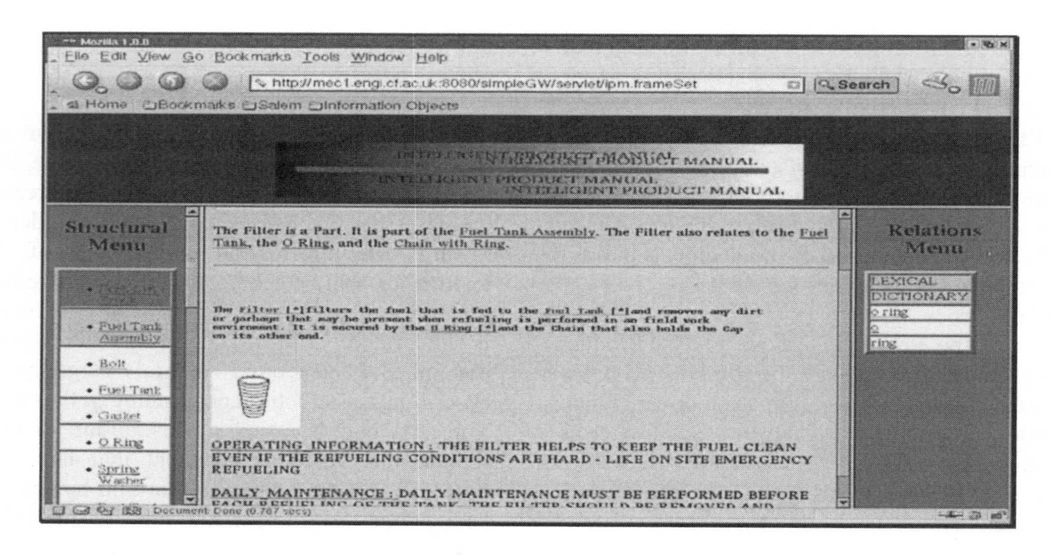

Fig. 6. Content Presentation.

7. Discussion & Conclusions

The framework that is presented in this paper models the main features of an application for documentation systems. The main entities involved in the process are identified and abstracted in a software framework.

The system as it is shown in the case study is capable of generating a fully functional Virtual Document. It allows dynamic updating of the data and offers a rich set of utilities such as the dynamic linking, personalised output based on the XML data, dictionary support, etc. The application can be executed as it is or extended to provide solutions customised to more specific needs of documentation systems.

Although the system is a valuable application it is slightly restrictive in the sense that the objects are functionally more effective when small and very well defined tasks are executed. The main task here is the selection and presentation of the data. This object-oriented framework does work satisfactorily but with limited flexibility on decision making and reaction to situations where the data might be in an unpredictable state (such as part of the data not being submitted to the PDM yet).

Another aspect that calls for improvement is that the system is centralised, and in the case of large products where the parts can be counted into several thousand components, it might prove to be computationally inefficient.

These problems are going to be addressed with future work that will extend the system in a distributed multiagent architecture.

Acknowledgements

The research described in this paper was performed within the I*PROMS Network of Excellence sponsored by FP6 of the European Community.

References

[1] Pham DT, Dimov SS and Setchi RM, Intelligent product manuals, Proc. IMechE. I-213 (1999) 65-76.

[2] Jorgensen EL and Fuller JJ, A web-based architecture for interactive electronic technical manuals (IETMs), ASNE Naval Logistics Conference. (1998).

[3] Bezanson WR, Performance support: online, integrated documentation and training, 13th ACM Annual Conference on Emerging from Chaos. Savannah. USA 1-10 (1995).

[4] Pham DT and Setchi RM, Adaptive product manuals, Proc. IMechE. C-214 (2000) 1013-1018.

[5] Rumbaugh J, (1991), "Object Oriented Modelling and Design", Prentice Hall.

[6] Maritn J and Odell JJ, (1998), Object-Oriented Methods, Prentice Hall.

[7] Schmid HA, Creating Applications from Components: A Manufacturing Framework Design, IEEE Software 6-13 (1996) 67-75.

[8] Perry BW, (2004), "Java servlet and JSP cookbook", O'Reilly.

[9] Bales DK, (2002), "Java Programming with Oracle JDBC", Thomson International.

[10] Goldfarb CF and Prescod P, (1999), "The XML Handbook", Prentice Hall.

[11] 2000, Xerces Java Parser, [WWW] URL: http://xml.apache.org/xerces-j [Accessed 2 December 2002].

[12] 2004, Jakarta-ORO, [WWW] URL: http://jakarta.apache.org./oro [Accessed 14 March 2003].

Intelligent Production Machines and Systems
D. T. Pham, E. E. Eldukhri and A. J. Soroka (eds)

Design and development of a decision support system for manufacturing systems

T.S. Mujber, T. Szecsi and M.S.J Hashmi

School of Mechanical and Manufacturing Engineering, Dublin City University, Dublin 9, Ireland

Abstract

To remain competitive industries must implement state-of-the-art technology to support the decision-making process. Industries may rely on these technologies to obtain benefits such as the improvement of process data management, better control over operations and improvement in both the planning and decision-making functions. In general, numerous modelling methodologies have been developed to support industries in gaining a better understanding of their system behaviours and processes that will enable them to make the kind of system designs and operational decisions to ensure market competitiveness. Examples of such methodologies are Process Modelling, Simulation Modelling, and Virtual Reality technology.

This paper presents an integrated decision support system that allows users who are not specialist in simulation modelling and virtual reality techniques to develop simulation models and virtual models automatically of manufacturing systems and addressing their behaviours.

1. Introduction

Manufacturing systems, processes and data are growing ever more complex. Product design, manufacturing engineering and production management decisions often involve the consideration of many interdependent variables, probably too many for the human mind to cope with at one time. These decisions often have a long-term impact on the success or failure of the manufacturing organization. It is extremely risky to make these major decisions based on "gut instinct" alone. Simulation provides a capability to rapidly conduct experiments to predict and evaluate the results of alternative manufacturing decisions. It has often been said that you do not really understand your industrial processes and systems until you try to simulate them [1].

Simulation modeling has been applied successfully for investigating current and anticipated planning strategies used to configure/reconfigure manufacturing systems in order to improve system performance. Such popularity of simulation has resulted in a large number of simulation packages available in the software market. Most of the simulation packages have the similar principles to design and model a manufacturing system. However, those packages have some limitations to fulfill the user requirements. Some of the simulation package limitations are listed below:

- Modelling and programming skills are required.
- Time consuming to develop simulation models.
- No real time interaction.
- Virtual models of many simulation packages do not support virtual reality devices e.g. Head Mounted Display (HMD).

This paper presents a new approach of developing a Decision Support System (DSS) that overcomes some of the above mentioned limitations of the presently available simulation packages.

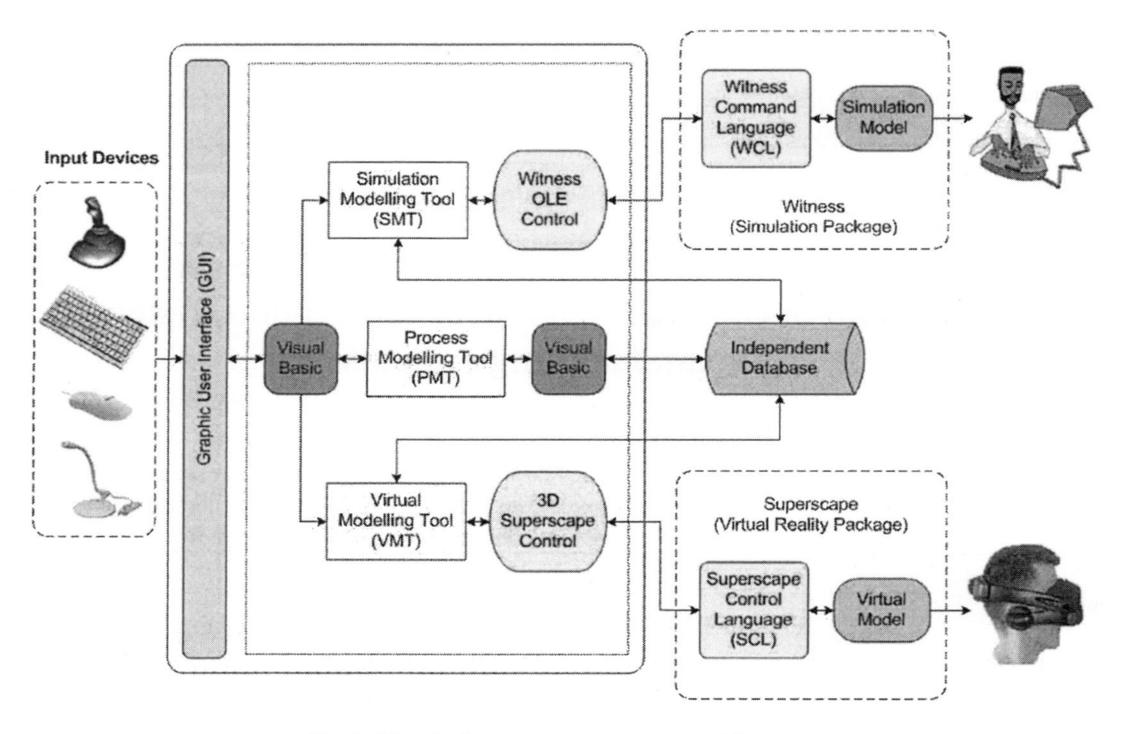

Fig.1. The decision support system architecture

2. The decision support system architecture

The architecture of the DSS is shown in Fig. 1. It presents the integration of the three main tools of the software and how the user can interact with them to develop a simulation model and virtual model of a manufacturing system.

As shown in Fig. 1. , the components of the decision support system include the following tools and devices:

- *Process Modelling Tool:*

The Process Modelling Tool (PMT) is the core of the DSS, which is used to obtain information regarding a manufacturing system to be stored in an independent database.

- *Simulation Modelling Tool:*

The Simulation Modelling Tool (SMT) is used to build a simulation model of a manufacturing system automatically.

- *Virtual Modelling Tool:*

The Virtual Modelling Tool (VMT) is used to create a virtual model of a manufacturing system automatically.

- *Independent Database:*

The independent database is the media to store the collected information regarding a manufacturing system that needs to be addressed.

- *Input devices:*

Input devices can be used to interact with the DSS which includes mouse, keyboard, joystick and microphone.

- *Output devices:*

A monitor or Head Mounted Display (HMD) can be used as output devices to view the generated models by the DSS.

3. Design and Development of the DSS

The DSS has been developed entirely using an object based programming language called Visual Basic. The development of the DSS incorporates the construction of Process Modelling Tool (PMT), Simulation Modelling Tool (SMT), and Virtual Modelling Tool (VMT) that can be used to generate simulation and virtual models for studying manufacturing systems. Using the DSS, one could:

- Describe the behaviour of manufacturing systems.
- Evaluate alternative hypotheses or theories based on the observed behaviour to find the optimum solutions.
- Use models to predict the future behaviour of a system.
- Measure the performance of real manufacturing systems.

The Graphic User Interface (GUI) of the DSS includes the following tools: PMT, SMT, VMT, Simulate and Output analyzing tools.

The DSS was designed to work in conjunction with Witness package and Superscape VRT software to allow non-expert users to develop simulation and virtual models automatically. The overall structural design of the DSS is shown in Fig. 3.

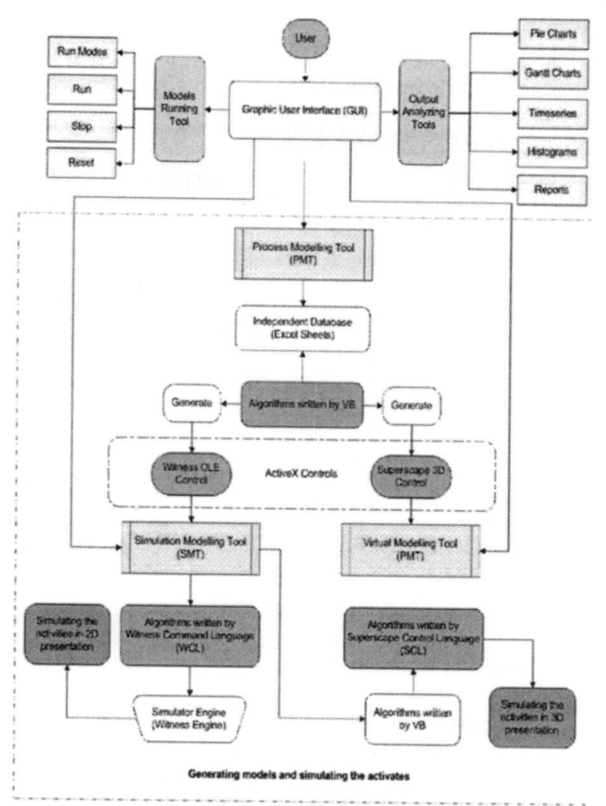

Fig. 3. Structural design of the DSS

3.1 Process Modelling Tool

The Process Modelling Tool (PMT) is the core component of the DSS, which is used to captures the operational logic and gather information regarding a manufacturing system and then it stores the gathered data in an independent database. The database is used to provide the simulation modelling tool and the virtual modelling tool with this information to be used to develop the simulation model and create the virtual model of a manufacturing system.

The PMT collects information from the user to define the resources and the process plan of the manufacturing system that needs to be studied. Fig. 4 shows the main elements of a manufacturing system that can be defined by the PMT.

The PMT has two options which are the Process Development and the Factory Layout.

Resources			Process Plan
Parts	Shifts	Buffers	Process Flows
Machines	Machine Tools	Labours	
AGVs	Transporters	Conveyors	Orders

Fig. 4: Main elements of a manufacturing system

- *Process Development:*

The function of this option is show the process modelling tool dialog pages interface that needs to be used to define all the resources and the process plan of a manufacturing system. Fig. 5 shows the Interface of the PMT that provides the user with several dialogs on a single form to define the resources and the process plan of a manufacturing system, where each resource has its own detail dialog page that can be used to specify its attributes. The process flows and the orders that represent the process plan of a manufacturing system can be defined using this interface as well to make it easier for the user than having more than one interface. As shown in Fig. 5, the user can access the detail dialog pages of each element of the PMT by clicking on the tab name of the element or the icon that represents the element.

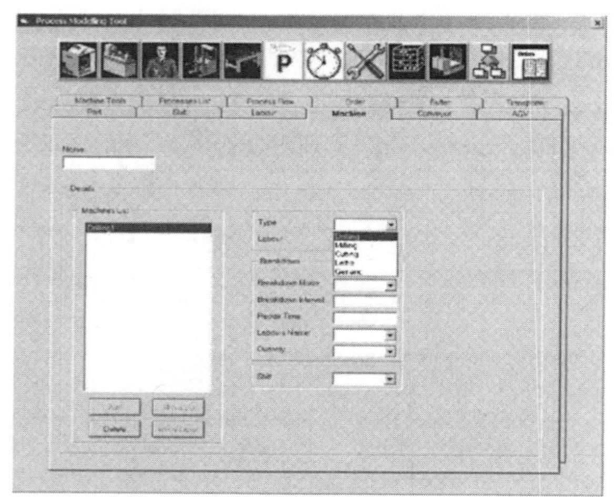

Fig. 5: PMT Interface

- *Factory Layout Window:*

Once all the resources of a manufacturing system have been defined using the PMT Interface, the factory layout window can be used to construct the layout of the system that needs to be studied. Fig. 6 shows the Factory Layout window that needs to be used to construct the factory layout of a manufacturing system.

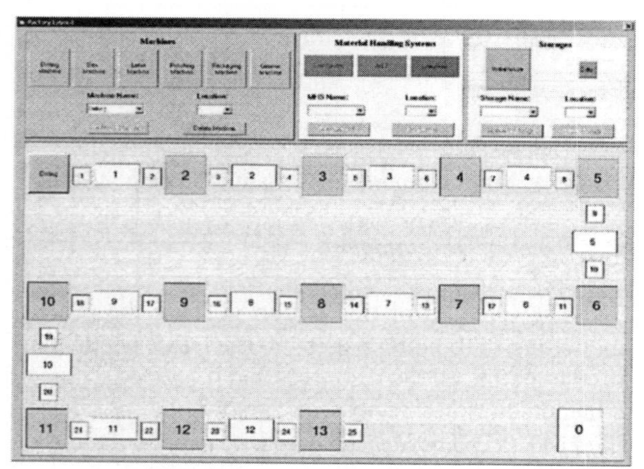

Fig. 6: Factory layout window

As shown on the window of the factory layout, there are three main types of resources that can be used to define the layout of a manufacturing system. These types include machines, material handling devices and storages. The user can define the position of each resource to be placed within the factory layout window, first by specifying the name of the resource from the list box which contains all the names of the resources that have been defined by the user and then specifying the location where the resource should be placed.

The main function of the Factory Layout window is to identify the x and y position of each resource which is required to feed the SMT and the VMT with the locations of the resources to generate the layout of the simulation model and the virtual model.

An Excel file needs to be created using the DSS to act as a database of the manufacturing system that needs to be studies. The data that has been defined by the user using the PMT and Factory Layout window is stored within the created Excel data sheets.

3.2 Simulation Modelling Tool

The Simulation Modelling Tool (SMT) allows users who are not very knowledgeable about simulation modelling techniques and programming to develop a simulation model of a manufacturing system automatically. It helps also expert users to shrink the time of developing complex simulation models.

The SMT applies the information that has been gathered about the resources using the PMT to develop a simulation model automatically. The gathered information is stored within a database in Excel file with multiple data sheets, where each resource has its own Excel data sheet for storing its relevant data.

The simulation software package that was chosen to be linked with the DSS to develop the simulation model is Witness from Lanner. Based on a survey conducted by Vlatka Hlupic [2], Witness package is the most used simulation software by industrial users and the second most used simulation software by educational users.

The database of a manufacturing system needs to be loaded first to the DSS before the user can generate the simulation model automatically. The database can be loaded by selecting *"Load Database"* option from the file menu options; an open window appears to allow the user to specify the name of the Excel file where the database of the manufacturing system is stored. When the database is loaded, the user can select the *"Generate Simulation Model"* command from the GUI of the DSS. The function of this command is to execute the algorithm in Fig. 7 to generate the simulation model automatically.

As shown in Fig. 7, the database includes information about the resources, which include machines, parts, conveyors, AGVs, labourers, machine tools, processes, transporters, buffers and shifts.

Witness OLE control is the mechanism that has been used to link the DSS software with Witness.

Algorithms developed by VB are used to read the Excel sheets database of the resources, and then this information is transferred into arrays that act as temporary database for the DSS. Algorithms written by VB linked with WCL commands are used to develop simulation models automatically in Witness in three phases as follows:

Define phase: Resources of a manufacturing system are defined to be included in the simulation model.
Display phase: Some of the resources are presented by icons in order to build up a pictorial representation of the manufacturing system.
Detail phase: The logic of the simulation model is specified in this phase, where the process plan and the orders have been used to detail the resources. Code is generated automatically to be attached to each resource for monitoring and performing different activities within the simulation model.

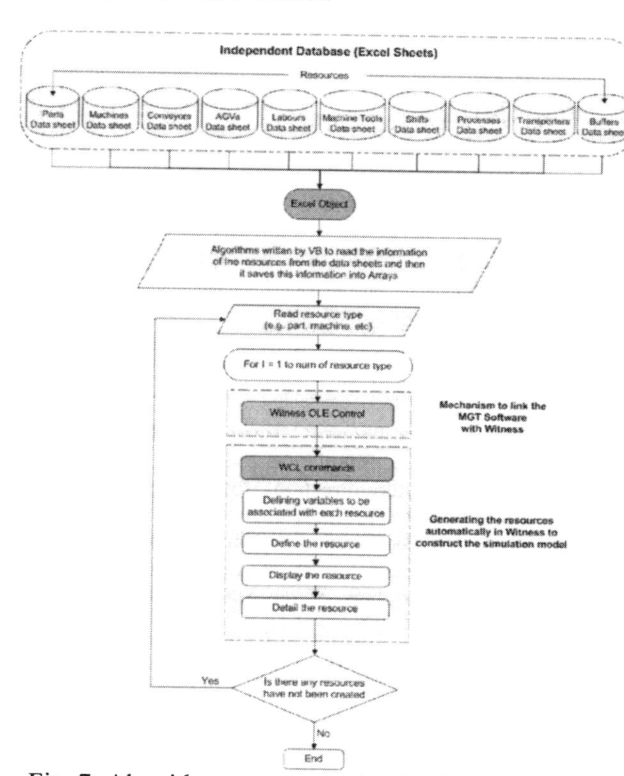

Fig. 7: Algorithm to generate the simulation model

Fig. 8 shows a snapshot of a simulation model generated automatically using the DSS.

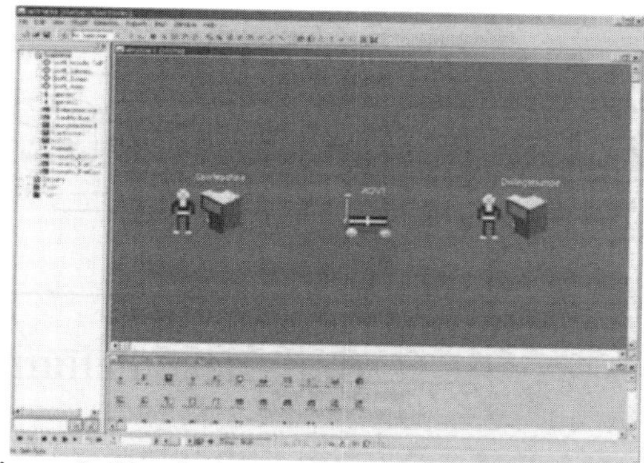

Figure 8: Simulation model generated automatically

3.3 Virtual Modelling Tool

The Virtual Modelling Tool (VMT) is applied to create virtual models automatically of manufacturing systems that represent simulation models in 3D scope. The virtual reality software that is chosen to be linked with the DSS to develop the virtual models is called Superscape VRT. The virtual reality package consists of an integrated suite of editors. These editors have been used to create a virtual environment of a manufacturing system. Virtual objects were also created that represent the resources of a manufacturing system which include a number of virtual machines, labourers and material handling systems. After the virtual objects of the resources were created, a number of attributes were applied such as size, position, textures, dynamics, colour and sound. Superscape Control Language (SCL) has been used to assign behaviour (coding) to the virtual resources. The purpose of the codes is to keep monitoring the values of properties on each resource and to command the resource to do different actions based on these values. to simulate the activities of the simulation model in 3D scope within the virtual model.

Figure 9 shows the algorithm that has been written by VB to generate the virtual model automatically of a manufacturing system. Fig. 10 shows a snapshot of a virtual model generated automatically using the VMT of the DSS.

2. Study approach

In practice, this case study has been undertaken within a Finnish steel manufacturing network in the period 2002-2004. For the pre-understanding, supply chain integration and ICT as an enabler for that integration task have been studied from a theoretical perspective. Also, the different e-supply chain designs and the main characteristics of them have been examined by reviewing the SCM and e-business literature. After theoretical investigation the authors have chosen a number of activities to analyze the state of art and state of practice in the ICT implementation in a case network.

The purpose was to describe and evaluate the development process towards an e-supply chain. The development process consists of developing an e-business roadmap based on chosen e-business framework, and the supporting ICT application. *Firstly,* unstructured interviews were conducted in order to fully understand the case companies' businesses and their activities and to select the business processes in which the new technical solution should concentrate. The first interviews were carried out in 12 companies, where the core competences, current level of co-operation and the current utilization of the ICT tools were pictured and evaluated. *Secondly*, based on these discussions, specific questionnaire for companies was designed in order to find out the critical issues in building a well-functioning supply network. The representatives of 6 companies answered 34 open-ended questions covering the following areas: the information sharing, process integration and collaborative relationships in the network. *Thirdly*, two sample order-delivery processes were modelled by following through their value chains. Theme interviews with the operational staff and key persons of the companies were conducted in order to identify the improving possibilities of the supply chains. The purpose of the interviews was to find out the critical issues in information flows. *Finally*, two workshops were arranged in order to start to develop the e-business road map for the case network. Totally 16 key persons from 9 different organizations (both companies and development organizations) gathered together and positioned today's state of art of the network in e-business, and also stated visions of the network in the short and long time period (years 2006 and 2010).

Parallel of the previous activities, representatives of 10 companies were interviewed and a technical questionnaire was carried out to gather requirements for an ICT prototype system to improve information flow in the case network. More interviews were held during the early development process to refine the requirements and at later stages of the development the prototype was demonstrated to the companies to further evolve the ICT system.

3. ICT in supply chain integration

3.1 Supply chain integration

Integration is a key concept in SCM and actually an essential prerequisite to SCM [6]. Getting the right product to the customer in the right place, at the right price, at the right time, and in the right conditions, is not only the lynchpin to competitive success but also the key to survival [7]. To stay competitive, companies strive to achieve greater coordination and collaboration among supply chain partners in an approach called supply chain integration [8].

[2] and [9] define supply chain integration as process integration upstream and downstream in the supply chain. According to [10], supply chain integration has three aspects: information integration, coordination and resource sharing, and organisational relationship linkages. The most important, information integration encompasses sharing of relevant knowledge and information among members of a supply chain. It may as well include sharing of design or manufacturing data and forecast and delivery scheduling data.

Significant investments in ICT generally, and Internet technologies in particular, are being targeted specifically at SCM, and ICT is seen as key to supply chain integration. E-business - the use of Internet-based computing and communications to execute both front-end and back-end business processes - allows partners redefine logistics flows so that the roles and responsibilities of members may change to improve overall supply chain efficiency.

3.2. Information integratiom with ICT

Information integration is the key starting point in supply chain integration. In a study focusing on e-procurement, [11] highlight several benefits of information integration: the enhancement of supply chain efficiency by providing real-time information regarding product availability, inventory level, shipment status, and production requirements, and the facilitation of collaborative planning among supply chain partners by sharing information on demand forecasts and production schedules that dictate supply chain activities. ICT reshapes business processes so that it has the potential to facilitate the flow of information between processes across the supply chain, and guarantees the accessibility of instantaneous and consistent information across all members of supply chain [12]. ICT reduces the product development time to market, reduces the product delivery time to the customer, improves workforce capabilities and flexibility, enhances the flexibility of the product facilities, and improves understanding and control of production processes [13]. ICT can lead to radical changes in business organizations, not only within the firm, but also between companies and with customers. ICT systems perform three vital roles in any type of organizations: 1) they support business operations (such as capturing point-of-sale data), 2) managerial decision making (such as choosing suppliers), and 3) strategic competitive advantage (for example, a firm's ability to integrate its entire supply chain) [14].

In particular, the question that has been much debated is what impact the ICT, particularly Internet, will have on intermediaries. While certain intermediaries add significant costs to the value chain, the aim of the manufacturer is to bypass them by dealing directly with customers [15]. [16] emphasize that by ICT, the companies in the value chain are creating internal and external connections to establish new capabilities within their markets, changing the way business will be conducted, and redefining roles and rules for how to direct a successful enterprise. This phenomenon is impacting every single part of business, all functions and departments, and will eventually affect every consumer worldwide as well. The effect is dramatically influencing how business transactions will be conducted in the future.

According to [17], ICT systems for SCM must incorporate six principles to adequately support enterprise planning and operations. The principles must be taken into account when designing or evaluating information systems as SCM applications. One of the evident reasons is the significance of information. 1) Information must be readily and consistently *available to all* that need it. Traditionally the information regarding SCM activities, such as order and inventory status, are often paper-based, which does not support the need to respond to customers and improve management decisions. Information availability can reduce operating and planning uncertainty. 2) Information must be *accurate*. Accuracy means that there is consistency between physical counts or status and information system reports. Increased information accuracy decreases uncertainty and reduces inventory requirements. 3) Information must be *timely*, meaning that the delay between when an activity occurs and when it is visible in the information system is minimized. The delay reduces planning effectiveness and increases inventory. 4) Information should be *appropriately formatted*. The right information must be in the right structure and sequence. Further, there are two principal expectations to the information systems. 5) Information systems must be *exception-based* in order to highlight problems and opportunities. If the system can identify the exception situations, which are not a part of a normal process, planners are then able to focus their attention on situations that require the most attention. 6) Information systems should be *flexible* to be able to meet both system users and customers' needs. Systems should be able to provide data tailored to specific customer requirements. [17]

3.3. e-supply design

[2] defines the supply chain as *"the network of organizations that are involved, through upstream and downstream linkages, in the different processes and activities that produce value in the form of products and services in the hands of the ultimate customer"*. Thus, instead of the term "*supply chain*", it would be more accurate to use the terms "*supply network*" or "*supply web*" to describe the net-structure of most supply chains.

Regarding the previous concepts, different, interrelated concepts can also be found in the literature, i.e., "*virtual supply chain*" [18], "*demand satisfaction community*" [19], "*value net*" [20], "*value web*" [21], "*virtual enterprise*" [22, "*virtual organisation*", [23] and "*extended enterprise*" [24]. All of these concepts present the new form of the digital business design, where the information flows play an integral role within the network. Especially the Internet enables the digital business design, which captures customer's real choices in real time and transmits them digitally to other participants of the network. These Internet enabled chains are frequently referred also to as "*e-commerce supply chains*" [25].

These different concepts of e-supply chains - chains that use e-business models and web technology - can be seen simply as processes necessary to transfer the goods sold over Internet to the customers, but widely e-supply chain is defined as wide-ranging topic related to the supply chain integration [16]. What ever the concept is termed, in the new digital business design, the customer dominates the supply chain and the chain is organized around the customer order.

This new structure of the supply highlights the simultaneous communication between different parties and the integration of the supply chain as a whole, not only the communication between the consecutive phases of the process. The common factors of all these concepts are that they are more or less [20], [25: focusing on the customer (producing more value to the customer and thus hooking the customer), dynamic designs that can be continuously developed and adjusted to changes, digital, agile and scalable, network-formed or "amoeba-like", and responding fast to changes.

4. Case study

4.1. Case network

The empirical study was carried out through the steel manufacturing industry network, which brings together a group of 19 companies in the steel sector in the Northern Finland. Engineering workshops, industrial service providers and the global steel producer as a focal company form the network. The focal company has a wide range of metal products and services like hot-rolled, cold-rolled and coated plates and sheets made of low alloy steels, cut steel products, steel tubes and pipes and long steel products. Engineering workshops have their own areas of expertise e.g. flame cutting, welding, machining, bending, forming, and shape cutting. Together with its suppliers, the focal company offers total solutions for construction and mechanical engineering and metal production industries.

4.2. The evolution of e-supply chains

In this study we have used the framework of [16] in order to develop the road map towards e-business. The work is based on the vision that the case network will have a common functioning ICT system in transferring logistics information between the network companies in the future. The whole development process of business network requires changes in way of action and improvements in using ICT more beneficial. In this study the authors are concentrating on examine what the ICT systems can offer to the development of e-business, even we understand that there are lot of many other business application that has to be improved, such as design and development, product and service introduction, purchase, procurement and sourcing, marketing, sales, customer service, engineering, planning, scheduling, manufacturing, logistics, customer care and human resources (table 1).

In the framework, there are four different e-business levels the network can use as a guide of understanding the position of the case network in the framework today, in the near future (in 2006), and in long-term future (in 2010). The levels are: 1) internal supply chain optimization, (level I/II) as a basic level, 2) network formation (level III), where the application of advanced supply chain management techniques is used to create the electronic network, 3) value chain constellation (level IV), where network members begin sharing resources and utilizing joint assets to enhance the network, and 4) full network connectivity (level V), which is achieved as use of the Internet is pervasive and the network is prepared to do business in a digital economy. Level V is not presented in the table 1, because the case network did not position itself so far up.

4.2.1. State of art – level I/II

The starting point of the development is the level I/II, which was set also today's position of the case network. The communication between case network companies today is based basically on mail, e-mail, phone calls or company visits that do not provide much visibility to the supply chain. Few companies have ERP-systems and if have, systems work independently as point solutions. The focal company has many ICT systems. It has recently invested to ERP system (SAP) that will replace most of the old ICT systems in three years.

Two of the smaller manufacturing suppliers have operation management systems that are easy to connect to common ICT system, which can help breaking down barriers between companies, speed up information flows and turn data into useful collaborative information. Small suppliers have problems in implementing ICT applications; they don't have enough knowledge and/or financial resources to invest.

However, there are some large and medium size of companies in the case network, which have ICT system implementation experience and knowledge. Furthermore, there are ICT software and hardware companies, which are closely cooperating with the case network companies. Two suppliers have already started to link electronically with their key customers, and can be positioned in the level III. One supplier has given access to the key customer into its ICT system, and tendering, ordering, and process control is managed electronically. Another supplier has web-based access with its key customer and manages reclamation there. The third example is a direct access between two ICT systems, where both companies can see the status of the production. Furthermore, there are other development processes concerning ICT systems, which are based on automatic message handling between two ERP-systems.

So, today the companies in the case network are in the different level in ICT applications and have different potential to adopt ICT, but they have common visions of e-commerce strategies and e-business models and the development work has been started. Also, the success examples from other industries, such as electronic industry, encourage the case network in their move towards e-business.

4.2.2. The first step to the level III

The first step in the development towards e-business, from basic level to the level of collaboration and cooperation, was positioned to be happened until the year 2006. This level III is actually starting level to build the network based on e-business. In this level, the commitments have been done in the high management level and the managers have decided of long-term development operations. Also, in this level, ICT knowledge in all companies is good, and the knowledge and resources are shared between the companies.

Table 1 Development steps of the case network towards e-business (based on [22])

Progression / Business Application	Level I/II Internal Supply Chain Optimization	Level III Network Formation	Level IV Value Chain Constellation
Information technology	2004 Point solutions	2006 Linked intranets	2010 Internet-based extranet
Design, development, product and service introduction	2004 Internal only	2006 Selected external assistance	2010 Collaborative design, enterprise integration
Purchase, procurement, sourcing	2004 2006 2010 Leverage business unit level		
Marketing, sales, customer service	2004 2006 Internally developed, programs, promotions	2010 Customer-focused, data-based initiatives	
Engineering, planning, scheduling, manufacturing	2004 MRP, MRPII, DRP	2006 ERP – Internal connectivity	2010 Collaborative network planning – best asset utilization
Logistics		2004 2006 Pull system through internal/ external providers	2010 Best constituent provider – dual channel
Customer care	2004 Customer service reaction	2006 Focused service – call centers	2010 Segmented response system, customer relationship management
Human resources	2004 Internal supply chain training	2006 Provide network resources, training	2010 Inter-enterprise resource utilization

The role of the focal company in developing its supply network is significant. In the case network, the focal company has to take an active role with its suppliers and start common internal ICT development projects. Also, the core competences of the companies have to be identified and strengthen by ICT applications. The case network will determine the used software solutions including the management and filing of mutual information and security issues. Furthermore the mutual information transmission formats needs to be decided together, these include the information transmission format, the used concepts and representation formats. Eventually, each company may have different ERP systems that can communicate with the common ICT system in order to handle mutual information. In case a company does not have own ERP system, it can transmit information to other companies by using the ICT system via web. The first elements of the ERP system of the focal company are planned to be in function in the year 2006. This is considered when developing the common ICT system. The objective is that once taped information is transferring in the same format along the whole supply chain.

4.2.3. The second step to the level IV

The second step in the development towards e-business was positioned to be happened until the year 2010. The focal company will have its own ERP-system as its main ICT system and there will be a common logistics system functioning between the network companies. Also the key customers and the logistics provider will be linked into net. Now, all the key suppliers are definitely connected, as well as distributors and customers. Furthermore, the case network will have common business processes and models. In this level IV, real-time information for engineering, production and transportation can be transferred along the whole supply chain. Also the customer's demands and orders are moving in the network transparently. The companies in this level understand the e-business methods and try suitable alternatives. Also mobile technology applications will be in function and they are integrated into common ICT system. That kind of example is information exchange between the transportation companies and logistics provider.

As it now seems, the development towards e-business will continue and the progress will lead to the full integrated network, where all the members are linked into common Internet-based extranet. Data will be transferred between the ICT systems creating visibility – best practises will arise and the superior competitive advantage will be possibility to gain.

4.3. SteelNet system

The case network adopt the modern business design and digitalize the logistics actions by developing the new open ICT application based on Internet and agent technology (later called SteelNet system). Agent technology is a new information technology solution in which there has been modest application in the logistics area so far. Agents are automated software entities operating through self-directed actions on behalf of the user –machines and humans. They have capability to interact with other agents reactively or proactively without constant human intervention. Agents can help to improve decision-making in collaborating companies by making the information exchange between companies possible. It is commonly accepted that agents can be described as software systems that are autonomous, co-operative (social behaviour), reactive and pro-active [27]. In the SteelNet system, agents are modelled based on the major functionalities of the companies (e.g. manufacturing, management and transportation) and they are able to communicate and collaborate with the agents in the other companies via Internet. The aforementioned division of agent responsibilities leads to a multi-agent system, where each company has several agents with different objectives and which communicate with their counterparts in the collaborating companies as well as with the agents in their own company.

The SteelNet system provides collaboration in information sharing, operational cooperation and dynamic supply chain configuration. In addition to real-time tracking of manufacturing-related information, the application includes the following basic services: a web application server to provide user interfaces, a user administration service, and an alarm service and an information management service. As the SteelNet system is a mutual system of the case network, it must be easily connected into companies' own ERP-systems. This requirement has been approached in three different ways: SteelNet system can communicate with company's own ERP-system by transmitting the mutual agreed information, SteelNet system can be used via web browser, while the agent container is situated into company's own server, and SteelNet system is used via web browser, while the agent container is situated in service provider's premises.

As [17] presented, ICT systems for SCM must incorporate six principles to adequately support enterprise planning and operations. Multi-agent systems seem to fill almost all of these principles. The SteelNet system makes the logistic information *available for all* the authorized users *accurate and timely* as agents deliver the information through the Internet. Agents can communicate with humans, with other information systems, and with other agents. Information is also *exception-based* in order to highlight problems and opportunities. Agents are monitoring their environments and reacting to changes. Intelligent agents are capable of reasoning based on the rules given by the user or knowledge learned from an open environment. Agent systems are also *flexible*, since agents can be organized according to different control and connection structures. Furthermore, agents can be created or discarded from a multi-agent system.

SteelNet business design is based on the digital supply chain concept – it represents one kind of e-supply solution. This new structure of the supply chain highlights the simultaneous communication between different parties and the integration of the supply chain as a whole, not only the communication between the consecutive phases of the process.

5. Conclusion

This paper has presented the development process, which consists of developing an e-business roadmap and supporting business design based on the agent software solution. Even though the case network seems to be in the very beginning in adopting the methods of the ICT, the need to improve the level of integration and information sharing is highly recognized in the networking companies. The companies involved in the case network have been enthusiastic to develop their business processes and business models towards e-business. Companies have taken a concrete step by taking part in development process of the SteelNet system. SteelNet system gives an opportunity for real-time and transparent information sharing. It is generated to fit for needs of small and big companies with different levels of ICT systems. Intensive co-operation between personnel of the companies and researchers has formed the system to answer for the requirements of every day duties. However, companies are not eager to share information unless there is a positive proof that sharing information is equally beneficial for all members of the supply chain. The withholding of information by even one member in the chain can lead to loss of trust and dysfunctional behaviour among all members despite the best technology facilitating information flow.

However, seamless integration with complete information sharing between all supply chain participants is still in the future. Since a number of autonomous companies belong to the supply chain network, it becomes imperative to develop a common mission, goals, and objectives for the group as a whole, while pursuing independent policies at an individual member's level. This scenario offers opportunities for design, modelling, and implementation of supply chain networks for maximum effectiveness, efficiency, and productivity in dynamic environments.

In the near future the SteelNet system will be tested for real. The purpose of this testing phase is to collect information for future development of the system. During this year the system is going to be extended to the real-time information management in the area of quotation process. The emphasis of the future work will be on importing business model into practice and developing semantics in the business network.

References

[1] Hamel, G; Prahalad, C., 1990, The Core Competence of the Corporation, Harvard Business Review, May-June.

[2] Christopher, M., 1998, Logistics and Supply Chain Management - Strategies for Reducing Cost and Improving Service, Prentice Hall.

[3] Handfield, R. B.; Nichols, E. L. Jr., 1999, Introduction to Supply Chain Management, Prentice Hall, Upper Saddle River, New Jersey.

[4] Maamar, Z.; Dorion, E.; Daigle, C., 2001, Toward Virtual Marketplaces for E-Commerce, Communications of the ACM, Vol. 44, No 12, pp. 35-38.

[5] Kalakota, R.; Robinson, M., 1999, e-Business. Roadmap for Success, Addison-Wesley

[6] Mouritsen, J.; Skjoett-Larsen, T.; Kotzab, H. 2003, Exploring the contours of supply chain management, Integrated Manufacturing Systems, Vol. 14, No 8, pp. 686-695.

[7] Mason-Jones R.; Naylor, B.; Towill, D. R., 2000, Lean, agile or leagile? Matching your

supply chain to the marketplace, International Journal of Production Research, Vol. 38, No. 17, pp. 4061-4070.

[8] Lee, H.; Whang, S., 2001, E-business and Supply Chain Integration, Stanford Global Supply Chain Management Forum, SGSCMF-W2-2001, November.

[9] Lambert, D. M.; Cooper, M.; Pagh, J. D., 1998, Supply Chain Management: Implementation Issues and Research Opportunities, The International Journal of Logistics Management, Vol. 9, No. 2, pp. 1-19.

[10] Lee, H. 2000, Creating value through supply chain integration, Supply Chain Management Review, Vol 4, No 4, pp.30-36.

[11] Min, H.; Galle, W.P., 2003, E-purchasing: profiles of adopters and nonadopters, Industrial Marketing Management, Vol. 32, pp. 227 – 233.

[12] Grieger, M., 2003, Information Integration in Supply Chain Management: Utilizing Internet Business Models, In Juga (Ed.) Proceedings of the 15th Annual Conference for Nordic Researchers in Logistics, NOFOMA 2003, 12-13 June, Oulu, Finland, pp. 26-41.

[13] Montgomery, J.; Levine, L. O., 1996, The Transition to Agile Manufacturing - Staying Flexible for Competitive Advantage, ASQC Quality Press, Milwaukee, Wisconsin.

[14] Haven, L.; Lewis, I., 2001, Logistics and electronic commerce: An interorganizational systems perspective, Transportation Journal, Vol. 40, No. 4, pp. 5-13.

[15] Timmers, P., 1999, Electronic Commerce – Strategies and Models for Busines-to-Business Trading, John Wiley & Sons, Ltd.

[16] Poirier, C. C.; Bauer, M., 2000, E-supply chain, Berrett-Koehler Publishers, Inc.

[17] Bowersox, D. J.; Closs, D. J., 1996, Logistical Management – The Integrated Supply Chain, McGraw-Hill International Editions, Singapore.

[18] Chandrashekar, A.; Schary, P. B., 1999, Toward the virtual supply chain: The convergence of IT and organization, International Journal of Logistics Management, Vol. 10, No. 2, pp. 27.39.

[19] Hewitt, F. 2000, Demand satisfaction communities: new operational relationships in the information age, The International Journal of Logistics Management, Vol. 11 No 2, pp.9-20.

[20] Bovet, D.; Martha, J., 2000, Value Nets – Breaking the Supply Chain to Unlook Hidden Profits, John Wiley & Sons, New York.

[21] Andrews, P. P.; Hahn, J., 1998, Transforming supply chains into value webs - Strategy and Leadership, Vol. 26, No. 3, pp. 6-11.

[22] Browne, J.; Zhang J., 1999, Extended and virtual enterprises – similarities and differences, International Journal of Agile Management Systems, Vol. 1, No. 1, pp. 30-36.

[23] Goldman, S. L.; Nagel, R. N.; Preiss, K. 1995, Agile competitors and virtual organisations. Strategies for enriching the customer, Van Nostrand Reinhold.

[24] Browne, J.; Sackett, B. J.; Wortmann, J. C., 1995. Future manufacturing systems - Towards the Extended Enterprise, Computers in Industry, Vol. 25, pp. 235-254.

[25] Gunasekaran, A.; Marri H. B.; McGaughey R. E.; Nebhwani M. D., 2002, E-Commerce and its Impact on Operations Management. International Journal of Production Economics, Vol. 75, No. 1-2, p. 185.

[26] Greis, N. P.; Kasarda, J. D., 1997, Enterprise logistics in the information era, California Management Review, Vol. 39, No. 4, pp. 55-79.

[27] Wooldridge, M.; Jennings, N. R., 1995, Intelligent agents: theory and practice, The Knowledge Engineering Review, Vol. 10, No. 2, pp. 115-152.

Intelligent Production Machines and Systems
D. T. Pham, E. E. Eldukhri and A. J. Soroka (eds)

Knowledge-based manufacturing within the extended enterprise

Berend Denkena, Peer-Oliver Woelk, René Apitz

Institute of Production Engineering and Machine Tools (IFW), Hannover Center of Production Technology, University of Hannover, Schoenebecker Allee 2, 30823 Garbsen, Germany

Abstract

Intelligent manufacturing techniques are in demand of several manifold types of information in order to achieve the goal of flexible, efficient, and competitive production. A lot of this information results from the product design process and can be used for manufacturing planning and execution provided that they are documented and managed in a structured way. This paper aims at taking advantage of well-structured information of the product design stage according to the STEP standard within the manufacturing domain. STEP is widely accepted for product data exchange and management. Its data models and methods provide a common basis for integrated collaboration processes in enterprises, allowing a holistic view that comprises areas like design, engineering, testing, manufacturing, and quality assurance. Thus, STEP can be regarded as an "enabling technology" for product-related IT infrastructures in enterprises. A promising approach is to use this methodology also within areas which are not currently associated with it, for example customer-supplier-collaboration, configurable machine tools, production management as well as knowledge management. In this paper some possible applications of STEP beyond the area of traditional product design will be discussed.

1. Introduction

Products and processes tend to increase in complexity. What is more, rapidly changing economical and technical environments call for collaboration between independent manufacturers. This trend is known as extended or concurrent enterprising. Successful concurrent enterprising requires an easy exchange of information between companies in order to efficiently develop new products. However, integrating several partners into a joint flow of information implies a number of challenges. These challenges are caused by different software platforms and systems or security questions, but may also concern questions of trust between independent companies.

Efficient support for collaboratively designing virtual products has to handle various processes in parallel in order to reduce time-to-market for new products. Interdisciplinary communication between the involved departments and companies becomes crucial. Virtual environments (in design) are used to optimise processes before they will be executed in reality (in manufacturing). Thus, the borderline between "product design" and "manufacturing" becomes indistinct and the need for a common basis for exchanging information arises, which is not limited to geometrical data only [1]. A topic that is often discussed within these trends is mass customisation, meaning small lot sizes in manufacturing and products that are tailored to individual customer demands.

From the point of view of Intelligent Manufacturing, it seems to be very promising to take advantage of well-structured information created within product design and used in manufacturing. One possible way to generate well-structured product information within the design process is the application of the STEP standard. The next section of this paper will outline its use for information structuring and knowledge representation. The subsequent sections will illustrate some applications of STEP-based information models in the manufacturing domain. Finally, the paper will give a short conclusion.

2. Knowledge representation

Knowledge and information are today widely accepted to be among the most relevant production factors of the 21st century. Owning information is no value in itself – it has to be used to prove its value. Product development and production have always been knowledge-driven processes. However, the economic relevance of the efficient use of knowledge in particular has been accepted with regard to engineering tasks within the 1980s and 1990s, when expert systems and knowledge-based systems were developed for supporting numerous well-defined tasks.

The term "knowledge" is often used without a clear definition. Knowledge is widely regarded to be fundamentally different from information and data. Knowledge resides in humans and contributes to decision-making, and it is transferred by interactions such as communication [2]. According to Harris, data can be defined as known facts, information as analysed data and knowledge as a combination of information, context, and experience [3]. Consequently, knowledge can only be built and used upon existing information and data. Knowledge can thus be seen as the top layer of a hierarchy of symbols, data, and information (see Fig. 1). It represents information in a personalised context that can e.g. be formed by experience.

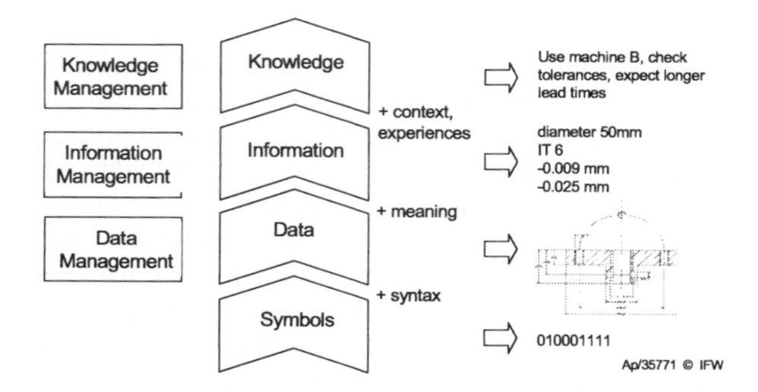

Fig.1. Data, information, and knowledge

Data and information have to be available in an easily accessible and well-structured way. Furthermore, the structure of the domain of application must be modelled.

Structured information is for example needed for purposes like multi-agent systems [4]. Within such applications, ontologies serve to model real-world entities and their relations as an environment for agent transaction. The question of efficiently building, maintaining and reengineering ontologies has not yet been thoroughly investigated with regard to the product development and manufacturing domain [1].

The question of computationally representing knowledge in a comprehensive and efficient way is a rather old one. In 1958, 'Advice Taker' as one of the first systems of its kind embodied knowledge representation and reasoning, and manipulated knowledge representation with deduction [6]. However, expressiveness and representation capabilities were still quite limited. Within the 1960s and 1970s a number of more powerful ways to computationally represent knowledge were developed within the information science community such as semantic networks [7] and frames [8]. Today`s discussion focuses on standards such as XML, RDF and their extension DAML/OIL [4]. To illustrate the wide range of information that is important for the manufacturing domain (and thus, the range of necessary information and data models), a brief overview is given in Fig. 2.

Still, the exchange of technical information and data also calls for less general concepts to store and exchange knowledge. One of these approaches is the STEP standard. Since its early days in the middle of the eighties, the importance of this standard has increased continuously. The STEP standard (as the abbreviation of "Standard for the exchange of Product Model Data" and defined by a set of documents of ISO 10303) is widely accepted for product data exchange and management. It is commonly used to handle and to document the results of the product design process. Its data models and methods provide a basis for integrated collaboration processes in enterprises, allowing a holistic view that comprises areas like design, engineering, testing, manufacturing and quality assurance. STEP can thus be regarded as an "enabling technology" for product-related IT infrastructures. Despite this fact, it is often only regarded as a standard for exchanging CAD data. However, future applications will extend this scope far beyond, as managing knowledge turns out to be as important as handling geometrical data, and collaboration during early design stages becomes vital for economic success. Within the next sections, some examples will be given.

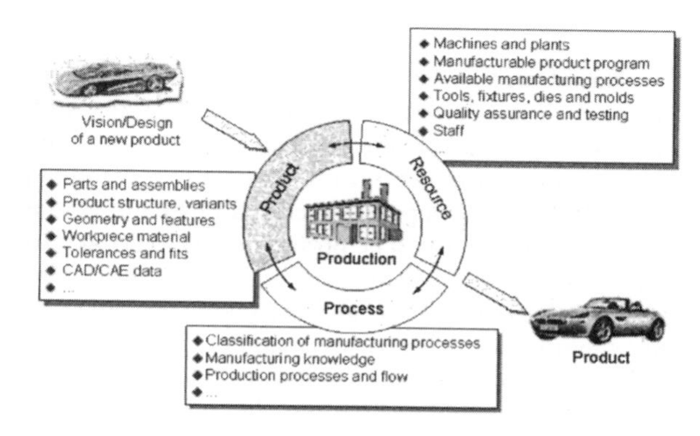

Fig. 2. Relevant information models in manufacturing (some examples)

3. Representation of information for reconfigurable machine tools

Beside the ability to design and to manufacture a specific product, it is important for enterprises to adapt their machine tools and manufacturing resources to changing requirements. Thus, not only modular products but modular manufacturing resources as well can be regarded as a key success factor. Conventional transfer lines and factories are missing this ability. Thus, a common description of machine tools and support of concurrent design and manufacturing of modules by different suppliers are needed. The use of the EXPRESS specification language (STEP part 11) for a modular machine tool model allows to extend and to reuse the model easily. This model is the basis for a configuration system for modular machine tools.

Any production process may be defined as a controlled transition of a work piece from an initial state to an intermediate or final state. For a cutting process, the change of a work piece state depends on the shape of the tool that performs the cutting process as well as on the relative movement between work piece and tool. A set of basic functions can be defined to describe for example a milling process. Each function is considered as an object or black box that has certain dimensions, kinematics and performances. These subsystems are connected to their environment by different interface types, which represent a set of predefined configuration patterns within a force flow matrix of the reference model. By means of this matrix and a set of rules, possible collisions of the configured model will be checked [10]. This approach allows to configure modular machine tools by using a STEP-compliant definition of modules and connections of the different suppliers.

4. Agent support for the extended enterprise

STEP-based information models may also be used for distributed artificial intelligence (in this case of intelligent software agents) in manufacturing execution. At a first glance, the relation between STEP and software agents in manufacturing is far from obvious. Nevertheless, in case of manufacturing of highly sophisticated products in small and medium batch-size, the relevance of manufacturing knowledge (for example available manufacturing processes at the shop floor, limiting constraints of specific machine tools, etc.) is more important than in serial production: Since a lot of simulation runs and process optimisations are carried out before a transfer line for serial production is commissioned, this procedure is often not suitable for job shop production. Thus, a MES (manufacturing execution system) for job shop production has to be able to deal with unforeseen disturbances. Current research activities deal with the application of software agents in this area, for example for integrated (in means of "parallel and adaptable") process planning and scheduling [11]. Since software agents implement a distributed process planning system closely related to the shop floor, knowledge about the product itself must be available.

Some of the relevant information models used for this purpose are influenced by STEP or are using it directly. For example, the geometrical representation of the product itself is based on CAD files. Assemblies are described according to the product structure specified by STEP part 42. The manufacturing tasks, which have to be performed in order to process the work piece, are defined by manufacturing features according to STEP-NC (ISO 14649) respectively STEP part 224. Figure 3 illustrates the application of a STEP-NC-based view on NC code for a particular manufacturing task, in this case the "manufacturing object" of a free-form pocket (some areas of the CAD model are removed for better perceptibility). Specific parameters of the tool path are revealed and appropriate milling strategies are applied to the tool path segments. Thus, necessary manufacturing knowledge is handled within the data model and can be taken into account for the agent's negotiations.

To describe the manufacturing abilities of individual resources, a similar approach was selected as described in the last section concerning modular machine tools. Relevant parts of these standards are used to create an ontology enabling the agents to perform communication as well as decision making in

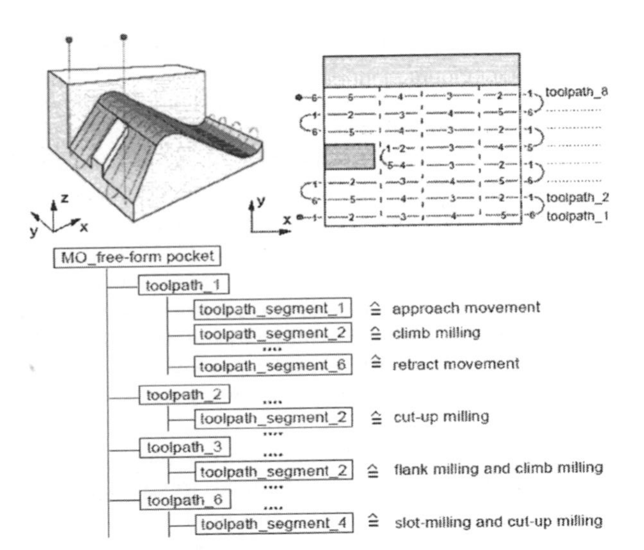

Fig. 3. STEP-NC-based object-oriented view on NC code

the distributed environment. Thus, software agents representing resources are able to determine if they are able to process a specific manufacturing feature of the product (requested by agents representing product orders) or not. Knowledge about the product and the resources use a joint process model.

5. Customer supplier collaboration

In order to improve cooperation between customer and supplier during product development, a description of product components and their component parts is necessary. This description may be based on constraints that define product requirements, enabling the supplier to use proprietary design knowledge. This methodology of variable descriptions of product components can replace the traditional delineation and supports the customer during the concept phase. Using constraints, the customer describes his requirements while being able to refer directly to already existing CAD geometry.

The supplier is able to use the description determined by the customer to perform the design autonomously. It is his task to work out the final product shape in detail by taking the given requirements into consideration. The description of requirements may be easily based on the formal description language EXPRESS (specified in part 11 of ISO 10303). If the constraints refer to product characteristics, for example geometry, the description will correspond to the level of the STEP standardisation (see Fig. 4).

Based on this concept, a JAVA-based software prototype has been realised during a joint research project [12], which enables the customer to describe the product requirements with the help of constraints and facilitates a company-specific adjustment of the descriptive structures. The prototype has interfaces to four commercial CAD systems, so that the customer can refer to existing CAD geometry. It is possible that the supplier simultaneously handles parts of the product component description in his own CAD System. The implementation features definition, application and communication modules. It can be used via the Internet. Thus, it is able to support distributed design teams.

6. Knowledge management within the extended enterprise

STEP may also be seen as bridge between manufacturing data and ontology-based knowledge management for extended enterprises [13]. Incorporating, storing, distributing, and finally using knowledge in a structured way by an appropriate software tool helps to avoid losing know-how. Furthermore, it allows to share the knowledge of experienced co-workers and may lead to an enhanced product quality and shorter development times.

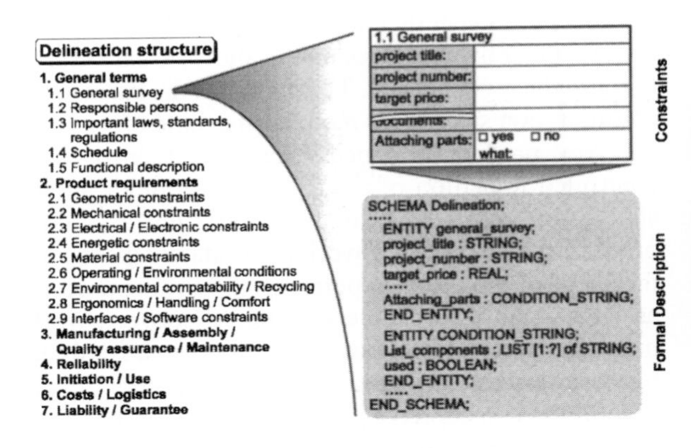

Fig. 4. Structure of a delineation for concurrent customer-supplier-cooperation

The foundations for a knowledge management system for department-spanning processes are adequate representation technologies, meta models and mechanisms for arbitrary knowledge items that allow to identify, classify, link, and retrieve different information and knowledge types from various sources, to provide it context-based, as well as to navigate over information items in an intelligent way

[14]. The management and retrieval of information is based on meta descriptions of information. Promising approaches towards knowledge management for extended enterprises integrate workflow and knowledge management.

7. Conclusion

STEP is a powerful basis for product data exchange and management in the virtual product creation process. It allows to integrate collaboration processes in enterprises and supports a holistic view on design, engineering, testing, manufacturing, and quality assurance. Methods and models provided by STEP can be seen as an enabling technology for product development within the extended enterprise.

Some possible applications have been presented in this paper, such as the use of data models influenced by STEP to describe modular machines. Furthermore, artificial intelligence systems for process planning can use information that is well-structured by STEP formalisms in order to negotiate about suitable operation sequences and resource allocation in job-shop production. Finally, collaborative product design and manufacturing management require appropriate means to store and access distributed knowledge. STEP may be seen as a tool to support knowledge representation and exchange as well as ontology-based domain modelling for knowledge management and machine learning strategies.

Acknowledgement

The Institute of Production Engineering and Machine Tools (IFW) is member of the I*PROMS Network of Excellence which is funded by the European Community.

References

[1] Krause, F.-L., Tang, T.; Ahle U. Abschlussbericht Integrierte Virtuelle Produktentstehung (Final Report, in German), Karl Hanser Verlag, München / Fraunhofer IRB Verlag, Stuttgart, 2002

[2] Shin, M.; Holden, T.; Schmidt, R. A. From knowledge theory to management practice: towards an integrated approach. Information Processing and Management, No.37, pp.335-355, ISSN 0306-4573, 2001

[3] Harris, D. B. Creating a knowledge centric information technology environment. http://www.htcs.com/ckc.htm, (Accessed 05-19-2005)

[4] Tönshoff H. K.; Woelk P.-O., Timm, I.; Herzog, O.: Flexible Process Planning and Production Control using Co-operative Agent Systems, Proceedings of the International Conference on competitive manufacturing COMA'01, pp. 442-449, Stellenbosch, Jan. 2001

[5] Apitz, R.; Brandes, A.; Woelk, P.-O.: Towards principles for designing ontologies for the production engineering domain. 13th Int. DAAAM Symposium "Intelligent Manufacturing and Automation: Learning from Nature". Vienna, Austria, pp. 17-18, October 23-26, 2002

[6] Mc Carthy, J.: Programs with Common Sense; reprinted in Semantic Information Processing; edited by M. Minsky; MIT Press; 1968

[7] Woods, W.A.: What's in a Link: Foundations for Semantic Nets ;In Representation and Understanding: Studies in Cognitive Science; edited by D. Bobrow and A. Collins; Academic Press; 1975

[8] Minsky, M.: A Framework for Representing Knowledge. The Psychology of Computer Vision. MIT-AI Laboratory Memo 306, June, 1974. (also: The Psychology of Computer Vision," P. Winston (Ed.), McGraw-Hill 1975, Mind Design," J. Haugeland, Ed., MIT Press, 1981, "Cognitive Science," Collins, Allan and Edward E. Smith (eds.), Morgan-Kaufmann 1992).

[9] Denker, G.; Hobbs, J. R.; Martin, D; Narayanan, S; Waldinger, R.: Accessing Information and Services on the DAML-Enabled Web. In: Proceedings of the Second International Workshop on the Semantic Web - SemWeb'2001, Hongkong, China, May 1, 2001

[10] Tönshoff, H.K.; Schnülle, A.; Drabow, G. Highly flexible and reconfigurable manufacturing systems due to a modular design. In: Proceedings of the CIRP 1st International Conference on Agile and Reconfigurable Manufacturing, Ann Arbor, USA, May 21-22, 2001

[11] Denkena, B.; Zwick, M.; Woelk, P.-O. Multiagent-Based Process Planning and Scheduling in Context of Supply Chains. In: HoloMAS - Holonic and Multi-Agent Systems for Manufacturing, Prague, Czech Republic, pp. 100–109, September 5-7, 2003

[12] Tönshoff, H.K., Uhlig, V. A Tool to Support Collaborative Product Development in Customer-Supplier Chains. In: International CIRP Design Seminar, Design in the New Economy, Stockholm, Sweden, pp. 85-89, June 6-8, 2001

[13] Denkena, B.; Apitz, R. Knowledge and Product Lifecycle Support for Extended Enterprises. 9th International Conference of Concurrent Enterprising, Espoo, Finland, June 16-18, 2003

[14] Tönshoff, H. K; Apitz, R.; Lattner, A. D.; Schlieder, C. KnowWork - An Approach to Co-ordinate Knowledge within Technical Sales, Design and Process Planning Departments. 7th International Conference on Concurrent Enterprising, Bremen, Germany, pp. 231-239, June 27–29, 2001

Intelligent Production Machines and Systems
D. T. Pham, E. E. Eldukhri and A. J. Soroka (eds)

Potentials of Voice Interaction in Industrial Environments

Matthias Peissner, Silke Lotterbach

Fraunhofer Institute for Industrial Engineering (IAO), Nobelstr. 12, 70569 Stuttgart, Germany

Abstract

This paper describes the range of aspects to be considered before implementing VUIs in industrial environments. It supports the decision process on whether or not a VUI is reasonable for a certain context of use. In order to create an overview of the relevant aspects, this paper refers to findings from the literature and integrates insights gained from interviews with carriers as well as suppliers of VUI applications.

1. Introduction

Voice user interfaces (VUIs) use speech technology to provide users with access to information, allow them to perform transactions, and support communications [1]. Today, advanced technology and software is available for a diversity of VUI applications. VUIs rely on automatic speech recognition (ASR) for user input and use speech, either recorded or synthesized, as their primary form of output. VUIs enable human-computer dialogues that cover only a limited spectrum of human conversation focussing on a predefined set of speech acts in a specified task domain. These restrictions reflect the limitations of today's speech recognition technology. Although the manageable dictionaries and grammars are getting larger and more complex, the full richness of human conversation will not be realistic in the foreseeable future [2].

Today, speech technology is widely used in call centre applications. In this domain, the usefulness of speech interaction is obvious. VUIs replace cumbersome touchtone systems or expensive human call agents and attend effectively upon the customers. They supply the caller with up-to-date information from huge databases or refer the caller to the competent customer consultant. Voice services for stock market information and rental car or ticket bookings are relatively widespread nowadays.

In other domains, however, speech interaction is currently only rarely used. Especially in the field of industrial production, the application of speech technology has still a great potential for growth. In the industrial domain, the requirements of the VUI are often even less complex than in other domains. Here, the context of use is well defined and often "known" by the system. Moreover, the user interface is controlled by a small number of recurrent users, which means that both interaction partners, the system and the user, can be trained for an efficient interaction.

This paper discusses the conditions for a beneficial application of VUIs in the industrial context. We describe important advantages and disadvantages of speech interaction in order to point out which contextual conditions have to be considered before employing speech technology.

2. Criteria for the implementation of voice user interfaces

2.1. Communication with computers is still restricted – yet effective

Many people argue that VUIs are ideal for human-computer interaction, as "Speech is a basic, universal mode of human communication" [3] – but this is only true for the communication between humans. Concerning the interaction between humans and technology, we are used to interact via displays, buttons or short commands. "While speech may be a natural form of communication for humans, speaking to computer artefacts or having them speak to us is a different activity" [4]. Therefore we cannot expect the same behaviour from systems as we do from other humans. Communication via voice is still far away from meeting these expectations. Today, communication with technology is task orientated and restricted, but if applications are well designed, VUIs offer a direct and efficient way of interaction.

Natural language has a huge advantage compared to fixed menu structures: users can compile a lot of information in short sentences, that otherwise would be hidden in deep hierarchic structures. The short sentence „I want to fly from Stuttgart to London tomorrow morning", contains the number of people travelling, date, estimated time, start and destination of the journey. Additionally, the user can describe in own terms what he wants, rather than adapt to the systems language. VUIs can – in contrary to graphical user interfaces (GUIs) – process various synonyms for a single command. VUIs can handle even complex tasks and are therefore ideal in many situations, but they are still far away from acting like the all-rounder systems known from science fiction movies.

2.2. Even large amounts of information can be easily input via voice, whereas output is limited

Most users can speak significantly faster than they can type or write [5]. Speech input can save time and avoid problems. Listening to speech output – on the other hand – is slower than accessing visual information. People can read faster than they can listen. Also, people can easily compare texts, pictures or figures visually. Comparable tasks are much more demanding when using a VUI and the users need to keep a lot of information in short time memory, which is not comfortable and may lead to mistakes. VUIs perform well when extensive user input is required, and the system reaction is either executing an action or giving short information.

Option menus can be displayed on GUIs very easily. Compared to this, VUIs often have difficulties in communicating all available options to the user. The auditory presentation of long menu lists is often not suitable for ergonomic reasons. This problem can be alleviated by experienced users who are acquainted with the functionality and necessary commands or by a visual aid (e.g. screen) that displays the possible options. Though the output of extensive information can be quite problematic, VUIs allow quick and easy access to a multitude of information. This is a clear advantage compared to GUIs, which can only display a certain amount of information. And "as the number of user-controllable functions on [...] devices increases, the user interface becomes overly complex and can lead to confusion over how to perform even simple tasks" [6].

Acoustic output allows vital information to be exposed to all users at a defined time. Users are alerted immediately and they do not have to watch a screen in order to catch the information. The opportunity of receiving real-time messages anywhere and anytime enables the operator to immediately react to critical events. Otherwise, the operator notices errors only when passing the affected machine by chance or when he gets back to his PC the next time. (cf. [7]).

2.3. Suitable information for VUIs is one that can easily be put in words

Spatial and visual information is difficult to be transferred into speech and is therefore less suitable for VUIs. The same applies to emotional information, as it is not only transmitted by words, but also by intonation, gesture and facial expression. Another problem is the acoustic presentation of structural information. Furthermore, it is not possible to present information in a parallel way, as it is possible with visual information (e.g. tables). Users can quickly switch between two pictures or pictures and text, but switching between two blocks of acoustic information is difficult. Also the manipulation or correction of visual data is possible, whereas the same action applied to acoustic information is time consuming and complicated, e.g. spelling mistakes (cf. [8]).

2.4. Dealing with linguistic diversity and overcoming physical handicaps

VUIs can be trained to understand different languages as well as accents or dialects. This may lead to a unique role of VUIs in the European Union due to the cultural and linguistic diversity: "There is no other advanced economic area that enjoys the cultural and linguistic diversity of Europe" [9].

For users with tendovaginitis, motorical impairments or just big hands it is often difficult to operate (mobile) computers manually. The implementation of a VUI can simplify this kind of work significantly. Also users with visual impairments can profit from VUIs.

2.5. Hands and eyes free, mobility provided

VUIs can be controlled via wireless headsets. This enables location-independent human-computer interaction. Users can walk around freely and do not have to come back to the PC terminal for receiving information or inputting data. VUIs economise not only the routes between workpiece and PC terminal but also the paths that the user's eyes and input devices (e.g. mouse, joystick, cursor, etc.) have to take on the screen. VUIs make it possible to directly access any item of an interactive application with a single voice command. Tedious navigation through extensive menu trees can be avoided by a flexible and intelligent VUI design. Moreover, wearing a headset allows the users to take a comfortable posture which helps to prevent postural deformities.

Speech applications provide a particular advantage when the user has to pay full visual attention to the primary work task while interacting with the system. The same holds for situations when the user's hands are busy with the primary task or not free for other reasons, e.g., wearing gloves, being soiled. In other words, conventional GUIs that demand visual attention and manual control interfere with most primary work tasks of industrial environments.

VUIs support efficiency by making it possible to accomplish two tasks simultaneously, e.g., operate a measuring device and input the displayed values into the system. This is not only beneficial in terms of time savings, but this is also supporting an ergonomic flow of work which is not disturbed by secondary tasks as keyboard entry or form-filling [10]. Thus, VUIs can improve productivity and precision [10] and help to economise inconvenient working procedures.

2.6. Straight usability – even in adverse conditions

Acoustical systems offer great advantages if the perception of visual information is impaired. This can be caused by either darkness or too much light, or by other circumstances, e.g. vibration. Situations like this occur for example during the maintenance of machines, when flaps or covers conceal sources of light.

VUIs provide special advantages "for work under uncomfortable circumstances – humidity, low temperatures, insufficient lighting, e.g., in deep freeze stores or during inspections outdoors" [11]. The only necessary equipment to be worn is a headset and radio receiver – and these are rather robust. In particularly hygienic environments a headset can easily be kept clean and is therefore more convenient than screen, mouse, touchpad or keyboard. VUIs are also beneficial, when there is no or only a very restricted keyboard or screen available (e.g. mobile phones, mobile devices).

Today's speech technology can handle noisy environments relatively well. Especially continuous noise is manageable, whereas unforeseeable and sudden peaks may lead to difficulties. As the user usually speaks through a microphone, the system can recognize commands very well and asks for repetition in case of problems. In contrary, the implementation of VUIs may not be favourable in very quiet environments or whenever human to human communication is vital.

2.7. Motivational aspects matter

Users may be not too happy at first, when VUIs are introduced into their working environment. Negative reservations are more widely spread than positive experiences. Therefore, a usable VUI is a very important precondition in order to create user acceptance for a new speech application. Only a highly acceptable VUI will be able to utilise the advantages of VUIs, e.g.

- Efficiency can be improved by omitting unpleasant and time-consuming tasks.
- Computer knowledge is not necessary for using a VUI.
- Users are often used to wearing earplugs and therefore accept wearing headsets very quickly, as this gives them the opportunity to communicate.
- VUIs can be adapted to individuals in a very short time. Therefore, recurrent users can choose the terms/language they

want to use and do not have to avoid foreign words or accents and dialects.

2.8. Economical factors

There are also economical aspects to be considered when employing speech applications (cf. [10]). At first, investments have to be made into hardware, software, installation and customisation of the system. Economical advantages are primarily due to improved man-machine interfaces which results in higher operator efficiency, as described in section 2.5, and in the prevention of operator errors. According to interviews with carriers of speech applications, a German company could avoid 50 % of all packaging errors by using a picking system controlled by voice. VUIs can also add to the company's image as they are seen "as modern and progressive technology that will be used by most companies in the future" (cf. [12]).

3 Conclusion

VUIs are not powerful enough to replace graphical user interfaces in the future. According to Shneiderman, "speech is the bicycle of user-interface design: it is great fun to use and has an important role but it can carry only a light load" [13]. However, under certain conditions and in certain contexts, speech – the same holds for bicycles – is the most comfortable and most efficient way to get things done.

In this paper, we have analysed the most important factors that influence the beneficial employment of VUIs in industrial environments. Further studies and interviews are planned in order to see how the specific requirements in individual production contexts can be met by a selective and purposeful application of speech technology. The here presented potentials of speech interaction make us expect significant market growths for VUIs in the production context. Special attention will be paid to multi-modal interfaces that combine the advantages of auditory and visual-manual interaction. According to the specific requirements of the current task and working context, the user can switch between interaction modalities or use a suitable combination of them.

In order to tap the full potential of these future user interfaces and in order to achieve a broad acceptance among the users and carriers of the emerging systems, a user-centred design approach will be indispensable. Especially innovative technologies require a high-level quality of use for their successful implementation.

References

[1] Cohen M and Giangola JP and Balogh J. Voice User Interface Design. Addison Wesley. 2004

[2] Allen JF and Byron DK and Dzikovska M and Ferguson G and Galescu L and Stent A. Towards Conversational Human-Computer Interaction. Dept. of Computer Science, University of Rochester, Rochester, NY 14627, AI Magazine 2001, retrieved 10.1.2005 at: www.cs.rochester.edu/research/cisd/pubs/2001/allen-et-al-aimag2001.pdf

[3] Markowitz JA. Using Speech Recognition, Prentice Hall PTR, 1996.

[4] Gardner-Bonneau D (Editor). Human Factors and Voice Interactive Systems, Kluwer Academic Publishers, p. 1, 1999.

[5] Zue V. Talking with Your Computer. Scientific American, August 1999, retrieved 10.1.2005 at: http://www.sciam.com/article.cfm?articleID=0009D2B7-F2E6-1C72-9B81809EC588EF21&catID=2

[6] Cohen PR and Oviatt SL. The role of voice in human-machine communication, in: Voice Communication between Humans and Machines, Herausgeber Roe D and Wilpon J. National Academy of Sciences Press, Washington, D. C., chapter 3, pp. 34-75, 1993.

[7] Janssen D and Peissner M and Schlegel T. FabSCORE – A framework for ergonomics in semiconductor productions. In: Khalid, H.M.: Work with computing systems: Proceedings of the 7th international conference on WWCS; Malaysia : Damai Sciences, 2004.

[8] Shneiderman B. The Limits of Speech Recognition, to improve speech recognition applications, designers must understand acoustic memory and prosody, Communications of the ACM, September 2000/Vol. 43, No. 9, pp. 63-65, 2000.

[9] Joscelyne A and Lockwood R. Benchmarking HLT progress in Europe – The EUROMAP Study. Euromap Language Technologies, Center for Sprogteknologi, Copenhagen 2003.

[10] Köppe C. Betriebswirtschaftliche Evaluation von Spracherkennungssystemen. Wirtschaftswissenschaftliche Fakultät der Universität Hannover, 2003.

[11] Helbig J and Schindler B. Speech-Controlled Human Machine Interaction. it – Information Technologie, issue 46 (6/2004), Oldenbourg Verlag, pp. 291-298, 2004.

[12] Michelson M and Dougherty M and Leppik P. Consumer Acceptance of Speech Technology, Phase 1: Focus Group and Usability Testing. Michelson & Associates Inc. & Vocal Laboratories Inc. & Voice Partners llc. Speech Technology Magazine, 2002.

[13] Shneiderman B. Designing the user interface, third edition, Addison-Wesley, 1998.

Intelligent Production Machines and Systems
D. T. Pham, E. E. Eldukhri and A. J. Soroka (eds)

Real Time Integrated Transportation System

S.G. Badiyani[a], V.H. Raja[a], B. Burgess[b], M. Ahmad[b]

[a] *Warwick Manufacturing Group, University of Warwick, Coventry CV4 7AL, UK*
[b] *B2B Manufacturing Centre, University of Teesside, Middlesbrough, Tees Valley, TS1 3BA, UK*

Abstract

The transportation industry is very competitive in our industrialised economy, with the pressure of fulfilling high quality service demands for efficient operations in supply chain and logistics management, and with this transport service providers are failing to achieve necessary high quality standards in their operations. This paper attempts to illustrate the integration opportunities of different upcoming technologies such as RFID, GPS, GPRS, e-Commerce concepts and high end computing, to improve the operations of transport service providers in meeting customer demands.

1. Introduction

In the present industrial world, companies are constantly deploying resources to shorten lead times and increase lead time reliability to the market. To achieve this milestone, management experts developed innovative management techniques like Just-in-Time, Supply Chain management and Logistics; where different firms collaborate and improve the operating efficiency to fulfil their customers demand and also provide rapid delivery of products [1]. Transportation plays important role in supply chain and logistics, as it is main operational area associated with the physical flow of inventory, right from raw material supply, to the distribution of finished products within supply chain.

Today transport service is a complex process comprising parcels sorting at the base station, route planning of parcel pickups and delivery, networking to achieve a wider coverage, and providing parcel tracking and tracing services to its customers [2]. With the rising cost of business operations and increasing competition in the transportation sector, service providers find it difficult to provide more efficient and higher quality services to their client, and have some issues undergoing the complex process mentioned earlier. The following issues were highlighted during the B2B Manufacturing Centre workshop at the University of Teesside.

- Service providers do not have any market place where they can find sufficient contract haulage work for any particular trip/route, as a consequence their operating cost is high and they find it difficult to provide cost effective service to their clients.
- Service providers find it difficult to plan the loading of the vehicles and also the routes, as at the initial stage they do not have sufficient and appropriate information about the parcels (size and weight) to be picked from their clients, and also about road congestion/traffic.
- Service providers currently do not have a tool for continuously monitoring the progress of each of their transporting vehicles, and so it is very difficult for them to have dynamic real-time routing for handling unforeseen events like popped in contract on existing route.

- Service providers often fail to notify their clients the time when a parcel pickup or delivery is about to take place, and hence there is often no-one immediately available at the clients end when goods need pickup or unloading; making the operations time consuming.
- Operators at the service provider's end often need to interrupt drivers and have conversations with them, to learn the load status, vehicle positions or giving instructions to the drivers for route alterations; which affects the drivers' performances and overall quality of the service.
- Service providers do not have any mechanism to record the time of the work for their drivers, which makes extremely difficult to comply with the new Working Time Directive (WTD) legislation.

In the transportation industry context, apart from the service providers, service consumers also have some issues related to the transportation which has an impact on their business operations. The issues that were highlighted by the service consumers were:

- Although companies tend to have preferred suppliers for repeat large contract haulage work; they also have a need for ad-hoc one off jobs which are usually sourced from small owner-driver haulage companies. There is currently difficulty communicating with small firms about opportunities because of a lack of a market place where they can find the appropriate cost effective services offered by these small firms.
- Consumers can not know the real time location of their goods on service providers' point-to-point tracking system, which sometimes in case of late delivery affects the planned operations of the consumers.
- Consumers often find service providers failing their time commitments, and usually without any notification from the service providers.
- In case of unforeseen events, operators at consumers end find extremely difficult to communicate with the transportation service providers for altering the planned pickup or delivery of parcels.

In order to overcome such difficulties and achieve improved operational efficiency, quality of customer services and business sustainability; an appropriate system needs to be developed an adopted by transportation service providers and consumers.

2. Proposed System

There has been a rapid development in various technologies like the introduction of GSM/GPRS wireless communication; from simple web-sites to complex business models like the virtual market place 'Amazon' and the virtual auction base 'EBay' in e-commerce field. For transportation industry, similar advanced solutions are available, like continuous tracking and tracing of vehicles/parcels, and software for route planning and scheduling. Also online up-to-date information about road traffic and congestion is easily available at any point of time. All these technologies are used separately. At present, the development of these technologies in an integrated fashion is very less, and there is no affordable commercial system available in the market. To make the working conditions more efficient in transportation services, a conceptual system called a Real Time Integrated Transportation System (RTITS) is developed based on an e-Business model and a modified and extended ParcelCall architecture [3]. The E-Business model comprises of a virtual (internet) market place similar to 'auction model' [4] for providing B2B services. In this case it is the transportation service for supply chain management, logistic and distribution of raw materials / manufactured products being offered. Whereas modified and extended ParcelCall architecture comprises of the real time tracking and tracing systems for transport vehicles (service provider's interest) and parcels (client's interest), with auto routing and electronic proof of delivery system.

RTITS provides personalized, cost effective and time saving transportation services to the client at one place; where clients can search for the service available in the virtual market (like auctions), as well as post their requirements on the market to be seen by the service providers (like reverse auctions). Apart from these basic services, client can also track their goods/parcels in real time, which helps in planning operations. For service providers, RTITS is a place for advertising/posting their services, and searching for clients and their requirements. RTITS also helps service providers in dynamic route planning, altering route plans and providing high quality customer service by tracing their vehicles in real time. RTITS improves the overall quality of the service as it automates the process and has less human intervention.

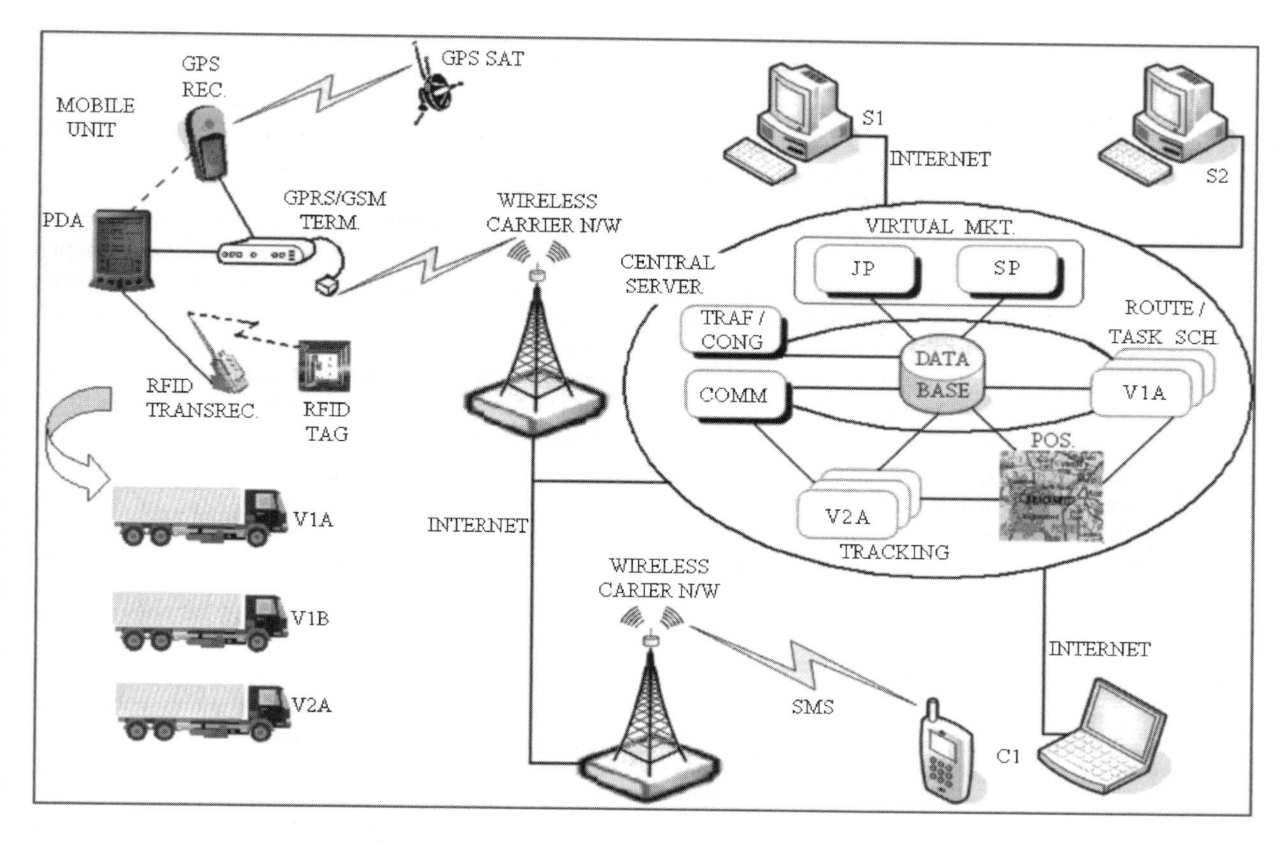

Fig. 1 Real Time Integrated Transportation System.

3. RTITS Components

Real time integrated transportation system is subdivided into two components; the Central Server and multiple Mobile Units, as shown in the Fig. 1. Each Mobile Unit is located within the transportation vehicle, and it continuously communicates with the Central Server via Public Wireless Carrier network.

3.1. Mobile Units

Mobile Units which are located within the transportation vehicle are the important part of the system, as the response sent from theses unit to the server are vital for decision making at the central server side. The mobile unit comprises different digital equipments interfaced via GPRS terminals as shown in the top left corner of Fig. 1. The interface between the central server and mobile unit is also through a GPRS terminal, where data is exchanged over the wireless mobile network. The GPRS terminal has a unique number used to identify the mobile unit in the GSM network and RTITS. The Global Positioning System Receiver (GPS REC) in the mobile unit receives the signals from four different GPS satellites and decodes it into meaningful signals (longitude, latitude, altitude and time) to represent the receiver's terrestrial position [5]; these signals are transmitted to Personal Digital Assistant's (PDA's) and to a central server via GPRS terminal. PDA executes two software modules; Route navigator, which accepts decoded signals from GPS receiver and navigates the driver on a route planned by the operator at the service providers' end. The other software module is designed for electronic proof of delivery processing; storing the information related to the task to be performed and current load (parcels) of the vehicle; recording Working Time Directives; and reading/writing Radio Frequency Identification (RFID) tag via RFID transceiver. RFID tags are used to label the parcels instead of traditional barcode labels, as RFID technology enables wireless electronic labelling and identification of the label using radio frequency over a short range. Another advantage of using RFID technology is that, it enables the presence of parcel at any point of time being detected, indeed real time tracking of parcels.

3.2. Central Server

The Central Server is the core of the entire system, which integrates different functional areas of RTITS. The central server comprises one central data base and six functional modules/work space, all operating in an integrated fashion. The central data base stores all the information about the system, and other modules pull the data from this central database; and have instances of relevant data which reduces the database interaction and data access time. A virtual market is a market place for transport service providers and service consumers for conducting business. Based on the trip/route, service providers post the details of the immediate services they can provide, in the service pool (SP) area. Similarly, service seekers post their requirements in the job pool (JP) area. There are multiple instances of Route/Task Schedule, each one for individual Module Units (different vehicles registered by service providers) which stores the route plan of that particular vehicle and tasks allocated to them, it also records all the parcel IDs and proof of delivery associated with the vehicle. Traffic/ Congestion module keeps the system updated of the current road traffic conditions through the operating region. Route planning and scheduling is based on route planning/ route finder software. The data representing the physical position of vehicles received from the mobile unit are stored in the Tracking instances as well as the central database. The positioning module pulls out maps from the data base and visually shows the positions of the vehicles based on the information from the tracking instances and route schedule. In this case the operator at the service providers' end gets access only to the visual information of the vehicles registered under their ownership. The communication module is like an information gateway of the system, which handles all the communication with the mobile units and the users of the system.

4. Working of Real Time Integrated Transportation System

4.1 Operations

The proposed real time integrated transportation system as shown in Fig. 1, is designed to work in the following manner. Both service providers and service consumer initially needs to register with the system, in addition to that service providers needs registering their vehicles with the system and installing mobile units. To illustrate all the possible combination of integrated operation, let us consider that at the start of the day, service provider 'S1' has some pickups and deliveries on a particular route. So in this situation service provider knows about the route and task to be performed, and enters the information (*Route, Pickup and Delivery Points, Load Capacity, Pricing*) per vehicle onto the database; depending upon the number of vehicles the services pool will be updated with the services entry that service provider can provide. In this case there are two entries one each for vehicle 'V1A' and 'V1B' for service provider S1. These entries display the different route and places where the service can be offered. While at the same time the route is planned and tasks are scheduled by the route/task scheduler based on the road traffic and congestion information; and such route and scheduled tasks are passed to the PDA in mobile unit, which will then use audio to instruct and guide the driver what task to perform next.

Also based on the schedule, the communication module will notify the clients of tentative times of pickups or deliveries, via e-mail or text message. At this time vehicles are already on the road performing the tasks that are directed by the PDA software. Whilst vehicles are on the move, the GPS receiver receives the signals and those signals navigate the driver on the pre defined route, also the same signals are passed to the central server where the tracking module records the coordination points for that particular vehicle. The positioning module pulls the map from the database and with the coordination points from tracking module shows the position of the vehicle on the map. This simplifies the information for the operator at the service provider's end to trace the vehicle on the planned route.

When the vehicle is navigated to the first destination, the PDA software instructs the driver on the task to be performed, say it instructs him to pick up a parcel. So for picking up parcels, the PDA will instruct the driver to activate one of the RFID labels with a predefined identification code (which was sent by the central server along with the task schedule), and attach it with the parcel. Now as soon as the parcel tag is activated, the message is sent back to the central server about the pickup, and this parcel is then associated with this unique mobile unit, and this information is stored in the database. From this point onwards continuous tracking of this parcel is possible.

One of the service consumers 'C1' needs to send a

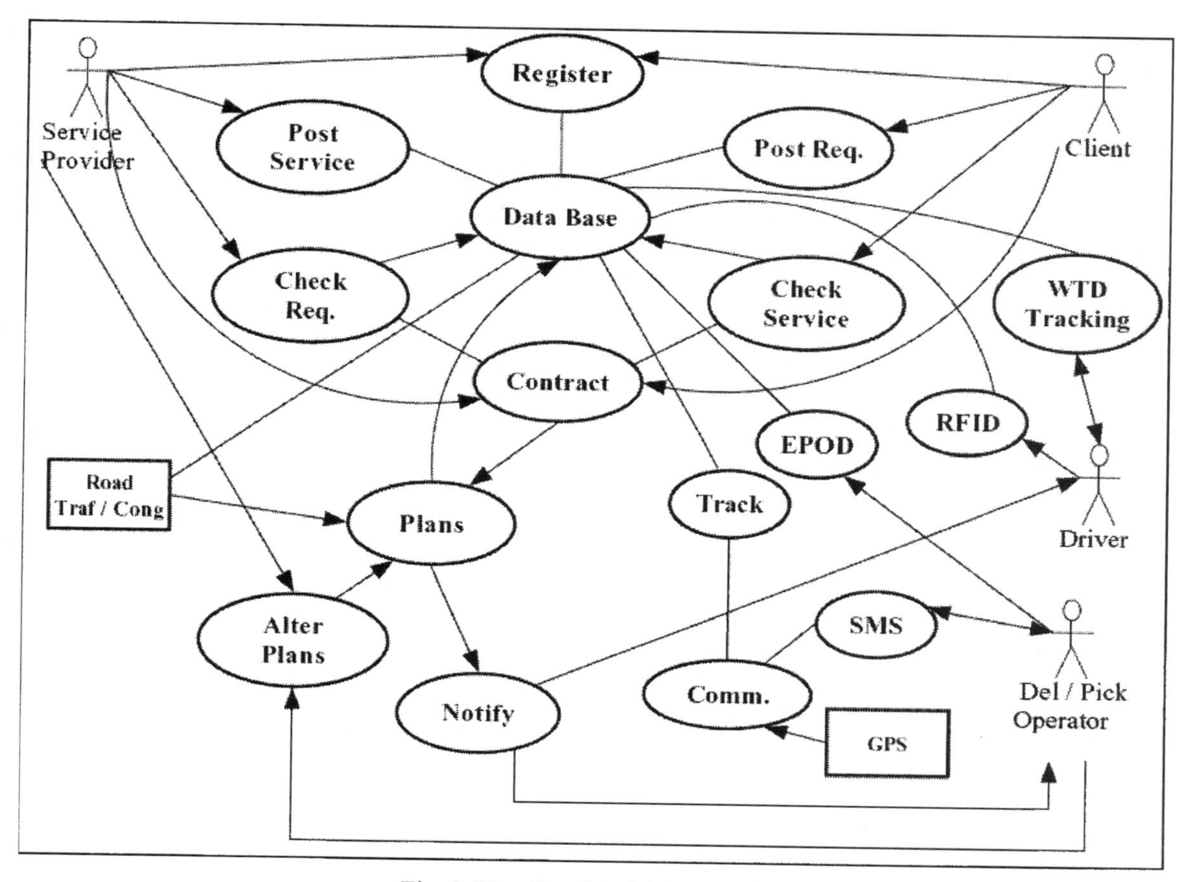

Fig: 2. Use Case Model for RTITS.

parcel from say pickup point 'A' and needs to deliver it to point 'B', so the operator logs on to the system and starts searching the service pool. In this instance the operator fails to find a particular match, and posts the requirements details (*Pickup Point, Delivery Point, Tentative Time of the Pickup and Delivery, Size and Weight of the Parcel*) into the task pool. The operator of service provider 'S1' finds the new entry into the task pool, and discovers that the destination point 'B' of the C1's delivery is on its route. The operator of 'S1' checks its route and realises that the vehicle just needs to alter the route for this pickup. To add a new client to its client base, the service provider takes up this contract and alters its route points in the database. Now the route / task scheduler modifies the route and sends the appropriate message to the mobile unit's PDA, without interrupting the driver and the driver is now directed to a different location for the new pickup. When the vehicle is routed for C1's task, the communication module sends notification to the client about the pickup time and also the RFID of the parcel, so that client can check the status/track the parcel once it has been picked up. With such dynamic route alterations, the service provider might fail his commitments given to their other clients, so for this the communication module at central server will send notification of new pickups/delivery time to them. In case of any unforeseen events at the pickup/delivery point, the system will allow operators to log on to the system and inform service providers about such events, which helps service providers in altering their work plan and avoid mismanagement in the service.

At the time of parcel delivery, the driver records electronic proof of delivery against the RFID of the parcel on its PDA, and this information is transferred to the central server at the same time and proof of delivery is recorded in the database. With this notification the load capacity of the vehicle is updated, which helps operator in taking other important real time decisions.

For tracking the status of the parcel in real time, it is not necessary for the clients to log on to the system.

This can be done by simply sending text message to the server with the *Client ID and RFID of the parcel*, and servers response will check the position of the parcel, and send a message back with the *distance of the parcel* and *time required* for parcel to reach at the client's destination.

At the beginning of work day/nights, drivers record their start time, break times and close time in to the PDA. These data along with the recorded activities of driver are passed to the central server. These recordings will help service providers to adopt and comply with the new Working Time Directive (WTD) legislation.

4.2 Use Case Model for RTITS

Based on the working of the integrated system, the Use Case Model for real time interactive transportation system is developed as shown in Fig. 2. The primary function of the system is to improve the efficiency of the transportation services in an integrated environment of e-commerce and real time tracking using GPS and RFID technologies. Primary actors in the use case model are those actors who are directly involved in providing services and the one who seek these services. So in this case they are Service Provider operators and operators at clients end. Where as the secondary actors in this model are drivers and operator at pickup/delivery point, who are not the decision making actors and who do not actually alter any of the system parameters.

Upon analysis of the working of the system and developed Use Case Model, it is clear that even service providers and service consumers are primary actors; their participation in the overall operation is limited to posting the service and availing a contract, they just enter the relevant information and then the system operates based on the preset rules or programmes. The system improves overall efficiency of the system, with less human intervention, which is time saving and cost effective in one way.

5. Conclusion

This paper has been a review of some of the critical issues in the current management of transportation services, like lack of a market place, difficulties in load planning and scheduling, lack of a continuous monitoring/tracking tool, failure of notification, and inefficient ways of dynamic route scheduling. To overcome these problems his paper proposes a Real Time Interactive Transportation System, which uses advanced technologies like GPS, RFID, Electronic Proof of Delivery, GPRS and e-Commerce concepts in a integrated fashion to provide virtual market place for conducting business; real time monitoring/tracking of parcels; dynamic route alterations, load planning and scheduling; and effective communication within the system.

Acknowledgement

The authors would like to thank the One North East RDA for their financial support for B2B Manufacturing Centre. This centre is in collaboration between the Warwick Manufacturing Group at the University of Warwick and the University of Teesside.

University of Warwick is partner of the EU-funded FP6 Innovative Production Machines and Systems (I*PROMS) Network of Excellence. http://www.iproms.org

References

[1] Bowersox DJ, Closs DJ and Cooper MB. Supply Chain Logistics management. McGraw Hill Irwin, New York, 2002.

[2] Jakobs K, Williams R, Grahm I and Lloyd A. Next generation Tracking & Tracing – A New Integrated Approach. Proceeding of 4th IEEE conference on Intelligent Transportation System (ITSC), Oakland, California, August 2001.

[3] Davie A. Intelligent tagging for transport and logistics: the ParcelCall approach. Electronics & Communication Journal. Jun 2002, pp 122-128.

[4] Deitel HM, Deitel PJ and Nieto TR. e-Business & e-Commerce HOW TO PROGRAM. Prentice-Hall Inc, New Jersey, 2001, pp 71-96.

[5] Brewer A, Sloan N and Landers TL. Intelligent tracking in manufacturing. Journal of Intelligent Manufacturing, 1999, 10, pp 245-250.

Intelligent Production Machines and Systems
D. T. Pham, E. E. Eldukhri and A. J. Soroka (eds)

Study On Manufacture Resource Evaluating System Of the ASP-based Platform

XIE Qing-sheng, YIN Jian

CAD/CIMS Institute of Guizhou University, 6 Guigong Road, Caijiaguan, Guiyang, China

Abstract

In the paper, the manufacture resource evaluating system of ASP-based platform was studied, and a two-stage decision model of selecting manufacture resources was presented. The linear mapping way was used to determine the marks of evaluating merits, and the analytic hierarchical process (AHP) was used to find the best manufacture resource. The decision model was very practical owing to citing the idea of some biding companies.

1. Introduction

The formation of network economic environment has transformed the traditional manufacture industry competition between individual enterprisers into the competition between supply chain and industry chain, in which enterprisers share complementary advantages and resources optimization [1，2]. Virtual enterprise is a temporary enterprise alliance which is leaded by the enterprise leader and joined by several alliance enterprises [3], when a market opportunity is appeared. .

Application Service Provider (ASP) is a kind of business operation mode and a service platform which provides third-party software and hardware application service via internet for customers. The successful operation of a virtual enterprise relies on the selection of alliance enterprises. On the ASP-based platform, the enterprise leader can fulfill the tasks of preliminary selection of alliance enterprises, information release, and optimal selection , by using the relative functions provided by the platform,.

The basic work of enterprise selection is evaluation. The most commonly-used evaluation methods are as follows: fuzzy evaluation method, analytic hierarchical process(AHP), multi-objective programming, neural network method etc. The efforts made in this paper lies on: establishing a manufacture resources evaluation system based on ASP platform; presenting two-stage decision model of selecting manufacture resource; making use of linear mapping to realize a quantification measurement of the corresponding relationship between detailed value and gradation level; realizing the optimal selection of manufacturing resources based on AHP.

2. The Synthesis Evaluation System of Manufacturing Resources

2.1. The Forming Process of Virtual Enterprise

A market opportunity may be a product, may also be a task of designing, or manufacturing or assembling. For a product, the alliance leader enterprise may invite public bidding as whole. And the whole task can also be divided into design, manufacture and assembling, each task can be further divided into several self-accomplished tasks(SAT) and several subcontract tasks. And then the alliance leader enterprise invites public bidding for the subcontract tasks. The product task decomposition tree is illustrated in Figure 1. Generally the smaller the granularity of the product task decomposed is, the shorter the accomplishing time will be, but with a higher cost. Therefore a reasonable breaking up the task has a great influence on final manufacturing cost and effect. For subcontract tasks, release the bidding invitation information to the potential alliance enterprises. And then make use of manufacture resources synthesis evaluation system to accomplish the selection of optimal alliance enterprises according to the tender information fed back in the stipulated time. The optimal alliance enterprises corresponding to respective subcontract tasks hence forming a virtual enterprise to face a certain market opportunity. Of course, the virtual enterprise formed by this way is not the optimal virtual enterprise in an overall sense.

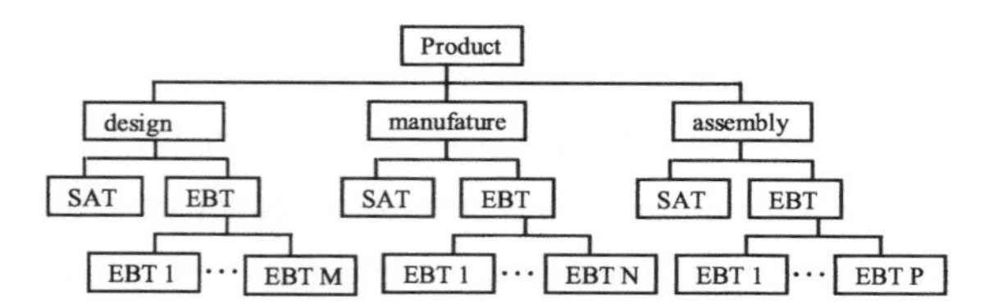

Fig1 decomposition tree of product task

With ASP platform, the alliance leader enterprise can use the functions provided by the platform such as alliance enterprise preliminary selection, information release, and optimal alliance enterprise selection to realize alliance enterprise preliminary selection, information release and weighed optimal alliance enterprise selection.

Worthy to be noticed, though the ASP platform provides algorithm and computer programme for optimal alliance enterprise selection, it makes determinations based on the information and data which is uploaded and transmitted to ASP platform by alliance enterprise and then stored in the data base of the platform. Therefore the truth of some crucial information such as the original annual examination of enterprise business license, the original accounting statement issued by accountant office etc. must be verified before signing a formal contract.

2.2. Two-stage Decision Model of Manufacture Resources

Theoretically, all the registered enterprises on the ASP platform are the potential alliance enterprises. As different task requires different manufacturer, there is a need to set up some qualification conditions for the manufacturers to meet before they become candidate enterprises and before the client extends bidding invitation to them. And the client, after receiving the tender information fed back, selects the optimal alliance enterprise based on the evaluation conditions. Because of this, the two-stage decision model is invested. The first-stage of decision model is for the preliminary selection of alliance enterprises, while the second-stage model is for the optimal selection of the alliance enterprises. The information used by the model are from the tender information of the candidate enterprises and manufacture resources collection module.

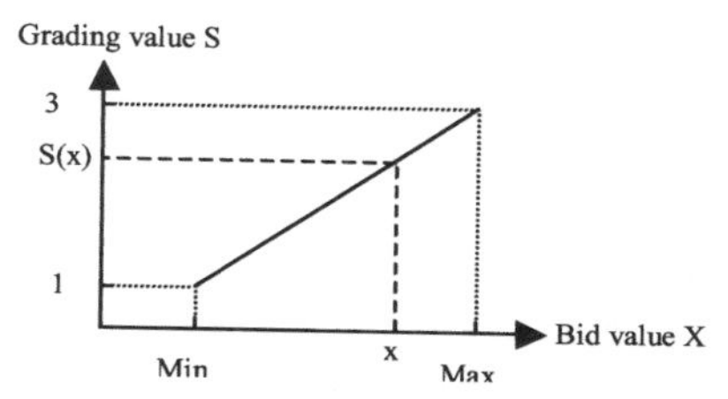

Fig2 Positive Index Linear

Table 1a

Qualification term for candidates

item	content
Juridical person and effective service	Having enterprise juridical person business license of last year
	Having tax register certificate
	Having P.R.C Organization Code Certificate
	Having manufacturing permit
Finance	Last two years auditing report issued by accountant office

Table 1b

Condition adjustable for candidates

item	content	Alternative item
Enterprise credit	Bank credit standing level	Non-required、A、AA、AAA
	Honor contract winner issued by government	Non-required、required
	Enterprise rank certificate	Non-required、required
Management	Quality guarantee system	Non-required、ISO9000、3C
Operation	Business scope	Machine processing, whole-set equip, other
	Net asset equal or larger than	(detailed amount)
	Business performance record of recent two years	Non-required、required
	Inspection or QC report of bid product	Non-required、required
Location	Location of enterprise	Non-required、Beijing, or other
Processing	Crucial equipment required	Machining center,electric spark,...

Table 2

Inspecting item of second-stage model

item	content	Grading method
Product design quality B1	C1 making quality of design drawing	Correct (1); standard &correct (2); standard & correct、CAD drawing (3)
	C2 design's satisfaction for overall function	Basic satifaction(1); satisfaction(2); better than requirement(3);
	C3 main technique parameter & index	Basic agreement(1); agreement(2); better than requirement(3);
	C4 design alternative	Basically reasonable (1); reasonable(2); excellent and creative(3)
	C5 structure design	Basically reasonable (1); fairly reasonable (2); reasonable, good strength, easy maintaining(3)
	C6 electrical control level	Low automation,hard for operation(1); moderateauotomation, convenient for operation (2); high automation, easy operation(3)
	C7 main material and spare ware	Using common product(1); partially using high quality product and name brand(2); completely using high quality products and name brands.(3)
Manufacture quality B2	C8 advantage of main processing equipment	Domestic, used for 10 years(1); domestic name brand, 5 years(2); international name brands, 5 years(3)
	C9 amount of main processing equipment	The least(1); the hightest(3); other determined by linear mapping
	C10 senior mechanic number for main processing equipment	
Cost B3	C11 bidding total price（FOB）	highest(1); lowest(3); other determined by linear mapping
After-sale B4	C12quality guarantee period	Half year(1); one year(2); two year(3);
	C13 after-sale quality responding speed	Organized, 3day arrival for emergency(1); Organized, 2day arrival for emergency (2); Organized, 1day arrival for emergency ((3)
	C14 spare articles and spare parts	Half year(1 分); 1 year (2 分); 2 years(3 分)
	C15 technical training	Yes, but rough(1); detailed training plan(2); perfect training plan(3)
Enterprise general information B5	C16 business performance	So-so market share(1); moderate (2); large marketshare, name brand product (3)
	C17 business status	Last year pay-off(1); last 2years pay-off(2); lasat 3year pay-off(3)
	C18 credit standing	Winning last-year honor-contract title(1); Winning last-2-year honor-contract title (2); Winning last-3-year honor-contract titl(3)
	C19 quality guarantee system	ISO9000 或 ISO14000(1 分)
	C20 cooperation	So-so(1 分); active、non beneficial suggestion(2 分); active、beneficial suggestion、helping to modifying relative articles.(3)
Design capability B6	C21 designer' number	smallest(1);largest(3); other determined by linear mapping
	C22 parallel design capability	Basically meeting requirement (1); meeting requirement(2); excellent(3);
	C23 coordinative design capability	Coordinative design capability under computer environment (3)
Aseemly B7	C24 number of mechanic equipped	smallest(1);largest(3); other determined by linear mapping
	C25 equipment in fitting shop	Basically meet the requirement (1); meet requirement(2); excellent (3);

Table 1 lists the inspecting items of the first-stage model. In it, Table 1a represents the qualification conditions for the candidate enterprises to meet. Table 1b deals with the conditions adjustable in accordance with actual situation. By using compound inquiry, the optimal alliance enterprise can be selected.

After receiving the bidding information from the candidate enterprises, using the second-stage model to conduct a synthesis evaluation to select the optimal alliance enterprise by their replies concerning processing time, price offer, technological condition deviation, after-sale service promise etc. Table 2 lists the reviewing items of the second-stage model. In Table 2, most of the indexes are belong to fixed quantitive index. And for those qualitative indexes, corresponding grades can be obtained by adopting certain grading rules. Here it adopts 4-level grading standard: lack of one item 0; optimal 3. As for the quantitative indexes, their grades are determined by means of “linear mapping approach.” Its principle is illustrated as in Fig 2. Among them, for the positive quantitative indexes (for example, equipment amounts), the lowest grade is 1, and the highest value is 3. The grade of other values for positive quantitative index is:

$S(x) = 2(x - Min)/(Max - Min) + 1$;

the grade for negative quantitative index is:

$S(x) = 2(Max - x)/(Max - Min) + 1$

Please notice that time is not listed in table 2 as an evaluation index. Here it adopts the thoughts of JIT (just in time), that is, the tender enterprise must provide service in accordance with the time requirement. Otherwise, it will be considered not responsive to the biding, therefore will be called off its evaluation qualification.

3. Synthesis Evaluation algorithm of Manufacture Resources

3.1. Weighing Measurement of Various Index

Table 2 lists out 25 evaluation indexes in 7 aspects and their grading methods. When the opportunity is a product and when it calls for bidding as a whole, adopt B1, B2, B3, B4, B5 to evaluate and select the optimal enterprise. When the market opportunity is a task of design, manufacturing, or assembling, or when a product is broken down into design, manufacturing, and assembling tasks, adopt B3, B5, B6 to give evaluation and select the optimal design enterprise, adopt B2, B3, B5 to select optimal processing enterprise, and adopt B3, B5, and B67to select optimal assembling enterprise.

Since different alliance leader enterprise has different concern about various indexes, it is necessary to determine the weight of various index according to detailed requirement.ASP platform provides the function of deciding weight by the users.

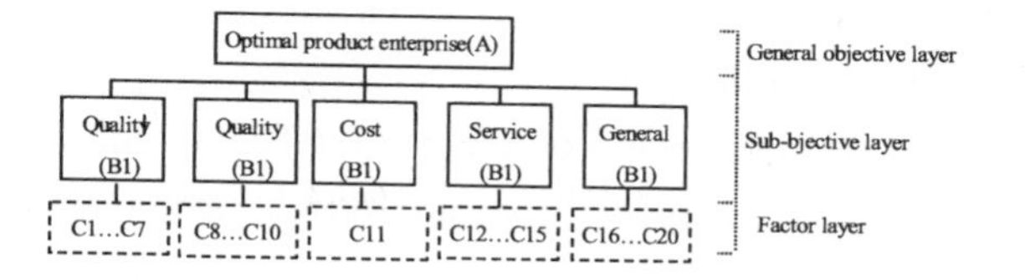

Fig3 optimal product enterprise evaluation hierarchical structure

Analytic hierarchical process AHP [6] is an effective tool for determining the weight of index. Its main thought is: break down a complicated multi-rule evaluation problem into an evaluating index with hierarchical structure. Make importance comparison between twain elements in the same level to form judgment matrix. Then make use of the method of working out matrix eigenvector obtain the weight of each level’s various element to one element of the higher level. And then adopting method ranking merging of weighed summation to work out the weight of the level’s various elements to overall objective.

Fig3 illustrates the step hierarchical structure of the optimal product enterprise, hereby, the judgment matrix of various layers elements to superstratum element and corresponding weight can be obtained. (table 3).

When forming the judgment matrix, the importance degrees of elements are indicted by number from 1 to 9. 1 represents the same importance of two elements, while 9 represents one element is extremely greater in importance than the other one. Evidently, to judge the importance degree between two elements is a rather fuzzy process and hard to represent this relationship with accurate data. Therefore to determine the graduation artificially is subjective and random. To avoid the confliction of importance degree between two elements, it is advisable to conduct a consistency test toward the matrix. When the random consistency rate Cr <0.1, the random of the matrix is acceptable, otherwise it needs a modification.

The eigenvector of the largest latent root for judging the matrix will become weight vector after going through normalization treatment. In table 3, W_{BA} is the weight of Layer B (sub-objective layer) in contrast to Layer A(overall objective layer), while W_{CB1}、W_{CB2}、W_{CB4}、W_{CB5} are the weights of layer C (element layer) in contrast to layer B (sub-objective layer). As there is only one element for B3, $W_{CB3}=1$.

Then synthesize the weights under one single sub-objective, and the weights of various elements in contrast to the overall objective will be obtained. The weight of layer C in contrast to layer A, W_{CA} will be obtained by following formula:

$W_{CA}=[W_{CA1},W_{CA2},W_{CA3},W_{CA4},W_{CA5}]$

$=[W_{BA}(1)*W_{CB1}, W_{BA}(2)*W_{CB2}, W_{BA}(3)*W_{CB3}, W_{BA}(4)*W_{CB4}, W_{BA}(5)*W_{CB5}]$

=[0.0157,0.0995,0.0651,0.0455,0.0257,0.0203,0.0268,0.1978,0.0432,0.0755,0.2158,0.0211,0.0297,0.0087,0.0174,0.0215,0.0341,0.0173,0.0120,0.0071]

Table 3a

A-B jugement matrix(Cr=0.045)

A	B1	B2	B3	B4	B5	W_{BA}
B1	1	1	2	3	3	0.2988
B2	1	1	2	3	4	0.3165
B3	1/2	1/2	1	3	4	0.2158
B4	1/3	1/3	1/3	1	1/2	0.0769
B5	1/3	1/4	1/4	2	1	0.0920

Table 3b

B1-C jugement matrix (Cr=0.0298))

B1	C1	C2	C3	C4	C5	C6	C7	W_{CB1}
C1	1	1/4	1/3	1/3	1/2	1/2	1/2	0.0526
C2	4	1	2	3	4	4	4	0.3331
C3	3	1/2	1	2	3	3	3	0.2180
C4	3	1/3	1/2	1	3	2	2	0.1524
C5	2	1/4	1/3	1/3	1	2	1	0.0861
C6	2	1/4	1/3	1/2	1/2	1	1/2	0.0681
C7	2	1/4	1/3	1/2	1	2	1	0.0896

Table 3c

B2-C jugement matrix (Cr=0.017)

B2	C8	C9	C10	W_{CB2}
C8	1	4	3	0.6251
C9	1/4	1	1/2	0.1365
C10	1/3	2	1	0.2384

Table 3d

B4-C jugement matrix (Cr=0.023)

B4	C12	C13	C14	C15	W_{CB4}
C12	1	1	2	1	0.2742
C13	1	1	4	2	0.3858
C14	1/2	1/4	1	1/2	0.1133
C15	1	1/2	2	1	0.2267

Table 3e

B5-C jugement matrix (Cr=0.0531)

B5	C16	C17	C18	C19	C20	W_{CB5}
C16	1	1/3	2	2	3	0.2340
C17	3	1	2	2	3	0.3706
C18	1/2	1/2	1	2	3	0.1877
C19	1/2	1/2	1/2	1	2	0.1301
C20	1/3	1/3	1/3	1/2	1	0.0776

3.2. Comprehensive Sequencing of Manufacture Resources

Table 4

Expert grading average value and comprehensive sequencing

	C1	C2	C3	C4	C5	C6	C7	C8	C9	C10	C11	C12	C13	C14	C15	C16	C17	C18	C19	C20	R
M1	3	2	2	3	2	2	3	1	3.0	3.0	1.0	2	2	2	2	2	2	1	1	2	1.7634
M2	2	2	2	1	2	1	2	2	1.8	1.0	2.4	1	1	1	1	1	2	0	0	2	1.7790
M3	1	2	1	1	2	2	1	1	1.0	2.1	3.0	1	1	1	1	1	1	0	0	2	1.6378

Suppose the alliance leader enterprise plans to call for a public bidding for one product as a whole, and there are three enterprises have replied to the bidding. Many experts are invited to grade them online by virtue of these 20 indexes from B1 to B5. And then average value of the experts grading values, $M=[m_1,m_2,\ldots,m_{20}]$, will be the final score.(table 4). Therefore, the comprehensive sequencing of manufacture resources is: $R=W_{CA}*M^T$

According to table 4, $R_2>R_1>R_3$, Enterprise 2 will be the optimal alliance enterprise for the whole product public bidding.

4. Conclusion

This paper has studied the comprehensive manufacture resources evaluation system of ASP-based platform. Since it has considered the operation reality of project contract of manufacture enterprise, and has borrowed idea from the bidding evaluation of bidding company, its decision-making model is more practical. It puts forward a two-stage decision model of manufacture resources selection. The first-stage model can realize the preliminary selection of manufacture resources, while the second-stage model can accomplish the optimal alliance enterprise selection faced with different market opportunities. It also puts forward a quantitative measurement of relationship between quantitative index value and grading ranks by using linear mapping method. It is able to realize the weight calculation of the evaluation index by using analytic hierarchical process (AHP), and further ale to adjust the weight distribution according to the real situation, make use of weighted summation to realize the optimum selection of manufacture resources.

References

[1] Xie, Qing-sheng. 2003. Mode and Strategy of ASP Development of Chinese Manufacture Industry[J], Manufacturing information realization of China, 32（1）:23-25(In Chinese)

[2] Xie, Qing-sheng. 2003. Research on Network Manufacture Mode and Developing Strategy of Chinese Enterprises [J], Digital Manufacture, 1(1): 142~152(In Chinese)

[3]Martinz M T, Fouletier P, Park K H. etc. Virtual enterprise-organization, evolution and control[J]. International Journal of Production Economics,2001, 74:225-238

[4]Camarinha Matos L M., Afarmanesh H, Garitac, etc. Towards an architecture for virtual enterprise[J]. Journal of Intelligent Manufacturing,1998, 9(2): 189-199

Intelligent Production Machines and Systems
D. T. Pham, E. E. Eldukhri and A. J. Soroka (eds)

User Interfaces for Service Oriented Architectures

W. Beinhauer, T. Schlegel

Fraunhofer IAO, Nobelstrasse 12, 70569 Stuttgart, Germany

Abstract

Service oriented architectures (SOAs) are developing more and more towards a fundamental pattern of information architectures within enterprises. The vision of loose coupling of coarse-grained components, dynamic interoperation of services with each other and semantic message routing over an enterprise service bus is particularly interesting for manufacturing enterprises that are following the approach of holonic production control. However, eventually software has to be fitted to human user's needs. Hence, composing processes or workflows dynamically implies that a user interface has to be generated dynamically as well. In this paper, we present preliminary results of deploying dynamic workflows to SOAs.

1. Motivation

Increasing business demand of the clients is the primary driver for developments in the manufacturing industry. Rapidly changing requirements, short predictability of order sizes and higher levels of individuality characterise the major challenges for manufacturing enterprises. An increasing level of individuality on the client side is linked to a vanishing level of product anticipation on the manufacturer's side. Efficiency and flexibility in production processes are the keys to the so-called mass customisation.

In terms of production organisation and supporting IT infrastructures, mass customisation essentially requires flexibility and reconfigurability as well – both on a computing level as well as in organisational aspects. With the approach of "holonic" production systems [6], the distribution of decision making over various, autonomous acting entities in production control has become a paradigm. This yields for computing systems as well as for organisational structures.

On the level of manufacturing control systems, the holonic approach entails:

- Flexible integration and reconfiguration of computing systems according to changing business needs
- "plug and play" integration and connectivity
- Distributed resources and orchestration
- Support for flexible production sequences determined by customers
- Distributed planning and control

Some of these issues have been picked up by newly developments towards service oriented architectures in distributed computing environments, enabling the "real time enterprise".

On the one hand, tightly integrated enterprise information systems accelerate the execution of business and reduce failures, on he other hand these linkages have to be flexible enough in order to be easily re-configurable in case of varying business needs. Hence, the coupling of enterprise software has to be loose in order to open the possibility of shorthand reactions on changing market conditions.

One of the most discussed solutions to this problem is service-oriented architecture (SOA). In SOAs, the formal description of business processes is separated from their projection to concrete software, giving rise for a better adaptability and re-configurability of the integrated software as a reaction on market changes. Moreover, flexible data exchange and automated data processing is eased. The business logic is kept in service-oriented process models, while the execution is performed by run-time engines. However, many business processes such a sequence of manufacturing manoeuvres still require manual user intervention. This yields particularly for business to business (b2b) processes between small and medium sized enterprises that do not rely on extended enterprise information systems. In order to keep track of the overall dynamic characteristics of such a SOA-based enterprise portal, user interfaces need to be included into the process models as well, as they reflect the business logic kept in these abstract models. Consequently, there is a need for generic adaptive user interfaces that support the workflow to be performed by human users within the framework of automated integration services.

In this paper, we point out the dependence of process modelling, workflow generation and their occurrence in automated user interfaces. Secondly, we present an approach of parameterized workflows that allow the pre-creation of process modules at design time, and their dynamic composition at runtime. Finally, we evaluate the first results among a b2b platform for dynamic supply chain management in manufacturing processes.

2. Approach

As stated above, a SOA needs a generic user interface that enables the performance of manual workflows in order to expand its full capability in terms of flexible business process integration. Generally speaking, a SOA consists of an enterprise service bus, a series of service providers and service consumers and a user interface that accepts user input such as triggering a specific action. A SOA does not essentially require deployment of web services, even if it is in most cases built on it. Service providers and service consumers act as sources and leaks for certain software modules. The business logic lying in between is also contained in one of the modules attached to the enterprise service bus. Such architecture is shown in Figure 1. The enterprise service bus interconnects all components of the IT infrastructure, including an orchestration engine that is able to execute static processes, a storage server that holds business data, an access control system, and usual encapsulated legacy systems.

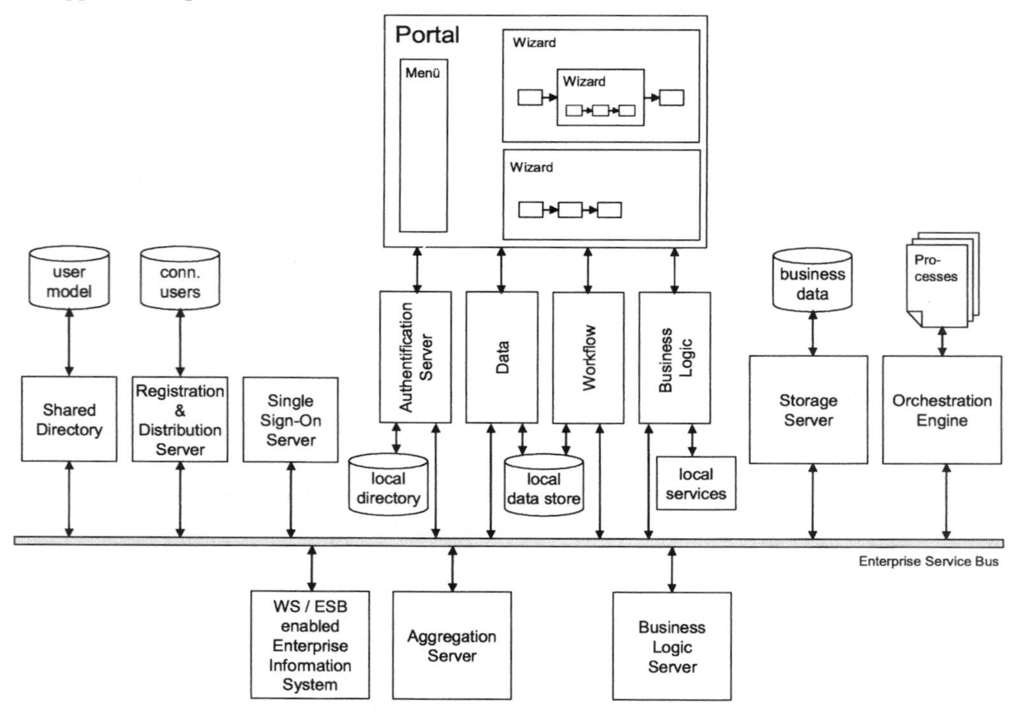

Fig. 1. A Service Oriented Architecture. The enterprise service bus interconnects all components of the IT infrastructure. Business logic and orchestration engine are kept apart.

2.1. Process

In the vast majority of cases, business and production processes need to be only slightly adapted – it is more an adjustment than a re-configuration. The coarse-grained processes, that describe the acting partners by their roles and mutual dependencies, basically remain constant, whereas the finer interaction patterns are subject to adjustments. For example, the procurement process in a supply chain application remains the same for every supplier, whereas some differences might occur concerning the detail of how an agreement becomes valid. In particular, this fine-grained step of a production process might require explicit user interaction on both sides of the supply chain.

The dynamic workflow composition may be effectuated by different ways: first, a steady dialog is held implicitly by a static process description, typically in a BPEL file, and is not altered during runtime.

2.2. Interface

Secondly, the dialog or at least some parts of it might be the result of a dynamic, ontology-based configuration. This composition might include the search and retrieval of a new component or business partner, so that new partners and new portlets are likely to join the process. Thirdly, the dynamic composition might be based upon a case base, where successful compositions are stored for later re-use. In this case, the execution of the process is deferred to a case based reasoning engine. The forth case of dynamic composition is the combination of the latter three. A case might include a static process part, which triggers maybe an ontology-based reasoning etc. Hence, nested process invocations require their user interactions to be performed within nested dialog wizards. This is illustrated in Figure 1.

A user front-end for a SOA-based production machine must be capable of displaying a variety of dialogues such as inlets, i.e. portlets within the user interface. This might be a simple browser window on a fixed screen, a touch screen or a browser window on a portable steering device. Hereby, the different altering types mentioned above suggest a two-phase generation of portlets for the tracking of the overall process on one side and the performance of explicit user interactions on the other. This reflects the two levels of business logic.

Our approach follows the correspondence of coarse-grained process tracking as well as the correspondence of fine-grained interaction patterns to specific dialogues. While the first workflows can be derived from the business logic, usually given as a BPEL description, the second dialogue patterns depend on parameterized functions. For the combination of these two, the dynamic user interface generator needs to be capable of crating nested dialogues.

Concepts for describing web-service user interfaces like Web Service User Interface (WSUI, [1]) have already been created. But so far none of these approaches has been used in practice, while current interfaces are developed purely manually, which harms the concept of service oriented architecture on the level of user interfaces.

3. User interfaces for services

3.1. Interfaces for process overview and control

As mentioned above, the coarse-grained process itself often remains static. For this reason, it is sufficient to provide a static description of the process sequence itself. This coordination may be executed in the form of

- a standard planner that executes the process with no user interaction;
- an interactive planner that plans with or without user input but always allows for interactively influencing the process flow during runtime;
- an interactive workflow that emphasizes the concept of controllability: the process is driven completely by the user, but may have an initial description to serve as basis for the mental model of the user, providing an overview on the possibilities while running a process instance (i.e. project).

Although much technology is needed for the planners, the user interface for process control can remain mainly static and is only parameterized by different process steps and process content.

3.2. Interfaces for interactive services

This is different for the services, which do not require a planning of services workflow connections like the coarse-grained process but require a differentiated user interface for input from, decisions by and output to the user. Depending on the complexity of this task, we differentiate between four types of service user interfaces, which we discuss in detail:

- Simple Service User Interfaces
- Complex Service User Interfaces
- Multi-Service User Interfaces
- Service Application User interfaces

Regardless, which type of user interfaces is needed, every dynamically executed interaction will need a proper description of the interaction required to the user. First, an initial classification will require to determine whether the task is an

- enter
- edit
- select
- write
- or any other specific interaction type.

Secondly, standardized data type classifications as given in UML will be required in order to define the data type that is handled in an interaction step. This is necessary for determining the classes of the interaction elements used. In this paper we do not discuss the specialties of e.g. voice user interfaces.

As there is no standard taxonomy of data types exact enough to allow for a dynamic user interface generation, there is a need for further standards in data and interface classification.

3.2.1 Simple Service User Interfaces

Simple service interfaces consist of only one parameter or a independent sequence of parameters with no transaction properties.

This simple type is rather easy to describe and transform. While modelling questions and complex structure also appear in the simple interfaces domain, no side effects occur in the user interface and once a transformation exists for a specific type, it is possible to use it with little parameterization in other contexts, too. The independent and sequential handling of more than one parameter allows for using single parameter interfaces that are only concatenated with no additional restrictions.

Problems are mainly on the level of how adequate elements can be found, like in the figure below.

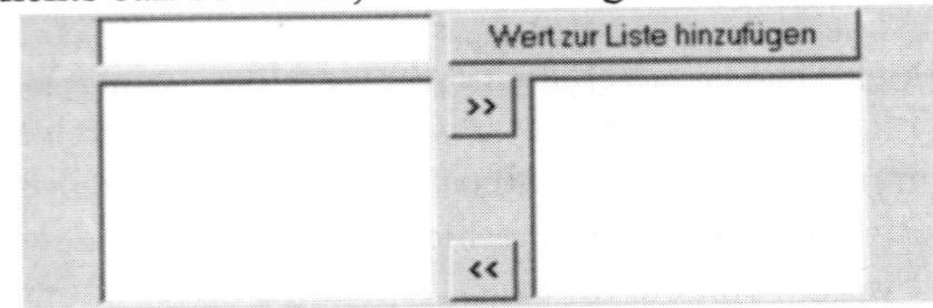

Figure 2: Possible representation of an expandable multi-selection (select m out of n) in a graphical UI).

3.2.2 Complex Service User Interfaces

Complex service interfaces require a description of elements or steps like in simple service interfaces but additionally need sequencing and structuring descriptions for these steps. While only one service is described, it can be very complex from an operational standpoint.

One modelling approach for such descriptions has been presented as InteractionCases in ([4]). InteractionCase allow for structuring interaction sequences into atomic clusters that can not be separated without disconnecting belonging information. They can be sub-structured using semantic groups and elements within these groups.

Describing interaction concerns this way for a service, it is possible to describe not only the data and technical requirements but also the user interface with the interactions necessary to use the service.

Complete service flows can be described this way. Where necessary, multiple InteractionCases can be attached to one complex service. Problems with complex service interface lie in the description of concatenation/sequencing of and element arrangement for the required interaction steps.

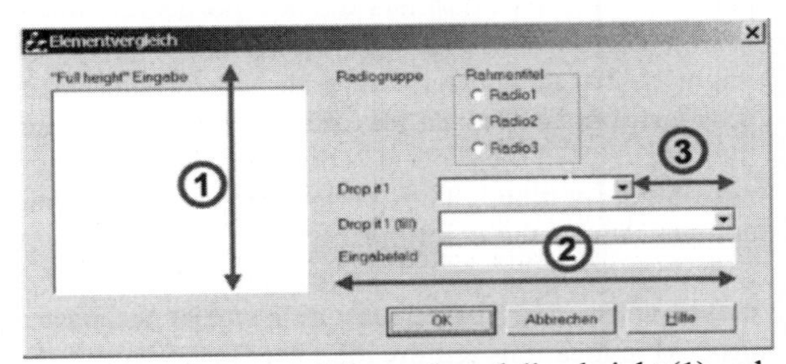

Figure 3: Determining and modeling height (1) and width (2) for complex interaction sequences in graphical UIs must deal with ranges for data (e.g. 1 to 200 characters) and elements (3)

3.2.3 Multi-Service User Interfaces

While independent complex services can be sequenced without side-effects, some services are dependent on others. Often the user interfaces has to consider dependencies between the services or parameters that are used for both – with consequences e.g. for their type and range.

This complexity requires additional provisions to ensure consistency with the service model. Options are the use of a semantic description of the service interfaces like with the Web Ontology Language OWL [5] together with a rule-base engine for composition or the use of process patterns enriched with interaction model descriptions dedicated to the connection of those services, providing process aspects and user interface description for ensuring proper work. The second approach is more static and model intensive while the first may produce unusable results, depending on the quality of the description and the composition engine. In our evaluation, we have chosen the second one due to more promising results and the broad use and existence of process descriptions in the logistics and production field of application.

Problems range from layout and element semantic questions as in figure to complex descriptions for the interconnection of services inadequately described at design time.

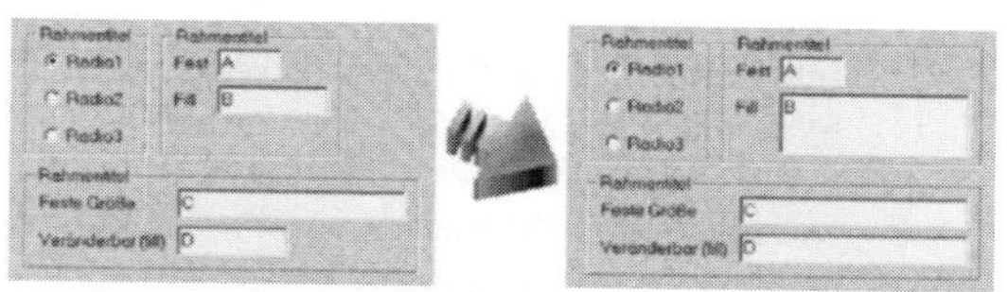

Figure 4: Rule-based element generation and adaptation to context and specified requirements

3.2.4Service Application User Interfaces

The composition of services to form a complete application – having the same problems as with other service interfaces – often profits from richer information at design-time of the interaction descriptions for the services and their interconnection.

This is the reason why for complex applications even with service oriented architecture many parts are described clearly and the environment is changing less dynamically, which even allows for generation of some interaction steps without a separate interaction description for the services – using only the context of the application and the capabilities of the services used. This context dependent description at design time in some cases allows for dynamic UI composition at runtime within the defined application context.

4. Interfaces for an SOA-based Portal

Building a dynamic portal for logistics services based upon a web-service architecture with automatic composition based on a semantic description, we have evaluated different approaches for constructing interfaces that provide the ability of dynamic adaptation in a limited portal environment.

Common web-service descriptions based on WSDL and BPEL descriptions proved insufficient for dynamic user interface generation of complex services. But standard conform descriptions combined with pattern descriptions and context information yielded enough information, to generate user interface without complex models and additional design steps.

Like in user interface generation approaches for non-SOA applications ([2],[4]), it is possible to transform a interaction specification into a generic element description, from which code can finally be generated (see figure below).

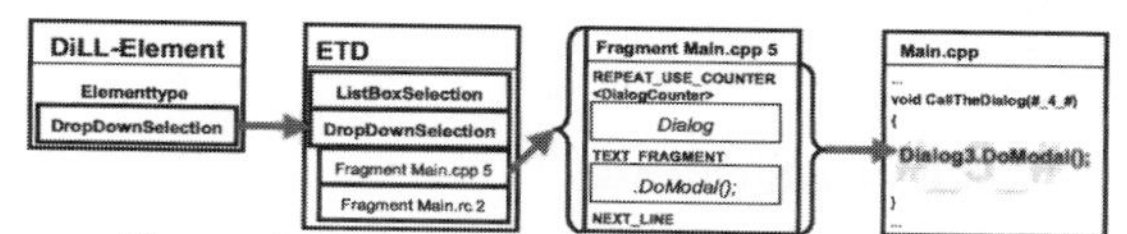

Figure 5: Schematic flow of design-time code generation for application user interfaces

In service oriented architecture applications, this process has to be done at run-time, because environment conditions are not completely clear at design-time. Only an attached user interface service could be used to provide an HTML-based interface that has been described when the service was generated.

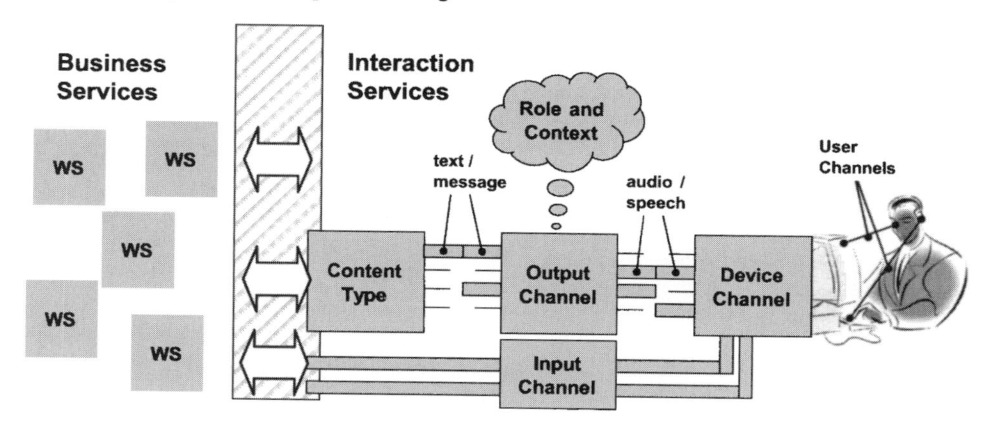

Figure 6: Semantic description of content, channels and device capabilities for routing service interface messages to the user in a distributed/ambient environment.

5. Future work

The proposed approach of generating dynamic dialogues for portals that rely on service-oriented architectures has been tested in a first step for a b2b business platform for dynamic supply chain management. It shows that the business process description designed for the steering of involved business partners and the services offered can serve as a basis for user interface generation, too. Future work tends towards a refinement of the existing infrastructure and the development of a library of pre-defined process patterns on the user interface level and an extension of the SOA user interface principle to more complex interaction environments like they occur in holonic production and ambient environments.

The results obtained need a further evaluation within a real manufacturing plant. Therefore, work in co-operation with the I*PROMS network is in progress. In this case, the dynamic creation of user interfaces will by coupled to the dynamics of ad-hoc networks. Therefore, the concept of self-describing services will be extended to self-describing control devices that form an intelligent network in their entirety.

6. Acknowledgements

This work has been carried out by the Fraunhofer Institute for Industrial Engineering (IAO), Stuttgart, Germany. Fraunhofer IAO is member of the I*PROMS network of excellence, supported by the European Commission. This work is related to KOMPASS, a project financed by the German Federal Ministry for Education and Research (BMBF).

References

[1] Anuff, E., Chaston, M., Moses, D., Kropp, A. (Eds.) (2002): Web Service User Interface (WSUI) 1.0. http://www.wsui.org/doc/20020726/WD-wsui-20020726.html

[2] Balzert, H. & Hofmann, F. & Kruschinski, V. & Niemann, Chr. (1996). The JANUS Application Development Environment-Generation – More than the User Interface. In Vanderdonckt J. (ed.), Proceedings of CADUI'96, Namur: Presses Universitaires de Namur, pp. 183-207.

[3] Endrei, M. et al. (2004). Patterns: Service Oriented Architecture and Web Services. IBM redbook SG24-6303-00.

[4] Schlegel, T., Burst, A., Ertl, T. (2004): A Flow Centric Interaction Model for Requirements Specification and User Interface Generation. In: Proceedings of the 7th International Conference on Work with Computing Systems, WWCS 2004.

[5] W3C (2004): OWL Web Ontology Language Reference. Retrieved February 28, 2005, from http://www.w3.org/TR/2004/REC-owl-ref-20040210/, W3C Recommendation 10 February 2004.

[6] Christensen, J.: Holonic Manufacturing Systems – Initial Architecture and Standards Directions. University of Hannover, 1994.

Intelligent Production Machines and Systems
D. T. Pham, E. E. Eldukhri and A. J. Soroka (eds)

A Roaming Vehicle for Entity Relocation

S. Kumar and G. C. Onwubolu

Department of Engineering, University of the South Pacific, Suva. Fiji Islands

Abstract

An automated guided vehicle (AGV) named ROVER, developed at the University of the South Pacific is described in this research paper. The vehicle is used to transfer material from a workstation to another in small and medium manufacturing enterprises. To achieve this, the AGV follows white markings on the factory floor. A microcontroller is used to control this direct current motor driven system. Sensors mounted on the AGV give information about its immediate surroundings. An outline of its control algorithm is provided. This paper also communicates results of experiments on braking distances, current consumption and velocity drop with incremental loading. The paper concludes with discussion on results that summarize the overall performance of ROVER.

1.0 Introduction

The advances in robotics can offer significant contribution to the enhancement of the capabilities of automated guided vehicles (AGVs). While robotics implement algorithms on mobile platforms that operate in strictly controlled and undisturbed environments, AGVs are built to operate in manufacturing environments that are noisy and constrained. An AGV shares its operating environment with factory workers and other expensive machinery. Nevertheless, AGVs can reap direct benefits in the form of more powerful and capable sensors and sophisticated algorithms developed from research in robotics.

Various researchers have made significant contributions in control systems, sensors, artificial intelligence, obstacle avoidance and navigational paradigms of robotic systems. These include the intelligent mobile platform [1], the mobile post system [2] and Omnimate [3].

This paper first discusses the mobile platform design of the AGV named ROVER – Roaming Vehicle for Entity Relocation [4]. It then presents the architecture of the control electronics and the control algorithm. Results from experiments dealing with station stopping distances and obstacle detection response distances are given along with results on its current consumption and velocity drop with incrementing load.

2.0 Platform design

The basic structure of ROVER is shown in Figure 1. The length, width and height are all 70 cm. Simple AGVs move backward, forward and are able to turn either side. Steering systems such as the differential steering and Ackerman steering can fulfill these demands. Ackerman steering system normally has two driving wheels at the rear or front and a single or a pair of other wheels dedicated to steering. The differential steering system utilizes two independently controlled driving wheels with other free wheeling wheels added for balance and support. Steering is achieved by independent control of the motors that drive the wheels. Differential steering offers the

Figure 1 Picture of ROVER

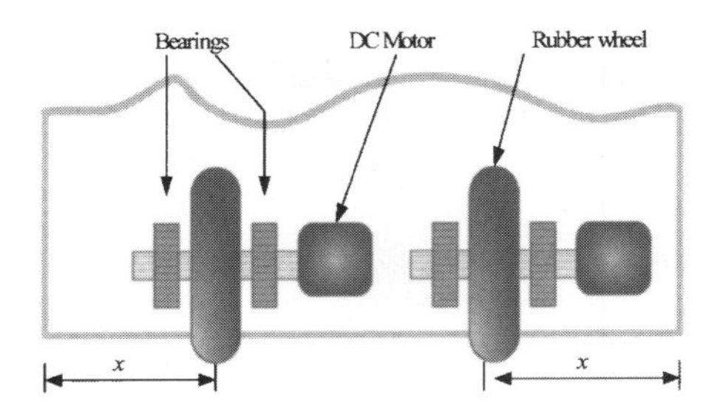

Figure 2 Positioning of the motors

advantage of rotation about the center of the mobile platform thus requiring significantly lesser space.

Hence, the differential steering system was implemented on this AGV using two independently controlled direct current (DC) motors with in-built gearboxes. These motors are normally used in automotive windshield wipers. Rubber wheels, 0.2 m in diameter, are attached to the motors via a steel shaft. Each wheel is pinned to the shaft while the internally threaded shafts are fastened to the armature of the motor. The shaft and the wheels are one unit and can be detached from the motor if need arises. Roller bearings on either side of the wheel support the shaft. Such a setup distributes the load directly to the chassis of the AGV instead of the armature of the motor. The motor therefore only provides tangential force to the shaft for movement of the wheels. Fabricated steel brackets house the bearings in position. Wheels have been placed along the central axis of the structure of the AGV.

Lead gel batteries with 12 volts and 4 Amp hour (4Ah) ratings are used to power the motors. With normal current draw, a single battery can supply power for approximately 3 hours.

Common in all differential drive systems is the addition of free wheeling castors for balance [5]. Four plastic castors have thus been mounted on each corner of the square structure. Fasteners link the castors to the chassis and that makes their height adjustable to a certain degree. Aluminum plates provide additional support to the castors. The wheels of the free wheeling castors have a 50mm diameter.

Positioning of the windshield wiper motors causes imbalance to the structure. These motors do not rotate at the same speed in both directions. Hence, they have been positioned such that they both operate at the same speed as shown by Figure 2. Weight is therefore concentrated towards the right of the AGV. In this setup, the battery compartment is placed towards the left. This reduces the intensity of the imbalance. The distance of both the wheels from the edge of the framework is the same as denoted by x in Figure 2. By ensuring that x is same on both sides, the center of the AGV also becomes the center of rotation.

A simple conveyor belt is used for loading and unloading entity on the AGV. The belt has a width of 45cm and extends over the width of the AGV. Being bi-directional, the unit picks-up and also drops-off entity from both sides of the AGV. A roller connected to an automotive wiper motor via a belt, drives the conveyor. This roller is used to adjust the tension of the conveyor belt. Smaller rollers on the either edges act as pulley while an aluminum sheet in the middle supports the belt. The belt slides over this sheet enabling the belt to handle larger weights.

3.0 Control circuit architecture

Figure 3 exhibits the operational framework of the electronics of the AGV. A common power supply unit meets the power demand for each component and therefore is the most central unit. The second most central unit is the PIC16F877 microcontroller. It is the hub of control for this AGV and transforms various data received from the sensors to appropriate output signals for the actuators. For intelligent interaction with its operating environment, the following sensors are utilized by the AGV.

1. Three Lynxmotion line tracing sensors mounted on an independent rolling unit attached to the mobile platform. These sensors track white lines on the black floor surface.
2. Eight infra-red (IR) sensors mounted on the vehicle to detect obstacles and provide information to perform obstacle avoidance.
3. Two IR sensors mounted on the side of the vehicle are used to detect stations where the AGV will stop during the line tracing process. It also determines the amount by which the AGV will turn to avoid an obstacle.

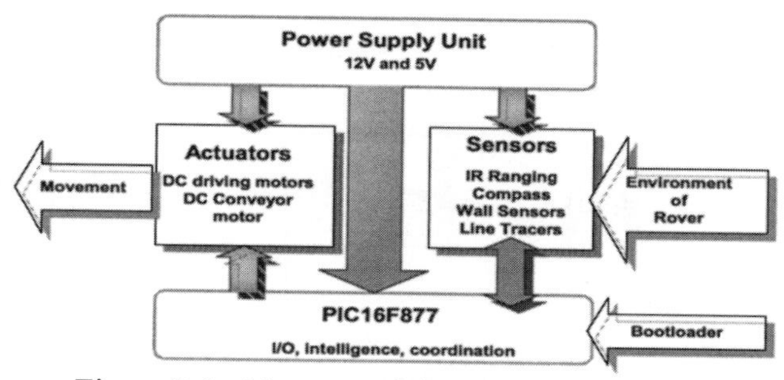

Figure 3 Architecture of the control circuit

4.0 Control algorithm

A PIC16F877 microcontroller coordinates the information from the sensors and initiates corresponding behaviors. These include getting data from line tracers and obstacle detecting sensors and appropriately determining the corresponding responsive actions of the motors.

The tasks that the AGV has to perform is vertically decomposed and is shown by Figure 4. Vertical decomposition deals with layering of intelligence. In the top most layer, the AGV exists with primitive survival behaviors such as movements in either direction using line tracing. New layers then evolve that use the upper layers. This includes navigation between nodes and stations and obstacle avoidance. Each task is mapped based on the primitives of sense-act that is unique to the reactive paradigm.

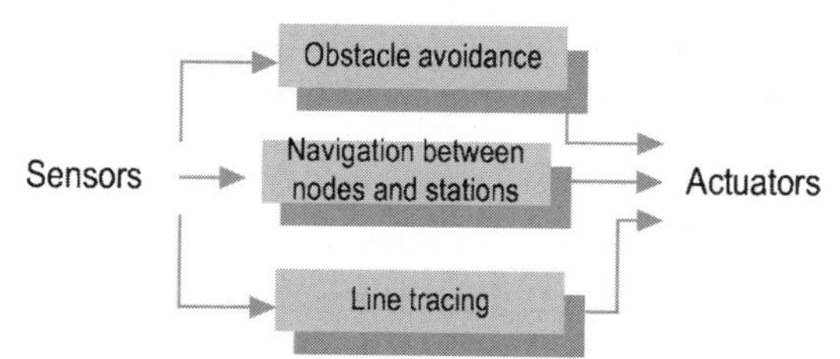

Figure 4 Vertical decomposition of tasks into Sense-Act organization.

It is important in vertical decomposition that each layer has access to sensors and actuators independently [6]. Moreover, if any of the new and advanced layers fail, the lower layer must still function. From Figure 4 it is clear that every layer has independent access to the sensors since each uses different sensors. While the actuators are common, the sensory inputs are designed as such that only one event will occur at a time. Also, if obstacle avoidance fails, the AGV will still be able to carry on with navigation between stations and nodes as long as there is no obstacle. In the case where the navigation layer fails, line tracing will still continue. The navigation layer is dependent on line tracing while obstacle avoidance is dependent on both.

Such a decomposition of tasks makes the execution more rapid. In addition, more behaviors can be added to make the AGV more intelligent. Since there is no form of memory mapping, all responses are analogous to biologically pure stimulus response reflexes [6]. Stimulus responses are programmed by behavior. For example if an obstacle is seen to the left the guidepath, the AGV without any planning will execute a stimulus response turning itself to the right. Programming by behavior makes the program codes modular and thus easy to test.

Finite state automata (FSA) is used to formally represent how the sequence of behaviors unfold. [6] describes FSA as a set of methodologies that indicates the output of a program at any instance. A state diagram, shown by Figure 5 presents the FSA for this AGV.

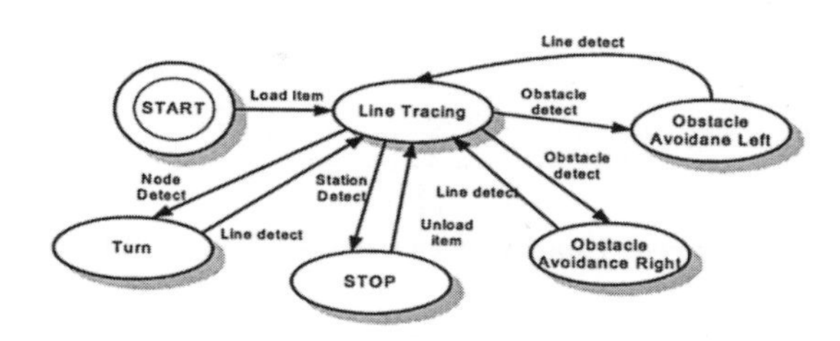

Figure 5 State transition diagram for AGV

As seen from Figure 5, the AGV can be in five states. These are line tracing, obstacle avoidance to the right, obstacle avoidance to the left , turns and stop. These five states are linked to each other by a set of stimulus. Arrows represent stimuli and a new behavior is triggered through it. From this state diagram, it is also noticed that the new state is always dependent on the current state. Also, when the AGV is in different states, sensory responses that are not relevant are ignored. For example while the AGV is in the obstacle avoidance to the right state, information from the third line-tracing sensor is ignored. When the AGV starts, it begins with line tracing. It then is able to detect obstacles, nodes and stations. The state diagram does not contain a final state of the AGV as it varies from one application to another.

5.0 Experimentation Results

Experiments were carried out to understand the basic behavior of the AGV. These experiments were carried out in an open space using the facilities at the University of the South Pacific.

5.1 Load versus time of travel

Load was gradually added to the AGV and the mean time taken to complete a 2.4 m path was recorded. The path was completed using line tracing. Figure 6 shows the results. It is noticed that the time to travel increases from 6 s to approximately 8 s with an increased load of 25 kg from no load (0 kg).

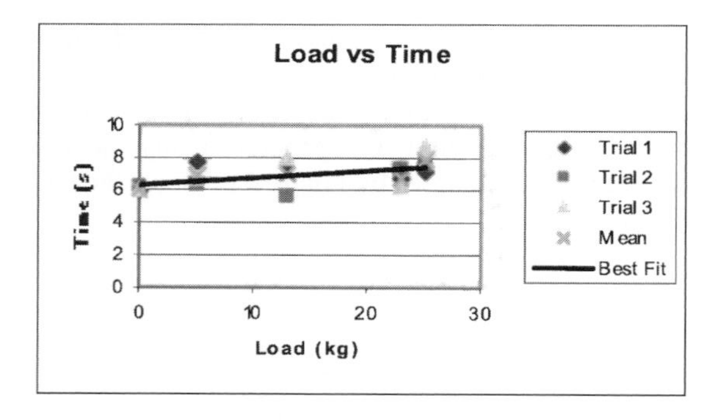

Figure 6 Load versus time for line tracing

This relates to a decrease in speed by 2 m/s with a 25 kg increase in load. It can therefore be inferred that there is an approximate decrease in speed of 0.08 m/s per a kilogram of added load.

5.2 Load versus current

The load versus current plot is shown by Figure 7. Increased loading affects the current consumption of DC motors. In this experiment, the average current draw was monitored with increased loading while the AGV was moving. There is a gradual increase of 1.5 A for a 25 kg increase in load. The difference is small courtesy of bearings mounted on either side of the wheels. The bearings absorb much of the load directed to the armature of the motor. This is noticed by the minimal change in current consumption of the two motors. A saturation in current draw is approached around 25 kg.

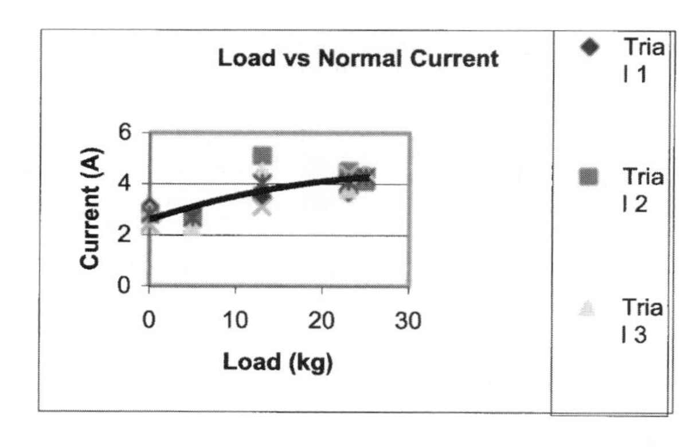

Figure 7 Load versus normal current

5.3 Station stopping response distances

Another experiment was carried out to observe the response distance of the AGV to stop upon detecting a station. The response distance consists of the extra distance the AGV travels once the sensors have detected the station. This response distance is caused by the time it takes the sensor to transfer information to the PIC16F877 and the time it takes the PIC16F877 to process the information and instigate a reactive response. The results of the experiment provide details on the distance over which the AGV can stop after a station is indicated.

Distances of the station were varied to notice if the stopping distance is affected by acceleration. The station was a white box with a width of 30cm. Table 1 presents the results which are an average of 4 trials.

Table 1 Distance to station and corresponding response distances

Distance to station (m)	Mean Response Distance (m)
0	0
0.5	0.19
1	0.24
1.5	0.23

From Table 1, it is noticed that the acceleration does affect the stopping distance. The AGV stopped within 0.2m when the station was 0.5m away and this value increased and remained almost the same for 1.5m and 2m.

6.0 Conclusion

Details of an AGV developed in-house at the University of the South Pacific have been presented. The AGV is capable of transporting load from one station to another using line tracing. Experimental results show key behavioral characteristics of the AGV with increased loading. It is noticed that current consumption approaches an asymptotic reference level with loading while the average speed of the AGV drops with loading. An asymptotic reference stopping distance is also approached when the station is more than 1 m away from the AGV start point.

Further work on the AGV will focus on minimizing the effect of loading on speed and thereby increasing the maximum permissible load the AGV can carry to more than 25 kg. It may be required later to increase the computing resources of the AGV so that much more capable control algorithms that can vary speeds and monitor position may be implemented.

Acknowledgements

The project was funded by the School of Pure and Applied Sciences research funds and the central pool funds of the University of the South Pacific.

References

[1] Crowley, J, 1984. Navigation for an intelligent mobile robot, *Technical Report, The Robotics Institute*, Carnegie Mellon University.

[2] Rafflin, C, 1996. Autonomous navigation of a post distributing mobile robot, *Advances in robotics : the ERNET perspective : proceedings of the Research Workshop of European Robotics Network, Darmstadt, Germany*, pp 137-144

[3] Borenstein, J, 2000. The omnimate: a guidewire free AGV for highly flexible automation, *Japan-USA Symposium on Flexible Automation*, University of Michigan, U.S.A.

[4] Kumar, S. 2003. Development of an automatic guided vehicle with the obstacle avoidance system, *Master of Science thesis,* Department of Engineering, University of the South Pacific.

[5] Borenstein, J and Koren, Y. 1989, Real time obstacle avoidance for fast mobile robots, *IEEE Transactions on Systems, Man and Cybernetics,* Vol 19, No. 5 , pp 1179-1187

[6] Murphy, R, *2000.* Introduction to AI Robotics, The MIT Press, U.S.A

Intelligent Production Machines and Systems
D. T. Pham, E. E. Eldukhri and A. J. Soroka (eds)

A system for automatic end-quality assessment of vacuum cleaner motors

D. Tinta[a], J. Petrovčič[a], B. Musizza[a], J. Tavčar[b], G. Dolanc[a], J. Koblar[b], Đ. Juričić[a]

[a] *Department of Systems and Control*
Jožef Stefan Institute, Jamova 39, 1000 Ljubljana, Slovenia
[b] *Domel Ltd., Otoki 21, 4228 Železniki, Slovenia*

Abstract

In this paper a diagnostic system for quality end-test of vacuum cleaner motors is presented. The system relies on innovative mechatronic solutions, which combine custom designed handling of units under test, laser vibrometry, electrical measurements and advanced signal processing. The result of processing of the measured signals are the so called features which serve to detect and localize even the most tiny faults either in electrical or mechanical part of the motor. Thus the accurate, reliable and sensitive diagnostic procedures allow for entirely fault-free final products.

---

1. Introduction

Continually increasing competition on the market of small electrical motors is pressing the manufacturers to decrease the production costs while increasing the quality of the final products. Domel Ltd Železniki ranks among the leading manufacturers of vacuum cleaner motors in Europe. As a way to cope with market competitors, particularly those from the far eastern countries that are invading the market by low prices, the major priorities of the company are improving quality and raising the efficiency of the production through lowering the production costs. A step towards this goal concerns the improvement of the process of quality assurance and, as part of that, fully automated end-quality assessment of the products. Hence maximal quality and almost 100% fault free final products are guaranteed.

There are several manufacturers in the market that offer automatic test stands for end-quality assessment of vacuum cleaner motors (e.g. Vogelsang & Benning [1], Schenck [2] and Artesis [3]). These systems, however, do fit only partly to the requirements set up by the company Domel. For example, the system of Vogelsang & Benning [1], as well as Schenk, allows the measurements of vibrations only at one point. According to the quality standards of the company as well as requirements set up by various customers vibration measurements are requested to be carried out at three different points on the motor's body. Therefore, the aforementioned systems, if installed, would only partly solve the problem of full quality assessment, thus additional (manual) measurements would still be needed.

The device to be tested is shown in Figure 1.

Fig. 1. Vacuum cleaner motor.

The solution presented below tends to provide fully automatic quality assessment tests which comprise

- determination of the key quality parameters of a vacuum cleaner motor,
- identification of any parameter of the motor violating the quality thresholds,
- fault localization in case that a motor of improper quality is identified.

These features lead to reduction of costs due to additional manual checkings.

The main purpose of the paper is to present a system for automatic end-quality assessment of vacuum cleaner motors at the end of the manufacturing line. The system has been in the regular use for half a year.

The paper is organized as follows. In the next section the structure of the diagnostic system is described first. After that the diagnostic cells are explained in more detail. In the fourth section the performance of the system is presented and, finally, concluding remarks are stressed.

2. System architecture

The system (c.f. Fig. 2) is composed of five major parts:

- 3 measurement and diagnostic cells,
- control unit for mechanical handling of the motors,
- data acquisition system.

The system operation relies on direct measurements of nine electrical and mechanical quantities i.e. power supply voltage, current, air pressure difference on the motor, revolution speed, electric power, vibrations and air temperature and humidity (needed for data conditioning) This set of measurements will be soon complemented by sound measurement.

Fig. 2. View at the end-quality assessment system installed on the assembly line.

The key assessment tasks are performed within three measurement modules running in parallel. Parallelism was needed in this case for two reasons:

a) relatively high number of different tests that need to be performed and
b) to comply with the 9 s production cycle.

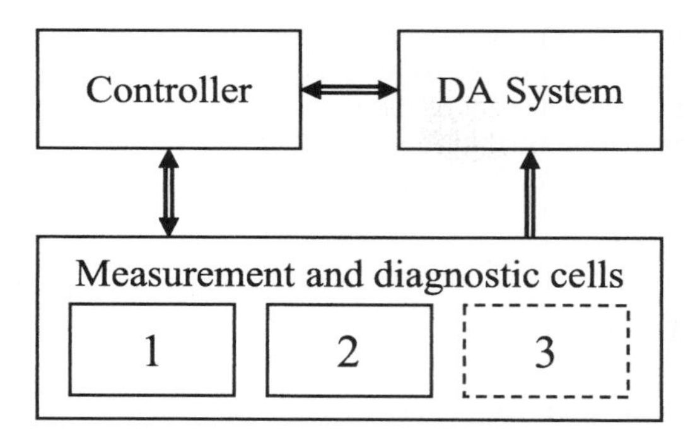

Fig. 3. Architecture of the diagnostic system.

The control unit is built around the Mitsubishi MELSEC-Q PLC which takes care of mechanical handling and tasks synchronization. The cycle of the diagnostic system conforms to the cycle of other parts of the assembly line. Based on information from position sensors located on the line and pneumatic actuators, the control system generates signals for positioning, start-up and shut-down of the motors during each test. In addition, it takes care of transporting motors through the diagnostic system.

The data acquisition system is based upon the National Instrument NI 6220 and NI 6221 modules.

The acquired data are first pre-processed and then delivered to the feature extraction algorithms which calculate the vector of features. Each feature (reflecting a particular aspect of quality) is checked in order to verify whether the device sticks to the quality requirements. If this is not so, detection and localization of tentative defects are performed. The entire application software is realized in LabVIEW.

Fig. 4. Control panel.

The data acquisition system is connected to the controller unit via RS 232 communication line.

3. Diagnostic modules

The first diagnostic module is aimed for testing the characteristic of the motors and the quality of commutation. The motor characteristic is defined by the following data: electrical current, electrical power, pressure difference and revolution speed. As it is very important that those data are obtained under nominal supply voltage, the measurement conditions are fully supervised in order to compensate for possible fluctuations. Based on sampled actual voltage as well as air temperature and humidity, the system is able to provide corrected values of the characteristic parameters i.e. normalized to the nominal supply voltage (230V) and standard atmospheric conditions. The quality of commutation is verified by calculating the root-mean-square (RMS) values of the current signal in 12 consecutive frequency bands each, being 2.5 kHz wide. The current signal is sampled with 60 kHz and Fourier spectrum is calculated from the samples (c.f. Fig. 5).

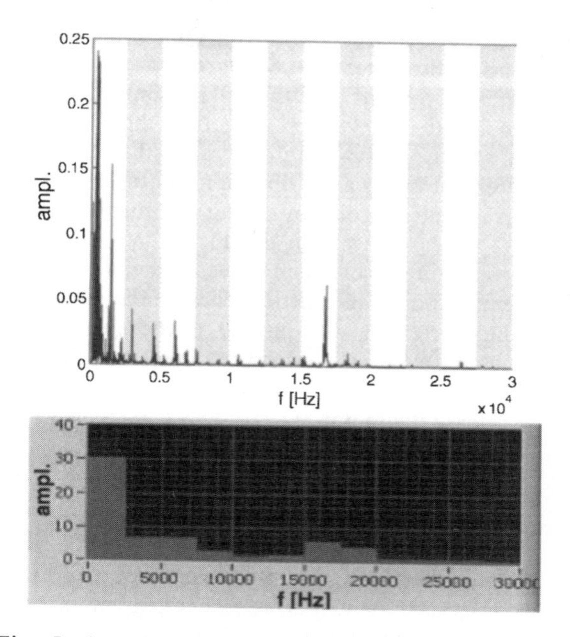

Fig. 5. An example of current frequency spectrum (above) and corresponding features i.e. RMS values of individual bands (below).

The second measurement and diagnostic cell shown in Figure 6 serves for vibration measurements. According to the customer's requirements the vibration must be measured at three points: on the cover in axial direction, on the cover in radial direction and on the housing of the vacuum cleaner motor in radial direction.

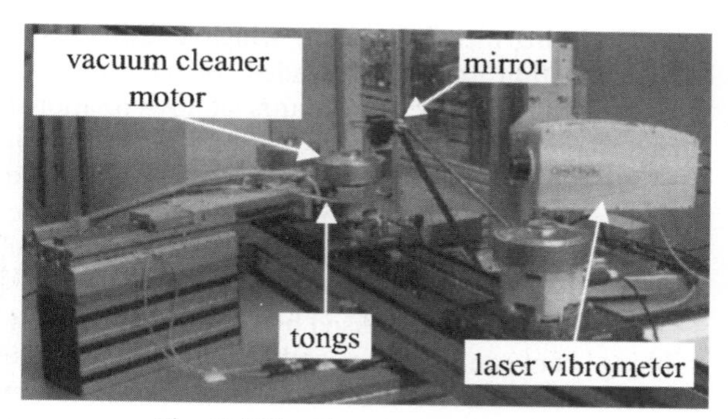

Fig. 6. Vibration measurement.

The vibration measurement is carried out as follows. First, a specially designed tongs grip the motor under test (see Fig. 7). In that way the motor is vibrationally isolated from the environment (i.e. the transporting pallet) and so the vibration originating from the assembly line cannot disturb the vibration measurement on the very motor. Vibrations are acquired by the laser vibrometer Ometron VQ-500-D. The vibrometer is fixed on a positioning mechanism, which allows for movement of the vibrometer in the vertical direction thus allowing measurement of vibration in the radial direction. The measurement in the axial direction (on top of the fan cover) is carried out by using a mirror, which redirect the laser beam in the appropriate direction. The positioning mechanism moves the vibrometer to another two points in order to measure the vibrations in radial direction, i.e. on the cover and on the housing. As soon as the measurements are finished the vibrometer returns to the initial position and the tongs release the motor.

Fig. 7. A motor gripped by tongs.

Vibrations of the vacuum cleaner motor are evaluated by RMS values calculated across 15 frequency bands, each 1 kHz wide, starting from 0 up to 15 kHz. For that purpose the vibration signal is sampled by 60 kHz. The RMS values of individual bands are calculated from the frequency spectrum of the vibrations signal. (Fig. 8).

The third measuring cell, which is still in the implementation phase, is intended for measuring of sound of vacuum cleaner motors at low rotational speed (approximately 30 revolutions per minute). Due to low sound intensity at this revolutions the measurements will be carried out in an anechoic chamber, in order to cut down the influence of disturbing noise from environment.

Mechanical faults caused by bearing faults or by rubbing between rotating and static parts of the motor become visible at low rotational speed. In that case typical bursts could be observed in the acoustic signal. The frequency of bursts depends on the geometry of bearing as well as on the rotational speed of the motor in the case of bearing fault. In case of rubbing it depends only on the rotational frequency of the motor [7]. The computational procedure starts with calculation of the envelope of sound signal by using Hilbert transform. Next, the frequency spectrum of the envelope is calculated [8]. Presence of certain fault is reflected in RMS value of the envelope in corresponding frequency bands, which are typical for individual fault. The procedure is illustrated in Figure 9.

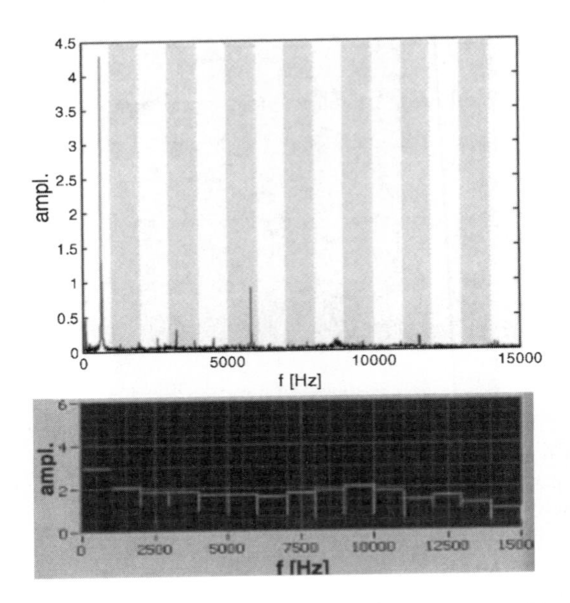

Fig. 8. Example of vibration frequency spectrum (above) and corresponding features i.e. RMS values of individual bands (below).

4. Diagnostic system performance

Each vacuum cleaner motor in the production line is associated a label. First, the code of the transporting pallet is read at the entrance of the diagnostic system. Then, the diagnostic results of the particular vacuum cleaner motor are being saved in computer under this code. In that way the tracking of the motor through individual measuring cells of the diagnostic system is ensured. At the exit of the system the pallet code is read again and is compared with that in the computer. If the codes are equal, it means that the testing was performed successfully. In the opposite case the adequate alarm is triggered. In the last operation of the diagnostic system the condition of the tested vacuum cleaner motor is recorded in the chip on the pallet. The record tells us if the motor is fault-free or the adequate code of the eventual fault.

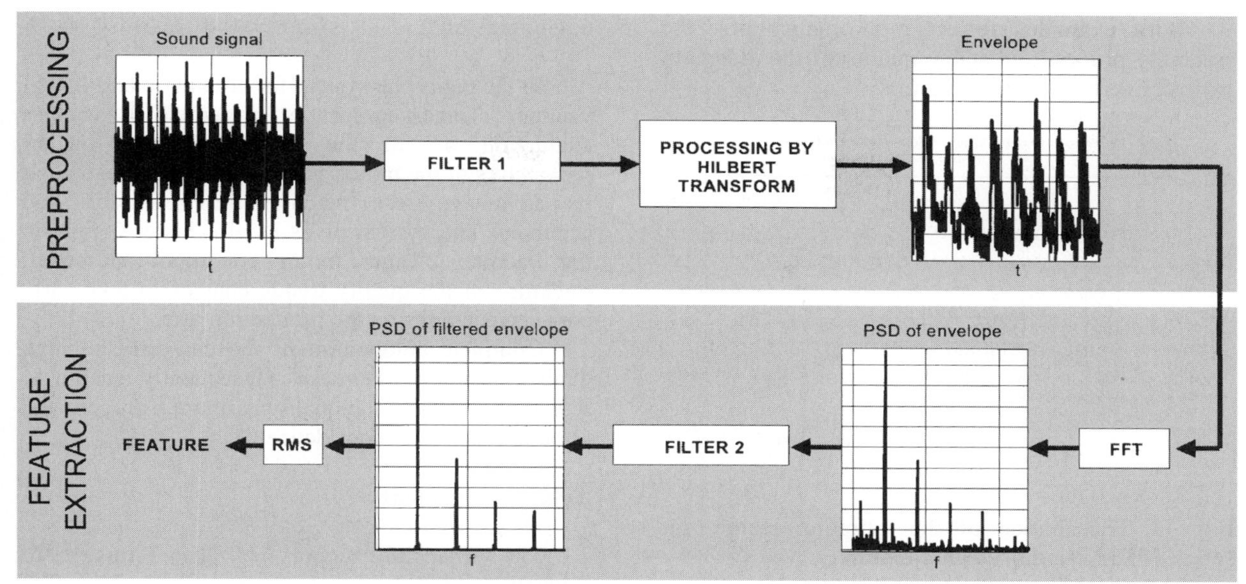

Fig. 9. Processing of sound signal.

The user interface of the system (Fig. 10) is used to display all the essential diagnostic results of the tested vacuum cleaner motors. Additionally it enables to set the thresholds of the features in both measuring cells, calculation of some statistical parameters of measured values and review of past testing results.

Fig. 10. User interface.

Figure 11. shows a distribution of RMS values of current signals taken in the frequency band from 0 to 2.5 kHz from a series of vacuum cleaner motors. It can be seen that the commutation quality of almost all of the motors falls within the required band (forbidden band is displayed in grey). Only in few cases the feature exceeds the threshold (shaded area in the graph).

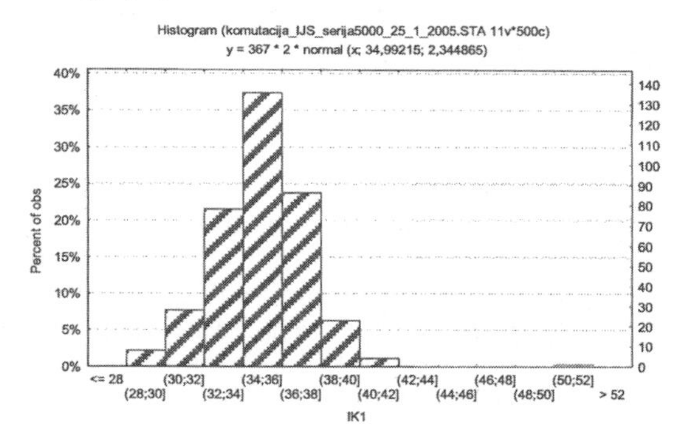

Fig. 11. Distribution of the commutation quality feature for series of 367 vacuum cleaner motors.

Figure 12 shows the distribution of vibration feature i.e. RMS values of vibration signals in frequency band around the rotational frequency of the vacuum cleaner motors (600-800 Hz) for a series of 368 motors. In this figure the vibrations on the housing in radial direction are presented. The vibrations of most of the motors are below the threshold (10 mm/s), just in some cases the threshold is exceeded (shaded area in the graph). It should also be noted that the mean value of the vibration features is quite below the threshold.

Both examples reflect high quality of the assembly process and stable quality of the assembly parts delivered by subcontractors.

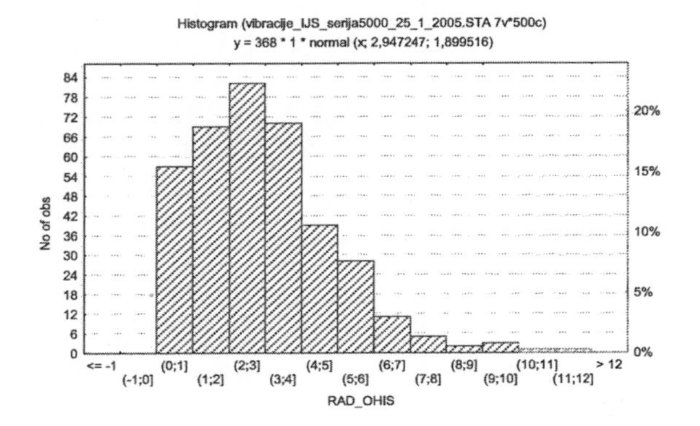

Fig. 12. Distribution of the vibration feature for series of 368 vacuum cleaner motors.

5. Diagnostic system evaluation

The performance of the diagnostic system bears a strong analogy with the performance of classical instrumentation. Indeed, the former can be viewed merely as a virtual instrument, which "senses" the motor condition. Therefore it could be evaluated in the same way like classical measuring instruments, i.e. according to the reference value from the ethalon device. In this case however, the ethalon device, which would give the reference value for motor condition, does not exist. The only available "reference" is operators' judgement. Unfortunately, the operators' judgements happen to be non-uniform in some "grey" cases in which a defect is not particularly obvious. The reason for this is operators' subjectiveness and the limitations of human perception. Therefore the diagnostic system was tuned and evaluated in an iterative procedure during the test production of the motors. In this phase the feature thresholds were set in deference to several aspects:

- full compliance with customers' requirements,
- consensus of opinion between operators responsible for quality control,
- past experience on quality assessment with similar motors,
- available statistics obtained during the test production.

6. Conclusion

In the paper the system for automatic end-test of vacuum cleaner motors is presented. It enables evaluation of all the most important quality parameters of the motor. Compared to manual testing its advantages are much better reliability and accuracy. The system gives us also better insight in the vacuum cleaner motor condition. Moreover, tracking of the quality trends enables to take early corrective actions on the production line.

Complete automation of the end-test and high quality standard assurance consequently enable to reduce the costs of the quality assurance process and to raise customers' confidence.

Acknowledgements

The financial support of the Ministry of Education, Science and Sport of Republic of Slovenia as well as of Domel Ltd. is gratefully acknowledged. The authors also thank to the reviewers for their fruitful comments, which have contributed to the improvement of the paper.

References

[1] Vogelsang & Benning (2005). Routine test systems, URL: www.vogelsangbenning.de/intro.html.

[2] Schenck (2005). Schenck Motor Test Systems, URL: www.schenck-usa.com/prod_motor_test.html.

[3] Artesis (2005). Motor Quality Monitor, URL: www.artesis.com/mqm.asp.

[4] Tinta, D., J. Petrovčič, U. Benko, Đ. Juričić, A. Rakar, M. Žele, J. Tavčar, J. Rejec in A. Stefanovska (2005). Fault diagnosis of vacuum cleaner motors. *Control Engineering Practice*, **13**, 177-187.

[5] Edwards, S., A.W. Lees in M.I. Friswell (1998). Fault diagnosis of rotating machinery. *Shock and Vibration Digest*, **30**, 1, 4-13.

[6] Yang, D. M. in J. Penman (2000). Intelligent detection of induction motor bearing faults using current and vibration monitoring. *Proceedings of COMADEM 2000*, Houston, 461-470.

[7] Benko, U., J. Petrovčič, Đ. Juričić, J. Tavčar in J. Rejec (2005). An approach to fault diagnosis of vacuum cleaner motors based on sound analysis. *Mech. syst. signal process*, **19**, 427-445.

[8] Randall, R.B. (2002). State of the art in monitoring rotating machinery. *Proceedings of ISMA 2002*, **4**, 1457-1477.

Intelligent Production Machines and Systems
D. T. Pham, E. E. Eldukhri and A. J. Soroka (eds)

Agent technology used for monitoring of automotive production

O. Sauer[a]

[a] *Department Production monitoring & control, Fraunhofer IITB, Karlsruhe, Germany*

Abstract

Today's automotive plants are equipped with heterogeneous software systems for different types of tasks, both factory planning and manufacturing operations. IT systems used for factory planning are summarized as 'digital factory tools'. On the operating level, software systems are not yet integrated and thus support separate tasks such as production order control, production monitoring, sequence planning, vehicle identification, quality management, maintenance management, material control and others.

A very promising technology for integrating existing software systems and their functionalities and to add assistant systems for the shop floor staff is to be found in software agents. In this paper the author describes the current use of software agents for today's production tasks and his vision for applying agent technology in future automotive production.

1. Current situation

1.1. Software support in automotive plants

Owing to the fact that the work load taken on by OEMs and suppliers has shifted to the benefit of the suppliers, OEMs have to focus increasingly on monitoring the production processes, on logistics and supply chain management, including the shop floor level. Today, shop floor staff use production monitoring tools that support manufacturing decisions which are merely based on production quantities. In the case of a facility breakdown or quality inspection results they only know that a certain number of vehicles is affected. However, they can neither identify the customer orders related to these vehicles nor their options, e.g. color, right or left hand drive, sun roof, etc. It would be a big step forward towards better and more transparent decisions if the shop floor people could base their decisions on identified vehicles/customer orders rather than on undefined production quantities. Production monitoring systems play a relevant role in supporting manufacturing operations; and, what is more, they can be seen as the operational part of the digital factory [1].

This means that in the years to come the above-mentioned production monitoring, maintenance management and logistics control systems will have to be integrated (see Figure 1) so as to allow for better and more transparent decisions and to recognize the impacts of decisions taken on the shop floor on just-in-sequence parts to be provided to the line, for instance. Another driving force behind new software technologies is the increasing number of vehicle models having ever shorter life cycles. IT systems must be more flexible with regard to changes and adaptations to the new model requirements. Some OEMs have app. 1.000 different IT-systems supporting the various steps within the order management business process including the above mentioned production

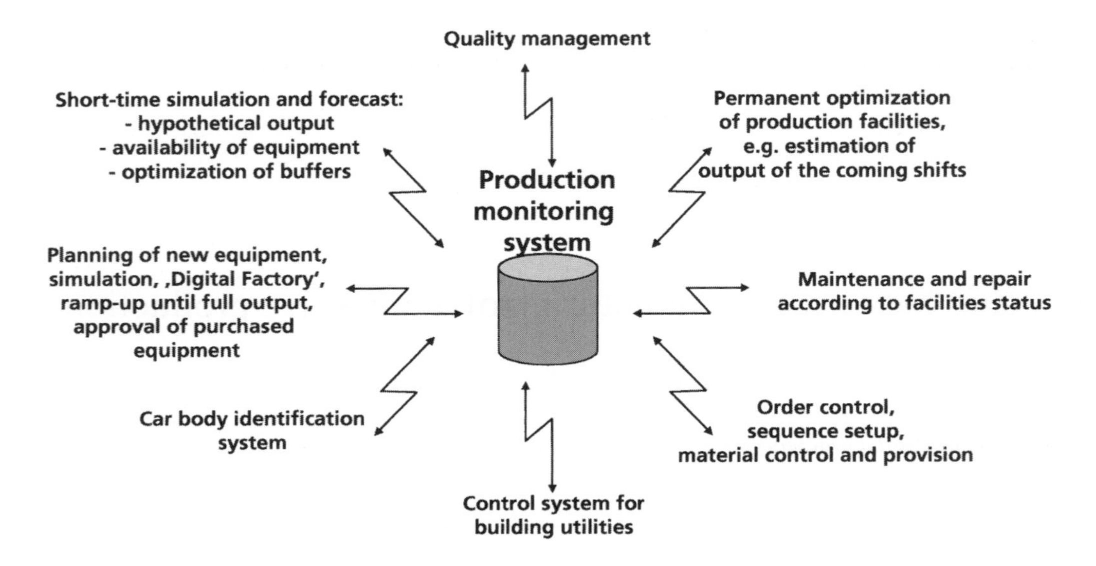

Figure 1 Examples of IT systems in today's automotive plants

related tasks.

Currently, a new generation of shop-floor related software systems is being developed - 'manufacturing execution systems' - to bridge the gap between today's stand-alone IT solutions. It is, however, evident that the software technologies applied today are not powerful enough to integrate existing software systems and deliver the required results, namely support of manufacturing decisions on the shop-floor level regarding all relevant aspects of production equipment, quality issues, provided parts and shift output. The integration aspect is in most cases supported by databases (Figure 2). It must be added that a couple of the above-mentioned systems, particularly those for production monitoring, require real-time data processing instead of database solutions because they receive a large number of signals from the programmable logistics controllers (PLCs) of the

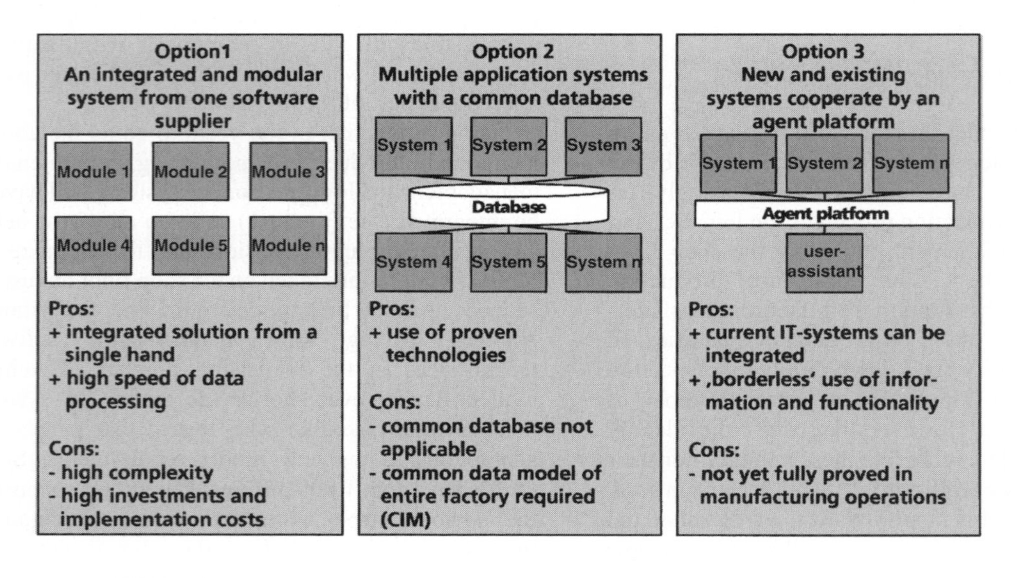

Figure 2: Possibilities for OEMs to integrate different software systems

production equipment. For real-time applications, database solutions are often not fast enough.

Another disadvantage of integration by means of a database is the required data model for all related applications. This data model is comparatively inflexible, particularly if changes or extension of functionalities are required or if new IT systems must be added. Additionally there is no software supplier (compared to commercial IT applications) who would be able to deliver all required modules for an automotive plant as a single-source supplier.

In the late 1990s a team of the Fraunhofer IPK developed a first concept for an Asian OEM allowing to integrate some of the above-mentioned IT systems, particularly those on the cell manufacturing level. Being ahead of their time, the Fraunhofer people designed a cell manufacturing system that combined production monitoring, quality control, maintenance / repair and tracking information for each shop - press, body, paint and trim shop. Applying the available software technology of that time, the plant in question went into operation in 1998.

Today it is evident that new software technologies have to be used to allow for a genuine integration of IT systems for production equipment, quality issues, provided parts and shift output and to preserve the existing software functionalities.

Let's assume it is useful to bring some of the existing operational IT systems of an automotive plant together - the user would be lost in view of the variety and abundance of functionalities. This means that a user-oriented assistant software that makes available the required functions and information for a specific working activity is essential for the users on the shop floor to take the required and expected decisions.

A more promising technology for integrating existing software systems and their functionalities and to add assistant systems for the shop floor staff is to be found in software agents.

1.2 State of the art for software agents in manufacturing environments

In the academic field, agent technologies have a long tradition, but their use in production and real-time applications has been very limited yet. Describing the state-of-the-art for agent-based systems we concentrate in the following on practical applications especially in automation applications.

The vision of a flexible and adaptive production is comprehensively elaborated in the vision paper ‚MANUFUTURE' published by the European Commission [2]. The application of Software Agents, however, especially for production monitoring tasks, has still not been described. Special issues, such as manufacturing order control, are very much focused on prototypical applications for single use, e.g. milling of power train parts. In this case agents representing work pieces, material flow devices and milling machines

One of the most prominent applications is described by authors from DaimlerChrysler [3] who have implemented a manufacturing cell for cylinder heads and other engine parts that is completely controlled by a software-agent-based system. The manufacturing cell works in serial production and has produced excellent results concerning capacity utilization and smoothed production. The control logic developed in this project by the PLC-supplier [4] is offered for further industrial applications. Most authors refer to this example to prove agent-technology being applicable on the shop floor.

Further work concerning decentralized manufacturing control with software agents was conducted in the PABADIS project [5]. Bioanalog behavior of software agents is currently studied in the CABDYN project [6]. A plug and produce approach was designed within a Plug-and-produce project funded by the German Federal Ministry of Education and Research [7]. Other aspects of software agents are described in I*PROMS [8], in Sonderforschungsbereich 627 [9] und a project conducted by the Institute for Process Control and Robotics (IPR), University of Karlsruhe, ESPRIT Projekt DIAMOND [10]. Finally the PAPAS project aims at transferring technologies for modelling, simulation and optimization to suppliers of production facilities, based on a plug-and-play concept for modular robots [11].

Hertzberg et.al describe results of the AgenTec project which have been applied to a prototype of a commissioning system to illustrate and test the integration of heterogeneous software and control systems [12]. Further examples for potential application in automation technologies are summarized by Urbano et.al. [13]. However, an industrial application is still missing.

2. Research and application development steps

2.1 Closed control loops and connection of monitoring and order management

The main ideas behind the closed loops as shown in Figure 3 are the following: there is in fact a connection between the concept of a factory and its control and monitoring during actual operation. The control philosophy, e.g. PUSH or PULL, has essential impacts on a factory's design and therefore on its profitability and flexibility. As a consequence the demands for future IT support of the enterprise are determined in a factory's initial conceptual phase. The tools of the 'digital factory' must in future integrate these linkages [14].

In detail the following three issues have to be tackled:

maintenance plans based on real and permanently updated statistical data of the production systems behaviour ('maintenance-on-demand').

3. The most interesting and most difficult point is to use production monitoring information for short term planning, e.g. daily sequence planning. To close this loop means to release manufacturing orders taking into account the actual status of the running production, e.g. actual and unexpected

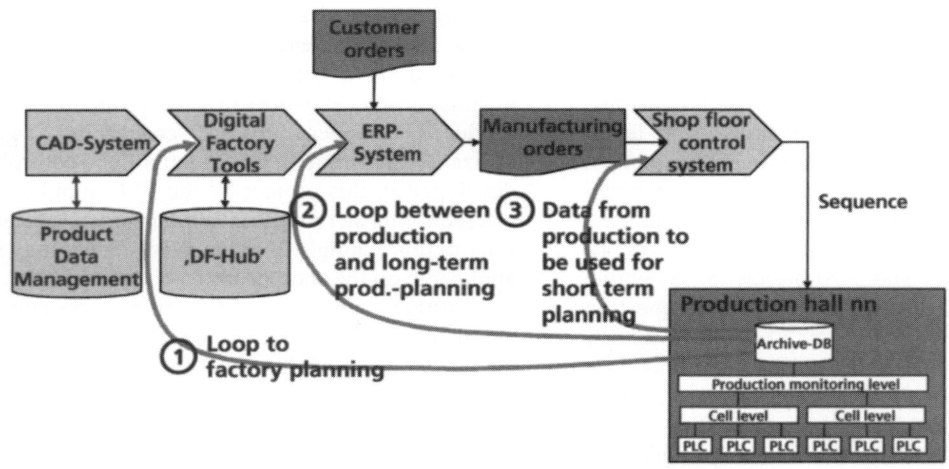

Figure 3: Establishing closed loops between production monitoring and control

downtimes of production equipment. Such a losed loop would also lead to the fact that OEM's could shorten their 'frozen zone' period. As a prerequisite the production monitoring system is combined with the car body identifying system, e.g. RFID-readers of 2D-bar code-readers.

1. The purpose of the 'digital factory', to begin with, is to minimize the investment risks with regard to the equipment (machines and installations) during the entire planning process of a factory. This results in new demands for planning tools, which must cope with imprecise and uncertain input data. The 'digital factory' can be compared to a factory-DMU (digital mock-up) with the difference that relevant simulators additionally allow predictions as to the dynamic behavior of the factory. Ideally, the input data used in the simulation are actual plant data recorded and processed by the production monitoring and control system. This feedback will lead to more reliable results of a planning process, because planning people can access real production data instead of estimated values.
2. Secondly, the information that is held in an archive database of a production monitoring tool might support the weekly or monthly production planning. The benefits are evident: not only will this result in more precise production plans, e.g. used as forecasts for suppliers, but also in better

2.2 Virtual Startup

When talking about manufacturing the buzzword 'digital factory' appears everywhere these days. DaimlerChrysler announced that they would in future not start any new work sites or production plants without first carrying out digital testing. An expert committee of the VDI (Society of German Engineers), in which the IITB is represented, recently defined the term as follows: "'Digital Factory' is a generic term for a comprehensive network of digital models and networks, which includes, among others, simulation and 3-D visualization. The purpose of the 'digital factory' is the overall planning, implementation, control, and continuous improvement of all factory processes and resources in connection with the product" [15]. In the future the 'Digital factory' will not only be used for factory planning. It is the vision to use the information stored in Digital factory databases to automatically configure and customize the production monitoring and control system's engineering part. It is already possible to automatically generate PLC programs from the virtual testing and commissioning (Figure 4).

3. Agent based production monitoring

In the field of production monitoring, maintenance management and cooperation with logistic-oriented IT systems, there still remains much room for manoeuvre for the application of software agent technology. To make it clear: today there doesn't exist any application of software agents in production monitoring and its real-time data processing - neither in the automotive industry nor in other branches. The model described in the following is therefore not just another agent based model but something completely different compared to today's production monitoring systems.

The development team at the Fraunhofer Institute for Information and Data Processing (Fraunhofer IITB) developed an agent-based production monitoring system and connected it to an assistant system which simulates the output of an automotive shop for the next three shifts on a short-term basis [16]. With the help of this assistant the shop floor personnel is able to foresee the impact of any kind of disturbance caused by unexpected changes in the production equipment, the buffers and the material flow systems, e.g. conveyors, skids, etc. This supports the manufacturing people intheir short-term decisions concerning shift capacities, increase or reduction in staff in the coming shifts, assignment of workers to the line, etc.

The agent-based production monitoring system is a new version of the Institute's existing system [17] in DaimlerChrysler's Bremen plant where it monitors and controls the shops for body, paint and assembly either from a central control room or from decentralized control panels on the shop floor. An overview over the system's architecture is given in Figure 5. The agent based system offers some new functionalities. Especially those that allow access and connection to other IT-systems, e.g. external data bases to control the sequence quality inside the monitoring system. The ProVis.Agent modules and those of related systems have been wrapped and enabled to communicate via a standard agent platform (JADE). Examples for these modules are

- the different input/output channels for the process data submitted from the PLCs,
- process variables such as production amount, status of line buffers, switches, etc., that are computed in the monitoring server,
- the interfaces between the monitoring server and its clients,
- archive functions that are conducted in a separate archive server and
- information clients with interfaces either to the above-mentioned assistant system or other services, e.g. radio/alarm server.

This architecture allows to cooperate and communicate with other existing IT systems and, what is more, enables their functions to be used by the

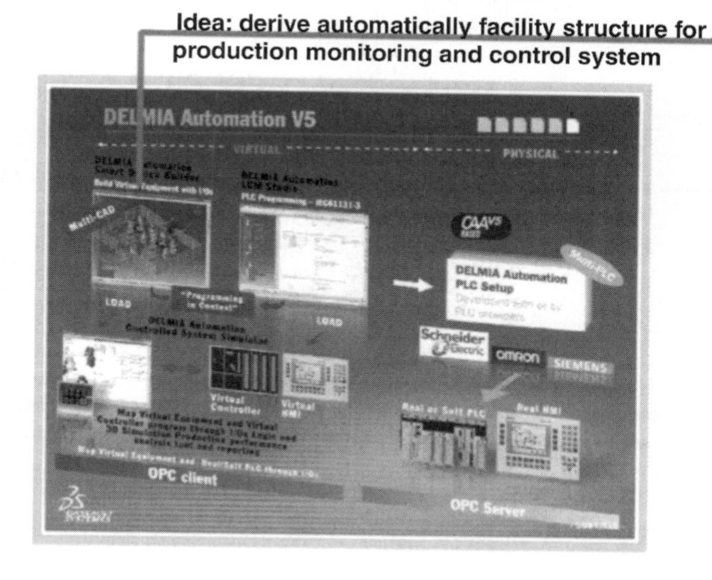

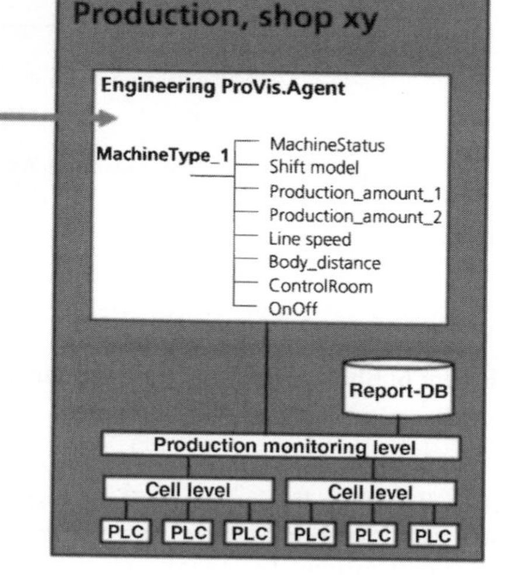

Figure 4: Virtual startup

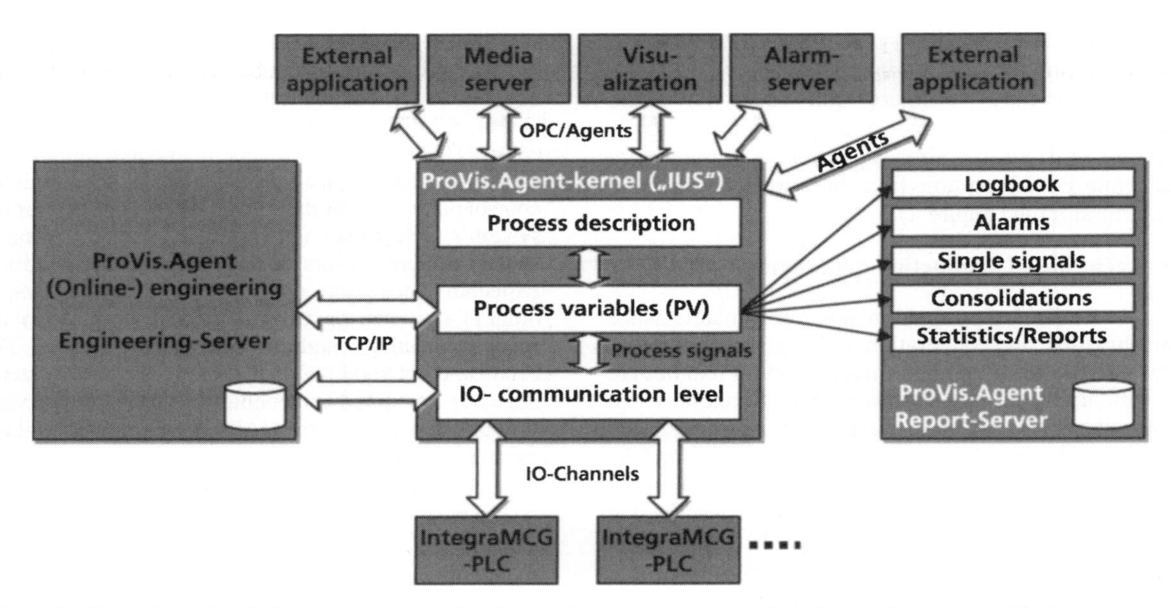

Figure 5: Short-time simulation system as production assistant tool connected to the production monitoring system

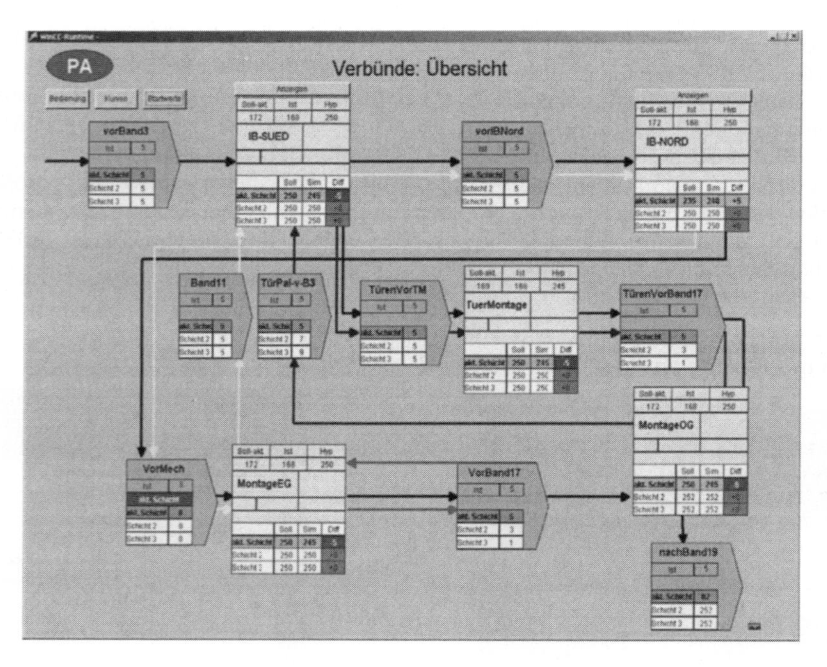

Figure 6: Short-time simulation system as production assistant tool connected to the production monitoring system

agents of the production monitoring system. Agent-based monitoring and control systems in automotive plants will allow to adopt 'plug and produce' procedures for more and more intelligent production systems in the coming years.

As one example of the above-mentioned assistant systems the IITB implemented a software solution that calculates for the 3 – 5 coming shifts the production output regarding all actual production breakdowns indicated by the production monitoring system as well as the buffer status calculated according to the current production output. In this case the interface shown in Figure 6 is WinCC. With the help of this assistant software, which is connected to the monitoring systems by a software agent, the shop floor people can react to failures on the shop floor level quickly and correctly. They are now able to adjust line speed and the work force for the coming shifts and to take efficient measures to keep the output as high as possible.

4. Conclusion

The proposed architecture for a production monitoring system which allows to monitor and control app. 300 PLC's per shop with real time requirements is a step forward to real time management promoted by [2]. Today systems like these can hardly be found in production environments because software agents still have to prove their stable real time operational availability. Secondly for a shared application like production monitoring, where process figures may be distributed to many clients all over the plant, it is impossible to install and maintain agent platforms on each client. With today's technologies this will result in
compromises between comprehensive use of agent technology and maintainability of IT-system. Next research steps are Ambient Intelligence architectures for production with decentralized intelligence.

References

[1] Sauer, O.: Einfluss der Digitalen Fabrik auf die Fabrikplanung, wt werkstattstechnik online 01/2004, pp. 31-34.

[2] European Commission: MANUFUTURE – a vision for 2020; Assuring the future of manufacturing in Europe; Luxembourg: Office for Official Publications of the European Communities. 2004 – 20 pp. ISBN 92-894-8322-9.

[3] Sundermeyer, K.; Bussmann, S.: Einführung der Agententechnologie in einem produzierenden Unternehmen – ein Erfahrungsbericht. Wirtschaftsinformatik, Vol. 43, No. 2, pp. 135-142, April 2001.

[4] Armando W. Colombo, A.W.; Neubert, R.; Schoop, R.: A Solution to Holonic Control Systems. Proceedings of the 8th IEEE International Conference on Emerging Technologies and Factory Automation, ETFA Oct. 18, 2001.

[5] EU Project PABADIS: Plant automation based on distributed systems. International IMS (Intelligent Manufacturing Systems) research project.

[6] CABDYN: Complex Agent based Dynamic Networks: Project Funding: Seed funding was provided as of 1st July 2003 for six months by the EPRSC under the Novel Computation Initiative (Grant No. GR/S63090/01).

[7] BMBF joint project ‚Plug and produce'; part of „research for tomorrow's prodcution' focussing on „flexible, temporary factories", funding number 02PD2001.

[8] Pham, D.T. et. al.: Innovative Production Machines and Systems (I*PROMS): A Network of Excellence for Innovative Production Machines and Systems. Proceedings of the 2004 2nd IEEE International Conference on Industrial Informatics, pp. 540-546.

[9] Research Area 3.4, project 3.4.1 smart factory im Sonderforschungsbereich 627: NEXUS – spatial world models for mobile context-aware applications, University of Stuttgart.

[10] DIAMOND: "Distributed Architecture for Monitoring and Diagnosis". EU Esprit Project No. 28735.

[11] Project ‚Plug and play Antriebs- und Steuerungskonzepte für die Produktion von morgen' (PAPAS). Joint project funded by German Federal Ministry of Education and Research

[12] Hertzberg, J. et al.: Agententechnologie zur Integration heterogener Systeme am Beispiel eines Prototypen für ein Kommissioniersystem. Proceedings zur SPS/IPC/DRIVES 2003, 25.-27. November 2003.

[13] Urbano, P.G.; Wagner, Th.; Göhrner, P.: Softwareagenten – Einführung und Überblick über eine alternative Art der Softwareentwicklung. Teil 2: Agentensysteme in der Automatisierungstechnik: Modellierung eines Anwendungsbeispiels. Automatisierungstechnische Praxis atp 45 (2003) Heft 11, pp. 57-65.

[14] Sauer, O.: Digital Factory – A comprehensive network of digital models and methods, visIT International 2004, pp. 12-13.

[15] Bracht, U.: Die Digitale Fabrik in der Automobil-Industrie – Vision und Realität. Proceedings der 5. Deutschen Fachkonferenz Fabrikplanung, 31.03.-01.04.2004. Landsberg: Verlag Moderne Industrie, 2004.

[16] Sutschet, G.: Störung im Griff. Ein Produktionsassistent für die Automobilfertigung. visIT 2 (2001), No. 2, pp. 6-7.

[17] Früchtenicht, H.W.; Sassenhof, A.: Werksweites Produktionswarten- und Anlageninformationssystem. In: annual report, Fraunhofer IITB, Karlsruhe : IITB, 2000, pp. 62-63.

Intelligent Production Machines and Systems
D. T. Pham, E. E. Eldukhri and A. J. Soroka (eds)

An Intelligent System for Manufacturing Fibre Optic Components

D.T. Pham[a], M. Castellani[b]

[a] *Manufacturing Engineering Centre, University of Cardiff, PO Box 688, Newport Road, Cardiff CF24 3TE, UK*
[b] *CENTRIA - Departamento Informática, Faculdade Ciências e Tecnologia, Universidade Nova Lisboa, Quinta da Torre, Caparica 2829-516, Portugal.*

Abstract

This paper presents an intelligent system for assembling optical fibre components. The system aligns the core of a fibre relative to a laser emitter to maximise the light output from the fibre. It searches for the location yielding the maximum light output by applying a gradient-based algorithm enhanced with momentum information. The parameters of the algorithm are adjusted on-line by a fuzzy controller according to the progress of the alignment operation. The control knowledge base is automatically generated using a new evolutionary algorithm. The fuzzy logic controller demonstrated a robust performance with alignment times and accuracies comparing favourably against those obtained using a manually designed controller.

Keywords: evolutionary algorithms, fuzzy logic, fibre optics, automatic assembly.

1. Introduction

A difficult task in the assembly of fibre optic transmitters is to align the laser emitter and fibre core. The active surface area of the former is in the range 1-2 μm^2 and the diameter of the latter is less than 10 μm. The aim of the alignment operation is to maximise the amount of light transmitted to the fibre core. Current manufacturing technology does not allow the required sub-micron fibre alignment to be achieved by precision fitting of pre-formed components. It is therefore necessary to use an active alignment technique, where the powered laser chip is moved relative to the fibre while the light output from the latter is monitored. When the maximum amount of light is obtained, the fibre assembly is permanently fixed relative to the laser chip using a fast setting adhesive or laser welding.

In the main, active alignment is performed manually at present. An operator controls a three-axis micrometer stage to effect the alignment by trying to maximise the fibre light output as monitored on an optical power meter display. Alignment times are typically long, ranging from 2 to 5 minutes depending on the product being manufactured. Automatic techniques under computer control use high-precision stepping motors to drive the stage supporting the fibre components. The search for the power peak is normally performed through a combination of area scans and gradient-based search routines. These automatic systems are slow compared to human operators. For this reason, the manufacturing process is still largely manual.

The main difficulty in aligning the components is caused by the irregularity of the power distribution in the laser beam. Rather than the ideal single Gaussian shape, the laser beam usually shows a maximum surrounded by a number of secondary peaks, plateaux and other irregularities. The alignment process is also complicated by positional inaccuracies of the stepping motors and mechanical vibrations of the frame. Simple gradient-based algorithms can converge to sub-optimal peaks and lengthy area scans and other techniques are necessary to ensure that the global power maximum is found.

Pham and Castellani [1] presented an automatic alignment algorithm based on "climbing" of the power slope supported by on-line momentum information. The parameters of the gradient-based search were adjusted via a fuzzy logic (FL) [2, 3] control module. The new automatic system allowed a considerable reduction in the laser-to-fibre alignment times.

Although the design of the knowledge base (KB) for the fuzzy controller was conceptually straightforward, the system still needed much manual effort to tune the rule base (RB) and membership functions (MFs) of the controller. This paper presents the application of a new evolutionary algorithm (EA) for automatically generating the KB for the laser-to-fibre alignment controller. Following a definition of the problem (section 2) and an overview of the alignment technique adopted (section 3), the EA is presented (section 4) and its implementation to the laser-to-fibre alignment problem is detailed together with the experimental results obtained (section 5).

2. Problem definition

Henceforth, a 3D cartesian reference frame *XYZ* will be assumed with *X* perpendicular to the surface of the transmitter and corresponding to the axis of the focusing lens. Z is taken to be vertical and points upwards. The laser-to-fibre coupling task requires the location of the maximum power value in the hypersurface formed by the power distribution in *XYZ* space.

All experiments in this study were carried out using the discrete lens alignment approach [4], where a separate lens is employed to couple the light into a cleaved fibre. To achieve optimum light coupling, the fibre must be positioned at a distance of a few mm from the laser chip and the components must be aligned with a precision better than 1 μm.

The devices used produced laser light of 1400μm in wavelength and maximum powers varying between 500μW and 2000μW. The power gradient along the focal axis was about two orders of magnitude smaller than along the other two coordinate axes and distinguished by a more regular shape.

To minimise the alignment time, the search algorithm has to find the power peak within the minimum number of cycles following the shortest path. A gradient-based algorithm looks attractive because of its ability to ascend quickly the optimisation surface. Unfortunately, such an algorithm is easily trapped by local minima, flat surfaces and other irregularities affecting real power distributions. Moreover, imprecise readings due to noise or mechanical vibrations of the frame holding the components can reduce its effectiveness.

The next section will present an enhanced gradient-based alignment technique modelled upon the dynamics of a body travelling through a physical medium and experiencing the attractive force of a potential field.

3. Laser-to-fibre coupling algorithm

The standard gradient-based search strategy has been enhanced with the ability to overcome some of the irregularities in the power distribution. In the following discussion, power readings are considered reversed in sign thus turning the problem into a minimisation task. In the case of an ideal laser emission with a Gaussian distribution, the power surface can now be thought of as an attractor basin centred at the power peak. For real power distributions, the areas of deviation from the Gaussian shape can be regarded as perturbations of the potential field.

The proposed search algorithm mimics the behaviour of an imaginary body attracted by a potential field and experiencing a friction force opposite to the direction of motion. The optical fibre is moved relative to the laser emitter according to the body's equations of motion. The example of a sphere falling down a valley can help visualisation of this strategy. Irregularities in real laser power distributions can be pictured as flat areas, "bumps" or "pot holes" along the hillside.

Consider the equations describing the motion of the body. A body of mass μ in an attractive potential field ε and travelling in a medium of friction coefficient κ experiences a total force $\mathbf{F}$ given by the composition of the attractive force $\mathbf{F}_A = -d\varepsilon/d\mathbf{r}$ and the friction force $\mathbf{F}_F = -\kappa \cdot \mathbf{v}$:

$$\mathbf{F} = \mathbf{F}_A + \mathbf{F}_F = -\frac{d\varepsilon}{d\mathbf{r}} - \kappa \cdot \mathbf{v} = \mu \cdot \mathbf{a} \tag{1}$$

where **a**, **v**, and **r** are respectively the acceleration, velocity and position of the body. Approximating the motion linearly for small movements yields the following equations:

$$\mathbf{a} = \frac{\mathbf{F}}{\mu} \qquad (2)$$

$$\mathbf{v} = \int \mathbf{a} \cdot d\tau + \mathbf{v}_0 \approx \mathbf{a} \cdot \Delta\tau + \mathbf{v}_0 \qquad (3)$$

$$\mathbf{r} = \int \mathbf{v} \cdot d\tau = \int (\mathbf{a} \cdot \Delta\tau + \mathbf{v}_0) \cdot d\tau + \mathbf{r}_0 \approx \approx \frac{1}{2} \mathbf{a} \cdot \Delta\tau^2 + \mathbf{v}_0 \cdot \Delta\tau + \mathbf{r}_0 \qquad (4)$$

where $\mathbf{v}_0$ and $\mathbf{r}_0$ are the initial velocity and position and $\Delta\tau$ is a small time interval.

Considering unitary time steps ($\Delta\tau = 1$) and approximating $d\varepsilon \approx \Delta\varepsilon_0$ and $\mathbf{dr} \approx \Delta\mathbf{r}_0$ enables eq. (2), eq. (3) and eq. (4) to be rewritten as:

$$\mathbf{a} = -\frac{\left(\frac{\Delta\varepsilon_0}{\Delta\mathbf{r}_0} + \kappa \cdot \mathbf{v}_0\right)}{\mu} \qquad (5)$$

$$\mathbf{v} = \mathbf{a} + \mathbf{v}_0 \qquad (6)$$

$$\Delta\mathbf{r} = \frac{\mathbf{a}}{2} + \mathbf{v}_0 \qquad (7)$$

Given μ and the ratio κ/μ, the displacement Δr of the body can be computed from the previous displacement and the associated power variation. Each Δx, Δy and Δz motion component can be calculated by resolving the vectorial equations (5), (6) and (7) along the desired direction.

Since the power gradient along the three coordinate directions is not known *a priori*, it is necessary to perform each motion component separately and measure the power at the end of each step.

Eq. (7) shows that the move is determined by two components, the first depending on the total force experienced by the imaginary body and the second relating to its momentum. The first component combines the action of the two forces acting on the body and changes its momentum according to the power gradient and the friction force. The magnitude of this change is affected by μ and the ratio κ/μ. μ represents the mass of the body and determines its sensitivity to the potential field, while the ratio κ/μ is related to the damping of the inertial motion. The momentum term produces an inertial motion giving the body a capability for overcoming obstacles like flat surfaces or local minima that would stop simple gradient-based algorithms. It also reduces the effect of inaccurate readings due to random noise or frame vibrations, as the errors are likely to cancel out in the momentum term.

The choice of the parameters μ and κ/μ defines the character of the search. Large values will make the descent of the power slope more conservative while small values will magnify the effect of the attraction force and let the body more freely descend the slope.

Different power distribution shapes and conditions require different settings and keeping constant values can be only a sub-optimal solution. The complexity of the problem and the interactions between the two parameters do not allow the derivation of a simple formula for the tuning of μ and κ/μ. The proposed algorithm employs a FL controller for the determination of μ and κ/μ in each direction of motion.

The FL system was implemented through the *Mamdani model* [5]. The core of the fuzzy system is constituted by the RB and the inference engine. The RB is composed of fuzzy if-then rules which describe qualitative non-linear relationships through antecedent-consequent pairs. The inference engine processes the rules and produces the output in the form of a fuzzy set. The defuzzifier converts the linguistic output into a crisp value. The RB and the MFs constitute the KB of the system. The FL controller has two inputs, respectively the measure of the laser beam power and its change with time, and two outputs, μ and κ/μ.

The search algorithm includes an initial scanning routine to locate a convenient starting point for the gradient descent. This point should be situated close enough to the main peak to minimise the alignment time and the risk of non-optimal convergence. On the other hand, the location of the starting point necessarily requires time consuming area scans. A trade-off is therefore necessary between search accuracy and scanning time.

The main limitation of the proposed algorithm was the need for effort and skills in the manual design of the fuzzy controller. The next section describes an EA developed by the authors to generate the FL module automatically.

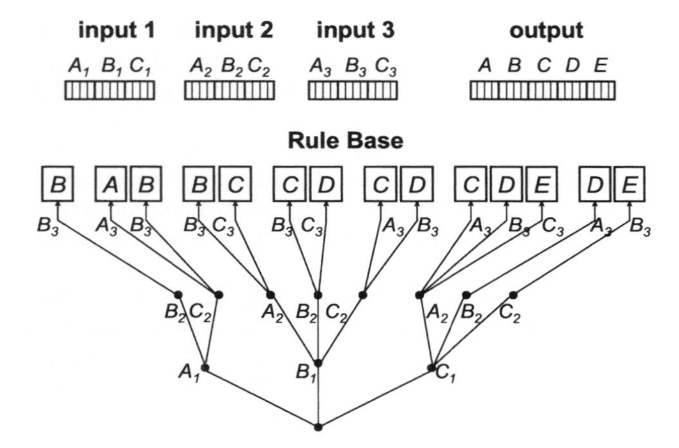

Fig. 1: Representation scheme

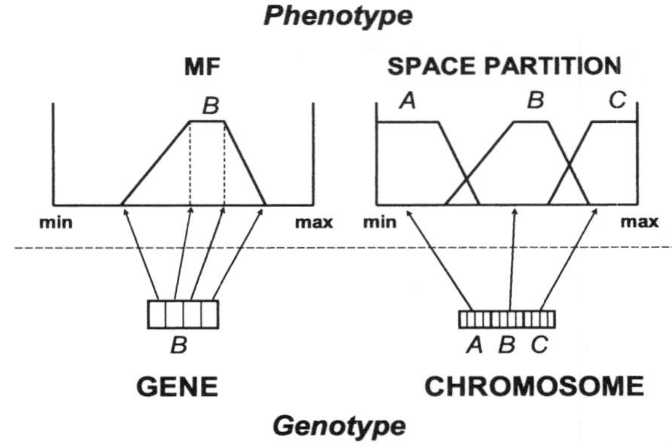

Fig. 2: Membership function encoding

4. Evolutionary algorithm

The evolutionary algorithm used in this work is described in Pham and Castellani [6]. For ease of referencing, this section gives a summary of the key features of the algorithm.

The proposed EA evolves FL systems through simultaneous learning of both the RB and the MF. The shape of MFs used for partitioning the input and output spaces is fixed prior to the learning process and is not subjected to evolution. As the significance of such a parameter is limited, it has not been included in the learning process.

The algorithm employs the generational replacement reproduction scheme [7], a new selection operator [8] and a set of crossover and mutation procedures [9] dealing with different elements of the fuzzy KB. The population is divided into three sub-groups to which different operators are applied. A specific integer-valued gene marks the species of each individual.

The genetic operators acting on the first sub-population work at the level of input and output fuzzy partitions. Crossover generates two new individuals by mixing the MFs of the two parents for each variable. Each parent transmits its RB to one of the offspring. Random chromosomic mutations can create new fuzzy terms, delete existing ones, or change the parameters defining the location and shape of a MF. Whenever genetic manipulations modify the set of MFs over which a RB is defined, a 'repair' algorithm translates the old rule conditions and actions into the new fuzzy terms. The sub-population of fuzzy systems undergoing this set of operations is called *species_1*.

The operators manipulating the second sub-population search for the optimal RB. Genetic crossover creates two individuals by exchanging sets of rules between the two parents. Each of the offspring inherits the input and the output space partitions from one of its parents. The mutation operator randomly changes the action of a fuzzy rule. To accommodate the new partitions, the conditions and actions of the swapped rules are translated into new linguistic terms. This group of solutions is named *species_2*.

The operators acting on the third sub-population deal with all the components of the fuzzy KB. Genetic recombination swaps all the MFs and the rules contained in a randomly selected portion of the input and output spaces. Mutation can take any of the forms defined for the modification of *species_1* and *species_2* genotypes. This third group of individuals is referred to as *species_3*.

The above mutation operators function at the KB level. There is also a mutation operator at the species level that transforms one species into another.

The algorithm represents candidate solutions using a variable-length encoding scheme. Fig. 1 gives an example of an encoded solution for a FL system having three input variables, each partitioned into three linguistic terms, and one output variable partitioned into five fuzzy terms.

A separate chromosome [10] is used to describe the partition of each input and output variable, each chromosome being composed of a number of genes equal to the number of fuzzy terms. Each gene is a real-valued string encoding the parameters defining the location and the shape of one MF. Fig. 2 details the MF encoding schemeThe fuzzy RB is represented as a multi-level decision tree, the depth of the tree

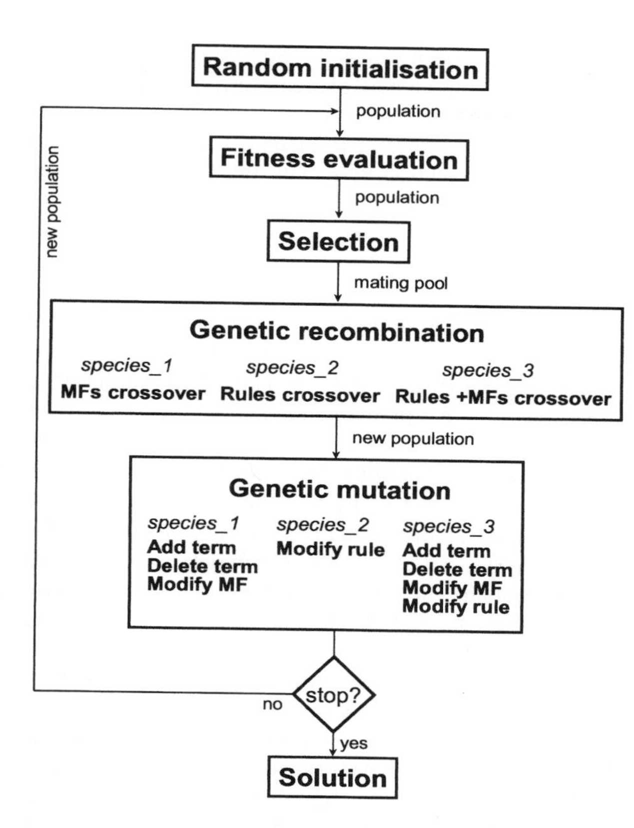

Fig. 3: Flowchart of proposed evolutionary algorithm

corresponding to the dimensionality of the input space. The rule antecedent is encoded in the full path leading to the consequent, each node being associated with a rule condition. The nodes at the last level of the decision tree represent the rule consequent and contain a fuzzy action for each output variable.

During the evolutionary search, the fitness of individual FL controllers was evaluated according to their precision and speed of locating the power peak. As it was not feasible to test the solutions on the real plant because of the long laser-to-fibre alignment times, it was decided to employ a simulated alignment task based on an empirical model of the laser beam. The model was built by fitting 3D power scans of a batch of laser emitters.

For each individual, the search algorithm was run until either the peak was localised or a pre-defined number of steps τ had elapsed. The search starting point and the value of the power peak were randomly varied every generation. The former was set to values normally guaranteed by the preliminary power scan. Once the algorithm had stopped, the final power

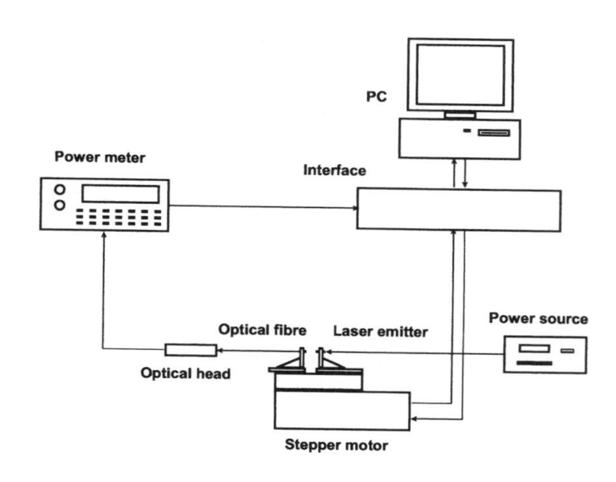

Fig. 4. Experimental system

reading and the number of search steps were recorded.

The fitness was then calculated as follows:

$$f(i) = \varepsilon_f + \frac{steps(fastest)}{steps(i)} \cdot \frac{\varepsilon_f}{10} \qquad (8)$$

where $f(i)$ is the fitness of individual i, ε_f is the final power reading, $steps(i)$ is the number of steps it took individual i to locate the peak and *fastest* is the solution that localised the peak most quickly during that evaluation trial. The contribution of the second term of eq. (8) has a magnitude 10 times smaller than the term expressing the precision. This was to bias the search first towards solutions able to locate the power peak and then solutions allowing fast alignment.

A flowchart of the algorithm is given in fig. 3.

5. Experimental settings and results

5.1. Experimental set up

The laser-to-fibre coupling algorithm was tested on a batch of optical fibres and laser emitter components. The mechanical unit consisted of two platforms each housing a clamp on which respectively the laser chip and the optical fibre were fitted. The platform carrying the optical fibre was fixed, while the other could be moved sequentially along each of the three axes by means of a piezoelectric stepper motor. Fig. 4 describes the experimental system.

The proposed search algorithm was used only for optimising the power in YZ planes (i.e. those normal to

Table 1. Evolutionary fuzzy controller – evolutionary algorithm settings

Parameters of evolutionary algorithm	
population size	60
number of generations	500
number of runs	3
mutation rate (KB)	0.15
mutation rate (species)	0.1
max number of terms per variable	10
Initialisation settings	
number of terms per variable	5
lower bound of space partition of ε	50 μW
upper bound of space partition of ε	2000 μW
lower bound of space partition of $d\varepsilon/dt$	[-450,-225] μW
upper bound of space partition of $d\varepsilon/dt$	[+225,450] μW
lower bound of space partition of μ	0
upper bound of space partition of μ	[3,6]
lower bound of space partition of κ/μ	0
upper bound of space partition of κ/μ	1
Rule Base	empty
Fitness function settings	
max number of evaluation steps	50
min power value at peak	500μW
min power reading at start of search	50 μW
max *X* dist. from peak at start of search	350μm
max *Y/Z* dist. from peak at start of search	8μm

the focal axis). Once the maximum power was found in a plane, the stage carrying the emitter was made to step a fixed distance along the X axis towards the fibre and the search was restarted in a new plane. The procedure was repeated until the YZ plane nearest to the focal point was found, that is the one where the peak value of the power was highest. A final tuning of the position was performed restarting the gradient-descent algorithm from the power peak location. The X step size was set experimentally to 80μm allowing the focal plane to be determined with a precision of 40μm, a value that did not substantially affect the peak power value.

For directions of motion perpendicular to the focal axis, a minimum step size was experimentally set to 0.3μm. When along a certain direction of motion, the value for the displacement Δr of the body was smaller than 0.3μm, the search along that direction was considered completed. Once on a given plane the motion had stopped in both the Y and Z directions, the power was considered optimised and the components were moved back to the position where maximum light coupling had been measured. Subsequently, the search was moved onto a new plane or terminated.

For each search direction, the fuzzy controller calculated the motion parameters. The EA outlined in the previous section generated the fuzzy KB for the control of μ and κ/μ.

5.2. EA settings

The sought solution had two inputs, the transmitted power ε and its time derivative, and two outputs, the control parameters μ and κ/μ. Table 1 summarises the setting of the EA parameters. Parameters defined over a range of values were randomly initialised in the given interval. The sought solution had two inputs, the transmitted power ε and its time derivative, and two outputs, the control parameters μ and κ/μ. If during the evolutionary process a solution experienced an input value outside the initialisation range, one of the two extreme terms had the support of its MF extended to include the new value.

To prevent the generation of infeasible fuzzy rules and minimise the size of the final solution, the population was initialised with a blank RB. Each time that, during the fitness evaluation phase, a candidate solution experienced a combination of input conditions which didn't lead to an existing rule action, a new rule was generated for that particular set of conditions with a randomly picked rule consequent

Once an individual was selected for mutation, the type of KB modification operator had to be chosen. For this purpose, a random procedure was used for *species_1* and *species_3* individuals, while solutions belonging to *species_2* could only undergo rule action modification. In the case of *species_1* individuals, the random procedure gave a 20% probability for the addition of one fuzzy term, a 25% chance for the deletion of one fuzzy term and the remaining 55% probability for the modification of one fuzzy MF. The same operators were assigned half of these probability values when applied to *species_3* individuals, whereas the modification of a rule action had a 50% chance. The allocation of these probabilities was according to practical experience which also indicated that the best rate for the species mutation operator was 0.1.

The evolutionary procedure was run three times and each time the best solution was saved. At the end, the three saved solutions underwent a further evaluation stage to select the final solution. This repetition of the evolution process was introduced to reduce further the

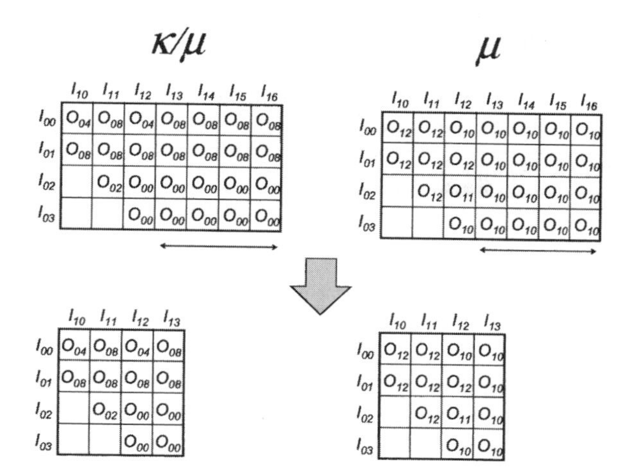

Fig. 5: Rule base simplification

risk of a sub-optimal result.

5.3. Results

Some first conclusions can be drawn by comparing the control KB produced by the proposed EA with the manually designed KB for the same control problem reported in [1].

In the automatically evolved KB, the input universes of power and change in power are respectively discretised into seven and four fuzzy terms, while the output universes of μ and κ/μ are respectively discretised into nine and three fuzzy terms. The fuzzy RB contains 25 production rules.

The manually created KB was slightly more compact in terms of fuzzy partitions and RB size. It divided the input space into 5×4 overlapping 'patches' each of them associated with a control response. The two output signals were created using respectively five and four output terms. On the other hand, the automatically evolved KB divided the input space into 7×4 overlapping patches, 25 of them associated with a control response. It therefore defined 8 more input sub-areas and 5 more control rules. Also the overall number of output fuzzy terms was greater, amounting to 9+3 terms.

Nevertheless, closer examination of the fuzzy response tables shows that, by applying some straightforward regrouping operations, the size of the learned KB could be greatly reduced. The upper part of fig. 5 depicts the fuzzy decision tables for the outputs of the two systems. As indicated by the arrows, both tables have the same entries for the last four columns. It is therefore possible to merge terms I_{13} to I_{16} to obtain a 4×4 decision matrix (see the lower part of fig. 5). Moreover, only 4 out of 9 elements of the term set for the first output are actually used in the control rules. If needed, the introduction of manual or automatic pruning should therefore noticeably reduce the KB complexity.

However, the complexity of the learned KB is acceptable, especially considering that the proposed EA does not contain any direct bias in favour of more compact solutions.

The evolutionary FL system was tested on the alignment of a batch of 40 different pairs of fibre optic components. For each alignment trial, a different laser emitter was used, while the optical fibre was changed every five trials. For each pair of components, the search algorithm was run until its completion or interrupted as soon as the power reading reached the maximum measurable value of 2000μW. Manual and conventional automated techniques were then used to search for further power improvements. If the final power peak value was not significantly different from the value obtained by the proposed automatic alignment system, the trial was considered to have succeeded.

The results obtained using the EA-generated fuzzy system were compared with those of an identical experiment carried out using the manually built FL controller described in [1]. The comparative data are presented in table 2.

Both automatic alignment systems successfully located the power peak in all the trials performed, suffering from slight occasional inaccuracies due to the imprecision of the stepping motors. The new fuzzy controller allowed better alignment times, giving an 8% reduction on the original gradient-based search time. However, such an improvement should be moderated by the magnitude of the standard deviation of the measurements.

6. Conclusions

A new EA was used to generate of the KB for a fuzzy controller in the assembly of fibre optic components.

Throughout the tests performed, the evolutionary fuzzy controller showed a remarkable robustness, producing a 100% success rate. In terms of alignment speed, the proposed technique obtained accuracy results that are comparable with those achievable by

Table 2: Comparison of manually built and evolutionary controllers

	Manually built fuzzy controller (40 alignment trials)			Evolutionary fuzzy controller (40 alignment trials)		
	Initial scan	**Peak search**	**Total**	**Initial scan**	**Peak search**	**Total**
average (secs)	17.65	15.20	32.85	17.00	13.95	30.95
worst (secs)	40	27	60	40	24	60
σ (secs)	8.36	5.91	10.93	8.75	5.88	11.29

manual design of the control policy. However, the evolutionary technique requires much less problem domain knowledge, is easily reconfigurable to new types of components and is fully automated.

As the only drawback, the generated KB was not as compact as a manually constructed KB as it included redundant fuzzy terms and rules. However, through straightforward manual tuning it was possible to reduce the size of the evolved solution.

Further work should address the creation of automatic pruning routines capable of detecting and suppressing redundant KB elements. It would also be worthwhile investigating the possibility of adding an evolutionary bias towards simpler solutions. The main advantages of producing more compact solutions would be an increase in transparency of the control policy and simpler hardware and software implementation.

Acknowledgements

This research was sponsored by Hewlett-Packard and the Welsh Assembly Government under the European Regional Development Fund programme and by the EC FP6 Network of Excellence on Innovative Production Machines and Systems (I*PROMS).

References

[1] Pham, D.T. and Castellani, M. (2001), Intelligent Control of Fibre Optic Components Assembly, Proc. IMechE, Part B, Vol. 215, No. 9, pp. 1177-1189.

[2] Zadeh, L. A. (1965), Fuzzy Sets, Information and Contr., Vol. 8, pp. 338-353.

[3] Ross, T. J. (2004), Fuzzy Logic with Engineering Applications, 2nd Ed., Wiley, New York.

[4] Gowar, J. (1993), Optical Communication Systems, 2nd ed., Prentice-Hall Int, New York.

[5] Mamdani, E. H. (1974), Application of Fuzzy Algorithms for Control of Simple Dynamic Plant, Proc. IEE, Vol. 121 No. 12, pp. 1585-1588.

[6] Pham, D.T. and Castellani, M. (2002), Outline of a New Evolutionary Algorithm for Fuzzy Systems Learning, Proc. IMechE, Part C, Vol. 216, No. 5, pp. 557-570.

[7] Davis, L. (1991), Handbook of Genetic Algorithms, Van Nostrand Reinhold, New York.

[8] Pham, D.T. and Castellani, M. (2005), A New Adaptive Selection Routine For Evolutionary Algorithms, submitted to Proc. IMechE, Part C.

[9] Fogel, D. B. (2000), Evolutionary Computation: Toward a New Philosophy of Machine Intelligence, 2nd ed., IEEE Press, New York.

[10] Eiben, A.E. and Smith, J.E. (2003), Introduction to evolutionary computing, Springer, New York.

Intelligent Production Machines and Systems
D. T. Pham, E. E. Eldukhri and A. J. Soroka (eds)

Automatic detection of Mediterranean water eddies from satellite imagery of the Atlantic Ocean

M. Castellani

CENTRIA - Departamento Informática, Faculdade Ciências e Tecnologia, Universidade Nova Lisboa, Quinta da Torre, Caparica 2829-516, Portugal.

Abstract

A new machine learning approach is presented for automatic detection of Mediterranean water eddies from sea surface temperature maps of the Atlantic Ocean. A pre-processing step calculates the direction of the temperature gradient at each point of the map. Given a map point, information on the surrounding thermal gradient field is extracted through a binary mask and organised as a numerical vector of gradient angles. This vector is forwarded to a multi-layer perceptron classifier that is trained to recognise patterns of gradient angles generated by positive and negative instances of meddy structures. The proposed system achieved high recognition accuracy with fast and robust learning results over a range of different binary masks. Results were also characterised by a very low detection rate of false positives. The simple and modular structure of the image processing routines reduce computational costs and make the system easily modifiable.

Keywords: oceanography, remote sensing, artificial neural networks.

1. Introduction

Due to its high salinity and high temperature, the presence of Mediterranean water strongly influences the hydrology and the dynamics of the Atlantic Ocean [1] and plays an important role in the transport of particles, suspended material and live organisms. Mediterranean water eddies (meddies) are of crucial importance in the process of, and account for, the diffusion of Mediterranean waters over thousands of kilometers with very little mixing [2].

Although the existence in the North Atlantic of a warm and salty tongue of water of Mediterranean origin has been known for a long time, the discovery of the existence of meddies is relatively recent. Meddies are mesoscale lens-like structures, in solid body cyclonic or anticyclonic rotation, with typical diameters of about 50 kms, periods of rotation of about 6 to 8 days and velocities of about 30 cm/s [3].

A decisive development in the study of marine phenomena came from the use of remote sensing data. In particular, recent studies indicated that mesoscale structures having a surface signature in temperature can be detected using satellite information. For this purpose, sea surface temperature (SST) maps created from satellite-borne infrared sensors are employed to visualise oceanic structures such as mushroom like dipoles, fronts and vortexes [4]. Despite current active research into the generation and circulation of meddies into the Atlantic (e.g., [2,5]), inspection of satellite maps still relies on visual interpretation of the images, which is labour intensive, subjective in nature and dependent on the interpreter's skills [6]. Given the large amount of satellite data which is collected daily about the Ocean, the development of automatic tools for analysis and mining of such imagery is nowadays of paramount importance.

This paper focuses on automatic detection of meddies from SST maps of the Atlantic Ocean. The study is part of a project aiming at remote identification of meddies with synergistic use of satellite thermal, colour and roughness images of the ocean (www.io.fc.ul.pt/fisica/research.htm - RENA project). The main problem in understanding SST maps is due to the strong morphological variations of oceanic phenomena that hinder the development of valid geometrical representations [7]. Structural patterns such as water vortexes are difficult to express into point-level constraints useful for shape recognition.

Two main approaches are customary for detection of oceanic structures. Both include a pre-processing stage where high-level features are extracted from the maps to highlight dynamic, structural or textural local properties. The first approach directly matches feature information with expert knowledge that describes the sought after patterns (e.g., [8,9,10]).

The second method relies on machine learning of the decision making policy that informs the recognition system (e.g., [11,12]). This approach requires the least design effort since it is much easier for experts to identify positive instances of the desired object than it is to specify precise domain knowledge. A learning classifier is also easier to reconfigure, other oceanic phenomena can be included in the identification algorithm simply by supplying the system with a new set of training examples.

This paper presents a new machine learning approach for automatic detection of meddies from SST maps of the North Atlantic. The algorithm uses an artificial neural network (ANN) [13] that is trained to recognise patterns of thermal gradient fields generated by meddy structures. In addition to built-in learning capabilities, the ANN ensures effective processing of large amounts of input data and a reasonable response to noisy or incomplete inputs.

The structure of the paper is as follows. Section 2 presents a review of the literature relevant to the subject. Section 3 details the problem domain. Section 4 presents the proposed algorithm. Section 5 describes implementation issues and application results of the proposed technique. Section 6 discusses the results and draws some comparisons with other published work. Section 7 concludes the paper and proposes areas for further investigation.

2. Automatic identification of oceanic eddies

Mesoscale structures and eddies in particular play an important role in the dynamics of oceans. A number of studies investigated the creation of automatic systems for detection and monitoring of such phenomena from remotely sensed SST maps.

Peckinpaugh et al. [11] used a two step procedure, namely a pre-processing stage for image segmentation and feature extraction, and the actual identification stage performed by an ANN classifier. SST images of the Gulf Stream region are segmented into sets of overlapping "tiles". For each sub-image, the magnitude and the direction of the maximum image "energy" is calculated from analysis of the local Fourier power spectrum [14]. These magnitudes and directions are fed as inputs into the ANN classifier. The authors reported encouraging accuracy results in discriminating warm eddies from cold eddies or open ocean areas.

Peckinpaugh and Holyer [15] also developed an eddy detection algorithm based on analytical description of eddy structures. An edge detection operator was initially used to outline the borders of oceanic phenomena. Several circle detectors were then compared to locate the centre and the radius of eddies. However the study was limited to a small sample of images and no evaluation was made on the possibility of detection of false positives.

Alexanin and Alexanina [9] regarded SST maps as oriented textures in the temperature field and calculated the dominant orientation of radiation contrast at each point. Since dominant orientations of thermal gradient are highly correlated to sea surface velocity directions, the centre of circular motion structures can be calculated from the orientation field via analytical models. Values of dominant gradient orientation values were averaged over one-two-day time intervals to reveal stable mesoscalar phenomena. The main drawbacks of this method are its computational complexity and the fact that the size of the target structures must be fixed beforehand.

Lemonnier et al. [6] used multiscale analysis of shapes to highlight areas of isotherm curvature which may contain eddy shapes. Analysis of oriented texture through phase portraits is used to classify the selected regions. The algorithm was tested on 100 SST images of the bay of Biscay. The main problem reported by the authors was a high rate of false detections [8] which required further manual processing of the image.

Cantón-Garbín et al. [12] first pre-processed SST

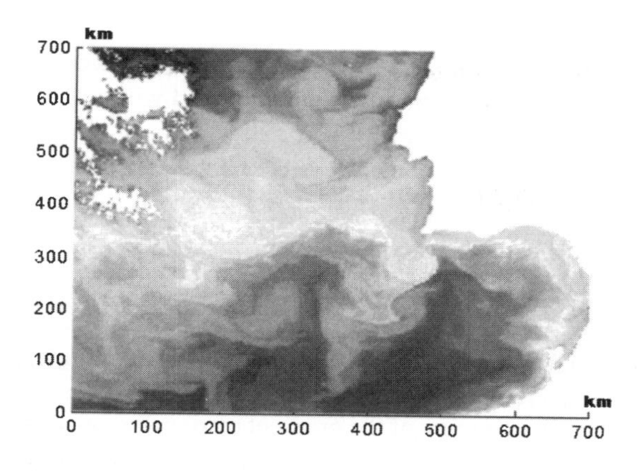

Fig. 1. SST map of the North Atlantic

images of the the Atlantic using an ANN module to mask cloudy areas. Images were then segmented using information on isothermal lines. Finally, features extracted from the segmented sub-areas were fed to a competitive layer of neuron-like processing elements for identification. A set of pre-classified examples of mesoscalar phenomena was used to train the processing elements. The authors also implemented a rule-based classification system based on experts' knowledge [7]. The paper reported good accuracy results for detection of upwellings and island wakes while the identification of warm and cold eddies resulted more problematic.

3. Problem domain

The aim of this study is the automatic detection of meddies from SST data collected via the Advanced Very-High Resolution Radiometer of the National Atlantic and Atmospheric Administration satellite. Temperature readings are arranged into SST maps. Each map covers an area of the Atlantic Ocean of 700x700 kms between 34-41°N and 6-15°W and contains one data point per square kilometer. The resolution of the temperature readings is within 0.1°C. Fig. 1 shows an example of SST map. White areas on the right part of the image correspond to land regions, namely the coastal areas of Portugal, Spain and Morocco. The white patches on the left of the image correspond to sea regions covered by clouds.

The survey of the previous section shows that automatic recognition of meddies from SST maps can be achieved following two different approaches. Due to the correlation between isotherm lines and water flow directions, both approaches usually exploit information on temperature gradient field.

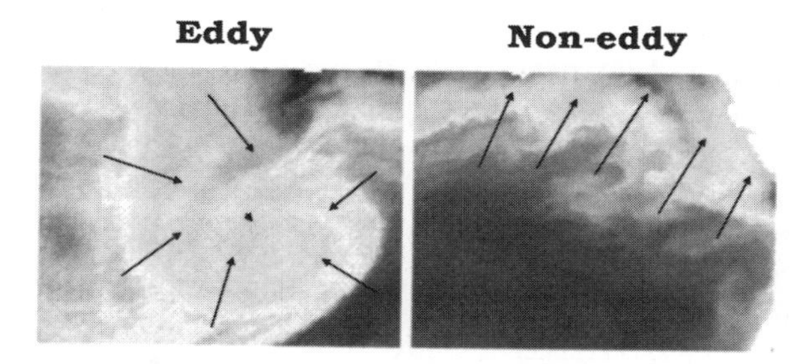

Fig. 2. Examples of likely gradient fields

The first approach usually matches thermal gradient maps with analytical descriptions of the sought after structures. A typical example is the work of Alexanin and Alexanina [9] which fitted dominant orientation maps of thermal contrast to elliptical curve models. The main limitation of this approach is the difficulty of complying with the high structural variability of eddies. Moreover, since temperature gradient traces in SST maps are often weak, image pre-processing is needed to enhance thermal contrast. Unfortunately, thermal contrast enhancement is likely to increase noise and spurious information as well, making recognition of structural patterns harder. Identification systems tend therefore to rely on rather complex algorithms which often show a certain degree of brittleness.

The second approach uses ANNs (e.g., [11,12]). ANNs are capable of learning arbitrarily complex non-linear mappings and can handle large amounts of sensory information. The approximate nature of their pattern matching and association processes makes them particularly suitable to deal with the high variability of mesoscale phenomena. Such robustness usually makes less critical the pre-processing stage, allowing faster and simpler procedures to be used. ANN learning capabilities also remove the need for time-consuming design of the identification knowledge.

The main drawback of ANN systems is that the decision making policy is usually not transparent to the user. For this reason, expertise can not be used to initialise the system and the learned knowledge can not be extracted for modelling purposes.

4 Proposed algorithm

The procedure starts with a very simple and coarse estimation of thermal gradient direction in SST maps. Temperature gradient direction fields around meddy centres are expected to be characterised by concentric patterns differing from fields created by other oceanic phenomena. Fig. 2 shows an example of likely gradient

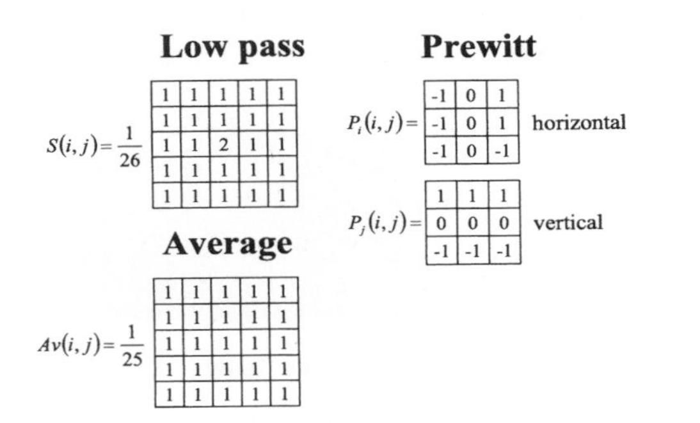

$$S(i,j)=\frac{1}{26}\begin{bmatrix}1&1&1&1&1\\1&1&1&1&1\\1&1&2&1&1\\1&1&1&1&1\\1&1&1&1&1\end{bmatrix}\quad P_i(i,j)=\begin{bmatrix}-1&0&1\\-1&0&1\\-1&0&-1\end{bmatrix}\quad P_j(i,j)=\begin{bmatrix}1&1&1\\0&0&0\\-1&-1&-1\end{bmatrix}$$

$$Av(i,j)=\frac{1}{25}\begin{bmatrix}1&1&1&1&1\\1&1&1&1&1\\1&1&1&1&1\\1&1&1&1&1\\1&1&1&1&1\end{bmatrix}$$

Fig. 3. Convolution matrices

fields associated to a meddy and a thermal front.

By sampling the thermal gradient direction at fixed steps around a point, it is possible to create a numerical vector that describes the gradient field in that region. Points at the centre of meddy structures will therefore be characterised by vectors that describe a surrounding area of roughly concentric outward or inward gradient directions. Such vectors, albeit noisy and irregular, are deemed to be separable from vectors generated by other oceanic phenomena.

Given an SST map point, the proposed method extracts a numerical vector describing the surrounding gradient field. This vector is forwarded to an ANN classifier for meddy detection. The ANN is trained to recognise patterns of gradient directions corresponding to meddy formations on a set of positive and negative instances. The ANN generalisation and noise rejection capabilities are expected to provide the required tolerance to noise and variability of meddy shapes.

4.1. SST image pre-processing

SST images are first processed using a low-pass filter to reduce noise and other irregularities, then the local vertical and horizontal temperature gradients are calculated using Prewitt edge detector [16]. The transformations follow the equations below:

$$smooth(i,j) = map(i,j) * S(i,j) \tag{1}$$

$$grad_i(i,j) = smooth(i,j) * P_i(i,j) \tag{2}$$

$$grad_j(i,j) = smooth(i,j) * P_j(i,j) \tag{3}$$

where *map* is the original 700x700 SST map, *smooth* is the low-passed image and $grad_i$ and $grad_j$ are respectively the horizontal and the vertical temperature gradient maps. S, P_i and P_j are the convolution matrices shown in fig. 3.

For each point $p(i,j)$, the horizontal and the vertical temperature gradient maps are averaged within a 5x5 window centred on $p(i,j)$.

$$X(i,j) = grad_i(i,j) * Av(i,j) \tag{4}$$

$$Y(i,j) = grad_j(i,j) * Av(i,j) \tag{5}$$

where X and Y are the averaged gradient maps and Av is the convolution matrix shown in fig. 3.

Local averaging of the gradient maps corresponds to a smoothing operation and has the dual purpose of reducing noise and highlighting areas of strong and coherent gradient field, revealing thus the boundaries of the main water currents.

Points marked as clouds or land are marked as "invalid" and are ignored in the convolution with matrices S and Av. In the convolution with matrices P_i and P_j, invalid points are given temperature values averaged from the values of the nearest neighbours. In the case all points in the region covered by the mask are labeled as invalid, the matching point of the new image is marked invalid. The importance of such points is however limited as they lie outside or at the very borders of visible ocean areas.

Finally, map θ of temperature gradient direction is calculated from temperature gradient maps X and Y. That is, for each point $\theta(i,j)$ of map θ, the thermal gradient vector is determined according to the parallelogram law from the horizontal and vertical gradient vector components $X(i,j)$ and $Y(i,j)$. The direction $\theta(i,j)$ ($0 \le \theta(i,j) \le 2\pi$) of the thermal gradient vector relative to a horizontal polar axis can be calculated as follows:

$$\begin{cases} X(i,j)>0 \,\&\, Y(i,j)\ge 0 \Rightarrow \theta(i,j) = arctg\left(\dfrac{Y(i,j)}{X(i,j)}\right) \\ X(i,j)=0 \,\&\, Y(i,j)>0 \Rightarrow \theta(i,j) = \dfrac{\pi}{2} \\ X(i,j)<0 \Rightarrow \theta(i,j) = arctg\left(\dfrac{Y(i,j)}{X(i,j)}\right)+\pi \\ X(i,j)=0 \,\&\, Y(i,j)<0 \Rightarrow \theta(i,j) = \dfrac{3}{2}\cdot\pi \\ X(i,j)>0 \,\&\, Y(i,j)<0 \Rightarrow \theta(i,j) = arctg\left(\dfrac{Y(i,j)}{X(i,j)}\right)+2\cdot\pi \end{cases} \tag{6}$$

In the case $X(i,j)$ and $Y(i,j)$ are tagged invalid, $\theta(i,j)$ is marked invalid too.

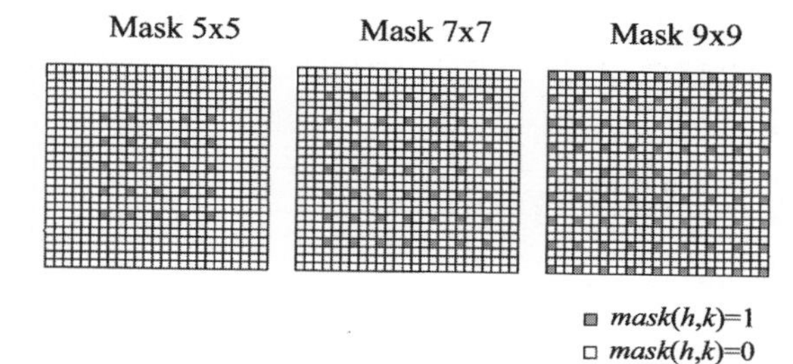

Fig. 4. Feature extraction masks

4.2. Numerical feature extraction

As previously stated, it is expected meddy structures to be characterised by more or less regular outward or inward concentric patterns of thermal gradient. By sampling the gradient direction field (i.e., map θ) of regions of SST images, such patterns can be revealed.

Given an SST map point $map(i,j)$, the proposed method scans at fixed steps a region around the corresponding point $\theta(i,j)$ of map θ. Each of the point values of thermal gradient direction represents a component of the feature vector describing the region surrounding point $\theta(i,j)$.

The sampling procedure is performed using a binary mask centred on $\theta(i,j)$ and reading gradient direction angles at '1' locations.

4.3. ANN meddy identification

Given an SST map point, the numerical vector describing the surrounding temperature gradient field is fed to a multi-layer perceptron (MLP) [17] classifier for meddy detection.

The MLP is perhaps the best known and most successful type of ANN. It is a fully connected feedforward ANN [13] composed of three or four layers of processing elements. The input layer gathers the incoming signals to the network and generally acts as a buffer. One or more *hidden layers* of neurons follow. Each hidden layer collects signals coming from the previous layer and processes them via a non-linear transformation function. It can be shown that no more than two layers of hidden neurons are required to form any arbitrarily complex decision region [17].

The ouput layer collects the signals from the last hidden layer and further processes them to give the final classification result.

Typically, the network undergoes a training phase where the weights of the connections between neurons are adjusted. Learning changes the system response by modifying the way incoming signals to the neurons are scaled.

Structure optimisation is of crucial importance for the performance of the ANN. Different topologies are usually trained and their merit assessed on the learning accuracy. The final configuration is often the result of a trial and error process, where several architectures are created and tested before the best solution is found.

Once the ANN architecture is optimised and the network is trained, the classifier is ready to operate.

5. Experimental settings and identification results

5.1. Experimental set up and algorithm implementation

Tests were carried out on a set of 32 SST maps of the Atlantic Ocean taken between March and July 2001 and made available by Instituto de Oceanografia, Faculdade de Ciências, Universidade de Lisboa, Portugal, (www.io.fc.ul.pt). Each map is organised as a 700x700 matrix of temperature readings with pre-marked land and cloud areas. A PentiumIII 1GHz processor with 512 MB of RAM was used for the tests.

Each map was pointwise pre-processed according to the algorithm described in the previos section. A set of functions was defined, each function implementing one step of the procedure. The processing of an image required about 30 minutes.

The distribution of meddies along the 32 maps is uneven due to different sea states and visibility conditions. IO experts manually identified a set of visible meddy structures, out of which 105 instances were selected for training and testing purposes. 445 examples of other sea structures (e.g., thermal fronts, open sea, etc.) were picked from the same maps as negative instances. A total of 550 data points were thus selected, each point corresponding to the centre of a meddy or a "non-meddy" region.

For each of the selected points, a numerical vector is generated from the map of gradient directions. Different masks were tested to sample the gradient field around the points. Experimental results refer to the three raster scan masks shown in fig. 4. *Mask 5x5* generates a 25-dimensional vector of gradient direction angles, *Mask 7x7* and *Mask 9x9* generate, respectively, a 49-dimensional and an 81-dimensional vector.

The MLP classifier takes as input the vector of angle measures generated via the masks of fig. 4 and produces a positive or negative identification response.

The network is trained via the Backpropagation (BP) algorithm [17] on a set of 440 randomly selected examples representing 80% of the total positive and negative meddy instances. The remaining 110 data points are kept for validation purposes.

The above partition generates a training set containing 84 instances of meddy patterns and 356 instances of non-meddy patterns. To improve learning, the size of the two classes was balanced by duplicating instances of meddy patterns until they reached 356.

5.2. Experimental results

Table 1 reports the learning results for the three feature masks of fig. 4. For each mask, two columns refer to the best MLP configuration respectively with one and two hidden layers. For each column, accuracy results are estimated on the average of 10 independent learning trials. The recognition accuracy is reported together with the standard deviation over the 10 trials. Fig. 5 shows the average accuracy within the span of plus or minus one standard deviation.

Table 1 also breaks down the classification accuracy on each of the two classes and reports the number of BP training cycles. The latter was optimised to maximise identification accuracy.

Experimental evidence shows that the proposed algorithm is able to recognise with high accuracy patterns of thermal gradient direction generated by meddy phenomena. Results showed robust performances with small standard deviations over the 10 learning trials. The very high recognition accuracy on negative instances of meddy patterns is particularly important since it assures a low rate of false detections.

Fig. 5 best shows the increase of the classifier accuracy with the size of the feature extraction mask. In particular, the classification accuracy on negative instances of meddy patterns seems to increase steadily with the span of the extraction mask, while the classification accuracy on positive examples of meddy patterns has a less linear behaviour. It should however be noted that such differences are comparable to the size of one or two standard deviations. It is therefore entirely possible that the observed variations are solely due to stochastic fluctuations.

A larger feature extraction mask seems also to allow the optimisation of less complex MLP structures. It is possible that larger masks convey a richer description of the temperature gradient field, making thus the recognition of different oceanic phenomena simpler. This would explain the possibility of training smaller

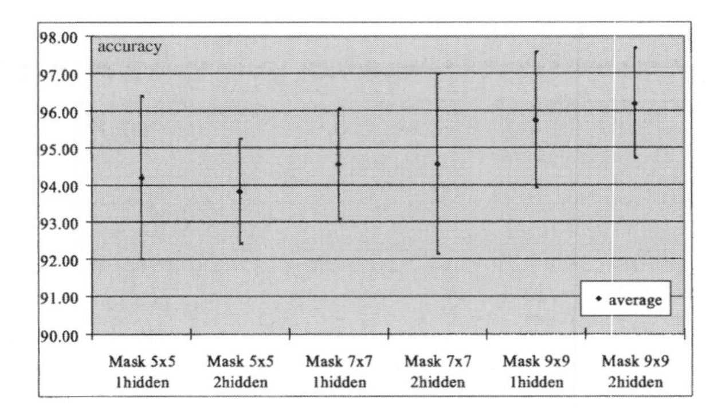

Fig. 5. Meddy identification accuracy results

ANN architectures when larger feature masks are used.

Differences in performance between structures composed of one or two hidden layers seemed to be modest. Four layered (two hidden layers plus input and output layer) MLPs required more training iterations, even though learning times were extremely reasonable.

The best performing classifier was achieved using *Mask 9x9* and used two hidden layers of 5 and 2 units. It required 3300 iterations of the BP algorithm to reach top performance for a learning time of 12 minutes circa.

6. Discussion

Experimental results proved that the proposed algorithm produces robust and high performing MLP solutions. These MLP solutions are able to successfully identify points in SST images that lie at the core of meddy structures. Identification is based on the temperature gradient field in a region around the point. The algorithm is fast, since the total processing time is mainly determined by the pre-processing stage that requires about 30 mins. The computational cost of the identification step is negligible due to the fast recall speed of the ANN classifier.

There are two ways the proposed method can be used for scanning SST maps. The first possibility is to scan pointwise the whole image and produce a classification result for each point. Identification results can be organised as a map, where each location corresponds to a positive or negative meddy detection response. Positive identifications are expected to be clustered in areas lying at the centre of meddy structures, while occasional false detections are likely to be scattered more evenly around the map. The application of a smoothing operator followed by thresholding may be sufficient to remove most of the

Table 1. Experimental results

	Mask 5x5	Mask 5x5	Mask 7x7	Mask 7x7	Mask 9x9	Mask 9x9
1st layer	15.00	25.00	20.00	15.00	10.00	5.00
2nd layer	-	15.00	-	5.00	-	2.00
features	25.00	25.00	49.00	49.00	81.00	81.00
accuracy	94.18	93.82	94.55	94.55	95.73	96.18
std. dva	2.19	1.41	1.48	2.42	1.82	1.47
meddies	88.10	87.62	87.62	87.14	88.10	90.48
non-meddies	95.62	95.28	96.18	96.29	97.53	97.53
iterations	8500	9500	1500	2500	500	3300

noise from the map of detection results.

A second possibility envisages the segmentation of the pre-processed SST image (i.e., the gradient direction map) into overlapping tiles of the kind used by Peckinpaugh et al. [11]. The size of each tile could for an example correspond to half the size of an extraction mask. Larger masks can be tested to further speed up the segmentation procedure. Positive detection results would then refer to the entire region covered by the mask.

Compared to the existing literature, the proposed algorithm shares the machine learning approach of Peckinpaugh et al. [11] and Cantón-Garbín et al. [12]. It is however conceptually simpler as image pre-processing is the result of the sequential application of straightforward convolution masks. Such a modular approach also allows easier modification, since masks can be quickly replaced, removed or introduced. Image pre-processing is the main difference between the proposed method and the one of Peckinpaugh et al. [11], while Cantón-Garbín et al. [12] also used a considerably more complex recognition system.

Regarding the pre-processing of SST images, the proposed method bears the most similarities with the analysis of oriented texture method of Lemonnier et al. [6] which is also used by Alexanin and Alexanina [9].

The method here presented also differs from the other approaches (e.g., [6,11]) since it uses gradient direction that is 2π periodical, instead of working with gradient orientation measures that are π periodical. It is believed the direction of the gradient vectors carries useful information for recognition of meddy structures.

As regards classification accuracy, it is difficult to draw a comparison between the proposed approach and other methods. Related work reported in this paper widely differ for geographic location, experimental settings and evaluation method. The recognition algorithm of several other models also targeted other mesoscalar structures such as upwellings and island wakes [7,8,12] and differentiated cold eddies from warm ones [11].

As a general remark, the proposed algorithm doesn't seem to suffer from the common problem of high detection of false positives [8,11,12]. Recognition rates vary widely in the literature. Thonet et al. [8] report 88% accuracy on identification of well formed eddies over 100 SST images of the Bay of Biscay. Cantón-Garbín et al., [12] obtained over 80% recognition accuracy for Atlantic eddies near the Canary Islands and 66% accuracy for eddies in the Mediterranean Sea. Guindos-Rojas et al. [7] report classification accuracies varying from 80% to 95% for the recognition of island wakes, cold and warm eddies and upwellings. Conservatively, it is possible to claim that the performance of the proposed algorithm is competitive with other results in the literature.

7. Conclusions and further work

This paper presented a new algorithm for automatic detection of meddies from SST maps of the Atlantic Ocean. The algorithm pre-processes the imagery via successive applications of simple convolution kernels. The direction of the thermal gradient is calculated for each SST map element.

Gradient field information in the neighbourhood of a map point is extracted through a binary mask and organised as a numerical vector of gradient directions. The numerical vector is forwarded to a MLP classifier for identification of meddy patterns. The network is trained on a set of pre-classified positive and negative examples of points belonging to meddy formations.

Experimental tests produced encouraging results and proved the efficacy of the proposed approach.

The proposed identification system achieved high accuracy with fast and robust learning results over a range of different extraction masks. Results were also characterised by very low rates of detection of false positives. This result is particularly important as meddies occupy only a small portion of SST map area.

Further work should focus on improvement of the image pre-processing procedure. Different convolution kernels should be tested and other image processing algorithms should be tried for generating descriptions of sea structures. Laws' texture energy measures [18], Fourier power spectrum analysis, Eulerian vector fields [19] are envisaged as possible alternatives for meddy characterisation. Different and more refined binary extraction masks should be tried as well.

Improvements should also focus on the design and training of the MLP classifier. Different classification algorithms could also be investigated.

Finally, software optimisation for compactness and speed may bring further enhancements of performance.

Acknowledgements

The author would like to thank Instituto de Oceanografia, Faculdade de Ciências, Universidade de Lisboa, for their expert technical support. The presented work is part of AI-MEDEA and RENA (http://www.io.fc.ul.pt/fisica/research.htm) research projects which are sponsored by Fundaçao Ciencia Tecnologia, Ministerio da Ciencia e Ensino Superior, Portugal.

References

[1] Reid, J. L. (1994), On the total geostrophic circulation of the north atlantic ocean: flow patterns, tracers, and transports, Progress in Oceanogr., Vol. 33, Pergamon, pp. 1–92.

[2] Oliveira, P.B., Serra, N., Fiúza, A.F.G. and Ambar I. (2000), A study of meddies using simultaneous in situ and satellite observations, Satellites, Oceanogr. and Society, ed. D. Halpern, Elsevier Science B.V..

[3] Armi, L. and Zenk, W. (1984), Large lenses of highly saline mediterranean water, Journal of Phys. Oceanogr., Vol. 14, 1560-1576.

[4] Parisi Baradad, V., Cabestany, J., García-Ladona, E. and Font, J. (2000), Modelling spatial structures in sst images through eulerian vector fields, Proc. SPIE, Vol. 4172. pp. 56-61.

[5] Bower A., Armi L. and Ambar, I. (1997), Lagrangian observations of meddy formation during a mediterranean undercurrent seeding experiment, Journal of Phys. Oceanogr., Vol. 27, No. 12, pp. 2545-2575.

[6] Lemonnier, B., Lopez, C., Duporte, E. and Delmas, R. (1994), Multiscale analysis of shapes applied to thermal infrared sea surface images, Proc. Intern. Geosc. and Remote Sensing Symp. 1994 (IGARSS'94), Pasadena - CA, Vol.1, pp. 479-481.

[7] Guindos-Rojas, F. Cantón-Garbín, M., Torres-Arriaza, J. A., Peralta-Lopez, M., Piedra-Fernandez, J. A. and Molina-Martinez, A. (2004), Automatic recognition of ocean structures from satellite images by means of neural nets and expert systems, Proc. of ESA-EUSC 2004, Madrid - Spain.

[8] Thonet, H., Lemonnier, B and Delmas, R. (1995), Automatic segmentation of oceanic eddies on AVHRR thermal infrared sea surface images, Challenges of Our Changing Global Environment, Conf. Proc. OCEANS '95, San Diego - CA, Vol. 2, pp. 1122-1127.

[9] Alexanin A. I. and Alexanina M.G. (2000), Quantitative analysis of thermal sea surface structures on NOAA IR-images, Proc. of CREAMS'2000 Intern. Symp. - Oceanogr. Of The Japan Sea, Vladivostok - Rus, pp. 158-165.

[10] Eugenio, F., Rovaris, E., Marcello, J. and Marques, F. (2003), Automatic structures detection and spatial registration using multisensor satellite imagery, Proc. Intern. Geosc. and Remote Sensing Symp. (IGARSS), Toulouse - France, Vol. 2, pp. 1038-1040

[11] Peckinpaugh, S. H., Chase, J. R. and Holyer R. J. (1993), Neural networks for eddy detection in satellite imagery, Proc. SPIE, Vol. 1965, pp.151-161.

[12] Cantón-Garbín, M., Torres-Arriaza, J. A., Guindos-Rojas, F. and Peralta-Lopez, M. (2003) , Towards an automatic interpretation and knowledge based search of satellite images in databases, Systems Analysis Modelling Simulation, Vol. 43, No 9, pp. 1249–1262.

[13] Pham, D. T. and Liu, X. (1995), Neural Networks for Identification, Prediction and Control, Springler-Verlag Ltd., London.

[14] Fitch, J. P., Lehmann, S. K., Dowla, F. U., Lu, S. Y., Johansson, E. M. and Goodman, D. M. (1991), Ship wake-detection procedure using conjugate gradient trained artificial neural networks, IEEE Trans. on Geosc. and Remote Sensing, Vol. 29, No. 5, pp. 718–726.

[15] Peckinpaugh, S. H., Holyer, R. J. (1994), Circle detection for extracting eddy size and position from satellite imagery of the ocean, IEEE Trans. on Geosc. and Remote Sensing, Vol. 32, No. 2, pp. 267-273.

[16] Matteson, R. G. (1995), Introduction to document image processing techniques, Artech House, Boston.

[17] Lippmann, R. P. (1987), An Introduction to Computing with Neural Nets, IEEE ASSP Magazine, pp. 4-22.

[18] Laws, K. I. (1980), Rapid texture identification, Proc. SPIE Vol. 238, pp. 376-380

[19] Parisi Baradad, V. (2000), Analysis of mesoscale structures though digital images techniques, PhD thesis, Univ. Politècnica Catalunya, Barcelona - Spain.

Intelligent Production Machines and Systems
D. T. Pham, E. E. Eldukhri and A. J. Soroka (eds)

Complementary agents for intelligent cluster sensors technology

S.Mekid

School of Mechanical, Aerospace and Civil Engineering, The University of Manchester, Sackville Street, Po Box 88, Manchester, UK.

Abstract

The purpose of this article is to discuss further expectations for a new concept of intelligent cluster sensors. New concepts of agents have to be implemented embedding new functionalities to dynamically manage cooperative agents for autonomous machines most required in manufacturing for zero defects production. The future trend is in intelligent clusters or distributed sensors in hybrid networks leading to a more global object control within a machine able to self organize into fault-tolerant and scalable systems.

1. A Challenging technology

There is a clear trend towards modularity of future holistic intelligent NC machines, PLCs and Robots based on distributed control design, which allows flexible control configuration and adaptation of systems. Such systems are governed by intelligent control systems consisting of a hierarchical structure of production control, machine control and drive control layers that have to implement open interfaces, learning capabilities, self-tuning mechanisms and sophisticated model-based prediction instruments in order to allow automated error-free machining for example. The topic discussed in this paper is applicable to PLCs and Robots but the author will emphasize the implementation in NC machines as a good example.

The multi process autonomous machine has to be equipped with new concepts of multi agent sensors cluster to measure different variables such as cutting parameters, in-process measurement of dimensions, geometric defects ...etc as shown in figure 1. After a static, dynamic and thermal analysis, a sensor mapping strategy on the machine is of paramount importance to ensure continuous feedback.

As an example, various wireless sensors are placed at high amplitudes locations of major modes of vibrations, while thermal sensors are located in critical locations subject to larger expansion. With such a sensor mapping, it is planned to implement compensation of errors and forces while machining to ensure zero defect workpiece in the autonomous mode.

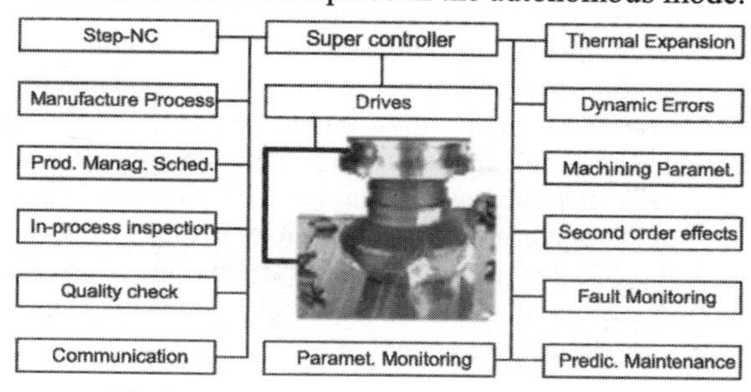

Fig.1: Error mapping in machine tools.

With the increasing number of sensors, the concern would be how to use reconfigurable sensors could be used for maintenance hence the required data will participate in feature recognition to help identifying the type of possible failure and immediately decide on the actions to be taken (self diagnostic and self service strategy, self healing requiring autonomous supervision. Finally, intelligent reconfigurable sensors are most needed. The potential advantages of such intelligent sensor concept includes:

- Reduced down time;
- Fault tolerant systems;
- Adaptability for self-calibration and compensation;
- Higher reliability;
- Master/Slave sensors mapping capability;
- Lower weight;
- Lower cost;
- Lower maintenance.

In the next part of the paper, the structure of the intelligent sensor and the different types of agents used to build the intelligence will be discussed.

2. Systems characteristics

Systems have different characteristics and requirements such as the case of the autonomous machine, the categories where agents as defined in cognitive systems are most suitable [1, 2] and defined hereafter:

- *Ill-structured:* usually designed systems do not have all the necessary structural information, they are ill-structured. Agents have a distinct notion about themselves and know that their environment can change, hence suitability for a dynamic environment.
- *Modular:* an agent applies very well to modular systems as it has dedicated tasks.
- *Decentralised:* an agent can easily be applied in decentralised system to ease self-decision making without interference with other processes.
- *Changeable:* an agent is very good for changeable systems. Change could be handled by both modular and decentralised systems. Decentralisation minimises the impact one module has on another module when it changes.
- *Complex:* with the increased complexity of systems, an agent is very well suited to handle such difficult task to model the behaviour of one or several components within its responsibility.
- *Real-Time:* most of the systems are currently operating in real time control where an agent is very suitable to achieve it.

3. Sensor Structure

More sophisticated sensors described, as intelligent or smart sensors have been known for more than two decades. These sensors are more complicated than traditional sensors as they simply gather, analysis and transmit data. The existing structure of intelligent sensors comprises the basic fundamental elements required in major manufacturing and other applications. And these are given as follows:

- **Sensing element**: type of material used to sense.
- **Signal conditioning and data conversion**. The signal obtained from the sensing element is modified, enhanced and converted to a discrete time digital data stream before passing through a processing element.
- **Processing element** that includes a microcontroller with an associated memory and software; this is the main component of the architecture where the incoming signal is processed.
- **Communication element**, which provides two-way communication between the processing element and users through, provided communication means such as wireless, optical fibres, serial buses, etc as well as interfacing to successful communication with outside world.
- **Power source**, energy required to power up the sensor.

4. Intelligent agents

The capabilities of the sensors are relatively limited by the degree of intelligence provided by the associated agents. Hence, it is important to identify the required agents as well as their structure to suit the above requirements. But according to [3] the definition of such agents is still not clear among the interested community. *"Some have tried to offer the general definition of agents as someone or something that act's one one's behalf, but that seems to cover all of computers and software. Other than such generalities, there has been no consensus on the essential nature of agents."* [4]

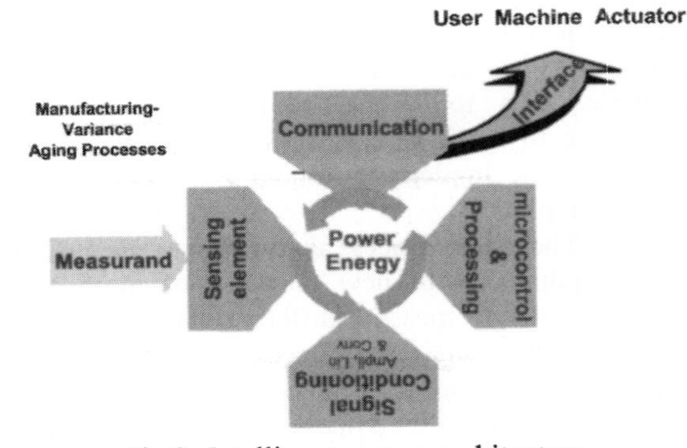

Fig.2. Intelligent sensor architecture.

A large list of definitions has already been proposed in the literature to include the following without privileging any of the definitions:

The Brustoloni Agent [5,6], *"Autonomous agents are systems capable of autonomous, purposeful action in the real world."*
The SodaBot Agent [7], *"Software agents are programs that engage in dialogs [and] negotiate and coordinate transfer of information."*
The Maes Agent [8], *"Autonomous agents are computational systems that inhabit some complex dynamic environment, sense and act autonomously in this environment, and by doing so realize a set of goals or tasks for which they are designed."* Maes also mentions other agent characteristic such as fast, reactive, adaptive, robust, autonomous and "lifelike".
The Nwana agent [9], to identify the emerging intelligence between agents, Nwana has emphasised autonomy, learning, and co-operation as characteristics that are very important in an intelligent agent but he did not mention the word 'intelligence' itself. In figure 3 shows the emerged agents from the interaction between the characteristics as defined by Nwana.

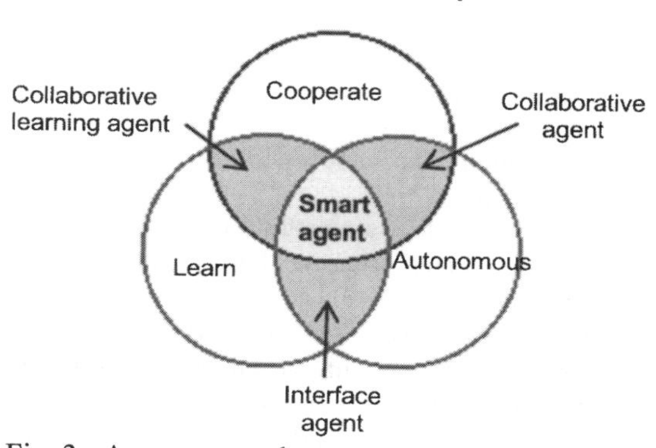

Fig. 3: An agent topology according to Nwana.

The Foner agent [10], according to Foner an agent must have certain characteristics that must be fulfilled in some way or another for an agent to be called an intelligent agent. The characteristics mentioned are autonomy,. personalization, risk and trust, domain, graceful degradation, co-operation, anthropomorphism and expectations.
The Petrie agent [4], a general description of an agent as someone or something that acts on ones behalf, is according to Petrie not a sufficient definition, since this description can be applied to all computers and software.
The Jennings and Wooldridge agent [11], according to Jennings and Wooldridge an agent must be *autonomous* an agent is autonomous if it is capable of acting without any direct guidelines from either humans or other agents. This means that the agent itself has control over its own actions and behaviour, i.e. the agent encapsulates its behaviour and internal state. If an agent is compared to an ordinary object that also has an internal state we can see an important difference; that there is at least one method in an object that can be invoked by another component. This implicates that an object is not autonomous.

The Hayes-Roth agent [12], finally the following definition of an intelligent agent is presented by Hayes-Roth: *"Intelligent agents continuously perform three functions: perception of dynamic conditions in the environment; action to affect conditions in the environment; and reasoning to interpret perceptions, solve problems, draw inferences, and determine actions."*

5. Single-agent and Multi or hybrid agent systems

In the application cited earlier where different types of measurements are required for different purposes, it is clear that agents have to either work alone or cooperate to achieve a task. The question could be whether a single agent could perform all the tasks or the concept of multi-agent has to be introduced. Communication between agents is important to transfer information between agents with a capability of prioritization. With multi-agent systems you can use parallelism, which speed up the system by parallel computation. You also gain robustness, scalability and the programming becomes easier [13].

Multi agent systems would not probably be of systematic use if the system is not suitable or of it is applied it may delay or will be an expensive solution. Further multi-agents are not suitable when used only to decentralise a system normally modelled as a centralised system. One should not try to provide multi-agent solutions to the wrong problems, it is foremost important to focus on the problems the multi-agents are meant to solve and not the possible benefits [14].

Today the focus of agent research is more on multi-agent systems than on single agent systems, since agents that co-operate and/or communicate can solve much more complex tasks than just one single agent is capable of [15]. Verification and validation of multi agents has already been addressed in [16]. Learning and autonomous capabilities should be part of the multi-agents strategy.
Nahm and Ishikawa [17] have proposed a hybrid network of agents as shown in Fig.4, they employ a lightweight middle agent, called ''interface agent'' for continuous or discrete interactions between local

agents (type 1), between a collection of local agents and a collection of remote agents (type 2), and between a local agent and a remote agent (type 3). Therefore, agent interactions are made only via the interface agent.

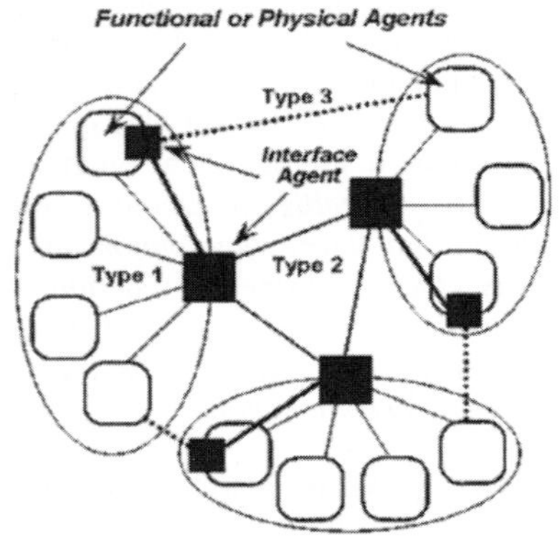

Fig.4: Proposed network application [17]

6.Example of agent's functionalities

A sensor agent can control and optimise the use of a sensor within the cluster or distributed sensors system. Among the tasks, it could decide which sensor has to measure according to its location and at an appropriate time. It could be replaced in case of failure and task immediately transferred to another equivalent sensor. The designed agent should have the ability to learn rapidly from the existing situation to decide which sensor should go for the current query. The agent could remember the more reliable sensors from low reliable ones and could learn as it progresses. With its local control this agent can determine the efficiency of the overall measurement, learn, cooperate and communicate with other agents.

An object agent, as another new concept dedicated to the object to be monitored or measured could be implemented. It conserves all information about it. The user could request at any time the current or history information on the object. If an object is moving in a fluid, a series of sensors could track it as it passes in their respective action area and the information is compiled by the sensor agent for the abject agent. Laser trackers for moving objects are very good examples.

7. Intelligent sensor functionalities

With an objective to emulate human capabilities in a fast process with parallel computation which may help in making an optimum decision, several functionalities were defined such as compensation, computing, validation, communication, integration and data fusion to ensure transfer of information between sensors.

7.1 Compensation functionality

The compensation is a method to improve measurements for better accuracy by considering the errors in the system. Time based or long term drift compensation due to degradation of the sensor elements.[18].

7.2 Computing functionality

Computing is the ability to provide the most relevant information in an efficient representation to the communication interface. This processing task involves signal conditioning, signal conversion; logic functions, data reduction and decision making that are used to enhance the signal received.

7.3 Communication functionality

The purpose of the communication functionality is to exchange the efficient information between the sensor and user and to allow users to reprogram for other measurements.

7.4 Validation functionality

Validation is a method to increase the overall reliability of the sensor. Faulty data can cause unexpected behaviors or system failure; Self-validation becomes more important when different data from multi-sensors or multi-measurements are sent to the system [18, 19].

7.5 Integration functionality

It concerns the integration of mechanical elements, sensors, actuators, and electronics on a single chip to eliminate wires connection among components, to reduce the overall size of the sensors, to optimally use the power energy and to reduce costs.

7.6 Data fusion functionality

This function ensures that only the most relevant information is transmitted between sensors.
According to the application reviewed above, development of the intelligent sensors has been progressing through improvement of the processing element. Using fuzzy or neural networks promotes sensor flexibility, self-compensation, self-validation, self organization functionalities [14]. As well as

integrating all elements into one single chip has contributed to the miniaturization of the sensors. Besides size-reduction and improving processing performance, other elements in the intelligent sensors have also been developing. For examples; measurement technology increases accuracy, stability and reliability of the sensing elements. Self-powered technology can enhance power consumption property. Communication technology such as wireless, bluetooth, infrared (IR), etc. extends communication range.

8. Practical expectations for intelligent sensors

The future complete structure of intelligent sensors depends on the development of agent philosophy as well as the hardware with smart sensing element. Intelligent sensors and actuators will form a network that is able to sense, process, communicate the information, act, and eventually modify system conditions based on this information. The intelligent network will mainly comprise intelligent sensors and actuators, communication channels and an advanced centralized processor.

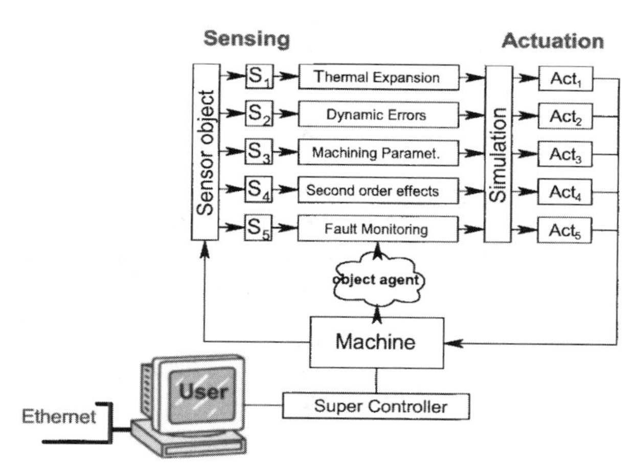

Fig. 6. Intelligent sensor cluster activity.

When the intelligent sensor network is becoming available, the ultimate goal to imitate human abilities may become a reality. Figure 6 present the overall system requirement with an example of intelligent sensors architecture in a manufacturing system using sensors clusters deployed in sensitive locations such as hazard warning applications with sensor agent as well as an object agent in closed loop structure.

8.1 Sensor Master/clusters set-up for single/hybrid processes

Reconfigurable and 3D measurement will be possible to track the existing phenomena. Self-calibrating will be easier because of the multiplicity of the sensors tracking the same parameter.

8.2 In process Parameter evaluation

In a autonomous process system, the sensing system should be able to track the failing process parameters in real time and apply self adjustment according to requirements. In more optimistic situations, the cluster could capture high order effects that may degrade the performance of the current process.

8.3 Extreme environment sensors for manufacturing processes and cutting interfaces

This is an extremely important issue required in manufacturing at standard level to monitor the cutting interfaces; e.g. laser processing, high speed machining. It is extremely important and challenging at nanoscale machining.

8.4 Plug and Play sensors with standardization of interfaces

Standardization of interfaces with rapid recognition of the host identity is of paramount importance to ease addition and mapping of different sensors and devices.

9. Conclusion

With the complexity trend in the new advanced systems requiring multiple variables closed loop control systems, multi agent in connexion with object agent in hybrid networks will constitute a fundamental structure for intelligent and autonomous systems but not a trivial solution. Various agents have been defined that could be implemented to build new intelligent cluster sensors. Robustness and accuracy of information are of paramount importance in such systems. The way forward is to define a strategy for self-learning capabilities within the local environment. Direct industrial benefits are zero-defects artefacts production, autonomous robots for complex and lengthy tasks and hybrid production lines with reconfigurable manufacturing systems.

Acknowledgment
The author gratefully acknowledges the support of the European Network of Excellence IPROMS for this work.

References

[1] Bradshaw, J. M., *Software Agents*, AAAI Press / MIT Press. Ed. (1997).

[2] Jennings, N. J. and Wooldridge, M.J. (1998), Applications of Intelligent Agents, In *Agent Technology - Foundations, Applications, and Markets*, Jennings, N. R. and Wooldridge, M. J., Ed., Springer-Verlag, Berlin, Germany, pp. 3-28.

[3] Andersson L.and Rönnbom Å., Intelligent Agents- *A New Technology for Future Distributed Sensor Systems?, MSc thesis,* Göteborg University, *1999*

[4] Petrie, C. J. (1996), "Agent-Based Engineering, the Web, and Intelligence", *IEEE Expert/Intelligent Systems & Their Applications 11,* 6, pp. 24-29.

[5] Brustoloni, Jose C. (1991), "Autonomous Agents: Characterization and Requirements," Carnegie Mellon Technical Report CMU-CS-91-204, Pittsburgh: Carnegie Mellon University

[6] Franklin, Stan (1995), Artificial Minds, Cambridge, MA: MIT Press.

[7] M.Coen http://www.ai.mit.edu/people/sodabot/slideshow/total/P001.html

[8] Maes, Pattie (1995), "Artificial Life Meets Entertainment: Life like Autonomous Agents," *Communications of the ACM,* 38, 11, 108-114

[9] Nwana, H.S., Software Agents: An Overview, *Knowledge Engineering Review,* 1996, 11, 3, 1-40.

[10] Foner, L.N. (1997), "What's an agent anyway?", To appear in *Proceedings of the First International Conference on Autonomous Agents (AA'97).*

[11] Jennings, N. J. and Wooldridge, M.J. (1998), "Applications of Intelligent Agents", In *Agent Technology - Foundations, Applications, and Markets*, Jennings, N. R. and Wooldridge, M. J., Ed., Springer-Verlag, Berlin, Germany, pp. 3-28.

[12] Hayes-Roth, B. (1995), "An Architecture for Adaptive Intelligent Systems", *ArtificialIntelligence: special issue on agents and interactivity,* 72, 329-365.

[13] Stone, P. and Veloso, M. (1997), "Multi-agent Systems: A survey from a machine learning perspective", Technical Report 193, Department of Computer Science, Carnegie Mellon University, Pittsburgh,.PA.

[14] Collier, T. C. and Taylor C., Self-organization in sensor networks, Journal of Parallel and Distributed Computing, 2004, vol. 64, pp. 866-873.

[15] Benoit, E. and Foulloy, L., High functionalities for intelligent sensors, application to fuzzy colour sensor, Measurement, 2001, vol. 30, pp. 161-170.

[16] M.Bourahla and M.Benmohamed Formal Specification and Verification of Multi-Agent Systems, *Electronic Notes in Theoretical Computer Science, Volume 123, 1 March 2005, Pages 5-17*

[17] Y.-E. Nahm and H. Ishikawa A hybrid multi-agent system architecture for enterprise integration using computer networks • *Robotics and Computer-Integrated Manufacturing, Volume 21, Issue 3, June 2005,Pages217-234*

[18] Tian, G. Y., Zhao, Z. X., and Baines, R. W., A fieldbus-based intelligent sensor, Mechatronics, 2000, vol. 10, pp. 835-849.

[19] Clarke, D. W., Intelligent instrumentation, Transactions of the Institute of Measurement and Control, 2000, vol. 22, pp. 3-27.

Intelligent Production Machines and Systems
D. T. Pham, E. E. Eldukhri and A. J. Soroka (eds)

Fuzzy Logic Modelling of a Powder Coating System

D. T. Pham, S.S. Dimov, O. A. Williams, N. B. Zlatov

Manufacturing Engineering Centre, Cardiff University, CF24 3AA, UK

Abstract

The successful application of powder coating technology depends on resolving practical issues such as reducing overspray and maximising transfer efficiency. Maximising transfer efficiency requires optimising application parameters that influence powder deposition efficiency. This paper presents the design and development of a predictive fuzzy-logic model for powder coating applications. To improve the performance of the fuzzy-logic model, the model was design based on optimised conditions as determined by statistical analysis. The fuzzy-logic based powder coating system has four application variables as inputs and one output, and gives excellent performance under the conditions investigated. The fuzzy-logic model can easily be implemented in the factory and has the ability significantly to reduce waste while maximising first-pass transfer efficiency.

1. Introduction

The concept of electrostatic powder coating has emerged as an attractive alternative to solvent based coating. The increasing popularity of this surface finishing technique can be attributed to a number of environmental and economic factors, and advancements in the use of the technology. The technology has been successfully deployed in special applications such as clear coating on brass fixtures and aluminium wheels while new applications are continually being developed [1]. However, users of the technology are still faced with challenges such as how to improve efficiency of the powder application process while achieving high coating quality.

The efficiency of a powder application process is frequently indicated by two common terms, material utilisation efficiency and first-pass transfer efficiency (FPTE). Material utilisation efficiency is the ratio of powder applied to powder used (including reclaim); on the other hand, first-pass transfer efficiency describes the productivity of the powder application process by measuring the ratio of applied powder to sprayed powder (without reclaim). Optimising powder deposited on the workpiece is important in achieving higher FPTE, which result in reduced waste and increased savings. This paper investigates the development of an intelligent powder coating system that can help to improve powder deposition on workpieces.

Figure 1 illustrates the powder application process. Uncoated parts are arranged on a conveyor. The operator sets up the process parameters for the particular coating job based on experience. Charged powder particles are sprayed onto the electrically grounded parts from powder coating guns which utilise corona charging technique. The charged powder particles deposited on the workpiece and adhere by means of electrostatic image forces. The powder-coated parts are then passed through and heated in a curing oven to ensure smooth coating.

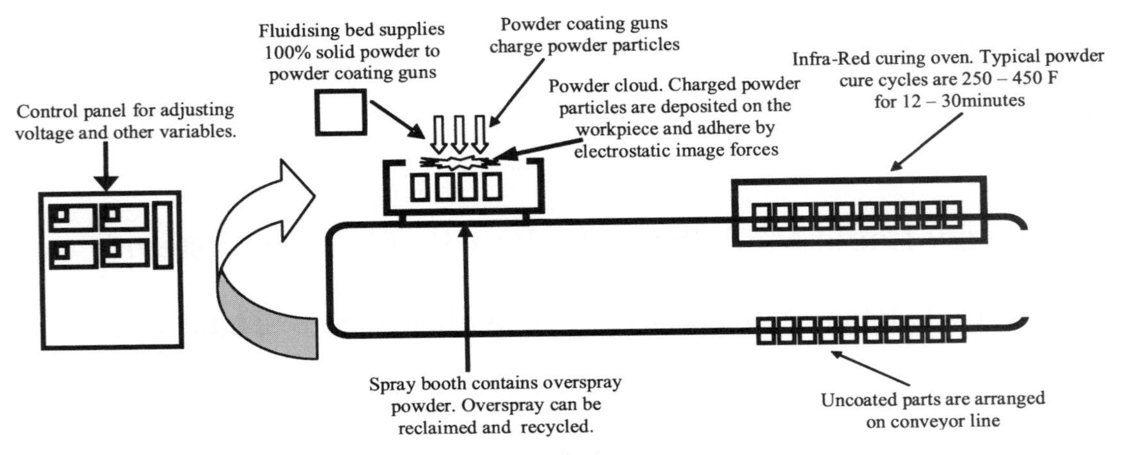

Figure 1: Powder application process

2. Fuzzy modelling and the design methodology for the fuzzy-logic based powder coating system

Fuzzy-logic modelling is a heuristic method and is suited to systems that require the ability to handle uncertainitiess and/or model imprecise reasoning procedures. A fuzzy-logic model uses IF-THEN rules in conjunction with logical operators. Membership functions map identified inputs of the system to membership values from 0 to 1, while the IF-THEN rules forms the rule base of the system. As inputs are received by the system, its inference mechanism evaluates all IF-THEN rules and determines their truth values. If a given input does not precisely correspond to an IF-THEN rule, then partial matching of the input data is used to produce an answer by interpolation. The process of combining all fuzzy conclusions obtained by inference into a single conclusion is called composition while defuzzification converts the fuzzy value obtained from composition into a "crisp" value. Defuzzification is necessary, since models of physical systems require discrete signals [2].

Figure 2 represents the schematic illustration of the methodology adopted for the design and development of the fuzzy-logic predictive model for a powder coating system. The design methodology utilised design of experiments to investigate different levels of application variables and to hunt for optimal values of the application variables. The fuzzy-logic powder coating system was designed based on the optimal conditions as determined by statistical analysis.

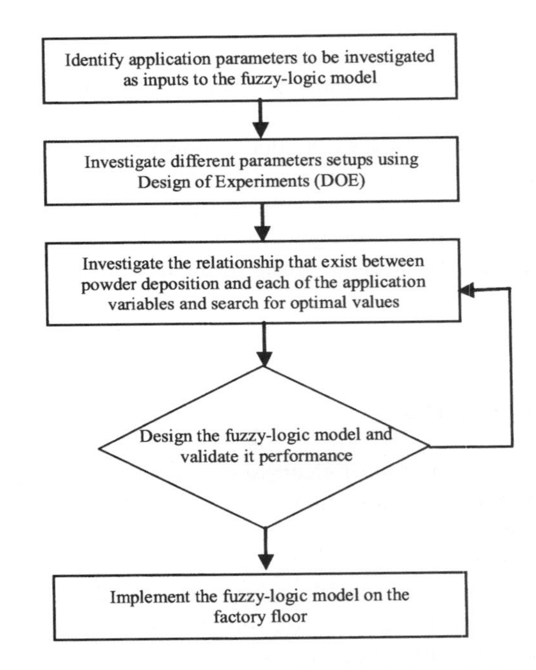

Figure 2: Design methodology for the fuzzy-logic predictive model

3. Experimental design and material setup

Based on practical considerations, design of experiments using orthogonal array was implemented to reduce the number of experimenatal runs required to investigate different combinations of application parameters. Detailed input setups for the L9 orthogonal array, single-level design with four factors at three levels, low, medium and high, with no interaction are

Table 1
Detailed parameter set ups

	Experimental setups								
Experimental factors	**1**	**2**	**3**	**4**	**5**	**6**	**7**	**8**	**9**
Powder-volume pressure (KPa)	117.21	117.21	117.21	137.90	137.90	137.90	158.58	158.58	158.58
Rotation speed	1.5	2	3	1.5	2	3	1.5	2	3
Gun-to-part distance (mm)	60	90	120	90	120	60	120	60	90
Gun current (μA)	40	50	60	60	40	50	50	60	40

reported in Table 1. The four application variables investigated in the study were: "powder-volume pressure (KPa)", "workpiece rotation speed", "gun-to-part distance (mm)", and "gun current (μA)". For each level of gun current investigated, the gun voltage was adjusted so that current limits at gun current setting.

The workpieces used for the study were 3-piece aerosol containers manufactured from large pre-decorated flat tinplate sheets. A black epoxy polyester gloss was chosen. This is able to produce 60-70microns coating thickness [3], and when cured gives a glossy surface finish. Figure 3 shows the 3-piece tinplate aerosol containers being powder coated in the pilot equipment which utilises corona charging.

Figure 3: 3-piece cans being powder coated in an electrostatic powder coating system

For the nine experimental conditions investigated in the study, 30 cans were sent through the powder booth while cans at positions 9 to 23 (15cans) were selected as samples. A total of 135 cans were taken as samples for data analysis. The weight of powder deposited on cans was obtained by measuring the weight of the cans using a microbalance before and after coating.

4. Assessing powder deposition using response surface method

Response surface method (RSM) is a composite technique that implements the steepest ascent method to search for optimum points. The method can be used to search for the optimal response within ranges of the experimental conditions and can model the change in the response variable without large changes in the operating parameters[4]. Using the experimental data the method was used to construct the response surface plots for the four application parameters. The fuzzy-logic model for powder coating was designed and developed to replicate the response surface plots given similar experimental conditions.

Figure 4 shows qualitatively the shape of the response plots given that powder-volume pressure is set at 137.90 KPa, can rotation speed at 2.25 turns in front of the powder coating gun, gun-to-part distance of 90mm and gun current of 50μamps. At these operational levels, predicted maximum powder deposition is 0.86grams.

Figure 4 also reveals that powder deposition increases initially with increase in the levels of can rotation speed, gun-to-part distance and gun current and decreases once these application parameters are set at higher levels. In addition the figure indicates that there may be a linear relationship between powder deposition and powder-volume pressure within the levels of powder-volume pressure (KPa) investigated in the study. However, it is impossible to say what happens at further levels of powder-volume pressure (KPa) outside those considered in the experiment.

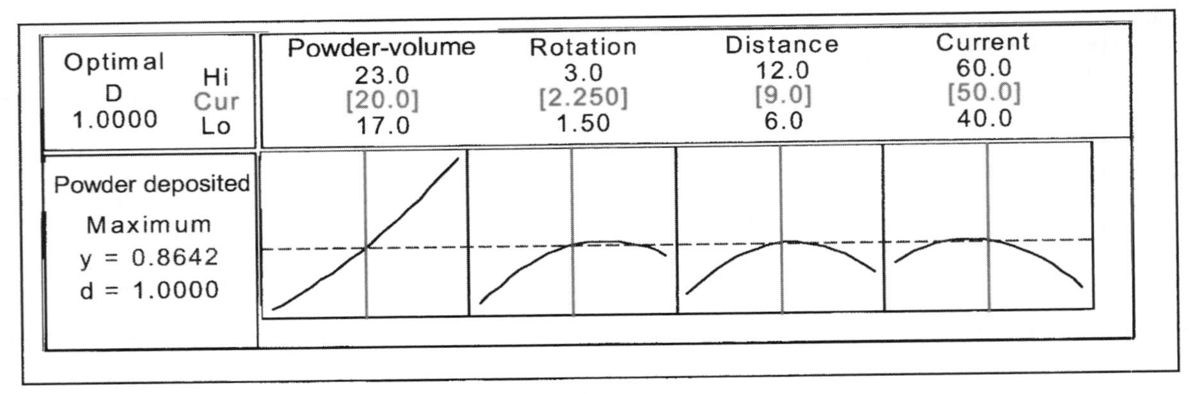

Figure 4: Response graphs of the application parameters

5. The fuzzy-logic based powder coating system

Figure 5 shows the components of a Mamdani type fuzzy-logic model. The figure depicts the triangular membership functions for the four input variables and one output variable to the fuzzy-logic model. The relation between the four inputs and one output of the fuzzy-logic model is represented by the rule matrix presented in Table 2. Various methods have being developed for the extraction of control rules for fuzzy-logic models [5], the approach adopted here was to extract the rules base of the fuzzy-logic model based on the behaviour of the process as depicted by the response surface plots. This approach is offer the advantage of modelling the behaviour of the fuzzy-logic model to resemble response surface plots of the process. The rule base of the heuristic based fuzzy-logic model is given in Table 2. A typical rule reads:

IF Powder-volume pressure is Medium,
Workpiece rotation speed is NOT High,
Gun-to-part distance is NOT High AND
Gun current is NOT Low,
THEN Powder deposited on the workpiece is
NOT Medium

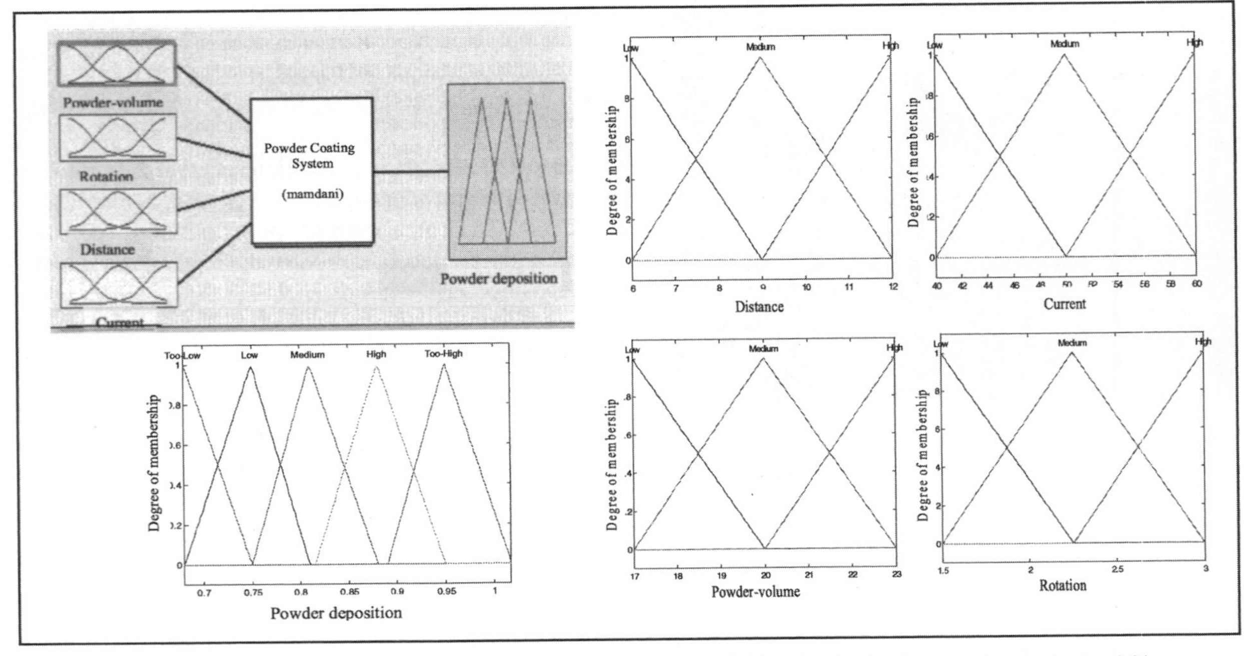

Figure 5: Components of the fuzzy-logic model and membership functions definitions for the four inputs and one output variables

Table 2
Rule matrix for the fuzzy-logic based model

Powder Deposited	Powder Volume	Rotation speed	Gun-to-part Distance	Gun current
Not Medium	Low	Not High	Not High	Not Low
Not Too Low	Not Low	Not Low	Not Low	Not Low
Not Too High	Medium	Not High	Not High	Not Low
Not Medium	Medium	Not Medium	Medium	Medium
Not Medium	Medium	Low	Not Medium	High
Not Medium	Not High	Low	Not Low	Not Low
Not Too Low	Not High	Low	Medium	Not Low

Figure 6 presents the output of the fuzzy-logic model corresponsing to inputs [137.90 KPa, 2.25 turns, 90mm, 50μA]. Given these levels of inputs the fuzzy-logic model predict an optimum "powder deposition" of 0.866 gram indicating 99% similarity to prediction shown in figure 4 from the response surface analysis. In figure 7 the response plots for the four application variables are presented showing close similarities to the plots presented in figure 4.

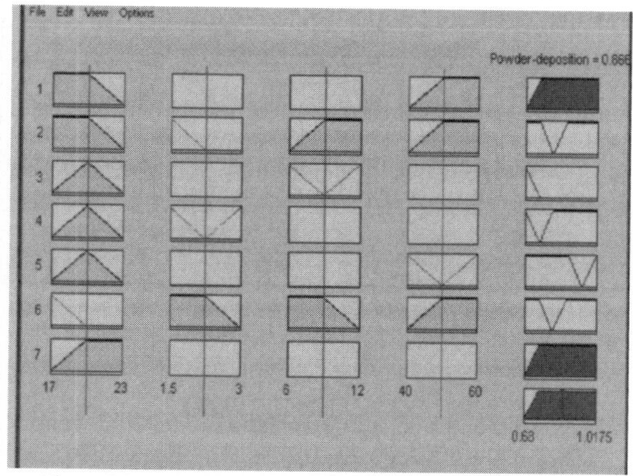

Figure 6: Matlab Fuzzy Logic Toolbox Rule viewer when inputs are [137.90KPa, 2.25, 90mm, 50 μA]

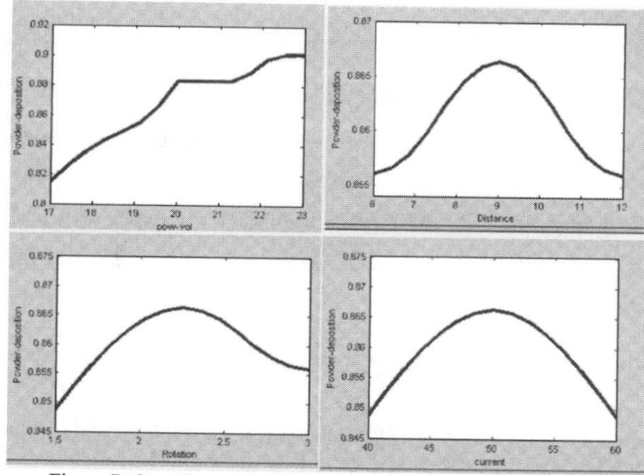

Figure 7: One-dimensional surface view of the fuzzy input variables

5.1. *Performance evaluation of the fuzzy-logic based powder coating system*

Figure 8 is a three-dimensional surface view indicating the shape of the response variable, powder deposition, given two inputs, rotation speed and powder-volume pressure (KPa). The figure indicates that an optimal powder deposition above 9.0 grams can be achieved if powder-volume pressure (KPa) is set at 158.58 KPa and workpiece rotation speed is at 2.25 turns while maintaining gun-to-part distance and gun current at the present levels. The fuzzy-logic based powder coating system thus has the capacity significantly to improve first-pass transfer efficiency by maximising powder deposition.

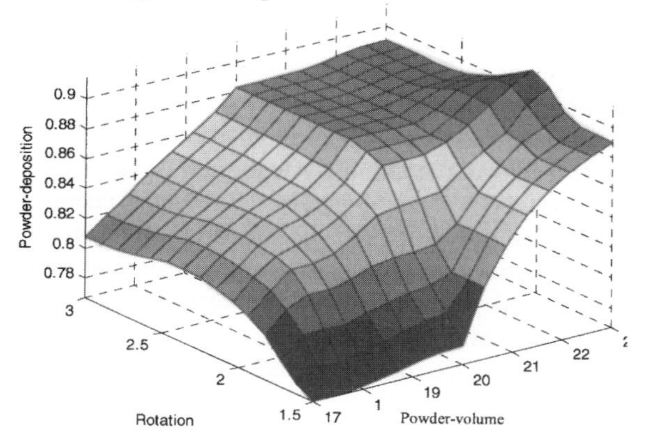

Figure 8: Three-dimensional surface view of the fuzzy-logic model

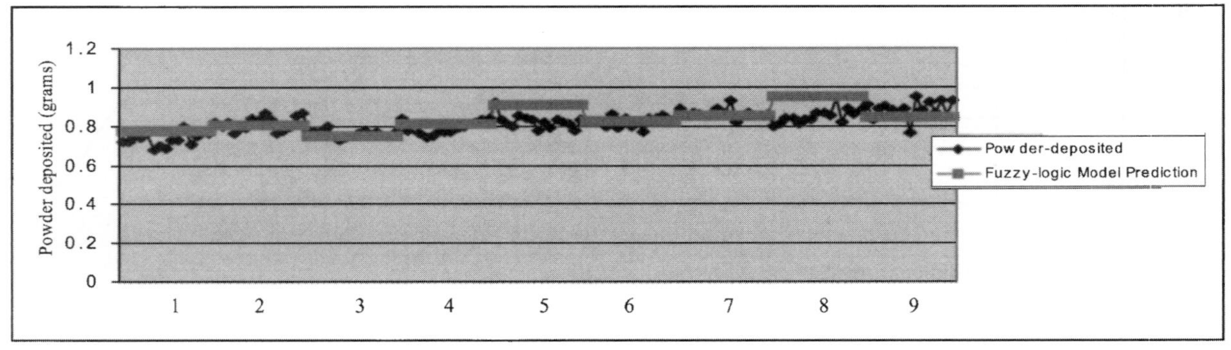

Figure 9: Experimental values and values predicted using the fuzzy-logic model

Figure 9 is a line graph comparison of the predicted outputs of the fuzzy-logic based model and the experimental data collected from 135 cans, 15 cans from each of the nine experimental setups. The plot shows qualitatively that similarities exit between the predicted values of "powder deposition" and the experimental data. Hence the fuzzy-logic based powder coating model can be said to perform well under the conditions investigated in the study.

6. Conclusion

Powder coating operatives frequently need to determine the input variable settings that optimise powder deposition. While each input variable is important in optimising powder deposition efficiency, it is important to understand how the combination of parameter settings affects coating jobs. Failure to understand the effects of input variable settings can result in reject and subsequent re-work which reduces the economic benefits that powder coating has over solvent-based coating. The economic benefits of an intelligent powder coating system that implements expert knowlegde (fuzzy-logic) include reduced energy requirements, low waste disposal costs, savings in labour and maintenance costs, increased production rates, reduced reject rates, and increased first-pass transfer efficiency.

This paper has demonstrated the application of fuzzy logic to powder coating. The fuzzy-logic model developed is based on the idea of designing the behaviour of the fuzz-logic model to be similar to the behaviour of a statistical representation of the process. Results obtained from the fuzzy-logic model for the powder coating system suggest the fuzzy-logic model has the ability to significant increase first-pass transfer efficiency, minimise the waste and reduce the time involved in selecting the best combination of application variables for coating jobs and improves savings.

Future work will include using the fuzzy model presented in this pa per in a control loop to realise an intelligent control system for the powder coating process.

Acknowledgement

The authors would like to thank the Engineering and Physical Sciences Research Council (EPSRC) / Department for Trade and Industry (DTI) for supporting this work as part of the LINK programme. The MEC is the co-ordinator of the EC-funded FP6 I*PROMS Network of Excellence.

References

[1] Gill D E. Powder coatimgs and their uses. Powder Coating Applications (1st edn), ed. Dawson S and Reddy V. Society of Manufacturing Engineers, Dearborn, Michigan,1990, pp 52 - 53.

[2] Lee C C. Fuzzy logic in control systems: fuzzy logic controller, IEEE Trans. Syst., Man. and Cybern., Vol.20, No.2, 1990, pp.404-417.

[3] Conesa C, Saleh K, Thomas A, Guigon A, and Guillot N. Characterization of flow properties of powder coatings used in the automotive industry. Hosokawa Powder Technology Foundation Scientific Journal, Japan, 22, 2004, pp 22 -106.

[4] Montgomery D C. Design and Analysis of Experiments (4th edn). John Wiley and Sons Inc., New York, 1996.

[5] Pham D T and Karaboga D. Intelligent optimisation techniques: genetic algorithms, tabu search, simulated annealing and neural networks. Springer-Verlag London, 2000, pp 249-258.

Intelligent Production Machines and Systems
D. T. Pham, E. E. Eldukhri and A. J. Soroka (eds)

Image segmentation using fuzzy min-max neural networks for wood defect detection

G.A. Ruz[a,b] and P.A. Estévez[a]

[a] *Department of Electrical Engineering, University of Chile, Casilla 412-3, Santiago, Chile*
[b] *SONDA S.A., Teatinos 500, Santiago, Chile*

Abstract

In this work a colour image segmentation method for wood surface defect detection is presented. In an automated visual inspection system for wood boards, the image segmentation task aims to obtain a high defect detection rate with a low false positive rate, i.e., clear wood areas identified as defect regions. The proposed method is called FMMIS (Fuzzy Min-Max neural network for Image Segmentation). The FMMIS method grows boxes from a set of seed pixels, yielding the minimum bounded rectangle (MBR) for each defect present in the wood board image. The FMMIS method was applied to a set of 900 colour images of radiata pine boards, which included 10 defect categories. The FMMIS achieved a defect detection rate of 95 percent on the test set, with only 6 percent of false positives. The area recognition rate (ARR) criterion was computed, to measure the segmentation quality, using as a reference the manually placed MBR for each defect. The ARR achieved 94.4 percent on the test set. The results show significant improvements compared with previous work and that the computational load of FMMIS is suitable for real-time segmentation tasks.

---

1. Introduction

Automated visual inspection (AVI) systems are an automated form of quality control normally achieved using a camera connected to a computer. The AVI framework includes five processing stages: image acquisition, image enhancement, image segmentation, feature extraction and classification [1]. A review of AVI research applied to the inspection of wood boards, concluded that segmentation is often the most time-consuming part of the process, and that usually does not locate all defects properly [2].

Colour image segmentation algorithms can be classified into one or more of the following techniques [3]: histogram thresholding, feature space clustering, region-based approaches, edge detection, fuzzy approaches, neural networks, physics-based approaches and hybrid techniques. The selection of a colour space is application dependent. Brunner et al. [4] found that, for wood images there is no advantage in transforming the red, green and blue (RGB) colour space into other colour spaces.

The flexibility of fuzzy sets and the computational efficiency of neural networks have caused a great amount of interest in the combination of both techniques. Among the neurofuzzy approaches, Simpson [5] introduced the fuzzy min-max (FMM) clustering neural network, where clusters are represented as hyperboxes in the *n*-dimensional pattern space. In [6], the FMM was extended and improved.

In this work , an image segmentation method based on the FMM neural networks is proposed. The new method is called fuzzy min-max neural network for image segmentation (FMMIS). The performance of this

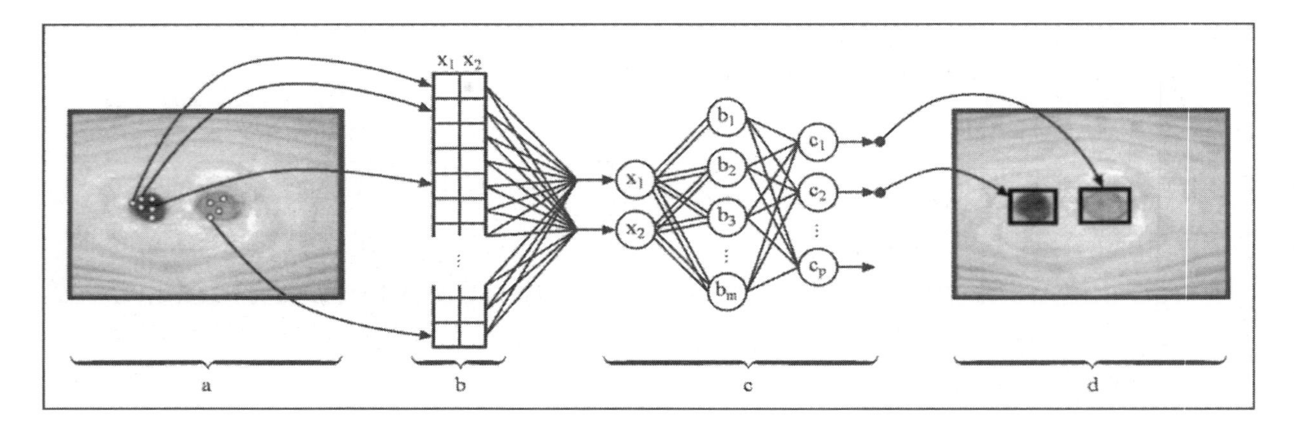

Fig.1. Different stages of the proposed image segmentation method. (a) seed selection process, (b) input patterns, (c) Fuzzy Min-Max neural network for Image Segmentation and (d) minimum bounding rectangles enclosing the defects, which correspond to the FMMIS output.

method is evaluated on the segmentation of wood (radiata pine) board images, which include 10 defect categories. The FMMIS method has been successfully applied to wood defect detection in [7] and to face detection in [8]. This paper is complementary to [7], where the FMMIS learning algorithm was not described in full, as is done here.

2. Image segmentation method

The proposed image segmentation method for wood defect detection consists of 4 stages, shown in Fig. 1. Each stage is described in the following subsections.

2.1. Seed selection process

To speed up the image segmentation process, the FMMIS does not use all the pixels from the image analysed. Instead, it only uses a few input pixels called seeds, to grow the hyperboxes. The seeds are automatically determined (located) by an ad-hoc procedure which is described in what follows.

Considering the great variability of colour of the wood boards, the seed selection is based on adaptive thresholding. For each board image, the mean colour intensity per channel, μ_t, and the minimum colour intensity per channel, κ_t, where t = R, G, B, are measured. A cumulative histogram per channel, H_t, is defined as:

$$H_t(n)=\sum_{i=\kappa_t}^{n} h_t(i) \tag{1}$$

where n is the colour intensity level ($n \leq 255$) and h_t is the histogram of the wood board image for channel t. In addition, a colour intensity level is selected, per channel, based on the cumulative histogram as follows:

$$\theta_t = \alpha H(\mu_t) \tag{2}$$

where α is a user-defined parameter, typically $\alpha \leq 0.01$. To detect defects that are brighter than those detected by using θ_t, an additional colour intensity is considered, and calculated as:

$$\rho_t = \frac{\theta_t + \mu_t}{2} \;. \tag{3}$$

For each wood board image, the seeds are the pixels belonging to the following intensity range:

$$I_t = \begin{cases} [\kappa_t \quad \theta_t] \wedge \rho_t & if\ \theta_t < \lambda_t \\ [\kappa_t \quad \theta_t] & if\ \theta_t \geq \lambda_t \end{cases} \tag{4}$$

where λ_t is a user-defined threshold for each channel, that allows to avoid false positives when all the defects on the wood board image are not too dark. In Fig. 1a. the seeds are represented as white circles.

2.2. Input patterns

The input patterns are the spatial coordinates of the seeds, determined in subsection 2.1. with each dimension normalized in the range [0,1]. Let X be a $S \times 2$ input matrix (see Fig. 1b.), where S is the number of seeds selected. The position of the hth seed in the image is represented by the vector $X_h = (x_{h1}, x_{h2}) \in I^2$, where the first coordinate indicates the column and the second coordinate the row of the image.

2.3. Fuzzy min-max neural network for image segmentation

The FMMIS is built using hyperbox fuzzy sets. A hyperbox defines a region in the *n*-dimensional geometric space by pairs of min-max points for each spatial coordinate of the image (rectangular boxes in the case of *2*-dimensional images).

Each hyperbox fuzzy set has associated a membership function which describes the degree of membership (spatial proximity) of a given pixel to a hyperbox in the [0,1] interval. Let each hyperbox fuzzy set, B_j, be defined by the ordered set

$$B_j = \{X_h, V_j, W_j, b_j(X_h, V_j, W_j)\} \tag{5}$$

where $V_j = (\upsilon_{j1}, \upsilon_{j2})$ is the min-point and $W_j = (\omega_{j1}, \omega_{j2})$ is the max-point. The membership function for the jth hyperbox is $0 \le b_j(X_h, V_j, W_j) \le 1$. Seeds contained within a hyperbox have full membership value equal to 1. The more distant the seeds are from the min-max bounds of the hyperbox the lower are their membership values. The membership function defined in [6] is used here:

$$b_j(X_h, V_j, W_j) = \min_{i=1,2}\left(\min\left([1 - f(x_{hi} - \omega_{ji}, \gamma)], [1 - f(\upsilon_{ji} - x_{hi}, \gamma)]\right)\right) \tag{6}$$

where f is the ramp function defined as,

$$f(g,\gamma) = \begin{cases} 1 & g\gamma > 1 \\ g\gamma & 0 \le g\gamma \le 1 \\ 0 & g\gamma < 0 \end{cases} \tag{7}$$

and γ is the sensitivity parameter, that controls how fast the membership value decreases when the seed is farther from the min-max bounds of the hyperbox. Fig. 2 shows the two-dimensional membership function of (6), with min-point V = [0.4 0.4] and max-point W = [0.7 0.7], and γ=1. The membership values ranging from 0 to 1 are represented by the gray-scale ranging

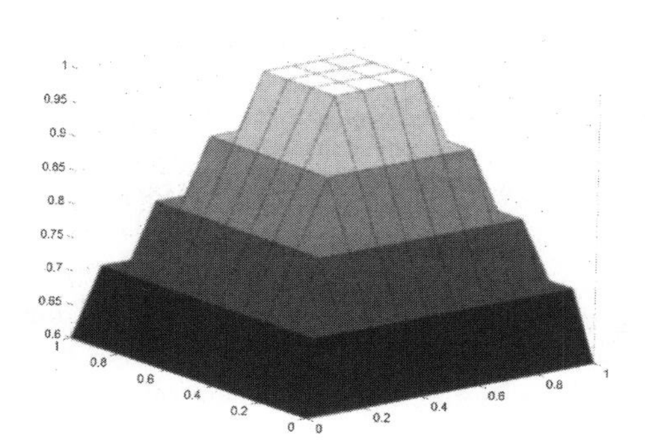

Fig.2. Example of the two-dimensional membership function associated to each hyperbox used by the FMMIS.

from black to white, respectively.

2.3.1. FMMIS learning algorithm

1. Initialisation: V_j and W_j points are initially set to 0. When a hyperbox is adjusted for the first time using the input pattern $X_h = (x_{h1}, x_{h2})$, the min and the max points of the hyperbox are made identically to this pattern,

$$V_j = W_j = X_h. \tag{8}$$

2. Hyperbox Expansion: When an input pattern (new seed pixel) is presented, the hyperbox with the highest degree of membership is found and expanded to enclose the input pattern. The hyperbox expansion is accepted only if the region contained by the expanded hyperbox is similar in colour to the region enclosed by the hyperbox before the expansion. A fuzzy colour homogeneity criterion based on the standard *Z*-function of the Euclidean distance between the mean colour intensities of the two hyperboxes is defined. In this way, the colour similarity, in the RGB space, between the hyperboxes before and after the expansion are compared. A user-defined parameter $\tau \in$ [0,1] is introduced to control the required degree of colour homogeneity for expanding hyperboxes. Formally, the following constraint must be satisfied to expand a hyperbox.,

$$Z(x,a,b,c) \ge \tau, \tag{9}$$

where *Z* is a fuzzy membership function defined as

$$Z(x,a,b,c)=\begin{cases}1 & 0\le x\le a\\ 1-2\left(\frac{x-a}{c-a}\right)^2 & a\le x\le b\\ 2\left(\frac{x-c}{c-a}\right)^2 & b\le x\le c\\ 0 & c\le x\le L\end{cases} \quad (10)$$

and x is the Euclidean distance between the mean colour intensities in the image region covered by the two hyperboxes (before and after expansion), measured in the RGB space. The parameters of (10) are set to: $a = 0$, $b=\frac{L}{2}$, $c = L$, and $L = 255\sqrt{3}$. If the expansion criterion for including the current seed is not satisfied, a new hyperbox is created starting with that seed as done in (8).

3. Hyperbox Overlap Test: Let us assume that the hyperbox B_j was expanded in the previous step. To test for overlapping, a dimension-by-dimension comparison is performed between B_j and all the rest B_k with $k\neq j$. Overlap exists between B_j and B_k, if one of the following four cases are met for dimensions i= 1,2:

Case 1: $\upsilon_{ji} < \upsilon_{ki} < \omega_{ji} < \omega_{ki}$

Case 2: $\upsilon_{ki} < \upsilon_{ji} < \omega_{ki} < \omega_{ji}$

Case 3: $\upsilon_{ji} < \upsilon_{ki} \le \omega_{ki} < \omega_{ji}$

Case 4: $\upsilon_{ki} < \upsilon_{ji} \le \omega_{ji} < \omega_{ki}$.

4. Hyperbox Contraction: If overlap exists between B_j and B_k, both hyperboxes begin to contract until the overlap is eliminated. The hyperbox contraction rules, that depend on the four cases previously described, are as follows:

Case 1: $\upsilon_{ji} < \upsilon_{ki} < \omega_{ji} < \omega_{ki}$,

$$\upsilon_{ki}^{new} = w_{ji}^{new} = \frac{\upsilon_{ki}^{old}+\omega_{ji}^{old}}{2} \quad (11)$$

Case 2: $\upsilon_{ki} < \upsilon_{ji} < \omega_{ki} < \omega_{ji}$,

$$\upsilon_{ji}^{new} = w_{ki}^{new} = \frac{\upsilon_{ji}^{old}+\omega_{ki}^{old}}{2} \quad (12)$$

Case 3: $\upsilon_{ji} < \upsilon_{ki} \le \omega_{ki} < \omega_{ji}$,

If $\omega_{ki} - \upsilon_{ji} < \omega_{ji} - \upsilon_{ki}$ then

$$\upsilon_{ji}^{new} = \omega_{ki}^{old} \quad (13)$$

otherwise,

$$\omega_{ji}^{new} = \upsilon_{ki}^{old} \quad (14)$$

Case 4: $\upsilon_{ki} < \upsilon_{ji} \le \omega_{ji} < \omega_{ki}$, by symmetry the same assignments as in Case 3.

5. Fine-tuning Hyperbox Expansion: After a single pass through all the seeds, there is a fine-tuning hyperbox expansion process, which allows the hyperbox to grow, if necessary, until the defect is completely enclosed. For 2D images, the hyperboxes are rectangles defined by four line segments. A horizontal line segment, hmin, and a vertical line segment, vmin, pass through the min-point of the hyperbox. Likewise, a horizontal line segment hmax, and a vertical line segment, vmax, pass through the max-point of the hyperbox. For the hyperbox B_j, and the colour channel t, the following notation is introduced: $h\max_{jt}(r)$, $v\max_{jt}(r)$, $h\min_{jt}(r)$ and $v\min_{jt}(r)$ are the colour intensities of the rth pixel belonging to the line segments hmax, vmax, hmin and vmin respectively. Let u_t be a colour intensity threshold defined for channels $t = R,G,B$. A line segment (any of the four) would be expanded if it contains pixels darker than u_t. The following conditions for the *while* cycles should be satisfied for each $t = R,G,B$.

Case 1: *while* $\left(\max_r\left(h\max_{jt}(r)\right)\le u_t\right)$

$$\left\{\omega_{j2}^{new} = \omega_{j2}^{old} + 1, \textit{update } h\max_{jt} \textit{ with } \omega_{j2}^{new}\right\} \quad (15)$$

Case 2: *while* $\left(\max_r\left(v\max_{jt}(r)\right)\le u_t\right)$

$$\left\{\omega_{j1}^{new} = \omega_{j1}^{old} + 1, \textit{update } v\max_{jt} \textit{ with } \omega_{j1}^{new}\right\} \quad (16)$$

Case 3: *while* $\left(\max_r\left(h\min_{jt}(r)\right)\le u_t\right)$

$$\left\{\upsilon_{j2}^{new} = \upsilon_{j2}^{old} - 1, \textit{update } h\min_{jt} \textit{ with } \upsilon_{j2}^{new}\right\} \quad (17)$$

Case 4: *while* $\left(\max_r\left(v\min_{jt}(r)\right)\le u_t\right)$

$$\left\{\upsilon_{j1}^{new} = \upsilon_{j1}^{old} - 1, \textit{update } v\min_{jt} \textit{ with } \upsilon_{j1}^{new}\right\} \quad (18)$$

6. Hyperbox Merging: After finishing the fine-tuning hyperbox expansion process (if necessary), a final step is added in order to merge hyperboxes belonging to the same defect. The number of hyperboxes constructed per defect depends on the value of τ in (9). Let c be the number of hyperboxes after the fine-tuning hyperbox expansion process. The centroid, CB_j (cb_{j1},cb_{j2}), of the hyperbox B_j is computed as:

$$cb_{ji} = \frac{\upsilon_{ji}+\omega_{ji}}{2} \quad (19)$$

for $i = 1,2$ and $j = 1,\ldots, c$. Let I_j be the image region contained within the limits of the hyperbox B_j. Let $d_E(I_j,I_k)$ be the Euclidean distance in the colour space between the mean intensities of the image regions I_j and I_k. The fuzzy membership function Δ_c, which measures the proximity in colour and space between two hyperboxes, is defined as:

$$\Delta_c = \min (Z(d_E(I_j,I_k),a,b,c), b_j(CB_j)) \quad (20)$$

where Z is the Z-function defined in (10), and b_j is the membership function defined in (6). The merging criterion is to merge hyperboxes whose proximity defined by (20) is greater than a given threshold.

The FMMIS representation as a neural network (see Fig. 1c.) makes possible to explore the parallelism of the algorithm. Although the learning algorithm is not necessarily neural (there is no biological principal underlying the expansion-contraction process), the execution of the network, once trained, fits in a neural scheme. For this case, a three-layered neural network was chosen to implement the FMMIS, as shown in Fig. 1c. The input layer, F_X, consists of two processing elements (PE), one for each dimension of the input pattern (seed) $\boldsymbol{X}_h = (x_{h1}, x_{h2})$. The second layer, F_B, consists of m PE (ideally one per each defect in the wood board image). Finally, the third layer , F_C, does the merging of the nodes of the second layer belonging to the same defect. There are two connections between each node of F_X and each node of F_B (see Fig. 1. c.). Each node of F_B in the three-layered neural network represents a hyperbox fuzzy set, where the connections from F_X to F_B are the min- and max- points. The transference function of F_B is the membership function defined in (6). The connections are adjusted using the learning algorithm described in this subsection. The connections between the nodes of F_B and F_C have binary values and they are stored in the $\boldsymbol{U}$ matrix, with value 1 if b_j follows the merging criterion giving origin to c_k , and with value 0 otherwise.

2.4. Output of the FMMIS

The last stage is to draw the rectangle (minimum bounding rectangle) on each defect using the min- and max-points of each hyperbox formed by the FMMIS algorithm, as shown in Fig. 1d.

3. Methods

A sample of 900 colour images (320×240 pixels) of wood boards was drawn from the University of Chile database [9]. Each image was manually labelled according to its largest defect, into one of the following 10 defect categories (see Ruz et al. [7] for details): birdseye & freckle, bark & pitch pockets, wane, split, stain, blue stain, pith, dead knot, live knot and hole. The data set, which corresponded to 90 images per category, was partitioned into two sets: 600 images for the training set and 300 images for the test set. The parameters of FMMIS using the training were set to, $\gamma=1$ (sensitivity parameter), $\tau = 0.99$ (degree of colour homogeneity used in hyperbox expansion), $u_R=195$ (fine-tuning hyperbox expansion parameter), and $D = 0.95$ (hyperbox merging parameter).

The performance of the FMMIS algorithm was measured on the test set using the following criteria: number of true positives TP (the number of defects contained by hyperboxes and correctly detected), number of false positives FP (the number of grain lines contained by hyperboxes, i.e., detected as defects), and the average processing time. The area recognition rate (ARR) criterion allows to compare the area of the hyperbox built automatically by FMMIS with the area of the manually placed minimum bounding rectangle. The ARR is defined as,

$$ARR = \left(1 - \frac{ukp}{tp} - \frac{unp}{tp}\right) \quad (21)$$

where tp is the total number of pixels in the MBR; ukp is the number of unrecognized defect pixels, i.e., the absolute difference between the defect pixels contained within the MBR and the defect pixels contained within the hyperbox determined by the FMMIS method; and unp is the number of unrecognized non-defective pixels, i.e., the absolute difference between the clear wood pixels contained within the MBR and the clear wood pixels contained within the hyperbox determined by the FMMIS method. The FMMIS algorithm was implemented in MATLAB 6.5 on a PC Pentium IV, 2.4 GHz, 512 MB RAM. The average processing time was measured for each image, from the seed selection process to the output of the FMMIS.

4. Experimental results

The FMMIS global performance on the test set, and its per-category disaggregation are shown in Table 1. The global TP rate achieved 95 percent of the total defects present in the test set, while the FP rate was 6 percent of the total grain lines (clear wood) present in the test set. The global area recognition rate achieved was 94.4 percent. In contrast, the global performance of the segmentation module of our previous AVI system [9], which used histogram-based multiple thresholding, obtained a TP rate of 94 percent with an FP rate of 32 percent on the same test set. None of the categories achieved an FP rate lower than 10 percent. On average, the number of selected seeds per image was about 100, i.e., 0.1 percent of the total number of pixels of an image. This fact makes the FMMIS algorithm very fast.

The last column of Table 1 shows the average processing time for FMMIS, including the seed selection process, which reached 0.11 ± 0.04 seconds per image. In comparison, the Seed Region Growing (SRG) algorithm of Adams and Bischof [8] obtained an average processing time per image of 3.03 ± 1.30 seconds (programmed in MATLAB and run on the same PC than the FMMIS algorithm) on the same test set, i.e., the FMMIS was 27 times faster than SRG. Moreover, the SRG algorithm obtained poor segmentation on the birdseye and freckle, stain, blue stain, and split categories (TP < 50 %).

Table 1
Performance of the FMMIS for wood defect detection, measured on the test set

Defect Category	TP %	FP%	ARR %	Time (sec)
Birdseye and freckle	91.6	10.8	90.3	0.12
Bark and pitch pockets	100.0	2.6	95.3	0.07
Wane	100.0	10.2	100.0	0.15
Splits	100.0	11.8	90.6	0.11
Stain	88.9	8.0	84.9	0.21
Blue stain	90.2	7.1	88.2	0.13
Pith	100.0	4.2	99.8	0.13
Dead knots	97.4	1.7	99.3	0.08
Live knots	100.0	4.1	97.1	0.08
Holes	100.0	0.0	98.3	0.06
Global (10 categories)	95	6	94.4	0.11± 0.04

5. Conclusions

The proposed colour image segmentation method achieved a high defect detection rate (95%) with a low false positive rate (6%) on images of wood boards. The FMMIS method is based on the original FMM, but with a new learning algorithm specially adapted for image segmentation tasks. The FMMIS method combines clustering with region-based techniques to obtain a substantially different method than the original Simpson's FMM. The results show that significant improvements have been obtained in comparison to previous work.

Acknowledgements

This work was supported in part by Conicyt-Chile, under grant Fondecyt 1050751.

References

[1] Pham DT and Alcock RJ. Smart Inspection Systems. Academic Press, London, 2003, 218 pp.

[2] Pham DT and Alcock RJ. Automated grading and defect detection: A review. Forest Prod. J. 48(4) (1998) 34-42.

[3] Cheng HD, Jiang XH, Sun Y, and Wang J. Color image segmentation: Advances and prospects. Pattern Recognition. 34 (2001) 2259-2281.

[4] Brunner CC, Maristany AG, Butler DA, VanLeeuwen D, and Funck JW. An evaluation of color spaces for detecting defects in Douglas-fir veneer. Industrial Metrology. 2 (1992) 169-184.

[5] Simpson PK. Fuzzy min-max neural networks. Part 2. Clustering. IEEE Trans. Fuzzy Sets. 1 (1993) 32-45.

[6] Gabris B and Bargiela A. General fuzzy min-max neural networks for clustering and classification. IEEE Trans. on Neural Networks. 11 (3) (2000) 769-783.

[7] Ruz GA, Estévez PA, Perez CA. A neurofuzzy color image segmentation method for wood surface defect detection. Forest Prod. J. 55 (4) (2005) 52-58.

[8] Estévez PA, Flores RJ, Perez CA. Color image segmentation using fuzzy min-max neural networks. Proceedings of the International Joint Conference on Neural Networks, 2005, Montreal, Canada.

[9] Estévez PA, Perez CA, Goles E. Genetic input selection to a neural classifier for defect classification of radiata pine boards. Forest Prod. J. 53 (7/8) (2003) 87-94.

[10] Adams R and Bischof L. Seeded region growing. IEEE Trans. Pattern Anal. Machine Intell. 16 (6) (1994) 641-646.

Intelligent Production Machines and Systems
D. T. Pham, E. E. Eldukhri and A. J. Soroka (eds)

Integrating human personnel, robots, and machines in manufacturing plants using ubiquitous augmented reality and smart agents

T. Schlegel[a], K. Müller[b]

[a] *Fraunhofer IAO, Stuttgart, Germany*
[b] *Danet GmbH, Weiterstadt, Germany*

Abstract

We describe a concept for an advanced manufacturing control system – the Smart Connected Control Platform. The approach enables a higher degree of automation, efficiency and control through a holistic model, decentralized production intelligence and modern sensor technology. Integrating ubiquitous and augmented reality interfaces supports human personnel in every aspect of the production process. To prevent information overload, intelligent agents paired with a semantic model help reduce the amount of information and deliver pre-interpreted and aggregated results to the manufacturing personnel. The advantages of this new approach are discussed in three example scenarios, which deal with advanced maintenance, customized production, and error diagnosis. The major contribution of our approach is the integration of machines, transportation robots, and human personnel using a decentralized smart agent platform and ubiquitous augmented reality technology. The platform described in this paper is currently under development.

1. Introduction

Today's manufacturing systems employ client-server networks and follow predefined deterministic programs. A plant supervisor schedules and downloads these programs form a central server over the network to the machines. Also Automatic Guided Vehicles (AGVs) are programmed to run on predefined routes through the factory and thereby transport production parts from one machine to another. All machine tools are supervised by human staff, who assist – for example – in positioning of individual parts. The work of the personnel at the machine tools themselves is oriented on details that the specific machine tools need, not on tasks in the production process.

In this paper we propose a novel approach to remedy the deficiencies that result from this inflexible and hierarchical setting. We introduce an innovative software platform, the Smart-Connected-Control Platform (SCC platform) for manufacturing enterprises. This platform currently under development allows controlling a production plant with help of an open distributed learning agent platform that integrates machines, robots, and – by using ubiquitous augmented reality – human personnel into a decentralized, networked control system.

Today, maintenance schedules for machines in manufacturing systems regard each machine isolated from the others. With the SCCP and its ability for point to point connection one machine can adapt to another's deficiencies, thereby extending maintenance periods.

In general the manufacturing of a single part consumes a considerable amount of time. Success in production can only be assessed after a part is produced and not during the production process itself. Though information is collected from places all over the factory the transmitted data is still barely interpreted and the data is used only locally. Hence automatic production process error recovery is hardly possible. The proposed SCC-platform integrates the various sources of data and correlates them in an intelligent way so that timely corrective actions are possible.

The production of customized products is still a challenge in current environments. It is usually accomplished by assembling a set of different, but similar standard components. The just-in-time production of customized basic parts is still unsolved. By using the Smart-Connected-Control Platform production of parts can be individually adjusted on the fly thus enabling a more customer-centered production process. A virtual reality interface allows (end)-customers the specification of their ideal product, which will subsequently be produced by the SCC-equipped manufacturing plant.

In the following sections we describe the advantages of our concept in supporting just-in-time maintenance, customized production, and error diagnosis in more detail. We also sketch a suitable architecture and show how integrated information sources and production tools could enhance current production with technology that will be available in the near future.

2. Related work

There exist many approaches to make manufacturing systems intelligent, faster, more flexible, and open to changes in vendor-specific definitions. This includes the application of artificial intelligence (AI) and smart agents as well as vendor neutral control architectures.

Albus [1] introduces the Intelligent Systems Architecture for Manufacturing (ISAM) as a hierarchical control architecture that allows modeling intelligent applications by assigning levels of control authority. The approach focuses on applications of artificial intelligence methods to the high level control of manufacturing systems. A reference model and initial implementation is demonstrated by Huang [2].

Musliner, Durfee, and Shin [3] focus in their definition of a Cooperative Intelligent Real-time Control Architecture on the separation of the AI technology from the real time control, with a structured interface between both parts of the system. The advantage of this approach lies in the usage of powerful AI methods without violating constraints applied by hard real-time conditions.

Shen and Norrie [4] apply an agent-based approach to cope with the challenges of manufacturing scheduling in open, dynamic environments. Pechoucek, Marík, and Stepankova [5] point to the advantages of using multi-agent systems in production processes and examine the coalition forming and social knowledge about collaborative agents. Recently, Walker, Brennan, and Norrie [6] presented a combination of AI techniques with heuristic job shop scheduling approaches and multi-agent systems to gain a dynamic and responsive scheduler.

One of the many advantages of using (multi-) agent systems is their openness. The idea of vendor independent systems has been examined by Lutz and Sperling [7,8] in the OSACA project.

With the rise of semantic web, semantic descriptions and matching have found their way into production applications, like semantic web-services for logistics [9]. Standards for description languages – necessary for service-oriented architectures (SOA) – have been created. Profiles of Web-Services can be described using DAML-S[1] or its successor OWL-S[2], based on the Resource description framework (RDF)[3]. While business processes can be described with languages like BPEL, holistic descriptions for more complex production processes are missing so far.

Approaches that use augmented reality applications on palmtop devices (PDAs) were presented by Benini et al. [10] in the context of virtual heritage, and by Wagner, Pintaric, Ledermann, and Schmalstieg [11] in the context of multi-user collaboration.

Virtual reality interfaces have been used in product design for a long time. Wesche and Seidel [12] developed an interface that allows freeform sketching, Wesche and Droske [13] developed a freeform styling application. Both systems are allow very advanced interaction styles for modeling a product in 3-D, without actually controlling the production process.

3. Semantic network for manufacturing plants

Approaches to intelligent manufacturing using AI methods tend to focus on control of machines or robots, sometimes also in combination. Most AI technology thereby concentrates on special tasks like logistics and scheduling. Human personnel are traditionally not integrated into these systems. With our approach, using handheld devices and advanced user interfaces, human-machine communication can be

[1] http://www.daml.org/services/daml-s/0.9/

[2] http://www.daml.org/services/owl-s/1.1/

[3] http://www.w3.org/RDF/

seamlessly integrated alongside with machine-machine communication. The advantages are obvious:

Wherever state-of-the-art technology is not sufficient for a task which is therefore performed by personnel the task can now be integrated into the control platform. Using ubiquitous augmented reality systems an operator can interact with the system on site, either to provide information that cannot be extracted by machine sensors or to carry out tasks that a human being can accomplish more efficient than a machine. Maintenance and repair crews as well as supervisors will have additional information and control through PDAs. Complex data can be overlaid with real parts and machines during the production process and thus support complex assignments.

Not only human operators can be seamlessly integrated in the maintenance and diagnose process. The SCC platform also allows customers to customize a product just-in-time prior to production. The design of a specific customized unique product can be performed in a virtual environment. The resulting specifications are then evaluated and implemented by the SCC platform, which programs the manufacturing plant and the production process just-in-time.

Even current operation procedures are enhanced by the highly networked SCC platform. Robots deliver goods tracked through smart tags in a more flexible way on routes that are determined just in time.

Furthermore the SCC platform – as a distributed agent system – can learn how to deal with error reporting and diagnosis. Operators on site can then use their advanced visualization technology and the SCC platform to diagnose problems in the production plant and react by flexibly reprogramming the production process.

3.1. The Smart-Connected-Control Platform

The SCC platform consists of infrastructure and semantic rules. The infrastructure consists mainly of network connections and distributed background knowledge for smart mobile agents. The network connections can be partially wired, partially wireless. Due to interferences using current wireless technology is not possible in all areas. As robots, especially Automatic Guided Vehicles, and human personnel can not drag along cables on their way but both are mobile they can change their location if they need to transfer data not having sufficient access. This, requires knowledge about hot-spots and suitable locations, which will be provided by the SCC-platform.

With this infrastructure in place communication paths between production machines and mobile entities, like humans or robots, can evolve to form dynamically changing networks. This allows point to point communication from human, robot or machine to each other, allowing every participant to influence each other's behavior. These ad-hoc networks provide the necessary operations layer for the smart mobile agents that form the other major part of the SCC-platform. The agents are distributed in nature to avoid bottlenecks in decision finding and increased flexibility. Decisions are negotiated among agents on different levels of abstraction with the well known AI approach of layering.

The semantic model in the Smart-Connected-Control platform bridges communication between heterogeneous production devices by ontology-mediated protocol adaptation. By this approach the system not only gains vendor independence. At the same time it supports formalized modeling of manufacturing processes and therefore automatic derivation of corrective actions in case of machine failure or human errors. The basis for this is constructed by a Unified Messaging Framework, able to transport any kind of production network data in a unified form, using different protocols but a standardized message and message description format.

The semantic message description can be interpreted by the intelligent platform in order to select, filter, aggregate and transform messages. Once this has been achieved, messages are re-encoded as far as necessary and sent to the derived information consumers, which can be devices connected to a machine or human personnel. As communication channels are part of the Smart Connected Control Platform, it is possible to adapt to new situations in real-time, for example by notifying a supervisor via his PDA instead of a voice message once he is in a loud area [14].

Using semantically described context information, it is possible to restructure and describe generated information as well as user input. A note left by a supervisor regarding problems with a special machine part in conjunction with parameters for an individually produced part, it is possible to improve knowledge management in and outside the factory: Platform and model deliver information in the right context, as platform state and platform model contain enough information for structuring, finding and applying knowledge items of the factory.

3.2. Integrating human personnel

Integration of human personnel into the new control structure is achieved by using mobile handheld

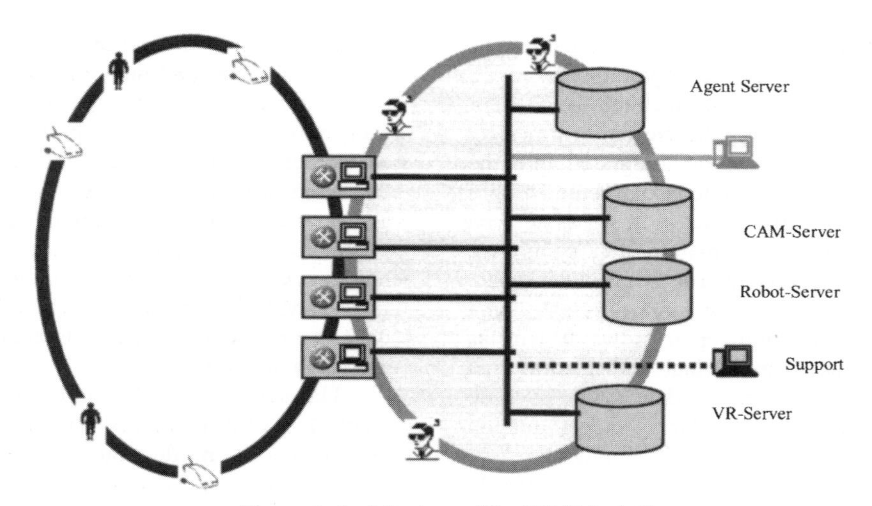

Figure 1: Architecture of the INMAS platform

devices (PDAs) in our prototype environment. Using ubiquitous computing and augmented reality technology personnel in the manufacturing plant has easy access to information and delivers input on site. In many situations this integration will be helpful: Human personnel can use its expertise in judging machine states and enter according data on site. Adversely, background information can be delivered on the fly to less knowledgeable personal. For example the PDA can be used as a navigation guide for external maintenance consultants – explaining to a repair crew where to find the machine in question, or as an interactive manual for a broader skilled crew, which explains what to do to open the machine or where to apply adjustments.

Also in the fully functional production process the on-site PDA is very useful. Personnel can use the PDA to overlay virtual information with the part in production to verify during the production process, if the part will be produced correctly. Therefore, it is possible to combine an ongoing quality control for each part with the dynamic reprogramming abilities of the SCC platform.

Moreover, with the possibility of controlling the actual production of a single part it is possible to guarantee first-part-correct production, even for unique, customized parts.

4. High-level architecture

The requirements and concepts presented above lead to a communication architecture that joins two levels: the factory floor level, where personnel, machines and robots are situated and the IT level, where the smart agents and the infrastructure govern the processes in the manufacturing plant. An overview over the architecture is shown in Figure 1.

The left circle shows the production plant floor and encompasses all participants located in the plant while the system operates. Robots (small carts) and personnel (humans) move in between the machine tools (shown in the middle, represented through computers and tool-icons).

The right side shows the network infrastructure with the participating servers. This is the infrastructure basis serving the agents (shown as heads) in the SCC platform to interact and control the system. So the agents are shown in a circle and touch all the hardware that can carry the mobile agents. The connection lines are drawn in solid for intra-net connections between the machine tools. The agent platform is a distributed system and all machine tools are equipped with numeric controls providing platforms for the mobile agents traveling on this network. The agents themselves have management overhead which will reside on the agent server in the system.

The CAM-server builds the repository for machine tool programs that can be downloaded to the machines as needed by the agents. A robot server builds the communication interface to the robots in the manufacturing plant. The connection to the support crew runs through a gateway. Therefore this line is dotted, as it is not part of the production plant intra-net. The VR server provides the interface to the customer, who will be able to order his customized product and – through the Smart-Connected-Control platform – will thereby program the manufacturing plant just-in-time according to his specific demands.

A CAM operator is involved in the beginning, when the system is set up. The system then works semi-automatically and the agents learn from the CAM operator. With time the agents become more and more autonomous in their decisions, so the CAM operator – who is now in a side position in the architecture – can step aside and leave the system most of the time in full-automatic mode. The learning process depends on actual run time of the fully implemented SCC platform in its final destination. Changes in processes (for example with relocations to another manufacturing plant or another company) will lead to the necessity of relearning.

5. Application areas

There are three scenarios that demonstrate advantages in applications of the Smart-Connected-Control platform to common manufacturing problems.

5.1. Advanced just-in-time maintenance

With help of the SCC platform pro-active maintenance is significantly improved. Machines and personnel enter machine states, necessary maintenance operations and their schedule into the SCC platform and the semantics enabled operating system reroutes parts, schedules maintenance operations or informs the production plant supervisor accordingly. The maintenance crew is notified of missing parts in case of an appointment on site.

The main advantages in this scenario lie in the direct communication from machine or personnel to machine and the corresponding self-calibration abilities. With the SCC platform it is no longer necessary to monitor single machines. Instead smart agents are able to monitor groups of machines. Furthermore, machines (or personnel with PDAs) are able to tell exactly when maintenance is necessary, not relying on statistically based estimations any more but on direct sensory data, models and events. Machines can go closer to limits for tolerances as one connected machine can compensate another's deficiencies and the platform can learn patterns of machine failure. Monitoring and supervising states of several machines together in an intelligent way (as done by higher level agents in the SCC platform) leads to just-in-time maintenance.

5.2. Customized production

A customer chooses and modifies a product in a virtual environment prior to its production. The production part's virtual representation will be adapted to the user input and once the user has finished the design phase, its semantic description will be analyzed and used as input for the SCC platform. The production process will then be adapted as needed for the customized product. This scenario uses the SCC platform with requirements generated on the fly.

Today, customization is implemented as assembling the 'customized' product from a variety of different basic parts, which were produced in series. So for example an individual car is built by choosing from a number of seats, wheels, colors, etc. Customization is therefore a question of logistics and keeping the right amount of varying basic parts in stock. MRP (Manufacturing Resource Planning) is thereby used to purchase components in time and solve the logistics problems. MRP systems do not aim at reprogramming machines in a production plant to produce a single unique basic part. Just-in-time production of customized basic parts is therefore not solved with the MRP approach.

The scenario demonstrates the SCC platform's flexible ability to build such a unique basic part just in time. Ideally, the semantic descriptions of machines in the manufacturing plant will be so strong, that the VR interface will automatically take restrictions from the production plant and process as well as for the resulting product into account and give a customer maximum tolerance in specifying his customized basic part. For example, this would allow the customer to specify his car seat by free form sketching and having it produced individually as a single unique part without impact to costs or prices.

5.3. Error reporting, diagnosis and self-healing

The SCC platform allows perception of machine states and errors for several machines together and interpreting these perceptions. These features can be used to derive overview states for the whole factory and therefore diagnose errors precise and on a high level. At the same time the system concludes possible counter measures and starts a self-healing process.

As a very advanced example, a system would not only make the perception that certain machines are malfunctioning. It would also find out, that the machines are installed in a certain corner of the production plant, that the temperature there is increasing and also smoke is detected and draw the intelligent conclusion, that there is fire in this corner. It would therefore automatically cut all power in this part of the manufacturing plant, call the fire department, and advise personnel how to leave the area of danger, guiding them e.g. through their PDAs.

With the systems ability to learn form the supervisor and the personnel in the manufacturing plant, its ability to derive the correct counter measures increases with time and lead to the functionality described above.

6. Conclusions

The Smart Connected Control Platform described in this paper is currently being developed. It will support higher automation capabilities in production, delivering a decentralized concept for connecting machines and interaction devices. Ubiquitous augmented reality technology will provide better control abilities for factories, starting with an enhanced communication infrastructure. This way it will enable a higher degree of automation, efficiency and control using a holistic model and decentralized production intelligence.

Integrating interfaces and support for human personnel in every aspect of the production process thereby leads to better control over the process. To prevent information overload, intelligent agents paired with a semantic model help reduce the amount of information and deliver pre-interpreted and aggregated results to the personnel.

Using virtual reality customers will be able to specify their ideal basic part. Its production description will be generated automatically by the SCCP and the part can be produced just-in-time.

Acknowledgements

The authors would like to thank the INMAS proposal core group for their discussions of above ideas. We would particularly like to thank Judith Mühl for her initiation of INMAS and her valuable contribution to this paper. Further work on this will be carried out in INMAS project, most of the partners being part of the I*PROMS network of excellence.

References

[1] Albus J. S. An Intelligent Systems Architecture for Manufacturing. Publication of the National Institute of Standards and Technology, USA, http://www.isd.mel.nist.gov/documents/publist.htm

[2] Huang H. M. Intelligent Manufacturing System Control: Reference Model and Initial Implementation. IEEE CDC, Kobe, Japan, 1996.

[3] Musliner D. J., Durfee H., and Shin K. G. CIRCA: A Cooperative Intelligent Real Time Control Architecture. IEEE Transactions on Systems, Man, and Cybernetics, Vol 23 (1993) 1561-1574.

[4] Shen W. and Norrie D. H. An Agent-Based Approach for Dynamic Manufacturing Scheduling. In Working Notes of the Agent-Based Manufacturing Workshop, Minneapolis, USA, 1998.

[5] Pechoucek M., Marík V., and Stepankova O. Coalition Formation in Manufacturing Multi-Agent Systems. 11th International Workshop on Database and Expert Systems Applications (DEXA'00), 2000.

[6] Walker S. S., Brennan R. W., Norrie D. H. Holonic Job Shop Scheduling Using a Multiagent System. IEEE Intelligent Systems, (2005) 50-57.

[7] Lutz P. and Sperling W. OSACA – the vendor neutral Control Architecture. Proceedings of the European Conference on Integration in Manufacturing IiM, Dresden, Germany, 1997.

[8] Sperling W. and Lutz P. Designing applications for an OSACA control. Proceedings of the International Mechanical Engineering Congress and Exposition, Dallas, USA, 1997.

[9] Janssen, D. and Lins, A. and Schlegel, T. and Kühner, M. and Wanner, G. (2004). Semantic Web Service Retrieval. In Proceedings of NCWS 2004, Mathematical Modeling in Physics Engineering and Cognitive Science, Vaxjo University Press, 2004

[10] Benini L., Bonfigli M. E., Calori L., Farella E., Ricco B. Palmtop Computers for managing Interaction with Immersive Virtual Heritage. Proceedings of EUROMEDIA, (2002) 183-189.

[11] *Wagner D., Pintaric T., Ledermann F., Schmalstieg D.* Towards Massively Multi-User Augmented Reality on Handheld Devices. To appear in: Proceedings of the Third International Conference on Pervasive Computing (Pervasive 2005), Munich, Germany.

[12] Wesche, G. and Seidel, H.-P.: FreeDrawer - A Free-Form Sketching System on the Responsive Workbench, VRST (2001)

[13] Wesche, G. and Droske, M. Conceptual Free-Form Styling on the Responsive Workbench, ERCIM News (2001) Nr. 44

[14] Beinhauer, W. and Schlegel, T.: User Interfaces for Service Oriented Architectures. To appear in: Proceedings of HCI International 2005, LEA, 2005

Intelligent Production Machines and Systems
D. T. Pham, E. E. Eldukhri and A. J. Soroka (eds)

Neural Networks in Automated Visual Inspection of Manufacturing Parts

P. Liatsis

School of Engineering and Mathematical Sciences, City University, Northampton Square, London EC1V 0HB, UK

Abstract

This paper describes the development of a visual inspection system for detection of blemishes in manufacturing parts. Initially, median filtering is used to improve the quality of the picture, next the Sobel operator is applied to locate the edges in the image, followed by Otsu's thresholding procedure that yields the binary edge image. The image is then decomposed into a number of non-overlapping, coarse resolution images, using the technique of coarse coding, and fed into a second-order neural network equipped with a built-in feature extraction mechanism. Finally, the performance of the system is evaluated and a study of its robustness to various types of noise is performed.

1. Introduction

The remarkable performance of humans in visual perception tasks such as the recognition of objects independently of geometric transformations is often taken for granted, until we try to develop systems with similar abilities. As a result, invariant object recognition is a major research area in computer vision. A number of approaches have been proposed to address the issue of image correspondence once geometric transformations are applied [1]-[3].

Higher-order neural networks have developed over the past decade as a powerful alternative to traditional multi-layered perceptrons [4]-[8]. The essence of HONNs is the expansion of the representation space using multi-linear input combinations. This concept is very attractive for use in recognition systems, since geometric feature extraction mechanisms can be incorporated within the structure of the neural network. This type of neural networks have also been shown to act as universal approximators [9].

Work on invariance theory has shown that features such as distances and line slopes defined by point pairs, angles of triangles defined by point triples, and cross ratios of lines defined by point quadruples are respectfully invariants to translation-rotation, translation-scale, translation-rotation-scale and perspective transformations. These invariants can be obtained by enriching the input representation space with all possible point combinations (up to the required order).

The practical use of HONNs is hindered by the combinatorial explosion of the higher-order terms. For instance, in the case of an MxN image and n-order combinations, the number of input terms will be augmented by (MxN)!/(MxN-n)!n!. Assuming an image with a spatial resolution of 256x256 pixels and second-order combinations, this translates to 2,147,450,880 additional inputs. To alleviate this problem, the technique of coarse coding has been proposed [10]. This decomposes the image into a set of non-overlapping, offset images of coarse resolution, such that the number of input combinations is reasonably bounded.

In this work, a visual inspection system is proposed for invariant blemish detection in axisymmetric industrial workpieces and is shown to exhibit 95% recognition accuracy. Finally, its performance is evaluated when the data have been corrupted with salt and pepper, structured noise, blurring and occlusion.

2. Higher-order neural networks

A major criticism of single-layered perceptrons [11] was that they were unable to perform non-linear separation, an example being the XOR problem, due to the simplicity of the resulting decision boundaries. A way of dealing with this problem was to generalise the perceptron architecture such that it accommodates intervening layers of neurons, capable of extracting abstract features, thereby resulting to networks that could solve reasonably well any given input-output problem. An alternative approach, based on recent studies of information processing exhibited by biological neural networks [12] as well as Group Method of Data Handling (GMDH) algorithms [13], was the expansion of input representation space by using multi-linear terms. This gave rise to a family of neural networks collectively known as higher-order neural networks.

In general, the output of a first-order neural network is defined by

$$y_i^1 = \begin{cases} f\left(\sum_{n=0}^{1} T_n^{hid}(i)\right) = f\left(T_0^{hid}(i) + T_1^{hid}(i)\right), hidden \\ f\left(\sum_{n=0}^{1} T_n^{out}(i)\right) = f\left(T_0^{out}(i) + T_1^{out}(i)\right), output \end{cases} \quad (1)$$

where f(net) is a nonlinear threshold function such as the sigmoid function and $T_0^k(i)$ is the bias term for output i of the k-th layer (where k takes the values hidden and output) given by

$$T_0^k = w_i^k \quad (2)$$

and $T_1^k(i)$ are the first-order terms for the i-th output unit of

$$T_1^k = \sum_j w_{ij}^k x_j^{k-1} \quad (3)$$

the k-th layer
where w_{ij}^k are the interconnection weights for each input x_j of layer $(k-1)$ and output node i of layer k.
Generalising to mixed n-th order networks gives [7]

$$y_i^n = f\left(\sum_n T_n^{hid}(i)\right) = f\left(T_0^{hid}(i) + T_1^{hid}(i) + \ldots + T_n^{hid}(i)\right) \quad (4)$$

for the nodes of the hidden layer, where $T_2^{hid}(i)$ and $T_n^{hid}(i)$ are given by

$$T_2^{hid}(i) = \sum_k \sum_j w_{ijk}^{hid} x_j x_k \text{ (2nd-order terms)} \quad (5)$$

$$T_n^{hid}(i) = \sum_n \cdots \sum_j w_{ij..(k-n)}^{hid} x_j \cdots x_n \text{ (nth-order terms)} \quad (6)$$

Consider an object and any two non-identical points A, B on the object. Next an arbitrary translation and/or rotation of the object within the image is applied and points A, B become A' and B'. Since the invariant under translation and/or rotation is the relative distance between any two points on the object, the output of the HONN can be hand-crafted to be invariant to this set of transformations by considering only the second-order terms [8]

$$y_i^{hid} = f\left(\sum_k \sum_j w_{ijk}^{hid} x_j x_k\right) \quad (7)$$

and by constraining the input-hidden weights to satisfy

$$w_{iAB}^{hid} = w_{iA'B'}^{hid} \text{ if } d_{AB} = d_{A'B'} \quad (8)$$

where d_{AB} and $d_{A'B'}$ are the Euclidean distances between points A, B and A', B' respectively.

The learning rule for HONNs is the backpropagation algorithm, appropriately modified to accommodate the inclusion of the higher-order terms in the hidden layer. The updating rule for the weights of the hidden layer is then given by [7]

$$\Delta w_{ijk}^{hid} = \eta\, \delta_i \sum_k \sum_j x_j x_k \quad (9)$$

where is the learning rate (typically set to 0.1), the 's are the error terms calculated as in the classical backpropagation, the x's correspond to the input pixel values, while k, j take values that satisfy the invariance constraints.

3. Coarse coding representation

As already pointed out, higher-order neural networks can suffer from a combinatorial explosion of the higher-order terms. The incorporation of the invariances in the network architecture partly alleviates this problem since it reduces the number of independent weights associated with the occurrence of particular distance. However, a great deal of storage is still necessary to associate each distance with the relevant pairs of pixels. A number of partial connectivity strategies and an alternative input representation scheme, namely coarse coding, have been recently proposed [6]. The approach pursued in this work uses coarse coding. The underlying concept in coarse coding is that it is possible to represent finely grained input fields using offset, overlaying fields of coarser pixels.

Coarse coding is a very important mechanism in biological systems. Humans and monkeys perceive binocular depth differences which are smaller than the resolving power of any single transducer (photoreceptor cell). This phenomenon called hyperacuity has motivated researchers to look more closely at coarse coding. A finely-tuned cell is responsive to a very narrow bandwidth of signal, e.g. bars of light that are exactly horizontal or vertical. In contrast, coarsely-tuned cells display a broad tuning curve, responding maximally to a rather narrow band of signals, and with gradually decreasing vigour as a function of the stimuli's similarity distance from the 'optimum' stimulus. Churchland and Sejnowski [12] suggested that coarse coding plays a significant role in colour perception, spatial hyperacuity, and stereoptic hyperacuity.

Sullins [10] first investigated the use of coarse coding in data structuring, where he showed that it reduced the number of units required to achieve a given level of accuracy. Let us consider the image of Fig. 1(a) which shows an input field of 8x8 pixels. This image can then be decomposed into a set of two overlapping but offset fields of coarser pixels, each of size 4x4 as shown in Fig. 1(b). The pixels in each coarse field are twice as large as the pixels in the original fine field. We require two sets of coordinates to refer to a fine field pixel, using the two coarse fields. For instance, pixel (x=5,y=5) in the original grid of Fig. 1(a) is represented by both pixel (x=C,y=C) in the first coarse field and pixel (x=II,y=II) on the second coarse field of Fig. 1(b). In this way each pixel in the original image is represented by a unique representation of pixels in the coarse fields.

The major advantage of coarse coding is that it allows the representation of a high resolution image by a combination of low resolution images. An analogy to coarse coding in the field of image analysis is the concept of scale-space representation. This method allows various characteristics of the images, such as edges, to appear at different levels of resolution.

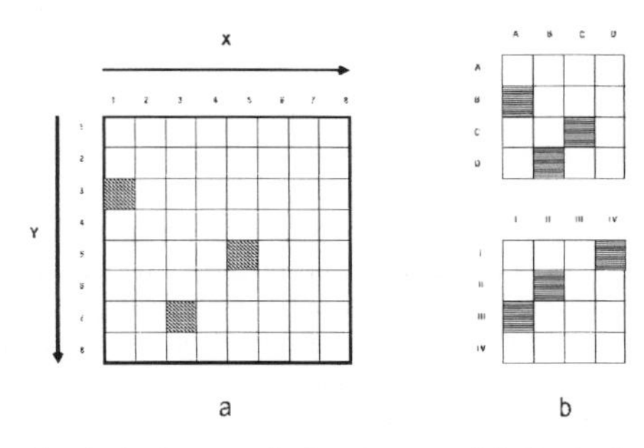

Fig. 1. a) A fine resolution image of 8x8 pixels. b) A possible decomposition of a) into a set of two offset, overlapping fields of 4x4 pixels.

A problem associated with coarse coding that needs to be addressed is how pixels that are not intersected by all the coarse grids are going to be handled, such as pixel (x=1,y=3) in Fig. 1(a). Spirkovska and Reid [6] have proposed two ways:
a) with wraparound, and
b) extending the original image field.
If coarse coding is implemented with wraparound, the pixel in question can be represented by pixel (x=A, y=B) and pixel (x=IV, y=I). In the case of b), we define the original field to be greater than the input field, e.g., we would use an input field of 9x9, instead of 8x8.

Using coarse coding, higher-order neural networks can be used to discriminate between objects in real-world images. First, the image is decomposed into coarse fields, each of sufficiently low resolution such that it would allow discrimination, while at the same time keeping at a minimum the number of possible pixel associations. Next, the network is applied to each coarse image in order to extract the necessary invariant features that will lead to the classification. Finally, the transfer function of the network is modified so that it takes into account the intersection of the coarse fields. Hence for a second-order neural network with built-in invariances to translation and rotation we have [8],

$$y^{hid} = f\left(\sum_{n}\sum_{j}\sum_{k} w_{ijk}^{hid} x_j x_k\right) \qquad (10)$$

where $f(.)$ is a non-linear transfer function (e.g. the sigmoid function), n ranges from 1 to the number of coarse fields used, j and k range from 1 to the coarse pixel grid size squared.

4. System structure

The proposed system is based on a IBM PC compatible (with a 80486DX Intel chip running at 25 MHz), equipped with a video digitizer board and a CCD camera. The data were on a 256x256 pixel 256 gray level matrix format suitable for further processing.

The core of the inspection system is a second-order neural network with built-in invariances to translation and rotation. However, some preprocessing was necessary before the images were presented to the network. In specific, median filtering was applied to enhance image quality, and then the Sobel operator was used to subdivide the image into distinct regions by detecting the discontinuities within the picture. Finally, the gray level edge images were thresholded using Otsu's method to provide the binary inputs for the network. Because of hardware limitations and to speed up the performance of the system, the resolution of the binary images was reduced to 64x64 pixels. The total processing time for the preprocessing steps was approximately 1 second.

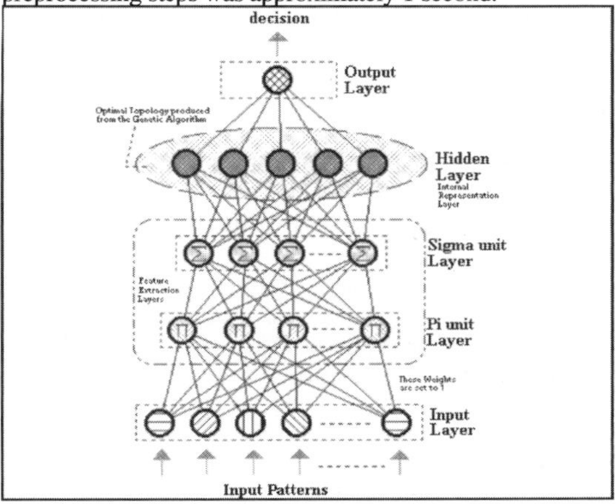

Fig. 2. 2nd-order neural network architecture.

Next the images were decomposed to a set of 5 coarse pixel images, each of 16x16 pixels. Parameters that play an important role in the selection of coarse grid sizes are the recognition problem in hand, and the available hardware. In the case of the 16x16 coarse grid images, 107 independent (input to hidden) weights were used for all possible pixel pair combinations. The training set for the network consisted of 12 satisfactory and 16 defective modules, presented in 20 random translations and orientations, thus giving a total of 560 training patterns. The optimal number of units in the hidden layer was determined using GAs [8], and was found to be 5 neurons. A schematic of the architecture is shown in Fig. 2. Training took nearly 7,000 epochs and next, the network was able to provide a classification decision in less than 0.1 seconds. In the next section, we will provide a brief description of the GA optimisation mechanism.

5. Genetic algorithms in HONN topology optimisation

Genetic algorithms (GAs) are suitable as tools for determining the optimal topology of a neural network since they have been repeatedly proven to overcome problems such as non-differentiable performance functions, and deceptive and multimodal topology spaces. Their operation is based on the doctrine of Darwinian evolution of the survival of the fittest [14]. The initial population of the GA (a set of potential solutions) is typically random, however with the use of genetic operators, such as reproduction, crossover and mutation, and a suitable fitness function, future populations learn accurately the problem at hand. In the current application, the learning accuracy, the training speed and the low topological structural complexity define the fitness of the population members.

The general GA adopted in this work is outlined below:

(a) Create randomly an initial population of likely solutions-topologies. Evaluate the objective values of all members. Calculate their representative fitness values.

(b) Use the mating roulette to select a pair of individuals with probabilities proportional to their fitness. Use the crossover operator to reproduce two children. Mutate them. Rectify their invalid alleles. Test them to ascertain their compliance with the population homogeneity constraints; if a child is diverse enough, evaluate its objective value, otherwise reject it.

(c) Repeat step (b) until a prespecified number of offspring has been produced.

(d) Remove an equal number of the worst members and insert the new ones.

(e) Repeat steps (b)-(d) until a total prespecified number of generations has elapsed or diversity has been expedited.

In order to assess the objective value of a member, a neural network simulator trains its corresponding structure for a representative number of epochs according to the classification attached to the individual so that a fitness can be assigned to it later on when the entire population is formed. The communications means between the neural and genetic simulators is a generator which performs a phenotype-to-genotype mapping. Various operators have been adapted to enable the GA to operate upon the topology population and evolve the optimal-minimal structures in a fast and efficient manner. These operators are especially chosen to fit the needs and the inherent characteristics of the optimisation requirements of the current application.

We make use of the Delete-Last technique [15] to replace the worst members of the population with new offspring. GAs have a tendency to create duplicate individuals, hence the No-Duplicates operator [16] is used to avoid premature convergence, but for very long chromosomes, this technique is still problematic. Hence in order to allow a more efficient usage of the allotted number of chromosomes, we improve the scheme by considering a safeguarded distance k such that the Hamming distance of a newly born child from its nearest population member is greater than or equal to k, or the child is discarded. Therefore, we linearly decrease k starting from d_U going down to d_L following the progression of the generations according to

$$k = round[d_L + (d_U - d_L)\,(G - g + 1)/G] \quad (11)$$

where g is the current generation and G is the maximum total number of generations. To alleviate the common problems of Super-individual and Close-race, we make use of Linear Ranking [14]. Its basic idea is to anticipate these problems by preallocating the probability of reproduction deterministically. Ranking disassociates the fitness from the computed objective value such that the best individuals are still selected more frequently but not as frequently as to dominate the population.

Since crossover operates more efficiently in the early generations when the population is sufficiently diverse and deteriorates towards the end when individuals become more similar, a kind of variable crossover is contrieved here; the number b of random points that break up the parental schemata starts from a predefined low bound b_L at the first generation and reaches b_U at the last generation G, by being increased linearly

$$b = round[b_L + (b_U - b_L)(g-1)G] \quad (12)$$

Mutation is used to compensate for premature loss of diversity but when increasing the probability Pm, the proliferation of good alleles is disrupted in a ratio equal to that of the poor ones. We use a diversity directed heuristic to accomplish an effective adaptive mutation. The probability interval is bounded by two predefined values P_{mL} and P_{mU} to restrict Pm from taking the extremes. The evaluation of Pm is done by

$$P_m = P_{m_L + (P_{m_U} - P_{m_L})(1 - H(a,b)/L)^2} \quad (13)$$

where H(a,b) is the Hamming distance of the two parents a and b and L is the chromosome length.

In order to encode a network topology to a chromosome, we use the binary connectivity matrix of the network, where an entry is 1 if the representative connection exists, and 0 otherwise. A maximum number H of hidden units has to be prespecified. Initially, the matrix is triangular to reflect the fact that only feedforward connections are allowed. The matrix is further pruned since input units cannot receive from any other input ones and output units cannot send to any other output ones. The residual trapezoidal matrix area is then encoded to a bit string in a row wise manner.

By placing a constraint that prohibits the outputs from being fixed, we alter the chromosome by picking up randomly a valid sender and then connecting it to it. Finally, any hidden unit is prohibited from being fixed or isolated, resulting to iterative pruning of all the fixed and isolated hidden units.

To calculate the raw fitness of a topology, we train it for a number of epochs T using the modified generalised delta rule for HONNs, as follows

$$T = T_0 + U_C \cdot T_0 \cdot R_C + U_U \cdot T_0 \cdot R_U \quad (14)$$

where T_0 is a small predefined baseline of epochs, U_C and U_U are respectively the fractions of th eunused connections and units of the network, and R_C and R_U are the predefined reward ratios that define the additional percentage of T_0 epochs to reward for each unused structural element. Finally, as the merit of the chromosome whose corresponding network topology is under evaluation, we take the average pattern classification error.

In this work, the following parameters were used: T_0=1000, R_C=60% and R_U=80%. The genetic evolution lasted G=300 generations, with a population size of 50 members, reproducing 10 offspring at each generation. The mutation bounds were P_{mL}=0.001 and P_{mU}=0.25, while the crossover lower and upper bounds were b_L=1 and b_U=50, respectively. The safeguard interval was (0,100). There were 107 input units, 1 output unitm and the maximum number of hidden units was set to 10, resulting to a chromosome length of 1232.

6. Results

This section will examine the performance of the system in unseen data and its tolerance to erroneous or missing data. Specifically, gray level salt and pepper noise, blurring, structured noise and occlusion will be examined.

Initially, the system was tested using 15 unseen components from each class (see Fig. 3), presented in 20 random translations and orientations. The corresponding confusion matrix is given in Table 1. It was able to identify the presence or the absence of the blemish with 95% accuracy. The misclassifications in this case affected the class of satisfactory modules and due to surface irregularities and the large size of the coarse pixels (or equivalently, the low resolution of the coarse grid images), which resulted to loss of significant discrimination information.

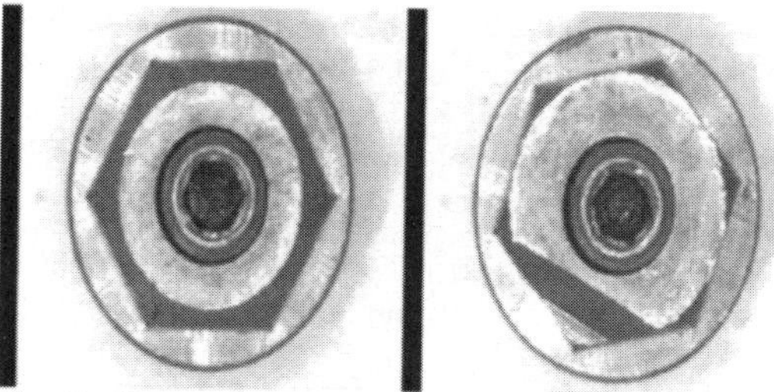

Fig. 3. Characteristic views of a) a satisfactory and b) a blemished module.

		Predicted	
		Satisfactory	Blemished
Actual	Satisfactory	272	28
	Blemished	0	300

Table 1. Confusion matrix for instances of unseen components.

Next, the testing data were corrupted with variable levels of gray level salt and pepper noise as shown in Fig. 4. The system maintained very good performance (~90% accurate classification) up to a noise level of 20%, and then its performance started to decrease to 85% for a noise level of 40%, while it was around 63% at 45% noise.

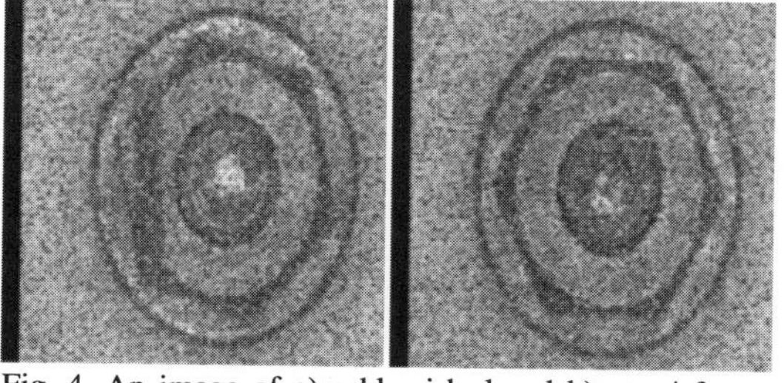

Fig. 4. An image of a) a blemished and b) a satisfactory module corrupted with 45% salt and pepper noise.

The next test involved artificial blurring of the images. Blurring occurs in images when we try to remove 'snow' noise or to reduce the effects of undersampling using smoothing filters. Its visual sensation is equivalent to the camera being out of focus. Fig. 5 shows blurred images of a blemished and a satisfactory module.

The simplest example of a smoothing filter is that employing spatial neighbourhood averaging using a rectangular neighbourhood of *nxn*. Then the smoothing operator is given by

$$g(x,y)=\sum_{i=\_n/2}^{i=n/2}\sum_{i=\_n/2}^{i=n/2} h_{sm}(i,j)\, f(x+i\,,y+j\,) \quad (15)$$

where $f(x,y)$ is the input image, n is the size of the neighbourhood and $h_{sm}(i,j)$ is the smoothing function, whose non-zero values in the neighbourhood are given by

$$h_{sm}=\frac{1}{n^2} \quad (16)$$

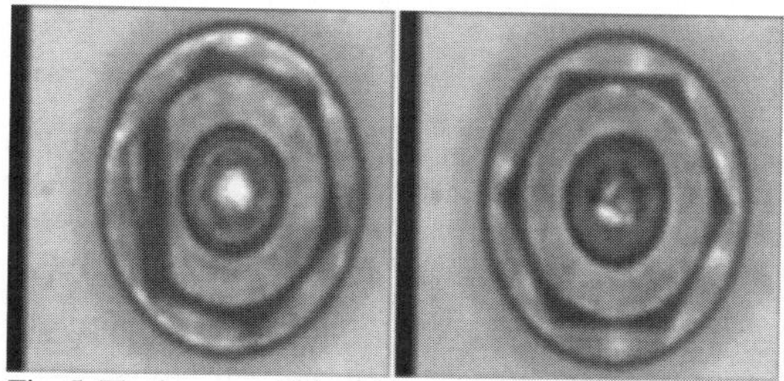

Fig. 5. The images of Fig. 3 blurred with a mask of 6x6.

When applying the system to the testing data, blurred using masks of size 2x2 to 11x11 pixels, it was found that its performance remained almost unaffected for smoothing masks up to size 4x4, while after that point it started to deteriorate, and in specific, for masks of 7x7 it was 50% (random).

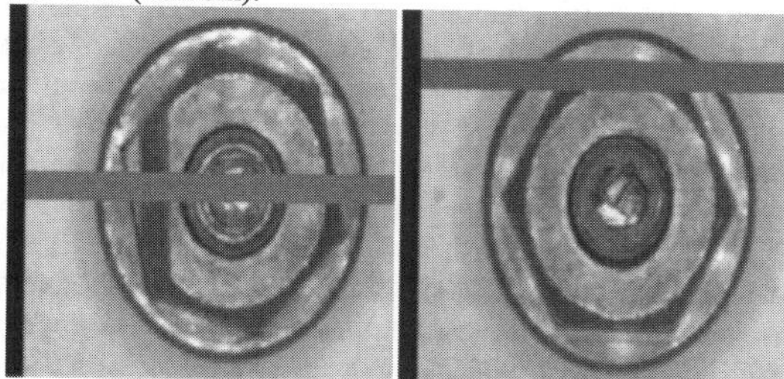

Fig. 6. The images of Fig. 3 with a superimposed line with width of 20 pixels.

In certain cases, such as in the transmission of TV images, structured noise rather than random noise is the predominant factor degrading image quality. The kind of noise examined here is the presence of a line pattern of random gray level, superimposed on the images. The experiments were carried out for line widths of 2 to 70 pixels (see Fig. 6). The results obtained with this kind of noise were quite encouraging. Although the slightest change (like the presence of a line with width of 2 pixels) in the image lowered the performance of the system, further increase of the width of the line pattern seemed to degrade its performance very slowly. More precisely, the system performed with acceptable recognition rate up to a line length of 20 pixels, while it decreased rapidly to 50% accuracy for a line pattern of 23 pixels width.

Finally, the system was tested in occluded scenes. The objects used for occlusion were squares with a linear dimension between 5 and 90 pixels. There was only one occluded object per image and the occluding square was always superimposed at random locations near the objects, as shown in Fig. 7.

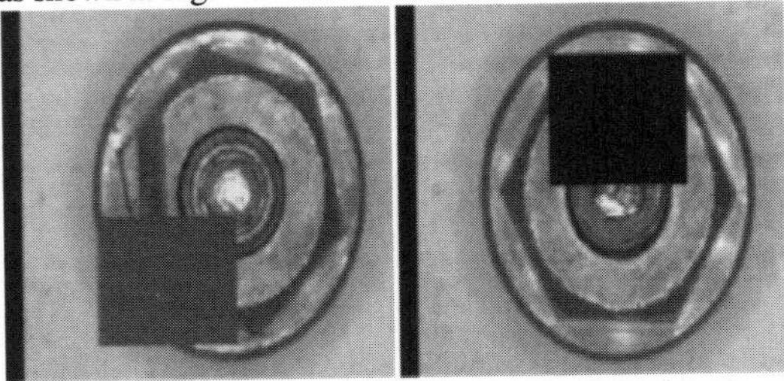

Fig. 7. Images of a) a blemished and b) a satisfactory module with a square of 90x90 pixels.

As expected, the performance of the system with occluded objects depended more on the size of the occluding objects and the size of the object in the image plane, rather than the similarity between the objects' classes. The performance of the system was nearly 95% for occluding objects of up to 10x10 pixels, then continued to be acceptable for sizes up to 40x40, and finally become random for squares of size 90x90.

7. Conclusions

This research involved the development of a neural network-based visual inspection system for invariant with respect to rotation and translation recognition of blemishes in axisymmetric components. In the preprocessing stage, impulse noise was removed from the picture, and the edges of the objects were located and provided in a binary format. Next, an image representation scheme, called coarse coding, allowed the representation of high-resolution images as a set of offset, overlaying coarser images. The core of the system was a second-order neural network with built-in invariances to translation and rotation, which employs simple geometrical equations to construct a feature extraction mechanism within the network architecture.

The performance of the system was evaluated in the inspection of industrial parts and tested for various types of noise. It was found that it performed with 95% recognition accuracy in unseen testing images, while it was quite robust to erroneous and incomplete data samples.

References

[1] Fukushima, K., A hierarchical neural network model for associative memory, Biol. Cybern., Vol. 50, pp 105-113, 1984.

[2] Troxel, S.E., Rogers, S.K., and Kabrinsky, M., The use of neural networks in PRSI recognition, Proc. IEEE ICNN, Vol. 1, pp 569-576, 1988.

[3] Barnard, E., and Casasent, D., Invariance and neural nets, IEEE Trans. Neural Networks, Vol. 2, No. 5, pp 498-508, 1991.

[4] Maxwell, T., Giles, C.L., Lee, Y.C., and Chen, H.H., Nonlinear dynamics of artificial neural systems, AIP Conf: Neural Networks for Computing, pp 299-304, 1986.

[5] Giles, C.L., and Maxwell, T., Learning, invariance and generalisation in higher-order neural networks, Applied Optics, Vol. 26, No. 23, pp 4972-4978, 1987.

[6] Reid, M.B., Spirkovska, L., and Ochoa, E., Simultaneous position, scale and rotation invariant pattern classification using third-order neural networks, Neural Networks, Vol. 1, No. 3, pp 154-159, 1989.

[7] Liatsis, P., Wellstead, P.E., Zarrop, M.B., and Prendergast, T., A versatile visual inspection tool for the manufacturing process, Proc. 3rd IEEE CCA, Vol. 3, pp 1505-1510, 1994.

[8] Liatsis, P., and Goulermas, Y.J.P., Minimal optimal strategies for invariant higher-order neural architectures, Proc. ISIE'95, Vol. 2, pp 792-797, 1995.

[9] Pao, Y.H., and Takefuzi, Y., Functional-link net computing: theory, system architecture and functionalities, IEEE Computer, Vol. 25, No. 2, pp. 76-79, 1992.

[10] Sullins, J., Value encoding strategies, TR-165, Comp. Sc. Dept., Rochester Univ., 1985.

[11] Minsky, M.L., and Papert, S., Perceptrons, MA: MIT Press, 1969.

[12] Churchland, J., and Sejnowski, T., The computational brain, MA: MIT Press, 1991.

[13] Farlow, S.J. (ed.), Self-organising methods in modeling: GMDH algorithms, Marcel Dekker Inc., 1984.

[14] Goldberg, D.E., Genetic Algorithms in search, optimization and machine learning, Addison-Wesley, 1989.

[15] Davis, L. (ed.), The handbook of Genetic Algorithms, Ch. 2, Pitman, London, 1989.

[16] Booker, J., Improving search in Genetic Algorithms, in Genetic Algorithms and Simulated Annealing, Pitman, London, 1987.

Intelligent Production Machines and Systems
D. T. Pham, E. E. Eldukhri and A. J. Soroka (eds)

Petri net controller synthesis using theory of regions

Z. Achour[a], A. Ghaffari[a], N. Rezg[a], X. Xie[a,b]

[a] *INRIA, ISGMP-Bat A, Ile du Saulcy, 57045 Metz cedex, France*
[b] *Ecole Nationale Supérieure des Mines de Saint-Etienne, 42023 Saint-Etienne cedex 2 France*
Email: xie@emse.fr

Abstract

Petri net controller synthesis is formally treated in this paper. Two supervisory control problems of plant Petri net models, forbidden state problem and forbidden state-transition problem, are defined. The theory of regions is used to provide algebraic characterizations of pure control places and impure control places for both problems. Thanks to Farkas-Minkowski's lemma, the algebraic characterizations lead to nice geometric characterization for the existence of control places for the two supervisory problems. A railway network application is presented.

1. Introduction

In our previous work [1], we proposed an approach for synthesizing a set of control places that solve a forbidden state problem of any bounded Petri net (PN) when such PN controller exists. The proposed approach is general. It is based on the computation of the maximally permissive and live using a Ramadge and Wonham-like approach [2]. It considers a general set of forbidden states in addition to liveness requirement and uncontrollable transitions. It establishes necessary and sufficient conditions for the existence of control places realizing the maximum permissive control. It then builds a compact PN representation of the closed-loop model respecting the control policy which can be efficiently used in the real-time control.

In this paper, a formal setting is given to the general PN controller design problem with respect to a desired behavior. Assumptions on the desired behavior that may be addressed by the synthesis approach are made clear. Two supervisory control problems, forbidden state-transition problems (FSTP) and forbidden state problems (FSP), are defined. Problems of the former class are given as a set of forbidden state-transitions, although some of them may lead to no forbidden markings. Problems of the latter class, subclass of the FSTP, are given as a set of forbidden markings. Any state-transition yielding a forbidden marking is forbidden. These problems are solved by the addition of control places. Control places may be with or without self-loops. They are called respectively pure and impure places. Necessary and sufficient conditions for the existence of such solution are established using the theory of regions. It is shown that there exist control problems that cannot be solved by adding pure control places. Impure control places have self-loops with controlled transitions which grant them with a higher power of control.

The paper is organized as follows. Section 2 defines the desired behavior and the PN controller design problems. Necessary and sufficient conditions for the existence of pure and impure control places are presented in Section 3. Section 4 presents a railway network application. Note however that applications to supervision of automated manufacturing systems can be found in [1].

2. PN controller design problem w.r.t. a desired behavior

We consider in this section the problem of designing control places to add to a given plant Petri net model (N, M_0), so that the resulting net (N_C, $M_{C,0}$) has a given desired behavior. The behavior of the Petri net model (N, M_0) can be represented by its reachability graph RG(N,M_0) or the set of its reachable markings R(N,M_0). We assume along this paper that the plant Petri net model (N, M_0) is bounded.

Two kinds of control specifications are addressed in this paper, forbidden state and forbidden state-transitions specifications. The PN controller design problem is defined for either specification.

2.1. The desired controlled behavior

Let R be the desired behavior of the controlled system (N_C, $M_{C,0}$). General approaches for computing this behavior are usually automata-oriented and follow the approach of Ramadge-Wonham [2] by taking into account safety requirements, liveness conditions and uncontrollable transitions.

Assumption 1: The desired behavior R of the controlled system (N_C, $M_{C,0}$) is a subgraph of the reachability graph RG(N, M_0) of the plant model.

In general, especially under specifications of forbidden transition sequences, R is not a subgraph of RG(N, M_0). The relaxation of this assumption is being investigated. Assumption 1 holds for a wide range of supervisory control problems including forbidden state problems, deadlock avoidance problems with/without uncontrollable transitions. In the following, nodes in R will be called *admissible* markings and arcs in R admissible state-transitions. According to whether the control requirements are expressed as a set of forbidden states or a set of forbidden state-transitions of RG(N,M_0), two classes of supervisory control problems can be defined.

Definition 1: A Forbidden State-Transition Problem, denoted FSTP for short, is a supervisory control problem such that the desired behavior R of a plant model (N, M_0) satisfies the following conditions: (i) $M_0 \in$ R; (ii) $\forall M \in$ R, $\exists \sigma \in$ R / ($M_0 \xrightarrow{\sigma} M$)$\in$R.

Condition (i) is obvious for any supervisory control problem. Condition (ii) implies that any marking M in R is reachable from M_0 by a sequence of transitions in R.

This class of problems encompasses a large number of real life situations, notably those where priority is given to some transitions over others. Such situation is very common at very large scale. For instance, routing problems in manufacturing systems may be seen as forbidden state-transition problems. Let us consider the small example of figure 1.

Transitions T1 and T2 are two tasks that lead to the same state. They may represent, for instance, the same operation performed by two different machines. For the sake of machine workload balancing, they have to be performed alternatively. The behavior of the plant model is given by its reachability graph (see Fig. 1.b). The desired behavior is given in figure 1.c. The set of forbidden state transitions includes ($M_0 \xrightarrow{T1} M_1$) and ($M_1 \xrightarrow{T2} M_2$), while M_1 and M_2 are admissible markings. This is an example of forbidden state-transition problems.

An important sub-class of FSTP is the forbidden state problem defined as follows.

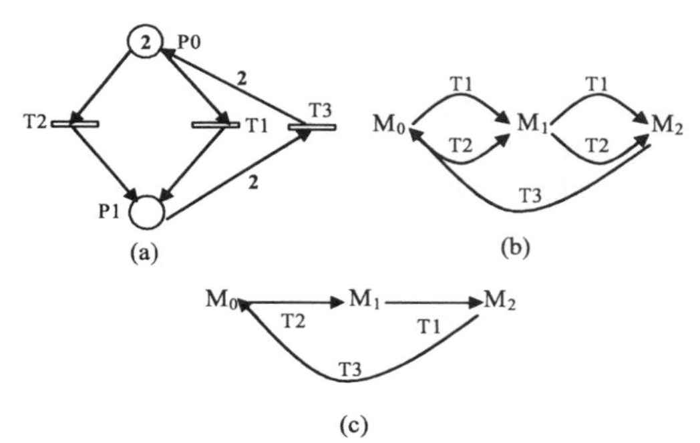

Fig. 1. A forbidden state-transition problem

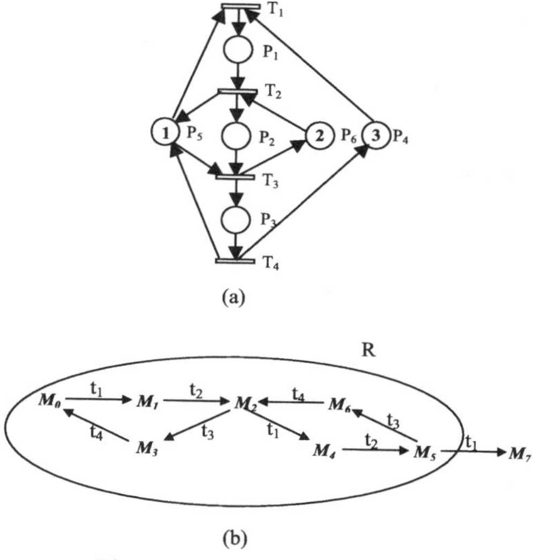

Fig. 2. A forbidden state problem

Definition 2: A Forbidden State Problem, denoted FSP for short, is a forbidden state-transition problem such that the desired behavior R of a plant model (N, M_0) satisfies conditions (i)-(ii) and $\forall(M \xrightarrow{t} M')\in$ RG(N,M_0), M$\in$R AND M'$\in$R $\Rightarrow (M \xrightarrow{t} M')\in$R.

Clearly, an FSP is an FSTP but the reverse is wrong. The FSTP defined in figure 1 is not an FSP. To summarize, in an FSP, any state-transition leaving the desired behavior R but feasible in the plant model leads to a non admissible marking. As a consequence of condition (iii), a FSP can also be defined by the set of admissible markings.

This class of problems covers a wide range of supervisory control problems including usual forbidden state problem and deadlock avoidance problem with/without uncontrollable transitions. Of course, an FSP cannot take into account specification such as priorities.

Consider the example of figure 2. According to its reachability graph (see Fig. 2.b), the modeled process may be in deadlock situation represented by the state M_7. Marking M_7 is not admissible and any state-transition yielding it is forbidden. The desired behavior associated to this forbidden state problem involves all markings apart M_7 and all state transitions apart $(M_5 \xrightarrow{t_1} M_7)$.

2.2. The PN controller design problem

Forbidden state problems and forbidden state-transition problems of bounded Petri nets may be solved by the addition of control places defined as follows:

Definition 3: A control place p_c of a plant Petri net model (N, M_0) is defined by ($M_0(p_c)$, $C^+(p_c,.)$, $C^-(p_c,.)$), where $M_0(p_c)$ is its initial marking and $C^+(p_c,.)$ (resp. $C^-(p_c,.)$) is its weighting vector of arcs from transitions of N to p_c (resp. from p_c to transitions of N). Note that the incidence vector of pc is C(pc, .) = C+(pc, .)–C-(pc, .).

Let $M_C^T = (M^T\ M_S^T)$ be any reachable marking of the controlled net (N_C,M_{C0}), where M is the marking of places of the plant model N, M_S is that of the added control places and superscript T denotes the transposition operator. The PN controller design problem may be defined as follows:

Definition 4: A PN controller design problem consists in determining a set of control places {p_{c1}, .., p_{cm}} to add to the PN plant model (N, M_0) such that, for any reachable marking $M_C^T = (M^T\ M_S^T)$ of the controlled net (N_C,M_{C0}), a transition t is firable in (N_C,M_{C0}) iff t corresponds to a state-transition from the marking M in R.

To reduce the behavior RG(N,M_0) of the plant model to R, state-transitions not belonging to R or yielding markings outside R from markings in R have to be prevented by the set of control places. Let Ω be the set of such state transitions denoted as *event separation instances*. For forbidden state problems, Ω is the set of all state transitions yielding outside R. It is given by: $\Omega = \{(M \xrightarrow{t} M')\in$ RG(N,M_0) / M$\in$R and M'$\notin$R}.

For forbidden state-transition problems, Ω is the set of all state transitions not belonging to R: $\Omega = \{(M \xrightarrow{t} M')\in$ RG(N,M_0) / M$\in$R and $(M \xrightarrow{t} M')\notin$R}.

3. Control places design

Given the PN plant model (N,M_0) and the desired behavior R defined in Section 2. This section presents the design of control places using the theory of regions.

3.1. Some basic relations

Consider any place p of the controlled net with initial marking M_0(p) and incidence vector C(p, .). Note that p can be either a place of the plant model or a control place.

For any transition *t* from any marking *M* in R, i.e., *t* is the label of an outgoing arc of the node *M* in *R* :

$$M'(p) = M(p) + C(p,t), \;\forall M[t > M' \in R \qquad (1)$$

where M' is the new marking or equivalently the destination node of arc t.

Consider now any non-oriented cycle γ of the desired behavior R. Applying the state equation to nodes in γ and summing them up gives the following *cycle equation*:

$$\sum_{t\in T} C(p,t)\cdot \vec{\gamma}[t] = 0, \quad \forall \gamma \in S \qquad (2)$$

where $\vec{\gamma}[t]$ denotes the algebraic sum of all occurrences of t in γ with a weight 1 for each transition in any given direction of the cycle and a weight of -1 for a transition in the opposite direction and S is the set of non-oriented cycles of the graph. $\vec{\gamma}$ will be called the counting vector of γ.

Consider now each node M of the desired behavior R. According to the definition of both FSTP and FSP, there exists a non-oriented path Γ_M from the initial state M_0 to M. Applying equation (1) along the path leads to: $M(p) = M_0(p) + C(p,\cdot)\vec{\Gamma}_M$ where $\vec{\Gamma}_M$ is the counting vector of the path Γ_M defined similarly as $\vec{\gamma}$. There may exist several paths from M_0 to M. Under the cycle equations, the product $C(p,\cdot)\vec{\Gamma}_M$ is the same for all these paths. As a result, the path Γ_M can be arbitrarily chosen. The reachability of any marking M in R implies that:

$$M_0(p) + C(p,\cdot)\vec{\Gamma}_M \geq 0, \quad \forall M \in \mathrm{R} \qquad (3)$$

which will be called the *reachability condition.*

3.1. Pure control places

Consider a new control place p_c and assume that it is pure, i.e. there does not exist any transition that is both input and output transition of p_c. A pure control place can be simply represented by its initial marking $M_0(pc)$ and its incidence vector $C(p_c, .)$.

According to Lemma 1, place pc verifies cycle equations (2) and reachability conditions (3). Unfortunately, these equations are not sufficient to obtain the desired behavior R. In order to obtain exactly R, for each event separation instance ($M \xrightarrow{t} M'$) in Ω, t should be prevented from happening by some place p_c. Since the control places are pure, t is prevented from happening at M by p_c if and only if:

$$M_0(p_c) + C(p_c,\cdot)\vec{\Gamma}_M + C(p_c,t) < 0, \quad \forall \left(M \xrightarrow{t} M'\right) \in \Omega \qquad (4)$$

Relation (4) will be called *event separation condition of* ($M \xrightarrow{t} M'$). Note that, for FSTP, M' of the plant model might be an admissible marking in R but the related marking M'_C = [M' M'_S] of the controlled net is a forbidden marking. This is equivalent to represent M' by two different copies: one as an admissible marking in R and the other one as a forbidden marking related to the event separation instance ($M \xrightarrow{t} M'$).

Theorem 1: A desired behavior R, subgraph of the reachability graph of a bounded Petri net (N,M_0), can be realized by adding pure control places to (N,M_0) iff there exists a solution ($M_0(p_c),C(p_c,.)$) satisfying conditions (2), (3) and (4) for any element of Ω.

Particularly, if R is the maximally permissive and live behavior of the controlled net, then Theorem 1 ensures that the synthesis of control places using the theory of regions allows the determination of the optimal control policy in the sense of maximum permissiveness, which the main result of [1,2].

In some problems, it is not possible to find a PN-based solution by means of only pure control places. Let us consider an example of these problems.

Consider the net of figure 3.a. The control specification is to not fire T2 after T1. The event separation instance to solve is ($M_3 \xrightarrow{T2} M_2$). Let p_c be a pure control place solving this problem. Then p_c satisfies the following system:

Reachability conditions of admissible markings:

i. $M_0(p_c) \geq 0$
ii. $M_3(p_c) = M_0(p_c) + C(p_c,T1) \geq 0$
iii. $M_1(p_c) = M_0(p_c) + C(p_c,T2) \geq 0$
iv. $M_2(p_c) = M_0(p_c) + C(p_c,T1) + C(p_c,T2) \geq 0$

Event separation instance :

v. $M_0(p_c) + C(p_c,T1) + C(p_c,T2) < 0$

Note that inequalities (iv) and (v) are contradictory. So, there is no pure control place to solve this problem. However, an impure control place

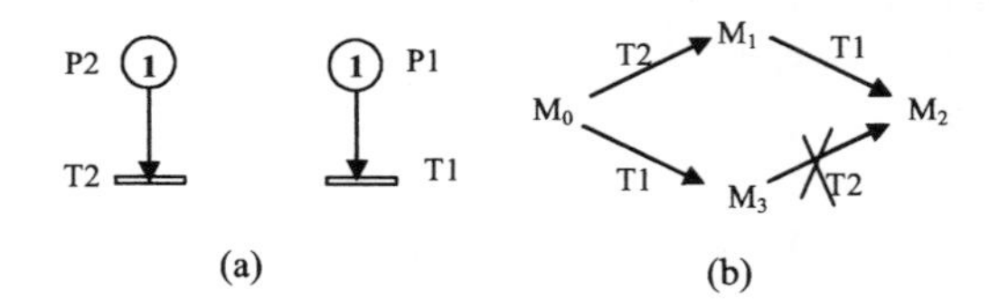

Fig. 3. A forbidden state-transition problem

exists as it will be shown in the next subsection.

3.3. Impure control places

An impure control place p_c is a place for which there exists at least one transition that is both input and output transition of p_c. The self-loops are introduced to increase the control power of control places. With respect to any event separation instance ($M \xrightarrow{t} M'$) in Ω, only control places p_c with self-loop connecting p_c and t need to be considered and the goal is to forbid t from firing at M. As a result, for each event separation instance ($M \xrightarrow{t} M'$), we look for a control place p_c defined by its initial marking $M_0(p_c)$, its incidence vector $C(p_c, .)$ and the weight $C^-(p_c, t)$ of the arc connecting p_c to t. First, according to Lemma 1, cycle equation (2) and reachability condition (3) hold.

Since the control places are impure, the reachability conditions (3) are no longer enough to guarantee the reachability of markings M in R. Additional conditions are needed to ensure firability of a transition t enabled at any markings M in R:

$$M(p_c) = M_0(p_c) + C(p_c, \cdot)\vec{\Gamma}_M \geq C(p_c, t)^-, \quad \forall M[t > \text{in R} \qquad (5)$$

The event separation condition related to ($M^* \xrightarrow{t} M'$) becomes:

$$M_0(p_c) + C(p_c, \cdot)\vec{\Gamma}_{M^*} < C(p_c, t)^- \qquad (6)$$

Relations (2), (3), (5) and (6) are necessary and sufficient conditions for the existence of impure control places as it is stated by the following theorem.

Theorem 2: A desired behavior R, subgraph of the reachability graph of a bounded Petri net (N,M_0), can be realized by adding impure control places to (N,M_0) iff there exists a solution $(M_0(p_c), C(p_c,.), C\text{-}(pc, t))$ satisfying conditions (2), (3) (5) and (6) for any element ($M \xrightarrow{t} M'$) of Ω.

Consider again the problem of figure 3. The separation instance may be solved by a impure control place p_c iff p_c satisfies the following relations.

i. $M_0(p_c) \geq 0$
ii. $M_0(p_c) + C(p_c,T1) \geq 0$
iii. $M_0(p_c) + C(p_c,T2) \geq 0$
iv. $M_0(p_c) + C(p_c,T1) + C(p_c,T2) \geq 0$
v. $M_0(p_c) \geq C^-(p_c,T2)$
vi. $M_0(p_c) + C(p_c,T1) < C^-(p_c,T2)$

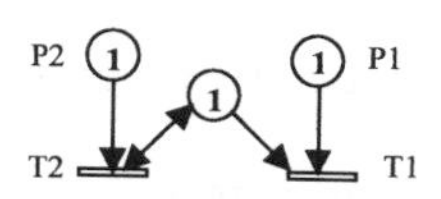

Fig. 4. The controlled net corresponding to problem of figure 3.

A control place defined by $M_0(p_c)=1$, $C(p_c,T1) = -1$, $C(p_c,T2) = 0$ and $C^-(p_c,T2) = 1$ is a solution of the above system (see fig. 4).

4. An actual deadlock problem in a railway network

The problem dealt with in this section concerns a railway network in Sardaigna (Italy) and was already addressed in [3]. In [3], the authors dealt with the problem of enforcing safety constraints (to avoid collision) inside and outside stations.

We are not considering the safety question here, but only the deadlock avoiding problem in the network part joining stations Chilivani and Olbia. So, we consider the high level model of the network part proposed in [3] (see fig. 5).

The railway network consists of 4 stations α, β, γ and δ. Except station α which contains 3 tracks, the other stations contain only two tracks. This capacity limitation is ensured by resources places p2, p8, p15,p21. Stations are linked by single tracks, except the network part between β and γ which is double-track. These capacity constraints are enforced by places p5, p11, p13, p18. At this abstraction level, all transitions in the model, representing departure and arrival of trains, may be considered as controllable.

Giua et al. proposed in [3] a siphon based solution method. It consists in identifying siphons that are likely to become empty and so entail a deadlock situation. For each siphon, a monitor place is added in such a way that there is always at least one token in the siphon's places. Monitors are added in an iterative way each time the number N of trains in incremented (marking of place p23).

The authors proved that if the PN model belongs to the particular class ES^2PR, which is the case for this example, then the siphon searching phase reduces to the solution of mixed linear program. As a result, the iterative procedure for monitor design remains valid only when the augmented model is still in the ES^2PR net class. Further, the convergence towards a live net cannot be guaranteed.

For this example, a solution was given in [3] for N such that $3 \leq N \leq 7$. For N<3 the controlled net is live. The procedure can no longer be applied for N>7 since the augmented net is not an ES^2PR net. Note that this synthesis method is not optimal, but for this particular example, the solution given in [3] is maximally permissive.

So, we applied the synthesis method proposed in this paper to the railway network problem. For N<7 we find the same solution as what is given in [3]. For N > 10, the size of the reachability graph tends towards the value 60000. Consequently, we limit ourselves to the instances corresponding to a number of trains varying from 7 to 10. Table 1 below recapitulates the obtained results.

Table 1.
Collision avoidance problem in a railway network.

N	Initial size	Size of R	Card(Ω)	Nbplace	CPU time (sec)
7	22800	20724	3228	9	100
8	35475	31121	6277	13	462
9	47350	40087	9681	18	882
10	55633	45698	12324	25	1045

5. Conclusion

We presented a rigorous setting for the general PN controller design problem w.r.t. a given desired behavior. Desired behavior is defined according to general forbidden state-transition specifications which include the case of forbidden state specifications. The closed-loop model consists of the original plant Petri net model and a set of control places. Control places computed using the theory of region based approach may be pure or impure.

The work pursues in several directions: (i) extending the proposed approach to take into account sequence dependent specifications and unobservable transitions, (ii) exploiting Petri net structure using formal analysis approach [4]. It might also be interesting to consider the use of constraint programming and intelligent search.

Acknowledgement

INRIA is partner of the EU-funded FP6 Innovative Production Machines and Systems (I*PROMS) Network of Excellence. http://www.iproms.org

References

[1] A. Ghaffari, N. Rezg and X. Xie, "Design of a Live and Maximally Permissive Petri Net Controller Using Theory of Regions," *IEEE Trans. on Robotics and Automation*, vol. 19/1, pp. 137-142, 2003.

[2] P.J. Ramadge and W.M. Wonham, "The Control of Discrete Event Systems" *Proceedings of IEEE*, vol. 77, pp. 81-98, 1989.

[3] A. Giua and C. Seatzu, "Supervisory control of railway networks withPetri nets," *Proc. IEEE Conf. Decision and Control*, p. 5004-5009, Orlando, Florida, 2001.

4] A. Giua and X. Xie, "Control of Safe Ordinary Petri Nets using Unfolding," *Journal of Discrete Event Dynamic Systems*, to appear.

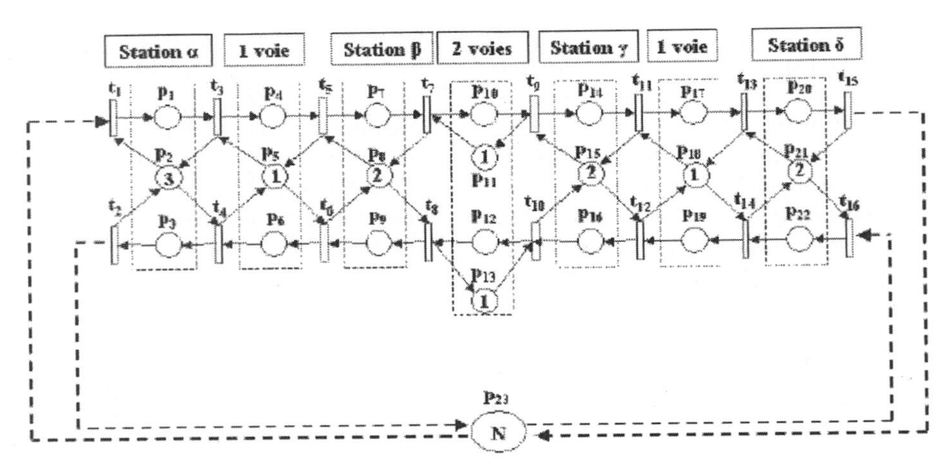

Fig. 5. A PN model of a railway network.

Intelligent Production Machines and Systems
D. T. Pham, E. E. Eldukhri and A. J. Soroka (eds)

Quality Control of High Pressure Die Casting using a Neural Network

M.I. Khan, Y. Frayman, and S. Nahavandi

Intelligent Systems Research Laboratory, Deakin University, Geelong VIC 3217, Australia

Abstract

This paper presents the application of a neural network to quality control of a High Pressure Die Casting (HPDC) process. HPDC is a complex process that is difficult to analyse and control by conventional methods which warrant the use of inductive methodologies such as neural networks. A feed forward neural network was used in this work to model the relationships between the process parameters and the quality indicators and a simple neural network based control framework is proposed to optimize the process parameters to produce castings with a desired quality. It should be noted that the predictive accuracy criterion is not enough alone to trust a neural network model.

1. Introduction

Modern manufacturing processes are generally complex and traditionally require thorough understanding of the process behaviour in order to control them adequately [1,2,3]. With an ever-decreasing workforce entering the shopfloor, it is quite likely that the specialized knowledge about such complex industrial processes may eventually disappear [4]. It is important to take measures to overcome the likely disappearance of this specialized knowledge. Computational intelligence techniques can help here if the manufacturing processes are adequately modelled and a required control strategy is further developed. In this research, a feed-forward neural network is used to model High Pressure Die Casting (HPDC) process in respect to castings quality.Then a possible control strategy to produce castings with a desired quality is proposed.

2. High Pressure Die Casting

The HPDC is an important manufacturing process used to produce a range of consumer products from small mobile phones to large automobile engine castings. The process begins by injecting liquid metal in the die by a moving plunger. The metal is solidified under high pressure and the part is extracted, generally by a robot, when the die is opened. The presence of around 150 process parameters with complex inter-relationships that impact the process behaviour makes it an inherently complex process [5]. The optimal settings of process parameters thus becomes a difficult task and can result in producing castings with undesirable defects like blistering, cold shuts and porosity. This work is concerned with the porosity defect in castings which is the highest occurring defect in the automobile component casting industry [6]. However, it is possible to apply the same framework for other casting defects.

The appearance of pores in castings due to improper process parameters settings is called porosity (Fig. 1). Two main types of porosity are gas and shrinkage porosity. Gas porosity appears due to the gas entrapped in the castings. It can be due to air injected in the die from shot cylinder (the cylinder which contains the molten metal before the plunger injects the metal in the die), the steam produced due to the high temperatures during the operation and the lubricant burnt during operating conditions due to elevated temperatures.

I.

II.

Figure 1: I) Gas porosity II) Shrinkage porosity from cross-sections of the castings.

The optimal settings of process parameters in HPDC are traditionally determined by using physical modelling techniques. These techniques have often resulted in conflicting results being reported by different authors [5]. Some of these traditional approaches suffer from the problem that they tend to concentrate on the material properties while ignoring the influence of the process parameters of the process itself [7]. Our hypothesis is that the HPDC process can be modelled more adequately by using inductive methods. In this work we were concentrating on modelling the relationship between the process parameters of the HPDC and the quality indicators as the first step. The next step was to design an inductive learning controller for controlling defects

3. Modelling the High Pressure Die Casting Process

As mentioned in the previous section the task of modeling HPDC is conducive to an inductive learning method. Out of several available inductive methods, we have decided to use a feed-forward neural network model, because:

- The data available in input/output format is suitable for supervised learning.
- Offline learning is more feasible at the moment, but in the future it may be required to learn online and adaptively for real time control (RTC). The feed-forward neural networks can handle adequately both the offline and online learning.
- Speed of learning is important for RTC, which rules out other supervised learning methods like Boltzmann machines.
- The HPDC process is highly nonlinear, which rules out linear and other simpler regression methods.
- Feed-forward neural networks have been used previously with considerable success in die casting [2, 3, 5, 7, 9].

The feed-forward neural network is represented by a structure consisting of weights and nodes. Weights are the learning parameters of a neural network similar to regression coefficients in linear regression. The nodes, each containing a transfer function are organized in layers. There are generally three types of layers in feed-forward neural networks. The input layer takes the inputs of the process to the hidden layer of the network for processing. After going through a transformation at the hidden layer, the transformed inputs are presented to the output layer. During the learning, the feed-forward neural network feeds the information about the difference in the desired and predicted outputs back to the hidden layer and a desired output is calculated for each neuron (node). This way of learning is called back propagation of errors [10].

Neural networks structure can be obtained for a given problem by either automatic network construction algorithms [11] or by trial and error. In the trial and error approach, different network structures are tested with a variable number of hidden neurons. The number of input and output neurons remains constant because the process being modeled has a specified number of inputs and outputs. The hidden neurons are the main processing units of the network and their number generally increases with increasing complexity of the problems.

The neural network model obtained in this work using a trial and error method consisted of four hidden nodes with tangent transfer function. The inputs consisted of eight process parameters:

- 1st stage velocity
- 2nd stage velocity
- Changeover position
- Intensity of tip pressure
- Cavity pressure
- Squeeze tip pressure
- Squeeze cavity pressure
- Biscuit thickness.

The outputs consisted of a four-level quality measure of the casting with 1 representing the best quality (minimum porosity) and 4 representing the worst quality (maximum porosity, basically a rejected casting). In between 1 and 4 there are two other acceptable levels of porosity. For a detailed description, please see [5].

4. Modelling Results and Discussion

After obtaining the four hidden nodes neural network predicting at a good accuracy, it was still necessary to find a way to find out whether the neural network model is conforming to the realities of HPDC process. It was decided to perform a sensitivity analysis around the mean to find out how the network has detected the influence of each process parameter on the quality of casting (occurrence of porosity) rather than only trusting a black box predicting with good accuracy. A comparison was carried out on the discovery made by neural network about the effects of process parameters on the quality of castings with the die casting research literature. The results obtained from sensitivity analysis are found to be generally in agreement with the die casting literature with a few exceptions. These exceptions nevertheless show that a neural network cannot be trusted one hundred percent only on the basis that it is predicting the quality of castings with good accuracy. It is unfortunately general practice to accept neural networks as the basis of good predictive accuracy. The results obtained from the neural network model are summarised in Table 1. The results which are in conflict with the literature are discussed in more detail.

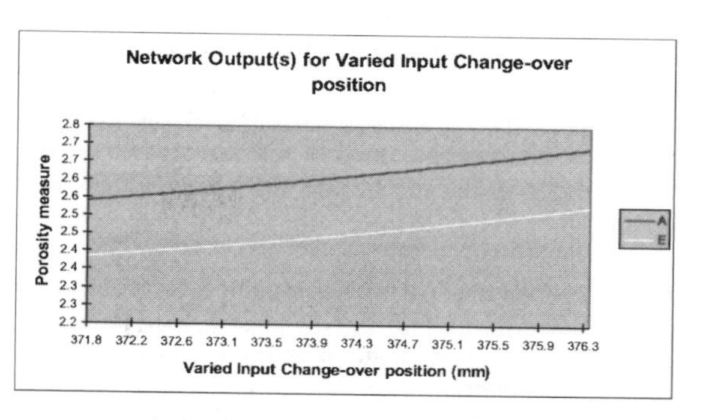

Figure 2: An increase in changeover position increases porosity.

Table 1: The results obtained from neural network model. The last column in the table compares obtained results with the existing knowledge of the influence of a specific process parameter on the level of porosity.

Process Parameters	Effect on Porosity	Literature Reference
1st stage velocity	Parameter increase causes a decrease in porosity level	In accordance with [12], [13]
2nd stage velocity	Parameter increase causes a decrease in porosity level	In accordance with [13]
Changeover position	Parameter increase causes a decrease in porosity level	In conflict with [14]
Tip & Cavity Pressures	Parameter increase causes both an increase and a decrease in porosity levels	Generally in accordance with [15], [16]
Biscuit size	Parameter increase causes an increase in porosity level Increase causes	In accordance with [14]

The first point of conflict regarding the influence of the changeover position on the level of porosity is of no surprise, since the HPDC process is a very complex one and further research in needed to find out if the results obtained in [14] are correct. It is shown in Fig. 2 that an increase in changeover position increases porosity. The changeover position is the process parameter measuring the point where the 1st stage velocity is accelerated to 2nd stage velocity. The later it is done, the better it should be according to the existing knowledge of the process, because it gives more time to the air present in the shot cylinder to escape before the plunger injects it into the die cavity. Ideally, the 1st stage of velocity should be changed to the 2nd stage (higher stage velocity) when the metal has reached at the end of the shot cylinder and all the air has escaped. It seems that the neural network's finding warrants further research to establish a change either in the network model or the established process knowledge. One way to do this is to design a neural network based controller of the process and find out whether it is able to reduce the porosity any better than the humans are doing with the existing process knowledge.

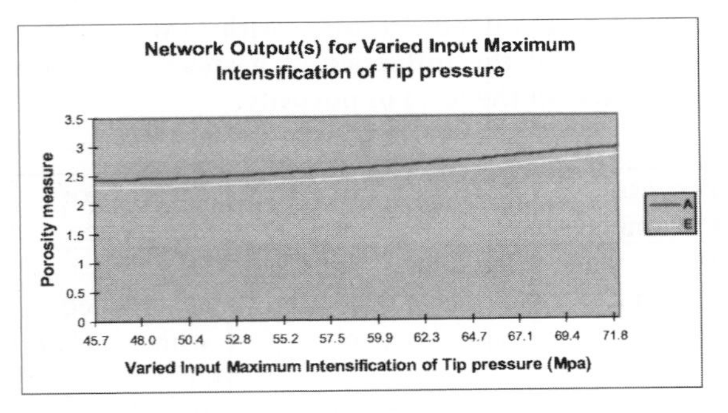

Figure 3: An increase in tip pressure seems to increase porosity.

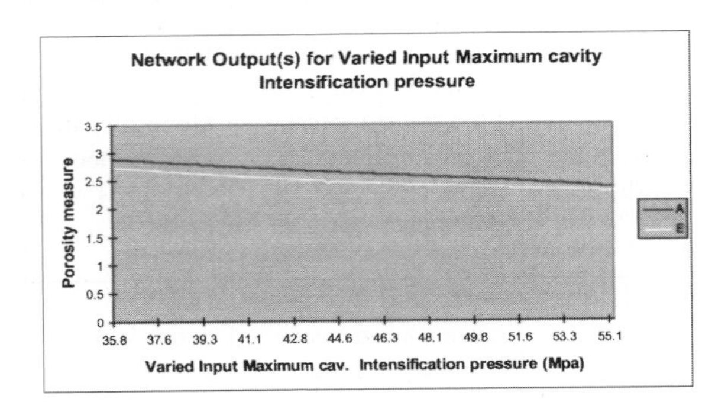

Figure 4: An increase in cavity pressure decreases porosity.

Fig. 3 shows an increase in porosity with increasing intensification of tip pressure, which seems to be in conflict with [15], [16]. Tip pressure is the pressure applied by the plunger when it comes to a halt to quickly solidify the casting. The pressure is applied in HPDC to minimize porosity as well to fill the pores due to shrinkage porosity and to bubble out any gases still remaining. This conflict can be explained in the light of Fig. 4. An increasing pressure developed inside the cavity decreases porosity which is in accordance to [15], [16]. The sharp contrast observed with respect to literature between Fig .3 and Fig. 4 can be explained when we analyze them together. The cavity pressure is the pressure imparted by the tip pressure to the die cavity. In an ideal world, it should be the same in magnitude, but the presence of biscuit acts as a bottleneck to the pressure. The biscuit is the part of casting which is formed at the end of the shot cylinder and at the beginning of the point of injection. It is a relatively small piece of metal which if solidified first (which generally happens) prevents the pressure applied to reach the solidus metal in the cavity. Consequently, only a fraction of the pressure applied at tip reaches cavity. The pressure that reaches the cavity decreases porosity since it reduces the shrinkage and allows the gases to escape. This capability of neural network to discern and discover the effect of each type of pressures further warrants a need to develop a neural network based controller.

The obtained results show that a neural network model is able to predict the level of porosity adequately.

5. Proposed Control Mechanism

The control strategy proposed in this work utilizes the obtained neural network model of the influence of process parameters on the level of porosity in accordance with commonly used approach to control [17].

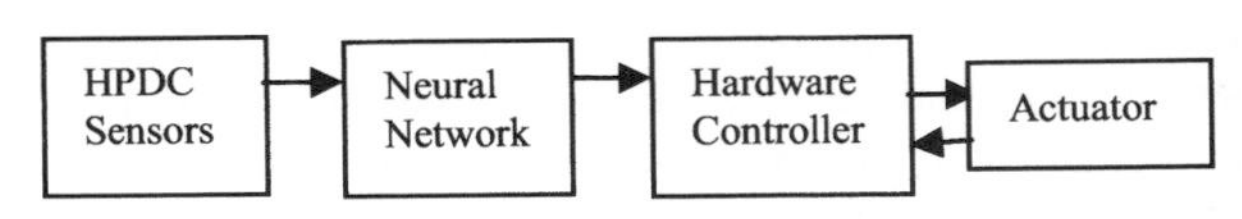

Figure 5: Proposed neural network control framework.

In the control framework of Figure 5, the obtained neural network model is used to select between the different HPDC process settings to the one that is likely to result in castings with acceptable levels of porosity. The process starts with the initial state and supplies its state (the values of process parameters) to the neural network model. The neural network model then verifies the optimality of the process state by predicting the likely level of porosity in the castings, and if the porosity level is predicted as acceptable, produced a signal to the hardware controller to maintain the current state of the process, otherwise it notifies the controller that the process has to be switched to another one of the available settings which is then evaluated again and if found adequate is used.

After every cycle of the casting production, the neural network can re-evaluate the likely effect of the particular process settings on the quality of the castings and generate required advise to the controller either to maintain the current process settings if the predicted level of porosity is adequate or to switch the process to different settings if the level of porosity is found to be inadequate.

6. Conclusion

Neural network can model a die casting process adequately and can be a reliable controller. It is important to note that a neural network predicting with enough accuracy may still not be a good model. It becomes more important to test a neural network using other criteria, especially when it is to be used in a controller. The industrial processes are very complex and the insufficiency of a neural network model may not be apparent if it is judged against the common criteria of predictive accuracy. One simple test in addition to good predictive accuracy can be sensitivity analysis. Sensitivity analysis helps to find out whether the neural network is able to learn the intricacies involved with each input (the process parameter in industrial process modelling), which is not otherwise clear from predictive accuracy alone.

7. Future Work

There is a need to further develop neural network's own reliability estimation techniques when it is used to model real world (industrial) systems. It becomes crucial in a neural network based controller because it is important to run the process in stable conditions. It is proposaed that to apply knowledge extraction techniques with specially designed neural networks [18] to trivialize knowledge gained by the neural network black box in a comprehensible form. This knowledge can be presented to the experts in the control domain (die casting in our case) and compared with the literature of the domain to find out how reliably the neural network is able to model the intricacies of the process. Sensitivity analysis as performed in this work is one of the ways to achieve this. Having more methods to verify the neural network is likely to get them accepted [19]..

7. Acknowledgment

This work was supported by the Cooperative Research Centre for Cast Metals Manufacturing (CAST). CAST was established and is funded in part by the Australian Government's Cooperative Research Centres Program.

References

[1] Rolfe B. Characterising Geometric Variation in Shape Manufacturing With A View To Process Parameter Feedback, PhD Thesis, The Australian National University, June 2002.

[2] Khan M.I, Frayman Y. and Nahavandi S. Modelling Of Porosity Defects In High Pressure Die Casting With A Neural Network. Proceedings of the Fourth International Conference on Intelligent Technologies 2003 (Intech'03), pp. 73-78, Chiang Mai University, Thailand.

[3] Khan M.I, Nahavandi S. and Frayman Y. Improving The Quality Of Die Casting By Using Artificial Neural Network For Porosity Defect Modelling. Proceedings Of The 1st International Light Metals Technology Conference 2003, pp. 243-245, CAST Centre Pty Ltd, Australia.

[4] Pattinson J. and Shirvani B. Skill Shortage In The UK Metalforming Industry. Proceedings Of The 7th International Conference on Sheet Metal 1999. pp. 45-52, Erlangen, Germany.

[5] Khan, M. I., Nahavandi, S. and Frayman, Y. High Pressure Die Casting Process Modelling Using Neural Network, in "ANN Applications In Finance and Manufacturing", Idea Group Publishing, USA, (accepted).

[6] Andresen W.T and Guthrie B. Using Taguchi and Meltflow To Determine Relationships Between Process Variables and Porosity. Transactions of The 15th International Die Casting Congress and Exposition, NADCA, Transaction No. G-T89-082, 1989, USA.

[7] Gordon A. Meszaros G. Naizer J. and Mobley C. Equations For Predicting The Percent Porosity In Die Castings. Transactions of The 17th International Die Casting Congress, NADCA, Transaction No. C-T93-024, 1999, USA.

[8] Huang J, Callau P and Conley J.G. A Study of Neural Networks For Porosity Prediction In Aluminium Alloy A356 Castings. In Thomas B.G. And Beckermann C. (Eds). Modelling of Casting, Welding, And Solidification Processes, VIII , TMS ,

June,1998, 1111-1118.

[9] Yarlagadda P.K.D.V and Chiang E.C.W. A Neural Network System for the Prediction of Process Parameters in Pressure Die Casting. Journal of Materials Processing Technology, 89-90, 583-590, 1999.

[10] Rumelhart D, Hinton G, and Williams R. Learning Internal Representations by Error Propagation, in D. Rumelhart et al (eds), Parallel Distributed Processing 1 , MIT Press, 318-362, 1986.

[11] Fahlman E. and Lebiere C. The cascade-correlation learning architecture, in D. S. Touretzky (Ed.), Advances in Neural Information Processing Systems, Morgan Kaufmann, San Mateo, CA, pp. 524-532, 1990.

[12] Garber L. W. Filling of Cold Chamber during Slow-Shot Travel. Die Casting Engineer, Issue Jul-Aug, 1981.

[13] Garber, L.W. Theoretical Analysis and Experimental Observation of Air Entrapment during Cold Chamber Filling. Die Casting Engineer, Issue May-June, 1982.

[14] Asquith B.M. The Use of Process Monitoring to Minimize Scrap in the Die Casting Process. Transactions of the International Die Casting Congress, NADCA, Transaction No. M-T97-063, 1997.

[15] Kay A, Wollenburg A, Brevick J, and Wronowicz J. The Effect of Solidification Pressure on the Porosity Distribution in Large Aluminium Die Casting. Transactions of The International Die Casting Congress, NADCA, Transaction # M-T97-094, 1987.

[16] Murray M. T, Chadwick G. A and Ghomashchi M. R Aluminium Alloy Solidification in High Pressure Die Casting. Materials Australasia, June 1990.

[17] Lee J. H. Modeling and Identification for Nonlinear Model Predictive Control: Requirements, Current Status and Future Research Needs, Proceedings of The International Symposium on Nonlinear Model Predictive Control: Assessment and Future Directions, Ascona, Switzerland, 1998.

[18] Khan M. I. Frayman, Y. and Nahavandi, S. (2004) Knowledge Extraction from a Mixed Transfer Function Artificial Neural Network, Fifth International Conference on Intelligent Technologies (InTech'04), University of Houston Downtown, Houston, TX (USA), 2-4 December 2004

[19] Gabriel I. And Dobnikar A. On-Line Identification And Reconstruction Of Finite Automata With Generalized Neural Networks, Neural Networks, Vol. 16, pp. 101-120, 2003

Intelligent Production Machines and Systems
D. T. Pham, E. E. Eldukhri and A. J. Soroka (eds)

Robot kinematic calibration using Genetic Algorithms

Kesheng Wang* and Jonathan Lienhardt

Knowledge Discovery Laboratory
Department of Production and Quality Engineering,
Norwegian University of Science and Technology,
N-7491 Trondheim, Norway

Abstract

Position and orientation accuracy of the end effector is affected by the precision of kinematic parameters of the robot manipulator. Thus good precision requires good knowledge of robot physical parameter values. However this condition can be difficult to meet in practice. Hence calibration techniques can be devised in order to improve the robot accuracy through estimation of these particular parameters. In this paper, the Genetic Algorithm is used to calibrate the robot kinematic accuracy. A kinematic model is formulated and conducted as an optimization problem for ABB robot manipulators. The objective is to analyze and evaluate the performance of the GA in optimizing such robot kinematic accuracy. In this algorithm, small changes in the kinematic parameters values represent the parent and offspring population and the end-effector error represents the fitness functions. A numerical example has been used to demonstrate the convergence and effectiveness of the given model.

Keywords: Genetic Algorithms, Robot manipulators, Robot kinematics, Robot calibration.

1. Introduction

A rather recent development in robotics is the usage of Computational Intelligence (CI) methods, especially Genetic Algorithms (GAs). Genetic Algorithm is a stochastic search approach in robotics which has shown high efficiency in certain multimodal and multidimensional domains. These algorithms imitate evolutionary procedures of biological systems to develop better and better solutions to the problem until one of them is satisfactory for the given purpose (even though it may not be the absolute optimum).

The manipulation and the offline coding of a robot is based on a kinematic model. This model is defined from the geometric characteristic of the robot, which is given by the constructor. Unfortunately, each robot is unique and those geometric parameters need to be readjusted for better to fit the reality. A robot pose can be defined in two ways. Either by giving the pose position and orientation values, or by giving the joint values of each link. The kinematic model is the mathematical tool that helps the transformation from the Cartesian space of the robot to the joint space (indirect kinematics) or the other way (direct kinematics).

* Corresponding author: E-mail: kesheng.wang@ntnu.no; Tel.: +4773597119; Fax:+4773597117

The calibration of robot manipulators using GAs has attracted many researchers. [6] [7] All these methods need complex inverse mathematical formulas to calculate the error values of joint variables. In this paper we introduce a new, low-cost, simple yet accurate approach using Genetic Algorithms for calibrating the robot kinematic accuracy. The calibration algorithm was implemented in the calibration of ABB robot manipulators.

2. Robot calibration problem

2.1. *Principle*

Position and orientation accuracy of the end effector of a robot manipulator is affected by the precision of kinematic parameters, such as machining tolerance, compliance and wear of the connecting mechanism and misalignment of the connected module components. Thus good precision requires good knowledge of robot physical parameter values. However this condition can be difficult to meet in practice. Hence calibration techniques can be devised in order to improve the robot accuracy through estimation of these particular parameters.

The transformation between position and orientation of the end effector and the joint variables is the most important effects for the absolute accuracy of a robot manipulator. Hence the main issue of robot kinematic calibration is to determine this transformation accurately. That means we can adjust the parameters describing the geometry of robot manipulators to improve the real pose of robot manipulators. This procedure is called the parametric calibration of robot manipulators.

The common procedure of the parametric calibration is divided into four major steps as follows:

1. Establishing the structure of kinematic model (both forward and inverse kinematics) of the robot for calibration.
2. Collecting a set of measurement data consisting of the end-effector pose and corresponding joint configuration.
3. Making the numeric identification of the model parameters. Generally this is based on non-linear least squares optimization.
4. Implementation of the identified model into the controller of the robot manipulator.

2.2. *The Denavit-Hartenberg model*

The Denavit-Hartenberg (DH) kinematic model [1] [3] [8] is based on four kinematic parameters per joint to describe the geometry of the robot (Figure 1). This model describes the geometric relationship between consecutive frames. A Cartesian coordinate is attached to each link in order to be able to define the position of the link at anytime. The transformation of rotation and translation of the fixed coordinates can be described in homogeneous matrices according to the kinematic parameters.

The aim of the identification process of the calibration algorithm is to readjust the four D-H parameters: a_i, d_i, α_i, θ_i by calculating correction values: Δa_i, Δd_i, $\Delta \alpha_i$, $\Delta \theta_i$. The forward kinematic equations are:

$$[X, Y, Z, \phi, \theta, \psi] = f(\theta_1, \theta_2, \ldots, \theta_N) \quad (1)$$

Where vector $[X, Y, Z, \phi, \theta, \psi]$ represents the position and orientation of the end-effector of a robot manipulator in a Cartesian coordinate, $\theta_1, \theta_2, \ldots, \theta_N$ are joint variables and f is a transformation matrix. The inverse kinematics is:

$$(\theta_1, \theta_2, \ldots, \theta_N) = f^{-1}[X, Y, Z, \phi, \theta, \psi] \quad (2)$$

3. Genetic Algorithms

A GA is a stochastic optimization method based on the mechanisms of natural selection and evolution [5] [9] [10]. In GAs, searches are performed based on a population of chromosomes represent solutions to the problem. A population starts from random values and then evolves through succeeding generations. During each generation a new population is generated by propagating a good solution to replace a bad one and by combining or mutating existing solutions to construct new solutions. GAs have been theoretically and empirically proven robust for identifying solutions to combinatorial optimization problems, as GAs conduct an efficient parallel exploration of the search space, while only requiring minimum information on the function to be optimized.

4. Robot error model

In a robot manipulator task, we want to achieve a defined position given a set of joint values. The actual position which is reached by the robot is always different for the model position which is calculated by the use of the forward kinematics. Theoretically the errors can be evaluated by equation (3) and (4). The forward and inverse error model can be expressed as the following two equations:

$$\begin{bmatrix} \Delta P \\ \Delta R \end{bmatrix} = J[\Delta\theta_i] \qquad i = 1,2,\ldots,N \qquad (3)$$

$$[\Delta\theta_i] = J^{-1}\begin{bmatrix} \Delta P \\ \Delta R \end{bmatrix} \qquad i = 1,2,\ldots,N \qquad (4)$$

Where J is Jacobian matrix and J^{1} is inverse Jacobian matrix.

The inverse Jacobean matrix is difficult to calculate, and it takes time. Because of the difference between the two sets of data, we can computer an enhanced model that will correspond better to the reality. We implement a GA search method to reduce the errors between the given and the actual pose of the end-effector of robot manipulator. The error between the given pose, and the actual pose achieved by the enhanced model is used as a fitness function of the GA approach.

This kinematic model will simply be created by adding correction values to the kinematic parameters of the robot manipulator. Those correction values will be used as the chromosomes of the GA, and their costs can be directly evaluated by the fitness function that we will try to minimize. For N measurements, we define the fitness function (FF), as

$$FF = \frac{\sum_{i=1}^{N}\sqrt{\Delta x_i^2 + \Delta y_i^2 + \Delta z_i^2 + \Delta\psi_i^2 + \Delta\theta_i^2 + \Delta\phi_i^2}}{N} \qquad (5)$$

5. Calibration algorithm

In the robot calibration algorithm, the kinematic parameter errors are used to represent the GA's individuals in population pool. The Mean Square Root of errors of absolute position and orientation of the end-effector of a robot manipulator represent the GA's fitness functions. The Algorithm could be described as the following steps:

1. Using the D-H parameters, $a_i, \alpha_i, \theta_i, d_i$ ($i = 1, 2, \ldots, N$), for a given configuration of the robot manipulator.
2. For a nominal position and orientation (P_N and R_N) of the end-effector, the corresponding joint variables, θ_i, are calculated by inverse kinematic equation (2).
3. The actual position P_c and orientation R_c are computed by a forward kinematics in equation (1).
4. Establish a set of measurement data consisting of the pose of the end-effector (position, P_m and orientations, R_m) and corresponding joint configuration, θ_i ($i = 1, 2. \ldots, N$) of the robot manipulator in a work space.
5. Calculating the errors of the end-effector, ΔP and ΔR.
6. The GA iterations start by generating an initial population pool with a number of individuals which contain the error variables of the joints: $\Delta a_i, \Delta\alpha_i, \Delta\theta_i, \Delta d_i$ ($i = 1, 2, \ldots, N$) randomly. These populations are used as parents from which the genetic operators are applied to produce offspring in a new generation.
7. Calculating the absolute position $P_c=[X, Y, Z]$ and orientation $R_c= [\phi, \theta, \psi]$ of the end-effector of the robot manipulator using the correction joint variables (by adding the joint error valuables which are generated in step 6).
8. Calculate the value of the fitness function using equation (5).
9. Reproduce the offspring in the next generation using a selection approach (roulette wheel or elitist strategy) and GA operators (crossover and mutation).
10. This process is iterated until criterion (such as a given number of iterations) is met.

6. Implementation

6.1. *Four different Types of configurations:*

In order to improve the accuracy of the robot manipulator, the kinematic parameters of a robot manipulator should be identified and readjusted by the use of robot kinematic calibration process. GA offers the possibility to optimize the kinematic parameters among a great number of different robot configurations. GA chromosomes can be easily used to represent kinematic parameters. For example, four different types of configurations (parameters) for a robot manipulator with six joint variables were selected:

Type 1: 24 chromosomes (Δa_i, $\Delta \alpha_i$, $\Delta \theta_i$, Δd_i) encoded each with 10 binary bits.
Type 2: 24 chromosomes (Δa_i, $\Delta \alpha_i$, $\Delta \theta_i$, Δd_i) represented by real numbers.
Type 3: 6 chromosomes ($\Delta \theta_i$) represented by real numbers.
Type 4: 18 chromosomes (Δa_i, $\Delta \alpha_i$, Δd_i) represented by real numbers.

In order to evaluate these 4 selections, the following GA parameters were chosen:

- Mutation rate: 0.05
- Crossover rate: 0.98
- Population Size: 50
- GA program stop if the best fitness value remains unchanged after 1000.

We obtain then the following results:

Type of Chrom.	Generation of converge	Mean Error	Error Ratio
Type 1	6035	265,5465114	20337,63%
Type 2	19131	5,057637714	289,26%
Type 3	6150	2,21727906	70,65%
Type 4	41285	0,830260798	-36,10%

Three indicators which were used to evaluate the efficiency of the algorithm are:

- The number of generations of convergence which was calculated to satisfy the convergence criterion.
- The mean error which is the value of the Fitness Function.
- The error ratio which represents the ratio of the error of the standard kinematic model and the enhanced one.

From the results, we can indicate that the best selection of the parameters is type 4 which only has three kinematic parameters, the Δa_i, $\Delta \alpha_i$, Δd_i.

In most cases, the GA can reach a stabile state (convergence) before the GA stops. When we run the GA with only three kinematic parameters, the error, after a 200 generations, is already very small.

6.2. Finding the best parameters

The best GA parameters should consist of (1) Mutation and crossover rate, (2) The size of population, and (3) Convergence criterion. In search for the best GA parameters, we programmed the GA to stop if, after a certain amount of generations, the best fitness remains unchanged. The bigger the amount of generation is, the longer the GA runs.

1. Mutation and crossover rate

The type 3 kinematic configuration was selected as a testing example. The size of population was 100 individuals and several combinations of mutation and crossover rate had been tested. We found that the solution is the one with a mutation rate of 0.1 and a crossover rate of 0.8. However if the computation time is not very important while running the program, a mutation rate of 0.05 and a crossover rate of 0.9 are recommended.

We also notice that the mutation rate has the most important impact to GA in this case and it must be kept under 0,1 to have satisfactory results.

2. Population size

The most interesting point here seems to be a population size of 175 individuals, which would offer the best compromise Generations/Error Ratio (See Fig. 1). Nevertheless, since the aim of the calibration process is to reduce the error, we will consider the best solution as being a Population Size of 200 individuals. The cost in terms of generations is higher, but we have a much better error reduction.

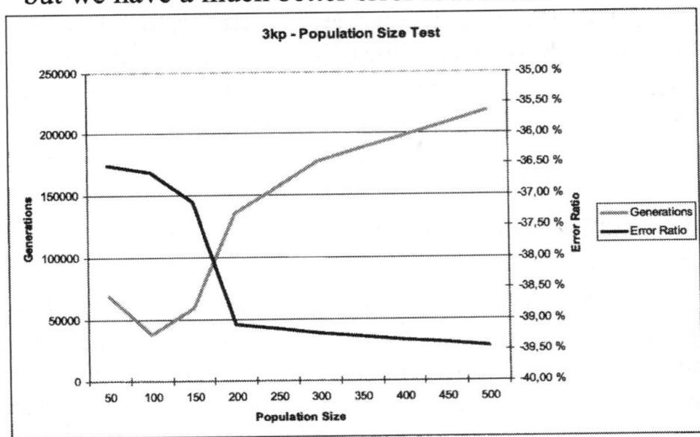

Fig. 1. Selection of the population size.

3. Convergence criterion

The convergence criterion used in the experiment is the number of generations where the best fitness remains unchanged. Practically, a counter is incrased when the best fitness remains the same, and reset to zero when reduced. The program stops when the counter reaches a certain level.

To find the optimum convergence criterion, different values for the maximum acceptable counter value were tested.

In Fig. 2, we can see that a maximum counter value of 5000 gives us good result by avoiding a too large amount of generations. We can also check the evolution of the best fitness value through the number of generations.

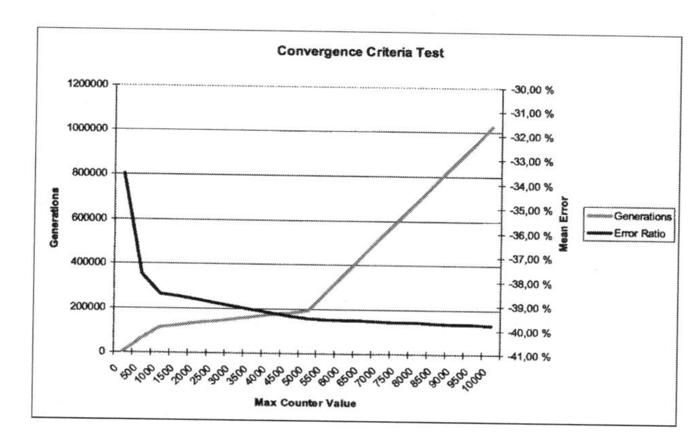

Fig. 2. Convergence testing.

After the convergence criterion's value of 5000 is passed, the number of generation increases dramatically as the error evolution is slow down. Before this value, the error reduction is acceptable but not sufficient. The convergence criterion needs to be high enough to reach a best fitness value. If we set the value of the counter to 100, we remain at the beginning of the convergence curve. The minimum acceptable value seems to be a counter of 500. Finally in order to have good results and to avoid a too long computation time, counter 5000 is recommended as a convergence criterion.

To implement the calibration algorithm described in section 5, a file containing the measurements results must be used. Joint values for the measured pose will be sent to the enhanced kinematic model, and the calculated pose will be compared with the measured one. The programming language for the implementation system is C^{++} [2] [4].

7. A Case Study

Applied the system to a 6 joints ABB Irb 6000 Robot, we found that the best results were obtained with the 18 real chromosomes arrangement. The GA optimum parameters were a value of 0.1 for the mutation rate and 0.8 for the crossover rate. A population contains 200 individuals (chromosomes). We stopped the algorithm after 5000 generations, which offered us an acceptable calculation time and an accuracy improvement ratio. Here is an example of the best result in which we used 42 000 generations and it took 281 seconds and the joint correction values are shown in Table 1.

Table 1. The best correction values.

Joint	Δa	$\Delta\alpha$	Δd
1	0.95761	0.000479141	-0.41084
2	-1.43162	5.79852e-005	-1.03977
3	1.20811	1.52593e-005	1.14438
4	0.186224	-0.00311594	-1.1102
5	0.497513	-0.00314035	-0.0293588
6	-0.925382	0.00036317	0.488601

The number of measurements is also important. Results show that we highly improve the accuracy on a few points in the robot workspace, and moderately on several points. Depending on the robot application, the desired accuracy can be different on many places of the workspace. But in each case, the calibration accuracy of the robot manipulator has been increased. The results we obtained are shown in Table 2 and Fig. 3.

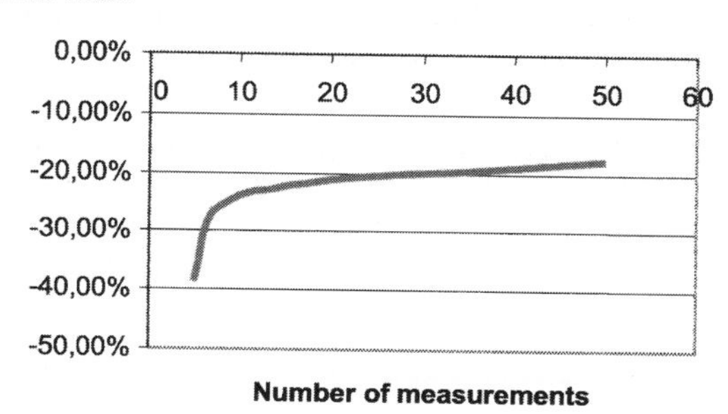

Fig. 3. The evolution of error ratio.

8. Conclusions

The robot kinematic calibration approach using Genetic Algorithm based on D-H convention has been presented. The effectiveness of the algorithm and its convergence in the presence of small joint errors and measurement errors has been demonstrated through a numerical implementation of ABB Irb 6000 Robot. The benefit of this approach is to avoid complex inverse Jacobean calculations in traditional robot calibration model.

In the experiment we used a basic Genetic Algorithm structure to enhance the accuracy of the robot manipulator, but there are many new techniques concerning Genetic Algorithms could be used, for example, ANN and Fuzzy logic technique could be integrated with GAs to build up hybrid computational system.

Acknowledgement

NTNU is a partner of EU-funded FP6 Innovative Production Machine and Systems (I*PROMS) Network of Excellence.

References

[1] Denavit, J. and Hartenberg, R. S., (1955), A Kinematic Notation for Lower-Pair Mechanisms Based on Matrices, Journal of Applied Mechanics, ASME, 22, pp.215-221.

[2] Gen, M., Cheng, R. (1997), Genetic Algorithm and Engineering Design, Wiley, New York.

[3] Goldberg, D. E., (1989), Genetic Algorithms in Search, Optimization and Machine Learning, Addison Wesley, MA.

[4] Kreyszig, E., (1988), Advanced Engineering Mathematics, 6th edition, John Wiley & Sons, Inc.

[5] Liberty, J. and Hord, J. M., (1996), Teach Yourself ANSI C++ in 21 Days, Premier Edition, Sams Publishing.

[6] McKerrow, P. J., (1991), Introduction to Robotics, Addison-Wesley Publishers Company.

[7] Mooring, B. W., et al, (1991) Fundamentals of Manipulator Calibration, John Wiley & Sons, Inc.

[8] Samak, S., (1999) Robot Kinematic Calibration Accuracy Using Genetic Algorithm, Proceedings of IMAC – XIX , A Conference on Structural Dynamics.

[9] Wang, K. (2005) Applied Computational Intelligence in Intelligent Manufacturing systems, International Seriess on Natural and Artificial Intelligence Vol. 2, Advanced Knowledge International, Pty Ltd, Australia.

[10] Zhuang, H., Wu, J. and Huang, W., (1996), Optimal Planning of Robot Calibration Experiments by Genetic Algorithms, Proceedings of the 1996 IEEE International Conference on Robotics and Automation.

Table 2. Influence of the number of measurements.

Number of Measurements	Generations	Mean Real Error	Mean New Error	Error Ratio
5	183150	1,299301968	0,796402695	-38,71%
10	173337	1,326358706	1,010070843	-23,85%
50	103166	1,376224662	1,131986938	-17,75%

Intelligent Production Machines and Systems
D. T. Pham, E. E. Eldukhri and A. J. Soroka (eds)

A performance evaluation approach of an inventory system with batch orders

Jie LI, Alexandru SAVA, Xiaolan XIE

INRIA I / LGIPM, ISGMP-Bat A, Ile du Saulcy, 57045 Metz, France
Tel: 33 3 87 54 73 64, Fax: 33 3 87 54 72 77, email: **li@loria.fr**, **sava@enim.fr**, xie@loria.fr

Abstract

This paper deals with an inventory system composed of a warehouse supplied from a manufacturing plant. Customer orders arrive randomly with random order size and the production capacity is finite. The inventory at the warehouse is controlled by a base-stock policy. Transportation time from plant to warehouse is a constant. An analytical approach is proposed to evaluate the order-to-delivery lead-time of the warehouse, the total inventory on order and the inventory holding and backlogging cost. Numerical comparisons with simulation show that the analytical approach is very efficient.

1. Introduction

Inventory systems are of great economical interest. The dynamic of such systems is triggered by customer orders and it is subject to constraints such as the production capacity, inventory costs and transportation lead time. Numerous performance evaluation and optimization approaches have been developed in order to evaluate these systems and help decision making. A presentation of the main inventory control policies can be found in [7].

Due to the page limitation, we limit ourselves to a review of papers most related to our work. An approach for performance evaluation and optimization of supply chains is introduced in [2]. The supply chain considered is composed of a set of inter-related stores and each store location is modeled by an $M^X/G/\infty$ queuing system. Inventory management is based on base-stock control policy. The authors derive analytical expressions for performance measures such as the store lead time, the filling rate, the service level and the inventory cost. However, it cannot take into account the production capacity of the plants, which is a key constraint in production-distribution systems. In [5] the authors developed a multi-stage inventory queue systems. A job-queue decomposition approach was proposed to evaluate the performance of serial manufacturing and supply systems with control at every stage. They proposed an efficient procedure to minimize the overall inventory cost. The model is based on *GI/G/1* queue, with products arriving one at a time. Batch sizes and transportation times are not considered. An optimal control policy for inventory systems with batch ordering is presented in [11]. However, the authors are taking into account only batches of fixed size.

The motivation of our work is to provide efficient methodologies for performance evaluation of realistic supply chains network models with features such as random batch orders, finite production capacities and transportation times. The work presented in this paper is the first step on the analysis of a basic supply network with a manufacturing plant and a warehouse. To the best of our knowledge, the work is new and there is no inventory policy evaluation approach in the literature which takes into account at the same time i) a random customer demand, ii) a finite production capacity; iii) a random batch size for orders, manufacturing and delivery quantities; iv) the transport time from the plant to the warehouse.

An analytical approach is proposed for performance evaluation based on several approximation techniques that are proved efficient for most manufacturing systems: (i) two-moment approximation, (ii) markovian approximation of general processes, (iii) approximation of dependent random processes by independent ones. More specifically, the analytical approach needs only the first two moments of random variables of the system to evaluate the order-to-delivery lead time of the warehouse, the total inventory on order and the inventory holding and backlogging cost. The production plant is modeled as a $M^X/G/1$ queueing system and the transportation system is represented as a $M^X/D/\infty$ queueing system. Approximate analytical expressions are derived for estimation of the first and/or second moments of order-to-delivery lead time and inventory on order. A log-normal distribution is then used to approximate the distribution of the inventory on order and to determine the total inventory holding/backlogging cost of the warehouse. Numerical comparisons with simulation show that the analytical approach is very efficient and it is not sensitive the workload of the production plant.

The reminder of the paper is organized as follows. Section 2 formally defines the two-stage production-distribution system considered in this paper. The analytical approach is presented in Section 3. In Section 4 we present some numerical results for comparison of the analytical approach and the simulation. Concluding remarks are given in Section 5.

2 The inventory system

This paper considers an inventory system with a warehouse supplied by a manufacturing plant. Only one type of products is considered. The warehouse faces batch customer arrival processes and generates orders to the plant. The plant processes orders from the warehouse according to its available capacity and on a FIFO basis. Any order filled at the plant is then send to warehouse and the transportation time is significant and cannot be neglected. Customer orders arrive randomly according to a Poisson process with arrival rate λ. The quantity of each order is a random positive integer variable X and the quantities of different orders are iid random variables. This implies that the arrival process is a compound Poisson process. Assume that the order quantity X has finite first two moments with mean μ_X and standard deviation σ_X. Upon the arrival of customer order of size X, if the on-hand inventory of the warehouse is enough, the order is totally filled. Otherwise, the customer order is only partially filled according to the on-hand inventory level and the remaining quantity is backlogged and will be filled by future product deliveries from the plant. This implies that customer orders can be split.

The inventory of the warehouse is managed following a base-stock control policy. This policy relies on the concept of inventory position which is equal to on-hand inventory at the warehouse minus the backlogged quantity plus the total pending order quantities. The base stock policy keeps the inventory position constant at the so-called base-stock level R. As a result, whenever a customer order of size X arrives, the warehouse immediately issues a replenishment order of size X to the plant. As a result, the replenishment orders arriving at the plant also form a compound Poisson process. The manufacturing plant is modeled by a single-server queueing system which serves each batch order on a unit basis. The processing time of each product unit is a random variable τ of general distribution with mean μ_τ and standard deviation σ_τ. The manufacturing plant can be modeled as an $M^x/G/1$ queueing system. The total time elapsed between the arrival of a replenishment order at the plant and the moment when the order is ready from delivery to the warehouse is called the plant lead time denoted as L_s. The delivery of products from plant to warehouse is made on the basis of replenishment orders. When a replenishment order of size X is ready at the plant, the batch of X product units is transported to the warehouse. The time needed to transport a batch from the plant to the warehouse is called transportation lead time L_t. We assume that the transportation time is a given constant. The order to delivery lead time of the warehouse, denoted by L, represents the time elapsed between sending an order to the plant and receiving at the warehouse the products of the order. It is defined by the sum of the plant and the transport lead time.

Another important issue in inventory management is the inventory holding cost and demand backlogging cost. Holding a unit of product in the warehouse incurs a cost of C_h per time unit. Demands waiting to be filled incur a cost of C_b per waiting time unit for each product unit of backlogged orders.

In the following, we propose an analytical approach for evaluating the first and second moments of the order-to-delivery lead-time of the warehouse and the average cost of the warehouse.

3 Performance evaluation

This section addresses the performance and cost evaluation of the production-distribution system. We first evaluate the first and/or second moments of the following performance variables:

L_s: lead-time at the plant for a typical order unit (not batch) issued by the warehouse

N_s: total number of units waiting to be or being processed at the plant

N_t: total number of units in transportation to the warehouse

L: order-to-delivery lead-time for a typical order unit issued by the warehouse

$N = N_s + N_t$: total number of units on order also called inventory on order.

These performance variables are then used in the evaluation of the inventory and backlogging cost.

3.1 The manufacturing plant model

As mentioned in the previous section, the plant can be modeled by a $M^X/G/1$ queuing system. As the warehouse is managed according to a base-stock policy, replenishment orders arriving at the manufacturing plant follows the same compound Poisson process as the customer orders. Let us first consider the service time T_B of a replenishment order:

$$T_B = \sum_{i=1}^{X} \tau_i \qquad (1)$$

where X is the size of the replenishment order and τ_i is the service time of i-th units of the order. By assumption, X and τ_i are mutually independent random variables. Hence,

$$m_{TB} = E[T_B] = m_\tau m_X \qquad (2)$$

where $m_\tau = E[\tau]$ and $m_X = E[X]$. By conditioning on X and by mutual independence of X and τ_i,

$$\begin{aligned} E[T_B^2] &= E[E[(\sum_{i=1}^{X} \tau_i)^2 \mid X]] \\ &= E[X.E[\tau^2] + X(X-1)(E[\tau])^2] \qquad (3) \\ &= m_x \sigma_\tau^2 + m_\tau^2(\sigma_x^2 + m_x^2) \end{aligned}$$

where, $\sigma_\tau^2 := Var[\tau]$ and $\sigma_x^2 := Var[X]$.

Combining relations (2) and (3),

$$\begin{aligned} \sigma_{TB}^2 &:= Var(T_B) \\ &= E[T_B^2] - E^2[T_B] \qquad (4) \\ &= m_x \sigma_\tau^2 + m_\tau^2 \sigma_x^2 \end{aligned}$$

Consider first replenishment orders as basic customers of the queueing system. The $M^X/G/1$ queuing system then becomes a $M/G/1$ queuing system with T_B as service time. The mean number N_B of batches in the plant can be computed by the Pollazek-Khinchin formula [1], where ρ is the traffic intensity of the quqLeueing system:

$$E(N_B) = \frac{\rho}{1-\rho} - \frac{\rho^2}{2(1-\rho)}(1 - \sigma_{TB}^2 / m_{TB}^2) \qquad (5)$$

$$\rho = \lambda m_{TB} \qquad (6)$$

From (2)-(6):

$$E(N_B) = \lambda \frac{\lambda(2m_x m_\tau - m_x^2 m_\tau^2 \lambda + \lambda m_x \sigma_\tau^2 + \lambda \sigma_x^2 m_\tau^2)}{2(1 - m_x m_\tau \lambda)} \qquad (7)$$

By Little's law, the mean of the total time L_B a replenishment order spends at the plant is as follows:

$$E[L_B] = E[N_B] / \lambda \qquad (8)$$

$E[L_B]$ is the average time a batch spends at the plant and $E[L_s]$ is a average time a product unit ordered by customers spends at the plant. These two concepts are different. Although the time of a product unit in a particular customer order (or batch) of size X spends in the pant is exactly the same as the time the related batch spends in the plant, $E[L_B]$ and $E[L_s]$ differ according to the distribution of the order quantity X. According to large number of numerical experiences, $E[L_B]$ and $E[L_s]$ are very close for most realistic distribution of X.

$$E[L_s] \approx E[L_B] \qquad (9)$$

Apply Little's law to product units, we have:

$$\begin{aligned} E[N_s] &= \lambda E[X].E[L_s] \\ &\approx \lambda m_x E[L_B] = m_x E[N_B] \end{aligned} \qquad (10)$$

Consider now the second moment of the performance indicators. First,

$$\sigma_s^2 := Var[N_s] = E[N_s^2] - E[N_s]^2 \qquad (11)$$

$$E[N_s^2] = E[E[(\sum_{i=1}^{N_B} X_i)^2 \mid N_B]]$$
$$= E[E[(\sum_{i=1}^{N_B} X_i^2 + \sum_{\substack{i,j\\ i\neq j}} X_i X_j) \mid N_B]]$$

where X_i is the number of units of replenishment order i. In general, N_B depends on the size X_1 of the batch being served and $X_1 \mid N_B$ does not have the same distribution as X_i with $i \neq 1$. By assuming the independence of N_B and X_1 and by assuming that $X_1 \mid N_B$ and X_i with $i \neq 1$ are iid, the following approximation is obtained:

$$\begin{aligned} E[N_s^2] &\approx E[N_B.E[X^2] + N_B(N_B-1).E^2[X]] \\ &= E[N_B].E[X^2] + (E[N_B^2] - E[N_B]).E^2[X] \end{aligned} \quad (12)$$

Given the service time distribution of the M/G/1 queue, generating function of the queue length of M/G/1 can be used to compute the second moment of N_B by the information of third moment of service time [8, 9]. Instead we propose to approximate the second moment of the queue length $E[N_B^2]$ using results of M/M/1 queuing systems. The approximation is based on observation of numerical experiences that suggests the ratio $E[N_B^2]$ / $E[N_{B,M/M/1}^2]$ is approximately the same as $(E[N_B])^2 / (E[N_{B,M/M/1}])^2$, i.e.

$$E[N_B^2] \approx \left(\frac{E[N_B]}{E[N_{B,M/M/1}]}\right)^2 E[N_{B,M/M/1}^2] \quad (13)$$

where $N_{B,M/M/1}$ is the queue length of $M/M/1$ queue with arrival rate λ and service rate $1/\mu_{TB}$.

The evaluation of the first two moments of $N_{B,M/M/1}$ can be obtained by its generating function $G(z)=\sum_{i=0}^{\infty} \pi_i z_i$,

where π_i is the probability that the queue length is i with $\pi_0 = 1 - \rho$, $\pi_i = \pi_0\ \rho^i$, for all i. As a result,

$$G'(1) = \frac{\rho}{1-\rho} = E[N_{B,M/M/1}] \quad (14)$$

$$G''(1) = \frac{2\rho^2}{(1-\rho)^2} = E(N_{B,M/M/1}^2) - E[N_{B,M/M/1}] \quad (15)$$

$$E(N_{B,M/M/1}^2) = \frac{2\rho^2}{(1-\rho)^2} + \frac{\rho}{1-\rho} \quad (16)$$

Combining (5), (13), (14) and (16),

$$E[N_B^2] \approx [1 - \frac{\rho}{2}(1 - \sigma_{TB}^2 / m_{TB}^2)]^2 (\frac{2\rho^2}{(1-\rho)^2} + \frac{\rho}{1-\rho}) \quad (17)$$

3.2 The transport system model

Remind that the transportation time L_t is assumed to be constant. The purpose of this subsection is to evaluate the distribution of the number N_t of product units in transportation. Notice that the arrival process at the transportation process is the departure process of replenishment orders from the plant and it is in general not a Poisson process. Let us approximate the arrival process of batches at the transportation system as a Poisson process. The transportation system can then be approximated by a $M^X/D/\infty$ queuing system [4, 6] with a fixed transportation time L_t. As the average arrival rate of product units at the transportation system is $\lambda E[X]$ and the transportation delay is L_t. From Little's Law,

$$m_t := E[N_t] = \lambda E[X].L_t = \lambda m_x L_t \quad (18)$$

The second moment of N_t is as follows, where N_{Bt} is the number of batches in transportation.

$$E[N_t^2] = E\left[\left(\sum_{i=1}^{N_{Bt}} X_i\right)^2\right]$$

Using the $M^X/D/\infty$ approximation which implies the mutual independence of N_{Bt} and X_i:

$$\begin{aligned} E[N_t^2] &= E\left[E\left[\left(\sum_{i=1}^{N_{Bt}} X_i\right)^2 \mid N_{Bt}\right]\right] \\ &= E\left[N_{Bt} E[X^2] + (N_{Bt}^2 - N_{Bt}) E^2[X]\right] \\ &= E[N_{Bt}] E[X^2] + \left(E[N_{Bt}^2] - E[N_{Bt}]\right) E^2[X] \\ &= E[N_{Bt}] Var(X) + E[N_{Bt}^2] E^2[X] \end{aligned}$$

N_{Bt} is the number of Poisson arrivals in a time interval of length Lt with $E[N_{Bt}] = \lambda L_t$ and $E[N_{Bt}^2] = (\lambda L_t)^2 + \lambda L_t$. Introducing these terms in the above equations leads to:

$$E[N_t^2] = \lambda L_t Var(X) + \left(\lambda L_t + (\lambda L_t)^2\right) E^2[X]$$
$$\sigma_t^2 := Var(N_t) = E[N_t^2] - E^2[N_t] = \lambda L_t (\sigma_X^2 + m_X^2) \quad .(19)$$

3.3. Order to delivery lead-time and inventory on order

We evaluate now the order to delivery lead-time $L = L_s + L_t$ and the total inventory on order $N = N_s + N_t$.

$$E[L] = E[L_s] + L_t \quad (20)$$

$$m_N := E[N] = E[N_s] + E[N_t] \quad (21)$$

Notice that N_s and N_t are in general dependent. An approximation of the second moment of N is obtained by considering N_s and N_t as two i.i.d random variables.

$$\sigma_N^2 := Var(N) = Var(N_s + N_t) \approx Var(N_s) + Var(N_t) = \sigma_s^2 + \sigma_t^2 \quad (22)$$

3.4 Total inventory cost estimation

By definition of the base stock policy, the following relation holds:

$$R = I - B + N \quad (23)$$

where R is the base stock level, I is the random inventory on hand, B is the backlogged quantities and N is the total inventory on order. Since customer orders can be partially filled, either I = 0 or B = 0. As a result,

$$I = (R-N)^+ \quad (24)$$

$$B = (N-R)^+ \quad (25)$$

where $(x)^+ = \max\{x, 0\}$. Note that the net inventory level $IN = I - B = R - N$.

The inventory holding cost for one product unit per time unit is C_h and the penalty per time unit generated by each product unit that the warehouse was not to deliver to the customer is C_b. The total inventory cost is:

$$C(R) = C_h E[I] + C_b E[B] \quad (26)$$

Combining (24) – (26),

$$C(R) = \int_0^\infty g(R-x) f_N(x) dx \quad (27)$$

where $f_N(x)$ is the pdf of the total inventory on order N and

$$g(x) = \begin{cases} C_h x, & \text{if } x \geq 0 \\ -C_b x, & \text{otherwise.} \end{cases}$$

Deriving the distribution function $f_N(x)$ of N is a very difficult task. In this paper, we propose to approximate $f_N(x)$ by a log-normal distribution [10] with mean equal to E[N] and variance equal to Var(N). That is:

$$f_N(x) \approx \frac{1}{Sx\sqrt{2\pi}} e^{-(\ln x - M)^2/(2S^2)} \quad (28)$$

$$S^2 = \ln\left(\frac{\sigma_N^2}{m_N^2} + 1\right) \qquad M = \ln m_N - S^2/2.$$

The use of log-normal approximation instead of usual normal approximation is mainly due to the large variance of N with respect of its mean. From (27), (28) and the definition of g(x),

$$C(R) = \int_{-\infty}^{\infty} g(R - e^{M+S.z}) \frac{1}{\sqrt{2\pi}} e^{-z^2/2} dz$$

$$C(R) = C_h \int_{-\infty}^{z^*} (R - e^{M+S.z}) \frac{1}{\sqrt{2\pi}} e^{-z^2/2} dz + C_b \int_{z^*}^{\infty} (e^{M+S.z} - R) \frac{1}{\sqrt{2\pi}} e^{-z^2/2} dz \quad (29)$$

where z^* is the solution of equation R – exp(M + Sz) = 0.

4. Numerical results

The aim of this section is to validate by simulation the analytical results presented in the previous section. We have implemented our analytical approach in C++. A discrete event system simulation tool-OMNeT++ (http://www.omnetpp.org) is used to model and simulate the inventory system. Numerical experimentation is done on the following example. The arrival rate λ of customer orders varies from 1 to 1.6. The batch size of each customer order X is an integer uniformly distributed in [3, 9]. The service time τ for each product unit is exponentially distributed with mean equal to 0.1. The transportation time is $L_t = 3$. Unit inventory holding cost C_h is 1 and unit backlogging cost C_b is 1. Base stock level is R = 50.

A very long simulation time of 10,000,000 time units is used. Results of both analytical approach and the simulation are given in Tables 2 and 3. The analytical approach provides very good fits moment estimation of order-to-delivery lead-time and the total inventory on order. Even though a number of approximations are made in the analytical approach, the second moment estimation of the inventory on order N is very good with error no more than 3.5%. Finally the log-normal approximation leads to good estimation of the total inventory cost. Moreover, the quality of the result seems to be rather insensitive to the traffic intensity ρ.

5. Conclusion

This paper proposes an analytical model to estimate the performances of a two stages production-distribution system. This system is characterized by i) a random client demand, ii) a finite manufacturing capacity; iii) a random batch size for orders, manufacturing and delivery quantities; iv) the transport time from the plant to the warehouse. We focused our attention on the lead time and the inventory cost. Our approach takes into account customer order size, finite production capacity of the plant, and transport time. Numerical results show that the analytical approach provides quite accurate performance/cost estimations.

An immediate direction is to optimize the inventory base-stock level in order to minimize the inventory cost. Another future direction for our research work consists in extending the approach to a inventory system with random transportation time.

6.Acknowledgement:

INRIA(Institut national de recherche en automatique et en automatique) is partner of the EU-funded FP6 Innovative Production Machines and Systems (I*PROMS) Network of Excellence. http://www.iproms.org

REFERENCES

[1] Cassandras, C. G. *"Discrete Event System: Modeling and Performance Analysis"*. Irwin, 1993.

[2] M. Ettl, G. E. Feigin, G. Y. Lin, D. D. Yao. "A Supply Network Model with Base-Stock Control and Service Requirements", *Operations Research,* Vol. 48/2, pp. 216-232, 2000.

[3] Kleinrock, L. *"Queueing Systems, Vol. I: Theory"*. John Wiley & Sons, New York, 1975.

[4] Liu, L., Kashyap, B.R.K., and Templeton, J.G.C, "On the $GI^x/G/\infty$ queueing System", *Journal of Applied Probability*, **27**, 671 – 683, 1990.

[5] L. Liu, X. Liu, D. D. Yao. "Analysis and Optimization of Multi-Stage Inventory-Queues", *Management Science* 50, 365-380 2004

[6] Hiroyuki MASUYAMA and Tetsuya TAKINE. "Analysis of an Infinite-Server Queue with Batch Markovian Arrival Streams", *Queueing Systems* 42 (3) p.269-296 November 2002

[7] Zipkin, P. H. *"Foundations of inventory management"*. Mc Graw Hill, 2000.

[8] Cooper, Robert B *"Introduction to Queuing Theory"*, Elseiver North Holland, 1981

[9] Buzacott, J., Shanthikumar, J., *"Stochastic Models of Manufacturing Systems"* . Prentice Hall, February 1993.

[10] Saporta, G., *"Probabilités Analyse des Données et Statistique"*, Technip, 1990

[11] Fangruo, C., "Optimal Policies for Multi-Echelon Inventory Problems with Batch Ordering", Operation Research, Vol. 48, No. 3, May-June 2000, pp. 376-389.

Table 1. given parameters (time unit: second)

	customer order frequency	Customer order size	Service time at plant	Transp. Time	C_h	C_b	R
mean	1 – 1.6	6	0.1	3.0	1.0	1.0	50
Std Dev		2	0.1	0			

Table 2 Simulation vs. approximation results for production lead-time Ls and order-to-delivery time L

Arrival rate	ρ	E[Ls] (sim)	E[Ls] (analyt)	Error (%)	E[L] (sim)	E[L] (analyt)	Error (%)
1.0	0.60	1.17551	1.17500	0.0433	4.17551	4.17500	0.0121
1.1	0.66	1.34188	1.34412	0.1665	4.34188	4.34412	0.0516
1.3	0.78	1.96235	1.95909	0.1665	4.96235	4.95909	0.0657
1.5	0.90	4.06190	4.05000	0.2940	7.06190	7.05000	0.1686
1.6	0.96	9.78959	9.80000	0.1062	12.7896	12.8000	0.0814

Table 3 Simulation vs. approximation results for inventory on order and inventory cost

Arrival rate	ρ	μ_N *(sim)*	μ_N *(an)*	*Error (%)*	σ_N *(sim)*	σ_N *(an)*	*Error (%)*	*Total Cost(sim)*	*Total Cost(an)*	*Error (%)*
1.0	0.60	25.4571	25.0500	1.625	14.1427	14.4062	1.828	25.564	26.6559	4.2712
1.1	0.66	29.0861	28.6712	1.447	15.6170	16.0360	2.613	23.0154	24.1640	4.9903
1.3	0.78	39.2413	38.6809	1.448	20.8461	21.5767	3.385	19.4695	20.3157	4.3461
1.5	0.90	64.1833	63.4500	1.155	39.6773	40.9942	3.212	28.8586	28.3903	1.6227
1.6	0.96	123.418	122.880	0.437	95.6309	97.3374	1.753	79.5283	77.7750	2.2046

Intelligent Production Machines and Systems
D. T. Pham, E. E. Eldukhri and A. J. Soroka (eds)

Applications of machine learning in manufacturing

D.T. Pham[a], A.A. Afify[a]

[a] *Manufacturing Engineering Centre, School of Engineering, Cardiff University, Cardiff CF24 0YF, UK*

Abstract

Machine learning is concerned with enabling computer programs automatically to improve their performance at some tasks through experience. Manufacturing is an area where the application of machine learning can be very fruitful. However, little has been published about the use of machine learning techniques in the manufacturing domain. This paper evaluates several machine learning techniques and examines applications in which they have been successfully deployed. Special attention is given to inductive learning which is among the most mature of the machine learning approaches currently available. The paper concludes with a summary of some of the key research issues in machine learning.

1. Introduction

Many knowledge-based systems have been employed to automate different operations in manufacturing. Examples are expert systems for decision support, intelligent scheduling systems for concurrent production and fuzzy controllers. A problem is the gathering of the required expert knowledge to implement these knowledge-based systems. Machine learning techniques can help in automating the time consuming process of knowledge acquisition that is essential to the development of a knowledge-based system. Automation would increase the speed and reduce the cost of development by decreasing the amount of time needed from experts and knowledge engineers. Automation could also uncover knowledge that might otherwise be overlooked by those involved in the knowledge acquisition process.

A great deal of research in machine learning has focused on classification, the task of developing a model, from a set of previously classified examples, that can correctly categorise new examples from the same population. Many manufacturing problems fall under the category of classification, where industrial domain experts are asked to assign a class label to an object or a situation based on the specific values of a set of parameters.

Machine learning approaches commonly used for classification include inductive learning algorithms such as decision tree induction and rule induction, instance-based learning, neural networks and genetic algorithms [1]. Among the various machine learning approaches developed for classification, inductive learning from instances may be the most commonly used in real world application domains. Inductive learning techniques are fast compared to other techniques. Another advantage is that inductive learning techniques are simple and their generated models are easy to understand. Finally, inductive learning classifiers obtain better accuracies compared with other classification techniques.

The paper is organised as follows: Section 2 presents the major techniques for machine learning that have been or could be applied in the manufacturing domain. Successful manufacturing applications of machine learning techniques are reviewed in section 3.

Section 4 concludes the paper with a summary of some of the key research issues in machine learning.

2. Machine learning approaches to classification learning

Several approaches to classification learning exist. This section will briefly review some of the main approaches: inductive learning, instance-based learning, neural networks and genetic algorithms.

2.1. Inductive learning

Inductive learning techniques can be divided into two main categories, namely, decision tree induction and rule induction. There are a variety of algorithms for building decision trees. The most popular are: CART [2] and ID3 and its descendants C4.5 and C5.0 [3,4]. These learning systems are categorised as "divide-and-conquer" inductive systems [3]. The knowledge induced by these systems is represented as decision trees. A decision tree consists of internal nodes and leaf nodes. Each Internal node represents a test on an attribute and each outgoing branch corresponds to a possible result of this test. Each leaf node represents a classification to be assigned to an example. To classify a new example, a path from the root of the decision tree to a leaf node is identified based on values of the attributes of the example. The class at the leaf node represents the predicted class for that example. Decision trees are generated from training data in a top-down, general-to-specific direction. The initial state of a decision tree is the root node that is assigned all the examples from the training set. If it is the case that all examples belong to the same class, then no further decisions need to be made to partition the examples, and the solution is complete. If examples at this node belong to two or more classes, then a test is made at the node, which will result in a split. The process is recursively repeated for each of the new intermediate nodes until a completely discriminating tree is obtained.

As with decision tree learning, there are many rule induction algorithms. Among them are AQ [5], CN2 [6], SRI [7] and RULES [8,9,10] which can all be categorised as "separate-and-conquer" inductive systems. In contrast to decision tree learning, rule induction directly generates IF-THEN rules. Rule induction systems produce either an unordered set of IF-THEN rules or an ordered set of IF-THEN rules, also known as decision lists [11], both including a default rule. To classify an instance in the case of ordered rules, the ordered list of rules is examined to find the first whose antecedent is satisfied by the instance. The predicted class is then the one nominated by this rule. If no rule antecedent is satisfied, the instance is predicted to belong to the default class. In the case of unordered rules, it is possible for some instances to be covered by more than one rule. To classify a new instance in this case, some conflict resolution approach must be employed. The general operation of separate-and-conquer rule induction algorithms is the same. They create the rule set one rule at a time. After a rule is generated, the instances covered by it are removed from the training data set and the same induction procedure is applied to the remaining data set until all the instances are covered by at least one rule in the rule set.

2.2. Instance-based learning

Instance-based learning is based upon the idea of letting the examples themselves form an implicit representation of the target concept [12]. In contrast to learning methods that construct a general, explicit description of the target concept when training instances are provided, instance-based learning methods, such as those using nearest-neighbour methods, simply store the training instances. Generalising beyond these instances is postponed until a new instance must be classified. Because of this, instance-based methods are sometimes referred to as "lazy" learning methods. A test instance is classified by finding the nearest stored instance according to some similarity function, and assigning the class of the latter to the former. Advantages of instance-based methods include the ability to model complex target concepts and the fact that information present in the training instances is never lost (because the instances themselves are stored explicitly). One disadvantage of instance-based approaches is that the cost of classifying new instances can be high. This is because nearly all the computation takes place at classification time rather than when the training instances are first encountered. Therefore, techniques for efficiently indexing training instances are a significant practical issue in reducing the computation required at classification time. A second disadvantage of many instance-based approaches, especially nearest-neighbour methods, is that they typically consider all attributes of the instances when attempting to retrieve similar training instances from the memory. If the target concept depends on only a few of the many available attributes, then the instances that are really most "similar" may be a long distance apart.

2.3. Neural networks

Neural networks provide a general practical method for learning real-valued and discrete-valued target concepts in a way that is robust to noise in the training data [13,14]. Neural network learning is well-suited to problems in which the training data corresponds to noisy, complex sensor data, such as inputs from cameras and microphones. The backpropagation algorithm is a common learning method adopted for multi-layer perceptrons, the most popular type of neural networks. The algorithm has been successfully applied to a variety of learning tasks, such as interpreting visual scenes, speech recognition, handwriting recognition and robot control. One of the chief advantages of neural networks is their wide applicability; however, they also have two particular drawbacks. The first is the difficulty in understanding the models they produce. The second is the often time-consuming training. Recent years have seen much research in developing new neural network methods that effectively address these comprehensibility and speed issues [15,16].

2.4. Genetic algorithms

Genetic algorithms are stochastic search techniques, which have been inspired by the process of biological evolution [17,18]. Several genetic algorithm-based systems for learning classification rules have been developed [19,20]. In these systems, rules are represented by bit strings whose particular interpretation depends on the application. The search for an appropriate rule begins with a population, or collection, of initial rules. Members of the current population give rise to the next-generation population by means of operations such as random mutation and crossover. At each step, the rules in the current population are evaluated relative to a given measure of fitness, with the fittest rules selected probabilistically as seeds for producing the next generation. The process performs generate-and-test beam-search of the rules, in which variants of the best current rules are most likely to be considered next. Genetic algorithms have been applied successfully to a variety of learning tasks and to other optimisation problems. For example, they have been used to learn collections of rules for robot control, to optimise the topology and learning parameters for neural networks and to solve job shop scheduling problems. Genetic algorithms have a potentially greater ability to avoid local minima than is possible with the simple greedy search employed by most learning techniques. On the other hand, they have a high computational cost.

3. Applications of machine learning techniques in manufacturing

This section discusses some applications of machine learning techniques in manufacturing with special emphasis on inductive learning. The focus is on inductive learning because detailed surveys concerning the other techniques can be found elsewhere (see for example [21]).

Machine learning techniques can be useful tools for discovering valuable patterns in data. As there are no universally applicable methods, it is important to have a clear understanding of the requirements of a particular problem and to choose the technique that best fits those requirements. In general, to be useful in a manufacturing application, a machine learning system should have the following abilities:

- dealing with different types of data (numerical, nominal, text and images),
- handling noise, outliers and fuzzy data,
- real-time processing,
- dealing with large data sets and data of high dimensions,
- producing results that are easy to understand,
- being simple to implement.

Machine learning algorithms are domain independent. In principle, they can be a very useful tool for the construction of knowledge-based systems. Efforts to apply machine learning methods follow a standard pattern. The main stages of the process are: formulating the problem, determining the representation, collecting the training data, evaluating the learned knowledge and fielding the knowledge base [22,23,24].

There is a wide range of manufacturing domains to which machine learning has been successfully applied.

In production operations management, it is increasingly important to develop automated scheduling systems as the production stages become more complicated. Learning-based scheduling, which involves automatic acquisition of dispatching rules, is one of the practical methods used to solve this problem. There have been various attempts to use learning in scheduling problems [25,26]. Proposed methods for acquisition of scheduling rules by using inductive learning techniques were applied to flow shop scheduling problems [27,28], job shop scheduling problems [29,30] and flexible manufacturing systems scheduling problems [31,32]. Experimental results indicated that effective scheduling can be achieved by applying the proposed methods.

Machine learning techniques have been applied to just-in-time (JIT) production systems. First, neural networks and decision trees were used to set the number of kanbans in a dynamic JIT factory [33]. The number of kanbans is important to the effective operation of a JIT production system. Results showed that neural networks and decision trees represent two practical approaches with special capabilities for dynamically adjusting the number of kanbans. Second, an approach based on inductive learning was used to predict JIT factory performance from past data that includes both good and poor factory performance [34]. In particular, the CART decision tree classifier was employed to generate rules automatically in a JIT manufacturing environment. The rules obtained can accurately classify and predict factory performance based on shop factors and can identify the important relationships between the shop factors that determine factory performance. Results indicated that inductive learning is a feasible technique for predicting JIT factory performance from dynamic shop floor data.

Semiconductor manufacturing is a complex operation that involves monitoring a large number of parameters from the early stages of production to the packaging of the end product. Improving the quality of the manufacturing process requires a great deal of data analysis work and is still accomplished by human engineers. The transaction volume in a typical wafer fabrication plant is as many as one million wafers a day [35]. The amounts of data involved make the data analysis task extremely time-consuming and difficult. Several authors proposed procedures for using machine learning techniques in semiconductor manufacturing [36,37]. Research results showed that machine learning techniques can be powerful tools for continuous quality improvement in a large and complex process such as semiconductor manufacturing.

Zhou et al. [38] developed an intelligent data mining system and applied it to drop testing analysis of portable electronic products to discover useful design knowledge. The rule induction method adopted is based on the C4.5 algorithm. Studies with the proposed system indicated that its approach is flexible and can be applied to a number of other design and manufacturing processes to reduce costs and improve productivity.

Aksoy et al. [39] presented an industrial visual inspection system that can be used to carry out quality control for mass production. The system employs the RULES-3 inductive learning algorithm.

Peng [40] developed a fuzzy inductive learning based intelligent monitoring system for improving the reliability of manufacturing processes. The method was successfully applied to diagnosing the conditions of tapping processes to ensure the quality of products.

Pham et al. [41] described an application of data mining and machine learning techniques in a steel bar manufacturing company. The application covered five areas, namely, make-to-order (MTO) versus make-to-stock (MTS) analysis, product sale profile analysis, rogue order determination, overdue order analysis and product combination analysis. The results demonstrated that the techniques employed can extract information to help intelligent decision making in industry.

Other applications include diagnosing motor pumps [42], accelerating rotogravure printing [43], determining suitable cutting conditions in operation planning [44], re-formulating and generalising the machining knowledge from a machining database [45], choosing sheet metal working conditions [46], discovering the laws governing metallic behaviour [47], modelling job complexity in clothing production systems [48], acquiring and refining operational knowledge in industrial processes [49] and identifying arbitrary geometric and manufacturing categories in CAD databases [50].

From the above-mentioned problems, features of successful applications of machine learning in manufacturing can be summarised as follows:

- the problem is of a sufficient degree of complexity,
- the problem can be formulated to match existing learning algorithms,
- ample training data are available in an appropriate format,
- the training data are representative,
- the data are free from noise or can be cleansed cost effectively,
- methods for learning and evaluating performance are suitable for the application under consideration.

4. Conclusions

Machine learning algorithms are demonstrably of significant value in a variety of real-world manufacturing applications. Dozens of companies around the world now provide commercial implementations of these algorithms (see www.kdnuggets.com), along with efficient links to commercial databases and well-designed user interfaces [51]. However, these algorithms also have some limitations. With the techniques described here, data sets having tens of thousands of training instances can be mined in relatively short times. However, many important data sets are significantly larger. To provide efficient machine learning methods for such large data sets requires additional research. Another problem is that most machine learning techniques to date handle only data with continuous and nominal values. A topic of considerable research interest is the development of algorithms that can learn regularities in other data types, such as text and images. Finally, although machine learning techniques can overcome a wide range of problems in manufacturing, they are still not applied on a large scale. Given these challenges, further developments in machine learning research can be expected in the future.

Acknowledgements

This work was carried out within the ERDF (Objective One) projects "Innovation in Manufacturing", "Innovative Technologies for Effective Enterprises" and "Supporting Innovative Product Engineering and Responsive Manufacturing" (SUPERMAN) and within the project "Innovative Production Machines and Systems" (I*PROMS).

An extended version of this paper will be published in the Proceedings of the Institution of Mechanical Engineers, Part B: Journal of Engineering Manufacture.

References

[1] Pham DT and Afify AA. Machine learning: Techniques and trends. Proc. of the 9th Int. Workshop on Systems, Signals and Image Processing (IWSSIP-02), Manchester Town Hall, UK, World Scientific, 2002, 12-36.

[2] Breiman L, Friedman JH, Olshen RA and Stone CJ. Classification and regression trees. Belmont, Wadsworth, 1984.

[3] Quinlan JR. C4.5: Programs for machine learning. Morgan Kaufmann, San Mateo, CA, 1993.

[4] RuleQuest. Data mining tools C5.0. Pty Ltd, 30 Athena Avenue, St Ives NSW 2075, Australia. Available from: http://www.rulequest.com/see5-info.html [Accessed: 1 February 2003].

[5] Michalski RS and Kaufman KA. The AQ19 system for machine learning and pattern discovery: A general description and user guide. Reports of the Machine Learning and Inference Laboratory, MLI 01-2, George Mason University, Fairfax, VA, USA, 2001.

[6] Clark P and Boswell R. Rule induction with CN2: Some recent improvements. Proc. of the 5th European Conf. on Artificial Intelligence, Porto, Portugal, 1991, 151-163.

[7] Pham DT and Afify AA. SRI: A scalable rule induction algorithm. Submitted to Proc. of the Institution of Mechanical Engineers, Part (C), 2004.

[8] Pham DT and Dimov SS. An efficient algorithm for automatic knowledge acquisition. Pattern Recognition, 1997, 30 (7), 1137-1143.

[9] Pham DT, Bigot S and Dimov SS. RULES-5: A rule induction algorithm for problems involving continuous attributes. Proc. of the Institution of Mechanical Engineers, Part (C), 2003, 217 (12), 1273-1286.

[10] Pham DT and Afify AA. RULES-6: A simple rule induction algorithm for handling large data sets. Submitted to Proc. of the Institution of Mechanical Engineers, Part (C), 2004.

[11] Rivest R. Learning decision lists. Machine Learning, 1987, 2, 229-246.

[12] Aha D, Kibler D and Albert M. Instance-based learning algorithms. Machine Learning, 1991, 6 (1), 37-66.

[13] Mitchell TM. Machine learning. McGraw Hill, New York, 1997.

[14] Pham DT and Liu X. Neural networks for identification, prediction and control. Springer-Verlag, London, 1999.

[15] Towell GG and Shavlik JW. The extraction of refined rules from knowledge-based neural networks. Machine Learning, 1993, 13 (1), 71-101.

[16] Jiang Y, Zhou ZH and Chen ZQ. Rule learning based on neural network ensemble. Proc. of the Int. Joint Conf. on Neural Networks, Honolulu, HI, 2002, 1416-1420.

[17] Pham DT and Karaboga D. Intelligent optimisation techniques: genetic algorithms, tabu search, simulated annealing and neural networks. Springer-Verlag, London Heidelberg, 2000.

[18] Freitas AA. Data mining and knowledge discovery with evolutionary algorithms. Springer-Verlag, Berlin, New York, 2002.

[19] Janikow CZ. A knowledge-intensive genetic algorithm for supervised learning. Machine Learning, 1993, 13, 189-228.

[20] Liu J and Kwok JT. An extended genetic rule induction algorithm. Proc. of the 2000 Congress on evolutionary computation, California, USA, 2000, 458-463.

[21] Monostori L. AI and machine learning techniques for managing complexity, changes and uncertainties in manufacturing. Engineering Applications of Artificial Intelligence, 2003, 16 (3), 227-291.

[22] Langley P and Simon HA. Applications of machine learning and rule induction. Communications of the ACM, 1995, 38 (11), 55-64.

[23] Braha D. Data mining for design and manufacturing: Methods and applications. Kluwer Academic Publishers, Boston, MA, 2001.
[24] Lavrač N, Motoda H, Fawcett T, Holte R, Langley P and Adriaans P. Introduction: Lessons learned from data mining applications and collaborative problem solving. Machine Learning, 2004, 57, 13-34.
[25] Priore P, Fuente D, Gomez A and Puente J. Dynamic scheduling of manufacturing systems with machine learning. Int. Journal of Foundations of Computer Science, 2001, 12 (6), 751-762.
[26] Akyol DE. Application of neural networks to heuristic scheduling algorithms. Computers and Industrial Engineering, 2004, 46, 679-696.
[27] Suwa H, Fujii S and Morita H. An acquisition of scheduling rules for flow-shop problems. Proc. of the 1996 Int. Conf. on Advances in Production Management System, 1996, 631-636.
[28] Murata T, Sugimoto T, Tsujimura T and Gen M. Rule conversion in knowledge acquisition for flow shop scheduling problems. Joint 9th IFSA World Congress and 20th NAFIPS Int. Conf., 2001, 4, 2417-2421.
[29] Suwa H, Fujii S and Morita H. Acquisition and refinement of scheduling rules for job-shop problems. IEEE Proc. of Int. Symposium on Industrial Electronics, 1998, 2, 720-725.
[30] Osisek V and Aytug H. Discovering subproblem prioritization rules for shifting bottleneck algorithms. Journal of Intelligent Manufacturing, 2004, 15, 55-67.
[31] Kim CO, Min YD and Yih Y. Integration of inductive learning and neural networks for multi-objective FMS scheduling. Int. Journal of Production Research, 1998, 36 (9), 2497-2509.
[32] Priore P, Fuente D, Pino R and Puente J. Dynamic scheduling of manufacturing systems using neural networks and inductive learning. Integrated Manufacturing Systems, 2003, 14 (2), 160-168.
[33] Markham IS, Mathieu RG and Wray BA. Kanban setting through artificial intelligence: A comparative study of artificial neural networks and decision trees. Integrated Manufacturing Systems, 2000, 11 (4), 239-246.
[34] Mathieu RG, Wray BA and Markham IS. An approach to learning from both good and poor factory performance in a kanban-based just-in-time production system. Production Planning & Control, 2002, 13 (8), 715-724.
[35] Shin CK and Park SC. A machine learning approach to yield management in semiconductor manufacturing. Int. Journal of Production Research, 2000, 38 (17), 4261-4271.
[36] Kang BS, Lee JH, Shin CK, Yu SJ and Park SC. Hybrid machine learning system for integrated yield management in semiconductor manufacturing. Expert Systems with Applications, 1998, 15, 123-132.
[37] Gardner M and Bieker J. Data mining solves tough semiconductor manufacturing problems. Proc. of the 6th ACM SIGKDD Int. Conf. on Knowledge Discovery and Data Mining, Boston, MA, 2000, 376-383.
[38] Zhou C, Nelson PC, Xiao W, Tirpak TM and Lane SA. An intelligent data mining system for drop test analysis of electronic products. IEEE Transactions on Electronics Packaging Manufacturing, 2001, 24 (3), 222-231.
[39] Aksoy MS, Torkul O and Cedimoglu IH. An industrial visual inspection system that uses inductive learning. Journal of Intelligent Manufacturing, 2004, 15, 569-574.
[40] Peng Y. Intelligent condition monitoring using fuzzy inductive learning. Journal of Intelligent Manufacturing, 2004, 15, 373-380.
[41] Pham DT, Packianather MS, Dimov S, Soroka AJ, Girard T, Bigot S and Salem Z. An application of data mining and machine learning techniques in the metal industry. Proc. of the 4th CIRP Int. Seminar on Intelligent Computation in Manufacturing Engineering (ICME-04), Sorrento (Naples), Italy, 2004.
[42] Giordana A, Neri F and Saitta L. Automated learning for industrial diagnosis. Proc. of the 3rd Int. Workshop on Multistrategy Learning, Harpers Ferry, West Virginia, WV, 1996, 125-134.
[43] Evans B and Fisher D. Using decision tree induction to minimize process delays in printing industry. In: Klösgen W and Zytkow JM (Ed.) Handbook of data mining and knowledge discovery, Oxford University Press, 2002.
[44] Hoffmann J. An intelligent tool for determination of cutting values based on neural networks. Proc. of the 2nd World Congress on Intelligent Manufacturing Processes and Systems, Budapest, Hungary, 1997, 66-71.
[45] Sluga A, Jermol M, Zupanič D and Mladenić D. Machine learning approach to machinability analysis. Computers in Industry, 1998, 37, 185-196.
[46] Lin Z-C and Chang D-Y. Application of a neural network machine learning model in the selection system of sheet metal bending tooling. Artificial Intelligence in Engineering, 1996, 10, 21-37.
[47] Mitchell F, Sleeman D, Duffy JA, Ingram MD and Young RW. Optical basicity of metallurgical slags: A new computer-based system for data visualisation and analysis. Ironmaking and Steelmaking, 1997, 24, 306-320.
[48] Hui PCL, Chan KCK and Yeung KW. Modelling job complexity in garment manufacturing by inductive learning. Int. Journal of Clothing Science and Technology, 1997, 9 (1), 34-44.
[49] Shigaki I and Narazaki H. An approximate summarization method of process data for acquiring knowledge to improve product quality. Production Planning & Control, 2001, 12 (4), 379-387.
[50] Ip CY, Regli WC, Sieger L and Shokoufandeh A. Automated learning of model classification. Proc. of the 8th ACM Symposium on Solid Modeling and Applications, Seattle, Washington, USA, ACM Press, 2003, 322-327.
[51] Mitchell TM. Machine learning and data mining. Communications of the ACM, 1999, 42 (11), 31-36.

Intelligent Production Machines and Systems
D. T. Pham, E. E. Eldukhri and A. J. Soroka (eds)

Applications of RULES-3 Induction System

M.S. Aksoy

Department of Information Systems, CCIS, King Saud University, Riyadh 11543, SA

ABSTRACT

In recent years, there has been a growing amount of research on inductive learning. Out of this research a number of promising algorithms have surfaced. Inductive learning algorithms are domain independent. In principle, they can be used in any task involving classification or pattern recognition. There have been several successful applications of inductive learning systems. In the paper a number of applications of RULES-3 induction algorithm are presented.

Keywords: Induction, Inductive Learning, RULES3, Expert Systems.

1. Introduction

In recent years, there has been a growing amount of research on inductive learning [1]. In its broadest sense, induction (or inductive inference) is *a method of moving from the particular to the general - from specific examples to general rules* [2]. Induction can be considered the process of generalizing a procedural description from presented or observed examples [3].

The purpose of inductive learning is to perform a synthesis of new knowledge, and this is independent of the form given to the input information [4]. In order to form a knowledge base using inductive learning, the first task is to collect a set of representative examples of expert decisions. Each example belongs to a known class and is described in terms of a number of attributes. These examples may be specified by an expert as a good tutorial set, or may come from some neutral source such as an archive. The induction process will attempt to find a method of classifying an example, again expressed as a function of the attributes, that explains the training examples and that may also be used to classify previously unseen cases [5].

Inductive learning algorithms are domain independent. In principle, they can be used in any task involving classification or pattern recognition [6]. There have been several successful applications of RULES-3 inductive learning system. Some of them are summarized in this paper.

The organization of the paper is as follows: In section 2, RULES3 inductive learning algorithm is outlined. The applications of RULES-3 algorithm are summarized in section 3, and section 4 is the conclusion.

2. RULES-3 Inductive Learning Algorithm

RULES-3 is a simple algorithm for extracting a set of classification rules from a collection of examples for objects belonging to one of a number of known classes [7]. An object must be described in terms of a fixed set of attributes, each with its own range of possible values, which could be nominal or numerical. For example, attribute "length" might have nominal values {short, medium, long} or numerical values in the range {-10, 10}.

An attribute-value pair constitutes a condition in a rule. If the number of attributes is Na, a rule may contain between one and Na conditions. Only conjunction of conditions is permitted in a rule and therefore the attributes must be all different if the rule comprises more than one condition.

RULES3 extracts rules by considering one example at a time. It forms an array consisting of all attribute-value pairs associated with the object in that example, the total number of elements in the array being equal to the number of attributes of the object. The rule forming procedure may require at most Na iterations per example. In the first iteration, rules may be produced with one element from the array. Each element is examined in turn to see if, for the complete example collection, it appears only in objects belonging to one class. If so, a candidate rule is obtained with that element as the condition. In either case, the next element is taken and the examination repeated until all elements in the array have been considered. At this stage, if no rules have been formed, the second iteration begins with two elements of the array being examined at a time. Rules formed in the second iteration therefore have two conditions. The procedure continues until iteration when one or more candidate rules can be extracted or the maximum number of iterations for the example is reached. In the latter case, the example itself is adopted as the rule. If more than one candidate rule is formed for an example, the rule that classifies the highest number of examples is selected and used to classify objects in the collection of examples. Examples of which objects are classified by the selected rule are removed from the collection. The next example remaining in the collection is then taken and rule extraction is carried out for that example. This procedure continues until there are no examples left in the collection and all objects have been classified. This algorithm can be summarized as follows:

Step 1. Define ranges for the attributes, which have numerical values and assign labels to those ranges. (Subtract the minimum values from the maximum values for those attributes and divide them to a number of intervals defined by the user).

Step 2. Set the minimum number of conditions (Ncmin) for each rule

Step 3. Take an unclassified example

Step 4. Nc=Ncmin-1

Step 5. If Nc<Na then Nc=Nc+1

Step 6. Take all values or labels contained in the example

Step 7. Form objects, which are combinations of Nc values or labels taken from the values or labels obtained in Step 6

Step 8. If at least one of the objects belongs to a unique class then form rules with those objects; ELSE go to Step 5

Step 9. Select the rule, which classifies the highest number of examples

Step 10. Remove examples classified by the selected rule

Step 11. If there are no more unclassified examples then STOP; ELSE go to Step 3

Here Nc is the number of condition(s) for each rule and Na is the number of attributes for each example.

3. Applications of RULES-3 Algorithm

Inductive learning algorithms are domain independent. In principle, they can be used in many tasks involving classification or pattern recognition [6]. There have been several successful applications of RULES-3 inductive learning system. Some of them are summarized below.

3.1. Industrial Visual Inspection

A system has been developed for industrial visual inspection. The system uses RULES-3 inductive learning algorithm to extract the necessary set of rules and template matching technique for feature extraction. In *Template Matching Technique*, the image to be inspected is scanned and the required features are extracted. These features are compared with those defined for the perfect pattern. This method greatly compresses the image data for storage and reduces the sensitivity of the input intensity data. A number of predefined binary templates can be used to extract the necessary features for images to be inspected. For this system 20, 3x3 masks are used to represent an image. Each example consists of 20 frequencies of each mask. The system was tested on five different types of tea or water cups in order to classify the good and bad items.

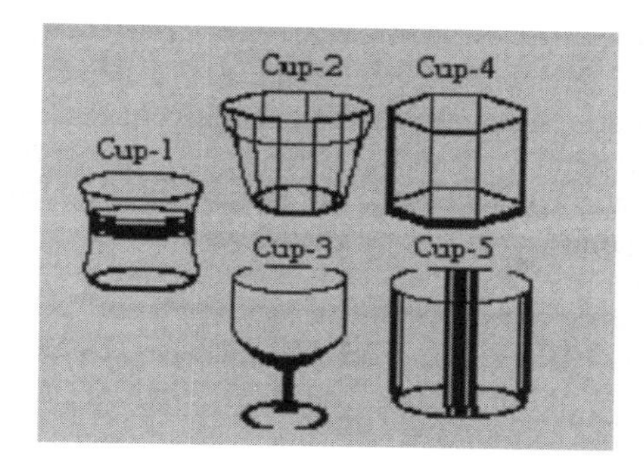

Figure 1. Five example good cups

It was trained using five good cups and then tested for 113 unseen examples (See Figure 1). The results obtained showed the high performance of the system: the efficiency of the system for correctly classifying unseen examples was %100. The system can also decide what the type of the cup being processed is [8].

3.2. Banknote Recognition

A system has been developed for this purpose. The system employs template matching for feature extraction and RULES-3 to extract the necessary set of rules and to recognize a banknote. 10 of 3x3 masks were used to represent a banknote. Each banknote (or example) is presented by the frequencies of the masks used. The system was trained for all back and front sides of different Turkish and Saudi banknotes. Ten images were used for each side of the banknotes for learning purpose. The system has been tested using many "unseen" examples and correctly classified all of them [9, 10].

3.3. Signature Recognition

An alternative technique for Signature Recognition has been developed. The technique employs template matching for feature extraction and RULES3 inductive learning algorithm to extract the necessary set of rules and to recognize a signature. 15 of 3x3 masks were used to represent a signature. Each signature (or pattern) is presented by the frequencies of the masks used. The system was trained using 144 signatures (16 signatures belonging to 9 different persons). An example signature is shown in Figure 2. The system has been tested using many unseen signatures and the ability to correctly classify them was found to be 97% [11].

3.4. Number plate Recognition

An alternative technique has been developed for number-plate recognition. The technique uses RULES-3 algorithm for learning and template matching technique for feature extraction. Each character (letter and number) that can be used in number-plates is considered as an example which consists of 20 frequencies of each 3x3 mask used for representation. In order to recognize a number-plate, all characters contained are recognized one by one. After bringing them together the number-plate being processed is recognized [12].

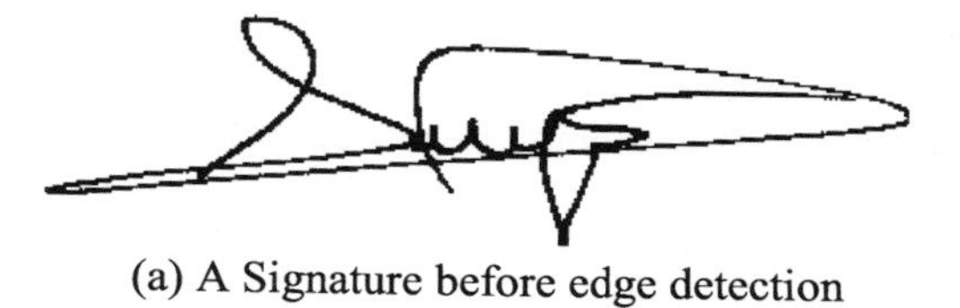

(a) A Signature before edge detection

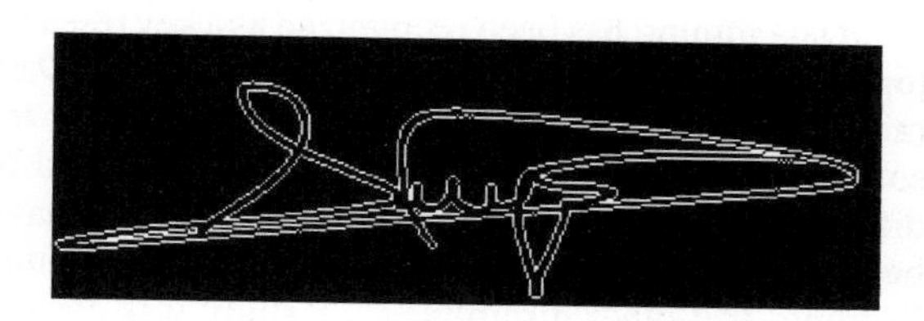

(b) A Signature after edge detection

Figure 2. An Example Signature before and after edge detection

3.5. Inspection of ceramic tiles

RULES-3 has been used for inspection and classification of ceramic tiles. RULES-3 is used for learning and template matching technique for feature extraction. 10 of 3x3 masks were used to represent a ceramic tile. More than 100 defective tiles (having crack and/or spot) and not defective tiles have been used for training. The efficiency of the system for correctly classifying many unseen examples was found to be 96% [13].

3.6. Barcode Recognition

A new technique has been developed for Barcode Recognition. The technique uses RULES-3 system. In the technique only the thickness of vertical lines are considered while the spaces in between are ignored. Each line is considered an attribute and its thickness is considered as its value. For each barcode a rule is extracted using RULES-3. No special hardware is required for the technique, only a PC and a barcode reader is enough. [14].

3.7. Dynamic System Identification

RULES-3 was employed for Dynamic System Identification. The two linear systems and two non-linear systems were modeled. The results obtained showed that induction-based modeling generally has the advantage of producing explicit models. In addition to this explicitness, the models induced by RULES-3, the algorithm adopted in this work, are more accurate than models constructed using neural networks. RULES-3 is easy to apply. It operates efficiently and does not require making assumptions about the structure of the system to be modeled [15].

3.8. Data Mining

Data mining has been recognized as a key research topic in database systems and machine learning. Data mining is one of the most important tools used for solving most of today's problems that are related to different sectors of our life. Different techniques have been developed for mining data in statistics, machine learning, and other disciplines. A study was carried out to show that RULES-3 is suitable for data mining. Table 1 shows a comparison of RULES-3 with three statistical, two Lazy, and six rule-based data mining algorithms on eleven real life data sets, it was realised that RULES-3 has many advantages over the other algorithms used in the study in terms of learning rate, accuracy and robustness to noisy and incomplete data [16].

3.9. Cell Formation Application

The design of cellular systems is a complex, multi-criteria and multi-step process which can have significant implications for the entire organization. Most research in this area focuses on the formation of the part-families and associated machine groups. Numerous quantitative techniques have been developed to address this part-family machine group formation problem. Existing approaches include mathematical programming algorithms for the matrix diagonalization, the application of network modeling, the use of similarity coefficient and rank order clustering. This study presents a new technique using reversed induction for cell formation in group technology. The starting point is the initial solution generated by traditional mathematical techniques. It creates cells with the greatest mean part numbers. The technique was applied to a manufacturing shop in TUVASAS (Turkish Wagon Industry Company) for 60 parts and 16 machines. The results show that it is at least comparable to other techniques[17].

3.10. Other Applications

Some other successful applications of RULES-3 are given below:
-The classification of well-known IRIS data [15],
-Parameters estimation in wastewater treatment [18].
-Application of RULES-3 to DNA Sequence [19].

Table 1. The overall performance of RULES3 compared to other algorithms used.

Algorithm	C1	C2	C3	C4	C5	Final Rate
RULES3	97.70	81.90	94.94	98.12	5.54	83.53
Neural Networks	94.53	83.27	93.73	96.35	10.43	83.51
K- nearest Neighbor	98.71	78.58	93.78	89.42	0.15	81.84
Logistic Model Tree	92.23	81.44	92.01	95.83	11.22	80.99
Ripper	85.03	84.55	91.92	98.84	0.36	80.82
C4.5	89.85	78.41	94.47	95.14	0.34	80.27
Naive Bayes	81.45	73.67	92.56	96.84	0.46	76.88
Multinomial logistic regression	85.41	77.66	92.73	96.64	12.15	78.65
Locally Weighted Learning	76.94	68.51	94.86	99.63	0.13	76.02
1-R	73.77	65.46	93.65	97.03	0.21	73.40
O-R	51.69	48.61	95.03	96.50	0.16	63.97
Weight (%)	25	25	20	20	10	

C1 : Learning Rate
C2 : Classification Accuracy
C3 : Robustness of the algorithm to noise
C4 : Robustness of the algorithm to missing values
C5 : Training time (sn)

4. Conclusion

The induction of rules from empirical data is a useful technique for automatic knowledge acquisition. It offers a modularized, clearly explained format for decision making which is compatible with human reasoning procedures. Also, the resulting rules are suitable for use in expert systems. In recent years, many task-oriented inductive learning systems have been developed that have demonstrated impressive performance in their specific domain of application.

This paper presents some applications of RULES-3 inductive learning system to different domains. However, some problems still remain. Most systems lack generality and extensibility.

References

[1] Nakakuki Y., Koseki Y. and Tanaka M. "Inductive learning in probabilistic domain", in proc. *Eighth National Conf. on AI*, Boston, July 29, August 3,1991, pp.809-814, 1990.

[2] Forsyth R., "Machine Learning principles and techniques", Ed: R. Forsyth, Chapman and Hall, London, 1989.

[3] Rubin S.H., "Expert systems for knowledge acquisition", in proc. *First World Congress on Expert Systems , Vol 3*, Orlando, Florida, December 16-19, pp.1793-1799, 1991.

[4] Kodratoff Y., "Introduction to machine learning", Pitman Publishing, London, 1988.

[5] Quinlan J.R., "Induction, knowledge and expert systems", in *Artificial Intelligence Developments and Applications*, Eds J.S. Gero and R. Stanton, Amsterdam, North-Holland, pp.253-271, 1988.

[6] Liu W.Z. and White A.P., "A review of inductive learning", in proc. *Research and Development in Expert Systems VIII.*, Cambridge, pp.112-126, 1991.

[7] Pham, D.T. and Aksoy, M.S., "A new algorithm for inductive learning", *Journal of Systems Eng.*, No.5, pp.115-122, 1995.

[8] Aksoy M.S., Torkul O. and Cedimoglu I.H., "An Industrial Visual Inspection System That Uses Inductive Learning", Journal of Intelligent Manufacturing, 15, pp:569-574, 2004.

[9] Sevkli M., Turkyilmaz A. and Aksoy M.S., "Banknote recognition using inductive learning", Int. Conf. On Fuzzy Syst. And Soft Computational Intelligence in Management and Industrial Eng., FSSCIMIE'02, İstanbul Technical Univ., pp.122-128, Turkey, May 29-31, 2002.

[10] Aksoy M.S., "Saudi Banknote Recognition Using Inductive Learning", Proc. of 2nd IIEC-2004, Riyadh, S.A, December 19-21, 2004.

[11] Al-Badri, T.M., "Signature Recognition Using Inductive Learning", Master Study, King Saud Univ. College of Comp. and Infor. Sciences, Riyadh, SA, 2004.

[12] Aksoy M.S., Cagil G. and Turker A.K., "Number plate recognition using inductive learning", Robotics and Autonomous Systems, (33), pp.149-153, Canada, 2000.

[13] Turkyilmaz A., Sevkli M. and Aksoy M.S., "Inspection of Ceramic Tiles Using Inductive Learning", Proc. 2nd Intern. Conf. Responsive Manuf., ICRM-2002, Univ. of Gaziantep, Gaziantep, Turkey, June 26-28, 2002.

[14] Aksoy M.S. and Bayram M., "A new technique to process and recognize barcodes using inductive learning", International Journal of Mathematical and Computational Applications, Vol.1, No.2, pp.1-6, Turkey, 1996.

[15] Aksoy M.S., "New Algorithms For Machine Learning", PhD. Thesis, University of Wales, College of Cardiff, U.K., 1994.

[16] Alasoos, B.A., "Data Mining Using Rule Extraction System- RULES", Master Study, King Saud Univ. College of Comp. and Infor. Sciences, Riyadh, SA, 2004.

[17] Col M., Torkul O. and Aksoy M.S., "A heuristic approach for machine cell formation", The 26th International Conference on Computers and Industrial Engineering, Melbourne, Australia, 1999.

[18] Sengorur B., Aksoy M.S. and Oz N., "Parameters estimation in wastewater treatment using inductive learning", Proc. 2nd. International Symp. on Intelligent Manufacturing Systems, Sakarya University, pp.81-87, Turkey, August 6-7, 1998.

[19] Karli K., "Application of Rule Induction Algorithm To DNA Sequence", Master Thesis, Fatih University, Institute of Science, Istanbul, Turkey, 2000.

Intelligent Production Machines and Systems
D. T. Pham, E. E. Eldukhri and A. J. Soroka (eds)

Correlation Analysis of Environmental Pollutants and Meteorological Variables Applying Neuronal Networks

A. Vega-Corona [a], Diego-Andina [b], F.S. Buendía-Buendía [b] J.M. Barrón-Adame [a]

[a] *Universidad de Guanajuato, Facultad de Ingeniería Mecánica, Eléctrica y Electrónica, México.*

[b] *Universidad Politécnica de Madrid, ETSI Telecomunicación, España.*

Abstract

In order to develop an environmental contingency forecasting tool for decision making, a Pattern Recognition method applying Neural Networks is presented. For this purpose, $\mathbf{SO_2}$ and $\mathbf{PM_{10}}$ time series concentrations are analyzed every hour and daily, as well as a variety of meteorological variables. These pollutant concentrations and meteorological variables are self-organized by means of a SOM Neural Network in different classes. Classes are used in training fase of a General Regression Neural classifier (GRNN) to provide an air quality forecast. In this case a time series set obtained from Environmental Monitoring Network (EMN) of the city of Salamanca, Guanajuato, México is used. Results verify the potential of this method versus other statistical classification methods and also variables correlation is solved.

1. Introduction

Air pollution is one of the most important environmental problems and is the result of human activities. Pollution has diverse causes and sources, such as industrial, commercial, agricultural and domestic activities. Combustion, used to generate heat, electricity or movement, is the process in which many pollutants are produced. Other activities like foundry and chemical production can induce to a deterioration of the air quality if it isn't controlled. Now a days, many first world countries make big efforts to minimize the effects of this activity (Kyoto) [1]. In polluted countries like México a continuos monitoring of the air quality to take forecast measures on possible negative effects in the population health is necessary. A special case with great pollution is the city of Salamanca Guanajuato in México (In México, this city occupies the fourth place in pollution). A great quantity of industries are located in Salamanca, in many cases are chemical industries and also of electricity generation. A pollution alert has been recorded in last years in Salamanca, when in several times the Ecological Mexican Standard NOM-085, has been surpassed [2]. Three years ago, an Environmental Monitoring Network (EMN) was installed in which time series about pollutant concentrations like Sulfur Dioxide ($\mathbf{SO_2}$) and particles $\mathbf{PM_{10}}$ among other meteorological variables are obtained (see Figure 1). In this research, a model that use a combination of Neural Networks (NNs) is applied. The model analyzes the correlation among different pollutant concentrations and meteorological variables and make an estimation of only one value known as Air Quality Index (AQI). Correlation among different concentrations and meteorological variables is a very important analysis for decisions making for a possible Environmental Contingency Forecasting (ECF). Analysis and evaluation about concentrations are given in the following sections.

Email addresses: `tono@salamanca.ugto.mx` (A. Vega-Corona), `andina@gc.ssr.upm.es` (Diego-Andina), `fulgencio@gc.ssr.upm.es` (F.S. Buendía-Buendía), `badamem@salamanca.ugto.mx` (J.M. Barrón-Adame).

1.1 *Principal air pollutants features*

Clean air is a gassy mixture composed by Nitrogen (78%), Oxygen (21%), Argon, Carbon Dioxide, Ozone and other gases in small quantities (1%). Therefore, the

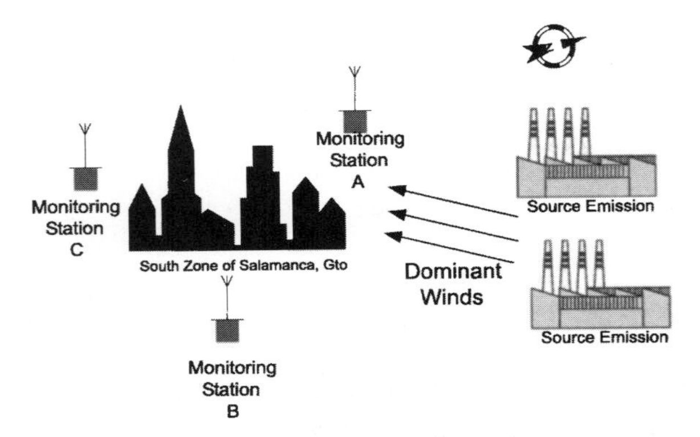

Figure 1. Sample points from Environmental Monitoring Network (EMN).

atmospheric pollution can be defined as the emission of great quantities of substances that perturb the physical and chemical air properties. Pollutants are classified in primary and secondary. Primary pollutants are in the atmosphere when they are originally emitted by the source. For evaluation purposes only Sulfur Dioxide and PM_10 particles are consider due to predominance in Salamanca. Secondary pollutants are those that experience chemical changes as a result of the meteorological effects or combination with other pollutants (as photochemical oxidizers) and some radicals like Ozone.

In the air quality evaluation for concentrations of Sulfur Dioxide ($\mathbf{SO_2}$),a daily mean estimation (24 hrs.) of 340 $\mu g/m^3$ (0.13 ppm) is considered and is equivalent to 100 AQI units. It doesn't almost have color and it has a spicy scent. When it is oxidized and it combines with water, it forms Sulfuric Acid (H_2SO_4), which is the main component of acid rain. For the case of suspended particles or breathable fragments with diameters among 0.3 to 10 microns ($\mathbf{PM_{10}}$), the air quality evaluation is a daily mean estimation (24 hrs.) of 150 $\mu g/m^3$, equivalent to 100 AQI units. The $\mathbf{PM_{10}}$ particles are solid or liquid particles in the atmosphere such as powder, metallic particles, cement, pollen, or compound organic. Breathable fragment of Total Suspended Particles (PST) is constituted by the particles with the diameters smaller than 10 microns, also well-known as $\mathbf{PM_{10}}$ particles.

Table 1
Health Levels and Air Quality Index (AQI)

Health Concern Levels	
Air Quality	AQI Values
Good	0 to 50
Moderate	51 to 100
Unhealthy for sensitive groups	101 to 150
Unhealthy	151 to 300
Dangerous	301 to 500

1.2 The Air Quality Index (AQI)

The AQI, provides daily information in a simple and uniform way on the air pollution concentration. The AQI is a value to inform at the population on the actions to reduce the air pollution or environmental forecasting. The AQI is a simple number into a scale from 0 to 500 [3]. Intervals in AQI scale are defined in Table 1, and are related to the health concerns on the population. Potential health concerns are due to a daily concentration of main pollutants like Sulfur Dioxide ($\mathbf{SO_2}$) and $\mathbf{PM_{10}}$ particles among others [1]. Intervals and conditions that describe air quality levels are also described in Table 1, and are explained as follow:

Good: AQI units between 0 and 50 are considered satisfactory and the air pollution possesses little or few risk.
Moderate: AQI units between 51 and 100 are acceptable; However, for some pollutants it can have a health concern for a small number of population.
Unhealthy for Sensitive Groups: When AQI units are between 101 and 150, members of sensitive groups can suffer effects in their health.
Unhealthy: All the population can suffer dangerous effects in the health when AQI values are between 151 and 300.
Dangerous: For a superior AQI values of 300 a warning alarm is emitted for health conditions, the whole population will be more probably affected.

In Figure 3, time series of pollution levels for $\mathbf{SO_2}$ and $\mathbf{PM_{10}}$ concentrations are shown and correspond to January 2005, and also health indicators accord Table 1, are shown. Also, in Figure 2, the one-dimensional estimation with the current method is shown.

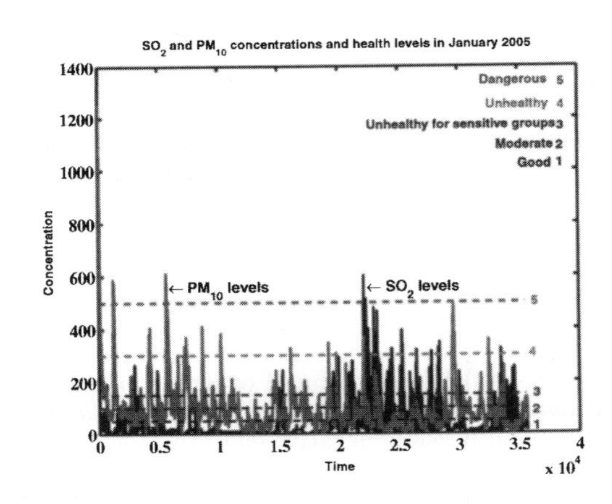

Figure 2. Concentration levels of $\mathbf{SO_2}$ and $\mathbf{PM_{10}}$ in January 2005.

The principal problem to solve in this research is to

consider more decision variables in the AQI estimation and also in the decision making process. Now a day, the AQI estimation in Salamanca is determined as it was explained in Subsection 1.1. An AQI estimation on pollutant concentration without considering other factors is computed with the current method. In this research an important factor to consider in AQI estimation are the variables consideration like wind speed in an automatic noise suppression method on the instrument measurements.

2. Model and Theoretical Fundament

The proposed method considers an automatic multivariable data analysis of time series obtained from EMN on monitoring points A, B and C geographically distributed in dominant winds direction and bigger population concentration (see Figure 1). The problem is to determine the correlation among all the variables involved in the decision making exercise on health risk for the population. Each monitoring point is considered like a sample point build with different sensors and also with its own perception field. Therefore, each point showed can be seen as a sensors fusion of where the time series are obtained. In problem solution a self-organized method that uses a Self-Organized Neuronal Network (SOM) has been proposed in order to build an automatic noise suppression method. Neural Networks (NNs) are computational structures and they can learn from examples [3]. In some multi-dimensional engineering problems (like air pollution) is necessary to recognize certain patterns without the necessity of knowing of data nature or their statistical distribution. Some patterns recognition techniques apply NNs to solve problems without the necessity of a prior data distribution knowledge or to make statistical suppositions. Other techniques in pattern recognition have the necessity to make statistical assumptions about data nature like Bayes theorem [4,5]. Consequently, NNs is an ideal tool to solve the problem here exposed due to their operation which is analyzed like a black box that minimizes the energy function [6,7].

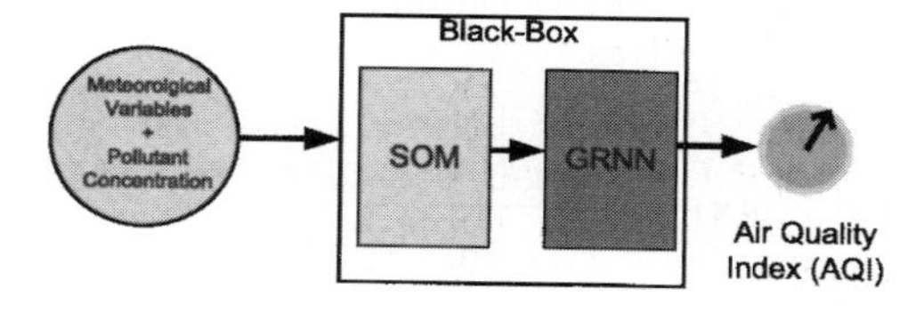

Figure 3. Proposed Model to estimate the AQI

2.1 *Variables definition*

Variables definition is considered like normalized concentration values about pollutants and normalized meteorological values (Wind Speed, Temperature and Relative Humidity). In Table 2, variables are defined in order to build a feature vector $\mathbf{x}_j$ and to define a pattern set $\mathbf{X}_* = \{\mathbf{x_1}, \mathbf{x_2}, .., \mathbf{x_j}, .., \mathbf{x_n}\}$. Let $\mathbf{X_{SO_2}}$ be a Sulfur Dioxide set and let $\mathbf{X_{PM_{10}}}$ be a particles concentration set and their corresponding pattern is defined as $\mathbf{x_j} = \{x_1, x_2, x_3, x_4\}$.

Table 2
Variables definition: $\mathbf{SO_2}$ concentration, $\mathbf{PM_{10}}$ concentration, **T**; Temperature, **RH**; Relative Humidity, **VV**; Wind speed.

	$\mathbf{X_{SO_2}}$	$\mathbf{X_{PM_{10}}}$
Variables x_i	$\mathbf{x}_j$	$\mathbf{x}_j$
x_1	SO_2	PM_{10}
x_2	T	T
x_3	HR	HR
x_4	VV	VV

2.2 *Proposed Model*

2.2.1 *Data Base and Pre-processing*

In this work, a real and historical time series database from the EMN has been used. Data series of three months from December to February in three years from 2002 to 2005 have been analyzed. During Winter the pollutant and the meteorological conditions have major health concern. Time series consider a total of 6,480 multidimensional patterns about pollutant and meteorological variables. In Figure 4, a typical day data concentrations for $\mathbf{SO_2}$ and its correlation with meteorological variables (Temperature and Relative Humidity) is shown. In this figure it is possible to appreciate the complicated nature of this problem.

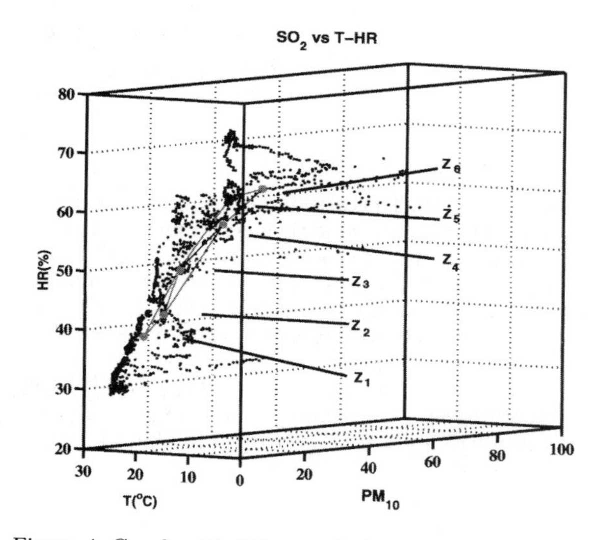

Figure 4. Graph with $\mathbf{SO_2}$ correlation and Meteorological variables, and Self-Organized Map.

2.2.2 Clustering Method

In this research a prior knowledge about patterns is unavailable. Therefore, in the classifier design, the pattern classes obtained applying the clustering method are used. Unsupervised learning is popularly adopted in data clustering where a prior class information is unavailable. SOM Neural Network is good for mapping similar patterns in a high dimension feature space to a much lower dimension output map while preserving the topological order.

Due to data nature is unknown a Self-Organized method has been proposed. In order to group the patterns in six classes according to health concern levels and noisy patterns (as shown in Table 1), a Self-Organized Maps Neural Network is apply as it is shown in Figure 5. The idea is to build different training pattern sets in order to design a classifier based on a NN. A SOM Neural Network structure with euclidian distance function and hexagonal topology and 3:2:1:1 structure has been proposed [8]. In order to have six clusters and therefore six prototypes or weights ($\mathbf{Z}_i$), one per class or AQI (as shown in Figure 4), which have a representation that involved variables in Neural Network training phase and which was mentioned in Section 2.2.1.

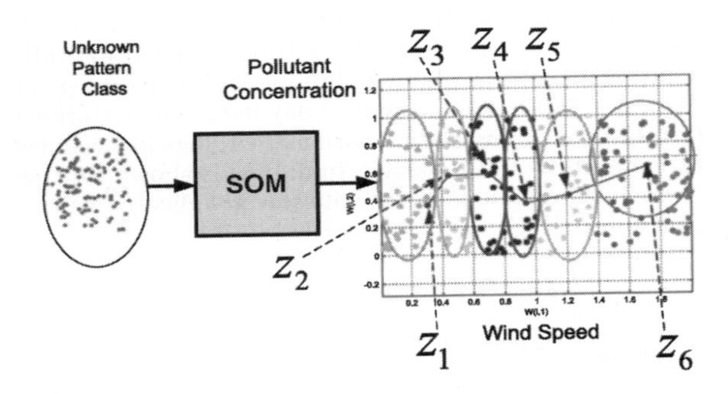

Figure 5. Self-Organized Neural Network

In Table 3, the classification for each health concern level of $\mathbf{SO_2}$ in function of their AQI intervals and their corresponding class $\mathbf{Z}_i$ (or category in a Self-Organized Map) are shown. Therefore, the center for each class is build as $\mathbf{Z}_i=\{\mu_{1i},\mu_{2i},\mu_{3i},\mu_{4i}\}$, where $\mathbf{Z_i}$ is the class center i of each type of AQI pattern, μ_{1i} is the pollutant concentration level prototype ($\mathbf{SO_2}$ concentration if the pattern belong to $\mathbf{X_{SO_2}}$ set and $\mathbf{PM_{10}}$ concentration if the pattern belong to $\mathbf{X_{PM_{10}}}$ set), μ_{2i} is the Temperature prototype, μ_{3i} is the Relative Humidity prototype and μ_{4i} is the Wind Speed prototype.

2.2.3 Classifier Design

In the classifier design, the pattern classes of the clusters with center $\mathbf{Z}_i$, obtained by means of clustering method are used for this purpose. A General Regresión Neural Network (GRNN) with clustering structure which is shown in Figure 6, is trained to obtain a continuos estimation for the AQI in two models. A first model is trained for $\mathbf{SO_2}$ concentrations set ($\mathbf{X_{SO_2}}$) and a second for $\mathbf{PM_{10}}$ concentrations ($\mathbf{X_{PM_{10}}}$). The GRNN was introduced by Donald F. Specht [9]. The main advantage of GRNN over a Multi-Layer Perceptron (MLP) is that, unlike the MLP which need a larger number of iterations to be performed in training phase to converge to a desired solution, the GRNN needs only a single learning pass to achieve optimal performance in classification. In general, the GRNN operation is described. Let $\mathbf{x}$ be a feature vector and y be a scalar and $f(\mathbf{x}, y)$ the joint probability density function (**pdf**) of $\mathbf{x}$ and y. The expected value of y given $\mathbf{x}$ is defined as

$$E[y \mid \mathbf{x}] = \frac{\int_{-\infty}^{\infty} y f(\mathbf{x}, y) dy}{\int_{-\infty}^{\infty} f(\mathbf{x}, y) dy} \tag{1}$$

the **pdf** is unknown, therefore it must be estimated from sample values of $\mathbf{x}_i$ and y_i from a kernel function estimator proposed by Parzen, see [9]. The estimator is defined as a reduced gaussian kernel $\exp(-\frac{D_i^2}{2\rho^2})$. Thus, is possible to obtain a discrete conditional mean of y given $\mathbf{x}$ or an estimation of $\hat{y}$ as,

$$E[y \mid \mathbf{x}] = \hat{y}(\mathbf{x}) = \frac{\sum_{i=1}^{n} y_i \exp(-\frac{D_i^2}{2\rho_i^2})}{\sum_{i=1}^{n} \exp(-\frac{D_i^2}{2\rho_i^2})} \tag{2}$$

where ρ_i is the kernel width, n is the number of all the patterns in the $\mathbf{Z}_i$ clusters and D_i is the euclidian distance among the input pattern and the $i-$th training pattern or ${D_i}^2 = (\mathbf{x} - \mathbf{x}_i)^T(\mathbf{x} - \mathbf{x}_i)$. The GRNN operation is simple, the input layer simply passes the patterns $\mathbf{x}$ to all units in the hidden layers composed by kernels functions $\exp(-\frac{D_i^2}{2\rho^2})$ and computes the squared distances among the new pattern $\mathbf{x}$ and $\mathbf{x}_i$ training samples; the hidden-to-output weights are just the targets

Table 3
Health concern levels respect to Air Quality Index and their category map representation for $\mathbf{SO_2}$ and $\mathbf{PM_{10}}$

Index Classification levels for $\mathbf{SO_2}$ and $\mathbf{PM_{10}}$		
Air Quality	AQI	Cluster Prototype
Good	0 to 50	$\mathbf{Z_1}$
Moderate	51 to 100	$\mathbf{Z_2}$
Unhealthy for sensitive groups	101 to 150	$\mathbf{Z_3}$
Unhealthy	151 to 300	$\mathbf{Z_4}$
Dangerous	301 to 500	$\mathbf{Z_5}$
* **Noise**	————	$\mathbf{Z_6}$

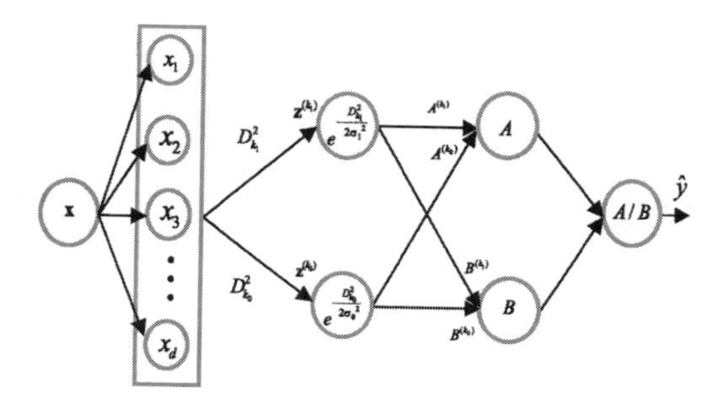

Figure 6. GRNN cluster structure

y_i, thus the output $\hat{y}(\mathbf{x})$, is simply a weighted average of the target values y_i of the training cases $\mathbf{x}_i$ close to the given input case $\mathbf{x}$. The only parameters ρ that need to be learned are adjusted using the algorithm proposed for us in [10].

In some problems like this research, the number of observations obtained can be sufficiently large that it is no longer practical to assign a separate node (or neuron) to each ith sample. Clustering method is used to group samples so that the group can be represented by only one node or prototype $\mathbf{Z}_i$ that measures distance of the input vector $\mathbf{x}$ from the cluster center $\mathbf{Z}_i$. However the cluster prototypes are determined, let us assign a new variable, m, to indicate the number of samples that are represented by the ith cluster center $\mathbf{Z}_i$. The estimation equation can then be rewritten as

$$\hat{y}(\mathbf{x}) = \frac{\sum_{i=1}^{m} A_i \exp(-\frac{D_i^2}{2\rho^2})}{\sum_{i=1}^{m} B_i \exp(-\frac{D_i^2}{2\rho^2})} \qquad (3)$$

and

$$\begin{cases} A_i(k) = A_i(k-1) + y_j \\ B_i(k) = B_i(k-1) + 1 \end{cases} \qquad (4)$$

where $m < n$ is the number of clusters, and $A_i(k)$ and $B_i(k)$ are the values of the coefficients for cluster i after k observations. $A_i(k)$ is the sum of the y_i values and $B_i(k)$ is the number of samples assigned to cluster i. The method of clustering can be as simple as establishing a single radius of influence, r. Starting with the first sample point $(\mathbf{x}_i, y_i)$, establish a cluster center, $\mathbf{x}_i$ at $\mathbf{x}$. All future samples for which the distance $|\mathbf{x} - \mathbf{x}_i|$, is less than the distance to any other cluster center and is also $\leq r$ would update equations (4) for this cluster. A sample for which the distance to the nearest cluster is $> r$ would become the center for a new cluster. Note that the numerator and denominator coefficients are completely determined in one pass through the data and no iteration is required to improve the coefficients [9].

2.2.4 Classifier Optimization

A problem in the estimations based in a GRNN is the adjustment of Perception Parameter (PP) of the neurons. Perception parameter is controlled by the ρ parameter to obtain a minimum classification error. It is solved using a multidimensional gradient algorithm and has been proposed by authors in [10].

3. Experimental Results

3.1 Data Clustering for $\mathbf{SO_2}$ and $\mathbf{PM_{10}}$ concentrations

In the experiments, a real time series data base for $\mathbf{SO_2}$ and $\mathbf{PM_{10}}$ concentrations have been analyzed every minute. The meteorological variables used were Wind Speed, Temperature and Relative Humidity, creating 6,480 four dimensional pattern vectors $\mathbf{x}_i$, one for each set of pollutant ($\mathbf{X_{SO_2}}$ and $\mathbf{X_{PM_{10}}}$), as is shown in Table 2. Both pollutant concentrations like meteorological variables are provided by the EMN from Salamanca. In the clustering method, six clusters have been performed from time series of years 2002, 2003, 2004. The clusters centers $\mathbf{Z}_i$ (or weights in the SOM) are analyzed according its features to determine the label or class of each prototype like in Table 3. Each prototype $\mathbf{Z}_i$, is used to build a GRNN in a cluster structure. Table 3, also shows the classification for each AQI level for $\mathbf{SO_2}$ and $\mathbf{PM_{10}}$ concentration in a Self-Organized Neural Network, where the center for each class is $\mathbf{Z}_i$ and as well is defined like $\mathbf{Z}_i=\{\mu_{1i},\mu_{2i},\mu_{3i},\mu_{4i}\}$, where, $\mathbf{Z}_i$ is the class center i, μ_{1i} is the $\mathbf{SO_2}$ or $\mathbf{PM_{10}}$ concentration level according set, μ_{2i} is the Temperature, μ_{3i} is the Relative Humidity and μ_{4i} is Wind Speed.

3.2 AQI estimation for $\mathbf{SO_2}$, $\mathbf{PM_{10}}$ and Meteorological variables

The complicated interpretation of the correlation among $\mathbf{SO_2}$ pollutant concentrations and meteorological variables like Temperature (T) and Relative Humidity (RH) is shown in Figure 4. In contrast, in Figure 8, is easy to appreciate and to classify the health concern levels of $\mathbf{SO_2}$ concentrations. Figure 8, shows that the complexity is minimized using the classes or categories related to each prototype $\mathbf{Z}_i$. In this case a GRNN estructure has been applied. GRNN is trained using the cluster prototypes $\mathbf{Z}_i$. Time series from 2005 in AQI estimation (direct mode of GRNN) have been used. When a pattern is presented in the GRNN input the AQI estimation it is immediate and is necessary only one pass. When a noisy pattern $\mathbf{x}$ is detected, the GRNN output is inhibited. Noise suppression is an innovation in this research because in the current method it is not considered.

Noisy pattern is an inconsistent element in the time series and it is caused by blasts of wind. Noisy elements

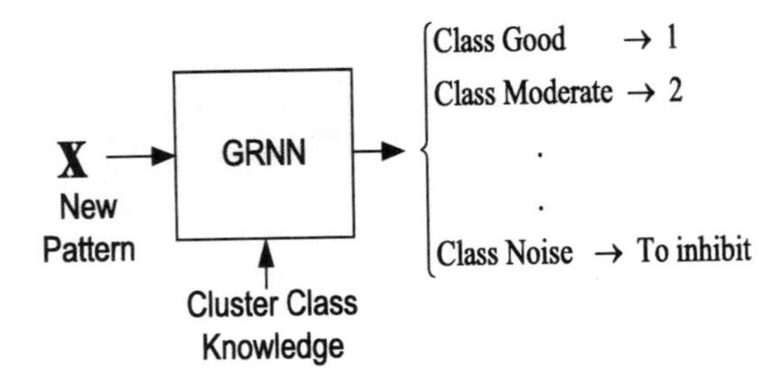

Figure 7. GRNN operation in direct phase

can cause bad estimates about AQI, so with this method a better estimate is obtained. The GRNN operation in direct phase is shown in Figure 7. The complicated in-

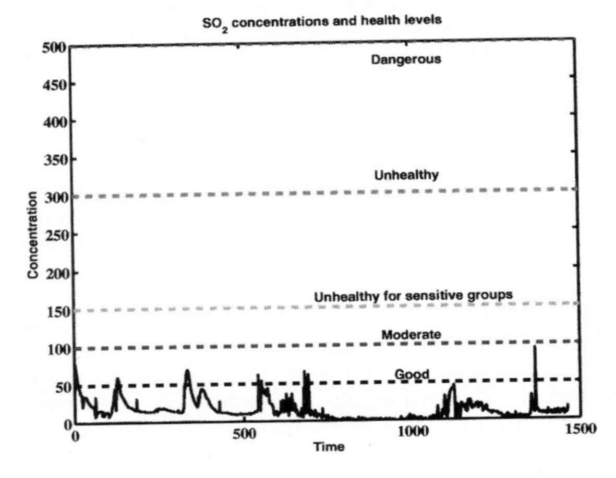

Figure 8. AQI estimation for $\mathbf{SO_2}$ concentrations.

terpretation of the correlation among $\mathbf{PM_{10}}$ pollutant concentrations and meteorological variables like Temperature and Relative Humidity is similar to $\mathbf{SO_2}$ concentration. Similarly, in Figure 9, the $\mathbf{PM_{10}}$ health concern levels are shown. The complexity is solved with this method using the classes corresponding to $\mathbf{Z}_i$.

4. Conclusions

In this work a vector set for $\mathbf{SO_2}$ and $\mathbf{PM_{10}}$ pollutants concentrations and meteorological variables has been built. In order to solve the AQI estimation considering $\mathbf{SO_2}$ and $\mathbf{PM_{10}}$ pollutants concentrations and meteorological variables a combination between Self-Organized Neuronal Network (SOM) and a General Regression Neural Network (GRNN) has been proposed. Both Neural Networks were trained and proven with multidimensional patterns of pollutants and meteorological variables. A correlation problem solution is given and results show the easy interpretation using the discrete classes for the AQI estimation. Concluding that the representation of values shown in Figures 9

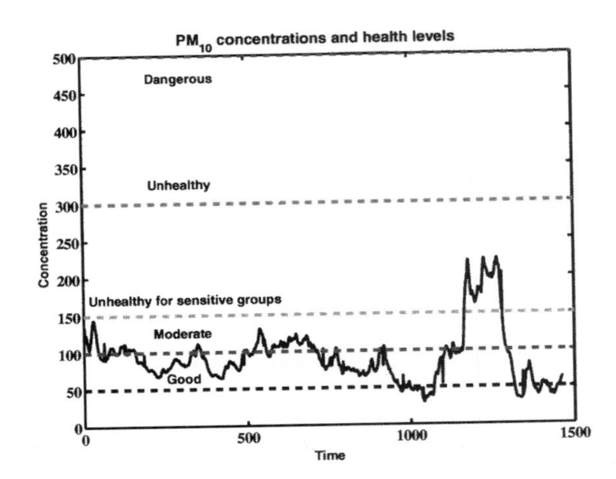

Figure 9. AQI estimation for $\mathbf{PM_{10}}$ concentrations.

and 8, allow to classify the patterns in function of their prototypes in a simple way than the multidimensional representation, therefore it is a good tool for making decisions in environmental forecasting.

Acknowledgements

Supported by City council of Salamanca, Gto. México

References

[1] Environmed Research Inc. Alpha Nutrition. Problem air pollution. http://www.nutramed.com/environment/particles.htm, 2004.

[2] Instituto Nacional de Ecología. Normas oficiales mexicanas para la protección ambiental. http://www.ine.gob.mx/ueajei/publicaciones/normas/, Diciembre 1994.

[3] S.Haykin. *Neural Networks.* Prentice Hall, 2nd edition, 1999.

[4] R.O.Duda and P.E. Hart. *Pattern Classification and Scene Analysis.* John Wiley and Sons, Inc., 2nd edition, 1973.

[5] Jain A.K. *Pattern Recognition.* John Wiley and Sons, Inc, 1988. pp.1052–1063.

[6] R.P.Lippmann. An introduction to computing with neural net. *IEEE ASSP Magazine*, pages 4–22, April 1987.

[7] D.Andina and A.Vega. Detection of microcalcifications in mammograms by the combination of a neural detector and multiscale feature enhancement. *Bio-Inspired Applications of Connectionism. Lecture Notes in Computer Science. Springer-Verlag*, 2085:385–392, 2001.

[8] T.Kohonen. The self organization map. *Proceedings. IEEE*, 78(9):1464–1480, September 1990.

[9] D.F.Specht. A general regression neural network. *IEEE Transactions on Neural Networks*, 2(6):568–576, 1991.

[10] Fulgencio S. Buendía Buendía, J. Miguel Barrón-Adame, Antonio Vega-Corona, and Diego Andina. Improving grnns in cad systems. *Proceedings on Independent Component Analysis and Blind Signal Separation: Fifth International Conference, ICA 2004. Granada, Spain. Springer-Verlag GmbH.*

Intelligent Production Machines and Systems
D. T. Pham, E. E. Eldukhri and A. J. Soroka (eds)

Early design analysis using rough set theory

Alisantoso D[a], Khoo L P[a] and Lu W F[b,†]

[a] School of Mechanical and Aerospace Engineering, Nanyang Technological University, Singapore
[b†]SIMTech Institute of Manufacturing Technology, Singapore

Abstract
This paper proposes a rough set based approach to early design analysis. The approach aims at providing a metrics to facilitate the evaluation of design concepts and enable early detection of inadequacy in design. Based on the rough set theory, an information system can be constructed using information such as design concepts and design capabilities gleaned from past designs. This information is then used to derive a set of design rules for the analysis of new design concepts. The proposed approach embodies a novel technique for the handling of unavailable information, which is a frequent occurrence in product design. The details of the approach, the novel technique and a case study are presented in this paper.

1. Introduction

Product design is an essential part of manufacturing that transforms a design concept into a complete execution plan for the manufacture of a product. As the specifications of the product need to comply with customer requirements, product design can be viewed as an iterative problem solving process that attempts to generate a design (solution) based on a set of product specifications (problem).

Design concept analysis concerns the mapping of design concepts to design capabilities [1]. It aims at assessing design concepts (condition attributes), in relation to product capabilities (decision attributes). One of the major problems in design concept analysis is the treatment of insufficient knowledge or incomplete information in design decision-making.

Many approaches to the handling of incomplete information had been developed. These include fuzzy set theory, rough set theory and Dempster-Shafer theory of evidence. An analysis of these approaches can be found in the work of Khoo and Zhai [2]. Of these, rough set theory, which was advocated by Pawlak [3], can be employed to extract concepts or decision rules from a given set of data and has been used successfully in many application domains [4]. In the domain of knowledge extraction, rough set theory offers the benefits of efficiency, ease of understanding and ability to generate results that can be interpreted directly [5].

The paper is organized as follows. Section 2 generally discusses the basic notions of design concepts analysis. Section 3 provides an overview of the rough sets theory. It provides a description of unavailable information, which is a special type of incomplete information associated with product design, and briefly discusses the effect of unavailable information on design concepts analysis. An approach to the handling unavailable information is explained in Section 4. A case study on the design of a dune buggy is presented in Section 5 to illustrate the approach developed. Section 6 summarizes the key achievements of this work.

2. Design Concepts Analysis

Product design comprises three main stages, namely conceptual design, embodiment design and detail design [6]. Among them, conceptual design is the most important stage where around 60% of the resources needed to manufacture a product are committed once the conceptual design is completed [7]. It is therefore prudent to minimize the design changes through proper analysis of the capabilities of design concepts with respect to product specifications as early as possible.

In this paper, design capability refers to the actual performance of the final product in relation to a design specification. The relationships between the design concept (DC) and design capability (CP) domains, and between the CP and the product specification (SP) domains are illustrated in Figure 1.

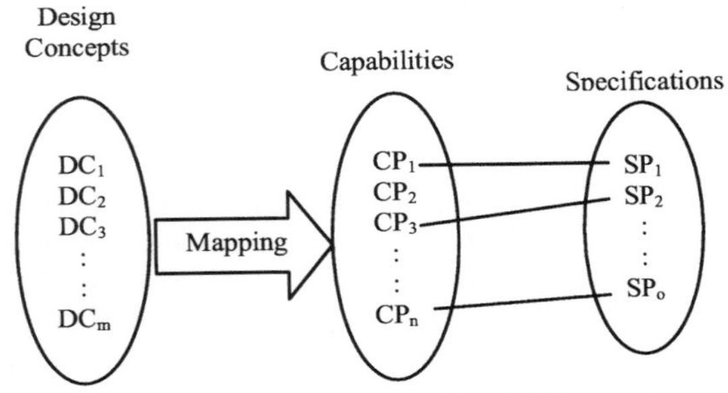

Fig 1. Design Concepts, Capabilities and Specifications Relationships

The concepts (DC_i) embodied in the design concept domain can be used to determine the capabilities (CP_j) of a product through a mapping process (Figure 1). The mapping between the design concept and design capability domains is highly convoluted. The design capability domain contains all the actual performance parameters that may or may not be required by the specifications (SP_k) denoted in the product specification domain.

Essentially, the purpose of product design is to realize a design that is able to satisfy a set of design specifications [6]. In practice, an experienced designer is able to evaluate and select suitable design concepts based on a combination of past design experience and engineering knowledge. This implies that past designs contain potentially useful knowledge for the design of similar products. Thus, it would be advantageous if this knowledge could be extracted to provide the basis for design concept analysis.

To realize such a mechanism, a past design database, which captures past design concepts and associated design capabilities, is necessary. In reality, it is highly probable that this database may contain imprecise, uncertain or missing design information, i.e. imperfect design information, for design concept analysis. For instance, in the design of a dune buggy, the design database may contain design information of dune buggies and dirt bikes. Information such as engine which is crucial for the design of dune buggies, is not applicable to the design of dirt bikes. Hence, this information is omitted in the design information of dirt bikes. As a result, when adapting the design information of dirt bikes for the design of dune buggies, missing or incomplete design information, i.e. unavailable information, will surface. This incomplete information can possibly be dealt with using the rough set theory [8][9]. In so doing, it is envisaged that useful design rules and knowledge for design concepts analysis can be extracted.

3. Rough Sets Theory and Unavailable Information

3.1 Basic notions

Rough set theory is a mathematical tool developed by Pawlak to synthesize the approximation of concepts from a given set of data, which is represented by an information system. Mathematically, an information system (I) can be depicted as follows.

$$I = \{U, A\}$$

where U is a non-empty finite set of objects and A is a non-empty finite set of attributes.

The set of attributes (A) can further be expressed as

$$A = \{C, D\}$$

where C is a non-empty set of condition attributes and D is a non-empty set of decision attributes.

The rough set theory has been applied in data driven decision-making [10][11][12] among others. Based on a set of conditions, a decision can be obtained. Many approaches to rules generation have been developed. They include the work of Mollestad and Komorowski [13], which derives decision rules using the notions of indiscernibility provided by the rough set theory, and the rough set based genetic algorithm approach proposed by Zhai et al. [9]. However, these approaches were developed to deal with rules extraction based on an information system with complete information and thus unable to handle unavailable information. In this paper, the work of Mollestad and Komorowski [13] has been extended to handle problems involving unavailable information. For clarity, the approach developed by Mollestad and Komorowski is summarized below.

Step 1. Based on the data registered in an information system, form a decision relative discernibility matrix comprising n columns and n rows, where n is the number of objects.

Step 2. Complete the $n \times n$ discernibility matrix. A discernibility matrix is a table that identifies the attributes that discern a pair of objects if the corresponding decision(s) of the pair of objects are different. More formally, for each row and column (i, j) of the discernibility matrix, insert the condition attributes that discern objects i and j where $i \leq n$; $j \leq n$, if the decision attribute for objects i and j are different.

Step 3. Form a rule for each row. For each entry in a cell, combine the conditional variables registered using OR Boolean relations to form a term. Subsequently, combine all the terms in a row with AND Boolean relations.

Step 4. Simplify the rules using Boolean algebra, if possible.

Step 5. Determine the value of the condition and decision variables for rule i based on the corresponding values of object i.

Table 1.
An Information System for Dune Buggies

Objects		Condition Attributes (Design Concept)			Decision Attributes (Capability)	
		Chassis	Wheel	Engine	Speed	Stability
Buggy A	X_1	Light	Narrow	Small	Slow	Stable
Buggy B	X_2	Light	Wide	Large	Fast	Stable
Buggy C	X_3	Heavy	Narrow	Large	Fast	Unstable

To illustrate, consider an information system with 3 objects and 5 attributes (3 condition and 2 decision attributes) gathered from a design database for dune buggies (Table 1).

Based on the information system depicted in Table 1, a decision relative discernibility matrix (Steps 1 and 2) was be derived. Subsequently, the attributes in the same cell are jointed using 'OR' Boolean operators, whereas the attributes in different cells are combined by 'AND' Boolean operators (Step 3 and 4). Based on Step 5, the final rule set is given as follows.

<u>*Rules set 1*</u>

Rule 1.1. IF ((Wheel is Narrow or Engine is Small) and (Chassis is Light or Engine is Small)) THEN (Speed is Slow and Stability is Stable).

Rule 1.2. IF ((Wheel is Wide or Engine is Large) and (Chassis is Light or Wheel is Wide)) THEN (Speed is Fast and Stability is Stable).

Rule 1.3. IF ((Chassis is Heavy or Engine is Large) and (Chassis is Heavy or Wheel is Narrow)) THEN (Speed is Fast and Stability is Unstable).

3.2 Unavailable information

In product design, analogy is frequently used to generate design concepts. New design concepts can be adapted from existing designs that are known to be able to support the required functions. In reality, some of the existing design concepts may not be directly relevant to the new design. Some of the design attributes that are relevant may not be available. This discrepancy, in effect, creates a different type of incomplete information known as unavailable information.

Grzymala and Hu [14] proposed a method that is based on probability theory for the repair of an incomplete information system. In their work, a value with the highest probability of occurrence is used to replace the incomplete value. However, such a method requires prior knowledge of data distribution, which may not be readily available for most data sets, small data sets in particular. Kryszkiewicz [8][15] used "null" value to replace the value incomplete attributes. "Null" value is employed to represents all the possible values attainable by the attribute with incomplete value. In essence, the value is assumed to be missing while in actuality it exists. Liang and Xu [16] used information entropy theory to deal with incomplete information. The incomplete variable was assumed to take any of the possible values. Latkowski [17] proposed a method to decompose an incomplete information system into a set of complete information systems. Subsequently, a set of rules was derived from each of the information systems and combined. Khoo and Zhai [2] employed pseudo-objects to replace objects with incomplete information. In their work, a pseudo-object was created for every combination of possible cases.

Other approaches are done by extending the operations of rough set theory. Felix and Ushio [18] solved the incomplete information problem using the discernibility relations of rough sets, i.e. surely discernible and possibly indiscernible. Wang [19] proposed a similar approach to the handling of incomplete information by extending the operators of rough sets. Interestingly, Pawlak [20] cautioned researchers not to complicate the operations of rough set theory unnecessarily as it may lead to mathematical complication and philosophical questions.

The approaches reviewed above appear to be able to handle incomplete information. However, they are based on the same basic assumption that the value of the attributes with incomplete information exist (in the information table) but is not known. They are not able to handle situations where the attributes are relevant but not available in the information table, i.e. unavailable information. The situation occurs when designers attempt to derive hybrid design rules (motorized bicycles, for example) based on two similar but different products (motorcycles and bicycles). Thus, no value can be assigned to these attributes with this type of incomplete information.

4. An Approach to the Handling of Non-Relevant Attributes in Design Concepts Analysis

This paper has extended the work of Mollestad and Komorowski [13] to create a new approach that facilitates the handling of unavailable information in an information system. Known as the Dissimilar Objects Algorithm, the proposed approach is summarized as follows:

1. Replace any unavailable information in the information system with "UA".
2. Derive the rules using the approach proposed by Mollestad and Komorowski by considering "UA" as a new value for the attributes.
3. Simplify the rules by eliminating rules containing "UA".

The rules derived from such an information system with objects having unavailable information based on the proposed Dissimilar Object Algorithm is more stringent compared to those obtained by the algorithm proposed by Mollestad and Komorowski, where unavailable information cannot be included [1] [21]. It is likely to be more robust (See Section 5.2 for a comparative study), more specific and potentially more useful for the evaluation of design capability. The reason being by including objects with some non-relevant attributes would result in a larger search space for possible solutions to a conceptual design.

5. A Case Study

5.1. Application of Dissimilar Objects Algorithm

The application of the proposed approach can be illustrated using the following case study. Assume that a company intends to design and develop a dune buggy. To realize this, the company carries out a preliminary study of dune buggies and dirt bikes as they are likely to be used in similar environment and shares many common features. For the purpose of illustration and ease of understanding, the information on 3 different types of dune buggies and 2 types of dirt bikes has been extracted from a past design database (Table 2). This is an extension of the information shown in Table 1 where additional 2 objects have been added.

Table 2.
An Information System with Incomplete Information

Objects		Condition Attributes (Design Concept)			Decision Attributes (Capability)	
		Chassis	Wheel	Engine	Speed	Stability
Buggy A	X_1	Light	Narrow	Small	Slow	Stable
Buggy B	X_2	Light	Wide	Large	Fast	Stable
Buggy C	X_3	Heavy	Narrow	Large	Fast	Unstable
Bike A	X_4	Light	Narrow	UA	UA	Stable
Bike B	X_5	Heavy	Narrow	UA	UA	Unstable

The condition attribute "Engine" of Objects X_4 and X_5, i.e. Dirt Bikes A and B, are unknown as dirt bikes are usually not built with engines. Accordingly, their values are denoted as "UA". Similarly, the decision attribute "Speed" for Objects X_4 and X_5 are unknown and their values denoted by "UA" since it is highly dependent upon the ability of the rider. Based on the proposed Dissimilar Objects Algorithm, the preliminary design rules have been extracted and the results are summarized as follows.

Rules Set 2

Rule 2.1. IF (Engine is Small) THEN (Speed is Slow and Stability is Stable)

Rule 2.2. IF ((Wheel is Wide or Engine is Large) and (Chassis is Light or Wheel is Wide)) THEN (Speed is Fast and Stability is Stable)

Rule 2.3. IF ((Engine is Large) and (Chassis is Heavy or Wheel is Narrow)) THEN (Speed is Fast and Stability is Unstable)

Rule 2.4. If (Engine is UA and Chassis is Light) THEN (Speed is UA and Stability is Stable)

Rule 2.5. If (Engine is UA and Chassis is Heavy) THEN (Speed is UA and Stability is Unstable)

As Rules 2.4 and 2.5 contain "UA", based on the Dissimilar Objects Algorithm, they are eliminated. Thus, the final rule set comprises 3 rules, Rules 2.1 – 2.3. These rules are checked against the original information system (Table 1). It can be readily shown that Rules 2.1, 2.2 and 2.3 satisfy Objects X_1, X_2 and X_3 respectively, implying that the rules are indeed reasonable.

5.2. A Comparative Study

It can be argued that since the rules containing "UA" entries are eliminated from the rule set eventually, those objects containing incomplete information such as Objects X_4 and X_5 in Table 2 should not have been considered in the generation of design rules. By removing Objects X_4 and X_5 from the information system, an information system identical to that of Table 1 can be obtained. Further analysis is needed to examine such a possibility.

Rules 1.1 and 1.3 are different from those derived by the proposed Dissimilar Objects Algorithm (i.e. Rules 2.1 and 2.3 respectively). It can be observed that Rules 1.1 and 2.1 draw the same conclusions. The premise of Rule 1.1, *'(Wheel is Narrow or Engine is Small) and (Chassis is Light or Engine is Small)'* can be simplified as *'(Engine is Small) or (Wheel is Narrow and Chassis is Light))*. Compared with that of Rule 2.1, which is *'(Engine is Small)'*, it is apparent that the premise of Rule 2.1 is a subset of that of Rule 1.1. Hence, the premise of Rule 2.1 is more stringent. It can also be readily shown that the premise of Rule 2.3 is indeed a subset of that of Rule 1.3.

6. Conclusions

In the research described in this paper, a novel approach based on rough sets theory to assist designers in design concepts analysis has been proposed. Based on the past design information registered in an information system, the approach is able to generate a set of design rules that describes the relationship between design concepts and design capabilities. As design capabilities are associated with design specifications, these design rules can be used to evaluate design capabilities of new or modified designs of similar products for compliance with design specifications. In so doing, an early design concepts analysis can be performed to detect design inadequacy during the conceptual design stage. In order to imbue the ability for handling of hybrid products, which are fairly common in product design, the Dissimilar Objects Algorithm has been proposed. Such an algorithm is able to handle unavailable information and explore a larger search space. It should be noted that in enlarging the search space, the selection of objects to be included must be done carefully. Careless inclusion of objects may yield unreasonable rules.

References

[1] Alisantoso D., Khoo L.P., Ivan Lee B.H., Fok S.C. (2004) "A Rough Set Approach to Design Concept Analysis in Design Chain", International Journal of Advanced Manufacturing Technology, ISSN: 0268-3768 (Paper) 1433-3015 (Online).

[2] Khoo, L.P. and Zhai, L.Y. (2001) "Multiconcept Classification of Diagnostic Knowledge to Manufacturing Systems: Analysis of Incomplete Data with Continuous-Valued Attributes", International Journal of Production Research, Volume 39, Number 17, pp. 3941-3957

[3] Pawlak, Z. (1982) "Rough Sets", International Journal of Computer and Information Sciences 11, pp. 341-356

[4] Komorowski, J.; Pawlak, Z.; Polkowski, L.; Skowron, A. (1999) "Rough Sets: A Tutorial", In Skowron, A., Editor, Rough Fuzzy Hybridization. pp. 3-98, Springer, New York

[5] Pawlak, Z. (1996) "Why Rough Sets?", Proceedings of the Fifth International Conference on Fuzzy Systems, Volume 2, pp. 738-743, New Orleans, Louisiana, September 8-11

[6] Ulrich, K.T.; Eppinger, S.D. (2000) Product Design and Development. Mc-Graw Hills Companies, Inc.

[7] Hundal, M.S. (1997) "Systematical Mechanical Designing: A Cost and Management Perspective", The American Society of Mechanical Engineers

[8] Kryszkiewicz, M. (1998) "Rough Set Approach to Incomplete Information Systems", Information Sciences, Volume 113, pp. 39-49

[9] Zhai, L.Y.; Khoo, L.P.; Fok, S.C. (2001) "Derivation of Decision Rules for the Evaluation of Product Performance Using Genetic Algorithms and Rough Set Theory", Data Mining for Design and Manufacturing, pp. 337-353

[10] Kusiak, A; Kern, J.A; Kernstine, K.H; Tseng, B.T.L. (2000) "Autonomous Decision-Making: a Data Mining Approach", IEEE Transactions on Information Technology in Biomedicine, Volume 4, Issue 4, pp. 274-284

[11] Nguyen, H.P.; Le, L.P.; Santiprabhob, P.; De Baets, B. (2001) "Approach to Generating Rules for Expert Systems using Rough Set Theory", Joint 9th IFSA World Congress and 20th NAFIPS International Congress, Volume 2, pp. 877-882, Vancouver, Canada, July 27

[12] Lee, S.; Vachtsevanos, G (2002) "An Application of Rough Set Theory to Defect Detection of Automotive Glass", Mathematics and Computers in Simulation, Volume 60, pp. 225-231

[13] Mollestad, T. and Komorowski, J. (1999) "A Rough Set Framework for Mining Propositional Default Rules", In Skowron, A., Editor, Rough Fuzzy Hybridization. pp. 233-262, Springer, New York

[14] Grzymala-Busse, J.W. and Hu, M. (2000) "A Comparison of Several Approaches to Missing Attribute Values in Data Mining", Second International Conference on Rough Sets and Current Trends in Computing, pp. 340-347, Banff, Canada, October 16-19

[15] Kryszkiewicz, M. (1999) "Rules in Incomplete Information Systems", Information Sciences, Volume 112, pp. 271-292

[16] Liang, J. and Xu, C. (2000) "Uncertainty Measures of Roughness of Knowledge and Rough Sets in Incomplete Information Systems", Proceedings of Third World Congress on Intelligent Control and Automation, pp. 2526-2529, Hefei, China, June 28-July 2

[17] Latkowski, R. (2003) "On Decomposition for Incomplete Data", Fundamenta Informaticae, Volume 54, pp. 1-16

[18] Felix, R. and Ushio, T. (1999) "Rules Induction from Inconsistent and Incomplete Data Using Rough Sets", Proceedings of IEEE International Conference on Systems, Man, and Cybernetics, Volume 5, pp. 154-158, Tokyo, Japan, October 12-15

[19] Wang, G. (2002) "Extension of Rough Set under Incomplete Information Systems", Proceedings of IEEE International Conference on Fuzzy Systems, Volume 2, pp. 1098-1103, Honolulu, Hawaii, May 12-17

[20] Pawlak, Z. (1998) "Granularity of Knowledge, Indiscernibility and Rough Sets", Proceedings of IEEE World Congress on Computational Intelligence and IEEE International Conference on Fuzzy Systems, Volume 1, pp. 106-110, Anchorage, Alaska, May 4-9

[21] Alisantoso D., Khoo L.P., Lee I.B.H. (2004) "An Approach to the Analysis of Design Concepts Using Rough Set Theory", Artificial Intelligence for Engineering Design, Analysis and Manufacturing (Special Issue), Volume 18, pp. 343-355

Intelligent Production Machines and Systems
D. T. Pham, E. E. Eldukhri and A. J. Soroka (eds)

Hypermedia-based adaptive expert system for advanced performance support

D. T. Pham[a], A. M. Huneiti[b]

[a]*Manufacturing Engineering Centre, Cardiff University, Wales, UK, E-mail: PhamDT@cardiff.ac.uk*
[b]*Information Technology Department, CIS division, Jordan University, Amman-Jordan, E-mail: A.huneiti@ju.edu.jo*

Abstract

This paper discusses the provision of advanced performance support through the integration of techniques used in developing knowledge-based systems, namely diagnostic expert systems and adaptive hypermedia systems. This integration is implemented by employing hypermedia which allows supporting content to be synchronized with the diagnostic expert system inference process. The paper proposes an integrated fault data model for diagnostic expert systems. It also introduces a technique for the adaptive retrieval of hypermedia-based technical information using data semantics. A special organisation of displays in the user interface of the adaptive diagnostic expert system allows operators, while applying the expert system for fault diagnosis, to request detailed information about a certain diagnosis procedure, and then return to the system to continue from where they left off. The techniques proposed in this paper are demonstrated through a prototype adaptive expert system for locating and correcting braking system faults in a forklift truck.

1. Introduction

Supporting the performance of workers in modern high technology job environments has become an increasingly complex, time consuming and costly task that requires advanced methods. Simple tasks of low information volume can be efficiently facilitated using traditional performance support methods such as paper documentation, electronic databases, lectures, instructor-led courses, and job aids. However, many problems are associated with these traditional methods especially when used to assist the operation of complex products or systems. These concern their portability, complexity, accuracy, reliability, and information maintainability [1]. There are also other problems associated with integrating and retrieving the information, the static structure of the presented material, the restricted support given to users, and the limitations of presentation methods [2].

Reliable Performance Support Systems (PSS) can enhance productivity and reduce training, time and costs and operating errors. Moreover, they can increase quality, task completion rate and worker autonomy [3, 4]. However, according to Fischer and Horn [5], without tools that are primarily PSS tools and with no clear methodology for building them or measuring their performance, PSSs will be limited to being just “an approach”.

This paper discusses the provision of advanced performance support through the utilisation and integration of technologies used in developing knowledge-based systems, namely diagnostic expert systems and adaptive hypermedia systems. Section 2 reviews advanced performance support systems. Section 3 introduces an integrated fault data model which enables the automatic generation of the knowledge base of a diagnostic expert system. Section 4 presents a methodology for the adaptive retrieval of technical information using data semantics. This methodology is supported by a stereotype user model and implemented using conditional semantic rules. Section 5 demonstrates

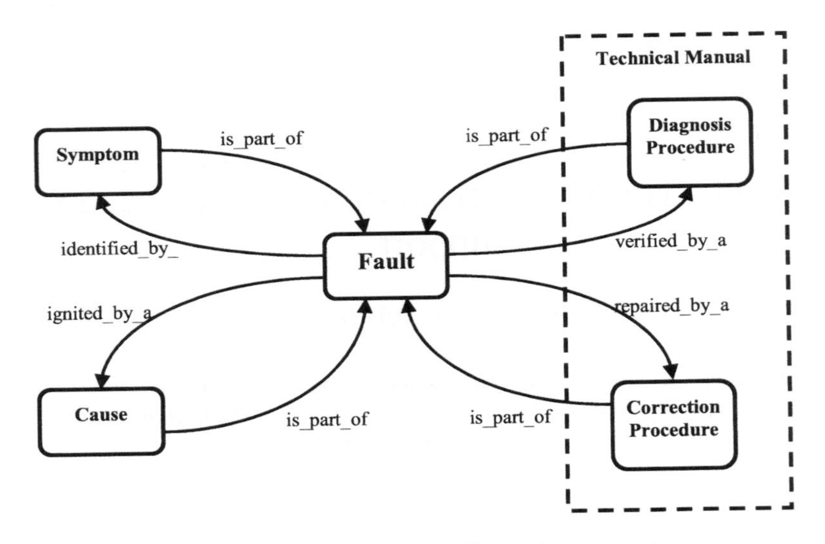

Fig. 1 Integrated fault data model for diagnostic expert systems

the user interface of the integrated adaptive diagnostic expert system for an all-terrain forklift truck. Finally, section 6 concludes the paper.

2. Advanced performance support systems

Electronic PSSs can range in complexity from a single help file for a specific task to a complete expert system for complex problem solving. These may include interactive task advising, tests for understanding, tutors, coaching, training, intervention, modular learning experiences, assessment feedback, and monitoring systems [6]. A specialised and more product-oriented type of performance support exists in the form of "product support". According to Pham et al. [7], product support consists of everything necessary to allow the continued use of a product, including user training, technical manuals, help lines, servicing, spare part ordering, and maintenance management.

Typically, a technician determines what to do by consulting with a diagnostic aid and then uses a fault code or a symptom tree to interpret the diagnostics output. The technician locates procedural information in a technical manual. In some cases, according to Cooper [8], the located information is sufficient, in other cases, different information is needed, or supplementary information may be required to bolster the understanding of the technician. Other information may be required to explain how to set up for and conduct a series of test until the fault is isolated and the correct procedure executed.

A common direction for advanced performance support is to supply task-specific and user-centred support in conjunction with information and problem solving capabilities. This direction is adopted by many recent advanced PSSs which include El-Tech [9], ADAPTS [8, 10, and 11], IPM [2, 7, and 12], and MMA [13]. The aforementioned PSSs consist mainly of two integrated "abstract" components that are designed and accessed in a task-based and user-centred manner. The first component is freely browsed technical information that supplies "how-to-do" type of information. The second component is the expert assistant that supplies guidance for more specific, complex, and difficult to learn tasks, in other words. "what-to-do" type of information.

Furthermore, users of PSSs have different levels of knowledge, expertise, and qualifications, and they also have different goals and objectives. Information quality and quantity can be controlled by managing the provision of the multimedia elements and/or associated hyperlinks. This can be achieved through adaptive hypermedia, for example to supply general information to novice operators and, in contrast, more detailed information to experienced operators. In adaptive hypermedia systems, the content, navigation, and presentation of information can be tailored to the needs of users by means of a user model [14, 15].

3. Integrated fault data model for diagnostic expert systems

As current products, equipment, and systems grow in size and complexity, it becomes more difficult to diagnose faults in them. Hence, the use of expert(s) knowledge to support user performance is increasingly necessary. Capturing and encapsulating expert diagnostic knowledge enables

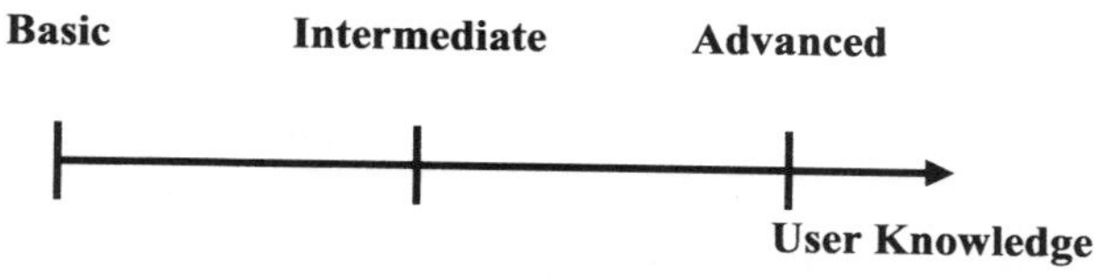

More detailed and complex information
More references
Less explanation
Less guidance
Less annotation

Fig. 2 Main principles of the adaptive strategy

permanent preservation of, and access to, this knowledge through special purpose expert systems.

Information used in diagnostic expert systems can be classified into "diagnosis" information and "technical" information. Diagnosis information is derived from specific data which identifies the symptoms and causes of faults and the diagnostic strategy that might be employed. This gives explicit fault-dependent information which is used in correcting faults and can be represented by heuristic rules. These rules are transformed into a set of formal rules to populate the knowledge base of the expert system. Although this type of information can be acquired from troubleshooting manuals, fault correction sheets, error codes and charts, the main source of this type of information is the human expert.

In contrast, technical information is obtained from detailed data associated with the steps of the diagnosis procedures, the structure, and the properties of the components of the supported product and their interaction. This type of information is used to enhance the understanding of the user by explaining, illustrating, and clarifying the performed diagnosis tasks, in other words "how-to-do" type of information. It includes a description of the work to be done, parts involved, precautions, warnings, and required instruments. The main source for this type of information is the technical documentation of the product itself.

Diagnosis information and technical information complement each other, and their integration gives more effective support for user performance. This integration is implemented through embedding references to technical information in the knowledge base of the expert system. These references are detailed descriptions of the diagnosis procedures which are available in the technical manual. Issues related to structured authoring and generation of hypermedia-based technical manuals are discussed in [16].

Diagnosis data is fault-oriented and can be modelled with regard to a specific set of faults. Figure 1 outlines an integrated fault data model. The model outlines the following hypotheses: (i) Every *fault* has one or more *symptoms* and *causes*, (ii) A *symptom* is an identifiable sign of existence of a *fault*, (iii) A *cause* is something that ignites a *fault*, (iv) Every *fault* has a *diagnosis* and a *correction* procedure, (v) A *diagnosis procedure* verifies a *fault*, (vi) A *correction procedure* repairs a *fault*.

Through the association of every fault with the corresponding diagnosis and correction procedure, the relationship between diagnosis information and technical information can be established. For implementation purposes, this integrated model is transformed into a database schema to be populated subsequently with diagnosis data. A user interface has been constructed on top of the database tables in order to facilitate the updating process of the structured diagnosis data in a graphical and user-friendly way. This structured set of data has been used "automatically" to generate the knowledge base for a prototype rule-based expert system that supports the location and the correction of braking system faults in a forklift truck.

4. Adaptive retrieval of hypermedia-based technical information using data semantics

4.1 Semantic data model for technical information

This semantic data model [16, 17] is extracted through a technical information usage analysis which abstracts a hierarchical representation of (i) the intended purpose for the technical information i.e. action vs. knowledge, (ii) the supported user tasks i.e. learning vs. performing, and (iii) the functional characteristics of the technical information elements i.e. fundamental, procedural, clarification, advice, etc. These abstract views are combined together into a semantic network and a set of rules and constraints, which are then mapped into a database schema. This mapping creates an ontology-like organisation which can then be used

Table 1
Relationship between adaptive support and user knowledge stereotypes

Adaptive Support	Basic	Intermediate	Advanced
Content:			
Main content	expanded	expanded	collapsed
Clarification information	yes	no	no
Advice information	yes	yes	no
Navigation:			
Fundamental hyperlinks	visible	visible	visible
Detailed Hyperlinks	none visible	some visible	all visible
Hot keywords	visible	not visible	not visible
Access method (guidance)	strict guided tour	expanded list	collapsed list
Presentation:			
Clarification icons	yes	yes	no

by authors in order to classify, organise, and build the structure of the product technical documentation. For further details, refer to [16].

4.2 Stereotype model for user knowledge

User models are often represented by either an "overlay" model or by a simpler "stereotype" model. The former is a representation of features of the user as an overlay on top of the domain model. The latter distinguishes several typical or stereotype users. The context of the work of the user is a deterministic factor in selecting the best type of user model to use and the features of the user to consider.

In the fault diagnosis domain, the high level goal of the user is "diagnosis" which is stable throughout the user interaction with the adaptive system. In addition, the low level goal of the user is the diagnosis or correction of a particular fault. These faults may change quite often during the work session. In addition, the required information is usually concise and precise for example, a single diagnosis or correction procedure, which enables the users themselves to estimate their knowledge of the task they are performing.

The suggested stereotype user model used to represent the knowledge of the user of the subject domain includes three stereotype values, namely basic, intermediate, and advanced. The goal or the task that the user wants to accomplish is identified through the provision of the identification code of the required diagnosis procedure supplied by the expert system. By adopting this explicit method for identifying the goals and knowledge of the users, a more accurate assessment will be yielded every time the system is invoked.

4.3 User knowledge-based strategy for adaptive support using conditional semantic rules

The proposed adaptive strategy can be perceived as a filtering mechanism for the technical information fragments retrieved and presented by the adaptive system. It determines which technical information is relevant and which is irrelevant, and how to present relevant information fragments by considering the knowledge and the goal of the user using conditional semantic rules. This filtering process accesses the semantically structured data of the technical documentation and work at the information fragment level. The main objective of the adaptive filter is to distinguish the "relevant" information fragments from irrelevant ones, depending on the user knowledge stereotypes, namely, basic, intermediate, or advanced. The relevant information fragments are then passed to be rendered and the irrelevant ones are discarded.

The main principles of the suggested adaptive strategy are outlined in Figure 2. The figure shows the relationship between the current user knowledge and the adaptive support features supplied to the user. As the knowledge of the user increases from "basic" through "intermediate" to "advanced", the complexity and detail of information and the number of visible reference links i.e. hyperlinks increases, and vice versa.

Table 1 depicts, in more detail, the relationship between the adaptive support given by the hypermedia system and the knowledge of the user. The adaptive support is categorised by content, navigation, and presentation. *Adaptive content support* is achieved by adapting the main content of the initial page accessed by a particular user to current user knowledge prototype. *Adaptive navigation support* is used to help users to find their way in the hypermedia space by adapting the page access methods and the provision of hyperlinks to

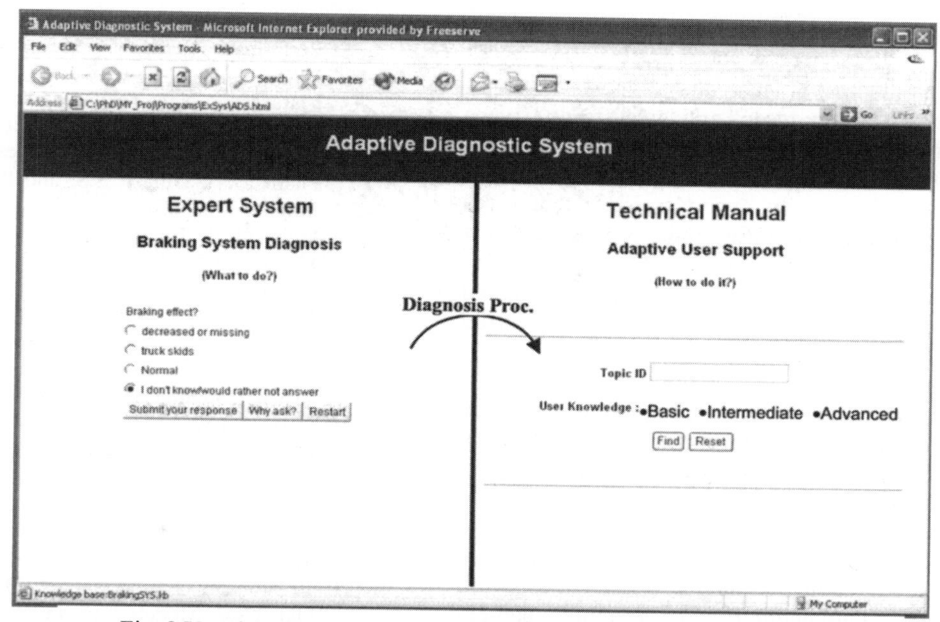

Fig. 3 User interface for the integrated adaptive diagnostic expert system

current user knowledge prototype. *Adaptive presentation support* is achieved by adapting the presentation of a page accessed by a particular user to current user knowledge prototype. For example, it can be implemented using visual icons which are applied to provide extra clarification for novice users. A distinct visual icon is associated with every information fragment and navigational hyperlink in order to clarify its functional characteristic, such as, fundamental, procedure, clarification, advice, or specification.

The values associated with every benchmark are an interpretation of the main principles of the adaptive strategy outlined in Figure 2. The adaptive support strategy is implemented by transforming the benchmarks and their values into conditional semantic rules which are applied on relationships between information fragments. These fragments are richly indexed in accordance with the semantic data model of the technical documentation described earlier.

5. Integrated adaptive diagnostic expert system for a forklift truck

A hypermedia-based adaptive expert system for locating and correcting braking system faults in a forklift truck was implemented in order to demonstrate the proposed techniques. The main objective of the application is to support the performance of relatively inexperienced technicians to perform at the level of more experienced and skilled technicians by encapsulating this expertise in a diagnostic expert system. The data used was extracted from the troubleshooting manual developed for a manufacturer of all-terrain forklift trucks as part of a project undertaken by the authors' laboratory within an EC-funded collaborative research program. The application works in conjunction with a Web-based prototype technical manual of the forklift truck, which supplies the technical information.

Figure 3 presents the user interface of the adaptive diagnostic expert system. The interface consists of two user displays which include the diagnostic expert system display (leftmost display) and the adaptive information retrieval display (rightmost display). These displays are combined into one display where the user goal in the form of an identification code for the requested diagnostic procedure is passed from left to right.

This organisation of displays allows operators, while using the system for fault diagnosis, to request information about a certain diagnosis procedure, and then return to the system to continue from where they left off. The required technical information is given in an adaptive manner considering the operator's knowledge of the requested diagnosis procedure. The integration of expert diagnostic assistance and technical documentation is achieved through the use of hypermedia which allows supporting content to be synchronized with the diagnostic ES inference process.

6. Conclusions

The main objective of an advanced PSS is to supply operators with "how-to-do" and "what-to-do" types of information. The former is achieved through a freely browsed technical documentation. The latter is achieved through an expert assistant that gives guidance for more specific, complex, and difficult to learn tasks. These services provide a better support when they are *inter-linked* and are designed to be accessed in a *task-based* and *user-centred* manner.

The quality of the support provided to improve the task performance can be substantially enhanced by integrating factual information and explanatory capabilities within knowledge-based expert assistant, adapted to the user knowledge of the performed task. It has been demonstrated that the advanced performance support concept can be implemented using techniques associated with developing knowledge-based systems namely diagnostic expert systems and adaptive hypermedia systems.

With regard to user modelling, further research can be conducted on investigating the application of implicit user models that are automatically detected by the system, in providing adaptive diagnosis support. Finally, a hybrid approach to user modelling which combines a stereotype and an overlay user model to represent the user knowledge about the technical documentation, might improve the accuracy of the adaptively delivered product information.

References

[1] Ventura, C. A. Why Switch From Paper to Electronic Manuals?. Proceedings of the ACM Conference on Document Processing Systems, January 2000, 111-116.

[2] Pham, D. T., Dimov S. S. and Setchi, R. M. Intelligent Product Manuals. Proceedings of the Institution of Mechanical Engineers, 1999, Part B, 213(I), 65-76.

[3] Desmarais, M. C., Leclair, R., Fiset J. and Talbi, H. Cost-Justifying Electronic Performance Support Systems. Communications of the. ACM, 1997, 40(7), 39 – 48.

[4] McGraw, K. L. Defining and Designing the Performance-Centered Interface, Moving Beyond the User-Centered Interface. ACM Interactions, 1997, 4(2), 19-26.

[5] Fischer, O. and Horn, R. Electronic Performance Support Systems. Communications of the ACM, 1997, 40(7).

[6] Cantando, M. Vision 2000: Multimedia Electronic Performance Support Systems. Proceedings of the 14th Annual International Conference on Marshalling New Technological Forces: Building a Corporate, Academic, and User Oriented Triangle, Research Triangle - United States, Oct 19-22, 1996.

[7] Pham, D.T., Dimov, S.S. and Peat, B.J. Intelligent Product Manuals. Proceedings of the Institution of Mechanical Engineers, Journal of Engineering Manufacture, 2000, 214(B5), 411-419.

[8] Cooper, D.W., Veitch, F. P., Anderson, M. M. and Clifford, M. J. Adaptive Diagnostic And Personalised Technical Support (ADAPTS). Proceedings of the IEEE Aerospace Conference, Aspen, Colorado, 1999, Paper No 4.602.

[9] Coffey, J.W., Canas, A. J., Hill, G., Carff, R., Reichherzer, T. and Suri, N. Knowledge Modelling and the Creation of El-Tech: a Performance Support and Training System for Electronic Technicians. Expert Systems with Applications, 2003, 25, 483-492.

[10] Brusilovsky, P. and Cooper, D. W. Domain, Task, and User Models for an Adaptive Hypermedia Performance Support System. Proceedings of the International Conference on Intelligent User Interfaces, 2002, San Francisco, CA, USA.

[11] Brusilovsky, P. and Cooper, D. W. ADAPTS: Adaptive Hypermedia for a Web-based Performance Support System. Proceedings of the 2nd Workshop on Adaptive Systems and User Modeling on the WWW, Toronto, Canada, May 11-14,1999, 41-47.

[12] Pham, D. T. and Setchi, R. M. Case-Based Generation of Adaptive Product Manuals. Proceedings of the Institution of Mechanical Engineers, 2003, 217 (B), 313-322.

[13] Francisco-Revilla, L. and Shipman F.M. Adaptive Medical Information Delivery Combining User, Task, and Situation Models. Intelligent User Interfaces, New Orleans LA USA, 2000, pp. 94-97.

[14] De Bra, P. Design Issues in Adaptive Web-Site Development. Proceedings of the 2nd Workshop on Adaptive Systems and User Modelling on the Web, 8th International World Wide Web Conference, 1999, Toronto, Canada.

[15] Brusilovsky, P. Methods and Techniques of Adaptive Hypermedia. User Modeling and User-Adapted Interaction, 1996, 6, 87-129.

[16] Pham, D. T., Dimov, S. S. and Huneiti, A. M. Semantic Data Model For Product Support Systems. Proceedings of the IEEE International Conference on Industrial Informatics INDIN 2003, August 21-24, 2003, Banff, Alberta, Canada.

[17] Pham, D. T. , Dimov, S. S., Setchi, R., Peat, B., Soroka, A., Brousseau, E., Huneiti, A. M., Lagos, N., Noyvirt, A., Pasantonopoulos, C, Tsaneva, D. and Tang Q. Product Lifecycle Management for Performance Support. Journal of Computing and Information Science in Engineering, 2004.

Intelligent Production Machines and Systems
D. T. Pham, E. E. Eldukhri and A. J. Soroka (eds)

Intelligent SERM (Strategic Enterprise Resource Management)

Ercan Oztemel, Tulay Korkusuz Polat and Cemalettin Kubat

Sakarya University, Engineering Faculty, Dept. of Industrial Engineering, 54187, Sakarya, TURKEY
eoztemel@sakarya.edu.tr, korkusuz@sakarya.edu.tr, kubat@sakarya.edu.tr

Abstract

Enterprise resource planning (ERP) is one of the driving factors for the success of the companies. However, it was realized in the recent developments that the companies need to extend the resource planning to resource management. SERM (Strategic Enterprise Resource Management) is specifically developed taking all management aspects of resource utilization (including planning) into account. SERM is composed of Strategic Management, CFA (Customer Focussed Activities), ERP, Technology Management, Performance Management allowing good management of complexities, dynamic changes and uncertainties. However, due to heavy complexities and dynamicity the support of computers and intelligent systems seem to be inevitable. This study defines a general framework for utilizing artificial intelligence especially agent technology supporting SERM.

1. Introduction

Today, it is clear that enterprises need to manage their business with very limited resources under complex, dynamic as well as uncertain situations. New applications, new studies, new productivity analyses are sought by the companies to overcome uncertainties and complexities. Note that, although, the Enterprise Resource Planning (ERP) is one of the driving factors for the success of the enterprises, the companies need to have management tools rather than planning tools due to complexities involved [1]. This study is an attempt to provide a general framework for such a management tool so called Strategic Enterprise Resource Management (SERM). The basic differences between management and planning tools are pointed out by Anderegg [2]. He mainly takes the attention of the readers the differences between ERP and Enterprise Resource Management (ERM). His study still lack of the strategic issues. SERM on the other hand is developed to make sure that the facilities for strategic issues together with the resource management are taken into account during a specific decision making [3]. SERM is developed in such a way that the all management activities and functional goals are designed in order to make sure that strategic objectives put forward by the management are (could be) fully satisfied. Today one of the main reasons behind unsuccessful organizations is either lack of strategic objectives or being unable to spread the objectives throughout the organizations. Each department or functional units, even each process, has its own objectives and goals. These should be contributing to satisfying overall company objectives defined at the strategic level. Moreover, the strategic objectives should be identified based on a serious analysis of the market, economic and political conditions, shareholders behaviors, customer attitudes, employee satisfaction etc. Strategic objectives are to be defined in such a way that the organization can be best managed obtaining competitive advantages. Performance indicators should be defined in order to monitor the level of satisfying or following the objectives. Furthermore, the organizations should define strategies and performance indicators in such a way that they manage their limited resources to generate better quality and better production environments for more productive results. By this way, organizations can update their strategies according to the current situations following the changes. SERM is a software based decision making environment ensuring that the corporate level strategic objectives are defined and then related functional goals are identified. The resources of the organization are then allocated based on the agreed objectives and goals.

2. A brief overview of SERM

SERM is a new approach for corporate level resource management. Figure 1 indicates basic components of SERM as well as the progress from MRP (Material Requirements Planning) to SERM.

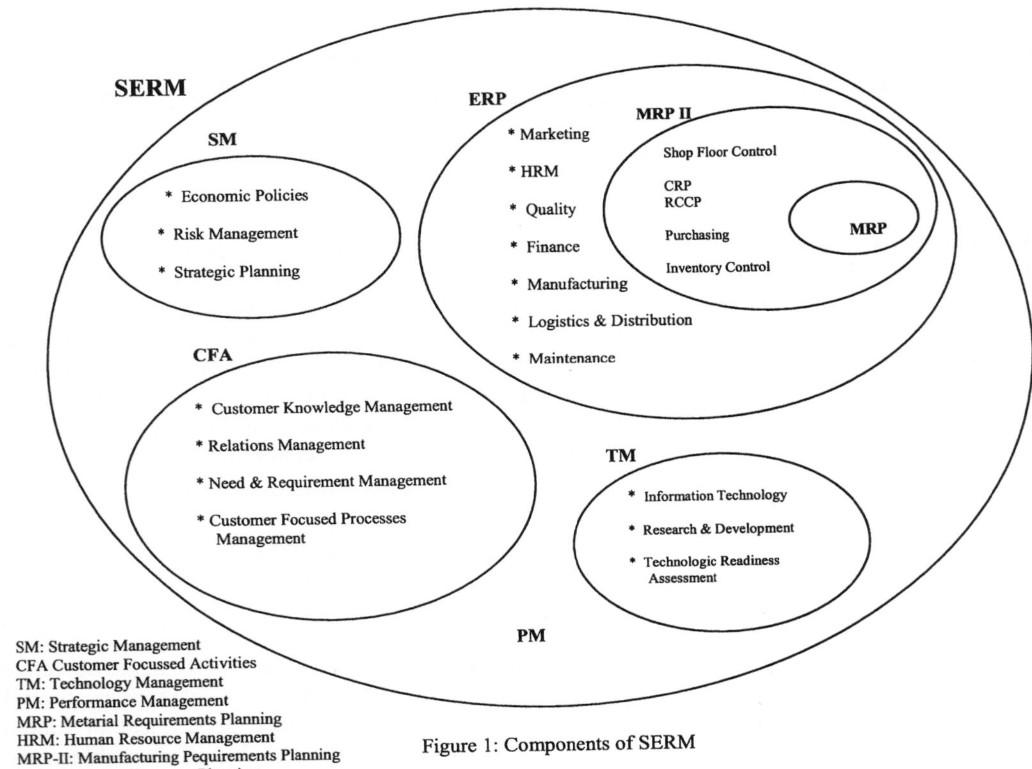

Figure 1: Components of SERM

As seen in the Figure, SERM consists of basic MRP [4] and MRP II (Manufacturing Resource Planning) [5] as well as ERP [6] capabilities. In addition to ERP, SERM provides a way to handle Strategic Management (SM) [7] that includes Strategic Planning, Economic Policies, Risk Management; Customer Focused Activities (CFA) including traditional Customer Relations Management, Customer Focused Process Management, Need&Requirement Management, Customer Knowledge Management; Technology Management (TM) that includes Research and Development (R&D), Information Technology and Technologic Readiness Assessment; and Performance Management (PM). Basic definitions of each component are given below. Detailed information and possible comparison with other existing planning systems can be found in Oztemel and Korkusuz [3].

This study provides a general framework for creating a fully automated SERM. The automation of SERM functions is intended to be performed by so called intelligent agents which are capable of reasoning and decision making using artificial intelligence technology. The following section first describes the intelligent agent technology briefly and then agents which are capable of performing SERM are explained.

3. Intelligent agents

Intelligent agents can be defined as autonomous systems perceiving the events from the environment, reasoning about those events and responding accordingly [8]. In order to perceive and act (reason) about the events, the agents should be equipped with suitable sensors and effectors as well as with the domain knowledge. That means, the agent should perform whatever action is expected from it to maximise its performance. The actions should be decided on the basis of the evidence provided by the percept sequence and whatever built in knowledge the agent possesses. Figure 2 provides a generic architecture for an intelligent agent.

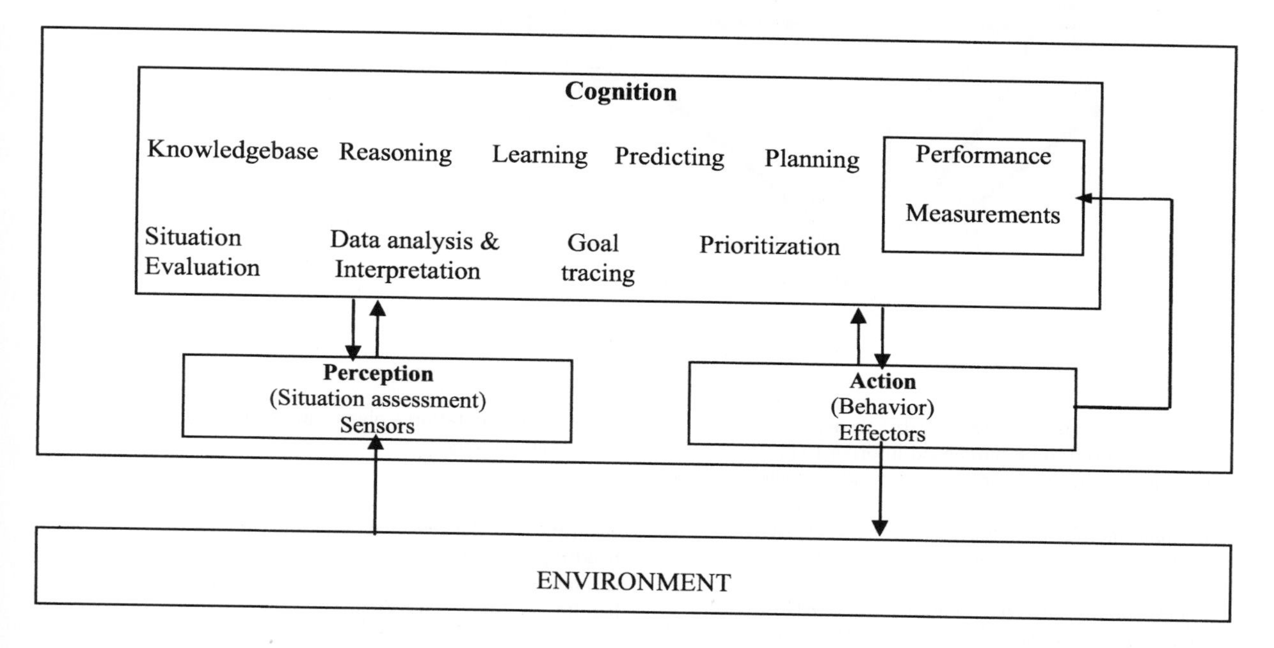

Figure 2: A generic architecture of an agent

As shown in the figure an agent may be comprised of;

- *Perception module* to perceive the events from the environment
- *Cognition module* to reason about perceived information
- *Action module* to act according to the response produced by the cognition module.

This generic architecture is found to be reasonable for SERM agents which are part of fully automated SERM environment. Agent based SERM environment is created using several agents following the architecture defined above. Note that each agent is responsible only for a specific functionality of SERM.

4. SERM agents

SERM agents can be defined as intelligent agents responsible for carrying out SERM related activities in an intelligent manner. Each agent is handles the activities it is assigned to. The agents can communicate knowledge to others to help them to make decisions. The knowledge protocols developed by Oztemel and Tekez [9] can be used to easy the communications.

Figure 3 shows SERM agents and their relations with each other.

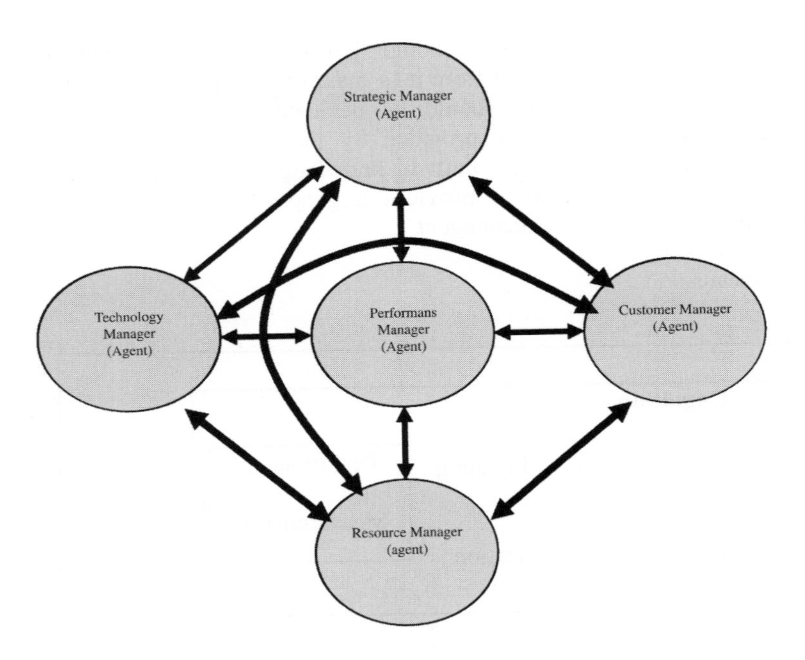

Figure 3: SERM Agents

- ***Resource Manager*** is responsible to define the list of material to be purchased in order to meet production orders defined through master production plan. It does not only define the list of material but also the amount of each item required. The agent is to handle the time of purchasing items obeying the lead times. This agent should also be responsible to perform human resource analysis, financial analysis as well as logistics, inventory and capacity analysis to make sure that the plant is able to respond the master production plan.

- ***Customer Manager*** is responsible for handling interaction with the customers and customer relations. It performs traditional CRM activities and more importantly manages customer focused processes. Main activities include;
  - Data collection, customer handling knowledge bank, knowledge elicitation and dissemination as well as some remedies for customer leverages for
  - Developing customer relations, recommends activities for customer recognition and involvement, performs customer classification and handles exchange of information as ell as customer satisfaction.
  - Identifies current and future customer needs to support strategic Manager
  - Sets up all processes towards customer requirements and needs. Process activities, performance indicators and implementation plans are organized accordingly.

- ***Technology Manager*** is responsible for ensuring that the organizations are capable of handling customer requests through technological readiness assessment. It will also handle the information technology developments in line with the corporate strategies and functional goals. It should handle;
  - ***Research and Development (R&D)*** capabilities such as generating new products, defining new processes, improving existing products with new features, improving exiting processes, producing new technologies and improving existing technologies to be taken into account and;
  - ***Technologic Readiness Assessment (TRA)*** to check-up the level of technology use and identify technological investments accordingly. This agent is based on so called capability maturity model [10].
  - ***Information Technology (IT)*** to provide effective information flow between managerial with. As complexity of products and processes are increasing everyday, it is vital importance for the enterprises to utilized information technology for efficient and effective information flow between product/service producing units as well

as managerial activities. Management information systems, Decision support systems and other similar information systems to be managed effectively. SERM requires internet or intranet based information management systems supported by efficient infrastructures. SERM based software should have distributed architecture and distributed database management systems. Artificial intelligence based technologies such as expert systems and neural networks could be utilized depending upon the application. Related software processes and procedure are to be defined for effective utilization of SERM.

- ***Strategic Manager (SM)*** is responsible to bring all information together to perform **Strategic Planning** in order to define corporate level strategies, related functional goals satisfying organizational mission and vision. It makes sure that the company performs market analysis, internal an external business analysis, organizational SWOT analysis, and takes customer, employee, supplier needs and requirements, as well as technological developments into account. The following is the basic activities of this agent. It should take **Economic Policies** of the government and market fluctuations effecting the organizations into account while forming the strategies. Possible risks also need to be taken into account. **Risk Management** module of SERM requires management to identify potential risks for each departmental unit (or processes for those enterprises implementing process management). This may involve managers to take immediate actions in case of a loss or risks occuring.

- ***Performance Manager (PM)*** is responsible for monitoring and managing the performance of the implementation of strategies and the success of achieving strategic goals. This should also define the performance of resource utilization and planning.

5. Conclusion

SERM is an enterprise level extended management tool for the managers. The agent technology is proven to be able to handle complex problems and generate autonomous decision making systems. The enterprise level decisions require all activities to be considered. Each activity may affect others both negatively and positively. The agents may easily handle the effects of each agent on others. This study proposes a multi agent architecture framework for Intelligent Strategic Enterprise Resource Management. The core of the proposed architecture is Performance Manager which is directed by the Strategic Manager. The knowledge exchange between agents can be carried out using a knowledge network which is able to transfer knowledge forms and messages using a specifically designed knowledge schema provided by Oztemel and Tekez [9].

Acknowledgement

The authors would like to thank Prof. Harun Taşkın for his valuable inputs during the discussion and also thanks I*PROMS network of excellence for creating such a discussion forum.

References

[1] www.imti21.org (Available on 10 April 2005)

[2] Anderegg T., 2004, "MRP/MRPII/ERP/ERM-Confusing terms and definitions for a Murky Alphabet Soup", http://www.cibres.com/articles/MRP-MRPII-ERP-ERM-Confusing-Terms-And-Definitions (available on 13 May 04)

[3] Oztemel E., Korkusuz Polat T., 2004, "A General Framework for SERM (Strategic Enterprise Resource Management)", In Proceedings of 4th International Symposium on Intelligent Manufacturing Systems, Sakarya, Turkey, pp. 714-719.

[4] Moustakis V., January 2000, "Material Requirement Planning MRP", http://www.adi.pt/docs/innoregio_MRP-en.pdf (available on 13 June 05)

[5] Ip W.H., Kam K.W., 1998, "An Education and Training Model for Manufacturing Resources Planning" Int. J. Engng Ed. Vol. 14, No. 4, p. 248±256, Printed in Great Britain. http://www.ijee.dit.ie/articles/Vol14-4/ijee1021.pdf (available on 01 April 05)

[6] Berchet C., Habchi G., 2005, "The implementation and deployment of an ERP system: An industrial case study", Computers in Industry, Article in Press, http://www.sciencedirect.com/science?_ob=MImg&_imagekey=B6V2D-4GDBT2B-1-Y&_cdi=5700&_user=746115&_orig=search&_coverDate=06%2F15%2F2005&_sk=999999999&view=c&wchp=dGLbVzz-zSkzV&md5=ee62081994d4c0593e0871c852c1200c&ie=/sdarticle.pdf (Available on 12 May 2005)

[7] Wheelen T.L., Hunger J.D., 2004, "Strategic Management and Business Policy", Ninth Edition, Pearson Education Inc., New Jersey

[8] Russell, S. J., Norvig, P., 1995, "Artificial Intelligence: A Modern Approach", Prentice-Hall International Inc., New Jersey

[9] Oztemel, E., Tekez, E. K. 2004 "A Reference Model for Intelligent Integrated Manufacturing Systems: REMIMS", In Proceedings of IMS International Forum on Global Challenges in Manufacturing, Cernobbio, Italy, pp. 1260-1267.

[10] Aouad G., Cooper R., Kagioglu M., Hinks J., Sexton M., 1998, "A Synchronized Process / IT model to support the co-maturation of process and IT in the construction sector", Construction Informatics Digital Library, http://itc.scix.net (Available on 15 May 2005).

Intelligent Production Machines and Systems
D. T. Pham, E. E. Eldukhri and A. J. Soroka (eds)

Performance analysis of two-machine continuous-flow model with delay

Iyad Mourani, Sophie Hennequin, Xiaolan Xie

INRIA/MACSI team and LGIPM
ISGMP-Bat. A. Ile du Saulcy
57045 Metz Cedex 1, France
{mourani, hennequin, xie@loria.fr}

Abstract.

This paper addresses the impact of transportation delays on the performance of single-product continuous flow production lines of two machines separated by one buffer. Operation-dependent failure model (ODF) is considered for the machines, i.e. a machine cannot fail if it is not working on a part. Volume to failure and machine repair times are exponentially distributed random variables. We prove that the high the delays, the less the throughput of the transfer line. Numerical results show that this result holds under more general conditions.

1. Introduction

To guide managerial decision making and choose between alternative courses of actions, the manager has to have suitable criteria. At the strategic level, these criteria are usually based on rather broad organizational goals. Nevertheless, for tactical and operational decisions, these general goals have to be converted into more tangible performance criteria that are susceptible to measurement and tracking over time. In addition, top management may set specific objectives for these performance measures that managers and workers have to aim at achieving. The usual performance measures in the manufacturing context relate to, throughput rate, capacity of a system, in-process inventory, cycle time, responsiveness to customer, quality of product and flexibility, which contains product flexibility and process flexibility. Usually, at least for low- and middle-level management, throughput rate is the dominant performance criterion.

Excellent literature surveys on performance evaluation of production lines can be found in Buzacott and Shanthikumar [1], and Dallery and Gershwin [2]. Two kinds of flow models are considered in the literature: discrete flow models and continuous flow models. The former are usually considered more realistic for discrete manufacturing, but the discrete processing of parts makes the performance analysis difficult particularly when using simulation. The latter provide a good approximation of material flows and makes the performance analysis more efficient. For the continuous-flow case, performance analysis of transfer lines has been largely studied (see [2, 3, 7, 10, 12]).

Though continuous flow models provide an interesting means to reduce the complexity inherent in traditional discrete parts modeling, existing continuous flow models do not take into account some important characteristics of manufacturing systems such as production lead-times and transportation delays. Many manufacturing processes have important delays in the material flow, such delays occur in oven processes (e.g. semiconductor diffusion), drying processes and testing. These delays usually have great impact on performance measures such as customer response time and work-in-process. Unfortunately, most existing continuous flow models ignore these delays.

There exist only two exceptions: *i*) The works of Van Ryzin et al [11], explicitly addressed the impact of delays for optimal flow control of job shops in order to minimize the discount and infinite-horizon average cost. A heuristic control policy for a flow shop with delay is derived using theoretical arguments and approximations. They considered a system composed of a set of flexible unreliable machines and neglected setup changes. They showed that the approach of Kimemia and Gershwin [6] for flow control of failure-prone job shops could be extended to systems with large processing times (i.g. delays), where by large they mean processing times that are not negligible relative to the interarrival time of parts (inter-loading time) or the relative frequencies of other events such as demand changes, failures and down-times. In the flow control model, delays are modeled using delay-differential equations in place of ordinary differential equations. The optimal control is determined by using the delay approximation technique due to Repin [9] and extended by Hess [4] and Hess and Hyde [5] to derive practical controls for linear-quadratic systems with delay. *ii*) The works of Mourani et al [8], which extend the model of van Ryzin et al [11] and propose a continuous Petri net model with delays for performance modeling and optimization of transfer lines. The new model is more realistic than classical continuous flow models yet keeps the simplicity and analyticity of continuous flow models.

In this paper we use the Petri net model proposed by Mourani et al [8] for the performance analysis and show that the throughput of the production line is a decreasing function of the delays. More precisely, we consider a continuous-flow model of transfer line composed of two machines and one buffer of finite capacity. Material transfer between machine M_1 and its downstream buffer B takes a finite amount of time . Machines are subject to operation-dependent failures and volume to failure and machine repair times are exponentially distributed random variables. Simulations show that this result stills hold for operating volumes to failure and times to repair with general distribution.

The rest of the paper is organized as follows. Section 2 addresses the continuous-flow production line including the notion of delays. Section 3 presents the performance analysis. Numerical results are given in Section 4. We conclude the paper in Section 5.

2. Production line with delay

Consider a continuous flow models of single-product transfer line constituted of two machines (M_1, M_2) separated by one buffer B of finite capacity H.

Van Ryzin et al [11] extended, by recognizing the importance of delays, the classical continuous flow models to take into account delay and considered the optimal flow control of a two-stage failure-prone system with transfer delay τ between stage 1 and input buffer of stage 2. The following differential equations describe the dynamic of the system:

$$dx(t)/dt = u_1(t) - u_2(t),$$

$$ds(t)/dt = u_2(t\text{- }0) - d(t).$$

where $d(t)$ represents the known demand rate at time t and $s(t)$ is the surplus at time t.

This paper uses the model of Mourani et al [8] and show the impact of transportation delays on the performance of two-machine failure-prone production lines.

Material flows continuously from outside world to the first machine, then waits a transportation delay τ, before arriving to the buffer, then the second machine, where it leaves the system (see Fig.1).

Thus, a transportation delay is considered between the machine M_1 and its downstream buffer B. That means that the parts produced on M_1 would not arrive immediately in the buffer B.

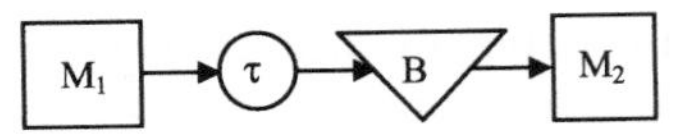

Fig. 1. Production line with a transportation delay between M_1 and B

Machines are subject to operation-dependant failures. This means that a machine can only fail while it is working on a part and so cannot fail if it is idle. Let $\alpha_i(t)$ as a failure state of machine M_i at time t with $\alpha_i(t) = 1$ if M_i is up and $\alpha_i(t) = 0$ otherwise. Each machine can be either up ($\alpha_i(t) = 1$) or down ($\alpha_i(t) = 0$). When it is up, it can be either working or idle (if it is starved or blocked). Machine M_i is starved at time t if one of the upstream machines is down and all buffers between this machine and machine M_i are empty. Machine M_i is blocked if one of its downstream machines is down and all the buffers between this machine and machine M_i are full. We assume that the first machine is never starved and the last machine is never blocked.

For machines subject to operation-dependent failures and for continuous flow model in which production rates can be reduced, we choose to use operating volumes to failure VTF_i to determine the failure times. Machine M_i fails when its cumulative production since last repair reaches VTF_i (see [10]). Time to repair is often used to characterize the reliability of a machine. Times to repair and operating volumes to failure (TTR_i resp. VTF_i) for each machine are generally distributed.

Further, in order to guarantee that the buffer capacity H of a buffer B is never exceeded, we restrict the total material in transit to B and in B to be smaller than H.

3. Performance analysis

In this section, we prove that the throughput of two-machine production line is a decreasing function of the delays.

Consider two transfer lines (system I and II) with two different transportation delays (τ resp. τ'). The same continuous Petri net of Mourani et al [8] is used here for representing transfer lines with transportation delays (see Fig. 2).

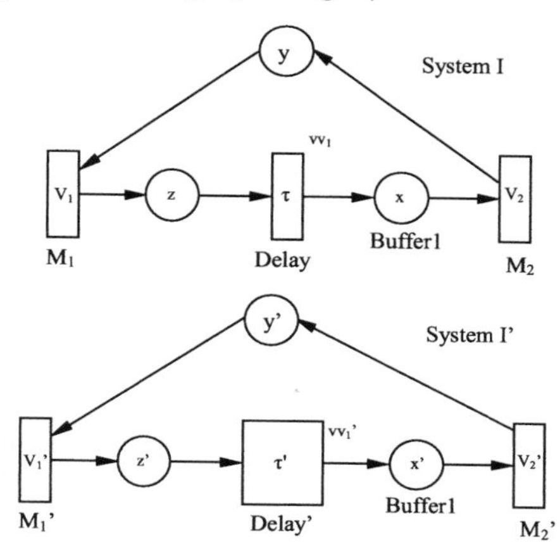

Fig. 2. Two-machine transfer lines with different transportation delays

The notations are:

- $v_i(t)$: the processing rate of machine M_i at time t, which have a constant maximal value U_i,
- $v_i'(t)$: the processing rate of M_i after the delay τ_i at time t,
- $x_i(t)$: the number of parts in the buffer B_i at time t for machine M_i,
- $y_i(t)$: the remaining capacity of buffer B_i at time t,
- $z_i(t)$: which represents the delayed part at time t,
- τ_i: the transportation delay, which the material flow waiting before arriving from the machine M_i to its downstream buffer

Initially the both systems I and I' are initially identical (i.e. $v_1(0)= v_1'(0)$, $v_2(0) = v_2'(0)$, $vv_1(0) = vv_1'(0)$ and $H(0)= H'(0)$), except the delays, where τ, τ' represent the transportation delay in system I (resp. system II) with $\tau, \tau' \geq 0$ and $\tau' \geq \tau$.

$pv_i(t)$: production volume produced by machine M_i at time t.
VTF_i: production volume to failure.

Theorem 1: if $\tau' \geq \tau$ so the throughput of system I is greater than or equal to that of system I', i.e., If $\tau' \geq \tau \Rightarrow Pr^I \geq Pr^{I'}$.
Where Pr is the total production volume manufactured by the line, or by the last machine in the line.

Proof:

We consider that both systems I and I' are coupled, thus the failures (resp. repairs) occur at the same production volume VTF_i (resp. same time TTR_i) for the both systems.

a) Let t_1 the time that the buffer x' becomes empty, i.e. $x'(t_1) = 0$, with hypothesis $v_i \geq v_{i+1}$, it is not possible to have: $y(t_2) = 0$, where $t_2 < t_1$, only if machine M_2 fails

1. $Pr^I = Pr^{I'}$, $\forall\ t \in [0, t_1)$ (coupling between systems I and I'),

2. At time t_1: $v_2' = vv_1' = 0$ (M_2' is starved) so $pv'_2(t_1) < pv_2(t_1)$ so M_2 fails before M_2' because ($pv_2(t_1) = VTF_2$) i.e. M_2 produces the production volume to failure before M_2'.

3. Let tf_2, tf_2' be the time when the production volume to the failure of M_2 (resp. M_2'); is achieved.
   So M_2, M_2' fail when producing the same production volume i.e., $pv'_2(tf'_2) = pv_2(tf_2)$ (coupling between systems I and I') and get repaired at the same time.
4. But $tf_2' > tf_2$ so at time tf_2 where M_2 breaks down, the production volume produced by M_2 is greater than it produced by M_2' i.e. at tf_2, $pv_2(tf_2) \geq pv_2'(tf_2)$
5. Let tr_2, tr_2' the time corresponding to the end of repair of M_2 (resp. M_2') after t_1.
   As $tf_2' > tf_2$, So at time tf_2 we have $Pr^I > Pr^{I'}$;
   If $tf_2' < tr_2 \Rightarrow Pr^I = Pr^{I'}$, otherwise $Pr^I > Pr^{I'}$.

b) Let t_2 the time that the remaining capacity in system I becomes empty, i.e. $y(t_2) = 0$, where $t_2 < tr_2$, then $v_1(t_2) = v_2(t_2) = 0$ (M_1 is starved and M_2 is down),
On the other hand, $0 < v'_1(t_2) \leq U_{1,max}$ (maximal production speed of machine M_1 and M_1').

For t_2 where $tf_2' \geq t_2$;

- At the worst of cases t_2 occurs at time tf_2 so the maximal quantity produced by M_1' is given as follows: $U_{1,max}(\tau' - \tau)$.

Consequently, at $t = \tau$, $0 < vv_1(\tau) \leq U_{1,max}$ but, $vv_1'(\tau) = 0$ and $0 < vv_1'(\tau') \leq U_{1,max}$.

So $\forall\, t \in [\tau, \tau']$ the difference in production volume between system I (with delay τ) and system II (with delay τ') is at the maximum $U_{1,max}(\tau' - \tau)$, thus at the worst of cases the delay of production in system II is given by $U_{1,max}(\tau' - \tau)$.

After t_2 the buffer capacity in system I would not become empty ($x = 0$) before that of system I' ($x' = 0$), except if M_1 fails, in this case and because of coupling M_1' fails too.
Consequently, $pv_1'(t) \geq pv_1(t) \Rightarrow v_1'(t) = v_1(t)$, $\forall\, t > t_2$
Then, $pv_1'(t) = VTF_1$, i.e. M_1' fails before M_1 hence it is not possible that the buffer becomes empty at time $t_3 > t_2$, i.e. $x(t_3) = 0$, where $t_3 > t_2$.
As a result, $Pr^I \geq Pr^{I'}$.

c) Let t_4 time where y' becomes empty, i.e. $y'(t_4) = 0$.
So $v_1'(t_4) = v_2'(t_4) = 0$ (M_1' is starved and M_2' is down), thus the repair of M_2' occurs at time $t = tr_2 + (\tau' - \tau) = tr_2' \Rightarrow t \in [tr_2, tr_2']$,

Hence, M_1 produces whenever M_1' is down
M_2 produces whenever M_2' is down

As a result, $Pr^I \geq Pr^{I'}$.

Let t_k time such as: $Pr^{II}(t_k) \geq Pr^I(t_k)$ so $pv_2'(t_k) > pv_2(t_k)$ (we have proved above that impossible to have $pv_2'(.) > pv_2(.)$), and that occurs if we have two conditions simultaneously:

1. There is no production delay, i.e. $x'(t_k) = x(t_k)$,
2. $v_2'(t_k) > v_2(t_k)$ and that could be happened if
   a) M_2 is down,
   b) The buffer is empty ($x = 0$) and the production speed after the delay is null ($vv = 0$) so M_2 is forced down i.e. $v_2 = 0$.

We will discus the two previous cases:
Case (a): if M_2 is down whenever M_2' is up, thus the production volume produced by M_2 is greater than it produced by M_2', i.e. $pv_2(.) \geq pv_2'(.)$ and so the production speed of M_2 is greater than it of M_2', which contradicts that is no production delay mentioned above.
Case (b): if $x = 0$ and $x' = 0$ (no production delay) so the production speed of M_2 is the same of M_2'.

Consequently, time t_k does not exist.

Q.E.D.

In the following section, we use the original Petri net given in [8] to show that the above results still hold for general distributions.

4. Numerical results

We consider in this section production lines with identical buffer and machines except the delays.

The parameters used in simulation are summarized in Table 1.

Table 1
Simulation data

$MVTF_i$	$MTTR_i$	U_i	τ	Simulation time
100 Parts	20 Time units	1 Part / time unit	1 Time unit	10^7 Time units

where:

- $MVTF_i$: Mean porduction volume to failure for M_i with $i = 1, 2$,
- $MTTR_i$: Mean time to repair for M_i with $i = 1,2$,
- U_i: Maximum processing rate for M_i with $i = 1,2$,
- τ: Transportation delay for machine M_1.

All random variables have the same type of distribution. Two distributions are considered.
Case (i): All random variables X follow an exponential distribution, i.e.

$$P[X \leq x] = 1 - exp(-x/m)$$

with $m = E[X]$ as defined in Table 1.
Case (ii): All random variables X follow a Weibull distribution with

$$P[X \leq x] = 1- exp(-x^2/\beta^2)$$

and β is chosen such that the means $E[X]$ are those of Table 1. More precisely, $\beta = 2m/\sqrt{22/7}$ where m is the related mean of Table 1.
We also vary the buffer size H_i to show the evolution of the impact of delay.

A. Case of exponential distributed random variables

In this case, operating volumes to failure and times to repair are exponentially distributed with rate w and μ respectively, i.e., mean volume to failure $MVTF = 1/w$ and mean time to repair $MTTR = 1/\mu$.

First, we vary the buffer capacity H and the delay τ to consider the evolution of the impact of delay. The throughput rate versus delay τ is plotted for two-machine line case (see Fig. 3).

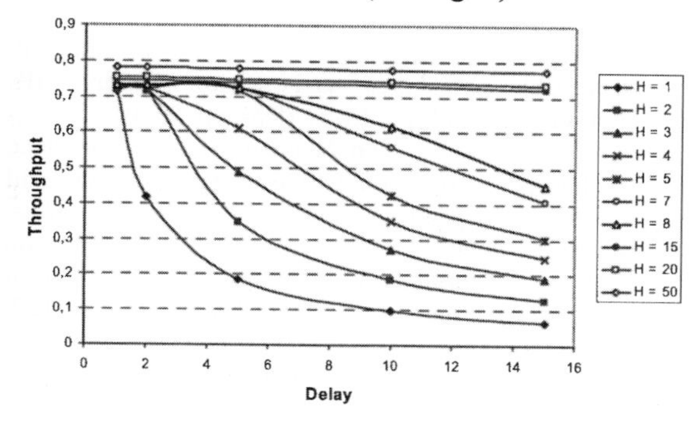

Fig. 3. Throughput rate versus delay τ_i for 2-machine case and exponential distribution

We notice from the figure presented above, that, in two-machine case, the impact of delay on the throughput rate of the production line is important for small values of buffer capacity H = 1, 2, 3 and 4 parts. For these values of H, increasing the delay significantly reduces the throughput rate. For medium buffer capacity such as H = 5, 7 and 8 parts, the throughput stays relatively constant until the delay reaches between 4 and 6 time units. When delay reaches these values the throughput begins to decreases. For large buffer capacity H = 15, 20 and 50 parts, the impact of delay on the throughput rate is very small, and the throughput stay nearly constant.

Further, it can be observed that the impact of delay is the greatest for H = 1, and it decreases as H increases and becomes null as H tends to infinity.

This result could be explained as follows. We consider the influence of both buffer capacity and delay on the throughput rate. On one hand, the impact of buffer capacity on the throughput strongly depends on the blocking/starving probability of machines. The higher the blocking/starving probability, the greater the impact of buffer capacity. Further the blocking/starving probability is a decreasing function of the buffer capacity H and becomes null when H tends to infinity (see Xie *et al* [13]). On the other hand, increasing the delay would delay the arrival of material flows at downstream machine and hence increases the starving probability of downstream machine, which again leads to lower throughput rate.

B. Case of random variables of Weibull distributions

Consider a two-machine line in which, operating volumes to failures and times to repair all follow a Weibull distribution.

We first vary the buffer capacity and then vary the delay. Simulation results for two-machine line are illustrated in Fig. 4 and 5.

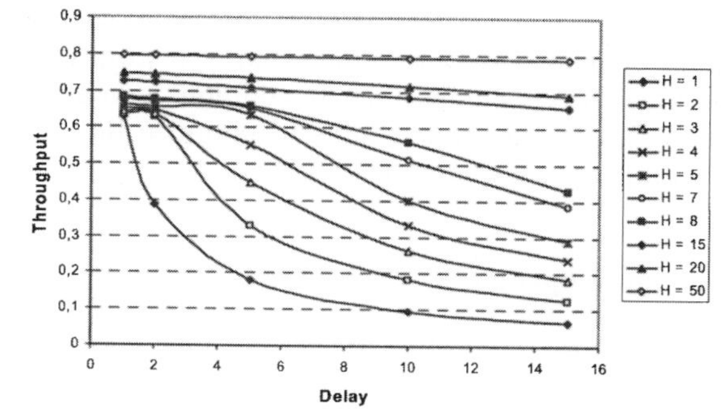

Fig. 4. Throughput rate versus delay for 2-machine case and Weibull distribution

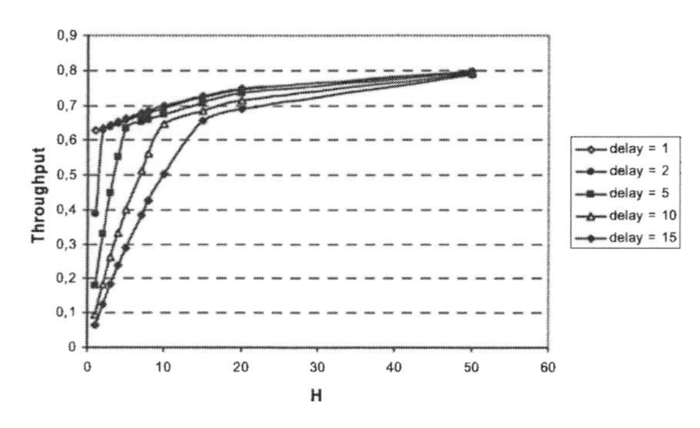

Fig. 5. Throughput rate versus buffer capacity for 2-machine case and Weibull distribution

The evolution of the impact of the delay is the same as that observed in the case of exponential distribution, i.e. the impact of delay on the throughput rate decreases as the buffer size increases and becomes null when the buffer size tends to infinity.

5. Conclusions

The performance analysis of single-product two-machine continuous-flow production line with transportation delays have been addressed in this paper. Machines are subject to operation dependant failures, i.e. a machine can only fail when it produces. Volumes to failure and times to repair are exponentially distributed. We have proved that the throughput of the transfer line is a decreasing function of the transportation delays. The simulations results show that this result stills hold using more general distributions for the volumes to failure and times to repair.

In further works, it would be interesting to analyse the performance evaluation in case of time dependant failures (machines can fail at any time) and more general manufacturing systems such as transfer lines with *N* machines and *N*-1 buffers and assembly/disassembly lines. In addition, it would be attractivé to use these results to optimize the lines considering transportation delays.

Acknowledgement

INRIA is partner of the EU-funded FP6 Innovative Production Machines and Systems (I*PROMS) Network of Excellence.
http://www.iproms.org

References

[1] Buzacott JA and Shanthikumar JG. Stochastic Models of Manufacturing Systems. Englewood Cliff, NJ: Prentice-Hall, 1992.

[2] Dallery Y and Gershwin SB. Manufacturing flow line systems: a review of models and analytical results. Queuing Systems, 1992, vol. 12, pp. 3-94.

[3] David R, Xie X and Dallery Y. Properties of Continuous Models of Transfer Lines with Unreliable Machines and Finite Buffers. IMA Journal of Mathematics Applied in Business and Industry, vol. 6, 1990, pp. 281-308.

[4] Hess RA. Optimal control approximation for time delay systems. AIAA Journal, 1972, 10 (11).

[5] Hess RA and Hyde JG. Suboptimal control of time-delay systems. IEEE Transaction on Automatic Control, 1973, 18, (6).

[6] Kimemia J and Gershwin SB. An algorithm for the computer control of a flexible manufacturing system. IIE Transactions, 1983, 15, (4).

[7] Kroese DP. On the decay rates of buffers in continuous flow lines. Methodology and Computing in Applied Probability, 2000, vol. 2, pp. 425-441.

[8] Mourani I., Hennequin S and Xie X. Continuous Petri nets with delays for performance evaluation of transfer lines. Proceedings of IEEE International Conference on Robotics and Automation (ICRA), Barcelona, Spain, pp. 3732-3737, April 18-22, 2005.

[9] Repin IM. On the approximate replacement of systems with lag by ordinary dynamic systems. Jouranl of Applied Mathematics and Mechanics, 1965, 29, pp. 254-264.

[10] Suri R and Fu BR. On using continuous flow line to model discrete production lines. Discrete Event Dynamic Systems: Theory and Applications, 1994, vol. 4, pp. 127-169.

[11] Van Ryzin GJ, Lou SXC and Gershwin SB. Scheduling job shops with delay. Int. J. Prod. Res., 1991, vol. 29, No. 7, pp. 1407-1422.

[12] Xie X. Performance Analysis of a Transfer Line with Unreliable Machines and Finite Buffers. IIE Transactions, vol. 25, No.1, 1993, pp. 99-108.

[13] Xie X, Mourani I and Hennequin S. Performance evaluation of production lines subject to time and operation dependent failures using Petri Nets. Proceedings of Eighth International Conference on Control Automation, Robotics and Vision (ICARCV), Kunming, China, pp. 2122-2127, December 2004.

Intelligent Production Machines and Systems
D. T. Pham, E. E. Eldukhri and A. J. Soroka (eds)

RULES-IS: An Immune-network inspired machine learning algorithm

D.T. Pham, A.J. Soroka

Manufacturing Engineering Centre, Cardiff University, PO Box 925, Cardiff, UK

Abstract

This paper proposes a novel algorithm for the extraction of rule sets. The extraction technique adopted is inspired by the immune system and its pattern recognition capabilities. The concept of an immune network is used which enables the algorithm to generate rule sets in an incremental manner. Such an immune network also enables the algorithm to learn from training examples that have not had classes specified. The algorithm developed has been tested on several different example data sets and has shown itself to be at least comparable to existing algorithms.

1. Introduction

This paper presents an algorithm for the extraction of rule sets that is inspired by the learning properties of the immune system.

The immune mechanism is a highly complex biological system for the identification and elimination of material that is foreign to the human body. When foreign organisms (antigens) invade the body they are identified by antibodies which bind with them, marking them for elimination. The agents involved in the elimination process include Lymphocyte cells that mediate the immune responses, Phagocytes which ingest these marked antigens and agents of the complement system - enzymes which attack the antigens [1].

To be able to perform such a task, it is considered that the immune system must possess the ability to learn about its surroundings, and recognise antigens it has encountered before [2].

Rather than being an exact model of the immune system, the proposed algorithm, RULES-IS (Immune-System Inspired Rule Extraction System), is based upon concepts related to the immune system. A widely adopted theory of how the immune system works is the Immune Network Theory presented in [3]. This has inspired work in the artificial intelligence domain [4, 5, 6, 7, 8].

The most significant difference between RULES-IS and other immune-system-based learning algorithms is that only the learning stage is related to the immune system, as opposed to both learning and identification as in [4, 6, 9, 10]. This opens the potential for more widespread exploitation of immune learning systems and techniques as the algorithm is not tied to a platform-dependent solution for classification.

2. Immune-System-inspired Rule Extraction Algorithm

2.1 Representation of Antibody and Antigen

Attributes within an example in a data set are represented by elements within the antibody and antigen. Due to the application area, additional types of elements are used within RULES-IS, namely, Range and Wild Card as shown in (Fig 1).

Antibody:

Leak	#1.1>9.1<	96.5	shade	*

Content:

class	range 1.1≤#≤9.1	number	string	wild card

Fig. 1 - Example Antibody

Range enables the antibody element to bind with an antigen element that is within a specified numerical range to deal with continuous-valued attributes. The antibody element in (Fig. 1) can bind with the antigen element if the value of the antigen is between 1.1 and 9.1. The Wild Card element enables that element of the antibody to bind with an antigen element of any type and value.

2.2 Immune Network

RULES-IS employs a form of Immune Network to store the antibodies and antigens presented to the system. The structure of the network (consisting of multiple layers) used is shown in (Fig. 2). The first layer (antibody layer) - similar to the Recognising Set in [3] - is occupied by what are termed active antibodies. They are used for matching with antigens and generating the final rule set. However, the network differs from that in [3] in that it only considers direct antigen-antibody relationships as opposed to antigen-antibody-antibody interdependencies within an immune network [3,11]. As using a full immune network can prove to be computationally expensive [9].

The other layers within the network consist of antibody-antigen pairings - the pairs used to form the antibody in the previous layer. Layers furthest away from the active antibodies are the oldest (new antigen-antibody pairs are appended to the head of the list).

The use of an immune network facilitates the implementation of incremental learning features as such algorithms need some method by which they can store their state between different learning sessions. By providing a data structure that can be saved and reloaded, it becomes possible to use the algorithm in an incremental manner.

In the immune network shown in (Fig. 2) when an antigen is presented to the system the antibody with the best match binds. Provided that the classes of the antibody and antigen are the same and that the quality of the bind exceeds a given threshold. If no such antibody exists then the antigen is used to form an antibody.

2.3 Matching Algorithm

The matching algorithm performs the task of calculating the quality of the match between an antibody and an antigen. The algorithm uses different methods for calculating the match quality depending on the type of data being represented by the element.

The matching algorithm uses direct bit-to-bit matching, as opposed to the bit-shifting approaches discussed in [2, 4]. It was found that for bit-shifting approaches discrete data sets could be a problem in that the algorithm could incorrectly identify examples presented to it [12].

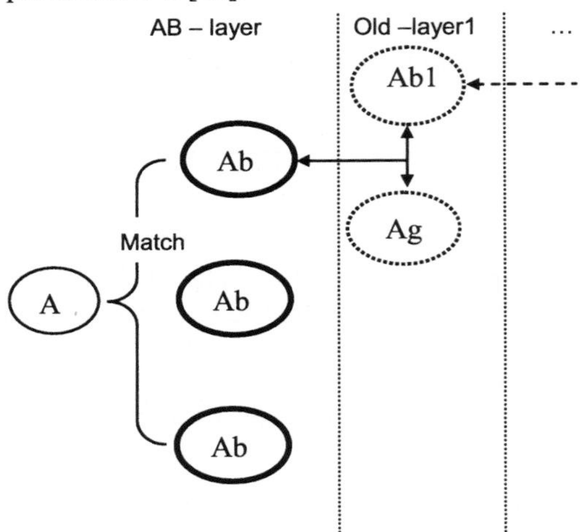

Fig. 2. Immune Network Structure

As in some other systems, such as [13], the antibodies in RULES-IS may contain wild-card elements. In RULES-IS a match between a wild-card element and another element is considered to be of equal quality to a match between two identical elements. The method employed in [13] (wild-card match is of lower quality) could have the side effect of generating more rules. Also it is possible that antigens which bind 100% with an appropriate antibody containing wild cards in RULES-IS would bind instead with another antibody or possibly not bind at all.

The matching algorithm presented in (Fig. 3) also takes into account the threshold so that, as more incorrect bindings occur, the threshold is raised and consequently the quality of match required for an antibody to bind with an antigen is increased.

2.4 Antibody Generation via Combination

This process generates a new antibody by combining an antigen and the antibody that binds with it. The process can be considered analogous to the primary response of the immune system which forms new antibodies in an attempt to deal with a previously unseen antigen. In RULES-IS, if the nth elements of the antibody and antigen are identical then, in the new antibody, the element will remain the same. If the nth elements of the antibody and antigen do not match and if they are both strings, a wild-card operator is used to replace them. This has proved to be useful in [13] for the recognition of promoter sequences. If the two elements are different numbers then a range is generated. This removes the need for quantisation levels to be specified by the user.

This new antibody is then checked against the other antibodies within the system to ensure that there is no inconsistency, thus preventing the antibody from classifying with a score of 100% an antigen that belongs to a different class. If there is no inconsistency then the test antibody which was generated replaces the original antibody, and the latter along with the antigen is placed in the previous layer. This process will reduce, if not eliminate altogether, the possibility of misclassification.

As can be seen in the algorithm shown in [Fig. 4], this system is capable of dealing with data instances that are not 'classified', i.e. the case where the class of the instance has not been specified. This allows the study of whether an algorithm based upon the immune system is capable of generating accurate rule sets when the classes of a proportion of the instances are not known.

The generation process involves several stages: generation of a new antibody, testing of this new antibody to ensure it does not misclassify any of the antigens already presented, and placing the antibody into the network if it is suitable.

Figure 4 illustrates the process for the generation of a new antibody. If, as in the first elements of the antibody Ab and antigen Ag, the two strings do not match a wild card is used in the resulting antibody. If, as in the second elements of the antibody and antigen the value of the antigen element is outside the limits of the range specified in the antibody element, then the range is expanded to encompass the value of the antigen element. The third elements illustrate that, when two numerical elements do not match, a range is generated which covers the two numbers. The fourth elements are two identical strings and so the element in the new antibody is the same string. The penultimate element in the antibody is a wild card and therefore the value of the corresponding element in the new antibody remains a wild card. The final elements are numbers of identical value and so the element in the new antibody takes the same value.

2.5 Antibody Generation from Antigen

Under certain circumstances antibodies cannot be generated using the process described in the previous section and have to be created directly from the antigen. Because RULES-IS does not perform mutation, it therefore cannot generate a new antibody from existing ones, as a real immune system would do.

The need to generate new antibodies directly might arise for different reasons. First, the combination procedure sometimes cannot produce an antibody because the new antibody will cause misclassification. Second, the are situations when no antibody matches the antigen sufficiently or no matching at all is achieved with the existing antibodies.

The process of generating the new antibody is straightforward in that the antibody is merely a clone of the antigen. The new antibody is therefore able to classify that particular antigen. The antibody is then placed in the network.

Ab

Val.A	#1.2>2.4<	3.1	X	*	1

Ag

Val.B	2.5	4.1	X	Z	1
⇓	⇓	⇓	⇓	⇓	⇓

New Ab

*	#1.2>2.5<	#3.1>4.1<	X	*	1

Fig. 4. - Creation of new Antibodies via Combination

However, if the antigen happens to have been for an example that has not had a class specified for it, then it is put back into the training data set, so that it can be reassessed when there are more antibodies in the system. This is only likely to happen when there are no antibodies in the system.

2.6 Decomposition

Antibody decomposing is a feature not present in a real immune system but it facilitates the generation of antibodies and helps to suppress antigens that misclassify. This technique also assists the algorithm in overcoming the problems mentioned in [13] concerning wild cards. It allows the match with a wild-card element to be equivalent to a full match and yet prevents the generation of antibodies that consist entirely of wild cards.

The primary factor contributing to the requirement for decomposition with RULES-IS is as with [13] in that the antibody contained too many wild cards caused either by many matches with different antigens, or by one very weak match that resulted in a significant proportion of the elements becoming wild cards. For this reason, a threshold is used during matching, but it alone does not guarantee to prevent this situation from arising.

If an antibody has a 100% match with an antigen, but the class of the antigen is not equal to the class of the antibody, this necessitates that the antibody be decomposed. The algorithm will essentially "drill-down" through the layers of the immune network until an antibody that does not produce an inconsistent classification is retrieved. The antigens that were bound to the antibody are then appended to the list of antigens that have yet to be presented to the system. The antibody is then placed in the active layer.

The decomposition feature is also very useful in incremental learning as it allows an antibody that misclassifies an antigen to be removed and a new antibody to be generated in its place. Once this has been done, the antigens that were appended to the list are re-presented to the system for re-incorporation into the network.

Following the decomposition, the threshold is incremented using (1). This equation will increase the threshold quickly when the current threshold ($trshld_{cu}$) is small relative to the number of elements (numEle), thus helping to prevent antibodies that will misclassify from being generated.

$$trshld_{new}=trshld_{cur}+(0.1*numElt - trshld_{cur}) \qquad (1)$$

2.7 Incremental Learning Memory

Various methods exist to perform incremental learning and therefore different approaches to the problem of memory - how to represent current knowledge so that it can be employed for future rule generation. Algorithms such as AQ15 [14] use a "perfect memory", where all training examples together with the rules generated are saved. Other algorithms, for example, RULES-4 [15], only a portion of the examples used to generate the rule set is saved.

Within the human immune system, approximately 5% of the least stimulated B-cells die off every day [2, 4]. This is the means by which the immune system controls the population of B-cells contained within the body. A system similar to this is therefore proposed for RULES-IS, in that 5% of the population of the immune network "dies" after rule set generation is completed. 5% of the Ab-Ag pairs are removed (if this proportion is sufficiently large) every time the algorithm is run incrementally. The pairs to be removed are always the oldest. Once the pairings have been removed, the network is saved to disk, to be reloaded the next time incremental learning needs to take place for this current rule set.

2.8 Overall Algorithm

First, the memory of the previous network, if it exists, is re-loaded. If not, the algorithm will work in batch mode. An antigen is then selected from the list of antigens, i.e. an example is chosen from the data set. If there are no antibodies then an antibody is generated from this antigen. The algorithm runs until there are no further antigens. If antibodies are present in the system, then the antigen will be compared against all of the antibodies.

Once this has been performed, the algorithm checks whether any of the best matches had a score of greater than zero. If not, then an antibody is generated from an antigen. As already mentioned this helps ensure that no antibodies consisting solely of wild cards can be generated. If the quality of the match is greater than zero, then the algorithm checks the number of antibodies that have effectively bound with the antigen to determine how many antibodies have a match score equal to the best match.

If only one antibody binds with an antigen, the algorithm checks to make sure that the classes of the antigen and antibody are not different. If they are identical then an antibody is generated via the combination process. After an antibody has been successfully created, the algorithm checks if there are any more antigens in the data set. If not, 5% of the older Antibody-Antigen pairs are removed, the trimmed network is saved and the process finishes.

When several antibodies have the same match score, they are arranged in a FIFO list. The first antibody in the list is taken and the classes of the antigen and antibody are compared. If they are identical or the antigen has no class specified for it then the algorithm again attempts to generate a new antibody from the antigen and antibody. If unsuccessful it then repeats the procedure until there are no further antibodies in the list. When all attempts have failed, a new antibody is generated from the antigen.

If the antibody and antigen classes are different,

then the antibody is decomposed as it has misclassified the antigen. The algorithm continues checking the matching antibodies against the antigen. When no valid antibodies can be found, the algorithm generates an antibody from the antigen.

Finally the rule set is extracted from the antibodies in the first or active layer.

3. Results and Discussion

3.1 Complete Data

In order to test the performance of RULES-IS, it was compared against other existing learning algorithms. Using criteria such as the number of rules generated, accuracy of the rules, average number of conditions, and ability to classify unseen data, the comparison was carried out on three standard data sets.

3.1.1Season Classification Problem

The season classification problem involves a small data set that contains eleven examples. This data set has been used by several researchers including [16, 17, 18] to examine the performance of various learning algorithms. The algorithms tested were C5.0, ID3, ILA, and RULES-3 and 4. Results for ID3 and ILA taken from [18].

As can be seen from the results shown in (Table 1) RULES-IS generates five rules which correctly classify the data set. This is identical to the results from the other algorithms tested with the exception of C5.0 which generates four rules plus one default rule (which can be considered as just another rule as it is the rule that fires if none of the other rules are true).

Table 1
Results for Season Classification Problem

Algorithm	Number of Rules	Accuracy of Rules
RULES-IS	5	100%
C5.0	4+1	100%
ID3	5	100%
ILA	5	100%
RULES-3	5	100%
RULES-4	5	100%

3.1.2 Flower Identification Problem

In the flower data set [16, 19], several classes are represented by only one example each, enabling assessment of the ability to deal with data sets with only a few examples of each class to learn from. The data set covers 25 classes. Therefore, the minimum possible number of 100% accurate rules is 25. (Table 2) shows some very interesting results. With respect to the number of rules, RULES-IS generates the same number as RULES-4 and one less than ID3 but four more than RULES-3. However, RULES-3 has an accuracy of 86.7% with the test data. The most striking result is in fact for C5.0, which generates one default rule and can only classify 6.9% of the examples.

Table 2
Results for Flower Identification Problem

Algorithm	Number of Rules	Accuracy of Rules
RULES-IS	27	100%
C5.0	0+1	6.9%
ID3	28	100%
RULES-3	23	86.7%
RULES-4	27	100%

3.1.3 Iris Data Set

The Iris data set [20] is often used for benchmarking machine learning and pattern classification algorithms [15] and in the evaluation of Immune System inspired algorithms [21]. The data set consists of 150 examples with 50 examples for each of the three Iris classes, divided into 70 examples for training and 80 examples used for testing. The 70 training examples were randomly picked from the 150 examples contained within the data set.

The results for are shown in (Table 3). In terms of the number of rules created, RULES-IS compares favourably with the other algorithms tested, generating fewer rules than even C5.0, which is considered to be amongst one of the best algorithms currently available. However, the accuracy of the four rules formed by RULES-IS is less than the 100% achieved by both C5.0 and RULES-3, but at least equivalent to the accuracy of RULES-4.

Table 3
Results for Iris Data Set

Algorithm	Number of Rules	Accuracy of Rules – Training Data	Accuracy of Rules – Unseen Data
RULES-IS	4	97.14%	95%
C5.0	5+1	100%	91.25%
RULES-3	14	100%	97.37%
RULES-4	10	97.14%	93.2%

When the accuracy of the rules generated by RULES-IS in classifying unseen data is examined it can be seen that of the examples classified an accuracy of 95% was achieved. This is comparable to the other algorithms lying approximately mid-way between the worst (C5.0) and the best (RULES-3).

3.2 Incomplete Data

To examine the ability of RULES-IS to deal with data sets where the classes of a proportion of the instances are unknown, a suitable data set had to be created. To do this the Iris data set was modified. The alterations were such that a proportion of the examples within a data set, from approx. 1% to approx. 95%, were randomly selected and had their class labels removed and replaced by the marker "#@#" which RULES-IS uses as an identifier for a missing class label. The rule sets generated are then tested and the results compared against the original data set which contains no missing class labels.

As most inductive learning algorithms are not able to handle the new data sets, a C5.0 based model was generated using the part of the data set where the classes are known. This model was then presented with the entire data set (containing both examples of known and unknown classes) and it then classifies these examples according to the rules previously generated. The result of this is then used to generate the final rule set, then this rule set is tested against the test data.

In (Fig. 5) the performance of RULES-IS initially is nearly identical to that of C5.0, until the point where the classes of more than 60% of the data set are unknown. When the proportion of unclassified data reaches such a level, the accuracy of RULES-IS is better than that of C5.0. Indeed, when 90% of the data sets are unclassified, the accuracy is still above 80%.

4. Conclusions and Further Work

RULES-IS has proved comparable to existing rule generation algorithms such as C5.0, ID3 and the RULES family in many respects, particularly the size and accuracy of the rule set created. RULES-IS has also demonstrated its ability to create rules using training data sets incorporating a proportion of unlabelled instances.

Future work could involve incorporate natural language processing techniques in antigen matching. This would allow linguistically sensitive matching between antigen and antibody based upon the similarity between words.

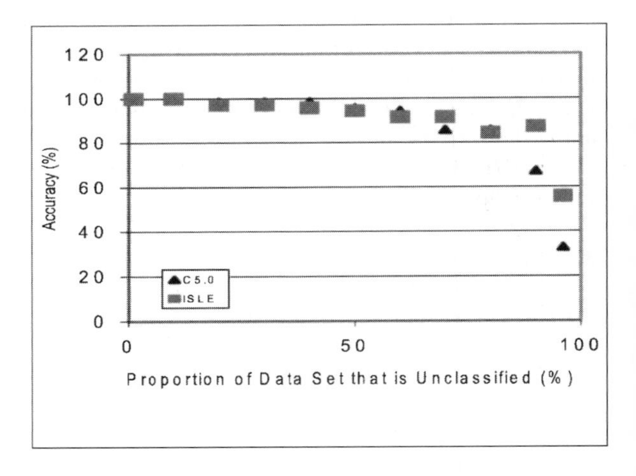

Fig. 5. Classification of unknown data

Acknowledgements

The research described in this paper was carried out within the INCO-COPERNICUS (EC) research project 96/0231 "Intelligent Product Manuals", the ERDF (Industrial South Wales) technology demonstration project "Intelligent Product Manuals for SMEs" the ERDF (Objective 1) project 'Supporting Innovative Product Engineering and Responsive Manufacturing' (SUPERMAN) project and the Innovative Production Machines and Systems (I*PROMS) Network of Excellence FP6-500273-2.

References

[1] Mayer M.M., (1973), The Complement System, Scientific American, 229, pp 54-66.

[2] Farmer J.D., Packard N.H., Perelson A.S., (1986), The Immune System, Adaptation and Machine Learning, Physica, Vol. 22, Part D, pp 187-204

[3] Jerne N.K., (1973), The Immune System, Scientific American, Vol. 229, Part 1, pp 52-60

[4] Hunt J.E. and Cooke D.E., (1995), An Adaptive, Distributed Learning System based on the Immune System, Intelligent Systems for the 21st Century, Proceedings International Conference on Systems, Man and Cybernetics, Vol. 3, pp 2494-2499

[5] Hunt J.E., Cooke D.E., Holstien H., (1995), Case Memory and Retrieval Based on the Immune System, IN Welose M. and Aomodt A., (Eds.) Lecture Notes in Artificial Intelligence 1010: Case-based Reasoning Research and Development, pp 205-216

[6] Hunt J.E. and Cooke D.E., (1996), Learning Using an Artificial Immune System, Journal of Network and Computer Applications: Special Issue on Intelligent Systems: Design and Application, Vol. 19, pp 189-212.

[7] Knight T. and Timmis J., (2002) A Multi-Layered Immune Inspired Approach to Data Mining. IN A Lotfi, J Garibaldi, and R John, editors, Proceedings of the 4th International Conference on Recent Advances in Soft Computing, pages 266-271, Nottingham, UK., December 2002.

[8] Knight T. and Timmis J. (2003) A Multi-layerd Immune Inspired Machine Learning Algorithm. In, *Applications and Science in Soft Computing*, (A. Lotfi and M. Garibaldi (Eds)) pp 195-202. Springer.

[9] Hunt J., King C., Cooke D., (1996), Immunising Against Fraud, Proceedings IEE Colloquium on Knowledge Discovery and Data Mining, London: UK

[10] Timmis J., Neal M., Hunt J., (2000) An Artificial Immune System for Data Analysis, Biosystems, Vol.55 (1/3), pp143-150

[11] Perelson A.S. (1989). Immune Network Theory. Immunological Reviews. No. 110 pp 5-36

[12] Soroka A.J., (2000), An Agent Based System for the Acquisition and Management of Fault Knowledge for Intelligent Product Manuals. PhD Thesis, Cardiff University, United Kingdom

[13] Cooke D.E. and Hunt J.E. (1995), Recognising Promoter Sequences Using an Artificial Immune System, Proceedings of the Third International Conference on Intelligent Systems for Molecular Biology, pp 89-97.

[14] Michalski R.S., Mozetic I., Hong J., Lavrac N., (1986), The Multi-purpose Incremental Learning Algorithm AQ15 and its Testing, Application in Three Medical Domains, Proceedings 5th National Conference on Artificial Intelligence, Philadelphia, PA: USA, pp 1041-1047.

[15] Pham D.T. and Dimov S.S., (1997), An Algorithm for Incremental Inductive Learning, Proceedings of the Institution of Mechanical Engineers Part B: Journal of Engineering Manufacture, Vol. 211, pp 239-249.

[16] Kottai R.M. and Bahill A.T., (1989), Expert Systems Made with Neural Networks. International Journal of Neural Networks, Vol. 1, No. 4, pp 211-226

[17] Pham D.T. and Aksoy M.S., (1995), RULES: a Simple Rule Extraction Algorithm. Expert Systems with Applications, Vol. 8, No. 1, pp. 59-65.

[18] Tolun M.R., and Abu-Soud S.M., (1998), ILA: an Inductive Learning Algorithm for Rule Extraction, Expert Systems with Applications, Vol. 14, pp 361-370.

[19] Aksoy M.S., (1993), New Algorithms for Machine Learning, PhD. Thesis, Cardiff University.

[20] Fisher R.A., (1936), The use of Multiple Measurements in Taxonomic Problems. Annals of Eugenics, Vol. 7, pp 179-188, 466-475

[21] Ji Z, Dasgupta D. (2004) Real-Valued Negative Selection using Variable-Sized Detectors. Proceedings of International Conference on Genetic and Evolutionary Computation (GECCO). Seattle, Washington USA, June 26-30, 2004.

Intelligent Production Machines and Systems
D. T. Pham, E. E. Eldukhri and A. J. Soroka (eds)

Semantic modelling of product support knowledge

R.M. Setchi, N. Lagos, S.S. Dimov

Manufacturing Engineering Centre, Cardiff University, Queen's Buildings, Cardiff CF24 3AA, UK

Abstract

This paper advocates the view that product support is a knowledge intensive process that should be modelled using knowledge engineering methodology. After highlighting the need for a new approach, the paper concentrates on defining the knowledge base of product support systems. The knowledge contained in the knowledge base is identified, and its modelling is described using notions introduced to capture the semantic complexity of the product support domain. The approach proposed is based on ontological formalisation that enables the creation of a knowledge model, which facilitates the sharing and reuse of product support knowledge. The applicability of the approach is illustrated using examples extracted from the knowledge based system developed.

1. Introduction

Product support is often described through the various forms of assistance that companies offer to their customers. Traditionally, it is associated with the provision of supplies, tools, equipment and facilities, as well as information. It may be in the form of installation instructions, user training, technical documentation, product manuals, help lines, servicing, spare parts, maintenance management, and product upgrades [1, 2].

Although, in general, access to knowledge is recognised as a critical part of any pro-active and creative business strategy, product support is rarely viewed as a knowledge engineering activity. In fact, as Kabel and Kiger [3] claim, job tasks involved in the creation of knowledge are perceived to be very different from those that have traditionally been expected of the technical writers and technical educators concerned with product support.

The aim of this paper is to demonstrate that product support is a knowledge intensive process that should be modelled using knowledge engineering (KE) methodology. The paper is organised as follows. Section 2 briefly reviews product support systems and highlights the need for adopting a KE approach. Section 3 discusses the knowledge contained within the knowledge base of a product support system and outlines its modelling process. Conclusions and directions for future research are discussed in the final section of the paper.

2. Background and related work

2.1. Product support

Pham et al. [1] define product support as everything necessary to allow the continued use of a product. It takes various forms, ranging from conventional paper-based technical manuals to more advanced interactive electronic technical manuals (IETMs), intelligent product manuals (IPMs) and electronic performance support systems (EPSSs). In particular, an IETM is defined as a technical manual authored in digital form and designed for electronic display by means of an automated authoring system [4]. An IPM, on the other hand, is referred to as a computerised interactive product support system that uses product life-cycle information, expert knowledge and hypermedia to provide just-in-time support to the user during the life of the product [1]. Finally, an EPSS is regarded as an integrated, readily available set of tools that help individuals to do their job and increase their productivity [5].

These definitions illustrate different approaches to product support systems, which are often described in terms of their purpose and features, enabling technology, development stages, and benefits. Most of the research in this area indicates the complexity of the domain, since knowledge is integrated from a number of engineering and non-engineering fields.

2.2. Product support and knowledge engineering

The importance of integrating knowledge engineering practices into the development of product support systems has been acknowledged by a number of researchers. Earlier studies focused on integrating reasoning mechanisms in product support systems, mainly as diagnostic tools. An example of such an approach is the work of Auriol et al. [6] who use case based reasoning to facilitate the troubleshooting of a welding robot.

Gradually, the focus in this area of research shifted from the reasoning mechanism to the knowledge base. Recent studies focus on user classifications, product and/or task structures, concentrating on how these are integrated. For example, the adaptive product manual developed by Pham and Setchi [7] is based on using product, user and task models. These models are integrated within a knowledge based system which uses cases that represent previously solved situations.

A similar approach is adopted by Brusilovsky and Cooper [8] who utilise integrated domain, task, and user models supporting the maintenance of equipment. The domain model represents the hierarchy of systems, subsystems and components, while the task model includes maintenance tasks, sub-tasks and steps. The components of the task hierarchy are connected to the components of the domain model using one relation only ("involve"), which limits the expressiveness of the representation. The three models are integrated within an expert system.

Latest research in product support indicates a trend towards semantic data modelling. For instance, Pham et al. [9] employ a semantic data model to generate virtual documentation. The model is based on data usage analysis, which abstracts the intended purpose of the product and task data elements, and their functional characteristics.

All these approaches are complementary rather than contradictory.

However, although these studies address the use of KE practices in product support, they do not follow a uniform approach towards the development of their knowledge bases. As a result, a major limitation of the previous work is the lack of design and knowledge reusability. This could become an obstacle in the nearest future when a new generation of much more complex and highly customised products emerges. The authors of this research share the vision that this challenge could be successfully addressed if product support modelling is unified, formalised and enriched by employing ontological principles.

3. Knowledge modelling approach

3.1. Product support knowledge sets

The goal of any product support system is to deliver knowledge that is accurate, applicable, reliable and user-tailored. To do that, the product support knowledge base should contain relevant knowledge about the products, users and their tasks (Fig. 1), as well as identifying the way in which these are linked to each other.

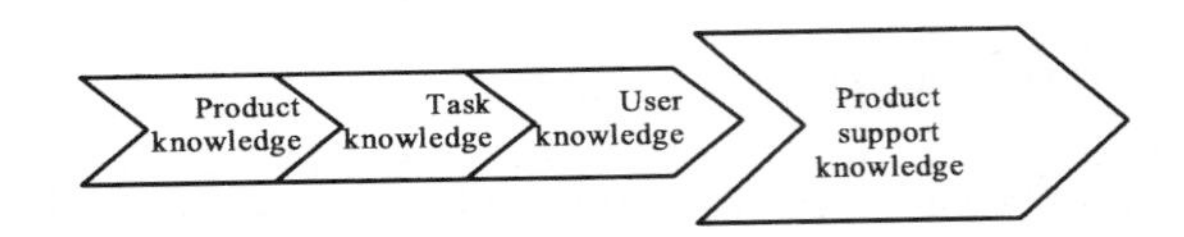

Fig. 1. Product support knowledge

If the knowledge available in a product support system is K_{PSS}, product knowledge is K_p, user knowledge is K_u and task knowledge is K_t, then for the product support system to be able to deliver optimal support, the following formal requirement must be satisfied.

$$K_p \cup K_u \cup K_t \subseteq K_{PSS} \quad \textbf{(1)}$$

3.2. From data to knowledge

The overall process of transforming data into knowledge has three stages (Fig. 2). First, relevant raw data of the products, tasks and users is gathered. This includes CAD models, bills of materials, drawings, assembly sequences, and user attributes. Next, this data is organised into information by identifying common relations between data items and employing them in accordance with ontological principles and semantics, as described in section 3.3.1. The resulting structures are further analysed and then used in model and case-based reasoning algorithms, to make them available for productive use, as explained in section 3.3.2. The links between the stages represent the two phases in the process of transforming data into knowledge.

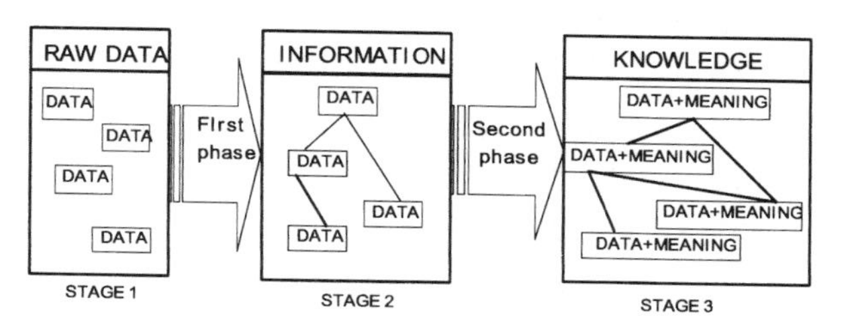

Fig. 2. Transforming data into knowledge

3.3. Conceptual modelling

Conceptual modelling involves the formalisation of domain knowledge in a high-level, application-independent, yet authoritative and rigorous way. The conceptual modelling aims the development of a knowledge base for product support (Fig. 3). The knowledge base contains two models: architectural and functional.

The *architectural model* is developed during the first phase. It is needed to represent real world information about the products, tasks, and users. Part of the architectural model are the squares named "Product", "Assembly", "Subassembly", "Part", and "Task", as well as their associations.

The square named "ProductType" and the lines that link it to the other squares, are a small fraction of the *functional model*, which is required for assigning meaning to the acquired information, according to the application context (i.e. product support). The functional model is created in the second phase.

Both models are further explained in the next two sub-sections. For simplicity, all references using capital letters in the rest of the paper relate to Fig. 3.

3.3.1. First phase

The objective of the first phase is to organise data in a way that ensures homogeneity and validity of the resulting information. The use of ontological principles is deemed appropriate because of the formality and richness of ontological knowledge modelling. This approach takes advantage of some widely accepted notions in KE, such as concept, instance, and relation.

Concept is a class of objects from the real world. Take the example of having several different cars such as Ford Mondeo, BMW 330i, and Peugeot 206 (see (A) in Fig. 3). All of them can be described by the general term "car" (B), which includes the characteristics that all of them share.

Instance is something that specifies a concept by illustrating a real world object that belongs to that concept, such as Peugeot 206 (A) for the notion of cars.

Relation is an attribute shared by objects in the subject domain that links and/or constrains them. One of the main ontological relations is generalisation. For example "car is-a vehicle" (C) is a generalisation type of relation.

In addition to these main ontological elements, this approach suggests the use of new structures, such as domain and connector. These are employed to enrich the representation ability of the system.

Domain is a hierarchy of concepts (i.e. concepts linked with "is-a" relation), which describes a part of the real world and is a concept itself. For example a domain "Product" (D) includes the hierarchy "car is-a vehicle is-a Product". Domains identify where each abstraction hierarchy is contained, thus clarifying the mapping to the different real world components.

Associating different concepts with the "is-a" relation and developing abstraction hierarchies (i.e. domains) is not enough in the case of a product support system. A relation that connects different domains is also needed. Take the example of "Product" (D) and "Assembly" (E) domains that are linked with the "Product has Assembly" relation (F). Relations that associate different domains are called *connectors*.

Connectors express "part-of", "is-composed-of", "is-realised-with", and aggregation relations. Connectors can also link instances that belong to different domains, if the concepts to which these instances belong are included within these domains and the domains are also linked with connectors. Connectors cannot link different knowledge sets.

The structural components described above, except for instances and their associations, enable the composition of the architectural model.

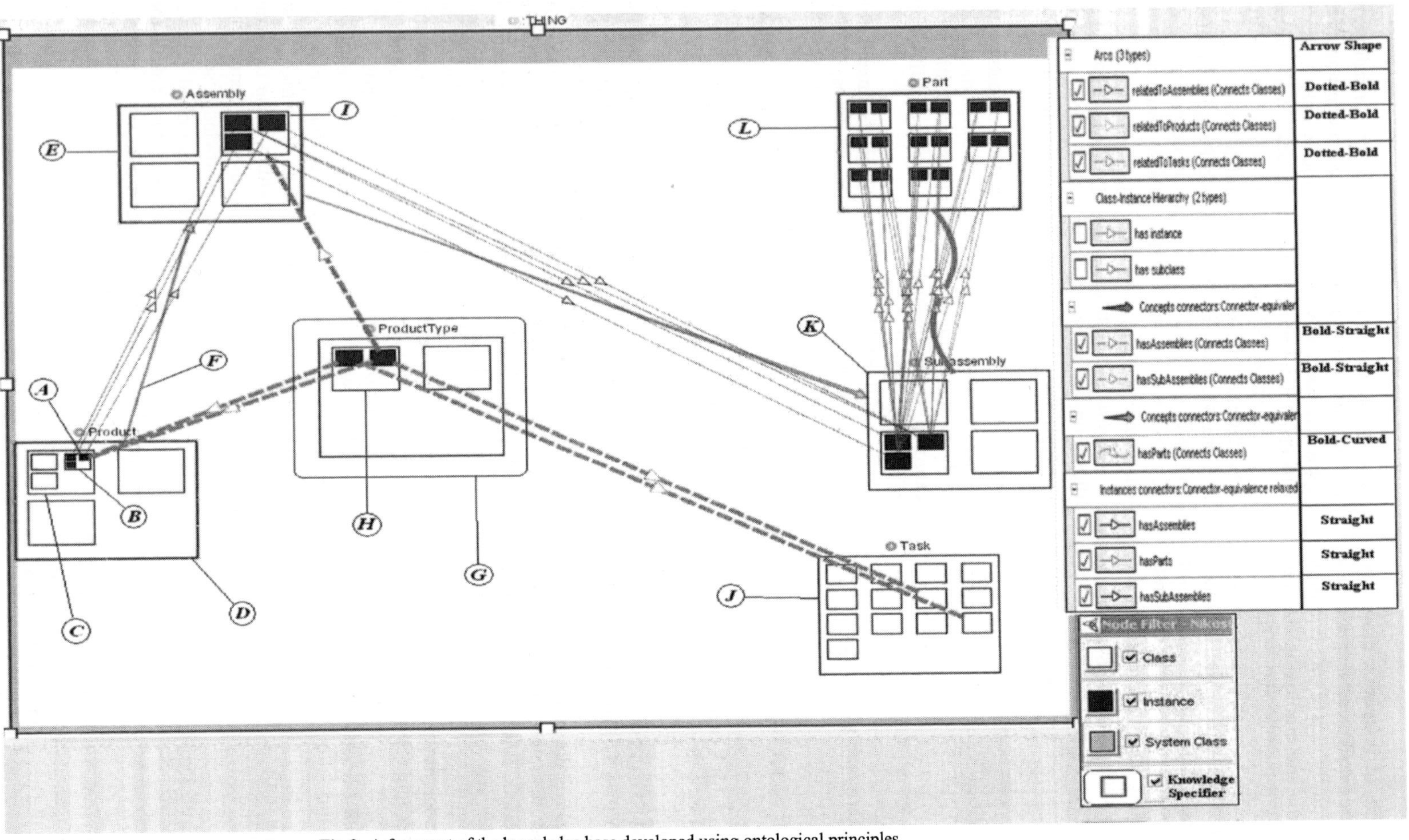

Fig.3. A fragment of the knowledge base developed using ontological principles

3.3.2. Second phase

The aim of the second phase is to put the structured information into productive use, within the context of product support. During this phase the architectural model is enriched by employing two new notions, *knowledge-specifier* and *arc*, which are further used in defining predicates and cases.

Knowledge-specifier is a property that is considered so significant within the application domain that it is represented as a different concept. For example, it may be useful to know whether a product is viewed as complex or not, since that can define the way in which the support provided is adapted. Therefore, a knowledge-specifier called "ProductType" (G), which defines product's complexity, is introduced. The level of the product support will be determined using mechanisms that include "ProductType". A simple example is the following predicate.

> *((ProductType = complex) AND (User = novice) AND (Task = complex))* ⇒ *ProductSupport = detailed*

Predicate 1. Knowledge-specifier within a predicate

Predicate 1 states that if the product is considered complex, the user is novice and the task is complex, then the product support solution should include details on all aspects of the query.

Arc is a relation that links the knowledge-specifiers with other concepts and/or domains. For example, "ComplexProduct" (H) can refer to concept "car" and "clutch" (I). Furthermore, knowledge-specifiers and arcs are used to relate different knowledge sets. In Fig.3, the task knowledge set (part of which is the task domain (J)) is related to the product knowledge set (composed by product, assembly, subassembly, and part domains) via "ProductType". The arcs in this case establish a constraint that is described by the following predicate.

> *Arc (ProductType, Task) = false* ⇒ *ProductSupport = null*

Predicate 2. Arc within a predicate

Predicate 2 states that if there is no arc between the knowledge-specifier and the task, then the product support system should not deliver a solution. This corresponds to the case when both the product and task are too complex for the user and (s)he would not be able to perform the task even with the help of the system (e.g. assembling a car). Such a situation is considered a safety hazard and should not be allowed. More complex predicates can be developed that take in consideration the status of the user and relate it to the above arc, which decide the kind of support given (see Predicate 3).

> *((Arc (ProductType, Task) =true) AND (ProductType = complex) AND (User = novice) AND (Task = complex))* ⇒ *ProductSupport = detailed*

Predicate 3. Arc and knowledge-specifier within a predicate

Connectors can be characterised as active or inactive depending on whether the real world components they describe exist or not. For example, if a product is very simple, then it may not have any assemblies and subassemblies, but only parts. The possibility of having such cases is represented by the *connector-dependence* (CD) measure. CD can take two values and be either true or false depending on whether the connector is always active or not. Take the example of the connectors shown in Fig. 3. The CD of the connectors between "Product" and "Assembly", and "Assembly" and "Subassembly" (K), is considered false because the existence of assemblies and subassemblies depends on the product's complexity (i.e. a simple product is one that has only part(s)), which is expressed by "ProductType". Therefore, only the connector that is directed to "Part" (L) has CD that equals to true, as it is always active. The connector-dependence measure establishes existential constraints on the architectural model (e.g. each "Product" should have at least one "Part"), which is a way to ensure the validity of the acquired knowledge.

The notions introduced enable the creation of the system's functional model. The combination of the architectural and functional models results in the development of the system's *knowledge model.*

3.4. Knowledge base and product support documents

The knowledge base of the system is developed using the knowledge model described above, populated with instances and their associations. Each attribute of each concept corresponds to a small fragment of a product support document, while the value of that attribute defines the way in which the content of that fragment changes (e.g. for Number of Disks=2, the double cutch disk document fragment is generated). Each concept is mapped to a part of the document that contains several of the aforementioned fragments that match the attributes that characterise each concept. The connectors and generalisation relations are used to search for part of documents that relate to the required concept, while the knowledge-specifiers and arcs are utilised to choose the presentation format of the provided document.

3.5. Implementation

The knowledge base is created using the Protégé ontology editor, and part of it is depicted in Fig. 3 with the Jambalaya plug-in. The squares in the figure represent concepts. An "is-a" relation is represented by placing a square within a bigger one. The connectors between different domains, with CD being false, are represented with bold straight lines, and with CD being true are shown as bold curved lines. The straight lines that link the black squares (instances) are connectors, with CD always false, linking different instances. The knowledge-specifier ("ProductType") is shown as a rounded square and the arcs that link it to the other concepts are represented by bold dotted lines. THING (the domain within which everything is contained) is the root concept of the ontology.

Currently, the knowledge base contains 278 frames, including 75 concepts, 80 slots, 22 facets, and 101 instances.

4. Conclusions and future work

This paper focuses on the representation of the knowledge about users, products, and tasks that a product support system should contain. Knowledge is acquired following a three stage process of transforming raw data into information and knowledge by the utilisation of ontological principles. New notions are introduced (domain, connector, knowledge-specifier, and arc) for formalising the acquired knowledge and applying it in the context of product support. A new measure (connector-dependence) is also proposed to ensure the completeness and validity of the developed structures. In addition, a knowledge base has been developed based on the knowledge model described.

This new approach for representing product support knowledge formalises the context within which a product support system operates and improves the product support responsiveness by prescribing the knowledge sets used and how these should be represented. The knowledge model provides an ontology that can be shared and reused by product support systems. It introduces the structural components needed, the links between them, and establishes their usability. The developed ontology advances the interoperability between product support and other knowledge intensive fields, by representing the elements of knowledge in a machine processable way. Furthermore, it facilitates the provision of adaptive and personalised support.

This work is a step towards formalising product support as a knowledge intensive process. Future work includes the development of a product support framework, where the developed knowledge base will be one of the major components.

Acknowledgements

The research described in this paper was performed within the I*PROMS Network of Excellence sponsored by FP6 of the European Community.

References

[1] Pham DT, Dimov SS and Setchi RM. Intelligent product manuals. Proc. IMechE. I-213 (1999) 65-76.

[2] Goffin K. Evaluating customer support during new product development-an exploratory study. Journal of Production Innov. Management. 1-15 (1998) 42-56.

[3] Kabel MA and Kiger R. Convergence of knowledge engineering and electronic performance support systems. IEEE Cross. in Comm. (1997) 45-51.

[4] Jorgensen EL and Fuller JJ. A web-based architecture for interactive electronic technical manuals (IETMs). ASNE Naval Logistics Conference. (1998).

[5] Bezanson WR. Performance support: online, integrated documentation and training. 13th ACM Annual Conference on Emerging from Chaos. Savannah. USA. 1-10 (1995).

[6] Auriol E, Crowder RM, McKendrick R, Rowe R and Knudsen T. Integrating case-based reasoning and hypermedia documentation: an application for the diagnosis of a welding robot at Odense steel shipyard. Eng. App. of AI. 12 (1999) 691-703.

[7] Pham DT and Setchi RM. Adaptive product manuals. Proc. IMechE. C-214 (2000) 1013-1018.

[8] Brusilovsky P and Cooper DW. Domain, task, and user models for an adaptive hypermedia performance support system. Proceedings of the IUI '02, ACM, San Fransisco, California, USA. (2002) 23-30.

[9] Pham DT, Dimov SS and Huneiti AM. Semantic data model for product support systems. IEEE International Con. Ind. Inf. (INDIN 2003). (2003) 279-285.

Intelligent Production Machines and Systems
D. T. Pham, E. E. Eldukhri and A. J. Soroka (eds)

SO_2 concentrations forecasting for different hours in advance for the city of Salamanca, Gto., Mexico.

U.S. Mendoza-Camarena, F. Ambriz-Colin, D.M. Arteaga-Jauregui, A. Vega-Corona

Faculty of Electrical, Mechanical and Electronics Engineering, University of Guanajuato, Tampico 912, Salamanca, Gto. Mexico.

Abstract

This work presents a method to predict Sulphur Dioxide (SO_2) concentrations for the city of Salamanca, Gto., based on real data provided by a Monitoring network during the months of December of years 2002, 2003 and 2004. A Generalized Regression Neural Network (GRNN) is used to perform the predictions. The GRNN was trained with the data obtained from years 2002 and 2003, and the data from year 2004 were the test group data. Predictions were made for 1, 12 and 24 hours ahead, and results have been promising. Root Mean Squared Error (RMSE) and Mean Absolute Error (MAE) are the performance measures for the GRNN.

1. Introduction

Prediction of atmospheric pollutant concentrations is of great importance for decision making, mainly for industrial urban areas. The purpose of this work is to develop a predictive model to forecast Sulphur Dioxide (SO_2) concentrations in the short term, for the city of Salamanca, Mexico, to warn over possible environmental contingencies.

Salamanca City, has a pollution problem due to the amount of factories that expel a number of pollutants. In an effort to reduce the levels of pollution in the city, the authorities established the "Patronato para el Monitoreo de la Calidad del Aire" (Patronage for Air Quality Monitoring), and the Atmospheric Monitoring Network (REDMAS) was created in 1999.

The Atmospheric Monitoring Network (REDMAS), consists of 3 Monitoring stations, their location was assigned according to meteorological studies, soil usage and population density.

Monitoring Stations consist of Measurement Instruments for five pollutants, meteorological sensors and data acquisition systems. Continuous analysis of air, allows for evaluating the Atmospheric Pollution and its distribution [1]. The Measured pollutants are: Sulphur Dioxide (SO_2), Nitrogen Dioxide (NO_2), Carbon Monoxide (CO) and Particulate Matter (PM_{10}), which are particles whose diameter is lesser than 10 micrometers.

Forecasting pollutant concentrations is a problem. The diffusion mechanism of air pollutants is quite complicated and it depends on several parameters of different nature, including topography of the area, and the energy balance that characterises urban environment [2]. Modelling atmospheric pollution phenomena is a complex task, influenced by a large number of external variables (e.g. Temperature, Relative Humidity, Wind Speed and Wind Direction).

Literature provides several approaches for time series modelling, being one of the most used the Neural Networks of the Multilayer Perceptron (MLP) type with a determined learning rule [2,3,4,5].

A Generalized Regression Neural Network (GRNN) is proposed in this work, as done in [6]. The inputs to the GRNN are the SO_2 concentrations time series data, and also the external variables Wind Speed (WS), Wind Direction (WD), Temperature (T) and Relative Humidity (HR) mentioned before.

Results for different number of external atmospheric variables are compared. Results for the Forecasting at different hours in advance are also compared, specifically for 1, 12 and 24 hours ahead. In the last part of this work, SO_2 concentrations are converted to an Air Quality Index (AQI), and estimated results versus real observations are compared.

2. Methodology

Figure 1 shows the flow diagram of the methodology that was followed for the completion of this work; it consists of 3 main phases: 1) Data Pre-processing, 2) Neural Network Training, 3) Neural Network Simulation and Results Evaluation.

2.1. Pre-processing

Provided data have flags and indicators, therefore, data must be debugged. Both pollutant concentrations and meteorological data are computed every minute by the Monitoring Network (REDMAS). The hourly average of this data is obtained to form the training and test data sets. Data was compensated because data from December 2002 had only measurements from days 1 to 28. December months of years 2002 and 2003 were used as the Training data set, and December 2004 was the Test data set.

2.2 General Description of GRNN

Giving a detailed explanation of the Generalized Regression Neural Networks is beyond the scope of this work, therefore, only a general description of how this type of Neural Networks work will be given. A deep study of the GRNN architecture and its applications can be found in [7].

GRNN algorithm is defined on equation (1).

$$E[y \mid \mathrm{x}] = \frac{\int y_i f_i(\mathrm{x}, y) dy}{\int f_i(\mathrm{x}, y) dy} \qquad (1)$$

where:

y_i is the estimator output corresponding to the pattern $\boldsymbol{x}_i$, $\boldsymbol{x}$ is the input pattern of the estimator, $E[y \mid \boldsymbol{x}]$ is the expected output value of the input pattern $\boldsymbol{x}$ and $f_i(\boldsymbol{x},y)$ is the joint probability density function (pdf) of x and y.

An approximation of (1) for N input patterns are estimated with

$$\hat{y}(\mathrm{x}) = \frac{\sum_{i=1}^{N} y_i f_i(\mathrm{x})}{\sum_{i=1}^{N} f_i(\mathrm{x})} \qquad (2)$$

where:

$$f_i(\mathrm{x}) = \exp\left(-\frac{(\mathrm{x}-\mathrm{x}_i)^T(\mathrm{x}-\mathrm{x}_i)}{2\sigma^2}\right) \qquad (3)$$

and:

$\mathbf{x}_i$ is the input training vector, y_i is the desired output corresponding to $\mathbf{x}_i$ and N is the number of training patterns for which $\{\mathbf{x}_i \rightarrow y_i\}$ (there is a correspondence between $\mathbf{x}_i$ and y_i).

Equation (3) is the most commonly used function in GRNN architecture, and it is the Radial Gaussian Base function, is the kernel's window or smoothing factor chosen for training the GRNN [8].

Given a training set and an independent test set, GRNN is trained by choosing the σ parameter, to obtain the best precision in the estimation. For most cases, this parameter is related to the error or noise distribution of the signal. In our structure we defined a single σ that produces the minimum MSE between the desired and the real output for the test group, which can be rapidly found using an iterative algorithm.

2.3 GRNN Training

Both, training and simulation of the GRNN are made using two different schemes. The first one considers only the current SO_2 concentration to predict the future concentration of SO_2. The second one, also considers different combinations of the environmental variables Wind Direction (WD), Wind Speed (WS), Temperature (T), and Relative Humidity (RH).

As shown in Figure 1, first step in training the GRNN is the initialization of the Kernel parameters. Conversely to other types of Neural Networks, in which number of neurons and hidden layers must be determined, the only parameter that needs to be modified in a GRNN to obtain a better approximation of the data to be modelled is the kernel's window σ. Right after, the GRNN is trained and σ is modified to get the minimum error. In this work the same σ was used for each neuron, and it is found iterating different values of σ.

Input neurons equal the number of observations

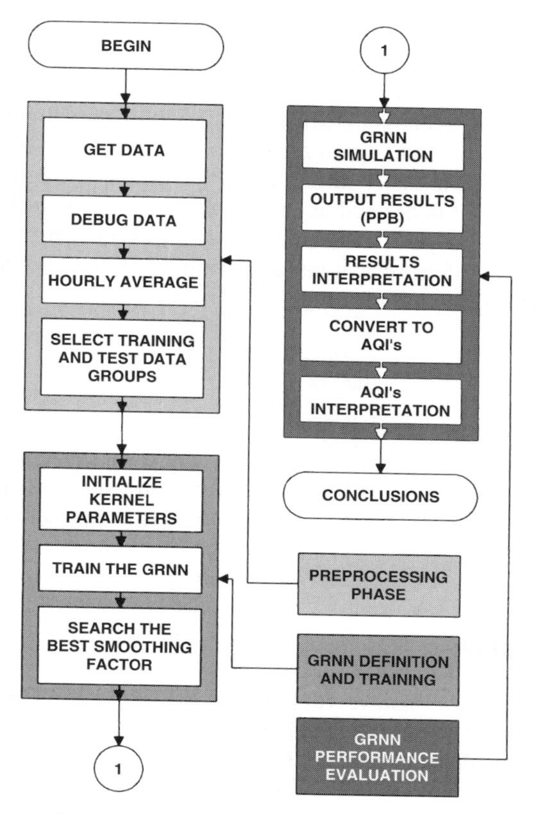

Figure 1. Flow Diagram for SO_2 Prediction.

the GRNN is trained with. In this work, input neurons are 1344 which is the total number of observations corresponding to the training period (Dec.2002 and Dec. 2003). Test group consists of 672 observations corresponding to Dec. 2004

2.4. GRNN Simulation

Prediction was made 1, 12 and 24 hours ahead, which are the periods mostly used when predicting time series.

Figure 2 shows how predictions are performed. Future concentrations are predicted taking into account current and previous concentrations. Vector ***x***, is the input vector at times $t=i-n-1$, ..., $t=i-1$, $t=i$, where *i*, is the current hour and *n* is the number of hours ahead with which prediction is done. Vector ***y***, is the output vector, whose elements correspond to the estimation of SO_2 levels at times $t=i+1$, $t=i+2$, ..., $t=n$, where *i*, is

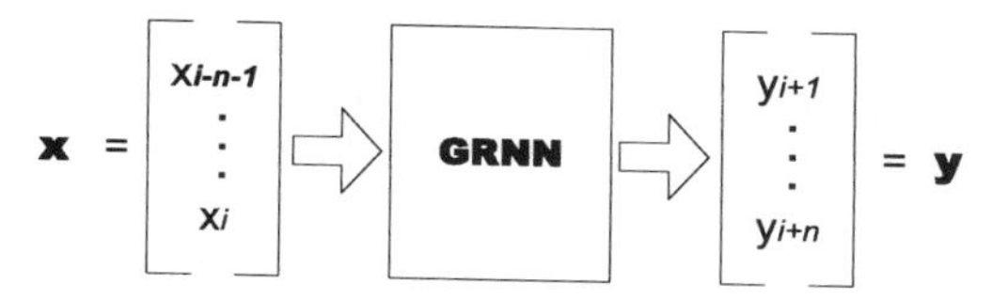

Figure 2. Prediction Scheme.

Table 1
Variables Definition

Variable Name	Convention
SO_2	x_1
Wind Speed (WS)	x_2
Wind Direction (WD)	x_3
Temperature (T)	x_4
Relative Humidity (RH)	x_5

the current hour and *n* is the number of hours ahead with which prediction is done.

Pollution levels are measured as an Air Quality Index (AQI) for authorities and common people to better understand or interpret data. AQI is defined as an overall scheme that transforms the weighted values of individual air pollution related parameters (e.g. SO_2, CO, visibility, humidity, etc.), into a single number or set of numbers [9]. In the case of Salamanca city, information provided by the Monitoring Network (REDMAS), is converted to IMECA's (Metropolitan Index of Air Quality). Based on the IMECA value for each pollutant, there are five categories to classify Air Quality, as it is shown in Table 2.

After getting and interpreting the results using SO_2 concentration expressed in Parts Per Billion (PPB), this values are converted to the Air Quality Index (AQI). Real and Estimated data are converted to AQI's, once

Table 2
Air Quality Index (AQI) Classification

IMECA's value	AQI Value	AQI Category
0-50	1	*Good*
51-100	2	*Satisfactory*
101-200	3	*Not Satisfactory*
201-350	4	*Bad*
>350	5	*Very Bad (Dangerous)*

this is done, real and predicted AQI's are compared.

Equations (4) and (5) show the formulas to convert SO_2 concentrations given in PPB to IMECA's.

$$CI = \left[\frac{PPB * 769.2307}{1000} \right]_{<130 PPB} \quad (4)$$

$$CI = \left[479.7701 * (\frac{PPB}{1000} - 0.13) + 100 \right]_{>130 PPB} \quad (5)$$

where:
CI is SO_2 concentration in IMECA's, and PPB is SO_2 concentration expressed in Parts Per Billion. If SO_2 concentration exceeds 100 IMECA's (equivalent to 130 PPB), SO_2 levels do not comply with Norm "Norma NOM-022-SSA1-1993" [10], therefore, formula (4) is used for values below the Norm and formula (5) is used for values exceeding the Norm.

Evaluation of the GRNN Performance was accomplished using the Root Mean Squared Error (RMSE), Mean Absolute Error (MAE) and the Effectiveness Percentage (EP).

Mean-squared error is the most commonly used measure of success of numeric prediction, and root mean-squared error is the square root of mean-squared-error, taken to give it the same dimensions as the predicted values themselves. This method exaggerates the prediction error - the difference between prediction value and actual value of a test case - of test cases in which the prediction error is larger than the others. If this number is significantly greater than the mean absolute error, it means that there are test cases in which the prediction error is significantly greater than the average prediction error [11]. Balaguer et al. [3], have used RMSE as an indicator of the relationship between predicted and observed data.

Root Mean Squared Error is computed according to equation (6)

$$RMSE = \sqrt{\frac{1}{n} \sum_{i=1}^{n} (\hat{y}_i - y_i)^2} \quad (6)$$

where $\hat{y}_i$, is the predicted value for a determined time $t=i$, y_i, is the real value for that same time and n is the number of observations.

Mean Absolute Error is the average of the difference between predicted and actual value in all test cases. Mean Absolute Error (MAE) is computed according to equation (7)

$$MAE = \frac{1}{n} \sum_{i=1}^{n} \left| \hat{y}_i - y_i \right| \quad (7)$$

where $\hat{y}_i$, is the predicted value for a given time $t=i$, y_i, is the real value for that same time, and n is the number of observations.

Effectiveness Percentage (EP), is used only for the case when AQI is predicted, since there are only five categories is easier for the GRNN to predict the exact value. EP is defined as the number of times that the GRNN succeeds to predict the AQI, divided by the number of observations and multiplied by 100 to express it as a Percentage.

$$EP = \frac{Na}{n} * 100 \quad (8)$$

EP is computed according to equation (8), where Na, is the number of times that GRNN succeeds to predict the AQI and n is the total number of observations.

3. Results

Table 3 shows values for RMSE and MAE for different combinations of external variables. External (atmospheric) variables considered are: Wind Direction (WD), Wind Speed (WS), Temperature (T) and Relative Humidity (RH).

The minimum error was obtained considering only one external variable (Wind Speed), although results obtained considering no external variables are very similar. This is probably because Wind Speed is a parameter that doesn't fluctuate much.

Table 4 and Figures 3, 4 and 5, show the prediction performance for 1, 12 and 24 hours ahead, without considering external variables. The best results were achieved for the prediction of 1-hour ahead SO_2 concentrations; the lowest MAE was for 12-hour ahead prediction though.

From Figures 4, 5 and 6 it can be observed that prediction is not very accurate when there are sudden variations in the plot. This agrees with data on Table 4, where it can be observed that RMSE is significantly greater than MAE, being an indication that there are values for which prediction error is much greater than the mean error.

Last part of this work is the conversion of SO_2

concentration levels to an Air Quality Index. This is done with real and predicted data.

Table 5 shows the results for the prediction of AQI's for 1, 12 and 24-hour ahead. RMSE and MAE are drastically reduced, and Effectiveness Percentage is introduced as a different effectiveness measure of the Prediction. The Best Results were achieved by the 12-hour ahead prediction. Figure 6, shows results of Table 5 in a graphical manner.

Table 3
Prediction performance considering different combinations of external variables.

No. of External Variables	Considered Variables	RMSE (PPB)	MAE (PPB)
0	None	28.41	17.28
1	WS	28.40	17.23
	WD	29.06	16.97
	T	28.62	17.22
	RH	29.36	17.88
2	WS, WD	29.1	16.97
	WS, T	28.58	17.24
	WS, RH	29.36	17.86
	WD, T	29.31	17.02
3	WS, WD, T	29.15	17.06
	WS,WD, RH	29.41	17.07
	WD, T, RH	29.35	17.17
4	WD, WS, T, RH	28.62	17.13

Table 4
Results for Predicting SO_2 concentrations at different hours ahead without considering external variables.

Hours Ahead	RMSE (PPB)	MAE (PPB)
1	28.41	17.28
12	31.96	16.95
24	37.27	22.26

Table 5
AQI Prediction results.

Ahead Hours	RMSE (AQI's)	MAE (AQI's)	EP
1	0.1728	0.1491	86.26%
12	0.1503	0.1205	89.43%
24	0.1949	0.1473	86.64%

4. Conclusions

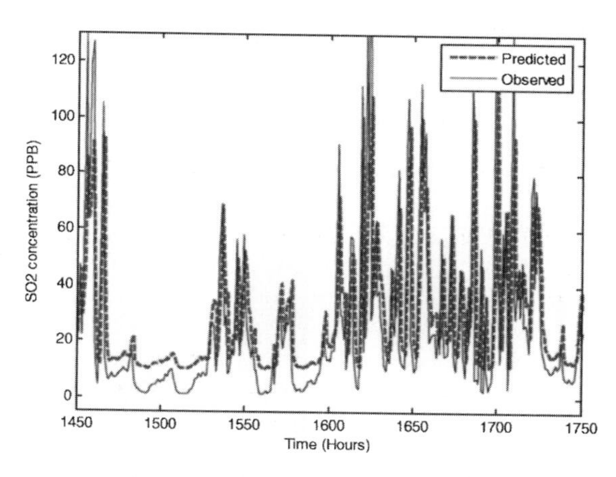

Figure 3. 1-hr. ahead Prediction without considering external variables.

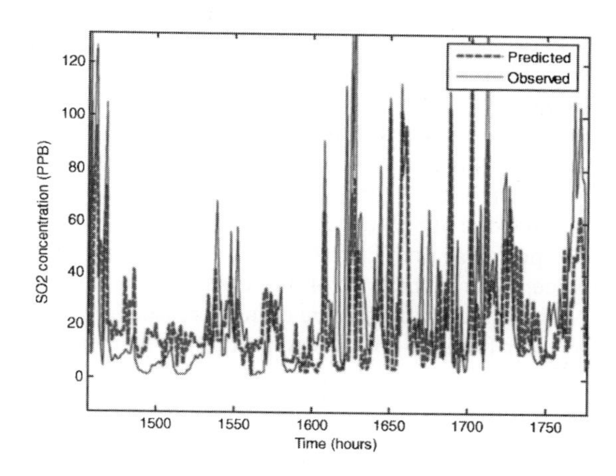

Figure 4. 12-hr. ahead Prediction without considering external variables.

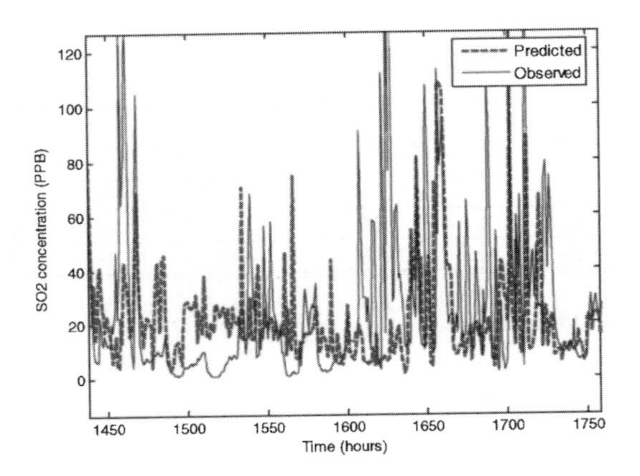

Figure 5. 24-hr. ahead Prediction without considering external variables.

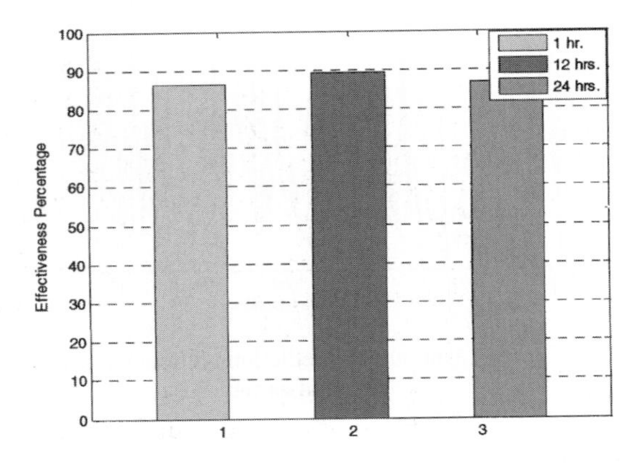

Figure 6. AQI Prediction at different hours ahead.

This work shows the utilization of a Generalized Regression Neural Network to predict pollution levels due to SO_2 concentrations. The model presented here has a low demand for input parameters it only needs one step for training. It has been shown that the knowledge of SO_2 concentrations alone can be enough to perform predictions of this pollutant with an acceptable error. Error increases as the time with which prediction must be done increases (e.g. error is greater for a 24-hour ahead prediction than error for a 1-hour ahead prediction), specially when sudden variations of SO_2 concentration occur. Prediction effectiveness is very good when Air Quality Index is predicted instead of concentration expressed in PPB, since it's easier to predict whether Air Quality Index will be Good, Satisfactory, Not satisfactory, Bad or Very Bad than predicting exact values of concentration given in PPB.

5. Further Work

The use of Artificial Neural Networks (ANN's) for Time Series Prediction has proven to be more effective than traditional Regression Techniques. The ANN's can also be used for Weather, Seismic and Volcanic Activity Forecasting.

References

[1] Patronato para el Monitoreo de la Calidad del Aire de Salamanca, A.C. http://www.prodigyweb.net.mx/redmas.

[2] I. Turias, F.J. González, P.L. Galindo. Application of neural techniques to the modelling of time-series of atmospheric pollution data in the 'Campo de Gibraltar' region.' Neural Network Engineering Experiences. Procc. of the 8° International Conference on Engineering Application of Neural Network (2003) 9-16.

[3] E. Balaguer-Ballester, E. Soria-Olivas, J.L. Carrasco-Rodriguez, S. Valle-Tascon. Forecasting of surface ozone concentrations 24 hours in advance using Neural Networks. http://citeseer.ist.psu.edu/566064.html.

[4] Davey, and S.P. Hunt. Time Series Prediction and Neural Networks. Journal of Intelligent and Robotic Systems, Vol. 31 (2000) 91-103.

[5] S.F. Crone. Prediction of White Noise Time Series using Artificial Neural Networks and Asymmetric Cost Functions. Proceedings Of The International Joint Conference On Neural Networks (2003) 2460- 2465.

[6] Chtioui, Y., Francl, L.J., and Panigrahi, S. Moisture prediction from simple micrometeorological data. Phytopathology Vol. 89 (1999) 668-672.

[7] D. F. Specht. A General Regression Neural Network. IEEE Transactions on Neural Networks vol. 2, No. 6 (1999) 568-576.

[8] Wasserman Philip D. Advanced Methods In Neural Computing. Van Nostrand Reinhold, 1993, pp. 147-158.

[9] M. Sharma, S. Aggarwal, P. Bose, A. Deshpande. Meteorology-based Forecasting of Air Quality Index Using Neural Network. Proceedings. IEEE International Conference on Industrial Informatics. (2003) 374 – 378.

[10] NORMA OFICIAL MEXICANA NOM-022-SSA1-1993. http://www.salud.gob.mx/unidades/cdi/nom/022ssa13.html.

[11]GRB Tool Shed. Help. Section 3.2 Error Measurements. http://grb.mnsu.edu/grbts/doc/manual/Error_Measurements.html.

[12] Fulgencio S. Buendía Buendía, J. Miguel Barrón-Adame, Antonio Vega-Corona, Diego Andina. Improving GRNNs in CAD Systems. Lecture Notes in Computer Science, Volume 3195, Springer, Oct 2004, pp. 160 - 167.

Intelligent Production Machines and Systems
D. T. Pham, E. E. Eldukhri and A. J. Soroka (eds)

Towards interactive analysis of virtual manufacturing simulations

K.R. Mahajan[a,b], W. Dangelmaier[b]

[a] *International Graduate School of Dynamic Intelligent Systems, University of Paderborn, Germany*
[b] *Heinz Nixdorf Institute, University of Paderborn, 11 Fürstenallee, 33102, Paderborn, Germany*

Abstract

Discrete-event material flow simulation tools are already offering possibilities to visualize factory operations in virtual worlds. This feature allows less experienced users to analyze the underlying system. Beyond this, visualization is not used to interact with the simulated (underlying) system to improve or control the material flow under disturbances. This paper presents a simulation-based two-tier framework, which seeks to control or improve material flow by means of user immersive visualization. The first tier uses optimization to compute the material flow based on deterministic disturbances. The second tier is a reactive algorithm, which computes solutions for probabilistic disturbances. The results of the two tiers are used for interacting with the underlying system using visualization. We show conceptually that it is possible to integrate real-time decision making in virtual simulation environments.

1. Introduction

These days, manufacturing systems are getting more and more complex. Simulation and visualization can help in understanding the underlying manufacturing system very well. However, it is not possible today, to control the material flow and results of the simulated system using bi-directional interactive analysis between simulation and visualization. An integration of real-time decision-making algorithms and virtual reality would provide advantages like the ability to visualize the entire plant (see Figure 3 for an example of a visualized factory), whilst still being able to control and analyze it real-time by interacting with the system in the event of disruptions or unforeseen events. This calls for an integration of existing and new work in areas like simulation and visualization, and real-time production scheduling.

There have been numerous works, which focus on real-time scheduling of the material flow based on simulation. Harmonosky et al. [4] present their work in the areas of real-time selective re-routing and scheduling algorithms based on simulation. They iteratively use simulation as a tool to find out the best policy from a set of alternative policies in real-time. Chong and Sivakumar [3] mention about reactive and predictive simulation based scheduling for dynamic manufacturing. They also use on-line simulation to evaluate the selected approaches and the corresponding schedules to determine the best solution. The selected solution was used until the deviation of actual performance from the estimated one exceeded a given limit. While the simulation-based approach for real-time scheduling seems very intuitive, more efforts are required to integrate these algorithms to provide decision-making abilities in 3D visualizations.

Using simulation as a tool for real-time scheduling presents several benefits and drawbacks. The most important benefit is that simulation can provide accurate information about a certain policy when the system behaviour is probabilistic. However, as the underlying manufacturing system gets bigger and as the number of "control points" (location where a decision to handle a part has to be taken) increase with each point providing several alternatives, the more difficult it is to employ simulation, especially in the real-time virtual world to obtain information about the best alternative policy. Secondly, the length of the look-ahead planning horizon, i.e. the length of the simulation run used to compare a set of alternatives, has little scientific backing. The more time each evaluative simulation runs, the longer it will take to make scheduling and control decisions, thus failing to provide decisions in "real-time".

In the areas of simulation and visualization, Mueck et al. [11] report the development of a real-time walk-through system coupled with a simulation system along with bi-directional information flow. The motive was to provide a completely immersive environment for the user, allowing him to interact with a certain process. However, the indications to lead the user to a significant process/object (a process that can recover the system from disturbances) were lacking. The user was not guided to the significant process automatically, based on critical events. Apart from the identification of the significant process, the user had no means to employ corrective policies in the virtual world to optimize the system.

In this paper, we describe a framework to integrate work in these areas and discuss related issues. In brief, the system will pre-calculate the material flow using best policies (from a range of alternative policies), based on deterministic input before hand by performing a static calculation (optimization). The material flow is then visualized in real-time and pre-simulated offline to handle disturbances. Section 2 describes the assumptions in our work. Section 3 shows the overall framework of our methodology. Section 4 and 5 describes how the manufacturing system will be modelled to perform the static optimization. Section 6 then describes how interactive analysis could take place in a virtual environment. The paper is concluded by mentioning new challenges and further research directions.

2. Assumptions

In this work, we do not yet consider the connection of the physical manufacturing system to the visualized simulation. This implies that we do not consider process and yield variations, etc originating from the physical manufacturing system and methods to bring back the deviations using control. All disturbances we consider result from either orders (jobs) and\or machine breakdowns in the virtual world. We specifically consider mass manufacturing environments with limited product variety.

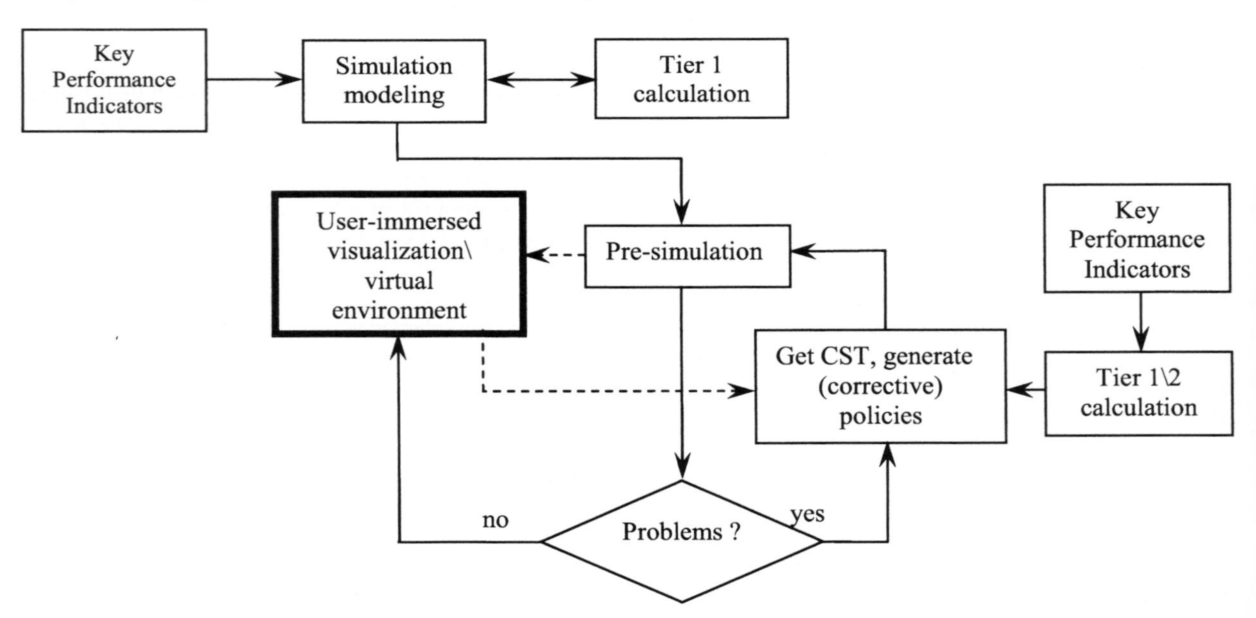

Fig. 1. Overall framework for interactive analysis in virtual environments

3. An overall framework

Figure 1 shows the overall framework of the envisaged system. The user has primarily two tasks. The first is to model the manufacturing system with all the (deterministic) parameters. Here, we let the user define the alternative control policies at so called "control points" in the simulation model. We also let the user decide which KPI's (Key Performance Indicators: make-span, costs, work in progress and equipment utilization) he wishes to achieve throughout the lifetime of a particular order. The user is typically given an option of setting priorities over the KPI's. All this information is modeled during simulation modeling and passed on to the tier 1 calculation. Based on this information, the tier 1 calculation will compute a sub-optimal material flow to be used in the system, which includes part routing decisions. The calculation will consider production rates, buffer sizes and current production loads (if any) to compute the flow. This flow is again given back to the simulation model (Figure 1). The tier 1 calculation primarily seeks to reduce complexity due to the alternative policies and to compute control policies for deterministic system disturbances.

Then, this material flow is pre-simulated in an offline fashion and visualized in a 3D environment. The visualization is the place where we seek to control and analyse the underlying system (bold box in Figure 1). The pre-simulation is done for three reasons. First, to "drive" the visualization (as is done traditionally), secondly, to validate tier 1 and tier 2 calculations, and thirdly, to simulate the effect of disturbances which might arise in due course of time. As seen in Figure 2, the pre-simulation, at time t_{Sim}, is a fast simulation, and leads the user-immersed real-time visualization, at time t_{VR}, (from now on real-time visualization is termed as virtual simulation) by time ΔT. Note however, that the pre-simulation is started from the same starting state as the virtual simulation (more details in next paragraph). The pre-simulation and the virtual simulation together handle disturbances in the system. The disturbances we consider are deterministic (arrival\cancellation of orders and machine maintenance (breakdown) s-chedules) and probabilistic (arrival\cancellation of new orders and machine breakdowns).

Assume, that at time ($t_{Sim} + x$) or ($t_{Sim} - x$), both > t_{VR}, information about a new order is available (deterministic disturbance). The pre-simulation stops after it goes into a complete irreversible damage state (due to the new order, assuming that the new order has completely different characteristics) at time (t_{Sim} + z). In other words, the pre-simulation determines the effect of the new order\disturbance. Meanwhile the virtual simulation continues, and tier 1 part of the calculation is then invoked which determines the material flow (resulting in corrective policies) based on the current (virtual) simulation state (CST), the KPI, the effect of the new order arrival (estimated by the pre-simulation), and the available alternative policies. On the other hand, when the disturbance is sudden machine failure (probabilistic disturbance, detected at, for instance, time t_{VR}), the virtual simulation continues, while the pre-simulation now comes back to the state at t_{VR} (=t_{VR}^1) (Figure 2) and rapidly simulates until the time where the pre-simulation goes into an irreversible damage state, due to the machine breakdown. The tier 2 calculation is then invoked which again considers the CST, KPI, the effect of the machine breakdown (obtained by the pre-simulation), and alternative policies to compute possible alternative part routes or corrective policies (refer Figure 1). An example of the

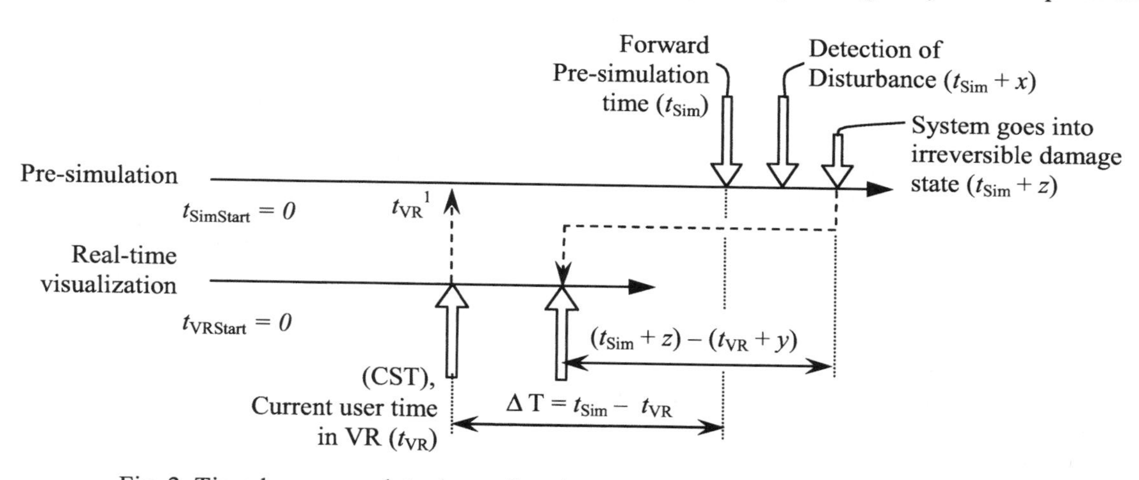

Fig. 2. Time-lag approach to detect disturbances and apply corrective policies

irreversible damage state is the state when a buffer is full and the preceding machine stops, because it can no longer send parts to the buffer. More on the issues related to considering the CST are discussed later. Now, we offer the following conjectures to the reader to help things fall in place: Conjecture 1: All deterministic disturbances will be detected by the pre-simulation mechanism, and will be handled by the tier 1 calculation phase. Conjecture 2: All probabilistic disturbances will be detected by the virtual simulation, and will be handled by the tier 2 calculation phase.

The computed corrective policies are then presented to the user. The sub-problems are to determine, by how much time does the pre-simulation lead the virtual simulation (ΔT), when does the user have to apply the corrective policy $((t_{\text{Sim}} + z) - (t_{\text{VR}} + y))$ (dotted lines with arrows in Figure 2) in the virtual world for all cases of disturbances, and for how long. It is also interesting to investigate if we could apply corrective policies as late as possible, with a view to keep the system as stable as possible, whilst still achieving the desired performance. The answer to when the user has to apply the corrective policies will have to be obtained not only by considering the CST, but also the information from the pre-simulation about how the virtual simulation will evolve over time.

When the user has these answers, his second task starts. A particular corrective policy will lead him to a certain location – a machine, a buffer, etc (this is explained in more details in the next section), where he navigates to in the virtual factory. This location can now be termed as a significant process that (Mueck et al. 2003 [11]) referred. We anticipate that when the number of corrective policies is too large, it may be wise to implement them at once, instead of navigating at numerous locations. The navigation leads to an immersive experience for the user. Situations where the user has to navigate through large physical distances between locations can also be handled in this setting, by stopping (and re-starting) the virtual simulation between navigating to the locations. After implementing the corrective policy (assume, at time $((t_{\text{Sim}} + z) - (t_{\text{VR}} + y)))$ (shown by dotted lines with arrows in Figure 2), the virtual simulation state and time is sent to the pre-simulation mechanism with the corrections (Figure 1). Again, both the virtual simulation and the pre-simulation mechanisms simulate the flow (starting both with state at time $((t_{\text{Sim}} + z) - (t_{\text{VR}} + y)))$, but with the addition that they now include the influence of the user's corrective policy. The time-lag mechanism repeats as explained earlier.
derivatives.

On the other hand, when there is no problem, the user continues to navigate in the system in a supervisory role using the walk-through system. In the following section methods to model the production system to optimize it are presented.

4. Modeling the manufacturing system

Control points will be implemented in the manufacturing system. The control points can be implemented as objects in the simulation system. Thus, a control point can be anything from a forklift to a machine. Alternative policies (resulting from decisions at control points) can be anything from an alternative material flow route within the factory or specific instructions for a specific part, to different steps of processing speeds for a machine. Alternative policies can be standard or special. Standard policies include, for instance, material flow routes, which cannot be changed, due to technological constraints. In this work special policies will be modeled separately, while standard policies will be "hard coded" during implementation. This means the system will be able to handle both these policy types simultaneously.

These policy definitions when expanded will result in a tree of alternative policies, which can be approximated as a network of alternative policies. The network will closely resembles the physical system configuration with the addition that if a machine\object has three policies to handle a part, then the machine will be replicated thrice in the form of nodes to reflect the three policies. Modeling the policies this way, makes it possible for the virtual user to relate a policy to one generic object (a machine, for instance). In the network, time, cost and production load derivatives required for handling one unit of flow in node j (which can, for instance be a machine) are placed on the preceding arc $i \rightarrow j$.

It is assumed that each policy will result in cost, time derivatives, while the production load derivatives will depend on number of parts assigned less the number of parts completed at any instant of time. Wherever it is not possible to place time and\or cost factors, either of these will be reduced to zero. If the underlying manufacturing system has more than one source and more than one sinks, then this can be modelled by placing dummy arcs from the super-source and super-sink to the several sources and sinks respectively and further adding zero costs, time and production load

5. Preliminary Tier 1 calculations

The tier 1 will compute a sub-optimal material flow based on the user KPI's like time (make-span) and costs. For the moment, no production loads on the factory are considered. The following are the known input about the system:

- time t required for each part a on machine j.
- number of parts n.
- cost k of operating each machine $j \longrightarrow$ hourly rate.
- number of buffers each with capacity B_{fi}.
- production rates $\xi_{in(i)}$ and $\xi_{out(j)}$ (measured in parts\min) before and after nodes i and j respectively.
- selective routing of parts n_p, $0 \leq n_p \leq n$, through the factory due to technological constraints of manufacturing n_p parts in the factory, if any.

The cost of operating each machine is obtained from the time product n took on machine j. Production rates can be obtained from the machine processing speeds. Then, the problem can be formulated as a network based LP, which will minimize make span and costs simultaneously. This concept is a combination of network theory and operations research (goal programming). We will make use of the nonpreemptive category where both functions are *of roughly comparable importance* (for detail treatment see Hillier, Lieberman 2001 [5]).

Note that it is feasible to have a large number of control points and the policies contained within the network. This is because the network simplex method can solve problems with several thousand nodes, having several thousand constraints in a matter of a few seconds.

6. Pre-simulation, real-time visualization, and interactive analysis

The tier 1 calculations can be solved using any of the commercial solvers. The material flow obtained, can then pre-simulated, and visualized, in our case, using *Technomatix eM-Plant*®. We will use the inbuilt time lag offered by *eM-Plant*® to model the time difference between pre-simulation and virtual simulation. Figure 3 shows visualization screen shots for a sample manufacturing system.

Assume that machine d randomly breaks down at some point in time during the virtual simulation. Because of the time-lag approach the virtual simulation will actually be behind the pre-simulator by time ΔT. Hence, we will parameterize the breakdown at the pre-simulation level, but ask it to occur at time (ΔT + time when breakdown occurs in the pre-simulation) during the virtual simulation. The pre-simulation will automatically detect a disturbance and will indicate where the disturbance occurred and when. Similarly, other disturbances can also be detected automatically by feedback from the pre-simulation or the virtual simulation.

As indicated earlier, the tier 2 calculations will be invoked to recover from the random machine breakdown. Tier 2 will consider alternative policies, KPI, etc., to compute corrective policies. Linking these corrective policies back to the objects will result in automatic detection of significant objects (processes). Tier 2 will mostly include algorithms which deal with reactive disturbances (several heuristic based algorithms have been developed, see Abumaizar and Svetska [1], Akturk and Gorgulu [2], Jain and Elmaraghy [6], Kim and Kim [7], Li et. al [8] and Yamamoto and Nof [12]). Because, the system immediately detects disturbances after they occur, it is possible that some time-lapse occurs before the system goes into a complete irreversible damage state. Hence, both tier 1 and tier 2 will have to provide control instructions which include decisions like when should the user be interacting with a significant process or processes.

If we assume that the result of the tier 2 calculation phase is to re-route the material flow (which was originally to be sent to machine d) to machine c, then our significant process would be object\buffer a. This is because, buffer a is the point where we defined the alternative control policies for the alternative material flow route. As discussed earlier the user may decide to navigate (for automatic motion planning algorithms see Mahajan et. al 2005 [9]) to the location and interact with the (significant process) object. The user will typically be given an option to implement the corrective policies from his current position, or to navigate to the significant process. Upon reaching the significant process, he will be able to select the process, and implement the corrective policy. This interaction could be done by sending messages to the pre-simulation mechanism, from the visualization. In the present case, the message would be instructions to change the control policy at buffer a. At this point, the loop is completed. Re-visiting Figure 1, if a new disturbance occurs, this time a deterministic order arrival, the tier 1 calculation would be activated, which would consider the algorithm described earlier.

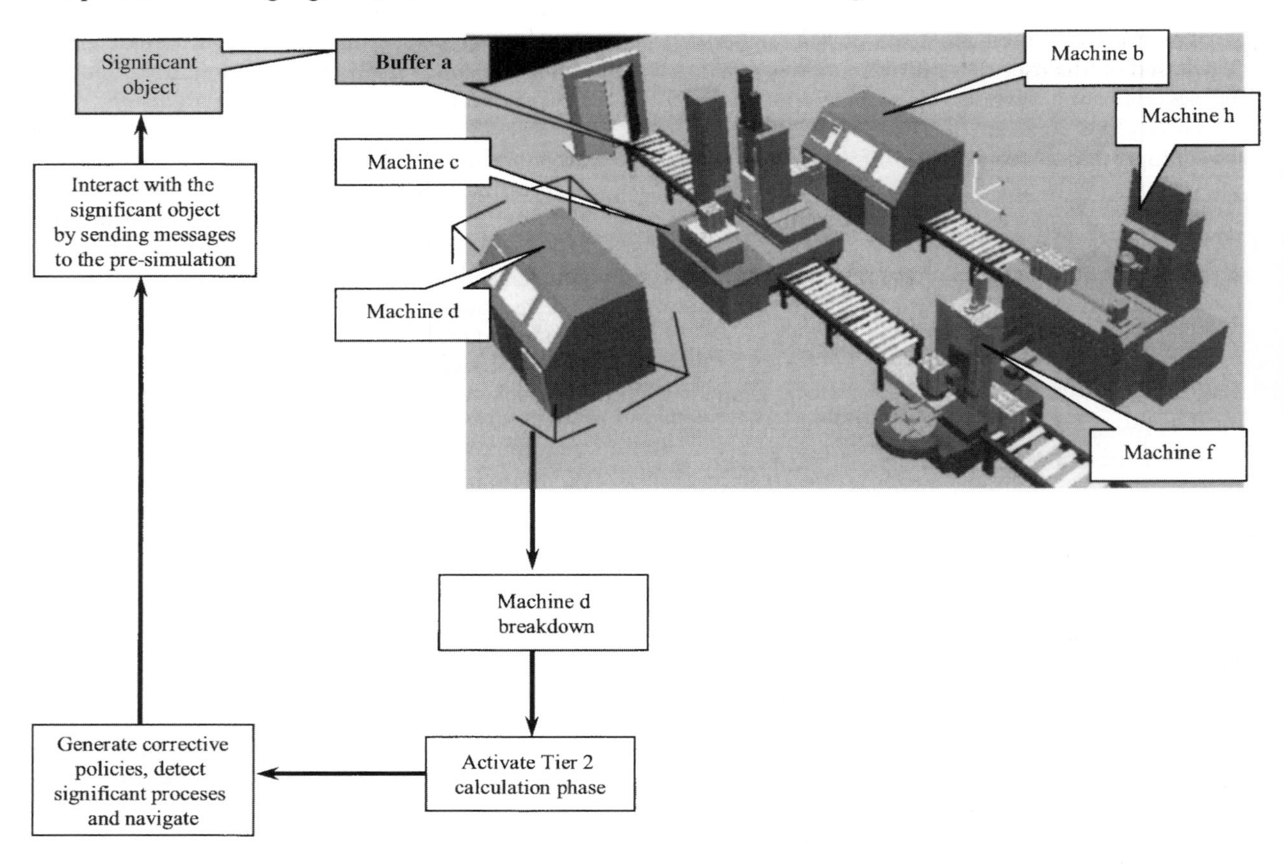

Fig. 3. Real-time visualization and control to support interactive analysis

7. Conclusions

Integration of visualization and real-time decision making algorithms discussed here serve as an inevitable starting point for the development of more sophisticated control algorithms in virtual environments. The real benefit from such an integration would be expected when the virtual simulation is made to emulate a real production facility. We have presented a theoretical framework to integrate work in the areas of visualization and real-time decision-making. We showed how problems can be automatically detected by the pre-simulation mechanism. Further, we discussed how significant processes could be identified, and how interaction between simulation and visualization could take place to correct a typical manufacturing problem. Much work remains to be done in the areas of tier 1 and tier 2 calculation phase, which will include algorithms for real-time decision-making based on the discussed disturbances. Work in these areas has been done in the past, but not in the context of virtual simulations and interactive analysis. In addition to this, the selection of the time difference between pre-simulation, and virtual simulation, and a basis for deciding when the virtual user should interact with the simulation, and for how long has to be established. Our future work will concentrate in these directions.

References

[1] Abumaizar, R. J., and J. A. Svestka. 1997. Rescheduling job shops under random disruptions. *International Journal of Production Research* 35 (7): 2065-2082.

[2] Akturk, M. S., and E. Gorgulu. 1999. Match-up scheduling under a machine breakdown. *European Journal of Operational Research* 112 (1): 81-97.

[3] Chong, C. S., A. I. Sivakumar, and R. Gay. 2003. Simulation-based scheduling for dynamic discrete manufacturing. In *Proceedings of the 2003 Winter Simulation Conference,* ed. S. Chick, P. J. Sanchez, D. Ferrin, and D. J. Morrice, 1465 - 1473. Piscataway, New Jersey: Institute of Electrical and Electronics Engineers.

[4] Harmonosky, C. M., R. H. Farr, and M. C. Ni. 1997. Selective rerouting using simulated steady state system data. In *Proceedings of the 1997 Winter Simulation Conference,* ed. S. Andradottir, K. J. Healy, D. H. Withers, and B. L. Nelson, 1293 -1298. Piscataway, New Jersey: Institute of Electrical and Electronics Engineers.

[5] Hillier, F. S., and G. J. Lieberman. 2001. Introduction to Operations Research. McGraw-Hill, ISBN 007232169.

[6] Jain, A. K., and H. A. Elmaraghy. 1997. Production scheduling\rescheduling in flexible manufacturing. *International Journal of Production Research* 35 (1): 281-309.

[7] Kim, M. H., and Y. Kim. 1994. Simulation based real time scheduling in a flexible manufacturing system. *Journal of Manufacturing Systems* 13 (2): 85-93.

[8] Li, R-K., Y.-T. Shyu, and S. Adiga. 1993. A heuristic re-scheduling algorithm for computer-based production systems. *International Journal of Production Research* 31 (1): 1815-1826.

[9] Mahajan, K. R., C. Laroque, W. Dangelmaier, S. Soltenborn, M. Kortenjan, and D. Kuntze. 2005. d^3 FACT insight: A motion-planning algorithm for material flow simulations in virtual environments. In *Proceedings of the 2005 Simulation and Visualization Conference*, ed. T. Schulze, G. Horton, B. Preim, S. Schlechtweg, 115 - 126. Erlangen Germany: SCS publishing house e.V.

[10] Minieka, E. 1978. Optimization algorithms for networks and graphs. Marcel Dekker, Inc. ISBN 0824766423.

[11] Mueck, B., Dangelmaier, W. and Fischer, M. 2003. Components for the active support of the analysis of material flow simulations in a virtual environment, In *Proceedings of the 15th European Simulation Symposium,* ed. A. Verbraeck and V. Hlupic, 367-371. Erlangen Germany: SCS publishing house e.V.

[12] Yamamoto, M., and S. Y. Nof. 1985. Scheduling\rescheduling in the manufacturing operation system environment. *International Journal of Production Research* 23 (4): 705-722.

Intelligent Production Machines and Systems
D. T. Pham, E. E. Eldukhri and A. J. Soroka (eds)

An integrated knowledge-based system for sheet metal processing

Y. Rao[a], J. Efstathiou[b]

[a]*School of Mech. Sci. & Eng., Huazhong University of Science & Technology, Wuhan 430074, China*
[b]*Department of Engineering Science, University of Oxford, Oxford OX1 3PJ, UK*

Abstract

This paper presents an integrated knowledge-based CAD/CAM system for sheet metal cutting-punching combination processing. The whole system consists of five functional modules, i.e. modeling, nesting, process-planning, NC-programming and simulation & reporting module, which are integrated seamlessly in terms of data processing. The knowledge base is the core of the integrated system, based on which, the system runs in an efficient and intelligent manner. This system has been successfully applied in industry, and an application example is demonstrated.

1. Introduction

Cutting and punching are two frequently applied processes in the sheet metal manufacturing industry. In general, the cutting process is conducive to manufacturing flexibility, whereas the punching process contributes to manufacturing efficiency and productivity. In order to combine the advantages of cutting and punching, some equipment manufacturers provide sheet metal producers with so-called combination machines, where the cutting and punching processes are combined. Namely, the complex-shaped-contour cutting and the small simple-shaped-hole punching can be carried out on one such machine without set-up. This kind of sheet metal processing is called combination processing. However, due to the process complexity of sheet metal components, the production preparation time for auxiliary activities (such as product design, optimal nesting, process planning and NC-code generation) is often remarkably longer than the sheet metal processing time on machine [1]. In order to make full use of such combination machines, a supporting CAD/CAM system is indispensable. In this paper, we present an integrated knowledge-based system for sheet metal cutting-punching combination manufacturing, which includes modeling and definition of sheet metal parts, optimal nesting, cutting/punching process planning, automatic NC programming, etc. This integrated CAD/CAM system has been successfully applied in several sheet metal production plants.

2. The modular structure of the system

The knowledge-based integrated system structure is depicted in Fig.1. The whole system consists of five functional modules, i.e. modeling, nesting, process-planning, NC-programming and simulation & reporting module. These modules share an unified objected-oriented data structure of sheet metal parts, based on which, the system is integrated seamlessly in terms of data processing. Furthermore, based on the knowledge base, the whole integrated system runs in an efficient and intelligent manner, starting with part modeling and ending with part processing on the machine.

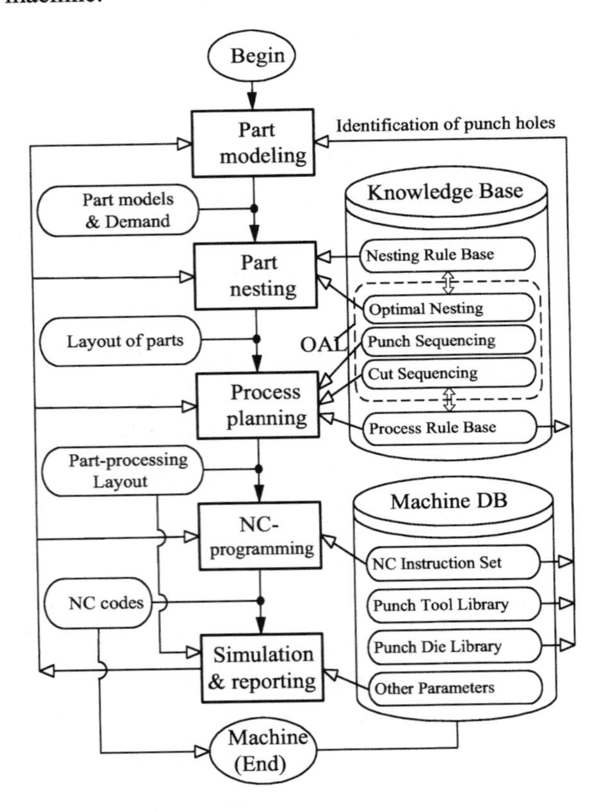

Fig. 1. The knowledge-based system structure

2.1. Modeling module

The modeling module is responsible for setting up the information models of sheet metal parts to be processed. There are three kinds of information in an established sheet metal part model, i.e. the geometrical information converted from outside CAD systems (e.g. AutoCAD) via DXF interface, the non-geometrical information (e.g. material, density) imported from outside databases (e.g. PDM system) or by manual input, as well as the information of the punching holes identified by the modeling module with the help of the process knowledge base and the machine database.

2.2. Nesting module

The task of nesting module is to decide the arrangement of the sheet metal parts onto the metal stock-sheet in such a way as to maximize the utilization ratio of the material whilst considering some sheet metal processing requirements. As showed in Fig.1, this module runs based on knowledge, which includes the nesting optimization algorithms and the nesting process rules. The nesting algorithms aim at high raw material utilization, whereas the process rules focus on sheet metal processing requirements and process optimization. The final optimal layout is a trade-off between the two objectives.

2.3. Process-planning module

The process-planning module takes responsibility of the cutting/punching process planning and optimization (such as punching tool sequencing and cutting path planning) for the sheet metal parts that have been laid out onto the stock-sheet by the nesting module. The knowledge for this module comprises the punching tool sequencing algorithm and cutting sequencing algorithm in the Optimization Algorithm Library and the cutting/punching process rules in the Process Rules Base. Some process rules are experiment-based (e.g. the rules for deciding cutting speed) while some are mainly empirical (e.g. the rules for deciding the type or length of an auxiliary cutting path).

2.4. NC-programming module

The NC-programming module automatically generates the NC codes suitable to the working of the machine tool. Firstly, according to the part-processing layout from the process-planning module, the NC-programming module generates the tool path data; secondly, in accordance with the NC instruction set of the machine tool, the tool path data are post-processed to generate the final NC codes able to carry out CNC working. In addition, the NC-programming module outputs some expected manufacturing data (e.g. the processing time and manufacturing cost) to the simulation & reporting module.

2.5. Simulation & reporting module

The simulation & reporting module is responsible for reporting the expected manufacturing results (such as processing time, manufacturing cost, material consumption and utilization ratio) according to the NC codes, part layout, some relevant processing parameters (e.g. process cost) stored in the machine database etc. There is a built-in simulation component in this module, which can simulate the sheet metal processing operation according to the NC codes and report such results as the punching sequence, cutting path and processing time. The report data can be used to improve the designing process within the system itself, as well as connecting the production planning and control system on shop floor.

3. The knowledge base of the system

The knowledge in the knowledge base (KB) is classified into three categories, i.e. the Nesting Rule Base (NRB), the Optimization Algorithm Library (OAL) and the Process Rule Base (PRB). The configuration of the KB and the KB-based system running information flows are depicted in Fig.1.

3.1. The Nesting Rule Base

The NRB is applied to the nesting module for considering the process constraints of sheet metal cutting-punching combination processing when generating the optimal layout. These nesting rules also work when the layout is improved via the human-machine interaction. Some examples of the nesting rules in NRB are listed below.

Rule 1: If there exists any punching hole (except circular shapes) in a part, then fix the orientation of the part or only allow it a 180 ° rotation.

Rule 2: If the shape of a part is long and narrow, then try to place the part close to the border of the stock-sheet to mitigate the heat distortion.

Rule 3: Try to lay out the small parts together (so that several adjoining parts may be continuously cut out, being pierced only once).

3.2. The Optimization Algorithm Library

The OAL comprises three categories of optimization algorithms for sheet metal cutting-punching combination processing: (1) sheet metal nesting algorithms, (2) punching tool sequencing algorithm and (3) cutting sequencing algorithm.

3.2.1. Sheet metal nesting algorithm

Generally, the objective of a nesting algorithm is

to minimize waste. So far, a great deal of work has been done on this well-known 'cutting and packing problem', and a lot of nesting algorithms have been developed [2]. In this study, the process requirements of cutting-punching combination manufacturing are fully considered. So, the nesting rules, as described in Section 3.1, are incorporated in the nesting algorithms. These nesting rules, however, may reduce the material utilization rate. Therefore, the trade-off between reducing waste and improving process is evaluated when generating a layout. We have developed several nesting algorithms [3] for different scenarios of the sheet metal part portfolio, and the rules for selecting appropriate algorithms are included in the NRB.

Algorithm 1: Nesting algorithm for identical complex shape parts.

Algorithm 2: Nesting algorithm for rectangular or close to rectangular shape parts.

Algorithm 3: Nesting algorithm for irregular shape parts.

Algorithm 4: Nesting algorithm in residual stock-sheets (with irregular borders).

Realizing that it is unlikely to generate purely automatic computerized solutions, we also develop a human-machine interaction module to improve the automatically generated layout.

3.2.2. Punching tool sequencing algorithm

The objective of punching tool sequencing in a layout is to save processing time in two aspects, i.e. the total tool-changing time and the total puncher-moving time. To save the tool-changing time, the frequency of tool change is kept to a minimum. First, those holes with the same shape and same size are classified into a certain type, which can be punched with a same tool. Second, the holes grouped in a same type are punched sequentially after one tool change. To save the puncher-moving time, the total puncher-moving distance should be the shortest, which can be formulated as a 'traveling salesman model'. As a typical NP-completeness problem, a lot of heuristic algorithms have been developed to provide a solution, such as the nearest neighbor algorithm, the shortest insert algorithm and the optimum weight spanning-trees algorithm. In this study, we adopt the nearest neighbor algorithm with the constraint of least frequency of changing tools for the punching tool sequencing.

3.2.3. Cutting sequencing algorithm

The most important objective of cutting sequencing is to ensure the parts' cutting quality (mainly by mitigating heat distortion) rather than minimize the total cutter-moving distance, because the benefits of the former obviously outweigh the loss associated with the latter. A rational objective of cutting sequencing is to minimize the total cutter-moving distance with the precondition of cutting process optimization. In this study, we develop a process-oriented cutting sequencing algorithm based on the following rules that ensure the cutting quality.

Rule 1: The inner contours of a part must be cut before the outer contour of the part.

Rule 2: The nesting parts must be cut before the part that contains them.

Rule 3: The part close to the border of the stock-sheet, which may be dynamically changed with part removals, should be cut first.

Rule4: With the precondition of Rule 3, the smaller parts take precedence over the bigger ones.

Rule 5: With the precondition of Rule 4, the nearest neighbor principle is adopted.

Based on these rules, a two-step cutting sequencing algorithm is developed as follows.

Step 1:

(1) Select a small part in a corner of the stock-sheet

as the first part (P1).

(2) If P1 contains some inner contours, then sequence these inner contours (using *Rule 4* and *Rule 5*) and add them before the P1's outer contour (*Rule 1*); otherwise, set the P1's outer contour as the first contour, and go to Step 2.

(3) If the P1's inner contours contain some nesting parts, then, according to the order of these inner contours, sequence the nesting parts of each inner contour (by calling this sequencing algorithm in a recursive manner) and add them before the inner contour (*Rule 2*).

Step 2:

(1) If all parts are sequenced, then the algorithm ends; otherwise select a neighbor of P1 as the next part (P2). This neighbor is chosen in compliance with the *Rule 3* and *Rule 4*.

(2) If P2 contains some inner contours, then sequence these inner contours and add them before the P2's outer contour; otherwise, set the P2's outer contour as the next contour, and let P1 = P2, go to Step 2.

(3) If the P2's inner contours contain some nesting parts, then, according to the order of these inner contours, sequence the nesting parts of each inner contour (by calling this sequencing algorithm in a recursive manner) and add them before the inner contour. Let P1 = P2, and go to Step 2.

After the cutting sequencing, the next step is to optimize the whole cutting path in accordance with the process knowledge in the PRB, which is described in the next section.

3.3. The Process Rule Base

The PRB is applied to the process-planning module for the sheet metal cutting/punching process planning in a layout. Some examples of the process rules in the PRB are described below.

Rule 1: Whether an inner contour of a part is eligible to be a punching hole, depends on the material and thickness of the part, the precision requirement of the contour, in addition to the shape of the contour and the available punching tools.

Rule 2: All punching holes should be processed before contour cutting.

Rule 3: If a contour is to be cut, the piercing point should be set on a waste area, and a 'cut-in' and a 'cut-out' auxiliary path should be appended to the cutting path of the contour. This rule is subdivided further according to different cases.

To ensure cutting quality, the cutting process of a contour usually begins and ends from a waste area nearby rather than from the contour itself. Starting from the piercing point, the cutting path leading to the contour is called 'cut-in' auxiliary path. After the pure contour cutting, the cutting path leaving from the contour is called 'cut-out' auxiliary path. The setting of the auxiliary paths of a contour, including their types (line or arc), positions and lengths, depends on the nature of the contour, the starting point, the surrounding area, as well as the material and thickness of the stock-sheet. In the PRB, there are a series of rules to ensure the auxiliary paths of a contour to be appended correctly.

4. Case study

The integrated system has been successfully applied in several sheet metal manufacturing plants. In this section, we introduce an application example in such a plant, where two punching-cutting combination machines (i.e. the 647ATC and 3500ATC both made in USA) are used to process sheet metal parts for the cabs of construction machinery.

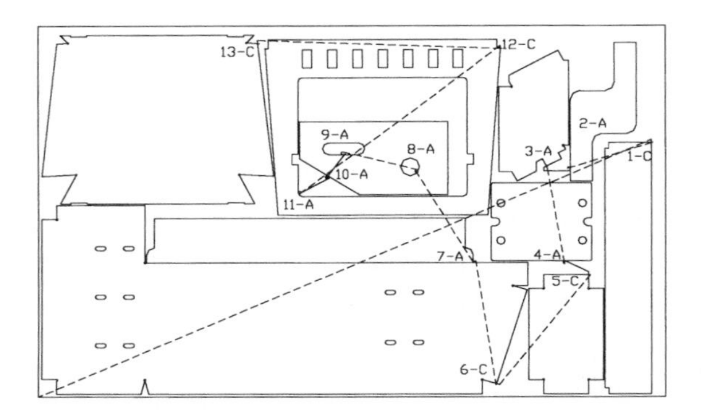

Fig.2. The optimized cutting path

Fig.2 shows an application example for processing 10 sheet metal parts (material = Q235, thickness = 2mm) in the stock-sheet sized 2000mm×1250mm. The applied optimal nesting algorithm is the Algorithm 3 in the OAL. In Fig.2, the cutting path planning result is described, where the numbers in order indicate the contour cutting sequence, and the capital 'C' or 'A' following a sequence number denotes the cutting direction of the corresponding contour: 'C' for clockwise and 'A' for anti-clockwise.

The machine employed in this case is the 3500ATC. The cutter's moving speed of the machine is 5000mm/min. The cutting speed is set at 2000mm/min, and the piercing time is set at 10seconds/time. In addition, the machine requires 50 seconds for a punching tool change, and 5 seconds for a hole punching. By processing simulation, the system reports the expected punching time 4.915min, the expected cutting time 14.399min and the expected total processing time 19.314min.

5. Conclusions

In order to make full use of the cutting-punching combination machines, a supporting CAD/CAM system, which integrates the functions of part modeling, nesting, process planning, NC-programming and simulation & reporting for sheet metal combination processing, has been developed. Based on the knowledge base, the system runs in an efficient and intelligent manner. The integrated system, which is developed in Visual C++ 6.0 and provided with a friendly human-machine interface, has been successfully applied in several sheet metal plants. Its applications, one of which is exemplified in this paper, have brought great benefits to the plants in terms of raw material utilization, manufacturing cost, productivity and processing flexibility.

Compared with other conventional approaches, the presented system is expandable without too much programming effort, thanks to its independent knowledge base. The further work is focused on the improvement and expansion of the KB so as to make the system more applicable.

Acknowledgement

University of Oxford is a partner of the EU-funded FP6 Innovative Production Machines and Systems (I*PROMS) Network of Excellence.

References

[1] Ehrismann R. and Ressner J. Intelligent manufacture of laser cutting, punching and bending parts, Robotics and Computer-Integrated Manufacturing, Vol.4, No.3/4, (1988) 511-515.

[2] Dowsland K. A. and Dowsland W. B. Solution approaches to irregular nesting problems, European Journal of Operational Research, 84 (1995) 506-521.

[3] Guo J. Sheet metal nesting algorithms considering the cutting process, Master Thesis, Huazhong University of Science and Technology, 2000 (in Chinese).

Intelligent Production Machines and Systems
D. T. Pham, E. E. Eldukhri and A. J. Soroka (eds)

Artificial intelligence methods a challenge for the modern polymer chemistry

Teodora Rusu [a], Oana Marlena Gogan[b]

[a] *"Petru Poni" Institute of Macromolecular Chemistry, Iasi, Romania, e-mail: teia@tuiasi.ro*
[b] *ARTI INFOMATICHE, ITALY, e-mail: ogogan@cs.tuiasi.ro*

Abstract

Solving today's challenges in polymer science and engineering will no doubt shape the future. The polymer science and materials engineering join forces with biology and medicine informatics to produce modern miracles and the interdisciplinary collaboration remain the true source of benefits.
The development of a macromolecular compound is an extremely laborious and time-consuming process. Molecular modeling and rational molecular design have become indispensable tools for the development of new compounds. Evolutionary techniques can help achieve the design of totally new molecules, some of them never even thought before. This paper presents the Artificial Intelligence approach as an alternative in the polymer designee.

---

1. Introduction

The chemical industry is sometimes referred to as the "keystone" industry because of the way the rest of the manufacturing sector relies on chemicals. Chemicals is nearly a $1.5 trillion global enterprise, and it is also the second largest consumer of energy in manufacturing and spends over $5 billion annually on pollution abatement.
In the development of the chemical industry a special role is held by the polymers industry. Polymers are macromolecules composed by repetition of a single unit (monomer). Some polymers occur naturally, such as starch, protein, and cellulose (cotton and wood), etc. Natural polymers have been used from long time in the manufacturing industry.
A real cultural and commercial revolution came when chemists began to synthesize polymers in the lab[1, 2]. These polymers, appropriately called synthetic polymers, can be found in everything from plastic bags to rubber tires, from water-proof fabrics to cell phones. Most polymer engineers feel the days of a market dominated by a few polymers are numbered. Many see the promise of the future in specialty polymers, which will be fashioned with complex sets of properties designed for specific needs but produced in relatively law volumes. Reinforcing polymer resins with graphite, boron, and super-strong amid fibers has already led to advanced composites that form the bodies of many of the latest spacecraft.
Reinforced plastics are economically competitive with metals in a business that must produce multiple makes of cars in a single year. Sheet-molded polyester and injection-molded polyurethanes reduce auto weight, increase fuel efficiency, and resist corrosion and dents Few parts of a car are incapable of being replaced by composites.

The phenomenal growth of electronics and computers in recent years has catalyzed research into the creation of conductive polymers. Plastics can be made electrically conductive through the addition of carbon or metal flakes. New research is aimed at making polymers intrinsically conductive by adding chemicals like iodine and sulfur trifluoride to the polymer chain. These altered polypyrroles, polyacetylenes, and polyquinolines are too susceptible to decay to be marketable yet. Still, these modified polymers and molded plastics have promising futures as static dissipaters in micro circuitry and as plastic batteries. Work in biopolymers also promises much for the years ahead. In the three decades since the discovery of the most complex nucleic acid structures, biochemists and molecular biologists have spliced genetic material, synthesized hormones, and created new strains of life. Polymer plastics and tissues work well together now, but synthetic biopolymers will far outshine today's techniques.

Production technologies of the future will combine tradition with innovation. Demand for composites in the auto industry has already led to a boom in the century-old process of compression molding. Scientists are utilizing new methods for combining previously incompatible ingredients, creating alloys and blends which often have properties better than any of their components. Commercialization is imminent for a new type of polymer factory inside the body of a living microbe. And astronauts have already formed the first polystyrene beads in space.

The first polymer inventors made progress by transforming natural materials in hit-or-miss fashion, but their work greatly accelerated after basic researchers clarified the fundamental characteristics, such as the relation between size or molecular weight and physical properties, that govern the behavior of polymers. In present the polymer science and materials engineering join forces with biology and medicine informatics to produce modern miracles and the interdisciplinary collaboration remain the true source of benefits. Solving today's challenges in polymer science and engineering will no doubt shape the future.

Modern chemistry has made important progresses in developing and testing theoretical concepts of new macromolecular compounds, with the structure related to impose properties for new areas of applications. The development of a macromolecular compound is an extremely laborious and time-consuming process. Molecular modeling and rational molecular design have become indispensable tools for the development of new compounds.

Evolutionary techniques can help achieve the design of totally new molecules, some of them never even thought before. This paper presents the Artificial Intelligence approach as an alternative in the polymer designee.

2. Polymer structures and properties

To apply the Artificial Intelligence techniques in computer-aided molecular design (CAMD), two things must be known to some extent: (a) the properties that are desired and how they relate to (b) the molecule's structure. The structure–activity relationship is needed to both determine the necessary properties and build a molecule that has those properties.

2.1 Polymer structure

Hybrid materials with different molecular architectures have attracted an increased interest in recent years.

Our interest focuses on copolymers containing hydrophilic and hydrophobic sequences that confer them special transport properties. The molecular ratio between the two sequences has an important role for the properties of such copolymers like shape-selective separation and water (small size particles) delivery.

It is known from literature that hydro gels reveal phase transitions making swelling-deflation changes in response to environmental modifications e.g. solvent composition [4, 5], ionic strength [6], temperature [7 - 10], electronic field [11] and light [12]. Such stimuli responsive polymers have been investigated with application to artificial muscles [13] immobilization of enzymes [14], concentration of dilute solutions [15, 16] and chemical valves [17].

Using stimuli responsive amphoteric gels did attempts for temporal control of water release. Thus, a swollen gel in ethanol immersed in water, undergoes immediately spontaneous translation and rotational motions with two-third of its volume immersed in water. In the course of its motion the gel gradually sinks and finally settles at the bottom of the vessel due to the contraction and cease of the motion [18]. The speed, duration and mode of gel motions are associated with its size, shape and chemical nature. Such polymer gels, exhibiting motion in water, have potential applications as soft-touch manipulators, target drug-delivery devices, micro-agitators, and micro generators.

Networks prepared by reacting functionalized polydimethylsiloxanes (PDMA) with different cross-linking agents are considered ideal models for studying

the network formation and properties [19]. PDMS is highly hydrophobic, very flexible and show a low variation of its properties with temperature. Since its mechanical properties are poor, for their improvement one should synthesize cross-linked copolymers based on PDMS [20].
In our group we have synthesized amphoteric networks (gels) based on PDMS – co – PMAA sequences. The polymer gels were obtained by radical copolymerization of polydimethilsiloxane macroinitiators with methacrylic acid in presence of ethileneglicol dimethacrilat as crosslinking agent[21]. The synthesis of the azoester macroinitiators (AzoPDMS), with different molecular weights of siloxane sequences and different contents of azo-groups, was realized according to Scheme 1.

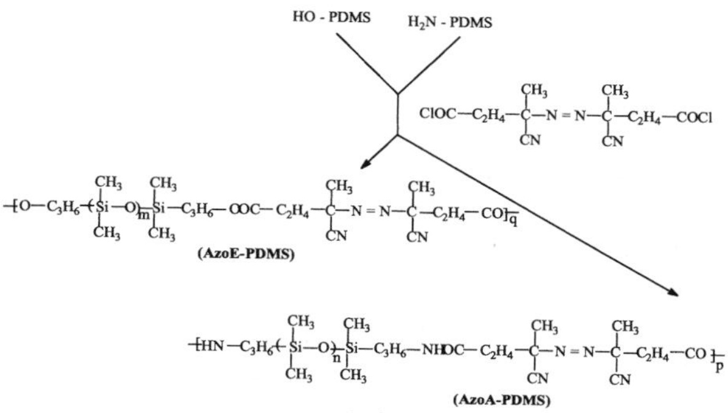

HO - PDMS H_2N - PDMS

(AzoE-PDMS)

(AzoA-PDMS)

Scheme 1

The synthesis of PDMS - poly(methacrylic acid) (PMAA) hydrophobic – hydrophilic gels was realised according to Scheme 2.

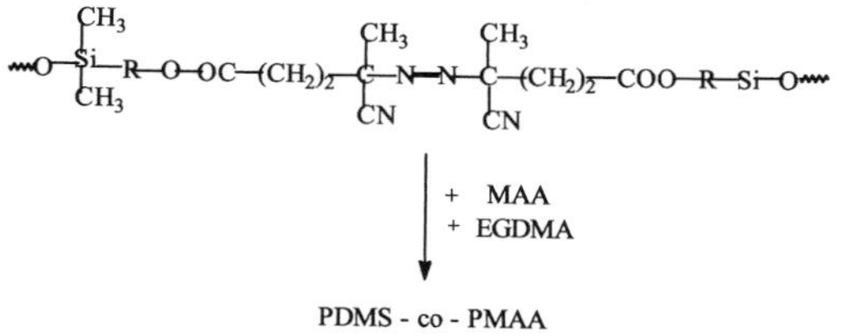

+ MAA
+ EGDMA

PDMS - co - PMAA

Scheme 2

2.2 Polymer properties

Physical properties of polymers are governed by three main factors:

- Number of monomer units in the chain, N
- Monomer units are connected in the chain. ⇒ They do not have the freedom of independent motion (unlike systems of disconnected particles, e.g. low molecular gases and liquids). ⇒ Polymer systems are poor in entropy.
- Flexibility: polymer chains are generally flexible.

Rectilinear conformation of a poly(ethylene) chain corresponding to the minimum of the energy. All the monomer units are in trans-position. This would be an equilibrium conformation at $T = 0$.

At $T = 0$ due to thermal motion the deviation from the minimum-energy conformation are possible. According to the Boltzmann law the probability of realization of the conformation with the excess energy U over the minimum-energy conformation is

$$pU \sim \exp(-\frac{U}{kT}) \qquad (1)$$

For carbon backbone the valence angle γ is fixed (for different chains $50° < \gamma < 80°$), however the rotation with fixed φ(changing the angle of internal rotation γ) is possible. Any value $\gamma \neq 0$ gives rise to the deviations from rectilinear conformation, i.e. to chain flexibility.
The flexibility is located in the freely-rotating junction points. This mechanism is normally not characteristic for real chains, but it is used for model theoretical calculations.
The linking of hydrophobic sequences, i.e., polydimethylsiloxanes (PDMS), to PMAA chains results in an amphiphilic material with solubility and solution behaviour depending on the molecular weights and molar ratio between the hydrophobic and hydrophilic segments. In solvents selective for one of the blocks, such copolymers form aggregates with a rather dense core issued of the insoluble blocks and surrounded by diffuse coronas formed from the soluble ones. Such micellar associations are of different types and, in many respects, resemble those of low molecular mass amphiphiles. However, the thermodynamic factors determining the association of block copolymers in organic solvents are different from those acting for amphiphiles in aqueous media; in the first case, the enthalpy contribution to the free energy change is responsible for the association phenomena, while for low molecular mass amphiphiles in water a positive standard entropy of micellization is the determining factor. Block copolymers containing hydrophobic and hydrophilic segments are excellent candidates for experimental investigations aimed at elucidating the driving forces determining the miscellization phenomena. Their solubility characteristics can be easily modified through changing the ratio between the hydrophobic to hydrophilic segments and additional parameters - such

as the chemical nature of the substitutions attached to the main chain, molecular weight a.s.o. - can be also varied and studied. Moreover, one can also consider polymeric structures containing not only neutral - neutral block, but also block copolymers composed of a neutral polymer linked to a polyelectrolyte.
The present approach aim to combine statistical mechanics of polymers with a coarse-grained simulation approach in order to develop a molecular-level understanding of the thermodynamics of such polymers.

3. Designee solution

In general, computer-aided molecular design requires the solution of two problems: the *forward* problem, which requires the computation of physical, chemical and biological properties from the molecular structure, and the *inverse* problem, which requires the identification of the appropriate molecular structure given the desired physicochemical properties [22, 23]. In the present study, we pursue two research themes in the genetic design framework. One is to investigate the efficacy of genetic design for problems with much larger and more complex design spaces. The second theme is to extend the original genetic algorithmic framework by incorporating higher-level chemical knowledge to better handle constraints such as chemical stability and molecular complexity by use of a *Multiobjective Tabu Search*.

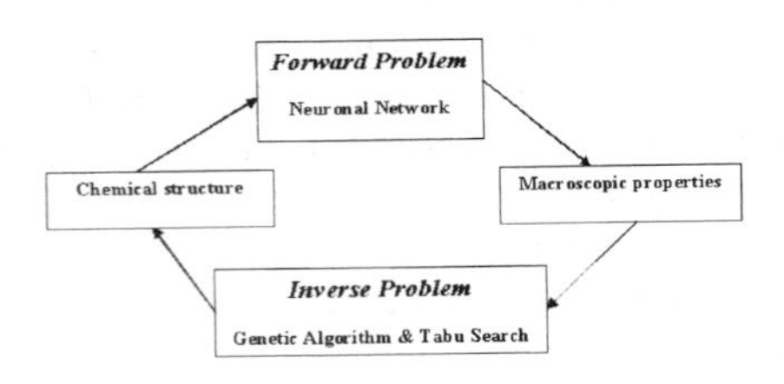

Scheme 3

The task of training sub-symbolic systems is considered as a combinatorial optimization problem and solvėd with the heuristic scheme of the Reactive Tabu Search (RTS)[24-26]. An iterative optimization process based on a ``modified greedy search" component is complemented with a meta-strategy to realize a discrete dynamical system that discourages limit cycles and the confinement of the search trajectory in a limited portion of the search space. The possible cycles are discouraged by prohibiting (i.e., making tabu) the execution of moves that reverse the ones applied in the most recent part of the search, for a prohibition period that is adapted in an automated way. The confinement is avoided and a proper exploration is obtained by activating a diversification strategy when too many configurations are repeated excessively often. The RTS method is applicable to non-differentiable functions, it is robust with respect to the random initialization and effective in continuing the search after local minima. The limited memory and processing required make RTS a competitive candidate for special-purpose implementations. The collection of theoretical data are then supplied to a neural network that is able to make the optimum selection according to the thermodynamic properties that it was trained to relate to a given structure[27, 28].

3.1. Genetic algorithm approach

Genetic techniques can use cyclic graphs to represent molecules. Vertices are typed by atomic elements. Edges can be single, double, or triple bonds (see fig. 1). Valence is enforced. The genetic graph software we propose evolves the population using crossover based on Tabu Search evaluator only; i.e.., mutation and reproduction are not implemented. These are trivial additions to the method, and we wanted to investigate the crossover operator first.

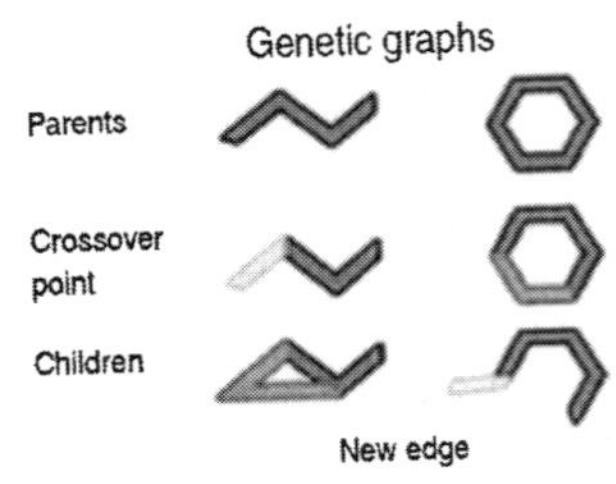

Figure 1. Genetic graphs manipulate molecules with two atoms represented as vertices, and the bond in between as an edge.

Each individual in the initial population is generated by choosing a random number of atoms. These atoms are randomly chosen from the elements present in the target molecule. Bonds are then added at random to construct a spanning tree; i.e., at this point all atoms are connected into a single molecule. Then a random number of additional bonds are added to create rings. The type of each bond is selected at random from the set of bond types present in the target molecule. The number of rings is chosen to be between half and twice the number of rings in the target molecule. The number of rings, by our definition, is always = bonds – atoms + 1. For this definition, single, double, and triple bonds are counted as one bond.

3.2. Multiobjective Tabu Search

The multiobjective Tabu Search algorithm is a population-based simulation for outline objectives combinatorial optimisation problems. Each solution maintains its own tabu list and the weights are adjusted in order to keep the solutions away from their neighbours and therefore attempt to cover the whole trade-off surface. The population of solutions explore their neighbourhood similarly to the classical simulated annealing, but weights for each objective are tuned in each iteration in order to assure a tendency to cover the trade-off surface. The weights for each solution are adjusted in order to increase the probability of moving away from its closest neighbourhood in a similar way as in the multiobjective tabu search algorithm [29]. From simulated annealing, this hybrid metaheuristic borrows the idea of neighbourhood search, probabilistic acceptance of candidate solutions and the dependence of this acceptance from a temperature parameter. From genetic algorithms, the approach incorporates the idea of using a sample population of interacting solutions.
The algorithms we used are: TS, simulated annealing and a GA (a population based approach with a local search). All the algorithms were searched for an ordering of polygons, that is the order in which they will be packed. The best results were obtained using a hibrid algorithm (genetic algorithm with TS algorithm).

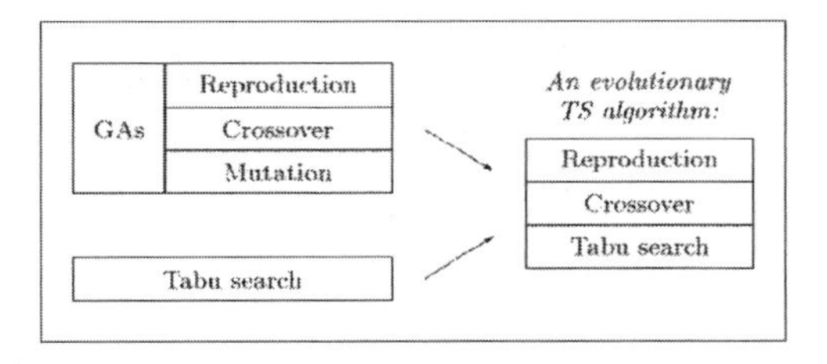

Figure 2. Schematic TS combination with genetic algorithm

For this work, the Tabu Search selection was used to choose parents in a steady state genetic system. The Tabu Search selection means that each parent is chosen by comparing to a Tabu List. Steady state means that new individuals (children) replace poor individuals in the population in agreement with the restrictions imposed by the Tabu List. By convention, after population-size individuals have been replaced, the generation is complete. The implementation follows this procedure:

```
1. Generate a random population of
      molecules
2. Repeat many times, gathering data
      periodically:
   1. Select two molecules from the
      population at random. Call the better
      molecule father.
   2. Select two molecules from the
      population at random. Call the better
      molecule mother.
   3. Make a copy of father and rip it into
      two fragments at random.
   4. Make a copy of mother and rip it into
      two fragments at random.
   5. Combine one fragment of the copy-of-
      father and one fragment of the copy-
      of-mother into a molecule called son.
   6. Combine the other fragment of the
      copy-of-father and the other fragment
      of the copy-of-mother into a molecule
      called daughter.
   7. Choose two molecules from the
      population at random. Replace the
      worst one with son.
   8. Choose two molecules from the
      population at random. Replace the
      worst one with daughter.
3. Repeat until satisfied
```

3.3. Neural network approach

Neural network (NN) applications to chemistry have been explored in recent years. The essential abilities and the flexibility of NN's are related to the interconnection of individual arithmetic units. Many kinds of networking strategies have been tested for chemistry application ad the one to have best result is the *"back-propagation"* (BP) algorithm. The BP network does not represent any particular architecture (a multi-layered net is generally used), but rather a special learning process.
The learning through back-propagation stems from that, the adjustments to the neural net's weights can be calculated on the basis of well-defined equations.
If $w_{ij}(n)$ represents weight of j neuron before fit and $w_{ij}(n+1)$ the weight after the fitting procedure, the Hebb's law for learning is:

$$w_{ij}(n+1) = w_{ij}(n) + c\, y_i\, y_j \quad (2)$$

Where y_i is output for i neuron (input for j) and y_j is the output for j, thus

$$w_{ij}(n+1) = w_{ij}(n) + c\, S(y_i)\, S(y_i) \quad (3)$$

If more than one neuron is activated, than the network forms local clusters, which depend on the input signals. The equation for the output of j will be:

$$y_j = \varphi\left[I_j + \sum_{k=-K}^{k=K} c_{j,k} y_{j+k} \right] \qquad (4)$$

Where I_j is the stimulus j: $I_j = \sum_{i=1}^{m} w_{ji} x_i$ and $\varphi(.)$ is a nonlinear function.

Nevertheless this procedure for correcting errors has very little in common with those processes responsible for the adjustment of synaptic weights in biological systems. The back-propagation algorithm may be used for one or multi-layered networks and is a supervised learning process[30].

4. Results and discussions

The tests were made by using experimental dates of thermodynamic properties for the PDMS-co-PMAA copolymers and carried out 8000 evaluations. Identical runs were carried out using the non-convex no fit polygon algorithm and the number of evaluations were adjusted so that the algorithms ran for about the same amount of time. To achieve this the searches were only allowed to carry out 300 evaluations.

Table 1:
Experimental dates fitness value for the PDMS-co-PMAA

Mating Pool after Reproduction with Crossover Site	Mating Parameter	Crossover Site	New Population	Integer Value	Fitness Value $f(x) = x^2$
010|10	3	3	01010	10	100
10|101	4	2	10010	18	324
110|10	1	3	11010	26	676
11|010	2	2	11101	29	841
Sum					**1941**
Average					**485.25**
Max					**841**
Schemata	**After Reproduction**		**After Crossover**		
Expected Count $\frac{\sum f(H)}{f}$	**Actual Count**	**Individual**	**New Count Expected**	**Actual Count**	**Individual**
0.01	0	-	0.00	0	-
1.83	3	2,3,4	2.54	3	2,3,4
0.33	3	1,3,4	1.60	2	1,3

The results for the different designed cases gives a percent of success rate of about 0.87 – 0.99 in achieving the design objective and the number of successful runs.

The average generation when the target was first located and the average number of distinct high-fitness solutions (see fig 3) at the end of the genetic design are in a reasonable limit and the error of the system is less than 3 % [29, 31]

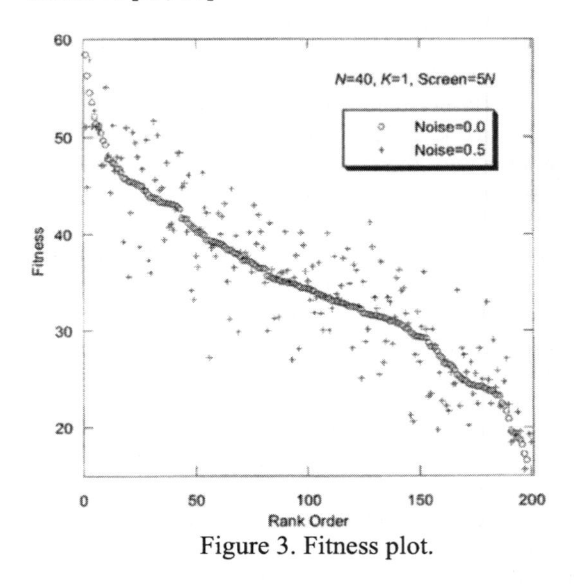

Figure 3. Fitness plot.

The abscissa is the rank order of fitness values based on the unperturbed fitness value. The ordinate is the fitness value before and after the addition of noise. Applying a random Gaussian error whose standard deviation is 50% (Noise = 0.5) of the standard deviation that generates the perturbed fitness values.

5. Conclusions

Genetic algorithms show the potential for the discovery of new structures that would normally not be considered by researchers. This advantage may increase the yield of discovery, dramatically influencing the development of chemistry in the coming decades. The advanced Tabu Search for the case matching will be a key point for future works.

Acknowledgement:

“Petru Poni” Institute of Macromolecular Chemistry, Iasi, Romania is associated partner of the EU-funded FP6 Innovative Production Machines and Systems (I*PROMS) Network of Excellence.

References

[1] Hermann Staudinger. From Organic Chemistry to Macromolecules: A Scientific Autobiography Based on Original Papers. New York: John Wiley, 1970.

[2] G. S. Whitby, ed. Synthetic Rubber. New York: John Wiley, 1954 .

[3] Ruck MJ. High-resolution spectroscopy of late-type stars. D.Phil. Thesis, University of Oxford, UK, 1994.

[4]Tanaka, F. and Ishida, M: *Physica A* 204 (1994), 660

[5] Katayama S., Hirokawa Y., and Tanaka T., Macromolecules, 17, 2462-2463,1984;

[6] Hirokawa Y., Tanaka T., J. Chem. Phys., 81, 6379-6380, 1984;

[7] Ohmine I., Tanaka T., J. Chem. Phys, 77, 5725-5729, 1982;

[8] Tanaka T., Fillmore D., Sun S. T..,Nishino I., Swislow G., Shah A., Phase Transition in ionic gels. Phys. Rev. Lett., 45,1636-1639, 1980;

[9] Tanaka T. Gel Sci. Am. , 244, 124-138, 1981;

[10] Tanaka T., Nishio I., Sun S. T., Ueno-Nishhio S., Collapse of Gels in an Electric-fielf Science, 218, 467-469, 1982;

[11] Irie M, Kunwatchakun D., Macromolecules, 19, 476-2480, 1986;

[12] Suzuki M., Polymer Gels,Ed. D. DeRossi, Plenum Press, New York, pp. 221-236, 1991

[13] Hoffman A. S., J. Control. Release, 6, 297-305, 1987;

[14] Freitas R. F. S., Cussler E. L., Chem. Eng. Sci, 42, 97-103, 1987;

[15] Trank S. J., Johnson D. W.,, Cussler E. L., Food Technol., 43, 78-83, 1989;

[16] Osada Y, Hasebe M., Chem. Lett., 9, 1285-1288, 1985.

[17] Hamurcu, E. E. and Bahattin, M.: Macromol. Chem. Phys. 196 (1995), 1261.

[18] He, X. W., Windmaier, J. M., Herz, J. E. and Magers, G. C.: Polymer 33 (1992), 866.

[19] Rusu, T., Pinteala, M., Iojoiu, C., Harabagiu, V., Cotzur, C., Simionescu, B. C., Blagodatskikh, I. and Shchegolikhina, O.: Synth. Polym. J. 5 (2) (1998), 29.

[20] Harabagiu V., Hamciuc V., Giurgiu D., Makromol. Chem., Rapid Commun., 11 (1990), 433.

[21] Rusu, T., S. Ioan, S. C. Buraga, Euro. Polym J, 37, 2005, 2001

[22] Venkatasubramanian, V.; Chan, K.; Caruthers, J.M. Designing Engineering Polymers: A Case Study in Product Design. *AIChE Annual Meeting*. Miami, FL, November, 1992, 140d

[23] J. Gasteiger, J, Zupan, , Angew. Chem. Int. Ed. Engl. 1993, 32, 503-527

[24] Hansen M.P., Tabu Search for Multiobjective Optimization: MOTS, Technical Report Presented at 13th International Conference on MCDM, Technical University of Denmark, 1997

[25] MM Mwembeshi, CA Kent and S Salhi, Chemical Engineering & Technology Journal , 2004, 27 (2) 130-138

[26] MM Mwembeshi, CA Kent and S Salhi , Computers and Chemical Engineering, 2004, 28 (9), 1743-1757

[27] Rusu, T, O. M. Gogan, Proceedings of Franco-Român Symposium on Applied Chemistry CoFrRoCA 2002, 129 – 132

[28] R Smithies, S Salhi and NM Queen, Neural Computation, 2004, 6, No 1. 139-157

[29] R Smithies, S Salhi and NM Queen (2005) "Predicting colorectal cancer recurrence: a hybrid neural network-based approach". in Metaheuristics: Progress as Real Problem Solvers (T. Ibaraki, K. Nonobe and M. Yagiura, Eds), Kluwer. Chap 12, 257-283

[30] T. Rusu, O. M. Gogan, Mol. Cryst. Liq. Cryst., Vol. 416, pp. 155–164, 2004

[31]T. Rusu, V. Bulacovschi, "Multiobjective Tabu Search Method Used in Chemistry ", Proceedings of The Second Humboldt Conference on Computational Chemistry, Nessebar, Bulgaria, September, 1-5, 2004

Intelligent Production Machines and Systems
D. T. Pham, E. E. Eldukhri and A. J. Soroka (eds)

Construction using combined materials – possibilities and limits

P. Dietz[a], T. Grünendick[a], A. Guthmann[a], T. Korte[a]

[a] *Institut für Maschinenwesen, TU Clausthal, Robert-Koch-Str. 32, 38678 Clausthal-Zellerfeld, Germany*

Abstract

Based on examples, the challenging task of construction using combined materials is presented in this article. First, recommendations for the correct design methodology are specified to meet the particular requirements for using combined materials. This is followed by examples out of the area of process engineering machines that represented the starting point for the specific methodology development. The advantage of using plastic deformation for an increasing load-bearing capacity is described in detail, it is proved that a combination of composites and metal leads to an optimisation of products, and the usage of ceramics for ventilators used under extreme conditions is presented.

1. Introduction

It is not sufficient anymore to focus only on approved processes and already existing, successful products to assure the competitiveness and existence of an enterprise on the market. It is indispensable to respond to customer and market needs with high innovation ability, new processes, products and approaches to assure the existence of the company.

Electronics and computer technology are meanwhile significant components of product development and construction. Whereas the possibility to increase the competitiveness of a company by usage of (new) materials and its innovation potential is noted so far only insufficiently.

The requirements of modern and market-driven products demand for a rethinking of the designer in this area. It is necessary not to insist furthermore to the well-known and approved materials and its conventional use, but rather to use and to combine materials in a functional way corresponding to the requirements.

In this connection, it is also necessary to consider the classic design methodology which promotes only insufficiently multi-functional materials and its use. Support within the product design process like "material-driven design" is provided for the phase of detailing, but not for the earlier phase of solution finding.

In the following, the material selection as a future task of the design engineer is formulated in detail, as well as the necessary interdisciplinary collaboration of the areas design, material and technology that only enables an integration of materials into the design process. Three constructive examples applying materials as a functional element are presented thereafter. They represented the basis for the current development of an enhanced design methodology approach that takes into account an early requirements-based, material-driven solution-finding [1].

2. Material choice – the design engineer's task

Thesis 1: From the designer's point of view, choosing a material represents a restriction that has nothing to do with the creative process of product development.

The basis of product design is always a number of demands that arise out of a number of areas and that are often far away from the actually achievable functionality. On the other hand, each material has a huge amount of properties that must be matched first to the demanded features.

Thesis 2: Choosing a construction material always represents a compromise, in that a material's specific properties favour the demanded properties, where others don't meet the demands.

One can conclude from this, that there is no such thing as an ideal material – also not for a specially selected design task. Nevertheless, one can also conclude that certain material characteristics can be promoted using material technology, e.g. by using alloys, or it is possible to shift the compromise in the direction of the product's demand by combining parts with different characteristics.

Thesis 3: By combining various materials when designing a product, this increases its chances of fulfilling the demands compared to using a single material in its design.

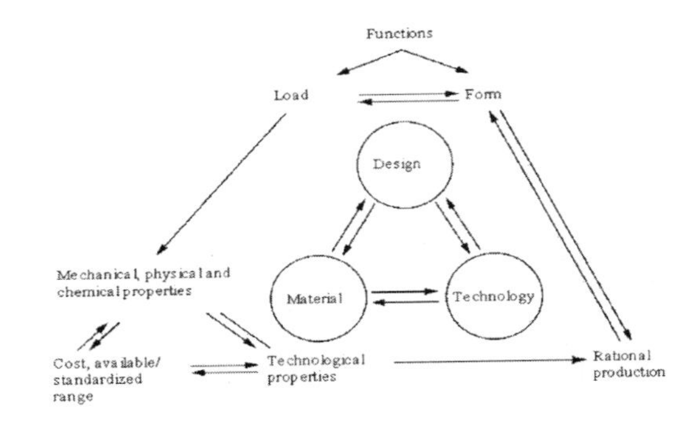

Fig. 1. Material choice with regard to the unity of design, technology and material [2]

Design engineers have been living from the design of surface protection structures, wear protection or the development of fibre-reinforced load carrying structures for generations. The essential point is that the designer's task is to construct product properties using material and production techniques, because the creation of the element itself is mostly bound to the creation of material properties and with this to the fulfilment of the demands by integrating material and production techniques. Economic viability – and often coupled directly to this are the amount of goods flowing into production - represents a feedback loop as shown in Fig. 1, that can influence the classical categories "design", "construction material" and "technology". In which way the designer chooses materials or combinations of these during the development process is demonstrated according to the guideline of the Association of German Engineers VDI 2221 [3] in Fig. 2. The right hand part shows the decision fields relevant to the materials.

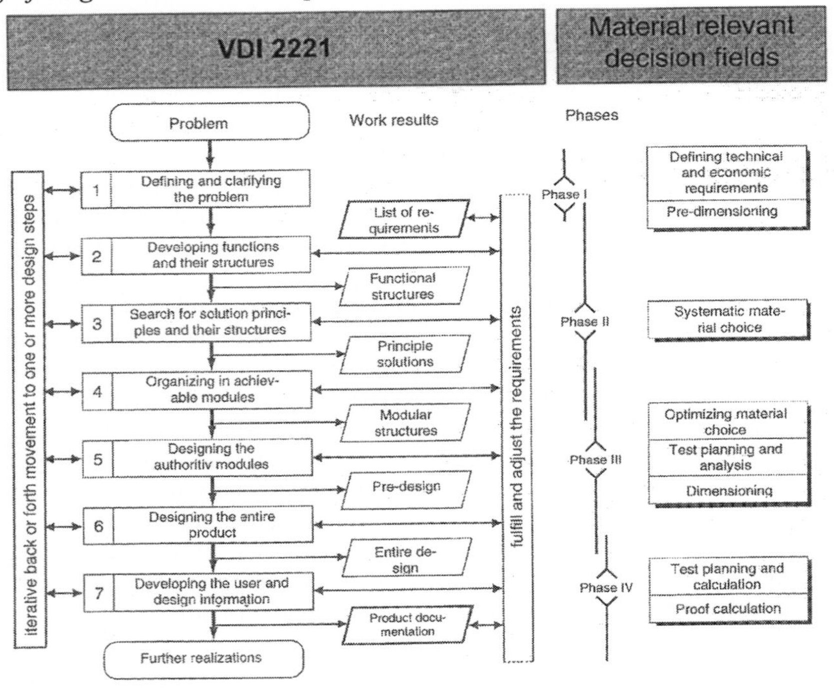

Fig. 2. General method of developing and designing in compliance to VDI 2221 and material relevant decision fields [3, 4]

Thesis 4: The economic success in using new materials to reach specific component properties depends on cost effective material production, machining and assembly of the to be developed parts.

Normally the construction

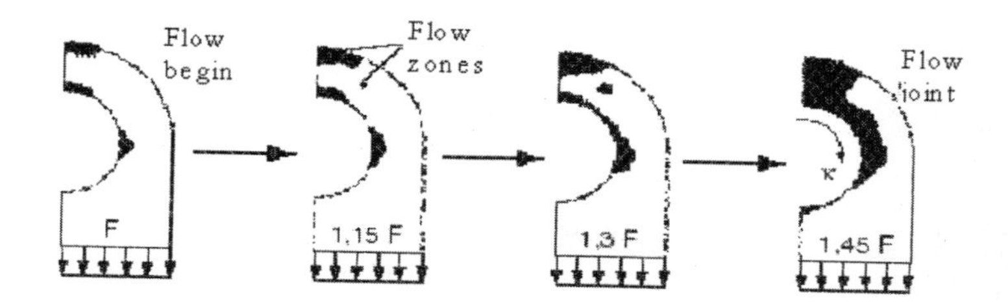

Fig. 3. Plastification of a bolt to link plate connection; κ contact-angle, F force [5]

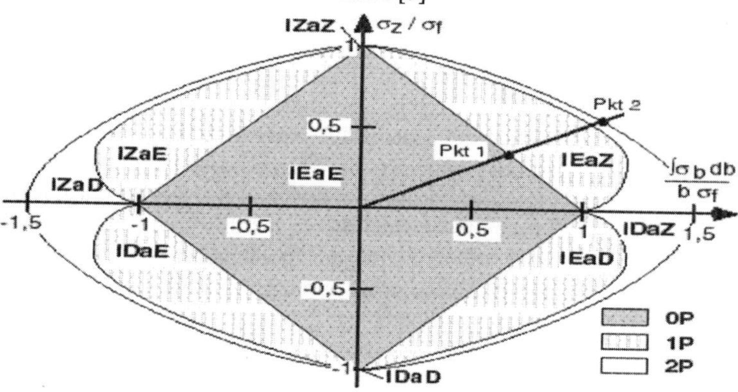

Fig. 4. Fields of solution for elastic and semi-plastic deformations due to different load combinations [5, 6]

material and manufacturing procedure are chosen in step 5: "developing the authority modules" and the definitive choice of material is taken in step 6 "developing the complete product". This late choice is basically caused by the up to this point very imprecise definition of physical dimensions, loads and other constraints of the design. The problem of choosing construction materials is similar to that of fulfilling demands like noise emission, ergonomic and economic restrictions, as a very detailed knowledge of the design's characteristics is required to come to an definitive rating of the parts properties. Therefore, an effective choice and optimisation procedure often requires a "feedback loop" in the design process to make sure that the information available gradually improves during the process of designing.

The following examples shall depict how design solutions for special development tasks were created by the dedicated use of composite material parts. By putting these examples into context with the mentioned theses it should also been shown up that, by accepting a compromise to reach design criteria, certain constraints are imposed on other properties. This can make the chosen solution questionable.

3. Using plastic deformation to increase load-bearing capacity

The stress limit of machine parts is usually determined by the criterion of linear elastic behaviour of the part and the material. The quite usable material behaviour in plastic stress regions is usually disregarded in calculations and treated as "extra safety". Fig. 3 shows that the true cause of part failure, the full plastification, holds an increase of load carrying capacity to 145% of the value determined by elastic calculation using a bolt to link plate connection as an example. The cause of this is that the plasticising area of the part has little effect on the stiffness of the connection. The interaction of stress distribution and material behaviour leads to stress distributions that show the possible increase of load bearing capacity by using a combination of two materials – one with ideal elastic and one with ideal plastic behaviour - if the distribution of bending stress allows such an equalisation (Fig. 4).

Referring to the theses set up in the introduction there can be no doubt that alone the change of characteristics of a known material leads to the focused-on-improvement of load carrying capability.

4. Combining composites and metal in the design of high speed air separators

Out of the optimisation process of a mechanical separation/ grading process with a separation limit of 1 μm particle diameter followed a design of rotor for a deflecting separator that has a peripheral velocity at the process area of at least 250 m/s. The consequent use of low weight principles to achieve high circumferal speeds leads to the use of hybrid construction incorporating composites to purposefully adjust the machine's properties towards load bearing capacity, distortion and dynamic behaviour. A prototype was developed whose process engineering elements were made out of composite material, the carrying structure from a combination of aluminium disks (because of the shaft-hub joint) and the collars from fibre reinforced composites (see Fig. 5).

Fig. 5. Hybrid rotor in two perspectives

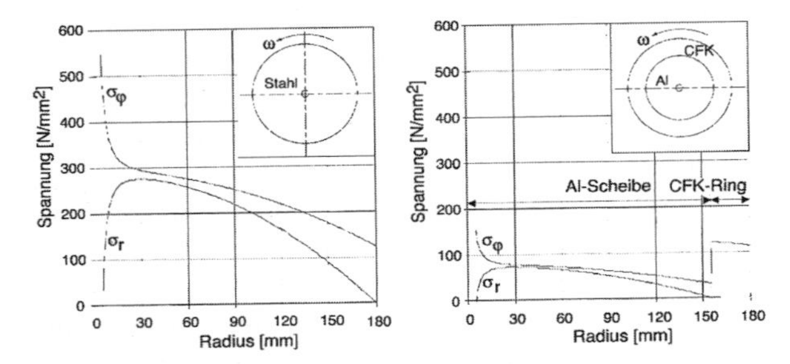

Fig. 6. Stress in disks of a high speed separator at a speed of 300 m/s (left: steel; right: aluminium disk with fibre reinforcement) [7]

Using the design method of function separation, the function of driving the fins and supports was realised with a metal disk, the centrifugal loads were taken up by the pressed on ring that reduces the centrifugal stresses in the disk by hindering its expansion. Because of this the outer ring should consist of material of low density, a high modulus of elasticity and high strength. The carbon fibre composite material (CFK-HM) has the best properties of the examined materials. Fig. 6 shows the tremendous stress differences compared to a steel disk in a calculated example.

The recited example illustrates that the optimisation of a structural part with special and higher than normal demands is possible by using materials specially suited to the design's demands, but that the detailed layout can only be achieved with close co-operation between material specialist, production technician and designer. As the demands placed on the separator were easy to isolate, an optimisation of the design was possible making use of material properties. The stumbling block mentioned in the introduction laid in a property of the separation process that was not foreseeable during the development: technical powder always contains a certain percentage of coarse grain that does not follow the airflow of the separator at these speeds and, through secondary turbulence, causes fin wear. For industrially successful use a protective coating against abrasive wear is necessary using this type of design.

5. Development of ceramic ventilators for hot gas circulation at temperatures up to 1350°C

High temperature processes with temperatures exceeding 1000°C play a big role in process engineering concerning energy consumption. This required the development and testing of a ventilator suitable for temperatures up to 1350°C. This goal can not be reached by further development of steel ventilators (max. 800°C, short term up to 1000°C). A solution was seen in a design with temperature resistant ceramics. SiSiC (silicon infiltrated silicon-carbide) has proven to be first choice for this application as it has high rate of thermal conduction and allows the production of large parts.

A significant problem with steel hot gas ventilators is heat flow via the shaft from the hot gas area to the surrounding environment. One should therefore prefer solutions that limit the heat flow out of the oven area. This lead to the design problem of the junction from hot to cold and the junction from steel to ceramic, i.e. the creation of a heat barrier on the one hand and on the other hand a suitable shaft-hub connection between the steel shaft and the ceramic wheel. A systematic examination resulted in the conclusion that, due to the notch susceptibility of ceramic materials and due to the high coefficient of expansion of steel compared to ceramics, a positive connection was unsuitable. A design in that the wheel was fully exposed to the gas temperature and the to be cooled area was moved outside the oven area was aimed for.

The experience gained with this wheel was used developing the final design: The wheel was mounted hanging, i.e. with a vertical shaft in a housing clad with fire resisting brick. A variable frequency motor drives the shaft via a belt. This creates defined loads on the two pedestal bearings, the mechanical load on the wheel itself is static during stationary operation. A non-positive system was developed for the wheel – shaft connection (Fig. 7), that uses the hollow shaft as a tension rod. On the cold, driven end there is a packet of disk springs that make sure that the pretensioning is large enough during different degrees of heat expansion. A cylinder of ODS (oxide-hardenend by dispersion superalloys) (9) is used in the hot gas area. The wheel (1) is aligned by the surfaces (2) and (3).

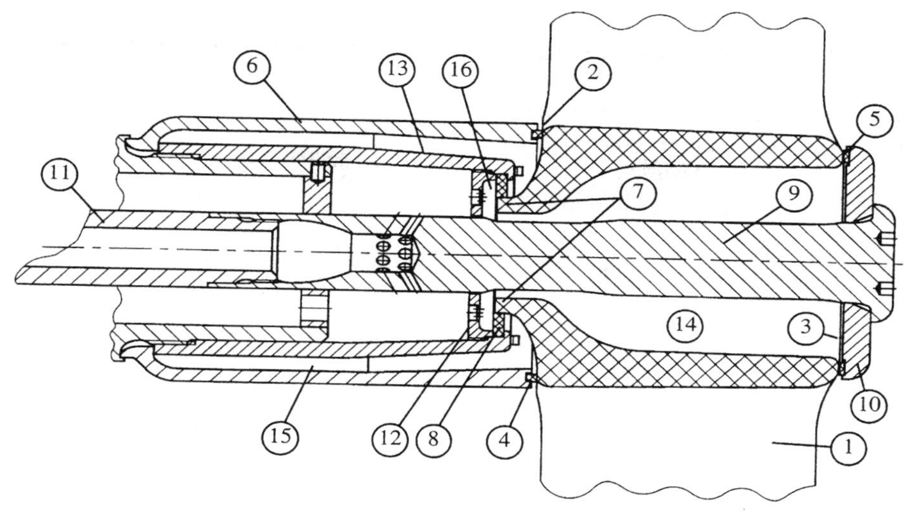

Fig. 7. Final design of the paddlewheel [8, 9]

This support against tilting is maintained even for differential expansion of the supports, as the parts can slide on each other. To avoid direct contact between metal and ceramic parts because of welding due to diffusion and to enable an easy dismantling, aluminium-oxide rings are inserted (4 and 8). They achieve an equalisation of radial expansion differences. It was noticed that these rings break up during operation. Their function is not compromised as they are trapped in grooves.

Torque is transmitted to the wheel via the ODS sleeve (6) and ring (4) as well as the tension rod (9), disk (10) and ring (5). On the ceramics side centering is taken care of by the small diameter offset ground in over the wheel's centering offset, so that during warming up only small diameter differences occur. The centering offset is surrounded by four zircon-oxide segments to avoid tensile stress in the ring (8). The observation, that thermal stressed ring shaped parts only have radial cracks after their unloading of a long operational period, led to this layout. Compressed air is blown against the metal plate (12) through the inner tube (11) to cool this and the tube (13). Hereby, the tube's (13) radial expansion is reduced and the centering is improved. The centering was so successful, that no heat induced imbalance occurred, even though play increased with high temperatures. The hot cooling air is led away through the outer tube. The cavities (14,15,16) are insulated with a fibrous ceramic material. A cooling air stream of 10 cubic meters per hour (induction state) was used, which is negligible compared to the delivery stream of around 500 m^3/h.

The detailed description of the development process of this ceramic-metal construction shall show, next to the positive effects of problem solving, which sorts of information exchange are required between mechanical engineering, material research, production engineering and process engineering, to create an engineering solution for an process engineering task. One of the most difficult problems for the designer was the unfamiliarity with the properties of ceramic construction materials: heat induced expansion, heat transfer, friction, physical properties etc. The co-operating firm from the ceramics industry had to get accustomed to terms like flux of force, notch effect, multidirectional stresses and integrate these into production.

6. Conclusion

By means of a number of examples it could be proved that, with the manifold demands placed on a modern and in line with the market product, the search for a construction material to fulfil these demands partly can cause large problems. In many cases an optimisation of a product can only be achieved by combining parts made from different, problem orientated material properties. Here, too, the choice of material always means a compromise between the product's demands and the material's properties. [10], [11], [12] present additional examples for a successful application of interdisciplinary solutions.

It could be shown that, on the one hand a different material behaviour in dependency of the loads - even without changing the material – contains optimisation potential (plastification capabilities of metal), on the other hand, that the combination of a number of materials leaves uncovered areas in the demands, as can be seen in the process engineering machine solutions.

All the above mentioned aspects and approaches are used for the current definition of an enhanced design methodology within the national-funded research project "Requirements-driven conceptual design (methodology) of constructions with incompatible materials (DFG 289/31-1)" [1].

Acknowledgement

The Institute of Mechanical Engineering, Clausthal University of Technology, is partner of the EU-funded FP6 Innovative Production Machines and Systems (I*PROMS) Network of Excellence.

7. References

[1] DFG 289/31-1: Anforderungsgetriebene Konzeption (Methodik) von Baukonstruktionen aus inkompatiblen Werkstoffen, technical report, publication foreseen, 2005.

[2] Grosse, A.: Interdisziplinäre Werkstoffauswahl durch Aufbau eines Material Data Mart. Diss. TU Clausthal 2000.

[3] VDI-Richtlinie 2221: Methodik zum Entwickeln und Konstruieren technischer Systeme und Produkte. VDI-Verlag Düsseldorf, 1993.

[4] Fischer, D.R.: Entwicklung eines objektorientierten Informationssystems zur optimierten Werkstoffauswahl. Springer, Berlin, 1995 (Diss.).

[5] Dietz, P.: Tragfähigkeitssteigerungen von Maschinenelementen durch teilplastische Verformungen. Konstruktion, October, 1999, Springer-VDI-Verlag, Düsseldorf, pp 27-35.

[6] Rothe, F.: Spielbehaftete Laschenverbindungen bei quasistatischer Belastung unter der Berücksichtigung nichtlinearer Randbedingungen. Diss. TU Clausthal 1994.

[7] Rübbelke, L.: Konstruktive Lösungen und Auslegemethoden für Hochgeschwindigkeitsabweise-radsichter aus Leichtbauwerkstoffen in der Verfahrenstechnik. Diss. TU Clausthal 1994.

[8] Dietz, P. (Ed.): Verfahrenstechnische Maschinen bei besonderen mechanischen, thermischen oder chemischen Belastungen. Springer, Berlin, 2000.

[9] Jakel, R.: Ein Beitrag zur Berechnung und konstruktiven Gestaltung keramischer Bauteile, dargestellt am Beispiel eines keramischen Heißgasventilators. Diss. TU Clausthal 1996.

[10] Dietz, P.; Grünendick, T.: Bauteilkonstruktion mit kombinierten Werkstoffen – Möglichkeiten und Grenzen am Beispiel verfahrenstechnischer Maschinen. Konstruktion 55 (2003), issue 9, Springer-VDI-Verlag, Düsseldorf

[11] P. Dietz; T. Grünendick: Interdisziplinäre Lösungsfindung am Beispiel der Welle-Faserverbund-Nabe-Verbindung. 9. Nationales Symposium der Sampe (The Society for the Advancement of Material and Process Engineering) an der TU Clausthal, 2003

[12] Dietz, P.: Auslegungskriterien für mehrlagig bewickelbare Seiltrommeln nach dem Prinzip des Leichtbaus. Konstruktion, November/December, 2004, Springer-VDI-Verlag, Düsseldorf, pp.75-80.

Intelligent Production Machines and Systems
D. T. Pham, E. E. Eldukhri and A. J. Soroka (eds)

Evolutionary Neuro-Fuzzy Modeling in Parametric Design

K.M. Saridakis[a], A.J. Dentsoras[a]

[a] *Dept. of Mechanical Engineering & Aeronautics, University of Patras, 26500 Rio Patras, Greece*

Abstract

A design approach is presented that confronts several issues of parametric design, by implementing soft computing techniques. The problem is stated in terms of design parameters and their associative relationships of different formalisms. Genetic algorithms are deployed in order to find the optimum solution according to custom optimization criteria. The best solutions of the genetic optimization are recorded, then a neuro-fuzzy procedure is used that limits the number of inputs and outputs and resolves problem's complexity by substituting existing associative relations with a fuzzy rule system. Redesign may be performed by searching the optimum solution under the same criteria but using the simplified fuzzy structure. A design of an oscillating conveyor is presented as an example case.

1. Introduction

Parametric design is performed through multi-dimensional solution spaces where design knowledge is usually structured in different formalisms. The consideration of the design as a combinatorial constraint satisfaction problem (CSP) may lead to a set of satisfactory but sub-optimal solutions that represent local extremes in the design space. In addition, the manipulation of both 'hard' (based on scientific principles) and 'soft' (based on designers' experience and judgment) design knowledge blocks and the discontinuity of the design space results in the inappropriateness of the common numerical and optimization methods in the case of design.

The current paper presents an approach where a design case is represented through generic design parameters and their associative relations. Both design parameters and their relations are classified into generic, problem-independent types. The underlying design knowledge representation and handling schemes are not associated with any specific formalism or abstraction level. By following specific algorithms, an hierarchical tree structure of the design parameters is formed. Searching the design space with genetic operations that are based on stochastic modules is an efficient way to cope with the strong non-linearity of most design problems and ensures that the optimization process would not be trapped in locally optimum solutions. The optimum solution is searched through a genetic algorithm (GA) based optimization. Fitness function is used by the GA for evaluating the generated populations of solutions depending on the problem under consideration.

In addition to the extraction of the optimum solution by means of genetic evolution and natural selection, records of the best individual solutions for each generation are kept until the final convergence. These records of elite solutions are then used by a neuro-fuzzy procedure in order to extract a simplified structure of the design problem. The designer has the flexibility to define which of the previously stated design parameters will be defined as inputs and outputs for the simplified problem's structure. For these parameters, values from the elite records are chosen and used for the training of a neural network thus constructing fuzzy sets and fuzzy rules that associate the defined inputs with the defined outputs. Then redesign can be performed through evolutionary optimization using the simplified fuzzy structure. Given that the simplified structure is based on best individuals and the structure is significantly simpler, the convergence to the optimum solution is achieved faster. Nevertheless the optimization criteria must be the same during all genetic optimization processes in order to maintain the consistency of solutions of the initial structure and of the fuzzy simplified structure.

The current approach is a step-wise process that possesses the required generality in order to be deployed systematically in domain-independent parametric design problems. The steps of the proposed methodology are described through an example of parametric design of oscillating conveyors and the results are discussed for efficiency. Finally the paper includes a short reference on future research work towards the enhancement of the proposed methodology.

2. Background issues

Design has been a subject of many research activities [1]. For these activities the common clarified requirement is to transform design into a robust and integrated scientific field. As a result, a series of techniques and methodologies have been deployed in the direction of establishing rules and scientific criteria in an effort to perform design tasks with systematic processes that incorporate the related available empirical knowledge.

The design information for each design problem is required to be abstracted and explored by formal methods [2] before the competing objectives are identified [3]. Tools and techniques have been proposed in order to model and utilize the abstract design knowledge [4] and to manipulate the design either as a constraint satisfaction problem (CSP) [5], or as a parametric multi-criteria optimization problem.

The establishment of design axioms [6] succeeded in formalization of the design problem. However the existence of successful past design solutions, which have been proven functionally insuperable while violating the design axioms, menace the validity of the axioms as far as their axiomatic nature is concerned. Many approaches that manage the design through design structure matrices (DSM) have been proposed [7]. Yasine et al. [8] in a DSM-based approach utilize connectivity maps for modeling and analyzing the design relationships in product development. Chen et al. [9] manipulates the decomposition of large scale design problems through a tree-based dependency analysis. Research has been deployed by Saridakis et al. [10] in the context of the complexity of the interdependencies of parameters in a large-scale design problem and the simplification of the generated tree-structures. However, many of the proposed approaches succeeded partially in expressing the design and manufacturing problem in an adequate mathematical form because of the diversity of the formalisms for design knowledge (rules, formulas, drawings, etc.), the uncertainty and the discontinuity that usually characterize design.

Latest advances in artificial intelligence (AI) tend to enhance the available efficient analytical evaluation tools and procedures in a variety of scientific domains. An artificial intelligence field of major importance is comprised by the so-called soft-computing techniques that include genetic algorithms (GA), fuzzy logic (FL) and artificial neural networks (ANN). Genetic algorithms (GAs) have been extensively used as optimization and machine learning methods [11]. Fuzzy logic and artificial neural networks are utilized in a variety of applications but their main characteristic is summarized in the possibility of management of uncertain and incomplete information in optimization, classification, modeling problems [12]. The combination of soft-computing techniques for resolving scientific problems has led in results that could not have been extracted with traditional methods. During the last two decades a huge growth has been reported in neuro-fuzzy modeling [13] that occasionally may include genetic algorithm deployment [14].

The possibility of diminishing the inherent difficulties of the design process have been incrementally researched [15] and intelligence techniques for automatic design synthesis have been proposed [16]. Genetic algorithms have been used in design problems either in conjunction with other methods [17] or as stand-alone design approaches [18]. Numerous methodologies have been based on fuzzy-logic and applied to design as integrated approaches or tagged with genetic algorithm processes for fuzzy multi-objective optimization [19]. Design approaches with the implementation of neural networks have also been proposed [20].

In the current paper an approach is proposed that formalizes design by the representation of the design knowledge in terms of design entities categorized into types and structured in hierarchical trees. The hierarchies of the design parameters are formed by utilizing the existing associative relations of different formalisms (computational formulas, empirical rules or lists of values) with design structure matrix (DSM). The search for the optimal solution is performed through genetic algorithms with custom optimization criteria. The best solutions through the populations of the genetic evolution are recorded and used as learning paradigms in a neural network that represents the input and output design parameters of the design problem under consideration. Then a fuzzy system is generated through fuzzy or subtractive clustering that represents the entire design problem in a simplified two-level fuzzy formulation. Other searches for optimal solution may proceed by applying genetic algorithms either to the initial multi-level hierarchical system or to the simplified fuzzy system based on elite solutions.

3. The Proposed Methodology

The current research work is elaborated under four basic guidelines: i) the elaboration of a systematic methodology for domain-independent parametric design, ii) the manipulation of available design knowledge of different formalisms, iii) the extraction of a globally optimal solution, and iv) evolution of the problem under consideration. The following paragraphs describe the deployment of the proposed methodology on the parametric design problem from the domain of oscillating conveyors preserving the four abovementioned guidelines.

3.1 Design entities and associative relations

A parametric design problem can be expressed in terms of design entities called design parameters (DP). A design parameter is a characteristic of the design case and its variance affects the final design outcome. A design parameter may refer to a physical attribute of the design object, to a function that the designed system should perform, or to a metric that characterizes the performance.

Firstly design parameters of the design problem and their associative relations must be stated. The design parameters may be either quantitative or qualitative. The relationships among the design parameters may be expressed in terms of computational formulas, empirical rules, selection matrices, experiment values etc. The expressed dependencies are registered in a DSM matrix [7] that performs partitioning and ordering of the design parameters. From the partitioned DSM a tree structure with the design parameters may be extracted that represents the design problem and ensures the existence of bottom-to-top solutions. A significant remark for this step is that design knowledge specific to the problem under consideration is required in order to tear interdependencies that form loops.

A tree structure that represents the design parameters from the design of oscillating conveyors is shown in figure 1. The design parameters are classified in two basic categories, the dependent and primary (non-dependent) design parameters. In the set of all design entities, there is always a subset with members that possess the unique property to be input design entities (primary or non-dependent) for the design problem under consideration and their instantiation determine, through the established associative relations, the rest (dependent) design entities. The primary design parameters may be characterized either as fixed or variable. The fixed primary design parameters (square boxes in fig.1) represent non-dependent design parameters used as inputs for the design problems whose values stay invariable during design cycles. An example of fixed primary design parameter is the '*particlesize*' that represents the diameter of a particle from the material to be conveyed. The material and its characteristics are specific and invariable for the parametric design of the conveyors. In the contrary the values of the variable design parameters (denoted as bold square boxes in fig.1) may change independently during design process. An example of variable primary parameter is '*troughtype*' that should take one of the three qualitative values (open, cylindrical, orthogonal) that affect in a certain way its dependent DPs (children). The values of the dependent design parameters (denoted with circles) are fully defined by their children DPs through their associative relations. An example of dependent parameter is '*convspeed*' that represents the conveying speed of the oscillating conveyor and is defined through a computational formula that relates seven children DPs. A special category of dependent design parameters is dictated by the variable dependent design parameters (denoted with bold circles) that represent dependent design parameters that are not fully defined by their children DPs. As far as the variable dependent DPs are concerned, their children DPs and the associative relations are only capable of reducing the design freedom to a smaller number of options or interval values but the final value must be user-defined.

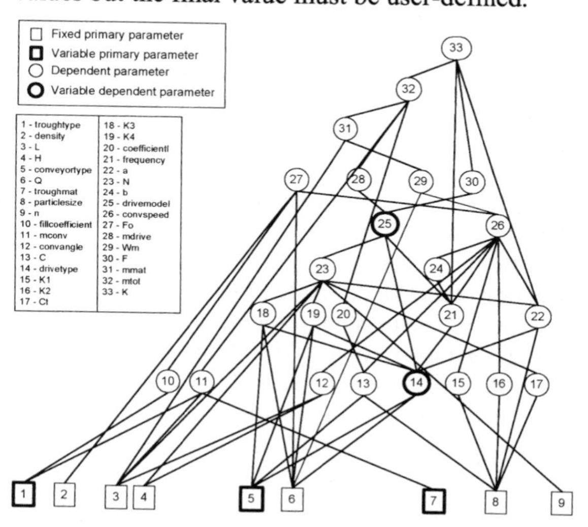

Fig. 1. Hierarchical tree of design entities.

The solution search is deployed in the design space formed by the variation of variable primary DPs and the variable dependent DPs. Two types of associative relations are used in the current approach: fuzzy rules and computational formulas. These relations are available or extracted from available design knowledge (selection tables, empirical rules etc.) in the field of oscillating conveyors.

3.2 Solution search with GAs

The optimal solution will be searched with criteria of convergence to pre-defined values for some critical DPs. The modeled design problem includes DPs both numerical and linguistic that are associated with non-linear and complex relations. Traditional optimization and numerical methods fail to cope with the aforementioned ascertainments while they may lead to a locally optimal solution. The utilization of genetic algorithm can surpass the posed difficulties but does not ensure that solution is the optimum in absolute measures or near optimum. Thus a combination of genetic algorithm with a traditional optimization method can be considered as a superior approach and most appropriate for the design problems. The present example is subsequent to small design space and only genetic algorithm is deployed.

The design problem of oscillating conveyors is expressed in terms of 33 DPs, of which 3 are variable primary DPs, 6 are fixed primary DPs, 2 are variable dependent DPs and 22 are dependent DPs. Their associative relations are translated into dependencies in a design structure matrix (DSM) and finally the tree of figure 1 is structured. After the statement of the design problem, the genetic algorithm is used (see fig.2). The GA affects the values of variable primary DPs 1, 5 and 7 and of the variable dependent DPs 14 and 25.

Prior to the initialization of the genetic optimization, three aspects need to be considered. Firstly, the criterion of optimization (genetic fitness function) must be defined. The optimization criterion depends on the design problem and relies on designer beliefs. In the example of oscillating conveyors 6 DPs (23, 25, 26, 27, 32, 33) have been considered as critical for the design outcome and characterized as performance variables (PVs). Preferable target values are defined for the abovementioned DPs and then GA tends to minimize the deviation from these target values. The following simple expression is used in the present example as optimization criterion:

$$\prod_{1}^{i}\left(\frac{\left|PV_i^{target} - PV_i^{current}\right|}{PV_i^{target}}\right) \quad (1)$$

where i is the number of performance variables, PV_i^{target} and $PV_i^{current}$ are the target and the current values of ith PV respectively. Other advanced optimization criteria may be used, for example the aggregation of weighted fuzzy preferences on values of the defined PVs.

Another aspect to be considered is the formal statement of the way that variable dependent DPs are partially valued from their children DPs and finally valued by the genetic algorithms processes. A characteristic example is the dependent variable design parameter *'drivemodel'* that depends on *'drivetype'*. The value for *'drivemodel'* can be selected among 34 different options that represent different drive unit models. Having defined the *'drivetype'* (e.g. drivetype = ''electromagnetic'') then only 5 options of *'drivemodel'* that represent electromagnetic drive units are valid for the genetic process.

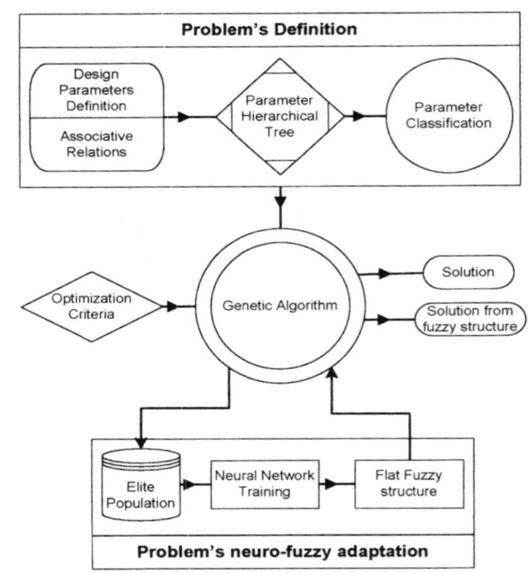

Fig. 2. Parametric design with soft-computing.

Finally the deployment of a genetic algorithm requires an adjustment of several features as population size, number of generations, selection and crossover properties, stopping criteria etc. Some general guidelines for tuning the aforementioned features may be found in the general GA theory but the optimal setting is problem specific and must be determined always in the context of the problem under consideration.

The genetic algorithm searches the design space for the optimal solution using stochastic processes. In the current approach not only the best solution is extracted but also the best solution in each optimization cycle is recorded as shown in figure 2. For the elaboration of the described soft-computing based design approach the respective toolboxes of MATLAB software [21] are used.

The genetic optimization is finally deployed in the aforementioned example for 30 generations. The best solution, the progress of crossover, the mean - worst - best solution scores are depicted in figure 3. After 10 runs, the average time needed to complete the solution search in the initial system is 94.2655 seconds.

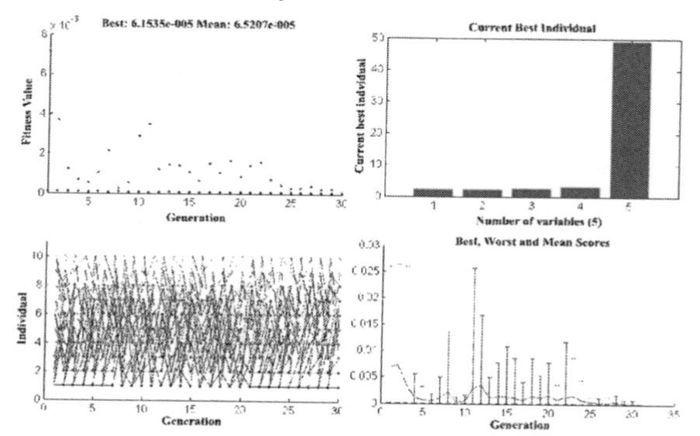

Fig. 3. Solution search with Genetic Algorithm.

3.3 *Transition to fuzzy model with ANN*

The genetic algorithm resulted not only to a final optimal solution but also to records containing the best (according to the posed criterion) solutions of each generation. These elite records contain values of inputs and outputs that can be used for the simplification of the initial design problem through neuro-fuzzy modeling. The simplified problem should have a flat structure of only two levels as shown in figure 4. The inputs of the simplified system are the variable primary DPs of the initial design problem and the outputs are the PVs of the initial problem.

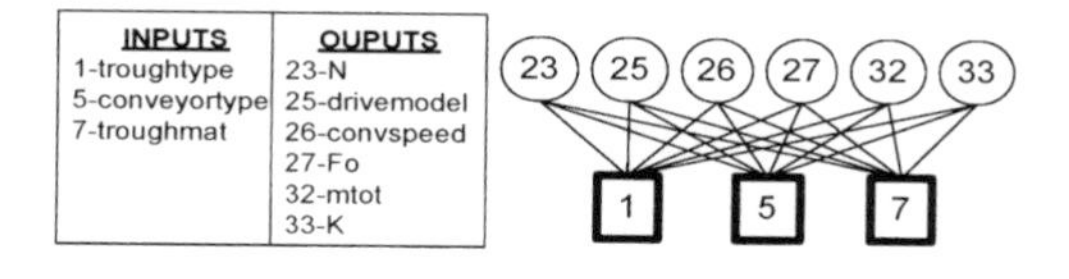

Fig. 4. Simplified design model.

A fuzzy structure that includes the abovementioned inputs and outputs must be generated. The next step is the training of a neural network using the 'elite' records provided from previous GA run. The number of the fuzzy membership functions selected for describing each input or output is custom and for the present example is set to 3. This is translated in 27 rules (regarding the 3 inputs) as shown in figure 5. In case of more complex problems subtractive clustering may be used in order to avoid the combinatorial explosion. The training of the neural network is deployed through a hybrid process that includes back-propagation and least-squares approximation. After the training of the ANNs, then the approximation through fuzzy structures is possible for all defined PVs. A representative fuzzy structure of PV '*convspeed*' is shown in figure 5 that contains the fuzzy rules and the outcome fuzzy surfaces for the input.

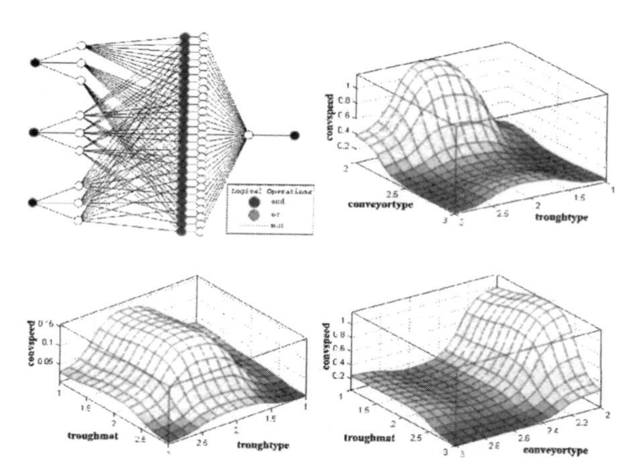

Fig. 5. Performance parameter 'convspeed' fuzzy rules and output space.

4. Conclusions

We have presented an approach that facilitates parametric design. The main advantages of the proposed methodology are: i) it is applicable to all parametric design problems in a formal and systematic step-wise process; ii) it copes with both qualitative and quantitative design knowledge; iii) many types of design relationships can be modeled; iv) the design problem is successfully approximated with simpler fuzzy structures; v) design objectives may be integrated in a unified optimization criterion. Future research work will be elaborated towards establishing more efficient optimization criteria and towards the automatic adjustment for choices concerning genetic algorithm's features.

Acknowledgements

The present research work has been done within the framework of the project 2991 funded by the Research Committee of the University of Patras, Greece.

University of Patras is a member of the EU-funded I*PROMS Network of Excellence.

References

[1] Regli W.C., Hu X., Atwood M., Sun W. (2000). A Survey of design rationale systems: Approaches, representation, capture and retrieval. Engineering with Computers (16): 209-235.

[2] Deng Y.M., Tor S.B., Britton G.A. (2000). Abstracting and exploring functional design information for conceptual mechanical product design. Engineering with Computers; 16:36-52.

[3] Otto K., Wood K. (2000). Product design. Prentice Hall.

[4] Wang L., Shen, W., Xie H., Neelamkavil J., Pardasani A. (2002). Collaborative conceptual design: state of the art and future trends. Computer-Aided Design (34): 981-996.

[5] Lottaz C., Smith I.F.C., Robert-Nicoud Y., Faltings B.V.(2000). Constraint-based support for negotiation in collaborative design. Artificial Intelligence in Engineering, Vol.14: 261-280.

[6] Suh N. (2001). Axiomatic design: Advances and applications. Oxford University Press.

[7] Browning Tyson R. (2001). Applying the design structure matrix to system decomposition and integration problems: A review and new directions. IEEE Transactions on Engineering Management, Vol. 48, No. 3, pp. 292-306.

[8] Yassine Ali, Whitney D., Daleiden S., Lavine J. (2003). Connectivity maps: Modeling and analyzing relationships in product development processes. Journal of Engineering Design, Vol. 14 (3).

[9] Li Chen, Zhendong Ding, Simon Li (2003). Tree-based dependency analysis in decomposition and re-decomposition of complex design problems. Journal of Mechanical Design, (in press).

[10] Saridakis K.M., Dentsoras A.J. (2004). Resolving the complexity of functional representation in design: A transition from computational relationships to performance hierarchical trees, 7th Conference on Engineering Systems Design and Analysis, UK.

[11] Goldberg, D. (1989). Genetic algorithms in search, optimization and machine learning, Addison-Wesley, Reading, Massachusetts.

[12] Kosko, B. (1992). Neural networks and fuzzy systems: A dynamical systems approach to machine intelligence, Prentice-Hall, Englewood Cliffs.

[13] Jantzen J. (1998). Neurofuzzy modeling, Tech report no 98-H-874, Technical University of Denmark: Oersted-DTU.

[14] Palade V., Bumbaru S, Negoita G. (1998). A method for compiling neural networks into fuzzy rules using genetic algorithms and hierarchical approach. 2nd Int. conference on knowledge-based intelligent electronic systems, Adelaide, Australia.

[15] S . Subba Rao, A. Nahm, Zhengzhong Shi, Xiaodong Deng, Ahmad Syamil (1999). Artificial intelligence and expert systems applications in new product development: A survey. Journal of Intelligent Manufacturing Vol.10: 231-244.

[16] F.J. Vico, F.J. Veredas, J.M. Bravo, J. Almaraz (1999). Automatic design synthesis with artificial intelligence techniques. Artificial Intelligence in Engineering Vol.13: 251–256.

[17] Yu, Tian-Li, Goldberg, D. E., Yassine, Ali & Chen, Y.-P. (2003). A genetic algorithm design inspired by organizational theory: A pilot study of a dependency structure matrix driven genetic algorithm. Artificial Neural Networks in Engineering, ANNIE.

[18] Goldberg, D. (2002). The design of innovation. Kluwer Academic Publishers, Massachusetts.

[19] Masato Sasaki, Mitsuo Gen (2003). Fuzzy multiple objective optimal system design by hybrid genetic algorithm. Applied Soft Computing Vol.3: 189–196.

[20] Shih-Wen Hsiao, H.C. Huang (2002). A neural network based approach for product form design. Design Studies Vol.23: 67-84.

[21] MATLAB © The Mathworks Inc, version 7.0.1.24704.

Intelligent Production Machines and Systems
D. T. Pham, E. E. Eldukhri and A. J. Soroka (eds)

Intelligent Conceptual Design of Robot Grippers for Assembly Tasks

D.T. Pham, N.S. Gourashi and E.E. Eldukhri

Manufacturing Engineering centre, Cardiff University, Newport Road, Cardiff CF24 3AA, UK

Abstract

The design of optimum gripping systems for robotic assembly tasks is non-trivial. Traditionally, production engineers have been in charge of this task. Their involvement, however, should be minimised in a modern computer-controlled manufacturing environment. This paper presents a fully automated gripper design system. The system receives the CAD models of the components to be assembled, identifies their geometric structures, determines their dimensions, groups them in families and then generates the best gripping system capable of handling them during the assembly. The paper also gives an example of the application of the system to a simple gripper design problem.

1. Introduction

A good gripper design is essential for the success of a robotic workcell [1]. Research in gripper design has two main directions. The first is influenced by the fact that the application of grippers is task-specific. Researchers adopting this approach attempt to create cost effective specific grippers to automate specific processes or versatile grippers for a wide range of applications. Conversely, researchers following the second direction try to develop methods and techniques to automate the design and selection of robot grippers for a given industrial use.

This paper presents a system that can automatically configure optimum gripper systems for robotic assembly tasks. Section 2 gives an overview of conceptual design and configuration problems. Section 3 describes the system and its five-step design procedure. The system receives the CAD models of the components to be assembled, identifies their geometric structures, determines their dimensions, groups them into families by considering the object-related factors listed above and then generates the best gripper system capable of handling them. An example of the application of the system to a simple end-effector design problem is given in section 4. Section 5 concludes the paper.

2. Conceptual Design and Configuration Problems

The design process consists of three main phases, namely, problem formulation, conceptual design and detailed design. Conceptual design is very complex as it requires a very high-level of mental activity. It is a decision-making process where complete and precise information about the problem is not available and yet the most important decisions have to be made. Above all, it is a creative activity by which new ideas and designs are conceived. French [2] describes it as "the phase that makes the greatest demands on the designer, and where there is the most scope for striking improvements. It is the phase where engineering science, practical knowledge, production methods, and commercial aspects need to be brought together".

Recent research has shown that up to 85% of the life-cycle costs of a product can be committed by the end of the this phase, while only about 5% of the actual life-cycle costs will have been spent [3].

Conceptual design takes the goal and specifications of the design problem, along with other constraints such as those imposed by materials, manufacturing, the environment, technology and economics as input and gives in return a scheme or a concept as output. This concept should satisfy all the design requirements and constraints, but does not have to contain much technical detail. However, all the means of performing each major function as well as the spatial and structural relationships should have been fixed [4].

The central activity in conceptual design is the generation of concepts. This is where possible design solutions are first envisioned [5]. It is the activity that leads to the success or failure of the product development mission. If the concept is poor, it can never be compensated for in later stages. It is during this activity that expertise and creativity are brought together so as to generate good concepts that can smoothly go through the consequent downstream activities and finally satisfy the needs of the design.

For the past two decades, researchers have been looking into the use of AI techniques, namely, expert systems and knowledge-based systems (KBS) for conceptual design. Krause and Schlingheider [6] give an overview of the application areas for KBSs in design and development. They identify activities which may be supported by KBSs and outline some industrial applications and research activities that have been undertaken.

In the application of KBSs to engineering design, the latter is sometimes considered a configuration problem. The design is achieved by configuring predefined components into a final solution that satisfies the design problem. Configuration has been a fruitful topic of research in AI in the past two decades, and in recent years it has also become a commercially successful and important application of AI techniques [7]. Many systems have been developed in the past for configuration; among the first generation systems is XCON [8]. XCON was developed to configure minicomputers. The VAX minicomputers were not delivered to customers as off-the-shelf packages but rather sold as a configuration of input, output and storage devices, supply power unit, processor board, main memory card and software. Within a period of five years XCON had processed nearly 100 000 customer orders [9]. Other examples of early generation configuration expert systems are Cossack [10] and MICON [11]. Some researchers have developed new configuration systems with new approaches [12; 13; 14].

3. Intelligent Gripper Design System

It is assumed that the 3-D CAD models for the components to be assembled will be available as inputs to the design system. This assumption is reasonable given the increasing use of 3-D CAD modelling in mechanical component design.

The output of the design system is an end-effector system with the minimum possible number of grippers needed to handle the various components to be assembled. Details of the steps followed by the design system are discussed in the following sub-sections.

3.1. Importing CAD Models

The first step is to import CAD models expressed in the STEP format which is common to most CAD systems. CAD models are loaded into the intelligent design software that forms the core of the automated design system.

3.2. Recognising Geometric Structures

Only the form (geometric structure) of the imported models needs to be recognised. This will be explained in the following sub-sections.

3.2.1. Reference Box

The imported solid component is bounded in what is called a reference box (see Fig. 1). This enables the general location and dimensions of the imported part to be determined.

3.2.2. Solid Object Checkers

Checkers are points on the vertices, edges and/or surfaces of a reference box. They form part of the standard templates for automatically identifying the geometric structure, location and dimension of the imported model. Their role is to determine whether or not the imported solid objects interfere with the reference box at those particular points where the checkers are located. This allows a comparison between the imported geometry and the reference box to be carried out so the imported geometry can be identified.

Fig. 2 shows checkers distributed on the vertices, edges and surfaces of a reference box. The number of checkers varies according to the number of geometric structures that the system is required to identify and the accuracy of that recognition. Since the initial concern is only with the simple geometric structures that most mechanical parts possess, a small number of checkers is sufficient.

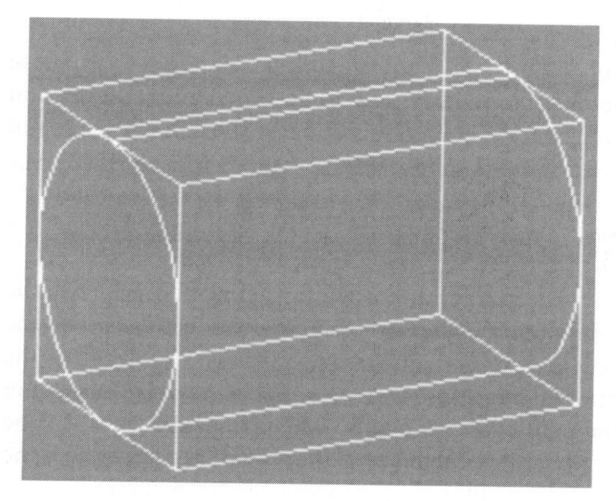

Fig. 1 An imported cylinder bounded by the reference box

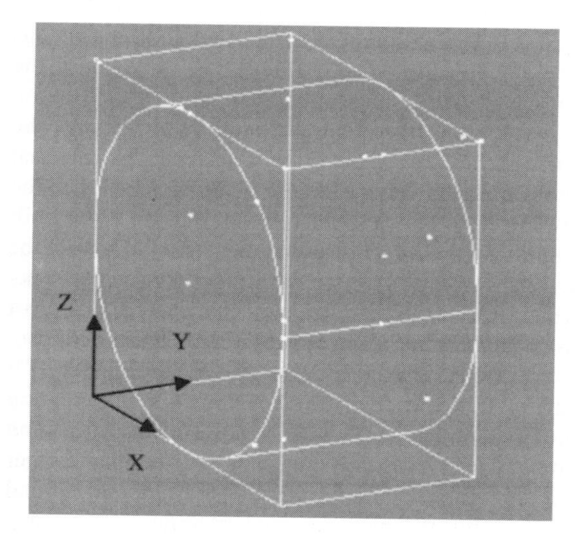

Fig. 2 Checkers distributed along the vertices, edges and surfaces of the reference box

The recognition of geometric structures is performed by applying rules. For example, the rule for recognising cylindrical structures operates as follows. Consider the imported solid depicted in Fig. 3. The figure shows the top view of an imported solid component bounded by the reference box. The checkers are located on the reference box edges, vertices and top surface at points 1-10 (note that checkers in the figure without a number displayed next to them are used for identifying other geometric structures and therefore should be ignored in this example).

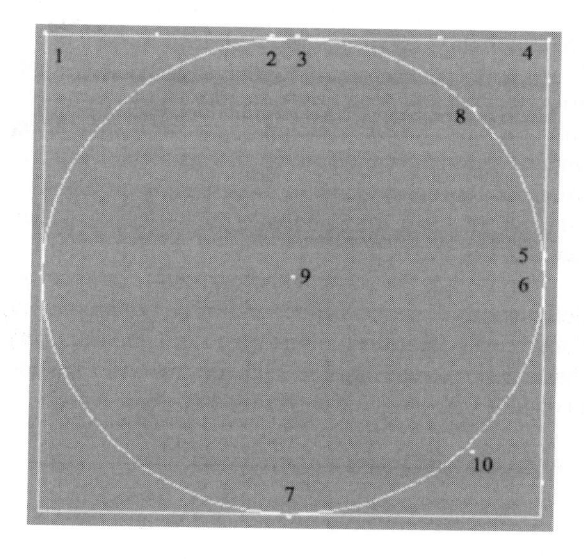

Figure 3: Position of some checkers on the reference box for identifying a cylindrical part

The rule collects the results of tests for interference between the checkers and the imported component. Based on that information, the rule infers the structure of the specific component. The tests reveal that the imported solid component does not interfere with the checker at point 1, which is located at the top left vertex of the reference box. Also, there is no interference at point 2. The tests find interference at point 3, which is located at the centre of the top edge. No interference is detected at points 4 and 5. Interference is revealed at points 6 and 7. At point 8, the checker interferes with the imported solid. This point is of particular importance because it reveals that the entity between points 3 and 6 is not a straight line. There is interference at point 9, which means that the component is not hollowed. No interference is detected at point 10. The overall conclusion is that the imported solid is a cylinder.

Currently, the system can recognise simple but commonly encountered cylindrical, prismatic and spherical parts presented in different orientations. Increasing the number of checkers and hence the number of rules will allow it to deal with more complex geometric structures.

3.3. Determining Information about the Imported Components

The third step is to determine the orientations and then all the relevant dimensions of the components to be assembled.

Again, the checkers provide information about the orientation of each component relative to its reference box. In the case of a cylinder, for example, three orientations are possible with respect to the X, Y and Z axes, namely, the cylinder may be lying vertically (along the Z axis), longitudinally (along the Y axis) or laterally (along the X axis).

Based on the orientation, dimensions can be determined. For the cylinder in Fig. 3, which is now known to be lying longitudinally, its length is equal to that of the reference box. The diameter of the cylinder is equal to the width or height of the reference box.

3.4. Grouping Components

Changing grippers to suit the requirements of different components during an assembly task is undesirable because it wastes production time. The versatility of the gripper, that is its ability to handle as many components as possible, is important for cost-effective assembly. A gripper designer should generate optimum systems so as to minimise, or ideally avoid, gripper changes during assembly. In the past, techniques such as changing the fingers and grooving the jaws (those parts of the gripper in contact with the component) of the gripper were often used.

Components tend to share some geometrical and other characteristics relevant to gripping. Grouping these components into families can facilitate the generation of optimum gripper systems.

The objective of this step is to group similar components into families based on the properties of individual components. A single gripper will then be generated to handle all the components of each group.

Grouping the components first requires them to be coded. Each component has attributes which are considered important to grasping. The weight, material, shape and size of a component are the most important attributes [15]. Therefore, the code for each component will hold information about these specific attributes. A four-digit code has been derived for this purpose. The structure of the code is shown in Fig. 4. For example, the character (:S) in the code indicates that the size of a component is (Small). This decision is made by the system according to a rule that classifies all the dimensions of a component into groups such as very small, small, medium and large. Another rule assigns a shape or geometric-structure class to each component according to the specific grasping requirements. Note that, for the prototype gripper design system developed, all components are assumed to be lightweight (:L) and made of rigid material (:R). These constraints were introduced to reduce domain complexity although they do not affect the generality of the proposed techniques.

3.5. Generating the Optimum Gripping Solution

The challenge for the designer is to generate the minimum number of grippers for handling all the components in an assembly task. The smaller the number of grippers involved in the assembly, the better the end-effector system. This is because there will be fewer gripper changes. Here, it will be shown how the minimum number of grippers can be generated.

All the information required by the design system has been determined: the geometric structure, the orientation and the dimensions of the components. Automatic grouping of the imported components has been achieved. What follows is the automatic configuration of the grippers.

The most important attribute is the shape of the component. The system initially searches for jaws that can handle the shape of the first component. Once the jaws are found, the rest of the gripper is built automatically to suit the component size [16]. Next, the system considers the second component and checks whether or not the first gripper can be used to handle it from the point of view of shape as well as size. If so, then there is no need to generate another gripper. If it cannot, however, then the same procedure for creating the first gripper is followed so as to construct a second one. A check is made at this point to determine whether the new gripper could hold both the first and second components, in which case the first gripper is discarded as it is redundant. The remaining grippers are generated in a similar fashion, with all the grippers created thus far being checked against the new component before a new gripper has to be designed and the new gripper being evaluated on all previously considered components to try and eliminate redundancies (multiple grippers for the same task). The process continues until all the components have been considered.

The best case is when the system generates a single gripper capable of handling all components. The worst situation is when a different gripper is required to handle each component.

Groups of components	Weight	Material	Shape	Size
	Code structure			
01	:L	:R	:CYL	:VS
02	:L	:R	:CYL	:S
03	:L	:R	:CYL	:M
04	:L	:R	:CYL	:L
05	:L	:R	:CYL	:VL
06	:L	:R	:PRI	:VS
07	:L	:R	:PRI	:S
08	:L	:R	:PRI	:M
09	:L	:R	:PRI	:L
10	:L	:R	:PRI	:VL
11	:L	:R	:SPH	:VS
12	:L	:R	:SPH	:S
13	:L	:R	:SPH	:M
14	:L	:R	:SPH	:L
15	:L	:R	:SPH	:VL

Key:
L: Light; R: Rigid; CYL: Cylinder; PRI: Prism;
SPH: Sphere; VS: Very small; S: Small;
M: Medium; L: Large; VL: Very large

Fig. 4 Code structure and the various groups of components

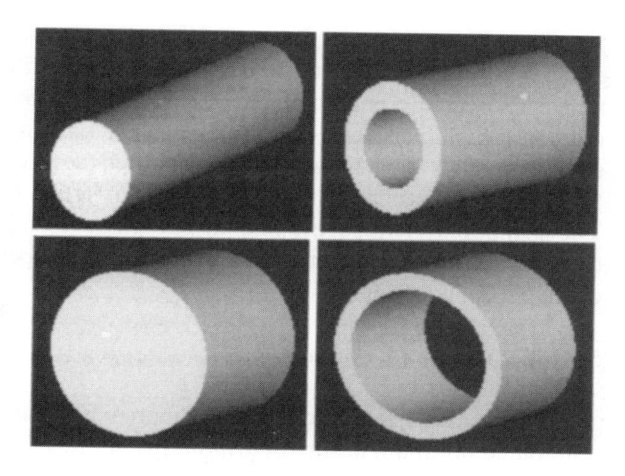

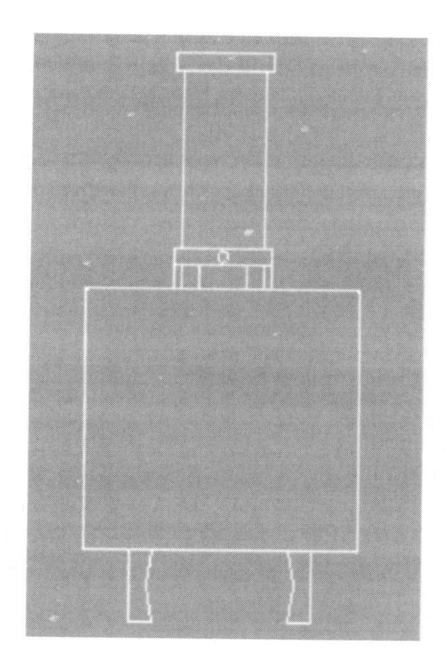

Fig. 6 A single-gripper system

4. Example Application

An example is presented in this section to illustrate the application of the design system to a simple gripper configuration problem. In the example, the system generates a single gripper capable of handling all the specified components

As the configuration process is fully automated, interaction with the system is minimised. The user is only required to specify the components that need to be assembled. The system then proposes the optimum set of grippers for the given component.

The design system first asks the user for the number of components to be assembled and the file name for each component. In the example, four components of the same geometric class, material type, size family and range of weights are specified (see Fig. 5). The system generates the solution shown in Fig. 6. The gripper system is optimal in the sense that it consists of a single gripper and thus requires no gripper change during assembly.

5. Conclusion and Further Work

This paper has described an approach to the automated optimum design of gripping systems. It has presented a five-step methodology and a system for fully automated design. Although the example application given in the paper is simple, the system is capable of handling more complex tasks.

A disadvantage of the current version of the gripper design system is that its operation could be slow when the number of components is large. This is because of the repetitive checking for redundancies after the generation of each new gripper. In future versions, it might be more expeditious to wait until the end of the gripper generation process before carrying out all checks at once. An additional benefit of this would be that potentially useful grippers would not be discarded prematurely.

As gripping involves surface contact, it might be more appropriate to consider the surfaces of components than their volumetric properties. Further work could focus on designing grippers using surface models of the components to be gripped. This is expected to facilitate the identification of common grippers for components that are different but have the same surface features.

Acknowledgements

The authors would like to thank the EPSRC (Innovative Manufacturing Research Centre, IMRC), the Welsh Assembly Government (Objective 1 and Objective 2 ERDF Programmes: SUPERMAN and ITEE) and the European Commission (I*PROMS FP6 Network of Excellence) for their support of this work.

References

[1] Causey, G. C & Quinn, R. D, 1998, "Gripper Design Guidelines for Modular Manufacturing", Department of Mechanical and Aerospace Engineering, Case Western Reserve University, http://dora.eeap.cwru.edu/agile/pubs.html

[2] French, M. J, 1971, "Engineering design: the conceptual stage", Heinemann Educational, London, UK.

[3] Kersten, T, 1995, "MODESSA, A Computer Based Conceptual Design Support System", Proceedings of the Lancaster International Workshop on Engineering Design, Lancaster, Springer-Verlag, London, UK, pp. 241-259.

[4] French, M. J, 1985, "Conceptual design for engineers", Design Council, London, UK.

[5] Hayman, B, 1998, "Fundamentals of Engineering Design", Prentice-Hall, Upper Saddle River, New Jersey, USA.

[6] Krause, F & Schlingheider, J, 1995, "Development and Design With Knowledge-Based Software Tools – An Overview", Expert Systems with Applications, Vol. 8, No. 2, pp. 233-248.

[7] Soininen, T, Tihonen, J, Mannisto, T & Sulomen, R, 1998, "Towards a general ontology of configuration", Artificial Intelligence for Engineering Design, Analysis and Manufacturing, Vol. 12, pp. 357-372.

[8] McDermott, J, 1982, "R1: A rule-based configurer of computer systems", Artificial Intelligence, Vol. 19, No. 1, pp. 39-88.

[9] Black, W. J, 1986, "Intelligent Knowledge-Based Systems: An introduction", Van Nostrand Reinhold, Wokingham, UK.

[10] Frayman, F & Mittal, S, 1987, "Cossack: A constraints-based expert system for configuration tasks". In Sriram, D & Adey, R (Eds.), Knowledge-Based Expert Systems in Engineering: Planning and Design, Computational Mechanics Publications, pp. 143-166.

[11] Birmingham, W, Brennan, A, Gupta, A & Siewiorek, D, 1988, "MICON: A single board computer synthesis tool", IEEE Circuits and Devices, pp. 37-46.

[12] Feldkamp, F, Heinrich, M & Meyer-Gramann, 1998, "SyDeR – System Design for Reusability", Artificial Intelligence for Engineering Design, Analysis and Manufacturing, Vol. 12, pp. 373-382.

[13] Myung, S & Han, S, 2001, "Knowledge-based parametric design of mechanical products based on configuration design method", Expert Systems with Applications, Vol. 21, pp. 99-107.

[14] Duric, M & Devedzic, V, 2002, "I-Promise – Intelligent protective material selection", Expert Systems with Applications, Vol. 23, No. 3, pp. 219-227.

[15] Pham, D. T & Yeo, S. H, 1988a, "A knowledge-based system for robot gripper selection: Criteria for choosing grippers and surfaces for gripping", International Journal of Machine Tools and Manufacture, Vol. 28, No. 4, pp. 301-313.

[16] Gourashi, N.S, 2003, PhD Thesis, "Knowledge-Based Conceptual Design of Robot Grippers", Cardiff University, UK.

Intelligent Production Machines and Systems
D. T. Pham, E. E. Eldukhri and A. J. Soroka (eds)

Knowledge-based and requirements-driven product development

D. Müller

Institut für Maschinenwesen, TU Clausthal, Robert-Koch-Str. 32, 38678 Clausthal-Zellerfeld, Germany

Abstract

The following paper discusses technologies and approaches for the optimisation of the product development and provides an outlook on future developments for an intelligent IT support for the design engineer. After highlighting the importance of the product design phase within the PLC and current application of PLM concepts and systems in the manufacturing industry, the paper concentrates on the integration of Knowledge Management techniques to the product development process. Additionally, the application of Requirement Engineering is introduced here to enable the specification of an optimised knowledge base. The implementation of these approaches is demonstrated on the basis of the European research projects PRIME and KARE, which feature both the concept of model-based data representation. Derived from their outcome, the future development of knowledge-based and requirements-driven intelligent software agents is motivated for a supported and integrated design process.

1. Introduction

The problems of the German automobile manufacturer Adam Opel AG domineered the headlines of European newspapers in autumn 2004. Not only these problems have pointed out again the continuous high cost pressure on the manufacturing industry. The global competition in this sector asks for high product quality at low costs and shortened times of development and production, expressed in a common manner. A company has to act effectively and flexibly if it wants to survive in the market these days. The introduction and integration of advanced IT-technologies to the business environment and processes is of particular importance here. Their rapid development has also a major influence on the structures of enterprises and the type of product development. Extended respectively virtual enterprises assemble as networked co-operation of (all) enterprises and persons that are connected to a product over its lifecycle, like for example manufacturer, supplier, service and customers. These collaborations are product related and they are dynamic. The application of Collaborative Engineering becomes more and more important in such distributed business environments. Suppliers and other partners of the product design will be integrated to the product development, that makes the existence of mechanisms for the exchange and control of product and process related data necessary. The inclusion of enterprises and persons over the entire product lifecycle from the first idea of a product up to its disposal represents another need for the implementation of methodologies and business processes for a supported management of product related or supportive data and their secured availability over the respective period and beyond. This is covered by the terms Product Lifecycle Management (PLM) and Product Lifecycle Support (PLCS).

The pressure of costs and time onto the companies is the reason for their extensive investigations on optimisation and rationalisation of business processes. Thereby, the cause and solution-finding for an improved added value is analysed in various ways. However, current measures for the reduction of costs can be found mainly in the area of production, namely labour costs. Considered under long-term aspects, this is not a solution for the actual problems of the enterprises. In fact, the biggest share of up to 85% of the accruing production costs comprises the production, but up to 80% of the total costs that have to be paid and 50% of the total turnover will be given and determined in the design and development phase [1]. An optimised product design process that takes exactly the relevant requirements into account holds accordingly a high potential to influence the costs.

This opinion is also supported by Dewhurst [2] in his open letter to the U.S. manufacturers. "Need to Cut Costs? Check your Design First" titles strikingly his position-paper. With reference to the year 1989 as new approaches in the product development have been promoted for once in the U.S. industry with immense savings (i. e. at Ford more than $1,2 Mrd), he compares these numbers of estimated savings with personnel costs in the course of production outsourcing to low wage countries. It is commonly known that personnel costs have a relative low influence to the total costs of a product. Dewhurst concludes that in case of savings these have to be initiated during the concept and design phase. He points out that the development of a new product requires creativity. There should be sufficient time available for the designers to find appropriate solutions, as today's design processes in industry are often too hasty and unstructured. Moreover, feasibility and refinement of multidisciplinary communication and exchange of product know-how and ideas is desperately needed to enable an optimised product specification.

However, just the availability of information and know-how in communication networks is no universal remedy, Brynjolfsson [3] accentuates in his strategy paper on the overcoming of the accumulating flood of information. Computers represent no substitute for the human capabilities in their information processing. They only provide assistance. Novel IT concepts and architectures should consider this aspect necessarily to generate quite applicable solutions. Brynjolfsson favours for that reason a double strategy of centralisation / decentralisation of information, depending on their kind. He differentiates between data easily to integrate of mostly numerical representation that can be processed by computers, and 'smooth' information like the type of customer relationship that are decentralised dispersed. This implies mostly a transfer of jurisdiction to the employees on site. "It is the idea to decentralise the decision making process to better utilise the capacity of the human information processing – not the capacities of the information processing by computers. It is the goal to free the employee's mind for advanced tasks." If this is referred to the product development, it could be derived that established basics corresponding to requirements are straight and clear at the developer's disposal, so that there is enough room for creative tasks and new ideas.

Quite mature systems for the management and organisation of product data are available on the market including also service-oriented architectures, as well as CAD systems are incorporating knowledge modules. The flood of data files accumulating more and more since the introduction of CAD systems in the 1980ies made the development of administrative systems necessary. Besides the functional enhancement of CAD systems (i. e. integration of management systems for drawings), the implementation of PDM systems has been of crucial relevance. PDM systems changed to information systems for an integrated data and process management with increasing product complexity and the introduction of engineering processes (i. e. Simultaneous or Concurrent Engineering). The continuous development of both kind of systems has been enormous over the past years. Advanced CAD systems manage no longer merely geometry and related attributes. Virtual test and knowledge modules replace the necessity of early prototypes and support the designer in his design task. The trend at the support of the entire product lifecycle including the need for corresponding data acquisition and availability over this period and across business networks as well as processes and workflows for organisation and control provided the basis for a functional extension of PDM systems, nowadays identifiable under the term PLM.

Bacheldor [4] shows that the PLM approach is more than just product design and not only relevant for the capital goods domain but affords also promising assistance for the domain of consumer goods in product development and production. The implementation of the concept at Karsten Manufacturing Co. maker of Ping golf clubs is like that "...taking the design and attaching all the various information related to that product – the market plan, design criteria, product specs, testing data, and other type of info...". Another example is provided by Grimes Industrial Design Co. in combination with the manufacture of shoe-skates. In detail, new designs are shared in time with the company divisions Marketing and Sales to offer the possibility of leverage and early development of sales campaigns and presentations.

Innovation and shortened time-to-market is the main interest here. Nevertheless, the complete integration and proper implementation of PLM concepts in the enterprises is mainly still in the fledgling stages.

2. The knowledge society

2.1. Innovation and vision

'Knowledge' is one slogan of these days. The capital of the companies is in their product know-how. Thereby, it has to be clearly distinguished that product knowledge is not only just data or information. The kind of knowledge has to be differentiated. Explicit knowledge is available in a formalised representation. Implicit knowledge is based on a person's experience, partly enhanced by the cognitive capabilities of a human being. There exist various investigations and approaches in the knowledge management domain to secure the availability of implicit knowledge in the company and that it is not lost anymore after a person's withdrawal, as this should be connected no longer to only the person as bearer of knowledge. Many actual applications or programme modules on the market are decorated with the attribute 'knowledge', but do not satisfy this according to the definition. For example, parameterisation has basically little in common with knowledge. The main problem lies in the representation of knowledge, respectively how it can be transformed into a formal model. A knowledge management system should promote the acquisition and utilisation of formalised know-how, and this also in a distributed environment.

Besides the technical problem to make knowledge processible, the question of 'how' the knowledge can be integrated in a supportive way to the enterprise's business processes is mostly neglected. It should be clear that i. e. the efficiency is definitely not only increased because of the provision of a filled database to the design engineer. "What follows 'Best Practices'?", the participants of the ProSTEP iViP Symposium 2004 wondered, summarised in [5]. For Weissflog [5] a reference to familiar and new facts is essential for a lasting success at the product development – the "ability to implement known issues efficiently and to design new things creatively out of the unknown". Kagermann [6] specifies: "Creative ideas arise from the ability to think productively against rules and to act. An innovation originates when these new ideas can be realised in an economic successful way".

Creativity is necessary for the generation of innovation, that is indisputable. Previous systems that own the attribute 'knowledge' can not achieve this yet. The functionality of corresponding systems is restricted to the pure capture and provision of knowledge. New knowledge is not created independently. Is a 'creative' knowledge management system possible at all? This problem can not quite be answered at the time.

2.2. PRIME

Within the scope of the European research project PRIME – Product Integrated Knowledge Management for the Extended Enterprise funded by the European Commission under the programme component Growth (GRD1-2001-40408), the Institute for Mechanical Engineering (IMW) as partner contributed substantially to the development of a practicable product knowledge management approach for the Extended Enterprise (EE), based on industrial requirements among consideration of the acceptance of all persons involved in the product lifecycle. The developed methodology has been written down in a public available workbook [7] that can be seen as guideline for the introduction of the PRIME concept. The main focus was on product related knowledge out of all phases of the product lifecycle which can be captured and evaluated within the framework of an EE, and this as well for all lifecycle phases. An overview of the PRIME knowledge management approach is provided in Fig. 1. The structure and functional components comprise the replenishment of knowledge in terms of capture and elicitation, its organisation and formalisation based on a model based representation, and the extraction by users via various browsing or report facilities. This provides the capability of an integrated solution for a productive application of knowledge. As this will go beyond the scope of this paper please refer to [8] for a detailed description of the PRIME concept.

A prototypical implementation of the concept has been realised in a web-based solution, which is currently under validation. Generally, it can be said that a basis for a distributed knowledge management is given with this concept. It features no 'creativity' yet, but because of its model-based layout it represents the knowledge basis for future developments.

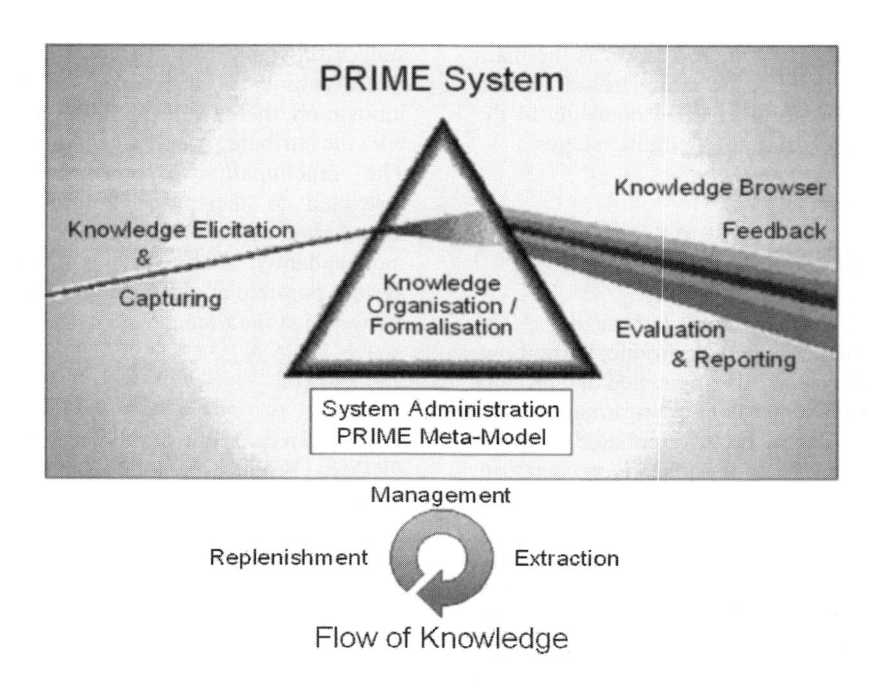

Fig. 1. PRIME Knowledge Management approach

3. Starting point: requirements

3.1 Requirement engineering

Whenever an engineer designs a new component or has to accomplish a design change, he has a certain idea of the product regarding functions, properties or structure. These result from the given requirements that will be interpreted individually based on the personal knowledge. According to the guideline of the Association of German Engineers VDI 2221 [9], the first design step consists of clarification and specification of the task designation to avoid i. e. pre-fixations and to identify only relevant items. The outcome of this procedure is a requirements document that may be adapted in further tasks. Customer orders defining expected specifications are mostly given in textual form, supplemented by verbal communication. However, the definition of requirements in a natural language based format goes along with problems like i. e. incompleteness, misinterpretation, inconsistency.

Consequently, a requirement process covers the three dimensions specification, representation and communication. The common activities involved in discovering, documenting, and maintaining a set of requirements for a computer-based system without reference to a specific discipline is known as requirement engineering [10].

Investigations in the domain of requirement engineering are in the focus of the IMW since mid of the 1990ies. A link of requirements engineering and knowledge management has been realised in the European research project KARE – Knowledge Acquisition and sharing for Requirement Engineering (EP 28916, [11]) which has been co-funded by the European Commission under the ESPRIT programme. A methodical approach and system prototype has been developed for the knowledge supported elicitation, analysis and negotiation of product requirements for complex systems. Natural language based requirements will be systematically captured and structured in the KARE requirement engineering approach. The transformation of requirements into a model-based representation, called formalisation, provides the fundamentals for an IT-based systematic evaluation and processing of them. Such processed requirements can be validated by means of a knowledge management system i. e. on inconsistency or completeness. The clear and explicit product specification as outcome of this procedure represents an improved basis for the product development. Integrated to the communication between customer and supplier, as one application scenario, the KARE approach enables a faster and more accurate bidding process.

The model-based approach of KARE, similar to the PRIME concept, has shown its potential based on a prototypical demonstrator, especially that the model-

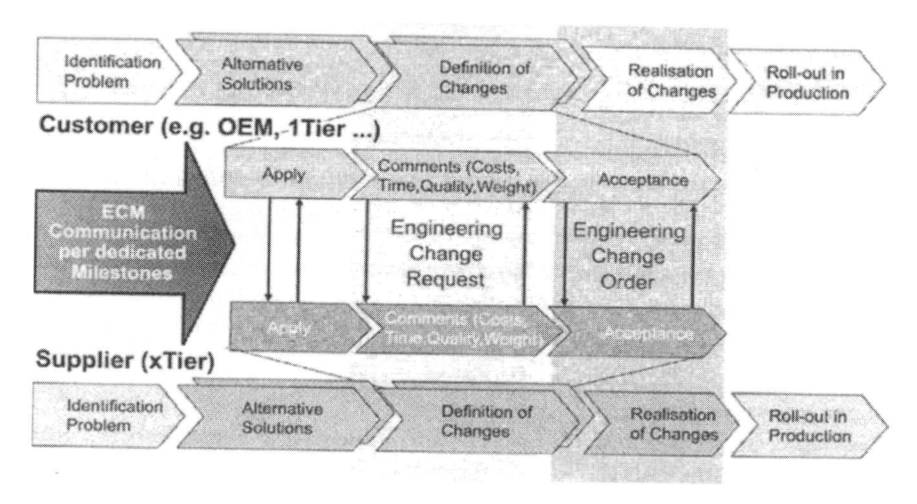

Fig. 2. General phases of the ECM and communication [12]

based representation of requirements enables the application of knowledge-based IT processes. Nevertheless, the validation showed also that this area should be subject of further research.

3.2 Configuration and change management

Requirements and product specifications are not inevitably constant during the design and development phase, especially for complex systems. Far from it, changes or adjustments are rather inevitable and an important component of the product development. The reasons for that are various: errors in development or planning, modified interface/ connecting parameters, updated customer requirements, modified market situation, uncertainties by interaction of different components etc.

Configuration management provides the possibility for a systematic identification of the product configuration to a discrete point in time during the product lifecycle. This enables to verify fulfilment, completeness and traceability of the current configuration. Emerging inconsistencies will initiate the necessity to start an engineering change process.

Any change can cause immense impact to the development and production process already. The engineering change management (ECM) controls, organises and documents the process. An engineering change cycle can be structured commonly in five phases (see Fig.2) according to Hirsch and Trautheim [12]. Especially the communication on negotiation of the engineering change request and order requires a clear structuring, control and synchronisation. The standardised implementation of corresponding procedures is supported by a workflow management system.

The common process of ECM shows a relationship to the requirements engineering processes. It can be also seen as embedded requirement process in certain circumstances, since in the end an alignment of requirements is also accomplished here.

4. Intelligent design support

In the discussion of the topics and problems in the previous sections, it was always possible to highlight a reference to the product requirements and their constructive realisation. A critical step in the requirement engineering domain is the identification and verification of requirements. This process is highly knowledge intensive. This process requires a detailed understanding of relevant key functions and properties of the future product specified by the requirements, and its system environment. The support by IT systems for an optimised alignment of design and requirements with respect to clearness, consistency and completeness is more or less insufficient at present. Consequently, there exists the need for the development of methods and applications that allow a structured, knowledge-based approach for specification, analysis and validation of requirements, supporting the virtual product development. It may be possible i. e. to run early test scenarios to identify and synchronise relevant design and interface parameters, to accomplish simulations or configuration analyses already during the development, and to be able to initiate necessary changes in time. This implicates a considerable reduction of costs by optimised, fully developed product specifications in shortened time of development.

Thinking one step further, this approach should not be restricted to a 'static' target/ actual comparison, but promote processes to support and guide the design engineer actively in his solution finding during the design process. Present knowledge and requirement systems are limited actually to the capturing and administration of corresponding objects, as mentioned before. Integrated expert systems, that are rule-based for the most part, reproduce the system contexts more or less exactly based on the conceptual, qualitative comprehension of the programmer. Knowledge is just reproduced in a sense, but not applied and extended. Intelligent software agents, so-called IT-applications inside a defined software environment with the goal of user assistance for specific tasks or problems, possess a certain autonomy in their behaviour. These agents own the basic quality of an orientation at the human behaviour and enable a natural interaction, therefore own a certain intelligence and should be acting like a personal assistant and process independently specified tasks. The field of application for such agents has been expanded considerably especially in the last years [13].

5. Conclusions

Model-based requirements and knowledge representations, as specified in the aforementioned research projects KARE and PRIME, could represent the information basis for a future development of particular intelligent software agents to execute i. e. solution finding, analyses and assessments upon that. A possible application scenario could be an autonomous requirements analysis for the extraction and generation of basic conditions and design alternatives, to provide exactly that knowledge and information to the design engineer that is necessary for his engineering task, no more nor less. The designer obtains more room for his creative work. Furthermore, it may be possible to implement an interactive consistency analysis related to the pre-determined requirements accompanying the design process. Consequently, this has the potential to reveal or avoid early erroneous design decisions, but could also propose or direct development trends.

A corresponding agent system could be integrated to the common design process and would be also applicable in distributed environments. How 'creative' such a system can be remains open so far, but it owns definitely high potentials to release the designer from lots of administrative tasks and to provide knowledge-based and requirements-driven support and guidance to him within an integrated, optimised design process.

Acknowledgement

The author gratefully acknowledges the financial support from the EC and the hard work of participants for the supporting projects KARE and PRIME.

The Institute of Mechanical Engineering, Clausthal University of Technology, is partner of the EU-funded FP6 Innovative Production Machines and Systems (I*PROMS) Network of Excellence.

References

[1] Heinrichs, W.: DAMA – the key to becoming a fast and flexible global electronics innovator, technical article, http://www.zuken.com, 2004

[2] Dewhurst, N.: Need to Cut Costs? Check Your Design First, letter to the U.S. Manufacturers, September 2004

[3] Brynjolfsson, E.; Den Kopf frei für anspruchsvolle Aufgaben, strategy paper, http://www.sapinfo.de/public/en/articlelist.php4/Author-1727740b6098533234, June 2004

[4] Bacheldor, B.: Deeper Than Designs, http://www.informationweek.com/story/showArticle.jhtml?articleID=26806195, Information Week, August 2004

[5] Weissflog, U.: Was folgt auf 'Best Practices'?, Produkt- Daten Journal Nr.2, November 2004

[6] Kagermann, H.: Editorial SAP INFO 119, August 2004

[7] The PRIME Consortium: PRIME Workbook, PRIME download area: http://www.prime-project.org/deliverables.html, 2003

[8] Düsing, C.; Müller, D.: A Knowledge Management approach for the Extended Enterprise, Institutsmitteilung Nr. 28, IMW Clausthal, 2003

[9] VDI Richtlinie 2221: Methodik zum Entwickeln und Konstruieren technischer Systeme and Produkte, VDI Verlag, Düsseldorf, 1993

[10] Kotonya, G.; Sommerville, I.: Requirements Engineering – processes and techniques, worldwide series in computer science, John Wiley & Sons Ltd., Chichester, England, 1998

[11] The KARE Consortium: KARE project website, http://www.imw.tu-clausthal.de/kare

[12] Hirsch, H.; Trautheim, A.: Engineering Change Management – abgestimmte, partnerübergreifende Prozesse, ProduktDaten Journal Nr.2, November 2004

[13] Keller, H.: Maschinelle Intelligenz – Grundlagen, Lernverfahren, Bausteine intelligenter Systeme, Friedr. Vieweg & Sohn Verlagsgesellschaft, Braunschweig, 2000

Intelligent Production Machines and Systems
D. T. Pham, E. E. Eldukhri and A. J. Soroka (eds)

Participatory innovative design technology tools for enhancing production systems and environment

T. Määttä, J. Viitaniemi, A. Säämänen, K. Helin, S.-P. Leino and K. Heinonen

VTT Industrial Systems, Tampere, Finland

Abstract

Production systems need to be improved in order to increase efficiency and environmental sustainability. There is a demand for a comprehensive method or principle to concurrently consider all the relevant factors affecting the whole environment of interest. This work is based on implementing virtual environment (VE) and augmented reality (AR) technologies in a number of case studies using a participatory approach. The findings indicate that it is not satisfactory to implement only one of these technology approaches in production systems and production environment development; it appears beneficial to integrate both types of technologies and corresponding procedures in the design process. Improvements should also focus on the design procedures in order to obtain the benefits of using these technologies in production systems and environment design.

1. Introduction

Manufacturing industry is involved with ever increasing global competition and requirements. Production systems must be improved to increase efficiency and environmental sustainability. A competitive manufacturing sector is vital for Europe to achieve its ambitious long-term economic, social and environmental targets [1] and to promote innovation in manufacturing.

One specific and widely studied approach to enhance production systems and environments is participatory design [2, 3]. The primary goal of participatory design is to gain better knowledge about the production system and environment being considered in the design process by involving different personnel from, for example, design, production, maintenance, safety or management in this process. This can be done, for example, by visualising the target system and its functions and effects and by sharing the knowledge between all participants [4]. The outcome of the design process may result in better quality of design according to the requirements, a more effective design and manufacturing process, and an increased added-value for the company, workers, and customers. One major effect of implementing a participatory design approach is the involvement of workers and their commitment to the design outcomes [5]. However participatory design needs efficient tools. Virtual Environment (VE) and other visualisation technologies have successfully been applied in participatory design [6, 7].

Visualisation technologies used in participatory design have several possible implementation levels ranging from simple tools describing reality to complex VE systems. Several concepts have been presented to describe these different levels of implementation [8-14]. For the purpose of this study the concepts of interest are Augmented Reality (AR) and Virtual Environments (VEs). AR technologies afford the capability to visualise virtual phenomena by adding information to the view of the real environment. VE technologies provide tools for a user to be immersed into the digital model of a real environment. The technologies were applied separately in the participatory design process due to the differences between them. There could be a great potential to integrate these methods so that they can be used concurrently in participatory design.

This paper presents work that was completed at VTT Technical Research Centre of Finland. A participatory approach and applications of both VE and AR technologies were implemented in industrial cases. The purpose of the work was to improve production systems and the production environment.

2. Technologies supporting participatory design

Depending on the maturity of the design stage we applied two different approaches when implementing participatory design in industrial cases. A Video Exposure Monitoring approach was utilised when defining and analysing the problem with the existing system. Virtual Environments were used to identify hazards and other targets of development from a digital system model.

2.1. Video Exposure Monitoring (VEM)

Applying methods and technology similar to AR can improve user comprehension of the real world environment as augmented images can provide information which cannot be seen in the real world. Exposure to some environmental factors is difficult to sense with human sense organs. Airborne contaminants can have a harmful effect on both products (contamination) and workers (occupational health). As airborne contaminants are normally invisible it is not easy to consider these factors during the development of the production environment.

The visualisation of invisible variables in the video picture can help to discover dependence between visible events and some easily measurable, but not visible variable. The PIMEX method (abbreviation for PIcture Mixed EXposure) is successfully used to visualise invisible physical variables. The basis of the PIMEX method was developed at the National Institute for Working Life in Sweden [15].

The PIMEX method is based on a combination of direct-reading instruments and video filming. The measurement instruments for environmental or other factors are connected to the equipment. Concurrently with the measurement, the work or other event under investigation is filmed. With the aid of PIMEX equipment, signals from the video camera and the measurement instrument are mixed to give information on the current level of the measured variable. The continuous signal from the measuring instrument is converted into a bar graph and displayed on the computer screen (Fig. 1). The height of the bar is proportional to the measured signal.

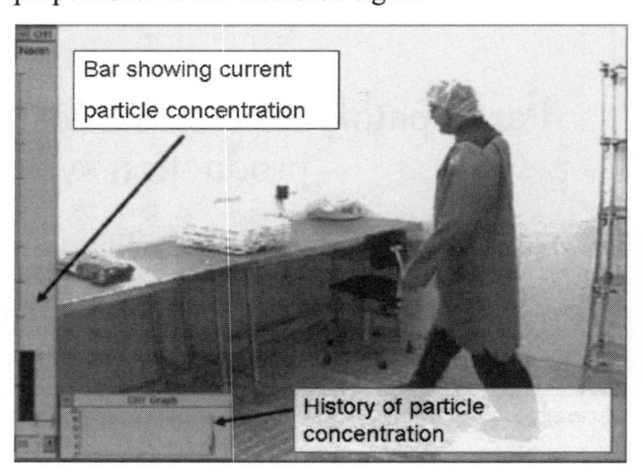

Fig. 1. Snapshot of PIMEX screen from measurements solving particle contamination in a clean room (readout descriptions added for clarification).

2.2. Virtual Environments (VEs)

Modern technologies such as VEs, Virtual Prototyping (VP) and Virtual Testing (VT) have proved to be useful tools in many fields of design and practice. For example, basic metal and parcelled goods production, electronics industry, food production, and the vehicle industry have utilised these tools to assess dependability and safety issues of products and production [4]. In contrast to previously time consuming prototyping and testing of plans, virtual prototyping and testing can be carried out in the early stages of the design process. VE systems enable design of the product, production and work tasks, and can also be used for training and practising skills before concretising the production lines or the product itself [16, 17].

Virtual prototyping involves creating computer-internal, virtual prototypes of systems and processes. In particular, virtual prototypes are used to verify such properties that would decrease both the time and cost associated with product or production. Human factors and manufacturability can be evaluated without fabrication of physical prototypes.

2.3. Participation

The main objective in our participatory design and development process was to bring together knowledge from the organisation and the experts [18, 19]. Visualisation methods such as VEs or VEM were applied in order to enhance communication. This process will facilitate the start of an internal development project in the company. The process is based on the use of methods that visualise the factors of interest, e.g. new production system layout, and the participation of different personnel groups in the company. A development group formed within the company utilises the visualisation methods to search for possibilities to enhance the production system and to evaluate the efficiency of alternative solutions. The process applies a general problem-solving model (or development cycle) consisting of the following phases: 1) Identification 2) Analysis 3) Solution development 4) Implementation and 5) Evaluation.

3. Application examples

3.1. Case 1: Steel factory

VEs were applied in a steel factory to evaluate safety and functions of new plant investments prior to implementation. The aim of the study was to investigate the advantages of implementing VE technology in a realistic industrial development project [20].

To use VEs in a safety analysis, the machines and environment were modelled with 3D simulation software based on photographs, video recordings and 2D drawings. The kinematics of the machines was also realised with the software so that the movement of different machines and their parts could be seen both with animation, and manually operated separately by the software. The project groups and the safety analysis and modelling expert made decisions on the level of accuracy and detail required in the model. Different colours were used for different parts. Digital human models were also used in the model in areas where humans would do the work or move. Cranes and vehicles were also modelled in the factory model and their ability to move was modelled as realistically as was necessary.

A workgroup was created for the safety analysis. The group consisted of a project manager, designers (mechanical, electronic, manufacturing engineers), the foreman, a safety specialist, a maintenance person and several workers. All these people were connected to the new department of the factory and had several years experience in the work conducted by that factory department. Most of them were also familiar with the safety analysis process.

Safety design and analysis is a systematic process in which the expertise of development groups and many different techniques are utilised. The goal is to systematically identify all the risk factors which may impact on the system. The risk assessments were applied to each hazard identified.

Several sessions were arranged during the analysis process. The hazards, danger zones and different tasks were analysed using 3D models and a VE. Animation of the machines was also used in the identification of hazards. 3D models were viewed at different angles and different distances during the sessions. Figure 2 shows an overview of one section of the factory. The tasks were analysed using animation.

Fig. 2. An overview of a 3D model from a section of the plant.

Certain parts of the machine and the structure of the factory building were analysed more precisely and the model was created with more details. Several safety factors in the process and faults in the drawings were detected by the implementation of VEs. According to the project manager the commission phase of the plant took place without any significant problems. One of the reasons for that was the use of virtual technology during the design phase.

3.2. Case 2: Medium size factories

There are several cases of investments representing middle term assets that have been evaluated by utilising a participatory approach and VEs, for example, in a food factory, in a factory for plastics injection moulded products, in a factory

5. Components of the case-based reasoning (CBR) system

This section covers the details of the CBR system developed. The main functions of the system are part input and its representation, sequencing, case retrieval, and case adaptation. These have been given in the following subsections. Machining sequence explained by algorithm is as follows:

5.1. Parts database and similarity system

The system consists of the following steps:

1. Creating the part on the CAD system (Solidworks).
2. Exporting the part data from the CAD system Using STEP format.
3. ISO 10303(STEP) is an international standard to provide an ambiguous representation and exchange mechanism of computer interpretable product information throughout the life cycle of a product. It is also provides a consistent data change format and application interfaces between different application systems. STEP format is useful in transferring product information from a CAD system to another, from CAD to CAE, from CAD to CAM, etc. The ultimate goal of STEP community is the manufacturing automation in a CIM environment.
4. Divided the STEP file to set of features and storing features in database.
5. parameters fall When facing a new part:
   - Create its feature form drawn it by solidworks and save it as a STEP format and flowing the step 2 and 3.
   - Retrieve a set of close parts (candidates for similarity assessment) from the database based on rough similarity criteria such as the total number of features and the types of these features.
   - Match the features of the input part against every feature in the set of candidates. The matching process is carried out using the type of features and dimensions of each entity.
   - Calculate the factor of similarity between the new part and every matched part. This factor expresses the percentage of identical feature among the two parts. The highest similarity factor indicates the part most similar to the new part.

The generation of the matching sequence explained by the algorithm is as follows:

1. Move to the top of the list of features of the part input.
2. Choose the first feature, call this the chosen feature.
3. Create a list called the precedence list and link it to the chosen feature.
4. Call upon the first sequencing rule.
5. Apply the rule to the case feature.
6. If any precedence feature is found, add it to precedence list of the chosen feature.
7. If all the sequencing rules have been applied, go to 8
   Else 7.1. Call upon the next sequencing rule.
   7.2. Go to step 5.
8. If the chosen feature is the last one in the feature list, go to 9, else move to the next feature. Now refer to this as the chosen feature, go to 3.
9. Stop.

These steps shown in fig.1

5.2. Case retrieval

The case retrieval module is the most important one in the system. The success of the system is governed by its ability to retrieve a case from the case library, which is in reality the "most similar one" to the part input by the user. As pointed out earlier, the system uses a case-matching methodology which is multi parametric and absolute.

5.3. Case adaptation

The unmatched features of the part are searched for in all the cases in the case library. If a match is found, its sub plan is extracted, adapted, and fitted into the process plan as explained in the previous sections. If a suitable match is not found, the system taken this features form other cases stored in the library. For the component shown in Fig. 2, the case study (Part5). This part stored as a STEP file format from the solidworks and contains three layers which are application layer, logical layer and the physical layer. The application layer contains application area, shape model, geometry, topology, form features and the dimensions and tolerances as specified by the designer. Topological is a another element in application layer and the basic topological entities are vertex, edge, face, loop and shell.

4. Conclusions

The proposed system incorporates several new features in various areas of process planning using case-based reasoning. In order to fulfill the requirements of rapidly changing product designs and configurations, the system uses several modifications to the modulus of CBR, namely, case matching, retrieval, and case adapter. A straight forward and realistic method for enumerating the similarity indices has been used which ensures that the case extracted as the most similar one is the closest to the input part and fulfils the process planning needs. The various modules receive assistance in their functioning from the process library which has been introduced in this research. Hence, it can be construed that the proposed system embraces the merits of both the variant and

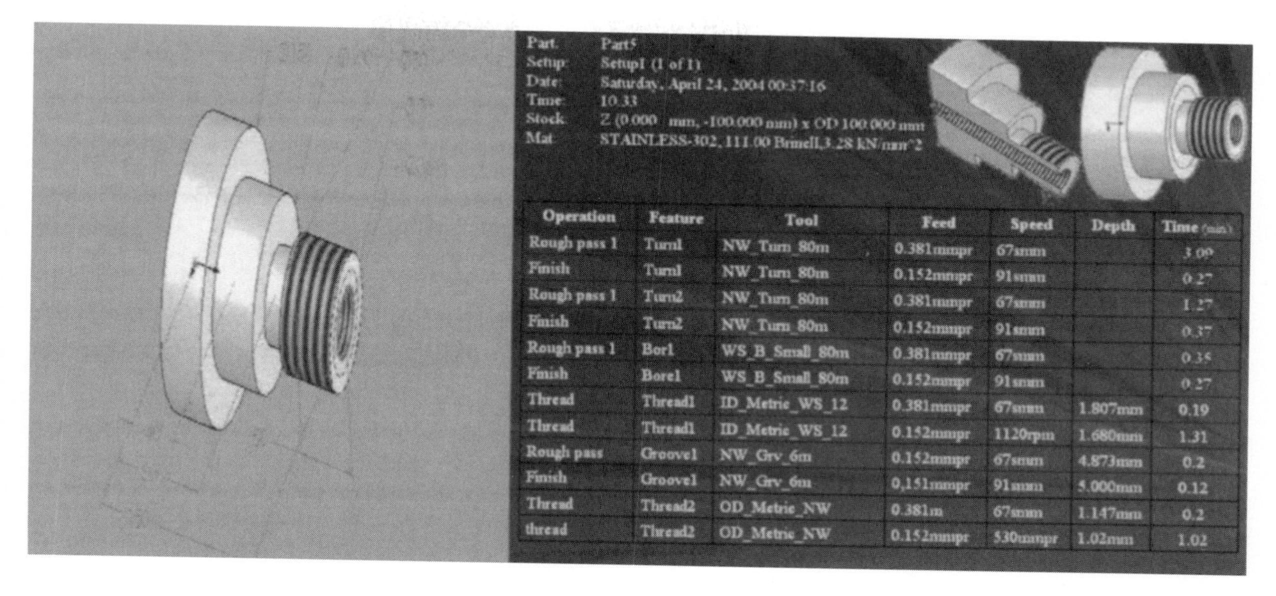

Operation	Feature	Tool	Feed	Speed	Depth	Time (min)
Rough pass 1	Turn1	NW_Turn_80m	0.381mmpr	67smm		3.09
Finish	Turn1	NW_Turn_80m	0.152mmpr	91smm		0.27
Rough pass 1	Turn2	NW_Turn_80m	0.381mmpr	67smm		1.27
Finish	Turn2	NW_Turn_80m	0.152mmpr	91smm		0.37
Rough pass 1	Bor1	WS_B_Small_80m	0.381mmpr	67smm		0.35
Finish	Bore1	WS_B_Small_80m	0.152mmpr	91smm		0.27
Thread	Thread1	ID_Metric_WS_12	0.381mmpr	67smm	1.807mm	0.19
Thread	Thread1	ID_Metric_WS_12	0.152mmpr	1120rpm	1.680mm	1.31
Rough pass	Groove1	NW_Grv_6m	0.152mmpr	67smm	4.873mm	0.2
Finish	Groove1	NW_Grv_6m	0,151mmpr	91smm	5.000mm	0.12
Thread	Thread2	OD_Metric_NW	0.381m	67smm	1.147mm	0.2
thread	Thread2	OD_Metric_NW	0.152mmpr	530mmpr	1.02mm	1.02

Fig.2. the case study.

the rule-based approaches of process planning. The performance of the proposed system is reasonably good. The efficiency and efficacy of the case retrieval and case adaptation modules have been verified. In such situations, the major role is played by the case adaptation module of the proposed system. The proposed system is robust in the sense that it easily takes care of the addition or deletion of extra features. The system exhibits a certain level of intelligence in developing the process plan.

Acknowledgements

This research was supported in Industrial engineering Depart. Faculty of engineering, Zagazig University. The authors would like to express their gratitude for the support.

References

[1] Sabourin, L. (1996), "an expert CAPP system", *Advances in Engineering Software*, 25, pp 51-59.

[2] Jiang, B., Lau, H., Chang, Felix T.S., Jiang H. (1999), "An automatic process planning system for the quick generation of manufacturing process plans directly from CAD drawings", Journal of Materials Processing Technology, 87, pp 97-106.

[3] Radwan, A. (2000), "A practical approach to a process planning expert system for manufacturing processes", Journal of Intelligent Manufacturing, 11, pp 75-84.

[4] Schank, R. C. and Riesbeck, C. K. (1989), "Inside Case Based Reasoning", Lawrence Erlbaum, Hillsdale, NJ.

[5] Schank, R. C. (1982), "Dynamic Memory: A Theory in Reminding and Learning in Computers", Cambridge University Press.

[6] Kolodner, J. L. (1988), "Extending problem solver capabilities through case-based inference", in Proceedings of a Workshop on Case-Based Reasoning, Clearwater Beach, FL, pp. 21–30.

[7] Hammond, K. J. (1989), Case Based Planning, Academic Press.

[8] Pu, P. and Reschberger, M. (1991), "Case-based assembly planning in Proceedings of Case-Based Reasoning Workshop", Washington, DC, Morgan Kaufmann, pp. 245–254.

[9] Yang, H., Lu, W. F. and Lin, A. C. (1994), "PROCASE: A case-based process planning system for machining of rotational parts", Journal of Intelligent Manufacturing, 5, pp. 411–430.

[10] Superatik, C. (1996), "Automated operation sequencing in intelligent process planning: a case based reasoning approach", International Journal of Advanced Manufacturing Technology, 12, pp. 21–36.

Intelligent Production Machines and Systems
D. T. Pham, E. E. Eldukhri and A. J. Soroka (eds)

A knowledge-based system to schedule multi-skilled labor with variable demand

S. Ibrahim, M. Khater, and H. Ghareeb.

Department of Industrial Engineering, University of Zagazig, Egypt

Abstract

Workforce is a required resource for any organization to achieve its goals. This resource should be managed well, and optimized to make the best use of it at minimum cost. The tour-scheduling problem is the problem of assigning labor to shifts for a given planning horizon.
We focus our work on solving the tour problem for organizations which operate continuously and comprise multi-skill labor classified hierarchically. The demand for labor is varying from shift to shift. The resulted schedule shows the on-off shifts for each employee. Moreover, we consider the problem of assigning tasks to employees according to their capabilities.
The problem is represented by integer-programming model which is mainly considered here as a criterion for evaluating the efficiency of the proposed knowledge-based system in terms of the number of surplus workdays. Also, the proposed technique has the advantage of distributing the surplus workdays evenly.

1. Introduction

Workforce allocation and personnel scheduling deal with the arrangement of work schedules and assignment of personnel to shifts to cover the demand for resources over time. This problem arises in manufacturing as well as in services, for example, telephone operators, hospital nurses, policemen, and air crews.

Manpower scheduling problem comprises four tasks as mentioned by Gary [1]. The first task is determining the quantity of work to be done. The goal of that task is to predict the characteristics of the system transactions that change over time. The second task is determining the required staffing to do the work for each time period. The outcome of that task is a decision of the number and skill levels of employees needed to meet the demand adequately throughout a planning cycle. The third task which we focus is developing a workforce schedule that supplies sufficient staffing while also accounting for employee requirement. Finally, the fourth task is controlling and administering the schedule as the work unfolds. This task involves the real-time control of the schedule, in which the manager assesses whether the schedule is ensuring that the work is actually done as planned.

Nysret [2] showed in his dissertation that the workforce scheduling problem is classified as NP-complete, which can not be solved in polynomial time but whether a given solution is right can be checked in polynomial. This is the nature of satisfiability problems which has two answers "Yes" or "No". The researchers studied this problem and developed different solution methods with three general approaches, optimization methods, heuristics, and search methods. Not all the researchers have dealt with the whole problem 'tour' but they have differed in the focused problem they have studied. Some of the researchers have focused their work on solving the day-off problem only. Others have considered the shift-scheduling problem which may include assigning tasks to employees. Few numbers of them have concentrated their work on the surplus dispersion, and how to reach an even distribution of the excess labor.

The first solution for this problem was given in 1954 by Dantzig [3] who proposed a mathematical programming model based on set-covering formulation to find the minimum tool booth attendants. Ahmed I.Z. Jarrah et al [4] presented a new methodology for solving complex tour scheduling problems. The methodology begins with an integer programming formulation, and then it introduces a set of aggregate variables and related cuts. These variables decompose the whole problem into seven subproblems (one for each day) that are much easier to be solved. Hoong Chuin [5] proved that a restricted version of the manpower shift scheduling problem is NP-hard, as it can not be solved in polynomial time. This observation allowed him to justify techniques which had been applied in manpower scheduling systems such as implicit enumeration and heuristic search. He focused his study on the changing shifts, and showed that the case of monotonic shift changes is polynomially solvable for both cyclic and non-cyclic schedules with the same computational time complexity. Amnon et al [6] combined the constraints network with knowledge based system to solve the timetabling problem. The knowledge-based techniques are utilized in the form of assignment rules, which assign a class of employees to a class of shifts. The other part of the method is a constraint heuristic to satisfy all the constraints of the problem. The mixed mode approach ends with a backtracking method to perform an exchange of assigned employees between competing shifts in case of a dead end. Alian Billonnet [7] used the mathematical programming models to solve the day-off problem for a hierarchical workforce. The formulation starts with an integer programming formulation to find the minimum workforce size and ends with another integer model to schedule the day-off. X. Cia, and K. N. Li [8] applied the genetic algorithm to solve the problem of multi skill labor in which there are three types of labor. Only one of them can do all the different jobs. The problem contains multi-criteria in which the parent selection is based on. Also, the problem was formulated as a multi-objective ILP model and a simple heuristic was provided to resolve the infeasibility created by the crossover if happened. Gert et al [9] used a simulation aided approach to solve the labor re-assignment problem. The proposed approach is based on Great Deluge Algorithm. The aim of work was to find an optimal combination of the existing and the improved personal structure by minimizing adjustments and personal costs and maximizing logistical goal achievements. Tijen et al [10] proposed a framework based on data envelopment analysis (DEA) for determining the optimum allocation of labor in cellular manufacturing systems (CMS). The study concentrated on the efficiency measurement and the determination of the number of operators in CMS under different labor assignment policies and when the demand rate and the transfer batch size were changed.

In this work, we study the tour-scheduling problem for continuous systems which operate for three daily shifts through the seven weekly days. The system includes homogenous hierarchical workforce classified according to their efficiency or experience. The proposed technique is to establish a knowledge-based system using Visual ProLog V5.2 [11]. The resulted schedule shows the off-shifts and the assigned shifts with specific tasks for each employee during a week.

The remainder of this paper is organized as follows. In the next section, we define the problem of interest and its assumptions. In section 3, we modify the model proposed by Sameh et al [12] to deal with our problem. The proposed knowledge-based system is summarized in section 4. In section 5, computational results obtained for a hypothesized data is taken as an example to illustrate the resulted schedule of the proposed technique. Finally, in the last two sections, the work is ended by the conclusion and proposal work that can be done in the future.

2. Problem definition

Sameh et al [12] dealt with the tour problem but for organizations which operate for a single shift per day. We presented the mathematical models and an expert system for this problem. But these techniques should be modified here to consider the new situation of the three daily shifts. A new constraint should be added to the day-off model of [12] for ensuring that each employee will work at most one shift per day and the difference in time between two successive assigned shifts is at least two off-shifts (16-hours) for each employee. Also, the expert system should be modified by adding the suitable rules that control the daily shift assignments. We assume that;

1. The system operates seven days per week.
2. There are three shifts per day; day-shift (D), evening-shift (E), and night-shift (N). The week-end days are the last two days.
3. Each shift is composed of number of tasks. Each task requires a number of labor with different capabilities.
4. Each labor will work five shifts per week.
5. Each labor will have two days off per week, not necessarily to be consecutive.
6. The system contains (s) different types of labor, classified hierarchically by their capabilities or experiences. A higher qualified labor can substitute a lower qualified one but not vice versa.
7. The schedule is rotating and is done for one week.
8. No employee absenteeism.

3. Shift-assignment model

This model is used for assigning on-off shifts to each employee. The output of the model shows when the worker is on and when he is off. Assuming there are (s) types of labor, each type contains (m_k) labor and the number of the weekly shifts is (n = 21).

The objective of the model is to find the minimum total number of assignments per week. The first constraint ensures that the number of labor required in shift (j) is assigned. The second constraint is to assign only five shifts (day-on) to each employee per week. The third constraint ensures that each employee will work at most one shift per day, and the difference in time between two successive assigned shifts is at least two off-shifts (16-hours) for each employee. The binary variable x_{kij} represents the assignment of the employees to shifts; $x_{kij} = 1$ when employee (i) of type (k) is assigned to work in shift (j), or $x_{kij} = 0$ when employee (i) of type (k) is not assigned to work in shift (j).

$$\text{Min } z = \sum_{k=1}^{s} \sum_{i=1}^{mk} \sum_{j=1}^{n} x_{kij} \qquad (1)$$

S.T

$$\sum_{k \leq K} \sum_{i=1}^{mk} x_{kij} \geq \sum_{k \leq K} R_{kj} \qquad K = 1,2,\ldots,s \quad j = 1,2,\ldots,n \qquad (2)$$

$$\sum_{j=1}^{n} x_{kij} = 5 \qquad k = 1,2,\ldots,s \quad i = 1,2,\ldots, m_k \qquad (3)$$

$$x_{kij} + x_{ki(j+1)} + x_{ki(j+2)} \leq 1 \qquad k = 1,2,..,s \quad i = 1,2,..,m_k \quad j = 1,2,..,n-2 \qquad (4)$$

$$x_{kij} = 0 \text{ or } 1 \qquad k = 1,2,\ldots,s \quad i = 1,2,\ldots, m_k \quad j = 1,2,\ldots,n \qquad (5)$$

4. The proposed knowledge-based system

Here, we establish a knowledge-based system to solve the manpower scheduling problem with the same predescribed conditions. The system is built using Visual Prolog V5.2. The methodology consists mainly of four steps:-

1. Feeding the system with the input data: The input data is the tasks requirements from each labor type (R_{tkj}) for each shift. These requirements are summed to form the shift requirements from each labor type (R_{kj}).
2. Building the facts of the system: The input data is transformed to facts. There are other facts that represent the status of the labor for each shift (on-off), the summation of his working shifts, shifts that contain surplus labor… etc.

3. Running the rules of the system: This step is the proposed algorithm which starts with calculating the workforce size, then assigns on-off shifts and a specific task simultaneously. The allocation process begins from satisfying the requirements from each labor type of first shift and goes on until it reach the last shift. After obtaining the weekly schedule the system distributes the surplus labor on shifts in even manner. Then, the weekly schedule of each labor is updated according to the surplus dispersion.
4. Printing results: Results are printed in a tabular format showing the required number from each labor type, the on-off shift schedule including the task assignment for each employee.

5. Computational results

The following data of Table 1 represents the task requirements for a given week. This data is hypothesized but it may fit the operating conditions of a system in continuous operation facilities like hospitals, call centers, police departments, and some manufacturing companies which can not be idle for any hour through the week. In this hypothesized system, labor is classified into three types (K1, K2, and K3) and their requirements are different form shift to shift and from day to day. The system comprises four different tasks (T1, T2, T3, and T4) which should be performed. Each entry in Table 1 represents the number of labor of type K required for doing task T in a weekly shift. It is required to find the minimum number of labor type to be available in the system to perform the required tasks. Also, it is required to find the weekly schedule which provides us with the tour of each labor; his off-shifts and his tasks in the working shifts.

After running the knowledge-based system, it provided us with the minimum number of labor from each type that should be available to meet the variable demand. These results are the same as the results obtained from solving the integer-programming model proposed by [12]. The results satisfy the maximum required number from each labor type and the five working-shifts constraints. So, the proposed algorithm has the same surplus labor of that of the integer-programming model.

Table 1
Labor requirements for task (T) during a week.
D/ day, E/ evening, N/ night shift.

Day	T	1			2			3			4		
	K	1	2	3	1	2	3	1	2	3	1	2	3
1	D	1	1	1	1	2	0	2	1	2	1	2	1
	E	2	1	2	1	1	1	0	1	1	2	1	1
	N	1	1	1	0	1	2	2	2	2	2	1	1
2	D	2	0	0	2	1	1	0	1	1	1	1	1
	E	1	1	1	1	2	2	2	1	1	2	1	1
	N	0	1	2	1	1	1	2	1	2	2	1	2
3	D	1	1	1	2	2	1	1	1	1	2	1	2
	E	0	0	1	1	1	1	2	1	1	1	1	1
	N	1	1	1	1	0	2	2	1	1	1	1	1
4	D	1	1	1	2	1	1	1	1	2	2	1	1
	E	1	1	2	1	1	1	2	1	2	2	2	1
	N	0	1	1	1	2	1	1	2	1	1	1	1
5	D	1	1	1	1	1	1	2	2	1	2	1	1
	E	1	1	1	1	2	0	1	1	1	1	0	1
	N	1	2	2	1	1	2	2	1	2	1	2	1
6	D	1	1	1	1	1	1	1	1	1	1	1	1
	E	1	0	1	1	1	1	1	1	1	1	1	0
	N	0	1	1	0	1	1	1	0	1	0	1	1
7	D	1	0	1	1	1	1	1	1	0	1	1	0
	E	1	1	1	0	1	1	1	1	1	0	1	1
	N	0	1	0	1	1	0	1	0	1	1	1	1

Table 2 illustrates the number of each labor type which should be available in the system to satisfy the demand. The required workdays is obtained from summing all the requirements of each labor type provided in Table 1. Also, the table provides us with surplus workdays of each labor type.

Table 2
Available labor in the system

Labor Type	Available number	Available workdays	Required workdays	Surplus workdays (Ava. –Req.)
K1	19	19 ×5 = 95	94	1
K2	18	18 ×5 = 90	89	1
K3	19	19 ×5 = 95	92	3

The tour-schedule is obtained as shown in Table 3. This table illustrates the scheduled shift (D, E, or N) with the specific assigned task (1, 2, 3, or 4), and the off-shifts (--). The advantage of the even distribution of the surplus workdays is cleared in this table. There are three surplus workdays which are assigned uniformly among shifts as one workday is assigned to the 1st shift, the second is assigned to the 20th shift, and the last workday is assigned to the 21st shift. Also, these workdays are assigned to different labor, they are the 17th, 18th, and the 19th labor of this type.

Table 3
The task-assignment schedule for type 3 labor. scheduled shift/assigned task, S: surplus workday.

	Day						
No.	1	2	3	4	5	6	7
1	D/1	E/2	E/4	N/3	--	D/4	--
2	D/3	E/2	N/1	N/4	--	E/1	--
3	D/3	E/3	N/2	--	D/1	E/2	--
4	D/4	E/4	N/2	--	D/2	E/3	--
5	E/1	N/1	N/3	--	D/3	N/1	--
6	E/1	N/1	N/4	--	D/4	N/2	--
7	E/2	N/2	--	D/1	E/1	N/3	--
8	E/3	N/3	--	D/2	E/3	N/4	--
9	E/4	N/3	--	D/3	E/4	--	D/1
10	N/1	N/4	--	D/3	N/1	--	D/2
11	N/2	N/4	--	D/4	N/1	--	E/1
12	N/2	--	D/1	E/1	N/2	--	E/2
13	N/3	--	D/2	E/1	N/2	--	E/3
14	N/3	--	D/3	E/2	N/3	--	E/4
15	N/4	--	D/4	E/3	N/3	--	N/3
16	--	D/2	D/4	E/3	N/4	--	N/4
17	--	D/3	E/1	E/4	--	D/1	E/S
18	--	D/4	E/2	N/1	--	D/2	N/S
19	D/S	E/1	E/3	N/2	--	D/3	--

6. Conclusion

In this work, we presented a knowledge-based system built using Visual Prolog V5.2 to solve the tour-scheduling problem in continuous-operating facilities which include hierarchical workforce. The proposed technique resulted in a number of labor equal that of the integer-programming models. Moreover, our proposed technique has the advantage of distributing the surplus workdays uniformly among the weekly shifts and the available labor.

7. Future work

In order to simplify our problem, we ignored some soft constraints, like the individual preferences, and allowance for part-time employees to share in satisfying the requirements. These aspects are intended to be done in the future. Also, the proposed knowledge-based system will be developed to deal with the sudden events that occur in the systems like the employee absenteeism which may make the current schedule outdated.

Acknowledgments

We are deeply indebted to everyone taught us to reach this scientific degree.

References

[1] Gary M. Thompson. Labor scheduling, part 3: developing a workforce schedule. ARTICLE, The Cornell Hotel and Restaurant Administration Quarterly. Vol. 40 Issue 1 (1999) 86-96.

[2] Nysret Musliu. Intelligent search methods for workforce scheduling: new idea and practical applications. Dissertation, Vienna University of technology, Austria, 2001.

[3] Dantzig G. A comment on Edie's traffic delay at tool booths. Operations Research Journal. 2(3) (1954) 339-341.

[4] Ahmed I.Z.Jarrrah, Jonathan F. Bard, Anura H. deSilva. Solving large-scale tour scheduling problems. Management Science. Vol.40 no.9 (1994)1124-1144.

[5] Hoong Chuin Lau. On the complexity of manpower shift scheduling. Computers operations research. Vol. 23 no. 1 (1996) 93-102.

[6] Amnon Mesels, Ehud Gudes and Gadi Solotrevcesky. Combining rules and constraints for employee timetabling. International Journal of Intelligent Systems 12(6) (1997) 419-439.

[7] Alian Billionnet. Integer programming to schedule a hierarchical work force with variable demand. European Journal of Operations Research 114, (1999) 105-114.

[8] X.Cai, K.N. Li. A genetic algorithm for scheduling staff of mixed skill under multi-criteria. European Journal of Operational Research. 125 (2000) 359-369.

[9] Gert Zulch, Sven Rottinger, and Thorsten Vollstedt, "A Simulation approach for planning and re-assigning personal in manufacturing", International Journal of Production Economics 90, pp 265-277, (2004).

[10] Tijen Ertay, and Da Ruan. Data envelopment analysis based decision model for optimal operator allocation in CMS. European Journal of Operational Research, In press.

[11] Prolog development center, "Visual prolog version 5.0", copyright 1986-1997.

[12] Sameh Ibrahim, Mahasen Khater, and Hani Ghareeb. Application of Manpower Scheduling in Manufacturing Processes. Al-Azhar University Engineering Journal, AUEJ. Vol. 8 no. 2 (2005) 24-34.

Intelligent Production Machines and Systems
D. T. Pham, E. E. Eldukhri and A. J. Soroka (eds)

Application of reinforcement learning to a mobile robot in reaching recharging station operation

L. Cragg and H. Hu

Department of Computer Science, University of Essex, Colchester CO4 3SQ, U.K.
Email: lmcrag@essex.ac.uk, hhu@essex.ac.uk

Abstract

Efficient control strategies for robot systems cannot always be developed by hand, especially when the robot system is operating in an unknown or uncertain environment. In this paper we show how Reinforcement Learning (RL) might be applied to improve the efficiency of a mobile robot in nuclear decommissioning characterisation, in particular allowing it to learn efficient routes back to a recharging station. We implement this learning functionality in a mobile agent (MA) environment. By doing so we can make use of the positive characteristics of MA mobility such as adaptability, fault tolerance and dynamic positioning of learning or control in a distributed system to supplement learning. Experimental results show how RL provides a more efficient method in this task than a non-AI control approach.

1. Introduction

If the task for a mobile robot is to navigate in an unknown or uncertain environment, it is difficult for a programmer to consider every eventuality when defining its behaviour. It is therefore beneficial to develop a mobile robot which can adapt to changes in the environment. One way to make a mobile robot more adaptable is to allow it to learn from experience. The Artificial Intelligence (AI) technique of RL allows a robot to learn from experience through interaction with an *environment*. The robot establishes the utility of taking an *action* from a range of actions when it finds itself in one of a range of environmental *states*. It learns which actions will lead over time to it realising optimum reward in pursuit of its goal.

In complex tasks such as search, rescue, space exploration or investigation in hazardous environments, mobile robots should be able to explore unknown environments and make decisions when encountering unexpected situations. In this paper we examine the use of RL to aid a mobile robot in the nuclear decommissioning task of characterisation (i.e. the exploration of unknown or uncertain hazardous environments to assess and map levels of radiation so that safe and cost effective decontamination can subsequently be applied).

Many existing robot nuclear characterisation systems are tethered, single robot systems [1], [2] and [3]. To speed up mapping and increase system fault tolerance it might be more appropriate to use multiple robots in which wireless network technologies rather than tethers are employed (despite interference or noise caused by radiation) as demonstrated in [4] and [5]. While a wireless system may allow increased system flexibility and reduce inter-robot space conflict issues, a wireless robot is constrained by the need to periodically return to a recharging station to recharge its battery. We employ RL to enable a mobile robot to learn efficient paths back to a recharging station through an environment.

The rest of this paper is structured as follows: Section 2 briefly outlines some background material. Section 3 describes how RL was implemented in a MA environment. Section 4 describes experiments which examine the rate at which a RL implementation can learn a stable policy, and compare RL against non-AI control in a number of environments. Section 5 analyses the results of our experiments and Section 6 provides some conclusions.

2. Background/literature review

As previously mentioned, RL is a control strategy in which an agent embedded in an environment attempts to maximise total reward (or return) in pursuit of a goal. The agent at any time step is in a particular state relative to the environment and can take one of a number of actions within that state to reach its goal. When the agent performs an action it receives feedback in the form of a reward from the environment which indicates if this is a good or bad action to take in attempting to achieve the goal.

The value of taking an action or being in any state can be defined using value functions e.g. the *Action-Value Function (or Q-Value)* $Q^{\pi}(s,a)$, the expected return when starting from state *s*, taking action *a*, and then following policy pi (π). The return can be defined as the sum of discounted rewards:

$$R_t = \sum_{k=0}^{T} \gamma^k r_{t+k+1}$$

where *t* is the current time step, gamma (γ) is the discount rate (a value between 0 and 1), *r* is reward and *T* the total number of rewards (in which $T = \infty$ or $\gamma = 1$ but not both). The policy is a method by which an action is selected.

In this paper an ε-greedy policy is employed. In ε-greedy action selection, the action with highest estimated reward is chosen most of the time (greediest action). However occasionally with a small probability, epsilon (ε), an action is selected at random. The aim of RL methods is to learn $Q^{\pi}(s,a)$ values so that the optimum action can be taken in any state. In this research we employ Q-learning, a temporal difference learning method which can learn from raw experience without a model of an environment's dynamics, and which updates estimates based in part on other learned estimates without waiting for a final outcome. It can be employed using simple algorithms which are not computationally expensive.

Fig. 1 presents the Q-learning algorithm: alpha (α) is the learning rate (a value between 0 and 1), which determines the rate at which Q-values are updated. γ is the discount rate, which causes future rewards to be worth less than immediate rewards, and $max_{a'}$ is the maximum reward attainable in the state following the current one.

```
initialise Q^π(s,a) arbitrarily
repeat (for each episode):
   initialise s
   repeat (for each step of episode):
      choose a from s using policy derived from Q
      take action a, observe r, s'
      Q^π(s,a) ← Q^π(s,a) + α [r + γ max_a' Q^π(s,a) − Q^π(s,a)]
      s ← s';
      until s is terminal
```

Fig. 1 Q-learning algorithm

Q-learning is proven to converge to an optimal action given that all state and action pairs are updated. By using an ε-greedy action selection mechanism all state and action pairs will be updated and the values of $Q^{\pi}(s,a)$ should converge given enough learning episodes. Q-learning and RL are described in more detail in [6].

3. System implementation

This section describes how RL was implemented in our system, as part of wider research into the use of MA mobility within multiple robot systems for nuclear decommissioning characterisation [7].

Our overall system is a MA based multi-robot environment [8]. A MA is a software entity which exists in an execution environment within which it has a clearly defined boundary. It can control to a large extent its own execution and interaction with other agents or computing components i.e. it is not purely reactive and it may be autonomous, asynchronous, dynamic and intelligent. A MA differs from a static software agent in the sense that it is mobile within its execution environment. When, as in this case, an agent's execution environment is composed of a number of distributed computers or robots, this allows it to move between computing nodes in order to perform tasks.

MAs have software engineering advantages over an amalgamation of other distributed computing architectures, can make our system more adaptable and robust [9], and allow us to dynamically position control or learning at the most appropriate location within a distributed environment [10]. By encapsulating learning in a MA we can apply these characteristics to this functionality, independent of the learning ability.

Our MA's main body is a live() function in which its main activity is located. We implement learning in this part of the agent. A flow chart showing its behaviour is shown in Fig. 2. As can be seen, the RL MA can operate autonomously to learn robot behaviours or be instructed via a user control GUI. A learnt policy can then be applied either to a simulated robot or a real Pioneer robot [11] in our lab, as shown in Fig. 3.

Once the RL MA is initialised it is provided with data about the environment in which the robot is located i.e. position of obstacles, the recharging station and the robot. A model of the environment is constructed based upon this information, which also describes the current state of the environment, potential moves the robot can take, and the reward which would be derived from taking a particular move or getting to a goal state (i.e. the recharger). A multi-dimensional array is initialised to store Q-values for all actions in all states along with the Q-learning algorithm.

The agent can enter an autonomous control mode in which Q-learning algorithm parameters are set, and the algorithm is applied to the model of the environment to simulate a number of learning episodes and in so doing update the values in the Q-value array. Alternatively these processes can be controlled by the user control GUI. The robot environment is an idealised grid based environment containing a number of obstacles in which each grid square represents a state. The robot can take one of eight actions in each state allowing it to move between states i.e. (North (N), South (S), East (E), West (W), NE, NW, SE, SW).

The mobile robot can learn to move through its environment towards a recharging station located in the corner of the environment. Moves outside the environment or through an obstacle are considered illegal and not permitted; the aim of the robot is to find the shortest legal path back to the recharging station. Using this system we conducted a number of experiments to test the ability of the system to learn in a number of environments, these are described in the next section.

4. Experiments

This section examines experiments, which we conducted with the RL system described previously. Two sets of experiments were conducted:

(1) *RL rate* – to assess the rate at which a stable policy could be learnt using RL in a variety of robot environments.

(2) *RL vs. non-AI control* – to compare RL against a non-AI control method.

A more detailed examination of set-ups employed for these two types of experiment is discussed in the following sections.

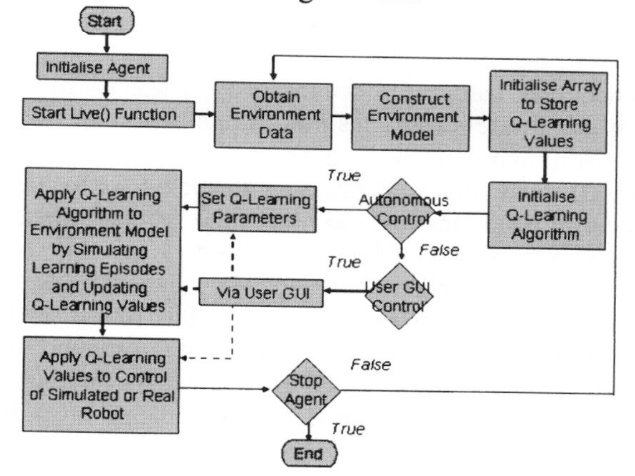

Fig. 2 Flow chart of a RL MA

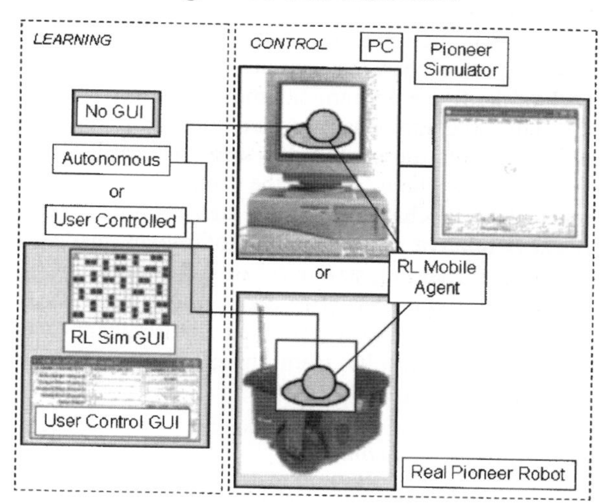

Fig. 3 RL based MA/mobile robot system

4.1. Reinforcement Learning Rate

A number of environments were selected to test the RL method as shown in Fig. 4 in which the triangle at the top left corner is the recharging station, a column is an obstacle, and the icon at the bottom right corner is the robot.

The aim of the RL rate experiments is to assess the rate at which a stable policy can be learnt using RL in a variety of robot environments. In these experiments the aim of RL is to guide the robot to the recharging station using the shortest route. In order to direct the RL learning process the following rewards were used: +100 if the robot reaches the battery recharging station, -1.0 if the robot makes a straight move (N, S, E, W), and -1.4 if the robot makes a diagonal move (NE, NW, SE, SW). These rewards encourage the robot to move to the recharging station in the least number of valid moves to achieve a maximum reward.

In the Q-learning algorithm α was set to 0.1 to encourage learning to occur slowly, γ was set to 0.9 so that future reward has more value than immediate reward, and ε in the ε-greedy algorithm was set to 0.001 to encourage exploration on average only once every 1000 moves, so that the robot utilises maximum reward actions more often.

The experiment set-up for the RL Rate experiments is shown in Fig. 5. In this set-up a RL MA is located on a PC on which it runs a number of simulations of the robot environments. By running a large number of simulations in which the robot starts in a variety of different locations, the RL MA can learn the optimum path the robot should follow given any start point in the environment. 50000 episodes (an episode being a successful run of the robot reaching the recharging station) were conducted for each of the five environments shown in Fig. 4.

The average positive and negative rewards obtained over these episodes were recorded; these would show how often the robot successfully reached the recharging station and obtained (+100 reward) relative to the number of moves made to reach the recharging station (-1 or -1.4 reward).

It was found that after about 40000 episodes a stable average reward for each environment was obtained. The results from these experiments are shown in Section 5.1.

4.2. RL vs. non-AI control

RL vs. non-AI control experiments were conducted in the same experimental environments used in the RL rate experiments. The aim of the RL vs. non-AI control experiments is to compare RL against a non-AI control method. As previously described, using RL the aim is to control the robot to move back to the recharging station using the shortest path possible. In order to direct the RL learning process, the same values for rewards, α, γ and ε were used as in the RL rate experiments.

Non-AI control in these experiments was implemented using a simple reactive control approach. Under the reactive control approach the robot attempts to find the shortest path back to the recharging station. It knows the position of the recharging station relative to its current location and attempts to take moves towards it. If it encounters an obstacle in its path it takes a move in a random direction before attempting once more to move towards the recharging station.

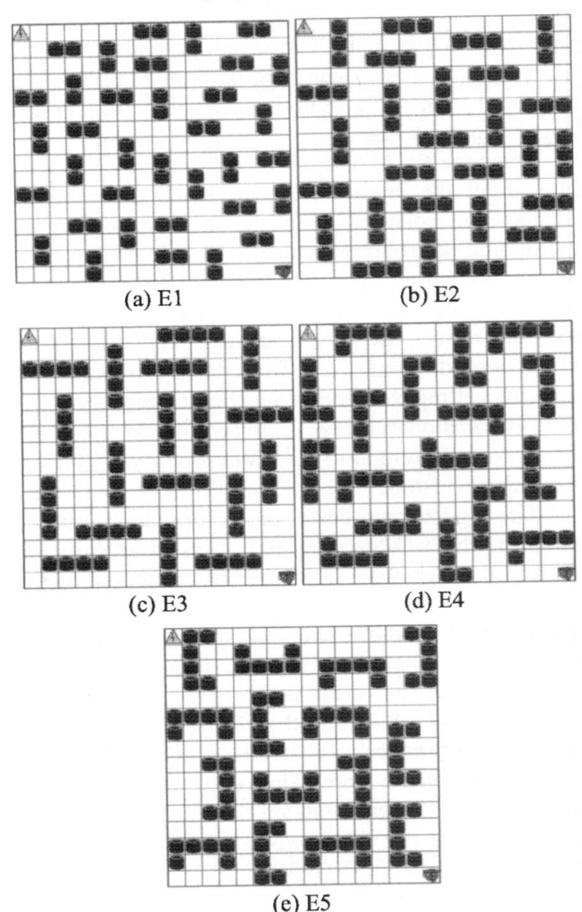

Fig. 4 Five Environment settings used in experiments

The experimental set-up for these experiments is the same as for the RL rate experiments as shown in Fig. 5. In addition to the RL MA a non-AI MA is also located on the PC and runs a number of simulations. By running a large number of simulations in which the robot starts in a variety of different starting locations the RL and non-AI MAs can be used to assess the overall effectiveness of both control methods to control a robot back to the recharging station.

The average positive and negative reward obtained by the RL and non-AI MAs in controlling the robot over 10 trials (1 trial = 10000 moves) was obtained to assess the relative percentage of positive to negative rewards obtained. As the non-AI method does not employ rewards, the type of move made and number of goals reached were recorded in order to calculate what the reward would be if measured with the RL implementation. 10000 moves for a trial acted as a timeout in the event that non-AI control failed to find a path to the charging station for the robot, and was also the figure at which the average positive reward obtained from both RL and non-AI control became stable. The results for these experiments are shown in Section 5.2.

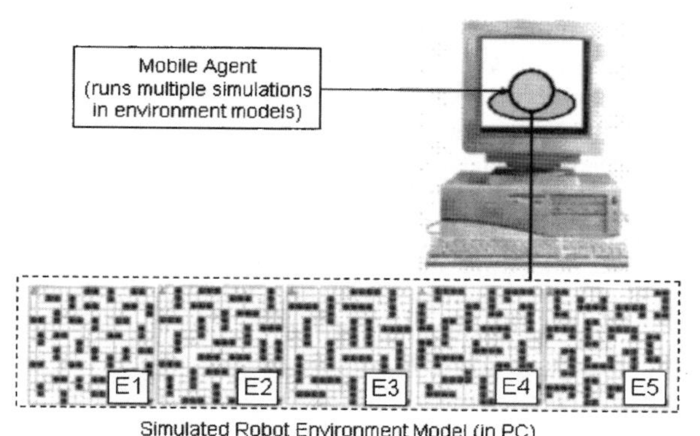

Fig. 5 Experiment set-up

5. Results and analysis

5.1 RL rate

RL rate experiment results are shown in Fig. 6, which include results for all environments from 0-50000 episodes. It shows that using this RL setup and environments, a stable policy can be learnt by approximately the 40000 episode mark, and that the most rapid learning occurs in the first 1000 episodes as shown by a rise in positive reward as an overall percentage of reward from 0% to 49-50%.

Learning continues to rise, but at a slower rate over the next 9000 episodes up to the 10000 episode mark from 49-50% to 79-82% for positive reward. Finally, over the final 40000 episodes up to the 50000 episode mark the learning rate continues at a slower rate and becomes stable at an 85-87% level. From Fig. 6 it can be seen that there is not a large difference between the learning rates or overall positive reward obtained in each of the five environments despite the difference in environment structure.

5.2. RL vs. non-AI control

The results for the RL vs. non-AI control experiments can be seen in Fig. 7. As can be seen, RL control outperforms non-AI control in each of the five environments. There is little difference in the overall performance of RL control in each of the five environments and it maintains a higher degree of positive reward relative to non-AI control.

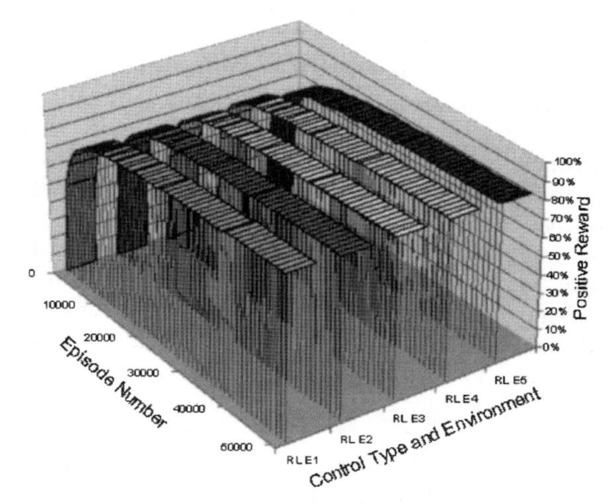

Fig. 6 RL rate experiment results 0-50000 episodes

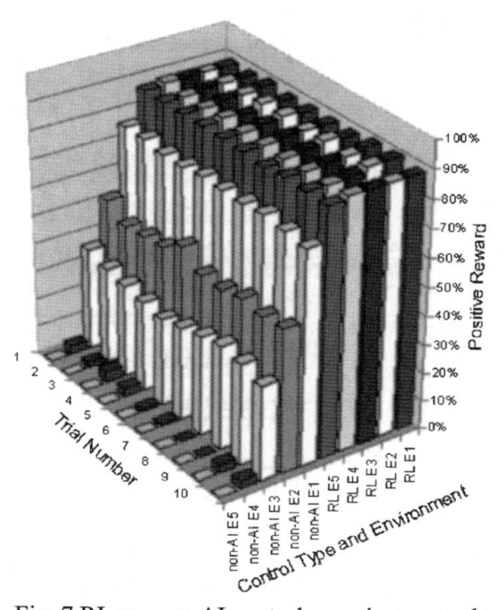

Fig. 7 RL vs. non-AI control experiment results

In some of the environments the non-AI control performance is very low, or fails completely because the robot cannot reach the recharging station because it becomes stuck in local minima and cannot escape e.g. non-AI E5 and non-AI E4.

While non-AI control can produce results that allow the robot to reach the recharging station successfully in some environments, RL control learns a path which is on average shorter. In other words, RL control provides a more efficient as well as more reliable means for directing the mobile robot back to the recharging station. Fig. 8 shows some paths taken using RL and non-AI control from seven random starting positions in environment E3. It demonstrates that when using non-AI control the robot always attempts to head directly towards the recharging station even if this means attempting to move through obstacles. This results in a larger average number of moves, and in some situations an inability to move round an obstacle. This is because it cannot learn a new strategy.

In contrast the paths for RL control show that RL can learn to find a path which navigates the robot around obstacles in order to reach the recharging station rather than attempting to move it through obstacles.

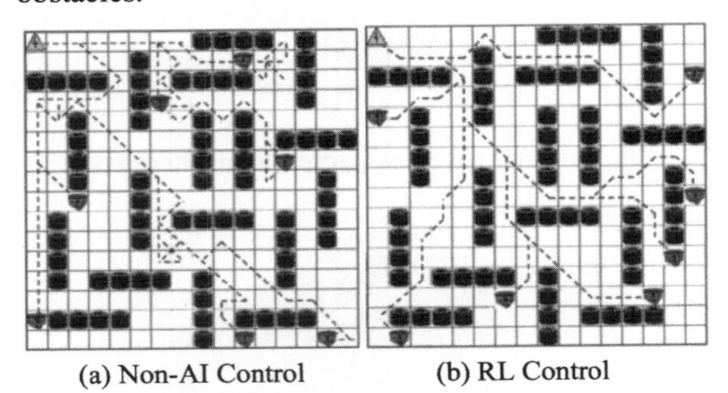

(a) Non-AI Control (b) RL Control

Fig. 8 Comparison of control from random start paths

6. Conclusions

In this paper, we have examined the use of the RL technique to provide a mobile robot with the capability to operate autonomously in hazardous environments in general, autonomously charging its batteries in particular. We have implemented the robot control architecture in a mobile agent environment, which provides software engineering advantages in comparison to an amalgamation of other distributed computing architectures. The proposed system is more adaptable and robust and allows a mobile robot to dynamically position control or learning to reach the most appropriate location within the environment.

Experimental results show the learning characteristics of RL control in five different simulated environments. It has been shown that RL can be used for a mobile robot to learn efficient control policies in a range of environments of varying complexity.

A comparison of the RL approach against a non-AI control method shows that RL provides a more efficient and safer method for a mobile robot to return to a recharging station. The RL method can be used to successfully guide the mobile robot back to the recharging station even in environments where the reactive method would have failed to find a path. This justifies the implementation of the AI technique in our wider multiple robot architecture.

References

[1]. US Department of Energy, Houdini II Remotely Operated Vehicle System – Innovative Technology Summary Report DOE/EM-0495, 2002.

[2]. Pacific Northwest National Laboratory, Technology Deployment Fact Sheet - Andros Robot for Canyon Disposition Initiative Remote Characterization, 2002.

[3]. Willis, W.D., Anderson, M.O., McKay, M.D., The Remote Underwater Characterization System, Proc. of the American Nuclear Societies 8th Topical Meeting on Robotics and Remote Systems, 1999.

[4]. Anderson M. et al., Demonstration of the Robotic Gamma Locating and Isotopic Identification Device, Proc. of the American Nuclear Soc. Spectrum, 2002.

[5]. US Department of Energy, Mobile Automated Characterisation System – Innovative Technology Summary Report DOE/EM-0413, 2002.

[6]. Sutton R. and Barto A.G., Reinforcement Learning: An Introduction, MIT Press, 1998.

[7]. Cragg L. and Hu H., Application of MAs to Robust Tele-operation of Internet Robots in Nuclear Decommissioning, Proceedings of IEEE International Conference on Industrial Technology–ICIT'03, pages 1214-1219, Maribor, Slovenia, 10-12 Dec. 2003.

[8]. Cragg L. and Hu H., Application of Mobile Agents in Multiple Online Robots, Proceedings of e-ENGDET 2004 (the 4th Int. Conf. on e-ENGineering & Digital Enterprise Tech.), Leeds, England, 1-3 Sept. 2004.

[9]. Cragg L. and Hu H., Implementing ALLIANCE in Networked Robots using Mobile Agents, Proc. of IAV 2004 5th IFAC Symposium on Intelligent Autonomous Vehicles, Instituto Superior Tecnico, Lisbon, Portugal, 5th-7th July, 2004.

[10]. Cragg L. and Hu H., Implementing Multiple Robot Architectures using Mobile Agents, Proc. of Control 2004, 6th-9th September, Bath, England, 2004.

[11]. ActivMedia Robotics, Software, Documentation and Tech. Support, http://robots.activmedia.com/, 2005.

Intelligent Production Machines and Systems
D. T. Pham, E. E. Eldukhri and A. J. Soroka (eds)

Intelligent manufacturing strategy selection

D.T. Pham, Y. Wang, S. Dimov

Manufacturing Engineering Centre, Cardiff University, CF24 3AA, Wales, UK

Abstract

This paper presents a new manufacturing strategy selection method. The idea is to regard a manufacturing decision making problem as a "*Game*" in a *Game Theoretic* sense, focus on causal relations between the different players in the *Game* and then use *Dynamic Causal Mining* (DCM) to extract their behaviour patterns. This combination of *Game Theory* and *Dynamic Causal Mining* enables the discovery of hidden strategies, which can be formed as rules governing the *Game* and to predict future outcomes, which could be significantly affected by present actions.

1. Introduction

Strategy selection implies making a choice between available opportunities with limited resources. Sometimes these opportunities are based on the behaviour of other competitors in the same system. Such decisions are common in daily life. The main goal of strategy selection is to make an optimal choice at a given time under a given circumstance. In manufacturing, there are two methods commonly used in making such choices, *System Dynamics* modelling and *Game Theory*. This paper proposes a combination of these two methods with data mining to produce a new way of extracting causal rules and making optimal selections. The proposed method hinges on regarding a manufacturing strategy selection problem as a "*Game*" involving several players. From a certain time point in the present to a certain time point in the future, any actions of any players in the "*Game*" at any moment may influence on the actions of others. Each player has a pattern of strategy selection, which is observable by other players, and to which other players have a counter pattern. A change in strategy pattern by a player might cause other players to change their selection pattern.

The paper is organised as follows. Section 2 briefly reviews the literature on manufacturing strategy selection. Section 3 gives details of the proposed method for manufacturing strategy selection. Section 4 gives an example illustrating an application. Section 5 concludes the paper.

2. Literature review

Game Theory provides a simple way of strategic analysis [1,2]. However, *Game Theory* is not the most efficient tool to solve problems involving causal dynamic decisions [3,4,5]. In areas where causal dynamics are involved, such as manufacturing strategy selection, *System Dynamics* can prove a better alternative. The main concept in *System Dynamics* is to understand how all the objects in a system interact with one another and then to create a model based on a dynamic hypothesis formulated for the system [6,7]. *System Dynamics* identifies the basic structure of a system, the behaviour of which can be understood through simulation or causal modelling [8,9]. Unfortunately, *System Dynamics* is based on individual understanding of the system. Each person may have a different opinion, and thus may create different results for the same problem.

In the literature, a *strategy* has been defined

as "the determination of the basic long-term goals and the objectives of an enterprise, and the adoption of courses of action and the allocation of resources necessary for carrying out these goals." [10]. Furthermore, a manufacturing strategy is "a pattern of decisions, both structural and infra-structural, which determine the capability of a manufacturing system and specify how it will operate in order to meet a set of manufacturing objectives which have been derived from business objectives" [11].

A manufacturing strategy comprises two parts, market oriented and operation oriented. There has been extensive research on both parts. Figure 1 illustrates the subjects involved in each part.

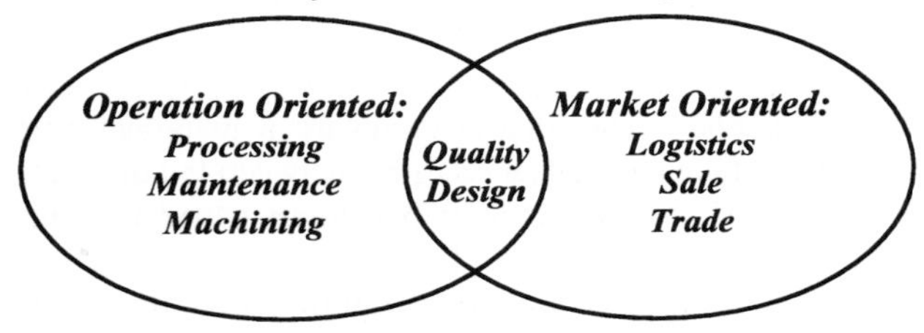

Fig. 1. Manufacturing strategy

Quality and design are in a grey area, because it involves both market and operation. It can be also said that the market part is the macro view and the operation part is the micro view of manufacturing.

Much research has focused mainly on the marketing part [12, 13], ignored the importance of operational process as part of manufacturing strategy, or concentrated on the operations part [14, 15] without giving any consideration to the business part. This paper emphasises that manufacturing strategy also involves a more detailed operational strategy and there should be a clear link between each of the market and the operational part.

3. Causal strategy selection

Figure 2 gives an overview of the proposed method of strategy selection. Enterprises have large amounts of stored data regarding previously selected strategies adopted by themselves and their competitors. In this paper, *Dynamic Causal Mining* (DCM) is proposed as a technique for mining these data and *Causal Game Theory* is the method used to select strategies. If the rules derived by DCM are not sufficient for strategy selection, the parameters of the DCM algorithm are changed in order to generate other rules to improve strategy selection. In the rest of the paper the term "*strategy*" is used to mean "*manufacturing strategy*".

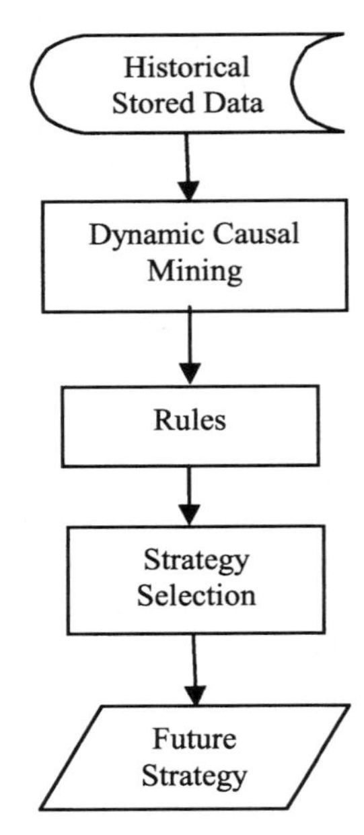

Fig. 2. The proposed method of manufacturing strategy selection

3.1 Game theory

A "*Game*" involves a group of players or participants whose payoffs depend both on how each player selects his own strategy as well as on how other players select their strategies. After all players have made a number of moves, they will try and predict what their opponents will do and select a strategy accordingly. The relevance of *Game Theory* to manufacturing is the conflict of interest and lack of resources among intelligent participants such as managers, engineers and designers. As long as there are more than one intelligent entities involved, *Game Theory* can be a useful tool.

The *Designer's Dilemma* referred to here is adapted from the "*Prisoner's Dilemma*" [1,2], which is a classical example to illustrate the fundamentals of *Game Theory*. Two players can choose between two moves, either "confess" or "deny". The idea is that each is better off by confessing than denying. However, the outcome obtained by both confessing is worse than by both denying. The whole *game* situation and its different outcomes are summarised in table 1.

Table 1
Prisoner's dilemma

Player2 / Player1	Confess	Deny
Confess	(-10,-10)	(0,-15)
Deny	(-15,0)	(-1,-1)

The table shows that for each possible pair of moves, cdgbdg c*sefv hgrsthe payoffs to Player 1 and Player 2 (in that order) are listed in the appropriate cells. Suppose Player 2 confesses his crime. Then Player 1 receives a 10-year prison sentence if he also confesses and a 15-year sentence if he denies his crime. Therefore Player 1 is better off confessing. Suppose Player 2 denies his crime. Then Player 1 receives no sentence for cooperating and confessing. However, if player 1 also denies his crime, both he and Player 2 receives a 1-year sentence. Finally, if both Players confirm their crime, they receive a 10-year sentence each.

The "*Prisoner's Dilemma*" can be easily applied to more realistic manufacturing problems. Imagine two designers available for design work. However, there is a manufacturer who does not have sufficient resources for both designers. The two designers are isolated from each other and the manufacturer visits each of them and offers them a deal. If only one designer accepts the deal, this designer will carry out the whole project and receive all the payment. If neither of them accepts the offer, the manufacturer will keep them for future projects, thus paying them a royalty. However, if one of them refuses the job, that designer takes no further part in the process and receives no payment. If both accept, and thus will be working together, each will receive less than if one had refused the job. The dilemma resides in the fact that each designer has a choice between only two options, but cannot make a good decision without knowing what the other one will do. This is shown in table 2.

Table 2
Designer's dilemma

Designer2 / Designer1	Accept	Refuse
Accept	(500,500)	(1000,0)
Refuse	(0,1000)	(100,100)

The example shows a one shot strategy profile setup according to *Game Theory*. As in the classical *Prisoner's Dilemma*, each of the players is tempted to deny (refuse). However, due to the uncertainty with the opponent's action, a rational player would realise that the opponent might confess (accept), thus leaving him with the worst payoff. To be on the safe side he will choose to confess (accept). If both players are rational, the solution to the *Prisoner's Dilemma* is that both should confess (accept).

In real life strategy selection is based on a player's knowledge of the opponent. If there is perfect knowledge, then the player has control of the *Game*. This knowledge is gathered from past records by applying data mining methods. As mentioned above, *Dynamic Causal Mining* (DCM) is applied in this work to extract knowledge to assist strategy selection in the form of causal rules among historical data of strategies.

3.2 Dynamic causal mining

A causality relation Caus(A_i ,B_k), indicates that an attribute A, adopted by player i influences attribute B, employed by player k. These attributes can represent characteristics related to a *manufacturing strategy*, such as reward payoff, inventory volume, production volume, etc. A causality relation from A to B implies that changes in the value of A would sometimes result in changes in the value of B. A polarity is introduced to indicate the direction of the change. There are four different combinations of polarity (+,-), (+,+), (-,+) and (-,-). The symbol "+" indicates increase. The symbol "–" indicates decrease. The combination (+,-) means if player i selects to increase the value of strategy A, player k will decrease the value of strategy B.

Let D denote a database which contains a set of *n* records with attributes *{X, Y, Z,....}*, where each attribute is of a unique type (sale price, production volume, inventory volume, etc). Each attribute is associated with a time stamp t. To apply DCM, the records are arranged in a temporal sequence (t = 1,2,...n). The causal relationships between attributes in D can be identified by examining the polarities of corresponding changes in attribute values. Let D_{new} be a new database constructed from D such that attribute ΔX_t in D_{new} is given by:

$$\Delta X_t = X_{t+1} - X_t \quad (1)$$

Three measures are required for DCM, *Simultaneous Positive Support, Simultaneous Negative Support and Balancing Support. Support* is defined as the ratio of the number of records of a certain polarity combination and the total number of records in the database. For database D_{new} and any two attributes ΔX and ΔY the three kinds of supports are defined as follows:

$$\textbf{Simultaneous Positive Support}\ (\Delta X, \Delta Y) = \frac{freq(+,+)}{n} \quad (2)$$

$$\textbf{Simultaneous Negative Support}\ (\Delta X, \Delta Y) = \frac{freq(-,-)}{n} \quad (3)$$

$$\textbf{Balancing Support}\ (\Delta X, \Delta Y) = \frac{freq(+,-)}{n} \quad (4.a)$$

$$\frac{freq(-,+)}{n} \quad (4.b)$$

where $freq(+,+)$ is a function giving the number of times an increase in X is associated with a simultaneous increase in Y, etc.

The simultaneous presence of combinations (+,+) and (-,-) indicates sympathetic changes and will produce a sympathetic rule. The simultaneous presence of combinations (+,-) and (-,+) indicates antipathetic changes and produces antipathetic rules. All supports relate to the frequencies of the occurring patterns. Given a user specified support, the problem of *DCM* is to find all rules where the support is larger than the user defined support.

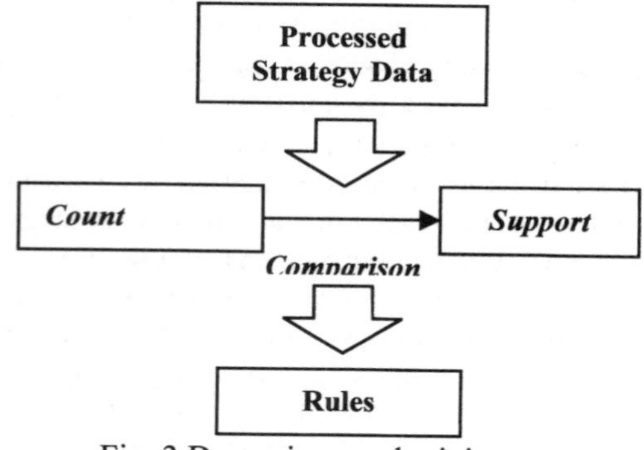

Fig. 3 Dynamic causal mining process

Figure 3 illustrates the DCM process. The process begins with counting the numbers of positive and negative values from incoming data and introduces measures to constrain the rule domain. The DCM is modified from *Association Mining Algorithm* [16] and can be decomposed into the following sub-problems.

(1) Identifying target attributes.

(2) Finding all set of attributes that have support above the predetermined minimum supports.

The second step determines the overall performance of mining dynamic causal rules. This is necessary to speed up the execution of the DCM process.

3.3 Causal game theory

The central idea of *Causal Game Theory* is for a player to make as many moves as possible to predict the opponent's selection pattern. The pattern should give information on how certain moves by one player will cause others to select their moves.

For a *Game* with N players, the normal form representation of *Game Theory* specifies for each player i a set of strategies S_i (with strategy $s_i \in S_i$) and a payoff function $u_i(s_1,....s_N)$ linked with the outcomes arising from each player i's selection of strategies$(s_1,....s_N)$. At each time step t, every player has to choose a strategy $s_{i,t}$. The payoff $u_i(s_{i,t})$ is the outcome of the *Game* after all players have made their selections.

Denote $s_{j-i} = (s_1,..., s_{i-1}, s_{i+1},....s_n)$ the list of strategies played by all the players' j other than i. The strategy profile $s^*_{i,t+1}$ for player i at time t+1 is selected by first identifying the causal rules between $s_{i,t}$ and $s_{j-i,t}$ from historical data.

A strategy profile $s^*_{i,t+1}$ is strictly dominant if the following is satisfied:

$$P_i(s^*_{i,t+1} = Caus\ (s_{i,t}, s_{j-i,t})) > P_i(s_{i,t+1}) \quad (5)$$

where P_i is a payoff function. This means the strategy profile has a higher payoff than any other profile at time t+1. A strategy profile $s^*_{i,t+1}$ is weakly dominant if the following is satisfied:

$$P_i(s^*_{i,t+1} = Caus\ (s_{i,t}, s_{j-i,t})) \geq P_i(s_{i,t+1}) \quad (6)$$

This means the strategy profile has a payoff higher than or equal to any other profile at time t+1.

A strategy profile $s^*_{i,t+1}$ is a Nash profile if the following is satisfied:

$$P_i(s^*_{i,t+1} = Caus(s_{i,t}, s_{j-i,t})) \geq P_i(s_{i,t+1}) \quad \&$$

$$P_i(s^*_{i,t+1} = Caus(s_{i,t}, s_{j-i,t})) \geq P_i(s_{i-j,t+1}) \quad (7)$$

This indicates the strategy profile $s_{i,t+1}$ depends on the strategy profile selected by any other player.

4. An illustrative example

Table 3 shows an original database, where the first column indicates the time instant, which could be hours, weeks, or years. The numbers in the other columns have the same units. They could, for example, be purchase prices or sales levels etc. The first row of table 3 can therefore be interpreted as in week 1, player 1 decides to produce 9 units; player 2 chooses to make 2 units etc.

Table 3
Input database

Time	P1	P2	P3	P4	P5	P6	P7
1	9	2	10	22	20	11	13
2	17	3	11	28	24	11	13
3	10	12	9	27	15	12	6
4	4	16	11	22	15	11	2
5	7	24	10	14	12	3	10
6	6	18	10	11	6	5	5
7	6	18	1	5	5	4	4
8	11	21	2	5	13	4	9
9	20	13	2	1	13	3	2
10	21	8	2	8	5	6	8

Step 1: Table 4 shows the 'dynamic' database that results after the difference calculation.

ΔP1	ΔP2	ΔP3	ΔP4	ΔP5	ΔP6	ΔP7
+8	+1	+1	+6	+4	0	0
-7	+9	-2	-1	-9	+1	-7
-6	+4	+2	-5	0	-1	-4
+3	+8	-1	-8	-3	-8	+8
-1	-6	0	-3	-6	+2	-5
0	0	-9	-7	-1	-1	-1
+4	+3	+1	0	+8	0	+5
+9	-9	0	-4	0	-1	-7
+1	-5	0	+7	-8	+3	+6

Table 4
Dynamic database

Step 2: Table 5 illustrates the 'pruned' dynamic database. Pruning is carried out to remove columns (attributes) where the level of support is below a set minimum. In this example, columns with two or more zeros (meaning two or more occasions when there are no changes to the values of the corresponding attributes) are removed.

Table 5
Pruned database

ΔP1	ΔP2	ΔP4	ΔP7
+8	+1	+6	0
-7	+9	-1	-7
-6	+4	-5	-4
+3	+8	-8	+8
-1	-6	-3	-5
0	0	-7	-1
+4	+3	0	+5
+9	-9	-4	-7
+1	-5	+7	+6

Step 3: Table 6 shows the different supports for the pairs of attributes in table 5. The supports are calculated according to equations (2) – (4).

Table 6
Counting results

	(+,+)	(-,-)	(-,+)	(+,-)
P1&P2	0.3	0.1	0.2	0.2
P1&P4	0.2	0.3	0	0.2
P1&P7	0.3	0.3	0	0.1
P2&P4	0.1	0.2	0.1	0.3
P2&P7	0.2	0.2	0.1	0.2
P4&P7	0.1	0.5	0.1	0

Suppose the threshold support is set to 0.3, which means any attribute pair with support with value larger than or equal to 0.3 is considered associated. The following association rules are thus obtained.

(+,+): (P1&P2), (P1&P7)
(-,-): (P1&P4), (P1&P7), (P4&P7)
(+,-): (P2&P4)

Note that (P1& P7) is a sympathetic rule because when one of the attributes increases its value, the other will automatically increase its value and vice versa. This kind of information can give a manager an insight into the strategy system being dealt with and thus enable better selection of strategy profiles.

5. Conclusion

This paper has proposed a new method of manufacturing strategy selection by combining *Game Theory* and *Dynamic Causal Mining*. To enable such a combination, a new form of the *Game Theory* is suggested, which focuses on the causal relationship between action patterns of each player, rather than the rationality of optimal selection. The paper has also suggested a new way of looking at manufacturing strategy. Players are not simply searching for the optimal payoff or the best preference. Often due to the lack of knowledge about the *Game*, players cannot identify the most suitable strategy. As a result, they need to understand what strategy is available and what causes it to change. If players can discover the underlying patterns, this can assist them in their decision making.

Acknowledgement
The authors are members of the EU-fund FP6 Network of Excellence for Innovative Production Machines and Systems (I*PROMS).

Reference

[1] Axelrod, R. The Evolution of Cooperation. New York: Basic Books. (1984).
[2] Dimand, R. W. and Dimand, M. The Early History of the Theory of Strategic Games. In Weintraub, E.R.(ed.) Duke University Press. (1992)
[3] Von Neumann, J. and Morganstern, O. The Theory of Games and Economic Behaviour. Princeton: Princeton University Press. (1944).
[4] Nash, J. Equilibrium Points in N-person Games. Proceedings of the National Academy of Sciences 36. (1950). pp.48 - 49.
[5] Maynard-Smith, J. Evolution and the Theory of Games. Cambridge: Cambridge University Press. (1982).
[6] Forrester, J. W. Industrial Dynamics. Cambridge, Mass: MIT Press. (1961)
[7] Forrester, J. W. System Dynamics and Learner-Centered-Learning in Kindergarten through 12th Grade Education. Paper D-4337, Cambridge, (1992)
[8] Senge, P. The Fifth Discipline. New York: Doubleday Mass. MIT. (1990).
[9] Sterman, J. Business Dynamics: Systems Thinking and Modelling for a complex World. Boston Irwin/McGraw-Hill. (2000).
[10] Chandler, J. Strategy and Structures: Chapters in the History of the Industrial Enterprise, MIT Press, Cambridge, Mass. (1962).
[11] Skinner, W. Manufacturing – the Formidable Competitive Weapon, Wiley, New York. (1985).
[12] Harrison, A. Manufacturing Strategy and the Concept of World Class Manufacturing. International Journal of Operations & Production Management, Vol.18, No.4, (1998). pp. 397-408.
[13] Gelders, L., Mannaerts P and Maes, J., Manufacturing Strategy, Performance Indicators and Improvement Programmes. International Journal of Production Research, Vol. 32, No. 4. (1994). pp. 797-805.
[14] Staelens F., Kruth, J.P. A Computer Integrated Machining Strategy for planetary EDM STC. (1989), pp. 187
[15] Kim, J. Optimal Model Building Strategy for Rapid Prototype Manufacturing of Sculpture Surface. University of Illinois at Chicago. (1997)
[16] Agrawal, R., Imielinski, T. and Swami, A. N. Mining Association Rules between Sets of Items in large Databases. In Proceedings of the 1993 ACM SIGMOD International Conference on Management of Data, P. Buneman and S. Jajodia, Eds. Washington, D.C. (1993). pp. 207-216

Intelligent Production Machines and Systems
D. T. Pham, E. E. Eldukhri and A. J. Soroka (eds)

Operating policy generation using a reinforcement learning agent in a melt facility

Doug Creighton and Saeid Nahavandi

Intelligent Systems Research Lab, Deakin University, Victoria 3217, Australia

---

Abstract

This study presents a methodology to allow a reinforcement learning agent to generate near-optimal policies for a melt facility. The application of the learning method to this industrial scale, dynamic, stochastic problem poses a number of challenges. The process is formulated as a semi-Markov Decision Problem. A novel method for application of RL agents to continuous state and action spaces, based on mapping continuous to discrete state and action spaces is developed. The agent successfully identified robust polices that improved on the best-practice of expert operators.

---

1. Overview

A reinforcement learning agent has been used to generate a robust operating policy for a melt deck facility. An initial parameter optimisation study, using methods based on established statistical and response surface techniques. The study highlighted the limitations of existing optimisation methods for policy optimisation in complex, dynamic systems. Through *what-if* analysis it was demonstrated that an increase in production output that can be achieved varying policy rules. An agent-based approach allows a robust operating policy to be automatically generated.

A set of trigger points or rules to control the timing of metal transfer from the melters resulted in sub-optimal performance, as the melt facility is a dynamic, stochastic system. An agent was used to identify a robust action-state policy, which is capable of maximising the long term average reward for a stochastic system instead. A strategy for controlling the production operation was learnt using these methods. This strategy has a lower long-term cost than current operating policies or those tested using *what-if* scenarios.

2. Problem Formulation

2.1 Overview

The facility studied included a semi-automated transfer device to load melters, two melters and an overhead telpher car used to transfer metal from the melter to the mould line. The melters are not completely unloaded as a base volume is required to expedite the preparation of the next melter load.

The state and action spaces can be formulated to define the system as a Semi-Markov Decision Process (SMDP), making it suitable for the application of a

Reinforcement Learning (RL) agent [1] to learn a robust operating policy. Process parameters, such as capacity and constraints, have been changed for commercial reasons.

The notation adopted uses a base and subscript to indicate data type and superscripts to indicate the object with which the data is associated. The base notation is as follows:

v_c	volume of a vessel, $v \in \mathbb{R}$ (kg)
v_{max}	maximum volume of the vessel, $v \in \mathbb{R}$ (kg)
v_{min}	minimum volume of the vessel, $v \in \mathbb{R}$ (kg)
v_d	discrete volume / number of moulds, $v \in \mathbb{Z}$ (kg)
t	cycle time (seconds)
s	state, $s \in S$

The superscript notation is listed below:

M_1	Melter 1
M_2	Melter 2
M	Mould line
L	Ladle
T	Total Transfer Volume (melter unload volume)
c	Cast or mould
w	Melter Charge

An initial simulation study identified the ladle transfer process as the bottleneck in the melt deck operation. This bottleneck cannot be removed by increasing the capacity of the ladle, as a non-linear relationship between melter preparation time and base volume in the melter.

At the commencement of the day shift all vessels are charged to their maximum capacity. The mould line can operate for several hours before being starved of melt. Once the buffered metal is consumed, metal shortages occur due to delays in melter preparation and metal transfer. At some points during the two shifts the mould line is not starved of metal and at other points it is. This suggested that identifying a transfer policy based on the state of the system might improve the production rate. Specifically, by varying the metal volumes removed from the melters, and the timing of the transfers, the productivity of the system could be improved. To avoid the need to generate a dynamic policy, and require continued learning by the agent, the reduction in ladle, melter and mould line capacities, due to slag build up, is neglected for this study.

Decision epochs are triggered by a combination of state transitions involving both the melters and the telpher car. A decision epoch occurs when the telpher car is available to commence unloading the melter and a melter is ready to be unloaded. The telpher car becomes available after the completion of the current melter unload procedure.

2.2 Metal demand

The goal of the agent is to maximise metal delivery to the mould line, but the mould line dictates the demand to the system. The mould line casts a mould every 30 seconds, t^M = 30 sec, consuming a metal volume of $v_d^c \pm 5\%$ kg. If the volume of metal held at the mould line falls to a level such that pouring another mould will cause the volume to drop below the minimum, or heal volume, the mould line is stopped. This type of interrupt to the casting process is known as a metal delay.

A delay restarting the mould line may occur after a metal delay caused by a shortage of metal at the pouring facility. Consequently, selecting the demand rate to drive the system is not straightforward. The rate of demand, or metal consumption, affects the speed of production, but demand rates much higher then capacity of the system will result in lower production rates. The policy learnt by the agent is only relevant for a particular demand level. A different policy is required for moulds of different volume and a transitional policy to optimise the change over between mould types.

The maximum demand level can be selected through a series of agent experiments. Rare events will trigger delays intermittently under all policies. A non-automated approach was used for this problem, due to the difficulty is accessing the reason for individual metal delays. A highest demand level without cyclic shortages was determined to be the maximum demand that could be met, using the agent's learnt operating policy. Metal delays still occur intermittently at this demand level due the stochastic nature of the system.

2.3 State definition

The metal volumes in each melter and the mould line are used to define the state of the system. The different locations of each melter affect the melt preparation time, due to different charge loading and unloading times, so it is not possible to treating the melters as identical processes and learn the same operating policy for both melters.

The number of moulds that can be poured, or the number of casts that can be made, from the in the vessel is measured using the integer unit *cast*. As an example, consider the mould line with a maximum volume of $v_{\min}^{M} = 14000\,kg$ and a minimum heal of $v_{\min}^{M} = 5000\,kg$. If $v_d^c = 100\,kg$, the maximum capacity of the mouldline is given by $v_d^M = \lfloor \frac{v_{\max}^M - v_{\min}^M}{v_d^c} \rfloor = 90\,cast$, where $\lfloor x \rfloor$ denotes the integer floor function.

The state space is approximated, by discretising the holding volumes of the melters, $v_d^{M_1}$ and $v_d^{M_2}$, and the mouldline, v_d^M. All metal volumes are discretised to converting their metal volume to the number of moulds that can be poured. This approximation is valid because as soon as the mouldline volume will fall below the heal, or minimum holding volume, the mould line is stopped. Metal volumes less than a mould volume will not affect production. As the demand pulls metal from the melters and it is difficult to transfer exact volumes, the transfer volume v_c^t is also discretised, approximated to v_d^t. The three full ladles, this approximation results in less then a 1% error in true volumes.

The set of states, S, is defined by $s = (s^M, s^{M_1}, s^{M_2})$, where $s^M \in S^M$, $s^{M_1} \in S^{M_1}$ and $s^{M_2} \in S^{M_2}$. The cardinality of the state space is given by $|S| = |S^M| \times |S^{M_1}| \times |S^{M_2}|$ where $|S^*| = \max(v_d^*)$. For the system studied, if $v_d^c = 100\,kg$ and no melter charging rules are enforced, the cardinality of the state space S is $|S| = 71 \times 71 \times 91 = 458\ 731$.

In reality, the melter charging rules ensure that the state space is much smaller. Melter charges are delivered in discrete batches, with delays between each batch. The charging rule employed resulted in the minimum final melter charge being 80% of $v_d^{M_1}$ or $v_d^{M_2}$. Consequently, $|S^{M_1}| = 17$, resulting in $|S| = 17 \times 71 \times 91 = 109\ 837$.

The state of the melter during preparation is estimated using a mathematical model. At the completion of the melter unload process the remaining melter charge is known. The mathematical model uses the unload time, original charge and other parameters to estimate the current temperature. At some stochastic time power is switched to this melter and the temperature is raised. At some time after the power switch metal charges will be delivered to the melter.

2.4 Action Selection

The agent selects the discretised metal transfer volume, v_d^t, at the decision epoch. This decision is based on the metal levels in each melter and the mould line. One, two or three full or partially full ladles can be removed from a melter during an unload operation.

The action space is discretised using the number of moulds that can be poured from a given volume, with units of *cast*. There may be a slight advantage of transferring a partial fraction of a *cast*, which would result to an extra one *cast* every few transfers. However, it has been neglected for this experiment in order to simplify the state space for the application of the RL agent.

The action space has also been reduced by applying common sense constraints onto the system. Given the transfer of metal is the bottle neck of the system, the transfer of low volumes is inefficient, as more metal is consumed during the transfer time than is delivered. Rather than letting the agent discover the inefficiency of these actions during its explorations, the agent's actions are restricted to volumes between the ladle's maximum capacity and half of the ladles capacity. An alternative minimum considered was the volume of metal consumed during the minimum cycle time of a ladle transfer,

$$v_{min}^{L} = \frac{t^{L} v_{c}^{c}}{t^{M}} \tag{1}$$

The less restrictive

$$v_{min}^{L} = \frac{v_{max}^{L}}{2} \tag{2}$$

was selected, as the smaller volumes might increase melter performance through alternate charge options. The agent was given the ultimate decision as to whether these less then efficient action were preferable.

The minimum and maximum volumes are then discretised, and the agent action set A is defined by

$$\begin{aligned} A = \Bigg\{ & \left(\left\lfloor \frac{v_{max}^{L}}{v_{d}^{c}} \right\rfloor \right), \\ & \left(\left\lceil \frac{\frac{3}{2} v_{max}^{L}}{v_{d}^{c}} \right\rceil, \left\lceil \frac{\frac{3}{2} v_{max}^{L}}{v_{d}^{c}} \right\rceil, \left\lceil \frac{\frac{3}{2} v_{max}^{L}}{v_{d}^{c}} \right\rceil + 1, \ldots, \left\lfloor \frac{2 v_{max}^{L}}{v_{d}^{c}} \right\rfloor \right), \\ & \left(\left\lceil \frac{\frac{5}{2} v_{max}^{L}}{v_{d}^{c}} \right\rceil, \left\lceil \frac{\frac{5}{2} v_{max}^{L}}{v_{d}^{c}} \right\rceil + 1, \ldots, \left\lfloor \frac{3 v_{max}^{L}}{v_{d}^{c}} \right\rfloor \right) \Bigg\} \end{aligned} \tag{3}$$

Where: $\lfloor x \rfloor$ denotes the integer floor function, and $\lceil x \rceil$ denotes the integer ceiling function.

To minimise the chances of metal shortage at the mould line, the maximum ladle volume, $\lfloor v_{max}^{L} / v_{d}^{c} \rfloor$ *cast*, is transferred whenever possible. If the remaining transfer volume is greater then the ladle capacity, a full ladle is transferred. The actual volumes transferred are subject to stochastic variation and are not rounded to *cast* volumes in the model.

A transfer cannot begin until there is spare capacity to accommodate the ladle of metal. Due to the highly stochastic nature of the mould line, if the transfer process was commenced before the space existed, then it is possible that insufficient capacity at the mould line would exist for the ladle volume.

When a decision epoch occurs and transfer is not immediately possible, it would be possible to postpone the unload decision until some future trigger point was reached in the modelled environment. This could improve the agent response, by allowing stochastic events between the initial decision epoch and the nominated unload trigger to affect the action selection. However, in the industrial setting the exact status information of equipment, such as the temperature, is not known. Also the communication between melt operators, mould line operators and the telpher car driver is limited.

3. RL agent design

3.1 General structure

At the completion of each production epoch, the time between decision events, the model outputs its current state and also evaluates and returns the cost function. The agent then uses this information when selecting future decisions, by aiming to satisfy a goal to maximise the long term, average reward.

The time between decision epochs is stochastic. The reward received for a state action selection is also stochastic, due to process variation. The Relaxed-Semi-Markovian Average Reward Technique (SMART) [2] is suitable for application to this system.

3.2 Agent reward

The aim of the agent is to maximise the metal production of the melt deck, without causing metal delays at the mould line. An appropriate measure of the agent's effectiveness is the number of moulds poured since last decision epoch divided by the time between epochs. This measure can be easily transformed to moulds per hours, or *casts per hour*, which is the standard shop floor measure of performance. This reward measure increases the communicability of results and agent performance.

If the mould line was modelled as a stochastic process, and line failures were included, it would not be possible to use just the number of moulds poured a

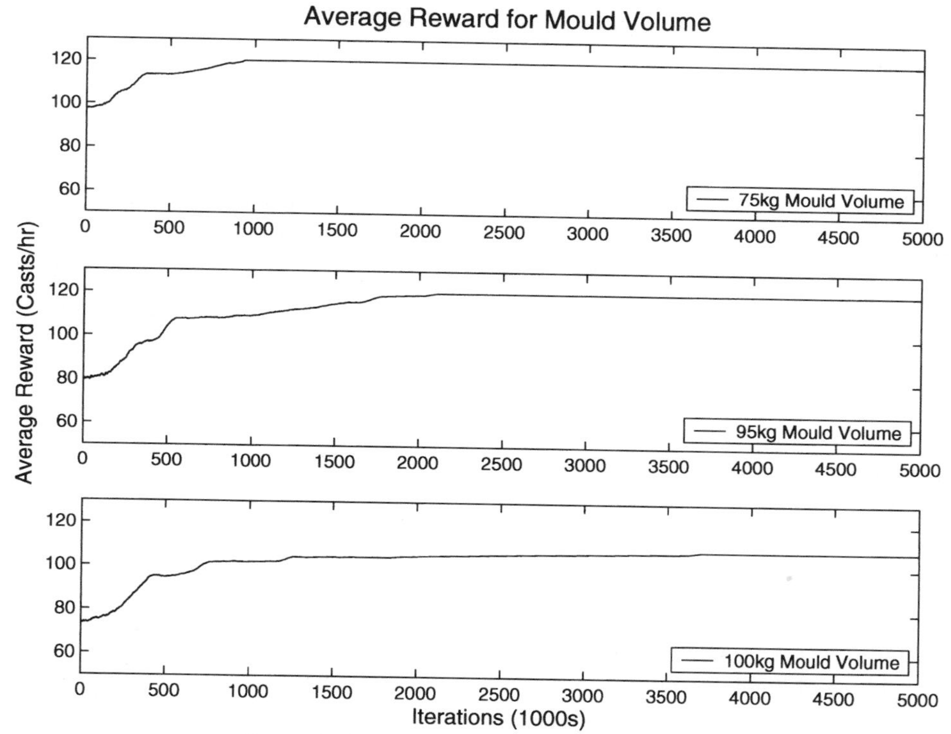

Figure 1: Average reward for mould volume

the reward function. Despite an optimal transfer policy, a mould line stoppage might result in no moulds being poured. Using the mould line uptime and downtime, rather then the total time, since the last decision epoch, a reward could be generated. However, due to the interdependence between the demand rate and the melt preparation rate, stochastic demand was not simulated.

4. Results and discussion

Initially the agent's actions were restricted to determining the number of full ladles to transfer from the melter to the mould line. The results of these experiments demonstrated that varying the transfer volume affected the melt deck production and delivery performance. To find a near optimal solution, the action set was then increased to consider a wider range of transfer options.

The agent could maximise its immediate term reward by always selecting a large metal transfer volume. Such an action also increases the preparation time available for the other melter, allowing an increased volume on the next transfer. However, when a large volume is removed from a melter, the time to prepare the subsequent charge also increases. Consequently, an eligibility trace [3, 4], with $\lambda = 3$ was used to increase the learning rate and correctly distribute reward to the actions. The Darken-Chang-Moody [5] 'search then converge' methodology was suitable for cooling the learning rate and exploration rate temperature.

The average reward received by the agent under different demand rates during the learn and converge process is illustrated in Figure 1. In the 75kg mould volume scenario the demand rate could be satisfied by a range of policies. As long as no metal delays occur at the mould line the agent's goal is achieved. Consequently, the agent converged to the maximum reward rate faster than for the higher demand policies. The demand rate for the 100kg mould volume

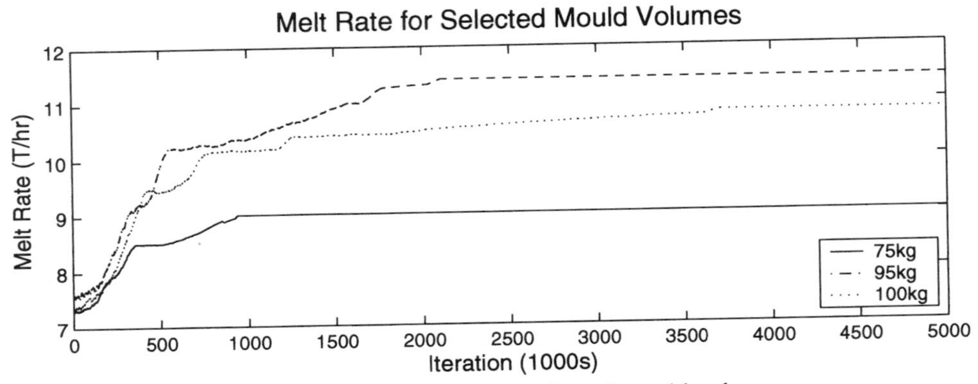

Figure 2: Hourly melt rate for selected mould volumes

scenario exceeds the maximum output rate achievable by the melt deck, resulting in mouldline shortages and a lower average reward achieved.

The melt rates of the optimal policy for each the demand rates illustrated in previous figures are compared in Figure 2. As a consequence of the restart delay on the mouldline, following a metal shortage, the average pour weight per hour for 100kg moulds was less then that achieved for 95kg moulds.

For the system studied, a maximum mould volume that could be produced without metal shortages was 95kg, corresponding to a melt rate of 11.4 Tonnes per hour. This is a 14% increase on the production rate that was achieved by expert operators.

Although the learnt operating policy was robust, severe interruptions in the production process will still result in shortages. The agent changes its production
policy to recover from metal delays as quickly as possible. Immediately following a metal delay the maximum volume transportable from the two melters is transported. This potentially could lead to further delays; due to longer preparation times for the proceeding melt preparations if further delays were experienced. This high risk strategy avoided short term shortages. Once the mould line volumes return to a higher level the agent begins transferring smaller volumes of metal again. This strategy allows the agent to hold a reserve in preparation for the next interruption, whilst slowly increasing the volume at the mould line.

5. Conclusion

A RL agent successfully found a near optimal operating policy for the metal transfer operation. The metal rate achieved is close to the upper bound determined by the power input. A policy enabling a 14% improvement was identified through a combination of sequential increases in demand and the application of the agent. This equates to an increase in productivity, or an equivalent value reduction in power costs. Importantly, this policy is robust to process variation, as it was learnt under stochastic conditions. The optimal operating policy can be varied and still achieve the agent's goal under most conditions, but the agent's policy is more robust to the random variations. Rare events will still result in metal delays, but the agent learnt a strategy to minimise mould line disruptions.

References

[1] Sutton R. S. and Barto A. G. Reinforcement learning: an introduction (MIT Press, Cambridge, MA, 1998).

[2] Gosavi A. Reinforcement learning for long-run average cost. European Journal of Operational Research 155: 654-674 (2004).

[3] Sutton R. S. Learning to predict by the methods of temporal differences. Machine Learning 3: 9-44 (1988).

[4] Tesauro G. Practical issues in temporal di®erence learning. Machine Learning 8(3/4): 257-279 (1992)

[5] Darken C., Chang J. and Moody J. Learning rate schedules for faster stochastic gradient search. In Proc. Neural Networks for Signal Processing 2 (IEEE Press, Piscataway, NJ, 1992).

Intelligent Production Machines and Systems
D. T. Pham, E. E. Eldukhri and A. J. Soroka (eds)

Quantitative utilities for the 'intelligent' definition of CNC programs for the high speed milling of complex parts

López de Lacalle, L.N., Lamikiz, A., Campa,F.J

Department of Mechanical Engineering, IPROMS associate members
Escuela Superior de Ingenieros Industriales, University of the Basque Country
c/ Alameda de Urquijo s/n, 48013 Bilbao, Spain, implomal@bi.ehu.es

Abstract

High speed multi-axis milling is a very complex process that requires a special care in the CNC program preparation in the CAM stage, which is critical for a successful process. A methodology to achieve high reliable machining programs (CN codes) is here presented. Two utilities have been developed to help CAM user to select the best toolpaths and cutting conditions. First, the lobe diagrams of the system material+tools+machine are ready to be used, for selection of the roughing conditions. Lobe diagrams are obtained via a theoretical approach, based on the frequency response of the tool/machine system.

Second an integrated utility to check the cutting force has been includes into software, to select those toolpaths which implies the minimum dimensional errors due to the tool deflection. In this way a more accurate, better-finished surface is machined, and a reduced tool wear is withstood. Two examples are discussed.

1. Introduction

Since the last nineties, high speed milling of complex parts such as moulds, dies, blades or blisks have become a sound technology in industry. Three axes milling canters have increased in number day by day. One of the successful factors is the use of CAM software with specially functions to make good use of the process, as is explained by López de lacalle et al. [1]. Most of forging dies and simple injection moulds are at date machined in three axes milling centres, but five axes centres have become a good solution for complex moulds or sculptured surface milling. The multi-axis (five axes) milling advantages can be pointed.

- The two additional orientation axes allow the machining of polyhedral parts, which cannot be machined using three axes machines.
- The length of tools, necessary large when deep cavities are machined, is reduced. Tool stiffness [2] is directly related with the L^3/D^4 tool slenderness factor, so a reduction in tool length dramatically reduces tool deflection and the lack of precision due to this effect.
- The tool orientation can be used to increase productivity by changing the type of machining strategy. Three examples can illustrative this. a) In the finishing of ruled surfaces, the flank milling strategy can be used [3], machining all inclined side faces with the cylindrical part of the tool, applying a big axial depth of cut. b) Other example is the machining of inclined planes: using the correct tool-axis orientation, a face milling operation can be carried out instead of a ball-end sculptured milling. c) The use of ball-end mills can

just be improved through a good selection of the orientation of the tool axis respect to surfaces. In this way cutting with the tool tip can be avoided.

In the same way that it is explained in the work [1] for the three axes machining, in 5 axes a new approach to the CAM stage can be applied, improving the reliability of the whole process. Definition of reliability for a machining process is 'achieving a good productivity with a low risk of wasted parts out of tolerances or with irrecoverable errors'. In five axes milling production, the CAM, and the CAM operator, is the centre of gravity of the planning process. Workshop workers can only change the actual values of cutting speed and feed rate making use of the machine dials (which modify the actual feed and spindle rotation speed respect to those programmed in the NC code), being impossible changes of the complex toolpath directly in the CNC interface. A new intelligent CAM procedure is here presented, shown in Figure 1, and some interesting examples are described. This production scheme includes a scientific model to evaluate the cutting forces.

2. CAM planning process

As may be seen in Figure 1, there are three stages in the generation of CAM cutting paths, according to the type of operation: roughing (*a*), semifinishing (*b*), and finishing (*c*).

Roughing (*a*) is of critical importance in HSM. The aim is to achieve not only productivity but also a highly uniform stock allowance, which will be removed in the course of finishing. In roughing there are three possibilities, each very much related to the size and hardness of the workpiece.

The object in finishing (*c*) is to achieve a roughness and tolerance specified by the client for each surface. The traditional strategy is *zig-zag*, but in this case the main drawback is that it intercalates two different cutting types, downmilling (also called climb milling) and upmilling. A solution may be to cut along one direction (zig), either downmilling (the most commonly used) or upmilling, but not along both of them. The best option is the use of milling strategies more closely adapted to each of the part zones and its geometry, depending on factors as control of the cusp, and as consequence the maximum roughness *Rt*, by varying the radial depth of cut in accordance with the workpiece slope. A check stage (*d*) is included for the NC programs, using an ad hoc software utility such as Vericut©, NC-Verify©, NC-Simul©. This software allows a virtual simulation previous to actual machining, in which different problems can be effectively detected, as collisions, machining outside of the machine workspace and problems due to tool gauging into the workpiece.

At the definition of cutting parameters, two developed utilities (*g,h*), are newly available to assist in the selection of toolpaths and cutting conditions. These utilities are applied after the selection of the recommended conditions given by toolmakers, directly obtained from links to the commercial databases (*e*) of these companies, usually calculated for suffering low tool wear.

2.1 Analysis of the stability of process

Milling can induce vibrations, resulting in machine tool excitation, or local dynamic phenomenon on the piece. In the case of self-excited vibration, it is the cutting process that generates oscillations. The origin of these oscillations, called chatter, is the dynamic excitation produced by the periodic irregular surfaces of the piece generated by the precedent tooth. The subsequent behaviour of these oscillations depends on the whole system machine-tool-piece. Parameters involving that phenomenon are the static and dynamic rigidity of the system, and its material mechanical properties.

The stability of chatter vibrations is analysed using the so-called *Lobe diagram*. These charts indicate the borderline between the stable and unstable cutting conditions attending to the axial depth of cut, a_p, and the spindle frequency in rpm. It is necessary to use different diagrams for each radial immersion, a_e, each machine-tool, each tool and for each part machined (geometry and material).A three dimensional model for the theoretical calculus of the lobe diagrams has been developed, presented in detail in Bravo et al., [4]. After calculation of the stability lobes for each combination of *machine+tool+raw material*, these are collected on-line with CAM software (step *g* in Figure 1), making possible the selection of cutting condition that ensue stability of milling process.

For low rigidity part pieces, these diagrams are calculated for each transitory stage, because modal features of the part change during milling. The aspect of one diagram for a thin wall (an aluminium wall of 150 x 50 x 0, 8 mm from an initial block measuring 150 x 50 x 4.8 mm) is shown in Figure 2.

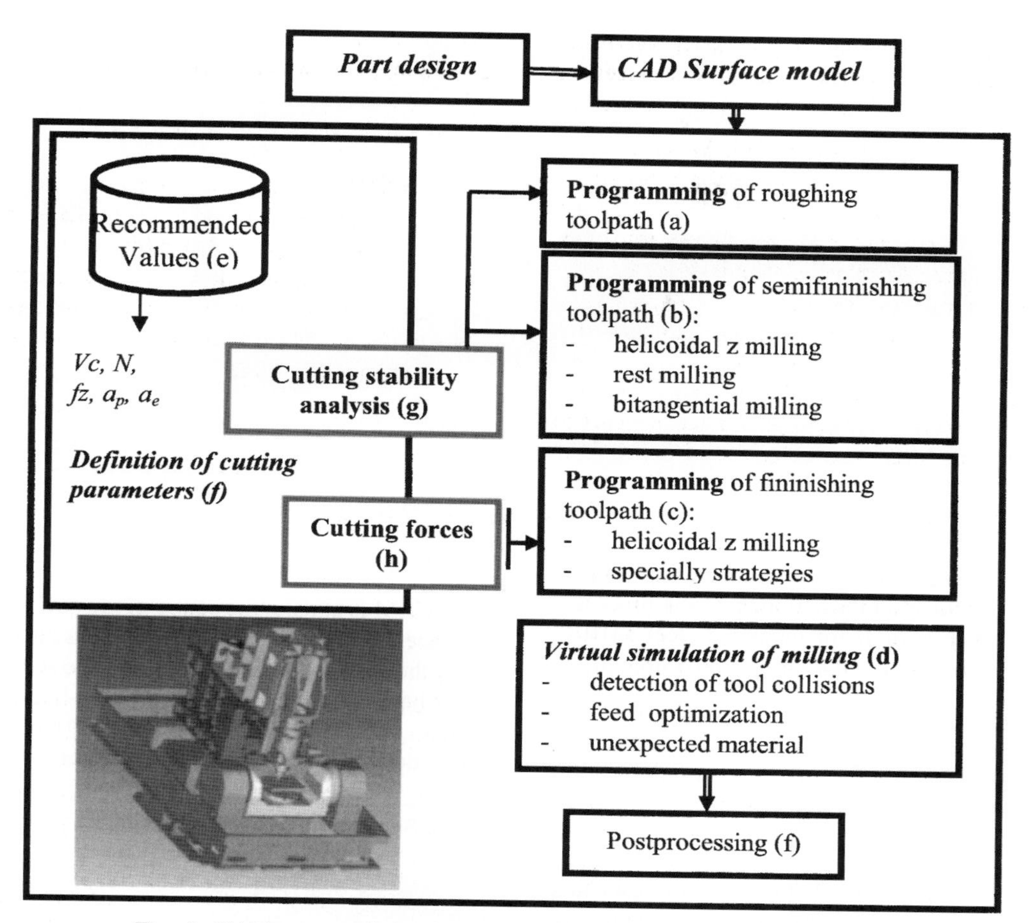

Fig. 1: CAM approach for the high speed milling of complex surfaces.

2.2 Force definition

For the optimisation of the ball-end high speed milling process of complex surfaces, a cutting force analysis utility has been developed. Sculptured surfaces are milling after a CAM toolpath optimisation stage, marked as (*h*) in Figure 1. The developed model is is explained in detail in [5,6]. In the model, the cutting force is divided in two components.

- On the one side the component from the shear force, inherent to the chip formation mechanism. This component is assumed to be proportional to the chip thickness (a_c).
- On the other side, the component from the friction forces of tool and chip. This component is assumed to be proportional to the cutting edge length

Resulting forces will be the addition of these two components; the contribution of each one is ruled by empirical coefficients. They depend on the workpiece material and the tool geometry. The three projections of the cutting force are in Eq.(1):

$$\begin{cases} dF_t(\theta,z) = K_{te}dS + K_{tc}\cdot a_c(\Psi,\theta,\kappa)db \\ dF_r(\theta,z) = K_{re}dS + K_{rc}\cdot a_c(\Psi,\theta,\kappa)db \\ dF_a(\theta,z) = K_{ae}dS + K_{ac}\cdot a_c(\Psi,\theta,\kappa)db \end{cases} \quad (1)$$

where *Krc, Ktc, Kac* , the so-called *shearing coefficients*, are in N/mm^2. *Kre, Kte, Kae,* the *friction coefficients* are in N/mm and $a_c$is in mm. Once the model was validated (Lamikiz et al. 2004),it has been integrated in a commercial CAM. Three approaches are possible at this point:

1. The model can be used *a posteriori*. A maximum admissible value for the cutting force is set; tool path is accepted if force is kept below this maximum value.
2. In three axes milling, the model can be used to evaluate *a priori* the performance of different possible machining strategies at some user-defined control points on the surface. The optimum tool-

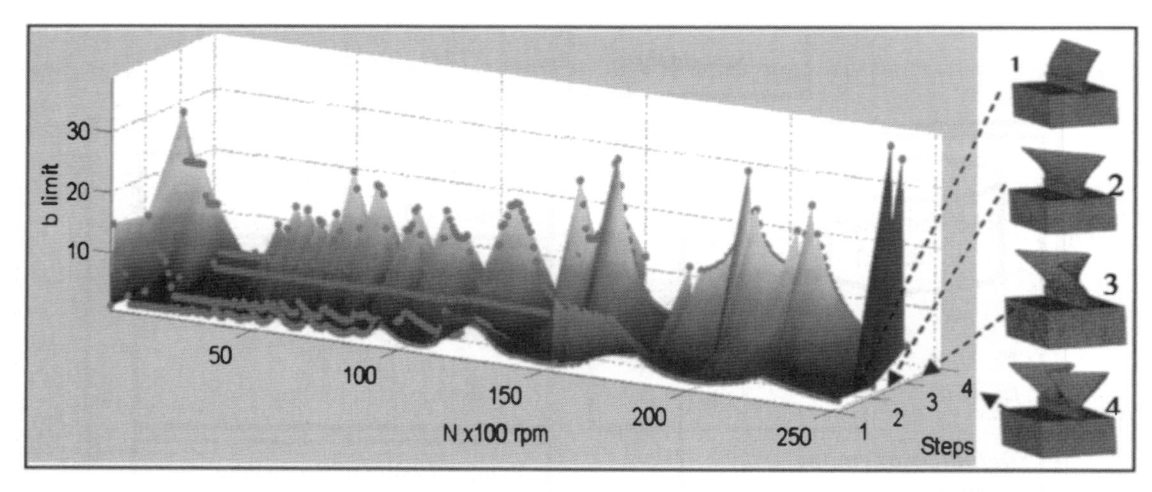

Figure 2: 3D Lobe diagram composed of spindle speed, part stages and maximum depth.

path will be the that which minimizes the value of the cutting force on those points. CAM process is detailed in Figure 3, for tempered steel 54HRC, tool ∅8 mm, $a_e = a_p = 0.4$mm,fz = 0.055mm/diente

3. In five axes milling, an on-line utility has been developed in order to assess the best orientation of the tool on each surface, milling surface with the tool orientation axes in that setting.

3. Example

A test part is in Figure 4. The first two tests have been performed in *zig-zag* strategy. In zig-zag one pass is in downmilling conditions whilst the return path is in upmilling, usually dimensional errors are a balance of those corresponding to each of the cases. The first row of figure shows the predicted cutting forces perpendicular to tool-axis and part; the second the real recorded ones along milling, and the third the accuracy by a coordinate measurement machine.

In test 1, the worst results are obtained using the zig-zag procedure, with errors bigger than 20µm in some points. The same time cutting forces have been recorded, mapping them in a colour diagram. As shown, only a small area was machined with forces near 90N. Case 2 shows better results, with a smaller area in which errors are more than 20 µm. Similarly to previous case, errors are located in zones where the force component perpendicular to tool axis is high. In test 3 the best results has been obtained, using only one milling sense, on that sense along where the cutting force is less than the others. The last figure contains results changing in all surface point milling sense to follow the minimum third case above.

4. Conclusions

The CAM is at date 'the centre of gravity' of complex surfaces milling. Two utilities are presented to make easy the planning of CAM process. The stability lobes have been previously calculated but on-line disposed to assist the definition of roughing, and finishing operations in low rigidity parts surfaces.

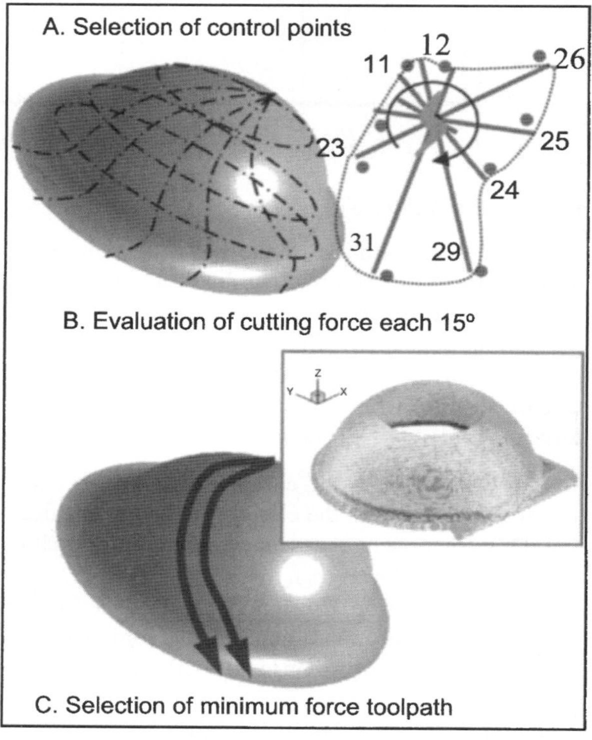

Figure 3: Integration of cutting forces in three axes.

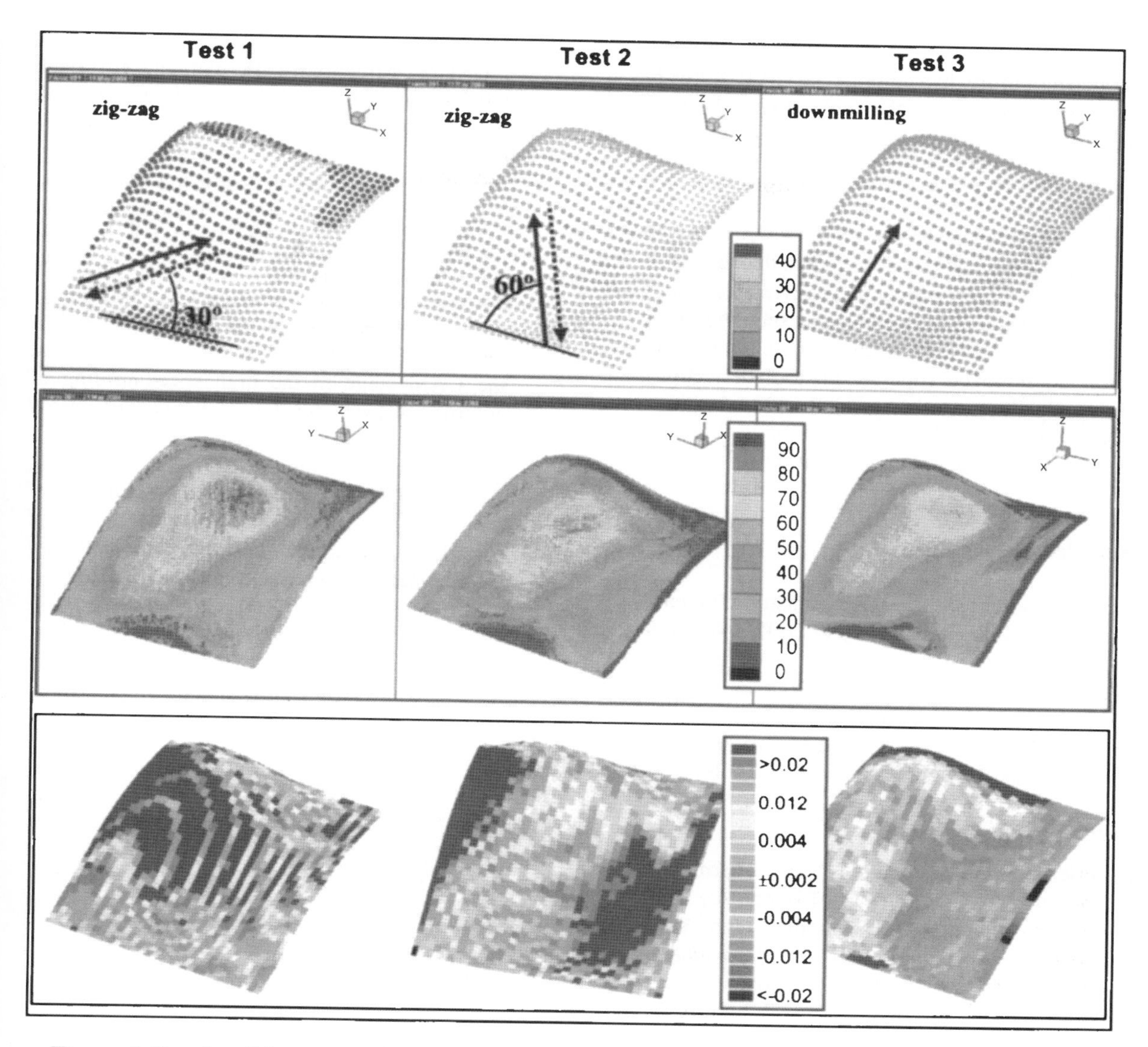

Figure 4: Results of the proposed methodology in three test parts. Up) Theoretical estimation of cutting forces (in N) perpendicular to tool axis and part. Middle) Real cutting forces (in N) monitored along milling. Down) Dimensional errors (in μm) measured in a CMM.

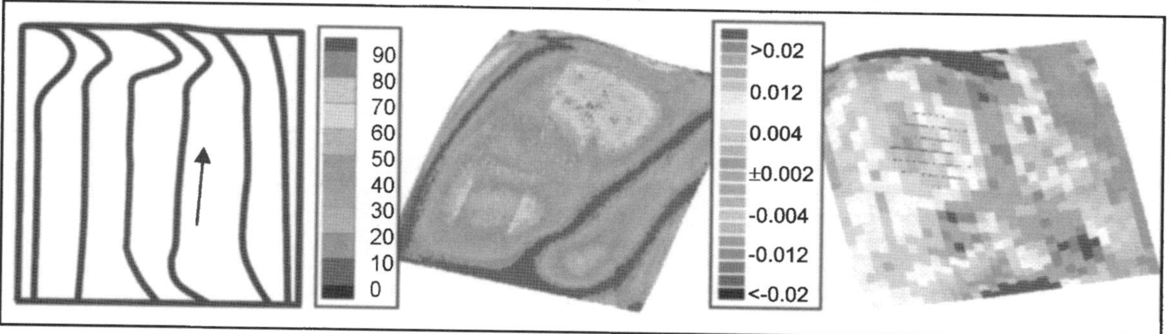

Figure 5: Left) Minimal cutting forces toolpaths, Middle) Monitored cutting forces, and Right) Results in a CMM (Coordinate Measurement Machine).

The estimation of cutting forces has also been integrated as a utility in a commercial CAM software for the best selection of the toolpaths in the high speed milling of complex surfaces.

Nowadays, the above approach is implemented in two companies. Hence, it is used by one die manufacture, who makes use of it for the five axes milling of insert blocks hardened more than 62 HRC, reporting a 12 % more production each month derived from the diminishing of once-again-executing of the CNC programs due to needed re-mill of that just material no removed (due to the tool deflection). Other application, in other general machining user, focused I n aeronautical components, stability of milling process is the common situation.

Acknowledgments

Thanks are addressed to Spanish Ministry of Science, for its financial support of project DPI2004-07569-C02-01 (sky-skin), and to the Basque Government for its support to Margune activity. Thanks are also addressed to the UE Feder funds, for its support in the Interregional Spanish and French Network Interreg for Aeronautical Manufacturing (AEROSFIN).

References

[1] López de Lacalle,L.N., Lamikiz,A. ,Salgado,M.A., Herranz,S., Rivero, A., 2002, Process planning for reliable high speed machining of moulds, *Int. J. of Production Research*, 40, n.12, pp. 2789 – 2809.

[2] López de Lacalle,L.N. , Lamikiz,A.; Sánchez,J.A. Salgado,M.A., 2004, Effects of tool deflection in the high speed milling of inclined surfaces, *Int. J. of Advanced Manufacturing Technology,* electronic reference DOI 10.1007/s00170-003-1723-x

[3] Jung-Jae Lee, Suk-Hwan Suh, 1998, Interference-free tool-path planning for flank milling of twisted ruled surfaces, *Int. J. Manufacturing Technology*. 14, n.11, pp. 795-805.

[4] Bravo, U., Altuzarra, O., López de Lacalle, L.N., Campa, F., Sanchez, J.A., Stability limits of milling considering the flexibility of the workpiece and the machine, *Int J. machine tool and manufacture*, to be published in 2005

[5] López de Lacalle,L.N., Lamikiz,A., Sánchez,J.A. Salgado,M.A., 2004, Cutting force integration at the CAM stage in the high speed milling of complex surfaces, *Int. J. of Computer Integrated Manufacturing*, to be published in 2005.

[6] Lamikiz,A., López de Lacalle,L.N., Salgado,M.A., Sánchez,J.A. 2004, Estimation of cutting forces in the ball-end machining of complex surfaces, *Int. Journal of Machine Tool & Manufacture,* 44, n. 14, pp.1511-1522.

Intelligent Production Machines and Systems
D. T. Pham, E. E. Eldukhri and A. J. Soroka (eds)

Using artificial neural networks for process planning of cylindrical machined components

H. El Awady[a]

[a] *Department of Systems Engineering, Faculty of Engineering, Zagazig University, Zagazig, Egypt*

Abstract

Computer-Aided Process Planning (CAPP) systems play an important role in an integrated CAD/CAM. One of the primary aims of a process planning system for machining is to interpret the design information and prescribe appropriate machining operations consistent with the requirements set forth by the designer. Over the years, there have been considerable efforts in developing CAPP systems using various approaches. A review of the reported literature in this area is presented. A CAPP methodology based on neural network is proposed. The detailed strategy for building the neural network model including designing, training and validation of the neural network is outlined. Illustrative examples are provided. The application of the proposed methodology is believed to contribute substantially in improved performance of CAPP systems.

1. Introduction

The automation of process planning activity by using Computer-Aided Process Planning (CAPP) has gathered much interest over the years. It is still a challenge for the engineers to bridge the gap created by the conflict between design demands and manufacturing requirements. As a bridge between Computer-Aided Design (CAD) and Computer-Aided Manufacturing (CAM), CAPP plays an important role in an integrated CAD/CAM system. One of the primary aims of a process planning system for machining is to interpret the design information and prescribe appropriate machining operations consistent with the requirements set forth by the designer. Given the complexity of the process planning activities, a direct mathematical formulation to meet the design specifications can be difficult or even formidable. This activity requires a significant amount of expertise and very often decisions have to be made based on intuition and understanding of the characteristics of the process planning in order to achieve the design specifications and meet the manufacturing constraints.

2. Background and previous related research

The following presents a survey on the application of various AI approaches in CAPP and their potential advantages and their shortcomings.

2.1. Applications of expert systems in CAPP

Several applications of expert systems in CAPP have been reported. A knowledge-based approach for part machining process selection has been reported by Edward G. [1]. It takes in manufacturing features as input and generates possible sequences of machining operations and their subdivisions. Younis [2] have used an expert system based approach that generates a sequence of machining operations using production rules, taking into account technological attributes of features of the part, modeled using a feature-based modeler. Jiang et al [3] have developed a CAPP system that extracts form features. Radwan [4] has reported development of an expert system based CAPP approach for machining of different kinds of surfaces and holes.

2.2. Applications of fuzzy reasoning in CAPP

There have been reports of applications of fuzzy reasoning in CAPP. Hashmi et al [5] have developed a fuzzy reasoning method for selection of machining speed for a depth of cut, material hardness. The necessary production rules were constructed based on knowledge extracted from data handbook.

2.3. Applications of neural networks in CAPP

Knapp et al [6] demonstrated the ability of neural network in the process selection and within feature process sequencing. In this work, two co-operating neural networks were utilized: the first one, a three layer back propagation neural network, takes in as input the attributes of a feature and proposes a set of machining alternatives; another fixed weight neural network selects exactly one of the alternatives. Parameters of the features are modified by the results of the operation until the final state of the feature has been reached. A neural network approach for automated selection of technological parameters of turning tools is reported by Santochi et al [7]. For each parameter, a neural network was designed, trained and validated. Marri [8] has developed a CAPP methodology that uses a neural network for selection and sequencing of machining operations for components with rotational symmetry. A neural network based CAPP approach for rotational components have been reported by Devireddy et al [9].

3. Proposed CAPP methodology for process selection.

The selection of machining operations is one of the important tasks in process planning. A process engineer (domain expert), who traditionally performs this task manually, routinely applies the knowledge that he acquired by learning the mappings between input patterns, consisting of features being machined and attributes (e.g. size, tolerance, surface finish) of a part and output patterns, consisting of machining operations to apply to these parts. Thus the general problem setting in machining operation selection and planning may be considered as a pattern classification task, which is ideally suited for application of neural networks.

3.1. Objectives and scope

The objective of the proposed methodology is automated selection of all feasible machining operation sequences and the preferred operation sequence after taking cost factor into consideration by using a neural network based approach. The scope of the present methodology is restricted to features commonly encountered in axisymmetrical rotational parts such as holes, external steps, etc. and the selection of operations is done for machining each feature of the part.

3.2. Strategy for development of the neural network architecture

The strategy followed for development of the neural network architecture is outlined as follows:

3.2.1. Identification of network input and output variables

The first step in development of the neural network architecture is deciding on the output process planning decision variables, followed by all related process planning input variables that could affect the selection of the output variables. It constitutes an important step as the optimal performance of the neural network depends on the chosen input patterns and output variables describing the problem. From our basic knowledge about metal cutting theory, the following factors influence one's decision for selection of machining operations:

(a) Type of the feature being machined
(b) Size of the feature
(c) Technological attributes required of the feature (e.g. tolerance, surface finish), and
(d) Machining processes and their capabilities.

The main factors to be considered in selecting a machining operation sequence are size of the feature to be machined, tolerance and surface finish specifications, which, therefore, constitute the process planning input variables. Having decided process planning decision variables, they are presented to the neural network as shown in Fig. 1.

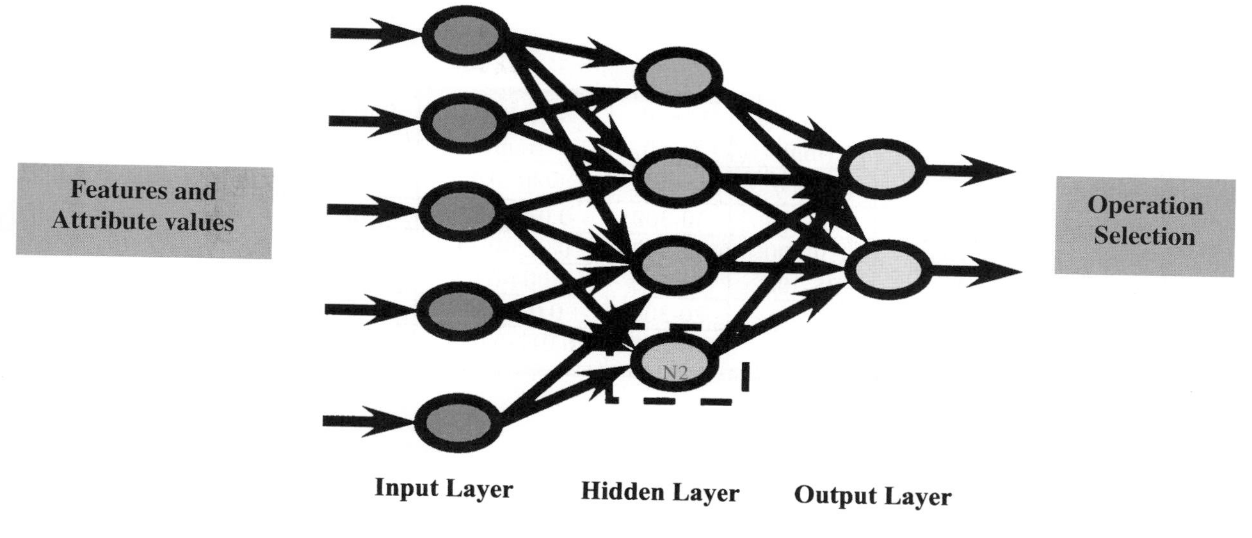

Fig. 1. General architecture of the Neural Network.

The input and output variables are organized into the input layer and the output layer nodes. Each input attribute of the feature is associated with a node in the input layer. The node takes on the value of the attribute, normalized to lie between 0 and 1. The aim of the network is to provide all the sequences of machining operations possible. Therefore, the network has a number of outputs equal to the machining operation sequences that is possible, each operation sequence being associated with a node in the output layer. Each output node can assume a value lying between 0 and 1 to indicate the order of preference. For example, for the 30+0.021 mm diameter hole with surface finish requirement of Ra=1.25μm, the operation sequence for machining the hole will be selected from amongst drilling-reaming, drilling boring- grinding, drilling-boring-honing.

3.2.2. Preparation of the set of training examples

In case of a neural network, the learning process or knowledge acquisition takes place by presenting the network with a set of training examples and the neural network through the learning algorithm implicitly derives the rules. Each training example represents a real machining case described by training inputs, comprising of feature attributes that correspond to the input nodes of the neural network and training targets, comprising of the alternative sequences of machining operations, which correspond to the output layer nodes of the network. Such sources included manufacturing handbooks, metal cutting textbooks and others. The scope of the present work has been restricted to axisymmetric rotational parts containing the following features, namely external step, taper, thread, hole. The machining operations and their capabilities, included in the present system, are given in Table 1.

Fig. 2 shows a set of rules developed by an expert system for generation of operation sequences. These rules are used to generate training examples of input-output pairs for presenting to the neural network. The relative costs between the alternative machining operation sequences are built into the outputs of the neural network while formulating the training examples by assigning values varying from 0 to 1 to the nodes.

3.2.3. Designing, training and validating of the neural network

All the neural networks developed for the study use a fully connected feed forward multilayer perceptron. The standard back propagation algorithm is used as the learning mechanism for the neural network. The sigmoid transfer function is used. In the absence of any theory or guide in the selection of the most appropriate configuration of the neural network, the optimum architecture can be only reached through trial and error with progressive adjustments of the weights by varying parameters such as number of hidden layers, number of nodes in each hidden layer, learning rate and momentum rate in order to obtain the minimum value of the network error. Accordingly, several experiments need to be carried out to identify

Table 1
Capabilities of typical machining operations

Operation	Diameter (mm)	Length (mm)	Tolerance (mm)	Surface finish (microns)	Precision attribute
Drilling	3-100	10-300	0.08-0.4	1.6-9.5	
Reaming	10-100	30-300	0.013-0.08	0.5-2.4	
Boring	19-300	60-900	0.025-0.13	0.4-6.3	Yes
Grinding (internal)	25-300	75-900	0.004-0.02	0.1-1.6	Yes
Turning/grooving/	5-100	20-300	0.03-.087	0.4-9.0	
Grinding (external)	5-100	20-300	0.012-0.035	0.07-1.6	Yes

- IF Diameter is 3-100 mm AND Tolerance is 0.08-0.4 mm AND Surface finish is 1.6-9.5 μm THEN Drilling is recommended
- IF Diameter is 10-100 AND Tolerance is 0.013-0.08 mm AND Surface finish is 0.5-2.4 μm THEN Reaming is recommended
- IF Diameter is 19-300 AND Tolerance is 0.025-0.13 mm AND Surface finish is 0.4-6.3 μm THEN Boring is recommended
- IF Diameter is 25-300 AND Tolerance is 0.004-0.02 mm AND Surface finish is 0.1-1.6 μm THEN Grinding(Internal) is recommended
- IF Diameter is 5-100 AND Tolerance is 0.03-0.087 mm AND Surface finish is 0. 4-9 μm THEN Turning/grooving is recommended
- IF Diameter is 5-100 AND Tolerance is 0.012-0.035 mm AND Surface finish is 0.07-1.6 μm THEN Grinding (External) is recommended.

Fig. 2. Excerpt from Expert System rules developed for generation of operation sequences.

the best architecture of the network having minimum error during training. At the end of training of the network, the generalization capability of the network is verified by validation tests presenting intermediate situations with respect to those proposed during training. Whenever the network fails in the validation set, the training set needs to be modified by adding to the training set the situations of the validation set, which have generated greater errors.

3.2.3. Designing, training and validating of the neural network

All the neural networks developed for the study use a fully connected feed forward multilayer perceptron. The standard back propagation algorithm is used as the learning mechanism for the neural network. The sigmoid transfer function is used. In the absence of any theory or guide in the selection of the most appropriate configuration of the neural network, the optimum architecture can be only reached through trial and error with progressive adjustments of the weights by varying parameters such as number of hidden layers, number of nodes in each hidden layer, learning rate and momentum rate in order to obtain the minimum value of the network error. Accordingly, several experiments need to be carried out to identify the best architecture of the network having minimum error during training. At the end of training of the network, the generalization capability of the network is verified by validation tests presenting pairs of inputs and outputs (validation set) describing intermediate situations with respect to those proposed during training. Whenever the network fails in the validation set, the training set needs to be modified by adding to the training set the situations of the validation set, which have generated greater errors.

3.3. Illustrative examples and results

In order to demonstrate the application of the proposed methodology, a neural architecture for automated selection of operations machining is shown in Fig. 3. The neural network was simulated using the EasyNN Plus ver. 2.od 2002-2003. After a number of experiments, an optimum structure of the neural network was obtained with 1 hidden layer and 9 neurons in the hidden layer. A total of 100 patterns were used in training the network. The learning rate and momentum rates were chosen as 0.9 and 0.83 respectively. The number of iterations required for training the network to the error level of 10% is 308. The training error graph is shown in Fig. 4. Consider an example of case study of dimensions D±0.02 mm and surface finish of Ra=1μm, for which the machining operation sequence is to be generated. The corresponding machining operation sequences generated by the neural network are: drilling, boring, Grinding, Turning, Taper turning, and grooving. Thus by interpreting the outputs of the neural network, the required operation sequence for the given operation feature is obtained. Upon examining the results from the neural network, they were found to be in agreement with the desired results. This suggests that the proposed neural network based approach holds a lot of potential for use in real manufacturing problems.

4. Conclusions

The article presents a detailed CAPP methodology for machining of rotational axisymmetric parts. It has been implemented using a neural network approach. The proposed CAPP methodology takes in as input the attributes of the features and automatically generates all the feasible alternatives for machining the feature in order of preference after taking their relative costs into consideration. Some advantageous features of the neural network based CAPP methodology as against the knowledge based approach are:

(a) Its efficient knowledge acquisition capability owing to its ability to implicitly derive the rules from sample machining cases presented to the neural network.

(b) Its capability to generalize beyond the original machining cases to which it is exposed during the training and face intermediate situations with reasonably good accuracy with respect to those proposed during the training.

(c) High processing speed once the neural network is trained.

Acknowledgements

This research was supported in Industrial engineering Depart. Faculty of engineering, Zagazig University. The authors would like to express their gratitude for the support.

References

[1] Edward G., 1998 "The analysis of methods for computer aided process planning", Provo, Utah, USA.

[2] Younis, M. A., Waheb, M. A. Abdel (2002), "A CAPP expert system for rotational components", Computers Ind. Eng., 33, 3-4, pp 509-512.

[3] Jiang, B., Lau, H., Chang, Felix T.S., Jiang H. (1999), " An automatic process planning system for the quick generation of manufacturing process plans directly from CAD drawings"; Journal of Materials Processing Technology, 87, pp 97-106.

[4] Radwan A. (2000), "A practical approach to a process planning expert system for manufacturing processes", Journal of Intelligent Manufacturing , 11, pp 75-84.

[5] Hashmi K., El Baradie M. A., Ryan M. (1998), "Fuzzy Logic Based Intelligent Selection of Machining parameters", Computers Ind. Eng., 35, pp 571-574.

[6] Knapp, G. M, Wang, H. (1992), "Acquiring, storing and utilizing process planning knowledge using neural networks", Journal of Intelligent manufacturing, 1, pp 1-15.

[7] Santochi M., Dini G. (1996), "Use of neural networks in automated selection of technological parameters of cutting tools", Computer Integrated Manufacturing Systems, 9, 3, pp 137-148.

[8] Marri, H. B., Gunasekaran, A., Grieve, R. J.(1998), "Computer-Aided Process Planning" A State of Art, International J. of Advanced Manufacturing Technology, 14, 4, pp 261.

[9] Devireddy Chandra R., Ghosh K. (1999), "Feature-based modeling and neural networks-based CAPP for integrated manufacturing", Int. J. Computer Integrated Manufacturing, 12, 1, pp 61-74.

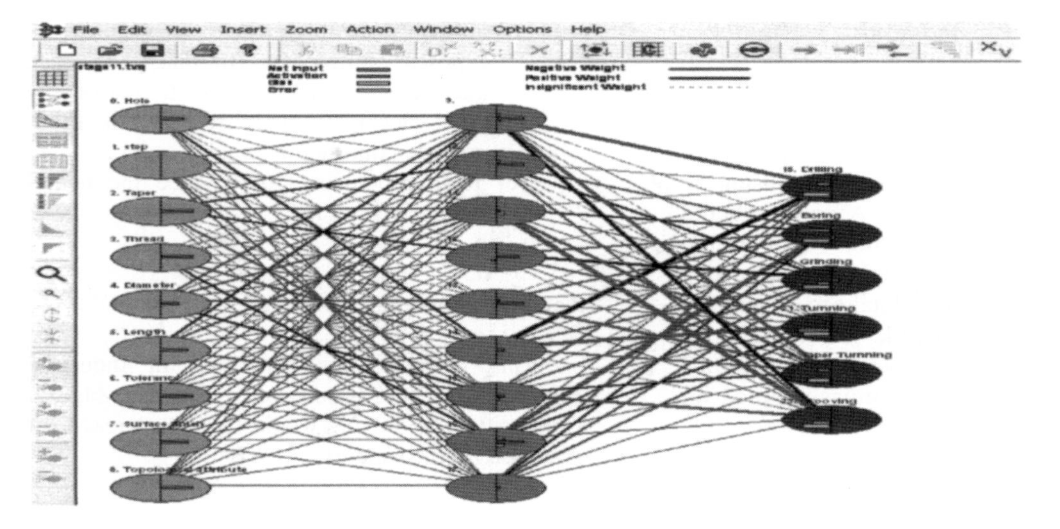

Fig. 3. Neural network architecture for automated selection of hole machining operations

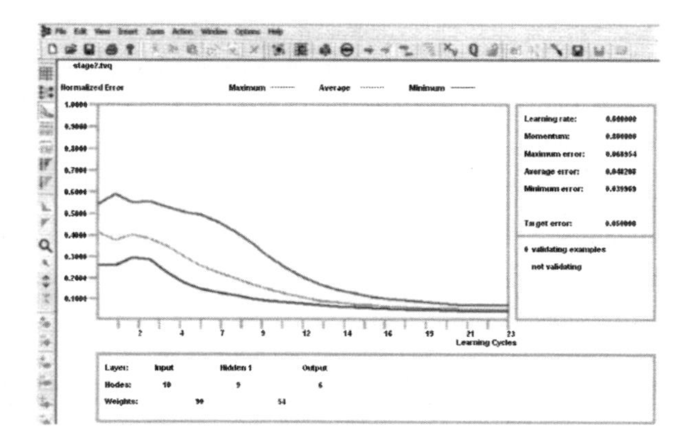

Fig. 4. Training error graph.

Intelligent Production Machines and Systems
D. T. Pham, E. E. Eldukhri and A. J. Soroka (eds)

Active Interpolated Surfaces for Robotic Arm Control: The Potential for High Speed Implementation in FPGA

D.Froud[a]

[a] *IAS Lab, University of the West of England, Bristol BS16 1QY, UK*

Abstract

The main contribution of this paper is to introduce a novel control system, formulated to take advantage of recently developed FPGA hardware in application to robot arm control, to demonstrate the potential of FPGA for manufacturing processes. Robotic Arms are commonplace in manufacturing. Active interpolated surfaces offer easy implementation in FPGA hardware and thus have fast processing as well as speedy programming and modification times, allowing for wide areas of applicability. High speed implementation potentially offers flexibility and fast adaptation. This paper reports on an active interpolated surface model applied, as an illustrative example application, to the task of removal of flawed coin templates in a mint. The task requires fast location of flawed items in a wide field of variably positioned coins, and the speedy determination of position and movement requirements to enable the location and extraction of the flawed coins using PUMA robotic arm.

1. Introduction

Robotic Arms are commonplace in manufacturing but suffer from some disadvantages such as process delay during installation and tuning, complex programming and minimal possibility for adaptation or reconfiguration without substantial further delay to the process of manufacture [1]. Real time learning and control has been postulated for several years [2], however the extended learning times for accurate positioning make them less attractive and costly in terms of process delay. A novel hardware implementation of an interpolated network, known as an active interpolated surface, offers easy implementation in hardware through Field Programmable Gate Arrays (FPGAs) [3] and addresses the problems associated with slow learning, and thus also adaptability and flexibility issues.

FPGAs are known for their massively parallel structure and thus fast processing times as well as ease and speed of programmability [4]. As the programming process is implemented through software control of the gate array, it allows for high speed development of the system through the software itself, giving large benefits of flexibility, and thus wide areas of applicability [5]. Although FPGAs have been identified as having massive potential for manufacturing control [5], industrial process control is clearly still the realm of custom design solutions indicated by the plethora of available reconfigurable boards. Although FPGAs have been used for robot control [6-8], a flexible robot movement optically determined error minimisation architecture for a manufacturing environment has not, to the author's knowledge, been addressed.

This paper reports on the simulated control of a robotic arm using an active interpolated surface. As an illustration of implementation, a robotic arm is modelled as a substitute for a human operator in the mundane task of quality control in a coin

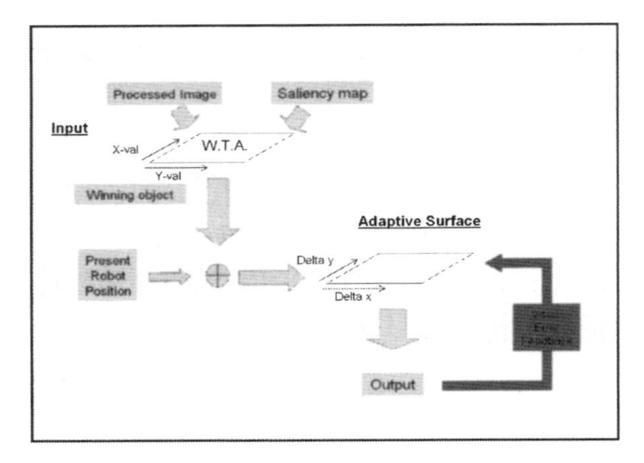

Fig. 1. System architecture.

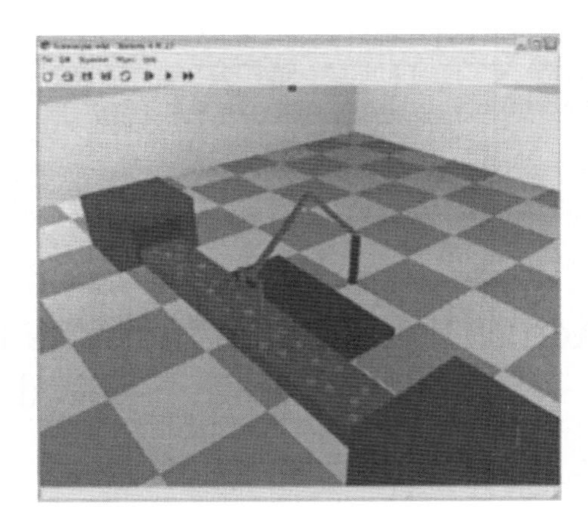

Fig. 2. Webots simulation.

manufacturing process, using data from a PUMA robotic arm. To avoid damage to machines during the forging process, coin templates must not be flawed at the edges on entry to forging machine and must therefore be removed prior to entry. This removal process has previously been carried out by human operators who must concentrate for long periods carrying out extremely repetitive work [9]. Implementing a fast learning robotic arm would offer a quick replacement possibility for the human operator by maintaining same conveyor system architecture. Minimal down time would be achieved due to fast learning compared with new human operators.

After detection the task requires a learning system to correctly position the robotic arm for efficiently removing the flawed coin templates. Coins are located in variable positions, with the most important criterion being that of 100% removal of the flawed specimens. The second criterion is one of speed. The ideal is that the operator – robotic or human – develops a rate of extraction such that the conveyor of coin templates never needs to be halted.

The active interpolated surface can be visualised as a collection of cells in a 2-dimensional array format that return an output in response to 2 inputs, similar to a look up table. The values in this array change according to experience from the process. Values also change after an experience to update all cells with no direct experience by interpolation around and between experienced cells. For implementation some work would be required on the image processing, but this could be carried out without interruption to the process, by using simulation of the robot arm and actual camera systems.

2. Method

2.1 Definition of the problem

The overall problem can be split into two main areas – detection and removal of flawed items. Flaw detection is not addressed in this paper. Removal of flawed items is addressed through the implementation of an active interpolated surface.

The removal task is broken down into 5 sections:-

1. Location and saliency of the flawed item
2. Location of the gripper/suction device
3. Error feedback for learning purposes
4. Movement algorithms
5. Learning algorithms

The flaw detection system was assumed to detect and colour-code the location of flawed templates. The model then locates the flawed items through colour extraction in the scene. Saliency is attributed to these items depending on the proximity to the end of the conveyor. Target items closer to the end of the conveyor must be removed first before they get out of reach of the robot and hence are attributed a higher saliency.

The robot arm tool head location was also colour-coded, allowing an easy error evaluation and feedback through distance error. Movement of the arm was modelled in accordance with a VAL 2 robotic arm controller.

Through the learning and movement algorithms the adaptive surface is warped to correct for movement error. This error comes about due to the time taken for the robot to move to the target location during which time the target has moved a distance down the conveyor. The surface quickly settles and becomes

Fig. 3. PUMA robot test rig setup.

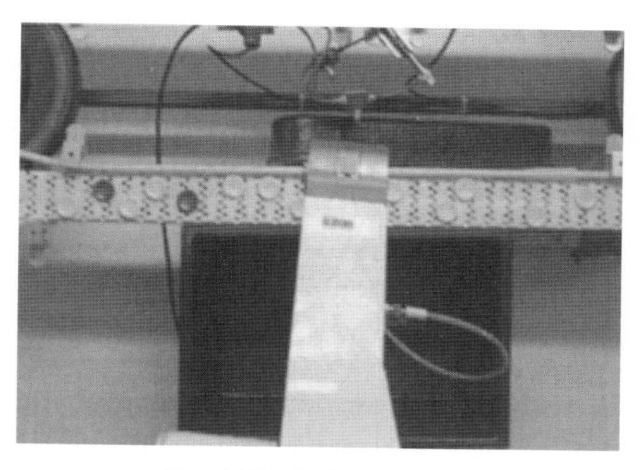

Fig. 4. Vertical camera view.

more and more static with experience. Interpolation still occurs until every possible movement has been carried out. A static surface is equivalent to a look-up table allowing speedy implementation of the relevant error correction. Interpolation of the surface allows good prediction for unknown locations, and sometimes successful movements without explicit knowledge of the movement required. Implementation using FPGAs allows rapid surface generalisation in a real time system.

2.2 Implementation of the interpolated surfaces

The system architecture consisted of a two connected surfaces (see Fig. 1) and included visual error feedback.

Visual input is combined with saliency (measured as downward position of the object on the conveyor) and fed into the input surface. This has fixed connections with a 'winner-take-all' (WTA) algorithm, and thus isolates the most salient object for removal. Having determined the position co-ordinates (x,y) of the most salient object, these co-ordinates are compared with the present position co-ordinates of the robot arm tool head. This provides the required change in x and y co-ordinates (dx, dy) for the robot to reach the target. As the robot arm takes time to move through the dx-dy transformation, the target on the conveyor will move also. The adaptive surface takes the dx and dy initial estimate and, through a trial and error process, determines and eliminates this error. As no the exact error is calculated, knowledge of the process is unnecessary. Non-lineararity in systems is observed and corrected for using the visual error feedback.

2.3 3-dimensional virtual model using Webots robot simulation package

A virtual 3-dimensional model was constructed using the Webots [10] robot simulation package (see Fig. 2). The active surface control system, outlined above, was implemented through a C++ controller program for the robot. A red coin indicated a flaw. Although it was not possible to add movement to the conveyor in this simulation package, the 3-dimensional visualisation was very helpful in the development of the surface for troubleshooting and fine tuning the active surface modification algorithm.

2.4 Modelling using a PUMA robot data

A real conveyor and robotic arm environment was constructed using a suction device instead of a gripper to lift and remove the simulated flawed coins (see Fig. 3). The suction device enabled a 2-dimensional robot arm movement system. The device could be moved in a horizontal plane to a position over any of the simulated coins on the conveyor, and suction used to pick up and remove the coin without any vertical robot arm movement. The PUMA robot was connected to a dedicated VAL 2 controller. Controlling the robot indirectly using a computer implementation of the surfaces and a serial link to the VAL 2 controller, or directly using a hardware FPGA interpolated surface configuration is considered as future work for the project.

Physical measurements such as robot and conveyor position and dimensions, acceleration and movement speed of the robotic arm, conveyor speed and allowable final position error were measured and included in a further C++ model. A vertical camera

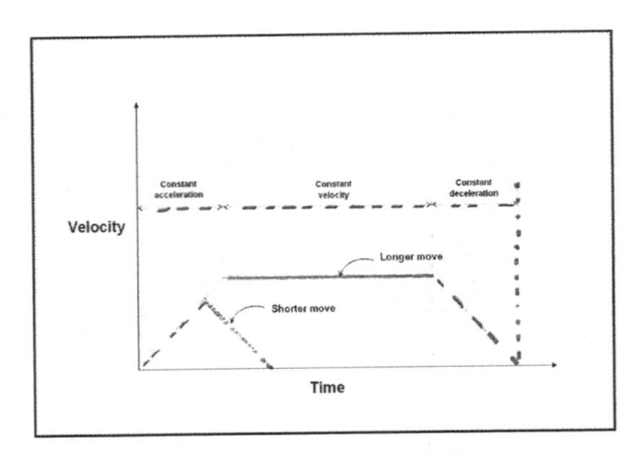

Fig. 5. Robot arm velocity profile.

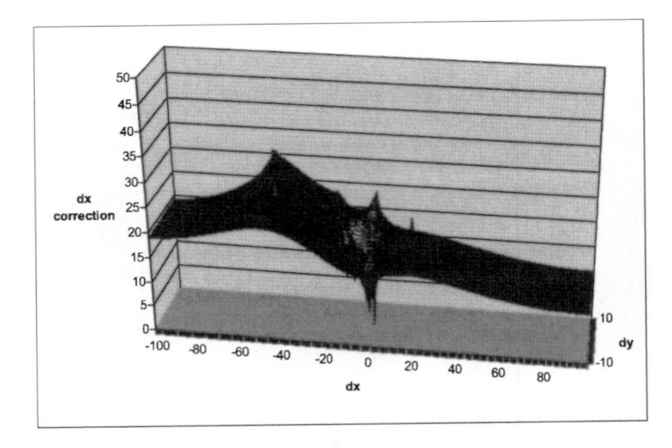

Fig. 6. Surface configuaration after 100 moves.

view of the PUMA robot shows the physical setup of the test rig (see fig. 4.) for visual processing.

In the simulation, the knowledge of the post movement position of the target was required for visual error feedback as an input for the active surface. To calculate this the time required for the robot to move to the determined location was evaluated, and the change in the target position thus calculated using the velocity of the conveyor. The velocity map for the robot arm movement is shown in fig. 5. Using an acceleration period in conjunction with a fixed maximum velocity for the robotic arm movement simulation, a non-linear adjustment output from the active surface was required for accurate position adjustment for robot movement in 2 dimensions. The VAL 2 controller takes the movement co-ordinates, and calculates and implements a straight line movement to the desired location with the movement profile shown in fig. 5.

2.5 Active Surface interpolation algorithm

With no initial knowledge, the surface makes no initial adjustment to the *dx* and *dy* movements determined from the visual input. This takes the robot arm to the exact location of the target before the movement. On reaching this position the error is visually determined, and this becomes the initial adjustment estimate. This is also incorrect as the new movement, with the inclusion of the position adjustment, will take a different time to perform, thus producing a different, but smaller error. This error is added to the previous adjustment and adopted as the fresh adjustment value, as shown in Eq. 1.

$$dx_{(n+1)} = dx_{(n)} + observed\ error \quad (1)$$
$$dx_{(n+1)} = \Sigma(nearest\ 4\ dx\ values)/4 \quad (2)$$
$$\Sigma\ (\Delta dx) < Threshold \quad (3)$$

Calculations have shown that for the maximum robot movement in this context (1 metre) with a conveyor speed of 0.1m/s, four trials of the same movement bring the error within the allowable error (±5mm).

Once an adjustment has been determined, the surface is generalised. If the adjustment is correct (i.e. the target was achieved) the position on the surface is fixed permanently. Otherwise the position determined is kept stationary for this generalisation procedure only, but is free to move for further generalising procedures. The method for generalising consists of each non-stationary point in the surface taking the value of the average of its neighbours as shown in Eq. 2. This process of averaging is repeated until the total change in values across the whole surface for that iteration is less than a threshold value (see Eq. 3). This threshold is important as it determines the speed of the generalisation process. The smaller the threshold, the more iterations are required in the generalising procedure. Usually this trades off against the number of moves of the robot arm required for success – a larger threshold leads to lower accuracy of movement.

2.6 Adaptation – switching sides

Finally to demonstrate and test the adaptability of the active interpolated surface system, the simulated robot was moved to the opposite side of the conveyor belt. The active surface was left as it had developed for the opposite side. Fixed positions were again made variable. As the expected results are that an inverse of the previous function is required, non-zeroing of the active surface would therefore extend the time required for settling the surface, at least for the initial move.

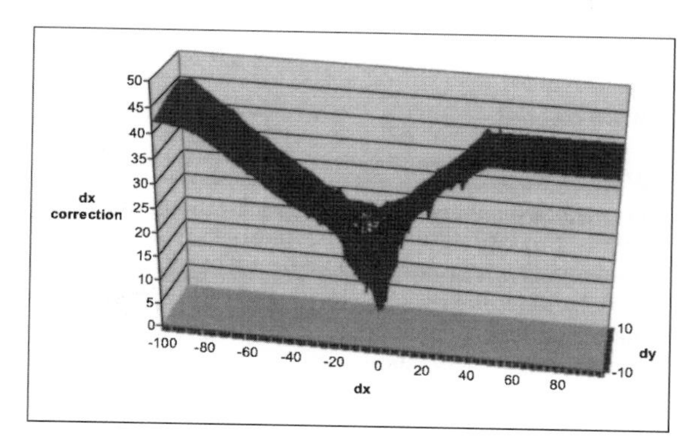

Fig. 7. Surface configuration after 1000 moves.

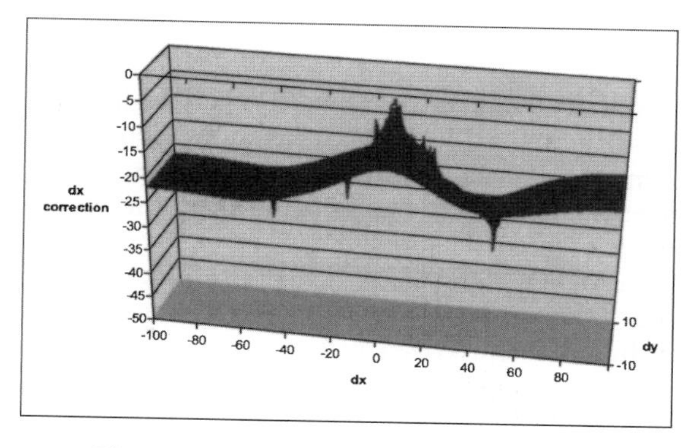

Fig. 8. Adapted surface after 100 moves

3. Results

For figs. 6-8 'dx' corresponds to the visually determined x-difference in position between the tool head and the target, 'dy' corresponds to the y-difference, and the 'dx correction' corresponds to the change in the x-difference to take into account the conveyor movement of the target during the arm move.

The active surface shown in fig. 6 was developed over 100 moves with an error correction threshold of 10. Correspondingly a further test was carried out using 1000 moves and an error correction threshold of 0.1. These results are shown in fig. 7. The central dip corresponds to the robot arm being exactly at the correct position for removal of the target during at the time of observation, and hence no correction factor is required. As the target distance from this point increases in any direction the correction factor required increases. The reason for this is that the time taken to move to the target position increases with distance, and thus the adjustment required to compensate for conveyor movement during this time also increases. The surface allows non-linear correction by having adjustment values unconstrained by equations. Instead they are determined using a trial-and-error approach and averaging between positions.

This non-linearity is illustrated in fig. 7 by the different zones. Near the central position the gradient is steep. Beyond this small zone there is a zone of reduced gradient. In fig.6 the surface decreases at high negative and positive x-values due to no relevant data existing in this area. The surface would be expected to continually increase with the gradient of the second zone as shown in fig 7. The flat zone on the right hand side of the graph in fig. 7 also indicates a lack of data. This lack of data is accounted for by the conveyor moving from left to right. Data for large moves to the right could not therefore be collected as the target would already have moved out of reach.

The corresponding numbers of iterations for the two examples illustrate the trade off between number of iterations and number of moves. To remove the first target, for a threshold of 10, the robot requires 10 moves and a total of 22529 iterations, while that with a threshold of 0.1 requires 484654 iterations, but in only 4 moves. Despite the large difference, after 100 moves, the comparison of number of items removed is 41:42 respectively. Although more moves are required for the initial learning in the lower threshold surface, overall performance is unaffected.

The number of iterations of the generalisation procedure per movement decreases rapidly. The total number of iteration procedures for 100 moves for the threshold of 10 is 45481. As can be seen from the previous paragraph, this is approximately double the number required for the first 10 moves, and less than 4 times the number required for the first move (13774). High levels of processing thus generally occur at the start of the adaptation process. The non-linearity of the process will dictate the rate of reduction in the number of attempts and iterations between attempts before successful removal for successive targets.

The adaptive process of the surface was then tested by switching the robot arm to the opposite side of the conveyor – effectively reversing the direction of movement. The surface was allowed to continue unchanged although fixed locations were again made variable. A plot of the new surface after 100 moves with a threshold of 10 is shown in fig. 8. The first target was successfully removed in 7 movements, compared to 10 for the original surface, with 37979 generalisations procedure iterations. This represents an increase in interations of 68%.

4. Discussion

An FPGA implementation offers very high speed processing in a parallel fashion. Taking the example of the smaller threshold surface shown in fig. 7, the number of iterations required to achieve successful removal of the first target totalled nearly 500,000 in 4 moves. An FPGA can theoretically run at clock speeds of 420 MHz or more, but 50 to 100 MHz is a more realistic maximum. Given 50 instructions for each cell in the surface to undertake per iteration, implementing a parallel system leads to iteration speeds of around 1 MHz. In terms of processing this translates into approximately half a second, with the 4 moves realistically taking the robot arm a further 2 or 3 seconds. With the conveyor moving at 0.1m/s and a maximum reach of the robot covering 1 metre of the conveyor, this represents less than half the length of the reachable area. Later removal attempts require less processing time. Although the number of moves is not guaranteed to decrease with further attempts, the error associated with each move is reduced, leading to faster overall success time compared with the original attempt. For the above simulation, the maximum number of moves observed for success was 7. Total distance moved by the arm, however, was reduced from 1.13 metres to 0.72 metres for the 7 move attempts. With the average movement success rate increasing to 60%, towards the end of the trial, attempts requiring more than 2 moves before success were unusual.

Adaptation to changed parameters in the process, whether position or speed, could easily be incorporated into this system. The learning and adaptation is, in theory at least, favourable comparable to a human operator faced with the same challenges.

The active surface technique can easily be extended to determine more than one parameter. For example if the robot required r and theta transformations instead of a single dx parameter, two surfaces could be implemented using a similar process.

The adaptability and speed of this process make implementation for non-linear systems particularly attractive where modelling and programming of a robotic arm could be costly or preventative.

Acknowledgements

I would like to acknowledge help from Farid Dailami, Tony Pipe and Chris Melhuish, as well as UWE CEMS department for use of the PUMA robot arm.

References

[1] Worn H, Wurll C and Henrich D. Automatic off-line programming and motion planning for industrial robots. Proceedings of the 29th International Symposium on Robotics 1998, pp 196-199.

[2] Ferreira A and Engel P. Positioning a robot arm: An adaptive neural approach. Proceedings of the 1996 International Workshop on Neural Networks for Identification, Control, Robotics, and Signal/Image Processing, pp 440-449.

[3] Bertin P, Roncin D and Vuillemin J. Programmable active memories: A performance assessment. In: G. Borriello and C. Ebeling, editors, Research on Integrated Systems: Proceedings of the 1993 Symposium, pages pp 88–102.

[4] Morris K. A bevy of boards: Selecting for success. J. FPGA and Programmable Logic, 2005, http://www.fpgajournal.com/articles_2005/20050412_board.htm.

[5] Welch J and Carletta J. A direct mapping FPGA architecture for industrial process control applications. Proceedings of the 2000 IEEE International Conference on Computer Design: VLSI in Computers & Processors, pp 595-599.

[6] Spenneberg D, Kirchner F, de Gea J. Ambulating Robots for Exploration in Rough Terrain on Future Extraterrestial Missions. In Proc. 8th ESA Workshop on Advanced Space Technologies for Robotics and Automation, ASTRA 2004.

[7] Jeon, J and Kim Y. FPGA based acceleration and deceleration circuit for industrial robots and CNC machine tools. *Mechatronics* 12, pp. 635–642.

[8] Alcaraz F. Adaptation of real-time software robot control units to generic hardware architectures. Dedicated Systems Magazine, 2003, http://www.omimo.be/magazine/emagazine/fulltext/2003q2_1.pdf

[9] Bjerknes J. Personal communication. Ex-employee of the Norwegian National Mint.

[10] Michel O. Webots: Professional mobile robot sim. J. Advanced Robotic Systems, 2004, 1, 1, pp 39-42.

Intelligent Production Machines and Systems
D. T. Pham, E. E. Eldukhri and A. J. Soroka (eds)

Design and Implementation of a Web Based Mobile Robot

S. Sagiroglu[a], N. Yilmaz[b], M. Bayrak[b]

[a] *Department of Computer Engineering, Faculty of Engineering-Architecture, Gazi University, Ankara, TURKEY*
[b] *Department of Electrical- Electronics Engineering, Faculty of Engineering-Architecture, Selcuk University, Konya, TURKEY*

Abstract

In this study, a web based mobile robot designed and implemented for long term and regular scientific purposes has been successfully presented. The mobile robot was controlled with portable PC on mobile robot. The PC can be connected to internet via Wi-Fi LAN adapter through wireless access point. A PC is used as a web server to provide control capability on the mobile robot through local network. The tests have shown that the designed robot is suitable for multi-purpose applications.

1. Introduction

Many types of mobile robots in practice have been successfully used in today's world such as wheeled, legged and flying used in a wide range of areas [1]. Legged mobile robots are used to lumber in forestry areas and to explore volcanoes. Wheeled mobile robots have been employed for cleaning, safety and/or carriage tool at home and offices. Flying mobile robots were used to explore in military and civilian areas in order to get precise information on atmospheric layers for weather forecast. Mobil robots have been also very popular for entertainment and movie sectors such as sumo wrestling, football matches and dangerous movie scenarios filmed. Figments of imagination creatures are also livened up with mobile robots too. Wheeled mobile robot designs are still popular because of their energy saving advantage and supports to different fields [1]-[6].

Most control systems on mobile robots have been also performed by computers to support complicated operations and tasks to be handled by human operators. Improving the comfort of human operator is always required. Connecting the mobile robots to computer system via wireless devices also helps human operator to handle and to control mobile robot efficiently and effectively within large distances [2]. This type of mobile robots is known as a tele-robot. This robot is usually used in unsafe and dangerous areas for human beings, for example in exploration of aerospace and volcanoes [3,4].

Recent developments in technologies help to improve mobile robots providing high speed and cost effective communication techniques. Web based mobile robots have been employed in robotic lectures [5], museums and art galleries [6].

Recent studies on mobile robots related to the study presented in this article have been reviewed in [3-9]. The works on remote experiments via internet based A four wheel tele-operated robot system [3], distance based mobile robot for obstacle avoidance [4], immobile cameras [5] and vision controlled autonomous mobile robots [8,9] have been achieved with the help of internet based control for distance and modern education.

A web based mobile robot designed for general purposes called SUNAR, (**S**elcuk **U**niversity **NA**vigation **R**obot) was presented in this study. The SUNAR is controlled via WAN and LAN communication.

In this work, to see the performance of SUNAR for two specific indoor and outdoor applications have been tested. In the indoor application, the robot was used for laboratory exercises in control, robotics, and assembly programming lectures. In the outdoor application, it is mainly used for observation the environments.

The mobile robot SUNAR designed and implemented in this work has been explained in the following section.

2. Developed Web Based Mobile Robot: SUNAR

The block diagram of the developed mobile robot is given in Fig.1. The mobile robot presented in this work was especially designed for web-based distance control for various applications. The robot system consists of the communication center and the robot vehicle. The communication center provides all connections of the mobile robot. The vehicle consists of the microcontroller units, the camera system, a portable PC with Wi-Fi LAN adapter, the motors and the motor drivers for movements of the vehicle and the camera, and the speed sensors. The robot is a three-wheeled robot. The two rear wheels are free. The front wheel steers and drives the mobile robot. For the simplicity and the cost, the body of SUNAR was constructed from PVC (Polyvinylchloride) and polyethylene materials.

Two microcontrollers (PIC 16F877) were used in hardware implementation. The slave microcontroller controls the camera system, the motors and the sensors. The controller provides the communication with the master microcontroller via I^2C serial bus. The master microcomputer manages the vehicle motion control, the communication with the portable PC and the commands of the slave microcontroller. The master microcontroller was connected to the portable PC via RS232 port. In addition, the four DC motors with gearbox, the four optical tachometers coupled to the dc motors, the four H-bridges for the DC motor driving and two LCDs were used in this implementation. The control center manages all activities in and around the robot vehicle.

The detail information about the vehicle motion system, the camera system and the communication system and the control center were given in following sections.

2.1. Vehicle Motion System

The drive system of SUNAR consists of a DC motor with gearbox and transmission of movement parts using with the gears and the shafts. There were two optic tachometers behind the DC motor with gearbox made from FAULHABER Company. The tachometers gave 15,000 pulses per cycle of output shaft.

Kinematics of SUNAR were given in equations (1)-(6). The kinematics equations for the rear wheel are:

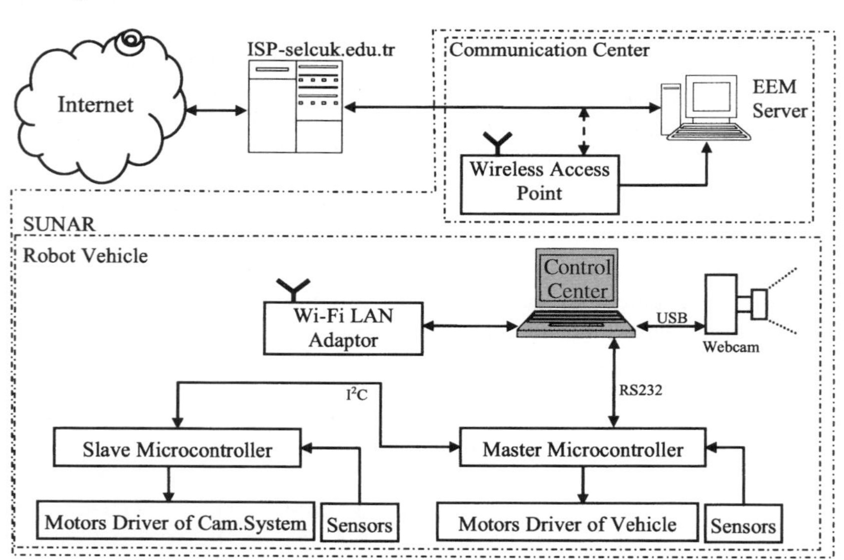

Fig. 1. Block diagram of SUNAR presented in this work

$$\dot{x}_{rear} = v_{front} \cos\theta \cos\alpha \quad (1)$$

$$\dot{y}_{rear} = v_{front} \sin\theta \cos\alpha \quad (2)$$

$$\dot{\theta} = v_{front} \frac{\sin\alpha}{\ell} \quad (3)$$

Kinematics equations for the front wheel are:

$$\dot{x}_{front} = v_{front} \cos(\theta + \alpha) \quad (4)$$

$$\dot{y}_{front} = v_{front} \sin(\theta + \alpha) \quad (5)$$

$$\dot{\theta} = v_{front} \frac{\sin \alpha}{\ell} \quad (6)$$

The positions of SUNAR were calculated by the master microcontroller using equations (1)-(6). The vehicle positions and the camera motion systems were achieved by PID controllers. These controllers are individually run on the master and the slave microcontrollers.

2.2. Camera Motion System

The camera motion system designed for this mobile system is used to achieve images around the mobile robot. The camera motion is achieved with two DC motors. The motors turn 360° horizontally (pan) and 270° vertically (tilt). The motion that comes from the first DC motor is transferred to the platform with a gear so that horizontal pan motion is fulfilled. The motion that comes from the second DC motor tilts the platform vertically via a gear set. The DC motors were driven with PWM based H-Bridge DC motor drivers controlled by PIC16F877.

2.3. Communication System

Three types of communications among the control center-master microcontroller, the wireless access point-Wi-Fi LAN adaptor and the master-slave microcontroller have been achieved. The first communication between control center and internet service provider (ISP) was acquired by a wireless access point and Wi-Fi LAN adapter. The communication range is expanded to 1 km using an external active antenna. The second communication between the control center and the master microcontroller was attained through RS232 port in the form of Modbus ASCII, 19200 Baud, 8 bit and the odd parity. The last communication between master and slave microcontroller was achieved through I^2C serial port.

2.4. Control Center

The control center on SUNAR consists of a portable PC compatible with IBM, a control program running on the computer and the communication units. The computer preferred was Intel Centrino 1.6 GHz microprocessor, 512 MB RAM and 30 GB HDD including an internal Wi-Fi LAN adapter.

The control center realizes the following four functions:

(1) capture the scenes from the web-cam and update the image on the web database.
(2) run the commands for the vehicle and the camera coming from the web or the LAN and update the system status on the web database.
(3) manage the communication with mobile robot and calculate the position of mobile robot from the information encoder.
(4) process the images and extract the knowledge from images for the operator.

2.5 Software Developed

In this study, software is developed to control all mobile robot system. This software consists of three programming parts including microcontrollers, control center, and the web interface. The microcontrollers were programmed using MPLAB assembler programming. The programming the web application was based on ASP for database and JavaScript for the timing and the refreshing functions. The control center was written in Delphi. The image capturing and the serial communication were done via the third party component. The screenshots for the control center and the web interface were given in Fig. 2.

As explained earlier, SUNAR mainly has two parts as shown in Fig. 1. The important controls were achieved in the control center. So there has been heavy communication between the master microcontroller and the control center. The control center designed achieves simple and complex missions.

The simple missions are to turn right or left, to act backward or forward, to adjust speed of acts of camera and the vehicle.

The complex missions are behind autonomous study such as wall and line tracking, object tracing, reaching desired positions and finding landmarks. These missions can be loaded to the control center via a web center of the robot. All mission commands are first stored to the system database and then executed. This reduces the process time. The time is important because several missions can be executed at the same time. While one mission continues the other starts so the time helps us to distinguish the missions. The status of missions is also stored to the database. It needs to be emphasized that all complex missions were realized by the control center.

The other part of SUNAR is the communication center. This center provides internet connection of the robot. It consists of a web server and a wireless access point. In this server, Windows XP based standard server programs are run.

As shown in Fig. 2, the program menu for the control center consists of the database, the status of UDP port, the serial port controls, the image capturing, the speed control of camera and the vehicle for manual usage, the calculated positions for the camera and the vehicle. In addition, two menu buttons ("Option", "Image Processing") also exist on the control center program. "Image Processing" menu is used for the image processing covering the filtering, segmentation, moment functions and recognition for the vision system "Option" menu provides a smooth control and communication utility covering the parameters.

The web interface has manual control option for the database, the captured images, the speed control of camera and the robot vehicle, the commands of semi-autonomous usage and the vehicle positions.

2.6. Vision System of SUNAR

This system was designed for further analysis such as navigation, recognition, observation, detection, path finding and estimation. Mobile robot navigation in a natural environment calls for a sophisticated sensory system. A method for visual landmark recognition was also used for both positioning and navigation in this paper. We define positioning as a process that uses the sensory input of the robot to derive the location and orientation of the robot with respect to a map of the environment. The image source is a webcam on the mobile robot. The webcam is connected to the portable PC via USB1.1 port. The captured image is the colour image in 320x240 resolutions. The images achieved from the webcam mounted on the SUNAR were filtered using median filter on 3x3 windowing.

For further image processing, median filtering and Sobel edge detection masks were used for filtering and segmentation. The details of these processes were not given here but will be explained in forthcoming articles.

3. Results and Conclusions

Experiments have been conducted on the designed mobile robot vehicle SUNAR as shown in Fig. 3. As mentioned earlier, SUNAR was especially designed for the multi-purposes including laboratory exercises, R&D, land marking and object recognition. The robot was especially designed to support virtual learning environment in control, robotics, image processing, vision and microcontroller laboratories. The mobile robot and its platform might also be used for the undergraduate and graduate R&D projects.

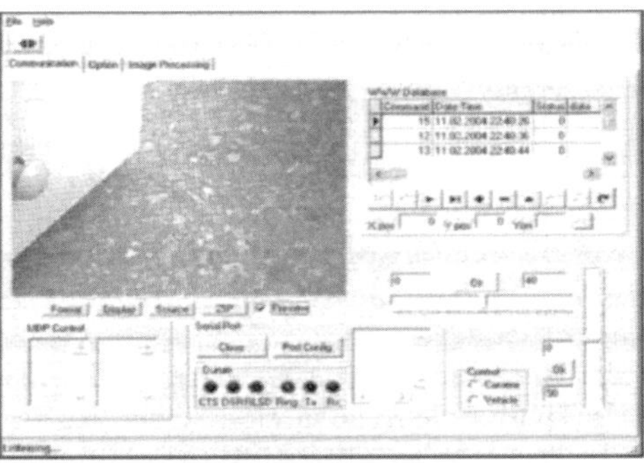

(a) Control center

(b) Web interface

Fig. 2. Screenshots for the developed software.

The robot was tested using two test environments as shown in Fig. 4. The first test, Test #1, was about the recognizing an object and avoiding an obstacle as shown in Fig. 4a. The environment used for Test #2 was for observing the growth of flowering plants as illustrated in Fig. 4b. In Test #1, SUNAR recognizes three different boxes. It then determines the position as required. So, the robot goes from one place to other without any crash. In Test #2, the flowerpots are recognized. Their positions are then determined. The pictures belonging to the plants (flowers) were captured and stored into a hard disk at regular time interval with particular distance by SUNAR. In this test, snapshot positions are defined by an operator to achieve the distance among images. These tests have been easily done on internet via the control center as illustrated in Fig. 2(a) or the web interface unit as demonstrated in Fig. 2(b).

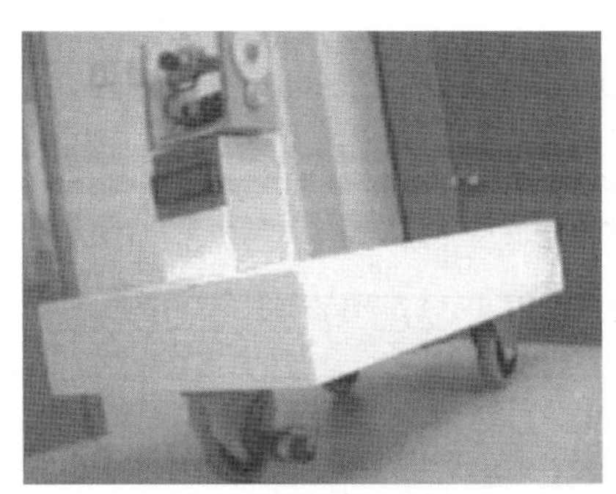

Fig. 3. Front view of SUNAR

(a) Test #1

(b) Test #2

Fig. 4. Test Scenarios of SUNAR

With the help of the designed and developed control centers on www services, faster communication is achieved for the mobile robot switching TCP/IP port on web or UDP port on LAN. The heavy communication between control center and client is also decreased with the help of the autonomous structure of the robot. It needs to be emphasized that the mobile robot system did not cost much as well.

In this study, a web based semi-autonomous mobile robot for multi-purposes was successfully designed and implemented. This paper shows that the designed robot can be used for multi-purposes easily and effectively. The only difficulty faced in the design and implementation was to lose wireless communication when the robot tested in different environments. Even though the work presented here is in its early stages, it still shows that it is feasible to carry out land marking and object recognition. The authors will focus on these applications in further studies.

References

[1] Asoh H, Motomura Y, Asano F, Hara I, Hayamizu S, Itou K, Kurita T, Matsui T, Vlassis N, Bunschoten R, Krose B: Jijo-2: An Office Robot that Communicates and Learns, Intelligent Systems, IEEE, Sept./Oct. 2001, Vol.16, 46-55.

[2] Diolaiti N, Melchiorri C.: Teleoperation of a Mobile Robot Through Haptic Feedback Haptic Virtual Environments and Their Applications, IEEE International Workshop, 2002 HAVE, 17-18 Nov. 2002, 67 -72.

[3] Schilling KJ., Roth H: Control Interfaces for Teleoperated Mobile Robots, Emerging Technologies and Factory Automation, Proceedings of ETFA'99, 7th IEEE International Conference, 1999, Barcelona Spain, 18-21 Oct. 1999, Vol.2, 1399 -1403.

[4] Park JB, Lee BH, Kim MS: Remote Control of a Mobile Robot Using Distance-Based Force, Proceedings of the 2003 IEEE International Conference on Robotics & Automation, , Taipei, Taiwan (14-19 Sept. 2003), Vol.3, 3415-3418.

[5] Schilling KJ., Vernet MP.: Remotely Controlled Experiments with Mobile Robots, System Theory, 2002. Proceedings of the Thirty-Fourth South Eastern Symposium, (18-19 March 2002) 71-74.

[6] Matsumaru K, Fhjimori A, Kotoku T, Komoriya K: Action Strategy for Remote Operation of Mobile Robot in Human Coexistence Environment, IECON'2000, 26th Annual Conference of the IEEE on Industrial Electronics Society, Nagoya Japan, (22-28 Oct. 2000), Vol.1, 1-6.

[7] Li THS, Chang SJ, Chen YX: Implementation of Human-Like Driving Skills by Autonomous Fuzzy Behaviour Control on An FPGA-Based Car-Like Mobile Robot, IEEE Transactions on Industrial Electronics, Vol.50 (Oct. 2003) 867-880.

[8] Vitabile S, Pilato G, Pullara F, Sorbello F: A Navigation System for Vision-Guided Mobile Robots, International Conference Proceedings on Image Analysis and Processing, Venice Italy (27-29 Sept.1999) 566 -571.

[9] Luo RC, Chen TM: Development of a Multibehaviour-based Mobile Robot for Remote Supervisory Control through the Internet, IEEE/ASME Transaction on Mechatronics, Vol.5 (Dec. 2000) 376-385.

Intelligent Production Machines and Systems
D. T. Pham, E. E. Eldukhri and A. J. Soroka (eds)

Docking of Autonomous Vehicles: A Comparison of Model-Independent Guidance Methods

G. Nejat and B. Benhabib

Computer Integrated Manufacturing Laboratory
Department of Mechanical and Industrial Engineering
University of Toronto, 5 King's College Road, Toronto, Ontario, Canada M5S 3G8

Abstract

In this Paper, two generic Line-of-Sight (LOS) sensing-based short-range guidance methodologies are presented for the docking of autonomous vehicles. The first method utilizes a passive LOS sensing scheme to provide vehicle corrective motions, while the latter method utilizes active sensing. The novelty of the proposed guidance methodologies is their applicability to situations that do not allow for direct proximity measurements of the vehicle. The objective of both proposed guidance methods is, thus, to successfully minimize the systematic errors of the vehicle, while allowing it to converge to its desired pose within random noise limits. Both techniques were successfully tested via simulations and are discussed herein in terms of convergence rate and accuracy, in addition to the types of localization problems that each method should be used in.

1. Introduction

In a typical two-stage autonomous vehicle motion execution, the first stage yields an initial approximate movement toward a desired goal (*long-range positioning*), whereas the second movement (*docking*) is a corrective fine-motion action based on high-precision feedback, [1]. During the latter stage, on-line motion planning would be necessary to achieve the desired (docking) pose (position and orientation) within required tolerances. The use of external non-contact task-space sensors, via passive or active sensing techniques, has been often advocated during this stage, [2]: (i) non-contact proximity sensors for on-line, navigation-based path-planning algorithms [3-5], or (ii) high-speed cameras in visual-servoing mode [6-8]. Furthermore, several early attempts in using active laser tracking systems have also taken place, consisting of retroreflectors placed on the bodies of mobile robots, for direct task-space motion perception [9-12].

All abovementioned methods, however, rely on relative or absolute vehicle-pose measurements. Frequently, even in the presence of task-space sensors, a vehicle's pose cannot be determined accurately due to the inability of the proximity sensors to measure orientation as precisely as position. This drawback may be overcome by utilizing a guidance-based method to *guide* the vehicle to its desired docking pose, within required tolerances, using indirect proximity measurements. We have, in the past, proposed several such generic guidance methodologies that utilize a Line-of-Sight (LOS) based task-space sensing system [2, 13]. These methods, however, require the use of the calibration model of the sensing system in order to guide the vehicle to its desired pose. In practice, though, situations may arise where this sensing-system calibration model may be unavailable to the guidance system, in which case a need for a model-independent guidance method rises.

In this Paper, we address the above localization problem by proposing and evaluating two different generic LOS-based guidance techniques that solely utilize the indirect proximity measurements to determine vehicle-motion commands. The first method utilizes a passive sensing scheme, while the latter method utilizes active sensing [14, 15]. The convergence rate and accuracy of the methods, in addition to the types of localization problems that each should be used in are also discussed.

2. Line-of-Sight Based Task-Space Sensing

External, non-contact LOS sensors can be used effectively to improve a vehicle's docking accuracy without undermining its performance. In our proposed system, a multi-LOS sensing system can be configured using several LOS sensing modules to provide sufficient and accurate sensory data for guidance-based motion planning of vehicles translating and/or rotating freely in multi-dimensional space. One individual module consists of a laser source, a galvanometer mirror, and a detector (e.g., PSD). The detector is mounted on the vehicle, while the laser beam, which defines the desired LOS, is aligned using a galvanometer mirror.

Both proposed methods incorporate LOS-based sensing with successive corrective motion actions that the vehicle needs to undertake in order to achieve its desired pose. The LOSs are aligned to hit the centers of targets (e.g., detectors) placed on the vehicle when it is at its desired pose, Figure 1a. However, in practice, after the vehicle is positioned through long-range positioning, the LOSs would actually be hitting their targets with offsets from their centers, due to systematic and random errors in the system, Figure 1b. These offsets are used to generate corrective actions to guide the vehicle from its actual pose to its desired pose via one of the two proposed methods.

3. Model-Independent Guidance Methodologies

The fundamental nature of the proposed generic guidance methods is that both *solely* use the detector offsets to generate motion commands of the vehicle and do not require the LOS sensing-system calibration model. Namely, the LOSs are pre-taught the desired angles, $\boldsymbol{\phi}_{desired}$, corresponding to the vehicle's desired pose by a "teaching by demonstration" method. Both methods are presented herein for the most generalized 6-dof vehicle-docking problem, in which the vehicle's shape is assumed to be a cube, with array type detectors on at least three of its orthogonal faces, Figure 1.

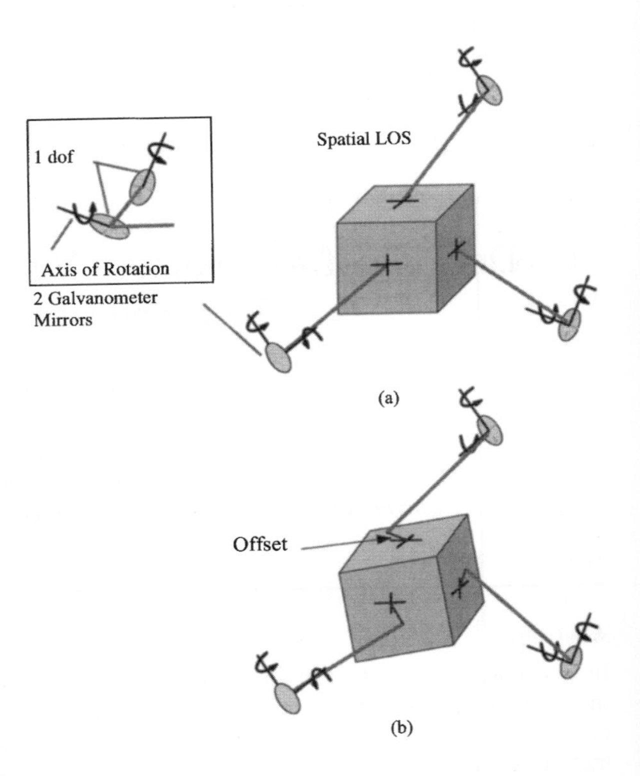

Figure 1: (a) Desired (theoretical) vehicle location and (b) actual vehicle relocation, respectively.

3.1 Guidance Based on Passive Sensing

While the vehicle is commanded to move to its desired pose ${}^{w}\mathbf{T}_{dv}$ $(x_d, y_d, z_d, \gamma_d, \beta_d, \alpha_d)$ with respect to the world coordinate frame, F_w, (i.e. long-range positioning), the galvanometer mirrors are in the process of aligning the three LOSs to hit the centers of the detectors. However, due to both systematic errors and random noise in vehicle motion, the vehicle only achieves an actual pose defined by ${}^{w}\mathbf{T}_{av}$ $(x_a, y_a, z_a, \gamma_a, \beta_a, \alpha_a)$. Offsets are measured in each detector's frame, F_{di}, and are used as feedback for the guidance of the vehicle to its desired pose via the *gradient-descent* method described below. It is important to note that once the LOSs are locked to the centers of the detectors at the vehicle's desired pose, they remain locked at this pose during the implementation of the guidance method.

Utilizing the detector offsets, the gradient-descent method determines the necessary incremental motion in all degrees of freedom of the vehicle, i.e., $(\Delta x, \Delta y, \Delta z, \Delta\gamma, \Delta\beta, \Delta\alpha)$. The gradient-descent method addresses rotational and translational error in separate sub-steps.

Rotation: The rotational motion commands are determined by utilizing all three PSD offsets:

PSD1: As can be noted in Figure 2, F_{d1} is perpendicular to the x-axis of the vehicle's (actual pose) frame, $^{a}F_{c}$, and, hence, only rotational errors about the other two axes (β, α) can be detected here. Utilizing the relationship $sin(\theta) \cong \theta$ for small angles in radians:

$$\beta_1 = \frac{e_{1z}}{r_{1z}} \tag{1}$$

$$\alpha_1 = \frac{e_{1y}}{r_{1y}} \tag{2}$$

where e_{1y} and e_{1z} are the y- and z-projections of e_1 on F_{d1}, respectively, and r_{1y} and r_{1z} are the distances from the projections of the measured detector offsets to the center of the vehicle at its actual pose, $^{a}F_{c}$, which can further be defined as:

$$\begin{aligned} r_{1x} &= h_1 \\ r_{1y} &= \sqrt{e_{1y}^2 + h_1^2} \\ r_{1z} &= \sqrt{e_{1z}^2 + h_1^2} \end{aligned} \tag{3}$$

where h_1 is the distance from the center of $^{a}F_{c}$ to the center of F_{d1}.

PSD2: As can be noted in Figure 2, F_{d2} is perpendicular to the y-axis of the vehicle's (actual-pose) frame, $^{a}F_{c}$, and, hence, only rotational errors about the other two axes (γ, α) can be detected here:

$$\gamma_2 = \frac{e_{2z}}{r_{2z}} \tag{4}$$

$$\alpha_2 = \frac{e_{2x}}{r_{2x}} \tag{5}$$

where

$$\begin{aligned} r_{2x} &= \sqrt{e_{2x}^2 + h_2^2} \\ r_{2y} &= h_2 \\ r_{2z} &= \sqrt{e_{2z}^2 + h_2^2} \end{aligned} \tag{6}$$

PSD3: As can be noted in Figure 2, F_{d3} is perpendicular to the z-axis of the vehicle's (actual-pose) frame, $^{a}F_{c}$, and hence, only rotational errors about the other two axes (γ, β) can be detected here:

$$\gamma_3 = \frac{e_{3y}}{r_{3y}} \tag{7}$$

$$\beta_3 = \frac{e_{3x}}{r_{3x}} \tag{8}$$

where,

$$\begin{aligned} r_{3x} &= \sqrt{e_{3x}^2 + h_3^2} \\ r_{3y} &= \sqrt{e_{3y}^2 + h_3^2} \\ r_{3z} &= h_3 \end{aligned} \tag{9}$$

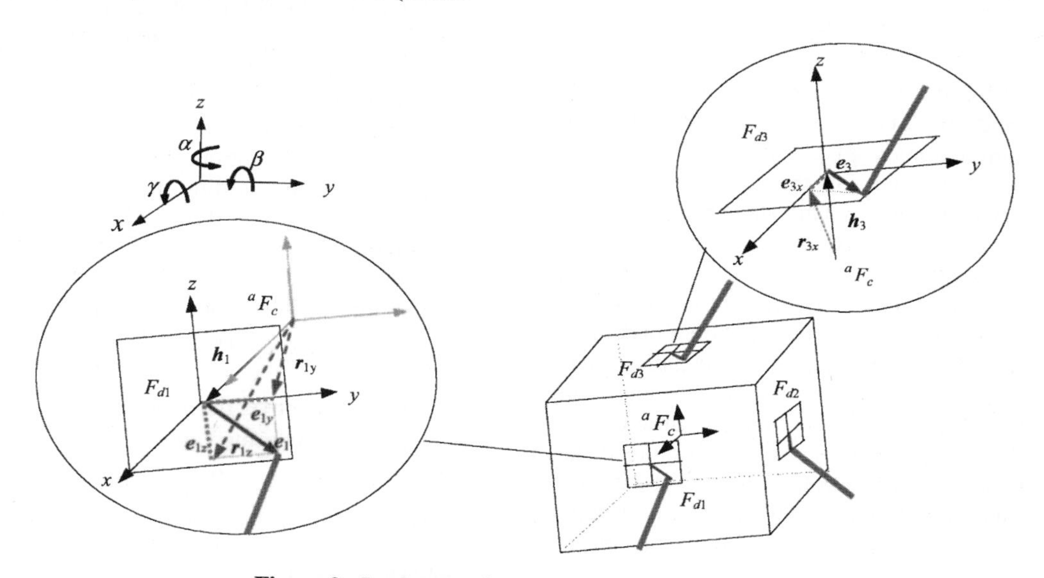

Figure 2: Implementation of Gradient-Descent Method.

The above rotational angles are averaged to determine the overall rotational movement required in (γ, β, α):

$$\Delta\gamma = (\gamma_2 + \gamma_3)/2 \quad (10)$$
$$\Delta\beta = (\beta_1 + \beta_3)/2 \quad (11)$$
$$\Delta\alpha = (\alpha_1 + \alpha_2)/2 \quad (12)$$

Translation: The translation motion commands are determined by minimizing the following cost function:

$$transerror = e_1^2 + e_2^2 + e_3^2, \quad (13)$$

where

$$\Delta x = 2(e_{2x} + e_{3x}) \quad (14)$$
$$\Delta y = 2(e_{1y} + e_{3y}) \quad (15)$$
$$\Delta z = 2(e_{1z} + e_{2z}) \quad (16)$$

In order to prevent excessive overshooting in the vehicle's motion and improve rate of convergence, the vehicle is commanded to move by amounts defined in Equations (10 to 12), and (14 to 16) based on a weighting scheme determined with respect to the overall vehicle motion. This procedure of localization is iterative in nature and stops when the vehicle is deemed to have converged to its desired pose, within acceptable tolerance levels.

The passive-sensing based guidance method is illustrated in Figure 3.

3.2 Guidance Based on Active Sensing

Stage 1: After the initial long-range positioning of the vehicle and the LOSs have been rotated to their corresponding angles, $\phi_{desired}$, at the vehicle's desired pose, the first stage of the guidance method is implemented, Figure 4. This stage requires that the sensing system determine the scanner angles corresponding to the vehicle's current (actual) location, ϕ_{actual}, by *active LOS-scanning,* via *an optimization process*. A search method (the Flexible Polyhedron in our work [16]) is used to determine the individual LOS orientations corresponding to the current vehicle location (within noise limits). Namely, an active search is carried out that minimizes the absolute distance between the point the LOS hits the PSD and its center. Each of the three LOSs is defined by two orthogonal angles, $\phi_{ijactual}$ (i = 1 to 3 and j = 1 to 2). Scanner

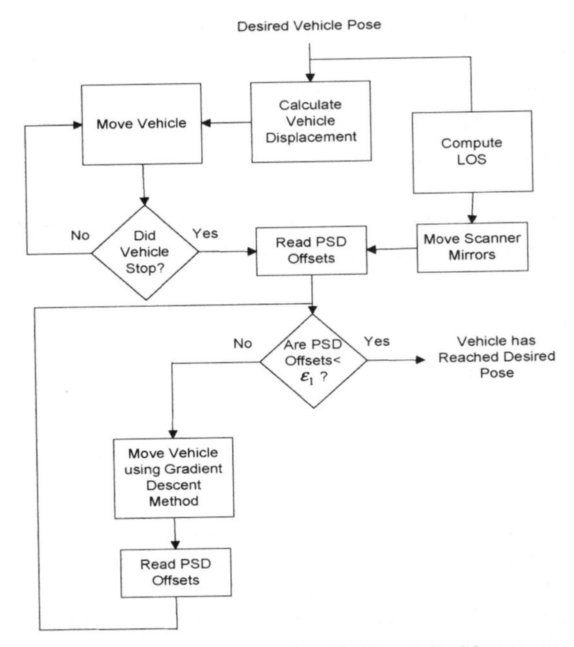

Figure 3: Passive-Sensing Guidance Architecture.

realignment is carried out independently and concurrently for each LOS.

Stage 2: The actual LOS angles, ϕ_{actual}, determined in Stage 1 are used along with the desired angles, $\phi_{desired}$, to determine a continuous guidance trajectory for the active control of the LOSs such that they effectively and efficiently guide the vehicle to its desired pose from its current actual pose, Figure 4. This trajectory is simply a collection of intermediate scanner angles, from ϕ_{actual} to $\phi_{desired}$. In order to minimize the systematic errors during the vehicle motion, the number of steps that the vehicle must take along this path can be optimized in terms of the smallest allowable vehicle motion. Herein, the number of steps is determined by dividing the initial detector offsets measured along the three PSDs (after long-range positioning) by twice the repeatability of the vehicle and, then, setting the number of steps to equal the maximum value determined among the three detectors, i.e., $\#\_Steps = \max\left({PSD\ offset}/{2 \times Vehicle\_repeatability} \right)$.

As noted above, the guidance trajectory is simply a collection of intermediate scanner angles, from ϕ_{actual} to $\phi_{desired}$, i.e., $\phi_{ijstepk} = \phi_{ijstepk-1} + \Delta\phi_{ij}$, where k = 1 to *#_Steps*, and $\Delta\phi_{ij} = {(\Delta\phi_{ijdesired} - \Delta\phi_{ijactual})}/{\#\_Steps}$. This stage is purely a calculation stage, neither the scanners nor the vehicle are physically moved.

Stage 3: The LOSs are used to guide the vehicle via the trajectory until it reaches its desired pose, Figure 4. First, the scanner angles (i.e., the LOSs) are oriented to the values corresponding to *Step 1* above (Stage 1), i.e., $\phi_{ijstep1} = \phi_{ijstep0} + \Delta\phi_{ij}$, and the PSD offsets, $\boldsymbol{e}_i$, are measured. These offsets are measured in each detector's frame, F_{di}, and used to guide the vehicle to its corresponding pose at this particular step via the described gradient-descent method, Equations (10 to 16). This procedure is then repeated for all subsequent Steps. The successive steps diminish the magnitude of the systematic errors in a continuous mode.

4. Simulations

4.1 Set-up and Procedure

The 6-dof-mobility vehicle's shape is a 0.1×0.1×0.1 m cube. The three spatial LOS sources are placed symmetrically around the perimeter of the (docking) workspace of the vehicle. The inaccuracy of the motion of the vehicle is represented twofold: systematic errors and random errors:

Systematic error = (inaccuracy/full_range) × Displacement_of_vehicle, (17)

where *(inaccuracy/full_range)* was chosen as 0.2 mm/m for translation and as 0.3 milli-deg/deg for rotation. Random noise was represented by a normal distribution; $N(\mu = 0.0, \sigma = 0.125\mu\text{m})$ for translation and $N(\mu = 0.0, \sigma = 0.8\,\mu\text{deg})$ for orientation.

In all the simulation tests, at the vehicle's *home* pose, the center of the vehicle was positioned at: $x =$ 0.0 mm, $y =$ 0.0 mm, $z =$ 0.0 mm, $\gamma = 0.0^\circ$, $\beta = 0.0^\circ$, and $\alpha = 0.0^\circ$ defined in the world frame, F_w, coordinates. In the specific simulation test presented herein, the vehicle was commanded to move from $x =$ 60 mm, $y =$ 60 mm, $z =$ 60 mm, $\gamma = 30.0^\circ$, $\beta = 30.0^\circ$, and $\alpha = 30.0^\circ$ to home. The vehicle's actual pose, after the initial (uncorrected) motion was $x = 8.1\ \mu\text{m}$, $y = 9.0\ \mu\text{m}$, $z = 8.8\ \mu\text{m}$, $\gamma = 8.5$ milli-degrees, $\beta = 8.4$ milli-degrees, and $\alpha = 8.4$ milli-degrees.

4.2 Examples

Figure 5 shows the offsets measured along each of the three PSDs versus the number steps (i.e. iterations) of the passive-sensing algorithm. The simulation results for the case of guidance using active sensing are shown in Figure 6. By examining the results, it can be deduced that the active guidance method has a tangibly faster convergence rate than the passive method.

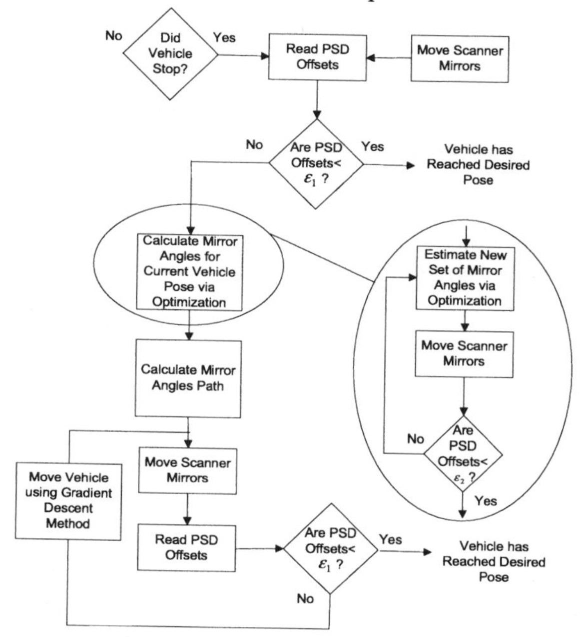

Figure 4: Active-Sensing Guidance Architecture.

5. Discussion and Conclusions

Two generic multi-LOS guidance-based methods are proposed for the localization of autonomous vehicles and robotic end-effectors, using passive sensing and active sensing, respectively. The methods are both invoked in the final stages of motion in order to accurately and effectively minimize the systematic errors of the vehicle, and allow it to converge to its desired pose within random noise limits. In terms of convergence rate and accuracy, the latter of the two methods is more desirable: It requires less vehicle movements, hence minimizing the addition of motion errors into the system, since every time the vehicle moves these errors are introduced. In addition, overshoots and jerky motions are, typically, not encountered due to the relatively small vehicle-motion corrections. The active localization algorithm, however, requires control over the scanner angles, e.g. LOSs. It is, thus, advisable that in situations where this type of control is not available to the guidance algorithm, i.e., where the LOSs are locked and cannot be moved, to use the passive guidance method instead. This method solely places emphasis on the magnitude and direction of the PSD offset measurements to determine vehicle motion.

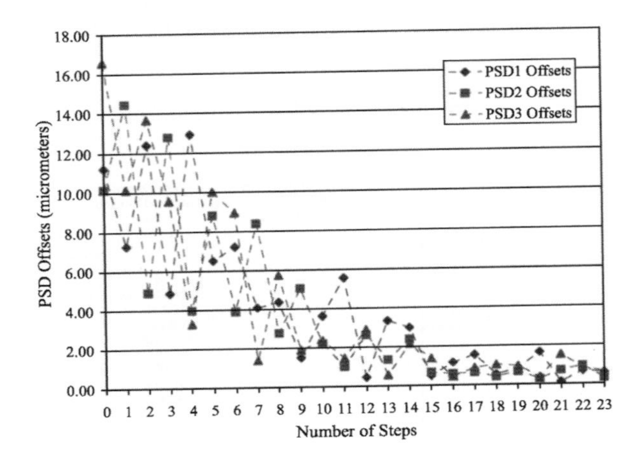

Figure 5: PSD Offsets for Passive-Sensing Method.

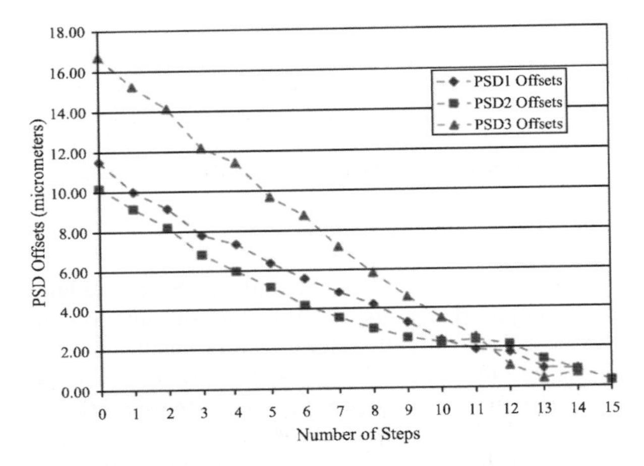

Figure 6: PSD Offsets for Active-Sensing Method.

Acknowledgements

This work has been supported by the Natural Sciences and Engineering Research Council of Canada (NSERC).

References

[1] R. Waarsing, M. Nuttin, and H. Van Brussel, "*Introducing Robots Into A Human-Centred Environment: The Behaviour-Based Approach*", International Conference on Climbing and Walking Robots; From Biology to Industrial Applications, Karlsruhe, Germany, pp. 465-470, 2001.

[2] G. Nejat, I. Heerah and B. Benhabib, "*Line-of-Sight Task-Space Sensing for the Localization of Autonomous Mobile Devices*," IEEE, Int. Conference on Intelligent Robots and Systems, Las Vegas, NV, pp. 968-973, 2003.

[3] H. Roth, and K. Schilling, "*Navigation and Docking Manoeuvres of Mobile Robots in Industrial Environments*," IEEE, Conference of Industrial Electronics Society, Aachen, Germany, pp. 2458-2462, 1998.

[4] P. Mira, R. Ferreira, V. Grossmann, and M.I. Ribeiro, "*Docking of A Mobile Platform Based on Infrared Sensors*," IEEE, Int. Symp. on Industrial Electronics, Guimaraes, Portugal, pp. 735-740, 1997.

[5] J. Vaganay, P. Baccou, and B. Jouvencel, "*Homing by Acoustic Ranging to a Single Beacon*," MTS/IEEE, Oceans Conference, Providence, RI, pp. 1457-1462, 2000.

[6] P. Roessler, S.A. Stoeter, P.E. Rybski, M. Gini, and N. Pananikolopoulos, "*Visual Servoing of A Miniature Robot Toward A Marked Target*," IEEE, International Conference on Digital Signal Processing, Santorini, Greece, pp. 1015-1018, 2002.

[7] C.J. Taylor, and J.P. Ostrowski, "*Robust Visual Servoing Based on Relative Orientation*," IEEE, Computer Vision and Pattern Recognition, Fort Collins, CO, pp. 574-580, 1999.

[8] D. Kragic, and H. Christensen, "*Cue Integration for Visual Servoing*," IEEE Trans. on Robotics and Automation, Vol. 17, No. 1, pp. 18-27, 2001.

[9] T.A.G. Heeren, and F.E. Veldpaus, "*An Optical System to Measure the End-Effector Position for On-line Control Purposes*," The International Journal of Robotics Research, Vol. 11, No. 1, pp. 53-63, 1992.

[10] J.R.R. Mayer, and G.A., "*A Portable Instrument for 3-D Dynamic Robot Masurements Using Triangulation and Laser Tracking*," IEEE Trans. on Robotics and Automation, Vol. 10, No. 4, pp. 504-516, 1994.

[11] Y. Koseki, T. Arai, K. Sugimoto, T. Takatuji, and M. Goto, "*Design and Accuracy Evaluation of High-Speed and High Precision Parallel Mechanism*," IEEE, Int. Conference on Robotics and Automation, Leuven, Belguim, pp. 1340-1345, 1998.

[12] C.J. Leigh-Lancaster, B. Shirinzadeh and Y.L. Koh, "*Development of a Laser Tracking System*," IEEE, Conference on Mechatronics and Machine Vision in Practice, Toowoomba, Australia, pp. 163-168, 1997.

[13] G. Nejat and B. Benhabib, "*High-Precision Task-Space Sensing and Guidance for Autonomous Robot Localization*," IEEE, Int. Conference on Robotics and Automation, Taipei, Taiwan, pp. 1527-1532, 2003.

[14] A. Bonen, R.E. Saad, K.C. Smith, and B. Benhabib, "*A Novel Electrooptical Proximity Sensor for Robotics: Calibration and Active Sensing*," IEEE Trans. on Robotics and Automation, Vol. 13, No. 3, pp. 377-386, 1997.

[15] P. Mowforth, "*Active Sensing For Mobile Robots*," IEEE International Conference on Control, Edinburgh, UK, pp. 1141-1146, 1991.

[16]J.A. Nelder and R. Mead, "*A Simplex Method for Function Minimization*," Computer Journal, Vol. 7, pp. 308-313, 1964.

Intelligent Production Machines and Systems
D. T. Pham, E. E. Eldukhri and A. J. Soroka (eds)

Simulation and Optimisation of a Tendon-Based Stewart Platform

Manfred Kraft[a], Erdmann Schäper[b]

[a]*Fraunhofer IPK, Pascalstr. 8-9, 10715 Berlin, Germany*
[b]*Technische Universität Berlin, IWF, Pascalstr. 8-9, 10715 Berlin, Germany*

Abstract

This paper deals with a special type of Stewart Platform. The usual rigid legs are replaced by tendons. By this means, larger scalability and a lighter-weight kinematic structure is made possible. The following disadvantages face this advantages: tendons can transfer only pulling forces and in general a redundant kinematic structure is required. In this paper a dedicated kinematic model is presented, as well as a tool for computing the controllable workspace. Various approaches are described, which yield a better relation of workspace to package space.

1. Introduction

More than 100 possible architectures have been proposed for parallel structures with two to six degrees of freedom (DOF). In the application for machine tools, very well known parallel structures are Gough-Stewart platforms, in which three types are most general, as shown in Fig. 1 [1, 2]. Fig. 1a illustrates the basic Stewart platform, called SSM (Simplified Symmetric Manipulator). The platform is connected to the base by six extensible legs that have distinct joints at both ends. Fig. 1b describes the TSSM (Triangular SSM) which has six joints on the base and only three on the platform. The structure of MSSM (Minimal SSM) is based of only three joints at both ends (Fig. 1c).

With a MSSM as an example, given the structure and constructional parameters, this paper will investigate the pose optimization of the moving platform of the MSSM in order to shorten the strut lengths, so that the workspace of the platform and the dynamic behavior will be improved.

For parallel kinematic machines (PKM) based on

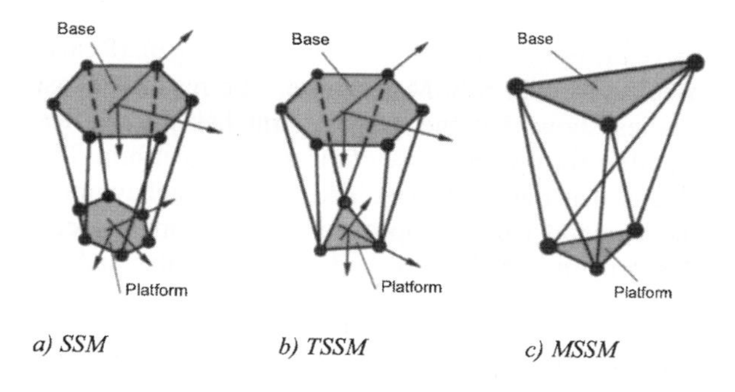

a) SSM *b) TSSM* *c) MSSM*

Fig. 1: Gough-Stewart platforms [3]

Stewart-platform with six DOF, since the spindle mounted on the platform is a rotational component, only five DOF are required to move the platform in the workspace. This means that one of the six DOF, namely the roll angle of the platform, is redundant for the spindle [4]. The use of redundant DOF allow to find an optimal pose of the platform, when the cutter location is given, which is resulted from NC programming.

The kinematic analysis of Stewart parallel manipulators (SPM) induces two main problems: the direct position analysis (DPA) and the inverse position analysis (IPA) of moving platform. The first problem can be solved by finding the pose, including position (x, y, z) and orientation (α, β, θ), of the platform when a set of strut lengths, i.e., a set of actuator displacements, is given. In the orientation (α, β, θ), the angles α and β denote the rotation of the platform about the x and y axes respectively, and θ denotes the roll angle of the platform [5, 6, 7]. The second problem, on the contrary, requires the strut lengths when the pose of the platform with respect to the base is given. Both problems are of practical interest for the control of the platform pose.

The DPA involves non-linear equations and admits many solutions, while the IPA has a unique and easy-to-be-found solution. For a SPM with non-geometric singularity, the platform poses corresponding to the DPA solutions in the workspace are not continuous. This property makes it possible to control the strut lengths of the SPM while moving the platform along a continuous trajectory [12, 13].

This means that only five degrees of freedom are required to move the platform in the workspace of hexapods with six degrees of freedom, since the spindle mounted on the moving platform is a rotational component and may be constrained for an optimal purpose.

The aim was to find an analytical algorithm to solve the optimal roll angle θ of the moving platform of a hexapod with MSSM structure. The results of case studies show that the longest strut has the shortest length in the optimal pose of the moving platform. This means that the workspace and dynamic behaviors of the hexapod will be improved in the optimal pose, if the structure of the hexapod is not singular and the structural parameters are determined.

1.1 The Challenge of a Tendon-based Stewart Platform

The main disadvantage of tendons, as compared to rigid legs, is the incapability to absorb compressive forces. Therefore, at least $n+1$ tendons are required for a n degree of freedom manipulator. In [8] the term 'Completely Restrained Positioning Mechanism' (CRPM) is used for $m = n + 1$ tendons and the term 'Redundantly Restrained Positioning Mechanism' (RRPM) is used for $m > n + 1$ tendons.

In this paper, we investigate kinematic structures

Fig. 2: Prototype of a 8 tendon kinematic structure

of the type RRPM. The RRPM algorithms represent a generalised approach and can easily adopted to the CRPM type.

A first prototype[1] of an 8 tendon structure (see Figure 2) is used to verify the results of simulation and optimisation.

The working prototype consists of a base frame with 8 linear drives, mechanical fasteners and guide rollers for the tendons. In order to keep the mechanical design simple, an irregular layout of the guide rollers was chosen. The platform in the centre is cuboid with mechanical fasteners in its 8 corners. A cuboid base frame and a 9-tendon-based structure have also been investigated for simulation and optimisation.

The aim of this paper is to explore the special issues associated with tendon-specific features. For general issues of parallel kinematic structures see [9].

[1] The mechanical design and realisation of the prototype was done by Technische Universität Berlin, Institut für Werkzeugmaschinen und Fabrikbetrieb, Fachgebiet Werkzeugmaschinen und Fertigungstechnik

2. Simulation of controllable workspace

2.1 The kinematic model

2.1.1. Assumptions

The following assumptions are postulated:

- Basis and platform are rigid bodies.
- Tendons are lines between platform and base.
- Mechanical fasteners are ideal ball joints.
- Roll-Pitch-Yaw angels describe the orientation of the platform.
- The drives yield a given maximum torque.
- A given minimum torque is necessary to tighten the tendons.

2.1.2. Force-Torque-Equilibrium

The following equations describe the static equilibrium for a given pose:

$$\begin{bmatrix} u_{B1} & \cdots & u_{Bm} \\ \hline p_{B1} \times u_{B1} & \cdots & p_{Bm} \times u_{Bm} \end{bmatrix} \begin{bmatrix} f_1 \\ \cdots \\ f_m \end{bmatrix} = -\begin{bmatrix} f_{ext} \\ m_{ext} \end{bmatrix}$$

$$[6 \times m] \qquad [m \times 1] = -[6 \times 1] \qquad (1)$$

with:

$u_{Bi} = f(pose, platform, base)$ (2)

$p_{Bi} = f(pose, platform)$ (3)

$fmin_i <= f_i <= fmax_i$ (4)

where:

m number of tendons

u_{Bi} unit vectors of tendons (base - platform) in base-coordinate-system

p_{Bi} direction vector (platform origin – platform corner) in base-coordinate-system

f_{ext} external force (gravitation, process forces, dynamic forces)

m_{ext} external torque (process torques, dynamic torques)

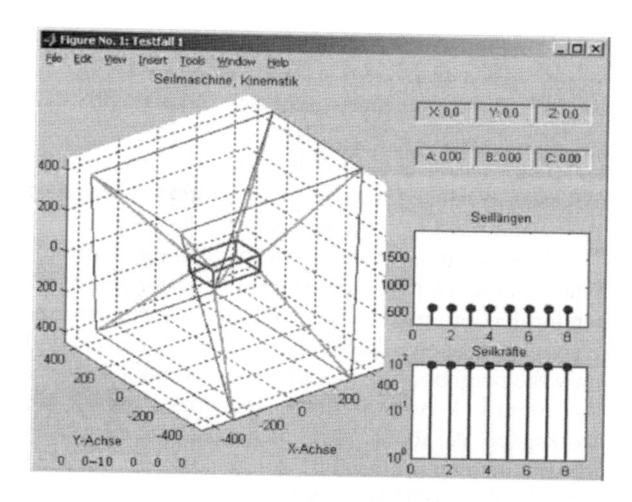

Fig 3: Tendon lengths and forces in dependency of platform pose and external forces / torques

2.2. The simulation tool

Because of the implicit non-linear dependencies of u and p on the current pose, it is not possible to give a closed form of the solution of equation (1) to (4) in terms of the current pose.

For CRPM and RRPM types, $m > 6$, for which the general solution is a convex subset of $(m - 6)$-dimensional vector space. This allows an optimal selection of one of the feasible solutions.

Different additional objective functions have been applied to find an optimal solution:

- Minimum sum of tendon forces, in order to optimise the energy consumption of the system
- Maximum sum of tendon forces, in order to optimise stiffness of the system.

Figure 3 shows the graphical interface of the simulation tool, which computes the inverse coordinate transformation, solves the force-torque-equilibrium, checks the auxiliary collision conditions (see 2.3) and looks for an optimal solution for tendon forces.

2.3. Controllable workspace of the prototype

The controllable workspace is the set of all poses for which the force-torque-equilibrium has a feasible solution and that are reachable without collision from a given zero position. The zero position is defined in the centre of the base frame.

Collisions are possible between:

- Base frame legs – Tendons
- Platform frame – Tendons
- Tendons – Tendons

The Base-Tendon collisions are not locally assessed, because they can be excluded by design of the base frame given the workspace is known.

There therefore remain 3x8 collision conditions for platform-tendon pairs, and 8x8 collision conditions for tendon-tendon pairs. 3x8 conditions for platform-tendon pairs are sufficient as the cuboid is convex and only the neighbouring faces need be examined.

Figure 4 and 5 show, that the controllable workspace is about 22% of package space. The maximum displacement of A- and B-axis (rotation

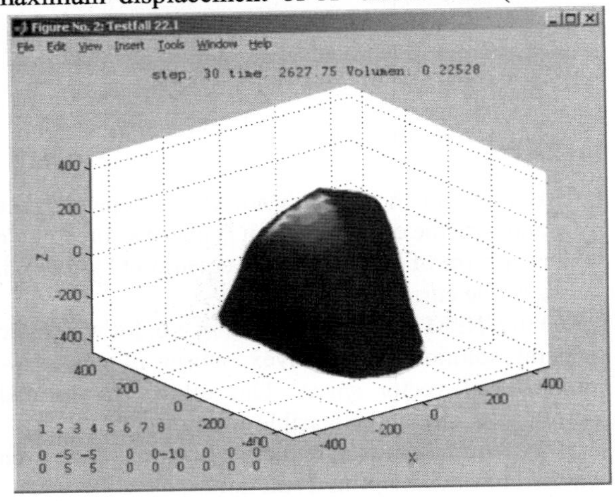

Fig. 4: Controllable workspace of prototype (any orientation)

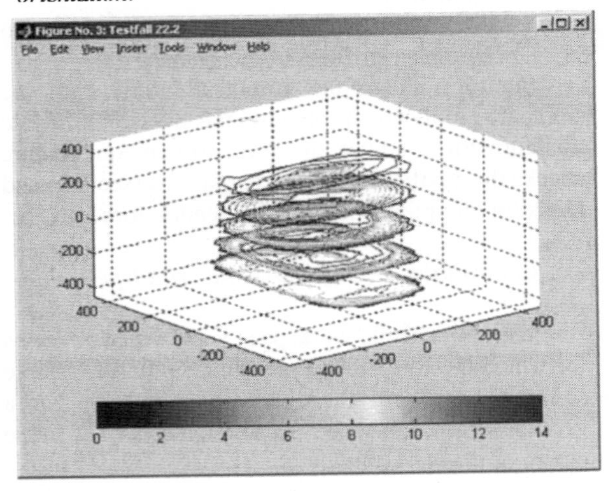

Fig. 5: Controllable workspace of prototype with possible displacement of A- and B-axis

around X- and Y-axis) is 14 degrees. The 14 degrees are reached only near the zero point, in the outer areas of workspace the possible displacement is obvious smaller.

The relatively small workspace and especially the low orientation displacement is the motive for investigation of several optimization approaches in the next chapter.

3. Optimisation

3.1. Reduction of mechanical fasteners of the platform

A design principle of most hexapods is the usage of twofold joints at the platform. The transfer of this principle to a 8-tendon platform is shown in figure 6.

The results of this approach are shown in Figure 7. The workspace now occupies about 26% of package space. But the most obvious change is the remarkable extension of rotational displacement.

The maximum displacement of the A and B-axes is now 45 degrees. In the main part of the workspace, the reachable displacement is 15 degrees and more.

These results confirm the assumption in [8] that the reduction of joint points by combination of mechanical fasteners substantially extends the controllable workspace.

In order to apply these findings to the prototype, the machine tool department of IWF has developed a new type of twofold universal joint with intersecting axis.

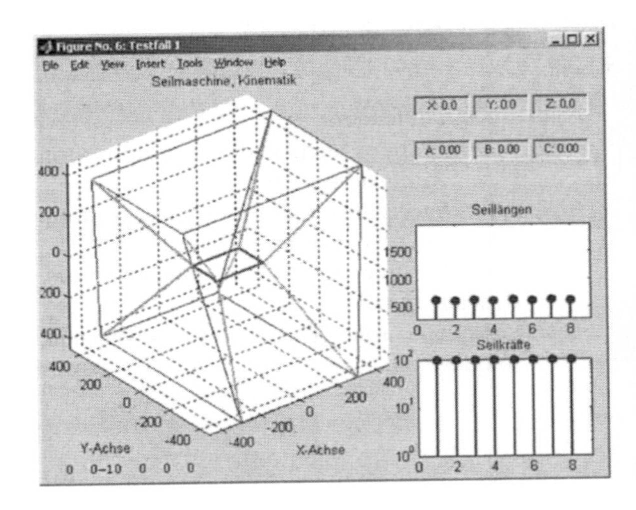

Fig. 6: Variation of the structure with twofold universal joints at the platform

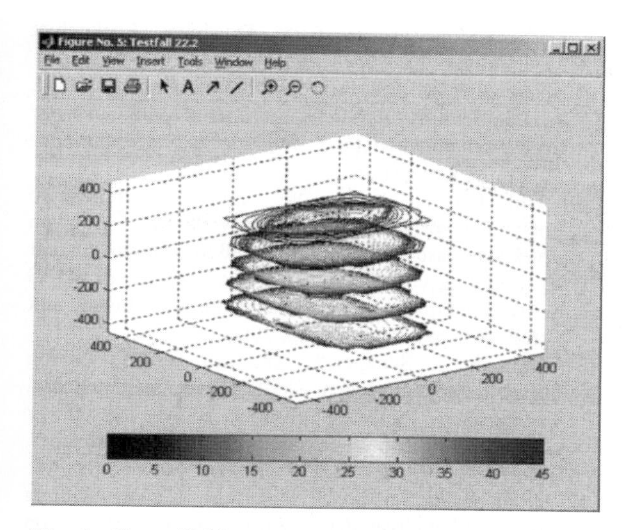

Fig. 7: Controllable workspace of a variant of prototype with possible displacement of A- and B-axis

3.2. *Tetraeder structure*

The basic idea of the tetraeder structure that one side of the frame is wide open, making it well suited for processing purposes. On the other hand, the base frame and platform should be kept as simple as possible. The outcome (see figure 8) of these design principles is a tetraeder frame and a equilateral triangle platform.

The wiring is an application of the results in chapter 3.1. There are twofold tendon guides on the 3 left-hand corners of the frame and a triple tendon guide on the right-hand side of the frame. The universal joints on the platform are all threefold.

Figure 9 shows the results of the tetraeder approach. This design yields a larger workspace and a

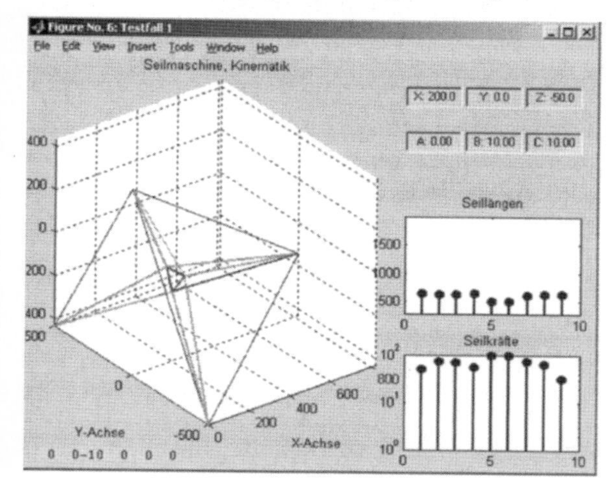

Fig. 8: A 9-tendon structure with a tetraeder base frame

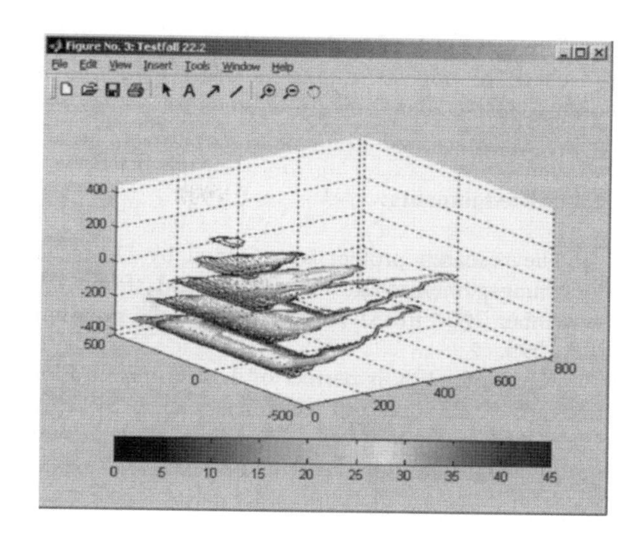

Fig. 9: Controllable workspace of the tetraeder structure

remarkably improved ability to make rotational displacements.

The rotational ability is limited only near the edges and the maximum displacement is larger than 45 degrees in the A and B-axes.

The controllable workspace is about 31% of package space.

Conclusions / Further Work

The results of simulation of the controllable workspace have been verified using the prototype (see Figure 2) of a tendon-based kinematic. In addition to the kinematic structure, it was necessary to design and implement a control system. Two basic tasks have been solved:

- Combined Position-Force-Control
- Self Calibration of the system

Both basic solutions require further improvement.

One of the possible strategies for the combined Position-Force-Control is intelligent decision making, which optimizes the real-time axis assignment to position control or to force control. This requires a model-based on-line estimation of the influence of the type of control on the overall accuracy and robustness of the coupled kinematic system.

The Self Calibration algorithm computes the zero points of the 8 axis quite accurately but requires about 15 minutes computing time. An intelligent choice of measurement positions based on learning techniques would reduce the required time significantly.

The main task of further work, however, will be the usage of process models in order to design a comprehensive adaptronic process control.

Acknowledgements

The research project was funded by Deutsche Forschungsgemeinschaft in the period from 01 September 2001 till 31 August 2002 in the program SPP 1099.

Fraunhofer IPK is partner of the EU-funded FP6 Innovative Production Machines and Systems (I*PROMS) Network of Excellence. http://www.iproms.org

References

[1] Merlet, J.-P.: The importance of optimal design for parallel structures. In: Boer, C. R.; Molinari-Tosatti, L.; Smith, K. S. (Eds): Parallel Kinematic Machines, Theoretical Aspects and Industrial Requirements. Springer-Verlag, London 1999, pp. 99-110.

[2] Negri, S.; Di Bernardo, G.; Fassi, I.; Molinari-Tosatti, L.; Bianchi, G.; Boer, C. R.: Kinematic analysis of parallel manipulators. In: Boer, C. R.; Molinari-Tosatti, L.; Smith, K. S. (Eds): Parallel Kinematic Machines, Theoretical Aspects and Industrial Requirements. Springer-Verlag, London 1999, pp. 69-84.

[3] Merlet, J.-P.: Efficient design of parallel robots. In: Neue Maschinenkonzepte mit parallelen Strukturen für Handhabung und Produktion. VDI-Bericht, Vol. 1427 (1998), pp. 1-13.

[4] Koren, Y.: Will PKM be adopt by industry? In: Boer, C. R.; Molinari-Tosatti, L.; Smith, K. S. (Eds): Parallel Kinematic Machines, Theoretical Aspects and Industrial Requirements. Springer-Verlag, London 1999, pp. 271-273.

[5] Castelli, V.–P.: Classification and Kinematic modeling of fully-parallel manipulators – A review. In: Boer, C. R.; Molinari-Tosatti, L.; Smith, K. S. (Eds): Parallel Kinematic Machines, Theoretical Aspects and Industrial Requirements. Springer-Verlag, London 1999, pp. 51-67

[6] Ma, O.; Angeles, J.: Architecture singularities of parallel manipulators. Int J Robotics and Automation, 7(1992)1, pp. 23-29.

[7] Wieland, F.: Entwicklungsplattform für Parallelkinematiken und Prototyp einer Werkzeugmaschine. Dissertation Universität Chemnitz 1999.

[8] Mielczarek, S.; Verhoeven, R; Hiller, M. "Seilgetriebene Stewart-Plattformen in Theorie und Anwendung (SEGESTA)", Zwischenbericht für den Zeitraum vom 1.1.1999 bis 1.8.2000 im DFG-Vorhaben HI 370/18-1, Gerhard-Mercator-Universität Duisburg, Fachgebiet Mechatronik.

[9] Neugebauer, R.; Harzbecker, C.; Drossel, W.-G.; Stoll, A.; Ihlenfeldt, S. "Parallel Kinematics in Manufacturing" in: Development Methods and Application Experience of Parallel Kinematics, Chemnitz, 2002.

[10] Uhlmann, E., Adam, W. "Zwischenbericht zum Forschungsvorhaben 'Entwicklung von Werkzeugmaschinen mit Parallelkinematik unter Verwendung von Seilantrieben' im Rahmen des DFG Schwerpunktprogrammes ‚Fertigungsmaschinen mit Parallelkinematiken", Berlin, 2002.

[11] Uhlmann, E.; Schäper, E.; Neumann, C.: Entwicklung einer hochdynamischen skalierbaren Seilkinematik, Tagungsband zum 3. Dresdner WZM-Fachseminar „Chancen für Parallekinematiken einfacher Bauart", 2001.

[12] Uhlmann, E.; Schäper, E.; Neumann, C.: Verwendung eines Freiheitsgrades zur Optimierung von Hexapodkinematiken, Fortschritt-Berichte VDI, Reihe 2, Nr. 620, S. 67 – 72, 2002.

[13] Uhlmann, E.; Liu, X.-W.; Schäper, E.; Neumann, C.: Pose Optimization of Parallel Kinematic Machines, Production Engineering Vol. IX/1, p. 113 – 116, 2002.

Intelligent Production Machines and Systems
D. T. Pham, E. E. Eldukhri and A. J. Soroka (eds)

Visual servoing of a robotic manipulator based on fuzzy logic control for handling fabric lying on a table

P.Th. Zacharia[a], I.G.Mariolis[b], N.A. Aspragathos[a], E.S. Dermatas[b]

[a] *Department of Mechanical & Aeronautics Engineering, Rion, Patras, Greece*
[b] *Department of Electrical & Computer Engineering, Rion, Patras, Greece*

Abstract

This paper introduces a visual servoed manipulator controller based on fuzzy logic to guide a fabric towards a sewing machine. The task of the end-effector is to handle a randomly located fabric on a table and feed it to the sewing needle along the desired seam. The proposed fuzzy controller determines the linear and angular velocity of the end-effector taking into account the current position and orientation of the fabric which derive from the vision system. The fuzzy rules are derived after studying the behavior of human workers during sewing and the membership functions are formed after simulation and extended experimentation. The experimental results demonstrate the efficiency of the system as well as the robustness of the controller performance.

1. Introduction

In apparel manufacturing the need for advanced automation requires that industrial robots should be capable of handling flexible materials with high efficiency. The up-to-date commercially available fabric handling systems are typically semi-automatic units, since they demand to a great extent the human intervention for performing tasks successfully.

On the other hand, fabric assembly involves complex operations in which human vision and manual dexterity are essential. Therefore, current industrial manipulators should be capable of manipulating flexible materials, since the application of flexible automation to the textile industry is extremely beneficial.

In contrast with the rigid materials, the highly flexible materials present additional problems since they distort and change their shape even under small-imposed forces. The difficulties related to handling flexible materials arise from the unpredictable, non-linear and complex mechanical behavior of flexible materials.

An automated machine system requires flexibility in order to handle various types of fabric and perform 'sensitive' operations such as sewing, where the fabric must be held taut and unwrinkled in order to obtain high seam quality.

Only a few researches [1,2,3,4,5] have worked on the automatic feeding of fabrics into the sewing machine. E. Torgerson et al. [3] introduced a method for the manipulation of various types of fabric shapes. The determination of robot motion paths is based on visual feedback defining the location of the fabric edges in world coordinates. The developed algorithm was proved to be effective for both polygonal and non-polygonal shapes, whereas the accuracy of the approximation to the desired seam line depends on the camera resolution.

D. Gershon and I.Porat [4] reported the FIGARO system, where an integrated robotic system for sewing has been developed. The system uses a PI controller, where the gains are selected by trial and error. The researchers used fabric tension control in order to prevent buckling and obtain good seam quality.

D. Gershon [5] proposed a parallel process decomposition of the robot sewing task. A robot arm manipulated the fabric to modify its orientation and control fabric tension during sewing. The sewing task was decomposed into four concurrent processes within a superposition parallel architecture. The robot's position control system is modeled as a mass-spring-damper system using the principle of mechanical equivalence, whereas the fabric is modeled as a nonlinear damped spring and the table friction acting on the fabric is included in the model. The complete system demonstrated robustness in the experiments.

Several industrial applications have addressed the manipulation problem where the location of the target is not a priori known. Using artificial vision to control the manipulator, the visual servoing systems are based on two approaches: position-based and image-based visual servo control [6].

In the position-based control systems, the error is computed in the 3D Cartesian space. The position error is computed using or not the model of the target depending on the visual features available in the image. The main advantage of this approach is that the camera trajectory is directly controlled in Cartesian space. However, this approach has the limitation of being sensitive to calibration errors that arise either from a coarse calibration of the camera or from errors appeared in the 3D model of the target.

In the image-based control systems, the error is directly computed in the 2D image space. The main advantage of this approach is that there is no need for the 3D model of the target and it is quite robust not only with respect to camera but also to robot calibration errors. However, the Jacobian matrix is a function of the distance between the camera and the target, which is not easily calculated and this is a serious a limitation of this model-free control approach.

In the sewing process, two robotic handling targets require sophisticated control. The first one deals with the control of the cloth tension and the second one with the path determination of the cloth. In the present paper, the second part is investigated, where a robot control system based on artificial intelligence and visual servoing is designed.

In this paper, a fuzzy controller is presented for handling and guiding a fabric to the sewing machine. The control law presented here is more effective than that of Cartesian systems, which suffers from calibration errors or from the numerical complexity of the Jacobian image, since the control function is based on fuzzy logic. The fuzzy rules derived after studying the behavior of the human workers for the sewing of a piece of fabric.

2. The robot sewing process

Fabrics are limp materials that present a rather unpredictable behavior during handling. The fabric bending rigidity is very small, so its shape is considerably changed when gravitational forces are applied to a piece of fabric. Therefore, sewing the fabric using a robot is not easy due to the wrinkling, folding and buckling.

Firstly, the features, namely the vertices of the fabric, are extracted from the image taken from the camera. After the edges of the fabric have been detected, the position-error (r) from the needle and the orientation error (θ) in relation to the seam line are computed (Fig.1). The linear and angular velocity of the fabric are derived from the position and orientation error through the designed fuzzy decision system. The position error (r) and the orientation error (θ) and their change with the time are the input data, whereas the linear and angular velocity of the end-effector are the output data.

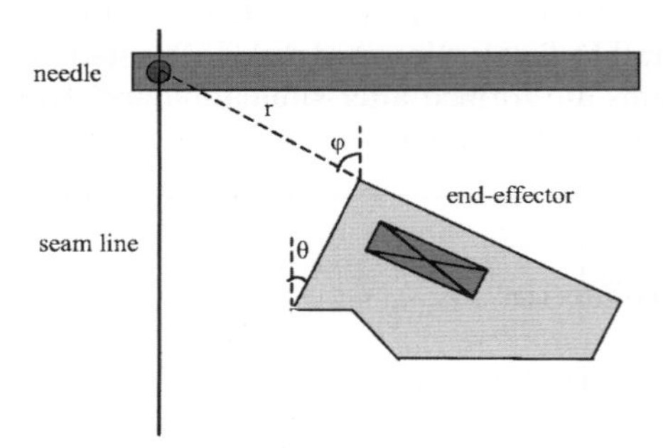

Fig. 1 Scene of the fabric lying on the table

Given the time step dt and the orientation angle φ, the new position of the end-effector can be computed. The fabric is transferred to a new position as a result of the movement of the end-effector, which is stuck on the fabric so as no slipping between the gripper and the fabric occurs. This process (phase 1) stops when the edge of the fabric reaches the needle with the desired orientation within an acceptable tolerance.

Next, the side of the fabric that coincides with the seam line is ready to be sewed. During sewing (phase 2), the fabric is guided along the seam line with a constant velocity, which should be the same with the velocity of the sewing machine, so that good seam quality is ensured. At each step, the direction of the sewed side is checked to be the same with that of the seam line. In the case that the side deviates from the seam line, the fabric is rotated around the needle until its side coincides with the seam line.

After one side of the fabric has been sewed, the fabric is rotated around the needle until the next side coincides with the seam line (phase 3). The orientation error (θ) of the next side in relation to the seam line and its time derivative are computed. These are the inputs of the fuzzy system that controls the rotation of the fabric around the needle, whereas the output is the angular velocity of the end-effector. When this side of the fabric approaches the seam line, it is ready for sewing.

The objective of our research work is to develop a robot controller capable of guiding the fabric towards the needle avoiding possible wrinkling, folding and buckling. Since the robot is in charge of handling the fabric lying on a table, our goal is to design a robust controller, which will automatically determine the path that the robot should follow.

2.1. The control system

In general, the control systems are based on the plant models and make use of geometrical parameters for robot path analysis. In our case, it is almost impossible to obtain a model running in real time. This is one of the reasons for selecting a fuzzy controller. In addition, the fuzzy logic controller is robust and quite fast in deriving the control outputs.

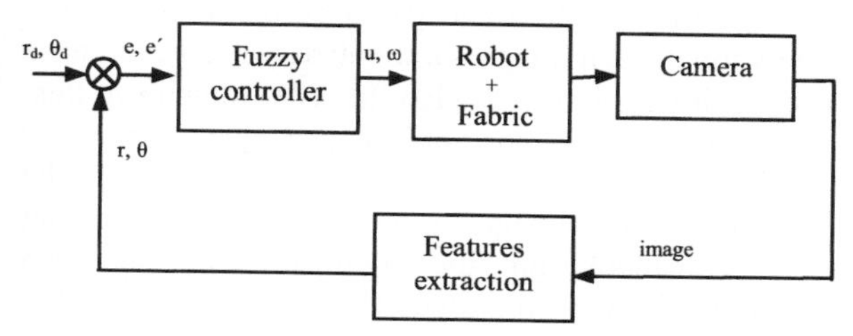

Fig. 2 The block diagram of the system

The proposed controller outputs the linear and angular velocity of the end-effector. The block diagram for the control system is shown in Fig.2. An analytical model of the robot and the fabric is not necessary. The knowledge of the behavior of the system has been incorporated into the controller and the system is able to respond to any position and orientation error regardless of the properties of the fabric.

The controller is able to manipulate a piece of fabric regardless of the shape and the type of the fabric. In addition, the controller is independent of the distance between the fabric and the sewing needle. It is also independent of the position and the orientation of the end-effector and there is no need for mathematical computations for different kinds of fabrics. Therefore, the proposed fuzzy controller is flexible and robust, because it is capable of handling various types and shapes of fabrics with efficiency.

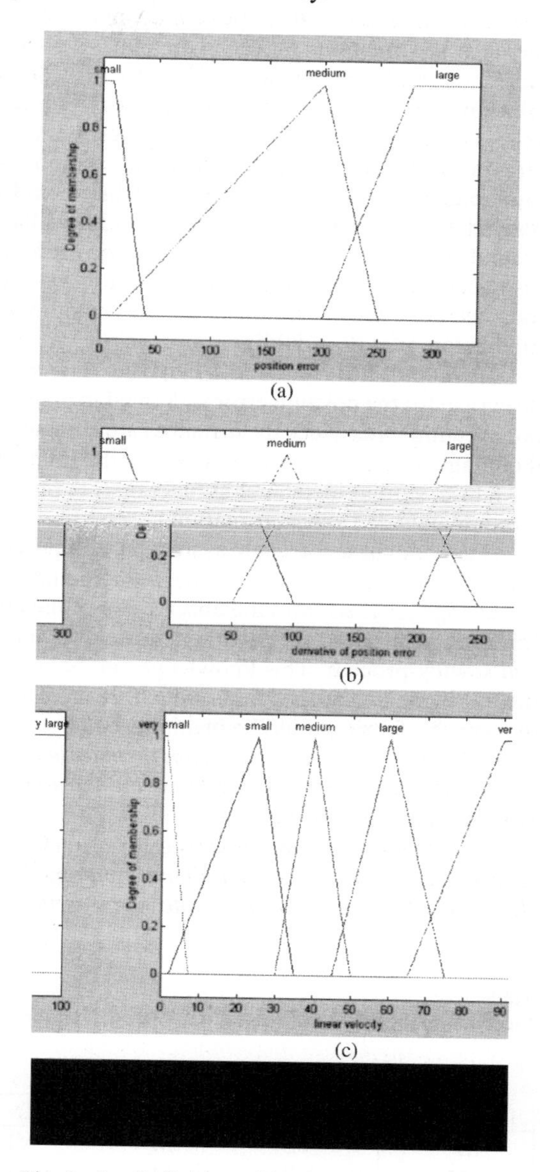

Fig. 3 a,b,c Definition of the membership functions

The construction of the fuzzy controller consists of defining the linguistic variables and the fuzzy terms associated with the numerical input and output data as well as the rules that govern the system.

Each one input variable is expressed by three linguistic terms: small (S), medium (M) and large (L), whereas each output variable is expressed by five linguistic terms: very small (VS), small (S), medium (M) and large (L) and very large (VL). The membership functions for the two inputs and the output of the system that controls the translation of the fabric are presented in Fig.3 a,b,c.

In the beginning, the type and the number of the membership functions for each variable as well as the range of each membership function are defined arbitrarily. After some experimental tests they have been altered in order to achieve the desired system efficiency. In some cases, where the membership functions were not well defined, the fabric could not reach the needle, because it was constantly oscillating around the needle. In other cases, the fabric made large displacements or large rotations when it was close to the needle, which resulted in an inappropriate orientation or even wrinkling.

It should be mentioned that the universe of discourse for the position error is defined in pixels and not in centimetres, since the distances are computed on the image. For the system that controls the rotation of the fabric, similar membership functions are provided through extended experimentation.

The rule base of each system is composed of 3×3=9 linguistic rules. The rule base includes the knowledge resulted from the observation of the fabric handling from the human workers during the feeding and sewing process. This knowledge can be given in the following two propositions, which are indicative of the way the fabric should be manipulated.

"The larger the distance from the needle and the smaller its change rate is, the faster the fabric should move towards the needle"

"The larger the angle between the fabric's side and the seam line and the smaller its change rate is, the faster the fabric should rotate towards the seam line."

Table 1
Fuzzy Associative Memory (FAM) of the fuzzy system

change rate	*position error* S	M	L
S	VS	M	VL
M	VS	M	VL
L	VS	M	L

Table 1 shows the Fuzzy Associative Memory of the system that controls the translation of the fabric, where the rule confidence is equal to one. The Fuzzy Associative Memory for the rotation is similar and is not presented due to space limits of this paper.

It should be mentioned that for the implication process the min operator is used, whereas for the aggregation process the max operator is used and the defuzzyfication process is performed using the centroid method.

2.2. Image features extraction

The basic idea as far as tracking is concerned is the use of a set of points called 'corners' that are actually the vertices of the tracked fabric in counter clock wise order defining its perimeter [7,8]. In our case, apart from the distortion by a similarity transformation (the product of a translation and a rotation), accessional distortion due to partial occlusion takes place [9]. In order to compensate for the loss of information, the tracking task demands more complex and robust algorithms. The 'corners' are used to define fragments of the fabric generating hypotheses and testing them in a two stages procedure.

In the initial stage a model description (MD) of the fabric is created [10]. At first the fabric is placed into the working area. The fabric's shape can be any arbitrary convex or not convex polygon. Using Harris corner detector [11] and post-processing the results, to eliminate noisy corners, the coordinates of its vertices ('corners') in the image plane are determined. Using that data certain geometric features of the polygon, namely the length of each side in pixels and the angle of each vertex in degrees, are automatically extracted.

In the second stage the position of the planar polygon in each image is extracted, which generates a scene description (SD) [10]. In order to achieve this, detection of the position of just two vertices is sufficient, since then the rotation and translation of the MD in the new image can be calculated. No scaling was considered but the algorithm that has been developed and is presented bellow can be modified to include scaling as well.

The acquired RGB image $I_{m \times n}$ is converted into grayscale $I1_{m \times n}$. Then the image histogram is calculated and the threshold's value is automatically set, in order to isolate the pixels that belong to the polygon. Afterwards, the binary image $BW1_{m \times n}$ is produced by thresholding the grayscale image $I1_{m \times n}$. and is segmented into binary objects by using the 8-connected components criterion [12]. Finally, the larger object assuming that corresponds to the part of the polygon that is not occluded by the robotic arm is selected creating a new binary image $BW2_{m \times n}$ (object binary image).

In the next step, the Harris corner detector is

applied to the object image $BW2_{m\times n}$ and to the intensity image $I1_{m\times n}$ as well. Afterwards, only the common 'corner points' are selected by combining those results.

In the last step, these 'corner points' are matched to the numbered 'corners' in the MD by using the same geometrical rules that were used in the extraction of the geometric features of the model. These rules match segments in the MD with segments in the SD generating N hypotheses consisting of N sets $\mathbf{V}_i$ (i=1,...N) of potential vertices of the polygon. The common area of the object in BW2 and the polygon defined in the image by a set $\mathbf{V}_i$ is calculated (in pixels) for every set $\mathbf{V}_i$. The ratio of the common area to the object's total area is defined as λ. If there is even one set $\mathbf{V}_i$ where λ is greater than a selected threshold the detection is considered successful (the hypothesis is valid) and the set that gives the maximum ratio is selected.

3. Experimental results

The experiments were carried out using a robotic manipulator with 6 rotational degrees of freedom and controlled by a PC Pentium IV 1.4MHz running under Windows 2000. The vision system consists of a Pulnix analog video camera at 768×576 pixels resolution RGB with focal capabilities ranging from 1m-∞. The camera is fixed above the working table in a vertical position, so that the fabric is kept in the field of view of the camera during the servoing (Fig.4). It is connected to Data Translation Mach 3153 frame grabber through a coaxial cable.

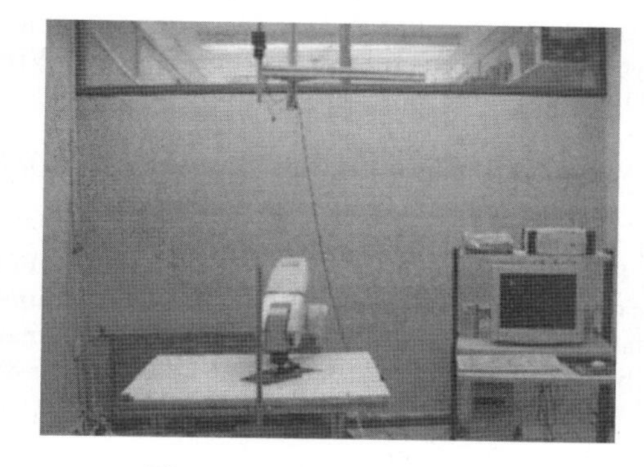

Fig. 4 The experimental stage

The shape of the fabric is a non-convex polygon, its colour is red (Fig.5a) and it has large resistance in bending so that its shape remains unchangeable and flat almost without folding or puckering during handling on the table. It should be mentioned that the handling of buckling is outside the scope of this paper. The location of the needle as well as the direction of the seam line are known in advance.

Furthermore, the location of the robot hand on the fabric has been selected, so that no wrinkles appear during the sewing process, according to the work [13] considering the type and the flexibility of the fabric. This makes the problem easier with respect to the image feature extraction.

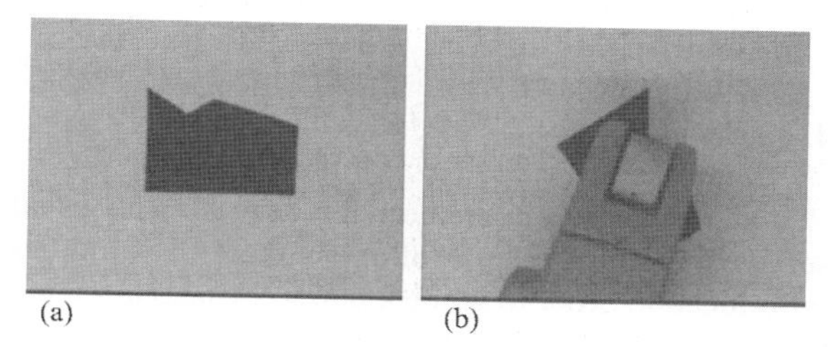

Fig. 5 (a) Image of the fabric on the working table (b) Image of the fabric handled by the robot arm

In the first stage of the experiments the tracking algorithm has been tested separately in a set of 100 RGB images (576x768x3 pixels) containing the fabric and the robotic arm in several positions and orientations (Fig.5b). Setting the threshold to 0.9, the detection success criterion ($\lambda > 0.9$) was satisfied for every image of the set.

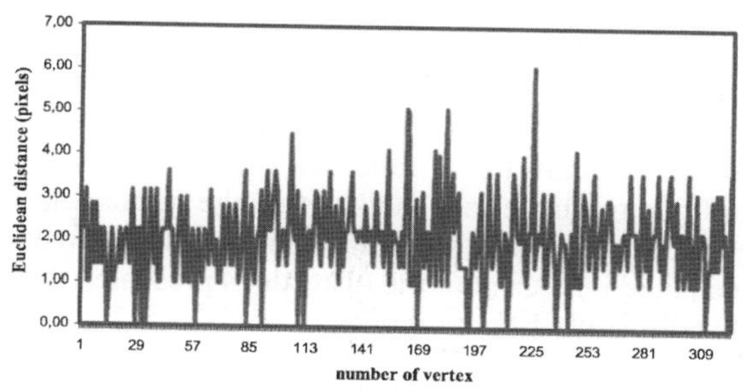

Fig. 6 The Euclidean distance in pixels of the detected vertices from their real position in the image

Moreover, the real position in this set of the 323 vertices that are not occluded has been manually determined and their Euclidean distance from the detected vertices has been calculated. The results are shown in Fig.6 where the diagram shows the Euclidean distance in pixels versus the number of the vertex (1 to 323). The mean Euclidean distance of all vertices is 2.0269 pixels while their standard deviation is 0.9876 pixels. Since the average length of the six sides of the polygon is about 164 pixels, the algorithm can be considered extremely accurate.

In the second stage of the experiments the complete system has been tested. The sewing process is repeated many times where each time the fabric starts from a different location on the table and the gripper is on a different location on the fabric. The accepted position and orientation error are set equal to 2 pixels and 0.5°, respectively. In all cases tested, the algorithms have been proved to be robust and efficient. When the accepted position and orientation error were set to lower values, the fabric made many oscillations around the needle until it reaches the needle.

4. Conclusions

In this paper, a method for designing a visual servoed manipulator controller based on fuzzy logic is introduced in order to guide a fabric towards a sewing machine and produce a good seam quality.

The experimental results show that the proposed approach is an effective and efficient method for guiding the fabric towards the sewing machine, sewing it and rotating it around the needle. Only in a few cases where the corners could not be detected, the robot failed to sew the fabric. Unlike some of the methods referred in the introduction, the proposed controller does not need many mathematical equations or calculations, but it proved to be rather simple and robust.

It should be stressed that the problem of sewing a piece of fabric using a robotic manipulator instead of a human expert poses additional problems, since it is more difficult for a robot to cope with buckling, folding or puckering. Considering the future research work, the proposed algorithm can be extended so that it can take into account the distortions presented during handling of fabric by robots.

Acknowledgements

This work is financed by the General Secretariat for Research and Technology of Greek Government as part of the project "XROMA-Handling of non-rigid materials with robots: application in robotic sewing" PENED 01. University of Patras is partner of the EU-funded FP6 Innovative Production Machines and Systems (I*PROMS) Network of Excellence.

References

[1] Bernardon E and Kondoleon T.S. Real time robotic control for apparel manufacturing. Conf. Proc. Robots 9(1) (1985), 4-46, Detroit.

[2] Gershon D. Strategies for robotic handling of flexible sheet material. Mechatronics 3(5) (1993), 611-623.

[3] Torgerson E. and Paul F.W. Vision-Guided Robotic Fabric Manipulation for Apparel Manufacturing. IEEE Control Systems Magazine (1988), 15-20.

[4] Gershon D. and Porat I. Robotic sewing using multi-sensory feedback. Proc. Of the 16th Internat. Symp. on Industrial Robots (1986), 823-834, Brussels.

[5] Gershon D. Parallel Process Decomposition of a Dynamic Manipulation Task: Robotic Sewing, IEEE Transactions on Robotics and Automation 6(1990), 357-366.

[6] Hutchinson S., Hager G.D. and Corke P.I. A tutorial on visual servo control. IEEE Transactions on Robotics and Automation 12(5) (1996).

[7] Paraschidis K., Fahantidis N., Petridis V., Doulgeri Z., Petrou L, Hasapis G. A robotic system for handling textile and non rigid flat materials. Proceedings of the Computer Integrated Manufacturing and Industrial Automation (CIMIA) (1994), 98-105, Athens.

[8] Fahantidis N., Paraschidis K., Petridis V., Doulgeri Z., Petrou L, Hasapis G. Robot handling of flat textile materials, IEEE Robotics & Automation Magazine 44(1) (1997), 34-41.

[9] Koch M.W. and Kashyap R.L. Using polygons to recognize and locate partially occluded objects. IEEE Trans. Patt. Anal Machine Intell. PAMI-9(4) (1987), 483-494.

[10] Ayach N. and Faugeras O. HYPER: a new approach for the recognition and positioning of two dimensional objects PAMI 8 (1986), pp. 44-54.

[11] Deriche R., Giraudon G. Accurate corner detection: An analytical study. Proc. 3rd Int. Conf. on Computer Vision (1990), 66-70.

[12] Vernon D. Machine Vision, Prentice-Hall (1991), 34 - 36.

[13] P.N. Koustoumpardis and N.A. Aspragathos, Fuzzy logic decision mechanism combined with a neuro-controller for fabric tension in robotized sewing process, Journal of Intelligent and Robotic Systems, no. 36 (2003), pp.65-88.

Intelligent Production Machines and Systems
D. T. Pham, E. E. Eldukhri and A. J. Soroka (eds)

A Computer Aided Design System for Product Oriented Manufacturing Systems Reconfiguration

S. Carmo_Silva[a], A.C. Alves[b], M. Costa[c]

[a] *Centre for Production Systems Engineering, University of Minho, Campus de Gualtar, 4700-057 Braga, Portugal (scarmo@dps.uminho.pt)*
[b] *Centre for Production Systems Engineering, University of Minho, Campus de Azurém, 4800-058 Guimarães, Portugal (anabela@dps.uminho.pt)*
[c] *Department of Production Systems, University of Minho, Campus de Gualtar, 4700-Braga, Portugal*

Abstract

Product Oriented Manufacturing Systems (POMS) are systems designed for variable product demand markets, and dynamically reconfigured for the manufacture of a single type of product or a family of similar products at a time. POMS systems can take several forms and result from exploring the use of flexible resources, a variety of new design philosophies, technologies and approaches to manufacture, including Cellular, Lean, Agile, Quick Response and Fit Manufacturing. In this paper a computer aided design system for the design and reconfiguration of POMS is addressed. A characterization and description of the structure and components of the design system, including a database, a user interface and a knowledge base and the description of some important data sets, are presented.

1. Introduction

It is recognized that the best performing manufacturing systems are those designed to suit specific manufacturing requirements of a product or a family of similar products, i.e. are Product Oriented Manufacturing Systems (POMS). The manufacturing system concept totally opposed to this is the one organized in functional departments, i.e. Functional Oriented Manufacturing Systems (FOMS) and, therefore theoretically able to manufacture any product variety. These systems tend to have poor system performance and provide poor customer service.

In the past, the decisions in favour of POMS were easier to take than today. Usually due to large production quantities, stable demand environment and reduced competition, POMS were established and remained unchanged for quite a few years. So, system reconfiguration was rarely required. Nowadays, due to high product demand variety and continuously increasing competition there is a need for fast and easy design of POMS to adapt them to meet customer changing requirements in an efficient and effective manner. This design process is oriented to the manufacture requirements of one product or a family of similar products, at a time. Unless POMS design is efficient and quick in achieving good configurations, which can be rapidly and easily implemented in practice, the advantages associated with POMS cannot be fully explored and, consequently, opportunities for maintaining a good customer service and competitive position in the market place can be lost. The use of FOMS could be sought as a good alternative to overcome such difficulties. However the requirements for fast delivery increased quality of goods and services and reduced cost of production do not favour FOMS either. This is why fast design or reconfiguration POMS is many times the way to follow. Moreover the POMS of today must be based on new technologies, reusable flexible resources and new approaches to manufacture, operation and control.

To be able to quickly attain good POMS designs and fast reconfiguration, computer aided design systems directly addressing POMS design must be used. Reported computer aided design systems (CADS) applicable to POMS tend to be restrictive, not focussed or unstructured. Thus they either implement a specific approach to POMS design, such as Production Flow Analysis [1], are developed for a wide spectrum of manufacturing systems without specifically focussing on POMS [2,3], or essentially are based on libraries of programmes to implement a given approach to design. Examples of CADS based on such approaches are those reported by Luong et al. [4] Mahadevan and Srinivasan [5] and Irani et al. [6].

In this paper a framework and an associated computer aided system directly addressing the designing of POMS, here called Computer Aided Design System for POMS (CADS_POMS), are proposed. They are based on a design methodology, referred to as the GCD (Generic-Conceptual-Detailed) methodology, developed by Silva and Alves [7]. The CADS_POMS eases data handling and the iterative design process and, at the same time, provides access to several methods and tools suitable for carrying out the design functions. This is done through a knowledge base and suitable interfaces. The knowledge base can be seen as repository of design and evaluation methods accessed and used at several design phases.

In the next sections a characterization of the Product Oriented Manufacturing System (POMS) concept is first presented and then the Computer Aided Design System for POMS and its framework are described. In the last section some concluding remarks are put forward.

2. Product Oriented Manufacturing Systems

A Product Oriented Manufacturing System (POMS) is defined as a set of interconnected flexible manufacturing workstations or cells, usually involving people, which simultaneously and in a coordinated manner address the manufacture of a product or a family of similar products, subject to frequent reconfiguration to be adapted to manufacturing requirements of different products or product families. A product may be simple, like a part, or complex, having a product structure with several levels. A set of cells that does not work under coordination towards synchronized production of end items, does not form a POMS. A paradigmatic example of Product Oriented Manufacturing System (POMS) is what Black [8] calls a linked-cell manufacturing system. Many manufacturing systems currently referred to as JIT, lean, flexible and virtual manufacturing systems may also be seen as POMS. Product Oriented Manufacturing (POM) can also be associated with concepts such as focused factory [9] and One-Product-Integrated-Manufacturing (OPIM) put forward by Putnik and Silva [10].

At a local scale a POMS can be seen as a network of balanced flow lines or manufacturing cells. This balancing explores flexibility of machines and enlarged skills of operators. These factors are considered by design methods as inputs to arrive to physical and operational systems configurations which are effective in achieving company objectives dependent on available manufacturing resources. The resources can be distributed in space and may be put together, in a localized site, or, alternatively, organized into virtual POMS. Today, these can benefit from intranet and internet based technologies, a prerequisite of the widely discussed Virtual Enterprise concept [11]. This approach to the virtual configuration of manufacturing systems was initially introduced in 1982, by McLean, Bloom and Hopp [12], and studied by several authors afterwards such as McLean and Brown [13], Drolet, Montreuil and Moodie [14] and Ratchev [15].

Although POMS lends itself to large quantities and small variety product requirements, it is particularly important in today's market environment to seek viable POMS for the "Make to Order" (MTO) and "Engineering to Order" (ETO) environments. This viability is ensured by exploring the reuse of flexible manufacturing resources and the organizational philosophies, techniques and tools associated with Lean Manufacturing (LM) [16], Agile Manufacturing (AM) [17], Quick Response Manufacturing (QRM) [18] and Fit Manufacturing [19]. Both LM and QRM favour production systems

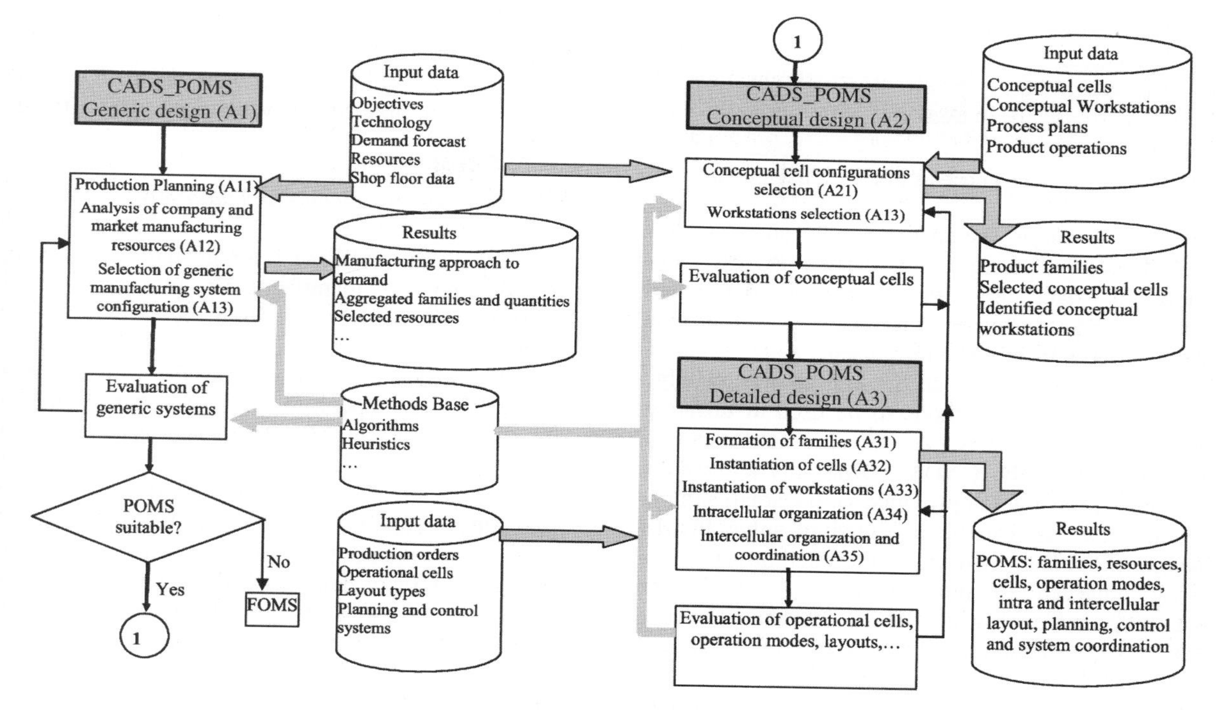

Fig. 1. CADS_POMS design Framework.

organization in multifunction autonomous units or cells working under integrated coordination for achieving production objectives. AM emphasizes the importance of rapidly changing system configuration for matching processing requirements from product demand changes.

3. Computer Aided Design System for POMS

3.1 CADS_POMS components and structure

The CADS_POMS framework is based on the Generic-Conceptual-Detailed (GCD) methodology for POMS design developed by the authors Silva and Alves [7]. A simplified representation of it is shown in Fig. 1.

The GCD methodology essentially puts forward a hierarchical multilevel and iterative design process for POMS and, at the same time, presents the designer with the set of design alternatives and parameters, which must be evaluated at each stage. This design process is extended to find production control and work coordination solutions, within and among cells, for complete product manufacture and assembly. Naturally, the design options and parameters are initially dependent on customer needs and derived functional requirements, as well as on the company objectives and restrictions.

In the design process aimed at reaching good solutions for both organizational and operational configurations of POMS several decisions at strategic, tactical and operational level must be taken. The first is to decide, based on market requirements and the company's internal and external environmental restrictions, if POMS are a viable alternative to manufacturing systems configuration. Only an affirmative answer to this question allows further POMS design.

In the GCD design methodology, all relevant data and restrictions are considered and a range of methods are used in the POMS design process. Under this methodology the design process is organized in three main phases, namely the generic, conceptual and the detailed one, and includes several design stages and activities. However important the GCD methodology may be, it can be of little use if not supported by a computer aided design system. This is done through the CADS_POMS that supports design activities from strategic planning to the POMS organization and workflow control mechanisms definition. The main components of CADS_POMS are a database, a methods base and a user interface with several menus and windows (Fig. 2). Software interfaces are also used to access and use several methods or algorithms for performing design functions and evaluating alternative design solutions.

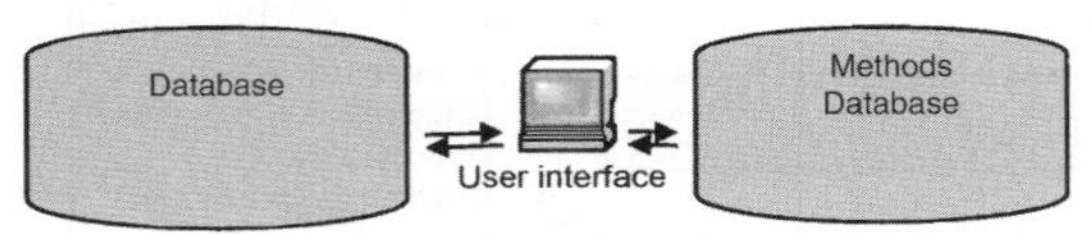

Fig. 2. Main components of the CADS_POMS

The database includes all relevant data for POMS system design. To design POMS very important data are the products data and manufacturing processes of products, as well as data about manufacturing resources used.

The methods base can be seen as a knowledge base providing the methods to be used at different POMS design phases. At the present stage of development of the CADS_POMS this knowledge base is centralized and provided with a small sample of methods that can be executed to solve POMS design problems. However, the methods base is being enlarged to provide the right tools for obtaining and evaluating POMS solutions at every design stage. Therefore, the CADS_POMS system will be able to quickly access a large variety of methods for efficiently solving the POMS design problems according to the GCD methodology and using data in the database. To further enhance this function a study is under course to evaluate and possibly to implement the methods base as a distributed knowledge base, updateable as new methods can be made available locally or remotely by a community of methods providers in a network of computing peers or servers. One important piece of software to implement this idea is a service interface for easy methods specification, access and local or remote execution through the Internet. This can be particularly important because the CADS_POMS could have several independent implementations, in companies for example, and all share the same distributed knowledge base of methods.

The user interface allows the user to perform several functions. An important one is the introduction of initial data for system design, both related with the underlined design philosophy and with objects, namely products, operations and resources of several types, including machines, which are available for configuration or reconfiguration of the manufacturing system of a company. This is important and instrumental to the main user function, which is to use the design methods in the design process having into account all restrictions and resources available. Additionally, the user plays a strong role in the selection of methods to use for supporting design functions. This means that the user must have a clear idea of the needs, purpose, role and usefulness of each method in order to be able to apply them according to the design process needs.

3.2 Fundamental concepts and data sets

Data in the database is organized for allowing specification of manufacturing processes in a manner that permits the user to fully explore alternative design solutions dependent on both available resources and processing flexibility. Thus a comprehensive specification of manufacturing processes is critical do the design success. The manufacturing process specification involves four levels. First, it is necessary to specify the *process plan* (PP) of each product. Process plans are defined at process planning level and are absolutely necessary as input data to POMS design. A process plan can be seen as a network of sub processes (SP). Each SP changes the processing state of a product and may involve a single or a set of manufacturing operations. An *operation* is an elementary conversion process performed in one product, component or workpiece. A process plan represents all theoretically possible alternative processes for manufacturing a product, i.e. the processes to take a product from an initial state of conversion to a final one.

The set of SP of the PP chosen for converting a product from an initial state to a final one is called an *operation plan* (OP). Therefore, usually a single process plan may specify or imbed several operation plans. So, at a second hierarchical level of process specification, the operation plan for a product has to be chosen or specified. Since, for arriving to an OP, it is usually necessary to choose one among a set of alternatives imbedded in the PP, a decision making process have to be carried out as referred above. In POMS design, this must take into account the design objectives, namely that of efficiently and dynamically reconfiguring POMS. For making a good choice, suitable methods for operation plan selection must be accessed and used, and a lot of user interaction is likely to be required.

Operation plans may provide alternatives for the sequence on which some operations can be carried out. Thus it is necessary, at a third level of manufacturing process specification to define or choose the *operations sequencing plan* (OSP), i.e. the order on which each operation of the OP of a product should be carried out. For an established manufacturing system configuration, this problem is more a problem of scheduling than a problem of system design. However, if efficient system operation is sought than the selection of an OSP for a product may be critical to POMS design.

The last manufacturing process specification level has to do with the choice of workstations to perform each operation of the operations sequencing plan. When a single workstation is available the choice is obvious. However, if more than one exists within a manufacturing system, then alternatives arise. This calls for a workstation selection process. This is also important for manufacturing design because it enables to finally settle the workstations to use in the manufacturing process of a product and, therefore, in the manufacturing system. This sequence of operations associated with the physical workstations that execute them is called *product routing* (PR).

Although a hierarchy of decision process steps can be envisaged from process plans to product routings, all, or at least some of the decision process steps may have to be integrated and solutions obtained in an interactive way. This is mainly performed during the Conceptual Design (A2) and Detailed Design (A3) of the GCD methodology.

To be able to specify PP and ultimately define PR, the processing operations of products must be specified. In the CADS_POMS, processing operations of a product result from the instantiation of generic operations, according to the physical transformation or assembly required to manufacture the product. This instantiation process is carried out by a process planner based on a number of parameters and operation attributes. The result is input to the database for POMS design. This data is likely to be reusable for several POMS design problems in the same manufacturing technological environment if production is to be repeated. This data can be used by the POMS designer who interacting with design methods can proceed to the choice of OP, OSP and ultimately PR. In a garment manufacturing environment, as an example, generic operations include cutting, sewing and attach zips. Operation attributes and parameters include the number of needles, number of threads, machine type, operator type and batch size.

To systematize the manufacturing process specification procedure and avoid data proliferation, in the Microsoft SQL relational database used, a table is defined for specifying operation attributes and parameters, referred to as the *characteristics* table. Another table contains the list of generic operations to be instantiated with data from the characteristics table for each product to manufacture in the system. A critical set of data in the database used by the CADS_POMS is shown in Fig. 3.

The machines table includes a list of all the machines that can be used for POMS design or reconfiguration. These include not only those available at the company but also those that can be acquired if necessary through buying, leasing or borrowing. Thus machines which are likely to be acquired in the market may also be listed. All these machines have attributes and processing characteristics which are listed in the characteristics table. Thus, both operations of products and machines in the system share the same characteristics table for its characterization or definition. This approach to machines and operations definition permits to identify, through a matching process, which machines can perform which operations. This matching process together with the already defined operation plans enables to choosing product routings. These are central to the specification and design of a POMS system. Whenever a POMS system has to be used to manufacture several products their processing requirements must be taken simultaneously into consideration for such choice.

4. Conclusion

Designing product oriented manufacturing systems (POMS) is a very complex task. Usually it cannot be carried out in an efficient way without computer aid. In this paper the framework of a computer aided system for POMS design, called CADS_POMS, is presented and briefly described. The framework is based on a POMS design methodology called GCD methodology and developed by the authors.

The main fundamental elements of the CADS_POMS system is a database a user interface and knowledge base that holds design methods for system design and evaluation at several design stages. The system design capability is both highly dependent on user interaction and on the availability of design methods. Apparently, it seems to exist advantageous that the knowledge base take a distributed form. This can be particularly important because the CADS_POMS could have several independent implementations, in companies for example, and all share the same distributed knowledge base of methods. Moreover it could be

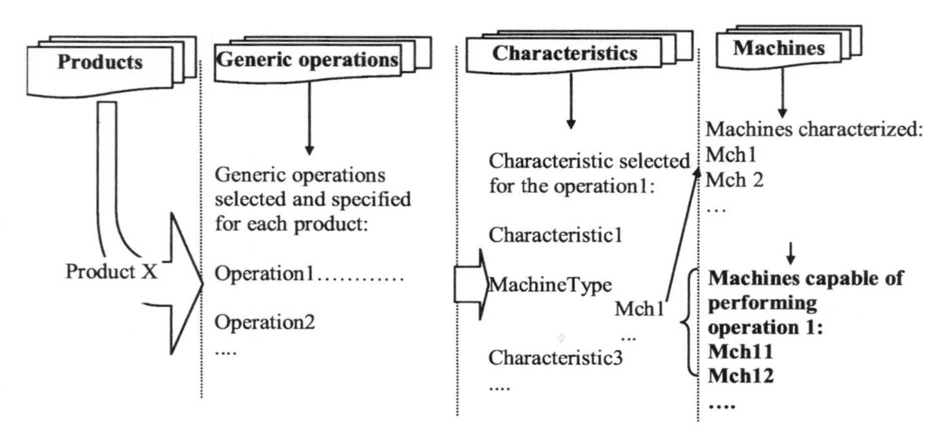

Fig. 3. Fundamental data sets for POMS design.

updateable as new methods could be made available local or remotely by a community of methods providers in a network of computing peers and servers. This idea is being validated and most probably will be implemented in the near future.

Acknowledgements

University of Minho is partner of the EU-funded FP6 Innovative Production Machines and Systems (I*PROMS) Network of Excellence. http://www.iproms.org

References

[1] Burbidge J L. Production Flow Analysis for planning Group Technology. Clarendon Press, Oxford, 1996.

[2] Cochran D S, Arinez JF, Duda J W and Linck, J. A decomposition approach for manufacturing system design. Journal of Manufacturing Systems, 20 (2001), pp 371-389.

[3] Suh NP, Cochran DS and Lima PC. Manufacturing Systems Design. Annals of the CIRP 47 (1998).

[4] Luong L, He J, Abhary K and Qiu L. A decision support system for cellular manufacturing system design. Computers and Industrial Engineering (2002)

[5] Mahadevan B and Srinivasan G. Software for manufacturing cell formation: issues and experiences. In: Proceedings of the Group Technology/Cellular Manufacturing Columbus, Ohio, 2003, pp 49-54

[6] Irani SA, Zhang H, Zhou J, Huang H, Udai TK and Subramanian S. Production Flow Analysis and Simplification Toolkit (PFAST). International Journal of Production Research 38(8) (2000) 1855-1874.

[7] Silva SC and Alves AC. In: V Marik, L Camarinha-Matos and H Afsarmanesh (Eds.) Design of Product Oriented Manufacturing Systems. Knowledge and Technology Integration in Production and Services, Kluwer Academic Publishers, 2002, pp 359-366.

[8] Black JT. The Design of the Factory with a Future McGraw-Hill

[9] Skinner W. The focused factory. Harvard Business Review (1974).

[10] Putnik G and Silva SC. In: L M Camarinha-Matos (Ed.) One-Product-Integrated-Manufacturing. Balanced Automation Systems I, Chapman and Hall, 1995.

[11] Camarinha-Matos LM and Afsarmanesh, H. The Virtual Enterprise Concept. In L M Camarinha-Matos and H Afsarmanesh, Infrastructures for Virtual Enterprises: Networking Industrial Enterprises Kluwer Academic Publishers, 1999, pp 3-14.

[12] McLean CR, Bloom HM and Hopp TH. The Virtual Manufacturing Cell. In: Proceedings of the 4th IFAC/IFIP Conference on Information Control Problems in Manufacturing Technology, 1982, 105-111.

[13] McLean CR and Brown PF. In H Yoshikawa and JL Burbidge (Eds.) The Automated Manufacturing Research Facility at the National Bureau of Standards. New Technologies for Production Management systems, North – Holland: Elsevier Science Publishers BV, 1987

[14] Drolet JR, Montreuil B and Moodie, CL. Empirical Investigation of Virtual Cellular Manufacturing System. In: Proceedings of the Symposium of Industrial Engineering-SIE'96, Belgrade, Serbia, 1996, pp323-326.

[15] Ratchev SM. Concurrent process and facility prototyping for formation of virtual manufacturing cells. Integrated Manufacturing Systems, 12 (2001) 306-315.

[16] Womack J, Jones DT and Roos D. The machine that changes the world. Rawson Associates, 1990.

[17] Kidd PT. Agile Manufacturing forging new frontiers. Addison Wesley Publishers, 1994.

[18] Suri R. Quick Response Manufacturing – A Companywide Approach to Reducing Lead Times. Productivity Press, 1998.

[19] Thomas AJ and Pham DT. In: R Schoop, A Colombo, R Bernhardt and G Schreck (Eds.) Making industry fit: the conceptualisation of a generic 'Fit' manufacturing strategy for industry. Proceedings 2nd IEEE Int Conf on Industrial Informatics, INDIN 2004, Berlin, 2004, pp 523-528.

Intelligent Production Machines and Systems
D. T. Pham, E. E. Eldukhri and A. J. Soroka (eds)

A Distributed Control Environment for Reconfigurable Manufacturing

T. Strasser[a], I. Müller[a], A. Zoitl[b], C. Sünder[b] and G. Grabmair[c]

[a] *Robotics and Adaptive Systems, PROFACTOR Research, 4407 Steyr-Gleink, Austria*
[b] *Automation and Control Institute, Vienna University of Technology, 1040 Vienna, Austria*
[c] *Industrial Informatics, University of Applied Sciences Wels, 4600 Wels, Austria*

Abstract

Future manufacturing is envisioned to be highly flexible and adaptable. New technologies for efficient engineering of safe, fault-tolerant and downtimeless systems and their adaptations are preconditions for this vision. Without such solutions, engineering adaptations of Industrial Automation and Control Systems (IACS) will exceed the costs of engineering the initial system by far and the reuse of equipment will become inefficient. In this work a new approach for a model driven, component based development of distributed, downtimeless real–time control, the controlled evolution of IACS and their execution during their adaptation based on the standard IEC 61499 is proposed. This new method increases engineering efficiency and reuse in component-based IACS significantly.

1. Introduction

The decisive factor for the market success of producing industries is a fast and flexible reaction to changing customer demands. New paradigms like "Flexible production up to lot size 1", "Mass Customization" or "Zero Downtime Production" will achieve these requirements but require completely new technologies for realisation [1].

In order to achieve the postulated flexibility of these paradigms technically, reconfiguration is needed at the physical and at the logical level. Therefore from the technological point of view a change from interlocked and rigidly coupled systems towards to functional complete, modular, if possible decoupled but closely cooperating and at best standardized components is in progress.

Future products, machines, plants and components are built up from flexible autonomous and intelligent mechatronic components to a distributed system. Compared to the higher complexity of such systems a number of advantages and new opportunities turn up: higher modularity, flexibility and (re-) configurability, scalability in functionality and computing performance, simpler system design and engineering by the use of functional complete mechatronic components, improved real time behaviour because of local computing performance, comprehensive real-time behaviour between components through deterministic and fault tolerant communication infrastructure, higher system availability through systematic distribution, and many more. Applications in automotive high-end products, modular manufacturing machines or highly flexible transportation facilities already show some of the demanded characteristics.

Distributed embedded real–time systems for industrial automation and control of plants that evolve towards downtimeless adaptable systems will play a key role to realize this vision [2]. Systems that can economically be changed by controlled evolution and which meet safety constraints and Quality of Service (QoS) stability will revolutionize manufacturing of goods, products and services. Production downtimes during system evolution need to be avoided. Controlled evolution of distributed embedded real–time IACS is therefore a prerequisite.

Component-based reconfiguration of control logic is technically hardly possible in current architectures of industrial automation and control systems. This hinders the users from getting the expected technical and economical advantages of reconfigurable systems. The component–based reference–architecture introduced by the IEC 61499 - "Function Blocks for Industrial Process Measurement and Control Systems" [3] for distributed industrial control systems features first basic concepts to reach that goal. It defines platform independent reconfiguration services at device level [4].

The challenge and aim of this paper is to present an approach to over–come limitations of IEC 61499 (e.g. methods for describing reconfiguration applications, real-time capability ...) methodologies and tools. Section 2 discusses the requirements for reconfiguration methods that are necessary for the proposed system evolution. Section 3 summarizes the main features and characteristics of IEC 61499 as a reference model for IACSs with special focus on its management capabilities for reconfiguration. Section 4 discusses an engineering approach for the controlled evolution of control applications and the regarding execution environment. First tests at the Automation and Control Institute of the Vienna University of Technology with a prototypic implementation of an execution environment for IEC 61499 will be presented in Section 5. The summary of this work concludes this paper.

2. Reconfiguration requirements on lower level control

This section identifies the basic requirements that have to be met in order to allow controlled evolution, i.e. reconfiguration of control applications, at the lower control level of automation and control systems. The requirements are introduced from three different viewpoints. The first is the domain of control and automation systems. The second deals with the motivations for reconfiguration processes on the device level. The last one introduces the aspect of engineering support necessary for adequate reconfiguration applications.

2.1. Requirements introduced through the domain

In order to provide a reconfiguration support for lower level control systems several requirements have to be met by the programming and modeling languages for control algorithms. Besides technical requirements, some of the requirements derive out of traditions of the control and automation domain. They all can be summarized to:

- *Predictability:* Current PLCs[1] show cyclic execution behavior, which is well known and widely spread. Distribution adds complexity here, which has to be hidden from the engineers.
- *Usability:* The control logic has to be as easy to understand and maintain as current PLC–based systems are.
- *Durability:* The system lifespan of IACSs is up to decades, which requires long–term upward and down–ward compatibility in case of replacement of embedded controllers.
- *Standards compliance:* In order to provide interoperability with other components, configurability with any standard–compliant tool and portability of software components are required.
- *Real time reconfiguration with guarantees:* Switching control applications or relocating them during full operation needs planned proceeding regarding reconfiguration and timing for changing from one to another system state.

2.2. Reconfiguration process requirements

In addition to the requirements introduced through the domain, other demands are posed from the appearance of normal and abnormal activities and conditions during execution of application programs. These circumstances may motivate reconfiguration processes. As a reaction to special situations of a control system (such as failures in hardware, mechanics and software) the control system needs to be changed and adopted in order to fulfill at least minimal functionality or just degrade to a safe state. This leads to the following requirements:

- Failure detection and system monitoring
- Failure diagnosis, selection and initialization of recovery applications

According to these requirements the systems have to be built of independent intelligent modules. The executed control application is composed of migratable components, which are executed independently of each other. But they are equipped with the ability to cooperate for solving control tasks if needed.

[1] Programmable Logical Controller

2.3. Requirements for engineering of reconfiguration applications

To keep the complex task of engineering reconfiguration applications manageable by control engineers the following requirements are introduced through the engineering process.

- *Application centred engineering:* Especially reconfiguration applications have to be considered from the whole applications' point of view. A device centred engineering approach will not be manageable and will lead to complex evolution scenarios.
- *Different engineering views:* Reconfiguration introduces additional views within the engineering. The application of the original system state has to be displayed in comparison to the new system state.
- *Reconfiguration modelling language:* According to the requirement of usability in Sec. 2.1 also the modelling language for reconfiguration has to be easy to understand and maintain. Therefore it is necessary to use a similar semantic.
- *Verification and validation:* Reconfiguration requires advanced methods for verification and validation to a bigger amount than "normal" applications. Critical constraints for reconfiguration are processing time and memory consumption according to execute the evolution without malfunction of running applications.
- *Version control:* The possibility of describing an evolution step of an IACS enables support of the whole system life cycle within engineering. The continuous system changes may be saved as different versions and enables for instance an undo–functionality.
- *Transition management:* During the process of reconfiguration the actual state of the system is a very important information to be introduced as transition condition within the evolution sequence. On the other hand the acknowledgement from the previous reconfiguration step has to influence the reconfiguration sequence.

3. IEC 61499 as a reference model for reconfigurable distributed control at real-time level

3.1. Main characteristics of IEC 61499

The programming constructs of today's existing systems, which are mainly based on the standard IEC 61131-3 [5] are too monolithic and closed to fulfil the requirements given above. The main restriction of current systems is the missing support for building control software out of independent components that can be added and removed at runtime. Therefore new ways for programming of lower level control systems have to be taken. The new standard IEC 61499 serves as a reference architecture that has been developed for distributed, modular, and flexible control systems. It specifies an architectural model for distributed applications in industrial–process measurement and control systems (IPMCS) in a very generic way and extends the function block model of its antecessor with additional event handling mechanisms and concepts for distributed systems.

Function blocks (FB) are an established concept for industrial applications to define robust and re–usable software components. The used network type is not in the scope of the specification, but a compliance profile for feasibility demonstrations, which is provided by the Holonic Manufacturing Systems Consortium, specifies its usage for Ethernet [6]. The following main features characterize this new standard [7,8]:

- Component oriented basic building blocks called FBs
- Graphical intuitive way of modeling control algorithms through the connection of FBs
- Direct support for distribution
- Definitions for the interaction between devices of different vendors
- Basic support for reconfiguration
- Based on existing standards of the domain

With the constructs and definitions of this standard most of the requirements introduced in the last section can be fulfilled. Because of these features IEC 61499 is suitable as a reference architecture for building new agile and adaptive lower level control of next generation automation and control systems.

3.2. Management interface of IEC 61499 devices

The configuration of a distributed automation and control system based on IEC 61499 can be enabled by the use of management functions which can be included in each device. For this purpose the standard defines a management application, represented by a device manager FB (depicted in Fig. 1). With the use of this FB combined with a remote application, which is described in the IEC 61499 Compliance Profile for Feasibility Demonstrations [6], access between different IEC 61499 compliant devices is allowed.

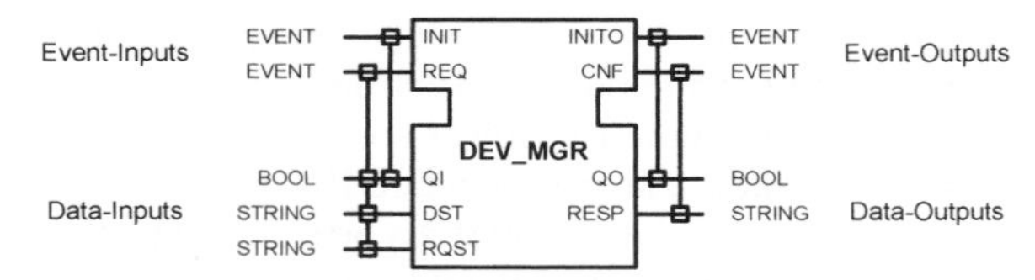

Fig. 1. Device Manager Function Block

This provides a very extensive facility to configure devices by sending XML[2] management commands from on to the other. The following standardized functions of the management application can be used to interact with a device:

- Creating FB instances,
- Creating event and data connections between instances of FBs,
- Downloading of new FB types and data types,
- Initiating the execution of FBs,
- Deleting FBs instances and connections,
- Supplying status information about the state of the device, the FB instances and their connections.

An example of this service interface is shown in Fig. 2 with the corresponding XML based configuration commands.

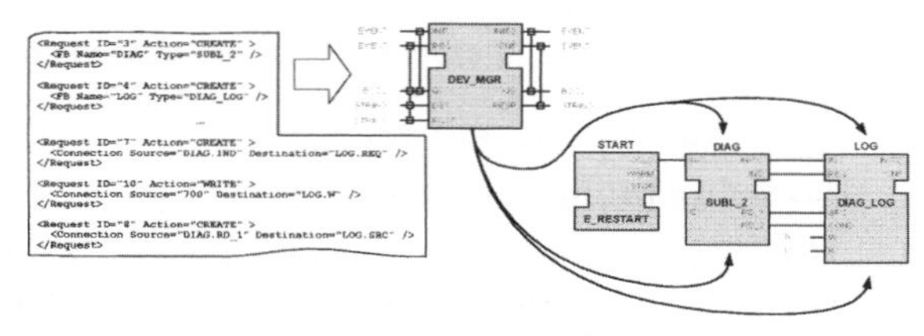

Fig. 2. Configuration example using the Device Manager concept

Through the use of XML commands for the (re)configuration communication, IEC 61499 provides a defined interface to other programs. Therefore generic tools can be developed to configure whole IEC 61499 compliant networks, independently from the vendor of the specific devices, although specific parameters have to be provided by the vendors of the devices. This could easily result in a variety of different tools, because no unitary interface is proposed. Probably resulting in increased effort of device reconfiguration, as the handling of different tools has to be learned and the appropriate selected.

2 eXtensible Markup Language

4. Engineering & runtime environment for distributed automation and control systems

To overcome the limitations of current embedded industrial automation and control engineering methods, we propose an application centred engineering method for efficient component based modeling of applications for controlled evolution of Industrial Automation and Control Systems (IACS) and its execution.

4.1. Approach for controlled evolution

The top–level approach focuses on replacing state–of–the–art "ramp down—stop—download—restart—ramp up" methods with a simple continuous system evolution, which is controlled by an evolution control application that is modelled with components in the same way as control applications. Evolution control can be either executed from an engineering environment or—if constraints regarding fault tolerance, real–time and safety have to be met—it can be distributed to different controllers. For safe and fault–tolerant control and evolution execution the reconfiguration application is verified [10] together with hardware capability descriptions of the different devices in the network leading to trustable QoS judgment.

4.2. Modelling of control and evolution control

The following engineering process provides a method for handling system evolution of Automation and Control Systems by an efficient support for engineering of control applications and evolution control applications. This engineering method consists of the following four major parts as depicted in Fig. 3:

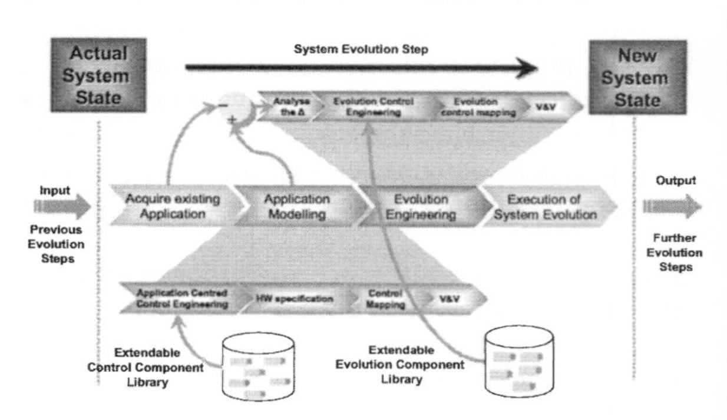

Fig. 3. Engineering process of a system evolution

1. **Acquire existing application**
   Collect all data necessary for describing the current system state and deliver it as input to the application modeling. The data consists of the system model in-

cluding applications currently running in the system, the hardware configuration of the system, the mapping of the applications to the different devices and the hardware capability descriptions.

2. **Application Modelling**
   Modelling the new control application is based on the existing ones by adding/removing components, their interconnections and specification of application properties (e.g. real–time constraints). The next step is the configuration of the hardware structure with devices and network connections. After this the modelled application parts are mapped to the according devices they should be executed on. The final step is the verification and validation of the control application in order to determine if the specified constraints will be met.
3. **Evolution Engineering**
   With the delta analysis differences between the current running control application and the newly modeled control application are determined. These differences serve as input and starting point for the modelling of the evolution control application that will change the existing application to the new one. The reconfiguration control properties and parameters are specified in the same way as control application properties (according to step 2). Similar to the mapping of the control application the reconfiguration application parts are mapped to the devices. The final step is the verification of the evolution control application.
4. **Execution of System Evolution**
   To execute evolution control applications the utilization of basic reconfiguration services at runtime platforms based on IEC 61499 management commands (cf. chapter 3) is necessary. The first step is to instantiate the evolution control application on the according devices. Next the execution of the evolution control application is done, i.e. the currently running control application is transformed into the new control application. This is achieved through basic reconfiguration services proving real–time constraint execution of reconfiguration processes at device level. To finish the evolution procedure the evolution control application will be removed after it has successfully executed all commands and all changes are finished.

Based on the basic reconfiguration services of IEC 61499 presented in Chapter 3 efficient engineering methods have to be provided to enable system evolution of IACSs.

4.3. Execution environment for control and evolution control applications

For the provision of evolution we propose an execution environment for small controller nodes of a distributed automation system based on the reference model of the standard IEC 61499. This execution environment provides an abstraction of the hardware, a safe encapsulated area for the execution of control algorithms, and a configuration management providing the basic services needed for the execution of "normal" control logic and of the evolution control application. Fig. 4 presents a schematic of the runtime [10].

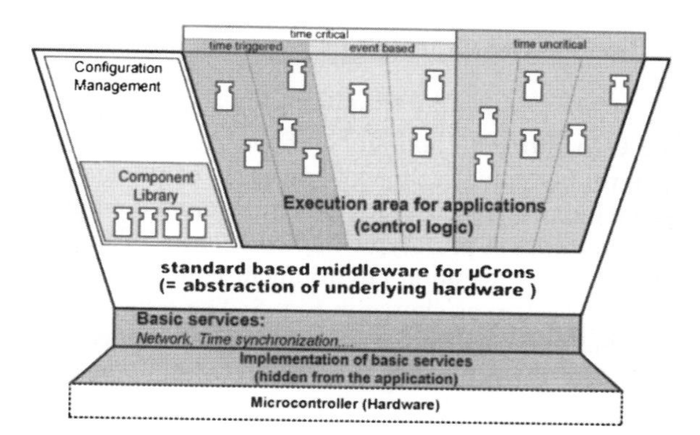

Fig. 4. Proposed execution environment for small intelligent controller nodes in a distributed automation system

4.3.1 Description of the execution environment

The **hardware abstraction** abstracts the hardware for the internal control applications in a device independent way. Typically, automation systems consist of many devices from different vendors. It provides interfaces to hardware specific features (communication and process interface).

The **execution area** provides the facilities for executing different applications in parallel without interfering each other. The applications can have different requirements regarding their quality of service. This means that time critical algorithms with time- or event triggered behaviour are executed in parallel to algorithms without any regards to timing. The algorithms itself are built out of IEC 61499 compliant FBs connected to each other.

The **configuration management** is responsible for controlling the life cycle of the applications in the execution area. Configuration commands [7,10] are send by an engineering tool or even by another execution environment of a device to create, destroy, start, stop, query and write components, variables, connections and even whole applications in predefined or calculated points in time.

The services it provides can be accessed remotely via the network to be used for example by engineering tools, or it can be accessed internally from applications controlling the reconfiguration processes. That allows special "reconfiguration applications", which have the knowledge to change themselves or other applications regarding to the system state. The real-time capabilities of the execution environment can be used to execute the configuration management like normal control applications. This allows providing real-time constrained execution also for the basic reconfiguration services.

5. Demonstration prototype

The "Testbed for Highly Distributed Control" at the Odo Struger Laboratory of the Automation and Control Institute, Technical University of Vienna provided the first test cases for validating our reconfiguration concepts. A prototypic implementation of an engineering and execution environment controls a pallet transfer system as shown in Fig. 5. The tests showed that the proposed basic reconfiguration services are sufficient for providing dynamic reconfiguration for lower level control.

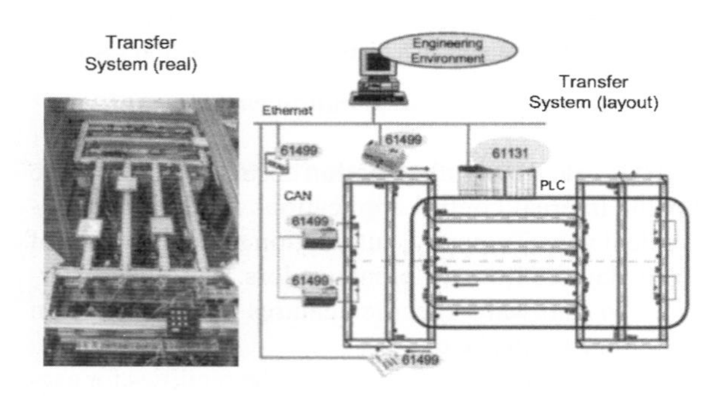

Fig. 5. Test-Bed

6. Summary and conclusions

The paper introduced a new engineering method and its execution on resource restricted embedded devices for controlled evolution of IACSs. Based on the requirements for reconfiguration methods we described a process for system evolution from an actual system state to a new system state. By aggregation of such single steps of system evolution, the whole life cycle of an IACS can be described. The reference model of IEC 61499 and its characteristics especially for management of applications have been demonstrated as basis support for this engineering method.

A new execution environment based on the standard IEC 61499 has been proposed as basis for the system evolution execution. First tests showed that the basic services are suitable for supporting reconfiguration at lower control levels.

Acknowledgements

This work is supported by the FIT-IT: Embedded System program, an initiative of the Austrian federal ministry of transport, innovation, and technology (bm:vit) under the FIT-IT contract FFG 808205 and 809447. Further information is available at: www.microns.org and www.easydac.org

PROFACTOR is partner of the EU-funded FP6 Innovative Production Machines and Systems (I*PROMS) Network of Excellence. Further information is available at: www.iproms.org

References

[1] ManuFuture: ManuFuture2003 / ManuFuture 2004.

[2] Koren, Y., Heisel, U., Jovane, F., Moriwaki, T., Pritschow, G., Ulsoy, G., Van Brussel, H.: Reconfigurable Manufacturing Systems. CIRP 48 (1999).

[3] International Electrotechnical Commision: Function blocks for industrial-process measurement and control systems. IEC Standard (2005).

[4] Brennan, R., Fletcher, M., Norrie, D., eds.: An agent-based approach to reconfiguration of real-time distributed control systems, IEEE Transactions on Robotics and Automation, Special Issue on Object-Oriented Distributed Control Architectures (2002).

[5] International Electrotechnical Commision: Programmable controllers - Part 3: Programming languages. IEC Standard (2003).

[6] Christensen, J.H.: HOLOBLOC.com - Function Block-Based, Holonic SystemsTechnology (Access Date March 2005).

[7] Christensen, J.H.: Basic Concepts of IEC 61499 (Access Date March 2005).

[8] Lewis, R.W.: Modeling control systems using IEC 61499. Number ISBN: 0 85296 796 9. IEE Publishing (2001).

[9] Vyatkin, V., Hanisch, H.M.: Formal-modelling and Verification in the Software Engineering Framework of IEC61499: a way to self-verifying systems. In: in Proceedings of the IEEE Conference on Emerging Technologies in Factory Automation (ETFA'01), Nice. (2001).

[10] Zoitl, Alois, Franz Auinger, Valeriy V. Vyatkin and Jose L. Martinez Lastra (2004).: Towards basic real-time reconfiguration services for next generation zero-downtime Automation systems. In: International IMS-Forum 2004.

Intelligent Production Machines and Systems
D. T. Pham, E. E. Eldukhri and A. J. Soroka (eds)

A multi-agent scheduler for reconfigurable manufacturing systems

S.P. Walsh[a]

[a] *School of Engineering & Technology, Deakin University, Geelong, Victoria 3217 ,Australia*

Abstract

The challenges of current and future manufacturing suggest that manufacturers will need to be more flexible and responsive if they are to thrive. Reconfigurable manufacturing systems are seen as one approach that could help with these challenges. However, many issues must be solved if a high degree of reconfigurability is to be achieved. Apart from the physical implementation issues that reconfigurable manufacturing systems present, they create significant issues that make effective scheduling more complex. This paper discusses the major issues of reconfigurable manufacturing systems from a scheduling perspective and describes a scheduler framework that has been developed, using multiple intelligent agents, that is capable of operating in a reconfigurable manufacturing environment. This multi-agent scheduler uses a task-resource heterarchical structure where agents represent the machines and the orders. The agents allocate tasks to machines by using an auction-based negotiation.

1. Introduction

If current trends are a guide, manufacturers will need to be increasingly flexible and responsive to cope with customer demands. There will be a need for flexibility beyond that which can be delivered by existing manufacturing system paradigms. These market conditions will require ever-quicker introduction of new products. The manufacturing system technologies available at present limit the level of agility that is possible in manufacturing. Adaptation of existing systems to new products is often slow (FMSs are often quite limited in their flexibility) and the launching of new systems can take a long time, limiting agility. The manufacturing systems that are used to achieve future manufacturing goals will require several key features. They must be [1]:

- Rapidly designed;
- Able to convert quickly to the production of new models;
- Able to adjust capacity quickly;
- Able to integrate technology and to produce an increased variety of products in unpredictable quantities.

One paradigm that has been conceived for tackling these problems is that of Reconfigurable Manufacturing Systems (RMS) [1,2]. An RMS is defined as follows. "A reconfigurable manufacturing system is designed for rapid adjustment of production capacity and functionality, in response to new circumstances, by rearrangement or change of its components." In practice such a system would consist of modular building blocks that are capable of being reconfigured as required.

Of course, there are many implementation issues that would need to be considered and dealt with before RMS or any other modular manufacturing system could become reality. Individual components would need to be designed to be part of reconfigurable manufacturing systems from the outset. Physical issue would also need to be considered. Also communication issues such as standard interfaces and protocols would be important. One important enabler that is required for RMS, which this paper focuses on, is an effective scheduling approach that is capable of operating with the unique challenges posed by reconfigurability.

2. Scheduling challenges in RMS

2.1 Layout

Apart from the major implementation issues that RMS exhibit, from a scheduling perspective, reconfigurability poses many challenges. With a traditional manufacturing system, the machines are arranged and configured in a fixed manner with a relatively fixed process plan for the orders. Even with a flexible manufacturing system the scope for reconfiguration is limited. The initial arrangement adopted could be a wide range of configurations but once it is implemented it is unlikely to be significantly altered. Reconfigurable manufacturing systems constructed from modular building blocks, on the other hand, would potentially be capable of being arranged, and then later re-arranged, into an unlimited number of configurations. Therefore, from a scheduling perspective, scheduling for an FMS would be different from that of an RMS. This is because in the FMS case the system configuration never changes and a scheduler specifically for that layout can be used, whereas in the RMS case the scheduler must be able to handle all configurations as the system may be radically altered at a later date.

2.2 Performance measures

With all manufacturing systems, performance measures are used to gauge the effectiveness of the system operation for a set of manufacturing objectives. These performance measures are based on the resources (e.g., Resource Utilisation, Throughput) or are based on the orders that are to be processed (e.g. Average Tardiness, Maximum Flowtime). Performance measures are sometimes a combination of different performance measures combined into one. The possibilities for performance measures were endless.

Traditionally, it was likely that during the design of a scheduler for a manufacturing system only a single performance objective, or a small set of objectives, is the focus of the scheduler. This is quite acceptable if the manufacturing system is always used for the same purpose and its structure never radically changed. However, with a reconfigurable manufacturing system, there is the possibility that the manufacturing objectives would change when the system was to be reconfigured. Hence, the scheduling system that is employed will require the capability to achieve 'good' performance for many, and ideally all, manufacturing objectives.

2.3 Distributed Scheduling

The scheduling approach that is most suitable for application to RMS (or other modular systems) is the use of some form of distributed scheduling. The reason for this is that it would allow the maximum degree of reconfigurability if the scheduling capabilities are built-in to the RMS constituent parts rather than requiring an extra module that would be wholly responsible for scheduling.

One form of distributed decision making is to use Multi-Agent Systems (MAS) [3,4,5]. In such systems decision-making is divided between a group of small local decision makers (i.e., agents) rather than being performed by a monolithic centralised scheduler. Scheduling with a multi-agent scheduler in a distributed scheduling environment provides new challenges that must be addressed.

The main issue to be addressed using this form of scheduler was for the agents to cooperate, as none will possess all the information necessary to solve the global problem. It is important to ensure that allocations are 'good' from the global sense rather than

Case 1 – No Cooperation

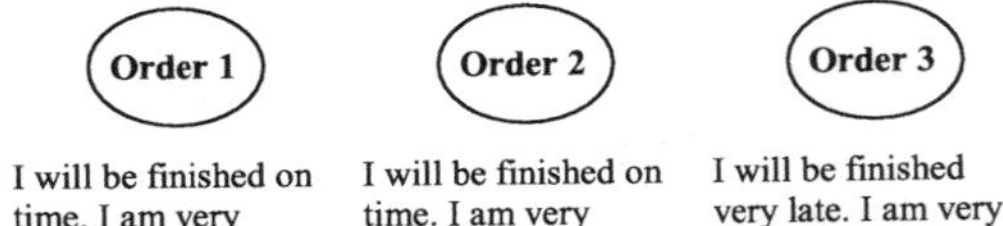

I will be finished on time. I am very happy. Tardiness = 0	I will be finished on time. I am very happy. Tardiness = 0	I will be finished very late. I am very unhappy. Tardiness = 90

Case 1 Mean Tardiness = 30

Case 2 – Cooperation

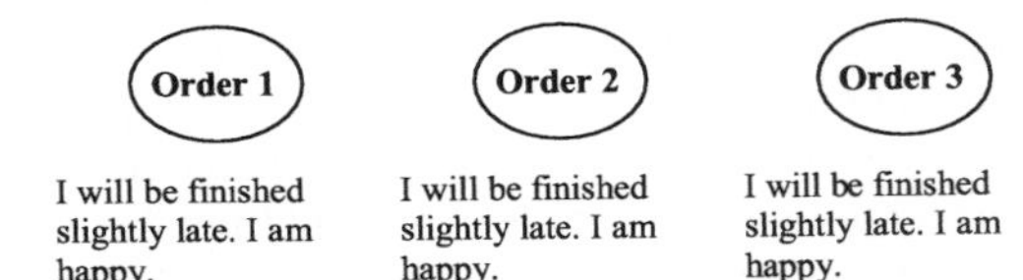

I will be finished slightly late. I am happy. Tardiness = 10	I will be finished slightly late. I am happy. Tardiness = 10	I will be finished slightly late. I am happy. Tardiness = 10

Case 2 Mean Tardiness = 10

Fig. 1. Cooperation vs No Cooperation

the point of view of the individual agents. Fig. 1 shows a simple illustrative example of the effects of cooperation.

In this example the three order agents have been

required to schedule their operations to a set of machines. In case 1 the order agents were greedy and seized the best allocations for themselves, achieving zero tardiness (minimizing tardiness being the objective). The third order then has to accept a poor allocation making it very late. In the second case orders 1 and 2 have cooperated, relinquishing their extremely good allocations to allow order 3 to be finished earlier. Even though they are less happy with their allocations than in case 1, the overall global performance (as measured by mean tardiness) is improved. None of the local schedulers (i.e., order and machine agents) have the perspective or the data necessary to achieve global goals. A method must be provided to force them to value global goals over their own local objectives.

2.4 RMS Scheduler Requirements

From this discussion of the problems and issues of a reconfigurable manufacturing system, as well as the problems inherent to multi-agent scheduling (and distributed scheduling in general), a list of objectives have been formed. In summary, the necessary features of a multi-agent scheduler for reconfigurable manufacturing systems are that it:

- Should be distributed;
- Must produce robust performance (i.e., must always work);
- Must produce effective performance;
- Must work for all manufacturing configurations;
- Must be able to achieve multiple performance objectives;
- Must seek to achieve global goals rather than local goals.

The following section provides a brief review of manufacturing scheduling techniques that use multi-agent systems or similar approaches.

3 Multi-agent scheduling

3.1. Multi-agent systems

A MAS can be defined as a loosely coupled network of problem solvers that worked together to solve problems that were beyond the individual capabilities or knowledge of each problem solver. These problem solvers (i.e., agents) are autonomous and can be heterogeneous in nature. The characteristics that are common to all MAS are [4]:

- Each agent has incomplete information, or capabilities for solving the problem, thus each agent has a limited viewpoint;
- There is no global system control;
- Data is decentralised;
- Computation is asynchronous.

Solutions to these problems are heavily intertwined and must be dealt with when designing a multi-agent system. Many broad ideas for the conceptual structure of multi-agent or other distributed manufacturing system scheduling have been developed and can usually be classified as hierarchical, heterarchical, or somewhere between.

3.2. Resource allocation

The key issue in the use of a multi-agent system to perform scheduling was how the tasks are allocated to resources. To achieve this, the autonomous entities must coordinate with each other to allocate tasks to resources, and they should cooperate with each other to ensure that those decisions lead to performance that is effective for the system as a whole. Cooperation and coordination are intertwined with each other and can not be readily separated.

A literature review by Tharumarajah and Bemelman [6] reviewed several approaches that had been used to solve distributed scheduling problems in both hierarchical and heterarchical control structures. Another review by Baker [7] looked at implementing commonly known manufacturing scheduling algorithms in heterarchical multi-agent schedulers. Several methods were reviewed and it was found that there was very little difficulty with implementing many methods in a multi-agent heterarchy. Very active research had used deterministic simulation (on-line simulation) algorithms [8] and near-optimal algorithms such as Lagrangian relaxation [9,10]. Other approaches have used dispatching rules [11,12]. Some of these methods are focused on seeking optimal or near optimal results. For many problems that MAS are suited to, seeking optimums is often impractical, as achieving optimum performance in a local problem provides no guarantee that the global performance will be improved [13].

Another common method for resource allocation that has developed is the idea of using some form of negotiation, such as auction-based negotiation, between agents, to allocate tasks. This idea was inspired by the common agent concept of bidding and was particularly well suited to heterarchical control structures. A very common auction-based negotiation method that was used was the Contract-Net method

[14]. There had been numerous uses of this method reported in the literature [9,14,15,16,17]. Parunak [18,19] noted several issues when using Contract-Net such as the need for excessive communication bandwidth if negotiations are not constrained and problems with temporal ignorance.

4. A Multi-Agent Scheduler for RMS

4.1. Overview

This section briefly describes a multi-agent scheduler that has been developed to handle the scheduling challenges imposed by RMS. In this scheduler each component in the manufacturing system has an agent assigned to handle all the scheduling and control issues relating to that component. These agents negotiate with each other to resolve the allocation of tasks to resources in a manner that is beneficial to the entire manufacturing system rather than to a few sections of the system. In the method that has been developed during this research the scheduling of tasks is handled in two separate methods. The first and most important of these methods is the initial allocation of tasks to resources. The second is the reallocation of tasks to improve production performance in the face of unexpected dynamic disturbances.

Initial Allocation uses an allocation-based negotiation method to determine the times that each task will be processed by the machines. A good example of an auction-based negotiation is Contract Net. A definition of this form of coordination and a discussion of its advantages and problems appeared earlier in this paper. Incorporated into the auction-based negotiation is a series of algorithms that use a system of 'penalties' to ascertain the effectiveness of its decisions in a 'global' rather than 'local' sense.

The reallocation of tasks is performed in response to dynamic events that cause a disruption in the operation of one or more parts of the manufacturing system. The best example of such a disruption is the breakdown of a Machine. Essentially, when an event occurs that will produce adverse 'global' effects, the reallocation algorithms of the affected agents are invoked. The agents involved calculate whether shifting tasks from one machine to another, or shifting the tasks to a later time, will produce more favorable 'global' performance than doing nothing. Generally this will require queries and negotiation between several agents to achieve.

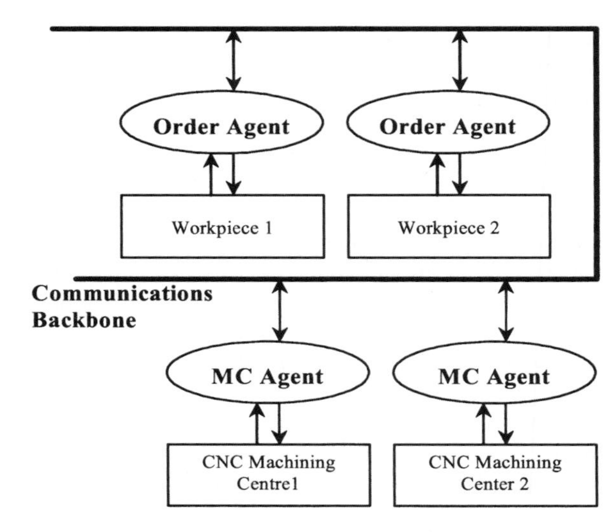

Fig. 2. Heterarchical structure

4.2. The agents

In this multi-agent approach, all physical components have an agent assigned to it that acts as its brain. For the purposes of this research there are two physical components that will be considered. These are the workpieces that are to be produced and the resources that process them. These are referred to as Machines and Orders in this paper. In the context of the manufacturing systems considered in this research an Order is a single discrete part that must have one or more tasks performed on it and a Machine is a mechanical device that performs the work on that part.

The agents are arranged in a heterarchical structure (as shown in Fig. 2) where no single agent is responsible for higher level decision making. All agents are responsible for their own components (nmachine agents being responsible for resources and order agents responsible for the products to be produced).

4.3 Allocation of tasks

The communication method that was selected to perform the initial allocation of tasks to resources can be classified as an Auction Based Negotiation. An auction-based negotiation requires communication between two parties, a manager and a set of bidders.

In the case of this scheduler the Orders Agents

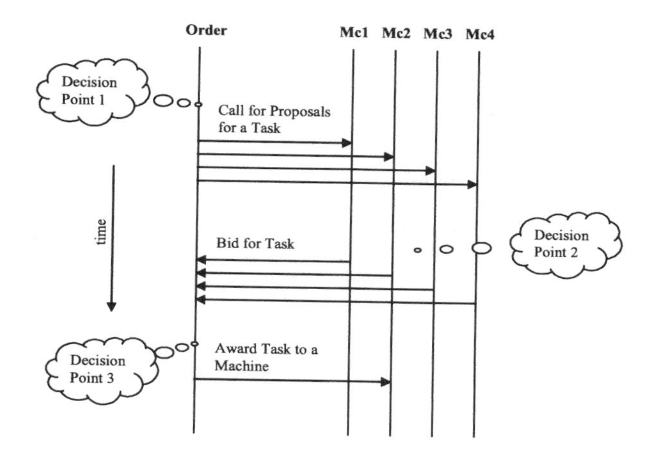

Fig. 3. Auction Based Negotiation

play the role of manager and Machine Agents play the role of bidders. In this situation the Order Agents have a list of tasks that must be performed by the Machines. It would be possible for the opposite (i.e. Machine Agents as managers and Order Agents as bidders) to have been used. That would have required to Machine agents to advertise their empty time slots to the orders to bid for. Such a method would have been equally valid but it seemed less intuitive than having the Orders request bids for their tasks.

However, just using an auction-based negotiation method alone is insufficient to achieve 'good' scheduling performance. This occurs at a series of decision-making points that occur throughout the negotiation. Fig. 3 shows the basic communication that occurs between an Order and Machines while initially allocating one of the Orders tasks, indicating three decision points. At these points scheduling algorithms that aim at improving 'global' performance are used. Fig. 3 only shows the communication at its most basic level. Further communication occurs during the second decision point stage.

The actions that occur at the three decision points are the Decision Point 1 is where the order agent determines the basic parameters (the time that it wishes a task to begin and the time that it wishes the task to be completed by) of when it wishes its task to be worked on. This information is passed to the relevant machines during the call for proposal stage. Decision point 2 is where the major work occurs. Using the task parameters each machine determines the best possible time (from its perspective) that the task can be inserted into its personal schedule. These solutions may require the shirting of already scheduled tasks. For each option that a machine trials it determines a penalty value. This penalty value is determined by asking all other agents that would be affected by the decision for their opinion of the proposed solution (a numeric value where low is good and high is bad). The solution that has the lowest penalty (i.e. the solution that disrupts all of the agents the least) is selected and presented as a proposal to the order agent.

Decision Point 3 is where the order agent selects the best proposal that it was offered. Unlike most auction based negotiations, in this approach the Order agent does not pick the bid that is immediately best for itself (in this case the bid that would result in the earliest completion time). It instead chooses the option that would have the least adverse effect on the system as a whole. This is done by judging bids on their penalty value.

5. Simulation tests

To test the scheduling performance of the multi-agent scheduler a simulation model of a reconfigurable manufacturing system was developed using the discrete event simulation modelling package Arena® by Rockwell Software [20]. Essentially, the simulation model consisted of groups of four machines and a shared buffer that were arranged into machine groups. The workpieces could move from any machine group to any other machine group through the transportation network. The simulation model could simulate different manufacturing layouts by disabling one or more machines or machine groups. For example a transfer line could be created by disabling all but one machine in each machine group, and a parallel machining situation could be created by disabling all but one machine group. A diagram of the simulation model layout is shown in Fig. 4.

Tests of the multi-agent scheduler with this simulation model have shown that the multi-agent scheduler is robust enough to operate under a wide

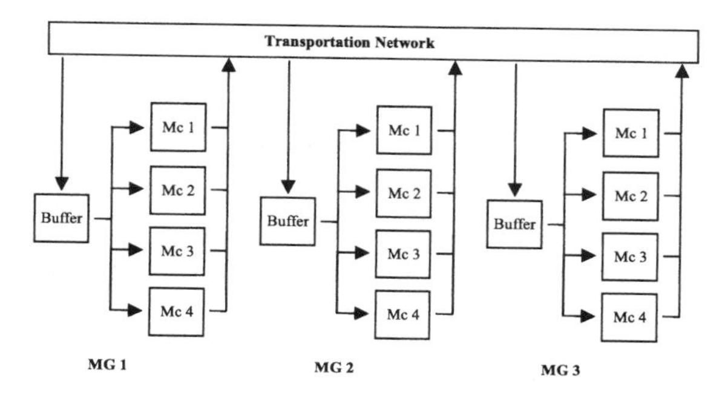

Fig. 4. – Simulation Model

variety of manufacturing layout. Simulations have included altering the manufacturing system from such diverse layouts as single machines, transfer lines, parallel machining, flow shop and job shop. The overall scheduling performance has been reasonably good in all these layouts but below the optimum performance that could have been achieved. Further work should be able to improve scheduling performance.

6. Conclusions

Simulation experiments demonstrated that the Multi-agent scheduler described in this paper way successfully able to achieve global goals under the challenging problems imposed by the nature of a Reconfigurable Manufacturing Systems. This was achieved without tailoring the multi-agent scheduler to each case in any way. Most of the multi-agent performance closely followed the performance of the benchmarks. At this stage the multi-agent scheduler should be considered a framework for shop floor scheduling in a RMS that can be further extended upon. While scheduling performance is adequate it could be improved by employing other innovative algorithms within this multi-agent framework.

Acknowledgements

I wish to acknowledge A. Tharumarajah for his many suggestions during this work.

References

[1] Mehrabi, M.G., Ulsoy, A.G. and Koren, Y. 2000. Reconfigurable Manufacturing Systems: Key to Future Manufacturing. *Journal of Intelligent Manufacturing,* 11 403-419.

[2] Koren, Y., Heisel, U., Jovane, F., Moriwoki, T., Pritschow, G., Ulsoy, A.G. and Van Brussel, H. 1999. Reconfigurable Manufacturing Systems. *Annals of the CIRP,* 2, 1-13.

[3] Lesser, V.R. 1999. Cooperative Multiagent Systems: A Personal View of the State of the Art. *IEEE Transactions on Knowledge and Data Engineering,* 11 (1), 133-142.

[4] Jennings, N.R., Sycara, K. and Wooldridge, M. 1998. A Roadmap of Agent Research and Development. *Autonomous Agents and Multi-Agent Systems,* 1 7-38.

[5] Sycara, K.P. 1998. Multiagent Systems. *AI Magazine,* (2), 79-92

[6] Tharumarajah, A. and Bemelman, R. 1997. Approaches and Issues in Scheduling a Distributed Shop-Floor Environment. *Computers in Industry,* 34 95-109.

[7] Baker, A.D. 1998. A Survey of Factory Control Algorithms That Can Be Implemented in a Multi-Agent Heterarchy: Dispatching, Scheduling, and Pull. *Journal of Manufacturing Systems,* 17 (4), 297-320.

[8] Duffie, N.A. and Prabhu, V.V. 1994. Real-Time Distributed Scheduling of Heterarchical Manufacturing Systems. *Journal of Manufacturing Systems,* 13 (2), 94-107.

[9] Owens, T.A. and Luh, P.B. 1991. An Integrated Optimization and Knowledge-Based Job Shop Scheduling System. *Proceedings of the 1991 IFAC Workshop on Discrete Event System Theory and Applications in Manufacturing and Social Phenomena,* Shenyang, China.

[10] Gou, L., Luh, P.B. and Kyoya, Y. 1998. Holonic Manufacturing Scheduling: Architecture, Cooperation Mechanism, and Implementation. *Computers in Industry,* 37, 213-231.

[11] Tsukada, T.K. and Shin, K.G. 1998. Distributed Tool Sharing in Flexible Manufacturing Systems. *IEEE Transactions on Robotics and Automation,* 14 (3), 379-389.

[12] Perkins, J.R., Humes, C.J. and Kumar, P.R. 1994. Distributed Scheduling of Flexible Manufacturing Systems: Stability and Performance. *IEEE Transactions on Robotics and Automation,* 10, 133-141.

[13] Tharumarajah, A. 2001. Survey of Resource Allocation Methods for Distributed Manufacturing Systems. *Production Planning & Control,* 12 (1), 58-68.

[14] Smith, R.G. 1980. The Contract Net Protocol: High-Level Communication and Control in a Distributed Problem Solver. *IEEE Transactions on Computers,* C-29 (12), 1104-1113.

[15] Maturana, F.P. and Norrie, D.H. 1997. Distributed Decision-Making Using the Contract Net Within a Mediator Architecture. *Decision Support Systems,* 20, 53-64.

[16] Bussmann, S. 1994. A Multi-Agent Approach to Dynamic, Adaptive Scheduling of Material Flow. *Distributed Software Agents and Applications: Proceedings of the 6th European Workshop on Modelling Autonomous Agents in a Multi-Agent World,* Odense, Denmark.

[17] Sousa, P. and Ramos, C. 1998. A Dynamic Scheduling Holon for Manufacturing Orders. *Journal of Intelligent Manufacturing,* 9 (107-112.

[18] Parunak, H.V.D. 1987. Manufacturing Experience With the Contract Net. In M.N. Huhns (ed.) *Distributed A.I.* Pitman.

[19]Parunak, H.V.D. 1989. Distributed AI and Manufacturing Control. *Decentralized A.I.: Proceedings of the First European Workshop on Modelling Autonomous Agents in a Multi-Agent World,* Cambridge.

[20] Walsh, S.P. and Nahavandi, S. 2002, Simulating a Reconfigurable Manufacturing System, Fourth International Conference on Modelling and Simulation, Melbourne

Intelligent Production Machines and Systems
D. T. Pham, E. E. Eldukhri and A. J. Soroka (eds)

Custom-Fit: A knowledge-based manufacturing system enabling the creation of Custom-Fit Products to improve the Quality of Life

A.Gerrits[a], C.L. Jones[b], R. Valero[c]

[a] *TNO Science and Industry, De Rondom 1, 5600 HE Eindhoven, The Netherlands*
[b] *Delcam InternationalPlc, Small Heath Business Park, Birmingham, B10 OHJ, U.K.*
[c] Aiju, Avd. de la Industria 23, Apartado 99, 03440 IBI Alicante, Spain

Abstract

A radical change [1] in manufacturing is starting to occur. Rapid Manufacturing [2] (RM) is based on completely new additive manufacturing techniques that produce fully functional parts directly from a 3D CAD model without the use of tooling. This offers the potential to change the paradigm of manufacturing, service and distribution with opportunities for producing highly complex and customised products.

The ambitious scope of the European Initiative CUSTOM-FIT, a Framework 6 Integrated Project, is to create a fully integrated system for the design, production and supply of individualised custom-products. Personalised and customised both to fit geometrically and functionally the requirements of the citizen for (initially) the medical & consumer goods sectors. This project is being funded by the EC over the next four years and a half and will become central to European research concerning RM. With a total budget of 16 Million Euros and with 55% industrial participation it will encompass the whole process from geometry capture through to the design of the product, the simulation of performance, production of parts and finally the distribution of the fully individualised products. This paper will describe the following subjects:

Introduction with the projects' background, output descriptions, technical, scientific, social & policy objectives core with a description of work and the enabling technologies. This Integrated Project is identified as highly ambitious with corresponding breakthrough and potential, and a carefully selected consortium provides the project a basis to become successful.

1. Introduction

CUSTOM-FIT is an industry led FP6 Integrated Project co-ordinated by Delcam plc, initiated by TNO Industrial Technology and Loughborough University. The partnership covers the complete vertical value chain from fundamental research, design and simulation up to the incorporation of a complete system, including end users.

The consortium is made up from broad base of some 32 partners (of which 35% SME) from 12 countries forming a complementary set of roles with their own individual expertise in the field. The proposed CUSTOM-FIT management includes a General Assembly, a Management Board, a Scientific and Industry Board and 8 Work Package Committees. The social aim of CUSTOM-FIT is to improve the quality of life by providing products to the citizen optimised to their individual geometrical shapes and requirements. This will improve the performance & comfort and at the same time reduce injuries [3]. The current production technologies are not able to automatically manufacture one-of-a-kind customised products and Custom-Fit is therefore the pioneer in using knowledge based (Rapid) manufacturing techniques. The following examples have been selected as test cases for the project and will illustrate its goal.

2. Applications

Prosthesis (see Figure 1)

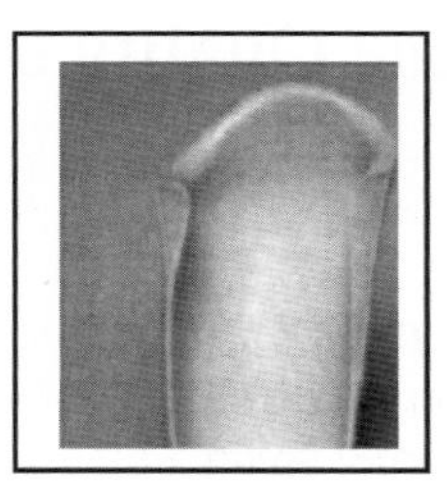

Fig. 1. Prosthesis socket

External attachments (nose, eye, ear, arm and leg, caused by birth defects, accident or cancer). For an amputee traditionally the socket is made using a negative and positive plaster model. The negative reproduces the stump topology and is produced by wrapping a plaster bandage around the stump. The positive is made pouring liquid plaster into the negative model. When the plaster becomes hard, it needs to be skilfully and correctly shaped for the socket's construction. Often several fittings are required before sufficient comfort is achieved. The new CUSTOM-FIT approach will be to scan the stump first, model the socket using new CAD techniques that will model variable graded materials and produce the socket with the LMT.

Implants (see Figure 2)

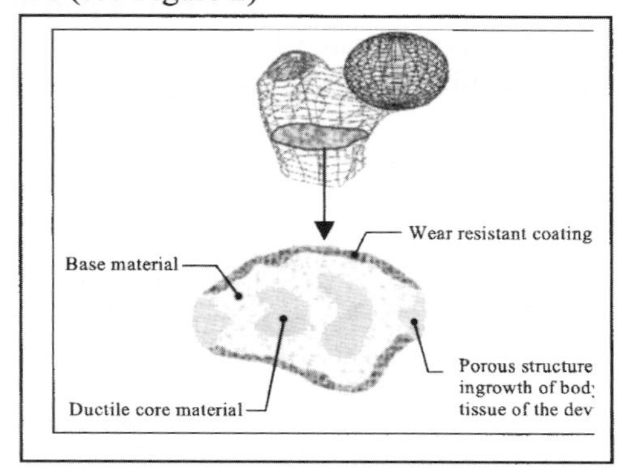

Fig. 2. Implant

This includes cleft pallet patients, trauma and amelioration (jaw bone fracture) of the dental base of the jaw, hip replacement and bone replacement for cancer patients. The aim within CUSTOM-FIT is to reduce the lead-time and investment in labour of conventional methods of manufacturing by using new knowledge based manufacturing methodology.

Helmets (see Figure 3)

Fig. 3. Complex Helmet

Protect the user against injuries by applying the RM technology to produce helmets that are custom made to improve fit and so reduces the risk of Traumatic Brain Injury (TBI). The inner cushioning will be designed to fit the rider's head. This optimised fit will also increase the comfort and therefore improve the safety by reducing tiredness and improve attention in traffic. Furthermore, a customised inner cushioning will allow the possibility of integrating spaces for integrated communication devices. This kind of integration of added functionality will be produced as a part of the signal component product.

Grips (see Figure 4)

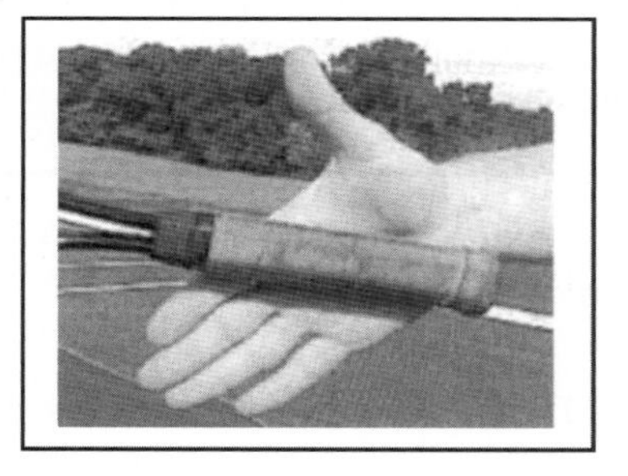

Fig. 4. Grip

Sport grips are used in different sports. In this project the targeted sports are racket sports, golf and exercise machines. In racket sports 50% of the players are affected by a tennis elbow. With a custom made grip the estimation on the number of injuries will be reduced by 30% at least. Today grips manufactured for exercise machines tend to be very basic in construction. This gives rise to a risk of injury and localised soreness/discomfort

The technological aim of CUSTOM-FIT will help transform the manufacturing sector of Europe into high-tech industries in order to regain competitiveness by:

- Moving from resource based manufacturing to knowledge based manufacturing
- Moving from mass produced single use products to new concepts of higher added value, custom-made, eco-efficient and sustainable products, processes and services

This will be achieved by developing and integrating a completely new and breakthrough manufacturing process delivering unrestricted geometrical freedom with graded structures (see Figure 5) of different material compositions.

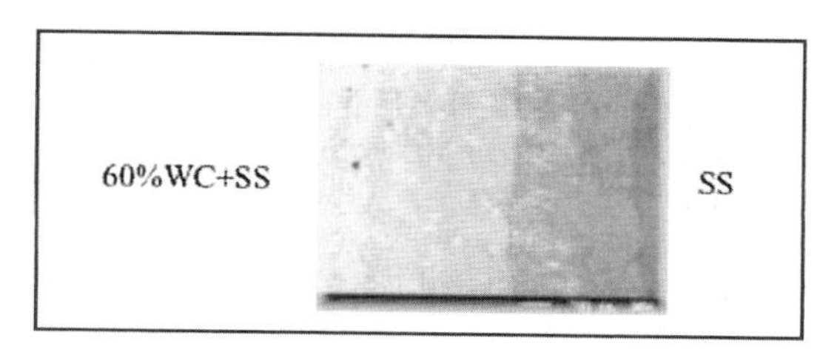

Fig. 5. Graded Material Wolfram Carbide-Steel

The products will inherently use less labour efforts since it will be based on Rapid Manufacturing (RM), IST and Material Science. This IP will give emphasis to products that will increase the quality of life of the European citizen, focussing on medical and consumer products that are customised to the shape, bio competitiveness and body contact properties of the human body. The new manufacturing system will produce high added value "on-demand" CUSTOM-FIT products, delivered to the citizens in hours rather than weeks.

3.Objectives

The specific objectives of the project are summarised below.

3.1 Technical & Scientific Objectives

- Capturing of data (f.i. Scanning, colour capturing)
- Processing of captured data (Input of all captured data, representation of graded structures, integration of non-geometrical requirements)
- Design of product (algorithms for neutral format and translator, mathematics and method to represent and design graded structures within a geometry and to transform subjective non-geometrical requirements into objective requirements)
- Manufacturing of the product (intended materials to be processed: Ti&Co alloys, engineering plastics, bio-compatible polymers, adjustment for processing other materials, manufacturing speed 25% faster than competitive manufacturing speeds)
- Final verification of the product (CEN, ISO)

3.2 Social & Policy Objectives

- Profound and credible know-how of the impact of the CUSTOM-FIT technology on supply chains, citizens, the economy and the environment [4]
- Several operational training programs on the implications of CUSTOM-FIT technology for the consumer
- A design philosophy
- Direct dissemination to 15% of all relevant bodies
- Successful commercial exploitation
- Four successful case studies by the industrial partners

Although customisation of products [5] has been recognised as having important commercial potential for many years it has generally been limited to relatively superficial cosmetic variations such as the choice of product colour, for example. The additional cost associated with customisation has been a major barrier to its wide spread adoption. The development of the fully integrated CUSTOM-FIT technology as a process will enable the determination of rapid product requirements, CUSTOM-FIT product design and tool-less rapid manufacturing [6].

This IP has the ambition to develop and integrate this system in order to enable complex customised products to be produced from unique combinations of (nano) materials without the need for dedicated production tooling. Products will be directly produced from 3D computer data without human interference. This will be undertaken in the fields of computer science & engineering (to capture the human geometry and to design the Custom Fit products), material science (invention of sustainable nano-materials & -processes), business re-engineering and logistics (brand development, supply and demand chains).

CUSTOM-FIT will drastically change how and where products will be designed and made, but there is still a long way to go; RM itself is not existing yet and furthermore a complete new manufacturing and supply system has to be developed [7].

In order to achieve the overall objectives the program will yield a number of radical innovations:

- New business process structure to accommodate and exploit the new process
- The ability to transform every kind of captured data into a standard neutral format enabling graded product design
- Design Modelling software that is suitable for functionally graded CUSTOM-FIT products and able to integrate non-geometrical requirements
- Simulation methods for checking the specifications of the designed functionally graded products
- A manufacturing process suitable for the rapid production of complex functionally graded products
- Development of novel material combinations for the rapid manufacture of functionally graded parts.

3.3 Core

Because the programme has adopted an integrated approach, covering a wide range of issues from consumer behaviour to materials testing, a multidisciplinary team will be required to undertake the programme:

- Engineers to develop the hardware
- Computer programmers/mathematicians to develop the product, process and performance modelling software
- Product designers to help develop the new design philosophy and devise the design rules for the Knowledge Based Engineering system
- Materials scientists to develop and test new materials
- Business process experts and managers from companies to assess the impact of RM on business and develop a strategy for business process reengineering to accommodate the new approach
- Training experts and educators to develop courses and teaching material to help retrain the existing and new companies

The project is split up in work packages logically enabling the creation of the complete manufacturing chain.

3.4 Implication on Business & Consumer

It is necessary to define the EU citizens' requirements for Custom-Fit products. All the requirements must be specified and the impact of the technology determined on the future social, economic and environmental landscape of the EU and also on supply, demand and logistics.

3.4.1 Capturing Technology

Within capturing technology different technologies (approx. 10 different approaches and technologies) are available to capture internal and external human geometry. Well-known technologies are body scanners, MRI and CT scanners. These technologies are outputting different formats that can not be directly used in Computer Aided Design environments.

In CUSTOM-FIT a translator will be developed to enable a translation of the different output formats of the capturing technologies to a neutral format.

3.4.2 Design technology and systems

Unfortunately, in addition to the lack of suitable hardware, currently there is no commercially available software that allows a designer to incorporate functional grading into a new product. Foremost no scientific know-how is available to represent, edit and add graded structures to (captured) geometrical data.

In CUSTOM-FIT a user-friendly system will be developed based on research for representing editing and added graded material structures. Further research will provide a methodology to transform subjective non-geometrical requirement into objective geometrical and physical requirements.

3.4.3 Rapid manufacturing processes, machines and materials

Rapid Manufacturing (RM) is defined as "the use of an additive manufacturing process to construct parts that are used directly as finished products or components". Since RM is not existing yet and currently the state-of-the-art involves only Rapid prototyping technologies, which are focussed on single pieces for evaluation purposes but not being suitable as end products. The intention is to make a step change with RM and overcome basic limitations as f.i. speed, accuracy, finish and costs all of which restrict present uptake up of the technology.

Four basic methods of production have been identified as potential machining techniques for Rapid Manufacturing purposes. All of them are potentially capable to depositing graded material to create single functionally graded components, all are currently in a development phase. In each case the research organisation is partner of CUSTOM-FIT, and will place their expertise at the consortium's disposal; then the most suitable parts will be used to establish a best of all worlds new CUSTOM-FIT process.

High Viscosity Jetting (Loughborough University, UK) is based on displacing a small drop of a high viscosity printable material to a desired location on a substrate. The HVJ process consists of a single jetting unit scaled to a block of multi-jets controlled in parallel to deposit the layer of a desired pattern. The jet units are controlled by the jet pressure, the distance from the substrate and the length of the jetting pulse. The system will have the potential to be the first true generic manufacturing process, enabling polymers, ceramic and metals to be processed with the same basic approach.

Metal Printing Process (SINTEF, Norway) is aimed at developing the equivalent of a high-speed photocopier that produces three-dimensional objects from powder material. Layers of powder are generated by attracting the metal or ceramic powder to a charged photoreceptor (PR) under the influence of an electrostatic field. The attracted layer is deposited on a building table where it is consolidated. More photoreceptors will be used to sequentially deposit different powders in the same layer. The process is repeated layer-by-layer until the three-dimensional object is formed and consolidated.

Multiple deflection continuous jet process (TNO Science and Industry, The Netherlands) is similar to a standard ink-jet system but it can currently use fluids with viscosities of approximately 300 mPa/s at jetting temperature. With this system a continuous stream of droplets is produced, the separate droplets are then electro-statically charged and deflected to form the desired pattern. Because of the higher achievable viscosities, stronger end products can be produced. Since it is a selective deposition technique it is possible to use several types of materials within the same product.

"Upside down" 3D printing process (De Montfort University, UK) is based in a polymer powder that is positioned precisely where it is required by the selective application of a temporary binder/electrostatic charge. Excess powder is removed and the material that has been deposited is fused together. This process is continuously repeated adding building and supporting materials until the desired functionally graded object has been completed.

As mentioned, in CUSTOM-FIT we face a big problem since RM does not really exist at the moment and RP processes are currently being used for Rapid Manufacturing with the problem that these machines have not been designed for these finished applications. Consequently they produce parts with poor or variable surface finish, tolerances that are difficult to maintain and the repeatability of the processes is often poor with very little real time adaptive control. Also, when manufacturing small volumes of prototypes the removal of support material is not a major issue but this will become so if there is a need to produce hundreds of thousands of parts in a short time. The speed of current layer manufacturing processes is a major issue with conventional manufacturing processes with injection moulding being 10 to 1000 times faster in material throughput.

The build envelope of current processes has been a problem but with the inclusion of Materialise as a partner who are producing very large stereolitography machines this problem is expected to be overcome.

RM offers a significant opportunity for new materials (especially nanomaterials and composites) to meet the demands for controlled porosity and graded materials. As a part of the project new nanomaterials and composites will be developed and tested to meet the requirements of functionally graded components identified in the proposed aforementioned test cases.

Nanocomposite materials consist of nanometer-sized inorganic particles dispersed in a polymer, and are an important upcoming new class of materials. The combination of nanocomposite materials with the use of additive manufacturing processes, such as the one proposed here, opens up a whole new horizon in component manufacture, as materials tailored to have different function can be deposited directly at the location where they are needed. In this Integrated Project, research will be performed in the application of nanomaterials needed for the specific products that will be made by the CUSTOM-FIT process.

The earlier described cases or "Test beds" will substantiate the CUSTOM-FIT approach allowing the partners to simulate the whole business process leading to the delivery of a fully customised product.

4. Conclusions

The ability to produce customised products, which are matched to the specific needs of an individual, are expected to have a major beneficial impact on the quality of life of European citizens. This Integrated Project is identified as highly ambitious with corresponding breakthrough and potential impact and therefore it is requiring an enormous effort and drive to achieve the objectives. The carefully selected consortium with a balanced partnership with complementary expertise in different sectors and geographical spread provides the project with a sound basis to be successful.

Acknowledgements

The authors, in particular, and CUSTOM-FIT consortium would like to express our gratitude to the European Commission for their financial support and assistance within the Sixth Framework Programme in developing this Integrated Project that would not have been possible without their help.

TNO is partner of the EU-funded FP6 Innovative Production Machines and Systems (I*PROMS) Network of Excellence. http://www.iproms.org

References

[1] Feenstra F., Holmer B., Tromans G., Moos N. and Mieritz B., "RP, RT, RM trends and developments / research survey" Proc. 4th Int. Conference on RP and Virtual Prototyping, 20 June 2003, London.

[2] Wohlers T., "The Wohlers Report 2003", USA 2003

[3] Alcañiz M., Montesa J., Lozano JA., Perpiñá C., Botella C., Baños R., Marco JH. A new realistic 3D body representation in virtual environments for the treatment of disturbed body image in eating disorders CyberPsychology and Behavior, 3:3, 433-441, Elsevier Science, 2000

[4] Porcar, R.; Such, M.J.; Alcántara, E.; García, A.C.; Page, A.; 2003; Applications of Kansei Engineering to Personalization (Practical ways to include consumer expectations into personalization); The Customer Centric Enterprise: Advances in Mass Customization and Personalization", edited by Mitchell Tseng and Frank Piller. Springer, New York / Berlin 2003; 301-313.

[5] Hague R., Campbell R.I. and Dickens, P.M., "Implications on Design of Rapid Manufacturing", Journal of Mechanical Engineering Science, January 2003 pp 25-30

[6] Wimpenny D.I, Hayes, Goodship V., Rapid Manufacturing – is it feasible?, International Journal of CAD/CAM & Computer Graphics, Vol15, N2-3-4, December 2000, pp 281-293, ISBN 2-7462-223-9

[7] Hague R., Dickens P.M., Mansour S., Saleh N. and Sun Z., "Design for Rapid Manufacturing", Rapid Prototyping and manufacturing Connference, Cincinatti, USA, April 2002, 10 pp

Intelligent Production Machines and Systems
D. T. Pham, E. E. Eldukhri and A. J. Soroka (eds)

Hybrid Manufacturing Technologies

E. Uhlmann, T. Hühns, C. Hübert

Technische Universität Berlin, IWF, Pascalstr. 8-9, 10715 Berlin, Germany

Abstract

In past few years the combination of different conventional machining processes to a hybrid manufacturing process has become an innovative approach for the machining of high performance materials. The aim in combining manufacturing processes is an increase in productivity, to broaden the limits of its application and an improvement of the product quality. In this connection a conventional manufacturing process is supported by an additional mechanical, thermal or chemical mechanism. This supporting mechanism either generates an independent contribution to the work results or it favourably alters the material properties or operational conditions during machining. At the Production Technology Centre (PTZ) of Berlin a number of hybrid manufacturing processes have been developed as well as known processes have been qualified and optimized for an application in an industrial environment.

1. Motivation

The use of new high-performance materials is often chosen to meet the demands on performance and quality of technical products. However, the excellent properties of these materials are accompanied by unfavourable processing characteristics. In order to resolve these new processing tasks, the combination of known manufacturing technologies to hybrid manufacturing methods is one alternative. The goal of the hybrid methods developed at IPK is a rise in productivity, the expansion of the spectrum of application to new materials groups as well as the increase in product quality via improved surface characteristics, minimised damage of subsurface or minimised form deviations.

2. Charakterisation of Hybrid Manufacturing Methods

Each hybrid method is characterised by one additional active principle, which is responsible for the improved performance compared to conventional methods. The various additional active principles may be categorised according to type: thermally, chemically as well as mechanically supported hybrid methods (see Fig. 1). With an additional thermal active principle, the processing of the component can be positively influenced with a specified change of the material properties or with the increase of the temperature gradients in the active zone. With an additional chemical active principle, on the other hand, passivating oxide layers from the work piece surface can be removed and the ductility of surface peripheral layers can be changed for subsequent processing. The mechanical support can occur via a superposition of the manufacturing processes with a directed or undirected secondary movement. There are two categories of hybrid methods if the relations in which the active principles stand to one another are taken into consideration:

1. Both manufacturing methods are cutting, joining or coating processes, whose active

Categorie 1 **Combination of two independent manufacturing methods**	Categorie 2 **Combination of an independent manufacturing method with a supporting process**
Thermally assisted hybrid processes • EDM-assisted grinding • Laser-assisted plasma spraying	**Thermally assisted hybrid processes** • Laser-assisted dry ice blasting • Laser-assisted cutting
Chemically assisted hybrid processes • Chemically-mechanical polishing • Electro-chemically-assisted EDM	**Mechanically-supported hybrid procedures** • Ultrasonic assisted die-sink EDM • Ultrasonic assisted profiling • Ultrasonic-assisted grinding

Fig. 1 Classification of the Hybrid Processes

principles allow an independent contribution with regard to process and work result.

2. One of the two methods is a cutting, joining or coating process. The other process supports the first during processing by changing the material properties or the processing conditions.

3. Thermally Assisted Hybrid Processes

Component processing via form changes is made difficult for many innovative materials due to their brittle-hard behaviour or their high strength. One approach to the increase in the productivity of the process as well as the quality of the work result is the change of the processing temperature via outside thermal exposure. In this way, the mechanical properties for, for example, advanced ceramics, are tend in the direction of more ductile cutting behaviour. A further increase in the local work piece temperature leads to a directed phase transition (melting, sublimating) of the material, and thereby also to a change in the form of the component in addition to the classical cutting process. Among the hybrid methods that are based on the support of a thermal active principle are:

- the EDM-assisted grinding[1, 2],
- the plasma- and laser-assisted cutting [7, 8, 9],
- the laser-assisted plasma-spraying [10] as well as
- the laser-assisted dry ice blasting.

The laser-assisted dry ice blasting is an exemplary

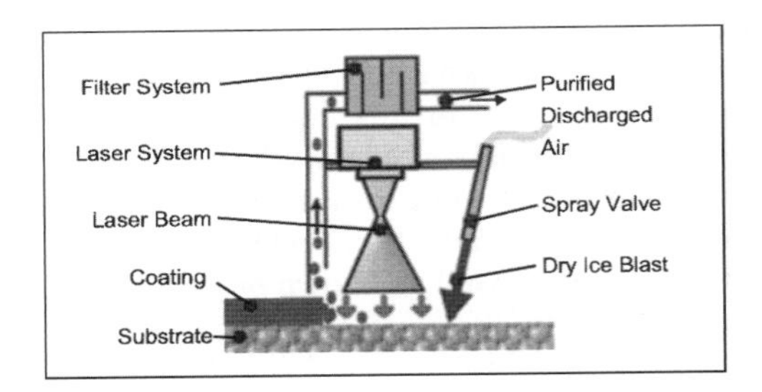

Fig. 2 Schematic Illustration of laser-assisted dry ice blasting

application of the combination of thermal affect of the laser beam with a conventional process in which the metal removal of the laser itself contributes to the cleaning result as well as strengthens the effect of the dry ice blasting (see Fig. 2). The dry ice blasting as well as the laser cleaning are environmentally-friendly cleaning methods. On the one hand, they are based on different mechanisms, on the other, they offer the same technological advantages. Neither procedure utilises secondary contamination in the form or a cleaning medium or blasting substances which require cost-intensive treatment or that have to be removed. Furthermore, both processes can be used to handle sensitive surfaces with non-conventional machining. It was against this background that the basis for the introduction of the hybrid process dry ice blasting and laser was worked out at the Production Technology Centre (PTZ) Berlin. Strongly sticking impurities should be removed via the mutually reinforcing active principles in which the use of dry ice blasting or laser alone is not possible, or the area to be treated is too

Fig. 3 Test station of the laser-assisted dry ice blasting

small. The starting point of the investigations was a modified CO_2-laser and a Nd:YAG-laser in combination with a dry ice pressure blasting system (see Fig. 3). Conventional dry ice blasting is based on the cleaning performance of three interacting mechanisms: chipping of material on the surface as a result of thermally induced stress and elastic gradients (thermo shock), the removal of particles using the impulses of the dry ice particles hitting onto the surface, and the pressure effect of the CO_2-gases which originates from the sublimating of the blasting material. During hybrid applications, the laser beam causes a rapid warming of the blasted material. This is immediately followed by the thermo shock effect of the dry ice blast, which significantly strengthens the effect. The hybrid use of both procedures allows an increase in the surface cleaning performance by up to 300 % [14, 15].

4. Chemically Assisted Hybrid Processes

A chemical material removal which supports the cutting processes is based on a chemical reaction of working materials with a fluid or gas medium. The reaction product is removed from the active zone, whereby the components are changed first on the microscopic level. The hybrid processes in this group are:

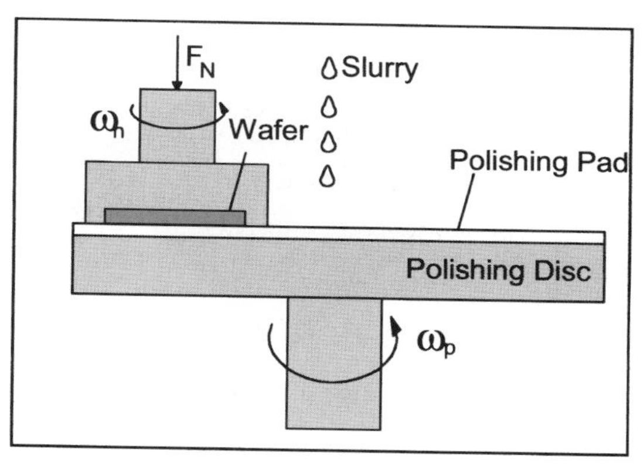

Fig. 4 Schematic composition of a wafer polishing machine for chemical-mechanical polishing

- electro-chemically-assisted EDM [3, 4] and
- chemical-mechanical polishing [5, 6].

In a row of industrial applications, especially in the semiconductor industry, there is a need for component surfaces of the highest quality. Meeting these quality demands requires further improvement of conventional lapping and polishing procedures. In chemical-mechanical polishing (CMP) two components work on the surface to be processed, which in combination make the desired processing possible. The mostly alkaline polishing solution

Fig. 5 Microchips on a wafer

(slurry) chemically attacks the surface of the component to be processed and reduces its resistance to changes of form. The mechanical components of the processing are realised via abrasive grains, which are floating in the polishing solution. During the processing, the polishing solution is continuously added and the work piece following a defined

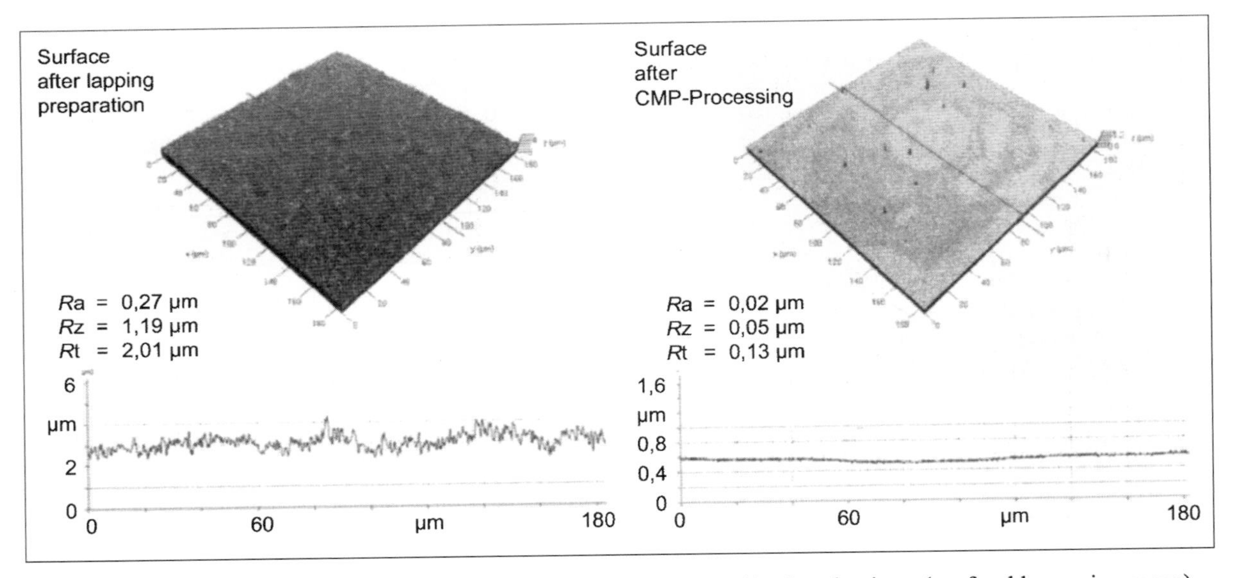

Fig. 6 Topography of the processed experimental work pieces made of lanthanaluminate (confocal laser microscopy)

kinematics is pressed onto a polishing pad (see Fig. 4). The process is industrially applied in the semiconductor branch as a super-finishing-process for single-crystal materials (see Fig. 5). The attainable surface roughness lies in the range of nanometers. With this method, essentially error-free work piece surfaces are possible, on which an undisturbed atom grid or crystal structure is exposed.

At the Production Technology Centre in Berlin investigations of wafers made of single-crystal lanthanaluminate ($LaAlO_3$) were carried out within the framework of a CMP processing research project. The wafers were used as a substrate for super-conducting coatings. The advantage of the use of CMP-processing lies in the production of surfaces, which are free of scratches. This is hardly possible with the use of conventional abrasive machining processes. The prerequisite for successful processing via CMP is an initial surface which already possesses a good surface roughness as well as exact knowledge of the material-dependent mechanical and chemical processing parameters. A multiple-stage lapping procedure for production of the initial surface was developed, which serves as the basis for the actual CMP-processing.

Within the framework of the technological investigations of the CMP process, an appropriate range of rotational speed of the polishing disc and a minimum required polishing pressure for an optimal processing result was found. Compared to traditional processing, a drastic shortening of the processing duration from 72 h to 13 h was recorded as well as a significant reduction in the observed surface roughness (see Fig. 6). [5, 6].

5. Mechanically Assisted Hybrid processes

The additional mechanical support of processing in the known hybrid processes occurs via a secondary movement, which is superimposed on the conventional tool kinematics. This brings about a change in the resulting kinematical contact conditions. Among the processes of this group are:

- ultrasonic-assisted die-sink EDM [11],
- ultrasonic-assisted profiling [12] and
- ultrasonic-assisted grinding [13].

For the cutting processing of brittle-hard materials, abrasive machining processes are called for. Whereas the grinding with bonded grains allows high metal removal rates of to be reached, a lapping process with loose abrasive grains yields high-quality surfaces with little damage of the subsurface. Thus, from machining of advanced ceramics arose the hybrid process of ultrasonic-assisted grinding. This is a combination of grinding and ultrasonic assisted lapping methods. The goal of this combination of methods was to bring the positive characteristics together in order to create an improvement in processing, via both bonded grains as

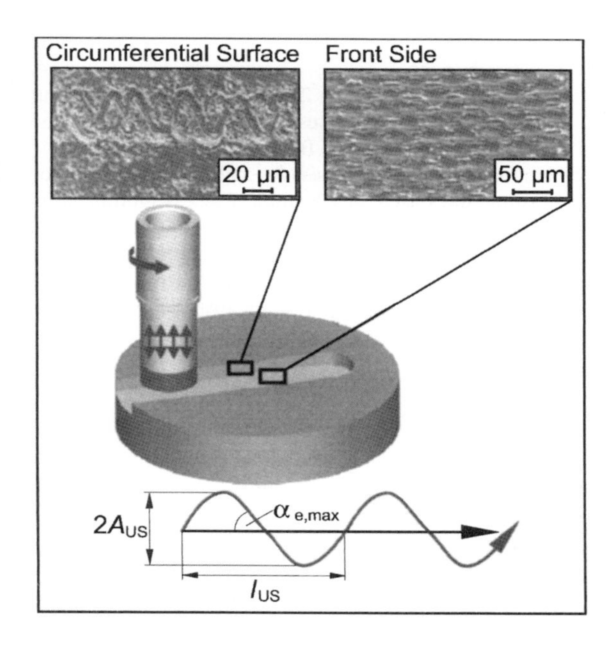

Fig. 7 Schematic illustration of the ultrasonic-assisted peripheral plunge grinding

well as ultrasonic assistance. In order to realise the vibration stimulus, the piezo-electrical principle is usually applied. The superposition of the face plunge grinding or the peripheral plunge grinding with a ultrasonic vibration leads to a fundamental change in the kinematical engagement conditions. On the one hand, as a result of the cutting movement they change to a scratching grain contact, and on the other as a result of the ultrasonic movement they change to an impulse-like discontinuous or sinus-shape engagement. On the surfaces, which have been processed by the tool on the front side, pocket like surface structures form due to the large engagement angles and short cutting length of the abrasive grains. The surfaces, which are processed through the circumferential side of the tool, are formed through sinus-shaped grinding score-marks (see Fig. 7).

Within the framework of the technological investigations, which were carried out at the Production Technology Centre Berlin, it was proven that the higher cutting speed and larger wave amplitudes of ultrasonic-assisted grinding lead to lower processing force and thereby to a more stable processing behaviour. In studies with work pieces made of zirconium oxide material removal rates were reached that lie over those of conventional grinding methods by a factor of 10 (see Fig. 8).

It was also determined that the abrasive

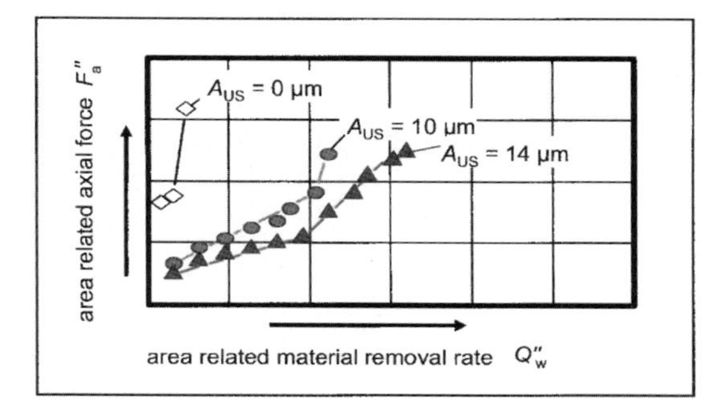

Fig. 8 Illustration of the course of axial processing force over the metal removal rate for the ultrasonic-assisted grinding of zircon oxide with different amplitudes A_{US}

mechanisms of conventional and ultrasonic assisted face plunge grinding differ in very fundamental ways. The continuous cutting contact in conventional processing leads to the rounding of edges and grain flattening, both of which indicate a high thermal load.

In ultrasonic-assisted grinding, the impulse-like impact in a grinding process characterised by its discontinuous engagement and high engagement angles leads to a micro chipping of the diamond grains. As a result of this micro chipping, sharpness of cutting is continuously generated during this process and the process force runs quasi-statically [13,15].

6. Conclusion

The overall aim in using hybrid manufacturing technologies is an increase in quality and productivity during machining of innovative materials. On a laboratory scale the presented processes exhibit a tremendous potential in fulfilling increasing machining demands. Until now very few hybrid technologies are used in an industrial application. Hence at the PTZ Berlin several R&D-projects are carried out to qualify and optimize hybrid manufacturing processes.

Appendix A. Authors

Prof. Dr.-Ing. Eckart Uhlmann is Managing Director of the Institute for Machine Tools and Factory Management (IWF) of the Technical University Berlin and Head of the Fraunhofer Institute for Production Systems and Design Technology (IPK)

Dipl.-Ing. Tom Hühns is Department Manager in

Manufacturing Technology at the IWF

Dipl.-Ing. Christoph Hübert is Head of Abrasive Machining in Manufacturing Technology at the IWF

References

[1] Zhang, J.H.; Ai, X.; Lee, K.W.; Wong, P.K.: Study on electro-discharge diamond wheel grinding (EDGM) of ceramic materials. Materials and Manufacturing Processes 11 (1996) 5, pp763-774

[2] Shu, K.M.; Tu, G.C.: Study of electrical discharge grinding using metal matrix composite electrodes. International Journal of Machine Tools and Manufacture 43 (2003) 8, pp 845-854

[3] Rajurkar, K. P.; Zhu, D.; Mcgeough, J.A.; Kozak, J.; De Silva, A.: New Developments in Electro-Chemical Machining. Annals of the CIRP 48/2 (1999), pp 559-579

[4] Schöpf, M.: Neue ECM/EDM Hybridtechnologie. Maschinenbau, Zürich 30 (2001) 3, pp 18-21

[5] Rieck, A.: Entwicklung einer Planarisierungstechnologie einschließlich Chemisch Mechanischen Polierens zur Fertigung hochauflösender Flächenlichtmodulatoren. Dissertation Gerhard-Mercator-Universität-Gesamthochschule Duisburg, 2000

[6] Uhlmann, E.; Kirchgatter, M.; Paesler, C.: Läpp- und Politurbearbeitung von einkristallinem Werkstoffen. 5. Seminar „Moderne Schleiftechnologie und Feinstbearbeitung", T.Tawakoli (Hrsg.), Stuttgart, 2004

[7] König, W.; Zaboklicki, A.K.: Laserunterstützte Drehbearbeitung von Siliziumnitrid-Keramik. VDI-Zeitschrift Band 135 (1993) 6, pp 34-39

[8] Bergs, T.: Analyse der Wirkmechanismen beim laserunterstützten Drehen von Siliziumnitidkeramik. Berichte aus dem Institut für Eisenhüttenkunde 13 (2002), pp 1-143

[9] Rozzi, J.C.; Pfefferkorn, F.E.; Shin, Y.C.; Incropera, F.P.: Experimental Evaluation of the Laser Assisted Machining of Silicon Nitride Ceramics. Journal of Manufacturing Science and Engineering (2000) 122, pp 666-670

[10] Nowotny, S.; Zieris, R.; Naumann, T.: Effiziente Flächenbeschichtung durch Laserunterstütztes Atmosphärisches Plasmaspritzen. Stahl (2002) 6, pp 43-45

[11] Huang, H.; Zheng, H.Y.: Ultrasonic Vibration Assisted Electro-Discharge Machining. SIM-Report (2003), pp1-8

[12] Liebe, I.: Auswahl und Konditionierung von Werkzeugen für das Außenrund-Profilschleifen technischer Keramiken. Dissertation TU Berlin, 1996

[13] Uhlmann, E.; Daus, N.-A.: Grinding of Structural Ceramic – Status and Perspectives for Industrial Applications. 10th International Ceramics Congress & 3rd Forum on New Materials. In: Proceedings of CIMTEC, Florence, Italy, 2002

[14] Uhlmann, E.; El Mernissi, A.: Arbeitsschutz und Sicherheitstechnik in der Demontage. Arbeits- und Ergebnisbericht, Teilprojekt A4, Sonderforschungsbereich 281 "Demontagefabriken zur Rückgewinnung von Ressourcen in Produkt- und Materialkreisläufen", TU Berlin, 2003, pp 119-160

[15] Uhlmann, E.; El Mernissi, A.; Dittberner, J.: Dry Ice Blasting and Laser for Cleaning. Process Optimization and Application. In: Proceedings of the Conference on Manufacturing, Modelling, Management and Control (IFAC-MIM'04), Athens, Greece, October 21 – 22, 2004

[16] Daus, N.-A.: Ultraschallunterstütztes Quer-Seiten-Schleifen. PhD-Thesis, TU Berlin, 2004

Intelligent Production Machines and Systems
D. T. Pham, E. E. Eldukhri and A. J. Soroka (eds)

Reconfigurable Robots towards the Manufacturing of the Future

N. A. Aspragathos

Department of Mechanical Engineering and Aeronautics, University of Patras, Rion 26500, Greece

---

Abstract

This work presents the research advances on reconfigurable robots for manufacturing applications. The differences between the fixed-anatomy and variable-anatomy robots and the promises for higher flexibility are presented. The main aspects of the reconfigurable robots are classified and discussed. The classification of the platforms and modules are presented and their characteristics are commented. Representative methods for the design, planning and control of reconfigurable robots and the relevant software for the simulation and operation of the reconfigurable robots are referenced.
The paper closes with the challenges and perspectives for the deployment of the reconfigurable robots in the manufacturing systems of the future. An integrated environment and a knowledge-base for rapid configuration of the reconfigurable robot in connection with an operating system for plug and play of the components are proposed as a perspective.

---

1. Introduction

Last decades, the robot is one of the main actors among the components of the modern manufacturing systems. A lot of research is devoted to the development of advanced control systems in order to enhance the flexibility of the robots and their capability to cope with uncertainty. However the current industrial robots have fixed anatomy corresponding to one or a limited number of tasks while the adaptation to new and more complex tasks is very difficult or impossible. Similarly the robotic manufacturing systems are designed and operate with few significant hardware modifications to process products of a limited variety. The flexibility of these systems relies mainly on the programmable controllers. The main barrier for the enhancement of the agility and flexibility of the manufacturing systems is considered the fixed physical architecture of the workcells, of the involved devices and particularly of the robots. The concept of reconfigurable manufacturing systems and robots is a promising direction towards a new era of agility and flexibility. A very intelligent brain has to be in synergy with a robust and versatile body to advance the efficiency of the system; therefore a mechatronic approach has to be followed for the next generation of robots.

A lot of research work has been devoted to the reconfigurable manufacturing systems and experts agree that this is the next step in the evolution of manufacturing industry [1]. However a little of research is carried out on the reconfigurable robotic workcells.

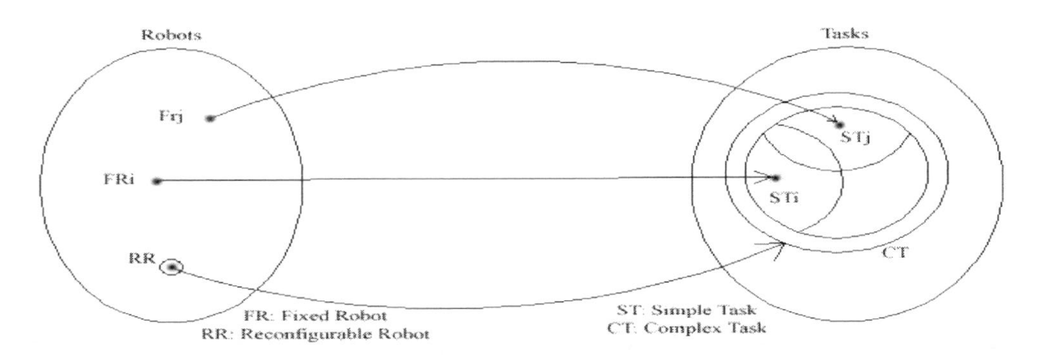

Fig. 1. Mapping of tasks to robots and vice versa.

I-Ming Chen [2] proposed an approach for the development of a component-based technology robot workcell, which can be rapidly reconfigured to perform a specific task. He claims that developed a unified control system and a platform for "plug and play" of reconfigurable robots in order to rapidly configure the optimal robotic workcell for the desired application. Other reconfigurable robotic workcells are reported for assembly automation [3] and reconfigurable fixturing and tooling [4,5].

The reconfigurable robots attracted the interest of a considerable number of researchers since the early 1980s. The concept of modular robot [6] can be considered the ancestor of the reconfigurable robot.

The usual industrial robots have fixed anatomy (topology) and change their shape by changing the coordinate variables. A reconfigurable robot is a structure of variable topology built by self-contained mechatronic modules, which can connect and disconnect easily to adjacent modules. The number, the type of modules and the way of connection vary from configuration to configuration. The modules are equipped with controllers, communications, actuators, sensors and connectors to join other modules and if they are identical then the reconfigurable robots are called homogeneous. The reconfigurable robot can change from one anatomy to another manually or by itself and in the second case is called self-reconfigurable. Particularly in space applications the robot should be able to diagnose and replace a failed module. The self-diagnosis and self-repair is a very important characteristic towards fault tolerant robots.

The ability of the reconfigurable robots to modify easily their anatomy can perform a remarkably wide variety of tasks that goes beyond the traditional robots of fixed anatomy. So a reconfigurable robot can be easily reconfigured for the manufacturing of quite different products, while the usual industrial robots can be used only for a very limited variation of the same product. In addition, it is expected that a reconfigurable robot can perform complicated task where more than one traditional robot is necessary. Fig. 1 illustrates the differences between reconfigurable and traditional robots in the mapping of the tasks to robots and vice versa.

The versatility and extensibility of reconfigurable robots promise a potential for a wide range of applications, low cost, greater autonomy and fault tolerance.

The purpose of this paper is to present a brief overview of the reported developments of reconfigurable robots for manufacturing applications accompanied by a discussion of some of the critical issues and the perspectives of their evolution. The presentation follows the main classes of the reconfigurable robots, modules, design aspects and control types. The paper concludes with the perspectives of the applications and the main challenges for the robotics research community towards the establishing a theoretical basis and practice for the deployment of the reconfigurable robots in the manufacturing industry.

2. Robot Platforms

For systematic study, the reconfigurable robots can be distinguished into two main categories namely the mobile robots and the stationary manipulators. Fig. 2 illustrates the main classification of the reconfigurable robots according to the basic structure. It is obvious that the boundaries between the classes are not clear since they overlap due to the versatility of the reconfigurable robots. For example a wheeled robot can be transformed to a walking robot or a caterpillar depending on the encountered terrain [7].

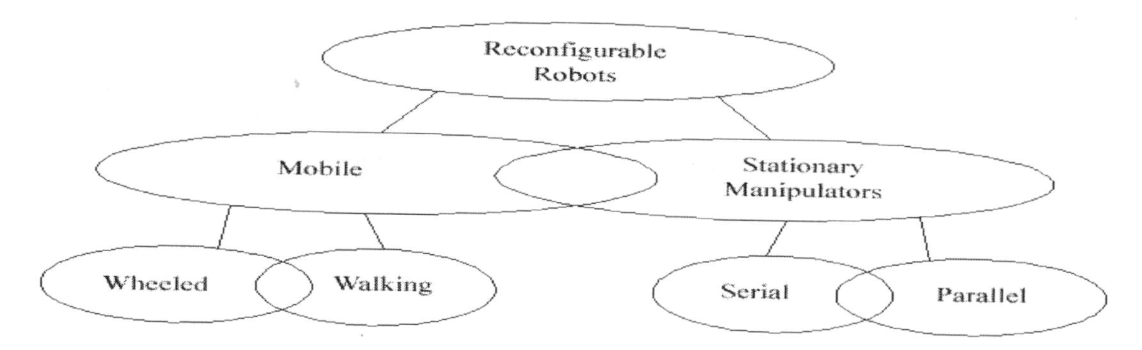

Fig. 2. A classification of the robot basic platforms.

"Tetrobot" built in Rensselear Polytechnic Institute can metamorphose from robotic arm to six-legged walker or octahedral platform [8]. The mobile robots are not in the core of the interest of this paper since at the present are not quite suitable for manufacturing applications like the reconfigurable manipulators. Most of the mobile reconfigurable robots are designed for non-manufacturing applications such as rescue robots or walking-wheeled robots for exploring irregular terrains [9]. This class is out of the interest of this review since limited to reconfigurable robots with applications close to manufacturing needs.

The stationary reconfigurable robots can be divided in two main classes the serial and parallel manipulators but hybrid serial/parallel manipulators appeared in the relevant literature eg. serial robots built from modules that are parallel robots. The "Puma" modular robotic manipulator introduced by Modular Motion Systems [10] and the Reconfigurable Modular Manipulator System (RMMS) developed at Carnegie Mellon [11] have similar conceptual design and can be considered as representatives of serial reconfigurable manipulators for industrial applications.

Parallel robots can apply high forces and present high accuracy. Recently modular design concepts are introduced to overcome the high kinematic complexity of parallel platforms [12], [13]. Hyper-Redundant reconfigurable Robotic Arms designed using as modules lower mobility parallel platforms [14], or compliant parallel mechanisms driven by optimized magnet-coil actuators [15]. Finally it is worth to notice the dual arm reconfigurable robot, which can adapt to a variety of tasks by modifying its different end-effector modules [16].

3. Mechatronic modules

The modules are mechatronic devices and can be distinguished in

- Passive or Active
- Fixed or Variable dimension and
- Connectors

The passive modules are building components of the reconfigurable robots without actuation ability but they can transfer motion, information and energy to the adjacent modules. On the other hand the active modules are really self-contained intelligent mechatronic devices including a build-in motor, sensors, a controller-driver, and the communication interface. Usually the active modules have fixed dimensions while the desired anatomy can be tuned with passive variable-dimension modules [12]. In some modules the connectors are integrated but special connectors are proposed to simplify the connection procedure [17]. A reconfigurable robot can be built by any number of modules, however the weight of the modules and particularly the weight of their actuators is a limiting factor. Karbasi et al [18] proposed a uni-drive system to regulate the motion of the modules using one motor, while Hafez et al [15] al proposed a lightweight, compliant mechanism driven by magnetic-coil actuators.

The module is the critical component and attracts the attention of the research community. Various types of modules are proposed with different actuators towards reducing the weight, the cost and the complexity while increasing the geometric, physical and mechanical compatibility. Reconfigurable robots with identical modules have several advantages: The modules can be mass produced decreasing the cost facilitate the repair and maintenance. However most of the proposed reconfigurable robots are built from two to four different modules. Unsal et all [19] propose the I-Cubes, which is a collection of two different modules namely links and cubes. On the other hand for the well-known serial robots PUMA and RMMS four different modules are designed.

4. Design, motion planning and control aspects

The design of the reconfigurable robots is a very complicate engineering task and the research and practice in this area is not mature enough for the mass deployment of the reconfigurable robots in the manufacturing industry. The development of reconfigurable robots includes design of standard modules, optimal anatomy adaptable to a diversity of task requirements, planning and control of the robot to accomplish the required task. In addition, the design approaches must take into account performance criteria and design constraints such as low payload-to-weight ratio, high speed and accuracy, isotropy, orientability, shape and volume of the workspace, scalability, dexterity and reduced cost and complexity. Part of these performance criteria and constraints are formulated in suitable indices for the analysis, design and motion planning of reconfigurable robots [14, 20]. Most of the criteria are used in design and analysis of the fixed-anatomy robots but in reconfigurable robots take a different meaning and definition [20].

The design and analysis for the reconfigurable robots is a quite complex task and the main effort and attention is devoted in the conceptual phase of the anatomy (topology) of the robot. In the case of fixed-anatomy robots, the designer has to select from a few well-established anatomies using plenty of expertise knowledge that maps the task requirements to the suitable robot anatomy. In the case of reconfigurable robots, the engineer faces a task oriented optimal anatomy problem. For simple anatomies with very few modules techniques based on enumeration of the possible combinations can be used. For generic complex highly versatile anatomies discrete optimization problem has to be solved to determine the best anatomy. Lemay and Notash [13] proposed a method for the synthesis of parallel manipulators using a combination of genetic algorithms and simulating annealing and a configuration engine for generating and designing the optimum parallel manipulator for a given task according to a set of criteria. An important and sensitive design task is the definition of the objective function and the constraints for the task where the robot has to be adapted. For motion planning and control of fixed-anatomy robots, the kinematic and dynamic equations and their inverse solutions are fixed so these are incorporated in the accompanying software. On the other hand for reconfigurable robots, techniques for the rapid automatic formulation of kinematic and dynamic equations are necessary [2]. Each module has its own motion controller and communication system. When the anatomy of the robot is determined the kinematic and dynamic model of the robot is formulated and the scalable controller is tuned.

The well known methods for robot control design has to be reconsidered and extended since the reconfigurable robots are highly complex structures and in general closed-form solution to their inverse kinematic problem can not be obtained. A promising approach for the planning and control of robots with large number of self-contained modules working in uncertain and changing environments cluttered with obstacles could be based on multi-agent control [21,22].

The software for design, planning and control of reconfigurable robots should be adapted to the requirements of scalability, complexity and highly reduced computational time. Pryor et all [23] proposed a generalized robotic software framework for kinematic control of reconfigurable robots, which includes generalized kinematics, dynamics, device interfacing and criteria-based decision making. A "software agent" approach is reported to reduce the computational time in the optimization of the robot anatomy [2].

5. Challenges and perspectives

The research and development of the reconfigurable robots it is divided to methods for design, planning, control and simulation as it is illustrated in Fig. 3. Despite the least overlapping the partitioned approach is a barrier for further advancement in the science base and practice for the deployment of reconfigurable robots in the manufacturing industry of the future. Therefore the research efforts have to focus towards the establishment of an integrated environment.

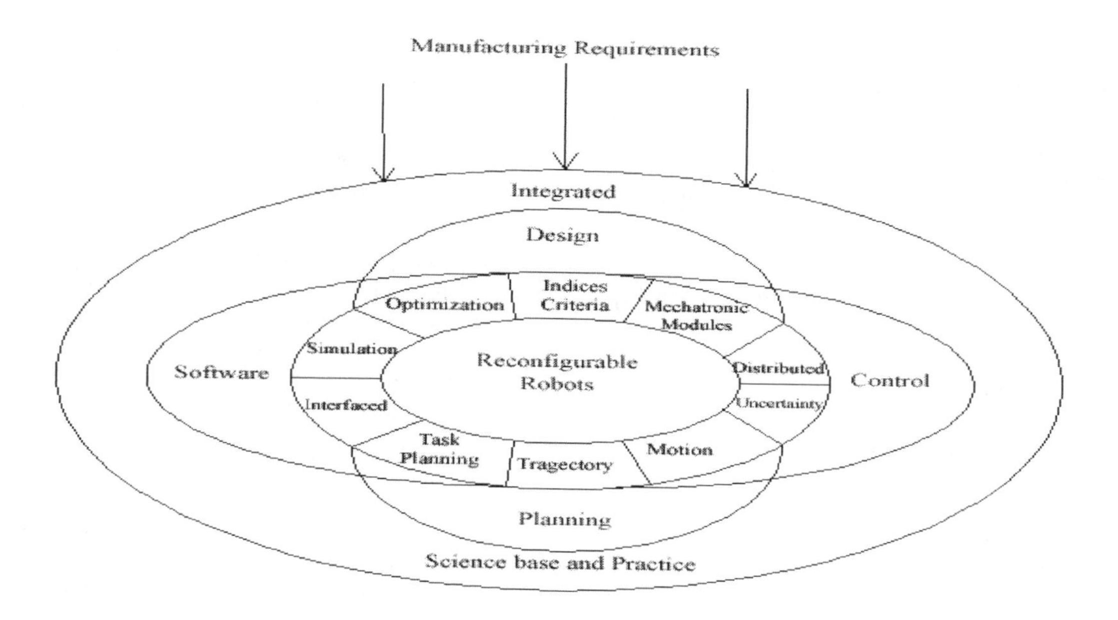

Fig. 3. An integrated environment for the design, planning and control of reconfigurable robots

A systematic investigation of the reconfigurable manufacturing systems in different levels of abstraction in order to determine the enhanced role and task requirements for the reconfigurable robots, since the applications of them lag behind the advancement of the reconfigurable manufacturing systems. Thus a systematic classification of tasks and an appreciation of their requirements are required. In this integrated science and practice environment a knowledge-based system for mapping the task requirements to the robot model and performance should be developed in order to rapidly determine the necessary configuration of the robot. This system should process simultaneously qualitative and quantitative knowledge for the conceptual design, planning and control of reconfigurable robots. In this environment the optimization of cost, performance, reliability, robustness, flexibility, and complexity should be considered [24]. For example a systematic design of a reconfigurable robot has to optimize the anatomy and the performance with the reliability and the cost of the robot, which can be estimated when, the planning and control is performed simultaneously with the design. An operating system for plug and play of the components of the reconfigurable robots has to be developed for rapid configuration of the required anatomy.

The main challenges in the advancement and deployment of the reconfigurable robots are the following:

- New anatomies and homogeneous mechatronic modules
- Optimal matching of robots structure to task requirements
- Intelligence for task planning and control with autonomous reconfiguration
- Integrated design, planning, control and simulation software

The great advances in Informatics and Artificial Intelligence could be exploited by the robotics and manufacturing research community in order to reach a level of science base and practice facilitating the deployment of the reconfigurable robots in the factory of the future for the freeing the workers from the unpleasant and hazardous working environments and for the benefit of the society.

6. Conclusions

This brief review of recent advances on the reconfigurable robots shows that this is a promising technology for increasing the flexibility and versatility of the robots. However the manufacturing applications of the reconfigurable robots are very limited. Therefore an integrated synergetic research and development effort is necessary to incorporate the reconfigurable robots into the reconfigurable manufacturing systems in the near future. On the other hand, the level of research on self- reconfigurable self-diagnostic and self-repair robots is in its infancy so it is difficult to estimate when this technology could be applied in manufacturing. For the deployment of these robots a great effort of basic research is still necessary.

7. Acknowledgements

University of Patras is partner of the EU-funded FP6 Innovative Production Machines and Systems (I*PROMS) Network of Excellence.

References

[1] M. G. Mehrabi, A. G. Ulsoy, Y. Koren, P. Heytler, "Trends and Perspectives in Flexible and Reconfigurable Manufacturing Systems." Journal of Intelligent Manufacturing, 13, pp 135-146, 2002.

[2] I-Ming Chen, "Rapid Response manufacturing through a rapidly reconfigurable robotic workcell", Robotics and Computer Integrated Manufacturing, 17, pp 1999-213, 2001.

[3] F. Yu, Y. Yin, X. Sheng, Z. Chen, "Modeling Strategies for reconfigurable assembly systems." Assembly Automation, 23, 3, pp 266-272, 2003.

[4] P. G. Ranky, "Reconfigurable robot tool designs and integration applications." Industrial Robot, 30, 4, pp 338-344, 2003.

[5] Y. Shen, B. Shinzadeh, "Dynamic Analysis of reconfigurable fixturing construction by a manipulator.", Robotics and Computer Integrated Manufacturing, 17, pp 367-377, 2001.

[6] R. Smith, R. Cazes, "Modularity in robotics, technical aspects and applications", IFS, Proceedings of International Conference in the Automotive Industry, pp 115-122, 1982

[7] S. Murata, E. Yoshida, H. Kurokawa, K. Tomita, S. Kokaji, Concept of Self-Reconfigurable Modular Robotic System", Artificial Intelligence in Engineering 15, pp383-387, 2001.

[8] G. Hamlin and A. Sanderson, "TETROBOT: A Modular Approach to Parallel Robotics", IEEE Robotics and Automation Magazine, 4(1), pp 42-49, 1997

[9] M. Yim, D. Duff, K. Roufas, "Polybot: a Modular Reconfigurable Robot", Proceedings of the International Conference on Robotics and Automation, pp 514-520, 2000.

[10] W. Schonlau, "MMS: A modular robotic system and model based control architecture.", Proceedings of the Conference on Sensors Fusion and Decentralized Control in Robotic Systems, Boston, MA, pp 289-296.

[11] C. Paredis, B. Brown, P. Khosla, "A rapidly deployable manipulator system", Proceedings of the International Conference on Robotics and Automation, pp. 1434-1439, 1996.

[12] G. Yang, I. Chen, W. Lim and S Yeo, "Kinematic Design of Modular Reconfigurable In-Parallel Robots", Autonomous Robots 10, pp 83-89, 2001.

[13] J. Lemay, L. Notash, "Configuration engine for Architecture Planning of Modular Parallel Robots", Mechanism and Machine Theory 39, pp 101-117, 2004.

[14] M. Badescu, C. Mavroidis, "New Performance Indices and Workspace Analysis of Reconfigurable Hyper-Redundant Robotic Arms" The International Journal of Robotics Research, 23, 6, pp. 643-659, 2004

[15] M. Hafez, M. Lichter, S. Dubowsky, "Optimized Binary Modular Reconfigurable Robotic Devices', IEEE/ASME Transaction on Mechatronics 8, 1, pp 18-25, 2003.

[16] Y. Fei, D. Fuqiang, Z. Xifang, "Collision-free Motion Planning of Dual-arm reconfigurable robots", Robotics and Computer-Integrated Manufacturing 20, pp 351-357, 2004.

[17] M. Nilsson, "Connectors for Self Reconfiguring Robots", IEEE/ASME Transactions on Mechatronics, 7, 4, pp 473-474, 2002.

[18] H. Karbasi, J. Huison, A. Khajepour, "Uni-drive modular robots: theory, design, and experiments", Mechanisms and Machine Theory, 39, pp 183-200, 2004.

[19] C. Unsal, H. Kiliccote, P. Khosla, "A Modular Self-Reconfigurable Bipartite Robotic System: Implementation and Motion Planning." Autonomous Robots, 10, 23-40, 2001.

[20] Pamecha, I. Ebert-Uphoff, G. Chirikjian, "Useful metrics for modular robot motion planning", IEEE Transactions on Robotics and Automation 13(4), pp 531-545, 1997.

[21] G. Birbilis, N. Aspragathos, "Multi-Agent Manipulator Control and Moving Obstacle Avoidance." ARK'04, Sestri Levante, 28-30 June 2004.

[22] H. Bojinov, A. Casal, T. Hogg, Multiagent control of self-reconfigurable robots.", Artificial Intelligence, 142, pp 99-120, 2002.

[23] M. Pryor, R. Taylor, C. Kapoor, D. Tesar "Generalized Software Components for Reconfigurable Hyper-Redundant Manipulators." IEEE/ASME Transactions on Mechatronics, 7, 4, pp 475-478, 2002

[24] V. Moulianitis, N. A. Aspragathos and A. J. Dentsoras. "A model for concept evaluation in design An application to mechatronics design of robot grippers." Mechatronics, Vol. 14, No 6, pp 599-622 July 2004.

Intelligent Production Machines and Systems
D. T. Pham, E. E. Eldukhri and A. J. Soroka (eds)

Robotised nano manufacturing: current developments and future trends

D.T. Pham[a], Z. Wang[a], S. Su[a], P.T.N. Pham[a], M. Yang[a], S. Fatikow[b]

[a] *Manufacturing Engineering Centre, Cardiff University, Cardiff, UK*
[b] *Abteilung für Mikrorobotik und Regelungstechnik, Universität Oldenburg, Oldenburg, Germany*

Abstract

This paper presents a general review of the area of nano manufacturing by robots considering current developments and future trends. The paper discusses nano motion and positioning, SPM-based nano vision and manipulation, and nano gripping and 3D handling, which are the main topics of interest in the field. Future trends are described in terms of expected applications of robotised nano manipulation and relevant requirements concerning the functions of robots. These include object-oriented nano manipulation, versatile 3D nano handling, automatic or semi-automatic 3D nano assembly, nano rapid prototyping and rapid tooling, and hybrid nano handling and manufacturing.

1. Introduction

Five years from now, nanotechnology will have an estimated global market of US $700 billion and this could escalate to US $1 trillion [1, 2, 3, 4] by 2015. With increased industrial demand for nano materials, structures, and devices, there is a growing need for robots to perform the handling, assembly, manufacturing, and testing of nano materials, components and systems.

Robotised nano manufacturing is an emerging field that deals with the controlled manipulation, handling and assembly of objects, and manufacture of nano materials, devices and systems with nanoscale accuracy [5, 6, 7, 8]. It is recognised as one of the key disciplines in nanotechnology, as the essence of nanotechnology is the ability to work at molecular level, atom by atom, to create large structures with fundamentally new molecular organisation [7, 8, 9]. By precise control of atoms, molecules or nano objects using nano manufacturing robots, new nano patterns, structures, devices and systems could be developed [5, 6, 10].

Fig. 1 illustrates a typical nano manipulation robot system structure currently under investigation by many researchers and scientists [5, 10]. It basically comprises five parts: (1) a human-machine interface for task or process control and communication between the operator and the system, (2) a controller for precise control of planned nano handling or manufacturing tasks or processes, (3) actuators for positioning the end-effectors / handling tools and nano objects to be manipulated, (4) end-effectors / handling tools for handling of nano objects, and (5) sensors for feedback control of nano operations such as imaging and force sensing. In a nano manipulation robot system, nano handling or nano manufacturing means that nano objects can be pushed or pulled, bent, twisted, cut, picked and placed, positioned, oriented, assembled, etc to build the desired nano patterns, structures, devices and systems [5].

Robotised nano manufacturing is a promising field that may lead to revolutionary new sciences and technologies, but the discipline is clearly in its infancy. It has the potential for major scientific and technological breakthroughs. However, the field also presents great challenges. This paper covers three challenging areas in the current development of robotised nano manufacturing including nano motion and positioning, SPM-based nano vision and manipulation, and nano gripping and 3D handling. This paper also discusses future trends in terms of expected applications of robotised nano manufacturing and requirements on the functions of robots. These include object-oriented nano manipulation, versatile 3D nano handling, automatic or semi-automatic 3D nano assembly, nano rapid prototyping and rapid tooling, and hybrid nano handling and manufacturing.

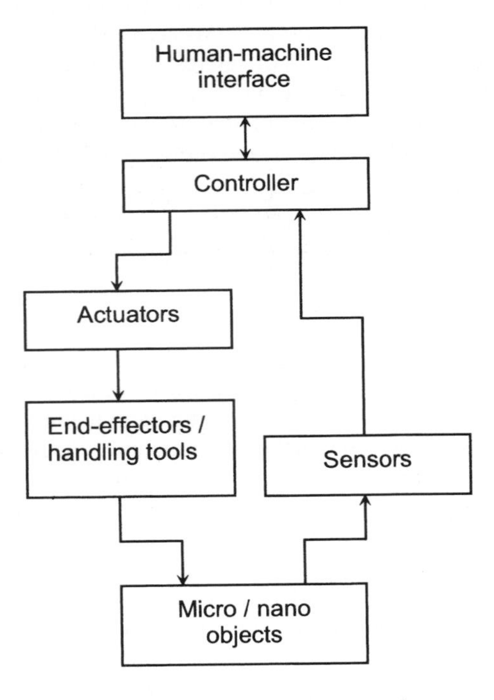

Fig. 1. Structure of a nano manipulation robot system.

2. Current developments

2.1 Nano motion and positioning

Nano motion and positioning can be regarded as early developments of nano manufacturing robotics leading to a variety of small micrometre scale objects being assembled with nanometre accuracy. This can be seen in many applications such as MEMS.

Nano resolutions for nano motion and positioning can be achieved by piezoelectric actuators. The piezoelectric actuator, in its most simple layout – a stack actuator, allows small deflections and provides high actuation forces. For some particular applications, piezoelectric actuators are built as bending actuators to increase deflections. This leads to a reduction in maximum achievable forces. The disadvantages common to piezoactuators are hysteresis, relaxation and drift effects. For instance, an scanning probe microscope (SPM) can move in x, y and z directions. The vertical displacement is controlled accurately by a feedback loop involving tunnelling currents or forces. However, nanoscale x, y motion is primarily open loop. Accurate horizontal motion relies on calibration of the piezoelectric actuators to circumvent creep and hysteresis problems [11, 12]. Thermal drift of the instrument is very significant. At room temperature, a drift of one atomic diameter per second is common. Thermal drift is negligible if the SPM is operated at very low temperatures such as 4K. This involves complex technology and is clearly undesirable. Therefore, for nano manipulation with SPMs at room temperature, drift, creep and hysteresis must be taken into account.

To reduce the influence of these effects, great effort has been put into the development of closed-loop control systems [13, 14] – either integrated into the piezo actuator or via an external deflection measurement system. However, as this principle is based on the voltage applied, the basic precision depends on the power source and the control system.

Apart from the problems with the actuators, the relative accuracy between the end-effector and the nano object to be manipulated remains a challenging topic for accurate handling of nano objects such as atoms, molecules and nano tubes.

2.2 SPM-based nano vision and manipulation

The invention of the scanning tunnelling

microscope (STM) was the first step along the road to technologies able to act at the level of individual atoms, which allows scientists to view and manipulate atoms directly [15, 16]. The success of the STM also led to the development of other scanning probe microscopes (SPMs), including the atomic force microscope (AFM) [17]. SPMs, as both nanoscale imaging sensors and manipulators, opened a new window into the nano world and have been a major force driving current developments in robotic nano manipulation.

In recent years, SPMs have been used for manipulating nano objects and building nano patterns [18, 19, 20]. One of the first demonstrations was by Eigler's group who used an SPM to position xenon atoms on nickel, iron atoms on copper, and carbon monoxide molecules on platinum [18]. A well-known example of their work was the IBM logo written with xenon atoms (see Fig. 2).

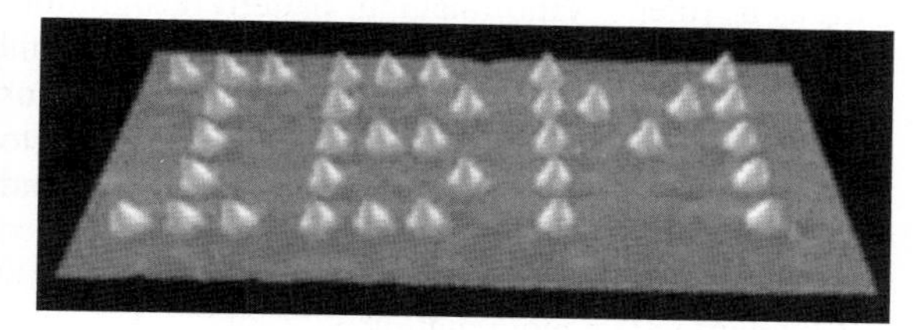

Fig. 2. IBM logo written with xenon atoms [18].

SPM-based nano manipulation is very important because manipulating arrangements between nano objects (or even atoms and molecules) provide a way to investigate and produce new generations of materials, devices, and systems with unique physical, chemical and biological properties. SPM-based nano manipulation enables breakthrough applications of nanotechnology [21]. However, there are currently two major problems with this manipulation technology: (1) SPMs are inherently not suitable for 3D operations due to their physical structures; and (2) the process is very slow as imaging and manipulation cannot be done at the same time.

2.3. Nano gripping and 3D handling

The manipulation of objects on a nanometre scale is a difficult research area that has been going on for many years. There are two main methods to pick and place nano objects. The first and most common method is "gripping" using surface forces. Compared with gravitational forces, adhesive forces are much stronger in the nano world. Therefore, an object, touched by a tip, adheres to that tip [22, 23]. In this case, the problem is not gripping, but releasing the object (the 'sticky' problem). Technologies for releasing nano objects are under investigation.

The second method to pick and place nano objects is to use specially designed grippers. In this case, the tips of the nano grippers have to be of similar dimensions as the nano objects to be able to grip them. Kim *et al* described a nano tweezer, based on two carbon nanotubes mounted on a small glass tip [24]. The gap between the tips of the nanotubes can be enlarged or closed by changing the electrical potential between the nanotubes. Another approach was made by Boggild *et al* (see Fig. 3) [25, 26, 27, 28] who used an electrostatic actuated gripper. At the end of each tip, smaller tips were grown, by using electron beam deposition. In this way, they were able to build a gripper with a gap as small as 25nm.

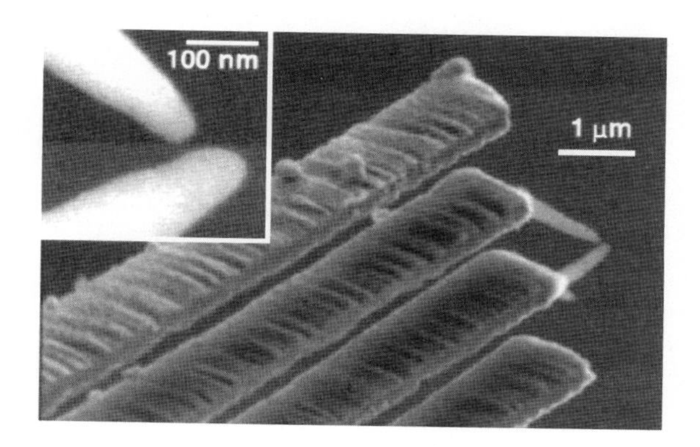

Fig. 3. A scanning electron micrograph of tweezers [25].

Nano handling has been an important field of research for more than ten years. Many manipulation tasks have been performed with SPMs [29, 30, 31]. The tip of an SPM can be used as a sensor or, as a tool pushing the nano object into the desired position. The most serious drawback of this technology is the need to work on a flat surface. To overcome this, more advanced set-ups used nano robots working in the chamber of a SEM [32, 33]. Work from the EU project ROBOSEM has contributed significantly to 3D nano handling in a SEM (see Fig. 4 and Fig. 5) [33, 34, 35, 36, 37, 38].

The development of nano tweezers and the use of SEMs are the two most promising approaches to nano gripping and 3D handling. However, further research into modelling of nano tweezers/grippers, imaging, pattern recognition and control techniques in the nano world is needed to develop robots capable of handling and assembling nano objects in 3D operations.

Fig. 4. ROBOSEM prototype robot [36].

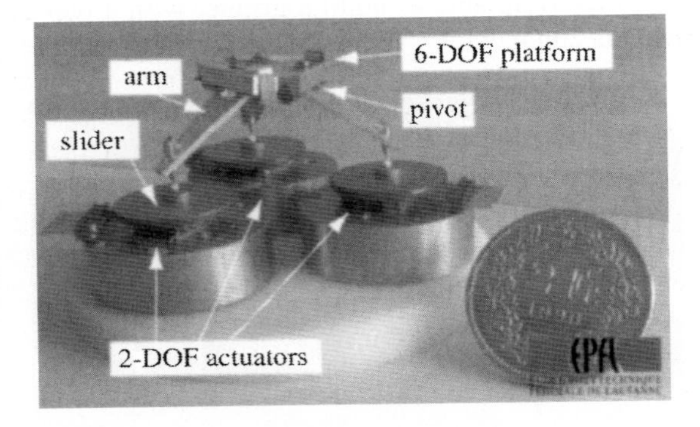

Fig. 5. ROBOSEM 6-DOF platform [37].

3. Future trends

Robotised nano manufacturing is based on nano handling and control techniques. It is clear that this technology is in its infancy, and there are currently four challenging aspects to be dealt with for applications in automatic object handling: (1) Accurate positioning of the end-effector relative to the object; (2) 3D operations; (3) smart sensor-based handling; and (4) automatic assembly or manufacturing.

New areas of development include object-oriented nano manipulation, versatile 3D nano handling, automatic or semi-automatic 3D nano assembly, nano rapid prototyping, nano rapid tooling and hybrid nano manufacturing. They are outlined in the following sub-sections.

3.1 Object-oriented nano manipulation

One of the challenging problems in nano manipulation is to ensure high relative positioning accuracy between the end-effector and the nano object to be manipulated. This topic is still under investigation for 3D nano manipulation. "Object-oriented" manipulation could provide a solution to this problem [39, 40].

Object-oriented nano manipulation has been proposed for accurate positioning of the end-effector relative to the object in an SPM or SEM [39]. This method uses real-time object pattern recognition to deal with the precise interactions between the end-effector of the robot and the object to be manipulated to remove the errors caused by environmental variations and instrument inaccuracy.

With this method, a reference pattern scale is used as a 'ruler'. Nano imaging, pattern recognition, detection and control techniques are key to this method. The interactions between the end-effector of the robot and the object could also be used to detect force and overload by nano force sensing and nano imaging techniques.

3.2 Versatile 3D nano handling

The need for versatile 3D nano handling systems will grow with their applications to the assembly and testing of nano materials, components and systems. The key problem remains the lack of versatile and smart nano handling systems, processes and techniques. Without them, nano handling tasks either have to be manually performed by well-trained technicians or are impossible to carry out.

Modular robotic manipulation systems could help fill the current technological gap in this area and thus make possible the large-scale exploitation of nanotechnology [10, 31]. They could provide scientists and engineers with useful tools for handling nano objects.

Apart from modularity, other expected features could be defined in terms of flexibility, adaptability, equipment compatibility, and ease of maintenance.

3.3 Automatic or semi-automatic 3D nano assembly

Automatic or semi-automatic 3D nano assembly is an important area in robotic nano assembly [10, 31]. New generations of materials, devices, and systems with unique physical, chemical and biological properties could be developed or produced by manipulating arrangements between nano objects (or even atoms and molecules). In recent years, research interest in robotic nano assembly has been growing and some progress has been made. However, the lack of efficient and effective techniques for assembling the nanoscale structures needed in many applications remains a major obstacle.

Typical nano assembly experiments involve a team of very skilled scientists working for many hours in a tightly controlled environment to build a pattern with tens of nano particles. Therefore, robotic nano assembly today is more of a demonstrative approach than a technique that can be routinely used.

It is clear that complex tasks cannot be accomplished unless the robot is programmed at a high level [9]. Also, nano assembly must be automated because the number of elementary operations required in most assemblies is often very large. Therefore, research is needed into programming languages suitable for assembly, and into automated techniques for nano gripping, optimal path planning, nano force detection, and overload protection.

3.4 Nano rapid prototyping and rapid tooling

A nano manufacturing robot could also provide a rapid prototyping or rapid tooling means for developing new nano products or tools. Nano rapid prototyping (NRP) is a technology for quickly creating nano patterns, structures and functional prototypes directly from design ideas or CAD models. Nano rapid tooling (NRT) concerns the rapid production of nano tools [41]. NRP and NRT are both means for compressing the time-to-market of nano products and, as such, are competitiveness enhancing technologies.

Building a nano stamper with an array of nano tubes could be an example of using a nano manufacturing robot for this application [42]. The robot could be programmed to pick and place nano tubes to form a desired pattern.

In many applications, scientists and engineers want to have a quick verification of their designs in building nano structures or tools. In such cases, a nano manufacturing robot could be used as a tool for nano rapid prototyping or tooling. This will significantly reduce the development cycle of new products.

3.5 Hybrid nano handling and manufacturing

There are problems when applying a robot system to the manipulation, handling and assembly of nano objects, or manufacture of nano products. The first problem is that of the size of the robot grippers relative to that of the objects to be manipulated (the 'fat finger' problem). The second problem is that of strong adhesive forces (the 'sticky' problem mentioned previously). Furthermore, there is a need for a robotised nano manufacturing system to carry out 'top-down' tasks such as nano lithography, cutting and electro-chemical machining. To meet these application requirements and overcome the above problems, a robotised nano manufacturing system should preferably be built with hybrid system technologies. The system could combine different nano manufacturing techniques. For instance, a robotised nano manufacturing system could be a platform for the development and use of self-assembly techniques in a more precisely controlled manner. Another example might be a system equipped with nano fabrication tools and other nano fabrication means for 'top-down' approaches.

4. Conclusion

The field of nanotechnology is generally advancing at a fast pace. A problem holding up the wide adoption of nanotechnology is the lack of tools, techniques and systems for mass manufacturing nano products. This paper has briefly reviewed the state of the art in robotised nano manufacturing. Successful developments in the areas of research discussed in this paper will enable the automation of nano manufacturing thus clearing the present bottleneck.

Acknowledgements

This work was part of the "Nano-Image Processing and Control Techniques in Nanoassembly" project funded by the British-German Academic Research Collaboration (ARC) Programme. The authors would like to thank the British Council for supporting the project. The MEC is the coordinator of the EC-FP6 I*PROMS Network of Excellence.

References

[1] Societal implications of nanoscience and nanotechnology. National Science Foundation, USA, 3, March 2001.

[2] New dimensions for manufacturing – A UK strategy for nanotechnology. DTI/Office of Sci. & Tech., 2002.

[3] Nanotechnology worth billions of Euros. Commission Press, IP/04/639, Brussels, 12 May 2004.

[4] Communication from the Commission: towards a European strategy for nanotechnology, COM (2004) 338 final, Brussels, 12 May 2004.

[5] Sitti M. Survey of Nanomanipulation Systems. IEEE-Nanotechnology Conference, 75-80, Maui, USA, 2001.

[6] Requicha AAG. Nanorobots, NEMS, and nanoassembly. Proc. IEEE, Vol.91, No.11, 2003, pp 1922-1933.

[7] Nanoscience and nanotechnologies: opportunities and uncertainties. The Royal Society & The Royal Academy of Engineering, July 2004.

[8] Nanotechnology Research Directions: IWGN Workshop Report. National Science and Technology Council / Committee on Technology / Interagency Working Group on Nanoscience, Engineering and Technology, September 1999.

[9] National Nanotechnology Initiative, National Science and Technology Council / Committee on Technology / Subcommittee on Nanoscale Science, Engineering and Technology, USA, July 2000.

[10] Smart nanorobots for automatic object handling in different application fields (NANOSMART). European project proposal, No. FP6-011862-2, 2004.

[11] Requicha AAG. Nanorobots, NEMS and nanoassembly. Proc. IEEE Nanoelectronics & Nanoprocessing, LMR, University of Southern California, 2002, pp. 1-12.

[12] Corbett J, McKeown PA, Peggs GN and R W Whatmore RW. Nanotechnology: international developments and emerging products", Annals of the CIRP, Vol. 49, No.2, 2000, pp523-545.

[13] Heerens WC, Holman AE and Tuinstra F. Using capacitive sensors for in situ calibration of displacements in a piezo-driven translation stage of an STM. Sensors and Actuators - A physical, Vol. 36, 1993, pp 37-42.

[14] Koops KR, Banning R, Scholte PMLO, Heerens WC, Adriaens JMTA, Koning WL. Design and construction of a high-resolution 3D translation stage for metrological applications. Applied Physics A, Vol. 66, 1998, pp 5857-5860.

[15] Binning G, Rohrer H, Gerber C, and Weibel E, Tunneling through a controllable vacuum gap. Applied Physics Letters, Vol. 40, No. 2, 1982, pp 178-180.

[16] Binning G, Rohrer H, Gerber C, and Weibel E. Surface studies by scanning tunneling microscopy. Phys. Rev. Lett., Vol. 49, No.1, 1982, pp 57–61.

[17] Binning G, Quate CF, and Gerber C. Atomic force microscope. Phys. Rev. Lett., Vol. 56, No.9, 1986, pp 930–933, 1986.

[18] Stroscio JA and Eigler DM. Atomic and molecular manipulation with the scanning tunnelling microscope. Science, Vol. 254, No. 5036, 1991, pp 1319-1326.

[19] Jung TA, Schlitter RR, Gimzewski JK, Tang H and Joachim C. Controlled room-temperature positioning of individual molecules: molecule flexure and motion. Science, Vol. 271 No. 5246, 1996, pp 181-184.

[20] Beton PH, Dunn AW and Moriarty P. Manipulation of C_{60} molecules on a Si surface. Applied Physics Letters, Vol. 67, No. 8, 1995, pp 1075-1077.

[21] The nano revolution. European Research, 2002, p 1.

[22] Fearing RS. Survey of sticking effects for micro parts handling. IEEE Int. Conf. Robotics and Intelligent Systems IROS '95, Pittsburgh, PA August 7-9, 1995.

[23] Fukuda F, Arai F, Dong L. Nano Robotic Would – From Micro to Nano. Proc. IEEE Int. Conf. on Robotics and Automation, 2001, Korea, pp 632- 637.

[24] Kim P and Lieber CM. Nanotube nanotweezer. Science, Vol. 286, 1999, pp 2148-2150.

[25] Boggild P, Hansen TM, Tanasa C, and Grey F. Fabrication and actuation of customized nanotweezers with a 25nm gap. Nanotechnology, Vol. 12, 2001, pp 331-335.

[26] Boggild P, Hansen TM, Molhave K, Hyldgard A, Jensen MO, Richter JJ, Montelius L ,and Grey F, Customizable nanotweezers for manipulation of free-standing nanostructures, IEEE-NANO 2001, S2.2 Nano-manipulations (Special session), 28 October 2001, pp 87-92.

[27] Petersen CL, Hansen TM, Boggild P, Bosisen A, Hansen O, Hassenkam T and Grey F, Scanning microscopic four-point conductivity probes, Sensors and Actuators A, Vol. 96, 2002, pp 53-58

[28] Lin R, Bøggild P, and Hansen O. Microcantilever equipped with nanowire template electrodes for multiprobe measurement on fragile nanostructures, Journal of Applied Physics, Vol. 96, No. 5, 1 September 2004, pp 2895-2900.

[29] Requicha AAG, Meltzer S, Terán Arce SFP, Makaliwe J H, Sikén H, Hsieh S, Lewis D, Koel BE and Thompson ME. Manipulation of nanoscale components with the AFM: principles and applications; IEEE Int. Conf. on Nanotechnology, Maui, 2001.

[30] Baur C, Bugacov A, Koel BE, Madhukar A, Montoya N, Ramachandran TR, Requicha AAG, Resch R, and Will P. Nanoparticle manipulation by mechanical pushing: underlying phenomena and real-time monitoring. Nanotechnology, Vol. 9, 1998, pp 360–364.

[31] Falvo MR, Taylor RM, Helser A, Chi V, Brooks FP

Jr, Washburn S, and Superfine R. Nanometre-scale rolling and sliding of carbon nanotubes. Nature, Vol. 397, No. 6716, 1999, pp 236-238.

[32] Yu M, Dyer MJ, Skidmore GD, Rohrs HW, Lu XK, Ausman KD, Von Ehr JR and Ruoff RS. Three-dimensional manipulation of carbon nanotubes under a scanning electron microscope. Nanotechnology, Vol. 10, 1999, pp 244-252.

[33] Garnica S, Hülsen H, Fatikow S. Development of a control system for a microrobot-based nanohandling station", Int. IFAC Symp. on Robot Control (SYROCO), Wroclaw, Poland, 2003, pp 631-636.

[34] Anon. Smart and small robots for the micro-world. (online). Available at <http://www.robosem.org/n_s_robosem_27052002.pdf >. (visited on 30 May 2005).

[35] Perez R, Agnus J, Clevy C, Hubert A and Chaillet N. Modeling, fabrication, and validation of a high-performance 2-DoF piezoactuator for micro-manipulation. IEEE/ASME Transactions on Mechatronics, Vol. 10, No. 2, April 2005, pp 161-171.

[36] Anon. Building the nanofuture: factory in a microscope. (online). Available at <http://europa.eu.int/comm/research/industrial_technologies/articles/article_370_en.html >. (visited on 30 May 2005).

[37] Anon. Smart and small robots for the micro-world: project images (online). Available at <http://www.cordis.lu/nanotechnology/src/pressroom-pub.htm>. (visited on 30 May 2005).

[38] Kleindiek S. Nanorobots for the SEM, Microsc Microanal 10 (Suppl 2), Microscopy Society of America 2004, pp 946-947

[39] Pham DT, Wang Z and Su S. Measurement technique and system based on sub-nano-scale reference patterns. UK Patent Application, No. 0303697.7, 2003.

[40] Knight W. Nanoscopic 'ruler' could provide microchip benchmark. New Scientist Breaking News, NewScientist.com, 25 February 2005.

[41] Pham DT and Dimov SS. Rapid Manufacturing – The technologies and applications of rapid prototyping and rapid tooling. Springer-Verlag London, 2001.

[42] Pham DT, Su S and Wang Z. Methods for making nanotube/nanoball-array composites, micro/nano-filters and tools. UK Patent Application, No. 0303696.9, 2003.

ning and simulation problems from small factories to large enterprises with hundreds of connected units.

It is obvious that the emulation provided by a multi-agent system can identify problems of the entire manufacturing process as well as it can help improve the performance of the process without influencing the real equipment that is actually not required for the purposes of testing. Modularity and easy reconfiguration of the multi-agent solution makes it possible to test several configurations and improve the efficiency of the process.

The proposed architecture is built on the multi-agent technology. The system is highly configurable and allows the user to model different configurations of the process. The simulation units (represented by the agents) contain particular parameters and constrains of the simulated units (e.g. production time, inputs, outputs, distribution of errors, etc.). The agent approach allows the integration of a new unit by simple means – the new agent is added. This makes it possible to change the configuration very easily, even in runtime. Each agent finds all its partners, which can provide him with the inputs or, more generally, with all agents that are able to provide inputs or to process its outputs. The configuration of the production process is self-organizing and it is defined by the parameters of the simulation units (agents). We recognize three main simulation units (agents): the processing unit, the stocks (buffers) and the transport units.

3 Time Synchronization

Time flow is one of the most complicated problems in simulation. Desired simulated time granularity usually goes to seconds or minutes, while the simulation period goes to weeks or months. The special *time synchronization agent* (TSA) is necessary. However, detailed simulation (e.g. with second or minute steps) of a long-time period can take a significant amount of real time. The synchronization agent has to be able to change the granularity of the simulated time in order to meet the best possible ratio between the required level of detail and the duration of the simulation.

The proposed system is based on *parallel discrete even simulation* paradigm. The problem is to keep the time flow in distributed agents environment synchronized and to guarantee proper causality of the events in the system. We propose the following time synchronization model for manufacturing simulation using multi-agent technology. The *time synchronization agent* is the centralizing point of the system and it is responsible for the time flow control in the entire system.

3.1 Confirmed Time Windows Synchronization

The time flow simulation consists of two synchronizing methods – *local synchronization* and *time windows* extension.

The local synchronization part is a non-central-oriented synchronization (it doesn't use TSA). It uses time stamps for each operation in the manufacturing process. Individual agents operate asynchronously and when a causal synchronizing event occurs (e.g. a low or a high amount of components in the buffers – input/output buffer under/overflow), they wait for the others. This method maximizes CPU utilization but running asynchronous agents may cause problems when the need of simulation interruption or parameter changes occurs (Figure 1-a).

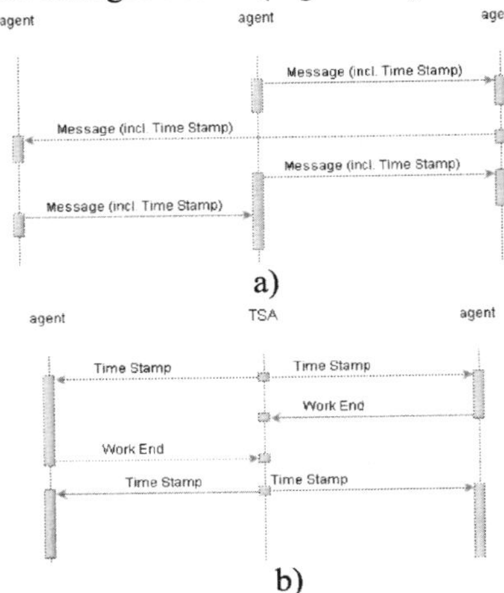

Figure 1: Confirmed time windows synchronization

The time windows extension of the previous method ensures the entire system to be synchronized at certain points of the simulation. Individual agents operate asynchronously using local synchronization within the time windows defined by the TSA. When an agent reaches the end of the time window, it sends a *confirmation* message to the TSA. TSA is responsible for collecting all the confirmation messages and subsequently it opens a new time window. This extension keeps the CPU load maximized. Due to the asynchronous nature of agents' operation taking place within the time windows, simulation can only be interrupted between two time windows (after collecting the confirmations and before opening of the new time window) (Figure 1-b).

The strong points of this approach are *fail-proof* operation (in terms of causality failure due to bad time-synchronization), possibility of *simulation interruption and continuation* between two time windows, very good *CPU utilization*, and the possibility of *time compression and dilatation* (one agent operates faster than the other, e.g. an idle machine consumes much less computation time than an operating machine). The weak points are *time non-uniformity* (each agent operates with different simulation time at the same moment) and *time non-linearity* (the speed of time flow is not constant).

In our case, the most important criterion is the fail-proof operation, the possibility of interruption and the CPU utilization. We accept message overhead of this method because confirmation messages can be extended by the information needed for other important components (see section 4.3).

4. System architecture

The architecture of the system can be divided into three functional parts:

i) simulation agents,
ii) time synchronization and control, and
iii) analytical and visualization part.

Simulation agents represent all the important production units. There is also a special set of agents for time simulation and control and for the visualization component. Together they form a manufacturing simulation multi-agent system.

The system is implemented in JAVA using the A-Globe agent platform [10]. This platform provides all the necessary support with minimum overheads and a very fast messaging system [9, 10].

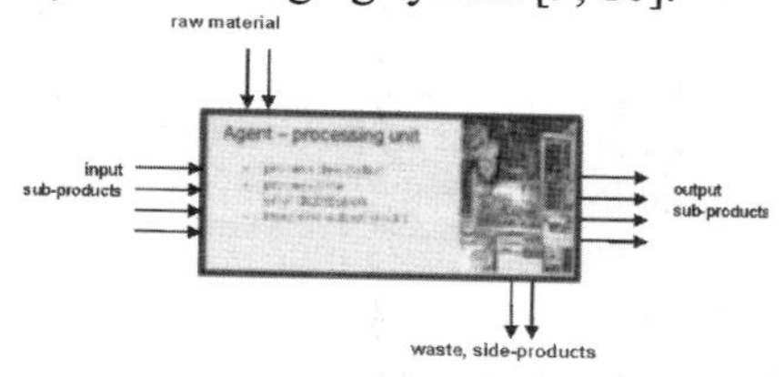

Figure 2: Agent as a general manufacturing unit.

4.1 Simulation agents

Simulation agents represent relevant manufacturing components. All the desired functionality is covered by an agent. We recognize three main classes of simulation agents – the processing unit, the stock, and the transport unit.

4.1.1. Agent structure

Each *manufacturing processing unit* is described by its main parameters (see Figure 2) such as inputs (raw material, sub-products), outputs (sub-products, waste), processing time, error distribution, size of input and output stocks, etc. The agent's behavior is to transform the inputs to the outputs upon the specified conditions.

Stocks (buffers) are a special case of processing units with reduced behavior. Stocks don't provide any transformation of the input sub-products to the output sub-products. They contain only parameters defining the size of the stock. The output product is the same as the input product.

Transport units are available for modeling time delays needed for transporting the sub-products (conveyor belts, etc.). The parameters of transport units are mainly time and the capacity. Once again, transport units can be modeled using reduced simulation agents.

Each simulation agent is equipped with a description of its desired role in the system. It is defined in the XML configuration file. The parameters of the agent are

- a set of input sub-products/raw materials
- a set of output sub-products/waste
- a set of production recipes
- input/output buffers capacities and critical limits
- failure conditions
- position of the machinery represented by the agent in the simulated world

Each *recipe* is defined by input components (with possible supplements), output components (with the error distribution of spoilage), and the processing time. During the simulation, an agent tries to keep its *input buffers* above critical limit and its *output buffers* below critical limit. When the agent's input buffers contain enough sub-products to start a certain receipt and the output buffers are able to accommodate the produced sub-products, the receipt can be executed. If more receipts can be executed at the same time, one of them is chosen randomly. The process of receipt execution is periodical as the simulation runs and any receipt can be carried out. When the *failure condition* is reached, the agent stops execution of the receipt and waits for the 'machine repairing'.

4.1.2. Contracting and Negotiation

Each transport of material or negotiation about sub-products availability is provided by inter-agent communication. The main goal of the agent's *con-*

tract module is to arrange the flow of material among agents. The *Contract-net-protocol* [11, 16] is used for contracting the producers and consumers of the agent's sub-products. The contracting behavior is triggered whenever the input or output buffer volume reaches its critical limit. After successful contracting, the agreed amount of sub-products is virtually transferred from the output buffer of the producer to the input buffer of the consumer. The duration of this transfer depends on the distance between agents in the simulated world.

Another negotiation protocol used is *query-inform* [12]. It is used for discovering the sub-products availability, respectively the time when a particular sub-product will be available in the producer's output buffer.

Every message contains a *time stamp* for time synchronization purposes. It is important for maintaining causality in the asynchronous time-flow environment. The contract module is aware of the last time stamp known to every co-operating agent and the future availability of the sub-products.

4.2 Time synchronization and control

Time synchronization of all simulation agents is provided by *confirmed time windows synchronization* method (see chapter 3.1).

Every agent is connected to the *time synchronization agent* (TSA). There is only one instance of this agent in the system. TSA is responsible for opening and closing the time windows. In fact, the TSA sends only a message responsible for opening a new time window. Information of the time window closing possibility is gathered from GEA (see bellow). This information is subscribed and when TSA receives confirmation of finishing all the activities within the current time window it then opens a new time window by broadcasting the opening message for the new time window (this message contains time stamps of the beginning and the end of this window). Just before this message is sent, it is the only moment when the simulation can be interrupted and TSA checks for requests for interruption provided by *simulation control module* (see below) before opening new time window.

The *Global Environment Agent (GEA)* is used for collecting all the distributed data of the entire system. It is responsible for manufacturing process visualization, simulation control, synchronization and visualization (see chapter 4.3).

Simulation Control Module (SCM) allows the user to control parameters of the simulation, such as to start, to pause and to stop the simulation, to adjust the time windows duration and to set the simulation starting and finishing time (see Figure 3). It also shows the current simulation time of the Logical Visio (see bellow). The time window settings consist of two parameters – *window size* (simulation time within one time window, e.g. 60s) and *window duration* (the corresponding minimal 'real' time, e.g. 3s). A combination of these two parameters defines the *speed factor* of the simulation in relation to the real time (simulation runs 20 times faster in this case). Once the user interrupts the simulation (by pressing the pause button) the time window parameters can be modified.

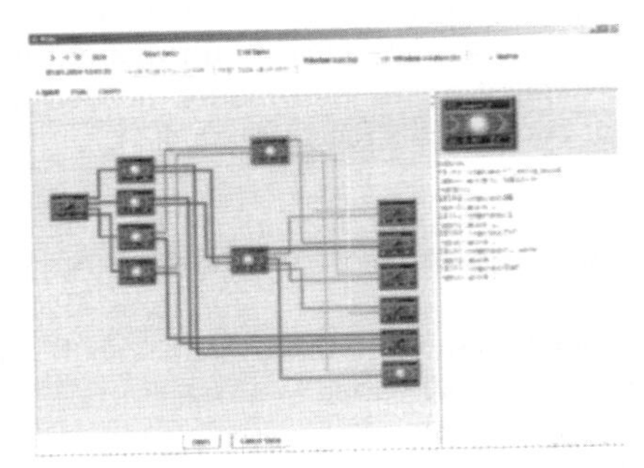

Figure 3: Logical Visio

4.3. Visualization and data evaluation

The simulation environment is presented to the user in a single compact application window. The time synchronization agent control is integrated in its bottom part (see Figure 3). It allows the user to control the necessary parameters of the simulation. All simulation units are clearly presented on the screen together with the inter-agent communication links and material flow. The simulation system presents internal data to the user in three different ways:

- schematic logical visualization,
- analytical data visualization,
- 3D simulation visualization.

Besides the 'on-screen presenting' features, all data can be stored in an external file for future analyses. All the visualization components are connected to the Global Environment Agent.

4.3.1 Logical Visio

The main screen of the simulation system contains three visualization tabs (see Figure 3). The first one ('Layout') belongs to the Visio that shows the logical structure of the simulation process. In the top-most part of the screen, there are simulation control components that are always accessible. The main part of the Visio (left) shows the logical structure of the system. On the right side of the screen the details of the selected agent are shown. Each agent is presented as a box with its name, the input/output buffer relative loads, machine load and the agent's current state (working, idle or broken). A more detailed description of an agent can be seen in the right-hand part of the screen. Agents are connected by lines that represent connections between machines as well as the communication or production flow (highlighted by different colors).

4.3.2. Analytical Visio

The other two tabs ('Plan' and 'Charts') belong to analytical visualization.

The *plan part* shows the processing activity of the simulation agents. It is able to show the load of the agents during the simulation in the time-table charts (see Figure 4-a). This Visio was adapted from the multi-agent planning system ExPlanTech developed by Gerstner Laboratory [5] and it is able to show not only the part of the production that has already been finished, but also the future plans (this can be used once the system is extended with the planning module). It is possible to show the overall time-table or inspect the individual, more detailed agent's time-tables.

The *charts part* shows several charts of the agents' activity and state (see Figure 4-b). For each individual agent, there are charts that show utilization of input/output buffers and machine load. Every chart shows the minimal, the maximal and the end value for each time window.

4.3.3. 3D Visio

The game-like visualization is provided for an in-depth inspection of the material flow and machines' activity in the manufacturing process (see Figure 5). It provides the user with intuitive overall information about the system configuration, together with detailed information about individual agents and sub-products. This component is based on Crystal Space game engine [14] and it was previously successfully deployed for several multi-agent systems [13].

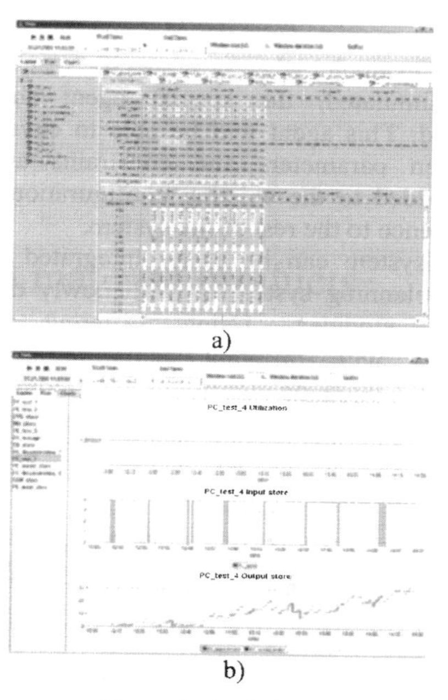

Figure 4: Analytical Visio.

4.3.4. Log Manager

The log manager collects all the information in the system within each time window. When all the information that belongs to the open time window is collected, the GEA is informed about the possibility of closing the current time window. All the information is stored and it is available for visualization components (i.e. immediately for the Analytical Visio and at the end of each time window for the others). The complete log is saved to a file for future use and further analyses.

5 Conclusions and future work

Agent technologies provide high flexibility and modular architecture with possibility of distributed processing. Simulation provides us with the possibility to test several configurations of production processes and to find out bottlenecks or possible problems. The system provides very fast and reliable experimental environment that makes it possible to choose the most expedient structure from the set of tested configurations. The agents running in the simulation define the structure of the process. The structure can be simply changed by adding or removing individual agents.

The natural features of multi-agent system such as openness guarantee easy extension in the future

Figure 7 : Cut through a CAD solid model showing the "empty" inside.

An extensive inventory of mathematical models to describe FGM structures (see e.g. [13 14, 15, 16, 17]) resulted in four potentially suitable options:

- Voxels have, like pixels, no dimension and are quantified by a scalar. Akin to the way that a digital image is represented by an array of pixels, a volumetric dataset is made of voxels laid out on a regular 3D grid [18, 19]. Besides the problem that a grid of voxels don't show mutual coherence, the geometric accuracy of a solid represented by a grid of voxels is determined by the number of voxels.
- FEM elements are used for Finite Element Analysis. Therefore a simplified model is used, consisting of many finite elements (tetrahedral, bricks, etc.) interpolating nodes in space.
- Particle system elements are objects that have mass, position, and velocity, and respond to forces, but that have no spatial extend. Because they are simple particles, by far the easiest objects to simulate. Despite there simplicity, particles can be made to exhibit the wide range of interesting behavior.
- Vague Discrete Modeling (VDM) elements have been developed so that the shape of a product doesn't have to be completely defined. A vague discrete model is vague in the sense that multiple objects are represented by one interval model, and that multiple shape instances can be generated based on certain rules.

All these types of entities do have the major disadvantage of using a lot of memory, some of them are rather complex or do not have any relation between the entities, badly influencing the editing of these kind of formats.

To overcome the disadvantages of above mentioned data formats for the descriptions of FGM structures, TNO developed an alternative approach. The next section describes how this approach is implemented in the software program Innerspace™. Basically the approach is based on the assignment of certain properties to specific areas within the enclosed volume of a solid. These arbitrary properties determine the characteristics of any point in the solid.

5 Innerspace principles

The best way of understanding what a modeling system for FGM structures should be capable to do is analyzing the situation from the point of view of a user.

Suppose a manufacturer of sweets developing a new kind of candy stick, see Figure 8. The instructions are:

The candy stick should be sour at one end and sweet at the other end

The candy stick should be red at the outside and white at the inside.

With these two instructions the model properties on each position have been described within the candy stick. The modeling system for FGM structures should now translate these desired model properties to material properties. The system must be able to give the value of the specified properties (sour, sweet, red, white) at any point so that the stick can be produced, realizing the correct material composition. This example shows it is a waste of effort to define the properties of every point of the solid representing the candy stick. To describe the structure of the candy stick, in this example the candy stick manufacturer started defining properties in some specific areas. The same procedure will be used to find an automatic way to calculate the properties of the stick at an arbitrary point (see Figure 9).

These areas, where the properties are equal will be named "props". The areas between these props are described by the transition function from one prop to the other, so-called "distribution function". Finally, when calculating the distribution of materials it will be clear that the distribution functions are limited by the boundary of the solid, so-called the domain. Summarizing, the modeling system for FGM structures is defined by:

- domains
- props
- distribution functions

5.1 Domains

Ultimate freedom of volumetric property distribution can be created with multiple domains. Every distribution function is valid in a specific domain. The props used by the distribution function, are not necessarily located in the domain.

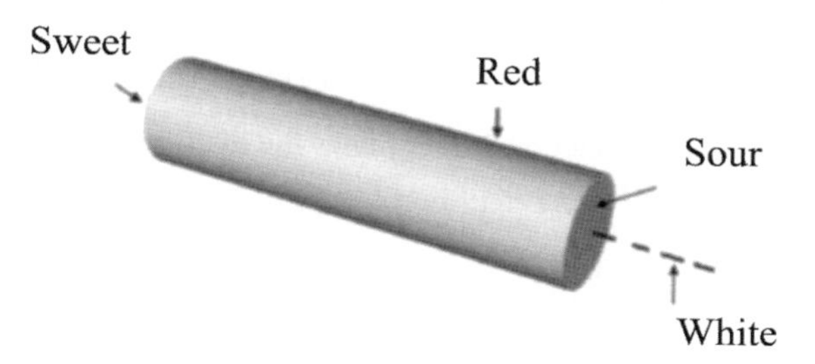

Figure 8 : Candy stick with desired properties.

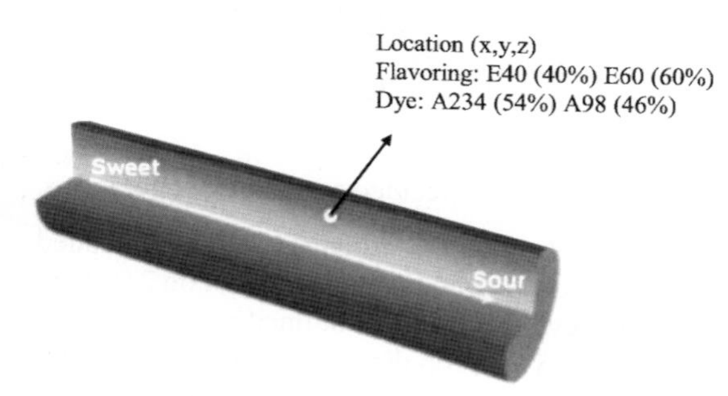

Figure 9 : Example of local properties in candy stick.

5.2 Props

For reasons of simplicity two types of props have been defined:

ISO prop

ISO props are areas with constant properties. They can be defined by a point, line or surface. ISO prop areas of the same type may not intersect or touch each other, since this causes an internal conflict.

Vari prop

For more advanced property distributions ISO props are not sufficient, therefore Vari props has been defined. Both props have the same behavior. The ISO prop however describes an area with constant properties, while the Vari prop describes an area with variable properties. An arbitrary Vari prop is dependent on one or more ISO or other Vari props.

Distribution functions

A distribution function describes the quantity (e.g. percentage sweetness) of a property as a function of the distance. In order to specify effective distributions two distribution functions are defined:

Absolute distribution function

The absolute distribution function is affected by the distance to a prop. This function gives the relation between a relative distance and a percentage of the property. The distance is a given value by the user, see Figure 10.

Relative distribution function

The relative distribution function is affected by the relative distance of two props. From a given point the shortest distance is calculated to both props. The relative position is used to calculate the values of the properties, see Figure 11.

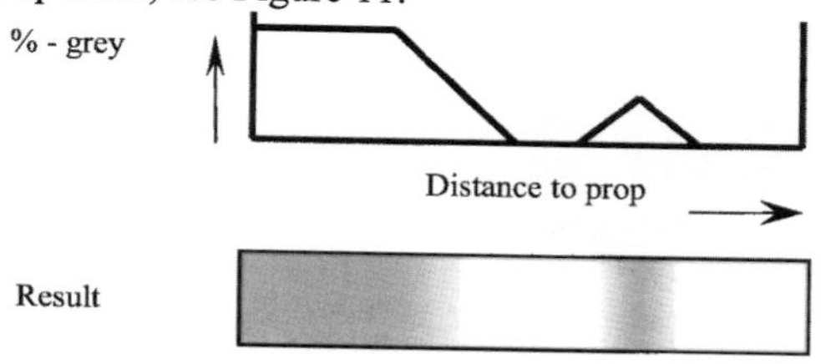

Figure 10 : Example of an absolute distribution function.

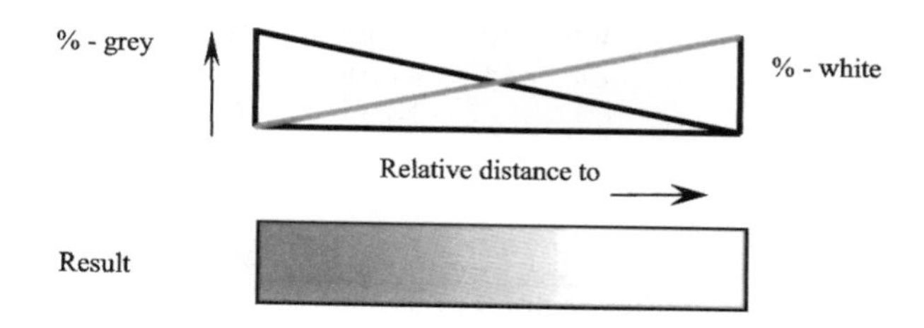

Figure 11 : Example of a relative distribution function.

Since all properties are defined by distribution functions and not by discrete elements like voxels, the resolution of the prop determination in theory is unlimited, and therefore can be tuned to any application.

An other example will be discussed now, in detail to demonstrate more extensively the capabilities of the developed method to define adequately a complex functionally graded structure. This graded structure can represent any type of product with graded material structures. An example is a dental crown with different properties, a hard wear resistant outside, graded to a stress absorbing ductile inside. For simplicity this example consists of simple cylinders demonstrating all the previously described available functionality.

Suppose a small cylindrical bar 1 (see Figure 12), completely enclosed by a bigger cylindrical bar 2. Both cylinders represent a domain. First will be focused on domain 1 represented by bar 1. In this domain two properties are created (as in the example with the candy stick) but with the distribution Function 1, displayed in Figure 13, a more advanced graded structure is generated (Figure 14).

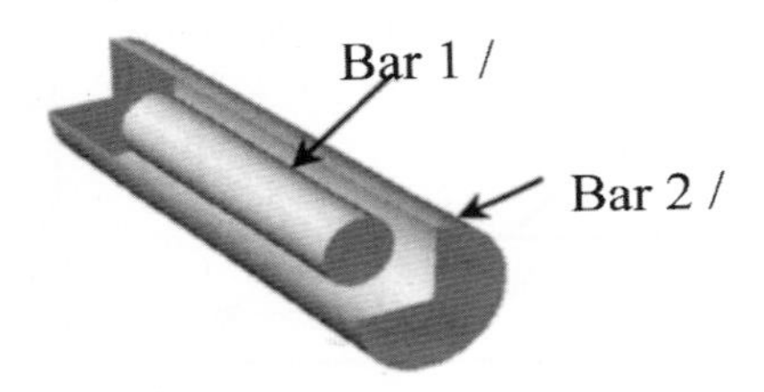

Figure 12 : Configuration of 2-bar example.

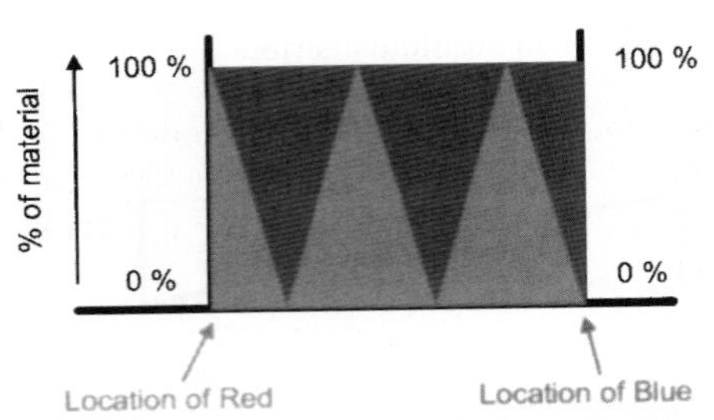

Figure 13 : Function 1 determines the distribution of the properties (colors) of domain 1.

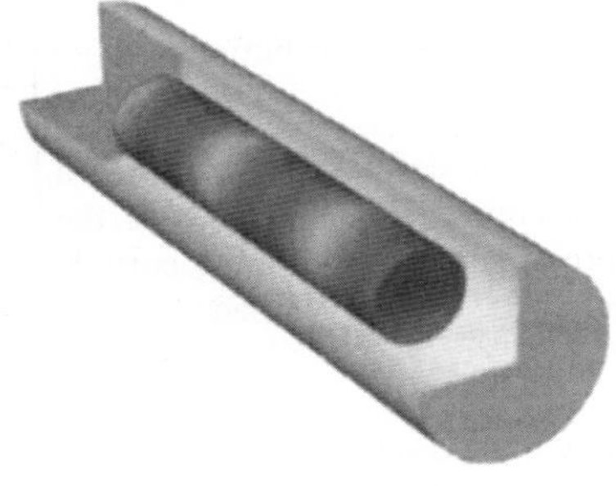

Figure 14 : Result of distributing the properties in bar 1.

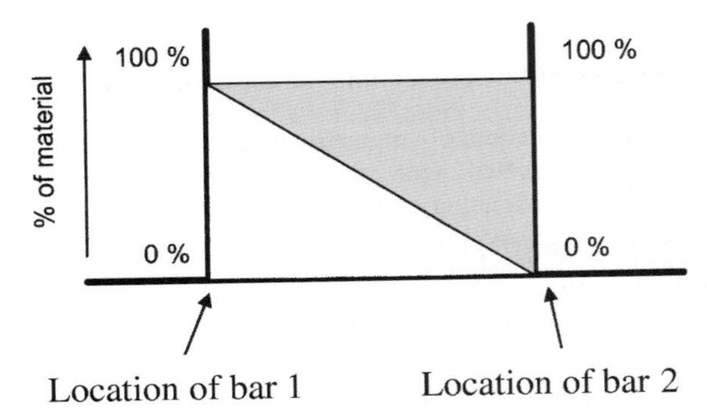

Figure 15 : Function 2 determines the distribution of properties of domain 2.

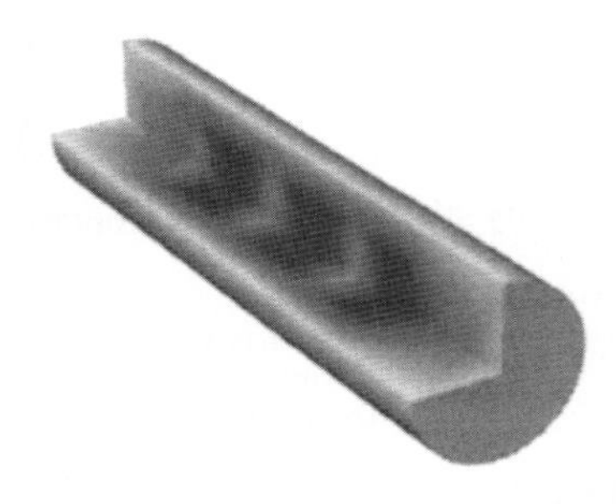

Figure 16 : Result of executed sequence of operations.

The properties created in domain 1 will be extended towards the outside of domain 2 by Function 2 (Figure 15). To achieve this, bar 2 is defined as an ISO-prop and bar 1 as a Vari-prop dependent on the same ISO-props as used to create the property distribution in domain 1. The result is displayed in Figure 16

6. Conclusion

This research, finding a solution to define graded material property distributions in a CAD like environment to be able to send graded material information to Rapid manufacturing machines for graded materials, succeeded.

The developed system covers the need of Research and Development centers in the field of experimental graded material and machines and is likely to cover both beginners as experienced user requirements

The creation of complex graded material distributions in complex geometries is far beyond the capabilities of the human brain and therefore can only be applied in full potential if it is supported or even driven by computerized optimizations in the future.

The design of Functionally Graded Material parts

(with Rapid Manufacturing technologies), and on a limited scale the production of these parts, has become possible. The creation of Functionally Graded Material products with Rapid Manufacturing therefore is now a reality.

TNO is partner of the EU-funded FP6 Innovative Production Machines and Systems (I*PROMS) Network of Excellence. http://www.iproms.org

References

[1] Wohlers, T., 2003, Wohlers Report 2003, Rapid Prototyping, Tooling & Manufacturing State of the Industry. Annual Worldwide Progress Report.

[2] Prinz, F., et al, 1997, Rapid Prototyping in Europe and Japan, World Technology Evaluation Center, published by Rapid Prototyping Association of the Society of Manufacturing Engineers.

[3] Jacobs, P.J., 2003, "Mobile Parts Hospital", Developed by Alion Now Operating in Kuwait to Support US-troops in Iraq, www.alienscience.com.

[4] Hopkinson, N., Dickens, P., 2003, Analysis of Rapid Manufacturing - using Layer Manufacturing Processes for Production, Proceedings of the Institution of Mechanical Engineers part C-journal of Mechanical Engineering Science 217.

[5] Masters, M., 2003, Digital Fabrication of Custom Hearing Instruments Worldwide, Proceedings International Conference of the Worldwide Advances and Setbacks in Rapid Prototyping, Tooling & Manufacturing.

[6] Griffith, M.L., et al, 1997, Multi-Material processing by LENS, Proceedings of the Solid Freeform Fabrication, Austin.

[7] Atwood, C., Griffith, M., et al, 1998, Laser Engineered Net Shaping (LENS™): A Tool for Direct Fabrication of Metal Parts, Proceedings of ICALEO, Orlando.

[8] Filser, F.T., 2001, Direct Ceramic Machining of Ceramic Dental Restorations, Ph.D. thesis, Swiss Federal Institute of Technology, Zŭrich.

[9] Mitsubishi Carbide, www.mitsubishicarbide.com.

[10] Jardini, A.L., Maciell, R., et al, 2003, The Development in Infrared Stereolithography using Thermosensitive Polymers, Proceedings of the 1st International Conference on Advanced Research in Virtual and Rapid Prototyping. Leiria, Portugal.

[11] Gothait, H., Even, R., Danai, D., 2003, Photopolymer Jetting Technology in Rapid Prototyping, Proceedings of the 1st International Conference on Advanced Research in Virtual and Rapid Prototyping. Leiria, Portugal.

[12] Höhne, K.H., et al, 1989, 3D-Visualization of Tomographic Volume Data using the Generalized Voxel-Model, Proceedings of the 1989 Chapel Hill workshop on visualization, NC.

[13] Knoppers, G.E., 2003, Research for VPD-Technology for Graded Materials: Basic architecture for an experimental VPD-system, TNO-report 42.03.006918.

[14] Gervasi, V.R., Crockett, R.S., 1998, Composites with Gradient Properties from Solid Freeform Fabrication, Proceedings of the Solid Freeform Fabrication, Austin.

[15] Kumar, A., Wood, A., 1999, Representation and Design of Heterogeneous Components, Proceedings of the Solid Freeform Fabrication, Austin.

[16] Jackson, T.R., et al, 1997, Modeling and Designing Components with Locally Controlled Composition, Proceedings of the Solid Freeform Fabrication, Austin.

[17] Jackson, T.R., 2000, Analysis of Functionally Graded Materials Object Representation Methods, Ph.D. thesis, Massachusetts Institute of Technology, Cambridge.

[18] Park, S.-M., Crawford, R.H., Beaman, J.J., 1999, Functionally Gradient Material Design and Modeling using Hypertexture for Solid Freeform Fabrication, Proceedings of the Solid Freeform Fabrication, Austin.

[19] Morvan, S., Fadel, G., 1999, Heterogeneous Solids: Possible Representation Schemes, Proceedings of the Solid Freeform Fabrication, Austin.

Intelligent Production Machines and Systems
D. T. Pham, E. E. Eldukhri and A. J. Soroka (eds)

Acoustic Holography in Solids for Computer-Human Interaction

W. Rolshofen[a], M. Yang[b], Z. Wang[b]

[a] *Institute for Mechanical Engineering (IMW), Technical University of Clausthal, Robert-Koch-Str.32, 38678 Clausthal-Zellerfeld, Germany*
[b] *Manufacturing Engineering Centre, Cardiff University, Newport Road, Cardiff CF24 3AA, Wales, UK*

Abstract

Computer-Human Interfaces for production machines can be generally categorised into two groups, tangible and intangible. The most commonly used computer input devices such as keyboards, touch-pads or touch-screens are tangible. In this paper, a new approach to the development of tangible acoustic interfaces based on acoustic holography is described.

The principle of acoustic holography in solids is described and the achieved theoretical results are presented in this paper. A prototype of such a tangible acoustic interface has been developed and will be further optimised.

1. Introduction

Following the use of visualisation and speech recognition for man-machine interaction interfaces, the next challenge for the future intelligent production machines and systems is haptic or contact technology which is based on acoustic source localisation and tracking. In this paper, an acoustic source localisation method based on the acoustic holography is described. This method is under development and will be implemented for various computer applications in the EU-project "Tangible Acoustic Interfaces for Computer-Human Interaction (TAI-CHI)" [1, 2].

2. Acoustic source localisation methods

Principally, there are two kinds of stimulation of physical objects: passive and active. In the passive mode any change in the acoustic properties of an object, due to its vibration as a consequence of interaction (knocking, tapping etc.), is detected and then used to estimate the location of the interaction. With regard to the active mode, the absorption of acoustic energy at the contact point of an object surface must be ascertained. Different types of impact can be identified like knocking, beating, scratching, rubbing, touching, as well as continuous motions.

Currently there are three passive methods under investigation for tangible acoustic interfaces: time delay of arrival (TDOA), time reversal [3] and acoustic holography. This paper focuses on acoustic holography in solids

3. Optic Holography

The idea of optical holography was formulated for the first time by Gabor [4] in 1948. There were many related inventions (e.g. lasers) before the principle was proved experimentally. A holographic picture is created by recording a hologram and then

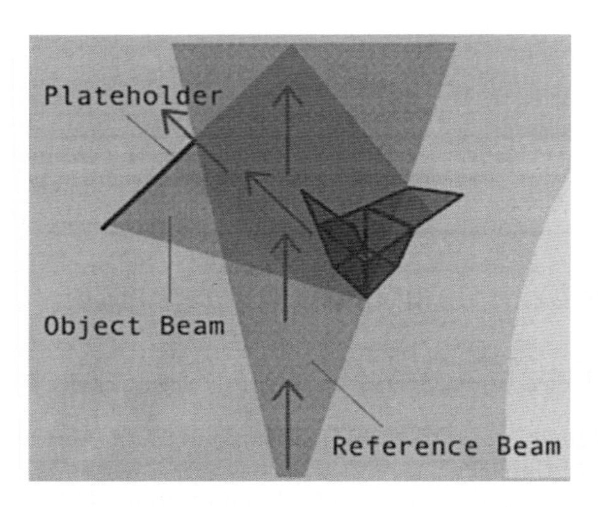

Fig. 1. Superposition of object- and reference beams on a photosensitive plateholder, where an optic hologram is created.

reconstruction of the receded hologram in three-dimensional space [5] (see Fig.1). If a wavefront propagates through an object point, a new spherical wave is built according to the Huygens-Fresnel's principle. This is called the object wave which is contrary to the first existing reference wave. The superposition of the amplitudes and the phase distribution can be recorded on a photosensitive plate (see Fig. 1).

The main advantage of holography is that it can reproduce three-dimensional information contents from the recorded two-dimensional holograms. This is because a three-dimensional field can be reconstructed from two-dimensional photosensitive surfaces using the saved phase information [6] [7].

4. Acoustic Holography

In acoustic holography, a discrete two-dimensional sound pressure field is stored and used to construct the three-dimensional sound pressure field, which can be the field of the particle velocity, the field of acoustic intensity vector, the field of surface velocity or the field of the intensity of a vibrating sonic source [8].

In general, acoustic holography is a reconstruction of either a three dimensional acoustic wave intensity field or a three dimensional acoustic wave phase field. Mathematically, a holographic reconstruction is just the convolution of the measured values with the values of the measured Green's function [8].

If a wave field $\Psi(r,t)$ is created by a sonic source, the following wave equation can be use to describe the wave propagation behaviour approximately:

$$\nabla^2\Psi(r,t) - \frac{1}{c^2}\frac{\partial^2\Psi(r,t)}{\partial t^2} = 0 \qquad (1)$$

where ∇^2 is the Laplace operator and "c" is a constant propagation velocity. In order to process the equation (1) more approximations are needed, such as an assumption of the existence of an infinitely dimensional surface surrounding the target area. Also, the Helmholtz equation with the wave number $k=\omega/c$ must be fulfilled.

Results in the complex wave field $\widetilde{\Psi}(r,\omega)$ can be obtained by applying Fourier transformation to the above equation. The values of the amplitude and phase in the new equation depend on the distance "r" between the position of the measurement to the source. For the spatial analysis of acoustic wave propagation behaviour, a constant frequency value ω is normally adopted in order to make a wave field meet the requirements of the Helmholtz equation.

$$\nabla^2\widetilde{\Psi}(r,\omega) + k^2\widetilde{\Psi}(r,\omega) = 0 \qquad (2)$$

5. Rayleigh-Sommerfeld Algorithm

With the increment of the distance between the source and the recording positions, further approximations become necessary. Concerning acoustic holography in solids, the Rayleigh-Sommerfeld approximation is promoted which is referred to Rayleigh's integrals and Sommerfeld's radiation condition. It is the most general and valid formula which can be applied through out in the entire space.

A mathematical model of Huygens-Fresnel principle can be deduced from the Rayleigh-Sommerfeld diffraction formula [9], because this corresponds to a convolution operation which, according to the convolution theorem, is just a multiplication in the Fourier space. From the analytical computation of the transfer-function (Green's Function), the wave propagation between hologram level and picture level can be calculated with the following formula (Rayleigh-Sommerfeld Algorithm) and visualized [7]. A visualisation can be realised by the holographic reconstruction of the complex wave field $\widetilde{\Psi}(x_B, z_B)$ with the hologram data $\widetilde{\Psi}(x_H, z_H)$, where F symbolises the Fourier transform and F^{-1} the inverse Fourier transform. Besides, the spatial co-ordinates are given with x and z, as well as the wave number k.

$$\widetilde{\Psi}(x_B,z_B)=F^{-1}\left\{F\left[\widetilde{\Psi}(x_H,z_H)\right]\cdot e^{ikz\sqrt{1-\lambda^2\left(\gamma^2+\delta^2\right)}}\right\} \quad (3)$$

In equation (3), the indices *B* and *H* indicate the source plane and the hologram plane respectively, λ is the wavelength, and γ and δ describe the associated local frequencies [7].

Based on the measured signals at the sensors' positions shown in Fig. 2, the amplitude and phase distributions of the acoustic waves are back-projected

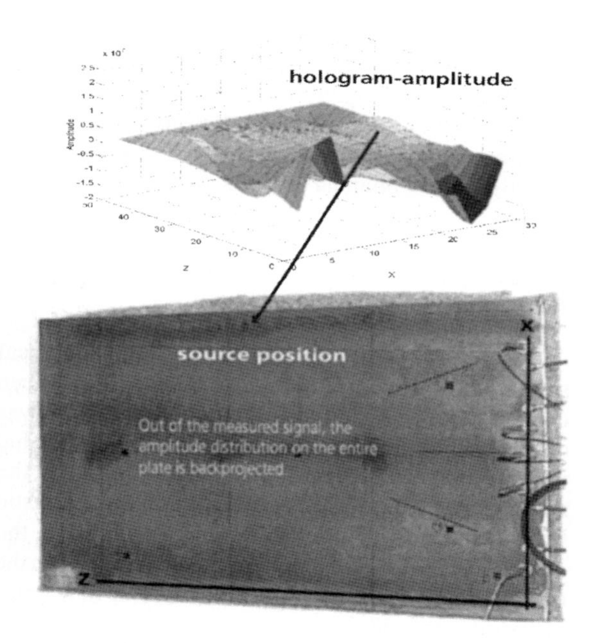

Fig. 2. Holographic description of the amplitude distribution after knocking at the source location on a metal plate [6].

onto a digital image plane from each sensor's position, in consequence of the superposition of the amplitudes and phases, two maxima are formed in the reconstruction amplitude and phase spaces respectively. Obviously the coordinates of the maxima correspond to geometrical location of the acoustic source.

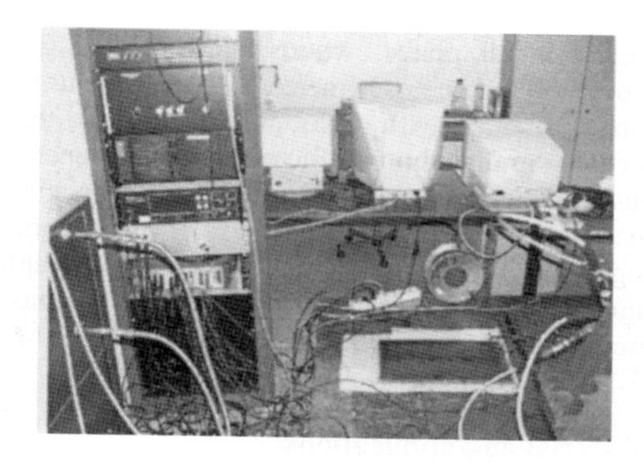

Fig. 3. Acoustic holography hardware and test rig.

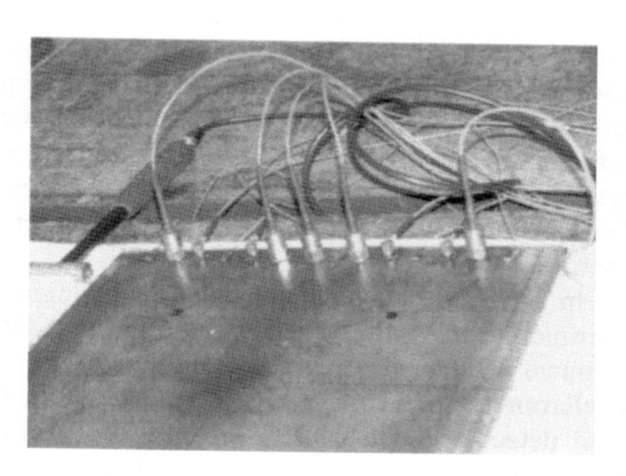

Fig. 4. A linear array of sensors mounted on a metal plate for acoustic holography experiments. Here two different kinds of sensors were tested: accelerometers and strain gauges.

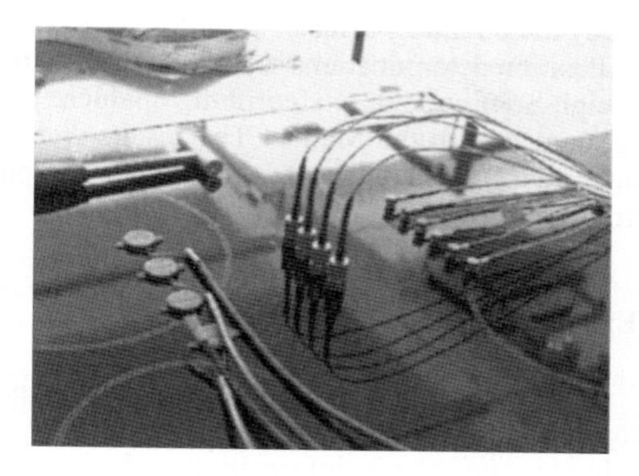

Fig. 5. A glass ceramic hob installed with three different kinds of sensors.

The theory and the developed algorithm of acoustic holography have been applied to various tests to validate their applicability. Therefore, different materials were selected for experiments, which included metal plates, wood boards and a glass-ceramic hob. In the experiments impulse stimulus were manually created with a hammer tapping the tested objects to simulate human finger taps. The generated acoustic signal was recorded at each sensor's position, then processed and finally backprojected to form a holographic image. From a reconstructed holographic image, a local maximum representing an acoustic source position can be identified (see Fig. 2).

6. Material and applications

The approximation of homogeneity and isotropy was used for simplify the analysis of behaviours of acoustic wave propagation in a physical object. However, based on this assumption, the developed theoretic model can only give approximate characterization of acoustic wave behaviours in real object. Therefore different materials have been tested to investigate how the approximations could affect the experimental results.

In order to carry out the acoustic holography experiments, a test rig has been established at IMW Technical University of Clausthal (see Fig. 3). Accelerometers and strain gauges were used in the test rig to detect acoustic signals (see Fig. 4). Various sensor arrangements have been designed and tested to optimize the quality of the reconstructed holographic images. An example of the sensor installation can be seen in Fig. 5, where the sensors were mounted in the middle of the plate instead of along the edges in order to study the boundary effects.

Position determinations were performed with the Rayleigh-Sommerfeld algorithm which was programmed with the Matlab. The results were 3-D images of acoustic holograms of amplitude and phase distributions presented in Figures 6, 7, and 8.

8. Experimental and simulation results

Basically, an acoustic holographic image is the reconstruction of acoustic wavefronts by their amplitude and phase information, which can be obtained from the signals detected by the sensors installed at different locations on the tested object. The first experimental setup and a reconstruction of a 3-D acoustic wave image are shown in Fig. 2. In the reconstructed digital holographic image, there existed several local maxima, however only one of them associated to the acoustic source. The others formed at the edges of the image were caused by the boundary effects.

In the experiments with a metal plate, three measurements have been carried out at a same position (co-ordinates x=15, z=30) on the tested object's surface. The results were presented in the Figures 6, 7 and 8. All the three reconstructed acoustic holograms have their locale maxima at the similar locations.

The in-solid wave propagation, like the used metal plate, was also simulated with the Finite Element Method (FEM). According to the elastic parameters of the tested plate as well as its size, the simulated displacements after a knock on the object surface have been calculated and shown in the Figures 9, 10 and 11.

After a certain amount of tapping force has been applied to a plate, the originating acoustic wave propagation due to the tap was simulated. In accordance with the idealised model of the theory, no captivation was applied to the boundary. Therefore, when displacements of all grid points have been calculated and assembled, strong wave reflections have been noticed occurring at the edges of the tested object. As a result of this, a conclusion can be made that the acoustic wave propagates faster in certain areas of the object (Fig. 11). A future step will be to set the plate in constraint condition during the simulation to investigate how boundary effects affect acoustic wave propagation at the sensors positions.

7. Conclusion

Laboratory experiments and mathematical simulations on in-solids acoustic holography have been conducted. The Rayleigh-Sommerfeld algorithm was developed for reconstruction of acoustic holographic images. At the moment, methods to improve the reliability of acoustic source localisation with acoustic holography are still under investigation. In future, the development of acoustic holography will involve in the development of new algorithms, the use of new types of transducers and searching for the best geometrical arrangement for acoustic sensors. Also the active method will be studied, where the medium is driven by a continuous acoustic energy source. Moreover, numerical simulations for the wave propagation insolids will be further strengthened.

Acknowledgements

This work was financed by the European FP6 IST Project "Tangible Acoustic Interfaces for Computer-

Human Interaction (TAI-CHI)". The support of the European Commission is gratefully acknowledged. IMW and the MEC are members of the I*PROMS Network of Excellence being funded by the EC under its FP6 Programme.

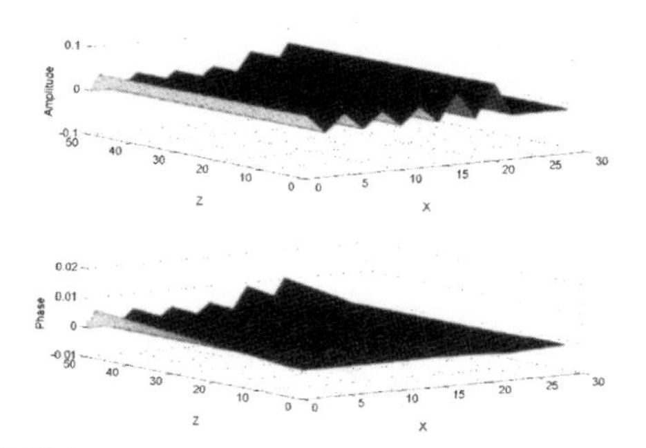

Fig. 6. Holographic result at source position (x=15, z=30), where the first of repeated measurements were done.

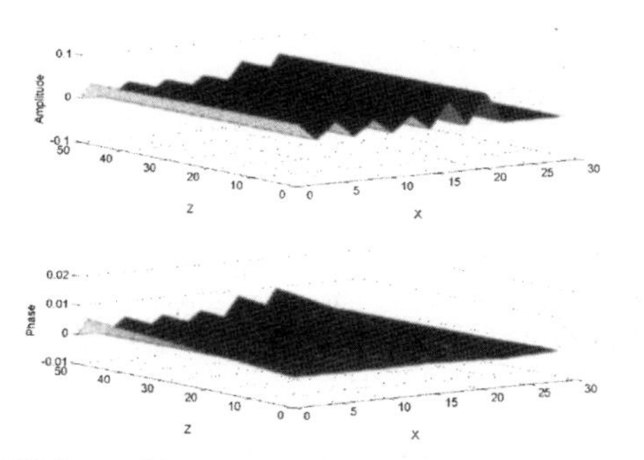

Fig. 7. Holographic result at source position (x=15, z=30), where the second of repeated measurements were done.

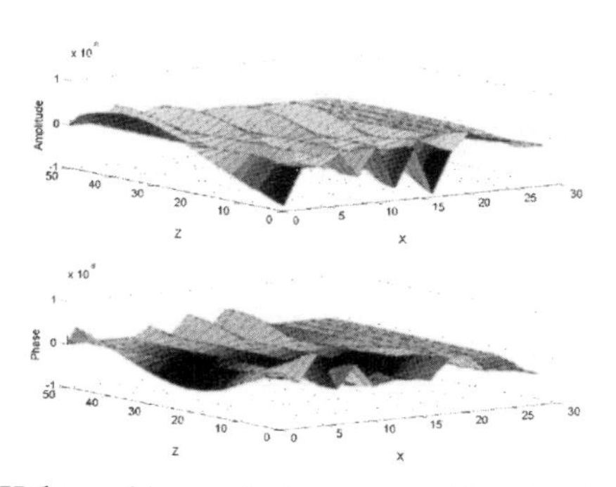

Fig. 8. Holographic result at source position (x=15, z=30), where the third of repeated measurements were done.

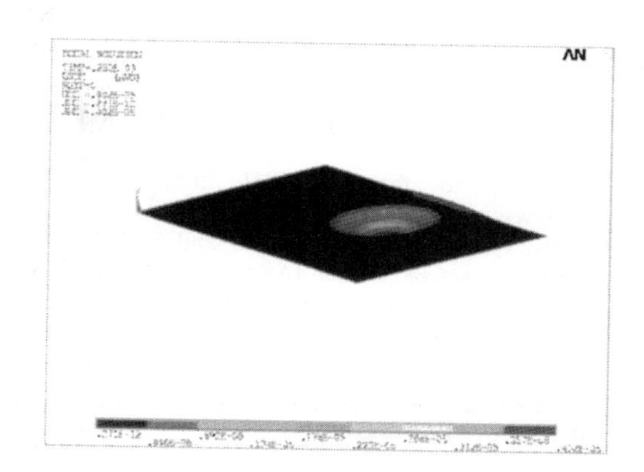

Fig. 9. First time step in FEM simulation. A certain force like a tap on the plate is simulated and the displacement of the material is approximated.

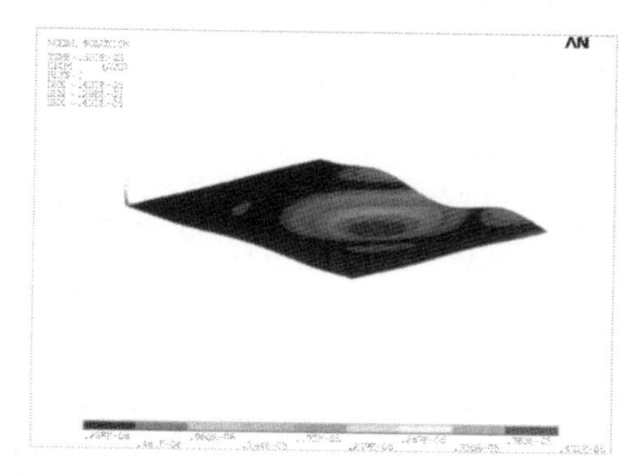

Fig. 10. Second time step in FEM simulation.

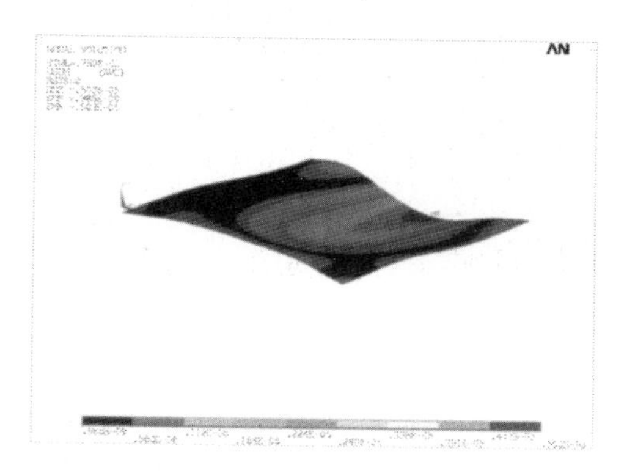

Fig. 11. Third time step in FEM simulation. A certain force like a tap on the plate is simulated and the displacement of the material is approximated. Instead of areas inside the plate, the boundary of the object can be seen in movement.

References

[1] TAI-CHI Consortium: Technical Annex of the Project, 2004.

[2] Reference to the TAI-CHI homepage. URL: http://www.taichi.cf.ac.uk/

[3] Ing RK, Quieffin N, Catheline S and Fink M. Time reversal interactive objects. JASA 115, 5 (2004), 2589.

[4] Gabor D. A new microscopic principle. Nature, 161 (1948), 777-778.

[5] MIT Museum´s Holography: eye on holography http://www.mit.edu/~sdh/holography/eoh/index.html.

[6] Düsing C, Rolshofen W. Berührbare akustische Benutzerschnittstellen. IMW Institutsmitteilungen, Nr. 29 (2004), 63-66.

[7] Roye W. Beitrag zur Weiterentwicklung der akustischen Holografie beim Einsatz in der Materialprüfung. VDI-Verlag, Reihe 5: Grund- und Werkstoffe, Nr. 117, 1987.

[8] Maynard J, Williams E and Lee Y. Nearfield acoustic holography: I. Theory of generalized holography and the development of NAH. J. Acoust. Soc. Am., 78 (1985), 4, 1395-1413.

[9] Williams E. Fourier Acoustics- Sound Radiation and Nearfield Acoustic Holography. Academic Press, 1999.

Intelligent Production Machines and Systems
D. T. Pham, E. E. Eldukhri and A. J. Soroka (eds)

Acoustic pattern registration for a new type of human-computer interface

D.T. Pham [a], Z. Wang [a], Z. Ji [a], M. Yang [a], M. Al-Kutubi [a], S. Catheline [b]

[a] *Manufacturing Engineering Centre, Cardiff University, Cardiff CF24 3AA, UK*
[b] *Laboratoire Ondes et Acoustique, ESPCI, 10 rue Vauquelin, 75231 Paris Cédex 05, France*

Abstract

This paper presents an analysis of acoustic pattern registration for a new means of human computer interaction. It discusses the relationship between variations in correlation coefficients and signal pattern registration parameters including signal sampling rates, amplitude digitisation resolutions and recorded signal lengths. Experimental results have confirmed that the accuracy of acoustic pattern registration is related to the data acquisition parameters adopted. Acoustic signal pattern registration with a single sensor could be improved using high sampling rates, high amplitude digitisation resolutions and long recorded signal lengths.

1. Introduction

Human-computer interfaces belong to one of two types, tangible (touchable) and intangible (or non-touchable, such as audio or video). The vast majority of input interfaces are tangible. Currently, the most important forms of such tangible interfaces include keyboards, mice, touch pads and touch screens. A common problem with these devices is that they restrict the mobility of users, constraining them to be in certain locations during interaction with the computer. To deal with this problem, tangible acoustic sensing technology has become an interesting subject for human computer interaction in recent years [1, 2, 3]. The main effort is to develop acoustics-based sensing technologies which can be adapted to virtually any physical object to create tangible interfaces, allowing the user to communicate freely with a computer, an interactive system or the cyber-world. Fig.1 illustrates a tangible acoustic interface (a large touch-screen) for human computer interaction developed by the authors' team at the MEC.

Fig. 1. A board on the wall acting as a tangible acoustic interface for human computer interaction at the MEC.

There have been two basic techniques developed for acoustic source localisation in this area: time delay of arrival (TDOA) and location pattern matching (LPM).

TDOA is a commonly used technique to locate the source by computing the differences in the arrival time of acoustic signals from a number of sensors. When an impact occurs on the surface of the structure, the time delay of arrival of the acoustic waves at each of the sensing elements will be recorded and the time delay can be easily determined by a computer. This technique has the advantages of quick response time and simple computation. The disadvantage is that it needs many sensors, resulting in complicated hardware configurations.

In contrast to the TDOA technique, LPM is an approach using one or a few acoustic sensors to record a signal pattern. In this technique, the acoustic pattern received from a sensor can be regarded as the superposition of different localised signal components with different frequencies, phases and reflections in an object. Compared with the TDOA technique, LPM has the advantage of simple hardware configurations, as usually only one sensor is needed [4, 5]. The drawback of the LPM technique is that it is slow, as it requires a relatively long recorded signal length compared with TDOA which uses a threshold crossing detection method [2, 6].

In addition, LPM has the feature of noise reduction due to the mean filter effect in pattern matching by the cross-correlation method [6, 7]. Cross-correlation is widely used in signal and image processing to recognise patterns in terms of similarity. It is a technique that gives the quality of a match with one or more templates or reference patterns [8, 9]. In the case of some TDOA applications, cross-correlation is also used to determine the relative time delay between the signals received from two or more sensors [6]. The maximum cross-correlation coefficient determines the time lag. In recent years, the dispersion of acoustic signal patterns produced by impacts has been discussed and analysed in a number of publications [10, 11]. However, the requirements for the accurate registration of acoustic signal patterns for human computer interaction have not been discussed. Such parameters as signal sampling rates, amplitude digitisation resolution levels, and recorded signal lengths are important for accurate signal pattern registration before computing correlation coefficients.

This paper will discuss the ways to acquire the acoustic signal patterns that are generated by physical contacts on a solid object surface, in terms of the sampling rate, amplitude digitisation resolution, and signal sample length. The accurate registration of signal patterns from one given position and the effect of these parameters on the cross-correlation function are the focus of this work.

2. Acoustic signal pattern registration and matching

A physical contact on a solid object or a surface (wall, table, etc.) will modify its acoustic properties by the way acoustic energy is distributed in the object and on its surfaces. When the surface of an object is touched, an acoustic vibration pattern will be generated at the point of contact and it can be detected using an acoustic sensor.

In the case of a single sensor-based system, as mentioned previously, the sensor output corresponding to an impact can be considered the superposition of many signal components with different frequencies, phases and reflections in an object. Fig. 2 shows a typical acoustic signal.

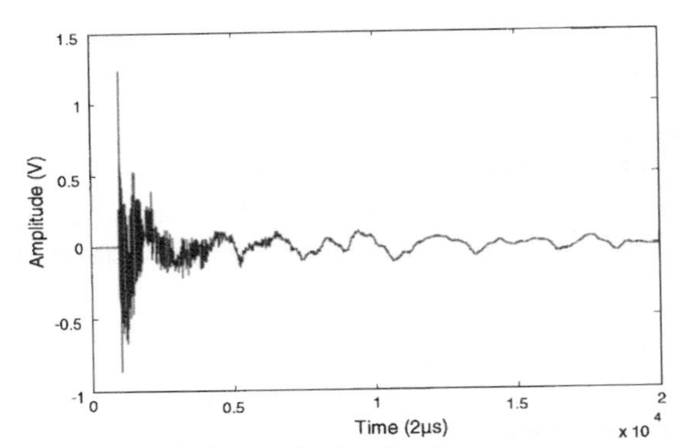

Fig. 2. A typical acoustic signal pattern.

The similarity or the degree of correlation between detected signal patterns can be decided by examining the correlation coefficient. Considering a fixed or reference pattern $s_0(t)$ and a pattern to be recognised $s_k(t)$, each with a population size of N samples, the correlation coefficient of signals s_0 and s_k can be written as

$$r_{s_0 s_k} = \frac{\mathrm{cov}(s_0, s_k)}{\sigma_{s_0}\sigma_{s_k}}, \quad k = 1, 2, 3, ... \tag{1}$$

where $\mathrm{cov}(s_0, s_k)$ is the covariance between signals s_0 and s_k, and σ_{s_0} and σ_{s_k} are the standard deviations of the signals s_0 and s_k.

When there is noise [12, 13] or a quantisation error [14], the signal patterns can be expressed as

$$x_0(t) = s_0(t) + n_0(t) \quad (2)$$

and

$$x_k(t) = s_k(t) + n_k(t) \quad (3)$$

where n_0 and n_k are noise or quantisation errors, and x_0 and x_k are the signal patterns with noise or quantisation errors. In this case, the correlation coefficient of signals x_0 and x_k can be written as

$$r_{x_0 x_k} = \frac{\text{cov}(x_0, x_k)}{\sigma_{x_0} \sigma_{x_k}}, \quad k = 1, 2, 3, ... \quad (4)$$

where $\text{cov}(x_0, x_k)$ is the covariance between signals x_0 and x_k, and σ_{x_0} and σ_{x_k} are the standard deviations of the signals x_0 and x_k.

In this work, the correlation coefficient $r_{x_0 x_k}$ is used to give the similarity degree of a match to one or more location templates or reference signal patterns. $r_{x_0 x_k}$ will be equal to 1 when two location acoustic signal patterns are identical. It is clear that the correlation coefficient will be affected by the quality of the recorded signal patterns.

In the registration of acoustic signal patterns, the accuracy is closely related to sampling rates, amplitude digitisation resolution levels and recorded signal lengths. The number of features in a pattern is directly proportional to its sampling rate and recorded signal length, and the signal amplitude accuracy is determined by its digitisation resolution levels or bits. In a particular application, the registration accuracy can be considered together with the sampling rate, amplitude digitisation, signal recorded length and object properties.

In order to express impact patterns precisely, high resolution and high speed analogue-to-digital converters are required for the registration of acoustic signal patterns. However, considering the registration accuracy and real-time processing requirements in a particular application, appropriate selection of data registration parameters is necessary. These aspects have been investigated experimentally and the results obtained are discussed in the following section.

3. Experimentation

The effect of the sampling rates, amplitude digitisation resolution levels and recorded signal lengths on the signal pattern registration has been investigated in a number of dedicated experiments. The experimental setup comprises a glass plate (length: 57 cm, width: 51 cm, thickness: 0.4 cm) providing the acoustic wave propagation medium, an accelerometer used as an acoustic sensor, an amplifier, an A/D converter and a computer (based on the Intel Pentium IV processor, 2.8 GHz). A block diagram of the signal pattern registration system is shown in Fig. 3. Fig. 4 shows the acoustic signal sensing circuit on the glass plate.

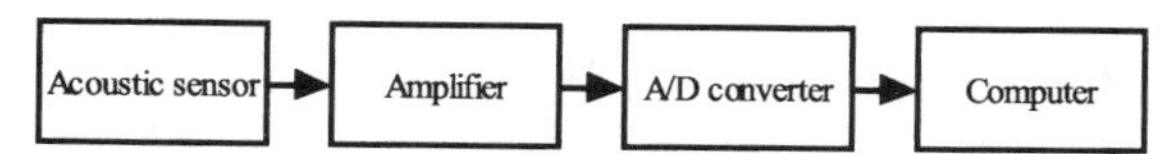

Fig. 3. An acoustic signal pattern registration system.

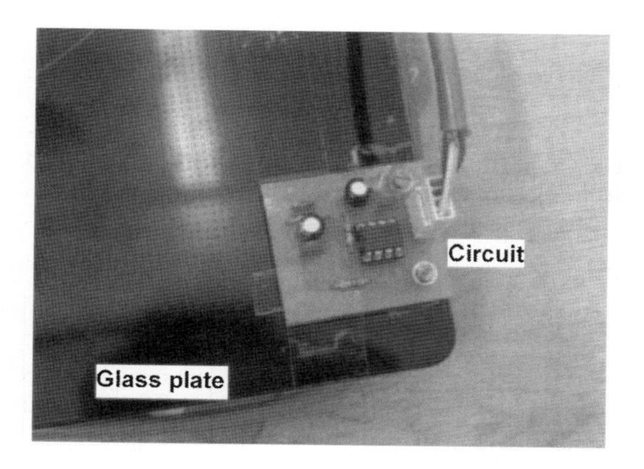

Fig. 4. Acoustic signal sensing circuit on a glass plate.

In the experiments, sampling rates between 1 KHz and 500 KHz, amplitude digitisation resolution levels between 8 and 16 bits, and recorded signal lengths between 5 ms and 60 ms were used to acquire the acoustic signals. The correlation coefficients and their deviations were obtained with 50 signal samples.

Fig. 5 shows three acoustic signal patterns recorded with sampling rates of 500 KHz, 31 KHz and 1 KHz. In Fig. 5(c), it can be seen clearly that the high frequency components were lost when the low sampling rate was used. The amplitude digitisation resolution employed in the experiments was 16 bits (the amplitude was between -1.25 V and 1.25 V). The recorded signal length was 60 ms. Fig. 6(a) shows the correlation coefficient distribution when the sampling rate is 1 KHz, and Fig. 6(b) depicts the standard deviations of correlation coefficients with sampling rates from 1 KHz to 500 KHz. It is clear that the standard deviation increases with a decrease in the sampling rate.

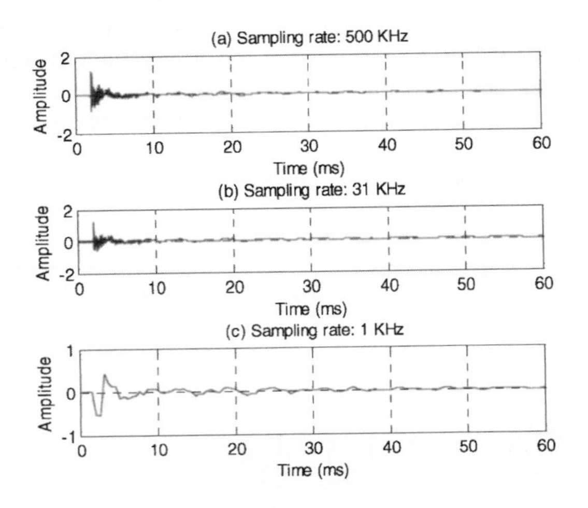

Fig. 5. Three recorded acoustic signal patterns with sampling rates of 500 KHz, 31 KHz and 1 KHz.

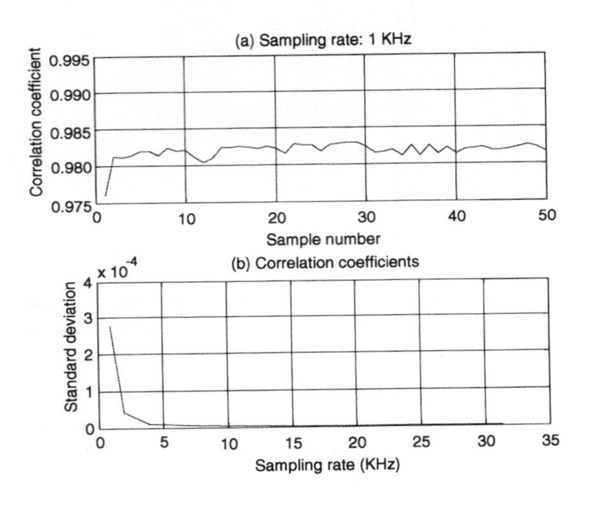

Fig. 6. (a) Correlation coefficient distribution when sampling rate = 1 KHz, and (b) Standard deviations of correlation coefficients with sampling rates from 1 KHz to 31 KHz.

Fig. 7 shows three recorded acoustic signal patterns with digitisation resolution levels of 16, 12 and 8 bits. The sampling rate used in the experiment was 500 KHz. The recorded signal length was 60 ms. Fig. 8(a) shows the correlation coefficient distribution when the amplitude digitisation resolution was 8 bits, and Fig. 8(b) plots the standard deviations of correlation coefficients with digitisation resolution levels from 8 to 16 bits. It can be seen that the standard deviation increases with a decrease in the amplitude digitisation resolution.

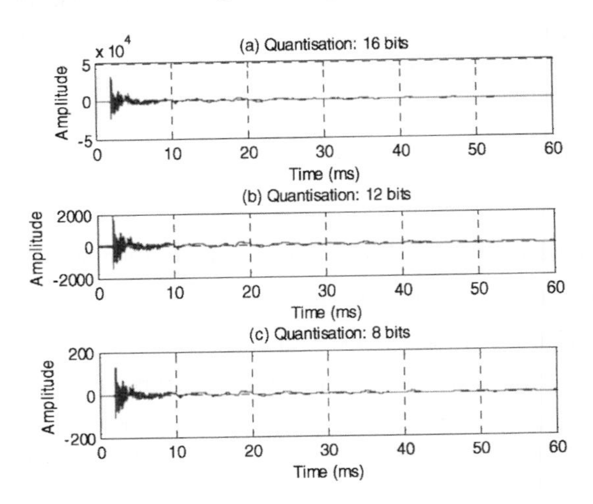

Fig. 7. Three recorded acoustic signal patterns with digitisation resolution levels 16, 12 and 8 bits.

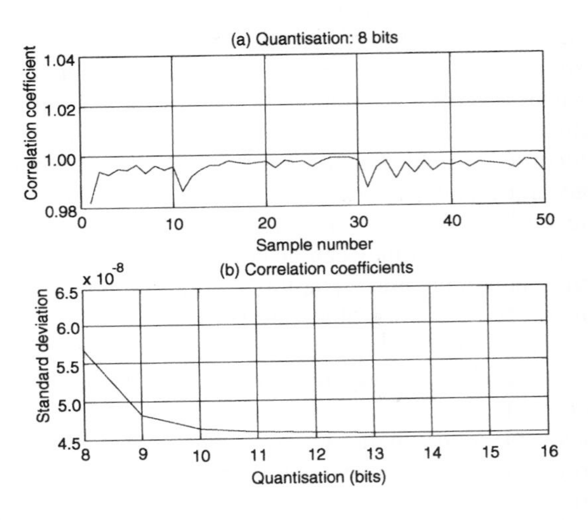

Fig. 8. (a) Correlation coefficient distribution when digitisation resolution = 8 bits, and (b) Standard deviation of correlation coefficients with digitisation resolution levels between 8 and 16 bits.

Fig. 9 shows three recorded acoustic signal patterns with signal lengths of 60 ms, 30 ms and 5 ms. The sampling rate used in the experiment was 500 KHz. The amplitude digitisation resolution was 16 bits. Fig. 10(a) shows the correlation coefficient distribution when sample length = 5 ms, and Fig. 10(b) shows the standard deviation of correlation coefficients with sample lengths from 5 ms to 60 ms. It can be seen, from Fig. 10(b), that the longer recorded signal length gives higher accuracy.

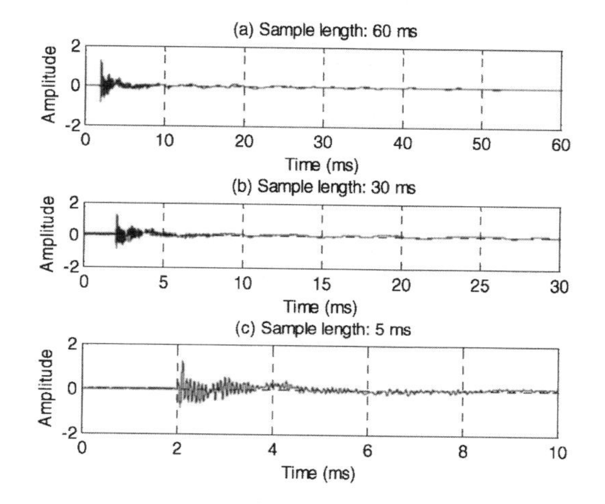

Fig. 9. Three recorded acoustic signal patterns with different signal lengths.

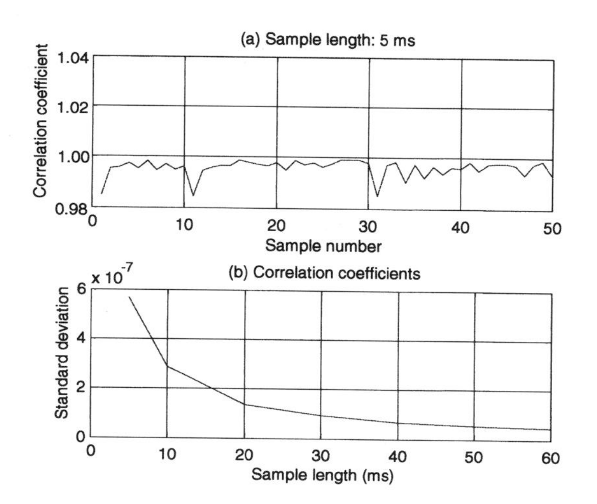

Fig. 10. (a) Correlation coefficient distribution when sample length = 5 ms, and (b) Standard deviation of correlation coefficients with sample lengths from 5 ms to 60 ms.

4. Conclusion

This work has shown that high accuracy of the registration of acoustic signal patterns can be achieved through correct selection of the sampling rates, amplitude digitisation resolution levels and recorded signal lengths. The experimental results have confirmed that the standard deviation of the correlation coefficient can be reduced by increasing the sampling rate, amplitude digitisation resolution and recorded signal length. Correct selection of data registration parameters can be made based on the registration accuracy and real-time processing requirements for a particular application.

Acknowledgements

This work was financed by the European FP6 IST Project "Tangible Acoustic Interfaces for Computer-Human Interaction (TAI-CHI)". The support of the European Commission is gratefully acknowledged. The MEC is the coordinator of TAI-CHI and the EC-funded FP6 Network of Excellence on Innovative Production Machines and Systems.

References

[1] Paradiso JA, Hsiao K, Strickon J, Lifton J and Adler A. Sensor systems for interactive surfaces. IBM Systems Journal, Vol.39, Nos.3&4, 2000, pp 892-914.

[2] Checka N. A system for tracking and characterising acoustic impacts on large interactive surfaces. MS Thesis, MIT, Cambridge, MA, 2001.

[3] Paradiso JA, Leo CK, Checka N and Hsiao K. Passive acoustic knock tacking for interactive windows. ACM CHI 2002 Conference, Minneapolis, Minnesota, 20-25 April 2002.

[4] Calson J, Sjöberg F, Ouieffin N, Ing RK, and Catheline S, Echo cancellation in a single-transducer ultrasonic imaging system. IEEE 5th Nordic Signal Processing Symposium Proceedings, 4-7 October 2002.

[5] Quieffin N, Ing RK, Catheline S, and Fink M, Real time focusing using an ultrasonic one channel time-reversal mirror coupled to a solid cavity. Journal of the Acoustical Society of America, 2004, Vol.115, No.5, 2004, pp 1955-1960.

[6] Ding Y, Reuben RL and Steel JA. A new method for waveform analysis for estimating AE wave arrival times using wavelet decomposition. NDT & E International, Vol.37, 2004, pp 279-290.

[7] Pham DT and Wang Z. Fringe pattern shift detection with nano resolution. IEEE 6th International

Conference on Signal Processing Proceedings, Vol.2, 2002, pp 1823-1826.
[8] Fisher RB and Oliver P. Multi-variate cross-correlation and image matching. The 6th British Conference on Machine Vision Proceedings, Vol.2, 1995, pp 623-632.
[9] Knapp CH and Carter GC. The generalized correlation method for estimation of time delay. IEEE Trans. Acoustic, Speech and Signal processing, Vol.24, 1976, pp 320-327.
[10] Prosser WH, Gorman MR and Humes DH. Acoustic emission signals in thin plates produced by impact damage. Journal of Acoustic Emission, Vol.17, Nos.1-2, 1999, pp 29-36.
[11] Ellwein C, Danaher S and Jäger U. A vibration signal family of impact events. European Symposium on Intelligent Techniques, Aachen, Germany, 14-15 September 2000.
[12] Gao Y, Brennan MJ, Joseph PF, Muggleton JM, and Hunaidi O. A model of the correlation function of leak noise in buried plastic pipes. Journal of Sound and Vibration, Vol.277, Nos.1&2, 2003, pp 133-148.
[13] Pham DT, Wang Z, Yang M, and Packianather MS. Statistical analysis of signal-to-noise ratios in fringe pattern matching. IEEE 6th International Conference on Signal Processing Proceedings, Vol.1, pp 626-639, 2002.
[14] Skydan OA, Lilley F, Lalor MJ, and Burton DR. Quantization error of CCD cameras and their influence on phase calculation in fringe pattern analysis. Applied Optics, Vol.42, No.26, 2003, pp 5302-5307.

Intelligent Production Machines and Systems
D. T. Pham, E. E. Eldukhri and A. J. Soroka (eds)

Expressive Gesture in Tangible Acoustic Interfaces

A. Camurri, C. Canepa, P. Coletta, A. Massari, G. Volpe

InfoMus Lab – DIST - University of Genova
Viale Causa 13, I-16145, Genova, Italy
www.infomus.dist.unige.it

Abstract

This paper presents a brief survey of currently ongoing research at DIST – InfoMus Lab on multimodal interfaces with a particular focus on Tangible Acoustic Interfaces (TAIs). A relevant aspect of our research is the multimodal analysis of the high-level expressive, emotional information, the involved non-verbal communication mechanisms, the role of such information and mechanisms in the design and development of expressive multimodal interactive systems based on TAIs. Research is carried out in the framework of the EU-IST project TAI-CHI (Tangible Acoustic Interfaces for Computer-Human Interaction), funded by the European Commission in the Sixth Framework Program. The paper provides a summary of the work performed in the project, including examples of ongoing experiments and prototypes developed with the new EyesWeb 4 open platform for multimodal processing (www.eyesweb.org).

---

1. Introduction

An important goal of our research is to explore paradigms of non-verbal interaction between humans and machines in the framework of multimodal environments. Music theatre, museum and science centre interactive exhibits, art installations, interactive Mixed Reality systems for therapy and rehabilitation, are key domains from which research takes useful inspiration as well as application scenarios where to exploit research results. A relevant aspect in such research is to investigate the role of expressive non-verbal communication mechanisms in interaction. A key focus of our work is on expressive gesture [1][2] i.e., on the high-level emotional, affective content that gesture conveys, on how to analyse and process this content in a multimodal perspective, on how to use it in the development of innovative multimodal interactive systems able to provide users with natural expressive interfaces [3].

Our research addresses these topics from multiple perspectives and with different objectives. From a methodological point of view, we aim at developing methodologies for scientific investigation and for evaluation and assessment of theories and models (see for example the subtractive approach described in [3]). Research focuses on the design and development of experiments aiming at empirically exploring specific aspects of expressive non-verbal communication deemed particularly important for developing expressive, multimodal interfaces (e.g., see the experiments on dance and music performances described in [2]). From the point of view of exploitation of research outcomes in concrete application scenarios, the focus is on the development of (i) a software platform for real-time multimodal processing and (ii) libraries enabling an immediate employment of the results in prototypes of innovative multimodal interactive systems (e.g., see the recent work on the new EyesWeb 4 open platform, www.eyesweb.org, and the multimodal interactive systems in [1]).

This paper provides a brief summary of the research activities at DIST – InfoMus Lab with reference to the EU-IST STREP project TAI-CHI

(Tangible Acoustic Interfaces for Computer-Human Interaction). Three main directions have been followed for research on Tangible Acoustic Interfaces (TAIs): (i) the development of an open platform for integrated multimodal processing of TAI data, (ii) the developments of models and algorithm for analysis of expressive gesture in TAI, and (iii) the development of early prototypes integrating algorithms for in-solid acoustic localisation systems.

2. The EyesWeb 4 platform

In research on TAIs the need for fully integrated and supported multimodal processing of data streams from several channels (e.g., visual, auditory) led us to a complete redesign of the EyesWeb open platform: the new EyesWeb version 4 [4]. A relevant aspect in the new EyesWeb version 4 platform is the explicit support to multimodality and cross-modality in the EyesWeb language. This is obtained through a new kernel, now providing (i) low-level scheduling mechanisms for managing different data streams (e.g., auditory and visual data) at different sampling rates, and (ii) high-level extensions toward integration of gesture and audio processing, aimed at real-time analysis of expressive information.

Concerning the low level support, besides the basic requirement to manage several datatypes (e.g., audiovisual and sensor systems streams) in a common environment, new features have been added.

A feature supports the automated transformation of datatypes of different, but compatible, domains. This is particularly useful to verify the effectiveness of an algorithm originally designed and implemented for a very specific domain. As an example, an FFT block working on matrices can be easily used to work on audio streams, since the conversion of audio buffers to matrices is automatically added by the system, without the need for any explicit action on the user side.

Another important feature is the possibility to design and develop blocks working on a whole family of datatypes. Where the previous version of EyesWeb could distinguish among specific blocks (working on a given datatype) or general purpose blocks (working on all datatypes), this new version supports the generalization of the block to the datatypes sharing a set of specified characteristics. This enables the development of blocks working on homogenous sets of datatypes, without the need to know them in advance, thus, it does not limit the expansibility of EyesWeb. Referring to the above FFT example, a better designed block could exploit such feature and work natively on both the audio buffer and the matrix datatypes, as they share core common characteristics (i.e., they both implement a common interface). This approach has the further advantage by a performance point of view, since it avoids consuming processing power for datatype conversion.

Moreover, it is under investigation the extension of the EyesWeb language toward the object-oriented programming paradigm. This consists in enabling the possibility of organizing EyesWeb blocks into hierarchies according to inheritance relationships. So, for example, all the blocks implementing a filter will derive from a generic filter block abstracting the common properties of filters. The introduction of dynamic binding will be also investigated: this will allow at the initialization of a patch the possible polymorphic instantiation of the correct derived block where an instance of the base block class is specified at design time.

Cross-modal and multimodal processing is supported by several features, including the timestamping of datatypes, which has been greatly enriched in this version of EyesWeb. At a low-level, each datatype is associated with a set of timestamps that the kernel can use to synchronize the data according to different needs.

However, support to multimodality and cross-modality does not concern low-level features only. It also includes high-level extensions toward integration of gesture and audio processing, aimed at music performance analysis and expressive information processing. Figures 4-6 show possible processing scenarios involving cross-modality and bimodality. Support to multimodal high-level processing allows the development of algorithms based on the integrated analysis of sound and gesture, e.g., the sound produced by hands (tapping on a table, handwriting, etc.) and the hand gesture. It also allows cross-modal processing such as video processing of visual representations of acoustic signals (e.g., auditory display) and acoustic-inspired analysis of movement, gesture, and texture. For example, the cochleogram of voice sound depicted in Figure 1 could be analyzed by means of suitable image processing algorithms in order to extract information that is not so directly accessible through audio analysis (e.g., activation of particular regions in the image, pattern matching with template images as in figure 4). Figure 6 shows an example of bimodal processing of hand gestures.

3. Expressive gesture in TAIs

Our research in the TAI-CHI project mainly concerns the multimodal and expressive aspects involved in interaction with TAIs. In particular, the work deals with the development of models and algorithms for multimodal high-level analysis and interpretation of integrated data extracted from video

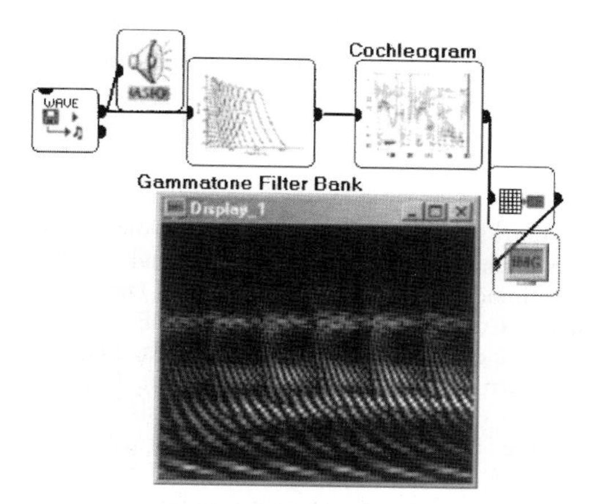

Fig 1. Extraction of a cochleogram in EyesWeb 4.

images, acoustic tangible interfaces, and acoustic localisation systems. A specific focus is on gestures performed by users and on the expressive content they convey.

The new EyesWeb 4 has been employed as basic platform for the design and development of a collection of software modules for high-level analysis of interaction with TAIs. Research has been devoted to both real-time analysis of expressive gesture in human-full body movement and analysis of expressive hand gestures. These analyses aim at detecting the way in which users approach the interface both with their hands and with the whole body. Concretely, we developed a collection of software modules for EyesWeb 4 for the interpretation of full-body gesture interaction. The software modules include low-level analysis of video images, e.g., enhanced background subtraction based on statistical models of the background, enhanced motion tracking based on Kalman filtering techniques and hybrid tracking techniques, enhanced colour blob analysis, contour extraction, convex hull, convexity defects, skeleton extraction. A further set of blocks is devoted to extraction of high-level expressive features, such as for example directness, impulsiveness, and fluentness [1]. The integration of these blocks with the blocks for hand motion tracking, finger localisation, and hand gesture analysis allows a broad analysis of how the user approaches a TAI, both at the level of global movement of the body and at the level of subtle hand movements.

Figure 2 shows an example from the first version of our s/w libraries for multimodal and cross-modal processing in EyesWeb 4. Hand gesture tracking and analysis is performed on a user interacting with a TAI. While the contact position is detected through an acoustic based localisation system, visual information is employed to get information on how the hand approaches and touches the interface (e.g., with a fluent movement, or in a hesitating way, or in a direct and quick way etc.). Moreover, the position and time of contact information obtained from audio analysis can be employed to trigger and control in a more precise way the video-based gesture analysis process: e.g., we are testing hi-speed and hi-res videocameras in EyesWeb 4 in which it is also possible to select the portion of the active ccd area using (x,y) information from a TAI interface.

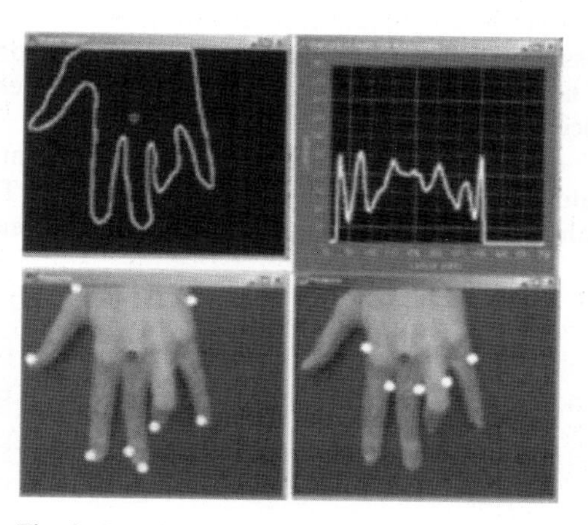

Fig. 2. Hand gesture analysis and tracking with EyesWeb 4 in tangible acoustic interfaces.

4. TAI interfaces in EyesWeb

We developed a first experimental TAI application integrated in an EyesWeb application (patch). It is based on two algorithms for in-solid localisation of touching positions developed by the TAI-CHI partners PoliMi (4 sensors, TDOA algorithm) and LOA (1 sensor). The patch (see Figure 3) integrates the two methods in order to increase the reliability of the detected touching position.

This EyesWeb application will be employed in a concert at La Biennale in Venice (composer Roberto Doati) in September 2005. The algorithms will be used for detecting the touching position on everyday life objects such as chairs and tables.

Another experimental TAI application we are developing consists in a method for performing acoustic handwriting, i.e., for recognizing a drawn letter from the sound produced by a pen on a TAI (Figure 6). This application integrates audio analysis with pattern matching and machine learning techniques.

Conclusions and future work

This paper provided a brief summary of our ongoing research on TAIs. In particular, it discussed the EyesWeb 4 platform for multimodal processing and the software libraries developed for multimodal analyisis of expressive gesture in TAIs.

Future work will concern a further extension of the EyesWeb 4 platform toward multimodal and cross-modal processing as well as the development of new software modules for expressive gesture processing.

A particular focus will be on the assessment and evaluation of the developed prototypes of TAIs. Such evaluation will be carried out in collaboration with psychologists in the framework of organized public events, such as for example the above-mentioned concert at La Biennale in Venice.

Acknowledgements

We thank our colleagues Anne-Marie Burns, Carlo Drioli, Barbara Mazzarino, Massimiliano Peri, Matteo Ricchetti, and Andrea Ricci for their contribute. We also thank the TAI-CHI partners PoliMi and LOA.

This research was partially supported by the EU-IST STREP Project TAI-CHI (Tangible Acoustic Interfaces for Computer-Human Interaction).

References

[1] Camurri A., De Poli G., Leman M., Volpe G. Toward Communicating Expressiveness and Affect in Multimodal Interactive Systems for Performing Art and Cultural Applications. IEEE Multimedia Magazine, 12(1): 43-53, IEEE CS Press, 2005.

[2] Camurri, A., Mazzarino, B., Ricchetti, M., Timmers, R., and G. Volpe. Multimodal analysis of expressive gesture in music and dance performances. In A. Camurri, G. Volpe (Eds.), "Gesture-based Communication in Human-Computer Interaction", LNAI 2915, Springer Verlag, 2004.

[3] Camurri A., Mazzarino B., Volpe G. Espressive interfaces. Cognition, Technology & Work, 6(1), pp. 15-22, Springer-Verlag, February 2004.

[4] Camurri A., Coletta P., Drioli C., Massari A., Volpe G. Audio processing in a multimodal framework. In Proc. 118th Convention, Barcelona, Spain, May 2005.

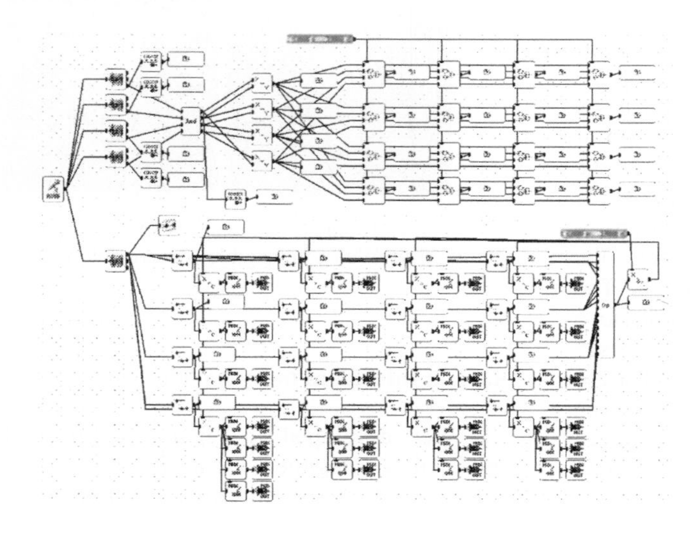

Fig. 3. An EyesWeb patch for in-solid localization of touching positions.

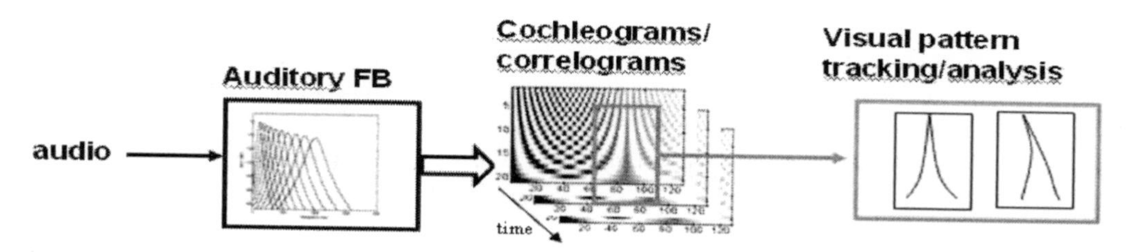

Fig. 4. An example of cross-modal processing: application of visual processing to cochleograms and correlograms.

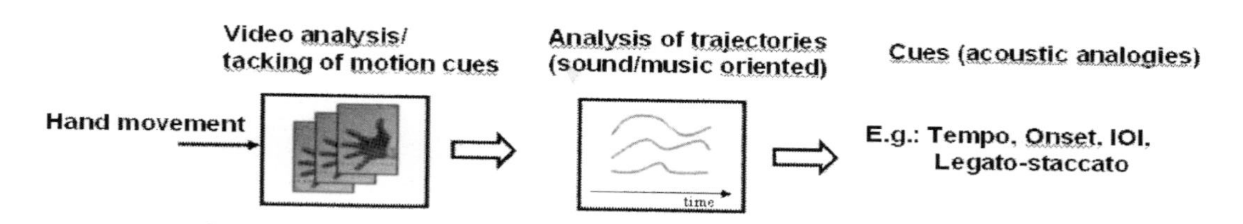

Fig. 5. An example of analysis of motion cues inspired to cues from auditory processing.

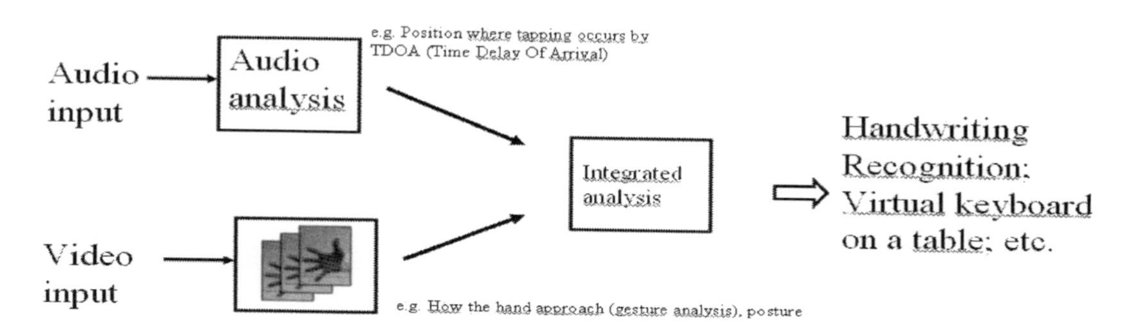

Fig. 6. An example of bimodal processing: algorithms based on the integrated analysis of the sound produced by hands (tapping on a table, handwriting) and the hand gesture (bimodal).

Intelligent Production Machines and Systems
D. T. Pham, E. E. Eldukhri and A. J. Soroka (eds)

Hardware structure of tangible acoustic interface systems

C. Bornand [a], M. Yang [b], Z. Wang [b], Z. Ji [b], M. Al-Kutubi [b]

[a] *eMbedded Information System, University of Applied Sciences, Y-Parc, CH-1400 Yverdon, Switzerland*
[b] *Manufacturing Engineering Centre, Cardiff University, Cardiff CF24 3AA, UK*

Abstract

Designing ergonomic interfaces for man machine interaction is a major task for today's computer design engineers. A new type of tangible man-machine interface which is based on acoustics is being developed. Such a tangible acoustic interface system usually consists of multiple acoustic sensors for detecting impact, a signal conditioning circuit for collecting the acoustic signals and a Digital Signal Processor (DSP) for acoustic signal processing and computing the location of the acoustic sources. This paper briefly outlines the operating principle of tangible acoustic interfaces and describes the hardware structures of such interfaces.

1. Introduction

The vast majority of computer input interfaces are tangible. Currently, the most widely used tangible interfaces are keyboards, mice, touch pads and touch screens. A common disadvantage with these input devices is the presence of mechanical and electronic parts at the point of interaction with the interface (switches, potentiometers, sensitive layers, force resistive sensors, etc). As a consequence direct contact with the intrusive mechanical parts inevitably affects the reliability of the input devices thus limiting their serviceable life.

In order to suppress the need for intrusive mechanical parts and sensing elements in the location of contact and interaction, an acoustics-based remote sensing technology has been developed and its hardware structure is described in this paper. This technology enables virtually any physical objects to be transformed into tangible interfaces, allowing the user to communicate freely with a computer without directly contacting with the mechanical or electrical sensing elements.

There are normally two acoustic source positioning methods being used for developing tangible acoustic interfaces. The first is called the Time Delay Of Arrival (TDOA) method. It uses multiple sensors for the detection of the time delay of arrival of the acoustic wave at several predetermined locations. After the extraction of the time arrival differences between the sensors, the location of the impact can be estimated from the calculation of hyperbolic intersections. The second is known as the Acoustic Pattern Matching method which normally utilizes a single sensor for the detection of the acoustic signal. With this method, samples of the acoustic patterns are recorded by tapping an object at pre-defined locations. Then the target location of a tap can be identified from its best match with all previous recorded samples of the acoustic patterns.

2. Operating principle of tangible acoustic interfaces

Any physical contact with a solid object or a surface (wall, table, etc.) will modify its acoustic properties by the way acoustic energy is distributed in the object and on its surfaces. Such perturbation of the acoustic pattern can be caused in two ways: first, by the acoustic vibration generated at the points of contact when tapping or moving a finger on the surface of an object (passive method); second, by the acoustic energy that is absorbed at the points of contact (proportional to contact pressure) when the object is activated with ultrasound (active method). As acoustic vibration propagates well in most materials, this means that the information about the interaction can be conveyed to a remote location, using the structure of the object itself as a transmission channel and suppressing the need for any overlay or any other intrusive devices over the area that one wishes to make sensitive.

3. Acoustic Sensors and signal conditioning

In order to find the proper sensor to be used for a tangible acoustic interface, various sensors have been tested including PVDF piezoelectric films, piezo disks, accelerometers, and normal miniature microphones (see Fig. 1). The experiments showed that piezo ceramic sensors have the highest sensitivity whilst PVDF sensors have the poorest sensitivity. The accelerometer gives the best signal-to-noise ratio (SNR) whilst the piezo sounder has the worst SNR. The experiments also showed that most of the commercially available piezo sensors have a relatively poor sensitivity to low frequency signals (a few hundreds of Hz).

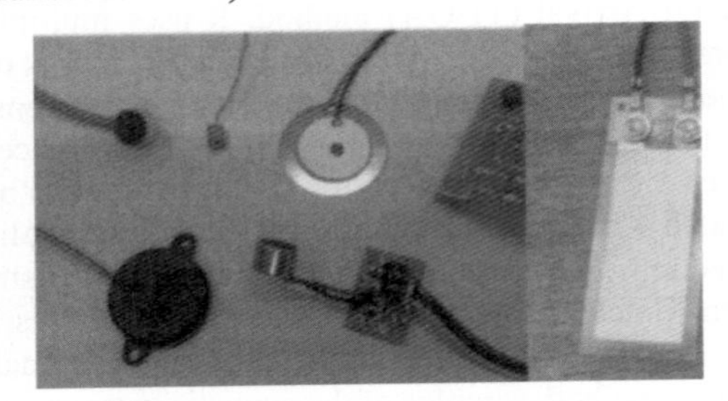

Fig. 1. A selection of the tested piezo sensors.

In order to enhance the sensor's sensitivity to low frequency acoustic signals, two piezo sensors have been developed and tested.

Fig.2 shows the two structures of cantilever beam that were employed for the new developed sensors. One has a free moving end and the other has a fixed end. Experiments show that the sensor with the fixed end has a better performance than that of sensors with a free end in terms of signal consistency and parasitic vibration, suggesting that more robust results can be expected.

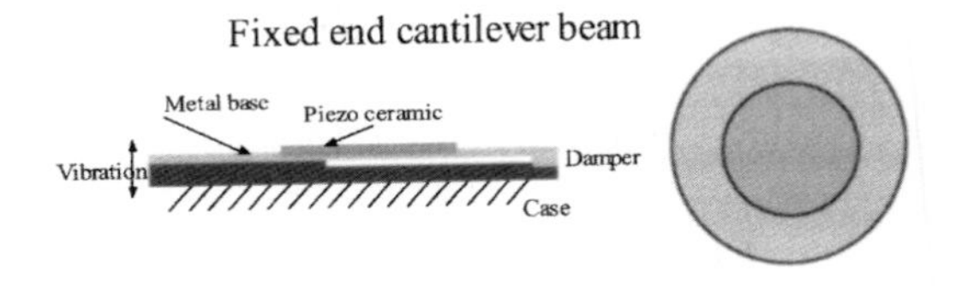

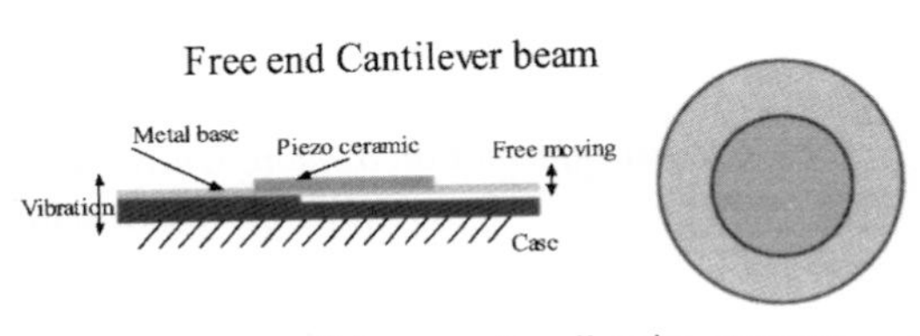

Fig.2. Structures of the acoustic vibration sensors developed for tangible man machine interfaces.

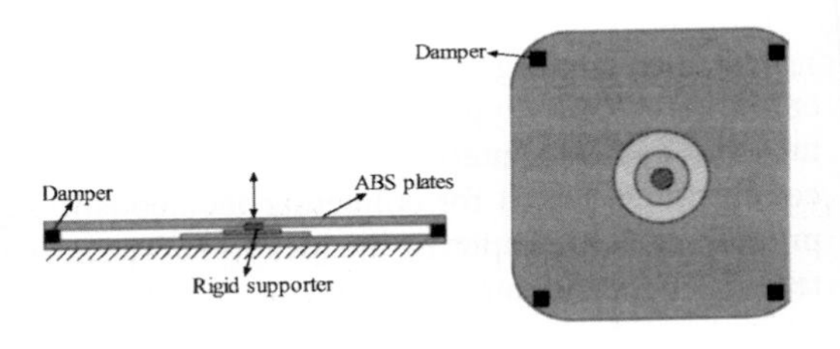

Fig. 3. A sensor structure for measuring low frequency vibration (sub-Hertz).

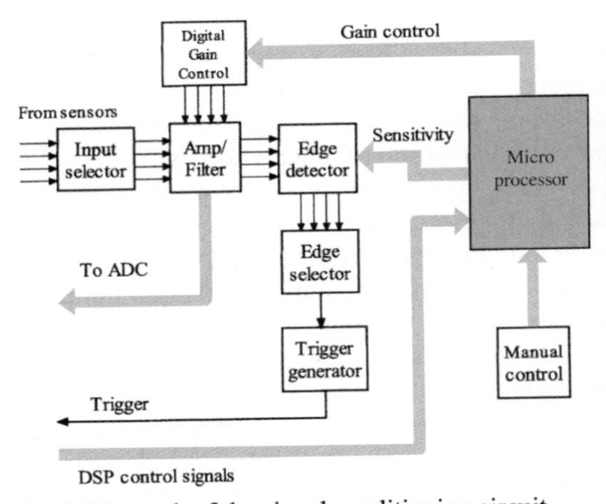

Fig. 4. Diagraph of the signal conditioning circuit.

The other sensor structure which has a high sensitivity to low frequency vibration signals is shown in Fig. 3.

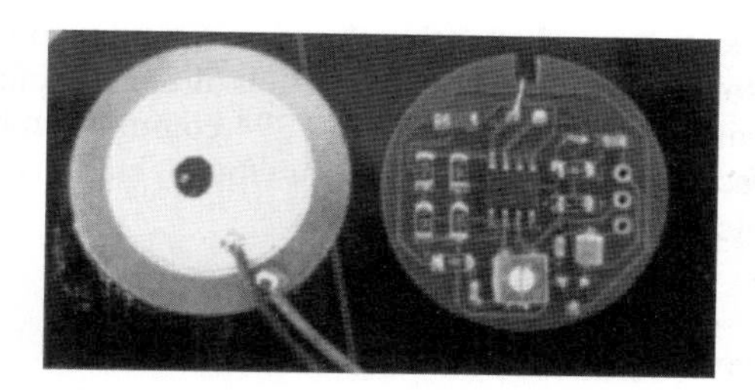

Fig. 5. A pre-amplifier was placed in the vicinity of the sensor in order to reduce the signal losses due to the high impedance of the piezo sensor.

A signal conditioning circuit (see Fig. 4) has been developed for the acoustic signal amplification, noise suppression and signal edge detection. In order to avoid signal loss and interference, the designed preamplifiers have high input impedance and have been placed in the vicinity of the sensor (see Fig. 5).

4. Embedded Digital Signal Processing (DSP) system

At the heart of the tangible acoustic interface system, a DSP system was used for the calculation of impact locations. It also performs the communication with other modality interface systems. The so-called PRESTO Embedded Digital Signal Processing System [5] shown in Fig. 6 has the features of open hardware architecture for single or dual processor applications using I/Os and communication, synchronization mechanisms for isochronous operations and data transfers for multimodal applications.

Fig. 6. The PRESTO Embedded Signal Processing System [5].

For the TDOA method multiple channel acoustic signals have to be acquired simultaneously at a high sampling rate for the reconstruction of the equivalent acoustic waveforms. After signal acquisition, the time delays between signals are extracted for the calculation of the acoustic source location. The computation results of the location are transferred to a PC for the visualization of the geometric location of the impact. If the result is a fully processed value expressing a location, then the transmission will occur only when such a location is detected, for instance at each tap given on the sensitive surface. In that case an asynchronous communication channel is sufficient. If the result is not a fully processed value, it might be necessary to use synchronous channels on which samples are transmitted at regular rates. In that situation it is preferable to establish a synchronization mechanism between the DSP and the isochronous channel. The information flow diagram is shown in Fig. 7.

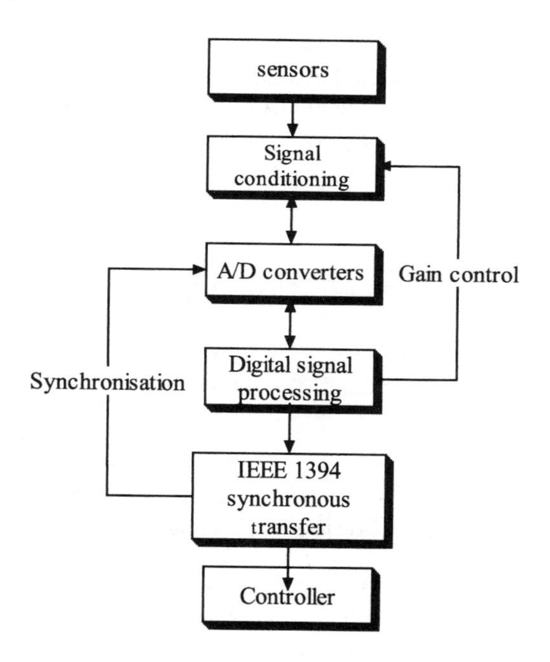

Fig. 7. Isochronous sampling and transfer over IEEE1394 (FireWire) bus.

5. Multi-channel AD converter and interface with DSP

At the output of the signal conditioning stage, the signals are sampled by a multi-channel A/D converter. Each channel has a maximum sampling rate of 1 MHz, and offers 16 unipolar channels with a resolution of 12 bits and a serial communication with a processor. In the chosen design, only two channels per A/D converter are used, among the 16 available. This gives globally 4 microphone inputs and 4 high impedance inputs. The four high impedance inputs are used for piezoelectric sensors.

The processor can be either a floating point SHARC ADSP21161 or a fixed point BlackFin ADSP-BF533 of Analog Devices. The ADSP-BF533 processor runs at a clock frequency up to 600 MHz with several interfaces such as 4 serial ports, a 16 bits asynchronous bus, an SPI and a 16-bit PPI and offers a very interesting power/price ratio.

At the centre of this classical architecture, an FPGA shown in Fig. 8 is used mainly to synchronize the signal acquisitions with one of the clocks: internal (quartz), real-time FireWire bus (1394b external clock), serial ports of the DSP or external word clock. Moreover, in order to avoid overloading the DSP unnecessarily, the localization algorithm is carried out only if a certain level of the analogical threshold is reached. The handling of the multiple interrupt signals is also done by the FPGA. Finally, 16 digital I/Os are also available on the FPGA for the configuration of the signal conditioning electronics.

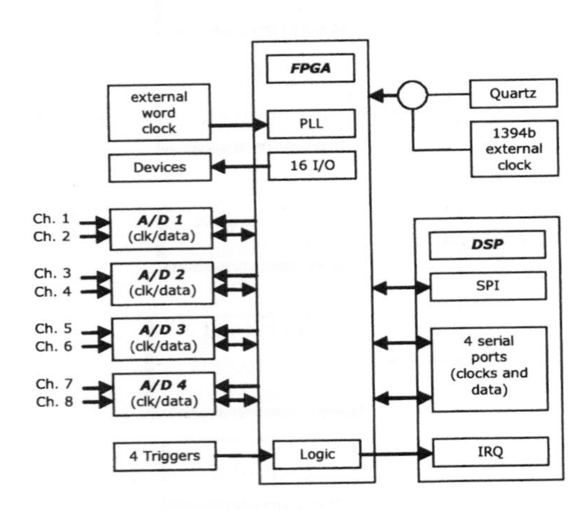

Fig. 8. A FPGA is designed for controlling and interfacing the ADC.

6. Results and conclusion

A TDOA algorithm has been implemented on the DSP system with BlackFin processor ADSP21533 @ 600 MHz. The results were comparable to the ones obtained on a 3 GHz Pentium 4 machine, with all the advantages of a compact embedded system compared to a PC with a big power supply and fan.

The selection of the sensors and the localization algorithms were directly responsible for the quality of the resulting information. In this work, the precision of the detection was better than 1 cm on a rigid wooden desk top. Due to the synchronization of the complete acquisition chain with the IEEE1394 communication interface, the data delivery was found stable and reliable.

Acknowledgements

This work was carried out as part of the Tai-Chi projects which is funded by the European Commission FP6 IST Programme. We acknowledge the members of the Tai-Chi consortium, for their help and stimulation. Particular thanks to the members of the Embedded Sensors Group of the MiS Institute for their support. The MEC is the coordinator of Tai-Chi and the EC-funded FP6 Network of Excellence on Innovative production Machines and Systems.

References

[1] Parisi R, Gazzetta R, Claudio E. Prefiltering approaches for time delay estimation in reverberant environments. 2002 IEEE International Conference on Acoustics, Speech, and Signal Processing ICASSP2002, Orlando, Florida, USA, May 13-17, 2002.

[2] Fröhlich B. and Plate J. The cubic mouse: a new device for three-dimensional input. In *Proceedings of the SIGCHI Conference on Human Factors in Computing Systems (CHI '00)* (The Hague, The Netherlands, April 1-6, 2000). ACM Press, New York, NY, 2000, 526-531.

[3] Note S, Lierop P. van Ginderdeuren, J.; Rapid Prototyping of DSP Systems: Requirements and Solutions; Proceedings of the Rapid System Prototyping, Sixth IEEE International Workshop, North California, 1995.

[4] http://www.acoustic-camera.com

[5] http://www.jdc.ch/en/dsp_index.html

Intelligent Production Machines and Systems
D. T. Pham, E. E. Eldukhri and A. J. Soroka (eds)

Impact Localisation Techniques for Tangible Acoustic Interfaces

D T Pham[a], M Al-Kutubi[a], Z Ji[a], M Yang[a], Z Wang[a], and S Catheline[b]

[a]Manufacturing Engineering Centre, Cardiff University, Cardiff CF24 3AA, UK
[b] Laboratoire Ondes et Acoustique, ESPCI, 10 rue Vauquelin, 75231 Paris Cédex 05, France

Abstract

This paper describes two impact localisation techniques for tangible acoustic interfaces. A rising-edge detection method was used for the time-delay-of-arrival (TDOA) approach and a cross-correlation method for the location-template-matching (LTM) approach. These methods were experimentally investigated using objects of different materials with different types of acoustic sensors. Experimental results have shown their potential for real-time localisation of impacts using common solid surfaces (a steel whiteboard and a hard wood board) and appropriate acoustic sensors for human computer interfaces.

1. Introduction

There are basically two types of human-computer interfaces that have been developed, tangible (touchable) and intangible (non-touchable). The vast majority of input interfaces are tangible. Currently tangible interfaces include keypads, mice and touch screens which are reliable and easy to use. However, a common problem with these devices is that they restrict the mobility of users, constraining them to be in certain locations during interactions with computers. A solution would be to convert virtually any tangible objects such as table tops, walls and windows into interactive surfaces.

In recent years, some attempts with visual and acoustic techniques have been made using tangible objects as natural alternatives to traditional computer interfaces for human computer interaction [1, 2]. The advantage of these techniques is that the interaction surface is no longer a delicate device with a specific size as is the case with touch screens.

A visual technique using a laser scanner was applied for accurately tracking the position of bare hands in a plane just above a large projection display [1]. It was very well suited to continuous tracking but less appropriate for mouse-click-like actions.

A passive acoustic technique was demonstrated by researchers at MIT on shop windows for public browsing [2]. Commercial tap-screens were promoted by I-Vibrations [3]. Those two applications were based on the time-delay-of-arrival method for localising the impact on the tangible surface.

In applications based on acoustic techniques for human computer interfaces [1, 3], glass was the only material considered for interaction with a PC. However, the techniques are not limited to glass.

The suitability of different materials of common use objects like steel whiteboards and hardboards for acoustic interfaces for human computer interaction has been investigated in this work. Two experiments corresponding to the TDOA and LTM approaches were carried out for passive impact localisation. In the experiments, the conventional TDOA approach was implemented using four acoustic sensors and the LTM approach using one acoustic sensor. Steel whiteboards and hardboards were chosen as acoustic wave propagation media as they are readily available, cheap and widely used for interior partitions and decorations. Projecting a display screen from a PC onto the surfaces of the boards also provides a freedom of mobility for public use. The acoustic signals generated by tapping on the object surfaces were first conditioned by an analogue electronic circuit and then digitised by a data acquisition card and fed into a PC to determine the acoustic source locations.

The main aim of the experiments was to demonstrate the feasibility of impact localisation techniques and not to compare them. The initial choice of a different material for each technique was made by taking into account the expected performance of the TDOA method for a homogeneous material (steel) and that of the LTM method for a non homogeneous material (wood).

The TDOA and LTM approaches and their implementations are discussed in the following sections.

2. TDOA Approach

The TDOA approach is popular in the area of acoustic source localisation [4]. The principle of TDOA localisation is to measure the time delays between the arrivals of the signals to the various sensors. These delays result from the differences in distances from the impact sources to the sensors at known locations.

Fig. 1 shows a block diagram of a TDOA system comprising a steel whiteboard, four sensors located at the four corners, signal conditioning hardware, a PCI data acquisition card and a PC. A knock on the whiteboard is detected by the sensors. The sensor signals are amplified and filtered by the signal conditioning hardware, digitised by the PCI data acquisition card and processed by the PC.

In the case of four sensors used for the localisation of an impact, as shown in Fig.1, the location (x, y) of the impact can be determined by the following equations

$$\sqrt{(x-x_1)^2+(y-y_1)^2}-\sqrt{(x-x_3)^2+(y-y_3)^2}=v\Delta t_{13} \quad (1)$$

and

$$\sqrt{(x-x_2)^2+(y-y_2)^2}-\sqrt{(x-x_4)^2+(y-y_4)^2}=v\Delta t_{24} \quad (2)$$

where (x_1, y_1), (x_2, y_2), (x_3, y_3) and (x_4, y_4) are the coordinates of the four sensors. v is the sound wave velocity in the object. Δt_{13} is the measured time difference between the sensor locations (x_1, y_1) and (x_3, y_3), and Δt_{24} is the measured time difference between the sensor locations (x_2, y_2) and (x_4, y_4).

In Eqs. 1 and 2, the wave velocity and the coordinates of the sensors are known, and the time differences between the paired sensors are to be measured so that an acoustic source location can be calculated using the two equations.

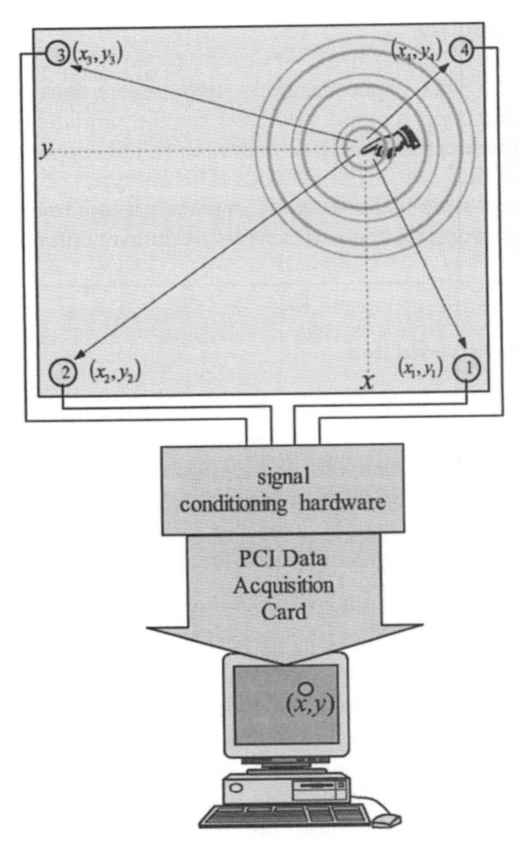

Fig. 1. TDOA system block diagram.

In this work, timing is determined by the rising-edge detection method. Rising-edge detection is a conventional method where the arrival time is counted from the time when the received signal amplitude first crosses a threshold [5]. It identifies the arrival of the fastest wave components and requires some filtering to distinguish it between wave modes. The triggered sensor normally is the one that is closest to the source and may include wave modes which do not propagate to other sensors.

The TDOA approach was implemented on a steel whiteboard with an area of 1.2×1.0 m^2, as shown in Fig. 2, using four accelerometers (as acoustic sensors) placed at the corners. Captured acoustic signals were sampled at a frequency of 100 KHz.

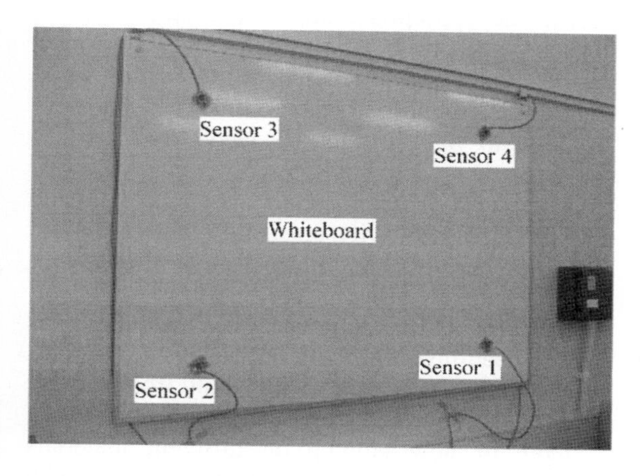

Fig. 2. TDOA experimental setup using a steel whiteboard and four acoustic sensors.

In the experiments, it was observed that error levels were not distributed evenly on the board. That may be caused by the differences in wave velocities as a result of the steel whiteboard surface being not entirely homogeneous. With digital conditioning of the raw acoustic signals, the achieved impact localisation error was about 30 mm on average. The maximum error was better than 50 mm.

3. LTM Approach

According to the time-reversal theory described by Fink, it is possible to reconstruct an acoustic signal in its original location in a scattering medium by recording the received signals and sending back the time reversed version of these signals through the medium [9]. This implies that the received signal carries its source location signature as a result of scattering in the transmission medium and reflections from complex boundaries.

With the Location Template Matching (LTM) approach, the uniqueness feature of the received acoustic signal, as a function of its source location, which is based on the time-reversal theory, is employed to localise an impact on an object. This is achieved by creating a template of received signals generated from taps mapped to their known corresponding locations. The source location can then be determined by cross-correlation analysis to find the best matching template.

Fig. 3 shows a block diagram of an LTM system comprising a hardboard, one sensor located at the bottom right of the board, signal conditioning hardware, a PCI data acquisition card and a PC. A knock on the hardboard is detected by the sensor. The sensor signal is amplified and filtered by the signal conditioning hardware, digitised by the PCI data acquisition card and processed by the PC. As mentioned above, the location of the sound source is determined by cross-correlation analysis.

Cross-correlation is a technique widely used to measure similarity between two signals. For example, cross-correlation has been applied to match unknown test ECG signals with template ECG signals in a database of known diagnoses to determine cardiac information [10]. In image processing applications, a finger has been tracked for real-time gesture interfaces by two-dimensional cross-correlation matches between the target image and the template images [11].

The similarity or the degree of correlation between detected acoustic signals can be decided by examining the correlation coefficient. Considering a template acoustic signal $s_0(t)$ and an acoustic signal to be recognised $s_k(t)$, each with a population size of N samples, the correlation coefficient of signals s_0 and s_k can be written as

$$r_{s_0 s_k} = \frac{\mathrm{cov}(s_0, s_k)}{\sigma_{s_0}\sigma_{s_k}}, \quad k = 1, 2, 3, \ldots \tag{3}$$

where $\mathrm{cov}(s_0, s_k)$ is the covariance between signals s_0 and s_k, and σ_{s_0} and σ_{s_k} are the standard deviations of the signals s_0 and s_k.

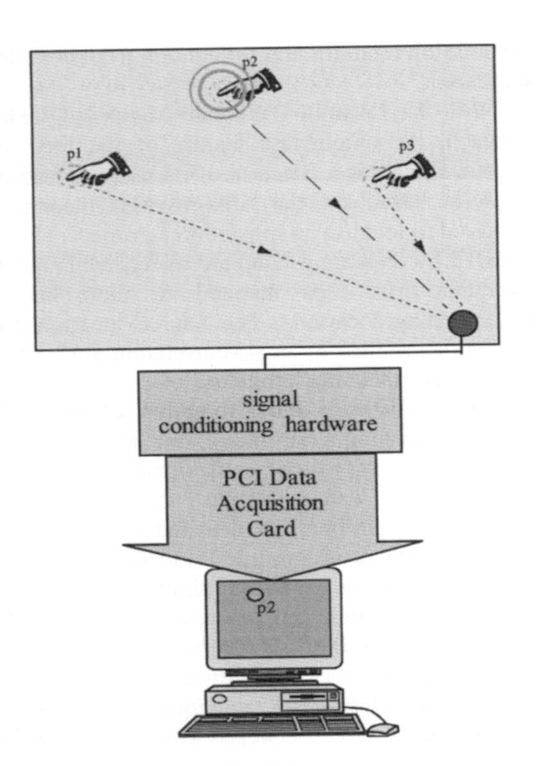

Fig. 3. LTM system layout. The impact is detected at location p2 after a template has been created for p1, p2 and p3.

Fig. 3 shows a hardboard with an area of 1.0×0.65 m^2, which was used in the experiment for the LTM approach. An electret microphone clamped at the bottom right of the board was used for recording the template signals and to receive the signals generated from tapping on the surface.

Fig. 4. LTM experiment using a hardboard.

The sampling rate here was reduced to 20 KHz because the location estimation is not related directly to the sound speed as in the TDOA method. A threshold was set to identify the closest match with the template using the cross-correlation coefficient as a measure for determination of the degree of similarity. A resolution of 100 mm was achieved in the experiment.

4. Implementation

Various factors need to be considered when implementing tangible interfaces including the surface material, sensor type, sampling frequency and the processing approach. It is important to choose the appropriate sensor for each material, impact type and application. For example, piezoelectric discs were used in [3]. In this work, electret microphones were found to be very sensitive to light impact on wood compared to piezoelectric sensors but less sensitive with glass or metal objects. Accelerometers on the other hand produce high sensitivity on metal surfaces but are more expensive than piezoelectric sensors or electret microphones.

Impact localisation on glass surfaces has already been investigated [2,3,4]. However, signals with different properties will result from similar impacts on different materials. Fig. 5 shows an example of how tapping on different materials produces signals having different properties with the sampling duration, frequency and smoothness. The signals shown in Fig. 5 were obtained from different types of acoustic sensors on different object surfaces. Fig. 5(a) depicts an acoustic signal from tapping on hardboard fitted with a microphone. Fig. 5(b) shows a signal from a steel whiteboard with an accelerometer. Fig. 5(c) displays a signal from a glass plate with a piezoelectric disk. The sampling frequencies for the three signals are 20 KHz, 100 KHz and 40 KHz respectively.

The accuracy of the impact localisation in the TDOA approach is directly related to the acoustic wave velocity in the medium. In a non-homogeneous medium, the wave velocity is not equal in all directions causing errors in the location detection. The concept of the template matching approach is based on the uniqueness of the signals received from different locations as a result of scattering. Thus, a heterogeneous medium may enhance the uniqueness of the received signal resulting in less ambiguity in recognising the location compared to the TDOA approach.

However, the reliability of this approach is strongly related to the particular template used in the calculation of correlation coefficients.

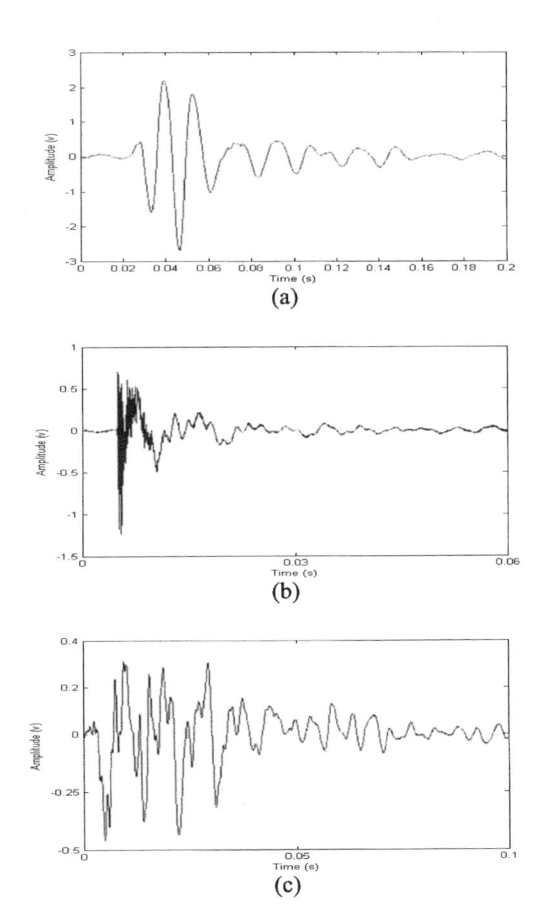

Fig. 5. Sample signals from tapping on (a) a hardboard, (b) a steel whiteboard, and (c) a glass plate.

Unlike the location template matching approach, the sampling rate in the TDOA approach needs to be high for fast wave velocities and for close impact locations. The TDOA approach is more suitable for identifying arbitrary locations within the working area defined by the sensors boundary, whereas the LTM approach suits the identification of specific locations predefined individually in the templates.

5. Conclusion

The TDOA and LTM approaches have been demonstrated for tangible acoustic interfaces. Different types of sensors were evaluated for different object materials. The chosen sensors used in the experiments that have demonstrated confident performance are accelerometers for metal sheets using the TDOA approach and microphones and piezoelectric discs for hardboard sheets using the LTM approach.

The experiments have shown that the TDOA approach is applicable to steel whiteboards for impact localisation provided that suitable sensors are used and appropriate digital filtering is applied. When the object becomes less homogeneous, the TDOA approach may fail while the LTM approach becomes more reliable. Promising results were achieved in the experiments in localising taps on the steel whiteboard and hardboard surfaces. The reliability and robustness of these approaches remain an issue under further investigation.

Acknowledgements

This work was financed by the European FP6 IST Project "Tangible Acoustic Interfaces for Computer-Human Interaction (TAI-CHI)". The support of the European Commission is gratefully acknowledged. The MEC is the coordinator of the EC-funded FP6 I*PROMS NoE.

References

[1] Paradiso JA, Hsiao K, Strickon J, Lifton J and Adler A. Sensor systems for interactive surfaces. IBM Systems Journal, Vol.39, Nos.3&4, 2000, pp 892-914.

[2] Paradiso JA, Leo CK, Checka N and Hsiao K. Passive acoustic knock tacking for interactive windows. ACM CHI 2002 Conference, Minneapolis, Minnesota, 20-25 April 2002.

[3] http://www.i-vibrations.com/

[4] Checka N. A system for tracking and characterising acoustic impacts on large interactive surfaces. MS Thesis, MIT, 2001.

[5] Ding Y, Reuben RL and Steel JA. A new method for waveform analysis for estimating AE wave arrival times using wavelet decomposition. NDT & E International, Vol.37, 2004, pp 279-290.

[6] Knapp CH and Carter GC. The generalized correlation method for estimation of time delay. IEEE Trans. Acoustic, Speech and Signal processing, Vol.24, 1976, pp 320-327.

[7] Prosser WH, Gorman MR and Humes DH. Acoustic emission signals in thin plates produced by impact damage. Journal of Acoustic Emission, Vol.17, Nos.1- 2, 1999, pp 29-36.

[8] Ellwein C, Danaher S and Jäger U. A vibration signal family of impact events. European Symposium on

Intelligent Techniques, Aachen, Germany, 14-15 September 2000.

[9] Fink M. Time-reversal mirrors. J. Phys. D: Appl. Phys. 26 (1993) UK, 1333-1350

[10]Bousseljot R and Kreiseler D. Waveform Recognition with 10,000 ECGs. IEEE Computers in Cardiology Proceedings, 24-27 Sept 2000, Cambridge, MA, 331-334.

[11]O'Hagan R and Zelinsky A. Finger Track – A Robust and Real-Time Gesture Interface. Advanced Topics in Artificial Intelligence, Tenth Australian Joint Conference on Artificial Intelligence Proceedings, 475-484, Dec. 1997.

Intelligent Production Machines and Systems
D. T. Pham, E. E. Eldukhri and A. J. Soroka (eds)

In-air passive acoustic source localization in reverberant environments

L. Xiao, T. Collins

Department of Electronic, Electrical and Computer Engineering, University of Birmingham, Edgbaston, Birmingham, B15 2TT, UK

Abstract

In-air passive acoustic source localization is important to the design and implementation of tangible acoustic interfaces for computer human interaction. The methods used for in-air passive acoustic source localization are discussed in this paper. A time delay estimation method based on general cross-correlation was chosen for the application. In addition, a phase transform and smoothed coherence transform are combined to suppress the reverberation and background noise from the environment. MATLAB simulations and practical tests were carried out under real reverberation conditions. Simulation and test results show that the method is effective and robust for single in-air passive acoustic source localization.

1. Introduction

In-air passive acoustic source localization is important in the design and implementation of tangible acoustic interfaces for computer-human interaction (TAI-CHI). Usually, the characteristics of the passive sources are not known and have a wide bandwidth (e.g. fingers snapping, human voice, percussive noise). In TAI-CHI systems, determination of the physical position of users is an essential capability. Without precise knowledge of the spatial location of users in an environment, the interface would not be able to react naturally to the needs and behaviour of the user. Generally, in a real acoustic environment, such as an office or meeting room, there are reflections from walls and other objects. Information used to separate and locate sources may include such multi-path characteristics, also known as reverberation. The presence of reverberation is probably the most challenging difficulty to be overcome when designing in-air passive acoustic source localization systems.

2. Methods used in passive source localization

Since the early 1990s, passive acoustic source localization applying microphone arrays has been an active research area. Many methods have been proposed to tackle the problem depending on the application and the scenario. These methods can be classified into five main categories: steered beamforming based method; spectral estimation based direction of arrival (DOA) method; near-field acoustic holographic (NAH) method; state-space approach based on particle filtering (PF); and time-difference-of-arrival (TDOA) estimation method.

Steered beamforming method is derived from a filtered weighted version of the signal data [1]. A maximum likelihood (ML) estimator steers the receiving beam of an array of microphones to various locations and searches for a peak in the output power in order to locate the source position. This method has rarely been applied to the localization problem because of its complexity. The ML estimator requires the solution of a set of nonlinear equations. Complexity prohibits its use in real-time applications.

Methods based on spectral estimation derive the source location from the spectral correlation matrix of the received signals. This category includes autoregressive (AR) modelling, minimum variance (MV) spectral estimation and Eigen-analysis techniques such as the multiple signal classification (MUSIC) method [1]. The MUSIC method can provide higher resolution to facilitate separating the DOAs of multiple sources that are located very close to each other. Thus, it is suitable for multiple source localization. These methods are proposed primarily for narrowband, far field source localization, although researches have tried to extend them to the near-field and wideband case.

NAH is a methodology that enables one to reconstruct acoustic quantities such as acoustic pressure, intensity, and power radiated from a finite object into 3D space based on the acoustic pressure measured in a 2D surface. The measured pressure is used as a boundary condition to solve the homogeneous acoustic wave equation in the region exterior to the sources [2]. Practical limitations of NAH are the large number of spatial sampling points (i.e. transducers) needed for accurate imaging and the fact that the algorithms are numerically intensive. Practical application of NAH to real-time passive acoustic source localization has seldom been seen so far.

The key to the particle filtering method is that the power from a true, continuous source follows a dynamical model from frame to frame, whereas there is no temporal consistency to the spurious peaks caused by noise, reverberation, etc. The sequential Monte Carlo method (particle filtering) is used to estimate the probability density function of the source location conditioned on all received data up to the current frame. PF based methods provide an efficient way of filtering the spurious peaks out of the cross-correlation function. The algorithm can be used for multiple source localization in a reverberant environment with known wave propagation speed and sensor positions [3]. One algorithm of this family provides not only the best performance in tracking an acoustic source in a reverberant room, but is also the most computationally efficient [4].

The method based on time delay estimation (TDE) firstly computes the TDOA between all pairs of microphones and then combines them, with the knowledge of the array geometry, to obtain the source location by solving the constrained quadratic optimization problem. Time delay between signals from any pair of microphones is obtained by first computing the cross-correlation function of the two signals, then the lag at which the cross-correlation function has its maximum is taken as the time delay. In terms of computational requirements, TDE based methods are the most efficient because they do not involve an exhaustive search over all possible angles as in DOA based method. Also, TDE can be directly employed for near-field and wideband source localization and has been extensively investigated. The majority of practical localization systems are TDE-based. However, TDE based methods are useful only for single source localization.

Many methods can be used for the time delay estimation such as the generalized cross correlation function (GCC) method [5], the adaptive method [6], the least square estimation method [7] and cepstral pre-filtering method [8]. Among them, GCC method is the most commonly used method for TDE and it is shown that GCC based on a phase transform (PHAT) is the most effective method on suppressing reverberation among a class of GCC based methods [5]. TDE based on PHAT was chosen for this application owing to its suitability for near-field and broadband applications, simplicity, as well as its modest computational requirements making it suitable for real-time implementation.

3. TDOA method based on PHAT

3.1 Mathematical model of TDOA based on cross correlation

Consider any two microphones, i and j. Let the signals received by these two microphones be $x_i(t)$ and $x_j(t)$, and they are represented as $X_i(f)$ and $X_j(f)$ in the frequency domain. Suppose the signals arriving at two microphones with a time delay of D between them.

$$x_i(t) = \alpha_i(t)s(t) + v_i(t)$$
$$x_j(t) = \alpha_j(t)s(t+D) + v_j(t) \tag{1}$$

where $s(t)$ is the source signal, $v_i(t)$ and $v_j(t)$ are the noise of microphones i and j at time t respectively, they are expressed as $V_i(f)$ and $V_j(f)$ in frequency domain; $\alpha_i(t)$ and $\alpha_j(t)$ are attenuation coefficients which are dependent on the distance between the source and microphone and the acoustic characteristics of the environment. Suppose $s(t)$, $v_i(t)$ and $v_j(t)$ are uncorrelated each other. Thus the cross-power spectrum with no pre-filtering is given by

$$\Phi_{x_i x_j}(f) = X_i(f)X_j^*(f) \tag{2}$$

The cross-correlation between these signals is given by the inverse Fourier transform of the cross power spectrum.

$$R_{x_i x_j}(\tau) = \int_{-\infty}^{\infty} \Phi_{x_i x_j}(f)e^{j2\pi f\tau}df \tag{3}$$

$R_{x_i x_j}(\tau)$ can be computed for the range of possible negative and positive lags. The lag at the maximum $R_{x_i x_j}(\tau)$ is the time delay between two signals. Hence it is given by

$$\tau_{ij} = \frac{1}{F_s}\arg\max(R_{x_i x_j}(\tau)) \tag{4}$$

where F_s is the sampling frequency. Substitute the Fourier transform of $x_i(t)$ and $x_j(t)$ into (2) and assume α_i and α_j are time independent, (2) can be transformed into

$$\Phi_{x_i x_j}(f) = \alpha\Phi_{ss}e^{-j\omega D} + \alpha_i S(f)V_j^*(f) + \alpha_j S^*(f)V_i e^{-j\omega D} + V_i(f)V_j^*(f) \tag{5}$$

where Φ_{ss} is the power spectrum of the source, $\alpha = \alpha_i\alpha_j$ and $\omega = 2\pi f$. Since the noise signals are assumed to be uncorrelated with the source signal and each other, the last three terms do not contribute towards the cross-correlation computation. Substituting (5) into (3), yields

$$R_{x_i x_j}(\tau) = F^{-1}[\alpha\Phi_{ss}(f)e^{-j\omega D}] = \alpha R_{ss}(\tau)\delta(\tau - D) = \alpha R_{ss}(\tau - D) \tag{6}$$

Thus we find that the cross-correlation function is just the incident signal's auto-correlation function time-shifted by the time delay *D*. Ideally, if the cross-corelation function is a delta function at *D*, the peak would be easily picked out. In reality, the auto-correlation of the incident signal ends up spreading the cross-correlation function around time delay *D*. When pre-filtering is used, as in GCC, the choice of the pre-filter must be geared towards minimizing this spread. If there happens to be multiple delays, as is the case in rooms with reverbration, the cross-corelation is given by

$$R_{x_i x_j}(\tau) = R_{ss}(\tau)\sum_m \alpha_m \delta(\tau - D_m) \tag{7}$$

If there is no noise and reverberation, the above method works well. However, if there are noise and reverberation, as in practical situations, it results in peaks in the cross-correlation function at each delay. If the delays are very close to each other and if the spread in the signal's auto-correlation function is very large, then the peaks may merge thus decreasing the resolution of the estimate. Improvement is necessary for the above cross-correlation method to get reliable TDE results in the presence of reverberation and noise. As a result, the method based on GCC is proposed.

3.2 The phase transform

PHAT is a special form of GCC based method proposed to suppress reverberation. For the TDE method based on GCC, a pre-filter is added to process $x_i(t)$ and $x_j(t)$ before they are cross-correlated. Suppose the transfer function of the pre-filter in the frequency domain is $H_i(f)$ for $x_i(t)$ and $H_j(f)$ for $x_j(t)$. The cross power spectrum between the filtered signal is

$$\Phi^g_{x_ix_j}(f) = H_i(f)H_j^*(f)X_i(f)X_j^*(f) = \Psi_g(f)\Phi_{x_ix_j}(f) \quad (8)$$

where

$$\Psi_g(f) = H_i(f)H_j^*(f) \quad (9)$$

The pre-filter used in PHAT is [5]

$$\Psi_g(f) = \frac{1}{|\Phi_{x_ix_j}(f)|} \quad (10)$$

where $\Phi_{x_ix_j}(f)$ is calculated from (5). Thus the estimated GCC with PHAT (GCC-PHAT) in the time-domain is given by

$$R^{PHAT}_{x_ix_j}(\tau) = \int_{-\infty}^{\infty} \frac{\alpha}{|\alpha\Phi_{ss}(f)e^{-j\omega D}|}\Phi_{ss}(f)e^{-j\omega D}e^{j\omega\tau}df = \int_{-\infty}^{\infty} e^{j\omega(\tau-D)}df \quad (11)$$

The integral on the left hand side reduces to a delta function at time delay D. Thus, we have

$$R^{PHAT}_{x_ix_j}(\tau) = \delta(\tau - D) \quad (12)$$

Therefore, under the assumption of completely un-correlated noise, the spread caused by the auto-correlation of incident signal disappears and a much sharper delta function for the cross correlation can be achieved. The time delay information is present in the phases of various frequencies and these are not affected by the transform. Because the transform tends to enhance the true delay and suppress all spurious delays. It turns out to be effective in moderate reverberation. One disadvantage of PHAT is that it tends to enhance the effect of frequencies that have low power when compared to noise power. This can cause the estimates to be corrupted by the effect of uncorrelated noise. To counter this problem, additional measures must be taken to suppress the background noise. An appropriate tool is the smoothed coherence transform (SCOT)[5]. The transfer function of SCOT is:

$$H_i^{SCOT}(f) = \frac{1}{\sqrt{\Phi_{x_ix_i}}} \quad (13)$$

$$H_j^{SCOT}(f) = \frac{1}{\sqrt{\Phi_{x_jx_j}}} \quad (14)$$

The effect of the SCOT is to suppress the energy in frequency bands containing a high noise level [5].

It should be noted that when calculating $\Phi_{x_ix_j}(f)$ in (5), it is assumed that α_1 and α_2 of (1) are time independent. This is not true if there is significant reverberation present. Therefore, ideal results of (12) cannot be achieved in a practical acoustic environment. However, in practice it takes a situation with relatively heavy reverberation and noise before this method tends to break down.

4. MATLAB simulation results

In order to verify the TDOA method based on PHAT and its application to in-air passive acoustic source localization in a real acoustic environment, MATLAB simulations were firstly carried out in a practical room. The dimension of the room is 7.4 m × 4.2 m × 3.1 m. The acoustic performance of the room and the reverberation are modelled by the acoustic image method [9]. The background noise is white noise, the signal to noise ratio (SNR) is between 4-20 dB depending on the source level and position of the microphone. The acoustic sources are some practically recorded passive acoustic sources such as human voice, hand clapping, and percussive noise (metal tapping wood). The simulation results are shown in Fig. 1. The practical coordinates of the acoustic source, represented by footnote 'p', and the estimated coordinates, represented by footnote 'est' are compared in Table 1. The errors between the practical value and the estimated value are also listed in the table as italics. The origin of the coordinate is chosen at one corner of the room as shown in Fig. 1, the x coordinate, y coordinate and z coordinate represent the length, width and height of the room respectively.

From Fig. 1 and Table 1, it can be observed that the estimated positions of the acoustic source match their practical counter part quite well. The errors between the practical value and estimated value are less than or equal to 5 cm. The simulation results prove that the TDOA method based on PHAT can be used for in-air passive acoustic source localization under reverberant conditions.

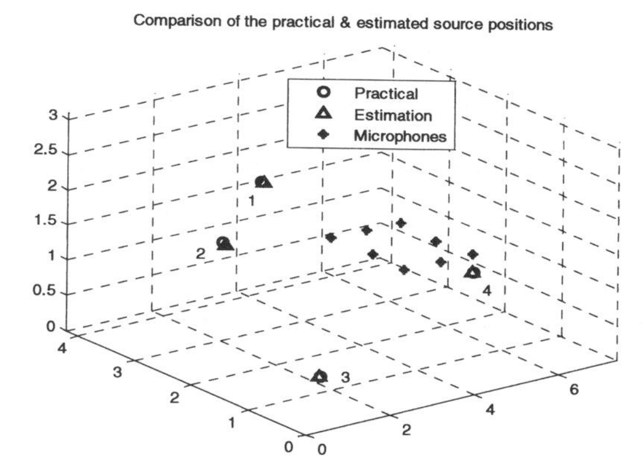

Fig. 1 MATLAB simulation results

5. Experimental tests

After the method was verified by the MATLAB simulation results, practical tests were carried out. The experimental setup is shown in Fig. 2(a) and the configuration of the microphone array is shown in Fig. 2(b). The SNR ranges from 4 dB to 14 dB in the practical tests. The test results are demonstrated in Fig. 3, and the practical and estimated results are compared in Table 2. Four kinds of practical passive acoustic sources are tested. These are human voice, hand clapping, wood tapping metal and metal tapping wood.

From Fig. 3 and Table 2, it can be observed that good agreements were found between the practical coordinates of the passive acoustic sources and those from estimation although the error between the estimated z coordinate and its practical counterpart of the 4th test is 15 cm, which is a little bit big. The reason is that for the 4th test, metal tapping wood, the metal tapped at the surface of a loud speaker and the surface has a small area. When the metal was tapping the surface, the whole area was vibrating and the acoustic pressure was transmitted from the whole surface to the microphone transducers, so the acoustic source is distributed on the whole vibrating surface. For the TDOA method based on PHAT, it can only localize a single acoustic source. If the acoustic source is distributed in a bigger area, the error between the practical time delay and that from estimation would be quite big or even it is totally wrong. Therefore, the method fails. We have tried tapping the surface of the central heating radiator in the room; the method did not work. Research is being carried out on in-air multiple passive source localization under noisy and reverberant conditions, using methods such as particle filtering.

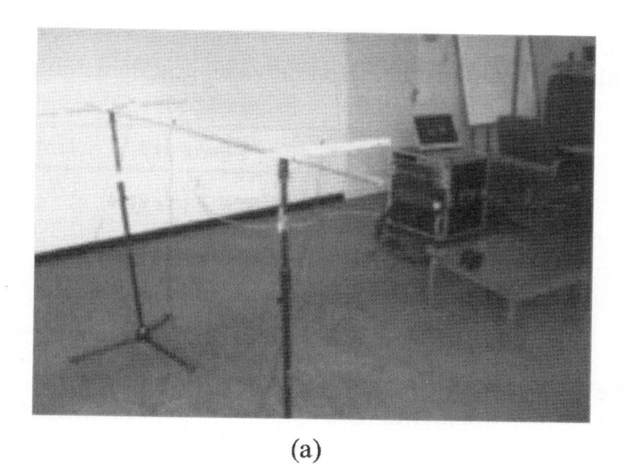

(a)

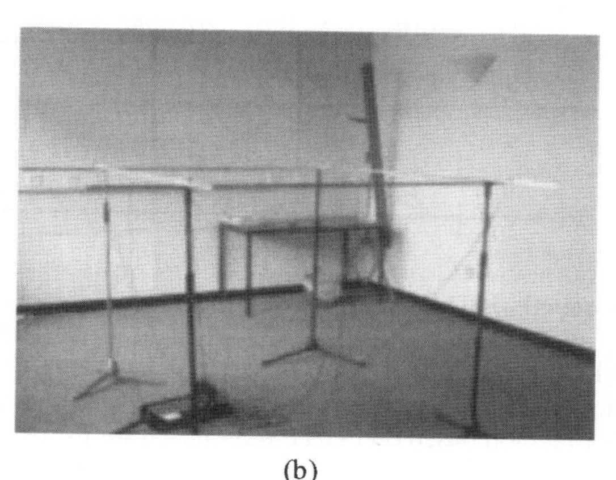

(b)

Fig. 2 Experimental setup (a) the whole test setup (b) the microphone array configuration

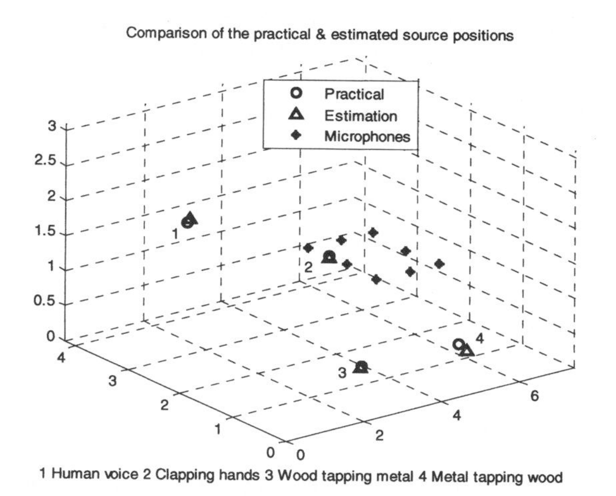

Fig. 3 Practical test results

Table 1
Simulation results comparison between practical and estimated source localizations

Tests	x_p(m)	x_{est}(m)	*error(m)*	y_p(m)	y_{est}(m)	*error(m)*	z_p(m)	z_{est}(m)	*error(m)*
1	3.	3.03	*0.03*	3.7	3.69	*-0.01*	1.	0.96	*-0.04*
2	2.	2.05	*0.05*	2.3	2.29	*-0.01*	2.5	2.47	*0.03*
3	1.	0.96	*-0.04*	0.5	0.49	*-0.01*	0.5	0.51	*0.01*
4	4.5	4.51	*0.01*	0.4	0.45	*0.05*	1.5	1.49	*-0.01*

Table 2
Test results comparison between practical & estimated passive source localizations

Tests	x_p(m)	x_{est}(m)	*error(m)*	y_p(m)	y_{est}(m)	*error(m)*	z_p(m)	z_{est}(m)	*error(m)*
1	2.46	2.47	*0.01*	3.72	3.67	*-0.05*	1.51	1.57	*0.06*
2	3.83	3.84	*0.01*	2.03	2.03	*0.*	1.42	1.38	*-0.04*
3	3.82	3.79	*-0.03*	1.4	1.4	*0.*	0.06	0.02	*-0.04*
4	5.25	5.25	*0.*	0.6	0.45	*-0.15*	0.44	0.4	*-0.04*

6. Conclusions

Different kinds of passive source localization method were discussed and compared in this paper. TDOA method based on PHAT was chosen for the in-air passive acoustic source localization because its simple algorithm, suitable for real-time implementation and easily applicability to broadband and near-field acoustic source localization. Results from MATLAB simulations and practical tests demonstrated the effectiveness of the method in practical applications. However, the method is only suitable for single source localization. Research is being carried out for multiple source localization in noisy and reverberant environment.

Acknowledgements

This work is funded by the EC 6th framework program Tai-Chi project.

References

[1] Chen JC, Yao K and Hudson RE. Source localization and beamforming. IEEE Signal Processing Magazine 19(2) (2002) 30-39.

[2] Hallman DL and Bolton JR. The application of nearfield acoustical holography to locate sources in enclosed space exhibiting acoustic model behavior. Proc. 12th IMAC (1994) 1076-1082.

[3] Larocque J, Reilly JP and Ng W. Particle filters for tracking an unknown number of sources. IEEE Trans. Signal processing 50(12) (2002) 2926-2937.

[4] Ward D and Williamson R. Particle filter beamforming for acoustic source localization in a reverberant environment. Proc. ICASSP-02 (2002) 1777-1780.

[5] Knapp CH and Carter GC. The generalized correlation method for estimation of time delay. IEEE Trans. ASSP 24(4) (1976) 320-327.

[6] Reed FA, Feintuch PL and Bershad NJ. Time delay estimation using the LMS adaptive filter – static behaviour. IEEE Trans. ASSP 29(3) (1981) 561 – 576

[7] Chan YT, Hattin RV and Plant JB. The least squares of estimation of time delay and its use in signal detection. IEEE Trans. ASSP 26(3) (1978) 217-222.

[8] Stephenne A and Champagne B. A new cepstral prefiltering technique for estimating time delay under reverberant conditions. Signal processing 59 (1997) 253-266.

[9] Allen JB and Berkley DA. Image method for efficiently simulating small room acoustics. JASA 65(4) (1979) 943-949.

Intelligent Production Machines and Systems
D. T. Pham, E. E. Eldukhri and A. J. Soroka (eds)

A framework for the integration of a trajectory pre-shaping technique into an advanced machine tool system

B. Jakovljevic[a], T. Kangsanant[b], A. Subic[a]

[a] *School of Aerospace, Mechanical & Manufacturing Engineering, RMIT University, Melbourne, Australia*
[b] *School of Electrical and Computer Engineering, RMIT University, Melbourne, Australia*

Abstract

Performance improvements are considered important for increased production capability and contouring accuracy of CNC machine tools. New generation machine tools look towards more advanced methods for controlling the drives to make the required improvements. The implementation of a previously developed analytical trajectory preshaping controller design technique into a new generation machine tool system is considered. A brief outline of the technique is provided, with a physical interpretation of key points. An overall layout and view of the integrated system is presented showing the main components, providing an overview of further automation and optimisation requirements. A number of issues are addressed as to the requirements for implementing the technique into an intelligent machining system environment.

1 Introduction

CNC Machine tools and the level of their technical development are a major factor in the production of goods through all aspects of industrial production affecting both the quality and the cost of manufacturing, as analysed by Weck and Bibring in [1]. This also implies ever increasing performance requirements for the drives. Consequently, they are increasingly operating closer to their physical limits in order to achieve increases in performance as determined in [2].

With increased performance the motion is increasingly in the transient. This leads to the requirement of machine tools operating with more advanced control techniques. These techniques need to address motion such as high speed cornering, circular profiles and straight line cutting at high speeds where much of the contouring is in the transient.

Monfared and Steiner in [3] give a perspective of what is to come in the near future for intelligent manufacturing systems, suggesting that a shift in paradigm is emerging which is transforming computerized manufacturing systems into intelligent manufacturing systems, with new techniques such as fuzzy logic and neural networks being implemented. In [4] it is shown that modern techniques such as neural networks can be efficiently used to control real-time operations of machine with accurate results.

The analytical finite settling time method which was developed in [5] is an advanced control technique for providing accurate transient response design. It is used for pre-shaping the trajectory signal to the machine tool axis controller. It can be implemented in varying degrees of complexity. In its more complex form a redesign of the controller is performed for each new motion profile. In [5] it is studied for implementing into an existing machine tool. However, due to complexity and processing and decision making requirements it is suited to the environment of a new generation advanced machine tool system. Through an overview of the technique this paper briefly provides a physical interpretation to some key equations and concepts which were not made in [5]. The paper then moves further on from [5] as it considers the issues of implementing the technique into a modern machining system.

2 Overview of the Analytical Technique

The technique was developed and shown in detail in [5]. This overview bridges a gap from [5] in that it provides a more detailed interpretation of the initial equations followed by an outline of design where the end tool position or velocity is available, rather than using the drive shaft measurement. This then aids the discussion of integration of the technique into an advanced machine tool system.

2.1 General characteristics of the technique

The analytical finite settling time method was developed to enable direct design of the velocity response profile in the discrete-time domain. This is the reverse of classical methods where the controller is tuned by adjusting the parameters until an adequate response is obtained. The analytical design method ensures that the following characteristics are met upon computation of the controller coefficients:

- No inter-sampling oscillation after settling
- Zero overshoot
- System stability

To obtain the desired characteristics, the same initial approach is used by Ogata in [6] for deadbeat design, but [5] then extends the development to arbitrary order of *F(z)* which sets the settling time and also providing flexibility in directly shaping the response profile. The derivation in [5] has been formalized for both *F(z)* and plant model of any order. Additional developments in [5] cater for type 0 models for use with velocity loop models identified with tachometer feedback, whereas in [6] only the design with a type 1 model is presented.

An outline is given for designing a trajectory pre-shaping controller, replacing the need for a position controller. This departs from [5] where the trajectory pre-shaping is outlined for adding to an existing machine with a tuned position controller. Another difference in this paper is that the compliance between the motor shaft and end tool is represented as inside the feedback loop. Measuring the end tool position is expected of an advanced new generation machine tool system.

2.2 Requirements

The design method requires accurate knowledge of the velocity control loop, compliance model and feedback device of each axis in the form of parametric transfer function models. These models can be obtained by experimental system identification to varying degrees of accuracy, which affects the contouring accuracy.

2.3 Outline of design with velocity loop model

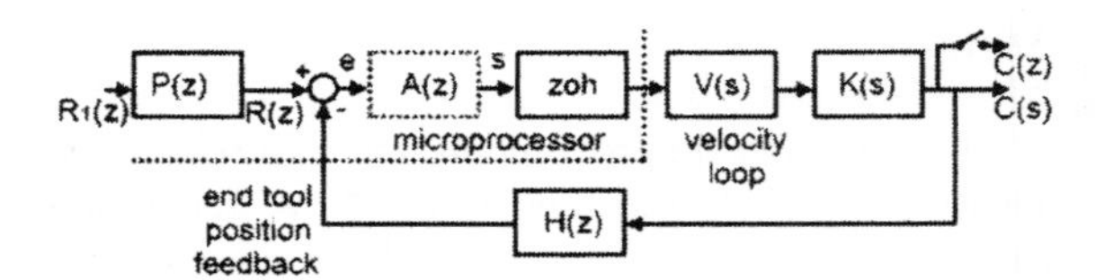

Fig. 1. Axis control system block diagram.

In fig. 1 the axis control system has an analog velocity control loop *V(s)* which uses an internal tachometer for velocity feedback. Externally, The end tool position signal is brought through the feedback device *H(z)* to the CPU. The input $R_1(z)$ would typically come from the interpolator. *P(z)* is the trajectory pre-shaping controller to be designed. *A(z)* is an intermediate controller obtained by the analytical technique in order to compute *P(z)*. The input to output transfer function for the system in fig. 1 can be expressed as:

$$C(z) = F(z)R(z) \qquad (1)$$

where *F(z)* denotes the overall transfer function. Imposing constraints on the initial equations is the key to obtaining the solution giving the characteristics as in section 2.1.

The deadbeat control method in [6] shows the design of a minimal settling time FIR controller in which the assumption is that the order of *F(z)* is equal to the order of the plant model. In this way the settling time cannot be specified nor can response sample values be pre selected. However, it does ensure a ripple free response after settling. Moving away from the deadbeat design method the analytical method from [5] uses higher order of *F(z)*. The order of *F(z)* determines the settling time. This approach gives more coefficients than equations, meaning that the user presets certain coefficients by setting sample points of the output response, then the rest are computed to fulfill the constraints set up in the equations based on the characteristics in section 2.1. The additional coefficients can then be used to shape the response profile with the above constraints forcing the remaining coefficients to fulfill the equations.

Designing the shape of the output response profile provides a means for shaping the acceleration profile indirectly through velocity profile design. This aids in avoiding jerks which can excite structural resonance of the machine tool causing vibration. A further cause of cutting accuracy problems is compliance between the drive shaft and the end tool, which can also be included in the computation when the machine tool is equipped for measuring the position or velocity of the end tool.

A brief outline is provided for designing a trajectory pre-shaping controller given a velocity control loop of a machine tool axis. This assumes a new controller design and with the trajectory pre-shaping technique implemented to replace any requirement for programming a position controller into the digital computer.

To achieve settling in a finite time, an FIR form is used for *F(z)* where the system poles are at the origin of the Z plane. From fig. 1 the error signal is:

$$E(z) = R(z) - C(z) \qquad (2)$$

For *R(z)* a step input is assumed for mathematical reasons, as it simplifies the equations compared with other inputs. A step input is not normally used in machine tool motion. At this stage *P(z)* in fig. 1 does not yet exist. When *P(z)* is finally determined and inserted it provides the actual reference signal *R(z)*. The response profile here is synthesized using a mathematical description of a step excitation to compute the desired response profile. Combining this with equation (1) and rearranging gives:

$$\frac{1 - F(z)H(z)}{1 - z^{-1}} = E(z) \qquad (3)$$

In the case of a unity feedback, the equation is:

$$\frac{1 - F(z)}{1 - z^{-1}} = E(z) \qquad (4)$$

It is required for *E(z)* to become zero after a finite settling time and of the same order as the overall transfer function *F(z)*. Then the expression for *E(z)* must have a limited number of coefficients after the division, and this is achieved by setting remainder of the division in (4) to zero. This is the approach taken in [6] also.

In fig. 2, the block diagram from fig. 1 is rearranged into an equivalent unity feedback loop. The feedback block *H(z)* is then equivalently placed before the speed loop.

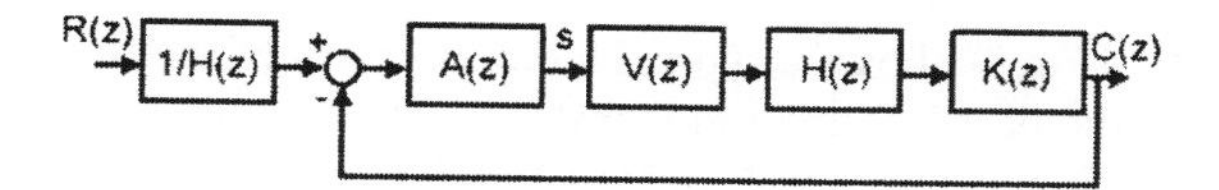

Fig. 2. Velocity loop and compliance model in equivalent unity feedback form.

The velocity loop and encoder and connection to digital computer through the DAC which is modeled as a zero order hold. The discrete-time transfer function of the velocity loop including the encoder feedback is then:

$$G(z) = Z\left\{\frac{1 - e^{-Ts}}{s} V(s)K(s)\right\} H(z) \qquad (5)$$

The velocity control signal *S(z)* also needs to be constant after settling time so that the velocity loop *V(s)* is not further excited, thus giving the ripple free response. Next, for the response to remain constant in the continuous time domain (not only the discrete domain) after settling time, the control signal *S(z)* must be constant, which is the same as in [5] as well as in [6] for the deadbeat design:

$$S(z) = \frac{C(z)}{G(z)} = F(z)\frac{R(z)}{G(z)} \qquad (6)$$

In this division, the remainder is also set to zero to achieve the requirement. *F(z)* can be set to any order and the additional coefficients are pre-selected to set the shape of the desired response. This leads into a complex derivation providing a solution for any settling time and useful with any model order.

Although this is a slightly different case than in [5], the equations and principle remain the same, and the details of the derivation can be found there. The remaining coefficients are then computed yielding *F(z)*. If the resulting profile is not suitable for the desired profile then re-iteration may be required. More on this in section 3.1.

The block diagram is then rearranged to include *P(z)* and remove *A(z)*. Then *P(z)* is calculated to achieve the equivalent result as:

$$P(z) = \frac{1 + V(z)H(z)K(z)}{1 + A(z)V(z)H(z)K(z)} \qquad (7)$$

3 Integrating Into a Machine Tool System

Fig. 3 shows a typical block diagram of the flow of the part information from the CAD drawing to the machine tool axis controller. The axis controller typically contains a position control loop with an inner velocity loop as in fig. 1 and a current controller internal to the velocity controller.

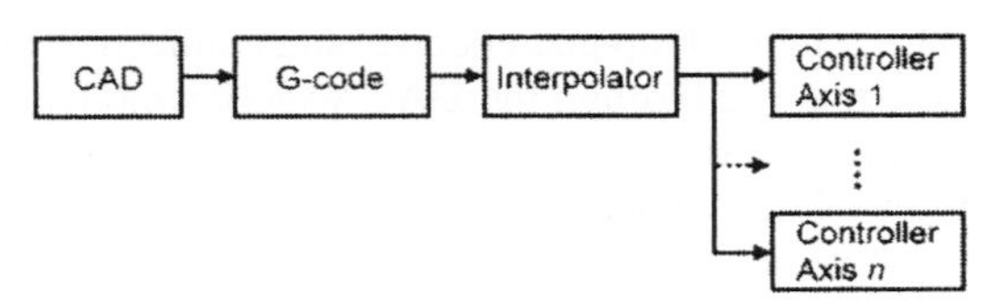

Fig 3. Typical flow of the part information to the position controller.

3.1 Preparing motion profiles

The next step is to determine what kind of profile is likely to give a useful result, after running through the analytical technique computation. A desired profile is input by inserting some of the sample points, and the analytical technique will determine the remaining points in order to maintain the characteristics in section 2.1. The result may be useful or it may not, in which case a reiteration is required. Although there can be any number of ways to suggest start profiles to help get to an optimal one, in [5] it was found useful to design a response based on a sine-exponential profile.

The multi-axis contouring is then achieved by the coordinated axis reference design using the concept of dynamical interpolation. In [5] it is shown how this can be done for high speed cornering, straight line motion where much of the motion is in the transient and for circular arcs as well as a brief mention of the approach required for arbitrary profiles. This is done mostly manually in [5], where now it becomes a question of how to implement it into an advanced machine tool system, which would to a large extent determine user interface operation, as the approach would perhaps still retain manual aspects. An intelligent system would be required to fully automate the design procedure.

3.2 Implementing trajectory pre-shaping

Pre-shaping assumes the trajectory signal comes from a block prior to the axis controller, as shown in fig. 1 as *P(z)* for a single axis. The *P(z)* block would be inserted in between the interpolator and the axis controller in fig. 3. This is the simplest way to implement the analytical design method. It involves generating the trajectory pre-shaping controllers once, with no further modification and is not modified for each motion of the machine tool. One way of using this approach is to design pre-shaping controllers which result in an improved dynamic matching of the axes. This yields a particularly useful result for multi-axis straight line motion at high speed when much of the motion is during the transient. The more complex approach is shown in fig. 4 with the same placement of the trajectory pre-shaping controller but with additional components which facilitate optimisation of the pre-shaper for each motion profile. The input to the system is information on the part derived from the CAD drawing. Fig. 4 shows the general flow of information from the part description to the generation of the pre-shaping controllers.

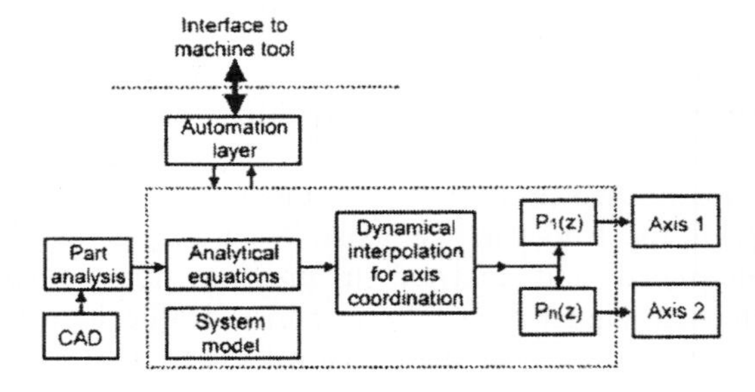

Fig. 4. Full utilization of the analytical design method.

It can be seen by comparing fig. 1 and fig. 4 that the interpolator is replaced in fig. 4, as the more complex approach provides the dynamical interpolation to coordinate the multi-axis motion taking into account the dynamics of each axis.

3.3 Obtaining parametric models

Fig. 5 shows the block diagram for the implementation of parametric system identification of the velocity control loop. The position loop is opened and the excitation signal is brought through the CPU and DAC to the velocity loop. The encoder feedback is connected to the data acquisition equipment. If measurement of the end tool can be obtained then the compliance model can be incorporated into the design as in section 2.3.

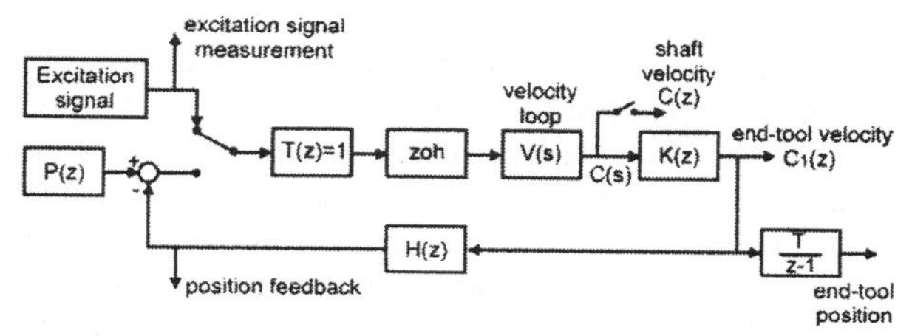

Fig. 5. System diagram for velocity loop system identification.

By implementing a trajectory pre-shaping approach the position reference signals are processed before they are fed to each axis of the CNC machine tool. If there is no position controller, then *T*(z) in fig. 5 can be stated to have a transfer function of 1/1, as the digital computer still exists and provides the signal to the DAC which is modelled as a zero order hold.

The actual procedures for system identification are outside the scope of this paper. A study has been made in [5] for a particular machine tool, however numerous literature is available on the topic.

3.4 The integration of main components

The block diagram in fig. 6 shows how the components of the analytical design method are interrelated and also the interface to the part information. The scheme is divided into three layers. At the core is the mathematical algorithm. The next level is the multi-axis coordination, and the top layer revolves around the interface between the analytical method and the machine tool. The part preparation from the CAD description is outside the scope of this research.

3.4.1 Core level

The mathematical algorithm of the analytical technique in the core layer computes the coefficients of the pre-shaping controller for any given settling time and any model order of the velocity loop. The solution also covers non-minimum phase models as shown in detail in [5]. The model may be type 1 or type 0 depending on whether the velocity or position signal was used in the modeling procedure. The settling time is the first parameter choice to be set.

3.4.2 Middle level

The middle level deals with the coordination of multi-axis motion. This is tightly integrated with the core layer. This stage aids the design of motion profiles for the axes then uses the analytical technique as the part which takes into account the system dynamics during the computation of desired profile feasibility.

Further iteration may be required between the two layers, and is where an intelligent software module can further be developed for searching for optimal solutions. Further questions can also be raised about optimisation issues. How to find a best solution, and what is the best solution for the criteria, which could be minimal contouring error, or minimum contouring time or other. In [5] the part profile is manually divided up into segments which can then be processed by the analytical design method. Regardless of the higher level optimization requirements, the analytical technique would remain the same, it would just require different response profile design approaches.

3.4.3 Top level

At the top level, the machine tool connects with the analytical design method. It is implemented through the main interface as the central component.

At this level the connection to the CAD and other parts is made. The part information from the CAD description requires segmenting into profiles for use by the analytical technique, see [5] for details. Implementing this segmentation requires further work in order to automate it for a machine tool system and to integrate it with the middle design layer especially for implementing an intelligent software module for optimal solution finding.

Automating this part assumes adopting an intelligent machining environment due to the complexity of the method. Further development is required for the integration into an advanced machining system with regards to the interface.

Optionally, a user interface can be implemented for manual assistance in the process of optimization of motion profile designs. In current machine tool environments, a certain level of automation exist in any case, here it is just being distinguished between fully automated where an intelligent system performs all operations, or the processes requiring intelligent decision making are performed by an operator or engineer. The level of automation for this section determines whether it is categorized into manually assisted by an operator or control engineer, or fully automated which assumes an intelligent machining system.

The interface could involve a database and software automation to aid the operator in the analysis of the part and the profile design process. Different levels of interaction between the operator and the machine tool may be required. The preparation of motion segments in [5] is performed manually, but the tasks can be automated, and it would be left to the operator to then choose what optimisation procedures to use and to validate the simulation or theoretical predictions prior to actual cutting.

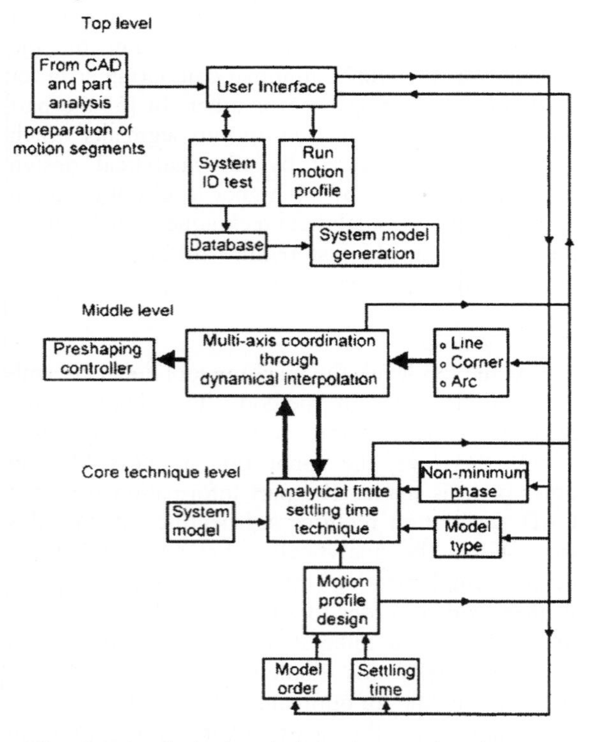

Fig. 6. Analytical technique integration diagram.

The machine tool axes can be tested periodically and obtained models can be compared with previous models which are stored in the database. This can also aid in machine tool monitoring by showing trends of parameter change with varying conditions and aging of the machine tool.

6 Conclusion

The analytical technique implements a complex computational algorithm requiring an iterative procedure for optimised profile generation. This requires significant computational resources for preparing a trajectory pre-shaping controller for each motion profile in the process of machining. It is therefore suited to an advanced machine tool system.

When using the analytical technique in the more complex implementation the standard method of G-code description of motion profiles in CNC machine tools is not adequately sophisticated to aid precise transient coordination of the axes with compensation of dynamical characteristics. A new module is required to replace the G-code description as well as the interpolator. The complexities involved around the analysis of a part and segmentation into components for preparation of motion profiles involves intelligent decision making, and to automate such procedures requires the adoption of an intelligent machining system environment.

References

[1] Weck M., Bibring H., (1984), "Handbook of Machine Tools", Volume 1, Types of Machines, Forms of Construction and Applications, John Wiley & Sons, New York.

[2] Weck M., Ye G., (1990), "Sharp Corner Tracking Using the IKF Control Strategy", Annals of the CIRP Vol. 39/1, pp. 437-441.

[3] Monfared M. A. S., Steiner S. J., (1997) "Emerging Intelligent Manufacturing Systems", International journal of Flexible Automation and Integrated Manufacturing, Vol. 5 (3&4), pp.151-170.

[4] Cohen G., (1998), "Neural networks implementations to control real-time manufacturing systems", Computer Integrated Manufacturing systems, Vol. 11, No. 4, p.243-251.

[5] Jakovljevic B., (2004), "A Methodology For The Improvement Of Contouring Accuracy In High Speed Machining", PhD Thesis, RMIT University, Melbourne, Australia.

[6] Ogata K., (1995), "Discrete-Time Control Systems", Prentice Hall, Englewood Cliffs, N.J., USA

Intelligent Production Machines and Systems
D. T. Pham, E. E. Eldukhri and A. J. Soroka (eds)

Accurate 3D modelling for automated inspection: A stereo vision approach

A. Bhatti[a,b] and S. Nahavandi[b]

[a] *CRC for CAST Metals Manufacturing, Deakin University, Vic 3217, Australia*
[b] *Intelligent Systems Research Lab., Deakin University, Vic 3217, Australia*

Abstract

A vision based approach for calculating accurate 3D models of the objects is presented. The strength of the work lies on multi-resolution based matching algorithm. The depth and dimensional accuracy of the produced 3D depth model lies on the existing reference model instead of the information such as laser signatures or different light pattern projections from extra hardware tools. It is a software based approach of finding accurate depth with minimal hardware involvement. The matching process uses the well-known coarse to fine strategy, involving the calculation of matching points at the coarsest level with consequent refinement up to the finest level. Vector coefficients of the wavelet transform-modulus are used as matching features, where modulus maxima defines the shift invariant high-level features (edges) with phase pointing to the normal of the feature surface. The technique addresses the estimation of optimal corresponding points and the corresponding 2D disparity maps leading to the creation of accurate depth perception model.

1. Introduction

The 3D reconstruction process can be categorized into three main categories: calibration (calculating the intrinsic and extrinsic parameters of the camera) [1, 2]; finding the corresponding pairs of points projected from the same 3D point on to the two perspective views [3, 4, 5, 6, 7, 8, 9]; and triangulation to project the 2D information back to the 3D space in order to create a 3D model.

Finding correct corresponding points from more than one perspective views in stereo vision [3] is subject to a number of potential problems such as: occlusion; ambiguity; illuminative variations; radial distortions, etc. A number of algorithms have been proposed to address at least some of these problems in stereo vision. The majority of these can be categorized into either area based or feature based algorithms. Area based approaches such as the ones used in [4, 5] are based on the correlation of two image functions over locally defined regions so the matching is performed on the basis of image pixel intensities. Dense depth maps can be achieved by correlating the grey intensities of image regions in the views being considered, assuming that they possess some similarities in the perspective views. The area based methods perform well in the presence of rich textured areas in the image but do not work well in featureless regions resulting in the correspondence of wrong image pixels. Another problem with area based algorithms is their handling of boundaries or edge information, which is very important in the context of generating accurate and dense depth maps. On the other hand, feature based algorithms such as approaches [6, 7] attempt to establish correspondences between the selected features, which are extracted from the images usually by using some explicit feature extraction algorithms. The main drawback in these feature based techniques is the retrieval of very sparse depth information; consequently, it is very hard to recover dense depth maps of the scene under consideration. The feature based algorithms also have the tendency of getting stuck in unnecessary features of the images and so they are quite dependent on the efficiency of the feature extraction algorithms.

Since this research was interested in obtaining dense three dimensional maps at a reasonable computational cost, this oriented the work, from the beginning, towards the first category. The algorithm described in this paper is a variant of this class of techniques. The original version was improved and eventually moved towards the concepts of multi-resolution, which involves the disparity estimation process at different resolutions, giving rise to the opportunities of correcting and upgrading the disparity maps as the process moves towards the higher resolutions. It can also be called an incremental process of creating dense disparity maps. Most famous and widely used multi-resolution analysis is wavelet analysis involving the decomposition of input signal into different resolution and sub-space. Their wide use is due to the fact that there exist a number of wavelet bases in the literature possessing the properties of short support, orthogonality and different approximation orders [10, 11]. In the context of stereo matching using multi-resolution analysis, some work has already been done using scalar wavelets [12] and quite promising results are achieved. However, the theory of multiwavelets evolved in early 1990s from wavelet theory and enhanced for more than a decade. Their success, over scalar wavelet bases, stems from the fact that they can simultaneously possess the good properties of orthogonality, symmetry, high approximation order and short support, which is not possible in the scalar case. In that work, multiwavelets based multiresolution approach is used to solve the problem of stereo image matching. Multiwavelets' coefficients are in fact image features which pass to detail space as couldn't be approximated by the approximation or scaling function. This is due to orthogonality that exists between scaling and wavelet functions.

The presented work is, in fact, a continuation of the work in [8, 9], where disparity maps are calculated by applying the correlation matching process directly on the wavelet coefficients at different resolutions. By using the concept of wavelet modulus maxima [10] the performance of the algorithm from [8] is improved as it involves the matching of coefficients representing the edges, with magnitude and phase, pointing to the normal edge surfaces.

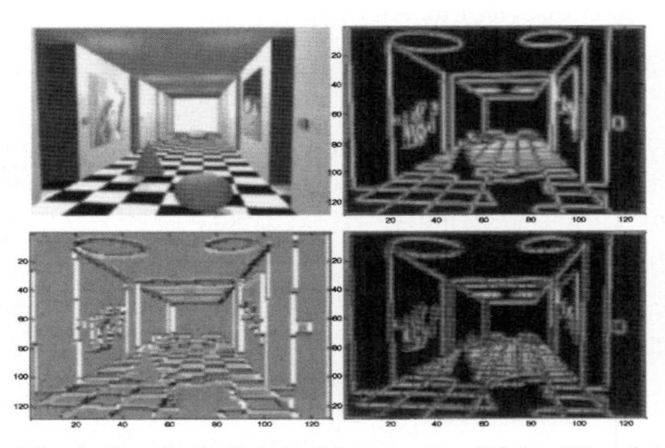

Fig. 1. Top Left: Original image, Top Right: Wavelet Transform Modulus, Bottom Left: wavelet transform modulus phase , Bottom Right: Wavelet Transform Modulus Maxima with Phase vectors.

2. Disparity estimation algorithms

The process of disparity estimation starts with the translation invariant wavelet decomposition [10, 19] of the original pair of images. Interested readers are referred, for any further information about the wavelet theory concepts, their generation and application, to [10, 11, 19].

The process starts with the matching of features at the coarsest resolution level. For each feature, in the first image, the best match is extracted from the other image using the correlation score as described below:

$$C_{x,y,d} = corr(W_{s,l}, W_{s,l,d}) \text{ for } d = 0 \quad d_{max} \qquad (1)$$

$$W_{s,k} = WTM_{s,k} \angle \Theta W_{s,k} \qquad (2)$$

where:

$W_{s,k}$ is wavelet modulus coefficients, s: scale or the resolution level, l: search space index out of r^2 search space and d: scalar disparity value.

As there are r^2 matrices, there must be less than or equal to r^2 candidate matches, which can be similar or different in terms of (x, y) coordinates. An example of wavelet modulus coefficients vector can be seen in Fig. 1. As we are dealing with the areas of image discontinuities of the surfaces, the concept of multi-window correlation [13] is used which proved to have better performance especially at image discontinuities.

1.1. *Probabilistic measure*

To keep the matching process consistent, a symbolic tagging procedure is introduced based on probability of occurrence and three different thresholds. Probability of occurrence (P_c), which is the probability of selection of any point as a candidate from any of the search space out of r^2 spaces and can be defined as

$$P_c(C_i) = \frac{l}{r^2} \text{ where } 1 \le l \le r^2 \quad (3)$$

where l: number of times a candidate match selected, and C_i: i[th] selected candidate out of r^2 search spaces. As all matching candidates have equal probability of being selected so the probability of occurrence for any candidate through one search space is $1/r^2$. The correlation score for each candidate is then weighted with the occurrence probability, which can be expressed as

$$CS_i = P_c(C_i)\sum_l C^i_{x,y,d} \text{ for } l \le r^2 \quad (4)$$

After that step, all candidate features points have a correlation score attached.

1.2. *Optimization and interpolation*

The candidate feature points are then divided into three pools based on three different thresholds T_i possessing the criteria $T_1 < T_2 < T_3$, where T_1 is the minimum criterion for the candidates to remain in the process and candidates violating that criterion are discarded. The candidates with the probability of consistency 1 are given the status of true candidates with no ambiguity whereas ambiguity does exist for the other candidates still in the process. Ambiguity is the phenomenon where there is more than one candidate available for a single point in the reference image. The problem of ambiguity is addresses with a geometric optimization procedure the details of which can be found in [9]. The matching process at the coarsest level ends up with a number of matching pairs, which needs to be interpolated to the finer level. The constellation relation between the coefficients at coarser and finer levels can be visualized by taking the decimation of factor 2 into consideration. The disparity from coarser disparity d_c to finer disparity d_F can be updated using the relation as follows:

$$d_F = d_f + 2d_c \quad (5)$$

where d_f is the local disparity obtained within the interpolated area. This process is repeated up to the finest resolution which in fact is the resolution of input image. A more detailed explanation of the matching algorithm can found in [8, 9].

3. Experimental outcomes

This research work is part of a major research program on a vision based automated inspection system capable of detecting faults in metallic objects and is the continuation of the work presented in [8, 9]. The motivation behind the idea is to extract accurate 3D depth of different surfaces in the object under consideration with minimal hardware involvement unlike many others' work which involves extra hardware such as laser scanners, laser and light pattern projectors, etc. [14, 15] to improve the accuracy of disparity maps. The reason and motivation behind the approach is to keep the process simple, as much as possible, from hardware complexity of apparatus so that it can be used in different environments without much change in the setup. As the stereo vision process solely is not mature enough to solve the problem to high accuracy, external information from the reference 3D model is used. Here a reference model of a simple object is produced in Matlab by manually measuring the physical dimensions. The next step of the work is to involve the reference models that can be produced using the common 3D drawing software such as AutoCAD, Solidworks etc. An example of the reference model is shown in Fig. 2.

The information that is extracted from the reference image model is the relative location of the discontinuities in between different surfaces and the dimensional range of the 3D depth. Relative locations of the different surface edges are most important as the stereo vision algorithms usually perform poorly in the areas where sharp depth changes occur.

The hardware setup of the presented work is

very simple and consists of stereo cameras headset from Videre Design and simple circular fluorescent lighting. In the presented work, objects under consideration are metallic plates which possess very shiny surfaces. To deal with this problem two pairs of images are taken for each object. One pair is taken at high gain of light intensity in order to dominate the surface discontinuities whereas the other is taken at lower gain which can dominate the surface features available and helps in finding the correct disparities by correlation measures. The former helps in extracting the overall structure of the object and later in detecting any abnormalities existing in different surfaces.

a)

b)

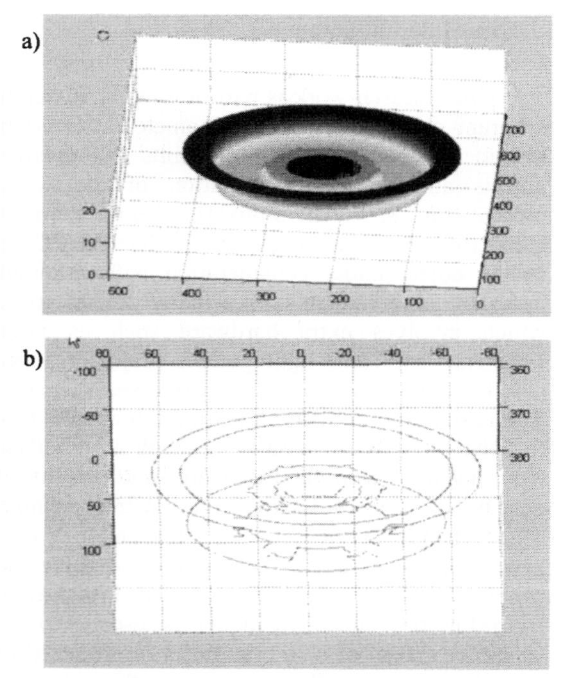

Fig. 2. a) Reference 3D depth model, b) Surface boundaries

An example of the outcome is shown in Fig. 3. It can be seen that a very smooth disparity map is obtained after it is optimized and refined using the information extracted from the reference model. One thing that needs to be clarified, is that the process is completely independent of the orientation or the location of the object on the image plane. An error graph is shown in Fig. 4, which in fact is the difference in the depth between the original measured and the recovered one with the aforementioned algorithm. The error is perhaps due to the error involved in the calibration process of the stereo cameras, which involves the recovery of internal parameters such as: focal length; optical centres; etc., and the external orientation of the two cameras: distance and rotation between the two cameras. The $[X\ Y\ Z]$ coordinates are extracted, from the calculated disparity map, using the equations as:

$$\begin{bmatrix} X' \\ Y' \\ Z' \\ W \end{bmatrix} \cong \begin{bmatrix} 1 & 0 & 0 & -C_x \\ 0 & 1 & 0 & -C_y \\ 0 & 0 & 0 & F_x \\ 0 & 0 & -1/T_x & 0 \end{bmatrix} \begin{bmatrix} u \\ v \\ d_{x,y} \\ 1 \end{bmatrix} \tag{5}$$

where:

$$\begin{bmatrix} X \\ Y \\ Z \end{bmatrix} = \begin{bmatrix} X'/W \\ Y'/W \\ Z'/W \end{bmatrix} \tag{6}$$

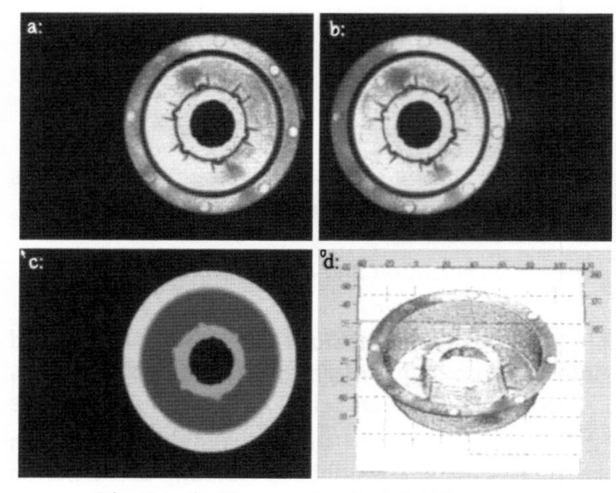

Fig. 3. a, b: Stereo pair of original images, c: Estimated 2D disparity map, d: 3D depth of model

Even though many sophisticated algorithms are available to calibrate the cameras and these techniques are quite optimized but still there is always room for small errors which accumulate with the error involved in the disparity estimations.

4. Conclusion

A multi-resolution disparity estimation

technique based on translation invariant wavelet and multi-wavelet transform modulus is presented. The technique addresses the estimation of optimal corresponding points and is capable of producing reasonably accurate 3D depth models.

The algorithm as a whole relies mostly on software rather than hardware. The optimization which involves the disparity errors correction is performed using an existing reference model of the object. Similar accuracy can be achieved using some extra hardware such as laser scanners, laser pattern projectors and different light pattern projectors, etc., which makes the process costly, complicated and needs special setup arrangements. Whereas the proposed algorithm uses minimal hardware equipment which comprises a stereo camera head set and a simple circular florescent lighting rod.

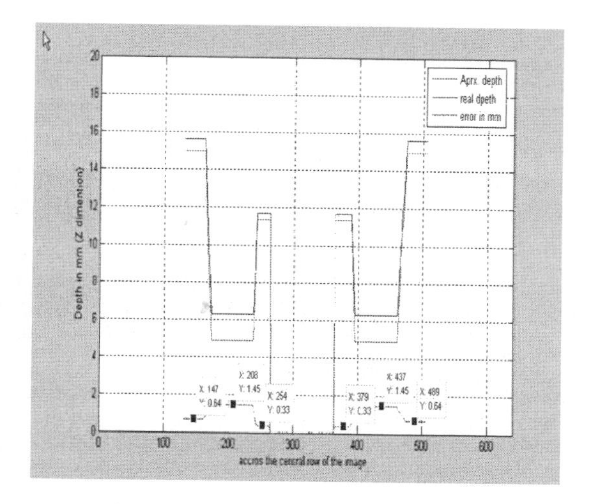

Fig. 4. Error plot between real 3D depth and the recovered one

5. Acknowledgment

This work was supported by the Cooperative Research Centre for Cast Metals Manufacturing (CAST). CAST was established and is funded in part by the Australian Government's Cooperative Research Centre Program.

References

[1] Z. Zhang. "A flexible new technique for camera calibration". IEEE Transactions on Pattern Analysis and Machine Intelligence, 22(11):1330-1334, 2000.

[2] J. Heikkilä, & O. Silvén, "A Four-step Camera Calibration Procedure with Implicit Image Correction". IEEE Computer Society Conference on Computer Vision and Pattern Recognition, San Juan, Puerto Rico, p. 1106-1112. 1997.

[3] O. Faugeras, Q-T. Luong The Geometry of Multiple Images. MIT Press, London, England, 2001.

[4] Z. Zhang, R. Deriche, O. Faugeras, and Q. T. Luong, A robust technique for matching two uncalibrated images through the recovery of the unknown epipolar geometry: Research Report 2273, INIRA, (1994).

[5] Olivier Faugeras, Bernard Hotz, Hervé Mathieu, Thierry Viéville, Zhengyou Zhang, Pascal Fua, Eric Théron, Laurent Moll, Gérard Berry, Jean Vuillemin, Patrice Bertin, Catherine Proy, Real time correlation-based stereo: algorithm, implementations and application, INRIA Reseach Report 2013 , (1993)

[6] H. Baker and T. Binford, Depth from Edge and Intensity Based Stereo: Int. Joint Conf. on Artificial Intelligence, (1981), pp 631-636

[7] Y. Ohta and T. Kanade, Stereo by Intra- and Inter-scanline Search Using Dynamic Programming: IEEE Trans. on Pattern Analysis and Machine Intelligence, 7(2), (1985), pp 139-154.

[8] Asim Bhatti, Saeid Nahavandi and Hong Zheng, Image Matching using TI Multi-Wavelet Transform: Digital Image Computing Techniques and Application, 2003, pp 937-946

[9] Asim Bhatti, Saeid Nahavandi, A Multi-Wavelet based Technique for Calculating Dense 2D Disparity Maps from Stereo: World Automation Congress, 2004

[10] S. Mallat, A Wavelet Tour of Signal Processing: Academic Press, Harcourt Place, London: 2001

[11] Özkaramanli H, Bhatti A., Bilgehan B., Multi wavelets from B-Spline Super Functions with Approximation Order: Signal Processing, Elsevier Science, 82(8), (2002), pp 1029-1046.

[12] He-Ping Pan, General Stereo Matching Using Symmetric Complex Wavelets: In Wavelet

Applications in Signal and Image Processing, Proceeding of SPIE, 2825, 1996, pp 697-721

[13] A. Fusiello, V. Roberto, and E. Trucco. Efficient stereo with multiple windowing. CVPR, 1997, pp 858–863.

[14] D. Scharstein and R. Szeliski, A taxonomy and evaluation of dense two-frame stereo correspondence algorithms: Intl. J.Comp. Vis., 47(1), 2002, pp 7–42.

[15] D. Scharstein and R. Szeliski. High-accuracy stereo depth maps using structured light. In IEEE Computer Society Conference on Computer Vision and Pattern Recognition (CVPR 2003), volume 1, 2003, pp 195-202.

[16] R. R. Coifman and D. L. Donoho, "Translation-invariant de-noising", in: A. Antoniadis and G. Oppenheim, ed., Wavelet and Statistics, Lecture Notes in Statistics, Springer-Verlag, 1995, pp 125-150.

Intelligent Production Machines and Systems
D. T. Pham, E. E. Eldukhri and A. J. Soroka (eds)

Best practice development within a manufacturing enterprise collaboration framework

Ove Rustung Hjelmervik and Kesheng Wang*

Knowledge Discovery Laboratory
Department of Production and Quality Engineering, NTNU,
N-7491 Trondheim, Norway

Abstract

Best Practice, representing value-creating processes, has been understood in terms of how teams learn, internalize, apply and improve experiential knowledge within an enterprise collaboration framework. In mature markets firms are dependent on knowledge development processes making best practice more efficient and effective for the purpose of competitiveness, without being copied by competitors. However, for the global firm to leverage best practice, ICT (Information and Communication Technology) needs to be applied both toward developing and transferring such knowledge. Thus, we need to understand if best practice can be developed through the supported of ICT-represented knowledge, and if so, how does it lead to best practice being improved/changed. We are reporting from research in progress. In our preliminary analysis we have discovered that, given certain organizational conditions, ICT-supported best practice representation may enhance best practice development.

Keywords: Knowledge Management, Best practice, manufacturing collaboration

1. Introduction

The business environment today is collaborative. In order to survive in a mature, global industry, such as the aluminum industry, manufacturers depend on enterprise-wide employee collaboration for their indigenous knowledge based productivity. By focusing on productivity rather than adding new capacity, one tries to prevent unwanted competitive actions. Productivity can result from reduced cost, or increased output, per unit of input. While the origin of tangible resources lies outside the firm, the competitive advantage of the enterprise is firm-specific knowledge. Best practice is a firm-specific knowledge, developed through collaboration between employees based on experience learned from such practice. Best practice business system is an ICT-based collaborative method for securing employee participation in the development of a firm. Best Practice development, transfer and application are based on iterative collaboration of experience. While collaborative ICT systems, such as holistic organized best practice business system[1], may be

[1] A holistic business system may consist of a schematic business process' value flow from input to output. The value flow is divided into work processes and activities. Each activity is described in form of best practice, and has a link to different procedural documents and support tools, such as customer relation management (CRM), enterprise resource planning (ERP), supply chain management (SCM), data warehouse and business intelligent tools. This way best practice guides employees

* Corresponding author: E-mail: Kesheng.wang@ntnu.no, Tel.:+4773597119, Fax: +4773597117.

able to transform firms' work practice and processes from deep hierarchies to lean production structures, we know little about whether or not they enhance best practice development. We want to find out if, and how, knowledge development is enhanced by computer-represented best practice. Our contribution is to explore this question using a case study from the aluminum industry where a best practices method newly was introduced to this global firm through an enterprise developed business system. Based on a preliminary analysis we found evidence of changes to best practice after the implementation of the new system. This can have relevance to "intelligent" production systems where organizational knowledge is represented through ICT, leading to combining experience from operative business processes. [9].

2. Best practice

Collaboration is often related to the nesting of heterogeneous information systems for the purpose of "sharing information between business processes across internal or external partners in the value chain network ... (which) supports the ... everyday business process that people work with as a matter of course" [12] (pp. 27, 28). Thus, intra-firm sharing of best practice is also a collaborative venture. Best practice (BP) is a recipe of knowing how to operate a process or a task [10]. Szulanski [15] describes best practice as "an internal practice that is performed in a superior way in some part of the organization and is deemed superior to internal alternate practices and known alternatives outside the company" (p. 28). Best practice, being embedded within the structure of the business model, is part of a complex set of interdependent routines (stock of knowledge) that is transferred, understood, applied, and improved or changed [7] [15] [17] [18]. It is complex [14] because it can involve structural changes to the organization as well as adoption of new technology and/or new work processes.

The purpose of BP is to operate a business process in the most effective and efficient manner known to the organization. This implies a process of leveraging current, and developing new, capabilities. Best practice of concern here is a firm's ability to secure a unified application, throughout its dispersed production sights, of the best way the firm knows how to produce its goods and services given its current stock of resources. Developing new knowledge requires the externalization of competitive experiential know-how. Externalizing experiential learning [1] requires the possibility for the team members to discuss experience made, structure the new knowledge and prepare it for electronic feedback to the process owner responsible for developing a best practice [18]. BP, representing the activities required for value-creating processes to run profitably, has been understood in terms of how teams learn, internalize, apply and improve experiential knowledge. In medium dynamic [2] markets [8] firms are dependent on collaborative development processes in order for externalized best practice to be inimitable for the competitors. However, BP, without the understanding of knowledge development and collaboration, may not cause superior performance [2]. Thus, we need to understand if best practice can be developed through the support of ICT-represented knowledge, and if so, how does it lead to best practice being improved/changed. On this issue the literature is scarce.

3. Best practice development and learning

Voss et al [14], claims that best-practice success depends on the link between best practice and learning, understanding, and performance. Learning is the process required for acquiring knowledge [1], while organizations learn "by encoding inferences from history into routines that guide behaviors" [11].

Experiential learning leads to knowledge development. This experience needs to be externalized. Such knowledge articulation is based on "a cognitive effort more or less explicitly directed at enhancing the (team's) understanding of these causal links" [18] (p. 431). Such articulation leads to a form of consensus. The next step in the cognitive process is to have the knowledge codified (p. 342) and transferred as experience to the manager responsible for the process (process owner, or gate keeper). The process owner (PO) will in turn convert it to new best practice. The PO

---

through relevant activities required for a safe and effective operation.

[2] A medium dynamic market can be equated to a mature market. We are in this article concerned with productivity gain based on indigenous improvements and innovation, rather than exogenous growth.

is responsible for incorporating experiential learning into the current practice before being transferred out to the users (See Fig. 1). Large companies are escalating their investments on ICT to the point of over-investing. Such over-investments do not result in added performance, a paradox attributed to failure of implementation [4] [6]. Such a failure is a result of lack of understanding, also in regard to applying ICT to the best practice development phenomenon. If ICT-supported knowledge representation enhances the development of best practice, how is this done, and what factors contribute to securing such process? Best-practice development typically comprises systematic collection, evaluation and refinements of experiential knowledge from operations applying similar technology for the purpose of delivering same processes or products. Based on this information, PO, being part of management, have the opportunity to do corrective actions before implementation.

4. ICT and collaborative manufacturing

An essential element in the application of ICT as a medium for improvements to their work processes is the employees' ability to *apply* the technology. Such capability is not "embodied within the technology" [13] (p. 420). ICT, or any other type of technology, must therefore be learned through practice within a "recurrent social practice of a community of users, (and) ... by the rules and resources implicated in their ongoing action" (p. 421). Only then can employees utilize ICT for the collaboration of best practice. For ICT to support the enhancement of new knowledge the technology must become second nature to the employees applying it. This can be achieved through the ongoing interaction with the technology so that "structures become routine (and) taken for granted" (p. 421). According to Orlikowski [13], "whether or not the technology or the work practices are changed, it is often an intended outcome of people's knowledgeable actions" (p. 421). This can result in an inertia reaction "where users choose to use technology to retain their existing way of doing things" (p. 421). In such a situation the users have "limited understanding and/or being skeptical of the technological properties available to them" (p. 421). Another result of people's knowledgeable action can be characterized by those employees who choose to "use the new technology to augment or refine their existing ways of doing things. ... resulting in ... noticeable improvements to work processes" (p. 422). All the users in Orlikowski research which fell into this second (and positive) category "used the technology with the intention of improving or enhancing their existing work process" (p. 423).

5. Findings and discussion

We are conducting a longitudinal, explorative, case study. A case study needs multiple sources of data and these studies typically combine data collection methods such as archives, interviews, and observation [8]. This preliminary analysis is based on semi-structured interviews, observations and participation, archives and documents applied for implementation of the best practice business system. We found that in our case firm the IT literacy was high, but that 30-40 percent of the employees have IT-phobia. It was pointed out to us that at the age of 35 years there was an increasing change from IT literacy to illiteracy at work (See Fig. 2). An interesting phenomenon was that while literate, some employees choose not to apply their IT knowledge at work. We also found that while the operating employees shared experience, staff and lower level management did not. Finally, we did a preliminary evaluation[3] of employees' feedback of experience, using the ICT's feedback function built into the system. We found seven experience transfers back to the process owners, of which three practices had been improved and five were pending. That is, knowledge development had been taking place. In addition, there had been three questions asked and response given.

ICT-literacy is a requirement for participating in the development of new best practice through the support of computer-knowledge. This includes understanding ICT as a tool rather then as a technology [13]. Our finding supports this. About 60-70% of our case firm's employees are IT literate, while some choose not to apply this literacy at work. Furthermore, whether or not the technology or the work processes are changed is often an intended outcome of people's

---

[3] The best practice business system started to be implemented on a roll-out basis with one plant at the time during 1. half of 2003. By spring of 2004, installation had been implemented in most plants. This preliminary evaluation was carried out first quarter 2004. This evaluation was done at one of the firm's business units, operating a number of dispersed production units.

knowledgeable actions, and depends on how the teams understand and *share meanings*, which *tools* are available and how the *social system* at the work place is structured [13] (p. 421). In our case firm, the purpose of the system was, according to the CEO, to improve current BP. We found that this was achieved by those capable of applying and understanding the system. We found evidence of Orlikowski's three organizational conditions present in our case firm, which requires teams to share a common meaning within a positive social system structure.

The end result of whether and how employees use the technology to enact different work processes or technology-in-practice, suggests "three distinct kinds of consequences: (1) consequences that represent no evident change in process, technology or structure; (2) consequences that represents some change in one or more of process, technology, and structure; and (3) consequences that represents significant changes in one or more of process, technology, and structure" (p. 421). We argue that the application of best practice system is meant to support condition two in Orlikowski's findings, that is, improvements in process, technology and/or structure. Therefore, provided employees have available to them, and understand, the ICT-tools at hand, and the knowledge represented by it, they can apply such knowledge. Furthermore, being able to apply the knowledge, teams have the ability to propose, or enact, changes to the work process, technology, or organizational structure. Such actions depend on satisfactory conditions at the work place. Such conditions are how the teams understand and share meanings; institutional conditions that include the rigidity of hierarchical structures, incentives and task assignments, and culture; and team competencies [13] (p. 421, 422). For employees demonstrating inertia, that is, do not go along with the application of a new system, negative conditions affect them adversely. At the same time, those willing to support change find these conditions amicable to the point that "users having moderate, competent or extensive understanding of their technology at hand ... (was) either moderately or highly motivated to use it to enhance their work practice" (p. 423). Our preliminary analysis supports the fact that those who applied BP through ICT-tools supported the sharing of experience.

For the organization, changes to the best practice imply often changes to technology and structure. Adoption of new collaborative ICT tools often requires structural changes or new ways to carry out work, such as new tasks and commando lines. For the firm to succeed with its ICT investment, however, it has to have ICT-competent employees [3] [5]. Through the application of ICT, employees can secure the organizations new ways of performing, resulting in an improved productivity and value creation for the business. Thus, ICT is a strategic necessity for firms. While ICT gives the organization the opportunity to play on a leveled playing field, it does not offer any competitive advantages. On the other hand, not controlling the right situation - processes, technology and structure, can reduce productivity. In relation to best practice, therefore, for multi-located large firms to gain competitive advantage within their industry, employees need to support the improvement of the current best practice processes and technologies applied in the value creating processes. We have in our research observed that IT-literate employees have supported the development of new best practice. Such capability supports enterprise collaboration, and enhances value creation.

6. Conclusion

Intrafirm best practice development is a collaborative manufacturing process. In order for the production company to improve current practice, one is dependent on the ability by the employees to develop new practice from experience. Such experience is based on ones ability to learn. In order for the organization to benefit from this experience, it is depending on the ability by the employees to externalize this new knowledge, and represent it onto ICT. Only through active participation by management will such development take place, be transferred, learned and applied. This development cycle is part of the collaborative manufacturing process. Its absence will lead to isomorphorism and lack of competitive capabilities. Our preliminary findings support the need for management's willingness to develop user-friendly collaborative systems encouraging employees to share experience and see the potentials for improving current practice, technology and systems.

Collaborative processes can lead to adjustments of work-processes and business model. This cyclical evolution of the business model, through the articulation and codification process of developing organizational knowledge, supports the dynamic capability building and hence improves the overall performance [18]. Our preliminary analysis supports our proposition that, given certain organizational conditions, ICT-supported knowledge representation may enhance knowledge development. However, our case demonstrates that without employees' willingness to share experience, firms will loose this knowledge, and loose its competitive advantage.

Acknowledgement

NTNU is a partner of EU-funded FP6 Innovative Production Machine and Systems (I*PROMS) Network of Excellence.

Reference

[1] Arrow, K. J., The economic implications of learning by doing. *The Review of Economic Studies*, Vol. 29, No. 3, June, 1962, pp. 155-173.

[2] Beaumont, N., *Technovation*, v 25, n. 11, November, 2005, pp. 1291-1297

[3] Brynjofsson, E., and L. Hitt, Paradox Lost? Firm-level evidence on the returns to information systems spending, *Management Science*, vol. 42, No. 4, April 1996.

[4] Brynjofsson, E., and L. M. Hitt, Beyond the productivity paradox: computers are the catalyst for bigger changes. *Communication of the ACM*, August 1998.

[5] Brynjofsson, E., The productivity paradox of information technology. *Communication of the ACM*, December 1993/Vol. 36, no. 12

[6] Davenport, T. H.; Process Innovation, Reengineering Work through Information Technology, Harvard Business School Press, Boston, 1993, chs. 3 and 10.

[7] Dierickx, I, and K. Cool; Asset stock accumulation and sustainability of competitive advantage, *Management Science*, Vol. 35, no. 12, December 1989

[8] Eisenhardt, K.M., and J.A. Martin; Dynamic Capabilities: what are they?, *Strategic Management Journal*, 21: 1105-1121, 2000.

[9] Kang, D., and R. Santhanam, A Longitudinal Field Study of Training Practices in a Collaborative application Environment, Journal of Management Information Systems, Winter 2003-4, Vol. 20, No. 3, pp. 257-281.

[10] Kogut, B., and U. Zander; Knowledge of the firm, combative capabilities, and the replication of technology; *Organization Science*, Vol. 3, No. 3, August 1992.

[11] March, J. G., and B. Levitt, Organizational learning, *Annual Review of Sociology*, 14, 1988, pp. 319-40.

[12] McClellan, M., Collaborative manufacturing: A strategy built on trust and cooperation. *Control Solution International*, 12.03, p. 27.

[13] Orlikowski, W. J., Using Technology and Constituting Structures: A Practice Lense for Studying Technology in Organizations; *Organization Science*, vol. 11, No. 4, July-August 2000, pp. 404-428.

[14] Rivkin, J. W., Imitation of complex strategies; *Management Science*, Vol. 46, No. 6, June 2000, pp. 824-844.

[15] Szulanski, G.; Exploring internal stickiness: Impediments to the transfer of best practice within the firm; *Strategic Management Journal*, Vol. 17 (Winter special Issue), 27-43, 1996.

[16] Voss, C. A., P. Åhlström, and K. Blackmon; Benchmarking and operational performance: some empirical results. *International Journal of Operations & Production Management*, Vol. 17, No. 10, 1997, pp. 1046-1058.

[17] Winter, S. W., and G. Szulanski; Replication as strategy; *Organization Science*, Vol. 12, No. 6, November-December 2001, pp. 730-743.

[18] Zollo, M., and S.G. Winter; Deliberate Learning and the Evolution of Dynamic Capability, *Organization Science*; May/June, 2002; 13, 3, p. 339.

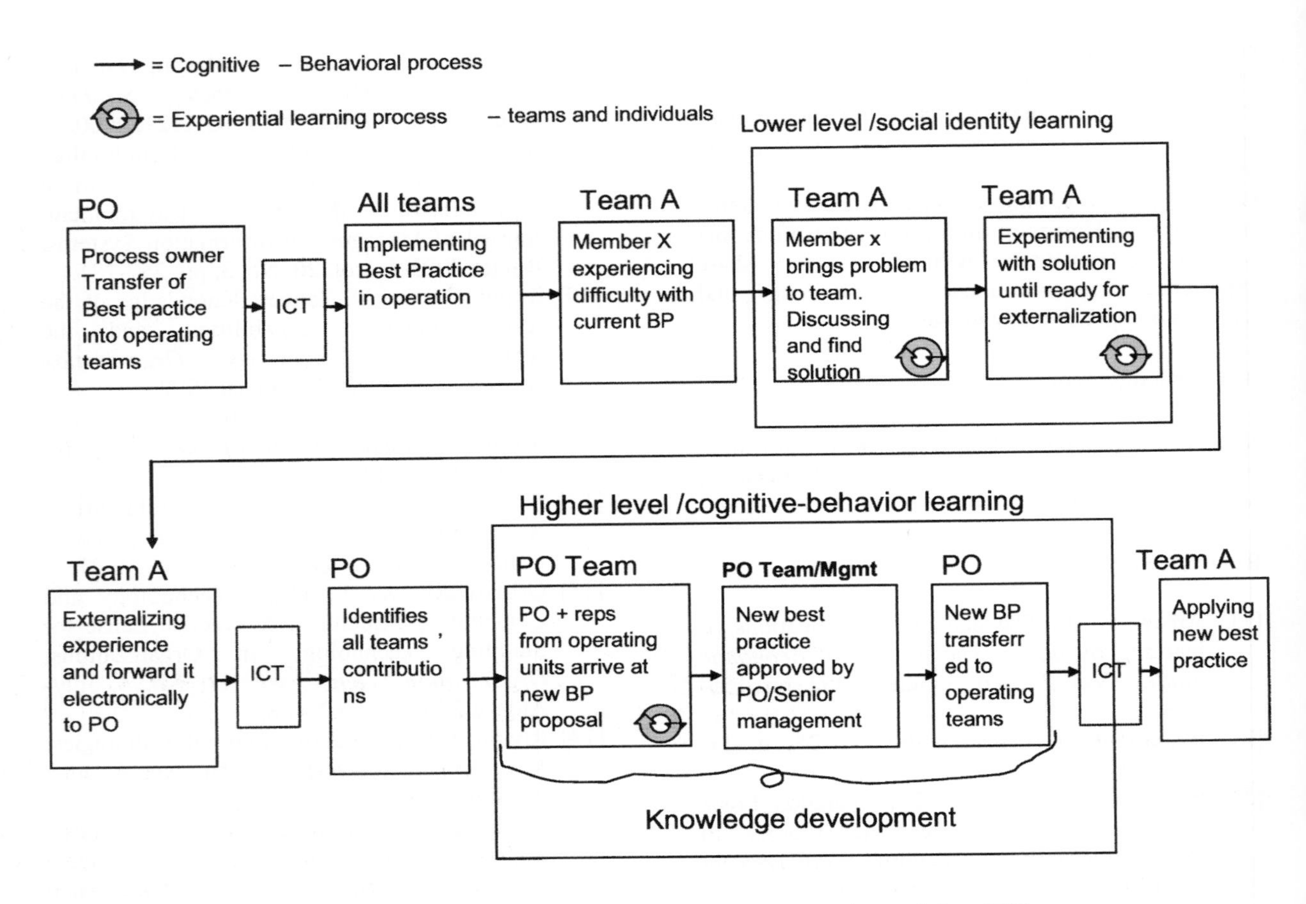

Fig. 1 Example of knowledge development cycle applying ICT.

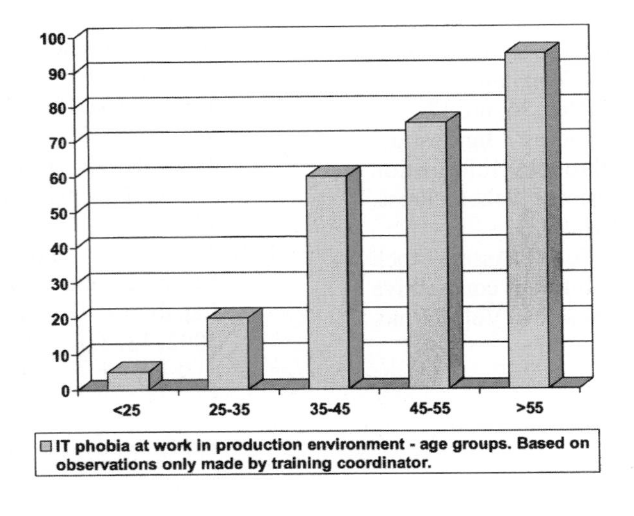

Fig. 2. An increasing change at the age of 35 years.

Intelligent Production Machines and Systems
D. T. Pham, E. E. Eldukhri and A. J. Soroka (eds)

Development of a Smart Supervisory Control System in a Sugar Mill Crystallisation Stage

R. Dodd[a], R. Broadfoot[b], X. Yu[a], A. Chiou[c]

[a]*School of Electrical & Computer Engineering, Royal Melbourne Institute of Technology University, Melbourne, Victoria, Australia*
[b] *Sugar Research Institute, Mackay, Queensland, Australia*
[c] *Faculty of Informatics and Communication, Central Queensland University, Rockhampton, Queensland, Australia*

Abstract

This paper discusses the development of an expert advisory system designed to provide expert knowledge in the control and management of a sugar mill crystallization stage. The smart supervisory control system is fundamentally a hybrid fuzzy expert system that incorporates fuzzy logic, relational databases and process models of the cystallisation stage. The primary topic of this paper will be a description of the framework of the smart supervisory control system with focus on: (1) system architecture, and (2) process models. The models presented are a core component in the expert system framework to predict of future pan stage operating conditions.

1. Introduction

Raw sugar production from cane is a nominally continuous operation, with 120-168 hours of processing per week, extending over 20-25 weeks of the harvest season. The crystallisation section, often loosely referred to as the pan stage, is the most complex part of the factory process where there are several batch wise or continuous crystallisation steps taking place concurrently [1].

In current Australian practice, two operators are normally employed on the pan stage and usually their duties extend no further than this section. There is considerable process interaction between the pan stage and centrifugal stage although management of the centrifugals is undertaken by different operators. The overall strategic management of the pan stage is quite difficult because of the very large number of process streams of varying compositions and crystal growth rate characteristics which must be managed [2]. Often the pan stage is managed in a sub-optimal manner because an overview of operations encompassing various sections - cane receival sections (cane quality), juice processing stations, the pan stage and centrifugal station - is not available.

The pressures on the Australian sugar industry to reduce the costs of sugar manufacture and increase the consistency of producing sugar of high quality require a smarter strategy for operation. This paper reports a development of a smart supervisory control system for sugar mill crystallization stage that has not been done in the world. It uses the advanced intelligent technologies to provide a standardised approach for pan operations by integrating key process models and combining them with the collective expertise and knowledge of pan operators [3]. It will

1. Result in a formal structure to the decision making procedure and reduce the number of *ad hoc* decisions and, consequently, the number of incorrect decisions;
2. Achieve increased productivity with the existing equipment, by employing strategies that are recommended by the smart supervisory control system;
3. Make improved use of the equipment capabilities to achieve sugar recovery, sugar quality and steam consumption targets while fulfilling the production rate requirements; and
4. Forewarn of potential problems with the current operating strategies.

This paper is organised as follows. Section 2 presents the smart supervisory control framework. Section 3 discusses the fundamental component of this system - a predictive mechanism to determine future pan stage operating conditions. The modelling integrates liquor and molasses feed rates to individual vacuum pans with information on the pan stage schedule and forward predictions of sucrose and impurity quantities from cane receival data to allow forecasting of future pan stage conditions. Section 4 presents some conclusions.

2. The Smart Supervisory Control Framework

The smart supervisory control system is essentially a hybrid fuzzy expert system incorporating fuzzy logic, explanatory capabilities and industrial process models of the crystallisation stage. The knowledge base is composed of human operator knowledge and mathematic dynamic models describing the crystallization process. The integration of such features leads to a challenge in the design and development of the smart supervisory control system. Previous research [3] acknowledges that no conventional software engineering methods exist to provide a solution to this problem.

2.1. Modular architecture

The smart supervisory control systems modular architecture is based upon conventional expert systems [4,5] and conventional If-Then fuzzy rule based systems design [6,7].

As depicted in Figure 1 the smart supervisory control system is designed as a modular architecture with clustered elements performing layered tasks. The partitioning of the system architecture aids in maintenance, accountability, upgradeability, adaptability and flexibility [8].

Compared to conventional expert system design the editor layer is unchanged. The data layer now also includes the dynamic interrelation models of the

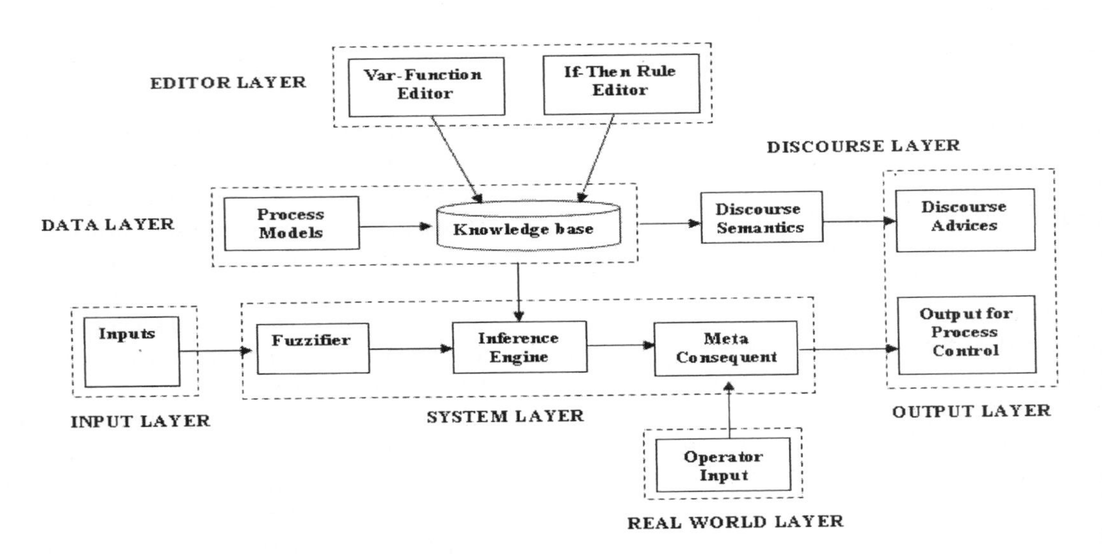

Fig.1 Proposed system architecture

crystallisation stage. The defuzzification component that is typical of fuzzy logic expert systems has been replaced by a meta-consequent function [8]. The discourse and real-world layers have also been added. The input and output modules have also been separated.

This unique modularity gives rise to the following layers:

- **Editor Layer** – Var function editor and If-Then rule editor component.
- **Data Layer** – Knowledge base and process model components.
- **System Layer** – Fuzzifier, inference engine and meta-consequent components.
- **Input layer** – Input user interface component.
- **Output layer** – Discourse advices and output for process control components
- **Discourse layer** – Discourse semantics and discourse advise components.
- **Real world layer** – Data from sources external to the data layer.

These layers can further be clustered depending upon functionality. These are (1) Standard layers – input, editor and data layers, (2) Discourse semantics – discourse and output layers and (3) Meta-consequent – system, real-world and output layers.

2.2. Standard layer

The input layer draws its information directly from the sugar mill online control system having been published as a series of relational databases. This information provides current data on the cane receival sections, juice processing stations, the pan stage and centrifugal station. Further information from the crystallisation stage operators, through the graphical user interface, can be provided to assist in determination of equipment performance ratings and operational problems and characteristics of the syrup, molasses and sugar process streams.

The editor layer provides the ability to modify the fuzzy membership function shapes and the rule associations with the dynamic process models. The final membership functions and rule base are stored in the knowledge base at the data layer. This layer also provides the facilities to make adjustments to the key process models that are associated with the rule base and explanations tagged to each of the rules.

2.3. Discourse semantics

Previous research [9] has show the need for expert system recommendations to be further accompanied by explanations to aid in the understanding and justification of presented advices to gain user acceptance. Limited research into explanatory capabilities for fuzzy expert systems has been carried out [10].

The system provides advice on:

- Recommended steam rate usage for the vacuums pans phase given current operating conditions
- Choice of A/B massecuite duties for "swing" vacuum pans
- Forecasting of stock tank levels and future disturbances
- Footing quantities to the vacuum pans; and
- Scheduling when pans should start and complete strikes

The discourse layer works to consolidate the inference process by formatting justifications for the presented advice in the most appropriate format. The method of presentation is an integral part of the output layer. The justification process is independent of the inference process which provides the final control output values that are recommended.

Explanations tagged to each rule in the If-Then fuzzy rule base are triggered and propagate through the inference mechanism upon a rule firing. This information is further passed to the discourse output component for further processing.

Aside from English based textual justification, justification for recommendation can be provided as a series of graphs or measures to show previous data trends on the pan stage schedule, productivity and steam rate usage. These can be presented against forward predictions of current pan stage operations and depict how the system recommendations can provide improvements.

2.4. Meta-consequent

The meta-consequent [8] layer provides an adjustment to the results of the inference process to adjust the process control models to information provided by the pan stage operators. This component replaces the defuzzifier found in traditional fuzzy logic based expert systems.

The meta-consequent layer is complimented by the real world layer to provide a method for mapping information on equipment performance ratings and operational problems and characteristics of the syrup, molasses and sugar process streams to the dynamic process models.

This provides meaningful interpretation of the inference process output by matching it to the current operating conditions at the crystallisation stage.

3. Predictive models

In order to predict future pan stage operating conditions a sequence of pan stage models have been developed. These models describe the operation of a portion of the overall pan stage process. By integrating these models into the expert systems knowledge base a predictive model for overall future pan stage operating conditions can be realised.

3.1. Crystallisation stage overview

The crystallisation section, as shown in Figure 2, is often referred to as the pan stage and is the most complex stage of the factory process for control purposes. Several batch wise and continuous crystallisation steps take place concurrently within this part of the process. Feed forward and feedback recycle streams are superimposed on this series of operations. The final stages in the raw sugar factory are the centrifuge station, which separates the sugar crystals from the mother liquor, and the sugar drying station [11].

Crystallisation takes place in large batch fed vessels known commercially as batch vacuum pans; in continuous vacuum pans; or in batch and continuous stirred cooling vessels referred to as crystallisers. Typical factories have between six and ten of these pans and about sixteen hours of residence time in the crystallisers for the low purity materials. The pan stage thus has an incoming stream of syrup and intermediate molasses recycle streams. Output products from the pan stage are raw sugar and final molasses.

3.2. Syrup rate prediction

The prediction of the quantities of sucrose and impurities in syrup from cane receival information allows a forward forecasting of the future pan stage loading of syrup. Liquor comprises the basic input to the pan stage.

Previous research into predicting the impurity loading to the pan stage was based on assigning impurity losses to factory products. These assessments demonstrated huge variability in the estimate of pan stage syrup impurities by +/- 20% during successive weeks of factory operation [12]. Conventional methods for factory sucrose and impurity balances [13] also exhibit large variations in the estimate of the sucrose and impurity quantities in syrup to the pan stage.

An empirical prediction model [14] has been developed using 2002 weekly crushing season data from Racecourse and Marian Sugar Mills.

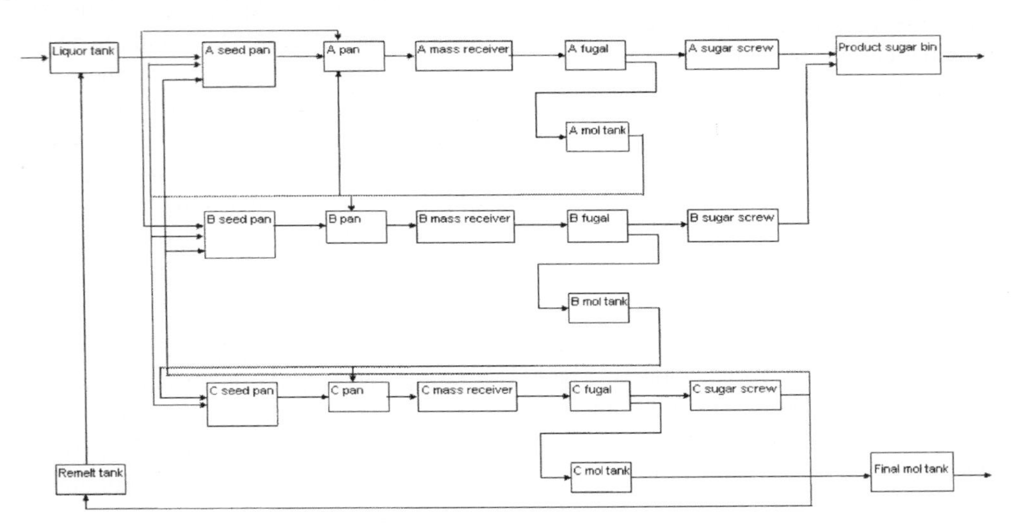

Fig.2 Pan stage operations

A model to forward predict the quantity of sucrose in syrup to the pan stage from cane is presented in equation (1). The relationship for impurities in syrup to the pan stage is presented in equation (2).

$$s = fqp/100 \quad (1)$$

where,

s is quantity of sucrose in syrup to the pan stage (t)
f is an empirical factory operational fraction
q is the quantity of cane crushed (t)
p is pol%cane of crushed cane (%)

$$i = fqp(100-t)/t \quad (2)$$

where,

i is quantity of impurities in syrup to the pan stage (t)
q is the quantity of cane crushed (t)
t is purity of syrup to the pan stage (%)

The empirical factory operational fraction can be calculated using data from the mill under consideration. Using 2002 weekly crushing season data from Racecourse and Marian Sugar Mills (excluding data for the first and final weeks of the season where potential errors in stock tank levels are substantial) the relationships presented in Table 1 were derived.

Syrup rates for the results presented in Table 1 were measured by magnetic flow meter. Cane quantities were measured at the cane receival station while pol%cane was calculated from juice laboratory analysis of the first expressed juice sample. Sucrose%syrup and dry substance%syrup values were measured in laboratory shift analyses.

Table 1. Seasonal empirical operational coefficients for Marian and Racecourse 2002 crushing season.

Sugar mill	Length of crushing season (weeks)	Average weekly empirical factory operational fraction	Seasonal standard deviation of empirical factory fraction
Racecourse	25	0.9725	0.0089
Marian	24	0.9767	0.0099

The empirical factory coefficients appear to be consistent among the various weeks for a particular factory (note the low values of standard deviation) and between factories (similar mean value). It may be considered that:

$$e = 1.0 - f \quad (3)$$

where,

e is a fractional estimate of impurity/sucrose loss between cane receival and the pan stage (i.e. impurity/sucrose losses to mud and bagasse).

Equation 3 indicates that the collective analytical errors in measurements and impurity/sucrose losses to bagasse and mud appear to be fairly consistent both within the mill crushing season and across mills.

The data required for the model represented by equations 1 and 2 are pol%cane, quantity of cane crushed and the purity of the syrup. The purity of syrup corresponding to the cane crushed in a shift (or shifts) will not be available until the lab analysis is performed later in the day. The previous day's information on syrup purity, or a rolling average value for syrup purity for several previous days would be used instead.

Although the variability of the test data for Racecourse and Marian 2002 season is small (as demonstrated by the standard deviation) it is still necessary to make feedback corrections based on actual pan stage data. Forward predictions can be made with the model and corrected using the error in prediction for pan stage quantities compared with actual production quantities. A rolling error average would be used to update predictions from the model.

By using rolling average error corrections for the prediction analytical errors in syrup and cane analysis, and even changes in the methods of cane analysis as may occur in the future, can be accommodated.

The prediction of the quantities of sucrose and impurities in syrup from cane receival information has been decomposed to a solution of only several variables while maintaining a high overall predictive capability.

3.3. Individual empirical vacuum pan models

Individual pan production rates have been modeled

by constructing empirical relationships for the rate at which each pan takes feed material (liquor, A molasses or B molasses) during the different phases of the pan's operation [15]. This boil-on rate for feed materials is a function of the massecuite level and phase of the pan, steam rate, head space pressure (vacuum), brix and purity of the feed liquor/molasses.

The approach has been applied to the pan stage of Racecourse Mill. An example is given in Figure 3 for pan 3 which is a small A strike pan at Racecourse Mill. The pan takes a footing of approximately 40 tonnes of high grade seed massecuite, boils on liquor initially and then A molasses to complete the A massecuite strike at a pan full level of approximately 60 tonnes. The liquor feed rate during different phases was determined from the change in level during the run up from the footing to the end of the liquor feeding period. To determine this rate, the brix of the massecuite in the pan, the brix of the starting seed footing and the brix of the liquor feed were measured. Similarly, massecuite samples at the start and end point of the A molasses feeding phase were taken and the brix measured to determine the A molasses feed rates achieved during pan operation.

Using this method it is possible to construct a piece-wise model of pan feed rate characteristics during each phase of the pan's operation for each of the vacuum pans. The current operating status of the vacuum pans can be determined using a simplified classification procedure based upon steam usage, level and change in level as presented in [16].

Racecourse Mill utilizes batch pans for all operations except for the C massecuite production. Thus, the model for the boil-on rate can be established

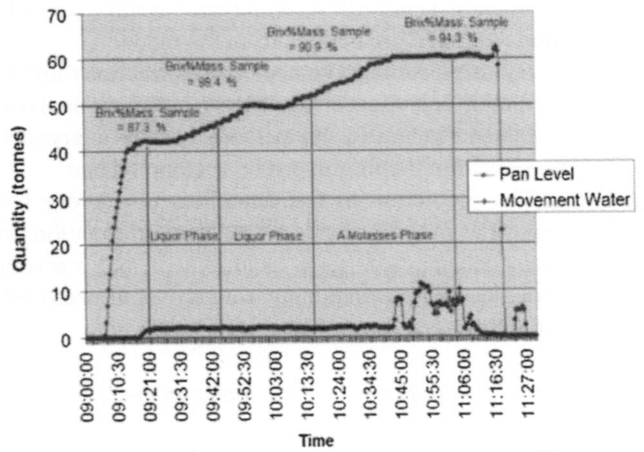

Fig. 3. Data for pan 3 at Racecourse sugar mill during an operating cycle on 04/09/2003

for each feed material during each phase (based on massecuite level) as the pan progresses to each target level where a cut out, change in feed material or pan discharge occurs. The boil-on rate models for each pan are determined by undertaking similar measurements through the different phases of each pan's operation.

The average pan feed consumption data gathered at Racecourse sugar mill during high grade strikes on 03/09/2003 and 04/09/2003 are presented in Table 2. These data are further subdivided depending on the massecuite level and steam rate. These data are not shown.

In order to determine a relationship between steam usage rates and the expected vacuum pan strike times a level based iterative computer simulation for a batch pan was also developed. A steam profile for all massecuite duties has been developed to assist in relating strike time to steam consumption of the vacuum pans. Given the expected schedule and operating status of vacuum pans, the strike times of pans within the schedule can be adjusted to avoid idling time and better improve the productivity on rate limiting pans. Further vacuum pan modeling information and results are presented in [15].

Combining projected vacuum pan feed rates into the pan stage schedule with predictions of sucrose and impurity quantities from cane receival data allows forward forecasting to ensure there are sufficient quantities of materials in stock during standard season operation, forewarn of potential problems with the current operating strategies and advise corrective procedures.

Table. 2. Average pan feed consumption data gathered at Racecourse sugar mill during high grade strikes on 03/09/2003 and 04/09/2003

Pan numbe r	Pan duty	Average liquor consumption (t/h)	Average A molasses consumption (t/h)	Average evaporatio n (t/h)
1	HG Seed	17	---	Not Available
2	B Mass.	27	63	12
3	A Mass.	15	17	8
4	A Mass.	111	52	24
6	B Mass.	17	90	27
9	A Mass.	34	33	4

3.4. Pan stage massecuite model

The steady-state model of the pan stage depicted in Figure 1 has been developed to calculate average flow rates of process streams using mass balances at each vacuum pan, stock tank and fugal. The model determines the average production rates of vacuum pan massecuite, C sugar remelt, molasses and sugar streams given the syrup purity and flow rate to the pan stage.

Crystal sizing determinations assist in calculating the necessary quantities of C sugar needed for the A and B pans to ensure final product sugar is of the required size. Typical conditions for crystal content, sugar size and coefficient of size variation have been assumed in this model and the purity rise of sugar at the fugals has also been taken into account.

3.5. Stock tank models

Once the empirical model for each pan has been established then the boil-on rates for each feed stream at the different stages of the pan stage schedule can be determined by summing the liquor, A molasses and B molasses feed rates for all the pans at that point in the schedule. Given the expected liquor production rate, C sugar remelt production rate and raw wash return from the refinery to the liquor tank during this interval, the predicted tank levels can be determined for the liquor tank. Similarly the predicted tank levels for the A and B molasses streams can be calculated from the production rates of the molasses at the centrifugal station and the sum of the consumption rates on the individual pans at a specific point in the pan stage schedule.

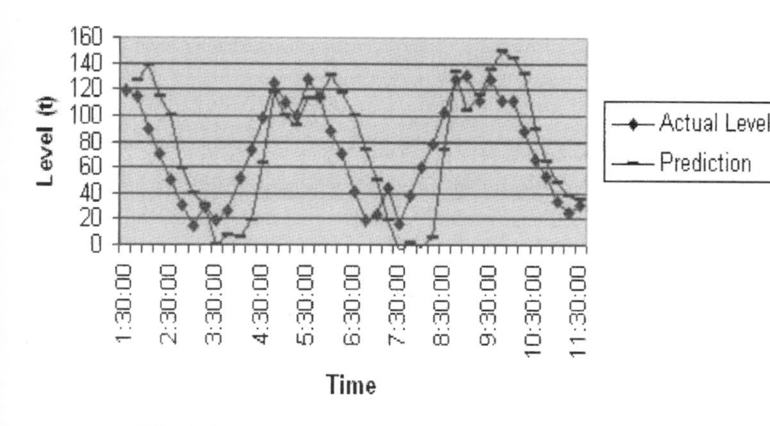

Fig 4. Syrup tank level prediction for 03/09/2003

Stock tank level prediction requires the utilization of the previously discussed models. A forward prediction of stock tank levels was made drawing from cane receival, stock tank and vacuum pan control system data for Racecourse mill on 03/09/2004. Using forward prediction of vacuum pan states and the key process models discussed, a 90 minute future forecast of stock tank levels was made utilizing a rolling error average over the three previous 15 minute intervals. Results for the syrup stock tank are presented in Fig. 4 and show a good correlation between predictions and actual stock tank levels.

4. Discussion and Conclusion

The overall strategic management of the pan stage is quite difficult because of the very large number of process streams of varying compositions and crystal growth rate characteristics which must be managed. Often the pan stage is managed in a sub-optimal manner because an overview of operations encompassing various sections - cane receival sections, juice processing stations, the pan stage and centrifugal station - is not available.

This paper reports a work in progress in the development of a smart supervisory control system. The system seeks to provide a unifying structure to assist pan stage operators in making early decisions, such as changes to steam rates, or allocation of pans to different duties (e.g. A or B massecuite) to avoid production rate difficulties, and maintain good operational performance with respect to sugar quality, sugar recovery and minimization of steam consumption on the total pan stage.

The smart supervisory control system is a hybrid fuzzy expert system incorporating fuzzy logic, explanatory capabilities and industrial process models of the crystallisation stage. Presently the most difficult part of process modelling has been completed and the construction of the advisory system is under way. The system hopes to contribute both to the field of fuzzy logic based expert system design and to the development of industrial process models for the crystallisation stage within a sugar factory.

Acknowledgements

The project contributors would like to thank the staff at Racecourse Sugar Mill for their cooperation and support. This project is supported by an Australian Research Council linkage grant. The funding assistance provided by the Sugar Research Institute is also acknowledged and appreciated.

References

[1] Broadfoot R, Beath AC, Wright PG and Miller KF. Modelling of batch sugar centrifugal performance. Proceedings of Chemeca 1998. (1988) 294-303

[2] Miller K.F and Broadfoot R. Crystal growth rates in high grade massecuite boilings. Proceedings of the Australian Society of Sugar Cane Technologists, Mackay, Australia. (1997) 441-447

[3] Dodd R, Yu X, Broadfoot R, and Chiou A. Development of a smart supervisory control system for pan stage operations in sugar factories. Proceedings of Information Technology in Regional Areas. (2002) 368-375

[4] Leung KS and Wong MH. An expert-system shell using structured knowledge: an object oriented approach. Computer (1990), 23(3), 38-46

[5] Gisolfi A and Balzano W. Constructing and consulting the knowledge base of an expert systems shell. Expert systems. (1993), 10(1), 27-37

[6] Goel S, Modi K, Shrivastava M, Chande PK and Gaiwak A. Design of a fuzzy expert system development shell. Proceedings of 1995 IEEE annual international engineering management conference. (1995) 343-346

[7] Berkan RC and Trubatch SL. Fuzzy Systems Design Principles: Building Fuzzy If-Then Rule Bases. IEEE Press, 1997

[8] Chiou A, Yu X and Lowry J. P-Expert: A Prototype Expert Advisory System in the Management and Control of Parthenium Weed in Central Queensland. In Dimitrov, V. and Korotkich, V (Eds), Fuzzy Logic: A Framework for the New Millennium, Heidelberg: Physica-Verlag. 2002

[9] Ye LR and Johnson PE. The impact of explanation facilities on user acceptance of expert system advice. Expert Systems with Applications. (1995), 9(4), 543-556

[10] Gregor S and Yu X. Exploring the explanatory capabilities of intelligent system technologies. Proceedings of Second International Discourse with Fuzzy Logic in the New Millennium. Physica-Verlag. 2000, pp 289-300.

[11] Broadfoot R and Beath A. Modelling of batch sugar centrifugal performance. Proceedings of 26th Australasian Chemical Engineering Conference. (1988) 6p.

[12] Broadfoot R and Miller K. Impurity transfer in raw sugar manufacture: impurity mass balance anomalies, Sugar Research Institute, 1990, internal report.

[13] The Standard Laboratory Manual for Australian Sugar Mills: Principles and Practices, Bureau of Sugar Experiment Stations Publications, Brisbane, Australia, 1984, pp 71-80.

[14] Dodd R, Broadfoot R, Yu X and Chiou A. Update on development of a smart supervisory control system for pan stage operations in sugar factories. Proceedings of Sugar Research Institute 2004 Regional Research Seminars (2004) Sugar Research Institute (CD-ROM).

[15] Dodd R, Broadfoot R, Yu X and Chiou A. Empirical pan modelling of vacuum pans for a sugar mill crystallization stage. Proceedings of the Australian Society of Sugar Cane Technologists, Mackay, Australia. (2005) 423-436

[16] Frew J and Wright P. Pan stage advisory scheme at Racecourse Mill. Commonwealth Scientific and Industrial Research Organisation Australia, 1977, report number CE/R-52

Intelligent Production Machines and Systems
D. T. Pham, E. E. Eldukhri and A. J. Soroka (eds)

From simulation to VR based simulation and testing

Pavel Ikonomov [a], Emiliya D. Milkova [a]

[a] *Industrial and Manufacturing. Engineering, Western Michigan University, Kalamazoo, MI 49024, USA*

Abstract

This papers presents, using Virtual Reality (VR), simulation, through product lifecycle that would provide design of better new products. Using comprehensive VR product simulation for new products, from design, manufacturing, and maintenance to the end-user and recycle state, allows not only makers but also the end-users to test the virtual products. By providing feedback, it also reduces design and production errors and contributes for better final products. In addition, the same VR product simulation can be used for training, virtual user manual, virtual repair maintenance, and for disassembly before recycle. A framework for building VR simulation for designed products is described and working application examples are provided. An implementation of the VR simulation applications for usage with extensive range of computation platforms and devices is presented.

1. Introduction

This paper describes an application of Virtual Reality (VR) for design, production, simulation and testing of the new products. The proposed VR simulation system allows new products to be designed and manufactured after being thoroughly tested not only at factory, but also by the consumers. It allows direct feedback to the designer from the factory floor and future users. This will allow an early discovery of design problems and errors, and faster acceptance of new products as well as capabilities for user to add new features for improvement in styling and quality of the product.

In general, the simulations attempt to represent certain features of the behavior of a physical or abstract system by the behavior of another system [1].

Virtual Reality provides more complex simulation technology than other simulations. Virtual Reality (VR) is the use of computer modeling and simulation to enable a person to interact with an artificial three-dimensional visual or other sensory environment. VR applications immerse the user in a computer-generated environment that simulates reality through the use of interactive devices, which send and receive information, that are worn as goggles, headsets, and gloves. [2].

To represent the real world, VR employs the power of the computers. While some of the human senses (e.g. smell, touch, feeling) cannot be represented or reproduced correctly with VR, additional cues, such as color changes, sound, and moving objects are used instead. One of the major problems of the VR system development is that in pursue for speed and visual effect, many features of the real physical system are omitted or simplified. While the cinematic effect can grab visual attention, the correct system simulation and behavior model are more important in scientific research. Additionally, accurate simulation system is useful only by a group of dedicated researchers and scientists that can operate the sophisticated equipment and software. This is the dilemma that needs to be resolved so that ordinary users can control highly sophisticated simulation system without extensive special training.

Here we propose a framework for building VR

simulation that is sophisticated to provide high quality immersive visualization, accuracy of the simulation and behavior model, and can be operated with natural human interaction.

We have built a VR system simulation that can be used both, industrial and end-user type of application. It can serve the both sides of products, engineers who design, manufacture, and test the products, and end-users that buy, use, and dispose those products.

1.1. VR for broad consumption

Films like 'Star Wars', 'Terminator', 'Star Trek', 'The Lord of the Rings', etc., has shown how Hollywood can produce VR like high quality visuals. Those types of application use simple rendering of moving objects, with fixed camera view. The computational time for several seconds of movie can run thousand of hours on numerous computers to reach the desired effect. While effects are marvellous, the visualization has limit or no relation with reality. It does not follow the rules of physics and cannot be applied for real simulation. At the same time, it enhances user expectation by showing how the VR simulation will look like in the future. Models and environments are highly accurate and give real life like perception to the user. Unsatisfactorily, the single camera viewpoint and motion are fixed and does not allow changes. Most of the pictures are not stereo and for those with stereo effect, special projection equipment is required.

Video games are another VR like application that provide cinematic like quality running in real time. Games use highly sophisticated software system, called 'game engine', that allow very fast real time simulation on ordinary computer with high-end graphics video card. In order to gain speed of rendering and movement, the models of objects and people are especially simplified with minimal use of polygons and creative usage of texture to represent details. Although they allow user to move freely in created environment, all paths and motions are still scripted as in the movies. User movements and actions are responded by highly sophisticated scripts using Artificial Intelligence techniques.

1.2. Simulation

There are many simulation systems for different purposes. Most of them represent physical phenomena accurately while allowing flexibility to the user to modify simulation properties at any time. The result can be dynamic or static. In most of the cases, there is time delay for the simulation software/hardware to respond to changes so it cannot be used in VR simulation. Some of the software allows limit applications to real time or near real time simulation that can be used in VR simulation. In such case, the computations for simulation can be done in real time or in advance (precompiled), or simulation model may be independently configured using generic parameters during its instantiation.

One of the most sophisticated simulation systems today is VHDL, used in electronic industry. VHDL was originally developed at the US Department of Defense in order to document the behaviour of the ASICs that supplier companies were including in the electronic equipments.

The idea of being able to simulate this behavior was so obviously attractive that logic simulators and logic synthesis tools, which read the VHDL and output a definition of the physical implementation of the circuit, were developed.

VHDL is, in fact, a fairly general-purpose programming language, provided that you have a simulator on which to run the code. It is possible to use VHDL to write a testbench that verifies the functionality of the design, using files on the host computer to define stimulus, interacts with the user, and compares results with those expected [3].

VHDL has proven itself as essential tool to design, test, and optimize the hardware in software simulation even before circuits were produced. Our goal with VR simulation is to play the same role in industry and everyday life while at the same time allowing interaction with the ordinary users without specific training

2. VR as simulation tool for design product testing

VR simulation systems have been proved to be very useful for industrial application using it with high-end computers, equipment, and peripherals. In the past, this setting was very expensive and its programming was so sophisticated and difficult to use that only few companies and research centers where able to design. With the advances of computers and software, it is now possible for most of the people to use VR simulation systems to make changes in the simulation using ordinary computers with generous requirements for processing power, memory and graphics. We would like to use the idea of a well-known software system beta testing practice to develop a beta testing production and maintenance system using VR simulation. Hereafter, we propose VR simulation application tools for usage at design, production, and management cycles of products life.

2.1. Beta testing

A common practice of software testing is that it is performed by an independent group of testers after finishing the software product and before it is shipped to the customer. Another practice is to start software testing at the same time the project starts and it is a continuous process until the project finishes.

In-house developers often test the software using what is known as 'ALPHA' testing which is often performed under a debugger or with hardware-assisted debugging to catch bugs quickly.It can be then handed over to the testing staff for additional inspection in an environment similar to how it is intended to be used. This technique is known as black box testing (or second stage of alpha testing) [4].

Following that, limited public tests known as beta-versions are often released to groups of people so that further testing can ensure the product has few faults or bugs. Sometimes, beta-versions are made available to the open public to increase the feedback field to a maximal number of future users.

In black box testing, the test engineer only accesses the software through the same interfaces that the customer or user would, or possibly through remotely controllable, automation interfaces that connect another computer or another processor into the target of the test [5].

2.2. Simulation as a tool of hardware/software testing

As it was described before, VHDL design is used for testing and optimizing the hardware in software simulation in electronics industry.

When electromechanical systems are designed, they are first tested with special software simulator. The design and joint simulation of the heterogeneous parts of a system in a unified environment offer many advantages, which include reduced system design and debug times, as well as increased simulation accuracy.

Flexible interfaces can be implemented for bridging VHDL with electromechanical systems simulators (EMSS) and VHDL simulators with the application programs that control the digital circuit simulated in VHDL correspondingly [5]. The design simulation can include motion simulation and FEA to provide testbed for various products from aircraft to micro-miniature switches. According to CAE engineer from Big Three automakers, they cut the cost at least in half using design simulation. Another company SiWave Inc., an optical switcher supplier, cut empirical testing by an order of magnitude. Using virtual prototyping, John Deer reduces the number of prototypes and saves time; from one to six months for each prototype [6].

2.3 What is missing from Virtual Reality simulation

Most of the existing VR simulation used is, in fact, 3-D visualization used to fly through the Virtual environment and play role for design product description and acceptance. Such VR simulation is still only a simple static 3-D visual presentation with limit interaction or animation that does not include the way real machines, objects, humans, and environment operate together in real life. In the past, the computational resources where the limitation factor, and the simulations were simplified for performance result. The computation power, however, is no longer obstacle for complete real time VR simulation.

VR-based applications are used to enable personnel to control and optimize the efficiency of manufacturing systems at different stages of the systems life cycle in real-time. They are able to integrate new core concepts within the design process [7].

As solution to the problems described above, Virtual Reality system that uses natural human interaction for control of the design products is proposed. While it works in a matter similar to beta testing and VHDL simulation, it allows not only testing of a new product, better interaction, and experience by user, but also natural interactions with virtual products and VR environment. By allowing feedback of data to the designer the virtual system can help him/her to improve product quality before going to actual production process.

The system proposed is not limited only to design/manufacturing engineers and maintenance. Ordinary people without training can test, modify, and provide feedback for the products that are still in design stage thus giving ultimate user an important tool to influence the design and manufacturing of the products that he/she desires.

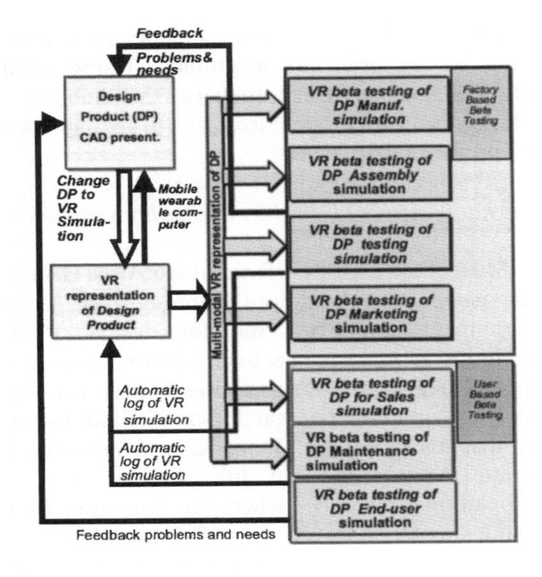

Fig. 1. Virtual Reality based beta testing system.

3. Design product VR simulation and testing

The Virtual Reality simulation world is build as representation of the real world. It is populated by machines, object and people. To be accessible to everyone, the VR system shall provide user with control and usage of machines and tools in natural matter without needs for special training. This requires that machines, tools, and humans (or avatars) to use a behavior model for the interaction with other objects, machines or humans. The behavior model can be as simple as predefined script or more complex one based on Artificial Intelligence.

The proposed system for VR beta testing works on the following way, see Fig. 1. First, designer produce Design Product (DP) CAD model following functional requirement, manufacturing restriction, and his/her understanding of the expectation of the end-user. In general, DP is a geometrical CAD model that includes constraints, physical properties, and tolerance requirements.

Next, in VR representation of Designer Product new information is added, such as how the product works, how it interacts with other products, how it is produced, and how it is used by the end-user. First, the designed product model is exported from CAD then simplified. In the following phase, additional custom designed models of parts, machines, tools, and VR environment are developed. This involves detailed application knowledge and implementation of the behavior of the actual product to be added to VR model and environment. Utilizing of such information allows general multi-modal interaction and visual models for VR product beta testing simulation to be build, which can be used by different groups of people.

Finally, the integration of this VR beta product models for each specific case would require minor changes of the visual data representation and behavior model, and adding of existing peripherals and visual hardware (head mounting display, wall type projection system-CAVE®, desktop or notebook computer, tablet, PDA or mobile device). In addition, control and feedback modules for each specific case can be added. The integration of those additional features allows simulation, control, and feedback of VR beta test product model to be performed.

During the usage of the VR system in Factory Base Beta Testing and User Base Beta Testing, an Automatic log of VR simulation can be loaded back to VR representation of DP. Designer can use this information to judge how well the system works, possible errors, or the usefulness of the designed functions. Then, if is needed, he/she can make changes of the DP to VR simulation. Through communication channels such as Instant Messaging, E-mail, and videoconference, direct feedback of problems and needs can be send to designer from production floor and end-user at any time. In such a case, a direct link from VR simulation can initiate the communication. Information from existing CAD product data management/product live-cycle management (PDM/PLM) system can be linked to close the cycle of VR simulation system.

4. Virtual Reality simulation implementation

4.1. Designing of the VR assembly simulation

One of the reasons that Virtual Reality simulations are not widely used is that Virtual Reality graphics required high-end powerful computers such as Silicon Graphics – SGI Reality Center, Onyx2, etc. On the other end are standard personal computers using some adequate hardware accelerated open GL video graphics board.

One of the benefits of using of the ordinary PC is reduced cost, but also very important is the possibility to use the available peripherals to set up a complete system that can be run anywhere. The system we have

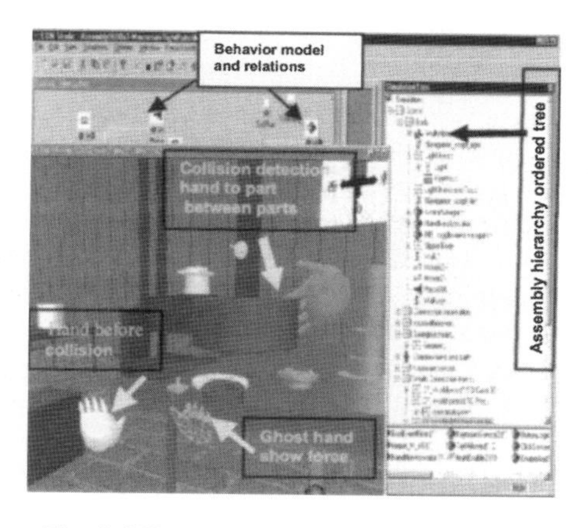

Fig. 2. VR assembly model and behavior.

used to develop and run the simulation, consist of computer with CPU Pentium 600Mhz, 256 RAM, and Open GL graphics card 32MB. For complete immersion and natural interaction, Head Mounted Display (HMD) Virtual research V8 (also Sony Glasstron-Stereo HMD, SVGA 800x600) and peripherals Polhemus Fastrack tracking system, and Fake Space pinch gloves are used. HMD is used to simulate immersive stereo visual feeling inside the Virtual Reality environment. Later, the same system with very minor modifications was tested on CAVE® type wall projection system. The elimination of HMD allowed collaborative VR simulation. The CrystallEyes Stereo Graphics shutter glasses allow accepting of stereo images projected on the walls by human eyes by subsequently switching on and off left and right eyes.

For rendering and behavior modeling, EON Studio Virtual Reality software was used, which supports most of the peripherals. Key parts of this VR software are: EON Studio -the main rendering and simulation engine and data exchange –most geometry was imported using data exchange from CAD software IGES file format. The rest of the geometry was developed in 3DS Max, Multigen Creator, VRML, and AutoCAD, and imported into the simulation.

4.2. Building behavior model of the VA simulation

VR Assembly simulation shall not only represent visually the assembly sequence, path, organization, etc, but also the relations between assembled parts or subassembly, tools and parts, machine operation, and control.

Constraint kinematics and connection relations are used by designer to specify the position and orientation of assembled parts. Those constraints can be applied to the axis, surface or be combined to simulate the assembly constraints between different parts. In the process of assembly, some of the constraints are released and some new ones are added, which requires special attention to maintaining them. To make assembly easy for the user, parts motion can be constraint to move on plane or axis to guide it to final position.

In the VR assembly different strength force between different assembled parts and subassemblies is used, to simulate how they are connected according to their weigh, tolerance fit or assembly relations [8].

One simple example is assembly of sequence of parts. Each part is assembled according to *hierarchical sequence* defined in assembly plan. For example part No. 3 cannot be assembled before part No1. is attached and then No. 2 is assembly, etc., see Fig. 2.

Position and strength of connection relations between each two parts or subassemblies shall represent the real one. For example if there are three parts stacked, different strengths shall be maintained between them so if we remove the lowest one the rest of the part shall move together with it. If we remove middle one, the top one shall stay with it or will not allow removal of the middle part if the top part is related to the lowest part. Finally, if we pull only the top part, the rest of them shall remain assembled. When there is fixed assembly, like hard-pressed one, then those parts shall remain connected even if we apply larger force while we disassemble the rest of the machine. This strength of the connections is not specified in the CAD design, so it is added during the building of the assembly model to make it more realistic. Information obtained from real assembly experts is added and represented.

In the Virtual Assembly, physical properties of the parts like weight, inertia mass, and spring constant and elasticity, need to be properly simulated. For example dropping a part will cause it to fall on the floor. Springs and other elastic parts can be pushed, pulled, bend, or twisted, which shall lead to proper physical simulation.

Collision detection was used not only to prevent interference of the parts and user, but also as an indicator. When part is touched/collides by the user hand it changes its color as indicator that the part can be grabbed, see Fig. 2. Designed assembly model flow allows straightforward process of the assembly operation, possibility to include hints, and help messages to support user during the VR assembly.

Fig. 3. Torque wrench bolt tightening

Fig.4. Crane and Press machine control

Application of the rigid body dynamics allows dynamic force simulation of moving parts. Inclusion of force feedback and gravity forces the simulation gives the feeling of touch and behavior of the assembled parts, and the sense of realism that increases the immerse feeling of the visual simulation [8].

When VR assembly is manual type, the user can see his hand, tools, and machines inside the Virtual world and work there by moving freely, grabbing, pushing/puling, and pressing buttons same way as in the real world. While this seems easy in the real life, any kind of interaction to and response from object/machines/equipment needs to be simulated, leading to complex behavior model that can slow the simulation.

This behavior model, extracted from real world, shall be kept in correct order. Force and relation in the virtual word shall be applied to avoid misinterpretation of the VR assembly.

For example when we assemble a cover to engine block with bolts, bolts are first assembled by hand in specified order, each bolt can be rotated and has one or two degree of freedom, translation on the axis direction and rotation around the same axis. Next, to tighten the bolts, tools are used. Torque wrench can be used, if necessary, to tight bolts with prescribed force, therefore scale for force simulation is displayed, see Fig. 3. After tightening, bolts become fixed to the engine with zero degree of freedom, and as result, the cover also become fixed; going from two degree – sliding on plane – to zero degree of freedom.

Machines/equipment control and usage in VR system is needed to resemble the natural matter of operation. For example let's have control of crane. User push buttons of the control box to move up/down, east/west, or south/north, and to control speed/acceleration, see Fig. 4. The user attaches the part to the crane on predefine positions as in the real case. At the destination position the part is detached from the crane and assembly connection forces between the parts or part and fixed place are added. As the crane manipulation is quite difficult process even in the real situation, these (snap) connections can be activated when the part is in close proximity within a specified tolerance to final fixed position.

When a press machine is used, the parts are assembled using press tool. User controls the press machine by pushing the controller holder to down position to run press down and up to revert it. During the press simulation, the press tool pushes the bearing inside the engine block, so the connections relation changes from simple assemble to tightening force between them and become press fixed type [9].

4.3. Input and control peripheral devices

Immersive type peripheral devices and ordinary, mouse and keyboard, input devices allow multi-modal VR system simulation to be used and tested on the ordinary computer, as well as on different type on computational devices such as notebook, PDA and tablets, when no special equipment is available.

Peripheral devices like 3D tracking sensors and grasp gloves are used to track the movement of the user's body, hands, and objects in Virtual Environment for tasks such as grasping and caring objects, using tools, and controlling machines. Tracking usually involves updating of the position and orientation of hands and head in real time.

Position/orientation tracking system with combination of grasping gloves allow user to move his own body and hands and see how their virtual representations move inside the Virtual Environment. User is able to grasp and interact with parts, machines, and tools in natural matter. User can grab and release the virtual objects, assemble and disassemble, and use tools and machines without requiring any preliminary learning or previous experience. Even though the tracking and grasping are two separate components of the VR simulation, the difference for the user is unnoticed and quite natural to use.

Tracking of the head is represented by a camera movement in VR closely follows the exact motion of human head from eye position. Hands and body tracking are employed not only to show the hands' positions and orientations, but also for tools used during the assembly. In some cases, tools can replace the hand for easy manipulation. To simplify the usage, grabbed parts or tools can be attached to hand in predefine position and orientation.

Due the problems of grabbing in VR, lack of touch sense and weight feeling, additional cues, such as hand color changing when collides or it is in close proximity with part, are introduced (See Fig. 2). Weight and connection strength is simulated by adding a ghost image of the hand. With increase in the force or weight the distance between the solid rendered image of the hand and ghost one increases, allowing the user to judge the magnitude of the connection force or weight. Sensor or simple on/of switch on each finger inside the gloves provide information changed of the position of each finger.

Use of tracking device and grasping gloves for natural interaction, in an immersive VR assembly, gives freedom to the users to operate in natural matter. It makes it possible for users to concentrate their attention to actual work on assembly, investigate assembly path accessibility, judge about the clearance between the parts, and how easy it is to make assembly.

To simplify the VR assembly tasks, a pre-run animation can be used to teach the user how to manipulate. This animation can also be used when we want to simplify the VR simulation, e.g. use for demonstration on computer that has lower computational capabilities like portable computer, tablet, or PDA. Such animation sequence simplification can be used if it is very difficult to manipulate the object. As an extension of the VR assembly, VR Disassembly simulation and simplified desktop assembly simulations for different specific tasks, such as bolt assembly sequence, machine operation, and control, were also developed.

5. Conclusion

We described usage of Virtual Reality simulation through product lifecycle. The proposed Virtual Reality simulation and test system allows complete geometrical representation and simulation of the functions of designed products. The proposed VR simulation system allows new products to be designed and manufactured after being thoroughly tested not only at factory but, also by end-user. There are no special requirements or knowledge of computer skills for the user to interact and control.

This development is continuing using multi-modal interaction and simulation tailored for each special computation platform like multi-wall projection-based Virtual Reality system, desktop computer, tablet, mobile wearable computers, and PDA. Interaction and representation are done in real time using same VR simulation tailored for different computational platform.

References

[1] Simulation, Wikipedia, the free encyclopedia http://en.wikipedia.org/wiki/Simulation

[2] Virtual Reality (VR), Britannica online http://www.britannica.com/search?query=virtual+reality&go_button.x=0&go_button.y=0&ct=

[3] V HSIC hardware description language, From Wikipedia, the free encyclopedia http://en.wikipedia.org/wiki/VHDL

[4], Software testing Alpha testing, Beta testing, Wikipedia, the free encyclopedia http://en.wikipedia.org/wiki/Software_testing

[5] N.C. Petrellis et al, Simulating hardware, software and electromechanical parts using communicating simulators, 7th IEEE International Workshop on Rapid System Prototyping (RSP '96), June 19 - 21, 1996,

[6] Louise Elliot, Cutting Development Costs with Design Simulation, Desktop Engineering, Vol.8, Issue 5 Jan 2003, pp.10-13.

[7] Ruy Mesquita et al, Usability Engineering: VR interfaces for the next generation of production planning and training systems, Computer Graphik topics 2000 5/2000, http://www.inigraphics.net/publications/topics/2000/issue5/5_00a07.pdf

[8] Pavel G. Ikonomov and Emiliya D. Milkova, Using Virtual Reality Simulation Trough Product Lifecycle, Proceedings of IMECE: International Mechanical Engineering Congress and R&D Expo Nov. 15 - 21, Washington DC.

Intelligent Production Machines and Systems
D. T. Pham, E. E. Eldukhri and A. J. Soroka (eds)

Identifying defects on plywood using a minimum distance classifier and a neural network

M.S. Packianather[a], P.R. Drake[b]

[a] *Manufacturing Engineering Centre, School of Engineering, Cardiff University, Queen's Buildings, The Parade, Newport Road, Cardiff CF24 3AA, UK*
[b] *Management School, University of Liverpool, Chatham Street, Liverpool L69 7ZH, UK*

Abstract

This paper describes the application of a minimum distance classifier and a neural network in identifying defects on plywood. The performance achieved by these two classifiers in this study has been used to compare the two methods for classification tasks. While the neural network misclassification fell to 13.5% that of the minimum distance classifier remained at 37% showing the superiority of the neural network. This is due to its inherent ability to deal with nonlinearity and create soft decision boundaries to separate pattern classes, thus making it a more efficient and intelligent classifier.

1. Introduction

Identifying defects on plywood which is the application considered in this paper is a problem suitable for some form of a pattern classifier. The classifiers chosen for this study are a minimum distance classifier (MDC) [1] and a neural network (NN) [2-5] which are examples of relatively simple and more intelligent classifiers respectively and are referred to as wood veneer inspection minimum distance classifier (WVIMDC) and wood veneer inspection neural network (WVINN) in this paper. The MDC and NN are described in sections 2 and 3. The wood veneer application considered in this study is explained in section 4, and the training of WVIMDC and WVINN are included in section 5. Finally, the conclusion is given in section 6.

2. The minimum distance classifier (MDC)

Any classifier can be viewed as a function which provides a mapping of the n -component features of a pattern into one of several pre-specified classes. This decision function is responsible for separating patterns which belong to different classes. The MDC operates on the discriminant function concept. This is explained here below. The basic building blocks of a generic classifier are shown in Figure 1.

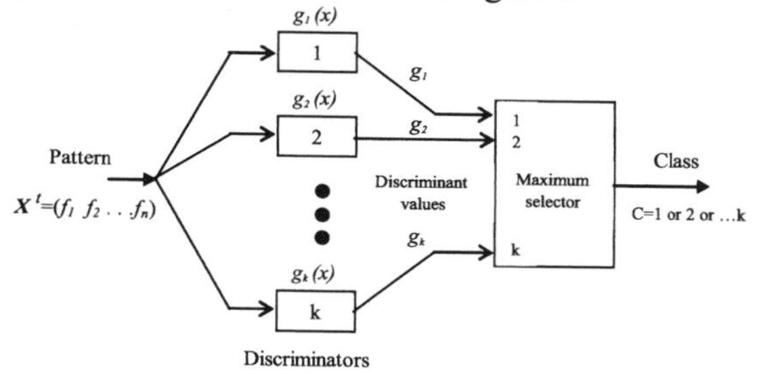

Fig. 1. A general classifier.

Building an MDC for a given set of patterns involves establishing the discriminant functions with which the classifier computes the discriminant values for a given pattern. These scalar values are passed on to the maximum selector for class assignment. The given pattern is classified as class 'i ' if and only if the i th discriminant function has the largest value. This is similar to the "winner takes all" method described in the literature [3].

2.1. Calculating the weights of the MDC

Since the discriminant function is of special importance in the building of a MDC, it is explained here in detail. It is assumed here that this function is linear. Let us assume that there are K distinct classes and each class is represented by a cluster of patterns. Each cluster is more simply represented by a "prototype" which is computed as the centre of gravity of the cluster. This computation is carried out using the patterns contained in the training set. The prototypes of each cluster are denoted by vectors $X_1, X_2, ..., X_k$. Let a given pattern, whose class is to be identified by the classifier, be denoted by vector X. The Euclidean distance between the given pattern and the i th class is expressed by the norm of the vector $X - X_i$ as follows:

$$\|X - X_i\| = \sqrt{(X - X_i)^t (X - X_i)}\text{, for } i=1,2,\ldots,k \quad (1)$$

The MDC computes the distance from the given pattern X to the prototype of each class. The unknown class of the given pattern will be decided by the closest class of the closest prototype. The equation above can be written as:

$$\|X - X_i\|^2 = X^t X - 2X_i^t X + X_i^t X_i \text{ ,for } i=1,2,\ldots,k \quad (2)$$

The term $X^t X$ in the above equation is independent of any class. Therefore the unknown class will be determined by the value of i for which the term $-2X_i^t X + X_i^t X_i$ is the smallest. This is redefined here as being determined by the value of i for which $2X_i^t X - X_i^t X_i$ or more simply $X_i^t X - 0.5X_i^t X_i$ is the largest. This property can now be used to describe the discriminant function:

$$g_i(x) = X_i^t X - 0.5X_i^t X_i\text{, for } i=1,2,\ldots,k \quad (3)$$

This equation shows that the discriminant function is of the general linear form which can be expressed as:

$$g_i(x) = D_i^t X + D_{i,n+1}\text{, for } i=1,2,\ldots,k \quad (4)$$

where $X^t = (f_1 f_2 \cdots f_n)$ contains n input features. The discriminant function coefficients, D_i, can be determined by comparing the two equations above as follows:

$$D_i = X_i \quad (5)$$

$$D_{i,n+1} = -0.5X_i^t X_i\text{, for } i=1,2,\ldots,k \quad (6)$$

Since there are $(n+1)$ components in total, by convention a '1' is appended as the $(n+1)$ th component of each pattern vector as they contain only n components. Figure 1 can be modified to produce the MDC as shown in Figure 2. More details on the MDC can be found in [1].

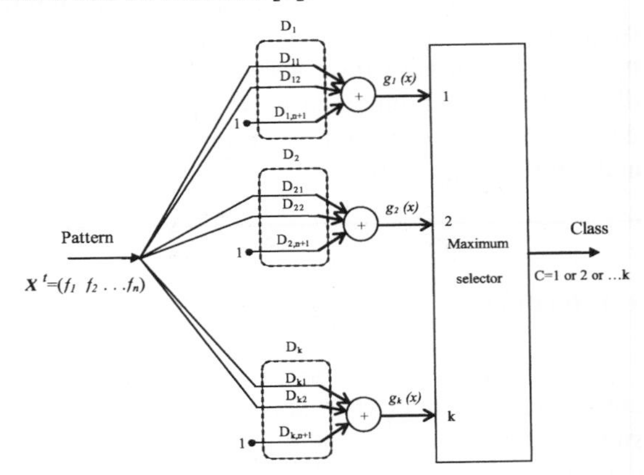

Fig. 2. Minimum distance classifier.

All of the $K*(n+1)$ weights required to build the MDC can be calculated as described above for the patterns contained in the training set.

3. Neural network (NN)

Neural networks are computational models inspired by the human brain. The multilayered feedforward neural network has an input layer to receive inputs from sensors or other sources, an output layer to communicate with the outside world and one or more hidden layers for data processing to transform the inputs into outputs. The architecture of a feedforward neural network with one hidden layer is shown in Figure 3. Each layer is made up of processing elements called neurons. Every neuron has a number of inputs, each of which must store a connection weight to indicate the strength of the connection. Connections are initially made with random weights. The neuron sums the weighted inputs and computes a single output using an activation function. A number of different activation functions can be used. In this study, a hyperbolic tangent function is used in order to increase the difference between the outputs. Each neuron in a layer is fully connected to every neuron in the subsequent layer forming a fully connected feedforward neural network. In a feedforward neural network, information flows from the input layer to the output layer without any feedback. Feedforward nets are guaranteed to reach stability and are faster than feedback nets [4]. There is one bias neuron for each hidden layer and the output layer, as illustrated in Figure 3, and they are connected to each neuron in their respective layer. These connections are treated as weights.

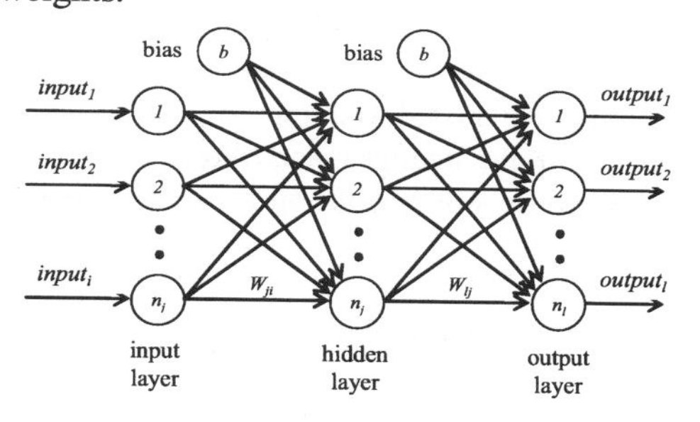

Fig. 3. Feedforward NN with one hidden layer.

The equations that describe the multilayered feedforward network training and operation are divided into two categories:

i. feedforward calculations;
ii. error back-propagation calculations.

3.1. Feedforward calculations

The feedforward calculations are used both in the training mode and in the recall operation of the trained neural network. The error back-propagation calculations are applied only during the training phase. The input neurons simply distribute the signal along the multiple paths to the hidden layer neurons. A weight is associated with each connection. All connections and data flow from left to right are shown in Figure 3. The way a neuron calculates its total input (*net_input*), and its output (as a function of its *net_input*), depends on the type of neuron and its activation function.

The feedforward values for a hidden layer neuron are calculated using the following equations:

$$net\_input_j = I_j = \sum_{i=1}^{n_i} W_{ji} O_i \tag{7}$$

$$output_j = O_j = f(net\_input_j) = f(I_j) = \frac{1}{1+e^{-I_j}} \tag{8}$$

where $i = 1...n_i$ defines an input neuron and $j = 1...n_j$ defines a hidden neuron.

The hyperbolic tangent function in (9) used in this study limits the output values between -1 and 1.

$$f(x) = \frac{e^x - e^{-x}}{e^x + e^{-x}} \tag{9}$$

Similarly, for a neuron in the output layer net_input_l and $output_l$ are calculated using the same equations as above:

$$output_l = O_l = f(I_l) \tag{10}$$

where $l = 1...n_l$ defines an output neuron.

The first derivative of this function $f'(I_l)$ is important in the error back-propagation algorithm. During the training phase, the feedforward calculation is combined with backward error propagation to adjust the weights. The error term for a given pattern is the difference between the desired output and the actual output (the output from feedforward calculation). The error calculation used in this study is as follows:

$$E_p = \sum_{l=1}^{n_l} \left(t_{pl} - O_{pl}\right)^2 \tag{11}$$

where E_p is the error of pattern **p**; $l = 1 \ldots n_l$ is the output neuron; t_{pl} is the desired (target) output for neuron l; O_{pl} is the actual output of output neuron l.

The *average sum of squared error (ASSE)* of all neurons in the output layer and over all the training patterns is given by:

$$ASSE = \frac{1}{P \times n_l} \sum_{p=1}^{P} E_p \qquad (12)$$

where P is the total number of patterns.

In this study the error was averaged after every 10 iterations and the learning was stopped when $ASSE < \varepsilon$, where a value of $\varepsilon = 0.03$ was deemed suitable for this work. This limiting value was used to stop the neural network from over-training, and was determined experimentally by monitoring the performance of three randomly selected neural network designs.

The error produced by the network for a pattern **p** is a function of weights alone since the input is fixed [5]. The back-propagation training algorithm aims to minimise the error (E_p) to an acceptable level by adjusting the weights so that a point on the error surface which represents the current state of the network moves in the steepest downward direction. A gradient descent based search method is explained in the following section.

3.2. Error back-propagation claculations

The most widely used method for training feedforward networks is gradient descent with momentum also known as the standard back-propagation algorithm [6]. The training algorithm calculates the direction in n-dimensions (where n is the number of weights that are to be adjusted) of the weight changes that will bring about the greatest reduction in the error function. The amount of the adjustment is controlled by two training algorithm control parameters namely:

i. η - learning rate (the step size along the error surface);

ii. α - momentum (a proportion of the previously calculated weight change).

The connection weights are changed according to this equation:

$$\Delta W_{ji}(new) = \eta * \delta_j * output_i + \alpha * \Delta W_{ji}(old) \qquad (13)$$

where δ_j for the neurons in the output layer and hidden layer is defined by:

$$\delta_j = f'(I_j)(t_j - O_j) \qquad (14)$$

$$\delta_j = f'(I_j) \sum_k \delta_k W_{kj} \qquad (15)$$

The smaller the learning rates, η, the smaller the weight changes from one iteration to the next will be and therefore the trajectory along the error surface will be smooth. Although this is desirable it is attained at the cost of a slower rate of convergence (learning). For high learning rates the weight changes are high and the network is able to train faster. However, if the learning rate is too large this will cause oscillation and the network will become unstable. An effective way of dealing with this problem is the addition of the momentum term. The momentum, α, is used to damp (stabilise) the high-frequency weight changes and reduce the risk of oscillation while still permitting fast convergence.

4. Wood veneer inspection application

Plywood is formed by bonding together a number of thin layers of wood, called veneers, using an adhesive. Veneer sheets can have defects which will produce poor quality plywood when the sheets are glued together. Identifying defects on surfaces such as wood has been pursued by researchers for many years. For example, Automated Visual Inspection (AVI) systems for identifying defects have been proposed by Pham and Alcock [7]. The generic AVI system is illustrated in Figure 4. The principles behind any such system include the use of image processing techniques, feature extraction to capture the essential characteristics of all defects and training a classifier to recognise these defects.

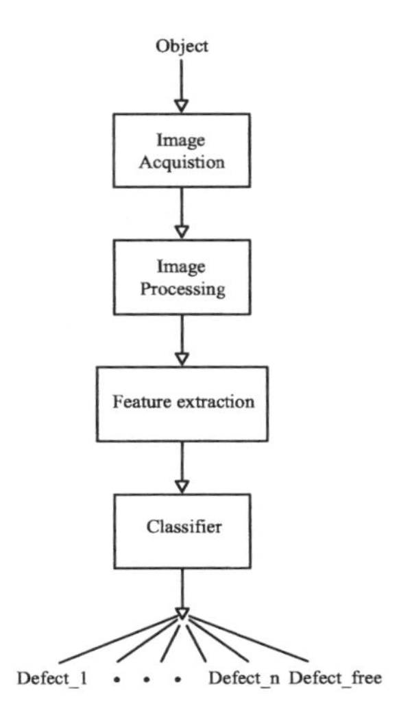

Fig. 4. Generic AVI system for defect identification.

In this study, using a charge-coupled device (CCD) matrix camera, the wood veneer defects were captured and stored on a digital computer. These images were converted into grey level histograms after applying segmentation and image processing algorithms. From the first and second order statistical features extracted from the histogram, 17 features were selected for training the WVIMDC and WVINN. All together 12 wood veneer defects and clear wood shown in Figure 5 were included in the examples used for training and testing the WVIMDC and WVINN.

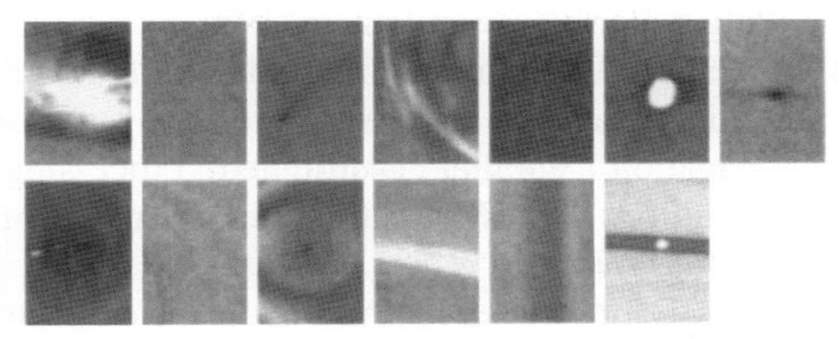

Fig. 5. Birch wood veneer samples – 1st row left to right: bark, clear wood, coloured streak, curly grain, discoloration, hole, pin knot; 2nd row left to right: rotten knot, roughness, sound knot, split, streak, worm hole.

5. Training the classifiers

For the application studied here, 232 examples (defects and clear wood) were employed, as detailed in Table 1. The classification of these examples had been performed by a human inspector. For subsequent classification experiments, for each class, 80% (185 in total) of the examples were selected at random to form the training set and the remaining 20% (47 in total) formed the test set. Three such sets of training and corresponding test data were created and are referred to as RS1, RS2 and RS3 in this paper.

Table 1
Pattern classes and the number of examples used for training and testing the WVINN.

Pattern class	Total	Used for training	Used for testing
bark	20	16	4
clear wood	20	16	4
coloured streaks	20	16	4
curly grain	16	13	3
discoloration	20	16	4
holes	8	6	2
pin knots	20	16	4
rotten knots	20	16	4
roughness	20	16	4
sound knots	20	16	4
splits	20	16	4
streaks	20	16	4
worm holes	8	6	2
Total	232	185	47

5.1 Training the wood veneer inspection minimum distance classifier (WVIMDC)

A Wood Veneer Inspection Minimum Distance Classifier (WVIMDC) was constructed in order to identify the defects. The MDC was built using all the wood veneer examples for the twelve defects and clear wood (given previously in Table 1).

The weights for the WVIMDC were calculated for the patterns in the training set in RS1, RS2 and RS3 as explained in section 2.1. A computer program was written in C to compute these weights and test the WVIMDC. The performance of the WVIMDC which was tested with the test examples in RS1, RS2 and RS3 is given in Table 2. The average misclassification of the WVIMDC is 36.88%.

5.2 Training the wood veneer inspection neural network (WVINN)

A suitable wood veneer inspection neural network was designed with seventeen neurons in the input layer, thirteen neurons in the output layer and one hidden layers with 51 neurons. The number of neurons in the input and output layers correspond to the number of features in the training data and the total number of classes which include twelve defects and clear wood. The number of hidden layers and the number of neurons in each hidden layer were determined by Taguchi Design of Experiments [8] and Response Surface Methodology [9]. Using these two methods the neural network topology was set to one hidden layer with 51 neurons, and the neural network control parameters which are the learning rate and momentum were optimised to 0.01 and 0.1 respectively.

The performance of the WVINN with its design parameters set at their best values described above are given in Table 2. Only the best of the three of the nine trials conducted for the WVINN with RS1, RS2 and RS3 are given in this table. The average misclassification of the WVINN is 13.48%.

Table 2
The performance of the WVIMDC compared with the WVINN for the test examples in RS1, RS2 and RS3.

	% Misclassification	
Test examples	WVIMDC	WVINN
RS1	36.17	10.64
RS2	31.91	8.51
RS3	42.55	21.28
Average	36.88	13.48

The results show that the NN based classifier has better generalisation capability compared to the MDC which is a "conventional classifier". The results obtained by both classifiers on RS1, RS2 and RS3 display the same trend in its performance.

6. Conclusion

The application of a minimum distance classifier and neural network for identifying defects on plywood has been presented in this paper. Firstly, a minimum distance classifier has been chosen to investigate the nature of the wood examples, and the construction of a WVIMDC has been explained in detail. Secondly, a neural network classifier has been selected to identify defects present in the wood veneer data, and a standard multilayered feedforward neural network back-propagation model has been described. Thirdly, the wood veneer application has been introduced. Fourthly, the performance of these two classifiers has been reported. Finally, the performance of these two classifiers has been compared and the results show that the efficiency of the WVINN is higher than that of the WVIMDC. This shows that the neural network behaves like a non-linear machine by having more than one layer of weights, and can be trained to learn the non-linear discriminating functions between the different classes.

Acknowledgement
The first author is a member of the I*PROMS Network of Excellence funded by the European Commission.

References

[1] Zurada JM. Introduction to Artificial Neural Systems. West Publishing Company, USA, 1992.

[2] Pham DT and Liu X. Neural Networks for Identification, Prediction and Control. Springer-Verlag, London, 1997.

[3] Haykin S. Neural networks: a comprehensive foundation. Prentice Hall, NJ, 1999.

[4] Nelson MM and Illingworth WT. A Practical Guide to Neural Nets. Addison-Wesley, Massachusetts, 1991.

[5] Taylor TG. The Promise of Neural Network. Springer-Verlag, Great Britain, 1993.

[6] Rumelhart DE Hinton GE and Williams RJ. Learning Internal Representations by Error Propagation. Parallel Distributed Processing: Explorations in the Microstructures of Cognition. 1 (1986), 318-362.

[7] Pham DT and Alcock RJ. Automatic Detection of Defects on Birch Wood Boards. Proc. I Mech E, Part E, J. of Process Mechanical Engineering. 210 (1996), 45-52.

[8] Packianather MS Drake PR and Rowlands HR. Optimizing the Parameters of Multilayered Feedforward Neural Networks through Taguchi Design of Experiments. Quality and Reliability Engineering International. 16(6) (2000), 461-473.

[9] Packianather MS and Drake PR. Modelling Neural Network Performance through Response Surface Methodology for Classifying Wood Veneer Defects. IMechE Part B, Journal of Engineering Manufacture. 218 (2004), 459-466.

Intelligent Production Machines and Systems
D. T. Pham, E. E. Eldukhri and A. J. Soroka (eds)

Issues concerned with the spatial optimisation of shapes

R. La Brooy[a], C. Jiang[b], W. Cheetham[c] Zhang, M[d],

[a]*PhD, FIEAust., Associate Professor of Advanced Manufacturing Engineering, RMIT, Australia,*
[b] *PhD, Research Engineer, Object Consulting, Australia.*
[c]*PhD, Research Engineer, Vision Systems, Australia.*
[d] *Ph,D Research Engineer, University of Newcastle, Australia.*

Abstract

Intelligent Manufacturing involves the use of enabling sciences to augment the manufacturing process. Nesting is a process where optimization of available materials can lead to millions of dollars in savings when the cost of the substrate is high. The paper addresses key issues when optimising shape placement on expensive parent substrates such as carbon fibre or leather. The outcomes attained, method of part and plate representation, objective function definition, overlap calculation and optimization dependence of initial parts organization in the Simulated Annealing optimization process are presented.

1.0 Introduction

There are many instances where the placement of shapes on laminae needs to be optimised. Often the issue is related to the broader requirement to minimise material wastage when placing these shapes on an expensive substrate. This was the case when the author directed a program to optimise shape placement on 2d surfaces for the aerospace industry. In order to so do, there were several issues identified as likely to affect the outcome of the work. The complexity of the problem can be depicted by a comprehensive nesting solution to the McDonnell Douglas Explorer (MDX) helicopter fuselage. The complete shipset comprised 1,326 shapes, requiring 22 nests on sheets of pre-impregnated carbon fibre. One such nest is shown in Fig. 1. Each nest was of dimensions 20mx2m and was produced for Boeing by the authors. Clearly such a process comprising thousands of parts had to be fully automated.

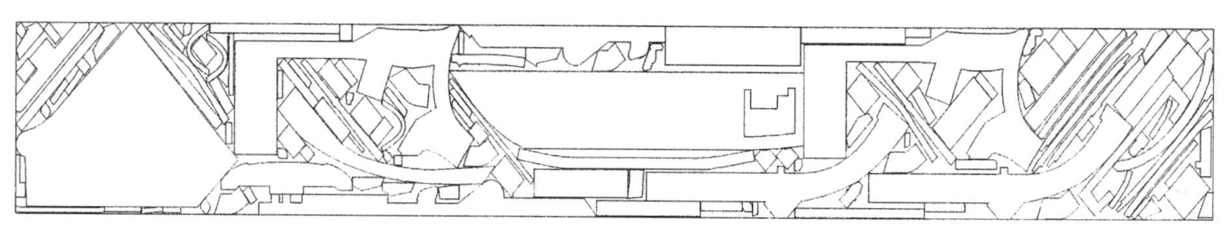

Fig 1. Example of McDonnell Douglas Explorer helicopter fuselage

2.0 Problem Definition

The authors have deemed it vital that every part be automatically enveloped by a polygon. The degree of approximation is user-defined and relies on an automatic generation of a net around the part. The number of nodes used is also automatically established for given levels of approximation. Fig. 2 shows a forbidden (no-go) zone where nesting must not proceed. Such instances occur when nesting on leather where hides are scratched or when manufacturing flaws in rolls of carbon fibre preclude the placement of shapes in the defective region.

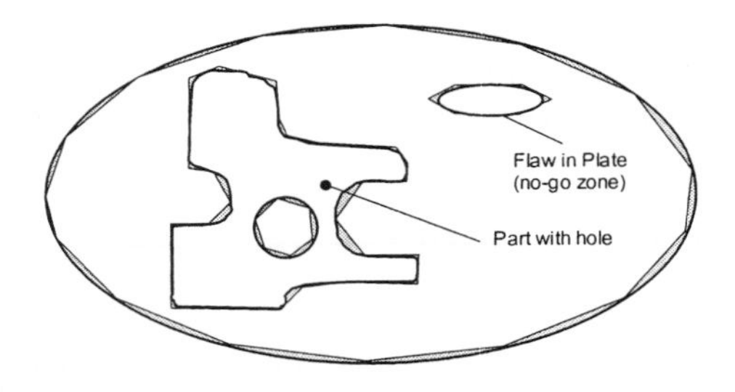

Fig. 2. Automatic Net Generation around shapes and plate.

Fig. 3 shows a case where the parent plate is highly irregular. A result of the authors' program for an elliptical nest is shown in Fig. 4.

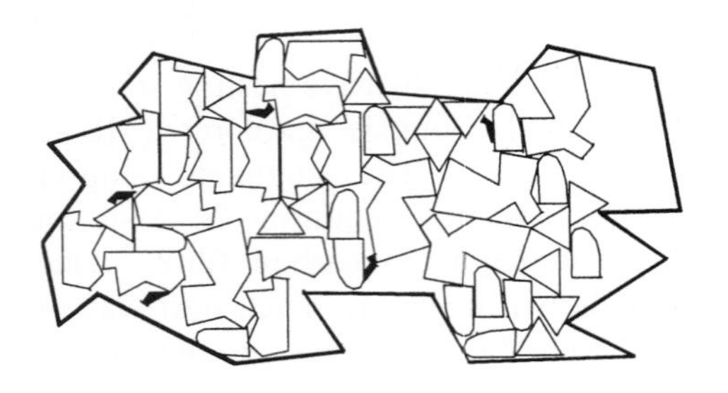

Fig. 3. Irregular plate

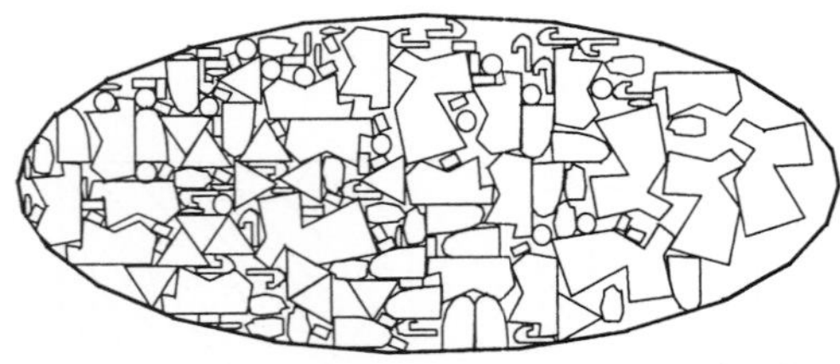

Fig. 4. Resulting authors' nest from fig. 2

When parts are laser-cut, the circumscribing part polygons may need to be nested with finite gaps between shapes. This occurs when common line cutting results in excess kerf. Here all polygons are spaced from each other and entities forming the polygon need to be *grown* from their original shape and gaps suitably filled. The situation is depicted in figure 5.

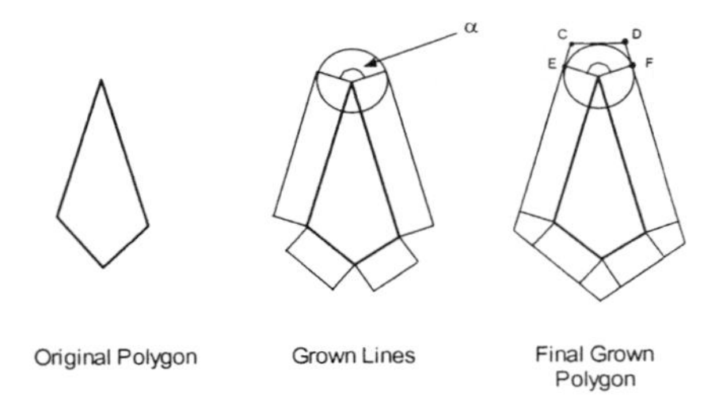

Fig. 5; Growing a polygon.

Additionally, part characteristics must encompass allowable degrees of rotation[1] and other orientation characteristics such as mirroring or flipping.

3.0 Objective Function

Clearly the objective function must include key characteristics of an optimum solution. A critical element is a dominant term that determines the nature of part distribution[2]. To illustrate the point, the authors determined that the second moment of area would be a key term in the objective function (Φ). In evaluating its efficacy, the objective function corresponding solely to this issue was:

$$\Phi = \Sigma A_i r_i^2 \quad (1)$$

Equation (1) was calculated for the nest. Here areas A and a centroidal distance from a focal point r from the top left side of the plate were dynamic part attributes. When invoked, the resulting nest for small parts is depicted in Fig. 6.

[1] Vital for nesting on carbon fibre or with patterned materials.
[2] Other terms can include overlap toleration with internal or external boundaries.

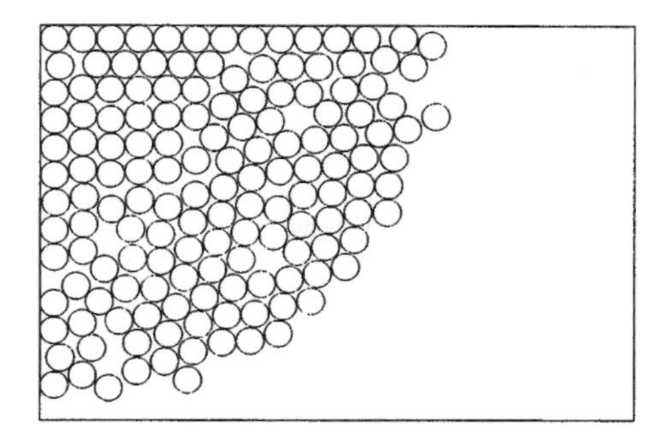

Fig. 6. Resulting optimum nest when minimizing $\Sigma A_i r_i^2$

It is apparent that when optimising $\Sigma A_i r_i^2$, a pre-disposition to a circular parts distribution results. This result was an impractical solution due to the remnant's shape. The authors then had to identify the inverse of the problem: i.e. what function needed to be integrated to produce a *rectangular* distribution of parts- and consequently, a rectangular, usable remnant. This led to an investigation into the use of even order terms in superquadrics. Fig. 7 shows how when the power n in the function $y=x^n + y^n$ is kept even and increases, a basic circle distorts to a square-like shape. It then follows that if a large value of n is used, the function would closely resemble a rectangular parts distribution.

The drawback with using higher order terms is the additional computational effort required, and the extremely high magnitudes of the term that results. These high values can however be normalised before use in the objective function, although more computational effort is needed.

4.0 Overlap between shapes

It is critical to determine rapidly, the degree of overlap between adjacent polygons on a two-dimensional plane[3] when the optimisation process drives parts to intermediate locations.

Consider a case where three parts comprise of polygons of l, m and n sides respectively. If overlap detection for all sides of all polygons was made, (lm+nm+ln) calculations are required to test whether every line comprising each of the three polygons intersects another. Most practical part enveloping polygons can have in excess of 50 sides. In order to reduce this complexity of calculation for every iteration of an optimisation process, the authors enveloped every part with a dynamic rectangular *bounding rectangle*.

If the general configuration of p polygons representing parts are now enveloped by individual bounding ***rectangles***, $[(p-1)+ (p-2)+ \ldots + 1)]x4^2 = \{(p-1)xp/2\}x4^2]$ calculations only are required, irrespective of the number of line segments comprising a contained part. In the 3 polygon example, $\{(3-1)x3/2\}x4^2$ will be << (lm+mn+ln). The authors then resorted to testing the intersection of bounding rectangles, determining whether these overlapped, then determined whether the shapes themselves overlapped, before resorting to checking individual line overlap when warranted.

5.0 Optimisation method

The authors used Simulated Annealing as the method of optimization. Here, the objective function Φ is evaluated for fitness according to the acceptance criterion:

$$e^{-\Delta\Phi/T} > \text{Random}[0,1) \qquad (2)$$

One part in the current nest was perturbated in a manner allowable for that part. The change in objective function for the entire nest was calculated and the acceptance criterion evaluated. The variable T was set high initially and reduced according to a *cooling schedule*. The process permitted worse, configurations to be accepted initially but automatically rejected these configurations as T was reduced.

6.0 Initial Configuration

The authors have found that the solution quality was dependent on the initial sequence by which parts were introduced to the plate.

[3] Warped planes in 3-d can be formed from 2-d shapes laid in3d forms.

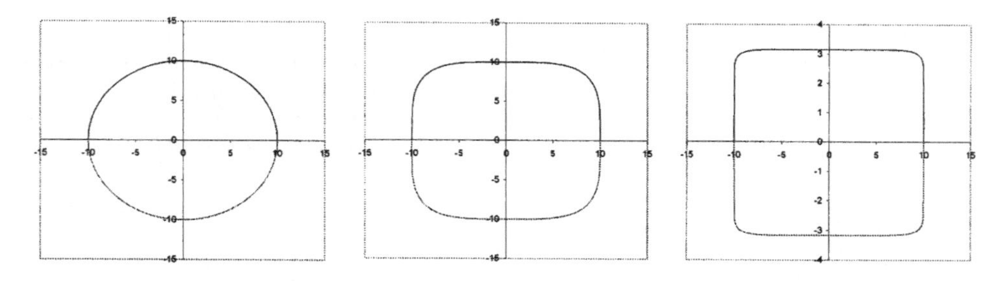

Fig. 7: Circular and Super-quadric distributions $x^2 + y^2$; $x^4 + y^4$ and $x^8 + y^8$ respectively for an ideal plate whose radius is 10.

Fig. 8 shows an example of the manner in which the authors automatically initiated the procedure. Note that parts are aligned, in sequence with large parts leading and located at the edge of the plate. In some instances it was beneficial to align parts based on their bounding rectangle widths.

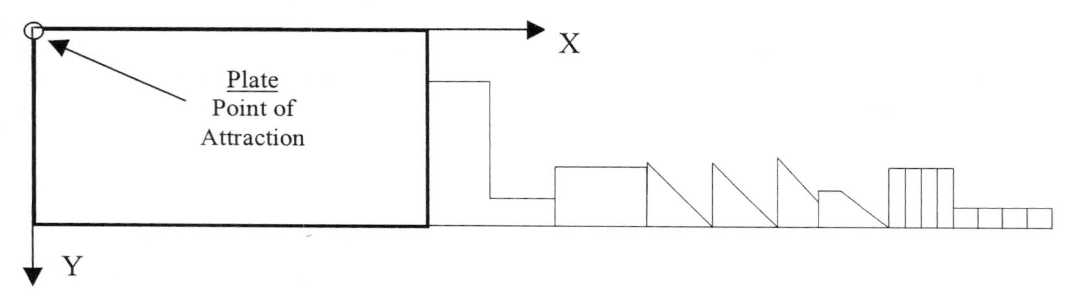

Fig. 8. Alignment of parts prior to initiating Simulated Annealing

7.0 Conclusion

The authors have outlined a highly successful method used to optimise parts layout for the aerospace industry. The manner by which parts are described and a rapid estimation of the degree of part overlap are critical to the convergence of the optimisation method. The authors have used Simulated Annealing as the preferred method of optimisation though the use of Genetic Algorithms was also tried. A description of Simulated Annealing in this context is the subject of a second paper in this series.

Acknowledgements

The authors acknowledge the $1.6 million research project, funded by the Australian Government's world-class CRC Program in Intelligent Manufacturing Systems & Technologies.

Bibliography

[1] AARTS & KORST, 1989, *Simulated Annealing and Boltzmann Machines*, John Wiley & Sons, Inc.; Brisbane

[2] AARTS, E.H.L., KORST, J.H.M. AND VAN LAARHOVEN, P.J.M., 1988, *A quantitative analysis of the simulated annealing: a case study for the travelling salesman problem*, J. Statistical Physics, Vol. 50, No. 1/2 pp 187-206.

[3] ADAMOWICZ, M., AND ALBANO, A., 1976, *Nesting two-dimensional shapes in rectangular modules*, Computer Aided Design, Vol. 8, No. 1 pp 27-33

[4] ALBANO, A., 1977, *A method to improve two-dimensional layout*, Computer Aided

Design, Vol. 9, No. 1, Jan-'77 pp 48-52

[5] ALBANO, A., AND SAPUPPO, G., 1980, *Optimal allocation of two-dimensional irregular shapes using heuristic search methods*, IEEE Transactions on Systems, Manufacturing and Cybernetics, Vol. 10, No. 5, May-'80 pp 242-248

[6] CHEETHAM, W, *On the Optimal Placement of 2S Shapes on 2D Laminae*, PhD, 1999, RMIT University, Australia

[7] DOWSLAND, K. A. AND DOWSLAND, W. A., 1992, *Packing problems*, European Journal of Operational Research, Vol. 56 pp 2-14

[8] DOWSLAND, K.A., 1996, *Genetic Algorithms - A tool for OR?*, Journal of Operational Research Society, Vol. 47, No. 4, Apr-'96 pp 550-561.

[9] DOWSLAND, KATHRYN A., 1993, *Some experiments with simulated annealing techniques for packing problems*, European Journal of Operational Research, Vol. 68 pp 389-399.

[10] DYCKHOFF, HAROLD, 1990, *A typology of cutting and packing problems*, European Journal of Operational Research, Vol. 44, pp145-159.

[11] GOLDBERG, D.E., 1989, *Genetic Algorithms*, Addison-Wesley, Reading, MA .

[12] ISMAIL, H.S., AND HON, K.K.B., 1992, *New approaches for the nesting of two-dimensional shapes for press tool design*, International Journal of Production

[13] AIN, P., FENYES,P. AND RICHTER, R., 1992, *Optimal blank nesting using Simulated Annealing*, Journal of Mechanical Design, Vol. 114, Mar-'92 pp 160-165.

[14] KIRKPATRICK, S., 1984, *Optimization by simulated annealing: Quantitative studies*, J. Statistical Physics, Vol. 34, No. 5/6 pp 975-986.

[15] KIRKPATRICK, S., GELATT, C.D. AND VECCHI, M.P., 1983, *Optimization by simulated annealing*, Science, Vol. 220, May-'83 pp 671-680.

[16] LAARHOVEN, P.J.M. & AARTS, E.H.L., 1987, *Simulated Annealing: Theory and Applications*, D. Reidal Publishing Company

[17] MARQUES, V.M.M, BISPO, C.F.G. AND SENTIEIRO, J.J.S., 1991, *A system fo the compaction of two-dimensional irregular shapes based on Simulated Annealing*, IEEE (IECON'91) CH2976-9 pp 1911-1916.

[18] METROPOLIS, N., ROSENBLUTH, A.W., ROSENBLUTH M.N., TELLER, A.H. AND TELLER, E., 1953, *Equation of state calculations by fast computing machines*, The Journal of Chemical Physics, Vol. 21, No. 6 pp 1087-109

[19] ZHANG, MM, 2004, *On the application of a generic nesting system with specific application to the aerospace industry*, PhD, RMIT University, Australia.

Intelligent Production Machines and Systems
D. T. Pham, E. E. Eldukhri and A. J. Soroka (eds)

Optimisation of nesting using Simulated Annealing

R. La Brooy[a], W. Cheetham[b], C. Jiang[c], Zhang, M[d]

[a]*PhD, FIEAust., Associate Professor of Advanced Manufacturing Engineering, RMIT, Australia,*
[b]*PhD, Research Engineer, Vision Systems, Australia.*
[c] *PhD, Research Engineer, Object Consulting, Australia.*
[d] *Ph,D Research Engineer, University of Newcastle, Australia.*

Abstract

Simulated Annealing (SA) is a recognized optimization process. The paper describes the process of Simulated Annealing that was used to optimise shape placement on expensive parent substrates such as carbon fibre or leather. This paper will also outline issues concerned with enabling SA to produce quality solutions in real time.

1.0 Introduction

Simulated Annealing is the optimising procedure mimicking the attainment of a low energy state of atoms caused by the slow annealing[1] of a metal. The optimisation algorithm simulates this physical procedure by reducing an analytic variable T, (directly analogous to temperature), systematically. There is then an expectation that the objective function (Φ) will reduce correspondingly. The scheme uses random numbers cleverly, to assist convergence of the algorithm.

Metropolis [1] introduced the criterion in accordance with the hill-climbing ethos of iterative improvement algorithms. It was stated that at a given Temperature T, if an objective function variable is perturbated causing a change in the function ΔΦ, then the new variable's value is accepted if:

$$e^{-\Delta\Phi/T} > \text{Random}[0,1) \qquad (1)$$

A scheme presented by Kirkpatrick et. al. [2, 3] blended the Metropolis criterion with extensive work and is now known as Simulated Annealing (SA). The authors have used this scheme to produce nests for aerospace applications. Fig. 1 is a nest produced for Boeing.

To demonstrate operation of the criterion, consider a perturbation in the early stages of a *minimisation*: T will be set high initially, (eg. T=2000). If a new objective function Φ_{k+1} is evaluated that is an improvement on the previous function Φ_k, the LHS of the acceptance criterion (1) will *always* be >1 and hence true. Consequently the new physical configuration[2] s_{k+1} creating the new objective function Φ_{k+1} will always be accepted.

[1] Metals are subject to annealing when heated to a high temperature and then cooled extremely slowly. The process allows strengthening of the metal by allowing movement across grain boundaries, or the very dissolution of these boundaries.

[2] For every physical configuration s_k, there will be a corresponding objective function Φ_k.

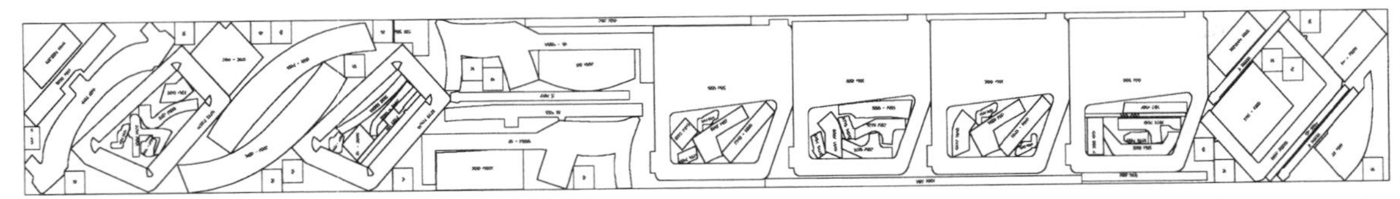

Fig. 1
The authors' use of Simulated Annealing to nest fuselage components of the *Mc Donnell Douglas Explorer* helicopter. Part identification markings are incorporated on each part.

Eg. If $\Phi_2 - \Phi_1 = \Delta\Phi = -100$, the LHS of (1) becomes:

$e^{100/2000} \cong 1.05$
which will always be true as Random[0,1) > 1

The criterion is intelligent in that if a configuration is produced with an increased objective function, it will have a high likelihood of acceptance provided *T is still large.* To demonstrate;
Now if $\Phi_2 - \Phi_1 = +100$, the LHS of (1) becomes;

$e^{-100/2000} \cong 0.95$

Unlike the first case, the probability of generating a random number less than 0.95 in the interval [0,1) is high but not certain, rendering the probability of accepting this new, non-optimal configuration at 95%. The probability of accepting such non-optimal configurations is high so long as T remains high. When T decreases, a very different picture emerges. Table 1 shows two new cases of bad configurations when T decreases to 100 and then 2:

Table 1
Key data from example

$\Delta\Phi = \Phi_2 - \Phi_1$	T	$e^{-\Delta\Phi/T}$
+100	100	≈0.37
+100	2	$\approx 2\times10^{-22}$

At T=2, the acceptance chance of a bad objective function is negligible. Hence the criterion encourages acceptance of both good and marginally adverse solutions in the early stages of optimisation when T is high. The situation is analogous to atoms in an annealing metal moving at random, settling temporarily and re-manoeuvring to reach an improved configuration.

Two key questions need to be resolved:

(a) How is the onset of a temperature reduction signalled?

(b) How then should the temperature reduce?

The objective function Φ_k must be quantified analytically by terms describing the present configuration s_k. A 'perturbation" results when a single variable has been changed, resulting in a new configuration s_{k+1}. Each accepted configuration, mapping to Φ_k, generally leads to an improved objective function. Each new perturbation is launched from the last accepted configuration.

2.0 The Markov chain length

A sequence of accepted transitions from a starting configuration can be connected through the search space until a rejection occurs. The *number* of *consecutive* transitions forming the sequence is termed the Markov chain length L_k at the temperature T_k. When the chain is long, there has been a large opportunity to search the solution space. The nomination of quasi-equilibria and hence the chain length can be specified by either:

- Fixing the total no. of perturbations at one temperature:
- Counting the number of *consecutive* rejected configurations.

The authors prefer specifying the onset of quasi-equilibria by detecting when the configuration becomes semi-stationary. The number of *consecutive* rejected configurations used is dependent on the complexity of the problem, the allowable move set and the objective function value as described by Cheetham [4] and Zhang [5].

3.0 The Cooling Schedule

The Cooling Schedule is the scheme by which the temperature is reduced during the execution of the SA algorithm. A cooling schedule in physical annealing dictates the rate of reduction of temperature directly affecting the grain structure of annealed material. If the material is cooled too quickly (quenching), then the material will contain dislocations (frozen in time) and grain boundary entanglements. Fast cooling therefore results in little dislocation removal, and results in a coarse, brittle product. In SA, quenching is akin to a fast iterative improvement algorithm that may only accept improvements in objective function to get a quick solution. When a material is cooled slowly, time is provided for the microstructure to realign itself and reduce the number of dislocations and entanglements. If enough time is allowed for cooling, residual stresses may be substantially removed achieving the lowest ground state (global minimum). SA is effective when made analogous to slow annealing and the cooling schedule is directly related to the manner by which stress relief is controlled.

Clearly, T is the driving potential and must be *reduced* in a systematic manner to encourage convergence appropriately. The authors have suggested that the temperature be reduced in a stepwise fashion, whenever the objective function reaches a *quasi-equilibrium*. The controlled reduction of temperature is termed a "cooling schedule" and an elementary schedule depicted in fig. 2 decrements temperature by:

$$T_{k+1} = 0.9^n T_i \qquad (2)$$

The schedule is plotted for the first 29 quasi-equilibria.

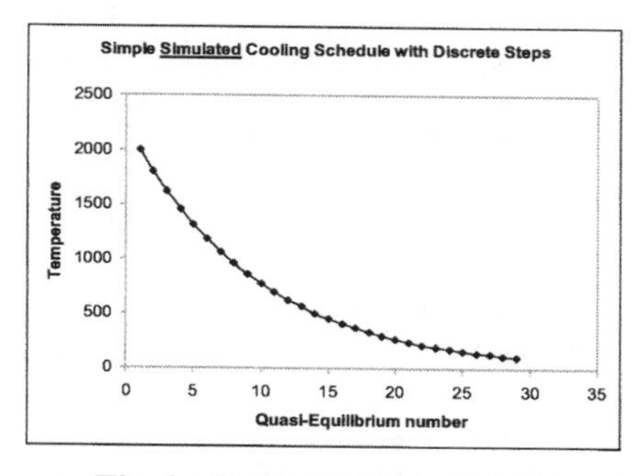

Fig. 2: Cooling schedule for T

The authors now show how the simple analytic function **$\Phi=x^2$** can be minimised over the range; **$-20 \le x \le 15$**. Figure 3 shows how SA and its Markov chains operate.

Particularly, Figure. 3b shows a map of the Markov chain lengths at the end of each quasi-equilibrium over the entire SA evaluation for this simple analytic function.

It should be noted that analytically, the solution is Φ_{min} occurs when $\partial\Phi/\partial x = 0$ at $x=0$ and is by far the simplest solution. This example is used to illustrate the relationship between formation path of Markov chains and cooling schedules using Simulated Annealing that can operate for non-analytic problems.

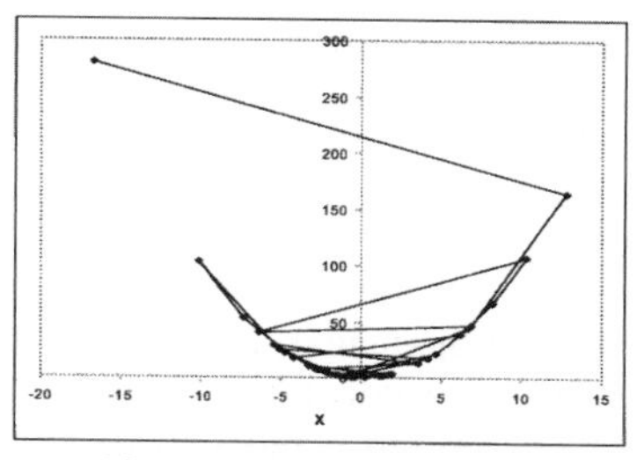

a) SA configuration path for entire SA run.

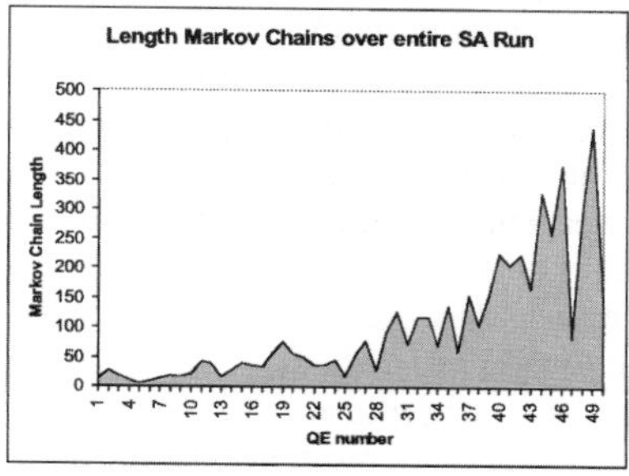

b) Objective function path for entire SA run.

Fig. 3: The path of SA and the corresponding Markov chains over the simulation

There are several factors that are inter-related in using simulated annealing effectively:

They are:

- The initial starting configuration of the parts:
- The initial temperature:

- The cooling schedule:
- Objective function-terms;
- Dominant objective function terms and the
- Allowable move set:

4.0 The move set

According to Aarts & Korst [6], theoretically, the starting configuration, or initial configuration, is independent of the global minimum. The authors have found that there is a dependency of solution on the initial configuration as did Marques et. al. [7].

Now the set of possible configurations that may be attained by perturbation of configuration s_i (to configuration s_j) will be defined by the *neighbourhood* $\boldsymbol{S_i}$ or *configuration space* as in Laarhoven & Aarts [8].

A neighbourhood r_i is a possible move or transition from one solution to another. The set of all neighbourhoods forms the move set $\boldsymbol{R}$.

Typically the algorithm randomly choses a move type to define a new iteration. The move set size will vary depending upon the application. By having fewer neighbourhoods, there is a greater probability of acceptance of a particular move *type* that may be needed to reach a better solution from the existing case. When there are too few neighbourhoods restricting the direction of the search space, the solution may become trapped in a local optimum.

To avoid solutions becoming 'stuck' whilst encouraging convergence, two different types of neighbourhoods are used;

- Convergent transitions - modifying the solution by small quantities promoting convergence, to solutions that iteratively reduce the objective function.
- Divergent transitions – adjusting the solution considerably, to promote search across a wider range of the solution space.

An effective move set will contain a balance of both types suited to a given application. For example, for the simple problem of minimising $y=x^2$ in the range $x_{min} < x < x_{max}$, a neighbour can be chosen randomly from the following move set of three moves:

$$R_i = \begin{cases} x_i + \delta_k & \text{or} \\ x_i - \delta_k & \text{or} \\ Rand\left(\dfrac{x_{\max} + x_{\min}}{2}\right) \end{cases}$$

where δ reduces each time a quasi-equilibrium k is reached.

In this example, the first two move types are convergent since they promote iterative traversing to a minimum. The third and last move type changes the current value of x to a random point in the solution space. Whilst the later could jump into less beneficial areas, it can also enable jumping closer to the optimum in x. Therefore this move type diversifies the search yet can allow faster convergence and the prospect of extrication from a local minimum. In the example there is no local minimum as the function produces a continuous, simple solution space. The effects can however be pronounced when the objective function is discontinuous as in the nesting of shapes.

5.0 The objective function

The key issue is to model the problem sufficiently well by an objective function, and define a move set promoting sufficient freedom to guide the solution set to a better configuration. The objective function is the principal means by which the quality of configurations is communicated to the SA algorithm. The authors use terms in the objective function that **drive** the solution towards a goal or optimum given that restrictions or constraints must not be violated. Such terms can be incorporated in Φ and can have relative weights assigned to them to balance the function. An alternative to including restriction or penalty terms in the cost function, is to structure a neighbourhood and move set that only allows feasible solutions (such moves are termed "greedy").

The use of penalty terms may permit infeasible or undesirable results to exist in the early stages of a search. Gradual elimination of such occurrences will occur as temperature is reduced. Allowing infeasible solutions is advantageous as it is possible that the global solution may only be visited in the solution space after traversing across a number of infeasible solutions outlined in Marques, Bispo & Sentieiro, [7], and shown in figure 4. The algorithm is therefore given flexibility to move through the search space hunting for high quality regions.

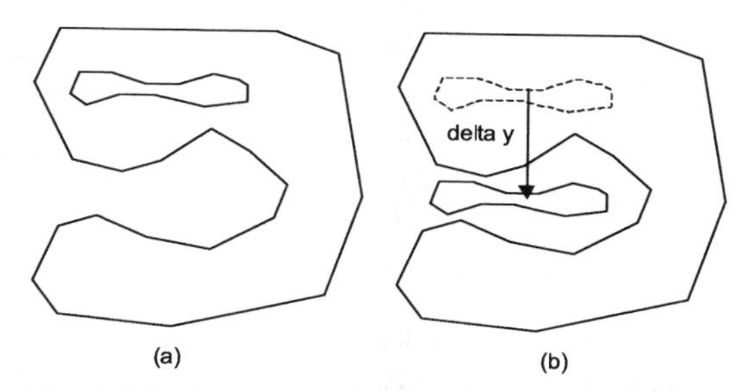

Fig. 4. Visiting the global solution part placement derived from a move from infeasible space.

6.0 Conclusion

In this paper, an overview of the Simulated Annealing algorithm was presented. A skeletal framework for the development of an automated nesting algorithm was outlined. The authors developed these ideas to produce nests of the quality of figure 1 using Simulated Annealing. T

Acknowledgements

The authors acknowledge the $1.6 million research project, funded by the Australian Government's world-class CRC Program in Intelligent Manufacturing Systems & Technologies.

References

[1] METROPOLIS, N., ROSENBLUTH, A.W., ROSENBLUTH M.N., TELLER, A.H. AND TELLER, E., 1953, *Equation of state calculations by fast computing machines*, The Journal of Chemical Physics, Vol. 21, No. 6 pp 1087-1092.

[2] KIRKPATRICK, S., 1984, *Optimization by simulated annealing: Quantitative studies*, J. Statistical Physics, Vol. 34, No. 5/6 pp 975-986.

[3] KIRKPATRICK, S., GELATT, C.D. AND VECCHI, M.P., 1983, *Optimization by simulated annealing*, Science, Vol. 220, May-'83 pp 671-680.

[4] CHEETHAM, W, 1999, *On the optimisation of 2D shapes on laminae*, PhD, RMIT University, Australia.

[5] ZHANG, MM, 2004, *On the application of a generic nesting system with specific application to the aerospace industry*, PhD, RMIT University, Australia.

[6] AARTS & KORST, 1989, *Simulated Annealing and Boltzmann Machines*, John Wiley & Sons, Inc., Brisbane

[7] MARQUES, V.M.M, BISPO, C.F.G. AND SENTIEIRO, J.J.S., 1991, *A system fo the compaction of two-dimensional irregular shapes based on Simulated Annealing*, IEEE (IECON'91) CH2976-9 pp 1911-1916.

[8] LAARHOVEN, P.J.M. & AARTS, E.H.L., 1987, *Simulated Annealing: Theory and Applications*, D. Reidal Publishing Company.

Bibliography

[1] AARTS, E.H.L., KORST, J.H.M. AND VAN LAARHOVEN, P.J.M., 1988, *A quantitative analysis of the simulated annealing: a case study for the travelling salesman problem*, J. Statistical Physics, Vol. 50, No. 1/2 pp 187

[2] ARBEL, AMI, 1993, *Large-scale optimization methods applied to the cutting stock problem of irregular shapes*, International Journal of Production Research, Vol. 31, No. 2 pp 483-500.

[3] ASKELAND, D.R., 1990, *The Science and Engineering of Materials*, Chapman and Hall.

[4] AZENCOTT, R., 1992, *Simulated Annealing: Parallelization Techniques*, John Wiley & Sons, Inc., Brisbane.

[5] BLAZEWICZ, J., HAWRYLUK, P., AND WALKOWIAK, R., 1993, *Using a tabu search approach for solving the two-dimensional irregular cutting problem*, Annals of OR, Vol. 41 pp 313-325.

[6] BROWN, D. E., AND HUNTLEY, C. L., 1992, *A practical application of Simulated Annealing to clustering*, Pattern Recognition, Vol. 25, No. 4 pp 401-412.

[7] CHAUNY, F., LOULOU, R., SADONES, S. AND SOUMIS, F., 1987, *A two-phase heuristic for*

strip packing: Algorithm and probablistic analysis, Operations Research Letters, Vol. 6 pp 1-15.

[8] CHUNG, J., SCOTT, D., AND HILLMAN, D.J., 1990, *An intelligent nesting system on 2-D highly irregular resources*, SPIE Vol 1293 Applications of Artificial Intelligence VIII pp 472-483.

[9] DOWSLAND, K.A., 1996, *Genetic Algorithms - A tool for OR?*, Journal of Operational Research Society, Vol. 47, No. 4, Apr-'96 pp 550-561.

[10] DOWSLAND, KATHRYN A., 1993, *Some experiments with simulated annealing techniques for packing problems*, European Journal of Operational Research, Vol. 68 pp 389-399.

[11] FILHO, J. L. R., TRELEAVEN, P. C., AND ALIPPI, C., 1994, *Genetic-Algorithms programming environments*, Computer, Vol. 27, No. 6, Jun-'94 pp 28-43.

[12] FOX, B.L., 1995, *Faster Simulated Annealing*, SIAM J. Optimization, Vol. 5, No. 3 pp 488-505.

[13] GLOVER, F., 1986, *Future paths for integer programming and links to artifial intelligence*, Comp. Oper. Res., Vol. 13 pp 533-549.

[14] HEISTERMANN, J., AND LENGAUER, T., 1995, *The nesting problem in the leather manufacturing industry*, Annals of OR, Vol. 57 pp 147-173.

[15] INGBER, L., 1989, *Very Fast Simulated Annealing*, Mathematical and Computer Modelling, Vol. 12, No. 8 pp 967-973.

[16] JAIN, P., FENYES,P. AND RICHTER, R., 1992, *Optimal blank nesting using Simulated Annealing*, Journal of Mechanical Design, Vol. 114, Mar-'92 pp 160-165.

[17] OLIVEIRA, J.F., AND FERREIRA, J.S., 1993, *Algorithms for nesting problems*, VIDAL, Rene V.V., "Applied Simulated Annealing",

[18] YAO, X., 1995, *A New Simulated Annealing Algorithm*, Intern. J. Computer Math., Vol 56 pp 161-168.

Intelligent Production Machines and Systems
D. T. Pham, E. E. Eldukhri and A. J. Soroka (eds)

Rapid Tooling Process Chain

D. Trenke[a], W. Rolshofen[a]

[a] *Institute for Mechanical Engineering (IMW), Technical University of Clausthal, Robert-Koch-Str.32, 38678 Clausthal-Zellerfeld, Germany*

Abstract

Methodologically, the Rapid Tooling Process Chain can be described with different steps in procedure. Every part of this technique is presented in this article, from the CAD construction to the final product. In the industrial use, the generating laying manufacturing process of rapid tooling is increasing its importance. This essentially depends on the opportunity to create complex structures very fast. Moreover, the quality of the produced components relies on every single process step and the performance of workmanship.

1. Introduction

The Rapid Tooling respectively Rapid Prototyping process chain (Fig. 1) starts from a 3D-CAD construction of the structural element which will be sintered. Resulting in a computer model, it is prepared with a specific Rapid Tooling software for the laser-sintering (placement, creation of support structures and information about layers). The edited data of the model is transmitted to the Rapid Tooling facility, where the structure is produced by a laser in layers. Finally, material textures can be improved through different steps in post-process like sealing with epoxy resin, sandblast, polishing and so on [3].

2. CAD construction

Prefacing mentioned, Rapid Tooling Process Chain starts with a CAD construction of the work piece which will be manufactured, e.g. a prototype of a mould or a functioning part. It should be kept in mind to build a 3D volume model instead of a 3D plane model, because in the following step the layer information would consist of just outlines of the model. But for laser sintering, planes are needed, which will be smelted during the Rapid Tooling process.

In the design step, several rules have to meet the requirements for Rapid Tooling, concerning this attention should be paid to height- and thickness ratio for root faces and spigots, as well as minimal sintering structures and normal construction rules. Most important is the fact that Rapid Tooling is a building production process layer by layer, that is contrary to other production techniques the complexity of the workpiece is no object. Therefore, the textured component the better, because this coincidence with shorter production times and lesser amount of used powder, which means lower expenses.

Moreover, draftsman should be aware of the advantage that any kind of complex structures can be created with laser sintering, which up to now could be produced only with big effort. An example for this is a coolant bore, which is a small three dimensional form.

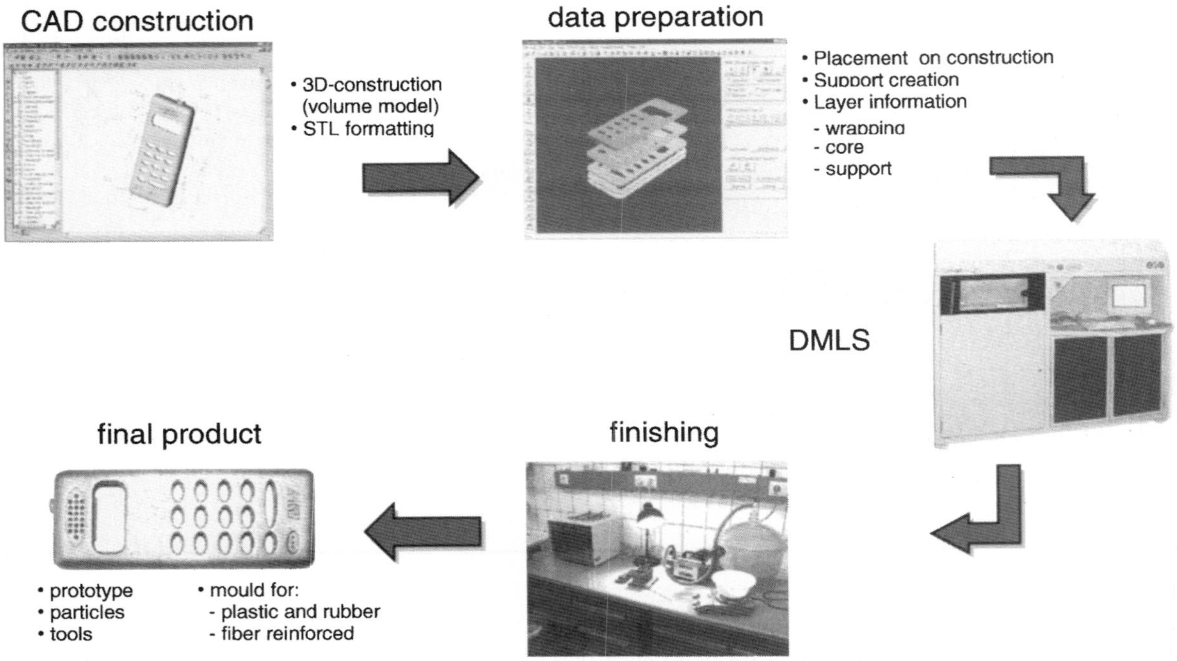

CAD construction

• 3D-construction (volume model)
• STL formatting

data preparation

• Placement on construction
• Support creation
• Layer information
- wrapping
- core
- support

DMLS

finishing

final product

• prototype
• particles
• tools

• mould for:
- plastic and rubber
- fiber reinforced

Figure 1: Rapid Tooling Process Chain.

Concluding, the constructed CAD model has to be merged in the STL format, which meant a special kind of triangulation, where the computerized model is built up in many small triangles. This result can be read by the Rapid Tooling Software. The following CAD programs are recommended for the model construction: ProE, CATIA or SolidWorks. In Fig. 2, for instance a mobile phone is designed with SolidWorks.

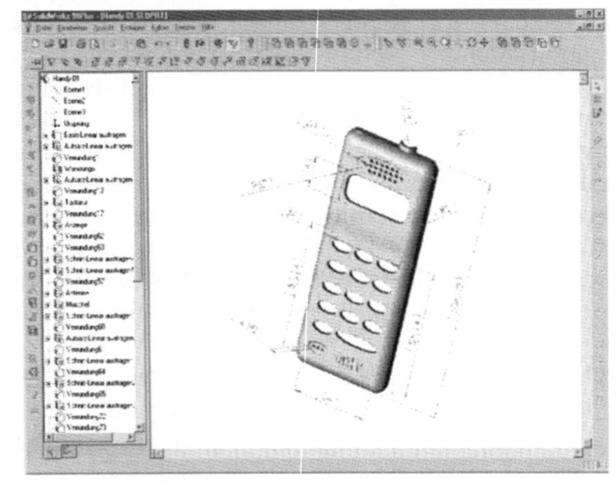

Figure 2: Model of a mobile phone designed with SolidWorks.

3. Data Preparation

The generated STL file is transferred to a special Rapid Tooling Software in the next step of data preparation. For this purpose, the program Magics RP of the enterprise Materialise is used at IMW. While using the direct metal laser sintering (DMLS), attention should be paid as follows, when a new "job" is going to be started:

Contrary to stereolithography e.g., the components cannot build levitating in the DMLS method, that implies that the components need always contact to a building platform, which will be laid also in the process chamber. To achieve this steady connection, support structures are created, like it is shown in Fig. 3.

On the one hand this support serves as brace for the overhang, which is bigger than 2mm, and on the other for abscission of the workpiece from the building platform. For an easier remove of this support structure, it has a comb like building and is exposed with special parameters.

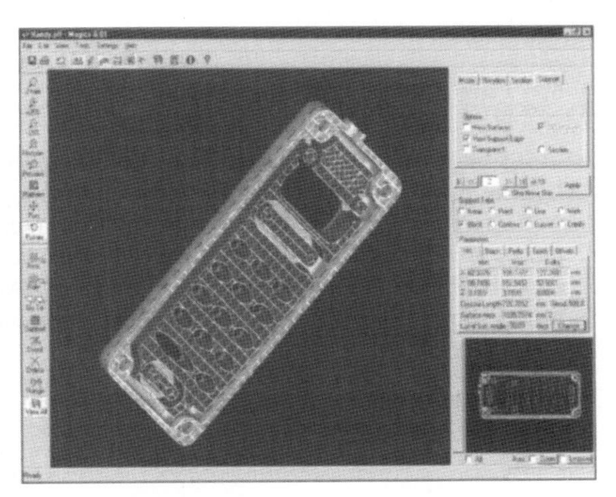

Figure 3: Mobile phone with support structure.

Form halves or functioning parts can be sintered direct on the platform, by what material can be saved and a steadfast base for the setting of the form in the tool arises.

In Fig. 4, the building platform, the workpiece and the support structure is pictured. This data leads to the information needed for each layer. The adjusted thickness of the layers depends on the particle size of the used metal powder. Sintered workpieces of DMLS have a typical thickness of layers from 0.05mm to 0.02mm.

Subsequently, the information for the layers are transferred to the Rapid Tooling facility, where model is generated by a laser.

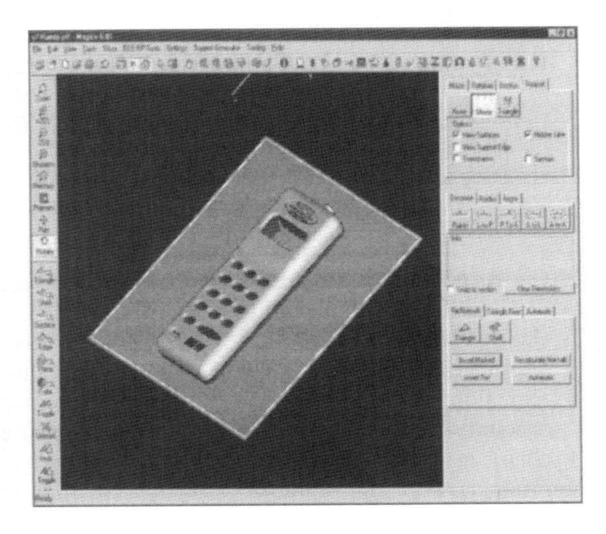

Figure 4: Workpiece, support- and building platform.

4. DMLS Building Process

The Institute for Mechanical Engineering uses a Rapid Tooling system called EOSINT M250 (s. Fig. 5), which consists mainly of the following components:

- CO_2–laser (200W) for the melting of the metal powder
- Support platform with mounted building platform
- Dosage facility; also for powder storage
- A wiper for the application of the metal powder on the building platform
- Process computer with software for system control

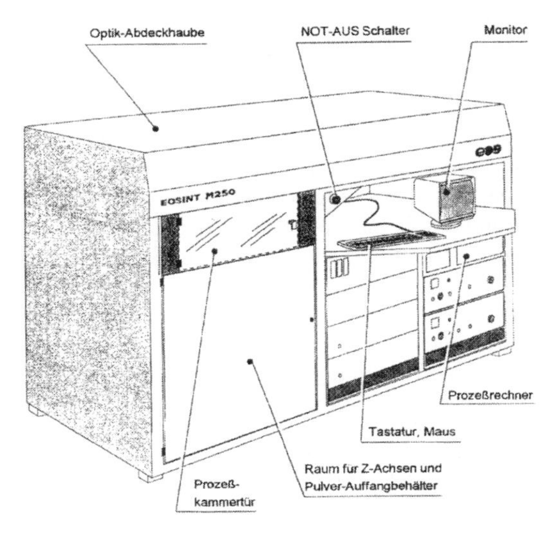

Figure 5: Rapid Tooling system EOSINT M250 [1].

At the beginning of the sintering, the building platform is covered with a thin metal powder layer for the first time. Then, this layer is molten by the CO_2–laser referring to the first layer geometry information.

After exposing the first layer, the building platform is lowered for the next layer thickness and the wiper is moved up to the stop on the right-hand side of the dosage platform. Necessary is the lowering, because after exposure the workpiece surface is irregular and rough, where the wiper could get caught.

When the wiper has reached his stop-position, the dosage platform is lifted until a sufficient amount of powder is allocated that the whole building platform can be fully covered again. The next time, the wiper is moved left and spreads so the next layer of powder, which is sintered with the second layer geometry data. Unnecessary powder falls in the collection cup on the left-hand side of the building platform. Sintering respectively exposing parameters are chosen to connect the new layer with the underlying. Every step is repeated until a complete, 3D workpiece is generated of the CAD informations.

Basis for the achieved quality of the sintered objects are the material properties of the used metal powder, particularly with regard to sintering behaviour and particle size. IMW uses DirectMetal 50, DirectMetal 20, DirectSteel 50 and DirectSteel 20, which are developed especially for DMLS.

A speciality of DMLS is that the workpieces are made in a skin-core structure. This means, that the volume of the object consists of different exposed regions. Therefore, the core and skin region have also an alternative layer thickness. But it must paid attention for the width of the laser beam, because he has a size of 0.6mm. The outer contour of the workpiece must be the same as the outer laser melting area. So the laser focus does not incident with the CAD boundary; it has to be shifted a little bit inside the object structure. It is called beam compensation.

Firstly, the contour is sintered and afterwards the inner area is made line by line where every single point is reached by the beam at least twofold. According to this double coverage, the temperature level can be guaranteed as high, what is necessary for a high quality. At last, the contour is exposed a second time, so the surface can be sintered more precisely and the surface roughness decreases, because the first melted area contains a higher heat conduction. With low scan velocity the workpiece density increases, in contrast to a high scan velocity, where the construction time decreases [4,5].

Besides selective laser sintering methods of the enterprises EOS and 3D Systems/DTM, there are further developments and modifications of these methods e.g. selective laser melting (SLM). Moreover, it is possible to use quite a number of materials for laser sintering, e.g. one component metal powder or ceramics.

5. Finishing

When the sintering is done, all workpieces and the building platforms are taken out of the Rapid Tooling system and possibly existing support structures are removed. The surface quality is improved by sandblasting. After sintering, the workpieces possess a final porosity of 30%, what could be infiltrated e.g. with epoxy casting resin. Moreover, the DMLS workpieces can be welded, soldered, eroded or shape cut during post-processing.

6. Final Sintered Products

Sorts of products which could be created with DMLS are forms and shapes of injection molding as well as fibre reinforced composite. The number of pieces is between some hundreds and thousands, but depend on the complexity of the workpiece. Also, functioning parts like gear and shaft could be produced directly.

Naturally, prototypes or display models are manufactured with Rapid Tooling (s. Fig. 6) [2].

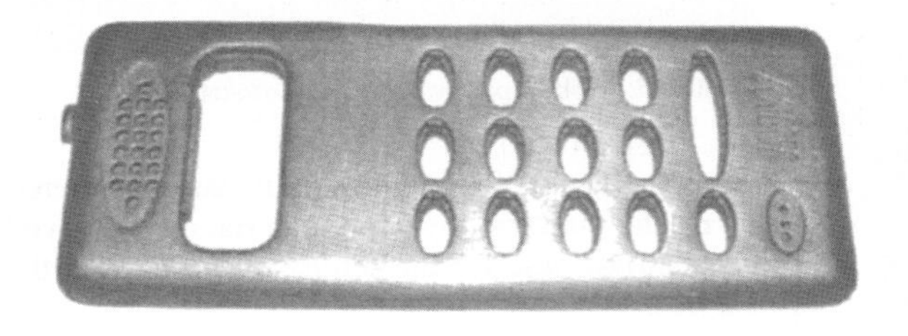

Figure 6: Mobile Phone jacket produced with DMLS.

The mobile shell was chosen for demonstration, because there differences could be made clear between Rapid Prototyping, Rapid Tooling and Rapid Manufacturing. During Rapid Prototyping, prototypes respectively models will be sintered, whereas Rapid Tooling generates mouldings for injection molding e.g. where then mobile shells in quite a number (> 100000) could be produced. Rapid Manufacturing means direct production of functioning parts. Therefore, Rapid Prototyping, Rapid Tooling as well Rapid Manufacturing describe each type of application of the sintered workpiece and they do not distinguish from the production process.

7. Conclusion

As it is shown, Rapid Tooling does not only meant the sintering in the proper meaning of the word, but also the whole process chain. With regard to quality improvements, new areas of application or just a reduction in construction time, this has to be taken into account. On the whole, the Rapid Tooling process chain in particular DMLS leads to high quality products with high number of pieces. On the other hand very complex prototypes and tools can easily be manufactured [6].

The intelligence consists of the possibilities, that through variations of the sintering parameters on the one hand different materials like metal, ceramic and synthetic could be processed and on the other components with specific properties like density, strength, stiffness etc. could be generated.

8. References

[1] EOS GmbH: Basis Training, Planegg, 1999.

[2] Trenke, D.: Anwendungen von Rapid Prototyping and Rapid Tooling in der Produktentwicklung, IMW Clausthal, 2001.

[3] Gebhardt, A.: Rapid Prototyping - Werkzeuge für die schnelle Produktentstehung, Hanser Verlag, Juli 2000

[4] Meiners, W.: Direktes Selektives Laser Sintern einkomponentiger metallischer Werkstoffe, Dissertation RWTH Aachen, 1999

[5] Song Y.-A.: Selektives Lasersintern metallischer Prototypen, Shaker Verlag, 1998

[6] Poprawe, R.: Lasertechnik für die Fertigung, Springer Verlag, 2005

Intelligent Production Machines and Systems
D. T. Pham, E. E. Eldukhri and A. J. Soroka (eds)

Reverse engineering and rapid prototyping for new orthotic devices

Y.E. Toshev[a], L.C.Hieu[b], L.P. Stefanova[a], E.Y. Tosheva[a],
N.B. Zlatov[b], S. Dimov[b],

[a] *Institute of Mechanics and Biomechanics, Acad. G. Bonchev Str., Building 4, Sofia 1113, Bulgaria*
[b] *MEC, Cardiff School of Engineering, Cardiff University, Queen's Buildings, Cardiff CF24 0YF, UK*

Abstract

The main aim of the study is to propose new personalized 3D designed elbow orthoses and new ankle-foot orthoses using reverse engineering and rapid prototyping. A general reverse engineering procedure was used including: data acquisition, preprocessing, contours and surface fitting and CAD model creation. Mituyoto CMM machine and Hymarc Laser Scanning system were used for scanning data acquisition. Different scanning angles were used to cover all the scanning points representing the human upper limb and the ankle-foot complex. The point cloud data were obtained. Preprocessing was used in order to remove the redundant point data and noises and to reconstruct the point cloud data into "optimal" data for CAD modelling. CopyCAD™ software was used for pre-processing the scanning data and for contours and surface fitting. Based on the contours representing arm and ankle-foot geometry, surface and solid models were constructed in CAD/CAM modelling packages ProEngineer™ and UG™. Error analysis was used to estimate the differences between the CAD models and the original point cloud data. ProEngineer™ was used to design the new orthotic devices. Wax print models of the upper limb and ankle-foot complex were obtained, laser scanned, CAD constructed and compared with the direct scanned CAD models. Different polymers versions of new devices were manufactured using rapid prototyping. Prototypes of new orthotic devices were obtained and analyzed.

1. Introduction

1.1. Reverse engineering

In reverse engineering (RE) of mechanical components, the shape of the existing component must be completely and accurately measured. The forces acting upon it must also be analysed in order to ensure that the new component can withstand the forces that the original withstood. In this study RE was used as an up-to-date methodology for constructing 3D CAD model of physical parts (human segments) by digitizing an existing part and then using that model to reproduce the part [1, 2]. The two basic tools needed for RE, are: (a) different tools "digitizing" an existing part - Coordinate Measuring machine (CMM), Computed tomography (CT) and MRI (Magnetic Resonance Imaging; (b) CAD software (modeller).

As well known, three dimensional scanning or "digitizing" is the process of capturing the geometry of existing physical objects through the use of "scanning devices". These non-contact devices, sometimes called 3D scanners, transmit various types of signals (laser, white light, radiation, sound waves, etc.) to determine distances. These devices collect an enormous amount of point data in a semi-random fashion, called "raw

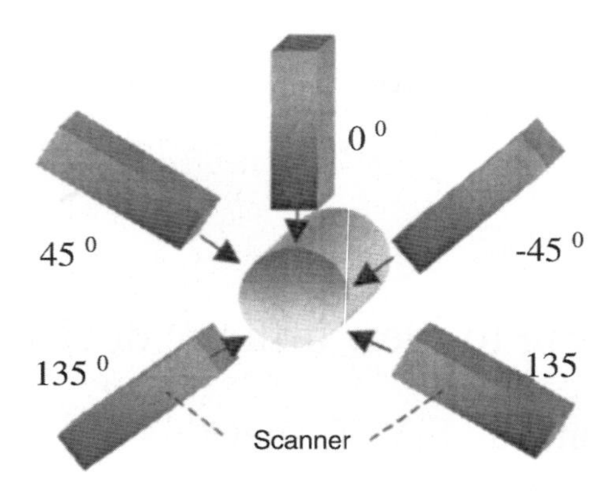

Fig. 1. Scanning angles used to scan human upper arm and ankle-foot complex

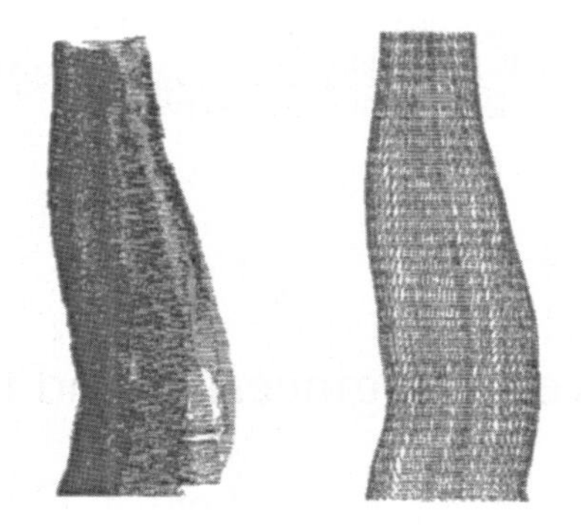

Fig. 2. "Original raw scanning data" of a human arm (left) and "approximate point cloud" after preprocessing (right).

point data". The point data could be organized in consecutive cross-sectional cuts or the point data in a fairly random form, called a "point cloud of data".

Based on laser scanning, CT or MRI scanning data and micro/milling tooling, 3D models of different anatomical segments of the human body have been obtained and many biomedical applications have been reported, mainly for: (a) making 3D physical biomodels (b) implant design (c) design for prosthetics and orthotics [1, 2, 3].

Special purpose reverse engineering programs may have many tools for performing general 3D shape manipulation, but their main focus is on the process of converting "raw point data" from the scanning devices into free-form surfaces as an accurate representation of the physical object, i.e. into "point cloud of data".

1.2. Biomedical applications of reverse engineering

The advancements in 3D scanning, computing power and specialized software applications have made possible some amazing processes in product design, competitive benchmarking, polymers micro tooling (rapid prototyping and rapid milling) and product verification.

2. Methods

2.1. Data Acquisition

Mituyoto CMM machine and Hymarc Laser Scanning System were used for scanning data acquisition. Different scanning angles were used to cover the surface points representing an arm or an ankle-foot complex (see Fig. 1). So called "original raw scanning data" was obtained (see Fig. 2, left).

2.2. Pre-processing

Because of using five scanning schemes for data acquisition and the involuntary movements of the arm during the scanning process, the resulted scanning point cloud contains redundant point data as well as noises.

Pre-processing was used in order to remove the redundant point data and noises and to reconstruct the point cloud data into "optimal" data for CAD modeling called "approximate point cloud". (see Fig. 2, right). CopyCAD™ software was used for preprocessing the scanning data.

2.3. Contours and surface fitting

Contours used to represent a human arm were then obtained based on the "approximate point cloud data". The surface model of the arm was obtained using CopyCAD™ software.

2.4. CAD Model creation and data analysis

Based on the contours representing the arm geometry, surface and solid models of an arm were constructed using CAD modeling packages such as UniGraphics™ and ProEngineer™.

Error analysis was accomplished between the CAD model of an arm and the "approximate point cloud" (See Fig. 1, right) extracted from the "raw

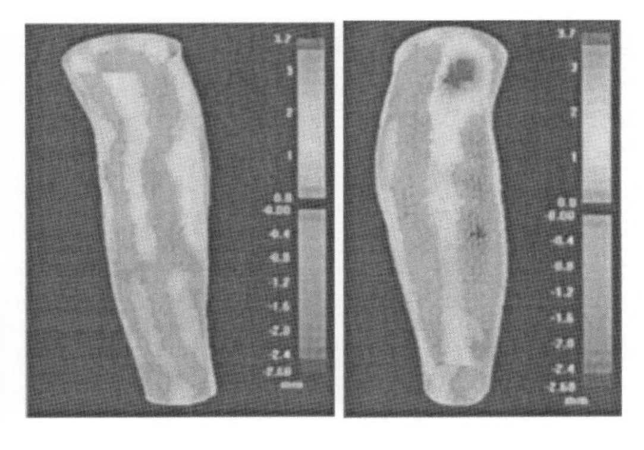

Fig. 3 Error analysis between the CAD model of an arm and the "approximate point cloud" extracted from the "raw scanning data".

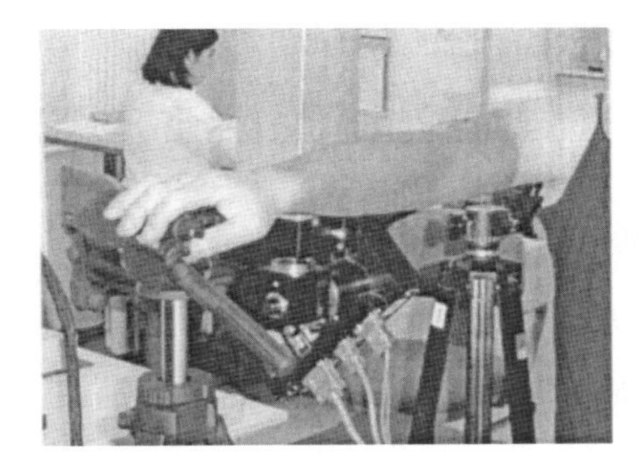

Fig. 4. Laser scanning of a human upper limb aimed to design a personalized elbow orthosis

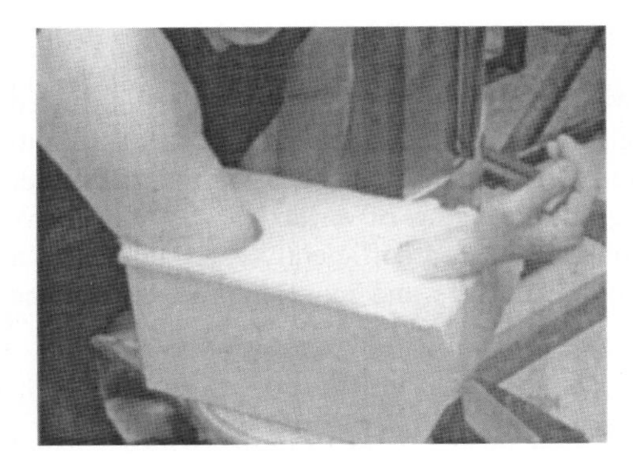

Fig. 5. Creation of a wax model of a human upper limb.

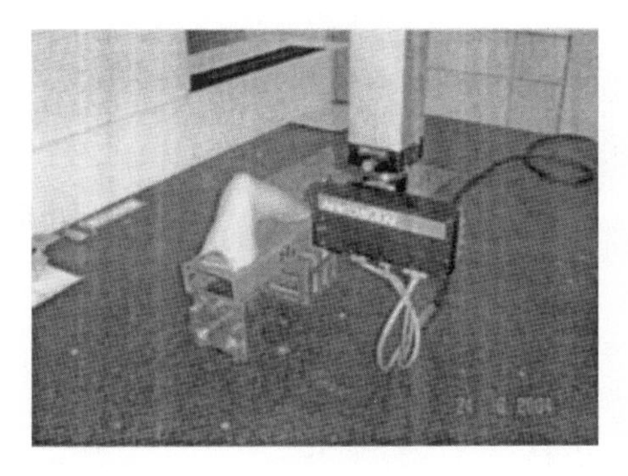

Fig. 6. Laser scanning of the wax model of a human upper limb aimed to design a personalized elbow orthosis

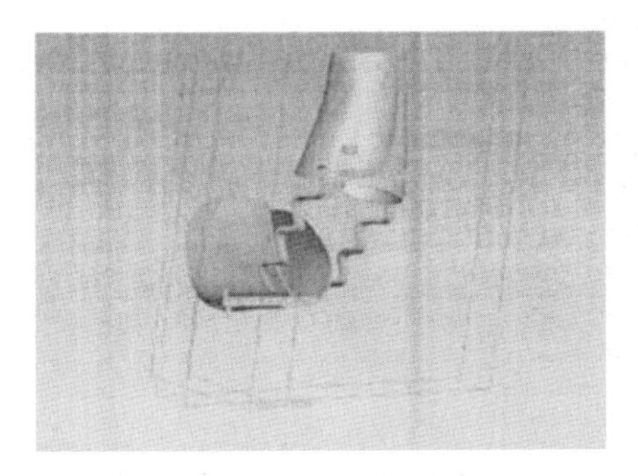

Fig. 7. CAD model of the new elbow orthosis using ProEngineer™

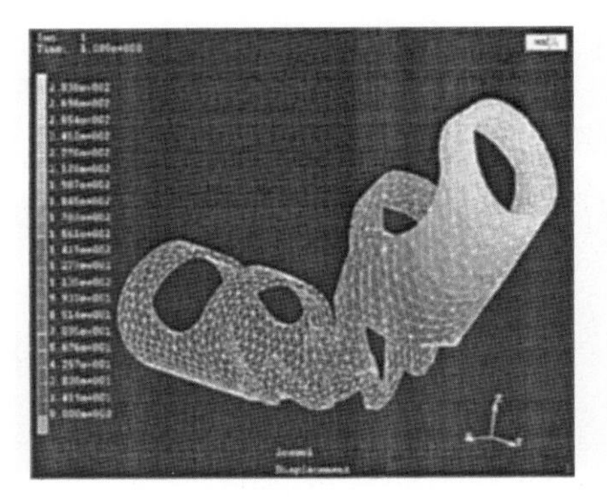

Fig. 8. FEA of the elbow Orthosis CAD model using Mentat FEA Solver™

Fig. 9. One of the prototypes of the new personalized elbow orthosis based on "in vivo" laser scanning.

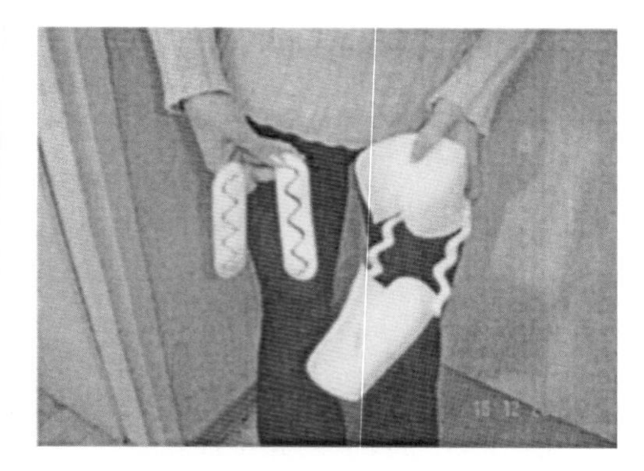

Fig. 10. The last prototype of the elbow orthosis was specially designed for double purpose: (a) to immobilize totally the elbow joint in case of ıma or disease; (b) to give possibility for elbow motion (Flexion/Extension) during rehabilitation.

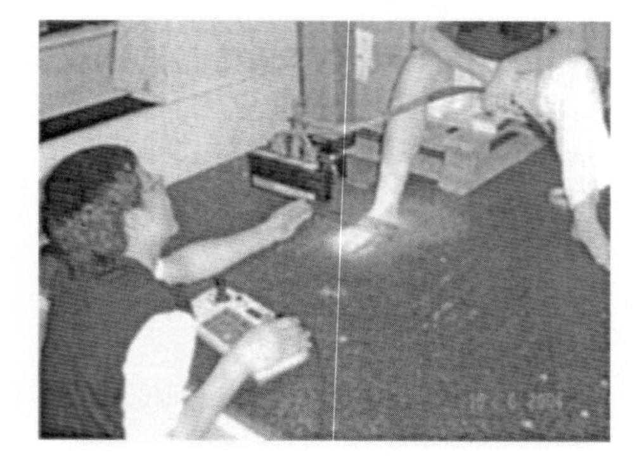

Fig. 11. Laser scanning "in vivo" of an ankle-foot complex.

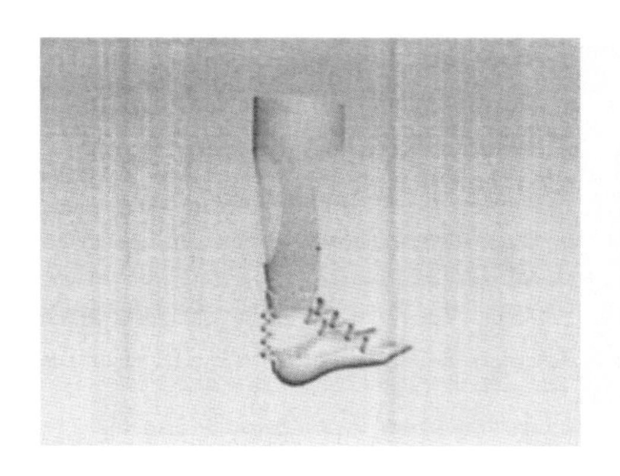

Fig. 12. CAD model of the new ankle-foot orthosis using ProEngineer^{TM}

scanning data" (see Fig. 3) and between the CAD model of an arm and the "original raw scanning data" (see Fig.2, on left) extracted from the "raw scanning data".

3. Results

3.1. New personalized elbow orthosis

3.1.1. Elbow CAD model based on laser scanning "in vivo"

A human upper limb was scanned *in vivo* using Coordinate Measuring Machine "Mitutoyo Euro Apex" and Laser Scanner "Hymare CTD" (see Fig. 4).

3.1.2. Elbow CAD model based on laser scanning of a wax model.

A wax model of a human upper limb was created (see Fig. 5) and laser scanned (see Fig. 6) in order to compare two basic approaches: "in vivo" laser scanning and scanning of a wax model. The wax model obtained was a sufficiently exact replica of a human elbow surface and more comfortable for the patient.

3.1.3. Manufacturing of the Elbow Orthosis using Rapid Prototyping

Different CAD models of the new personalized Elbow Orthoses were FEA analyzed (see Fig. 8) and manufactured (see Fig. 9).

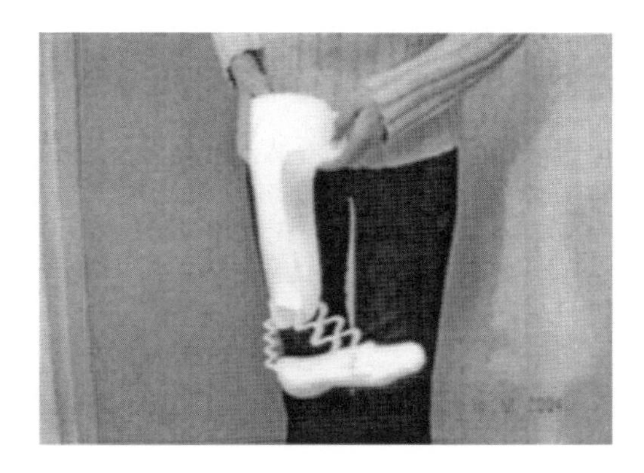

Fig. 13. Prototype of new personalized ankle-foot orthosis based on wax model and laser scanning

The last prototype (see Fig. 10) was specially designed for double purpose: (a) to immobilize totally the elbow joint after trauma or disease; (b) to give possibility for elbow motion (Flexion/Extension) during rehabilitation.

3.2. New personalized ankle-foot orthosis

The new personalized ankle-foot orthosis was designed using the same procedure as described above for the elbow orthosis. The same two initial approaches have been used: laser scanning "in vivo" (see Fig. 11) and wax modelling plus laser scanning. CAD solid models were obtained using UniGraphics™ and ProEngineer™ (see Fig. 12). Rapid Prototyping was used to manufacture different polymers prototypes of the personalized ankle-foot orthosis (see Fig. 13).

4. Conclusion

To achieve the main goal (design and manufacturing of new personalized orthoses) two initial approaches have been used: (a) a human upper limb and a human ankle-foot complex were scanned *in vivo* using Coordinate Measuring Machine "Mitutoyo Euro Apex" and Laser Scanner "Hymare CTD; (b) wax models of a human upper limb and a human ankle-foot complex were obtained and after that laser scanned.

The wax modelling presented sufficient surface accuracy and was more comfortable for the patient due to the shorter time for the procedure than in the case of "in vivo" scanning.

On the base of the scanned data 3D geometrical models of a human upper limb were created and different polymers prototypes were designed in two steps using reverse engineering: (a) pre-processing and contours/surfaces fitting (Surfacer™, CopyCAD™ and GeomagicStudio™); (b) CAD model reconstruction and evaluation (UniGraphics™ and ProEngineer™.

The CAD models of the orthoses were analyzed using Finite Element Analysis. Layer Manufacturing (Rapid Prototyping) was used to manufacture different polymers prototypes of new Elbow Orthoses and Ankle-Foot Orthoses.

The preliminary survey show good perspectives for implementation of the mew Orthoses in the industry.

Acknowledgements

The study was supported by the Royal Society (joint project of the Bulgarian Academy of Sciences and MEC, Cardiff University) and by the Bulgarian Research Found (Grant No TN-1407/2004).

References

[1] Hieu LC, Bohez E, Vander-Sloten J, Oris P, Phiena HN, Vatcharaporn E and Binhc PH. Design and manufacturing of cranioplasty implants by 3-axis CNC milling. Technology and Health Care, 8, 2000, pp 85-93.

[2] Popov E and Chan FM. The effective way of doing computer-aided reverse engineering. Proceedings of the International Conference on Manufacturing Automation. Hong Kong, 2002, pp 137-145.

[3] Stefanova, L., Le, C.H., Zlatov, N., Taiar, R., Tosheva, E., Toshev, Y. (2004) 3D computer modeling of elbow orthosis using reverse engineering. In: Proceedings of the International Conference on Bionics, Biomechanics and Mechatronics, Varna, pp. 267-270.

Intelligent Production Machines and Systems
D. T. Pham, E. E. Eldukhri and A. J. Soroka (eds)

Towards rapid sheet metal forming

O. Owodunni, S. Hinduja and S. Mekid

School of Mechanical, Aerospace and Civil Engineering, The University of Manchester, UK

Abstract

This paper examines two techniques which have potentials for rapid sheet metal forming. The two techniques are dieless sheet metal forming and sheetmetal forming using reconfigurable tooling. For each technology, the mechanics of the manufacturing process, the equipment required, Computer-aided manufacturing (CAM) system, accuracy achieved and the economics are discussed. The research issues are also given consideration.

---

1. Introduction

The factory of the future requires characteristics which include responsiveness, environmental sustainability, reduced cost and lead time. Rapid prototyping and manufacturing meets some of these requirements. Current approaches of applying rapid prototyping/manufacturing to sheet metal forming adopt the direct or indirect tooling approaches. Indirect rapid tooling uses a rapid prototyping technique to produce a pattern which is then used to cast the tool in the required metallic material or as a substrate to which a metal coating is applied. On the other hand, the direct tooling method fabricates the tool directly in a metallic material by using a high energy beam such a laser to melt a metallic powder or wire layer-by layer. While these methods are often faster than conventional tooling methods, the tools fabricated are specific to the shapes for which they are produced and cannot be reused. Also, for the direct tooling method there is a limit to the size that can be produced. There is therefore a need to develop a flexible and sustainable technology for rapid sheetmetal forming. This paper examines some techniques which have great potentials to meet this important need. These techniques are dieless sheet metal forming and the use of reconfigurable tooling. Dieless sheetmetal forming, also referred to as Incremental Sheetmetal Forming (ISF), manufactures sheetmetal parts without using conventional special-purpose tools. Sheet metal forming using reconfigurable tooling (or discrete dies), on the other hand uses tools of similar complexity to the conventional one but made of discrete elements, making it possible to rearrange them into a new shape.

2. Dieless forming of sheet metals

ISF is used to form sheet metal into complicated shapes without the use of conventional punch and die. In the process local micro-deformations incrementally occur on the blank through a rotating and translating single point tool. Deformation of the sheet can be achieved with or without a supporting, fixed tool about which the single point forming tool moves (see Figure 1). The research of Powell and Andrew at Cambridge University in the UK [1, 2] is perhaps the earliest reported literature that can be most closely linked with the ISF process. They developed an incremental forming process for forming flanged sheet metal parts without dedicated tooling by repeated localized deformation. The examination of various flanges showed that the most significant reactions are limited

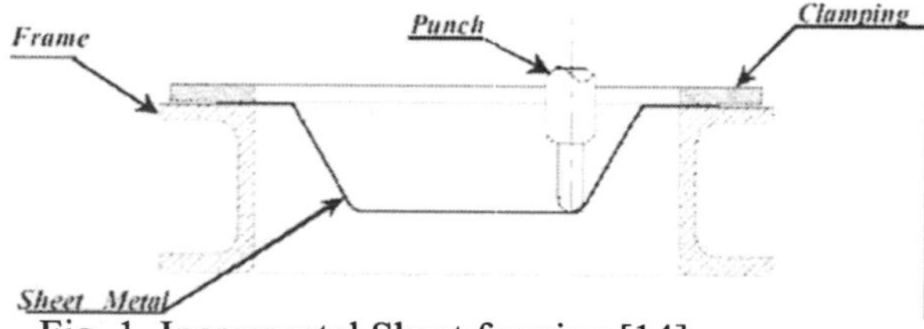

Fig. 1. Incremental Sheet forming [14].

to the vicinity of the current position of the tool. This result suggested the possibility of using non-dedicated tooling which only provided support for the blank near a current position of the forming tool. Since that initial research, several other groups from various parts of the world have also contributed to this technology. As reported by Matsubara [3], in the early part of the 1990s several Japanese researchers investigated the incremental sheet metal forming process. Contributions have also been made from China [4, 5, 6], South Korea [7, 8], Italy [9, 10] and Germany [11].

2.1 Mechanics of ISF

The mechanics of ISFis similar to that of conventional conventional spinning, shear forming and flow forming in that the deformation mechanism is localized to a small region under the forming tool. The deformation occurring in forming products is due to local stretching. Bending occurs at clamping position or as a result of process design to achieve it. The forming limit curve contrasts with that of conventional sheet metal forming. The strain conditions ranged from pure uni-axial stretching to bi-axial stretching. The dominant mode of failure is due to excessive thinning of the sheet in areas overstretched. Researchers such as Kim, Yoon, and Yang [7, 8], Filice et al [9], Fratini et al [10] and Hirt et al [11] have made contributions to the understanding of the mechanics of incremental sheet forming and overcome existing limitations.

Kim and Yang [7] carried out experiments on computer controlled incremental ISF process. The results show the deformation in the blank to be shear-dominant enabling the final sheet thickness strain to be predicted. To improve the formability of the sheet metal forming process through the achievement of uniformity of thickness in the final formed product, a double-pass forming method was suggested and found to be very effective. The intermediate geometry in this 2-pass process was determined by using the predicted thickness strain distribution to ensure that shear deformation is reduced in the highly deformed region and increased in the less deformed.

Yoon and Yang [8] attempted to overcome the disadvantage of previous ISF process which limits the deformation occurring in forming products to local stretching alone. The process suggested to overcome this disadvantage did not require any dedicated die, minimized loss of material and thickness strains through adoption of unconstrained forming and replacing stretching with bending deformation respectively.

Filice et al [9] examined the material formability in ISF by the determination of the forming limit diagram from results of tests developed to produce different strain conditions in a sheet. The strain conditions ranged from pure uni-axial stretching to bi-axial stretching. The result obtained showed that the forming limit curve contrasted with that from conventional sheet metal forming. Fratini et al [10] examined the effects of material properties on formability in ISF using typical sheet metal forming materials with different hardening coefficient, anisotropy, ultimate tensile strength and percentage elongation. The result showed that the formability is highly dependent on the strain hardening coefficient of the material and the percentage elongation. An empirical expression for the forming limit in terms of the material properties was obtained and suggested as a tool for predicting formability from material properties alone. To overcome the problems of excessive thinning in ISF, Hirt et al [11] proposed multistage forming and the use of tailor rolled blanks.

2.2 Equipment and software for ISF

The equipment required for ISF can be standard CNC milling machines, though specialized machines have also been developed [12]. In addition, some minimal toolings are required to offer non-specialised support and clamping to the sheer metal during the forming process. A simple cylindrical forming tool is also required. The tooling (shown in Figure 2) employed by Matsubura [3] for forming convex shapes consists of the upper plate, which can slide vertically and on which the blank is placed and clamped by another plate. Instead of a punch and die, the technique employs a simple forming tool and a fixed support tool. The forming process is accomplished by the movement of the forming tool around a contour corresponding to a cross section of the intended shape followed by downward movement to the next cross-sectional plane.

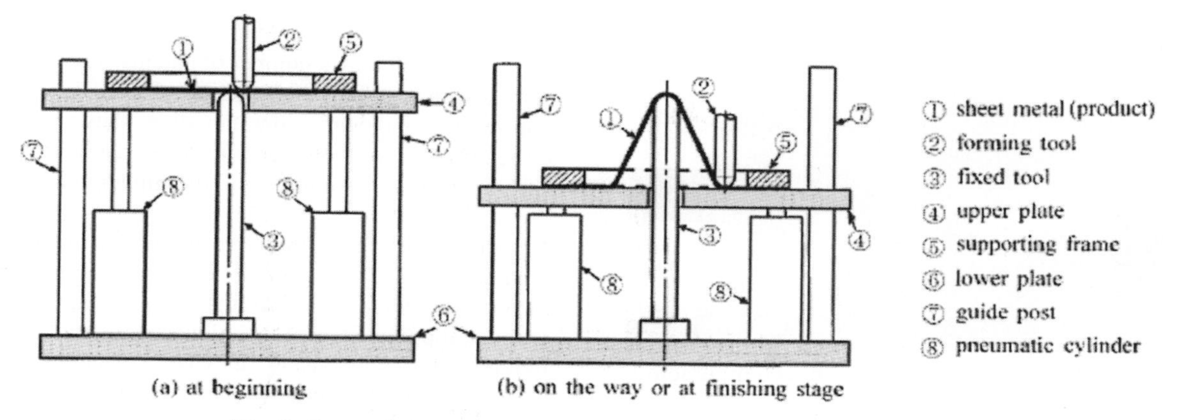

Fig. 2. Set up for supported incremental forming [3].

Standard CAM systems for generating toolpath for machining are often employed in incremental forming, the toolpath strategy being the contour parallel type.

2.3 Accuracy of incremental ISF

The accuracy of the formed part is affected adversely by spring back arising from bending that occurs, as against pure stretching, in the deformation of the sheet metal. Accuracies of between 1.5mm and 2 mm have been achieved [13] depending on the geometry of the formed part. Factors that affect the accuracy includes the shape of the formed part, the shape and size of the tool, the forming toolpath, material and thickenss of the blank etc. some of these factors affecting accuracy have started to be investigated and techniques for addressing them are being suggested.

Ambrogio et al [14] noted that the inaccuracy arising from springback arise from bending zones close to the clamping fixture. To understand this, an investigation of the effect of the tool diameter and toolpath offset on the accuracy was carried out using experimental and an explicit finite element technique. 3D laser scanning and surface reconstruction was used to obtain the geometry of the formed part. Good agreement between the numerical prediction and the actual geometry was obtained and used to suggest a strategy for error reduction incremental forming. A methodology was proposed [15] for reducing the geometric error by modifying the ideal path of the forming tool using the ideal and actual positions of points formed in an earlier increment. The actual positions are obtained by using a 3D digitizing system. The method allows the forming error to be reduced by approximately 70%.
Hirt et al [11] proposed that to overcome the problem of deviations from specified geometry, a first component is formed and measured. The deviation is inverted, scaled by a factor and applied to the required shape to obtain a modified shape which when used for the forming gives an improvement in the accuracy.

2.4 Economics of ISF

While the process can form almost as intricate shapes, compared to the conventional sheet metal forming process, the tooling costs can be as low as 5-10%, shorter lead-time for prototyping obtains and forming forces are considerably lower. Thus a standard CNC machine which is quieter and occupies less floor space can be employed.

2.5 Research issues in ISF

There has not been a systematic investigation of the mechanics of incremental sheet metal forming which consider the process variables such as shape, tool shape and size, blank material and toolpath strategy/parameters. A validated finite element analysis of the process also needs to be addressed as well as the issues involved in designing products for incremental sheet metal forming. Rather than using CAM systems developed for machining toolpath generation, software dedicated to incremental sheet metal forming needs to be developed. Also, while 3 axes CNC machines have been used, 4 and 5 axes ones have not been investigated.

3 Forming Of Sheetmetal using Reconfigurable Tool

Conventional sheetmetal forming uses tools which have to be fabricated each time a new shape is required. This

Fig. 3. Sheet metal forming with reconfigurable punch and die [17].

results in long lead time, high tooling cost and waste of material. Thus the replacement of such conventional tool with one that can be reconfigured each time a new product is required offers considerable advantage. Ongoing contributions to the understanding of this technology and research issues are discussed in sections 3.1 to 3.8.

3.1 Mechanics of sheetmetal forming using reconfigurable tooling

The mechanism of sheetmetal forming over discrete elements has not been given much attention by the research community. One reason for this is because, in practice, the discontinuous surface of the die is usually smoothed resulting in similar situation as conventional die or dies made of compliant materials. Thus, the mechanism of forming is similar to that of conventional dies. Since application of discrete dies has been to stretch forming and other sheetmetal forming cases where a blankholder is nor used.

3.2 Equipment for sheetmetal forming using reconfigurable tooling

Conventional sheet metal press is used for reconfigurable dies. Thus, the main difference in equipment is toolings. As for conventional sheet metal forming, only one tool is required for operations such as stretch forming. The use of reconfigurable punch and die have also been investigated by researchers [16, 17] as illustrated in figure 3. There is a similarity in the technology of reconfigurable tool proposed by various researchers. Differences arise in the method of positioning and clamping the discrete elements and the method of smoothening the discontinuous surface of the tool. Walczyk [16] reviewed the history of advances in the development of reconfigurable tools from 1923 including a number of patents. The history, started with two-dimensional array of pins, varying in applications and degree of automation, shows that the basic principle for reconfiguration was established very early in the history of this rapid tooling approach.

Nakajima [18] made the first attempt to create an automatically reconfigurable tool and tested some methods to increase the load that the pins can support without slipping Several techniques were studied for moving the pins using a CNC machine. The shortcoming of the types of control was the difficulty to move one pin without dragging the adjacent pins.

In 1980, the first self-contained automatically controlled 3D reconfigurable tool was patented by Pinson [19]. Although it was similar to earlier developed ones, it had the capability of being set automatically by computer-controlled servo-actuators connected to each pin.

Investigation carried out by Walczyk and Hardt [16], took the idea of Nakajima further and solved the problems related with his investigation. There in the MIT laboratories, a reconfigurable tool was designed, which was capable of setting the pins automatically and served as a die for a forming press. The solution to Nakajima's problem of the dragging between pins was to separate each row of pins using a sheet metal spacer, rigidly attached to the die frame. This way, the maximum forming load per pin is limited by the failure of the row divider material.

The MIT research, proposed that all pins of a densely packed die should be square in cross section to ensure *load path isolation* and the pin tips be rounded to avoid piercing and enable a smooth surface to be obtained. To ensure the frame is stiff enough for uniform distribution of the clamping load over the pin matrix a maximum allowable deflection is set, and the structure is solved using analytical methods or FEA analysis. The uniformity of the clamping force is also improved using a thin interpolating material layer between the wall that compresses the pins and the first row of pins.

Recently, Papazian and co-workers [20] developed a reconfigurable tooling concept for stretch forming of sheet metal in aerospace applications. The die (figure 4), consisting of 1120 pins, is reconfigured using servo-actuators at each pin. However, the die costing over a million pounds would not be affordable to most small and medium enterprises.

Fig. 4. A large scale reconfigurable tool [20]

Researchers at the university of Manchester have also carried out some preliminary investigations on the use of reconfigurable tooling [17]. A low-cost rapid tool that is easily reconfigurable by a single computer-controlled actuator or conventional CNC machine was developed and tested. The bed-of-pins concepts was enhanced with a novel approach of positioning and clamping the pins in a simple, low-cost and time-saving way (see figure 5). Tool path strategies, from simplest to the most complex, for positioning the pins by a single actuator or CNC machine were proposed and evaluated to obtain an optimal strategy. The concept was tested for sheet metal forming as shown in figures 3 and 6.

3.5 CAD/CAM system for sheetmetal forming using reconfigurable tooling

To reconfigure the tool, software to derive the position of the discrete elements and the control instruction for the actuator device or CNC machine used for positioning the discrete element is required. As shown in the contributions of various researchers, the positions of the discrete elements are determined by intersecting the surface model of the die/punch with a line which represents the centre line of the discrete element. The positions of the tip of the discrete

Fig. 5. Positioning the pins for reconfigurable tool using a CNC machine [17].

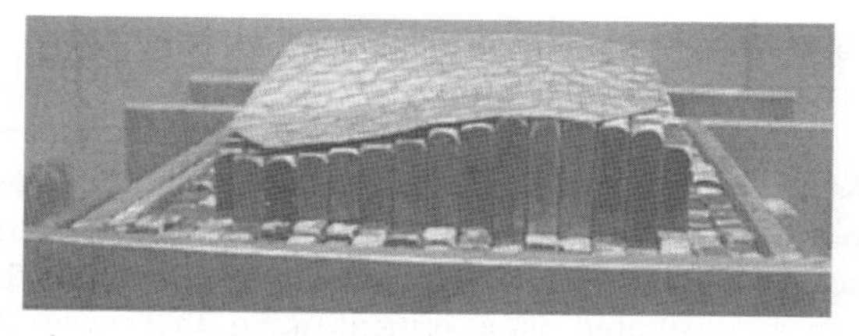

Fig.6. Forming results [17].

elements are then used as an input to the program which generates the instructions for the actuator device or CNC machine. For the CNC machine, it possible as demonstrated in the contributions of owodunni et al [17] to generate efficient toolpath by minimizing idle time of the positioning tool.

3.6 Accuracy of sheetmetal forming using reconfigurable tooling

The accuracy of the sheetmetal formed with this technology depends, amongst other factors, on the accuracy with which the discrete elements are positioned and the accuracy after the discontinuities on the surface of the discrete elements are smoothened. It is possible to achieve accuracy close to the few microns achievable by most positioning devices such as CNC actuators. In addition to the accuracy of positioning the discrete elements, the accuracy of the formed product is affected by factors such as springback as occurs in conventional sheetmetal forming. As demonstrated in the contributions of Papazian et al [20], it is possible to compensate for such spring back and other size variations such as the effects of the compliant interpolator and release of the residual stresses using techniques such as experimental determination of the deformation transfer function and FEA method.

3.7 Economics and sustainability issues in sheetmetal forming using reconfigurable tooling

Considerable savings in cost, time and material is achieved using reconfigurable tooling. For example the contributions of papazian et al [20] reports that a reduction to $1/8^{th}$ and $1/3^{rd}$ of the current tool fabrication cycle time and labour hour respectively is obtained. From the contributions of owodunni et al [14], it was shown that it possible to also achieve great savings in the capital investment to the extent that the cost of a reconfigurable tool can be comparable to that of a conventional die.

3.8 Research issues in sheetmetal forming using reconfigurable tooling

More investigations are needed to achieve low-cost positioning mechanism that are faster and for large scale tools. The limit on the size of small features that can be captured is a critical issue that need the consideration of the research community. Efficient CAM system that will give more consideration to optimal planning is another important research issue. Experimental and numerical analysis of the mechanics of sheet metal forming using discrete elements also need to be investigated.

4 Conclusions

This paper has reviewed contributions in the research which have great potential for rapid manufacturing of sheetmetal parts. Reduction cycle time and labour is possible for these developments. Also the possibility of having a capital cost for the technology comparable to conventional tooling has been demonstrated for prototypes. Several research issues that will address the mechanics, the equipment, software, accuracy and economics of rapid sheetmetal forming need to be given consideration.

Acknowledgement

The University of Manchester is a partner of the EU-funded FP6 Innovative Production Machines and Systems (I*PROMS) Network of Excellence.

References

[1] Powell NN. Incremental of sheet metal forming: flanged components, Proc. of the 28th MATADOR Conference, UMIST, April 1990, 359-365.

[2] Powell N N and Andrew C. Incremental forming of flanged sheet metal components without dedicated dies. Proc. of the Inst. of Mech. Engineers, Journal of Engineering manufacture, 1992, 206, 41-47.

[3] Matsubara S A computer numerically controlled dieless incremental forming of a sheet metal. In: Proc. of the Inst. of Mech. Engineers, part B, J Eng Manuf, 215, 7, 2001, 959–966.

[4] Jie L, Jianhua M and Shuhuai H. Sheet metal dieless forming and its tool path generation based on STL files. Int J Adv Manuf Technol, 2004, 23, 696–699.

[5] Dai K, Wang ZR and Fang Y.CNC incremental sheet forming of an axially symmetric specimen and the locus of optimization, Journal of Materials Processing Technology 151, 2004, 63–69.

[6] Yang H, Zhan M, Liu YL, Xian FJ, Sun ZC, Lin Y, and Zhang XG, Some advanced plastic processing technologies and their numerical simulation, J. Materials Processing Tech., 151, 1-3, 2004, 63-69.

[7] Kim T. J. and Yang D. Y. Improvement of formability for the incremental sheet metal forming process. *Int. J. Mech. Sci.*, 2001, 42, 1271-1286.

[8] Yoon SJ, Yang DY Investigation into a new incremental forming process using an adjustable punch set for the manufacture of a doubly curved sheet metal, Proc. Inst. of Mech. Engineers, part B, J Eng Manuf 215, 7, 2001, 991–1004.

[9] Filice L. Fratini L and Micari F. Analysis of material formability in incremental forming, Annals of the CIRP, 51, 1, 2002, 199-202.

[10] Fratini L, Ambrogio G, Di Lorenzo R, Filice L and Micari F. Influence of Mechanical Properties of the sheet Material on formability in single point incremental forming. Annals of the CIRP, 53, 1, 2004, 207-209.

[11] Hirt G, Ames J, Bambach M and Kopp R. Forming strategies and Process modelling for CNC Incremental sheet forming. Annals of the CIRP, 53, 1, 2004, 203-206.

[12] http://www.nextfactory.com manufacturer of Amino dieless CNC sheet metal forming system.

[13] Leach D, Green AJ and Bramley AN, Sheet production without dies or special punches. *Metallurgia*, 68, 2, 2001, pFT12.

[14] Ambrogio G, Costantino I, De Napoli L, Filice L, Fratini L and Muzzupappa M. Influence of some relevant process parameters on the dimensional accuracy in incremental forming: a numerical and experimental investigation. Journal of Materials Processing Technology 153–154, 2004, 501–507.

[15] Ambrogio G, De Napoli , Filice L and Muzzupappa M. Improvement Geometrical Precision In Sheet Incremental Forming Processes. Proc. of ESDA04, 7th Biennial conf. on Engrg Systems Design and Analysis, July 19-22, 2004 Manchester, United Kingdom.

[16] Walczyk D F and Hardt DE. Design and analysis of reconfigurable discrete dies for sheet metal forming, J of Manufacturing Systems, 17, 6, 1998, 436-454.

[17] Owodunni OO, Diaz-Rozo J, and Hinduja S. Development and Evaluation of a Low-Cost Computer Controlled Reconfigurable Rapid Tool, Computer-Aided Design and Applications, 1, 1-4, 2004, 101-108.

[18] Nakajima, N., A newly developed technique to fabricate dies and electrodes with wires,Bull of JSME, Vol. 12 No. 54, 1969, 1546-1554.

[19] Pinson, G. T., Apparatus for forming sheet metal, US patent no. 4212188, July 1980.

[20] John M. Papazian, Elias L. Anagnostou, Robert J. Christ, Jr., David Hoitsma, Patricia Ogilvie, and Robert C. Schwarz, Tooling For Rapid Sheet Metal Parts Production, 6th Joint FAA/DoD/NASA Conf. on Aging Aircraft, San Francisco, CA, USA, September 16-19, 2002.

Intelligent Production Machines and Systems
D. T. Pham, E. E. Eldukhri and A. J. Soroka (eds)

Agent-based multiple supplier tool management system

R. Teti

Department of Materials and Production Engineering, University of Naples Federico II, Piazzale Tecchio, Naples, 80125, Italy

Abstract

A multi-agent tool management system (MATMS) was developed in the framework of a negotiation-based multiple-supplier system for automatic tool management and procurement. The MATMS operation is illustrated for the real industrial case of a turbine blade producer (customer) requiring dressing jobs on worn out CBN grinding wheels from different tool manufacturers (suppliers) in a supply network. The MATMS was implemented by making use of agent development tools integrated in the Foundation for Intelligent Physical Agents (FIPA) Agent Management Reference Model and the fundamental aspect of agent communication was realised through the FIPA Agent Communication Language (ACL).

1. Introduction

Recently, a new software architecture for managing supply networks at the tactical and operational levels has emerged. It views the supply network as composed of a set of intelligent (software) agents, each responsible for one or more activities in the supply network and interacting with other agents in planning and executing their responsibilities [1].

An agent is an autonomous, goal-oriented software paradigm that operates asynchronously, communicating and coordinating with other agents as needed [2].

A multi-agent system consists of a group of different types of agents that can take on specific roles within an organizational structure [3]. It can be defined as a loosely coupled network of problem solvers that interact to solve problems that are beyond the individual capabilities or knowledge of each problem solver [4]. Important characteristics of a multi-agent system are [5]:

- each agent has incomplete information or capabilities for solving the problem (limited viewpoint)
- there is no system global control;
- data are decentralised;
- computation is asynchronous.

Most research on multi-agent systems focuses on the coordinative intelligent behaviour among a collection of autonomous intelligent agents considering that the group of agents provides more than the sum of the capabilities of its members [6].

The adoption of multi-agent technology is based on 3 fundamental system domain characteristics [7]:

- data, control, expertise or resources are inherently distributed;
- the system is naturally regarded as a society of autonomous cooperating components;
- the system contains legacy components that must interact with other, possibly new software components.

Supply network management by its very nature has all the above domain characteristics [3]. A supply network consists of suppliers, factories, warehouses, etc., working together to fabricate products and deliver them to customers. Parties involved in the supply network have their own resources, capabilities, tasks, and objectives. They cooperate with each other autonomously to serve common goals but also have their own interests. A supply network is dynamic and involves the constant flows of information and materials across multiple functional areas both within and between network members. Multi-agent technology, therefore, appears particularly suitable to support collaboration in supply network management.

In this paper, the development of a multi-agent tool management system (MATMS), operating in the framework of a multiple-supplier network, is presented. The aim of the MATMS is the optimum tool inventory sizing and control, including on-time delivery and cost and stock-out risk minimisation, of CBN grinding wheels for Nickel base alloy turbine blade fabrication.

2. Turbine blade manufacturing

Turbine blades are manufactured along several production lines, each for one aircraft engine model requiring a set of CBN grinding wheel types (part-numbers). Each part-number is planned to work a maximum number of blades: when it reaches its end of life, it is sent for dressing to an external supplier in a supply network and remains unavailable for a time defined as dressing cycle time. For each part-number, a sufficient number of CBN grinding wheels (serial-numbers) must be always available (on-hand inventory) to prevent production breakage due to tool run-out. The part-number on-hand inventory size, I, depends on:

- number of pieces per month, P;
- number of pieces per wheel, G;
- number of months required without new or dressed wheel supply, C (coverage period) heuristically selected.

The wheel demand, D, for each part-number is given by:

$$D = (P/G) * C - I_0 \quad (1)$$

where: P/G = tool demand rate (number of wheels per month); I_0 = initial part-number inventory size.

Traditional tool management consists in the strategic planning of the wheel inventory size based on the selection of a coverage on-hand inventory for each part-number (number of wheels for production needs in the coverage period C). This procedure does not always prove adequate; the producer, aware of this drawback, increases or reduces the part-number inventory level on the basis of experience. The results of this policy, founded on skilled staff knowledge, are the historical inventory size trends: in some cases, the expected trend matches the historical one; in other cases, it is underestimated, with risk of stock-out, or overestimated, with excessive capital investment [8].

The historical trend is a fit solution that prevents tool run-out and useless investment: it can be used as a reference for assessing alternative tool management strategies.

3. Generic agent shell

The MATMS was developed by making use of agent development tools integrated in the FIPA Agent Management Reference Model that provides for the creation, registration, location, communication, migration and retirement of agents [9]. The entities contained in the reference model (Figure 1) are logical capability sets (i.e. services) and do not imply any physical configuration. These can be combined in the physical implementation of Agent Platforms (AP) defined by FIPA as the environment where agents can physically exist and operate [10]. FIPA agents located on an AP utilise the facilities offered by the AP for realising their functionalities. In this context, an agent, as a physical software process, has a physical life that has to be managed by the AP. The implementation details of individual AP and agents are the design choices of the individual agent system developers.

The Generic Agent Shell (Figure 1) provides for several layers of reusable services and languages. They are concerned with agent communication services to exchange messages with other agents, specification of coordination mechanisms (shared conventions about exchanged messages during cooperative action with other agents), services for conflict management and information distribution (voluntary or at request information of interest to other agents), reasoning and integration of purpose built or legacy application programs. The glue that keeps all layers together is a common knowledge and data management system on top of which these layers are built. The approach allows for a clear distinction between an agent's social know-how (communication services, coordination mechanisms, information distribution services and other) and its domain level problem solving capability.

Purpose built application programs can make use of this agent architecture to enhance their problem solving capabilities and to improve their robustness through coordination with other agent-based applications. Pre-existing (legacy) application programs can also be incorporated with little adaptation and can experience similar benefits. This latter point is important because in many cases developing the entire application afresh would be considered too expensive or too large a change away from proven technology.

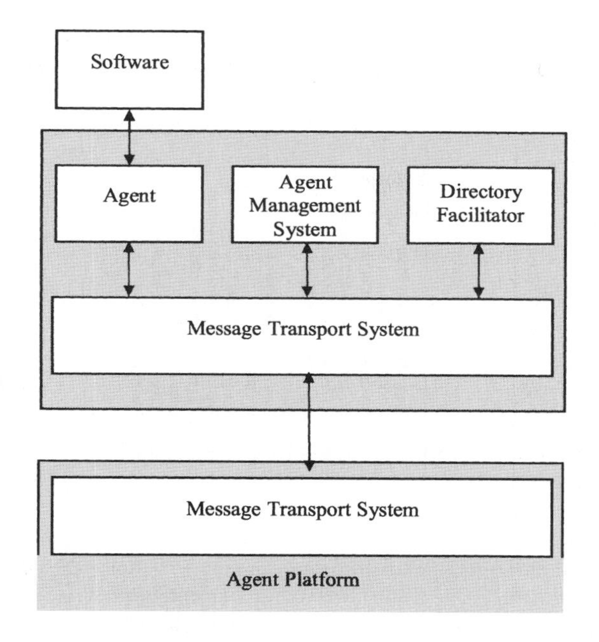

Fig. 1. Agent management reference model.

4. Multi-agent tool management system

The MATMS activities are carried out according to the multi-agent interaction and cooperation protocols described below.

In Figure 2, the block scheme of the developed MATMS, subdivided into three functional levels is reported:

- the Supplier Network Level, including the external tool manufacturers in the supply network;
- the Enterprise Level, including the logistics of the turbine blade producer;
- the Plant Level, including the production lines of the turbine blade producer.

The Supplier Network Level is responsible for carrying out the dressing jobs on worn out CBN grinding wheels. It comprises only one type of agents, the Supplier Order Acquisition Agents (SOAA_i), that represent the external tool manufacturers (suppliers) activity of acquiring dressing job orders from the turbine blade producer (client).

The Enterprise Level is responsible for coordinating the MATMS activities to achieve the best possible results in terms of its goals, including on-time delivery, cost minimization, and so forth. It comprises three different intelligent agents performing the fundamental tool management activities:

- the Resource Agent (Ra) that merges the functions of inventory management, resource demand estimation and determination of resource order quantities; the Ra domain specific problem solving function is the Flexible Tool Management Strategy (FTMS) [11].
- the Order Distribution Agent (ODA) that selects the external supplier to which the CBN grinding wheel dressing job orders should be allocated on the basis of negotiations and constraints; the ODA domain specific problem solving function for order allocation is implemented in ILOG OPL Studio 3.5 [12].
- the Dressing Time Prediction Agent (Dtpa) that carries out the predictions of CBN grinding wheel dressing cycle times, founded on historical data, under both a supplier-independent and a supplier-dependent basis; the DTPA employs as domain specific problem solving function a neuro-fuzzy paradigm known as Adaptive Neuro-Fuzzy Inference System (ANFIS) [13, 14].

The Enterprise Level comprises also a Knowledge & Data Base Agent (K&DBA) that handles all the information relevant for tool management activities, including the updating of historical data on CBN grinding wheel dressing cycles, and a Warehouse Timer Agent (WTA) that takes care of the incoming and outgoing CBN grinding wheels, including the evaluation of actual dressing cycle times.

The Plant Level is responsible for the manufacturing of turbine blades. It comprises only one type of agents, the Production Agents (PA_j), representing the various production lines of the turbine blade producer factory.

4.1. MATMS agent communication

Communication is a fundamental aspect of a multi-agent system activity and takes place through exchange of messages between agents. The latter use a common language, the Agent Communication Language (ACL), to transfer information, share knowledge and negotiate with each other.

The most widely used ACL is the one developed by the Foundation for Intelligent Physical Agents (FIPA), FIPA ACL [15], based on the "Speech Act Theory" by J.R. Searle [16] derived from the linguistic analysis of human communication and founded on the idea that speech does not constitute only communications but also real actions.

The unit for communication analysis is the message, called communicative act or performative. In FIPA ACL, the types of communicative acts that constitute the language basis are identified through the analysis of communication processes of interest for the world of artificial agents.

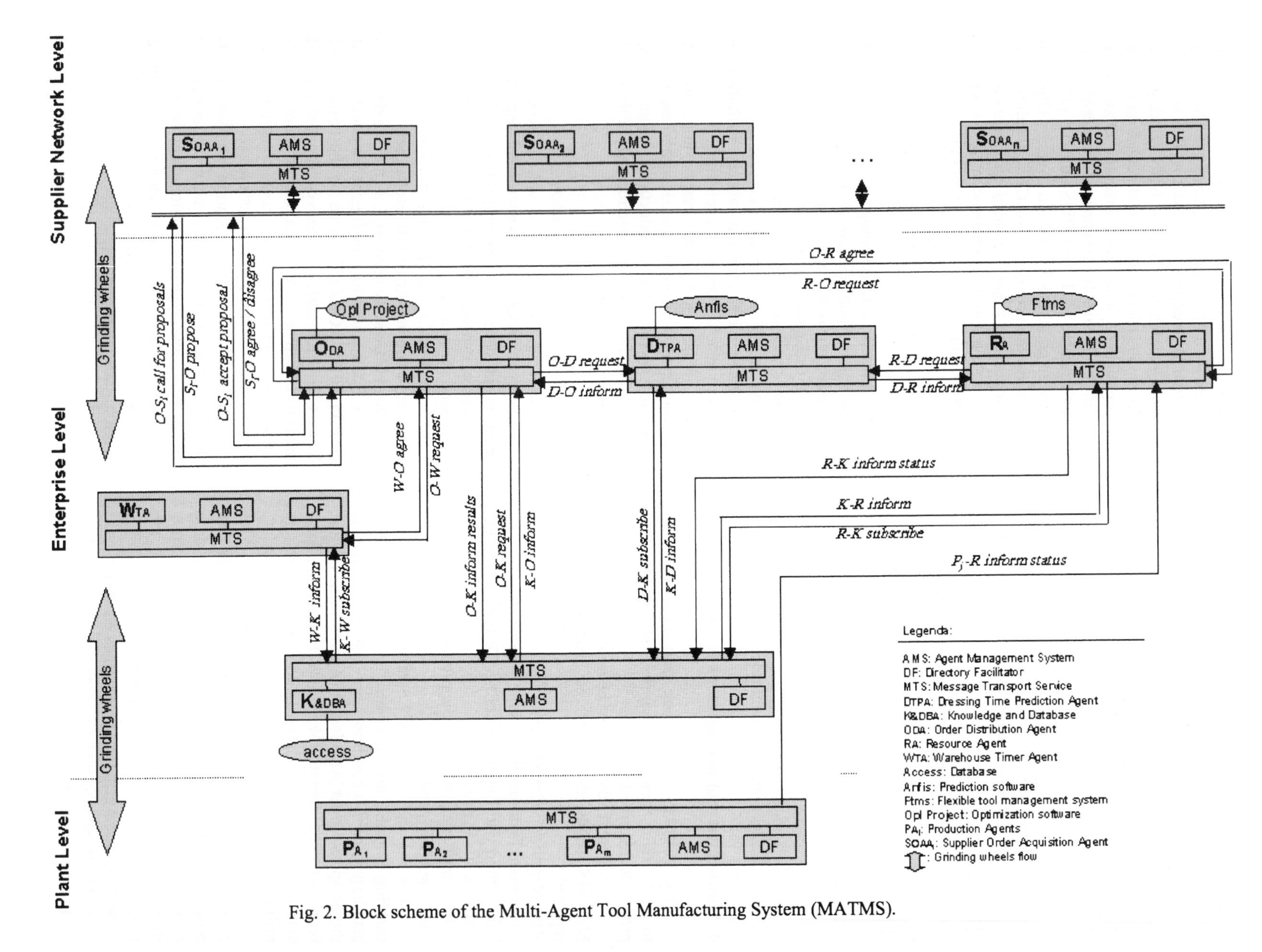

Fig. 2. Block scheme of the Multi-Agent Tool Manufacturing System (MATMS).

Table 1
FIPA ACL performatives utilised in the MATMS

PERFORMATIVE	DESCRIPTION
Inform	the sender informs the receiver that a given proposition is true
Subscribe	the action of requesting a persistent intention to notify the sender of the value of the reference, and to notify again whenever the object identified by the reference changes
Request	the sender requests the receiver to perform some action
Agree	the action of agreeing to perform some action, possibly in the future
Refuse	the action of refusing to perform a given action and explaining the reason for the refusal
Propose	the action of submitting a proposal to perform a certain action, given certain preconditions
Call for proposals	the action of calling for proposals to perform a given action
Accept proposal	the action of accepting a previously submitted proposal to perform an action

FIPA ACL offers a set of standard communication acts or performatives to describe agent actions: e.g. inform and confirm, request and query, agree and accept, propose, etc. It also allows users to extend them if the new defined ones conform to the rules of the ACL syntax and semantics.

In Table 1, the communication acts or performatives utilized in the MATMS are summarized.

4.2. MATMS functioning

In this section, the functioning of the MATMS is described in terms of agent communications and activities in the three system levels (see Fig. 2).

Initially, the RA in the Enterprise Level receives information on CBN grinding wheel end-of-life events from the PA_i in the Plant Level through P_j-R inform communicative acts containing the part-number and serial-number of the worn out CBN grinding wheels.

The RA starts collecting the necessary information to decide whether or not the worn out CBN grinding wheels should be delivered to an external supplier for dressing.

For each worn out CBN grinding wheel, the RA obtains information on current monthly production, P, part-number tool life, G, and part-number inventory level, I, by regularly interrogating the K&DBA through a R-K subscribe act, followed by a K-R inform reply.

Moreover, the RA asks the DTPA, through a R-D request act followed by a D-R inform reply, for a supplier-independent dressing cycle time prediction.

To issue a dressing cycle time prediction, the DTPA needs the updated historical dressing cycle times that it obtains from the K&DBA through a D-K subscribe act, followed by a K-D inform reply.

On the basis of the supplier-independent dressing cycle time prediction and the values of P and G for the relevant part-number, the RA evaluates the demand for CBN grinding wheel dressing to provide for part-number inventory sizing and control through its domain problem solving function: the FTMS.

If the RA does not consider necessary the dressing operation for a certain part-number, it informs the K&DBA through a R-K inform act that the worn out CBN grinding wheel serial-number is kept on-hold in the enterprise warehouse.

If the RA deems necessary to issue a dressing job order, the RA sends a R-O request act to the ODA asking to start a procedure for dressing job order allocation; this request is followed by an O-R agree reply.

The task of the ODA consists in allocating the required dressing job order to one of the external tool manufacturers in the supply network on the basis of negotiations and constraints. To this purpose, the ODA needs to gather information necessary for supplier selection from the relevant agents: DTPA, K&DBA, $SOAA_i$.

The ODA starts negotiating with the $SOAA_i$ in the Supply Network Level to obtain from each of them the dressing price and time (offers) for the required worn out CBN grinding wheel. The negotiation is initiated by an O-S_i call for proposals act to which the $SOAA_i$ respond with their offers through S_i-O propose acts.

Simultaneously, the ODA obtains from the DTPA the supplier-dependent dressing cycle time prediction (one for each supplier in the network) and from the K&DBA the dressing job reference price, due-time, etc., through O-D and O-K request acts followed by D-O and K-O inform replies.

On this basis, the ODA ranks the responding suppliers and allocates the dressing job order to the first supplier in the rank through an O-S_i accept proposal. The selected $SOAA_i$ accepts (refuses) the dressing job order through a S_i-O agree (refuse) act. In case of order refusal, the ODA contacts the second supplier in the rank list, and so on.

At the end of this procedure, the ODA informs the K&DBA about the order allocation results through an O-K inform act and requests the WTA to send the worn out CBN grinding wheel for dressing to the selected supplier using an O-W request act, followed by an W-O agree reply.

The WTA records the delivery and reception dates of each CBN grinding wheel for actual dressing cycle time evaluation. These data are regularly fed to the K&DBA that makes them available for further requests and interrogations by the relevant agents; this is obtained through a K-W subscribe act, followed by W-K inform replies.

5. Conclusions

The development of a multi-agent tool management system (MATMS) for automatic tool procurement, operating in the framework of a negotiation-based multiple-supplier network, was presented.

The MATMS was developed by making use of agent development tools integrated in the FIPA Agent Management Reference Model that provides for the creation, registration, location, communication, migration, and retirement of agents. The fundamental aspect of agent communication for information transfer, knowledge share, and negotiation was realised in the MATMS through the FIPA Agent Communication Language (ACL).

The MATMS functioning was illustrated with reference to the real industrial case of a turbine blade producer (customer) requiring dressing jobs on worn out CBN grinding wheels from different tool manufacturers (suppliers) in a supply network.

Acknowledgements

This research work was carried out with support from the EC FP6 NoE on Innovative Productions Machines and Systems – I*PROMS.

References

[1] Fox, M.S., Barbuceanu, M., Teigen, R. Agent-Oriented Supply-Chain Management, Int. J. of Flexible Manufacturing Systems, 200, 12, pp 165-188.

[2] Chen, Y., Peng, Y., Finin, T., Labrou, Y., Cost, R., Chu, B., Sun, R., Willhelm, R. A Negotiation-Based Multi-Agent System for Supply Chain Management, Working Notes of the ACM Autonomous Agents Workshop on Agent-Based Decision-Support for Managing the Internet-Enabled Supply-Chain, May 1999.

[3] Yuan, Y., Liang, T.P., Zhang, J.J. Using Agent Technology to Support Supply Chain Management: Potentials and Challenges, M. G. De Groote School of Business, 2001, Working Paper N. 453

[4] Durfee, E.H., Lesser, V. Negotiating, Task Decomposition and Allocation Using Partial Global Planning. In Distributed Artificial Intelligence, Eds. Gasser, L., and Huhns, M., San Francisco, CA, Morgan Kaufmann, 2, 1989, pp 229-244.

[5] Sycara, K.P. Multi-Agent Systems, AI Magazine, Summer 1998, pp 79-92.

[6] Fischer, K., Müller, I.P., Scheer, A.-W. Intelligent Agents in Virtual Enterprises, Proc. 1st Int. Conf. on Practical Appl. of Intelligent Agents & Multi-Agent Technology (PAAM '96), London, 1996, pp 205-223.

[7] Bond, A.H., Gasser, L. Readings in Distributed Artificial Intelligence, Morgan Kaufmann,1988.

[8] D'Addona, D., Teti, R. Intelligent Tool Management in a Multiple Supplier Network, Annals of the CIRP, 2005,Vol. 54/1

[9] FIPA Agent Management Specification, SC00023K, 2002/12/03, Foundation for Intelligent Physical Agents, Geneva, Switzerland, 2002, http://www.fipa.org.

[10] FIPA Agent Management Specification, SC00023J, 2002/12/03, Foundation for Intelligent Physical Agents, Geneva, Switzerland, 2002, http://www.fipa.org.

[11] Teti, R., D'Addona, D. Emergent Synthesis Approach to Tool Management in a Multiple Supplier Network, 5th Int. Workshop on Emergent Synthesis – IWES '04, Budapest, 24-25 May, 2004, pp 41-50.

[12] Teti, R., D'Addona, D. Agent-Based Mulptiple Supplier Tool Management System, 36th CIRP Int. Sem. on Manufacturing Systems - ISMS 2003, Saarbrücken, 3-5 June, 2003a, pp. 39-45.

[13] Teti, R., D'Addona, D. Grinding Wheel Management through N-F Forecasting of Dressing Cycle Time, Annals of CIRP, 52/1, 2003b, pp. 407-410.

[14] Teti, R., D'Addona, D. Multiple Supplier Neuro-Fuzzy Reliable Delivery Forecasting for Tool Management in a Supply Network, 6th AITEM Int. Conf., Gaeta, 8-10 Sept., paper 10_4, 2003c, pp. 127-128.

[15] FIPA97 Specification part 2, Agent communication language. Foundation for intelligent physical agents, October 1997.

[16] Searle, J.R. Speech acts. Cambridge University Press, Cambridge, England, 1969.

Intelligent Production Machines and Systems
D. T. Pham, E. E. Eldukhri and A. J. Soroka (eds)

A correlation between the friction coefficient and the roughness of rolls in a hot flat rolling

M. Durante, F. Memola Capece Minutolo, A. Langella

Department of Materials and Production Engineering
University of Naples Federico II,
P.Tecchio 80, 80125 Naples, Italy

Abstract

The principal factors affecting spread in hot flat rolling are geometric (initial width and height, roll radius) and tribological (roll roughness, friction coefficient, rolling temperature).

The factors affecting the frictional conditions in the contact zone are very important also for the design of the rolling mill. In particular, in order to calculate, by analytical formulae and FEM simulation, the rolling power and the maximum contact angle, the value of the friction coefficient is necessary.

The aim of this work is the evaluation of the friction coefficient starting from the knowledge of the values of rolls roughness. The influence of this parameter on lateral spread in flat rolling has been evaluated by experimental tests.

The tests have been conducted at high temperature (1000°C) using steel and cast iron rolls, for different values of the area reduction and of the rolls roughness.

In order to determine the value of the frictional coefficient for different conditions, a series of FEM simulations has been conducted. By FEM results, obtained setting the value of friction coefficient, the spread has been evaluated and compared with the values measured in experimental tests. In this way, the value of the friction coefficient has been correlated to the value of the rolls roughness.

1. Introduction

Many parameters affect the rolling processes and complex formulae must be considered for the roll mill design. So, in the last years, the FE analysis is very effective in calculating the surface profile and the cross sectional area of workpiece after a roll pass. Several numerical studies that simulate rolling passes have been presented on the basis of three dimensional finite element analysis [1-4]. Carlsson and Iankov [5-6] used FE simulation to investigate the effects of friction, back and front tension, on contact pressure distribution and lateral spread in flat rolling of wire.

In particular, the 3-D FE method [7-9] has been also used in the analysis of strip rolling, shape rolling, slab rolling and to solve problems associated with special shaped strip rolling, also with friction variation models.

Friction is arguably the most important parameter in a rolling process. Essentially, there are three friction models that concern metal forming and therefore rolling [10]. They can be defined as the Amont's law, the law of constant friction and the theory of adhesion.

In the rolling process a method to determine the friction coefficient, assumed constant along the contact arc, is based on the maximum angle of bite.

The friction coefficient, in this case, is determined

only in the moment when the outer edge of the workpiece comes in contact with the roll surface and the value is affected by the force with which the metal is pushed into the rolls.

Generally, in order to increase the friction coefficient and consequently the maximum angle of bite, the roughness of roll surface is increased. In Ekelund's formula, the friction coefficient is expressed as a function of rolling temperature, considering the material of the rolls and the surface roughness.

As reported in Wusatowsky [11]:

for cast iron and rough steel rolls

$$f= 1.05-0,0005\ T \quad (1)$$

for ground steel rolls

$$f= 0,8\ (1,05-0,0005\ T) \quad (2)$$

for chilled and smooth steel rolls

$$f= 0,55\ (1,05-0,0005\ T) \quad (3)$$

where T is the rolling temperature in °C.

In hot rod rolling another parameter, that can be easily measured, affected by the friction, is the spread. There are many formulae to calculate the spread, but the ones where the roughness of rolls is considered are empirical, and have a limited field of validity.

In this work, the evaluation of friction coefficient as a function of roll roughness has been studied with 3-D FEM method. The analysis has been conducted on the basis of spread values measured in rolling experimental tests, using different values of roughness and materials for the rolls.

2. Experimental tests

In order to evaluate the influence of the rolls roughness on the workpiece spread, some tests have been carried out using a two high reversing mill with rolls having a diameter of 480 mm.

The initial billet was 1200mmx125mmx125mm in dimensions, and the material a low carbon steel, whose chemical composition is reported in table1.

Table 1
Steel chemical composition

Element	C	Mn	Si	P
%	0.23	0.5	0.27	0.045

The billets were heated in a gas furnace up to a temperature of 1075°C before the rolling process. The temperature was then measured with an optical pyrometer before the entrance in the first stand. In all the test carried out, this value was of 1070°C. Two couples of rolls were used, the first ones were steel rolls, the second cast iron rolls. On each roll a turning operation was executed, after which the values of centre-line average height (R_a) were measured.

Rolling tests were then carried out using different values of surface roughness. In tables 2 and 3 these values, utilized respectively for steel rolls and cast iron rolls, are reported. In particular, in the tables, the values of R_{al} and R_{at}, measured with a Rank Taylor Hobson profilometer, in parallel and transversal directions with respect to the rolls axis, are related. In a first series of tests, the spread of the rolled section for the different roughness and for both couples of rolls, was investigated. In these tests the final height of the workpiece was 104 mm.

A second series of tests was carried out on the steel rolls using the highest value of roughness (R_{al} = 10.30μm), in order to evaluate the greatest reduction that is possible to obtain in these operating conditions without pushing the billet.

Table 2
Values of roughness for steel rolls

	R_{al}(μm)	R_{at}(μm)
1	10,30	3,90
2	5,10	2,30
3	2,20	0,95
4	0,19	0,07

Table 3
Values of roughness for cast iron rolls

	R_{al}(μm)	R_{at}(μm)
1	8,10	2,90
2	4,20	1,60
3	1,80	0,70
4	0,16	0,05

3. Experimental results

The measurement of the spread was executed considering the mean between the maximum and the minimum value of the width, measured on the rolled section.

In tables 3 and 4 the values of the spread for the different values of the roughness, both for steel rolls and for cast iron rolls, are reported.

From the second series of tests , it was possible to

find that the greatest reduction that we can obtain, starting from an initial height of 125 mm, takes to a final height of 41 mm. The spread of the workpiece in this tests was of 23 mm.

Table 4
Final width (b_2) measured for different values of roughness of steel rolls

R_{al} (μm)	R_{at} (μm)	b_2 (mm)
10,30	3,90	128
5,10	2,30	129,5
2,20	0,95	131
0,19	0,07	131,5

Table 5
Final width (b_2) measured for different values of roughness of cast iron rolls

R_{al} (μm)	R_{at} (μm)	b_2 (mm)
8,10	2,90	128,5
4,10	1,60	129,8
1,80	0,70	131,0
0,16	0,05	131,8

4. FEM analysis

A 3-D numerical model was developed, based on Marc Autoforge calculation code, using linear solid tetrahedral elements with 8 nodes; the roll was assumed to be a rotating rigid surface, so it is not necessary to assign the material properties. From the scheme reported in fig.1 it is possible to note the surface of symmetry and the rigid surface that brings into contact the workpiece with the roll; as soon as the workpiece touches the roll, this rigid surface breaks off from the workpiece.

The simulation is developed under isothermal conditions, at a temperature of 1070°C.

The material used for the workpiece is a C22 steel, set from the library of the program.

In fig. 2 and 3 some steps of the numerical simulation are reported.

The simulations were conducted setting different values of the friction coefficient between roll and workpiece; after calculation, the spread of the workpiece was measured.

From the comparison between the spread values obtained by experimental data and by simulation, a correlation between the parameter R_a and the friction coefficient is possible, both for steel and for cast iron rolls.

In fig.4 and fig.5 these correlations are reported.

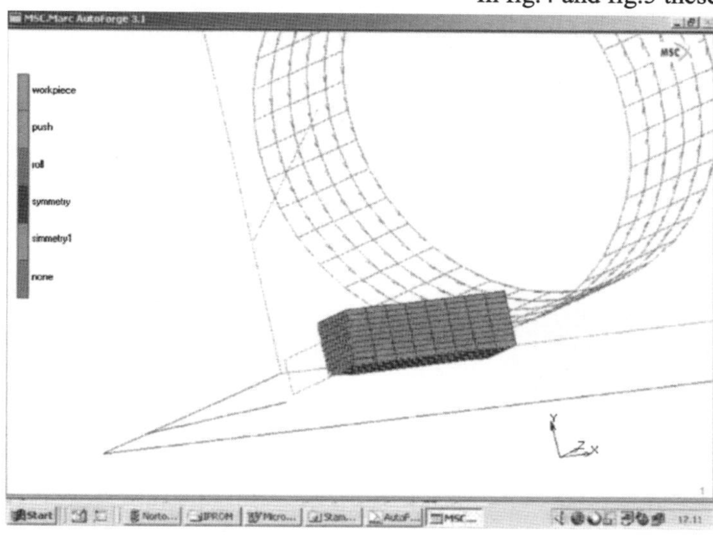

Fig.1 Scheme of the simulation

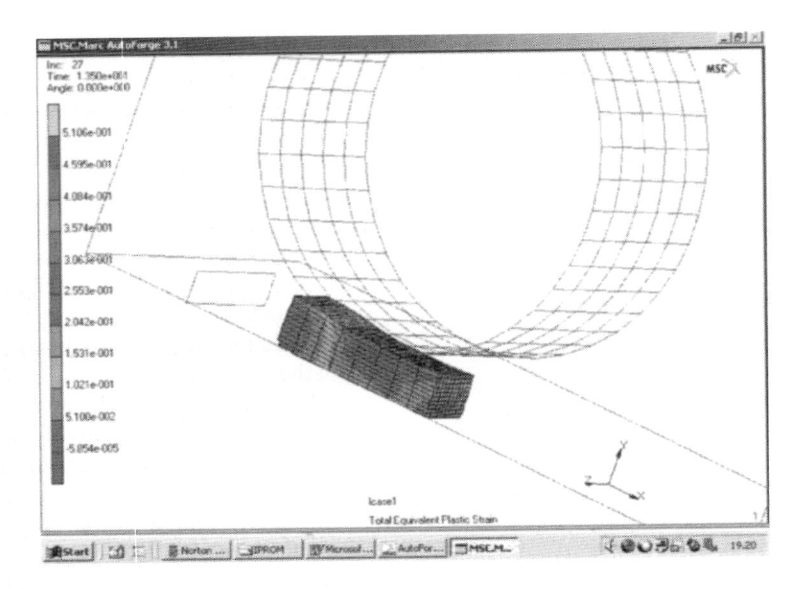

Fig. 2 Rolling process simulated by FEM

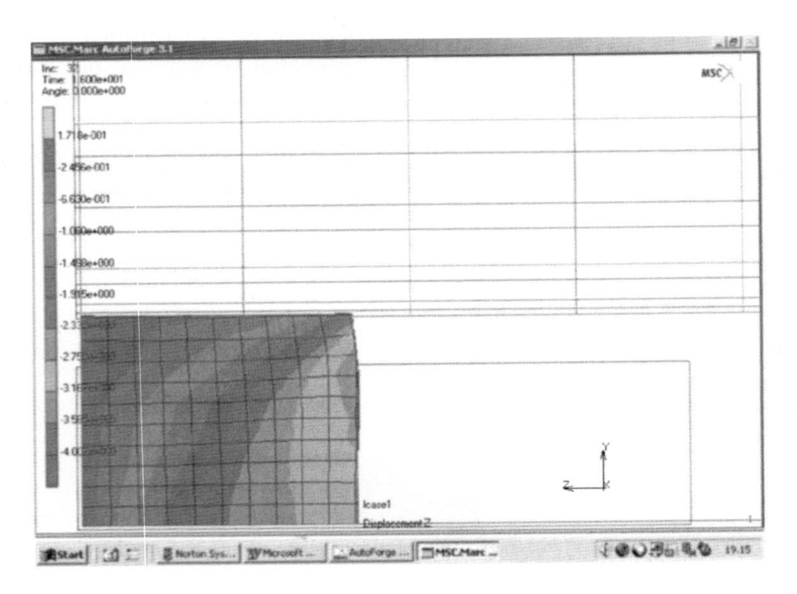

Fig.3 Spread calculated by FEM simulation

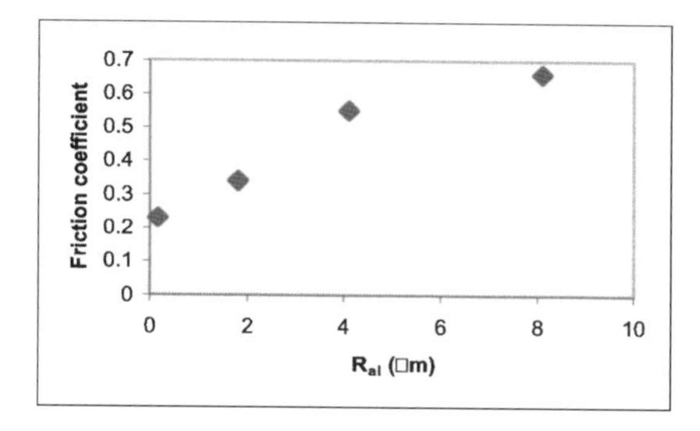

Fig. 4 Correlation between friction coefficient and roughness for steel rolls

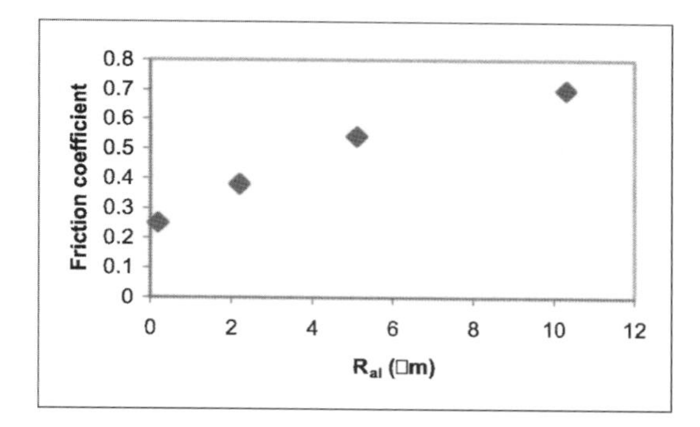

Fig. 5 Correlation between friction coefficient and roughness for cast iron rolls

5. Conclusion

In this work, the friction coefficient between rolls and workpiece in a hot rolling process has been evaluated, starting from the values of rolls roughness.

In particular, the friction coefficient and the rolls roughness (R_a) have been correlated comparing the values of workpiece spread, measured in experimental tests and obtained varying the roughness of rolls, with the ones assessed by FEM setting the value of friction coefficient.

The friction coefficient appears to increase with the roughness for both types of rolls material. In particular, the values of friction coefficient estimated for cast iron rolls are lightly higher than the ones obtained for steel rolls.

In order to validate the FE method, a direct comparison between simulation results and experimental data has been carried out by means of the formula concerning the greatest height reduction.

In fact, a value of the friction coefficient equal to 0.69 has been calculated, putting in the following formula (4) the data obtained from the results of the second series of experimental tests:

$$\Delta H = 2r\,(1-\cos\rho) \tag{4}$$

where r is the rolls radius,
ρ is the friction angle
ΔH is the maximum height reduction

For this value of friction coefficient a value of the workpiece width results from the second series of experimental tests.

In fig. 6 the values of the workpiece width as a function of the friction coefficient values, obtained from the FEM results with steel rolls and from the data of the second series of experimental tests, are reported.

Furthermore, in fig. 6 the final workpiece width, obtained with a rolls roughness R_a equal to 0,19 μm, is reported for a value of friction coefficient of 0,3.

This one was calculated by the formula (3), assuming just so smooth steel rolls with a value of R_a equal to the above mentioned and a rolling temperature of 1070°C.

Formulae (1), (2) and (3) present in [11], supply approximate values of friction coefficient, while adopting a FE method it is possible to establish an exact correspondence between the roughness (R_a) and the friction coefficient.

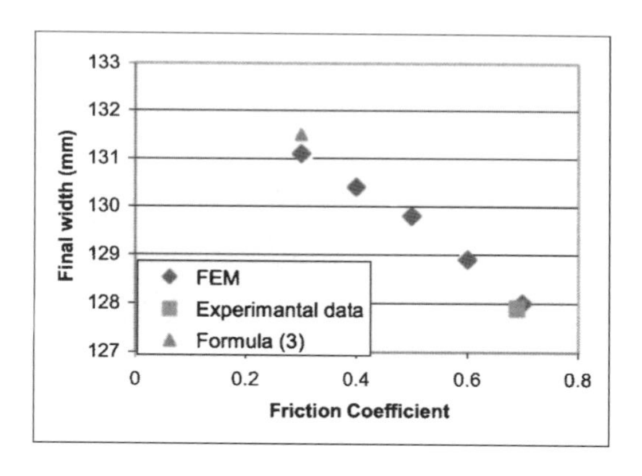

Fig.6 Comparison between FEM simulation, experimental and analytical results

The knowledge of the friction coefficient, starting from the rolls roughness, can be employed not only for an evaluation of the rolled section spread, but also for an assessment of the power necessary for the manufacturing.

Acknowledgements

This research work was carried out with support from the EC FP6 NoE on Innovative Production Machines and Systems – I*PROMS.

References

[1] W. Shin, S.M. Lee, R. Shivpuri, T. Altan . Finite-slab element investigation of square to round multipass shape rolling. Journal Materials Processing Technology, 33(1992) 141-154.

[2] K. Komori, Simulation of deformation and temperature in multipass calibre rolling. Journal Materials Processing Technology, 71(1997) 329-336.

[3] Z.Y. Jiang, S.W. Xiong, Liu X.H. GD Wang, Q. Zhang. 3-D rigid-plastic FEM analysis of the rolling of a strip with local residual deformation. Journal Materials Processing Technology, 79 (1988) 109-112.

[4] ZY. Jiang, A.K. Tieu. A method to analyse the rolling of strip with ribs by 3-D rigid visco-plastic finite element method. Journal Materials Processing Technology, 117 (2001) 146-152.

[5] B. Carlsson . The contact pressure distribution in flat rolling of wire. Journal Materials Processing Technology, 73 (1998) 1-6.

[6] R. Iankov, Finite element simulation of profile rolling of wire. Journal Materials Processing Technology, 142 (2003) 355-361.

[7] K. Mori, K. Osakada, T. Oda. Simulation of plane-strain rolling by the rigid-plastic finite element method. International Journal of Mechanical Science, 24 (1982) 459-468.

[8] C. Liu, P. Hartley, C.E.N. Sturgess, G.W. Rowe. Finite-element modelling of deformation and spread in slab rolling. International Journal of Mechanical Science, 29 (1987) 271-283.

[9] P. Hartley, S.W. Wen, I. Pillinger, C.E.N. Sturgess, D. Petty. Finite element modelling of section rolling. Ironmaking and steelmaking, 20 (1993) 261-263.

[10] D.R. Durhan, B. von Turkovich. Modelling of frictional boundary condition in metal deformation. Annals CIRP, 40 (1991) 235-238.

[11] Z. Wusatowsky, Fundamentals of rolling, Pergamon Press, London-New York, 1969.

Intelligent Production Machines and Systems
D. T. Pham, E. E. Eldukhri and A. J. Soroka (eds)

Advanced Hybrid Mechatronic Materials for ultra precise and high performance machining systems design

Fabrizio Meo[a], Angelo Merlo[b], Maria de la O Rodriguez[c], Bernhard Brunner[d], Massimo Ippolito[e]

[a] *Fidia S.p.A.,Corso Lombardia 11, 10099 San Mauro Torinese, Italy*
[b] *Ce.S.I. Centro Studi Industriali, via Tintoretto 10, 20093 Cologno Monzese, Italy*
[c] *Fundacion Fatronik, Ibaitarte 1, 20870 Elgoibar, Gipuzkoa, Spain*
[d] *Fraunhofer Institut Silicatforschung, Neunerplatz 2, 97082 Wuerzburg, Germany*
[e] *Sequoia Automation srl, via XXV Aprile 8, 10023 Chieri, Italy*

Abstract

The machinery field is traditionally a sector poor in term of structural material solutions, but where machine performances are increasing so sharply that new materials are urgently required. Getting more intelligent and integrated structural solutions is essential to really introduce enormous benefits in machine tool design and development. The future technological challenge will be the availability of a new class of smart composite materials whose elasto-dynamic response can be adapted in real time in order to significantly enhance the performance of structural and mechanical systems like machine tools under a wide range of operating conditions. The project "Advanced Hybrid Mechatronic Materials for ultra precise and high performance machining systems design" (HYMM) is trying to answer these issues.

1. Introduction

The ever growing competition on the international markets pushes manufacturers towards shorter design cycles and decreasing manufacturing times and costs for their products. This trend generates a demand for smart, flexible and faster machining systems, easy to set up and configure, which are able to drastically reduce machining time, while improving the final accuracy. Machine axis acceleration of speed 3-5 times higher than conventional ones together with machining accuracy in the range of 0.5-1 μm (considering workpiece size of 1500-2500 mm) will be the most probable targets of the new generation of machining systems inside manufacturing shops. Inertial forces, dynamic vibration and stability problems arising from such accelerations will be so big that, if no suitable solutions are provided, the precision and machining quality will be definitely endangered.

Strong mass reduction of mobile machine parts together with the increasing of their stiffness and damping to get excellent static, dynamic and thermal stability of the structures is becoming a 'must', to ensure a technological and cost-effective achievement of such ambitious goals.

Conventional materials for building machine tools are commonly cast iron, welded steel and in some case aluminum-alloy. Recently, especially in some running or recently finished EU projects (ULMAT, VINO, PRIMA) studies have been carried out to introduce

Polymeric Matrix Composites (PMC) fibers reinforced materials, aluminum Honeycomb and glued structures. Even if good results in term of mass reduction and damping increase seem to have been achieved within these projects, this is absolutely not enough to get the declared excellent technological machine performances.

Moreover thermal stability, in such a range of expected machine accuracy, is so important that a gradient of temperature in the machine structure of few degrees can drastically compromise the machining precision.

The machinery field is traditionally a sector poor in term of structural material solutions, but where machine performances are increasing so sharply that new materials are urgently required. Getting more intelligent and integrated structural solutions is essential to really introduce enormous benefits in machine tool design and development.

The future technological challenge will be the availability of a new class of smart composite materials whose elasto-dynamic response can be adapted in real time in order to significantly enhance the performance of structural and mechanical systems like machine tools under a wide range of operating conditions.

This paper first describes the main problems to be dealt with, then discusses some possible passive and active solutions for the damping of vibrations and possible solution for the compensation of thermal deformations.

2. Requirements and possible solutions

Requirements have been defined in terms of which improvements are significant and can be expected, relatively to critical areas that have been identified in the design of machine tools. The source for this are both the experience of the machine tool builders and the results of tests that were performed during the first phase of the project.

2.1 Reduction of Vibration in a low frequency domain (0-50 Hz)

The vibrations in this range are due to inertial effect induced by forces generated by motors during acceleration phase (in positioning or following curved profile while machining). The result is a bad surface quality and accuracy, noise, and sometimes the breaking of the tool or of its inserts. The workaround usually used to counterbalance this vibration is to decrease the dynamic parameters of the system (axes speed and acceleration, gain of the position control loop, gain of the velocity control loop inside the servo-drives). In the end the result is either a decrease of the quality of machining, or an increase in the time needed (Numerical Controls usually provide parameters to customize the compromise between quality and time), which in general means a degradation of the system performance. The best quantitative way to evaluate the dynamic quality of the system is measuring the bandwidth of the electro-mechanical system constituted by the mechanical structure plus the servo drive. The typical bandwidth of a low quality axis would be about 10 Hz, where a high quality axis can reach about 100 Hz, basically depending on its mass. Moreover in some cases these frequencies may change significantly according to the machine position.

Possible ways to counterbalance this effect are implementing very light, stiff and intrinsically damped material/structure for the moving parts (passive solutions) and possibly acting at level of drive technology and CNC control algorithms.

2.2 Reduction of Vibration in a high frequency domain (50-1000 Hz)

The vibration in this range are induced by the cutting process, and cause bad surface quality and roughness, noise, and sometimes the breaking of the tool or of its inserts.

The tests confirmed what was already expected: the main critical frequency can be calculated as critical Frequency = Spindle Speed * number of cutters. Higher harmonics, and in some cases other harmonics are also present.

Variability of critical frequency in this case must be taken into account: both spindle speed and number of cutters can change in a wide range according to the kind of machine and the kind of machining operation, but they change every now and then during the operation, due to tool change and to the performing of a different machining operation. The most critical case occurs when the spindle rotates at speeds which excite the resonant modes of the machine, which happens when the shaft rate or its harmonics lie in the vicinity of the resonant peaks of the machine. In some cases multiple frequencies need to be damped. In a general case a wide frequency range (50-1000 Hz) must be considered.

A significant damping of the amplitude of this vibration would allow a much higher speed during the machining operations, or a higher removal rate of the material, both resulting in an extreme saving of time for the machining operation and as a consequence higher productivity. Since this vibration usually arises from the ram of the machine, the target will very likely be achieved through innovative active devices for the damping of vibrations generated around the spindle unit.

2.3 Reduction of structural displacements due to thermal effects or elastic deformations

Thermal aspects have a big importance since they represent one of the main causes of lack of accuracy in machine tools (40 – 70 % of the overall inaccuracy is due to thermal causes).

When a machine tool is working, there are many heat sources. Some of them are internal sources such as the heat originated in different elements of the machine, by motors, spindles, gears, rolling elements, ball screw, guideways, etc. Other heat sources come from the machining process itself, such as the heating of the tool, workpieces, chips, or the heat (or cooling) provided by the lubrication and cooling systems. There are also external heat sources such as variations in environment temperature, radiation, air stream, etc. Some of these sources are more or less well known and controllable, but others are difficult to control.

Several attempts were already made to compensate the negative effects of at least the dominant disturbances, like for instance in the projects HIQU (High Quality in Moulds and Dies Production, EP 6293; 1993÷1996) and MEDECOTHER (Measurement and compensation of Thermal error in machine tool, BE 3545, 1996÷1999). Being a physical phenomenon that is affected by no linearity, hysteresis and a multi variable dependence, the problem has been approached most of the times either in a too simplified way or adopting a too complex procedure. The conclusion was always that achieved results could be applied to individual applications only, with almost no possibility of transferring them to another application or machine tool.

Many steps can be taken in order to decrease the lack of accuracy of machine tools due to thermal deformations: on one hand there are the steps that can be taken into account during the design phase, to diminish the heat source effects, acting on heat transfer across the machine and to the environment, modifying the geometric architecture of the machine. On the other hand it is possible to use rough thermo-mechanical compensation methods as well as compensation by means of numerical control. Moreover warm-up cycles can be performed every time the machine is switched on.

In this case the problem is basically the measurement of deformations at Tool Center Point (TCP), since these phenomena are usually slow enough to allow an effective compensation to be performed by the Numerical Control. The main problems is then how to measure in real-time this drift by means of smart sensors.

3. Passive solutions for the reduction of vibrations

Light damped solutions are taken into consideration. Passive dampers tuned at one frequency do not represent an innovative solution: they are state of the art and above all they work at just one frequency, while a wide bandwidth in the frequency domain is needed.

Two different kinds of solutions are being analysed in terms of achievable stiffness, weight, cost, and production technology:

- polymeric composite materials fibers reinforced (CFRP) material, using carbon fiber and epoxy matrix;
- sandwich (pyramidal core) construction consisting of prismatic core and steel skins (see Figure 1 and 2).

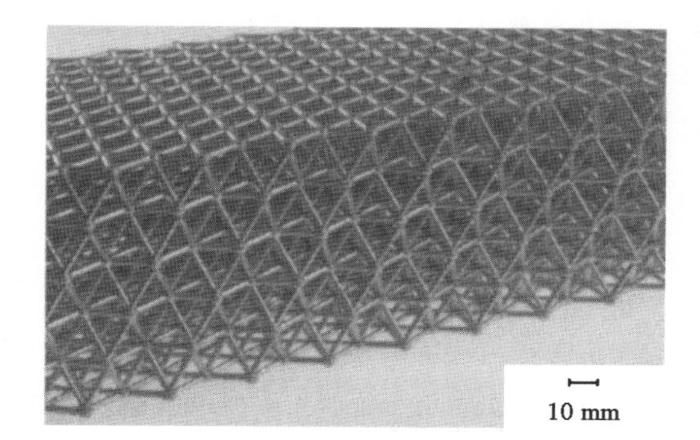

Figure 1: example of pyramidal core construction

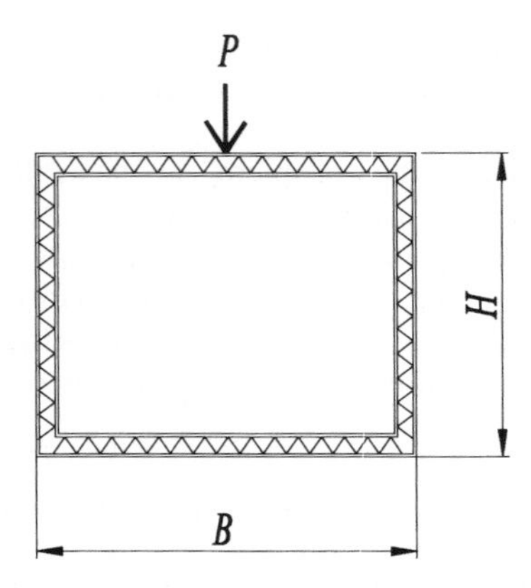

Figure 2: Cross-section of the designed Ram with sandwich walls

4. Active solutions for the reduction of vibrations

4.1 Tuned Absorbers

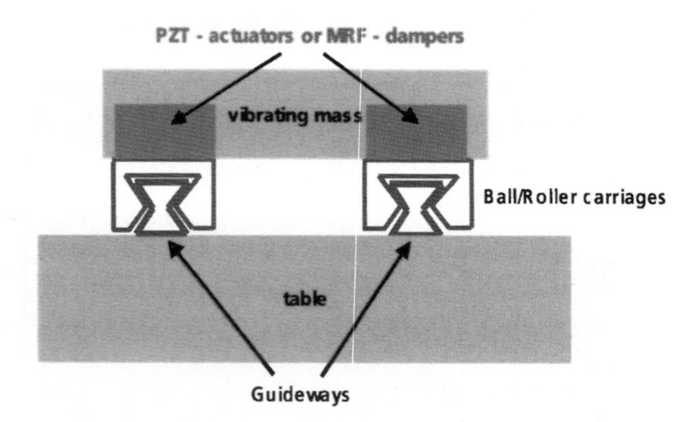

Figure 3: tuned absorber

This solution is based on piezo-electric (PZT) actuators or magnetorheological fluid (MRF) dampers located close to the guideways to obtain smart active damped guideways. It may be used to damp the so-called "cross-effect", consisting in vibrations being transmitted from one axis of the machine to another one, and also vibrations transmitted from the external.

4.2 Smart Platform

This is an active vibration spindle control with actuators working in series with the flux forces; it is constituted by a platform with four degrees of freedom (three rotational and one translational). The system consists of a "Fixed platform", a "Mobile platform" and piezo-actuators.

Figure 4: Smart Platform

The system uses three piezo stack actuators having the longitudinal axis parallel with the spindle axis. When the length of the piezo actuators is modified in a suitable way (with opportune algorithms) the spindle is rotated around a virtual axis lying in a plane normal to the axis of the ram.

The main advantage of this system is the better control of spindle position to be used for active vibration control. Some disadvantages could be related to the loss of stiffness (due to the 'in-series' arrangement of the piezo-actuators) and to the size of actuators which can cause a big size of the device.

5. Solutions for thermal deformation problems

One topic in the project is the integration of smart sensors for measuring thermal deformations in the machine structure. This integrated measurement can simplify the mathematical model to predict the tool center point (TCP) movement caused by thermal expansion or vibration. For the control algorithms and the working of the machine tool, reliable sensor sytems are crucial. Optical sensor systems have been analysed, because they are not subject to electromagnetic interference effects, can be easily embedded in the machine structure, have a long lifetime and can measure temperature and strain at same time.

5.1 Fiber Bragg Gratings

Fiber Bragg grating (FBG) sensors (see Figure 5) have been subject to continuous and rapid development since they were first demonstrated for strain /deformation and temperature measurements in 1989.

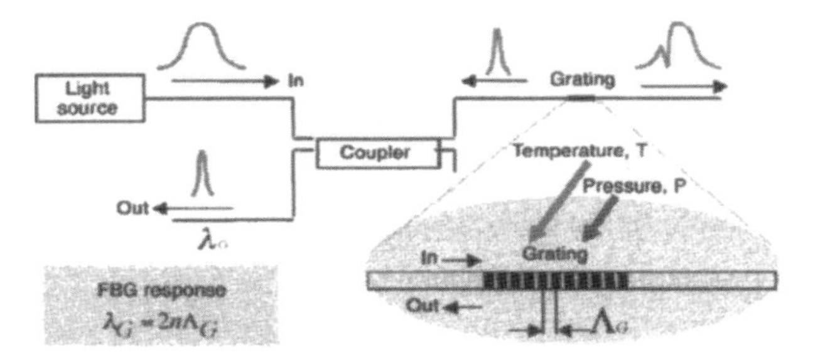

Figure 5: Principle of operation of fiber Bragg gratings

The main reason for this is because FBG sensors have a number of distinguishing advantages [1],[2], [3]:

- high sensitivity (1 με = 1μm / m = 0.0001% in strain; 0.5°C in temperature)
- maximum strain 1%
- small size (diameter ~ 0.5 mm) and light weight
- immunity to electromagnetic interference (radiation, radio frequency interference), no electrical contacts
- immunity to chemical agents, insensitive to corrosion
- temperature range –40°C – 300°C
- possibility of measuring at multiple points at arbitrary spacing with a single optical fiber ⇒ "multiplexing

All these advantages lead to different technical applications for measuring of strain, displacement, pressure, vibration, deformation, temperature.

5.2 Extrinsic Fabry Perot Interferometer Sensors

The basic principle of extrinsic Fabry Perot interferometer (EFPI) sensors is based on the multi-reflection Fabry-Perot interference between the two reflecting mirrors. The EFPI sensor is made using a single-mode optical fiber and a multi-mode fiber as the two reflectors. The two fibers were inserted and fused by epoxy into a quartz capillary tube with a larger diameter. The cavity length between the two surfaces of the optical fibers is changed when an external load is applied onto the sensor area. The cavity length can be determined by using a CCD spectrometer. So the strain of the sensor can be calculated.

Advantages are similar to the FBGs:

- absolute measurements of strain, displacement and temperature providing consistence time after time
- insensitivity to light loss due to fiber bending, cable length or light source fluctuations
- precision better than 0.05 % (∓ 1 °C, resolution 0.1 °C; 10 μm, resolution 2 μm)
- frequency range up to 1 kHz
- embeddable in laminates or bondable on surfaces (diameter of the glass fiber ~200 μm)

The conclusion is that optical fiber sensors are suitable for thermal and vibration measurements of machine tool parts, and will be tested for this goal.

6. Conclusions

The project, that will end in mid-2007, has now entered the phase of building a set of samples that will allow to verify the effectiveness of the described solutions.

Even a partial success of these solutions, that is even though only some of them demonstrate to be effective, will however represent a significant breakthrough in the area of high-performance machine tools.

Acknowledgements

The project "Advanced Hybrid Mechatronic Materials for ultra precise and high performance machining systems design" (HYMM) is a project funded by the European Commission under the NMP priority (NMP3-CT-2003-505206).

The author wish to thank the other partners of the consortium: Ce.S.I. (Centro Studi Industriali) (I), Sequoia Automation (I), Cambridge University – Engineering Department (UK), MS-Composites (F), Fundacion Fatronik (E), Czech Technical University (CZ), Fraunhofer ISC (D).

Fidia S.p.A., Cambridge University, Fundacion Fatronik and the Czech Technical University are partners of the EU-funded FP6 Innovative Production Machines and Systems (I*PROMS) Network of Excellence. http://www.iproms.org

References

[1] Y. Rao, Optics and Lasers in Engineering 31 (1999), 297 – 324

[2] M. Wippich et.al., The industrial Physicist 6/2003, 24-27

[3] J. Degrieck et. al., NDT.net 4, (1999)

[4] S. Takeda et.al., Composites A 33 (2002), 971 – 980

[5] N. Theune et.al., Proceedings of OPTO 2000 Conference

[6] Kin-tak Lau, et.al., Composites B 32 (2001), 33 – 45

[7] V. Dewyaters-Marty et.al., J. Intell Mater Syst Struct 9 (1998), 785 – 787

[8] J. Leng et.al., NDT&E International 35 (2002), 273 – 276

Intelligent Production Machines and Systems
D. T. Pham, E. E. Eldukhri and A. J. Soroka (eds)

Advanced ultrasonic system for high performance composite laminate nondestructive evaluation

G. Caprino, R. Teti, V. Lopresto, I.L. Baciu, T. Segreto, A. Busco

*Department of Materials and Production Engineering,
University of Naples Federico II, Ple. Tecchio 80, 80125 Italy*

Abstract

A quantitative non-destructive evaluation (NDE) of impact damaged biaxial and quadriaxial non-crimp fabric (NCF) composite laminates made by resin infusion under flexible tooling (RIFT) technology using a volumetric ultrasonic (UT) pulse-echo immersion testing system was carried out. NCF composites laminates made by RIFT technology were subjected to drop weight low-velocity dynamic tests to generate impact damage in the composite material. The main scope of the UT NDE based on UT image analysis is the characterisation, measurement and comparison of the damage development in the entire laminate material volume.

1. Introduction

The evaluation of the damage after impact load can be useful in the determination of the residual properties of advanced composite materials. The interaction between failure modes is also critical to understand damage mode initiation and propagation in these new generation materials [1].

Since it has been found that relatively low impact energy levels can determine significant strength losses in composite laminates, extensive research efforts have been devoted to the relationships between impact parameters, failure modes, damage extension and residual strength retention after impact.

Due to the increasing interest of the aerospace industry, non-crimp fabric (NCF) composite laminates will be considered in this paper.

In comparison with classical laminates introduced in aeronautical structures with the primary goal to reduce weight, NCF composite present lower fabrication costs. Moreover, this composite material type has shown a great potential where high impact resistance is required.

The aim of the present research is to validate what has been found on classical composite laminates in terms of impact damage.

In order to verify the influence of different material design configurations, two different NCF laminate lay-ups were considered.

The ultrasonic (UT) nondestructive evaluation (NDE) system utilized in this work for consists of a purposely designed hardware configuration and a custom made software code, Robotest©, developed in the Lab View environment [2].

2. Ultrasonic Nondestructive Evaluation System

The UT NDE system hardware configuration (see Fig. 1) is composed of a number of functional elements:

- Oscillator/detector, generating the electrical pulses for the UT probe and receiving the returning pulses;

- Transmitter/receiver UT probe;
- Digital oscilloscope connected to the oscillator/detector allowing for the acquisition, visualization and digitalization of the UT pulses; through a GPIB interface, the data are transferred to a PC for UT signals post-processing;
- PC, carrying out the UT data acquisition and processing, and controlling the displacement of the mechanical system;
- Mechanical system, consisting of a 6-axis Staubli RX 60 L robotic arm displacing the UT probe.

The software for UT NDE is a custom made software, Robotest©, developed in the LabView environment, with the purpose to control the UT NDE system displacement with 6 degrees of freedom and provide for the complete UT signal detection, storage and analysis (3D UT NDE) [2].

The Robotest© software contains various options for different test procedures. In this work, UT tests were carried out using the Full Volume Scan (FV-Scan) procedure [3, 4].

This kind of scan consists in the detection and digitization of the whole UT waveform for each position of the transducer during scanning. At the end of the scan, UT data are organized in a volumetric file containing the whole set of complete digitized UT waveforms.

From the UT volumetric file, UT images corresponding to any segment of the UT signal, i.e. to any portion of the material thickness, can be obtained and analyzed. The software also allows to retrieve the single UT waveform corresponding to any given in-plane location, identified by mouse clicking on the UT image of interest.

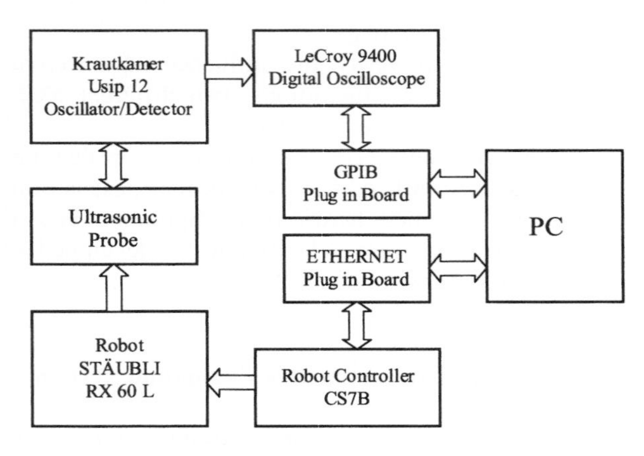

Fig. 1. Ultrasonic NDE System.

2. Materials and experimental work

2.1 Materials

Low velocity impact tests were carried out on NCF composite laminates 4 mm in thickness, made by the RIFT technology.

The matrix was an Hexcel HexPly®M36 resin and the carbon fibres were type Tenax HTS 5632 12k. Polyester stitching yarn (50 dtex) was used for connection with stitch length 2.5 mm. The final fibre volume fraction was 62%.

Biaxial laminates were constructed with stacking sequence $[(+45°/-45°), (0°/90°), (+45°/-45°), (0°/90°), (90°/0°), (+45°/-45°), (90°/0°), (+45°/-45°)]_s$ whereas quadriaxial laminates were manufactured with stacking sequence [+45°/0°/-45°/90°, +45°/0°/-45°/90°, 90°/-45°/0°/+45°,90°/-45°/0°/+45°].

In both cases, a final fibre aerial weight of 267 g/m^2 per layer was obtained.

2.2 Mechanical testing

The mechanical tests were carried out on an instrumented falling weight machine type Ceast Fractovis (Fig. 2) connected to a Das 4000 program for data recording and analysis. Rectangular specimens 100 mm x 150 mm were cut from the laminates and impacted with different impact energies.

A cylindrical indenter with hemispherical nose was used for impact tests. The fixture in figure 3 was used for NCF laminate specimen clamping.

Fig. 2. Ceast Fractovis falling weight machine.

Figure 3. NCF laminate clamping fixture

Table 1
Impact testing program (pen = penetration)

Laminate configuration	Impact energy (J)
Quadriaxial	9-12-16-20-25-30-40-pen
Biaxial	9-16-20-30-40-pen

Table 2
Test conditions for drop weight low-velocity impact tests

Sample id.	Height (m)	Mass (kg)	U (J)	Speed m/s
QHL3IM2	0.255	3.6	9	2.24
QHL3IM3	0.34	3.6	12	2.58
QHL3IM4	0.453	3.6	16	2.98
QHL3IM5	0.567	3.6	20	3.33
QHL3IM2	0.255	3.6	9	2.24
QLL3IM6	0.708	3.6	25	3.73
QLL3IM7	0.546	5.6	30	3.27
QLL3IM8	0.728	5.6	40	3.78
QLL3IM9	0.411	5.6	22.58	2.84
BHLIM2	0.254	3.6	9	2.24
BHLIM3	0.453	3.6	16	2.98
BHLIM4	0.566	3.6	20	3.33
BHLIM5	0.546	5.6	30	3.27

Table 1 and 2 summarize the test parameters utilised in the impact testing program. The analysis of the preliminary test data allowed for the selection of suitable energy levels that were adopted in the continuation of the experimental program to obtain controlled damage levels in the specimens (Table 2).

The force-displacement curves each impact test were recorded and post-analysed.

2.3 Ultrasonic testing

After mechanical testing, the specimens were subjected to UT volumetric scanning using the custom made software code Robotest© for FV-Scan UT NDE .

Pulse-echo immersion volumetric UT scans were carried out using a focused (49.6 mm focal length), high frequency (15 MHz) transducer for maximum resolution. The oscillator/detector was set up at 90 dB gain and medium damping. The digital oscilloscope was set up at 1 V/div, 0.5 μs/div, and sampling frequency 100 MHz, resulting in 500 samplings detected for each UT waveform.

Each specimen was scanned over an area of 110 x 155 mm with scan step 1 mm. After UT NDE the delaminated area was measured.

3. Results and discussion

3.1 Mechanical testing results

In Fig. 4, the delaminated area, measured through UT NDE, is plotted vs. impact energy, U. All experimental points approx. follow a straight line. Interestingly, the biaxial laminates show a smaller damaged area.

In comparison with classical laminates, a larger delaminated area was noted for the NCF material. For the impact energy necessary to obtain the barely visible impact damage (BVID), a delaminated area of 4000 mm^2 was observed, much higher than the correspondent value for classical laminates (2500 mm^2 for the 4 mm thick laminates) [5]

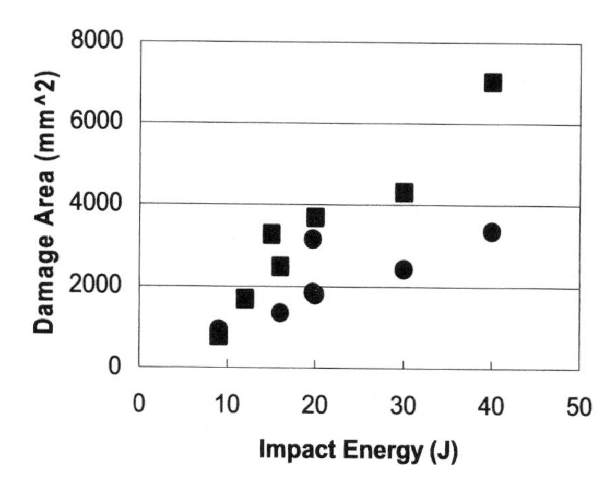

Fig. 4. Delaminated area vs. impact energy.
■ = quadriaxial laminates, ● = biaxial laminates.

The inspection of thick laminates results, therefore, quite critical because very small impact energies may produce large delaminated areas within the material structure whereas the laminate external surface only shows a small damaged zone. The problem is less critical for biaxial laminates that showed smaller delaminated areas under the same test conditions as for the quadriaxial laminates.

3.2 Ultrasonic non-destructive evaluation results

3.2.1. UT image generation

As previously illustrated, the Robotest© software code allows for the generation of single or multiple UT images of any thickness portion of the internal structure of the material under examination by processing the UT volumetric file containing the complete UT waveforms detected during the FV-Scan.

In order to obtain UT images, it is first necessary to retrieve from the volumetric file a typical UT waveform. Then, a time gate is set on this waveform to identify the material portion(s) to be represented. The time gate can be divided into a desired number of equal sub-gates to obtain multiple images. One image is generated for each sub-gate and each image represents the internal structure of the corresponding portion of laminate thickness.

In this work, a multiple image generation procedure was utilized by setting a time gate on the whole UT waveform and dividing it into four parts to generate four UT images, each corresponding to 1/4 of the laminate thickness (about 1 mm).

This choice was justified by the laminate stacking sequence and the typical behavior of composite laminates, where delamination develops preferentially at the interfaces between layers having different orientations [6].

In order to generate a UT image of the material thickness portion corresponding to the gated region, different algorithms can be applied to the UT waveform segment.

In this work, the peak amplitude value of the segmented signal was calculated. A 2D matrix of size equal to the number of scan steps in the x and y directions, containing the peak amplitude values of the gated signal segment for each material UT interrogation point, was obtained. This matrix was mapped to form 2D images using grey tones or pseudo colors (Figure 5).

3.2.2. UT image analysis

In Figure 5, the four UT images obtained from the volumetric UT files of a quadriaxial and biaxial laminate are reported for drop weight low-velocity impact tests with same impact energy (9 J). Each image in the series represents the laminate internal structure of 1/4 of the material thickness, starting from the upper surface (first image on the left) down to the opposite lower surface (last image on the right).

By looking at the UT images in Figure 5, it can be seen that the damage develops along the interfaces between layers having different orientations. Moreover, the delamination extension increases with increasing distance (depth) from the impact surface. This results in the well known hat-shaped configuration of the delamination damage [7].

The UT analysis also reveals an absence of delamination in a small zone directly below the impactor-material surface contact point.

The presence of a delamination at a given interlayer within the material results in a "shadow" in the subsequent UT images. Consequently, the (d) UT images in Figure 5 represent the in-plane projection of the damage developed at different depths in the whole laminate thickness, rather than the damage in the fourth portion of the laminate thickness. Their examination allows for the assessment of the global damage induced into the material according to the traditional UT scanning techniques (C-Scan) and was used for global delaminated area measurements.

By comparing the UT images for the quadriaxial NCF specimen (see Fig. 5 upper image series) and for the biaxial NCF specimens (see Fig. 5 lower image series) impacted under the same testing conditions (see Table 2), it can be seen that the internal damage is always larger for the quadriaxial laminate, as stated in section 3.1 (see Fig.4).

In Fig. 6, the four UT images obtained from the volumetric UT file of a classical quasi-isotropic carbon fiber resin reinforced laminate tested under similar conditions as the NCF laminates (UT probe 15 MHz, step 0.5 mm, scanning area 75 mm x 75 mm, impact energy 8.43 J) [8].

By comparing the UT image series of the classical and the NCF laminates, it can be clearly seen that the latte advanced composite material type is characterized by a much more reduced and circumscribed impact induced delamination damage, confirming the great potential of NCF composite laminates when high impact resistance is a critical material property.

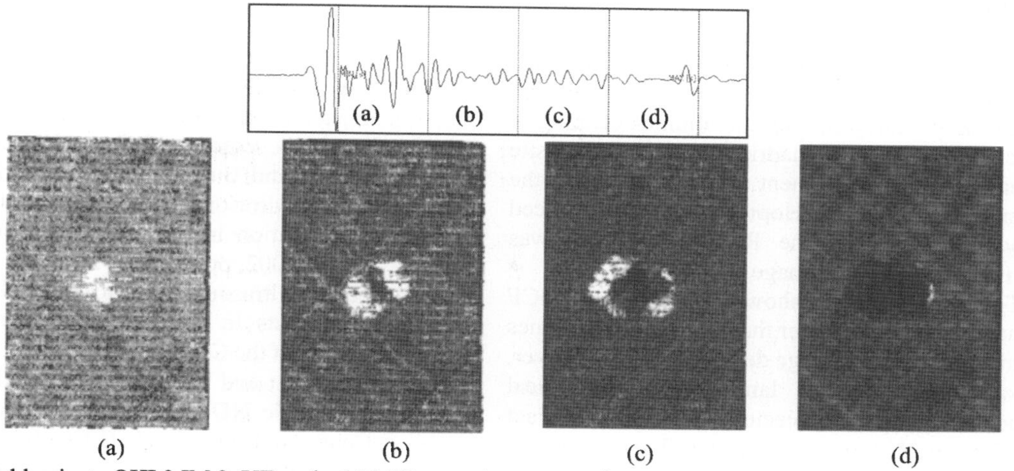

Quadriaxial laminate QHL3 IM 2: UT probe 15 MHz, step 1 mm, scanning area 110 mm x 155 mm, impact energy 9 J

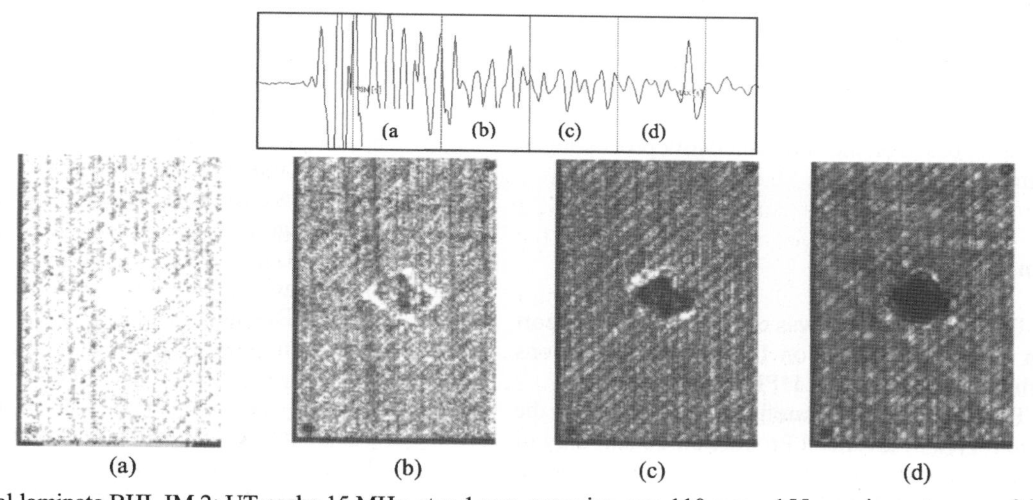

Biaxial laminate BHL IM 2: UT probe 15 MHz, step 1 mm, scanning area 110 mm x 155 mm, impact energy 9 J

Figure 5. UT images from drop weight low-velocity impact test specimens QHL3IM2 (upper) and BHLIM2. (lower) (a) surface damage, (b) and (c) internal damage, (d) in-plane projection of the damage

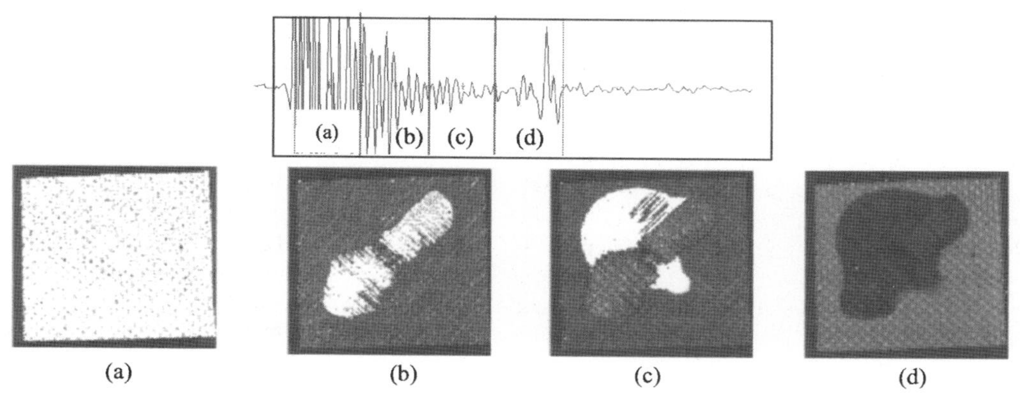

Figure 6. UT images from drop weight low-velocity impact test samples for carbon fiber resin reinforced laminate Ind3d1: UT probe 15 MHz, step 0.5 mm, scanning area 75 mm x 75 mm, impact energy 8.43 J.
(a) surface damage, (b) and (c) internal damage, (d) in-plane projection of the damage.

4. Conclusions

An innovative 3D UT NDE system, based on complete UT waveform detection and processing, was utilized to carry out the full volume-scanning of impacted biaxial and quadriaxial NCF composite laminates. The measurement and comparison of the volumetric damage development in the advanced composites made by the RIFT technology was performed through UT image analysis.

The results obtained showed that the biaxial NCF laminates performed better than the quadriaxial ones in terms of internal damage development. Moreover, by comparing the NCF laminates with classical composite laminates subjected to the same impact conditions, in NCF material it is possible to observe a much more reduced and circumscribed delamination damage.

The further developments of the research work will include the automatic characterization and measurement of volumetric damage in advanced composite components on the basis of UT image evaluation based on the neural-network processing of UT image data [9].

Acknowledgements

This research work was carried out with support from the EC FP6 NoE on Innovative Productions Machines and Systems – I*PROMS.

The authors are grateful to the staff of the Laboratory for Advanced Production Technology of the University of Naples Federico II for their help and support and to P. De Santo for the realisation of the UT FV-scan tests as part of his thesis work.

References

[1] Liu D. and Malvern L.E. Matrix cracking in impacted glass/epoxy plates, J. of Composite Materials, Vol. 21, 1987, pp. 594-609.

[2] Teti R. and Buonadona P. 3D surface profiling through ultrasonic reverse engineering, Intelligent Computation in Manufacturing Engineering 3, Ischia, 2002, pp. 387-394.

[3] Teti R. Ultrasonic Identification and Measurement of Defects in Composite Material Laminates, Annals of the CIRP, 1990, vol. 39/1: pp. 527-530.

[4] Teti, R. and Buonadonna P. Full Volume Ultrasonic NDE of CFRP Laminates, 8th Eur. Conf. on Comp. Mat. (ECCM-8), Naples, 1-5 June, 1998, Vol. 3: pp.317-324.

[5] Caprino G. and Lopresto V. Fracture surfaces of CFRP laminates statically loaded at the centre", Proc. of ECCM11, 31 May- 3 June Rhodes, Greece, 2004.

[6] Liu, D. Impact induced delamination: a view of bending stiffness mismatching, J. Compos. Mater., 1988, vol. 22: pp. 674-92.

[7] Abrate, S., Impact on Composite Structures, Cambridge University Press, 1998.

[8] Teti R., Lopresto V, Buonadonna P and Caprino G. Ultrasonic Non-destructive Evaluation of Impact Damaged CFRP Laminates, 10th Eur. Conf. on Comp. Mat. (ECCM-10), Bruges, Belgium, 3 – 7 June, 2002, paper n. 346.

[9] Teshima T., Shibasaka T., Takuma M., Yamamoto A. Estimation of Cutting Tool Life by Processing Tool Image Data with Neural Network, Annals of the CIRP, 1993, vol. 42/1: pp. 59-62.

Intelligent Production Machines and Systems
D. T. Pham, E. E. Eldukhri and A. J. Soroka (eds)

Application of Artificial Neural Networks in the prediction of quality of wastewater treated by a biological plant

C. Leone, G. Caprino

Department of Materials and Production Engineering, University of Naples "Federico II"
Piazzale Tecchio, 80, 80125 Naples, Italy

Abstract

Industrial processes generate large quantities of waste, resulting in health problems and adverse environmental impact. In particular, the treatment and reconditioning of wastewater is a complex problem, due to the existence of strong non-linearity effects, time variant parameters and multivariable coupling not allowing the adoption of simple models to predict the process efficiency and the output water quality.

In this paper, the ability of Artificial Neural Networks (ANNs) to predict the quality (pH, electrical conductivity, chemical oxygen demand (COD)) of the wastewater coming from a pharmaceutical industry after treatment in a biological plant was verified. Using a commercial ANN software, various network architectures, differing in the number of hidden layers and nodes, were tested, in order to find an optimised solution in terms of both precision and learning time. The effectiveness of each ANN configuration was verified by the "leave-k-out" method. Even the simplest ANNs tested were able to correctly describe the pH, due to the relative insensitivity of this parameter to the process conditions. Matching the actual variation of the electrical conductivity proved harder, this task being achieved at the expense of a complication in the network architecture. However, the parameter most difficult to reproduce was the COD, which underwent considerable oscillations within the time window considered. The best ANN architecture was made of seven nodes in the input layer, two hidden layers of fifty nodes each, and three nodes in the output layer. By this solution, reasonable predictions were obtained, provided the input parameters were appropriately selected.

1. Introduction

Industrial processes result in large quantities of waste, generating health problems and adverse environmental impact. In the last decades, the increasing social awareness of this phenomenon has highlighted the need for a sustainable production, pushing the national and international legislations through more and more stringent regulations [1]. In this scenario, the development of optimised systems for the waste treatment and recycling has become crucial.

Food and pharmaceutical industry discharge highly concentred effluents (waste water), mainly containing organic components, which can be processed by means of biological treatments as anaerobic digestion in order to decrease pollution and reduce environmental impact [2-4].

The water treatment is a complex process where the existence of strong non-linearities, time variant parameters and multivariable coupling do not permit the use of simple models to reliably predict the efficiency of the treatment plants or the output water conditions [5]. In principle, Artificial Neural Networks (ANNs) seem to be promising to this task, since their main advantages have been experienced when multivariate problems, for which no analytical solutions exist, but a sufficiently large data-base is available, have been approached [6, 7]. In fact, ANNs have found quite recently widespread application in many engineering fields, among which pattern recognition, failure analysis, non-destructive evaluation, as well as process control [8-10]. Nevertheless, developing an efficient ANN for each single problem generally requires considerable efforts to correctly select the optimum network architecture, type of input parameters, and training algorithms [11].

In this work, ANNs were used to predict the water characteristics (pH, electrical conductivity (C), chemical oxygen demand (COD)) of a wastewater produced by a pharmaceutical plant after biological treatment. Different network configurations were developed and tested, changing the organization of the input data. An optimised network architecture, consisting of seven nodes in the input layer, two hidden layers of fifty nodes each, and three nodes in the output layer, was individuated and its predictive ability was ascertained. The most difficult parameter to reproduce was the COD. However, a reasonable estimate of this parameter was achieved by appropriately selecting the input parameters.

2. The biological plant

The biological plant considered is made of two tanks (Fig. 1), labelled as "equalization stage" and "biological stage", respectively. In the first tank, water volumes deriving from different production lines are collected and mixed. The liquid is then conveyed by a hydraulic pump to the second stage, where a biological mass digests the pollutant components.

In predetermined days, at 8 am, a water sample is extracted from both the equalization and the biological stage, and the following parameters are measured:

- pH;
- C;
- COD;

these quantities are affected by the indexes "e" and "b" hereafter, to distinguish the equalization and the biological stage, respectively.

In addition to the previous parameters, the water volume V in the equalization stage and the concentration MLSS of the biological mass in the biological tank are determined. From the values measured, a skilled operator suitably sets the flow rate P of the hydraulic pump, to successfully perform the treatment process. Due to the flow rates usually adopted, up to three days can be necessary to completely evacuate the water volume contained in the equalization tank.

The flow rate of the water outcoming from the biological stage is adjusted in such a way that the liquid volume within this tank is held constant.

For control purposes only, the COD_e is also measured periodically, at 8 pm.

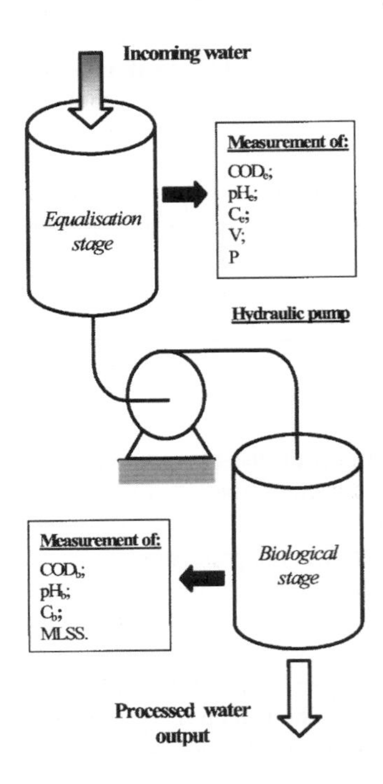

Fig. 1. Schematic of the biological processor.

3. Scope and ANN architectures

The scope of this work was the prediction of the parameters correlated with the water quality in output (pH_b, C_b, COD_b) at 8 am, knowing the input parameters (pH_e, C_e, COD_e, V, P, MLSS). The software used to this aim was Neudesk version 2.11 by Neural Computer Sciences, running in Windows environment, which allows many types of architectures and training algorithms to be selected by the operator.

The possibility to freely configure the architecture of an ANN, although useful in many instances, also raises the problem of the network optimisation. The latter was carried out suitably changing the typology and number of input data, the number of hidden layers, and the nodes characterizing each of them. In what follows, the symbol $n_1/n_2/\ldots/n_k/\ldots n_n$, where n_k

indicates the number of nodes in the k-th layer, will be used to designate the generic ANN architecture. In almost all the cases examined, the number of nodes in the output layer, n_n, was fixed, putting n_n=3 (i.e. pH_b, C_b, COD_b).

The degree of accuracy achieved by ANN during the training process was evaluated through the root-mean-square (RMS) error, calculated by comparing the network outputs with the actual output values. Generally, the RMS error progressively decays, until an asymptotic value is obtained, which represents the likely accuracy intrinsic in the architecture considered.

In the training stage, the allowable error was 0.01, the momentum 0.7, the step size 1 and the maximum iteration number (epochs) 1000. In all cases, the allowable error was achieved before the maximum epochs would be reached.

To test the effectiveness of the ANNs selected, the "leave-k-out" method was adopted: all the input data sets available, except one in turn, were used to build the training data set; the latter was utilised to train the network; then, the trained network was called to predict the pH_b, the conductivity, C_b, and the COD_b of the unknown specimen.

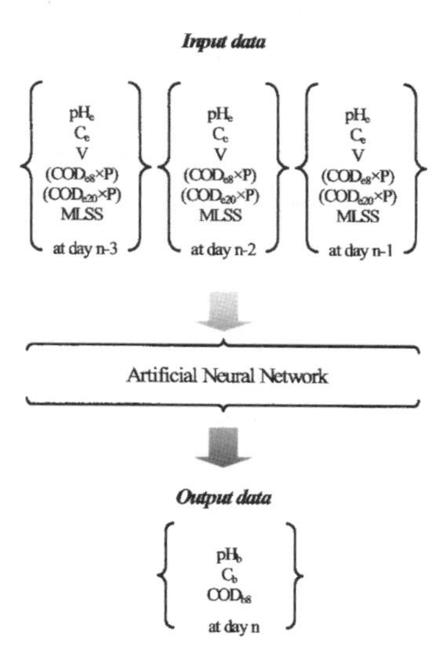

Fig. 2. General architecture of the first ANNs tested.

4. Results and discussion

As previously specified, the maximum time required to process the entire content of the equalization tank is 3 approximately days. Therefore, it was speculated that, if a correct evaluation of the output water quality at the generic day n is wanted, information about the conditions of the water at days n-3, n-2, and n-1 should be given to ANNs. With this in mind, the first ANN structures used (indicated as ANN1 in Table 1) were built according to the scheme in Fig. 2, holding constant the number of input layer (n_l=18). The symbols COD_{e8} and COD_{e20} in the figure and in Table 1 indicate the COD measured in the equalization stage at 8 am and 8 pm, respectively.

Table 1
Experimental matrix of the input and output data used in the networks organization

Input data			
Input parameter	ANN1	ANN2	ANN3
pH_e	$\{n-i\}_{i=1}^{i=3}$	-	n-1
C_e	$\{n-i\}_{i=1}^{i=3}$	$\{n-i\}_{i=1}^{i=3}$	n-1
(COD_{e8}×P)	$\{n-i\}_{i=1}^{i=3}$	$\{n-i\}_{i=1}^{i=3}$	n-1
(COD_{e20}×P)	$\{n-i\}_{i=1}^{i=3}$	$\{n-i\}_{i=1}^{i=3}$	n-1
V	$\{n-i\}_{i=1}^{i=3}$	$\{n-i\}_{i=1}^{i=3}$	n-1
MLSS	$\{n-i\}_{i=1}^{i=3}$	$\{n-i\}_{i=1}^{i=3}$	n-1
COD_b	-	$\{n-i\}_{i=1}^{i=3}$	n-1
Output data			
Output parameter	ANN1	ANN2	ANN3
pH_b	n	n	n
C_b	n	n	n
COD_b	n	n	n

n: number individuating the day.

From Fig. 2, the product (COD_e×P), representing the total oxygen demand transferred from the equalization to the biological tank, was adopted.

The ineffectiveness of the general architecture depicted in Fig. 2 was revealed even in the training phase. It was observed that, irrespective of the complexity of the structure of hidden layers, the RMS value remained quite high, although, as expected, the time to converge was lower, the lower was the number of layers and nodes.

Figs. 3 to 5 compare the experimental data with the predictions of the best ANN configuration complying with the scheme in Fig. 2. Clearly, the pH_b is well described. However, this is somehow anticipated, depending on the relative insensitivity of this parameter to the process conditions. Instead, large inaccuracies are found in the evaluation of C_b and COD_b, which undergo large oscillations during time. Nevertheless, some similarities between the measured and predicted general trends can be also noticed.

In an attempt to improve the ANN accuracy, in a

second step the input data set was partially changed in nature, and further variations in the ANNs configuration were also considered. The main features of these tests can be seen in Table 1: in the case of ANN2 architectures, the information concerning the pH was eliminated, and was substituted by the corresponding COD_b values; in the case of ANN3, all the input parameters listed in Table 1 were utilised, with reference uniquely to the day before the one for which the prediction was wanted. Further, in ANN2 the Only the best results achieved by these routes are presented and discussed in the following.

In figures 6, 7 and 8 the prediction resulting from ANN2 are graphically shown, and compared with the observed data.

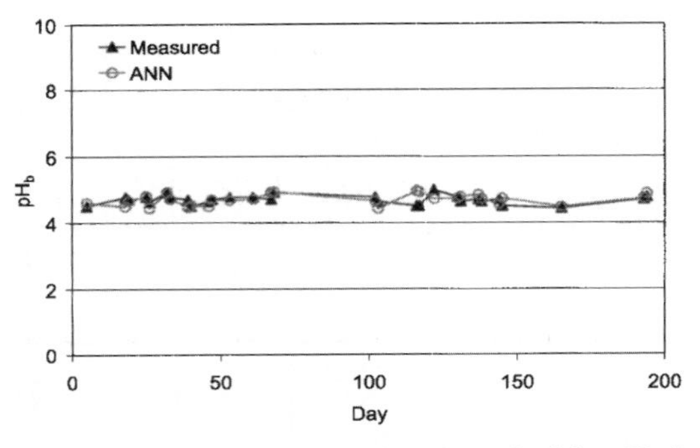

Fig. 3. Experimental and predicted trend of the pH_b (in arbitrary units) during time. ANN1 architecture: 18/50/50/3.

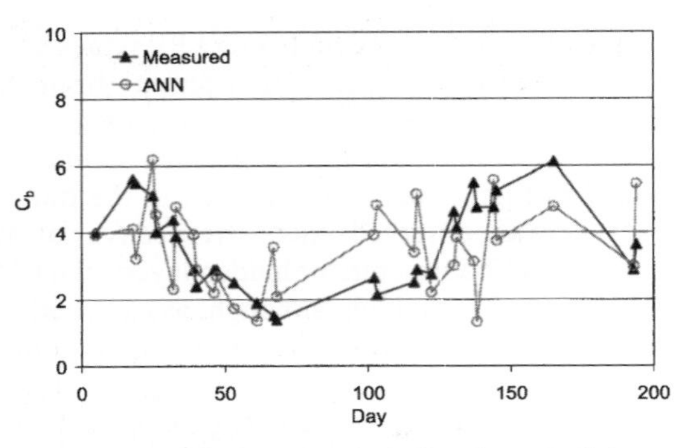

Fig. 4. Experimental and predicted trend of the conductivity C_b (in arbitrary units) during time. ANN1 architecture: 18/50/50/3.

Singularly, despite the elimination of pH_e as input datum, the network preserves its ability to predict the pH_b (Fig. 6). This is not surprising, simply highlighting the negligible significance of pH in assessing the ANN efficiency.

For what concerns C_b and COD_b (Figs. 7 and 8), little improvement is found in the behaviour of ANN2 compared to ANN1. Indeed, the general trend of the measured data seems to be slightly better shaped in ANN2. However, oscillations even larger than verified in the actual behaviour are predicted from the latter.

As previously noted, the number of input data was dramatically reduced to eight in ANN3 (Table 1). In practice, all the types of experimental data available were used, but the information given was limited to the last day before the prediction day. The best results were obtained with a 7/50/50/3 network, and are graphically represented in Figs. 9 to 11.

Looking at conductivity (Fig. 10) and COD (Fig. 11), ANN3 clearly performs better than the previous networks.

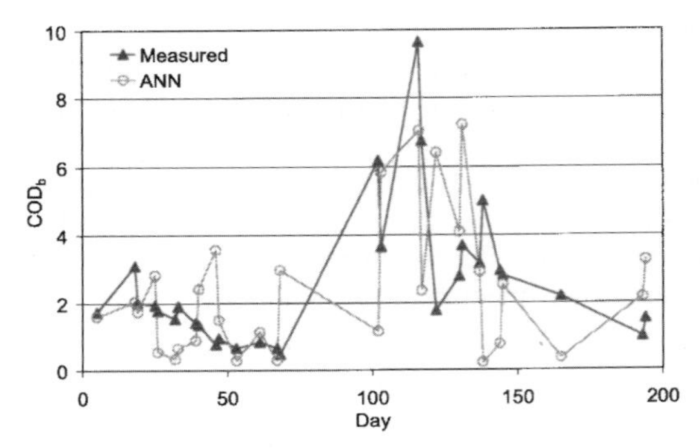

Fig. 5. Experimental and predicted trend of the COD_b (in arbitrary units) during time. ANN1 architecture: 18/50/50/3.

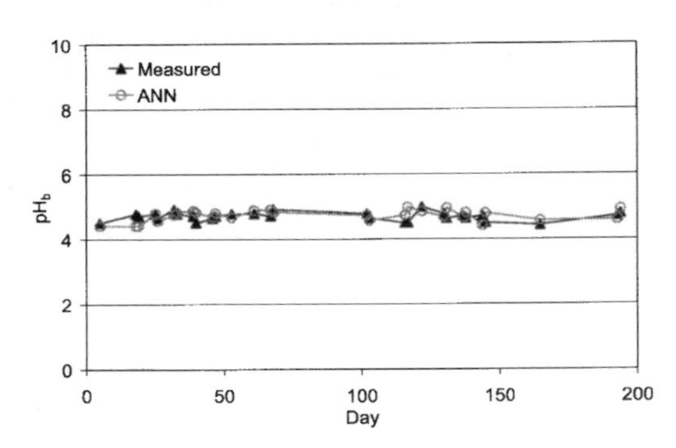

Fig. 6. Experimental and predicted trend of the pH_b (in arbitrary units) during time. ANN2 architecture: 18/50/50/3.

The agreement between predicted and experiments is reasonable, though the oscillations of the theoretical curve is larger than actually found. This suggests that the output water quality in a given day is correlated with the equalization tank conditions during the day before (ANN3 network), rather than during the three previous days (ANN1 and ANN2). Of course, the convergence of ANN3 to the precision set was quicker than verified for ANN1 and ANN2, due to its simpler architecture.

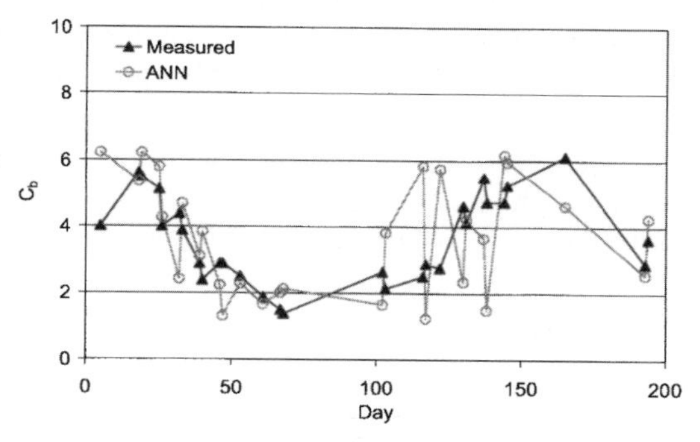

Fig. 7. Experimental and predicted trend of the conductivity C_b (in arbitrary units) during time. ANN2 architecture: 18/50/50/3.

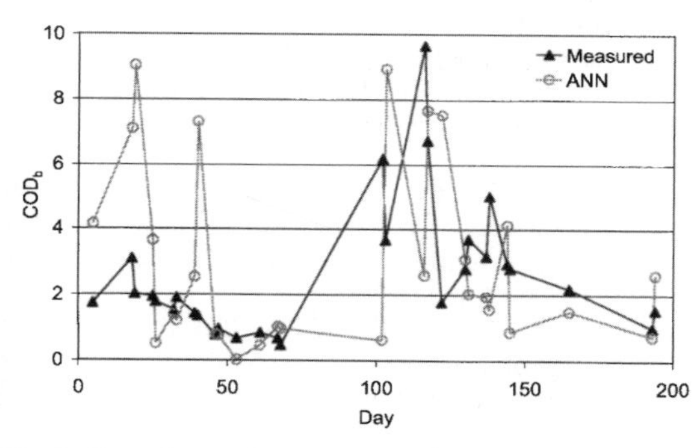

Fig. 8. Experimental and predicted trend of the COD_b (in arbitrary units) during time. ANN2 architecture: 18/50/50/3.

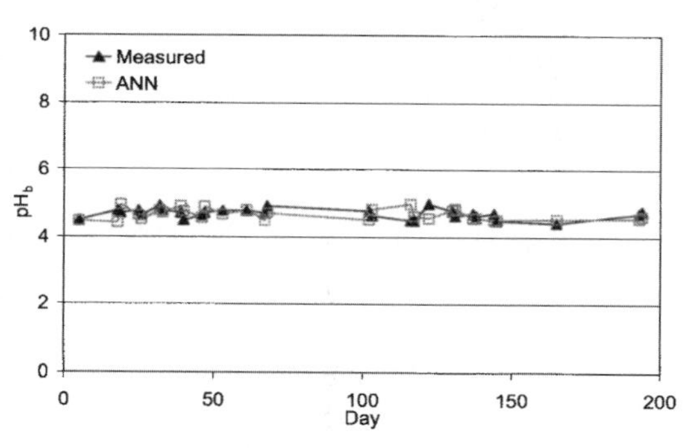

Fig. 9. Experimental and predicted trend of the pH_b (in arbitrary units) during time. ANN3 architecture: 7/50/50/3.

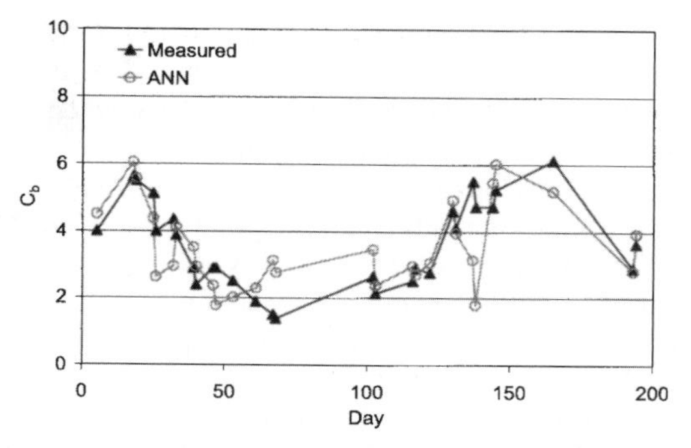

Fig. 10. Experimental and predicted trend of the conductivity C_b (in arbitrary units) during time. ANN3 architecture: 7/50/50/3.

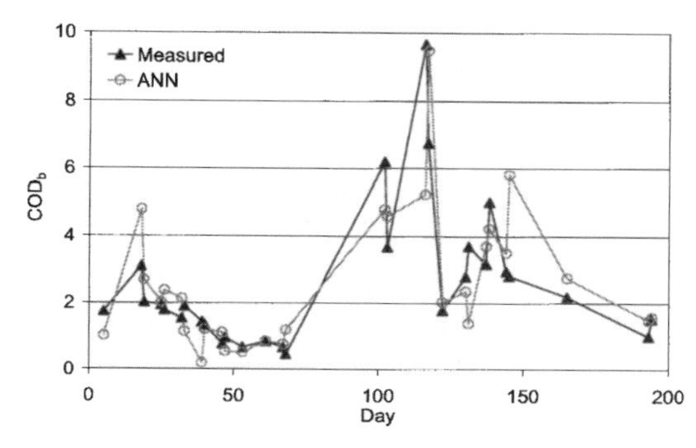

Fig. 11. Experimental and predicted trend of the COD_b (in arbitrary units) during time. ANN3 architecture: 7/50/50/3.

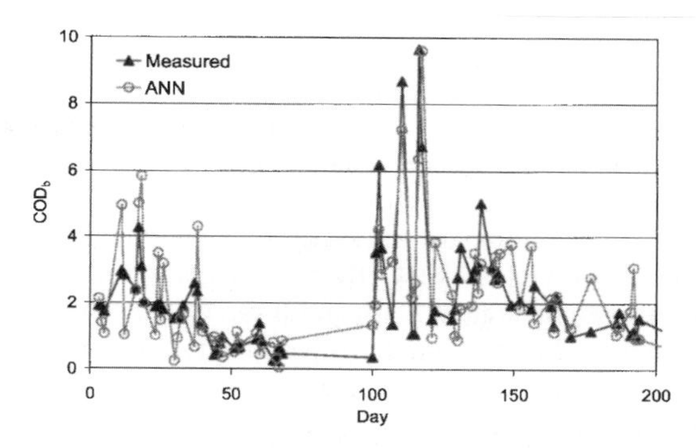

Fig. 12. Experimental and predicted trend of the COD_b (in arbitrary units) during time. ANN3 architecture: 7/50/50/3. Complete set of experimental data.

Since the measurement of the characteristic parameters of the chemical plant was not carried out daily, not all the data collected could be used in the training stage. In fact, indicating by 1/2/3/4/5/6 six consecutive days, in which the measurement was not performed in days 1 and 6, only the data of day 5 could be used to train ANN1 and ANN2, because of the need (Table 1) to know what occurred during the three previous days (2, 3, and 4). In principle, the same data set could provide three valid points (days 3, 4, and 5) for the training of ANN3. To allow for a comparison of the networks on a common basis, the training stage of the ANNs previously discussed was performed using the same data set, made of all the points (Figs. 3 to 11) satisfying the needs of ANN1 and ANN2. After ANN3 was trained, the remaining experimental points (not used in training) were added to the diagrams in Fig. 9-11, and ANN3 was called for a complete prediction of all the results available. The results of this analysis are shown in Fig. 12 with reference to COD_b, which was apparently the most difficult parameter to describe. In this case also, quite good agreement is found between prediction and experiments.

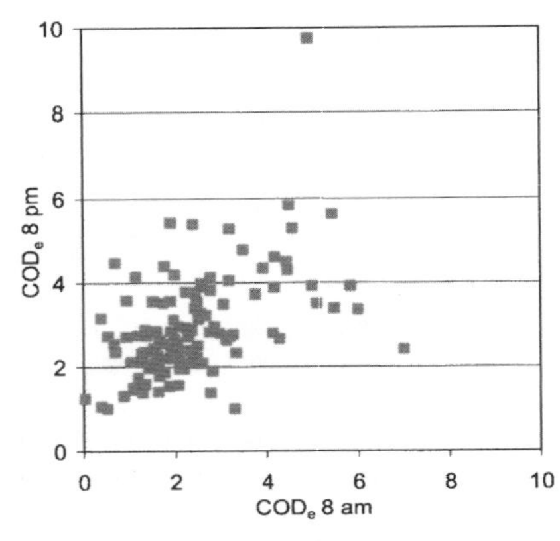

Fig. 13. Comparisons of the COD_e values (in arbitrary units) measured at 8 am and 8 pm of the same day.

Probably, rather than depending on the architectures tested, the inaccuracies noted in the ANNs behaviour are intrinsic to the problem faced. Figure 13 yields the COD values measured in the equalization stage at 8 am and 8 pm of the same day. Evidently, large fluctuations within short time periods can occur in the incoming water composition, because of new liquid volumes continuously flowing into the tank. Of course, this aspect cannot be captured by single COD values, as was implicitly assumed in using COD_{e8} and COD_{e20} in training the ANNs.

5. Conclusions

In this work, the ability of artificial neural network (ANN) technology to predict the quality of industrial wastewater after a biological treatment was investigated. From the results presented and discussed in this work, the main conclusions are as follows:

the pH of the outcoming water undergoes small variations, and is easily predicted, whichever the ANN used;

the electrical conductivity, and especially the chemical oxygen demand (COD), suffer large fluctuations within quite small time periods; despite this, a reasonably good prediction of these quantities can be achieved, provided the ANN is suitably optimized;

rather than with the ANN architecture, the inaccuracies noted are probably correlated with the input parameters, which are intrinsically unable to properly describe the actual behaviour of the biological process.

References

[1] Italian law D. LGS n°152, may 11, 1999, Anti-pollution regulation about water and reception of European Community directive n° 91/271/CEE and 91/676/CEE, pub. S.U. n°101/L, G.U. n° 124, may 29, 1999.

[2] Cheremisinoff P. N., Water and Wastewater Treatment Guidebooks. Prentice Hall Inc, Englewood Cliffs, NJ; 1994.

[3] Advances in Water and Wastewater Treatment Technology, Matsuo T., Hanaki K., Takizawa S., Satoh H., Eds., Elsevier Publ., B.V. ISBN:0-444-50563-6, 2001.

[4] Mulligan C. N., Environmental Biotreatment Technologies for Air, Water, Soil, and Wastes, ABS Consulting/Government Institutes, Rockville, MD, ISBN: 0-86587-890-0, 2002,

[5] Miller R.M., Itoyoma K, Uda A., Takada H, Bhat N., Modelling and control of a chemical waste water treatment plants, Comput. Chem. Engag, 21 pp 947-952, 1997.

[6] Beale R., Jackson T., Neural Computing: an Introduction, Institute of Physics Publ., Bristol, 1992.

[7] Bishop C.M., Neural networks and their applications. Rev. Scl. Instrum. Vol. 65, n°6, pp. 1803-1832, 1994.

[8] Hamed M.M., Khalafallah M.G., Hassanien E.A., "Prediction of wastewater treatment plant performance using artificial neural networks" *Environmental Modelling and Software*, 2004, Vol. 19, Issue 10, pp. 919-928.

[9] Onkal-Engin G., Demir I., Engin S.N., "Determination of the relationship between sewage odour and BOD by neural networks" *Environmental Modelling and Software*, 2005, Vol. 20, Issue 7 , pp. 843-850.

[10] Grieu S., Traore A., Polit M., Colprim J., "Prediction of parameters characterizing the state of a pollution removal biologic process" *Engineering Applications of Artificial Intelligence*, 2005, Vol. 18, Issue 5, pp. 559-573.

[11] Touche R. and Co., Best practice guidelines for developing neural computing application - an overview, Ministry of Defence Procurement Executive, UK, 1993.

Intelligent Production Machines and Systems
D. T. Pham, E. E. Eldukhri and A. J. Soroka (eds)

Chip form monitoring in turning based on neural network processing of cutting force sensor data

T. Segreto[a], J. L. Andreasen[b], L. De Chiffre[b], R. Teti[a]

[a]*Department of Materials and Production Engineering, University of Naples Federico II*
Piazzale Tecchio 80, 80125 Naples, Italy
[b]*Department of Manufacturing Engineering and Management, Technical University of Denmark*
Bldg. 425, 2800 KSG. Lingby, Denmark

Abstract

Sensor monitoring of chip form during longitudinal turning of carbon steel Ck45 was carried out through the detection and analysis of cutting force sensor signals. Both single chip form classification and favourable/unfavourable chip form identification were considered. Signal processing for feature extraction was carried out through a parametric method of spectral estimation. Decision making on chip form typology was performed through a supervised neural network (NN) approach, using diverse back-propagation feed-forward NN configurations, in view of the development of an on-line and real time chip form control procedure.

1. Introduction

Efficient chip control in operations which tend to produce long continuous chips like turning, drilling and tapping is associated with problems due to a general lack of rules for predicting chip breaking and to changes in chip breakability because of variations of the processing conditions [1, 2].

Process variations resulting from tool wear, non uniformity of workpiece material, thermal effects, etc., cause variations in the chip form during a machining operation [3]. Therefore, continuous monitoring of the produced chips is needed to avoid unacceptable chip forms and the risk of product or machine damage.

In a previous work [3], an intelligent sensor monitoring technique for chip form control during turning, based on cutting force signal detection and neural network (NN) sensor data analysis, was established and verified using the first experimental results from turning tests carried out with selected cutting conditions, yielding well defined chip forms.

In this paper, sensor monitoring of chip form during longitudinal turning of C steel was applied to experimental conditions characterised by a wide range of process parameters, generating different chip form typologies under more realistic production situations. The intelligent methodology utilised was drawn from the one in [3] and developed in view of an on-line chip form identification procedure.

2. Experimental procedures

Machining tests were carried out by longitudinal turning on a CNC-lathe using standard throw-away inserts. The inserts (Seco TP301) were triangular coated carbides. In all tests, a standard 25 mm square tool holder was used with a cutting edge angle of 90°. Carbon steel Ck45 cylindrical bars with 245 mm diameter were used as work material. These bars were held in a 3-jaw chuck and supported by a live centre at the other end. For each cutting test, the feed force component was measured by the use of a specially designed tool force Kistler dynamometer built to be mounted on the turret of a CNC lathe. The amplified feed force signal was band-pass filtered using a fourth order Butterworth filter in order to increase the resolution of the A/D conversion by suppressing the DC and resonant components. The signal was input into a Brüel & Kjær signal analyzer to be A/D converted and Fast Fourier transformed. The duration of the cutting test and the force signal acquisition was carried out with sampling frequency 2.5 kHz. The total number of samples for the feed force component in a cutting test was 2.5kS.

To reduce the influence of random events, the feed force spectrum was averaged using 20 instant spectra. The averaging process was performed with a max overlap, thus reducing the time for averaging to about 800 ms. Once created, the averaged spectrum was saved on PC for post-analysis.

In Table 1, the cutting test conditions are summarised. The chip form for each test was classified according to the ISO standard 3685 (see Fig. 1) [4].

Cutting		Favourable	Unfavourable	
Straight	1 Ribbon Chips	1.2 Short		1.1 Long/1.3 Snarled
Mainly up curling	2 Tubular Chips	2.2 Short	2.1 Long	2.3 Snarled
	3 Spiral Chips		3.1 Flat/3.2 Conical	
Mainly Side curling	4 Washer type Chips	4.2 Short	4.1 Long	4.3 Snarled
Up and Side Curling	5 Conical Helical Chips	5.2 Short	5.1 Long	5.3 Snarled
	6 Arc Chips	6.2 Loose/6.1 Conn		
	7-8 Natural Broken Chips	7 Elemental		8 Needle

Fig. 1. Chip form classification.

Table 1
Summary of machining tests
D = bar diameter; a = depth of cut; f = feed rate

Signal Id.	D (mm)	a (mm)	f (mm/rev)	Chip form	Chip type
1	240	0.5	0.1	Snarled 2.3	Unfav.
2	240	0.5	0.2	Snarled 2.3	Unfav.
3	240	0.5	0.3	Snarled 2.3	Unfav.
4	240	0.5	0.4	Snarled 2.3	Unfav.
5	239	0.5	0.5	Snarled 2.3	Unfav.
6	245	2	0.1	Snarled 2.3	Unfav.
7	245	3	0.1	Snarled 2.3	Unfav.
8	245	3	0.3	Snarled 2.3	Unfav.
9	245	3	0.4	Snarled 2.3	Unfav.
10	241	1	0.1	Long 2.1	Unfav.
11	241	1	0.15	Long 2.1	Unfav.
12	245	2	0.05	Long 2.1	Unfav.
13	245	3	0.05	Long 2.1	Unfav.
14	241	1	0.2	Short 5.2	Fav.
15	241	1	0.3	Short 5.2	Fav.
16	245	2	0.15	Loose 6.2	Fav.
17	245	2	0.2	Loose 6.2	Fav.
18	245	2	0.25	Loose 6.2	Fav.
19	245	2	0.3	Loose 6.2	Fav.
20	245	2	0.4	Loose 6.2	Fav.
21	245	3	0.025	Loose 6.2	Fav.
22	245	3	0.15	Loose 6.2	Fav.
23	245	3	0.2	Loose 6.2	Fav.
24	245	3	0.25	Loose 6.2	Fav.
25	245	3	0.4	Loose 6.2	Fav.
26	245	3	0.5	Loose 6.2	Fav.
27	245	3	0.6	Loose 6.2	Fav.
28	245	3	0.75	Loose 6.2	Fav.

3. Sensor signal analysis

Sensor signal analysis was performed to discriminate chip form on the basis of cutting force data. The 2.5 kS feed force file from each cutting test was subdivided into 5 sub-files, each representing a 0.5 kS signal specimen.

Feed force signal specimens were processed in order to achieve the spectral estimation through a parametric method [5, 6].

In this procedure, the signal spectrum is assumed

to take on a specific functional form, the parameters of which are unknown. The spectral estimation problem, therefore, becomes one of estimating these unknown parameters of the spectrum model rather than the spectrum itself.

From the signal specimen (measurement vector), p features or predictor coefficients $\{a_1, \ldots, a_p\}$ (feature vector), characteristic of the spectrum model, are obtained through linear predictive analysis (LPA) [7]. Feature extraction was implemented through the application of Durbin's algorithm [8] with p = 4, 8, 16.

4. Neural network data processing

Neural network (NN) sensor data analysis under supervised training was utilised to carry out a pattern recognition procedure in high dimensions feature spaces [9] using the 4-, 8-, and 16-component feature vectors extracted from the 0.5 kS feed force sub-files.

Two different pattern recognition procedures were applied: one for individual chip form classification and the other for favourable/unfavourable chip form identification.

Three-layer feed-forward back-propagation NN were built with the following configurations (see Fig. 2): the input layer had a number of processing elements (nodes) equal to 4, 8 or 16, according to the number of input feature vector components. The number of nodes at the hidden layer was 4, 8, 16, 32 or 64, depending on the number of input nodes. The output layer contained only one node, yielding a coded value associated with the chip form.

In Table 2, the NN output coded values and graph symbols for each chip form category are reported together with the number of corresponding available training cases.

Training and testing of the NN was performed using training sets made of the 4-, 8-, and 16-component feature vectors, respectively.

The leave-k-out method [10], particularly useful when dealing with relatively small training sets was applied: one homogeneous group of k patterns, extracted from the training set, was held back in turn for testing and the rest of the patterns was utilized for training. In this application, a k = 1 value was used.

The available training set is made of a total of 140 training cases and was subdivided in 2 different ways:

- 45 Snarled, 20 Long, 65 Loose, and 10 Short chip form cases, for single chip form classification;

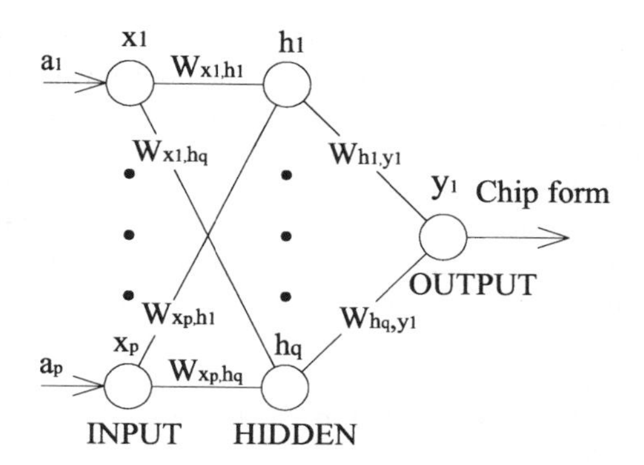

Fig. 2. NN configuration for chip form classification

Table 2
NN output coded values, graph symbols and number of training cases for each chip form category

Chip form	Coded value	Graph symbol	# of training cases
Snarled	0	♦/◊	45
Long	1	●/○	20
Loose	2	▼/▽	65
Short	3	▲/△	10
Favourable	0	▲/△	75
Unfavourable	1	●/○	65

- 75 favourable (Loose and Short) and 65 unfavourable (Snarled and Long) chip form category cases, for favourable/unfavourable chip form identification.

5. Result and discussion

In Tables 3 and 4, the success rate (SR) values of the NN processing are reported for each NN configuration and for the two classification procedures.

As regards the single chip form classification, all NN total SR values are low (see Table 3): the highest value, obtained with the 4-4-1 NN, is only 60%.

This poor result is due to the intrinsic difficulty in identifying four distinct chip forms on the basis of a relatively small number of training cases (see Table 2).

Moreover, a negative influence on the NN total SR is exerted by the Short chip form SR range (0% – 50%), consequence of the very low number of Short chip form training pairs (10 cases over 140 total cases).

Finally, the four chip forms under examination present similarities of shape (in particular, the couples given by the Short and Loose forms, on the one hand,

and the Snarled and Long forms, on the other hand) responsible for similarities in the training pairs that can generate confusion in the NN during the knowledge acquisition phase.

As regards the favourable/unfavourable chip form identification, all NN configurations display significantly higher total SR values (see Table 4) in comparison with the previous classification procedure. The highest NN total SR, obtained with the 4-4-1 NN, was 87%.

This is the consequence of a rather well balanced training set (75 favourable and 65 unfavourable chip form cases) made of decidedly distinct chip types: Loose and Short for the favourable chip form category, on the one hand, and Snarled and Long for the unfavourable chip form category, on the other hand.

Table 3
NN SR for single chip form classification

NN Configuration	NN SR				
	Snarled	Long	Loose	Short	Total
4 - 4 - 1	33	55	86	20	60
4 - 8 - 1	37	35	75	20	53
4 - 16 - 1	28	40	69	0	47
8 - 8 - 1	37	50	72	20	54
8 - 16 - 1	22	30	41	0	30
8 - 32 - 1	28	40	33	50	34
16 - 16 - 1	35	40	36	20	35
16 - 32 - 1	31	55	35	10	35
16 - 64 - 1	28	30	33	30	31

Table 4
NN SR for favourable/unfavourable chip form identification

NN Configuration	NN SR		
	Favourable (Loose/Short)	Unfavourable (Snarled/Long)	Total
4 - 4 - 1	84	90	87
4 - 8 - 1	84	86	85
4 - 16 - 1	72	67	70
8 - 8 - 1	82	70	77
8 - 16 - 1	56	56	56
8 - 32 - 1	53	55	54
16 - 16 - 1	57	73	65
16 - 32 - 1	54	64	59
16 - 64 - 1	49	56	52

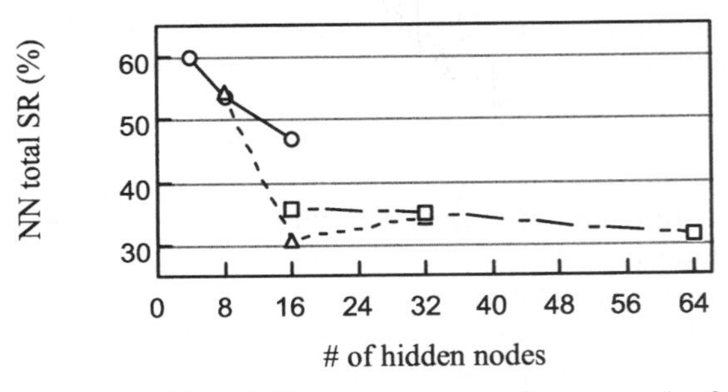

Fig. 3. NN total SR versus number of hidden nodes for single chip form classification.
○ 4 inputs nodes, △ 8 input nodes, □ 16 inputs nodes.

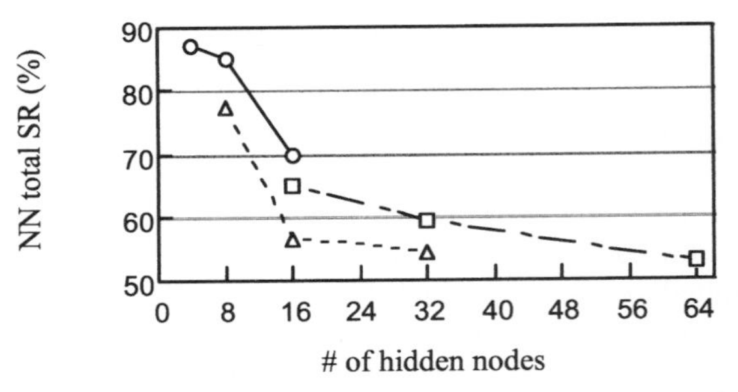

Fig. 4. NN total SR versus number of hidden nodes for favourable/unfavourable chip form identification.
○ 4 inputs nodes, △ 8 input nodes, □ 16 inputs nodes.

This satisfactory result is reinforced by the fact that the NN SR for each category is equivalently high (84% for favourable, 90% for unfavourable chip form).

In Figures 3 and 4, the NN total SR values for the two classification procedures are plotted versus number of hidden nodes for different numbers of input features.

From the figures, it can be seen that the NN total SR decreases with increasing number of hidden nodes as well as with increasing number of input features for both classification procedures.

In fact, the NN architecture characterised by the lowest number of input nodes and hidden nodes (4-4-1 NN configuration) provided for the best identification results in both data processing paradigms.

In Figures 5 and 6, the NN output values (see Fig. 5a and 6a) and the NN classification error given by E = (actual output – desired output) (see Fig. 5b and 6b) for the 4-4-1 NN configuration are reported versus number of training pairs for both classification procedures.

Though Figure 5 is for the highest NN total SR in single chip form identification, many misclassification cases (black symbols) can be seen, particularly for the Short chip form, showing deficient learning conditions for this type of NN knowledge acquisition.

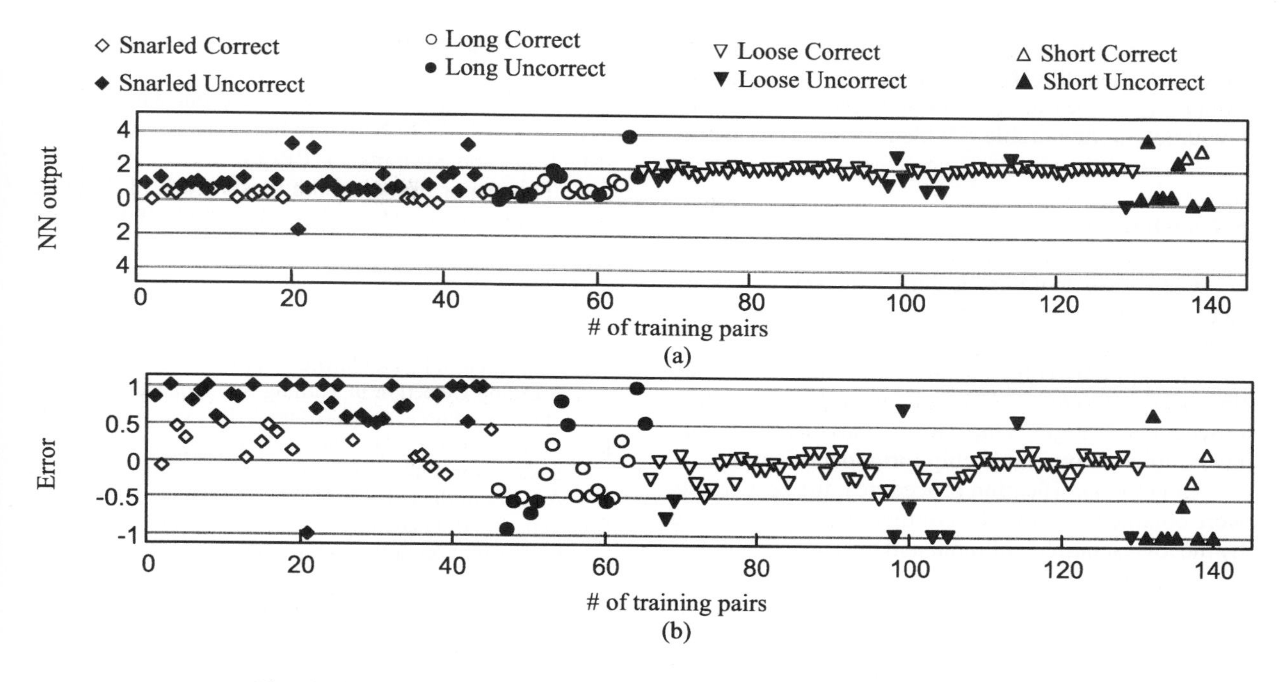

Fig. 4. Single chip form classification: highest SR = 60% (NN configuration 4-4-1).
(a) NN output and (b) NN classification error vs. number of training pairs.

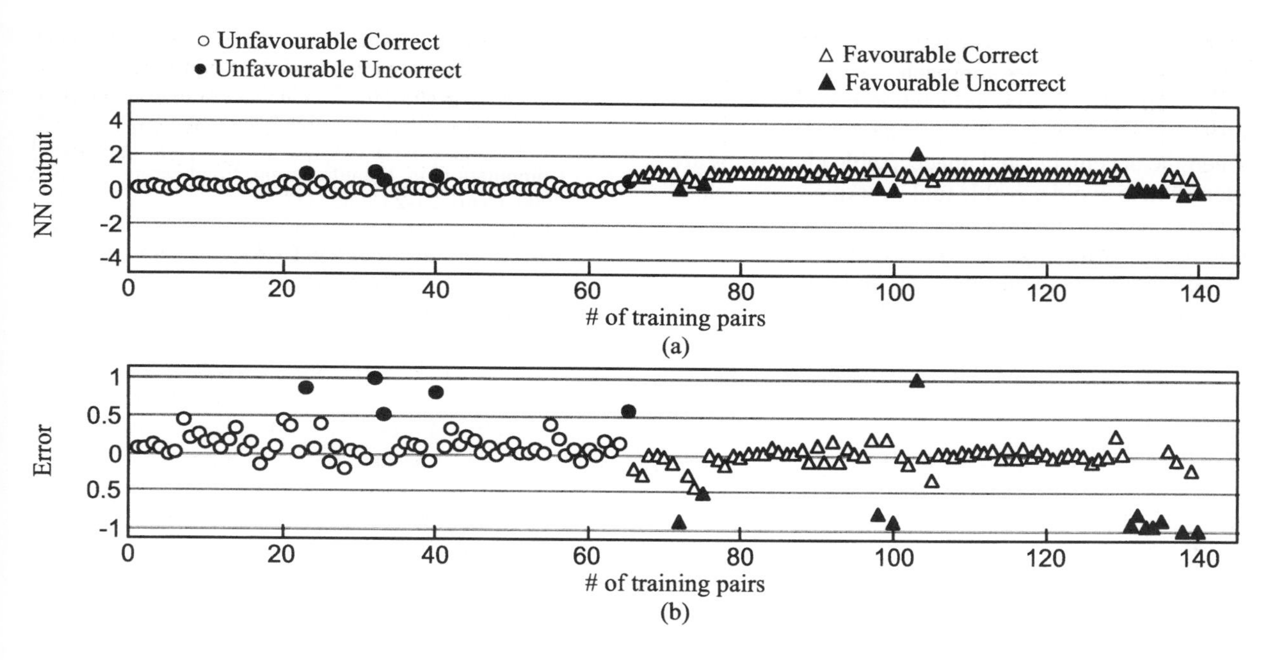

Fig. 5. Favourable/Unfavourable chip form identification: highest SR = 87% (NN configuration 4-4-1).
(a) NN output and (b) NN classification error vs. number of training pairs.

On the contrary, Figure 6 displays only few uncorrect classification cases, revealing more exhaustive learning conditions for the favourable/unfavourable NN identification procedure.

6. Conclusions

Chip form monitoring during turning of carbon steel Ck45 was carried out using cutting force sensor signal detection and analysis.

Decision making on chip form was performed through different NN based pattern recognition procedures: single chip form classification and favourable/unfavourable chip form identification.

In both classification procedures, the best results were obtained with the NN configuration characterized by the lowest number of input and hidden nodes: 4-4-1.

The single chip form classification procedure presented a low performance, the highest NN total SR being only 60%. This was due to severe classification requirements (four distinct chip forms) coupled with deficient learning conditions for this NN approach.

The favourable/unfavourable chip form identification procedure provided for a high performance, yielding a NN total SR as high as 87%.

This attractive result is the consequence of more reasonable identification requirements (two distinct chip categories) together with more thorough learning conditions for this NN approach.

From an industrial application point of view, an efficient chip form monitoring and control is achieved when the acceptable and unacceptable chip forms are identified to avoid possible tool, workpiece or machine damage during machining. This was shown to be obtainable with the favourable/unfavourable NN classification procedure that can be implemented for on-line process monitoring applications with high identification SR.

Acknowledgements

This research work was carried out with support from the EC FP6 NoE on Innovative Productions Machines and Systems – I*PROMS.

This paper draws on the thesis work of Antonio Cuscianna who is acknowledged for his collaboration.

References

[1] Andreasen, J. L., De Chiffre, L. (1993). Automatic Chip-Breaking Detection in Turning by Frequency Analysis of Cutting Force, Annals of the CIRP, Vol. 42/1: 45-48.

[2] Andreasen, J. L., De Chiffre, L. (1998). An Automatic System for Elaboration of Chip Breaking Diagrams, Annales of the CIRP, Vol. 47/1 : 35-40.

[3] Teti, R., Buonadonna, P., D'Addona, D., 2001, Chip Form Monitoring through Cutting Force Sensor Signal Processing, V AITEM Conf., Bari, 18-20 Sept.: 21-30.

[4] ISO 3685: Tool-life testing with single-point turning tools. First ed. 1977, Annex G: 41.

[5] Teti, R., Buonadonna, P. (1999). Round Robin on Acoustic Emission Monitoring of Machining, Annals of CIRP, Part 3, Int. Doc. & Rep.: 47-69.

[6] Jawahir, I.S., Van Luttervelt, C.A. (1993). Recent Developments in Chip Control Research and Applications, Annals of the CIRP, Vol. 42/2 : 659-693.

[7] Rabiner, R.L., Shafer, R.W. (1978). Digital Processing of Speech Signals, Prentice-Hall, Englewood Cliffs, NJ.

[8] Stearns, S.D., Hush, D.R. (1990). Digital Signal Analysis, Prentice-Hall, Englewood Cliffs, NJ.

[9] Teti, R., Buonadonna, P., 2000, Process Monitoring of Composite Materials Machining through Neural Networks, 2nd CIRP Int. Sem. on Intelligent Computation in Manufacturing Engineering - ICME 2000, Capri, 21-23 June: 377-382.

[10] Masters, T. (1993), Practical Neural Networks Recipes in C++, Academic Press, San Diego, CA, USA.

Intelligent Production Machines and Systems
D. T. Pham, E. E. Eldukhri and A. J. Soroka (eds)

Critical aspects in the application of acoustic emission sensing techniques for robust and reliable tool condition monitoring

E.M.Rubio[a], I.L. Baciu[b], R. Teti[b]

[a] *Department of Construction and Manufacturing Engineering, Distance Learning University of Spain (UNED) Juan del Rosal, 12, 28040-Madrid, Spain*
[b] *Department of Materials and Production Engineering, University of Naples Federico II, P.le Tecchio 80, 80125 Italy*

Abstract

Acoustic emission sensing techniques for tool condition monitoring (TCM) in machining processes have been widely studied in recent years. Some approaches have been developed and successfully implemented in industry. However, standard solutions allowing for a quick and easy application of this sensor monitoring method have not been found yet. This is basically due to the great complexity associated with tool wear mechanisms together with critical aspects in the design of cutting tests for tool wear monitoring experimentation and in the set-up of the measurement chain. In this work, the main critical points associated with the definition of the cutting tests for TCM have been analysed. The study has revealed that it is necessary to pay special attention to the previous wear level in the tool and to the way it was achieved. Besides, the need for a standardised methodology for cutting test design, allowing for a better comparison of results obtained through different TCM test procedures, has been evaluated. As a first approach, the minimum set of elements and characteristics that should be included in such standardised methodology has been established.

1. Introduction

Turning, drilling, milling and grinding are material removal processes widely used in many industrial sectors. In some of them, such as the car or the aerospace industry, the geometrical accuracy and surface finish requirements have been strongly increased in recent years. To meet these new needs, conventional machine tools have been replaced by numerical control machine tools (NCMT) to eliminate the variability introduced by the operator and to obtain better quality parts [1-4].

On the other hand, in order to increase productivity and reduce manufacturing costs, NCMT have been associated with other components, mainly robots and computers, in flexible and computer-integrated manufacturing systems able to operate automatically during long periods of time. Such advanced manufacturing systems demand an optimal performance at all machining stages. This includes just in time tool change when the tool has reached a certain wear level. However, the predetermined moment to perform the tool change is, in general, not easy to predetermine. To reduce costs, it is desirable to utilize the tool life as thoroughly as possible but this depends, case by case, on the required geometrical and surface accuracy levels.

The availability of a tool condition monitoring (TCM) techniques allowing for the on-line control of tool wear state and proper actions to counteract any decrease in machining performance caused by worn out tools or unexpected catastrophic tool failure, is becoming a critical requirement in order to optimise and improve the manufacturing process efficiency [1-3, 5-7].

Development and implementation of tool condition monitoring systems have been the study subject of many researchers during the last years [1, 2, 8-19]. Particularly, acoustic emission (AE) generated by friction between tool and work piece during the cutting process has been analysed by a large number of authors and employed, in different ways, to develop systems allowing for the on-line tool wear monitoring [7, 20, 22-32]

However, standard solutions for industrial applications have not been found yet. This is mainly due to the complexity of tool wear mechanisms and the possibility to select among different alternatives in each step of the establishment of a tool wear monitoring procedure, so that not only one solution is possible.

In the following, the main critical points found in the definition of both the cutting tests and the measurement chain set-up to carry out an AE based tool wear monitoring procedure are illustrated and analysed.

2. Background

2.1. Friction and tool wear

During a cutting process, friction takes place at the workpiece-tool interface. This phenomenon is very complex and involves combined effects of adhesion, erosion, abrasion, fatigue, plastic deformation, diffusion and corrosion [21]. The contribution of each effect to friction depends on different aspects such as the specific properties of the workpiece and the tool, the surface roughness and the operating and environment conditions [2, 11, 18, 20, 21]; in any case, the adhesion mechanism operates in the widest range of cutting processes [21].

Friction produces a gradual wear of the cutting tools causing a negative influence on the quality of the machined surface and on the workpiece geometry and dimensions specified in the design. In the worst case, breakdown of the tool and damage to the workpiece or the machine can occur [13, 23, 28]. If the machine tool operates with other machines tools and robots under the control of a computer, the breakdown of the machine generates a critical problem for the whole system.

In order to prevent such problems, various methods for tool wear monitoring have been proposed, but none of these has been universally accepted due to the complex nature of this type of procedures [1, 8].

Tool wear monitoring methods have been classified traditionally into direct and indirect ones, according to the sensors used. In direct methods, the tool wear is measured directly using different types of sensors such as optical, radioactive or electrical resistance sensors. In indirect methods, the tool conditions are evaluated on the basis of parameters measured during the cutting operation such as cutting forces, acoustic emission, or vibrations [1].

These methods can be performed on-line, while the cutting process is being performed, or off-line, interrupting the process to carry out the control either on the machine or away from it. Off-line methods are not very practical, especially in automated manufacturing systems.

The development and implementation of on-line direct monitoring methods is quite difficult to carry out since the tool is not clearly visible during the cutting operation because of tool movement (in the rotating tool case) and the presence of chips and cutting fluids. For these reasons, methodologies explored for on-line direct tool condition monitoring have not had a good acceptation in shop floors.

The last trends in tool condition monitoring are focused on the development of on-line indirect methods. There are several in-process indirect measurement systems based on different types of sensors or a combination of them. The most common sensors are vibration and acceleration, load and power or acoustic emission and force [24]. Among them, acoustic emission and force sensors are the most widely employed due to their interesting results [9, 24]. However, there are different critical points in the set up of these systems that can be improved in order to obtain better results.

This work is focussed on the analysis of some of the most critical problems involved in the experimental set up of tool condition monitoring systems based on acoustic emission detection and analysis.

2.2. Friction and acoustic emission

The friction phenomenon produces elastic stress waves. These waves, generated by the release of energy stored within a material and called acoustic emission (AE), propagate through the material and can be detected by adequate AE sensors when they arrive at the material surface, provoking small displacements.

The mechanical signal can be turned into an electrical signal of low amplitude and high frequency by a piezoelectric AE sensor. The electronically processed signals can be later analysed with different mathematical and computational techniques [1, 12, 18, 29, 30]. AE signals can be classified into two types: continuous and burst. The continuous type AE signals are associated with plastic deformation in ductile materials and the burst type ones are observed during crack growth, impact or breakage.

It is widely accepted that the main sources of AE stress waves (see Fig. 1) during cutting are associated with [1, 12, 18, 30]: the primary (1) and secondary (2) shear zones, particularly in the case of fresh tools; the tertiary shear zone or the frictional contact at the interface between tool flank and workpiece, in case of flank wear (3); the frictional contact between the tool rake face and the chip, in case of crater wear (2); the crack growth at the contact zone between tool tip and workpiece (4); the plastic deformation of the chip (5); the collision between chip and tool (6); the chip breakage and catastrophic tool fracture.

As it can be easily realised, the friction phenomenon is not easy to deal with due to the great number of effects involved. However, AE signals provide for the possibility of identifying, by means of signal parameter changes, the tool wear state, essential for predicting tool life and detecting malfunctions in the cutting process such as chip tangling, chatter vibrations and cutting edge breakage.

Obtaining information on tool wear by means of AE detection and analysis generally includes the following steps:

- define the cutting tests;
- define the tool condition monitoring tests;
- select/propose the method(s) for signal processing;
- select/propose the process modeller(s);
- set-up the experimental layout;
- perform the cutting and tool condition monitoring tests;
- process the signals;
- model the tool wear process.

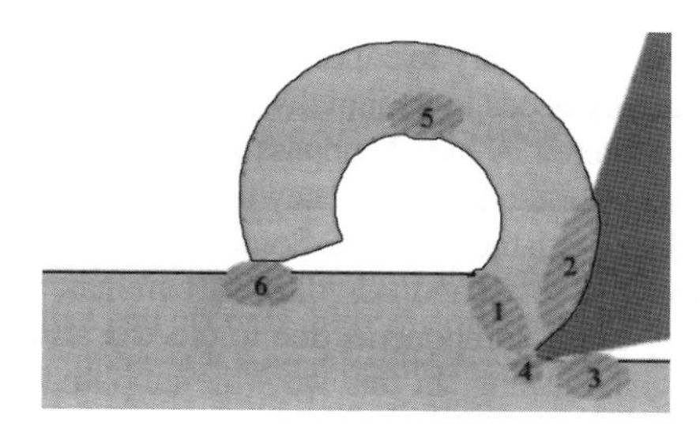

Fig. 1. Main sources of AE stress waves associated with chip formation: 1) primary shear zone, 2) secondary shear zone and craterization by friction, 3) tertiary shear zone and flank wear by friction, 4) crack growth at the tool tip-workpiece contact, 5) chip plastic deformation and 6) chip-tool collision.

By reviewing the literature, it can be verified that the information about some of these steps is often incomplete or absent and, therefore, the whole analysis (cutting/tool wear monitoring tests, signal processing and so on) does not provide for reliable results because some factors have not been taken into proper account.

In the following, the main problems found in the first two steps in the AE-based TCM procedure establishment are going to be analysed along with the solutions most commonly utilised in practice or proposed as the best alternatives.

3. Problems and solutions

3.1. Definition of the cutting tests

The correct definition of the cutting tests can be considered as the first point to improve in order to develop and implement an adequate on-line AE-based TCM system.

Although in the literature there are papers presenting an accurate description of the cutting tests [26], most of the authors do not say, or at least not with the desired detail, how they obtained the wear of the tools used in the tests: the time of use, the used workpiece material or under what cutting conditions the wear was achieved. This is a critical issue since this data is needed to complete the information about the experiment and to have a better knowledge about the process, allowing for more reliable conclusions. For example, it is important to know the time of use to carry out a comparison of the obtained results when the cutting conditions, workpiece material or type of tool are changed.

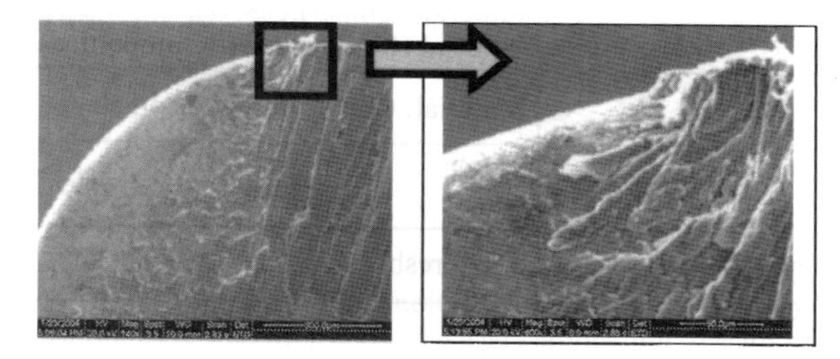

Fig. 2. BUE and BUL obtained by adhesion mechanism.

On the other hand, it is important to have information about the workpiece material used for previous tool wear generation. In general, nothing is specified about the used workpiece material. For example, if it is the same as the one for the AE tests or if the tool wear was achieved by machining different types of workpiece materials. This fact is important since, depending on the used material and the cutting conditions, different variations of the tool geometry can occur. As mentioned above, the main cause of wear development is the adhesion mechanism. This kind of friction consists basically in the transfer of small particles from the tool to the chips [33]. But, sometimes, the incorporation of macroscopic fragments from the workpiece material to the tool surface occurs [34]. These

investigates the powder-coating process as employed in the existing plant and presents a methodology for designing gun nozzles and evaluating their efficiencies. It discusses the application of Computational Fluid Dynamics (CFD) to validate gun nozzle designs and the experimental results obtained. Figure 2 illustrates the design methodology implemented in this work.

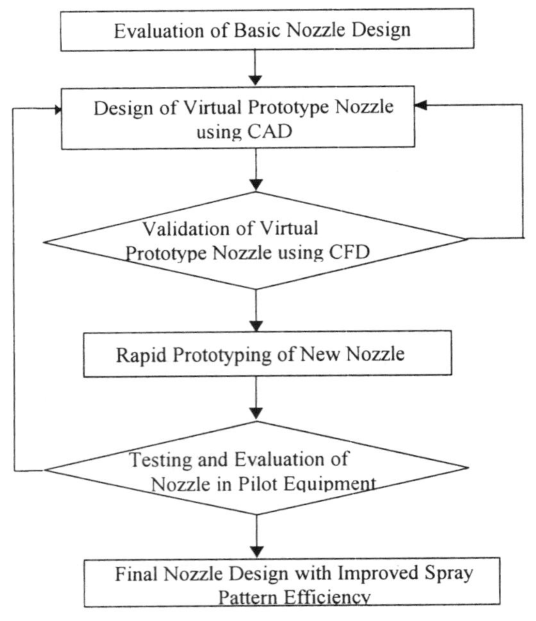

Figure 2: Methodology for developing efficient nozzles

2. Evaluation of the efficiency of the basic powder coating gun nozzle

The methodology implemented in this paper makes use of reverse engineering and solid modelling to copy the existing basic gun nozzle design and then modify it to generate new efficient nozzles capable of achieving increased powder deposition efficiency. Figure 3 shows a 3D model and a CAD drawing of the existing nozzle. A powder deposition test was conducted for that nozzle spraying on a flat plate with constant gun pressure and voltage. Figure 4 depicts the spray pattern produced. The distance of the gun nozzle to the target surface was maintained at 90mm. The choice of gun-to-part distance was guided by results obtained from preliminary experimental work. The area covered by the existing basic nozzle was approximately 16000mm2 (see figure 4). The estimated required coating areas for the new nozzle designs are shown in table 1.

Table 1: Required coating areas for each gun

New nozzle design	Required coating area
Top gun nozzle	5025mm2
Middle gun nozzle	8000mm2
Bottom gun nozzle	4000mm2

The pilot plant had the basic nozzle fitted to all the three guns. The first (top) gun produced 50% more than the required coating area, the second (middle) gun gave in excess of 68% and the third (bottom) gun oversprayed by 75%.

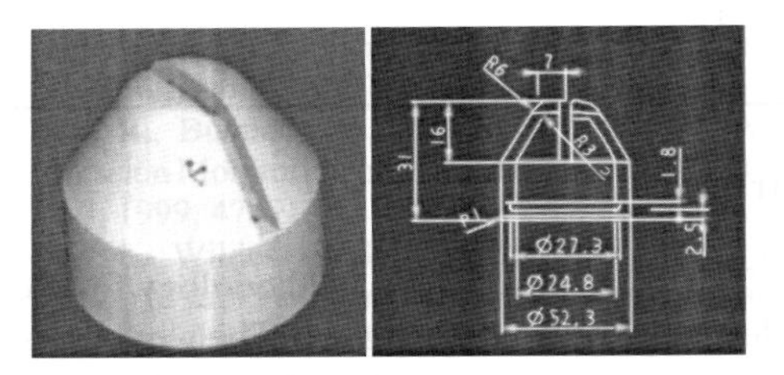

Fig.3: 3D model and CAD drawing of the existing nozzle

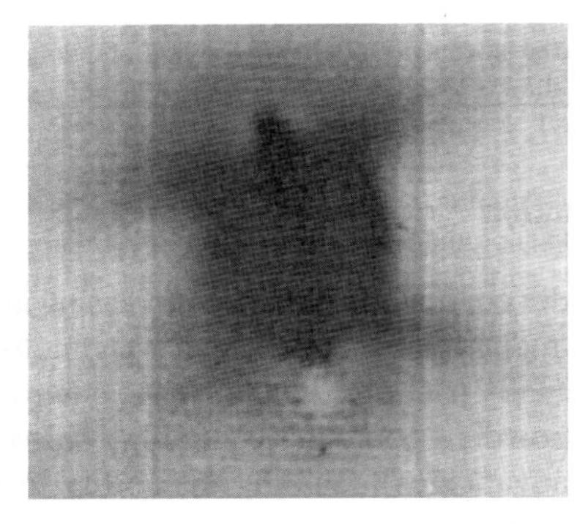

Fig.4: Powder deposition test using the basic nozzle on a flat plate (300mm x 300mm)

3. Creation of Virtual Prototype Powder Coating Nozzle and CFD Validation

In sections 3.1 and 3.2, a brief description is given of the computer-aided design of a new prototype nozzle and its validation using the FLUENT CFD software. The results of computer simulations of the spray patterns of the prototype nozzle under different inlet and static pressures are also presented. The simulations were conducted to determine the static pressure distribution between the coating gun and the workpiece.

3.1 *CAD Design of Virtual Prototype Nozzle*

Figure 5 shows the CAD model of the prototype nozzle constructed using the Pro/Engineer 3D solid modelling software. The nozzle has a variable prismatic aperture of dimensions 2.5mm x 15mm to 5mm x 17mm.

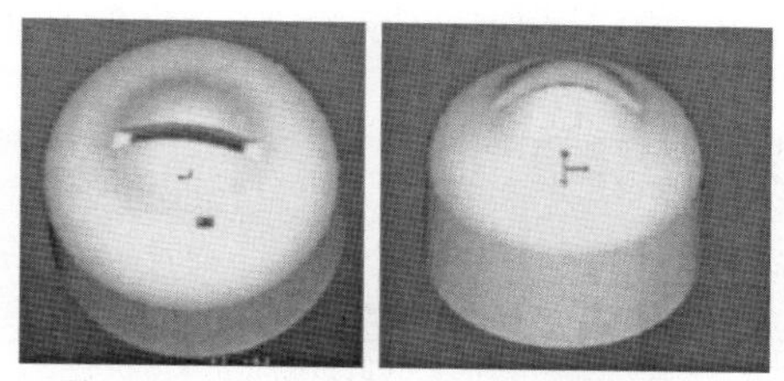

Fig.5: Prototype nozzle with prismatic aperture

3.2 *CFD Validation of Virtual Prototype Nozzle*

3.2.1 *Simulation of Inlet Nozzle Pressure*

Fluidised powder travels from the fluidised bed through the Venturi pipe to the gun nozzle (see Figure 6). The Venturi pipe ensures that fluidised powder is supplied to the powder feed hose at a consistent rate. The pressure of the powder feed from the fluidising bed and Venturi pipe plays a critical role in deciding the flow rate of the fluidised powder to the gun. To monitor the flow condition in the system, the powder feed pressure was simulated using FLUENT. Figure 7 presents the finite element mesh constructed for the analysis by FLUENT. Figure 8 shows the speed distribution of the fluidised powder particles. From figure 9, it can be seen that the pressure at the outlet is higher compared to other areas of the model.

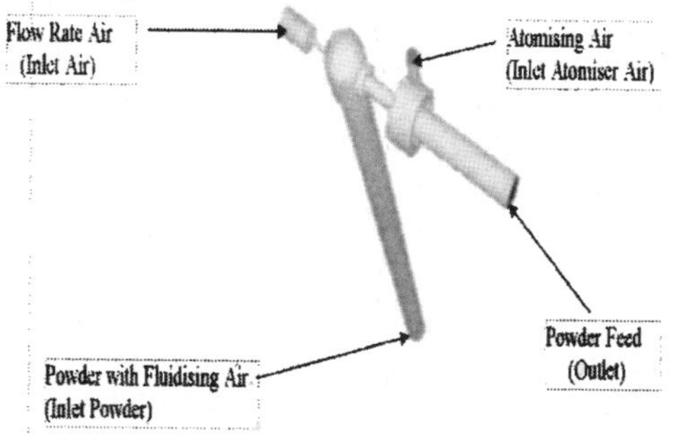

Fig.6: Model of Venturi pipe and atomising air inlet

Simulation results from FLUENT revealed that the outlet pressure varied between 10^3 and 10^5 Pa with different input and atomiser air flow rates. The outlet pressure from the Venturi pipe is the inlet pressure at the powder feed hose and therefore the inlet pressure of the nozzle.

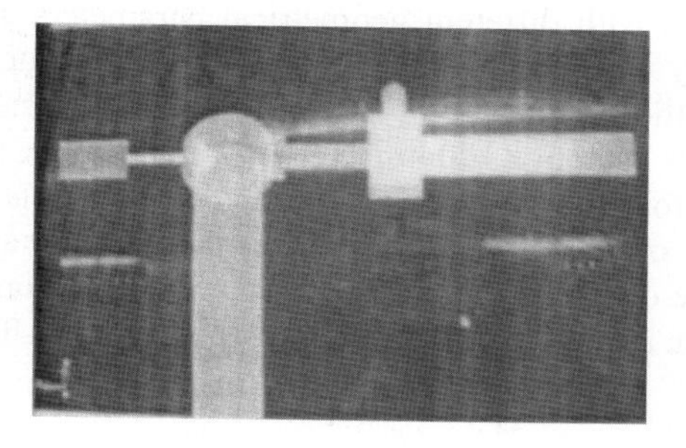

Fig.7: Finite-element mesh for FLUENT analysis

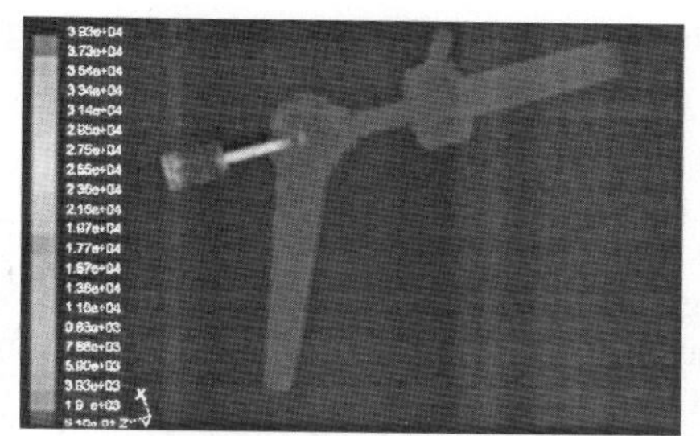

Fig.8: Powder particle speeds computed by FLUENT

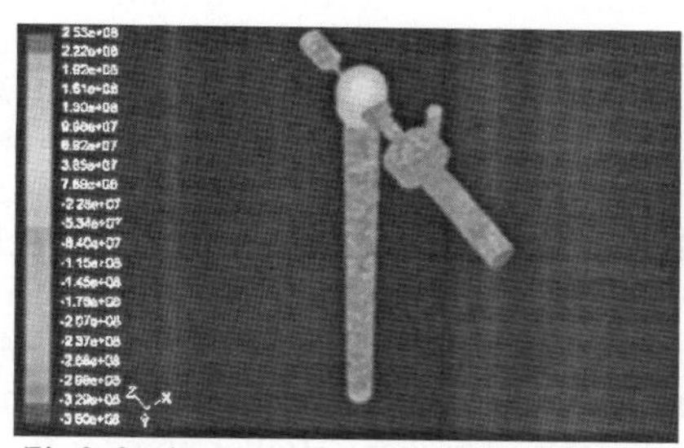

Fig.9: Static pressure computed by FLUENT

3.2.2 *CFD Simulation of Spray Patterns Using Virtual Prototype Nozzle*

The motion of the electrically charged powder particles from the corona spray gun is governed by a combination of electrical and aerodynamic forces [2]. Electro-charging of the powder is aimed at improving powder deposition on the grounded workpiece. The spray gun determines the trajectories of the charged particles and hence plays a significant role in the final transfer efficiency. Particle field charging can be calculated according to the method of Pauthenier and Moreau-Hanot [3], [4].

To determine the requirements for efficient gun nozzles, the dynamic behaviour of five prototype nozzles with different geometrical parameters and a variety of inlet gun pressures was investigated in simulation. The prototype nozzles were directed to spray a 100mm x 100mm flat plate. In equation 1, the body force acting on a charged powder particle as a result of the attraction between the particle and a charge on a flat plate is presented. The equation was written as a user defined function (UDF) in FLUENT:

$$F = \frac{1}{4\pi\varepsilon_0} \cdot \frac{Q_1 Q_2}{r^2} \qquad (1)$$

where:

F is the body force (N)

r is the distance between two attractive charge (m)

ε_0 is the permittivity constant of free space

Q_1 is the charge on particle (Coulomb)

Q_2 is the charge on the flat plate (Coulomb)

Figure 10 depicts a prototype nozzle with a rectangular aperture of $80mm^2$ positioned in front of the flat plate to be sprayed. The results of the FLUENT simulation of the flow of charged particles from the gun nozzle to the target show that, when the prototype nozzle was subjected to an inlet pressure of 10^3 Pa, the flow pattern was more concentrated and better directed towards the targeted surface, with less dispersion and overspray than when the inlet pressure was increased to 10^5 Pa. Thus, the prototype nozzle when subjected to an inlet pressure of 10^3 Pa achieved a higher powder deposition efficiency compared to the case when the inlet pressure was 10^5 Pa. This result indicates that, with a higher inlet pressure, the total powder flow rate is higher. Higher powder flow rates accelerate the build up of powder particles on the target, which increases the possibility of back ionisation and poor coating quality. Figures 11 and 12 depict the flow of the powder particles from the prototype nozzle when subjected to inlet pressures of 10^3 Pa and 10^5 Pa respectively.

Figures 13 and 14 show the static pressure distributions inside the nozzle and around the target flat plate with inlet pressures set at 10^3 Pa and 10^5 Pa respectively. In figure 13, reasonably even distributions of high static pressure in the nozzle and the target can be observed. Figure 14 reveals that the static pressures on the target were low even though the

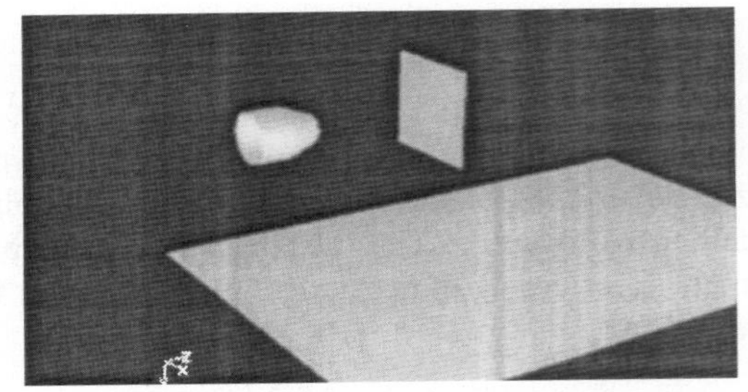

Fig.10: CFD model of prototype nozzle and a flat target

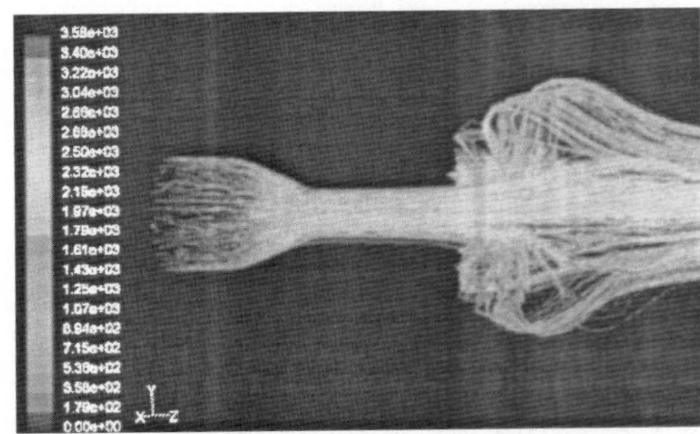

Fig.11: Flow of powder particles with inlet pressure at 10^3 Pa

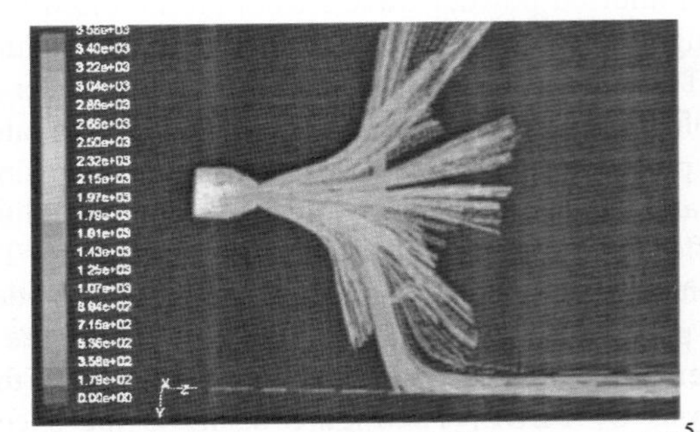

Fig.12: Flow of powder particles with inlet pressure at 10^5 Pa

pressure inside the nozzle was high.

The results of the CFD modelling of static pressure distributions is consistent with the results obtained from the simulation of charged particles flow from the nozzle to the target, with inlet pressures set at 10^3 Pa and 10^5 Pa. High inlet pressures of 10^5 Pa and above do not necessarily translate into improved powder deposition efficiencies because the flow of charged powder particles emerging from the nozzle at high inlet pressures is divergent and not concentrated on the target. The simulation results for the prototype nozzle indicate that increased efficiency is achieved when the inlet pressure is set at 10^3 Pa, with more than 90% of sprayed charged powder particles deposited on the target surface.

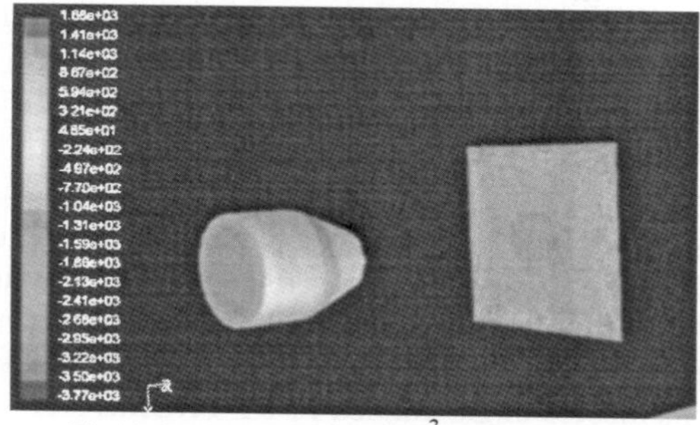

Fig.13: Static pressure with 10^3 Pa inlet pressure

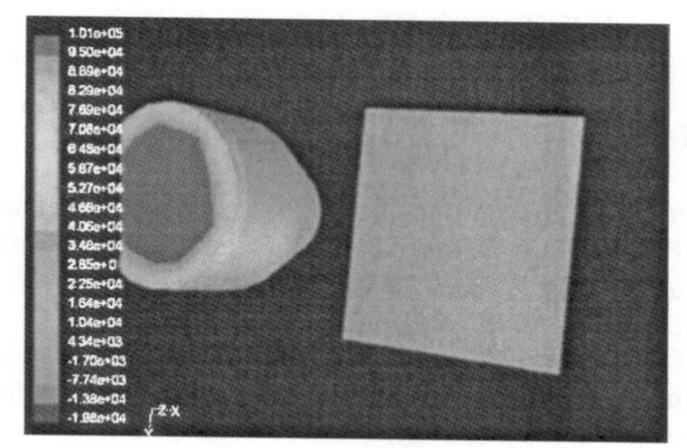

Fig.14: Static pressure with 10^5 Pa inlet pressure

4. Rapid Prototyping of New Nozzle Designs

Rapid prototyping (RP) is a term that describes the technology used for quickly fabricating parts and functional prototypes from CAD models [5]. In this work, a Selective Laser Sintering (SLS) RP machine [5] was employed to produce six different nozzles. The nozzles were manufactured in order to identify appropriate designs for the three powder coating guns. Results from the CFD simulation of the prototype nozzle were used to choose the parameters for the new nozzles. The nozzles were produced from Duraform material [5] and tested under normal operating conditions. Figure 15 and Table 2 present the six nozzles and their design parameters.

5. Nozzles Testing and Evaluation

All the six SLS nozzles were tested in the pilot plant with the inlet pressure set to 10^3 Pa, the spray gun voltage to 40 KV and the gun-to-target distance to 900mm. The target flat plate had dimensions of 300mm x 400mm. The spray efficiency of each nozzle was assessed by measuring the surface area covered during a given period. Figure 16 shows the spray patterns of the six candidate nozzles while table 3 presents the approximate coated surface areas.

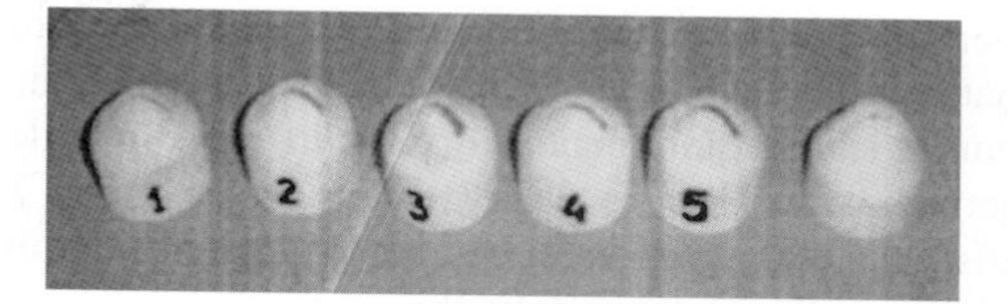

Fig.15. Six SLS prototyped nozzles

Table 2: Characteristics of prototype nozzles

Nozzle	Prismatic aperture	Effective surface
N1	2.5mm X 15mm	$S_{N1}=37.5mm^2$
N2	4mm X 15 mm	$S_{N2}=60mm^2$
N3	5mm X 15mm	$S_{N3}=75mm^2$
N4	4mm X 16mm	$S_{N4}=64mm^2$
N5	5mm X 16mm	$S_{N5}=80mm^2$
N6	5mm X 17mm	$S_{N6}=85mm^2$

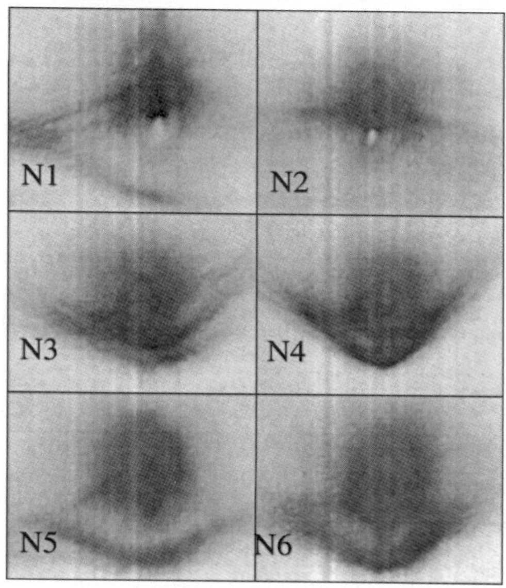

Fig.16. Spray patterns of candidate nozzles

Table 3: Estimated coating area of candidate nozzles

Nozzle	Coating area
N1	$CA_{N1}=5100mm^2$
N2	$CA_{N2}=4800mm^2$
N3	$CA_{N3}=8666mm^2$
N4	$CA_{N4}=6933mm^2$
N5	$CA_{N5}=9098mm^2$
N6	$CA_{N6}=1280mm^2$

Close examination of the spray pattern of nozzle 5 and comparison of the coating area obtained indicate that nozzle 5 shares similar properties with the simulated prototype nozzle. Further analysis of the operational efficiencies of the other nozzles in terms of coating area also indicates that nozzle 3 satisfies the geometrical requirements for coating the middle section of the 3-piece cans, while nozzles 1 and 2 satisfy the requirements for coating the top and bottom sections of the cans respectively.

Finally, all the three selected nozzles, N1, N2 and N3, were tested in the pilot plant on cans with height 132mm and diameter 40mm. Nozzle 3 was fitted to the gun responsible for coating the middle part of the cans, nozzle 1 to the gun for coating the top part of the cans, and nozzle 2 to the third gun. Figure 17 shows a 3-piece can being powder-coated using the prototype nozzles.

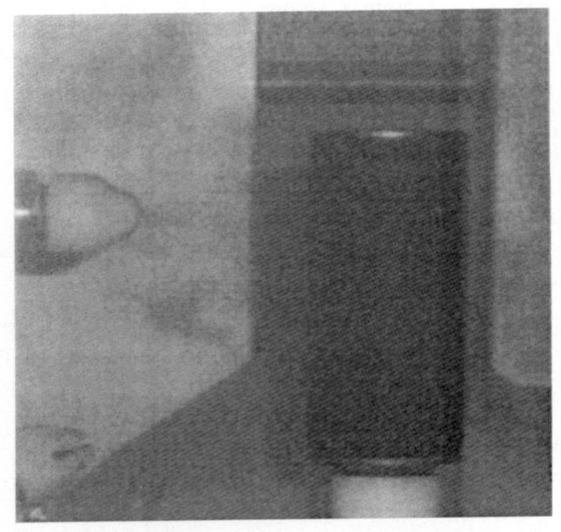

Fig.17: A 3-piece can being powder-coated in the pilot plant

The results of the tests in the pilot plant indicate a 15% increase in powder deposition efficiency. The methodology implemented in this work has therefore led to a systematic efficiency improvement.

6. Conclusions

This paper has presented a methodology for increasing powder deposition efficiency in electrostatic powder coating. The methodology gives an approach to nozzle design to reduce overspray. A standard computational fluid dynamics system, FLUENT, was used to calculate the inlet pressure at the gun nozzle. The CFD simulation offered an opportunity to investigate a wide range of nozzle parameters with a view to finding optimal candidate nozzle designs for the powder application process. Experimental results obtained with the candidate nozzles indicated that powder deposition efficiency improved significantly.

Acknowledgement

The research described in this paper was carried out with funding provided by Engineering and Physical Sciences Research Council (EPSRC-LINK) and the Department of Trade and Industry (DTI). The authors are memebers of EC-funded FP6 I*PROMS Network of Excellence.

References:

[1] Reddy V. Powder spray technologies and their selection. Powder Coatings Applications(1st edn), ed. Dawson S and Reddy V. Society of Manufacturing Engineers, Dearborn, Michigan,1990, pp 111-115.
[2] Guskov S. Electrostatic phenomena in powder coating: new methods of improving Faraday-cage coating, finish quality and uniformity, and recoating operations", Nordson Corporation, Ohio, 1996.
[3] Ye Q and Domnick J. On the simulation of space charge in electrostatic powder coating with a corona spray gun, powder technology, 2003, pp 250-260.
[4] Ye Q, Steigleder T, Scheibe A and Domnick J. Numerical simulation of the electrostatic powder coating process with a corona spray gun. Journal of Electrostatics, 54 , 2002, pp 189-205.
[5] Pham D T and Dimov S S. Rapid Manufacturing: the Technologies and Applications of Rapid Prototyping and Rapid Tooling. Springer-Verlag London 2001.

Intelligent Production Machines and Systems
D. T. Pham, E. E. Eldukhri and A. J. Soroka (eds)

Development of an artificial neural network for defects prediction in metal forming

V. Corona, A. Maniscalco, R. Di Lorenzo

Dipartimento di Tecnologia Meccanica, Produzione e Ingegneria Gestionale
Università di Palermo, Viale delle Scienze 90128 Palermo, Italy

Abstract

In bulk forming processes the prediction of ductile fractures and flow defects is a very important task. Generally the ductile fracture criteria are utilized for the estimation of damage accumulated by the material during the deformation. The principal defect of this approach is that each criterion gives a good result for some processes, instead not a good performance is obtained in relation to other processes. Thus, a considerable advantage is obtained by the implementation of a tool able to predict ductile fracture occurrence independently by the particular process analyzed. The aim of this paper is the development of an artificial neural network which can predict the occurrence of fracture with no dependence by a particular criterion. The paper presents the procedure leading to the choice of the most performing network. In particular, the impact of different network architectures (number of neurons and layers), transfer functions, input vectors, data structures, learning rules was analyzed. The implemented neural network can recognize fracture occurrence in a wide range of bulk forming processes and gives very performing results.

1. Introduction

The goals of forming process design mainly concern the production of defect free products. Different approaches to predict fracture occurrence have been presented in the literature mainly based on the application of fracture criteria [1]. Generally, damage is identified as a variable related to the stress and strain conditions: fracture occurrence is predicted according to a critical value of damage for the given material. The state of the art in this field is represented by several criteria that use different critical values such as the tensile strain energy per unit of volume or taking into account also the hydrostatic stress [2,3].

Other approaches refer to voids growth model and are founded on the idea that cracks occur after void nucleation, growth and coalescence; thus, these criteria take into account macro variables related to fracture mechanics [3,4,5].

The weakness of such techniques concerns the need of a critical damage value to identify fracture occurrence. Actually, to identify this critical value it is necessary to know the specific process conditions and to have experimental observations.

It must be also observed that commercial numerical codes utilize ductile fracture criteria and are able to predict fracture when the damage factor reaches its critical value.

The most relevant drawbacks of approaches based on ductile fracture criteria are related to the consideration that each criterion usually has good performance in fracture prediction for particular stress - strain paths, but may fail for different processes.

In this way, an artificial intelligence tool for ductile fracture prediction with no dependence from a particular criterion is proposed.

Actually, a great advantage in process design may be obtained by implementing a tool able to predict fracture occurrence for a wide range of cold forming process without the use of ductile fracture criteria.

In the approach here addressed, an artificial neural network was developed in order to classify stress – strain paths [6]: the network is able to identify process conditions leading to fracture as well as to recognize safe paths (i.e. paths which do not determine fracture).

2. The artificial neural network approach

As mentioned, the aim of this paper is the implementation of a tool able to predict ductile fracture occurrence independently by ductile fracture criteria. In particular the analysed processes are cold bulk forming applications. The variables taken into account influencing fracture occurrence were: the effective strain, effective stress, maximum principal stress, mean stress and tangential stress (ε_{eff}, σ_{eff}, $\sigma_{maxprinc}$, σ_{mean}, $\sigma_{XY/RZ}$). The complexity of the relation between these variables and the fracture occurrence and the lack of an explicit linkage interpreting the phenomena leading to fracture, led to the utilization of a black-box approach.

Actually, the utilization of a neural network is the most suitable approach when an implicit knowledge is available, thus neural networks represent the black box able to receive as an input the process variables and to provide as an output the occurrence of fractures.

The problem here addressed is a classification problem, in fact the aim is to explore the different combinations between the input variables and properly classify them in the categories fracture occurrence or fracture absence. One of the main advantages in the application of such tool consists in its generalization ability: the network was trained on data determined from different processes thus it proved its capability to deal with data sets affected by noise. Indeed the data used to train the network were obtained from numerical simulations and were affected by the noise related to the sensibility of the specific utilised numerical code.

The implemented approach consists in the former analysis of experimental tests from the technical literature in which forging processes presenting fracture occurrence are reported [1,3,7-10]. Thus, these processes were simulated utilising DEFORM 2D 8.1 code and the stress-strain paths was tracked in the workpiece region where fracture was expected. Furthermore, it was necessary to collect data also for processes in which fracture does not occur.

As the investigated processes are concerned, several axi-symmetric test cases were took into account: extrusion processes with central bursting occurrence, upsetting of billets with different geometry presenting superficial cracks, blanking of thick sheet. The material utilized in all the analyzed processes was AISI 1040 steel. Thus, the training set was built by the collected stress-strain data. A value equal to 1 was associated to the output variable corresponding to stress-strain paths leading to fracture occurrence; on the contrary, a value of the output variable equal to 0 was associated to stress-strain paths determining fracture absence. In this way, the training set was composed by more than two thousand data deriving by the numerical simulations and containing both paths determining fracture and paths avoiding fracture. Furthermore, about 400 data records (extracted from the training set) were used for the network validation. Once the network was trained and validated, it was tested utilizing about 200 data records (related to forming processes not included in the training data) in order to evaluate its predictive capability. The network was developed within a MATLAB environment.

In the following paragraphs, the evolution of the network that led to the best performing network architecture is analyzed.

3. Network morphology

A Multi Layer Perceptron feed-forward back-propagation network was used for the analyzed classification problem.

The multilayered feedforward neural networks (MFNN) are one of the main classes of static neural networks and their architecture includes a large class of network types with many different topologies and training methods.

In particular they are indicated for non-linear solutions and for ill-defined problems.

The morphology of static multilayered feedforward neural networks consists of many interconnected neurons. These neurons form layered network configurations through only feedforward interlayered synaptic connections in terms of the neural signal flow.

In general, an individual neuron aggregates its weighed inputs $w_{ik}\phi_k$ and yields outputs y_i through a nonlinear activation function F (see equation 1).

Figure 1 shows the structure of a neuron highlighting the input signals and their processing leading to the neuron output.

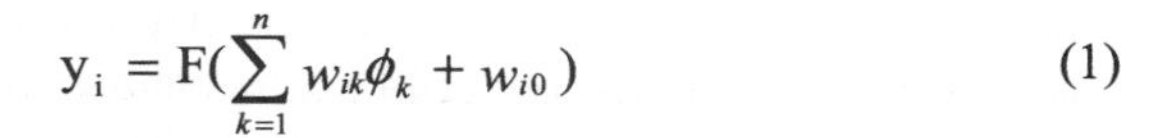

$$y_i = F(\sum_{k=1}^{n} w_{ik}\phi_k + w_{i0}) \qquad (1)$$

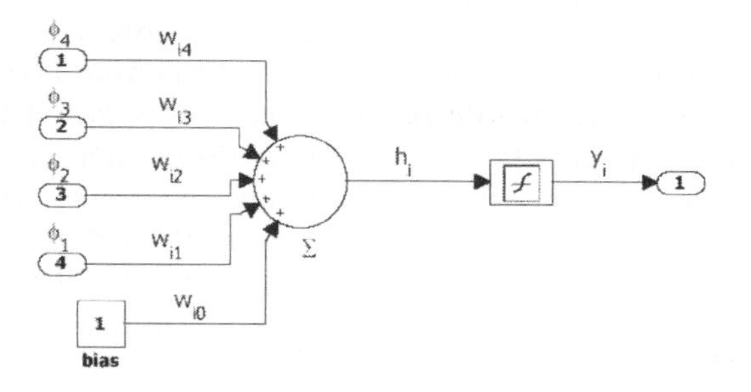

Fig. 1 Neuron structure

As the transfer function is regarded, it was necessary to utilize a non-linear function to transfer the information between the neurons, the Tan-Sigmoid transfer function was chosen since it provided the best training performance. On the contrary, the Satlin transfer function was used for the output neuron because of the necessity of an output variable that varies between 0 and 1 (see Fig. 2).

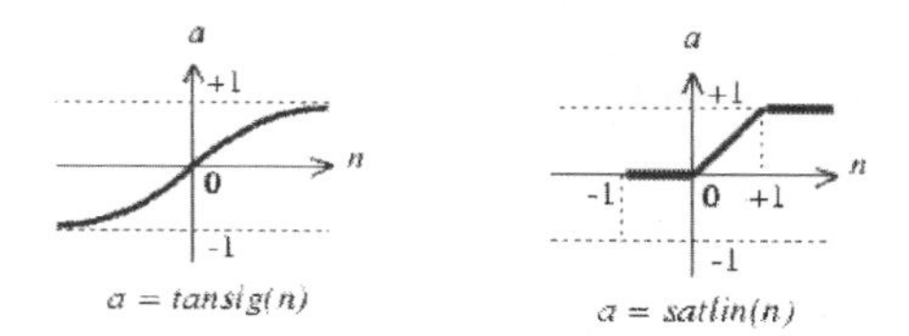

Fig. 2 Transfer Functions

4. Number of hidden layers and neurons

The typical back-propagation network consists of an input layer, an output layer, and at least one hidden layer. With a minimum of three hidden layers it is able to solve problems of any complexity; in fact the choice of the number of hidden layers is related to the possibility to classify correctly the outputs in a complex space of solutions.

The number of layers has a strong influence on the obtained performance. On the other hand the optimum layout is strongly dependent from the particular application and a generic rule to determine the layer's number doesn't exist. Instead, it is possible to determine the minimum number of hidden layers L_H with the following empiric relation:

$$\min(L_H) = \log_2(n_{input} + n_{output}) \qquad (2)$$

where n_{input} and n_{output} are the number of input and output neurons respectively.

For the investigated problem the number of input variables was equal to five and it corresponds to the number of input layer neurons. The output layer consisted of one neuron associated to a variable that assume the value 1 if the stress-strain data conduce to fracture, 0 otherwise.

Thus, the minimum number of hidden layer taken into account is equal to 3 and networks with three and four hidden layer were developed in order to determine the best performing one.

As the number of hidden neurons N_{HN} is regarded, the lack of rigid rules for the determination of such number led to take into account the following empiric equation which gives an indicative value for the number of hidden neurons:

$$N_{HN} = \frac{TS \times e\%}{(n_{input} + n_{output})} \qquad (3)$$

where *TS* indicates the training set dimension (i.e. the number of input data records); *e%* is the acceptable percentage error (in the present application such acceptable level was fixed to 3%).

In order to improve the performance of the network, the following procedure was applied: starting from the number of hidden neurons calculated from equation 3, such number was increased if the obtained network was not able to learn from the training set, while it was reduced if no generalization was provided (i.e. if test error increased).

On the basis of the previous considerations, the configurations taken into account were composed by a number of hidden neurons that varied from 9 to 13 for the network with 3 hidden layers and from 10 to 14 for the network with 4 hidden layers.

It has to be observed that the distribution of the hidden neurons over the hidden layers was widely investigated and the best performing architecture was determined. In particular, Fig. 3 and 4 show the network performance in terms of absolute percentage error at the varying of neurons number in the case of 3 and 4 hidden layers respectively.

Figure 5 shows how the best network configuration with 4 hidden layers (14 hidden neurons) provides (in the phases of training, validation and test) a prediction error smaller than the one provided by the best 3 hidden layers configuration (10 hidden neurons).

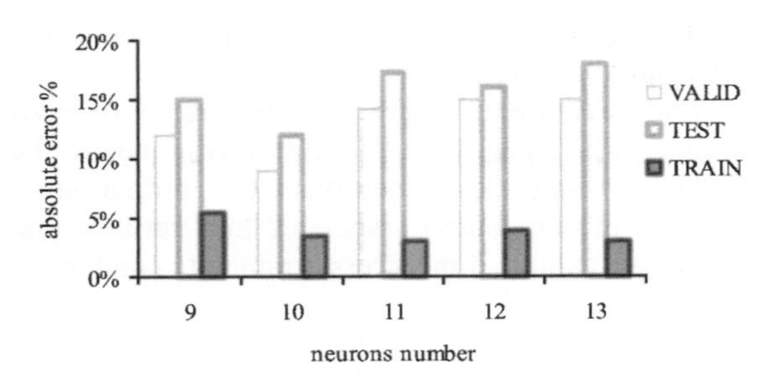

Fig. 3 Absolute % error vs. neurons number for the network with 3 hidden layers

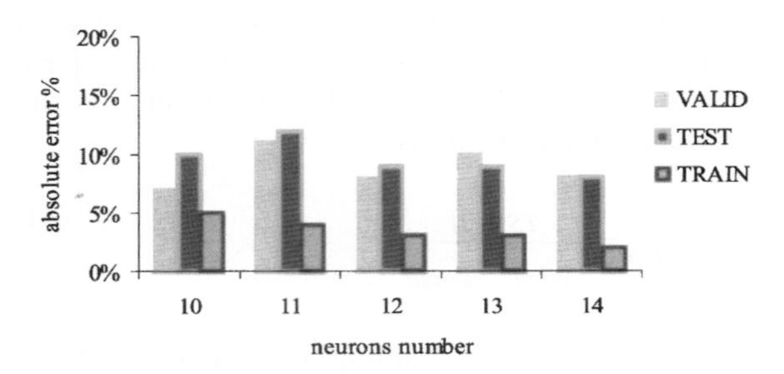

Fig. 4 Absolute % error vs. neurons number for the network with 4 hidden layers

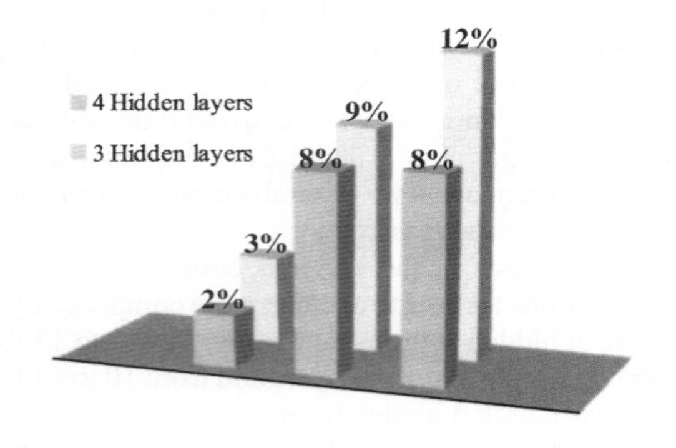

Fig. 5 Comparison of the performances of the two investigated networks

5. The training process

The training procedure used to solve the analyzed problem is the back-propagation method.

It was created for multiple-layer networks and nonlinear differentiable transfer functions. Generally in such procedure the training phase is carried out by providing to the network a set of input vectors associated to the expected output vectors and the training is performed until the network is able to classify input vectors correctly. In other words the network has to be able to associate to a certain input the expected output value.

According with this procedure the difference between the output of the final layer and the desired output is back-propagated to the previous layers. On the basis of such back propagation the connection weights are adjusted in order to reduce such difference Generally, such adjustment are developed according to the well known "Delta Rule". Nevertheless, the principal drawback of such learning in feedforward, back-propagation architecture is that the learning can converges in a local minima.

Actually, there are many variations to the learning rules for back-propagation networks and in the analyzed application Batch Gradient Descent with momentum, Quasi-Newton and Levenberg-Marquardt Methods were taken into account as learning algorithms.

The Levenberg-Marquardt learning algorithm was selected because it guarantees the convergence to the best solution with the smaller computational time.

6. Input vectors and data structures

The performance of the network is strongly influenced by the input variables and by the choices about the training set structure. Indeed the input variables must be characterized by a small correlation (preferably a correlation index lower than 70%).

In the investigated case the five variables were accurately chosen on the basis of the technological knowledge available on the process, also with the aim to avoid redundancy of information.

As the data structure is regarded, the training set dimension must be appropriate for the problem complexity and network size in terms of neurons number.

The minimum training set dimension was chosen according to the following general empiric rule ($n_{weigths}$ indicating the total number of networks weights):

$$TS >> 5 \times n_{output} \times n_{weights} \quad (4)$$

The data collected in the training set must be chosen in order to include complete information about the investigated phenomenon.

An increase of the network performance can be obtained by a normalization of the utilised data. Furthermore, a clustering of the data is very useful since the networks performance strongly increases when it can learn from properly structured data subsets.

As mentioned, for the application here addressed 2000 data records for the training was chosen, normalized data were provided to the network for training, validation and testing phases and a clustering of the data was applied.

As the latter aspect is concerned, the clustering was performed by an analytical approach based on data clustering analysis and also by creating data subsets on the basis of two criteria: process kind and fracture occurrence/absence.

6. Analysis of the results

The result of the optimization process above described, led to the best performing network architecture which consists of four hidden layers with 14 hidden neurons and 57 links representing the weights for the transfer of the information among the neurons.

The network structure is reported in Fig. 6.

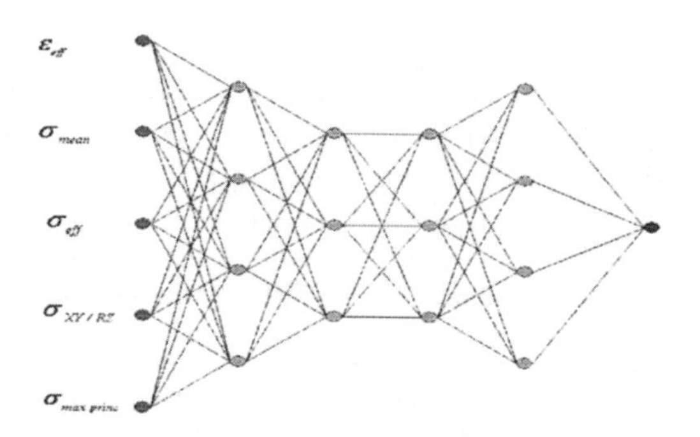

Fig. 6 The best performing network architecture

The developed network was trained, validated and tested through the data available on the analyzed problem, i.e. by the data obtained from the numerical simulations of the chosen processes. In particular, the data were distinguished in three sets to be utilized for training, validation and test respectively.

As the performance of the network is concerned, it was measured through the mean square error (MSE) calculated on the basis of the difference between the expected output and the calculated one.

The evolution of the performance of the network during the training, validation and test epochs is reported in the Figure 7.

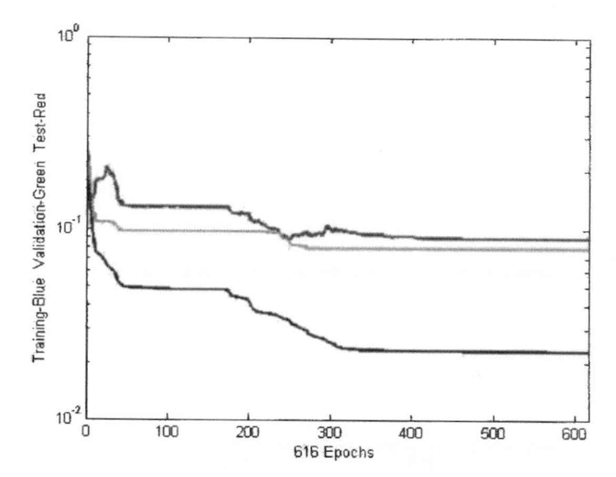

Fig. 7 The ANN performance

Furthermore, the results related to the three phases of training, validation and test were also evaluated by calculating an global percentage error over the data which gives a more accurate indication of the correctness of the predictions supplied by the network. The obtained errors in the three phases are summarised in Table 1

Table 1
Percentage error in the training, validation and test phases.

	Global percentage error
Training	2%
Validation	7%
Test	8%

The results of the network development are summarised in the following figures 8-9-10.

In particular Fig. 8 reports the training performances of the network. In all the cases corresponding to fracture absence (i.e. safe parts) the network has to provide a value equal to 0 of the output variable; actually, as it can be observed in Fig. 8, the response of the network is 0 in most of the "safe" cases during training while few points are incorrectly predicted (points not equal to 0).

As well Fig. 9 reports the predictions of fracture for the training points in which fracture is effectively expected.

As it can be observed, again few points have wrong predictions.

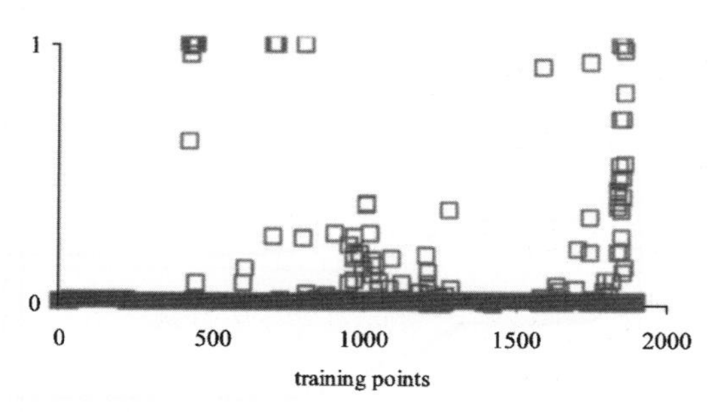

Fig. 8 Training phase: identification of the fracture absence

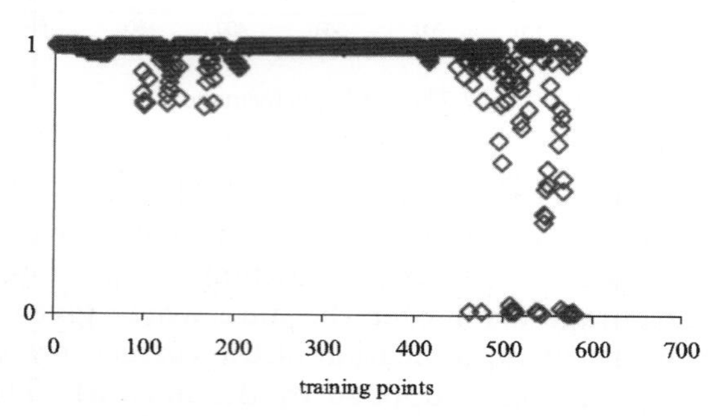

Fig. 9 Training phase: identification of the fracture occurrence

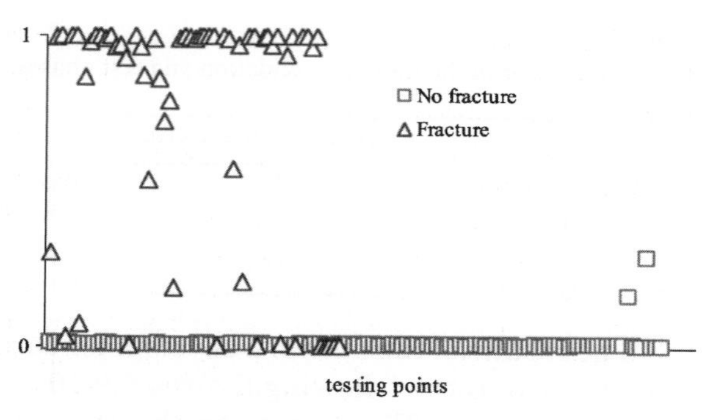

Fig. 10 The testing results

The results of the network testing phase are reported in Fig. 10 where the high predictive performance of the network is shown.

The network is able to recognize the fracture occurrence and absence, in fact most of the points indicating fracture (triangle symbol) are closer to 1 which means that fracture is carefully predicted; as well almost all the points indicating no fracture (square symbol) rely on the 0-axis.

7. Conclusions

According to the described results it can be assessed that the predictive capability of the obtained network and its generalization power are very useful in order to implement a valid tool in forming processes design.

Further developments of such tool will consist in an extension of the forming process data utilised for the network training in order to develop a predictive tool valid for a wider range of processes.

Moreover, an integration of the neural network with the numerical code will be implemented in order to include network predictions among the numerical output in a finite element environment.

References

[1] Landre J. Pertence A. Cetlin P.R. Rodrigues J.M.C. and Martins P.A.F. On the utilization of ductile fracture criteria in cold forging. Finite elements analysis and design Vol. 39, 2003: 175-186.

[2] Cockcroft M.G. Latham D.J. Ductility and the workability of metals. J. Inst. Metals, V..96, 1968: 33.

[3] Gouveia B.P.P.A. Rodrigues J.M.C. Martins P.A.F. Ductile fracture in metalworking: an experimental and theoretical research. J. Materials Processing Technology, V. 101, 2000: 52.

[4] Rice J. R. and Tracey D. M.. On the Ductile Enlargement of Voids in triaxial Stress Fields. Journal of the Mechanics and Physics of Solids, V.17, 1969: 201-217.

[5] McClintock F.A. A criterion for ductile fracture by the growth of holes. Trans. ASME J. Appl. Mech., V. 17, 1968, 363.

[6] Di Lorenzo R., Maniscalco A., Corona V., Micari F. Prediction of ductile fracture in bulk metal forming: an artificial neural networks based approach, submitted to 8th ICTP Conference, 2005.

[7] Ragab A.R. Fracture limit curve in upset forging of cylinders. Materials Science and Engineering, V. A334 2002: 114–119.

[8] Koa D.C. Kimb B.M. The prediction of central burst defects in extrusion and wire drawing. Journal of Materials Processing Technology, V. 102, 2000:19-24.

[9] Borsellino C. Micari F. Ruisi V.F. Central bursts prediction in extrusion. Proc. of the 1st Conference and Exhibition on Design and Production of Dies and Molds, 1997:341-346.

[10] Borsellino C. Micari F. Ruisi V.F. A new shear energy criterion for ductile fracture prediction in cold forming. Proc. of the 4th A.I.TE.M. Conf., 1999:379-386.

Intelligent Production Machines and Systems
D. T. Pham, E. E. Eldukhri and A. J. Soroka (eds)

FEM analysis of advanced high strength steel car body drawing process

S. Al Azraq[a], R. Teti[a], S. Ardolino[b], G. Monacelli[b]

[a] *Dept. of Materials and Production Engineering, University of Naples Federico II, Italy*
[b] *Elasis SCpA, Pomigliano d'Arco, Italy*

Abstract

Advanced High Strength Steels (AHSS) have been intensively applied to automobile components to improve crashworthiness, without increasing the body weight, under a strong pressure of the requirements for fuel consumption, durability and energy absorption. Some of the problems encountered in the forming of AHSS, due to lack of knowledge about the material behaviour, are presented and described with the purpose of showing the necessity of sheet metal simulation to save cost and time by replacing physical tryouts with virtual tryouts. The codes and elements normally used in these types of simulation programs, as well as their common approaches, are presented and evaluated. Finally, a test case of an automobile component (hood structure) was analysed using the incremental approach of the AutoForm FEM code.

1. Introduction

Automobiles play an important role in our life. This makes it necessary to incessantly try to reduce their production cost in line with the innovation of production technology. At the same time, measures are being taken toward the reduction of fuel cost and the improvement of safety so that eagerly-pursued harmony with the social and natural environment can be established. It is safe to say that thin steel sheets for automobiles have made progress in responding to the market needs. In recent years, it has become one of the most important tasks for automobiles to make the reduction in weight of auto bodies compatible with the improvement of crashworthiness, particularly with the aim of reducing CO_2 gas emissions by improving fuel consumption [1]. To satisfy these demands, a new steel grade, the Advance High Strength Steel (AHSS), has been developed, pushing the boundaries of what was previously possible with conventional steel grades. A major negative consequence of using these new materials is the absence of knowledge about their behaviour during both the drawing process and future physical applications. Engineers turned to forming simulations in order to closely investigate and try to minimize this problem. The aim of industrial application of drawing process simulation is to replace the physical tryout by computer tryout to reduce time and cost and improve quality in the die design/manufacturing cycle [2]. In the past, a number of analyzing methods have been developed and applied to various forming processes. Some of these methods are the slab method, the slip-line field method, the viscoplasticity method, upper and lower bound techniques and Hills general method. These methods have been useful in qualitatively predicting forming loads, overall geometry changes of the deformed blank and material flow, and approximate optimum process conditions. However, a more accurate determination of the effects of various process parameters in the drawing process has become possible only recently, when the finite element method (FEM) was developed

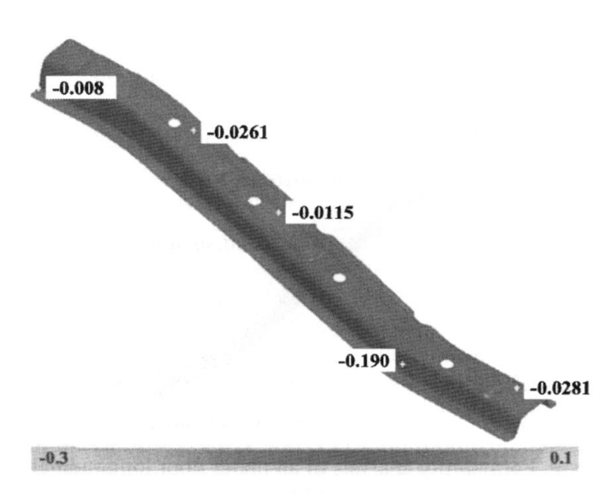

Fig. 5. Thinning analysis of an under floor longitudinal member of Fiat Stilo.

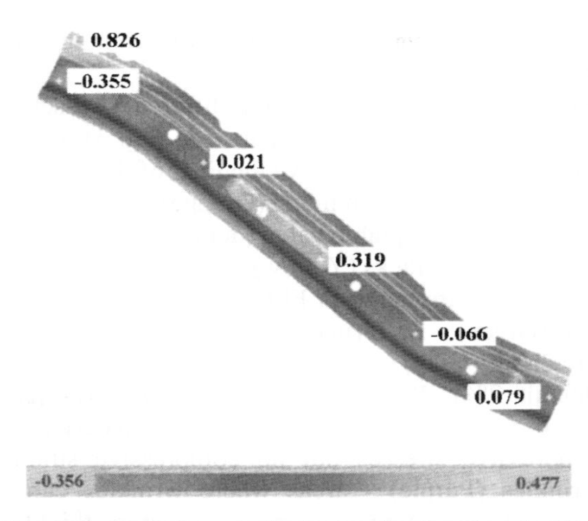

Fig. 6. Springback analysis of an under floor longitudinal member of Fiat Stilo

These simulation programs are used to replace experimental tryouts by virtual tryouts, saving time, money and resources. Another significant advantage of its employment is the improvement, simplification, speed up and enhancement of the die design process.

The further developments of the research will include the use of a neural network approach to the modeling of the behaviour of the AHSS work material under variable strain rate conditions by reconstructing the stress-strain curves from tensile test experimental data obtained using different strain rate values [15].

Acknowledgements

This research work was carried out with partial support from EC FP6 NoE on Innovative Production Machines and Systems – I*PROMS.

References

[1] Takahashi M. Development of high strength steels for automobiles. Nippon Steel Tech. Rpt. N. 88 July 2003.

[2] Makinouchi A, Teodosiu C and Nakagawa T. Advance in FEM Simulation and its Related Technologies in Sheet Metal Forming. SME, 1998.

[3] Boogaard A.H. Van Den, MeindersT., Huetink J. Efficient implicit finite element analysis of sheet forming processes. Int. J. Numer. Meth. Eng., 56, 2003: 1083-1107.

[4] American Iron and Steel Institute & Auto/Steel Partnership, Automotive Steel Design Manual, Revision 6.1, August 2002. http://www.steel.org.

[5] International Iron Steel Institute, Advanced High Strength Steel Application Guidelines, March 2005. http://www.worldsteel.org.

[6] American Public Works Association, Vehicle Corrosion Caused by De-icing Salts, Special Report #34, September 1970. http://www.apwa.net.

[7] Baboian, R. Environmental aspects of automotive corrosion control, in Corrosion and Corrosion Control of Aluminum and Steel in Lightweight Automotive Applications, NACE Intern., Houston, TX, 1995.

[8] Dust Control, Road Maintenance Costs, Cut with Calcium Chloride, Public Works, Vol. 121, No. 6, May 1990: 83-84.

[9] Shaw, R., Zuidema, B. K. New high strength steels help automakers reach future goals for safety, affordability, fuel efficiency, and environmental responsibility SAE Paper 2001-01-3041.

[10] Miller B and Bond R. The practical use of simulation in the sheet metal forming industry. Joint paper, Confed. of British Metalforming Technical Conf. 2001

[11] Reddy, J.N. An introduction to finite element method, Mc Graw Hill, Inc., 2nd ed., 1993.

[12] Hoon Huh, Tae Hoon Choi. Modified membrane finite element formulation for sheet metal forming analysis of planar anisotropic materials. Int. J. of Mech. Sci., 42, 1999: 1623-1643.

[13] Kawka, M., Makinouchi; A. Shell-element formulation in the static explicit FEM code for simulation of sheet stamping. J. of Mat. Proc. Tech., 50, 1995: 105-115.

[14] Sester M. Implementation of the selected elastoplastic models in autoForm-incremental. Workshop 3DS, Villetaneuse, June 2002.

[15] Giorleo G., Teti R., Prosco U. and D'Addona D. Integration of neural network material modelling into the FEM simulation of metal cutting, 3rd CIRP Int. Sem.on ICME, Ischia, 3-5 July 2002: 335-342.

Intelligent Production Machines and Systems
D. T. Pham, E. E. Eldukhri and A. J. Soroka (eds)

Flexible tool management strategy for optimum tool inventory control

D. D'Addona, R. Teti

Department of Materials and Production Engineering,
University of Naples Federico II, Piazzale Tecchio 80, 80125 Naples, Italy

Abstract

The design and functioning of a Flexible Tool Management Strategy (FTMS) paradigm for optimum tool inventory sizing of CBN grinding wheels for nickel base alloy turbine blade fabrication is presented. The FTMS is integrated in an agent-based tool management system as a domain specific problem solving function of the intelligent agent responsible for tool inventory sizing and control. The evaluation and comparison of the performance of the FTMS paradigm was carried out with reference to two real industrial cases of CBN grinding wheel inventory management.

1. Introduction

This work is part of a wider scope research concerned with the development and implementation of a Multi-Agent Tool Management System (MATMS) for automatic tool procurement [1-4]. The MATMS operates in the frame of a negotiation based multiple-supplier network where a turbine blade producer (customer) requires from external tool manufacturers (suppliers) the performance of dressing operations on worn-out CBN grinding wheels for nickel base alloy turbine blade fabrication [5, 6].

In this paper, the development and implementation of a novel Flexible Tool Management Strategy (FTMS) paradigms, integrated in the MATMS as domain specific problem solving function of the intelligent agent responsible for optimal tool inventory sizing and control, is presented as an alternative to traditional tool management.

2. CBN grinding wheel tool management

Turbine blades are manufactured along several production lines, each for one aircraft engine model requiring a set of CBN grinding wheel types (part-numbers). Each part-number is planned to work a maximum number of blades: when it reaches its end of life, it is sent for dressing to an external supplier in a supply network and remains unavailable for a time defined as dressing cycle time. For each part-number, a sufficient number of wheels must be always available (on-hand inventory) to prevent production breakage due to tool run-out. The part-number on-hand inventory size, I, depends on: # of pieces/month, P; # of pieces/wheel, G; # of months required without new or dressed wheel supply, C (coverage period) heuristically selected. The wheel demand, D, for each part-number is given by $D = (P/G) * C - I_0$, where: P/G = tool demand rate (# of wheels/month); I_0 = initial part-number inventory size.

3.2 Design, implementation and validation of a Distributed Control System Real-time Framework (DCRF)

3.2.1 Implementation of the DCRF

The first step for this implementation is the porting of TAO to RTAI, done by the consortium and OCI. OCI (Object Computing, Inc), a USA company, are among the implementers and the best world experts in TAO, while the consortium itself includes some of the best European experts about RTAI. Nevertheless this step proved to be a very complex one, and is still currently in progress, although some good results have already been achieved. The main one was the verification of direct and normal collocation working in real-time.

A code release was published on the SourceForge site and mailed to the ACE/TAO and RTAI mailing lists. A full-time hardware platform was constructed for testing the port. Anyway at the same time the consortium elaborated a contingency plan for the case that the TAO port to RTAI proves to be unfeasible.

3.2.2 Configuration system

The requirements point out that configuration of the OCEAN control system and in general the configuration of a component based system is a complex problem and it is not solvable by ad hoc programming meaning that configuration data and application functionality are not clearly separated. This leads to tight-coupled applications, where (re)configuration, modification, validation and reuse are difficult to achieve. The CORBA Component Model (CCM) addresses interesting aspects about these issues.

CCM is a framework that supports the definition, code generation, packaging, assembly and deployment of CORBA components.

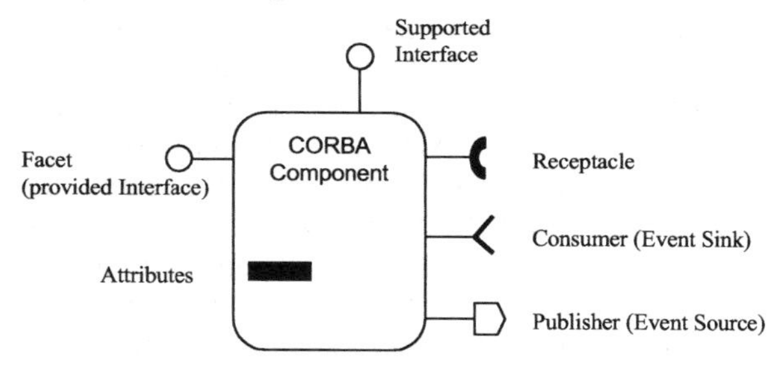

Figure 3: Structure of a CORBA Component

CCM components are the basic building blocks in a CCM system. A major contribution of CCM derives from standardizing the component development cycle using CORBA as its middleware infrastructure. Component developers using CCM define the IDL interfaces that component implementations will support. Next, they implement components using tools supplied by CCM providers. The resulting component implementations can then be packaged into an assembly file, such as a shared library, or a DLL, and be linked dynamically. Finally, a deployment mechanism supplied by a CCM provider is used to deploy the component in a *component server* that hosts component implementations by loading their assembly files. Thus, components execute in component servers and are available to process client requests.

Looking at the mentioned requirements for OCEAN's configuration, and taking into account the CCM features, designing and realizing a proprietary configuration approach is not advisable. This would lead to a low chance of general acceptance outside the consortium and to a lack of manpower for implementation and maintenance.

For a basic realization of the example scenario, two different CCM implementations have been considered: K2 and CIAO. K2 is a commercial CCM implementation from the company iCMG. It has been tested in a non commercial evaluation version which can be downloaded for free. This version is based on TAO. CIAO is TAO's freely available implementation of the CCM.

In the moment K2 is more advanced than CIAO, offering a CIDL compiler and graphical tools to build component skeletons, descriptors and assemblies.

In CIAO the CIDL compiler so far does not create component and assembly descriptions. There are so far also no graphical tools to build component assemblies.

The porting of ACE/TAO to RTAI so far leads to encouraging results, but CIAO has been considered not yet mature enough to be used at the moment for the implementation of the OCEAN demonstrators. The consortium decided to use a temporary configuration system that WZL (University of Aachen) is currently developing. In this system, parameters are included in XML files. The CCM approach should be adopted as soon as possible, but out of the timeframe of the project.

3.3 Design, implementation and validation of Motion Control base and extension components for an open numerical control system

The detailed functionality of the components is another focus area. The aim here is to achieve a definition of a reference architecture for the motion control components of the open control systems by extending existing architectures (OSACA, OMAC, JOP) and new proposals. The goal is to provide standardized and published interface descriptions for components which are up to now part of monolithic control blocks and which will be extracted from there; this way machine tool builders or end users will be able to integrate additional functionalities independently of control manufacturers.

This part of the project started with a pre-analysis of the major user requirements for the Motion Control Base (MCB) components (HMI Server, PLC, Kinematics, Motion Control Kernel) taking into account the different actors involved in the use and development of NC systems, i.e. control vendors, control developers, machine tool builders and industrial end users. It started from an analysis of the current market and industrial context, including the use of open source software and open control architectures, the hardware platforms and operating systems, the PLC functionalities currently used, and then, through an analysis of the needs of the consortium partners, papers and the answers to a questionnaire specifically created, performed a preliminary definition of functionalities and interfaces for these components. The final result of the task was a preliminary description of the components through several kinds of UML diagrams.

The following activities aimed at extracting from the main control block (the so-called Motion Control kernel) the following functionalities:

- HMI (Human Machine Interface) server component: it distributes data from all active components to the client applications; it is not only a server used to supply data, but also represents the entry point to gain access to the whole run-time machine functionality; through this, it is possible to know the status of the machine, to set up and control the part-program execution phase, to read/write parameters and data of the various NC sections, while the client applications are the front end to communicate with the user which make use of the HMI component to access the NC. In some cases the client applications can operate locally, on the same hardware hosting the real time NC structure, in other cases they operate remotely, on a second hardware platform using a network connection for the communication.
- PLC component: this is a general purpose tool, an application-independent logic sequencer which behaves in a different way according to its functionalities (the programming languages handled) and the programmable logic code downloaded into it. The programmable logic allows to customise the component through the insertion of the machine-dependent part (tool change, palette management, machine supervision, control of the tool loading system, of the part loading system, of auxiliary axes, of lube oil and coolant system, of other sensors and so on).
- Kinematics component: this is an essential component for a NC and handles coordinate transformations from the user reference system to the physical axes reference system and vice versa. It is always mandatory in case of having more than 3 axes or non-orthogonal axes relations. As the demand to develop or to adapt the kinematics equations appears very often, it is necessary to standardise a kinematics component with a well-defined interface. Thus it becomes feasible to develop new kinematics by creating a new instance of such an interface.
- The Motion Control kernel itself can be defined as a component including the command generation for a single channel of a NC. Such a complex functionality includes: interpretation of the part program, motion preparation (look-ahead, tool compensation, etc.), path interpolation and axis control, while it excludes the above mentioned kinematics functionalities.
- Process Control component: This component provides different functionalities depending on the considered application and user needs, like: chatter control, thermal compensation, and tool wear/breakage control. Also kinematic compensation (screw compensation) and compensation of structural deformations may be treated by such a component. The process control component offers standard interfaces for the control of specific processes, e.g. EDM, ultrasonic, Plasma/Laser/Waterjet cutting, Welding, Polishing, etc.
- Safety Component: This component has to guarantee a fail-safe behaviour of the control system and the machine tool. Fail-safe means that when a failure occurs the whole systems has to

where ε_1 and ε_2 represent translations along axis parallel to the working plane while φ_x and φ_y represent rotations along the same axis. The effect of the errors on the shape of the guide has been analyzed using the Mathematica package (version 4.0) on a Pentium III, 128 MB, 700 MHz computer. Two views of the simulated guide are in Fig.6 and 7. For the simulation the following machining parameters were chosen: feed velocity = 500 mm/min and RPM = 1200. The simulated errors were $\varphi_x = 2°$ and $\varphi_y = 3°$.

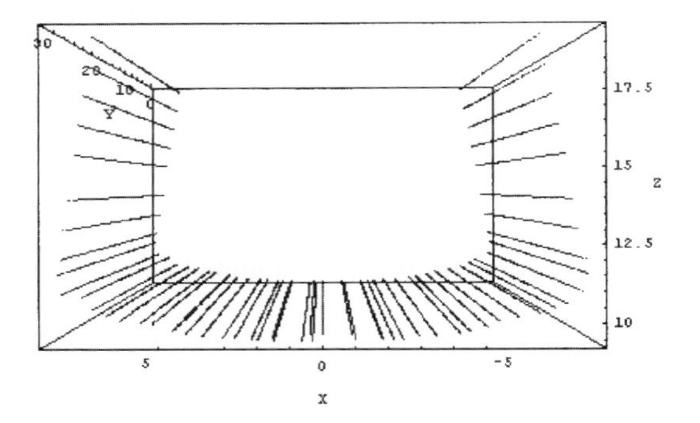

Fig. 6. Transversal view of the milled guide.

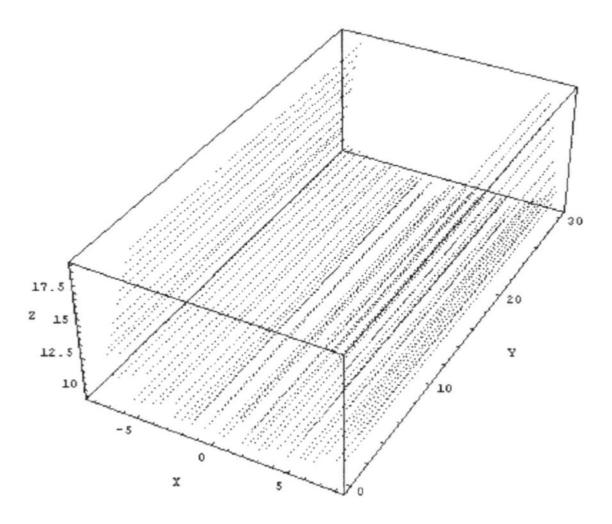

Fig. 7. Perspective view of the milled guide.

6. Future developments

In the prosecution of the research work, it is envisaged to generate, on the basis of the available points cloud a b-spline surface by a real-coded-genetic algorithm [9]. The availability of an optimized continuous surface will make it possible to realize an adaptative control of the too trajectory to minimize the machining errors.

Acknowledgements

This work was carried out with support from the EC FP6 NoE on Innovative Productions Machines and Systems – I*PROMS.

Reference

1 van Luttervelt C.A., Childs T.H.C., Jawahir I.S., Klocke F. and Venuvinod P.K. Present Situation and future Trends in Modeling of Machining operations: Progress Report of the CIRP Working Group 'Modeling of Machining Operations'. Keynote Paper, Annals of the CIRP, 47(2) (1998) 587-626.

2 Hong M.S. and Ehmann K.F. Generation of Engineered Surfaces by the Surface Generation System. International Journal of Machine Tools and Manufacture, 35 (1995) 1269-1290.

3 Prisco U. Prediction of workpiece accuracy in machining. D.Phil. Thesis, University of Napoli Federico II, Italy, 2003.

4 Portman V., Inasaki I., Sakakura M. and Iwatate M. Form-Shaping Systems of Machine Tools: Theory and Application. Annals of CIRP, 47 (1998) 329-322.

5 Giorleo G., Carrino L., Polini W. and Prisco U. Topography of Turned Surfaces. Proceeding of the 6th International Conference on Material Forming ESAFORM (2003). Salerno, Italy, April 28-30, 551-554.

6 Portman V.T. and Weil R.D. Higher Order Approximation in Accuracy Computations for Complex Mechanical Systems. Proc. of the 3rd CIRP Seminar on Computer Aided Tolerancing, Chacan (1993) 195-211.

7 Reshetov D.N. and Portman V.T. Accuracy of Machine Tools. ASME PRESS, New York 1988.

8 Wang W.P. and Wang K.K. Geometric Modelling for swept Volume of Moving Solids. IEEE Computer Graphics Applications 6 (1986), 8-17.

9 Capello F., Nigrielli V. and Ruggirello A. Data Fitting with B-Spline Surfaces by a Real-Coded Genetic Algorithm. Proc. of the 4th CIRP Seminar on Intelligent Computation in Manufacturing Engineering, Sorrento (2004) 669-674.

Intelligent Production Machines and Systems
D. T. Pham, E. E. Eldukhri and A. J. Soroka (eds)

Reconstruction of 3-D Relief Surfaces from 2-D Images

D. T. Pham [a], M. Yang[a], F.M. Chan

[a] *Manufacturing Engineering Centre, Cardiff University, Cardiff CF24 3AA UK*

Abstract

Various techniques have been employed to obtain 3-D surface models in the redesign of products and in the building of customized designs. However, these 3-D models have to be either painstakingly designed through an elaborate CAD process or reverse engineered from existing sculpted prototypes using multi-dimensional metrology. This paper presents a grey-level-based approach to 3-D relief surface reconstruction from a photographic 2-D image. A software package has been developed to convert a captured image into a 3-D data format for the interactive manipulation, visualization and subsequent reconstruction of the 3-D surface model.

1. Introduction

To perform reverse engineering, many digitizing techniques have been developed ranging from the simple touch comparator to sophisticated laser triangulation techniques [1,2]. There are also techniques based on tactile sensors [3,4], laser range sensors [5,6] and vision systems [7,8] developed to extract information from a physical part. Each technique has its own merits and demerits.

Less attention has been paid to the potential use of computer vision and image processing technology as a means of generating CAD models with 3-D relief surfaces from photographic 2-D images. This is important in circumstances where a particular product no longer exists, nor the drawings for it, but it is still necessary to recover details of the product. Such a process may also be important due to increasing demands for visually realistic three-dimensional model regeneration for the art, multimedia and virtual reality environments.

The proposed method does not require the labour intensive process of designing complex objects on a CAD system if their photographic images are available. It is suitable for those applications which do not require the final relief models to be as accurate as the models created with traditional reverse engineering techniques. Also, if the constructed 3-D model is mainly used for a visualization of the corresponding 2-D image, the proposed method is more easily realizable than those based on accurate 3-D geometric measurements.

Computer vision has proved to be of value in component recognition and inspection due to its low cost, ease of operation and real-time capacity [9, 10]. Al-Kinidi et al. [11] showed that the data extracted from photographic images could have potential for surface texture assessment while Peng et al. [12] proposed a photometric stereo approach to 3-D shape recovery based on multiple-image capturing and processing. In the work reported here only a single photographic image was acquired, either from a digital camera or from a scan, for the recovery of the surface texture. The objective of the paper is to show that the grey-level information extracted from captured 2-D images can be converted to a data format suitable for the subsequent 3-D modeling and rapid prototyping of the surface texture.

3. Multiplication coefficients were applied to the values of x, y, z of each individual point to allow the manipulation of the shape of the 3-D model before the rapid prototyping process (see Fig. 3).
4. Commercial reverse engineering software was used to assess the quality of the image-processed data.
5. The CAD model with three-dimensional relief was reconstructed.
6. The surface model was sent to a rapid prototyping machine to manufacture the part.

Two case studies have been carried out to validate this approach which included 3-D representation of 2-D pictures.

1. Reverse engineering design of a paperclip

This is a situation where a company no longer supports a particular product, but it is still necessary to recover details of the product for which a drawing does not exist. This case involved 3-D remodeling from a binary image in which only the contour of the object was extracted for the reconstruction (see Fig. 5). This is particularly useful for reconstructing a 2-D object without or with little surface texture but having complex contours which would be time-consuming to create using design software.

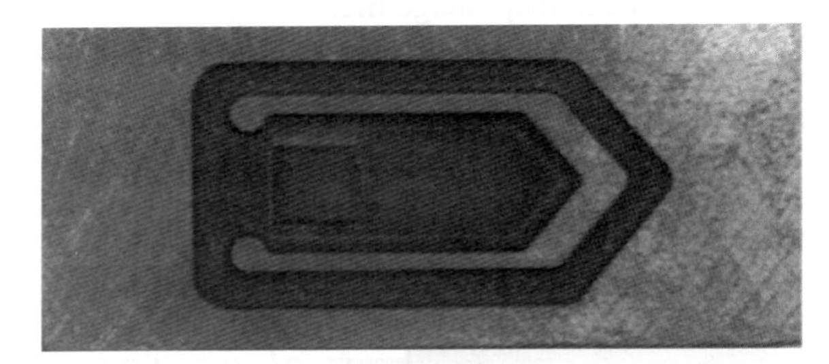

Fig. 5 A 3-D relief surface of a paper clip created from the 2-D image.

2. Recreation of 3-D relief surface from a 2-D visual image

The user interface of the software used to extract the grey information of the image is shown in Fig. 6 (b). The conversion was made pixel-by-pixel into a 3-D data model for CAD modeling. A rapid prototyping system then built the 3-D relief surface based on the digitised grey intensity of the 2-D image. This is useful for producing a replica of a 3-D work of art from a 2-D photographic image.

6. Conclusions

The ability to convert photographic images automatically into 3-D relief models can find wide applications in reverse engineering. The proposed technique offers a simple and useful method to produce virtual stereo images which are same as the original photographs. Software was developed as an "easy-to-Use visual tool to convert the captured images into the required CAD/CAM data format. After the creation of the 3-D data model based on the grey level of a 2-D image, a surface profile of the object derived from the 3-D data can be viewed and verified using reverse engineering software. Finally, a product model is prototyped.

The proposed approach is still under development and more experimentation is required to evaluate its efficiency and accuracy.

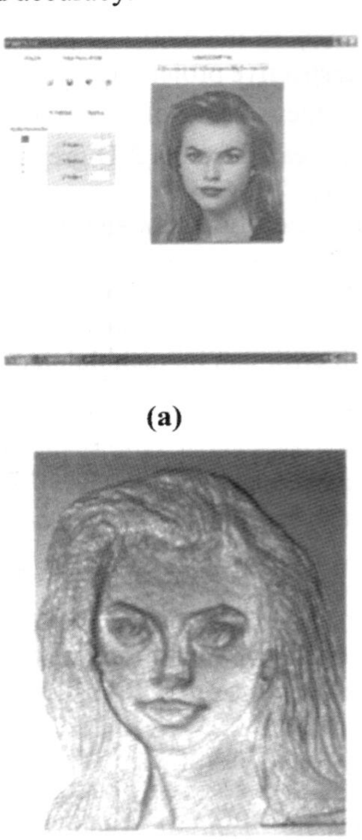

(a)

(b)

Fig. 6 (a) data conversion software user interface; (b) a 3-D relief surface created from its 2-D image.

Acknowledgements

This work was carried out as part of the SUPERMAN project funded by the European Commission and the Welsh Assembly Government under the European Regional Development Fund program. The authors gratefully acknowledge the support of I*PROMS which is financially funded by the European FP6 Program.

References

[1] C. Bonivento, A. Eusebi, C. Melchiorri, M. Montanari and G. Vassura, "WireMan: A Portable Wire Manipulator for Touch-Rendering of Bas-Relief Virtual Surfaces", 8th International Conference on Advanced Robotics, ICAR'97, Monterey, California, USA, July 7-9, 1997.

[2] J. Moos, A. Linney, S. Grindord, S. Arridge and J. Clifton, "Three dimensional visualisation of the face and skull using computerised tomography and laser scanning techniques", Eur J Orthod, pp. 9:247-53, 1987.

[3] M. Moll, "Shape Reconstruction Using Active Tactile Sensors", Ph.D. Thesis, Computer Science Department, Carnegie Mellon University, Pittsburgh, PA, 2002.

[4] A. Bicchi, J. K. Salisbury, and D. L. Brock, "Contact Sensing from Force and Torque Measurements", The Int. J. of Robotics Research, 12(3): pp. 249-262, 1993.

[5] C. K. Chua, S. M. Chou, W. S. Ng, K. Y. Chow, S. T. Lee, S.C. Aung and C. S Seah, "An integrated experimental approach to link a laser digitiser, a CAD/CAM system and a rapid prototyping system for biomedical applications" Int J. Adv. Manuf Techn., vol. 14, pp110-115, 1998.

[6] K. H. Lee, H. Park and S. Son, "A framework for laser scan planning of freeform surfaces", Int J. Adv. Manuf Techn., vol. 17, pp171-180, 2001.

[7] P. J. Armstrong and J Antonis, "The development of an active computer vision system for reverse engineering", Proc. Instn. Mech. Engrs, Part B, 214, pp 615-618, 2000.

[8] D. W. Manthey, K. N. Knapp and D. Lee, "Calibration of a laser range-finding coordinate-measuring machine", Optical Engineering, Vol 33, no 10, pp 3372-3379, 1994.

[9] S. J. Sternberg, "High-performance cameras cut cost of real-time imaging", Laser Focus World, pp. 33-38, March 1994.

[10] D. O'Neill and C. Lewis, "Slow-scan CCDs simplify real-time plasma spectroscopy", Laser Focus World, pp. 103-109, October 1994.

[11] G. A. H. Al-Kinidi, R. M. Baul and K. F. Gill, "Experimental evaluation of shape from shading for engineering component profile measurement", Proc. Instn. Mech. Engrs, Journal of Engineering Manufacture, Part B, 203(B4), pp 211-216, 1989.

[12] Q. Peng and M. Loftus, "Using image processing based on neural networks in reverse engineering", Int. J. of Machine Tools & Manufacture, vol 41, pp 625-640, 2001.

[13] R. Hall, Illumination and Color in Computer Generated Imagery, Springer-Verlag, New York, 1989.

Author Index

CHECK FOR 1 PARTS